ASTRONOMY AND ASTROPHYSICS ABSTRACTS

A Publication of the Astronomisches Rechen-Institut Heidelberg
Member of the Abstracting Board of the International
Council of Scientific Unions

Volume 7
Literature 1972, Part 1

Edited by
S. Böhme · W. Fricke · U. Güntzel-Lingner
F. Henn · D. Krahn · U. Scheffer · G. Zech

Springer-Verlag Berlin Heidelberg GmbH 1972

Astronomisches Rechen-Institut
Heidelberg
Director: Prof. Dr. W. Fricke

Astronomy and Astrophysics Abstracts
Editor-in-Chief: F. Henn

Astronomy and Astrophysics Abstracts
is prepared under the auspices
of the International Astronomical Union

ISBN 978-3-662-12283-9   ISBN 978-3-662-12281-5 (eBook)
DOI 10.1007/978-3-662-12281-5

Preface

Astronomy and Astrophysics Abstracts, which has appeared in semi-annual volumes since 1969, is devoted to the recording, summarizing and indexing of astronomical publications throughout the world. It is prepared under the auspices of the International Astronomical Union (according to a resolution adopted at the 14th General Assembly in 1970).

Astronomy and Astrophysics Abstracts aims to present a comprehensive documentation of literature in all fields of astronomy and astrophysics. Every effort will be made to ensure that the average time interval between the date of receipt of the original literature and publication of the abstracts will not exceed eight months. This time interval is near to that achieved by monthly abstracting journals, compared to which our system of accumulating abstracts for about six months offers the advantage of greater convenience for the user.

Volume 7 contains literature published in 1972 and received before August 15, 1972; some older literature which was received late and which is not recorded in earlier volumes is also included.

The authors of papers who have sent us abstracts on request have effectively contributed to the success of our service. We should like to express our gratitude to them. We acknowledge with thanks contributions to this volume by Dr. J. Bouška, who surveyed journals and publications in the Czech language and supplied us with abstracts in English, and by the Commonwealth Scientific and Industrial Research Organization (C.S.I.R.O.), Sydney, for providing titles and abstracts of papers on radio astronomy.

We also extend our warmest thanks to Mrs Monika Betz, Miss Helga Ballmann, Mrs Karola Gudé, and Mrs Brigitte Furian, who typed the text of this volume on IBM 72 Composers and compiled the pages from abstract slips in a perfect form for offset reproduction, to Miss Gisela Nollert, for punching material for the author index and the subject index, and to Mr Hartmut Jahreiß and Mr Heiner Schwan for helping with the layout of the text.

Heidelberg, September 1972

Siegfried Böhme
Walter Fricke
Ulrich Güntzel-Lingner
Frieda Henn
Dietlinde Krahn
Ute Scheffer
Gert Zech

Contents

Positional Astronomy, Celestial Mechanics

Space Research

Theoretical Astrophysics

Sun

Earth

Planetary System

Stars

Interstellar Matter, Gaseous Nebulae, Planetary Nebulae

Radio Sources, Quasars, Pulsars, X Ray-, Gamma Ray-Sources, Cosmic Radiation

Stellar Systems

Introduction

Astronomical bibliographies

Astronomy and Astrophysics Abstracts begins documentation and abstracting as from the year 1969. For information on astronomical literature before this date consultation of one of the following bibliographies is suggested:
(1) J. J. de Lalande, Bibliographie Astronomique, Paris 1803 (this work covers the time from 480 B. C. to the year 1803, VIII + 966 pages).
(2) J. C. Houzeau, A. Lancaster, Bibliographie générale de l'astronomie, Volume I (in two parts), Bruxelles 1882, 1887, Volume II, Bruxelles 1889. The complete title of Volume II is "Bibliographie générale de l'astronomie ou catalogue méthodique des ouvrages, des mémoires et des observations astronomiques, publiés depuis l'origine de l'imprimerie jusqu'en 1880". A new edition of these volumes was prepared by D. W. Dewhirst in 1964.
(3) Bibliography of Astronomy, 1881 - 1898. The literature of this period was recorded on standard slips by the Observatoire Royal de Belgique. From the material (some 52.000 items) a microfilm version was produced by University Microfilms Limited, Tylers Green, High Wycombe, Buckinghamshire, England, in 1970.
(4) Astronomischer Jahresbericht, 1899 gegründet von Walter Wislicenus, herausgegeben vom Astronomischen Rechen-Institut in Heidelberg (formerly in Berlin), Verlag W. de Gruyter, Berlin. For the period from 1899 to 1968 sixty-eight volumes were published, each of which, in general, covers the literature of one year.
(5) Bulletin Signalétique – Section Astronomie, Astrophysique, Physique du Globe. Published by Centre de Documentation du Centre National de la Recherche Scientifique, Paris. This publication is a continuation of "Bibliographie Mensuelle de l'Astronomie" founded in 1933 by the Société Astronomique de France. The publication is continued.
(6) Referativnyj Zhurnal. Founded in 1953 and published by Vsesoyuznyj Institut Nauchnoj i Tekhnicheskoj Informatsii, Akademiya Nauk, Moskva. The publication is continued.

Concept of Astronomy and Astrophysics Abstracts

This abstracting service aims to present a comprehensive documentation of the literature in all fields of astronomy and astrophysics. It appears in semi-annual volumes, two of which cover the literature of a calendar year. The half-yearly period of issue is regarded as an optimal period of time for summarizing papers into subject categories and for the presentation of abstracts as quickly as possible after the publication of the original literature. The time limits at which the documentation begins and ends for a volume are not sharply defined, except in the sense that all literature will be covered which was received by the editors within these limits.
Vol. 7 is devoted to the recording summarizing and indexing of astronomical publications of the year 1972 received from January 1, 1972 to August 15, 1972; it also records a number of papers issued before 1972 but received within the given period of time.

The main characteristics of the concept of Astronomy and Astrophysics Abstracts may be summarized briefly.

(1) Titles of papers are given in the language of their authors whenever possible. If they are not in English but supplied with English translations they will be given in English. Abstracts are presented in English, French or German. Titles of papers in Russian are given in English.
(2) Authors' abstracts are used whenever possible. As a rule, popular articles were not abstracted; however their titles are usually given with the notation "Popular article".
(3) As a rule, each paper has been classified into one of 108 numbered subject categories and allocated a serial number within the category. In this way each item is numbered by six figures, the first three of which indicate the number of the category. Three further figures indicate the serial number within the category, which was allocated in the order of the receipt of the abstract. Reference to an abstract in Volume 1 is indicated by "01" before the number of the category; for example, 01.074.028, denotes Volume 1, category 074, abstract 028. Vol. 2 is indicated by "02", etc., Vol. 7 by "07".
A paper may have been classified into more than one category. Then its abstract has been allocated a number in one of the categories involved, and in the other category (or categories) the paper has been indicated by the title and a reference to the abstract number.
Papers whose authors are not named were treated like those with authors' names, with one exception: reports from correspondents of journals whose names were unknown were not numbered.

(4) There are categories which suggest the presentation of the material in subject groups. For instance, a subject group may be formed by all information received on the same solar eclipse, comet, nova, etc. The unsorted presentation of such material in a subject category would be inconvenient for the user, even if the individual comet, etc. were included in the subject index.

The following subject categories are subdivided into subject groups:
008 Observatories, Institutes. The publications of observatories and astronomical institutes are listed in alphabetical order of the towns of the institutions, each town forming a numbered subject group. For each publication a reference to an abstract number is made.
010 Societies, Associations, Organizations. The publications of each one form a subject group. The groups are presented in alphabetical order.
079 Solar eclipses. All publications related to one solar eclipse form a subject group.
103 Comets: Listed Objects. All publications related to the same comet form a numbered group.
124 Novae. All publications related to one nova form a subject group.
125 Supernovae. All publications related to one supernova form a subject group.

(5) Border fields of astronomy and astrophysics have been taken into account by presenting titles of papers occasionally without abstracts. The selection of papers for inclusion has been made according to the degree of relevance to astronomical research.

1

Transliteration of the Russian alphabet

The transliteration of the Russian alphabet in use in Astronomy and Astrophysics Abstracts is presented here.

А	а	a	Р	р	r
Б	б	b	С	с	s
В	в	v	Т	т	t
Г	г	g	У	у	u
Д	д	d	Ф	ф	f
Е	е	e	Х	х	kh
Ё	ё	e	Ц	ц	ts
Ж	ж	zh	Ч	ч	ch
З	з	z	Ш	ш	sh
И	и	i	Щ	щ	shch
Й	й	j	Ъ	ъ	''
К	к	k	Ы	ы	y
Л	л	l	Ь	ь	'
М	м	m	Э	э	eh
Н	н	n	Ю	ю	yu
О	о	o	Я	я	ya
П	п	p			

This transliteration was recommended by the Abstracting Board of the International Council of Scientific Unions in 1969. It is essentially the same as the transliteration proposed by the Academy of Sciences, Moscow, and used by the Referativnyj Zhurnal (see Referativnyj Zhurnal, 51. Astronomiya, 1969 No. 1). It may be noted that the letters can be read and printed by usual data processing machines.
In the literature however the names of Russian authors can be found transliterated in different ways. We present the names in the form in which they are given in the references cited.

Sources of information

The majority of sources of information for this volume are given in section **001 Periodicals** and in section **008 Observatories, Institutes**. The term "periodical" has been used in its widest sense for publications in a sequence of undetermined duration, even if the intervals of appearance are not regular. Section 001 records 288 periodicals with their full titles and with abbreviations which are in use in Astronomy and Astrophysics Abstracts. It may be noted that the titles of the periodicals are given in their original languages, and that Russian titles have been transliterated applying the transliteration given above. Section 008 records 150 periodicals; these are publication series of observatories and astronomical institutes which have not been included in section 001. The abbreviations of the titles of the periodicals have been given so that in most cases they permit recognition of the full title without recourse to the key in section 001. The steadily growing number of periodicals makes it necessary to use more extensive abbreviations and to abandon the use of very condensed ones.
Other abstracting journals have been consulted in order to examine the degree of completeness of our service. Occasionally, in particular in Physics Abstracts, Referativnyj Zhurnal, and Bulletin Signalétique abstracts of papers were found which had not come to our attention. In such cases Astronomy and Astrophysics Abstracts cites these papers, but also gives reference to the abstracting service which acted as the source.

Classification into a scheme of subject categories

The subdivision of astronomy and its border fields into sub-ject categories is facilitated by the fact that the astronomical objects appear to be particularly well suited for the formation of categories. Sun, moon, earth, planets, comets, and meteorites, the various kinds of stars, galaxies, radio sources, quasars, and pulsars etc. suggest natural subdivisions. It may be assumed that such subdivisions can be maintained for long periods of time. Experience shows, however, that progress in research may imply changes in the classification scheme, in particular, in fields where the expansion of knowledge is explosive.
A few explanatory remarks may be in order on some of the subject categories. Section 002 includes short news notes whose titles and authors are given, but the authors of the notes have not been included in the author index. In section 003 books on astronomy and astrophysics and its border fields are listed which came to our notice from January to August 1972. References to book reviews are given if the review appeared quickly.
For completeness of documentation, personal notes (section 006) and obituaries (section 007) are listed. In section 012 (Proceedings of Colloquia, Congresses, Meetings, and Symposia) the proceedings etc. are listed with titles and editors. The individual papers are classified into their corresponding subject categories, but not included in the subject index. The main subjects of these symposia are cited in the index under section 012.
Errata to papers communicated by the authors are listed at the end of the corresponding subject categories.

Author index and subject index

The subject category and the serial number forming six figures for each abstract have been used as a means of reference in the author index and the subject index. These references are more precise than page references. They offer considerable advantages in indexing by means of data processing machines, and they are more convenient for the user.
The author index of this volume contains 6771 names. A complete reference comprises six figures, three for the subject category and three for the serial number within the category. In the case of more than one reference to abstracts in one category, the number of the category is given only once and not repeated in the immediately following references. The total number of papers (some do not give names of authors) recorded in this volume is about 6300
We consider the subject index as only a first approximation to an optimal index covering all fields of astronomy and astrophysics and their border fields. Several iterative steps appear to be necessary until an index has been compiled for one of the subsequent volumes which may then serve as a kind of standard for the near future. The assigning of one or more key words to a paper is undoubtedly a difficult task. Some journals have started giving key words together with the titles of papers. These key words are chosen by the authors themselves and are in many cases identical with our designations of subject categories with no additional specification. In fact, in some cases it may be more useful to refer to a subject category as a whole than to an item number, in particular, if the total number of abstracts in a category is very small, and if more specific key words do not provide a proper description of the paper.
While each volume is scheduled to contain an author index and a subject index, the magnetic tapes containing the index information will be used to produce separate index volumes (authors and subjects) at intervals of a few years.
The text of the publication was typed on IBM 72 Composers in the editorial office, and it was given to the printer in a form ready for offset reproduction. The author index and the subject index were compiled and printed by means of electronic computer (Siemens 2002).

Abbreviations

AAS	American Astronomical Society
AAVSO	American Association of Variable Star Observers
Abh.	Abhandlungen
Abstr.	Abstract
Abt.	Abteilung
Acad.	Academy, etc.
Accad.	Accademia
Adv.	Advances
AG	Astronomische Gesellschaft
AIAA	American Institute of Aeronautics and Astronautics
AJB	Astronomischer Jahresbericht
Akad.	Akademie
An.	Anales, etc.
Ann.	Annals, etc.
Arch.	Archiv, etc.
Ark.	Arkiv
ASA	Astronomical Society of Australia
Asoc.	Asociación
ASP	Astronomical Society of the Pacific
Ass.	Association
ASSA	Astronomical Society of Southern Africa
Astrofis.	Astrofisica, etc.
Astrofiz.	Astrofizika, etc.
Astron.	Astronomy, etc.
Astronaut.	Astronautics, etc.
Astrophys.	Astrophysics, etc.
ASV	Astronomical Society of Victoria
ASWA	Astronomical Society of Western Australia
Atmosph.	Atmosphere, etc.
BA	Bulletin Astronomique
BAA	British Astronomical Association
BAN	Bulletin of the Astronomical Institutes of the Netherlands
Ber.	Berichte
BIH	Bureau International de l'Heure (Paris)
Bol.	Boletin
Boll.	Bolletino
Bull.	Bulletin
Byull.	Byulleten' (Bulletin)
Circ.	Circular
Cl.	Classe
Coll.	Collection
Commun.	Communication
Comun.	Comunicazioni
Contr.	Contributions, etc.
COSPAR	Committee on Space Research
C.S.I.R.O.	Commonwealth Scientific Industrial Research Organization
Dep.	Department
Diss.	Dissertation
Div.	Division
Dokl.	Doklady (Reports)
ESO	European Southern Observatory
ESRO	European Space Research Organization
Fis.	Fisica, etc.
Fiz.	Fizika, etc.
Fys.	Fysica, etc.
Géod.	Géodésie, etc.
Geod.	Geodesy, etc.
Geofis.	Geofisica, etc.
Geofiz.	Geofizika, etc.
Geofys.	Geofysik, etc.
Geol.	Geology, etc.
Geogr.	Geography, etc.
Geophys.	Geophysics, etc.
Ges.	Gesellschaft
Glav.	Glavnyj (Main)
Gos.	Gosudarstvennyj (State)
HRD	Herzsprung-Russell diagram
Hydrogr.	Hydrography, etc.
IAF	International Astronautical Federation
IAU	International Astronomical Union
ICSU	International Council of Scientific Unions
IEEE	Institute of Electrical and Electronics Engineers
Industr.	Industry, etc.
Inform.	Information
Inst.	Institute, etc.
Instn.	Institution
Ionosph.	Ionosphere, etc.
Issled.	Issledovaniya (Research)
Ist.	Istituto
Izv.	Izvestiya (News)
Jb.	Jahrbuch
JO	Journal des Observateurs
Journ.	Journal
Kl.	Klasse
Lab.	Laboratory
Mag.	Magazine
Mat.	Matematica, etc.
Math.	Mathematics, etc.
Mech.	Mechanics, etc.
Med.	Mededelingen
Medd.	Meddelande, Meddelser
Mekhan.	Mekhanika, etc.
Mém.	Mémoires
Mem.	Memoirs, Memorandum, etc.
Meteorol.	Meteorology, etc.
MIT	Massachusetts Institute of Technology
Mitt.	Mitteilungen
MVS Sonneberg	Mitteilungen über Veränderliche Sterne, Sonneberg
Nachr.	Nachrichten
NASA	National Aeronautics and Space Administration
Nat.	Naturwissenschaftlich, etc.
Naut.	Nautics, etc.
NBS	National Bureau of Standards
NRAO	National Radio Astronomy Observatory (Green Bank)
NRL	Naval Research Laboratory (Washington)
Obs.	Observatory, etc.
OSA	Optical Society of America
Oss.	Osservatorio, Osservazioni, etc.
Ped.	Pedagogika, etc. (Pedagogics)
Phil.	Philosophical
Phys.	Physics, etc.
Planet.	Planetary
Priklad.	Prikladnoj (Applied)
Proc.	Proceedings
Progr.	Progress, etc.
Pubbl.	Pubblicazioni
Publ.	Publications
Rap.	Raportoj
RAS	Royal Astronomical Society
RAS Canada	Royal Astronomical Society of Canada
Rech.	Recherches
Rend.	Rendiconti

Abbreviations

Rep.	Report	Techn.	Technics, etc.
Repr.	Reprint	Tekhn.	Tekhnika, etc.
Res.	Research	Teor.	Teoreticheskij
Rev.	Review, etc.	Terr.	Terrestrial, etc.
Ric.	Ricerche	TH	Technische Hochschule
Roy.	Royal, etc.	Theor.	Theoretical
SAF	Société Astronomique de France	Tidssk.	Tidsskrift
SAI	Società Astronomica Italiana	Trans.	Transactions
SAO	Smithsonian Astrophysical Observatory	Trudy	Trudy (Publications)
SAS	Société Astronomique de Suisse	Tsentr.	Tsentral'nyj (Central)
Sci.	Science, etc.	Tsirk.	Tsirkulyar (Circular)
Sect.	Section	TU	Technical University
Ser.	Series, etc.	Uch. Zap.	Uchenye Zapiski (Treatise)
S. I. R.	Service International Rapide des Latitudes	Univ.	University, etc.
Sitz.-Ber.	Sitzungsberichte	URSI	Union Radio Scientifique Internationale
Soc.	Society	Verh.	Verhandlungen
Soobshch.	Soobshcheniya (Communications)	Veröff.	Veröffentlichungen
Sternw.	Sternwarte	Wet.	Wetenschappen
Stud. Cerc.	Studii şi Cercetari	Wiss.	Wissenschaften, etc.
Supl.	Suplemento	Zeitschr.	Zeitschrift
Suppl.	Supplement	ZfA	Zeitschrift für Astrophysik
SuW	Sterne und Weltraum	Zhurn.	Zhurnal (Journal)

Periodicals, Proceedings, Books, Activities

001 Periodicals

AAS Photo-Bull.
AAS (American Astronomical Society) Photo-Bulletin. Published by Eastman Kodak Company, Rochester, N.Y., for the Working Group on Photographic Materials.

Abh. Deutsch. Akad. Wiss. Berlin
Abhandlungen der Deutschen Akademie der Wissenschaften zu Berlin. Klasse für Mathematik, Physik und Technik. Publisher: Akademie-Verlag, Berlin.

Acad. Roy. Belgique, Bull. Cl. Sci.
Académie Royale de Belgique, Bulletin de la Classe des Sciences (Koninklijke Academie van België, Mededelingen van de Klasse der Wetenschappen). 5ᵉ Série. Palais des Académies, Bruxelles.

Acta Astron.
Acta Astronomica. Publisher: Komitet Astronomii, Polskiej Akademii Nauk, Warszawa - Kraków.

Acta Phys. Austriaca
Acta Physica Austriaca. Publisher: Springer-Verlag, Wien.

Acta Univ. Carolinae Math. Phys.
Acta Universitatis Carolinae, Mathematica et Physica. Administrace: Matematicko-fyzikální fakulta University Karlovy, Praha.

Actas Acad. Nacional Cienc. Lima
Actas de la Academia Nacional de Ciencias Exactas, Fisicas y Naturales de Lima. Lima - Peru.

Adv. Astron. Astrophys.
Advances in Astronomy and Astrophysics. Publisher: Academic Press, New York — London.

AIAA Journ.
AIAA Journal. A Publication of the American Institute of Aeronautics and Astronautics devoted to Aerospace Research and Development. Published by the American Institute of Aeronautics and Astronautics, New York, N.Y

Am. Scient.
American Scientist. Society of Sigma Xi, New Haven, Conn.

Ann. d'Astrophys.
Annales d'Astrophysique. Revue internationale bimestrielle publiée par le Centre National de la Recherche Scientifique et éditée par son Service d'Astrophysique, Paris. After Vol. 31 replaced by "Astronomy and Astrophysics".

Ann. Françaises Chronométrie Micromécanique
Annales Françaises de Chronométrie et de Micromécanique, publication annuelle de l'Observatoire de Besançon, du Centre Technique de l'Industrie Horlogère et de la Société Française de Chronométrie et de Micromécanique. Rédaction et administration: Observatoire de Besançon. Publiées avec le concours du Centre National de

la Recherche Scientifique et des organismes corporatifs.

Ann. Géophys.
Annales de Géophysique. Revue Internationale trimestrielle, publiée par le Centre National de la Recherche Scientifique, Paris.

Ann. Obs. Astron. Météorol. Toulouse
Annales de l'Observatoire Astronomique et Météorologique de Toulouse. Publisher: Gauthier-Villars, Paris.

Ann. Physics
Annals of Physics. Publisher: Academic Press Inc., New York, N.Y.

Ann. Physik
Annalen der Physik. 7. Folge. Publisher: Johann Ambrosius Barth, Leipzig.

Ann. Physique
Annales de Physique. Publisher: Masson et Cie., Paris.

Ann. Soc. Sci. Bruxelles
Annales de la Société Scientifique de Bruxelles. Série I: Sciences Mathématiques, Astronomiques et Physiques. Published by Institut de Physique, Heverlé-Louvain.

Annual Rep. Astron. Inst. Greece
Annual Reports of the Astronomical Institutes of Greece. Published by the Greek National Committee for Astronomy. Academy of Athens, Research Center for Astronomy and Applied Mathematics.

Annual Rev. Astron. Astrophys.
Annual Review of Astronomy and Astrophysics. Publisher: Annual Reviews Inc., Palo Alto, California.

Ann. Univ.-Sternw. Wien
Annalen der Universitäts-Sternwarte Wien. In Kommission bei Ferd. Dümmlers Verlag, Bonn.

Anzeiger. Österreich. Akad. Wiss. Math.-Nat. Kl.
Anzeiger. Österreichische Akademie der Wissenschaften. Mathematisch-Naturwissenschaftliche Klasse. Publisher: Springer-Verlag, Wien.

Applied Optics
Applied Optics. A monthly publication of the Optical Society of America. Published for the Optical Society of America by the American Institute of Physics, New York, N. Y.

Arch. Sci. Genève
Archives des Sciences, éditées par la Société de Physique et d'Histoire Naturelle de Genève. Publisher: Imprimerie Kundig, Genève. Subscription address: Librairie Payot, Genève.

Ark. Astron.
Arkiv för Astronomi. Utgivet av Kungliga Svenska Vetens-

kapsakademien, Stockholm. Printed by Almqvist & Wiksell, Stockholm.

Ark. Geofys.
Arkiv för Geofysik. Kungliga Svenska Vetenskapsakademien, Stockholm.Printed by Almqvist & Wiksell, Stockholm.

Artificial Satellites
Artificial Satellites. Publication of Polish Scientific Institutions. Polish Academy of Sciences, National Committee of Geophysics and Geodesy, National Committee for Space Research, Warsaw. Publishing Office: Palac Kultury i Nauki, Warszawa.

Asoc. Argentina Astron. Bol.
Asociación Argentina de Astronomía. Boletin. Editor: Instituto Argentino de Radioastronomía, Provincia de Buenos Aires, Argentina. Printer: Talleres Gráficos "Renovación", La Plata, República Argentina.

Astrofizika
Astrofizika. Izdatel'stvo Akademii Nauk Armyanskoj SSR, Erevan. [An English translation is published in "Astrophysics".]

Astrofiz. Issled. Izv. Spets. Astrofiz. Obs.
Astrofizicheskie Issledovaniya. Izvestiya Spetsial'noj Astrofizicheskoj Observatorii. Akademiya Nauk SSSR. Publishers: Izdatel'stvo "Nauka", Leningradskoe Otdelenie, Leningrad.

Astron. Astrophys.
Astronomy and Astrophysics. A European Journal. Published by Springer-Verlag, Berlin – Heidelberg– New York.

Astron. Astrophys. Suppl. Ser.
Astronomy and Astrophysics. Supplement Series. A European Journal. Published by the Astronomical Institute Lausanne and Geneva Observatory, Switzerland, on behalf of the Board of Directors.

Astronaut. Acta
Astronautica Acta. An Archive Journal of the International Academy of Astronautics. Published by Pergamon Press, New York – Oxford.

Astronaut. Aeronaut.
Astronautics & Aeronautics. A Publication of the American Institute of Aeronautics and Astronautics. Published monthly by the American Institute of Aeronautics and Astronautics, Easton, Pennsylvania.

Astron. in der Schule
Astronomie in der Schule. Zeitschrift für die Hand des Astronomielehrers. Herausgegeben vom Verlag Volk und Wissen, Berlin. Redaktion: Sternwarte Bautzen.

Astron. Journ.
The Astronomical Journal. Published for the American Astronomical Society by the American Institute of Physics, New York, N. Y. Editorial Office: Department of Astronomy, Columbia University, New York, N. Y.

Astron. Nachr.
Astronomische Nachrichten. Publisher: Akademie-Verlag, Berlin.

Astron. Soc. Pacific Leaflet
Astronomical Society of the Pacific. Leaflet. Edited by the Astronomical Society of the Pacific, San Francisco, California.

Astron. Tidssk.
Astronomisk Tidsskrift. Edited by Astronomisk Selskab, København; Norsk Astronomisk Selskap, Oslo; Svenska Astronomiska Sällskapet, Stockholm. Printed by John Griegs Boktrykkeri, Bergen.

Astron. Tsirk.
Astronomicheskij Tsirkulyar, izdavaemyj Byuro Astronomicheskikh Soobshchenij Akademii Nauk SSSR. Tipografiya Astrosoveta AN SSSR, Moskva.

Astron. Vestn.
Astronomicheskij Vestnik. Publishers: Izdatel'stvo "Nauka", Moskva.

Astron. Zhurn. Akad. Nauk SSSR
Astronomicheskij Zhurnal. Akademiya Nauk SSSR. Publishers: Izdatel'stvo "Nauka", Moskva. [An English translation is published in "Soviet Astronomy AJ"].

Astrophysics
Astrophysics. The Faraday Press cover-to-cover translation of Astrofizika. The Faraday Press, Inc., New York, N. Y.

Astrophys. Journ.
The Astrophysical Journal. Published in collaboration with the American Astronomical Society by the University of Chicago Press, Chicago, Illinois.

Astrophys. Journ. Suppl. Ser.
The Astrophysical Journal. Supplement Series. Published in collaboration with the American Astronomical Society by the University of Chicago Press, Chicago, Illinois.

Astrophys. Letters
Astrophysical Letters. An International *EXPRESS* Journal. Published monthly by Gordon and Breach Science Publishers Ltd., New York – London – Paris.

Astrophys. Norvegica
Astrophysica Norvegica. Edited by The Institute of Theoretical Astrophysics, University of Oslo (Det Norske Videnskaps-Akademi i Oslo). Universitets-forlaget, Oslo.

Astrophys. Space Sci.
Astrophysics and Space Science. An International Journal of Cosmic Physics. Published by D. Reidel Publishing Company, Dordrecht – Holland.

Atti Accad. Nazionale Lincei. Mem.
Atti della Accademia Nazionale dei Lincei. Serie Ottava. Memorie. Classe di Scienze fisiche, matematiche e naturali. Sezione I: Matematica, Meccanica, Astronomia, Geodesia e Geofisica. Published by Accademia Nazionale dei Lincei, Roma.

Atti Accad. Nazionale Lincei. Rend.
Atti della Accademia Nazionale dei Lincei. Serie Ottava. Rendiconti. Classe di Scienze fisiche, matematiche e naturali. Published by Accademia Nazionale dei Lincei, Roma.

Australian Journ. Phys.
Australian Journal of Physics. Published by the Commonwealth Scientific and Industrial Research Organization, East Melbourne, Victoria.

Australian Journ. Phys. Astrophys. Suppl.
Australian Journal of Physics, Astrophysical Supplement. Published by Commonwealth Scientific and Industrial Research Organization, East Melbourne, Victoria.

BAV Rundbrief
BAV Rundbrief. Mitteilungsblatt der Berliner Arbeitsgemeinschaft für Veränderliche Sterne. Editor: BAV Berliner Arbeitsgemeinschaft für Veränderliche Sterne eV., Berlin.

BBSAG Bull.
Bedeckungsveränderlichen Beobachter der Schweizerischen Astronomischen Gesellschaft, [Swiss Astronomical Society's Eclipsing Variable Observers], Bulletin. To be obtained from R. Diethelm, Winterthur, Switzerland.

Bol. Inst. Mat., Astron., Fis. Univ. Nacional Córdoba
Boletin del Instituto de Matematica, Astronomia y Fisica, Universidad Nacional de Córdoba (R. A.).Dirección General de Publicaciones, Córdoba (Argentina).

Bol. Liga Latinoamericana Astron.
Boletin de la Liga Latinoamericana de Astronomia. Publicado por la Asociacion Argentina Amigos de la Astronomia, Buenos Aires, Argentina.

Boll. Geod. Sci. Affini
Bolletino di Geodesia e Scienze Affini. Pubblicazione dell'Istituto Geografico Militare, Firenze.

Boundary-Layer Meteorology
Boundary-Layer Meteorology. An International Journal of Physical and Biological Processes in the Atmospheric Boundary Layer. Published by D. Reidel Publishing Company, Dordrecht—Holland.

British Astron. Ass. Circ.
British Astronomical Association, Circular. Editorial Office: 97 Hawkswood Drive, Hailsham, Sussex.

Bull. American Astron. Soc.
Bulletin of the American Astronomical Society. Published for the American Astronomical Society by the American Institute of Physics Inc., New York, N. Y.

Bull. Astron. (BA)
Bulletin Astronomique. 3e Série. Publié par le Centre National de la Recherche Scientifique, Paris. After Vol. 3 (1968) replaced by "Astronomy and Astrophysics".

Bull. Astron. Inst. Czechoslovakia (BAC)
Bulletin of the Astronomical Institutes of Czechoslovakia. Published under the auspices of the Czechoslovak Academy of Sciences by Academia, Praha. Editor: Astronomical Institutes of the Czechoslovak Academy of Sciences, Praha.

Bull. Astron. Inst. Netherlands (BAN)
Bulletin of the Astronomical Institutes of the Netherlands. Publisher: North-Holland Publishing Company, Amsterdam. After Vol. 20 replaced by "Astronomy and Astrophysics".

Bull. Géod.
Bulletin Géodésique, being the Journal of the International Association of Geodesy. Nouvelle Série. Publié par le Bureau Central de l'Association Internationale de Géodésie, Paris.

Bull. Geograph. Survey Inst.
Bulletin of the Geographical Survey Institute. Published by the Geographical Survey Institute, Ministry of Construction, Tokyo, Japan.

Bull. Obs. Astron. Beograd
Bulletin de l'Observatoire Astronomique de Béograd. Editor: Observatoire Astronomique de Béograd. Printed by Naucna delo, Béograd.

Bull. Sci. Yougoslavie
Bulletin Scientifique. Conseil des Academies des Sciences et des Arts de la RSF de Yougoslavie. Section A: Sciences Naturelles, Techniques et Médicales. Redaction et Administration: Opatička ul. 18/II, Zagreb (Yougoslavie).

Bull. Signal.
Bulletin Signalétique. Section 120: Astronomie, Physique spatiale, Géophysique. Centre de Documentation du Centre Nationale de la Recherche Scientifique, Paris.

Bull. Signal.
Bulletin Signalétique. Bibliographie des Sciences de la Terre. Section 220, Cahier A: Minéralogie, Géochimie, Géologie extraterrestre. Centre de Documentation du C.N.R.S., Paris; Département Documentation du B.R. G.M., Orléans.

Bull. Soc. Roy. Sci. Liège
Bulletin de la Société Royale des Sciences de Liège. L'Université, Liège.

Byull. Abastuman. Astrofiz. Obs.
Abastumanskaya Astrofizicheskaya Observatoriya, Gora Kanobili. Byulleten'. Akademiya Nauk Gruzinskoj SSR. Publishers: Izdatel'stvo "Metsniereba", Tbilisi.

Byull. Stantsij Optichesk. Nablyud. Iskusstv. Sputnikov Zemli
Byulleten' Stantsij Opticheskogo Nablyudeniya Iskusstvennykh Sputnikov Zemli. Published by Astronomicheskij Sovet Akademii Nauk SSSR, Moskva. Beginning with number 60 (1971) the title of the publication changed in Nablyudeniya Iskusstvennykh Nebesnykh Tel.

Canadian Journ. Phys.
Canadian Journal of Physics. Published by the National Research Council of Canada, Ottawa. Printed in Canada by the University of Toronto Press, Toronto, Ont.

Celestial Mechanics
Celestial Mechanics. An International Journal of Space Dynamics. Publishers: D. Reidel Publishing Company, Dordrecht—Holland.

Ciel et Terre
Ciel et Terre. Bulletin de la Société Belge d'Astronomie, de Météorologie et de Physique du Globe. Administration: Avenue Circulaire, 3, Bruxelles. Printed by Imprimerie R. Louis, Bruxelles.

Circ. d'Information
Circulaire d'Information. Union Astronomique Internationale. Commission des Etoiles Doubles. Address: Observatoire de Meudon, Meudon, France.

Coelum
Coelum. Periodico bimestrale per la Divulgazione dell' Astronomia. Editor: Osservatorio Astronomico Universitario di Bologna.

Comments Astrophys. Space Phys.
Comments on Astrophysics and Space Physics. A Journal of Critical Discussion of the Current Literature. Publishers: Gordon and Breach Science Publishers, Inc., New York – London.

Comptes Rendus Acad. Bulg. Sci.
Comptes Rendus de l'Académie bulgare des Sciences. (Doklady Bolgarskoj Akademii Nauk). Sofia.

Comptes Rendus Acad. Sci. Paris
Comptes Rendus hebdomadaires des Séances de l'Académie des Sciences, publié avec le concours du Centre National de la Recherche Scientifique. Imprimerie: Gauthier-Villars, Paris.

Contr. Atmosph. Phys.
Contributions to Atmospheric Physics – Beiträge zur Physik der Atmosphäre. Publisher: Friedrich Vieweg & Sohn, Braunschweig.

Cosmic Electrodynamics
Cosmic Electrodynamics. An International Journal devoted to Geophysical and Astrophysical Plasmas. Printed in The Netherlands by D. Reidel Publishing Company, Dordrecht–Holland.

COSPAR Inform. Bull.
COSPAR. Information Bulletin. Address: COSPAR Secretariat, Paris.

Deutsche Geod. Kommission Bayer. Akad. Wiss.
Deutsche Geodätische Kommission bei der Bayerischen Akademie der Wissenschaften. Reihe A: Höhere Geodäsie; Reihe B: Angewandte Geodäsie; Reihe C: Dissertationen; Reihe D: Tafelwerke; Reihe E: Geschichte und Entwicklung der Geodäsie. Published by Verlag der Bayerischen Akademie der Wissenschaften, München.

Documentat. Observateurs
Documentation des Observateurs. Rédaction: Station d'Astrophysique de Forcalquier.

Documentat. Observateurs Circ.
Documentation des Observateurs. Circulaire. Rédaction: Station d'Astrophysique de Forcalquier.

Dokl. Akad. Nauk
Doklady Akademii Nauk SSSR. Seriya Matematika, Fizika. Publishers: Izdatel'stvo "Nauka", Moskva.

Dunsink Obs. Publ.
Dunsink Observatory Publications. The Observatory of the School of Cosmic Physics, Dublin Institute for Advanced Studies, Dublin.

Earth Extraterr. Sci.
Earth and Extraterrestrial Sciences. Published by Gordon and Breach Science Publishers, London.

Earth Planet. Sci. Letters
Earth and Planetary Science Letters. A Letter Journal devoted to the Development in Time of the Earth and Planetary System. Publisher: North-Holland Publishing Company, Amsterdam.

El Universo
El Universo. Organo de la Sociedad Astronomica de Mexico, Mexico, D. F.

Endeavour
Endeavour. A review of the progress of science, published in four languages by Imperial Chemical Industries Limited, London.

ESO Bull.
European Southern Observatory, Bulletin. Edited by European Southern Observatory. Office of the Director: Hamburg.

Fortschritte Phys.
Fortschritte der Physik. Publisher: Akademie-Verlag, Berlin.

Gaz. Astron. Mém.
Gazette Astronomique. Mémoires van het Sterrenkundig Genootschap van Antwerpen, (de la Société d'Astronomie d'Anvers), Antwerpen. Printer: «De Voorzorg», A. Van Leuvenhaege, Antwerpen.

Geochim. Cosmochim. Acta
Geochimica et Cosmochimica Acta. Journal of the Geochemical Society. Publishing House: Pergamon Press, Ltd., Oxford.

Geodezja Kartografia
Geodezja i Kartografia. Komitet Geodezji Polskiej Akademii Nauk. Publisher: Państwowe Wydawnictwo Naukowe, Warszawa.

Geomagn. Aeronom.
Geomagnetizm i Aehronomiya. Akademiya Nauk SSSR. Izdatel'stvo "Nauka", Moskva [An English translation is published in "Geomagnetism and Aeronomy".]

Geophys. Journ.
The Geophysical Journal of the Royal Astronomical Society. Published for the Royal Astronomical Society by Blackwell Scientific Publications, Oxford – Edinburgh.

Gerlands Beiträge Geophys.
Gerlands Beiträge zur Geophysik. Publisher: Akademische Verlagsgesellschaft Geest & Portig K.-G., Leipzig.

Glasnik Mat.
Glasnik Matematicki. Published by the Society of Mathematicians and Physicists of the S. R. of Croatia. Publisher: Drustvo Matematicara i Fizicara S. R. Hrvatske, Zagreb.

Helvetica Phys. Acta
Helvetica Physica Acta. Schweizerische Physikalische Gesellschaft. Publisher: E. Birkhäuser, Basel.

Hemel en Dampkring
Maandblad van de Nederlandse Vereniging voor Weer-en Sterrenkunde en van de Vereniging voor Sterrenkunde, Meteorologie, Geophysica en Aanverwante Wetenschappen in Belgie. Publisher: Wolters-Noordhoff N. V., Groningen.

IAU Circ.
International Astronomical Union, Circular. Central Bureau for Astronomical Telegrams, Smithsonian Astrophysical Observatory, Cambridge, Mass.

IBM Journ. Res. Development
IBM Journal of Research and Development. Published bimonthly by International Business Machines Corporation, Armonk, New York.

Icarus
Icarus. International Journal of Solar System Studies. Publisher: Academic Press, New York – London.

ICSU Bull.
ICSU Bulletin. International Council of Scientific Unions. Secretariat: 7, Via Cornelio Celso, Rome, Italy.

IEEE Spectrum
IEEE Spectrum. Published monthly by the Institute of Electrical and Electronics Engineers, Inc., New York, N. Y.

Inform. Bull. Southern Hemisphere
Information Bulletin of the Southern Hemisphere. Editorial Office: Observatorio Astronómico, La Plata, Argentina.

Inform. Bull. Variable Stars
Commission 27 of the I.A.U. Information Bulletin on Variable Stars. Konkoly Observatory, Budapest.

Infrared Physics
An International Research Journal. Publisher: Pergamon Press Ltd., Oxford – London – New York.

International Journ. Theor. Phys.
International Journal of Theoretical Physics. Publisher: Plenum Publishing Company, Donington House, London.

Irish Astron. Journ.
The Irish Astronomical Journal. A Quarterly Publication under the auspices of the Observatories of Armagh and Dunsink. Subscription address: Managing Editor, Irish Astronomical Journal, Armagh Observatory, Northern Ireland.

Izv. Akad. Nauk Armyan. SSR
Izvestiya Akademii Nauk Armyanskoj SSR. Fizika. Publisher: Izdatel'stvo AN Armyanskoj SSR, Erevan.

Izv. Glav. Astron. Obs. Pulkovo
Izvestiya Glavnoj Astronomicheskoj Observatorii v Pulkove. Akademiya Nauk SSSR. Izdanie Glavnoj astronomicheskoj observatorii v Pulkove, Leningrad.

Izv. Komissii Fiz. Planet
Izvestiya Komissii po Fizike Planet. Akademiya Nauk SSSR. Astronomicheskij Sovet. Moskva.

Izv. Krymskoj Astrofiz. Obs.
Izvestiya Krymskoj Astrofizicheskoj Observatorii. Akademiya Nauk SSR. Publishers: Izdatel'stvo "Nauka", Moskva.

Jenaer Rundschau (Jena Review)
Jenaer Rundschau (Jena Review). Publisher: VEB Verlag Technik, Berlin.

JETP Letters
JETP Letters. A translation of JETP Pis'ma v Redaktsiyu of the Academy of Sciences in the USSR. Published semimonthly by the American Institute of Physics, Lancaster, Pennsylvania.

Journ. Astronaut. Sci.
The Journal of the Astronautical Sciences. Published by the American Astronautical Society Inc., Baltimore, Md.

Journ. Astron. Soc. Victoria
The Journal of the Astronomical Society of Victoria.

Printed by D. Buscombe Printers, Glen Waverley, Victoria.

Journ. Astron. Soc. Western Australia
The Journal of the Astronomical Society of Western Australia. Edited by the Astronomical Society of Western Australia, Perth, W. A.

Journ. Atmosph. Sci.
Journal of the Atmospheric Sciences. Published by the American Meteorological Society, Boston, Mass.

Journ. Atmosph. Terr. Phys.
Journal of Atmospheric and Terrestrial Physics. Publishers: Pergamon Press, Oxford – London – New York.

Journ. British Astron. Ass.
Journal of the British Astronomical Association. Subscription address: British Astronomical Association, Burlington House, Piccadilly, London.

Journ. British Interplanet. Soc.
Journal of the British Interplanetary Society. Printed in Great Britain by Unwin Brothers Ltd., The Gresham Press, Old Woking, Surrey, and published by The British Interplanetary Society, London.

Journ. Fluid Mechanics
Journal of Fluid Mechanics. Published by Cambridge University Press, London – New York.

Journ. Geophys. Res.
Journal of Geophysical Research. An International Scientific Publication. Published three times a month by the American Geophysical Union, Washington, D. C. First section: Space Physics; Second section: Physics and chemistry of the solid earth, planetology, geodesy; Third section: Oceans and atmospheres.

Journ. History Astron.
Journal for the History of Astronomy. Publisher: Science History Publications Ltd., Cambridge, England. American Representative: Neale Watson Academic Publications, Inc., New York City, U.S.A.

Journ. Math. Phys.
Journal of Mathematical Physics. Published by the American Institute of Physics, New York, N. Y.

Journ. Navigation
The Journal of Navigation. Published quarterly by The Royal Institute of Navigation at the Royal Geographical Society, London.

Journ. Optical Soc. America
Journal of the Optical Society of America. Publisher: American Institute of Physics, New York.

Journ. Phys. A. General Phys.
Journal of Physics A. General Physics. Europhysics Journal. Published by the Institute of Physics and the Physical Society, London, England, in association with the American Institute of Physics, New York.

Journ. Physique
Journal de Physique. Publication de la Société Française de Physique, Paris.

Journ. Plasma Phys.
Journal of Plasma Physics. Publishers: Cambridge University Press, London.

Journ. Proc. Roy. Soc. New South Wales
Journal and Proceedings of the Royal Society of New South Wales. Published by the Society, Science House, Sydney.

Journ. Quant. Spectrosc. Radiat. Transfer
Journal of Quantitative Spectroscopy & Radiative Transfer. Publisher: Pergamon Press, Oxford — New York.

Journ. Roy. Astron. Soc. Canada
The Journal of the Royal Astronomical Society of Canada, devoted to the advancement of astronomy and allied sciences. Printed by the University of Toronto Press, Toronto, Ontario, Canada.

Kometn. Tsirk. *Kiev*
Kometnyj Tsirkulyar. Gruppa po Issledovaniyu Komet Astrosoveta i Mezhduvedomstvennyj Geofizicheskij Komitet Akademii Nauk SSSR. Kievskij Universitet im. T. G. Shevchenko.

Komety i Meteory
Komety i Meteory. Akademiya Nauk Tadzhikskoj SSR. Astronomicheskij Sovet Akademii Nauk SSSR. Publishers: Izdatel'stvo "Donish", Dushanbe.

Kosmich. Issled.
Kosmicheskie Issledovaniya. Akademiya Nauk SSSR. Publishers: Izdatel'stvo "Nauka", Moskva.

Kozmos
Kozmos. Popular Astronomical Journal of the Slovak Central Observatory in Hurbanovo. Publisher: Slovenská ústredná hvezdáren v Hurbanove.

L'Astronomie
L'Astronomie et Bulletin de la Société Astronomique de France. Revue mensuelle. Rédaction: Société Astronomique de France, Paris.

L'Universo
L'Universo. Rivista dell'Instituto Geografico Militare. Direzione, Redazione e Amministrazione: Istituto Geografico Militare, Firenze.

Magnitnye Polya Solnech. Pyaten
Magnitnye Polya Solnechnykh Pyaten. (Supplements to Solnechnye Dannye. Byulleten' (*Solar Data*)). Publishers: Izdatel'stvo "Nauka", Leningrad.

Math. Rev.
Mathematical Reviews. Published by the American Mathematical Society, Providence, R. I.

Mem. Fac. Sci. Kyoto Univ.
Memoirs of the Faculty of Science, Kyoto University. Series of Physics, Astrophysics, Geophysics, and Chemistry. Printed by Yamashiro Printing Publishing Co. Ltd., Kamigyo, Kyoto.

Mem. Roy. Astron. Soc.
Memoirs of the Royal Astronomical Society. Published for the Royal Astronomical Society by Blackwell Scientific Publications, Oxford — Edinburgh.

Mem. Soc. Astron. Italiana
Memorie della Società Astronomica Italiana. Nuova Serie. Pubblicate sotto gli auspici del Consiglio Nazionale dell Ricerche. Publisher: Tipografia Baccini & Chiappi, Firenze.

Mercury
Mercury. The Journal of the Astronomical Society of the Pacific. Published by the Astronomical Society of the Pacific, San Francisco, California.

Messtechnik
Messtechnik (Zeitschrift für Instrumentenkunde). Publishers: Verlag Friedrich Vieweg & Sohn GmbH, Braunschweig.

Meteoritics
Meteoritics. The Journal of the Meteoritical Society. Published quarterly by The Meteoritical Society and Arizona State University Bureau of Publications. Editorial address: Center for Meteorite Studies, The Arizona State University, Tempe, Arizona.

Meteoritika
Akademiya Nauk SSSR. Komitet po Meteoritam. Publishers: Izdatel'stvo "Nauka", Moskva.

Mitt. Astron. Ges.
Mitteilungen der Astronomischen Gesellschaft, Hamburg. Printed by G. Braun, GmbH, Karlsruhe.

Monatsber. Deutsch. Akad. Wiss. Berlin
Monatsberichte der Deutschen Akademie der Wissenschaften zu Berlin. Mitteilungen aus Mathematik, Naturwissenschaft, Medizin und Technik. Publisher: Akademie-Verlag, Berlin.

Monthly Notes Astron. Soc. Southern Africa
Monthly Notes of the Royal Astronomical Society of Southern Africa. Published by the Astronomical Society of Southern Africa, Royal Observatory, Cape Province, South Africa.

Monthly Notes International Polar Motion Service
Monthly Notes of the International Polar Motion Service. Published by the Central Bureau, International Latitude Observatory of Mizusawa, Mizusawa-shi, Iwate-ken, Japan.

Monthly Notices Roy. Astron. Soc.
Monthly Notices of the Royal Astronomical Society. Published for the Royal Astronomical Society by Blackwell Scientific Publications, Oxford — Edinburgh.

Moon
The Moon. An International Journal of Lunar Studies. Publisher: D. Reidel Publishing Company, Dordrecht — Holland.

MVS Sonneberg
Mitteilungen über Veränderliche Sterne. Edited by Sternwarte Sonneberg (Zentralinstitut für Astrophysik, Bereich Sternphysik) der Deutschen Akademie der Wissenschaften zu Berlin.

Nablyud. Iskusstv. Nebesn. Tel
Nablyudeniya Iskusstvennykh Nebesnykh Tel. Published by Astronomicheskij Sovet Akademii Nauk SSSR, Moskva.

Nachr. Akad. Wiss. Göttingen
Nachrichten der Akademie der Wissenschaften in Göttingen. II. Mathematisch-Physikalische Klasse. Vandenhoeck & Ruprecht, Göttingen.

Nachr. Karten-, Vermessungswesen
Nachrichten aus dem Karten- und Vermessungswesen. Editor: Institut für Angewandte Geodäsie (Abt. II des

Deutschen Geodätischen Forschungsinstituts). Published by Verlag des Instituts für Angewandte Geodäsie, Frankfurt a. M.

Nature
Nature. Editorial and Publishing Offices: Macmillan Journals Limited, 4 Little Essex Street, London; 711 National Press Building, Washington, D. C.

Nature, Phys. Sci.
Nature, Physical Science. Editorial and Publishing Offices: Macmillan Journals Limited, London – Washington.

Naturwissenschaften
Die Naturwissenschaften. Publisher: Springer-Verlag, Berlin – Heidelberg – New York.

Nauchn. Informatsii
Nauchnye Informatsii. Astronomicheskij Sovet Akademii Nauk SSSR, Moskva.

Nuovo Cimento
Il Nuovo Cimento. Rivista Internazionale e Organo della Società Italiana di Fisica, Series A, B. Publisher: Nicola Zanichelli, Editore, Bologna.

Nuovo Cimento Lettere
Lettere al Nuovo Cimento, a Cura della Società Italiana di Fisica. Editrice Compositori, Bologna.

Nuovo Cimento Rivista
Rivista del Nuovo Cimento a cura della Società Italiana di Fisica. Editrice Compositori, Bologna.

Nuovo Cimento Suppl.
Supplemento al Nuovo Cimento. Publisher: Nicola Zanichelli, Editore, Bologna.

Observations Artificial Earth Satellites
Observations of Artificial Satellites of the Earth (Nablyudeniya Iskusstvennykh Sputnikov Zemli). Magyar Tudományos Akadémia Csillagvizsgáló Intézete, Budapest.

Observatory
The Observatory. A Review of Astronomy. Publishers: The Editors of "The Observatory", Royal Greenwich Observatory, Herstmonceaux Castle, Hailsham, Sussex, England.

Optik
Optik. Zeitschrift für das gesamte Gebiet der Licht- und Elektronenoptik. Publishers: Wissenschaftliche Verlagsgesellschaft mbH., Stuttgart.

Orion Schaffhausen
Orion. Zeitschrift der Schweizerischen Astronomischen Gesellschaft (SAG). Bulletin de la Société Astronomique de Suisse (SAS). Administration: Generalsekretariat der SAG, Schaffhausen.

Österreich. Zeitschr. Vermessungswesen
Österreichische Zeitschrift für Vermessungswesen. Editor and Publisher: Österreichischer Verein für Vermessungswesen, Wien.

Peremennye Zvezdy, Byull.
Peremennye Zvezdy, Byulleten', izdavaemyj Astronomicheskim Sovetom Akademii Nauk SSSR. Published by Astronomicheskij Sovet Akademii Nauk SSSR, Moskva.

Peremennye Zvezdy, Prilozhenie
Peremennye Zvezdy, Prilozhenie (The Variable Stars, Supplement). Astronomicheskij Sovet Akademii Nauk SSSR, Moskva.

Phil. Mag.
The Philosophical Magazine. A Journal of Theoretical, Experimental and Applied Physics. Eighth Series. Publisher: Taylor & Francis, Ltd., London.

Phil. Trans. Roy. Soc. London
Philosophical Transactions of the Royal Society of London. Series A, Mathematical and Physical Sciences. Published by the Royal Society, London.

Phys. Abstr.
Physics Abstracts. Science Abstracts, Series A. An INSPEC Publication, published by The Institution of Electrical Engineers, London.

Phys. Ber.
Physikalische Berichte. Herausgegeben von der Deutschen Physikalischen Gesellschaft e. V.und von der Deutschen Akademie der Wissenschaften zu Berlin. Friedrich Vieweg & Sohn, Braunschweig.

Phys. Blätter
Physikalische Blätter. Physik-Verlag, Mosbach/Baden.

Phys. Bull.
Physics Bulletin. Published by the Institute of Physics and the Physical Society, London, England.

Phys. Earth Planet. Interiors
Physics of the Earth and Planetary Interiors. A journal devoted to observational and experimental studies of the Earth and Planetary interiors and their theoretical interpretation by the physical sciences. Publisher: North-Holland Publishing Company, Amsterdam, Netherlands.

Phys. Fluids
The Physics of Fluids. Published by the American Institute of Physics, New York, N.Y.

Phys. Letters
Physics Letters. Volumes A and B. Publisher: North-Holland Publishing Company, Amsterdam.

Phys. Rev. A
Physical Review A, General Physics. Published for the American Physical Society by the American Institute of Physics, Lancaster, Pa., and New York, N.Y.

Phys. Rev. B
Physical Review B, Solid State. Published for the American Physical Society by the American Institute of Physics, Lancaster, Pa., and New York, N. Y.

Phys. Rev. C
Physical Review C, Nuclear Physics. Published for the American Physical Society by the American Institute of Physics, Lancaster, Pa., and New York, N.Y.

Phys. Rev. D
Physical Review D, Particles and Fields. Published for the American Physical Society by the American Institute of Physics, Lancaster, Pa., and New York, N.Y.

Phys. Rev. Letters
Physical Review Letters. Published weekly by The Amer-

ican Physical Society, New York, N. Y.

Phys. Today
Physics Today. Published by the American Institute of Physics, New York, N.Y.

Physica
Physica. Publishers: North-Holland Publishing Company, Amsterdam, The Netherlands, on request of the Foundation "Physica", Utrecht.

Planet. Space Sci.
Planetary and Space Science. Pergamon Press, Oxford – London – New York.

Plasma Physics
Plasma Physics. Publisher: Pergamon Press, Oxford, England.

Pokroky
Pokroky matematiky, fyziky a astronomie. Editor: Jednota čs. matematiků a fyziků. Publisher: Academia, Praha.

Postępy Astron.
Postępy Astronomii. Czasopismo Poświecone Upowszechnianiu Wiedzy Astronomicznej. Polskie Towarzystwo Astronomiczne, Warszawa. Printed in Poland by Pánstwowe Wydawnictwo Naukowe, Lódź.

Priroda
Priroda. Publishers: Izdatel'stvo "Nauka", Moskva.

Proc. Astron. Soc. Australia
Proceedings of the Astronomical Society of Australia. Published for the Society by Sydney University Press, Sydney.

Proc. Cambridge Phil. Soc.
Proceedings of the Cambridge Philosophical Society (Mathematical and Physical Sciences). Publishers: Cambridge University Press, London.

Proc. IEEE
Proceedings of the IEEE. Published monthly by the Institute of Electrical and Electronics Engineers, Inc., New York, N. Y.

Proc. Koninkl. Nederl. Akad. Wet.
Koninklijke Nederlandse Akademie van Wetenschappen. Proceedings. Series B, Physical Sciences. Publishers: North-Holland Publishing Company, Amsterdam.

Proc. National Acad. Sci. U.S.A.
Proceedings of the National Academy of Sciences of the United States of America. Published monthly by the National Academy of Sciences, Washington, D.C.

Proc. Roy. Irish Acad.
Proceedings of the Royal Irish Academy, Section A: Mathematical, Astronomical and Physical Science. Published by the Royal Irish Academy, Dublin.

Proc. Roy. Soc. London
Proceedings of the Royal Society of London. Series A: Mathematical and Physical Sciences. Published by the Royal Society, London.

Progr. Theor. Phys. Japan
Progress of Theoretical Physics. Published for the Research Institute for Fundamental Physics and the Physic-

al Society of Japan. Publication Office: Progress of Theoretical Physics, Yukawa Hall, Kyoto University, Kyoto, Japan.

Progr. Theor. Phys. Suppl.
Supplement of the Progress of Theoretical Physics. Published for the Research Institute for Fundamental Physics and The Physical Society of Japan. Publication Office: Progress of Theoretical Physics, Yukawa Hall, Kyoto University, Kyoto, Japan.

PTB Mitt.
PTB Mitteilungen. Amts- und Mitteilungsblatt der Physikalisch-Technischen Bundesanstalt, Braunschweig — Berlin.

Publ. Astron. Soc. Japan
Publications of the Astronomical Society of Japan. Published by the Astronomical Society of Japan. Office of the Society: Tokyo Astronomical Observatory, Mitaka, Tokyo. Agent: Maruzen Co. Ltd. (Export Department), Nihonbashi, Tokyo, Japan.

Publ. Astron. Soc. Pacific
Publications of the Astronomical Society of the Pacific. Published in Provo, Utah, by the Astronomical Society of the Pacific, San Francisco, California. Printed by Brigham Young University Press, Provo, Utah.

Publ. Roy. Obs. Edinburgh
Publications of the Royal Observatory, Edinburgh. Published by The Royal Observatory, Edinburgh, Scotland.

Publ. Tartu Astrofiz. Obs.
W. Struve nimelise Tartu Astrofüüsika Observatooriumi, Publikatsioonid. Eesti NSV Teaduste Akadeemia, Tartu.

Quarterly Journ. Roy. Astron. Soc.
Quarterly Journal of the Royal Astronomical Society. Published for the Royal Astronomical Society by Blackwell Scientific Publications, Oxford.

Radio Sci.
Radio Science. Published by the American Geophysical Union, Richmond, Virginia.

Referativ. Zhurn. 51. Astron.
Referativnyj Zhurnal. 51. Astronomiya. Vsesoyuznyj Institut Nachnoj i Tekhnicheskoj Informatsii. Moskva.

Referativ. Zhurn. 52. Geod. i Aehros"emka
Referativnyj Zhurnal. 52. Geodeziya i Aehros"emka. Vsesoyuznyj Institut Nauchnoj i Tekhnicheskoj Informatsii. Moskva.

Referativ. Zhurn. 62. Issled. kosm. prostranstva
Referativnyj Zhurnal. 62. Issledovanie Kosmicheskogo Prostranstva. Vsesoyuznyj Institut Nauchnoj i Tekhnicheskoj Informatsii. Moskva.

Rep. Progr. Phys.
Reports on Progress in Physics. Published by The Institute of Physics and the Physical Society, London.

Rev. Geophys. Space Phys.
Reviews of Geophysics and Space Physics (formerly Reviews of Geophysics). Published by the American Geophysical Union, Richmond, Virginia.

Revista Astron.
Revista Astronomica. Organo de la Asociación Argentina Amigos de la Astronomia, Buenos Aires.

Rev. Modern Phys.
Reviews of Modern Physics. Published for The American Physical Society by the American Institute of Physics, Lancaster, Pa., and New York, N.Y.

Rev. Sci. Instruments
Reviews of Scientific Instruments. Published by the American Institute of Physics, Lancaster, Pa., and New York, N.Y.

Rezul'taty Nablyud. Sovet. Iskusstv. Sputnikov Zemli
Rezul'taty Nablyudenij Sovetskikh Iskusstvennykh Sputnikov Zemli. Published by Astronomicheskij Sovet Akademii Nauk SSSR, Moskva.

Ric. Astron.
Ricerche Astronomiche. Specola Vaticana, Città del Vaticano.

Ric. Sci.
La Ricerca Scientifica. Serie Seconda. Rivista del Consiglio Nazionale delle Ricerche. Consiglio Nazionale delle Ricerche, Roma.

Ric. Spettrosc.
Ricerche Spettroscopiche. Laboratorio Astrofisico della Specola Vaticana. Specola Vaticana, Città del Vaticano.

Říše hvězd
Říše hvězd. Czechoslovak popular astronomical journal. Publisher: Orbis, Praha.

Roy. Astron. Soc. New Zealand Circ.
Royal Astronomical Society of New Zealand, Variable Star Section, Circular. Publication Office: Greerton, Tauranga, New Zealand.

Roy. Astron. Soc. New Zealand Variable Star Sect. Repr.
Royal Astronomical Society of New Zealand, Variable Star Section, Reprint. Publication Office: Greerton, Tauranga, New Zealand.

Rumanian Sci. Abstr.
Rumanian Scientific Abstracts. Natural Sciences. Publishers: The Scientific Documentation Centre of the Academy of the Socialist Republic of Romania, Bucureşti.

Sci. American
Scientific American. Published monthly by Scientific American, Inc., New York, N.Y.

Science
Science. American Association for the Advancement of Science, Washington, D.C.

Sci. Progr. Découverte
Science Progrès Découverte (formerly Science Progrès, La Nature). Revue publiée avec la participation du Palais de la Découverte. Published by Dunod, Editeur, Paris. Imprimerie Bayeusaine, Bayeux.

Sci. Rep. Tôhoku Univ.
The Science Reports of the Tôhoku University. First Series (Physics, Chemistry, Astronomy). Published by the Faculty of Science, Tôhoku University, Sendai, Japan.

Sitz.-Ber. Bayer. Akad. Wiss.
Bayerische Akademie der Wissenschaften. Mathematisch-Naturwissenschaftliche Klasse. Sitzungsberichte. Publisher: Verlag der Bayerischen Akademie der Wissenschaften, München.

Sitz.-Ber. Deutsch. Akad. Wiss. Berlin
Sitzungsberichte der Deutschen Akademie der Wissenschaften zu Berlin. Klasse für Mathematik, Physik und Technik. Publisher: Akademie-Verlag, Berlin.

Sitz.-Ber. Heidelberger Akad. Wiss.
Sitzungsberichte der Heidelberger Akademie der Wissenschaften. Mathematisch-Naturwissenschaftliche Klasse. Publisher: Springer-Verlag, Heidelberg.

Sitz.-Ber. Österreich. Akad. Wiss.
Sitzungsberichte. Österreichische Akademie der Wissenschaften. Mathematisch-Naturwissenschaftliche Klasse. Abteilung II: Mathematik, Astronomie, Meteorologie und Technik. Publisher: Springer-Verlag, Wien.

Sky Telescope
Sky and Telescope. Published by Sky Publishing Corporation, Cambridge, Mass.

Smithsonian Contr. Astrophys.
Smithsonian Contributions to Astrophysics. Smithsonian Institution Astrophysical Observatory, Cambridge, Mass. Printed by Smithsonian Institution Press, City of Washington. For sale by the Superintendent of Documents, U. S. Government Printing Office, Washington, D. C.

Smithsonian Year
Smithsonian Year. Annual Report of the Smithsonian Institution, including the financial report of the Executive Committee of the Boards of Regents. Published by the Smithsonian Institution, Washington, D.C.

Solar Physics
Solar Physics. A Journal for Solar Research and the Study of Solar Terrestrial Physics. Publishers: D. Reidel Publishing Company, Dordrecht–Holland.

Solnechnye Dannye Byull.
Solnechnye Dannye. Byulleten'. *(Solar Data)*. Publishers: Izdatel'stvo "Nauka", Leningradskoe Otdelenie, Leningrad.

Soobshch. Byurakan. Obs.
Soobshcheniya Byurakanskoj Observatorii. Akademiya Nauk Armyanskoj SSR, Erevan.

Soobshch. Gos. Astron. Inst. Shternberg
Soobshcheniya Gosudarstvennogo Astronomicheskogo Instituta im P.K. Shternberga. Publishers: Izdatel'stvo Moskovskogo Universiteta, Moskva.

Southern Stars
Southern Stars. The Journal of the Royal Astronomical Society of New Zealand (Inc.). Address of the Society: P.O. Box 3181, Wellington C1, New Zealand.

Soviet Astron. AJ
Soviet Astronomy AJ. A translation of the Astronomical Journal of the Academy of Sciences of the USSR. Published by the American Institute of Physics, Inc., New York, N.Y.

Spaceflight
Spaceflight. A Publication of the British Interplanetary

Society. Printed by Eyre & Spottiswoode Limited at Grosvenor Press, Portsmouth, and published by the British Interplanetary Society, London.

Space Science Rev.
Space Science Reviews. Publishers: D. Reidel Publishing Company, Dordrecht–Holland.

Springer Tracts Modern Phys.
Springer Tracts in Modern Physics. (Ergebnisse der exakten Naturwissenschaften). Springer-Verlag, Berlin–Heidelberg–New York.

Sterne
Die Sterne. Zeitschrift für alle Gebiete der Himmelskunde. Johann Ambrosius Barth, Leipzig.

Sternenbote
Sternenbote. Monatsschrift für Österreichs Amateurastronomen. Publisher: Astronomisches Büro, Hermann Mucke, Wien.

Stockholms Obs. Ann.
Stockholms Observatoriums Annaler. Printed by Almquist & Wiksell, Stockholm.

Strolling Astronomer
The Strolling Astronomer. The Journal of The Association of Lunar and Planetary Observers, Publication Office: The Strolling Astronomer, Box 3 AZ, University Park, New Mexico.

Stud. Cerc. Astron.
Studii şi Cercetări de Astronomie. Editura Academiei Republicii Socialiste România. Editorial Office: Observatorul Astronomic, Bucureşti.

Stud. Geophys. Geod.
Studia geophysica et geodaetica. Published for the Geophysical Institute of the Czechoslovak Academy of Sciences by Academia, Praha.

Stud. Soc. Sci. Torunensis
Studia Societatis Scientiarum Torunensis, Toruń–Polonia. Sectio F (Astronomia).

Stud. Univ. Babeş-Bolyai
Studia Universitatis Babeş-Bolyai. Series Mathematica-Physica. Publishers: Intreprinderea Poligrafica, Cluj.

SuW
Sterne und Weltraum. Astronomische Monatsschrift. Publisher: Verlag Sterne und Weltraum Dr. Vehrenberg, Düsseldorf, Germany.

Tellus
Tellus, a bi-monthly Journal of Geophysics. Svenska Geofysiska Foreningen. Printed in Sweden by Almqvist & Wiksells Boktryckeri AB, Uppsala.

Trans. Astron. Obs. Yale Univ.
Transactions of the Astronomical Observatory of Yale University. Published by the Observatory, New Haven.

Trans. Roy. Soc. Canada
Transactions of the Royal Society of Canada. Published by the Royal Society of Canada, National Research Building, Ottawa.

Trudy Astrofiz. Inst. Alma-Ata
Trudy Astrofizicheskogo Instituta, Alma-Ata. Akademiya

Nauk Kazakhskoj SSR. Publishers: Izdatel'stvo "Nauka" Kazakhskoj SSR, Alma-Ata.

Trudy Glav. Astron. Obs. Pulkovo
Trudy Glavnoj Astronomicheskoj Observatorii v Pulkove. Akademiya Nauk SSSR. Izdanie Glavnoj astronomicheskoj observatorii v Pulkove, Leningrad.

Trudy Inst. Teor. Astron.,Leningrad
Trudy Instituta Teoreticheskoj Astronomii. Akademiya Nauk SSSR. Publishers: Izdatel'stvo "Nauka", Leningrad.

Trudy Tashkent. Astron. Obs.
Trudy Tashkentskoj Astronomicheskoj Observatorii. Akademiya Nauk Uzbekskoj SSR. Publishers: Izdatel'stvo "FAN" Uzbekskoj SSR, Tashkent.

Tsirk. Astron. Inst. Tashkent
Tsirkulyar Astronomicheskogo Instituta. Akademiya Nauk Uzbekskoj SSR. Izdatel'stvo "FAN" Uzbekskoj SSR, Tashkent.

Tsirk. Astron. Obs. L'vov
Tsirkulyar. Astronomicheskaya Observatoriya. L'vovskij Ordena Lenina Gosudarstvennyj Universitet imeni Ivana Franko. Publisher: Izdatel'stvo L'vovskogo Universiteta, L'vov.

Umschau
Umschau in Wissenschaft und Technik. Umschau-Verlag Frankfurt a. M.

Urania Barcelona
Urania. Revista de Astronomia y Ciencias Afines. Organo de la Sociedad Astronómica de España y América, Barcelona; Unión Nacional de Astronomia y Ciencias Afines, Madrid.

Urania Kraków
Urania. Miesiecznik Polskiego Towarzystwa Miłośników Astronomii, Kraków. Publisher: Krakowska Drukarnia Prasowa, Kraków.

Vasiona
Vasiona. Revue d'Astronomie et d'Astronautique. Bulletin de la Société Astronomique "R. Bosković", Beograd.

VdS Nachrichtenblatt
Nachrichtenblatt der Vereinigung der Sternfreunde e.V. After Vol. 18, No. 3 published in combination with "Sterne und Weltraum". Verlag Sterne und Weltraum Dr. Vehrenberg, Düsseldorf, Germany.

Veröff. Astron. Rechen-Inst. Heidelberg
Veröffentlichungen des Astronomischen Rechen-Instituts Heidelberg. Verlag G. Braun, Karlsruhe.

Veröff. Sternw. Sonneberg
Deutsche Akademie der Wissenschaften zu Berlin. Institut für Sternphysik. Veröffentlichungen der Sternwarte in Sonneberg. Publisher: Akademie-Verlag, Berlin.

Vesmír
Vesmír. Přírodovědecky časopis Čs. akadmie věd. Publisher: Academia, Praha.

Vestn. Khar'kov. Univ.
Vestnik Khar'kovskogo Universiteta. Seriya Astronomicheskaya. Publishers: Izdatel'stvo Khar'kovskogo Universiteta, Khar'kov.

Vestn. Kiev. Univ.
Vestnik Kievskogo Universiteta. Seriya Astronomii.
Publishers: Izdatel'stvo Kievskogo Universiteta, Kiev.

VJS Naturforsch. Ges. Zürich
Vierteljahresschrift der Naturforschenden Gesellschaft
in Zürich. Printer and Publisher: Leeman AG, Zürich.

Weltraumfahrt
Weltraumfahrt, Raketentechnik. Publisher: Umschau-
Verlag, Frankfurt a/Main.

Wiss. Zeitschr. Friedrich-Schiller Univ. Jena
Wissenschaftliche Zeitschrift der Friedrich-Schiller-Uni-
versität. Jena. Mathematisch-Naturwissenschaftliche
Reihe. Edited by the Rektor der Friedrich-Schiller-Uni-
versität Jena.

Wiss. Zeitschr. Humboldt-Univ. Berlin
Wissenschaftliche Zeitschrift der Humboldt-Universität
zu Berlin. Mathematisch-Naturwissenschaftliche Reihe.
Edited by the Rektor der Humboldt-Universität, Berlin.

Yamamoto Circ.
Yamamoto Circular. Published by the Yamamoto Observatory, Kamitanakami – Kiryutyo, Otu, Siga-ken, Japan.

Zeitschr. Angew. Physik
Zeitschrift für Angewandte Physik. Publisher: Springer-
Verlag, Berlin–Heidelberg–New York.

Zeitschr. Astrophys. (ZfA)

Zeitschrift für Astrophysik. Publisher: Springer-Verlag,
Berlin–Heidelberg–New York. After Vol. 69 (1968)
replaced by "Astronomy and Astrophysics".

Zeitschr. Geophys.
Zeitschrift für Geophysik. Publisher: Physica-Verlag,
Würzburg, Germany.

Zeitschr. Naturforschung
Zeitschrift für Naturforschung. Europhysics Journal.
Teil a: Astrophysik, Physik, Physikalische Chemie.
Published by Verlag der Zeitschrift für Naturforschung,
Tübingen, Germany.

Zeitschr. Physik
Zeitschrift für Physik. Publisher: Springer-Verlag, Berlin-
Heidelberg–New York.

Zemlya i Vselennaya
Zemlya i Vselennaya. Astronomiya, Geofizika, Issledovaniya Kosmicheskogo Prostranstva. Nauchno-Populyar-
nyj Zhurnal Akademii Nauk SSSR. Publishers: Izdatel'-
stvo "Nauka", Moskva.

Zentralblatt Math. Grenzgebiete
Zentralblatt für Mathematik und ihre Grenzgebiete. Publisher: Springer-Verlag, Berlin–Heidelberg–New York.

Zvaigžņota Debess
Latvijas PSR Zinātņu Akadēmijas Radioastrofizikas
Observatorijas Populārzinatniks Gadalaiku Izdevums.
Izdevnieciba "Zinātne", Riga.

002 Bibliographical Publications

002.001 News and views.
Nature, Vol. 235, 9 - 16, 74 - 82, 125 - 136, 193 -
202, 247 - 256, 360 - 368, 417 - 422 (1972). − Tachyons and
gravitons, p. 10 - 11; Helium in globular cluster stars, p. 14 -
15; Separate big-bang for QSOs? , p. 16; Two families of X-ray
variables, p. 76; Maffei 2, an Sb spiral galaxy, p. 76; Neutral
hydrogen near the sun, p. 78 - 79; Gravitation: Waves from
black holes, p. 80; Understanding the hadron era, p. 81 - 82;
More X-rays from NP 0532, p. 132; Blast waves and the Galaxy, p. 134; Another QSO-galaxy pair, p. 134 - 135; A record
cosmic ray primary? , p. 193; Gravitational radiation: Too
much too soon? , p. 199; Visible solar features and interferometry, p. 201 - 202; Binary stars as astronomical accelerators,
p. 247 - 248; Mars: Red planet is not dead, p. 251 - 252; Carbon chemistry of the lunar surface, p. 252 - 253; Imploding-
exploding waves, p. 361 - 362; Radio observations of Cygnus
X-1, p. 364; Jupiter: Probing by Pioneer, p. 419; Cosmology:
Gravity and Fred Hoyle, p. 421 - 422.

**002.002 Annotations on the papers on geomagnetism and
aeronomy in "News of higher educational establishments. Radiophysics", 1970, Vol. 13, No. 6, 7, 9, 10.**
Geomagn. Aeronom., Vol. 12, 171 - 173 (1972). In Russian.

002.003 News notes.
Sky Telescope, Vol. 43, 10 - 13, 93 - 94, 158 - 159,
229 - 231 (1972).
(1) Radio emission from Antares B; Finnish meteorite studied;
Further renovation at Greenwich Observatory; Planetariums in
North America; New nearby galaxies; Possible neutron star;
New work on Sirius B; Time signal experiment; Boston Planetarium developer; Alert to meteor watchers; Flare star mono-
graph. (2) Minor planet 1971 UA; Popigay depression: A Siberian astrobleme? ; Catalogues of supernovae and quasars; Sunrise USA; Solar system experts; Moonquakes; The Milky Way
before Galileo. (3) IC 10 and the Maffei galaxies; Eruption in
solar corona observed by OSO 7; AAS meeting in Seattle. (4)
Hubble constant revised; Radio photograph of Maffei 2; The
largest asteroids; Carbon dioxide in the Martian polar caps; Algol as a radio source.

002.004 News.
Nature, Phys. Sci., Vol. 235, 1 - 2, 21 - 22, 41 - 42,
61 - 62, 81 - 82, 101 - 102, 121 - 122, 141 - 142, 161 - 162
(1972). − Redshift selection effects, p. 1 - 2; White dwarf
polarization, p. 21; Uhuru: More X-ray sources, p. 62; Planets:
Search for Jovian X-rays, p. 82; Cosmic gamma rays, p. 101 -
102; Hills and vales on Mars and Venus, p. 121; HEOS-A2 working well, p. 122; Astronomy by the sea, p. 141 - 142; Variability of 3C 120, p. 142; Pulsars probe the galactic field, p. 161;
Quasar variability: Doubts about 3C 273, p. 162; Most systems
go, p. 162; TD 1A: ESRO comes of age, p. 162.

002.005 Science news.
Priroda, No. 1.72, p. 94 - 103; No. 2.72, p. 103 -
113; No. 3.72, p. 100 - 109 (1972). In Russian.
(1) Is the hypothesis of Low valid? Tectonic map of the
moon (*N. B. Semikhatova*); Similarity in the structure of Mars
and the earth. (2) Soviet stations on the orbit of Mars; Do
microorganisms exist on the moon? New 2-m telescope at the
Shemakha Astrophysical Observatory. (3) M. V. Lomonosov
Gold Medals 1971 for V. A. Ambartsumyan and H. Alfvén;
The relief of Mars (*Yu. M. Pushcharovskij*); Three stages of
the earth's formation; Astronomical rock painting in Armenia
(*B. E. Tumanyan, V. M. Masson*).

002.006 News from science and other informations.
Zemlya i Vselennaya, 1972, No. 2. In Russian.
X-ray pulsar (*N. N. Chugaj*), p. 4; Model of an X-ray source,
p. 29; Circular polarization of the light of planets, p. 35; Mass
of Pluto, p. 40; Radio map of Venus, p. 41; Professor Zwicky's
lectures (*B. A. Vorontsov-Vel'yaminov, B. V. Komberg*), p. 42;
Molecules in interstellar gas (*V. A. Bronshtehn*), p. 60; New
methods for measuring the diameters of stars, p. 61; Rotation
of Saturn's satellites, p. 61; Encounter between the earth and
asteroids, p. 61; Supernovae service (*P. G. Kulikovskij*), p. 61;
Astronomy at schools in the Vinnitsa region (*I. D. Il'evskij*),
p. 80.

002.007 News and views.
Nature, Vol. 236, 9 - 14, 55 - 60, 95 - 100, 139 -
146, 197 - 204, 325 - 330, 369 - 376, 425 - 432 (1972).
Astrophysical abundances under fire, p. 9; Cosmic rays and
scaling, p. 59; Variations in radio stars β Lyrae and β Persei, p.
99; Cosmology: Conformally invariant theory of gravity, p.
100; Sco X-1 at optical and X-ray frequencies, p. 144 - 145;
Anatomy of lunar magic, p. 197; Cen X-3: Peculiar binary, p.
201; Crab pulsar speeds up again, p. 204; Selenology: Surface
structure, p. 328; More variations from Sco X-1, p. 328;
TD-1A looks at stellar atmospheres, p. 372 - 373; Solar system:
Too many cooks, p. 374 - 375; Association of pulsars with
supernova remnants, p. 375 - 376; Magnetosphere: Data from
IMP 5, p. 427; 30 Doradus observed in the infrared, p. 427 -
428.

002.008 News.
Nature, Phys. Sci., Vol. 236, 1 - 2, 17 - 18, 49 - 50,
81 - 82, 97 - 98, 113 - 114 (1972). − ESO blank polished, p. 2;
NP 0532: Giant pulses, p. 18; Cooperation in Astronomy at
Meudon, p. 18; Interstellar molecules: Probing the background,
p. 50; Pulsars: X-rays from Vela, p. 82; Interplanetary space:
Solar system plasmas, p. 98; Thermal models of the moon, p.
113 - 114.

002.009 News notes.
Sky Telescope, Vol. 43, 282 - 284, 351 - 352 (1972).
(5) Popular astronomy in East Germany; Measurements of
star diameters; Intergalactic matter and the Milky Way; Comet
Bradfield; Solar neutrino dilemma; Evolution in very close
binaries; Gamma Geminorum as a binary; (6) Occultations of
stars by Jupiter and Ganymede; Acoustical waves on the sun;
Formaldehyde in Allende; Centaurus X-3 as a binary system.

002.010 News from science and other informations.
Zemlya i Vselennaya, No. 1 (1972). In Russian.
Law on meteorites, p. 53; At the explosion location of
the supernova of 1184, p. 53; Hydroxyl in other galaxies, p. 55.

**002.011 Annotations on the papers on geomagnetism and
aeronomy published in "News of higher education-
al establishments, Radiophysics", 1971, Vol. 14, Nos. 1 - 5.**
Geomagn. Aeronom., Vol. 12, 377 - 380 (1972). In Russian.

**002.012 Nuclear and relativistic astrophysics and nuclidic
cosmochemistry: 1963−1967. Volume IV.**
B. Kuchowicz.
Nuclear Energy Information Center, Warsaw. Rev. Report
No. 45, 209 pp. (1971).

002.013 Science news.
Priroda, No. 4.72, p. 101 - 112; No. 5.72, p. 104 -
112 (1972). In Russian.
(4) New value for the mass of Pluto; Results of measurements of γ-radiation; Extraordinary velocity of the motion of

galaxies; Faster than the velocity of light? ! (5) Possible source of gravitational waves; Registration of oscillations in the moon's interior; On the nature of the Hellas region on Mars; On the long way to Jupiter (*S. A. Nikitin*).

002.014   Chronicle.
Urania Kraków, Vol. 43, 24 - 27, 50 - 53, 87 - 89, 115 - 121, 144 - 148, 176 - 177 (1972). In Polish.
(1) The dimensions of the radio source on Jupiter (*A. Marks*); Astronomical observations of polarized light (*B. Kuchowicz*); Radio source 3C58, a supernova remnant (*M. Pańków*); Structure of quasars (*B. Kuchowicz*); The earth captures cosmic matter (*A. Marks*); Storms on the Black Sea and solar activity (*A. Marks*). (2) Phobos (*L. Zajdler*); The brightest star in our Galaxy (*B. Kuchowicz*); Observations of changing geographical longitudes of continents (*M. Pańków*); Radio observations of supernova remnants (*M. Pańków*). (3) Did the flux of solar protons change over the last million years? (*B. Kuchowicz*); A Seyfert galaxy regularly changes its optical brightness (*B. Kuchowicz*); Microwaves from Mars (*B. Kuchowicz*). (4) Do the galactic clusters originate in an explosion?(*B. Kuchowicz*); Observations of supernova 1572 (*B. Kuchowicz*); Discovery of the gravitational waves?! (*A. Marks*); Craters on the satellites of Mars (*A. Marks*); Does life exist on the planets? (*B. Kuchowicz*); Transit of the earth on the sun's disk (*M. Pańków*). (5) First molecular lines beyond our Galaxy (*B. Kuchowicz*); Detection of promethium in the star HR 465 (*A. Marks*); First extragalactic source of gamma radiation (*B. Kuchowicz*); Characteristics of Jupiter and stability of the solar system (*B. Kuchowicz*); Did Mars possess a more dense atmosphere? (*A. Marks*); First results of the telescope observations of Mars during the last great opposition (*A. Marks*); Nicolaus Copernicus exhibition at the Silesian Planetarium (*C. Ługowski*). (6) Do we see the black holes? (*A. Marks*); New theory connected with the Cassini division (*A. Marks*); Is Pluto an iron-rich planet? (*A. Marks*).

002.015   Science news.
Priroda, No. 6.72, p. 88 - 97 (1972). In Russian.
New start to Venus (*S. A. Nikitin*); Hurricanes on Mars; "Export" of lunar glass; Porous snow on Ganymede? Variability of emission lines in Seyfert nuclei (*I. I. Pronik*); How to investigate black holes (*M. A. Korets*); Magnetospheric emission of the earth.

002.016   News from science and other informations.
Zemlya i Vselennaya, 1972, No. 3. In Russian.
The flight of Venera 8, p. 19; Globular cluster or a most nearby galaxy? (*M. S. Frolov*), p. 20; How many novae flare-ups in the Andromeda nebula and the Galaxy? p. 28; Radio nuclei of galaxies, p. 28; What is the reason for the sudden change of the period of pulsars? p. 28; News of solar bursts, p. 40; Lectures for astrophysicists (*L. P. Grishchuk*), p. 78; Program "Eole", p. 78.

002.017   Astronomy and Astrophysics Abstracts. Vol. 6, Literature 1971, Part II.
S. Böhme, W. Fricke, U. Güntzel-Lingner, F. Henn, D. Krahn, U. Scheffer, G. Zech (Editors).
Published for Astronomisches Rechen-Institut, Heidelberg by Springer-Verlag, Berlin – Heidelberg – New York. 10 + 560 pp. Price DM 72.00 [Subscription price per volume DM 57.60] (1972).

002.018   Kurzberichte aus der Forschung.
SuW, Vol. 11, 12 - 15, 39 - 40, 104 - 107, 137 - 140, 164 - 165, 188 - 190 (1972).
(1) Gab es wirklich einen Urknall? ; Solarkonstante; Staub in der Erdatmosphäre und Oberflächentemperatur der Erde; Schwarze Löcher in Doppelsternsystemen; Anlaufschwierig-

keiten beim Effelsberger Radioteleskop; Ballonteleskop THISBE: Erfolgreiche Startkampagne in Texas/USA; Die Ablenkung des Lichtes am Sonnenrand im Radiowellengebiet gemessen; Möglicher neuer Begleiter von M31. (2) Zweifel am kosmologischen Ursprung der Rotverschiebung von Quasaren; Das Absorptionslinienspektrum von 4C 05.34; Eindeutigkeits-Satz des Sternaufbaus nur eingeschränkt gültig; Helligkeit des Nachthimmels im fernen Infrarot; Mariner 10; Spiralstruktur des Andromeda-Nebels bis 23 kpc vom Zentrum; Radio-Photographie des Himmels; 1971 UA (Objekt Kohoutek), ein Planetoid mit außergewöhnlicher Bahn. (3) Erste Ergebnisse des Infrarot-Spektrometers von Mariner 9; Kern von Cen A entdeckt; Eigenbewegungsmessungen mit Galaxien als Bezugssystem; 91,5 – cm–Flugzeug–Teleskop; Radiokarte von Maffei 2; Hochenergetische Gammastrahlung aus dem galaktischen Zentrum; Skylab-Mannschaft im Zeiss Planetarium; Sonnenwind und interplanetarer Staub. (4) Das menschliche Auge als Detektor für kosmische Strahlung; Neuer Röntgenpulsar in der Nähe des Crab-Nebels; Großes Antennensystem für amerikanische Radioastronomie; Topographie der Venus; Das neue sowjetische Radioteleskop; Neue Strahlungsquellen im fernen Infrarot; Apollo-17-Landeplatz bekanntgegeben; Apollo 15: Photographien der Korona und des Zodiakallichtes. (5) Pulsare und galaktisches Magnetfeld; Optische Identifikation von Cyg X-1; Fernseh-Technik in der Astronomie; System der Planetenmassen. (6) Suche nach Antimaterie; Radar-Echos von Planeten; Nochmals: Solare Neutrinos; Struktur von VY CMa; Kosmische Staubschicht bedeckt die Erde; Eine neue Hypothese über Jupiter; Marsmond Phobos hat zwei Ebenbilder auf der Erde (*J. Classen*).

002.019   Die Entstehung der astronomischen Fachzeitschriften in Deutschland (1798 - 1821).   D. B. Herrmann.
Veröff. Archenhold-Sternw. Berlin-Treptow, No. 5, 158 pp. (1972).

002.020   Centre de Données Stellaires. Inform. Bull. No. 3.
J. Jung (Editor).
Compiled at Observatoire de Strasbourg. 24 pp. (1972). – The individual contributions are induced in their corresponding subject categories – see abstracts 002.901, 041.033 - 041.037, 113.039, 152.012.

002.021   News and views.
Nature, Vol. 237, 9 - 16, 69 - 76, 193 - 200, 249 - 256, 367 - 374, 425 - 432, 481 - 488 (1972).
Origin of X-ray background radiation, p. 12 - 13; Solar magnetism: Sunspot variations, p. 13; Mars: Recent Mariner results, p. 70 - 71; Radical view of anomalous redshifts, p. 193; Poor man's rocket, p. 193 - 194; Time variation of gravity constant, p. 198; Optics: Diffraction gratings, p. 200; Interacting radio galaxies or radio trails? p. 249; Two kinds of pulsar-quake? p. 251; Luna 20 samples, p. 256; Low frequency radio signals from extensive air showers, p. 367; Inhomogeneous earth, p. 367 - 368; X-ray astronomy: Uhuru dominates Madrid meeting, p. 369 - 370; Age determinations and lunar evolution, p. 425 - 426; Model for type I supernovae, p. 430; Why Venus is cloudy, p. 488.

002.022   News and views.
Nature, Vol. 238, 9 - 16, 69 - 76 (1972).
Intensity of solar neutrinos, p. 10; Hard X-ray flare in Cygnus X-1, p. 13; Elementary particles and cosmology, p. 69 - 70; Stellar radiation: Infrared excesses, p. 72 - 74.

002.023   News.
Nature, Phys. Sci., Vol. 237, 1 - 2, 17 - 18, 33 - 34, 49 - 50, 81 - 82, 97 - 98, 113 - 114, Vol. 238, 17 (1972).
Nonthermal causes of continental drift, p. 1; Quasars and galaxies: Looking for correlations, p. 2; Supernova remnants:

X-rays from Aquila, p. 18; X-ray astronomy: Pinning down sources, p. 34; Interstellar hydrogen: Unusual cloud, p. 50; Space: The ultraviolet earth, p. 81; Prognoz in orbit, p. 82; Background radiation no longer anomalous, p. 97; Radio astronomy: Southern hemisphere microwaves, p. 98; Magnetic reversals: Length of the Kiaman, p. 114; 150 years of Scotish Royal astronomy, Vol. 238, p. 17.

002.024 **News notes.**
Sky Telescope, Vol. 44, 16 - 17 (1972).
(1) Bright supernova in NGC 5253; Polarized X-rays from the Crab nebula; Harvard Observatory opens new building.

002.025 **Nouvelles brèves.**
Ciel et Terre, Vol. 88, 70 - 71, 138 - 143, 225 - 228 (1972).
(1) Catalogue général des étoiles variables — Tome III. (2) Entrée du soleil dans les constellations; Les dimensions des mers lunaires; La petite planète exceptionnelle 1971 UA; L'occultation de Beta Scorpion C par Io; Mesures infrarouges de Ganymède; L'orbite de la comète Ikeya-Seki (1968 I); Désignations définitives des comètes de 1970; Nova Cephei 1971. (3) Les petites planètes exceptionnelles en 1972; La comète périodique Shajn-Schaldach; Les comètes Tsuchinshan; Etoiles invisibles ou «trous noirs»; La masse de l'atmosphère.

002.026 **Rassegna delle riviste e notizie brevi.**
P. Maffei.
Coelum, Vol. 40, 17 - 21, 62 - 66, 109 - 113 (1972).

002.027 **Theses and dissertations on the history of astronomy.**
Journ. History Astron., Vol. 3, 73, 149 (1972).

002.028 **Bibliography.** Z. Kopal, M. Moutsoulas, J. W. Salisbury (Editors).
The Moon, Vol. 3, 472 - 491; Vol. 4, 250 - 268, 507 - 535 (1972). — Current critical bibliography of the entire field of lunar studies.

002.029 **Forschung und Technik.**
Phys. Blätter, 28. Jahrgang, p. 35 - 38, 181 - 183, 224 - 228, 278 - 280 (1972).
Kosmisches Primärteilchen von ca. 4×10^{21} eV — und was es lehrt, p. 35 - 38; Die Polkappen des Mars, p. 183; Die Landungen von Mars 2 und Mars 3, p. 224 - 225; Englische Astronomen identifizieren eine Röntgenstrahlquelle, p. 225; Rot-Verschiebung einer Kaliumlinie im Schwerefeld der Sonne gemessen, p. 226; Jupiter und Io bedeckten Beta Scorpii, p. 227 - 228; Heftige Explosionsvorgänge im Herzen der Milchstraße, p. 278; Neue Erkenntnisse über weitentfernten Quasar, p. 279 - 280; Entstehung der Spiralarme von Galaxien, p. 280.

002.030 **Science and the citizen.**
Sci. American, Vol. 226, No. 1, p. 44 - 47, 50 - 52; No. 2, p. 40 - 42, 44; No. 3, p. 40 - 44; No. 4, p. 54 - 57; No. 5, p. 48 - 50, 52, 54; No. 6, p. 51 - 53 (1972). — The sirens of Titan, No. 1, p. 46; Moonglow, No. 1, p. 47; Coming of age, No. 2, p. 41; Amazing Mars, No. 3, p. 40; Through a gas darkly, No. 4, p. 57; The Mariner becomes a Viking, No. 5, p. 49 - 50; The mysterious moon, No. 6, p. 51; The case of the missing neutrinos, No. 6, p. 53.

002.031 **Mitteilungen aus Wissenschaft und Literatur.**
Sterne, 48. Jahrgang, p. 51 - 52, 120 - 122 (1972).
(1) Jahreszeitliche Schwankungen des Schwerepotentials der Erde (*H.-U. Sandig*); AP Librae ist eine extragalaktische Radioquelle (*J. Gürtler*); Gebäudeschaden durch Meteoritenfall (*J. Classen*). (2) Wechselwirkung zwischen Radiogalaxien (*J. Gürtler*); Nichtvariable Sterne im Zustandsbereich der Cepheiden (*F. Schmeidler*).

002.032 **Nouvelles de la science.**
L'Astronomie, 86ᵉ année, p. 117 - 118 (1972).
Occultations de petites planètes par la lune; Une pluie d'étoiles filantes en avril 1972? ; Vesta et les Hyades.

002.033 **Astronomical notebook.** J. S. Griffith.
Spaceflight, Vol. 14, 32 - 33, 71 - 73 (1972).
(1) The atmospheres of Mars and Mercury; Radio observations of Venus; The satellites of Saturn; The Neptune — Pluto system; The origin of small interplanetary particles; Disk shaped stars; Technetium stars; The origin of the galactic halo; (2) The ice caps of Mercury!; A contribution towards the theory of the formation of the planets; Cepheid variables; Low mass objects that have gravitationally collapsed; Water in stars; Stellar collisions; Rapidly varying blue stars; Galactic wakes; The timing of light flashes from the Crab nebula pulsar.

002.034 **Astronomical notebook.** J. S. Griffith.
Spaceflight, Vol. 14, 192 - 194, 231 - 233, 272 - 274 (1972).
(5) Model of Jupiter; Practical Jovian photochemistry; Solar wind; Icarus and general relativity; Red objects in Sagittarius; The outer appearance of a black hole; An intriguing hydrogen cloud; Maffei 1 not a nearby galaxy? (6) Planetary nebulae; Radio observations of planetary nebulae; The Gum nebula; Intergalactic stellar 'breeding grounds'; Snowplows and supernovae; Presupernovae; Outer regions of elliptical galaxies; Quasars, black holes and the galactic centre; Galactic winds; Expansion of the optical remnant of a supernova. (7) A new massive member of the local group of galaxies? The origin of spiral waves; Pulsars, quasars and supernovae; The Crab pulsar; Extragalactic pulsars; The expansion of clusters of galaxies; Gravitational radiation from pulsars; Search for optical pulsars; A new hydrogen cloud.

002.035 **Film reviews.** J. C. Gilbert.
Spaceflight, Vol. 14, 198 - 199 (1972).

002.036 **The AIP program for physics information. A national information system for physics and astronomy, 1972-1976.**
American Inst. Phys., *New York.* 6 + 67 pp. (1971).

Errata

002.901 **Errata: Centre de Données Stellaires.** Inform. Bull. No. 2 [see 06.002.026]. M. Bischoff.
Centre de Données Stellaires, Inform. Bull. No. 3, (see 002.020), p. 23 - 24 (1972).

003 Books (Astronomy and Astrophysics)

003.001 **Vistas in astronomy, Vol. 13.**
A. Beer (Editor).
Pergamon Press, Oxford – New York – Toronto – Sydney – Braunschweig. 8 + 296 pp. Price £ 9.00 (1972). – The individual contributions are included in their corresponding subject categories – see abstracts 041.008, 042.021, 044.011, 045.008, 064.022, 080.015, 102.009, 113.014, 114.057, 114.058, 122.050 - 122.053, 125.013, 131.068, 162.020.

003.002 **The environment of the earth.** F. Delobeau.
Astrophys. Space Sci. Library, Vol. 28.
D. Reidel Publishing Company, Dordrecht – Holland. 9 + 113 pp. Price hfl. 30.00 (1972). – Contents: (I) The earth in space; (II) The terrestrial magnetic field; (III) The terrestrial ionosphere; (IV) The outer ionosphere and the van Allen belts; (V) The borders of the terrestrial environment; (VI) Disturbances in the atmosphere; (VII) Micrometeorites.

003.003 **Astrophysik I. Eine Einführung.** H. Sautter.
Gustav Fischer Verlag, Stuttgart. Uni-Taschenbücher UTB-No. 107, 6 + 115 pp. Price DM 11.80 (1972). – Contents: 1.) Optische Grundlagen; 2.) Optische Beobachtungsinstrumente; 3.) Sternparallaxen; 4.) Sternspektren; 5.) Die Physik der Sternatmosphären; 6.) Die Sonne.

003.004 **Astrophysik II. Eine Einführung.** H. Sautter.
Gustav Fischer Verlag, Stuttgart. Uni-Taschenbücher UTB-No. 108, 6 + 143 pp. Price DM 12.80 (1972). – Contents: 1.) Die radioastronomische Empfangstechnik; 2.) Sternentwicklung; 3.) Galaxien; 4.) Kosmologie.

003.005 **Physics and astronomy of the moon.**
Z. Kopal (Editor).
Second edition. Academic Press, New York – London. 11 + 303 pp. Price $ 15.00 (1971). – The individual contributions are included in their corresponding subject categories – see abstracts 094.188 - 094.193.

003.006 **Advances in Astronomy and Astrophysics, Vol. 8.**
Z. Kopal (Editor).
Academic Press, New York – London. 11 + 348 pp. Price $ 19.50 (1971). – The individual contributions are included in their corresponding subject categories – see abstracts 094.194 - 094.196.

003.007 **Advances in Astronomy and Astrophysics, Vol. 9.**
Z. Kopal (Editor).
Academic Press, New York – London. 12 + 252 pp. Price $ 16.50 (1972). – The individual contributions are included in their corresponding subject categories – see abstracts 094.197, 094.198, 095.012, 117.031.

003.008 **Die Relativität der Trägheit.** H.-J. Treder.
Akademie-Verlag, Berlin. 6 + 119 pp. Price DM 19.80 (1972). – Contents: Teil I: Zur Frage der Modellierbarkeit der Einsteinschen Effekte in Galilei-invarianten Gravitationstheorien gemäß Weber und Riemann; Teil II: Riemannsche Mechanik und Mach-Einstein-Doktrin; Anhang I: Die Induktion von träger Masse in einer relativistischen Mechanik; Anhang II: Einige Konsequenzen der Mach-Einstein-Doktrin für die Himmelsmechanik, Geophysik und Kosmogonie.

003.009 **Astrodynamics. Orbit correction, perturbation theory, integration. Volume 2.** S. Herrick.
Van Nostrand Reinhold Company, London – New York – Cincinnati – Toronto – Melbourne. 14 + 348 pp. Price £ 9.75 (1972). – Contents: Numerical integration and Cowell's and Encke's methods; Differential processes (second chapter); The method of variation of parameters; Hamiltonian mechanics; Spheroids; Introduction to general perturbations; Appendices: Numerical analysis and Gaussian summation integration; Astrodynamical values of the physical constants.

003.010 **Scattering and absorption of light in the atmosphere.**
G. Sh. Livshits (Editor).
Trudy Astrofiz. Inst., *Alma-Ata*, Vol. 18. Izdatel'stvo "Nauka" Kazakhskoj SSR, Alma-Ata. 154 pp. Price 1 Rbl. 36 Kop. (1971). In Russian.

003.011 **Theory and experiment in exobiology, Volume 1.**
A. W. Schwartz (Editor).
Wolters-Noordhoff Publishing, Groningen, The Netherlands. 4 + 160 pp. Price Dfl. 36.50 (1971). – The contributions within the subject scope of Astronomy and Astrophysics Abstracts are included in their corresponding subject categories – see abstracts 061.048, 094.234, 099.061, 091.024.

003.012 **The vortex theory of planetary motions.**
E. J. Aiton.
MacDonald, London; American Elsevier Inc., New York. 9 + 282 pp. Price hfl. 68.00 (1972). – Contents: Introduction; The theories of Kepler and Galileo; The vortex theory of Descartes; The influence of the Cartesians, 1637 - 1700; Attraction theories: Newton and his precursors; The harmonic vortex of Leibniz; The Cartesian vortex theory, 1700 - 1729; The introduction of Newton's theories in France; Attempts to reconcile the Cartesian and Newtonian theories, 1728 - 1734; The last days of the vortex theory; Conclusion.

003.013 **Geology and physics of the moon.** A study of some fundamental problems.
G. Fielder (Editor).
Elsevier Publishing Company, Amsterdam – London – New York. 8 + 159 pp. Price hfl. 75.00 (1971). – The individual contributions are included in their corresponding subject categories – see abstracts 094.235 - 094.245.

003.014 **Astronomy. Volume 7. Acceleration processes in space.** L. I. Dorman.
Itogi nauki. VINITI Akad. Nauk SSSR, Moskva. 233 pp. Price 1 Rbl. 68 Kop. (1972). In Russian. – Contents: 1. Statistical acceleration mechanisms in a turbulent plasma; 2. Regular acceleration mechanisms; 3. Gas of accelerated particles and its interaction with plasma and electromagnetic radiation; 4. Acceleration processes in the magnetosphere of the earth; 5. Acceleration of particles on the sun; 6. Acceleration and deceleration of particles in the interplanetary space; 7. Properties of accelerated particles of extrasolar origin; 8. Acceleration and propagation of fast particles in the Galaxy; 9. Acceleration processes in stars and galaxies of different type; 10. Acceleration of particles in the metagalaxy; 11. Composition, spectrum and sidereal variation and the problem of the origin of cosmic rays; 12. Acceleration processes in cosmos and generation of electromagnetic radiation.

003.015 **The physics of cosmic X-ray, γ-ray, and particle sources.** K. Greisen.
Gordon and Breach Science Publishers, New York – London – Paris. 8 + 115 pp. Price DM 38.40 (1971). – This book originally appeared as a section of 'Astrophysics and General Relativity', 1968 Brandeis University Summer Institute in Theoretical Physics, Volume 2, published by Gordon and Breach in 1971 = 06.143.032.

003.016 Relativistische Astrophysik. G. Dautcourt.
Akademie-Verlag, Berlin; Pergamon Press, Oxford; Vieweg + Sohn, Braunschweig. Wissenschaftliche Taschenbücher, WTB Vol. 86, 157 pp. Price DM 9.80 (1972). — Contents: Relativistic processes in the Galaxy; Quasars and extragalactic radio sources; Evolution of the metagalaxy; Appendices.

003.017 General relativity. Papers in honour of J. L. Synge. Edited for the Royal Irish Academy by L. O'Raifeartaigh.
Clarendon Press, Oxford. 9 + 277 pp. Price approx. £ 8.00 (1972). — The contributions within the subject scope of Astronomy and Astrophysics Abstracts are included in their corresponding categories — see abstracts 062.037, 066.087 - 066.090, 162.057.

003.018 New techniques in astronomy. H. C. Ingrao (Editor).
Gordon and Breach Science Publishers, New York – London – Paris. 12 + 446 pp. Price £ 12.30 (1971). — This book is an English version of the original Russian "Novaya Tekhnika v Astronomii" (Volumes 1 & 2), published by the USSR Academy of Sciences in 1963 and 1965, respectively. It contains a collection of conference papers on astronomical techniques and instrumentation.

003.019 Plasma, the fourth state of matter. D. A. Frank-Kamenetskii.
Translated from Russian by J. Norwood, Jr.
Macmillan, London; Plenum Press, New York, 8 + 159 pp. Price £ 6.00 (1972).

003.020 Astronomisch-chronologische Tafeln für Sonne, Mond und Planeten. P. Ahnert.
Verlag Johann Ambrosius Barth, Leipzig. 5th edition, 47 pp. Price DM 10.20 (1971). — Reviews in Astron. in der Schule, 9. Jahrgang, p. 47; 1972 (*H. J. Nitschmann*); SuW, Vol. 11, 85; 1972 (*A. Kunert*).

003.021 Solar eclipses and the ionosphere. M. Anastassiades (Editor).
Plenum Press, New York. 14 + 309 pp. Price $ 18.50 (1970). — Review in Journ. Atmosph. Terr. Phys., Vol. 34, 171; 1972 (*J. A. Ratcliffe*).

003.022 The life of Benjamin Banneker. S. A. Bedini.
Charles Scribner's Sons, New York. 434 pp. Price $ 14.95 (1972). — Review in Sky Telescope, Vol. 43, 182 (1972).

003.023 Mathematical astronomy for amateurs. E. A. Beet.
W. W. Norton & Company, Inc., New York, N.Y. 143 pp. Price $ 7.95 (1972). — Review in Sky Telescope, Vol. 44, 44 - 45 (1972).

003.024 Johannes Kepler, 1571–1630. Yu. A. Belyj.
"Nauka", seriya "Nauchno-biograficheskaya literatura", Moskva. 296 pp. Price 95 Kop. (1971). In Russian. — Review in Priroda, No. 3.72, p. 118 (1972).

003.025 Campanus of Novara and medieval planetary theory. Theorica Planetarum.
F. S. Benjamin, Jr., G. J. Toomer (Editors).
University of Wisconsin Press, Madison. 16 + 492 pp. Price $ 20.00 (1971).

003.026 Nowe materiały do działalności publicznej Mikołaja Kopernika z lat 1512–1537 (New materials to Nicolas Copernicus's public activities from 1512 to 1537).
M. Biskup.
Państwowe Wydawnictwo Naukowe, Warsaw – Toruń, Poland.

96 pp. Price zł 12.00 (1971). — Reviews in Journ. History Astron., Vol. 3, 148 (1972); Urania Kraków, Vol. 43, 152 (1972).

003.027 Geometric optics. J. W. Blaker.
Marcel Dekker, Inc., New York. 127 pp. Price $ 9.50 (1971). — Review in Sky Telescope, Vol. 43, 46 (1972).

003.028 Les radiotélescopes. E.-J. Blum.
Presses Universitaires de France, Paris. 128 pp. Price F. 3.95 (1972). — Review in Sky Telescope, Vol. 43, 318 (1972).

003.029 Lunar orbiter photographic atlas of the moon. D. E. Bowker, J. K. Hughes.
National Aeronautics and Space Administration, NASA-SP-206. [Available from Superintendent of Documents, U.S. Government Printing Office, Washington, D.C.]. 41 pp., 675 plates. Price $ 19.95 (1971). — Review in Sky Telescope, Vol. 43, 250 (1972).

003.030 New horizons in astronomy. J. C. Brandt, S. P. Maran.
W. H. Freeman and Company, San Francisco. 496 pp. Price $ 12.50 (1972). — Review in Sci. American, Vol. 226, No. 6, p. 126 (1972).

003.031 Stars for the space age. J. B. Breed III.
World Publishing Company, New York, N.Y. 64 pp. Price $ 5.95 (1971). — Review in Sky Telescope, Vol. 43, 46 (1972).

003.032 'Augen in das All' (Eyes into space). G. Breuer.
L. Schwann Verlag, Düsseldorf. 114 pp. Price DM 12.80 (1970). — Review in Spaceflight, Vol. 14, 235; 1972 (*H. O. Ruppe*).

003.033 What star is that? P. Lancaster Brown.
Thames & Hudson, London; Viking Press Inc., New York. 224 pp. Price £ 3.50, $ 12.95 respectively (1971). — Reviews in Journ. British Astron. Ass., Vol. 82, 146 - 147; 1972 (*A. C. Curtis*); p. 228 - 230; 1972 (*P. Lancaster Brown*); Sky Telescope, Vol. 43, 45 - 46 (1972).

003.034 The books of Autolykos: On a moving sphere and on risings and settings.
F. Bruin, A. Vondjidis (Editors, Translators).
American University of Beirut, Beirut. 10 + 83 + 70 pp. Price $ 8.00 (1971). — Review in Journ. History Astron., Vol. 3, 147; 1972 (*W. D. Stahlman*).

003.035 Activation analysis in geochemistry and cosmochemistry. Proceedings of the NATA Advanced Study Institute, Kjeller, Norway, 1970 September.
A. O. Brunfelt, E. Steiness (Editors).
468 pp. Price N.Kr. 144.00 (1971). — Review in Geochim. Cosmochim Acta, Vol. 36, 507 - 509; 1972 (*R. R. Keays*).

003.036 Mikołaj Kopernik jako ekonomista. S. Cackowski.
Państwowe Wydawnictwo Naukowe, Toruń, Poland. 73 pp. Price zł 10.00 (1970). — Review in Urania Kraków, Vol. 43, 152 (1972).

003.037 Atlas optischer Erscheinungen. Ergänzungsband. M. Cagnet, M. Françon, S. Mallik.
Springer-Verlag, Berlin – Heidelberg – New York. 35 pp. and 15 plates. Price DM 88.00; $ 25.50 respectively (1971). — Review in SuW, Vol. 11, 172; 1972 (*K. Bahner*).

003.038 Comparative guide to programs in earth sciences,

physics and astronomy. J. Cass, M. Birnbaum. Harper and Row, New York, 18 + 246 pp. Price $ 4.95 (1972). [Reprinted from "Comparative Guide to Science and Engineering Programs"].

003.039 **Submillimetre spectroscopy.** G. W. Chantry. Academic Press, London. 10 + 386 pp. Price £ 6.00 (1971).

003.040 **Infinity and intelligence.** G. V. Chefranov. Rostov. universitet, Rostov n/Donu. 176 pp. Price 74 Kop. (1971). In Russian. – Review in Referativ. Zhurn. 51. Astron., 7.51.2 (1972).

003.041 **Beginnings of lunar soil science. Physico-mechanical properties of the lunar soil.** I. I. Cherkasov, V. V. Shvarev. Nauka, Moskva. 200 pp. Price 1 Rbl. 14 Kop. (1971). In Russian. – Review in Referativ. Zhurn. 62. Issled. kosm. prostranstva, 6.62.200 (1972).

003.042 **Radio telescopes.** W. Christiansen, I. Högbom. Translated from the English edition. Mir, Moskva. 238 pp. Price 1 Rbl. 60 Kop. (1972). In Russian. – Review in Referativ. Zhurn. 51. Astron., 6.51.122 (1972).

003.043 **Report on planet three and other speculations.** A. C. Clarke. Harper and Row Publishers, New York, N.Y. 250 pp. Price $ 6.95 (1972). – Review in Sky Telescope, Vol. 43, 182 (1972).

003.044 **The Mariner 6 and 7 pictures of Mars.** S. A. Collins. Superintendent of Documents, U.S. Government Printing Office, Washington, D.C. 159 pp. Price $ 4.25 (1971). – Review in Sky Telescope, Vol. 43, 245, 246; 1972 (*C. A. Cross*).

003.045 **Eyewitness to space.** H. L. Cooke, J. D. Dean. Harry N. Abrams, Inc., New York, N.Y. 227 pp. Price $ 35.00 (1971). – Review in Sky Telescope, Vol. 43, 112 (1972).

003.046 **Données spectroscopiques relatives aux molécules diatomiques.** L. Denis-Gausset (Editor). Pergamon Press, London–New York. 7 + 515 pp. Price F 350.00 (1970). – Review in Planet. Space Sci., Vol. 20, 955 - 956; 1972 (*P. K. Carroll*).

003.047 **Water in the universe.** V. F. Derpgol'ts. Nedra, Leningrad. 224 pp. Price 54 Kop. (1971). In Russian. – Review in Referativ. Zhurn. 51. Astron., 6.51.51 (1972).

003.048 **Gravitation and universe.** R. H. Dicke. Translated from the English edition. Mir, Moskva. 102 pp. Price 34 Kop. (1972). In Russian. – Review in Priroda, No. 5.72, p. 121 - 122 (1972).

003.049 **Exploring the moon and the solar system.** T. Dickinson. Copp Clark Publishing Co., Toronto, Ont. 72 pp. Price Canadian $ 6.50 (1971). – Review in Sky Telescope, Vol. 44, 44 (1972).

003.050 **Astronomia przedkopernikowska.** J. Dobrzycki. Państwowe Wydawnictwo Naukowe, Toruń, Poland. 58 pp. Price zł 10.00 (1971). – Review in Urania Kraków, Vol. 43, 153 (1972).

003.051 **A guide to earth satellites.** D. Fishlock (Editor). MacDonald, London; Elsevier, New York. 12 + 160 pp. Price £ 2.50; $ 7.95 respectively (1971/72). – Review in Nature, Vol. 236, 38; 1972 (*W. J. G. Beynon*).

003.052 **The Euler–Mayer correspondence (1751–1755).** E. G. Forbes. American Elsevier Publishing Company Inc., New York, N.Y. 118 pp. Price $ 11.00 (1971). – Review in Sky Telescope, Vol. 43, 385 - 386 (1972).

003.053 **Tobias Mayer's "Opera Inedita".** E. G. Forbes. American Elsevier Publishing Company Inc., New York, N.Y. 166 pp. Price $ 14.75 (1971). – Review in Sky Telescope, Vol. 43, 386 (1972).

003.054 **Introduction to geophysics. Mantle, core and crust.** G. D. Garland. W. B. Saunders Company, Philadelphia, Pa. 14 + 420 pp. Price $ 14.50 (1971).

003.055 **Understanding the earth.** I. G. Gass, P. J. Smith, R. C. L. Wilson (Editors). M.I.T. Press, Cambridge, Mass. 355 pp. Price $ 9.25 (1971). – Review in Spaceflight, Vol. 14, 78; 1972 (*J. S. Griffith*).

003.056 **Galileo reappraised.** C. L. Golino (Editor). University of California Press, Berkeley – Los Angeles. 8 + 110 pp. Price $ 5.00 (1971). – Review in Journ. Roy. Astron. Soc. Canada, Vol. 66, 173; 1972 (*J. MacLachlan*).

003.057 **Dom i środowisko rodzinne Mikołaja Kopernika.** K. Górski. Państwowe Wydawnictwo Naukowe, Toruń, Poland. 54 pp. Price zł 6.00 (1968). – Review in Urania Kraków, Vol. 43, 151 (1972).

003.058 **Plasma physics, Vol. 9, Part B.** H. R. Griem, R. H. Lovberg (Editors). Academic Press, New York. 324 pp. Price £ 7.45 (1971).

003.059 **In the world of gravitational forces.** N. P. Grushinskij, A. N. Grushinskij. Nedra, Moskva. 160 pp. Price 27 Kop. (1971). In Russian. – Review in Referativ. Zhurn. 52. Geod. Aehros"emka, 4.52.104 (1972).

003.060 **Atlas and gazetteer of the near side of the moon.** G. L. Gutschewski, D. C. Kinsler, E. Whitaker. National Aeronautics and Space Administration, Washington, D.C. 538 pp. Price $ 15.00 (1971). – Review in Sky Telescope, Vol. 43, 318 (1972).

003.061 **Time and space. Measuring instruments from the fifteenth to the nineteenth century.** S. Guye, H. Michel. Translated by D. Dolan in collaboration with S. W. Mitchell. Pall Mall Press, London. 289 pp. Price £ 10.50 (1971). Review in Journ. History Astron., Vol. 3, 65 - 66; 1972 (*G. L' E. Turner*).

003.062 **Our blue planet. The story of the earth's evolution.** H. Haber. Translated from the German edition by E. Stuhlinger. Charles Scribner's Sons, New York, N. Y., 8 + 88 pp. Price $ 1.95 (1972).

003.063 **Celestial mechanics. Vol. 2.** Y. Hagihara. M. I. T. Press, Cambridge, Mass. Part 1, 2. 18 + 16 + 920 pp. Price $ 30 each part. (1971/1972). – Review in Sky Telescope, Vol. 43, 317 (1972).

003.064 **Moons and planets: An introduction to planetary science.** W. K. Hartmann.
Bogden and Quigley, Inc., Tarrytown-on-Hudson – New York. 11 + 404 pp. Price $12.95 (1972). – Review in Sky Telescope, Vol. 44, 44 (1972).

003.065 **The structure and physical properties of the earth's crust.** J. G. Heacock (Editor).
American Geophysical Union, Washington, D. C. 10 + 348 pp. Price $19.00 (1971). – Review in Nature, Vol. 238, 53; 1972 (*S. A. F. Murrell*).

003.066 **Naar de sterren kijken.** R. van Helden.
Gemeente Utrecht, Bureau Culturele Zaken. 60 pp. (1971). – Review in Hemel en Dampkring, Vol. 70, 61; 1972 (*T. de Vries*).

003.067 **The mind of the scientist.** M. Hoskin.
Taplinger Publishing Co., Inc., New York, N. Y. 128 pp. Price $5.95 (1972). – Review in Sky Telescope, Vol. 43, 386 (1972).

003.068 **Astronomy one.** J. A. Hynek, N. H. Apfel.
Benjamin, Menlo Park, Calif. 12 + 402 pp. Price $10.50 (1972).

003.069 **Planets, stars, and galaxies. An introduction to astronomy.** S. J. Inglis.
John Wiley & Sons, Inc., New York – London – Sydney – Toronto. Third edition. 19 + 498 pp. Price $10.95 (1972). Review in Sky Telescope, Vol. 43, 317 (1972).

003.070 **Informatie in woord en beeld over Sterrekunde.** F. P. Israel.
Moussault's Uitgeverij, Amsterdam. 60 pp. Price f 5.90 (1970). Review in Hemel en Dampkring, Vol. 70, 63; 1972 (*J. W. Wijbenga*).

003.071 **Biblioteka Mikołaja Kopernika.** L. Jarzębowski.
Państwowe Wydawnictwo Naukowe, Toruń, Poland. 86 pp. Price zł12.00 (1971). – Review in Urania Kraków, Vol. 43, 152 (1972).

003.072 **Astronomia Mikołaja Kopernika.** C. Iwaniszewska.
Państwowe Wydawnictwo Naukowe, Toruń, Poland. 74 pp. Price zł 12.00 (1971). – Review in Urania Kraków, Vol. 43, 153; 1972 (*S. R. Brzostkiewicz*).

003.073 **Astronomy: Fundamentals and frontiers.** R. Jastrow, M. H. Thompson.
John Wiley & Sons, New York – London –Sydney – Toronto. XIV + 35 + 404 pp. Price £6.50 (1972).

003.074 **Mimozemské civilizace.** S. A. Kaplan et al.
Academia, nakladatelství Československé akademie věd, Praha. 312 pp. Price Kčs. 36.00 (1972).

003.075 **Spherical astronomy. Textbook for college students specialized in astronomical geodesy.** V. Z. Khalkhunov.
Nedra, Moskva. 304 pp. Price 76 Kop. (1972). In Russian.

003.076 **Geological time.** J. S. Kirkaldy.
Contemporary Science Paperback No. 46. Oliver & Boyd, Edinburgh. 8 + 133 pp. Price 37.5 p. (1971). – Review in Journ. British Astron. Ass., Vol. 82, 235; 1972 (*G. E. Satterthwaite*).

003.077 **American astronauts and spacecraft.** D. C. Knight (Editor).

Franklin Watts, Inc., New York, N. Y. 176 pp. Price $7.95 (1972). – Review in Sky Telescope, Vol. 44, 44 (1972).

003.078 **The time.** A. I. Konstantinov, A. G. Fleer.
Standartizdat, Moskva. 367 pp. Price 1 Rbl. 22 Kop. (1971). In Russian. – Review in Referativ. Zhurn. 51. Astron., 7.51.154 (1972).

003.079 **Man and his universe.** Z. Kopal.
William Morrow, New York, N. Y. 314 pp. Price $7.95 (1972). – Review in Sky Telescope, Vol. 43, 386 (1972).

003.080 **Mars – the mysterious planet.** K. Zh. Kovachev, B. Zh. Kovachev.
Nauka i izkustvo, Sofiya. 304 pp. Price 1.24 Lv. (1971). In Bulgarian

003.081 **Nuclear and relativistic astrophysics and nuclidic cosmochemistry.** Vol. 1. B. Kuchowicz.
Gordon and Breach Science Publishers, London. 380 pp. Price £7.50 (1972). – Review in Journ. British Interplanet. Soc., Vol. 25, 315 (1972).

003.082 **Fundamentals of lunar astrometry.** K. A. Kulikov, V. B. Gurevich.
Izdatel'stvo "Nauka". Glavnaya Redaktsiya Fiziko-Matematicheskoj Literatury, Moskva. 392 pp. Price 1 Rbl. 99 Kop. (1972). In Russian. – Review in Referativ. Zhurn. 51. Astron., 6.51.44 (1972).

003.083 **The planet earth.** K. A. Kulikov, N. S. Sidorenkov.
Nauka, Moskva. 184 pp. Price 56 Kop. (1972). In Russian. Review in Referativ. Zhurn. 51. Astron., 7.51.41 (1972).

003.084 **Space, time, motion.** I. V. Kuznetsov (Editor).
"Nauka", Moskva. 623 pp. Price 2 Rbl. 20 Kop. (1971). In Russian. – Review in Priroda, No. 3.72, p. 116 - 117 (1972).

003.085 **Astronomi och astrofysik.** G. Larsson-Leander.
C. W. K. Gleerup Bokförlag, Lund. 353pp. Price S. Kr. 56.85 (1971). – Review in Astron. Tidssk., Årg. 5, p. 46 - 47; 1972 (*P. O. Lindblad*).

003.086 **Two planets.** K. Lasswitz. Translated from the German edition (Leipzig, 1897) by H. Rudnick.
Southern Illinois University Press, Carbondale. 8 + 406 pp. Price $10.00 (1971). – Review in Science, Vol. 175,567(1972).

003.087 **Earthbound astronauts.** B. Lay, Jr.
Prentice-Hall, Inc., Englewood Cliffs, N. J. 198 pp. Price $6.95 (1971). – Review in Sky Telescope, Vol. 43, 310 - 311; 1972 (*R. Hillenbrand*).

003.088 **The physics of pulsars.** A. M. Lenchek (Editor).
Topics in Astrophysics and Space Physics. Gordon and Breach Science Publishers, New York. 10 + 174 pp. Price $14.50 (1972). – Reviews in Science, Vol. 176, 1230 - 1231; 1972 (*F. D. Drake*); Sky Telescope, Vol. 43, 317 (1972).

003.089 **Thermal characteristics of the moon.** (Progress in Astrophysics and Aeronautics, Vol. 28).
J. W. Lucas (Editor).
MIT Press, Cambridge Mass. - London, 13 + 340 pp. Price $14.95 (1972). – Review in Sky Telescope, Vol. 44, 45 (1972)

003.090 **Basic dynamics of some moon effects.** G. L. Luebbers.
Published by the author, P. O. Box 1034, La Costa Station, Malibu, Calif. 4 + 76 pp. Price $4.00 (1972).

003.091 Hypothesen zur Planetentheorie des 17. Jahrhunderts.
Y. Maeyama.
Institut für Geschichte der Naturwissenschaft, Johann Wolfgang Goethe-Universität, Frankfurt am Main, 133 pp. (1971).
Review in Journ. History Astron., Vol. 3, 66 - 67; 1972 (*D. T. Whiteside*).

003.092 Of a fire on the moon. N. Mailer.
Little, Brown & Co., Boston, MA. 472 pp. Price $ 7.95 (1970). – Review in Spaceflight, Vol. 14, 38 - 39; 1972 (*W. I. McLaughlin*).

003.093 Catadioptric imaging systems. J. Maxwell.
American Elsevier Publ. Company, New York, N. Y. 102 pp. Price $ 17.00 (1972). – Review in Sky Telescope, Vol. 43, 386 (1972).

003.094 Physique et dynamique planétaires. P. Melchior.
Vamder – éditeur, Louvain – Bruxelles. Vol. 1: Géodésie et astronomie géodésiques, 8 + 247 pp.; Vol. 2: Gravimétrie, potentiel gravitationnel de la terre et de la lune, 6 + 311 pp.; Vol. 3: Géodynamique, 6 + 268 pp. Subscription Price: Vol. 1 - 4: FB 1400, US $28.00 respectively (1971/ 1972).

003.095 Astronomie, een nieuw profiel van het heelal.
D. H. Menzel.
Elsevier, Amsterdam – Brussel. 320 pp. Price f 69.00, Belg. F 1150 respectively (1971). – Review in Hemel en Dampkring, Vol. 70, 62 - 63; 1972 (*T. Dethier*).

003.096 Survey of the universe. D. H. Menzel, F. L. Whipple, G. de Vaucouleurs.
Prentice Hall International Inc., London. 860 pp. Price £ 8.25 (1971).

003.097 Introduction to mathematical fluid mechanics.
R. E. Meyer.
Wiley and Sons, New York – London. 11 + 185 pp. Price £ 6.15 (1972). – Review in Nature, Phys. Sci., Vol. 238, 16; 1972 (*A. G. Mackie*).

003.098 Introduction to classical and modern optics.
J. R. Meyer-Arendt.
Prentice–Hall, Inc., New York, N. Y. – Englewood Cliffs, N. J. 558 pp. Price $ 15.95 (1972). – Review in Sky Telescope, Vol. 43, 317 - 318 (1972).

003.099 The theory of relativity. C. Møller.
Oxford University Press, New York. Second edition. 14 + 558 pp. Price $ 44.25 (1972).

003.100 Rotation of the earth and climate. A. S. Monin.
Gidrometeoizdat, Leningrad. 112 pp. Price 84 Kop. (1972). In Russian. – Review in Referativ. Zhurn. 51. Astron., 5.51.41 (1972).

003.101 1972 yearbook of astronomy.
P. Moore (Editor).
Norton and Company Inc., New York. 227 pp. Price $ 5.95 (1971). – Review in Sky Telescope, Vol. 43, 182 (1972).

003.102 Medieval chronicles and the rotation of the earth.
R. R. Newton.
Johns Hopkins University Press, Baltimore – London. 825 pp. Price $ 15.00 (1972). – Review in Sky Telescope, Vol. 44, 45 (1972).

003.103 Venus and Mercury. A. E. Nourse.
Franklin Watts, London – New York. 68 pp. Price $ 3.75 (1972). – Review in Sky Telescope, Vol. 44, 45 (1972).

003.104 Dividends from space.
F. I. Ordway III, C. C. Adams, M. R. Sharpe.
Thomas Y. Crowell Co., New York. 309 pp. Price $ 10.00 (1971). – Review in Sky Telescope, Vol. 43, 248 - 250; 1972 (*E. M. Brooks*).

003.105 Menschen messen Zeit und Raum. E. Padelt.
VEB Verlag Technik, Berlin. 168 pp. (1971). – Review in Urania Kraków, Vol. 43, 153; 1972 (*L. Zajdler*).

003.106 Physical cosmology. P. J. E. Peebles.
Princeton University Press, Princeton, N.J. 16 + 282 pp. Price $ 9.00 (1971). – Reviews in Science, Vol. 176, 650 - 652; 1972 (*G. B. Field*); Sky Telescope, Vol. 43, 250 (1972).

003.107 The origin of the universe. J. A. Piasecki.
Philosophical Library, New York. 8 + 56 pp. Price $ 3.75 (1972).

003.108 Les sciences de la terre à l'heure des satellites.
J. Pouquet.
Presses Universitaires de France, Paris. 259 pp. Price Fr. 22.00 (1971). – Review in Sky Telescope, Vol. 43, 250 (1972).

003.109 Star myths and stories. P. M. Proctor.
Exposition Press, Jericho, N.Y. 183 pp. Price $ 6.00 (1972). – Review in Sky Telescope, Vol. 44, 45 (1972).

003.110 Nuclear reactions in stellar surfaces and their relations with stellar evolution. H. Reeves.
Gordon and Breach, New York – London – Paris. 10 + 87 pp. Price £ 4.00, $ 9.60 respectively, cloth; £ 1.95, $ 4.75 respectively, paperback (1971). – Review in Nature, Vol. 236, 284 - 285; 1972 (*B. Pagel*).

003.111 Precision astrolabe, Portuguese navigators and transoceanic aviation. F. M. Rogers.
Academia Internacional da Cultura Portuguesa, Lisboa. Distributed in U.S.A. by Wm. S. Sullwood, Publishing, Taunton, Mass. 397 pp. Price $ 8.00 (1971). – Review in Journ. Navigation, *London,* Vol. 25, 135 - 137; 1972 (*D. H. Sadler*).

003.112 The beauty of the universe. H. Rohr.
Viking Press, Inc., New York, N.Y. 87 pp. Price $ 10.00 (1972). – Review in Sky Telescope, Vol. 43, 318 (1972).

003.113 Les cadrans solaires anciens d'Alsace.
R. R. J. Rohr.
Editions Alsatia, Colmar, France. Collection 'Richesses de l'Alsace'. 272 pp. Price Trade edition F 115.00; Numbered edition, full leather F 320.00 (1971). – Reviews in L'Astronomie, 86e année, p. 202 - 204; 1972 (*R. Sagot*); Journ. Roy. Astron. Soc. Canada, Vol. 66, 129 - 130; 1972 (*F. A. Stebbins*); Orion Schaffhausen, 30. Jahrgang, p. 30 - 31; 1972 (*E. Antonini, E. Wiedemann*).

003.114 The amateur astronomer and his telescope.
G. D. Roth.
Faber & Faber, London. 179 pp. Price 75 p. (1972). – Review in Journ. British Astron. Ass., Vol. 82, 312; 1972 (*A. C. Curtis*).

003.115 Introduction into astronautics. Volume 2.
G. M. Ruppe.
Translated from the English edition. Nauka, Moskva. 575 pp. Price 2 Rbl. 94 Kop. (1971). In Russian. – Review in

Referativ. Zhurn. 62. Issled. kosm. prostranstva, 5.62.66 (1972).

003.116 **Hamiltonian cosmology.** M. Ryan.
Lecture Notes in Physics, Vol. 13. Springer-Verlag, Berlin – Heidelberg – New York. 7 + 169 pp. Price DM 18.00 (1972).

003.117 **Toruń w czasach Kopernika – Urbanistyka, architektura, sztuka.** B. Rymaszewski.
Państwowe Wydawnictwo Naukowe, Toruń, Poland. 73 pp. Price zł 10.00 (1969). – Review in Urania Kraków, Vol. 43, 151 (1972).

003.118 **Från jorden till solen.** A. E. Sandström.
Bokförlaget Aldus/Bonniers, Stockholm. 116 pp. Price S.kr. 13.50 (1970). – Review in Astron. Tidssk., Årg. 5, p. 49 - 50; 1972 (*T. Elvius*).

003.119 **Från Thales till Einstein.** A. E. Sandström.
Bokförlaget Aldus/Bonniers, Stockholm. 213 pp. Price S.kr. 20.50 (1971). – Review in Astron. Tidssk., Årg. 5, p. 49 - 50; 1972 (*T. Elvius*).

003.120 **Beyond the Observatory.** H. Shapley.
Scribner's Sons, New York. 224 pp. Price $ 2.65 (1972). – [Reprint of the 1967 edition].

003.121 **Cosmic rays and nuclear interactions at high energies.** D. V. Skobel'tsyn (Editor).
Translated from Russian by F. L. Sinclair. Lebedev Physics Institute Series. Consultants Bureau, New York. 6 + 228 pp. Price $ 29.50 (1971).

003.122 **Basic physics of stellar atmospheres.** T. L. Swihart.
Pachart, Tucson, Ariz. 12 + 86 pp. Price $ 7.95 (1971). – Review in Science, Vol. 175, 292 (1972).

003.123 **Fernrohrmontierungen und ihre Schutzbauten für Sternfreunde.** A. Staus.
Verlag Uni-Druck, München. 3rd revised edition. 84 pp. Price DM 24.00 (1971). – Review in SuW, Vol. 11, 58; 1972 (*G. D. Roth*).

003.124 **Astronomische Navigation.** Eine Einführung in die astronomische Navigation für Sportschiffer und Sternfreunde. W. Stein.
Verlag Klasing u. Co. GmbH, Bielefeld – Berlin. 208 pp. Price DM 13.80 (1971). – Review in SuW, Vol. 11, 85; 1972 (*A. Kunert*).

003.125 **Soviet rocketry: The first decade of achievement.** M. Stoiko.
David and Charles Ltd., Newton Abbot, Great Britain. 11 + 272 pp. Price £ 3.25 (1971). – Review in Planet. Space Sci., Vol. 20, 955; 1972 (*D. G. King-Hele*).

003.126 **Planet earth. Its physical systems through geological time.** A. N. Strahler.
Harper and Row, New York. 10 + 438 pp. Price $ 11.00 (1972).

003.127 **Principles and applications of palaeomagnetism.** D. H. Tarling.
Chapman and Hall, London. 164 pp. Price £ 1.75 (1972). – Review in Nature, Vol. 238, 53; 1972 (*S. A. F. Murrell*).

003.128 **Continental drift.** D. H. Tarling, M. P. Tarling.
G. Bell and Sons Ltd., London. 122 pp. Price £ 1.50 (1971). – Review in Journ. British Interplanet. Soc.,

Vol. 25, 316 (1972).

003.129 **Les fondements de la mécanique céleste.** Y. Thiry.
Gordon and Breach, Science Publishers, Inc., New York. 198 pp. (1970).

003.130 **Geometrical methods in cosmic geodesy.** A. P. Tishchenko.
Nauka, Moskva. 114 pp. Price 62 Kop. (1971). In Russian. – Review in Referativ. Zhurn. 62. Issled. kosm. prostranstva, 4.62.267 (1972).

003.131 **Atlas of variable star identification charts.** V. P. Tsesevich, M. S. Kazanasmas.
Nauka, Moskva. 350 charts. Price 3 Rbl. 2 Kop. (1971). In Russian.

003.132 **Sterne und Menschen.** Aufsätze und Vorträge. A. Unsöld.
Springer-Verlag, Berlin – Heidelberg – New York. 8 + 170 pp. Price DM 29.80 (1972). – Review in Phys. Blätter, 28. Jahrgang, p. 238; 1972 (*H. Scheffler*).

003.133 **The sun in H-alpha light with a spectrohelioscope.** F. N. Veio.
Adams Press, Chicago, Ill., 56 pp. Price $ 1.25 (1972). – Review in Sky Telescope, Vol. 43, 317 (1972).

003.134 **Life sciences and space research IX.** Proceedings of the open meeting of Working Group 5 at thirteenth plenary meeting of COSPAR, Leningrad, 1970 May 20–29. W. Vishniac (Editor).
Akademie-Verlag, Berlin. 190 pp. Price M 50.00, $ 11.93 respectively (1971).

003.135 **Mikołaj Kopernik – dzieje jednego odkrycia.** W. Voisé.
Państwowe Wydawnictwo Naukowe, Toruń, Poland. 76 pp. Price zł 10.00 (1970). – Review in Urania Kraków, Vol. 43, 152 (1972).

003.136 **Johannes Hevelius and his catalog of stars.** I. Volkoff, E. Franzgrote, A. D. Larsen.
Brigham Young University Press, Provo, Utah. 89 pp. Price $ 8.00 (1971). – Review in Sky Telescope, Vol. 43, 182 (1972).

003.137 **Extragalactic astronomy.** B. A. Vorontsov-Vel'yaminov.
Textbook for students at universities. Nauka, Moskva. 464 pp. Price 1 Rbl. 32 Kop. (1972). In Russian.

003.138 **Thermospheric circulation.** W. L. Webb (Editor).
MIT Press, Cambridge, Mass. – London. 373 pp. Price $ 14.95 (1972). – Review in Sky Telescope, Vol. 44, 44 (1972).

003.139 **The discovery of our Galaxy.** C. A. Whitney.
Alfred A. Knopf, New York. 308 pp. Price $ 10.00 (1971). – Reviews in Sky Telescope, Vol. 43, 46 (1972); p. 243, 244; 1972 (*M. S. Roberts*).

003.140 **Kant's cosmogony.** Translated by W. Hastie, with a new introduction by G. J. Whitrow.
Johnson Reprint Corporation, New York – London. 40 + 205 pp. Price $ 12.50 (1970). – Review in Journ. History Astron., Vol. 3, 67 - 68; 1972 (*M. A. Hoskin*).

003.141 **Prediction and analysis of solar eclipse circum-**

stances. W. Williams, Jr.
AFCRL-71-0049, Final Report Contract F19628-70-C-0087.
Arthur D. Little, Inc., Cambridge, Mass. (1971). – Review in
Applied Optics, Vol. 11, 486; 1972 (*J. N. Howard*).

**003.142 Progress in elementary particle and cosmic ray
physics, Vol. 10.**
J. G. Wilson, S. A. Wouthuysen (Editors).
North-Holland Publishing Co., Amsterdam. 325 pp. Price
Dfl. 78.00, $ 21.75 respectively (1971).

003.143 The earth's age and geochronology.
D. York, R. M. Farquhar.
Pergamon, Oxford – New York. 8 + 178 pp. Price £ 2.50
(1972). – Review in Nature, Vol. 238, 53; 1972 (*S. A. F.
Murrell*).

**003.144 Physics of the earth and planets. Figures and inner
structure.**
V. N. Zharkov, V. P. Trubitsyn, L. V. Samsonenko.
Nauka, Moskva. 384 pp. Price 1 Rbl. 99 Kop. (1971). In Russian. – Review in Referativ. Zhurn. 51. Astron., 5.51.242
(1972).

003.145 Theory of gravitation and the evolution of stars.
Ya. B. Zel'dovich, I. D. Novikov.
Nauka, Moskva. 484 pp. Price 2 Rbl. 14 Kop. (1971). In Russian. – Review in Referativ. Zhurn. 51. Astron., 4.51.775
(1972).

**003.146 Calendar of the Tartu Observatory for the year
1972.**
Tallinn. 81 pp. (1971). In Estonian. – Review in Referativ.
Zhurn. 51. Astron., 5.51.44 (1972).

003.147 Bridge into space.
Izvestiya, Moskva. 623 pp. Price 3 Rbl. 23 Kop.
(1971). In Russian. – Review in Referativ. Zhurn. 62. Issled.
kosm. prostranstva, 6.62.64 (1972).

003.148 Mariner–Venus 1967: Final project report.
National Aeronautics and Space Administration,
Washington, D.C. NASA SP-190, 301 pp. $ 3.00 (1971). – Re-
view in Sky Telescope, Vol. 44, 45 (1972).

003.149 Apollo 15: Preliminary science report.
Published by NASA Manned Spacecraft Center.
National Aeronautics and Space Administration, Washington,
D.C. NASA SP-289, 536 pp. Price $ 8.00 (1972). – Review in
Sky Telescope, Vol. 44, 45 (1972).

003.150 Problems of cosmochemistry and meteoritics.
Akad. Nauk USSR. Kom. po meteoritam, Komis. po
kosmokhimii i meteoritike pri otd. nauk o Zemle i Kosmose,
I-nt geokhimii i fiz. mineralov. Nauk. dumka, Kiev. 221 pp.
Price 1 Rbl. 34 Kop. (1971). In Russian.

003.151 Skies of the Andes.
Available from John H. Lutnes, Photographic De-
partment, Kitt Peak National Observatory, Tucson, Ariz. –
Review in Sky Telescope, Vol. 43, 316 - 317; 1972 (*D. di
Cicco*).

003.152 Radio frequency interference handbook.
R. E. Taylor (Editor).
National Aeronautics and Space Administration. NASA
SP-3067. [Available from U.S. Government Printing Office,
Washington, D. C.], 264 pp. (1971). – *MWS*

003.153 Atoms and molecules in astrophysics.
T. R. Carson, M. J. Roberts (Editors).

Academic Press, London – New York. 10 + 370 pp. Price
£ 7.50 (1972).

**003.154 First supplement to the third edition of the general
catalogue of variable stars.** B. V. Kukarkin, P. N.
Kholopov, Yu. N. Efremov, N. P. Kukarkina, N. E. Kurochkin,
G. I. Medvedeva, N. B. Perova, Yu. P. Pskovsky, V. P. Fedoro-
vich, M. S. Frolov.
Astronomical Council of the Academy of Sciences of the
USSR. Sternberg State Astronomical Institute of the Moscow
State University, Moscow. 324 pp. (1971). In Russian and
English. – Containing information on 2216 variable stars desig-
nated in 1970 and improved information on 1907 previously
designated variable stars.

003.155 Des astres, de la vie et des hommes. R. Jastrow.
Translated from English by C. de Richemont.
Éditions du Seuil, Paris. 199 pp. Price F 21.00 (1972).
Review in Bull. Signal., Vol. 33, Section 120, No. 6898 (1972).

003.156 Origin of chemical elements. R. J. Tayler.
The Wykeham Science Series. Wykeham Publications Ltd., London – Winchester. 9 + 169 pp. (1972).

**003.157 Evolution of the protoplanetary cloud and forma-
tion of the earth and the planets.**
V. S. Safronov.
Translated from Russian. Israel Program for Scientific Trans-
lations, Keter Publishing House Ltd., Jerusalem, Israel. 212 pp.
Price $ 18.00 (1972).

003.158 1972 celestial calendar and handbook.
C. F. Johnson, Jr.
To be obtained from C. F. Johnson, Jr., 48 Roberts Street,
Watertown, Conn. 06795, U. S. A. 38 pp. Price $ 1.50 (1971).
Reviews in Sky Telescope, Vol. 43, 112 (1972); Strolling
Astronomer, Vol. 23, 187; 1972 (*J. R. Smith*).

**003.159 I. Kant: Universal natural history and theory of the
heavens,** with a new introduction by M. K. Munitz.
Ann Arbor Paperbacks, University of Michigan Press, Ann
Arbor, Mich. 24 + 180 pp. Price $ 2.45 (1969). – Review in
Journ. History Astron., Vol. 3, 67 - 68; 1972 (*M. A. Hoskin*).

**003.160 Extra-terrestrial civilizations – Problems of inter-
stellar communication.** S. A. Kaplan (Editor).
Translated from Russian. Israel Program for Scientific Trans-
lations, Keter Publishing House Ltd., Jerusalem, Israel. 272 pp.
Price £ 11.70 (1971). – Review in Journ. British Interplanet.
Soc., Vol. 25, 309 - 312; 1972 (*A. T. Lawton*).

003.161 Maps and man. N. J. W. Thrower.
Prentice-Hall Inc., Englewood Cliffs, New Jersey.
184 pp. Price $ 5.95, $ 2.95, respectively (1972). – Review
in Sky Telescope, Vol. 43, 386 (1972).

003.162 Fundamentals of astro-dynamics.
R. R. Bate, D. D. Mueller, J. E. White.
Dover Publications, New York; Constable, London. 12 + 455
pp. Price £ 2.25 (1972). – Review in Nature, Vol. 238, 471 -
472; 1972 (*D. G. King-Hele*).

003.163 The solar system. F. W. Cousins.
John Baker, Ltd., London. 300 pp. + 138 plates.
Price £ 5.00 (1972).

003.164 Geological problems in lunar and planetary research.
J. Green (Editor).
American Astronautical Society Publications, Tarzana, California. 14 + 736 pp.(1971). – Review in Nature, Vol. 238,
295; 1972 (*S. Banerjee*).

003.165   **Living on the third planet.**
H. Alfvén, K. Alfvén. Translated by E. Johnson.
W. H. Freeman and Company, San Francisco – Reading. 8 +
187 pp. Price $ 4.95, £ 2.20 respectively (1972).

003.166   **Physics of the solar continuum radio bursts.**
A. Krüger.
Akademie-Verlag, Berlin. 206 pp. Price DM 39.00 (1972).

003.167   **The rush toward the stars.**   T. Logsdon.
Franklin Watts Ltd., London. 224 pp. Price $ 6.80
(1970). – Review in Spaceflight, Vol. 14, 39; 1972 (*I. Graham*).

003.168   **Physics of the upper atmosphere.** Proceedings of
the International School of Atmospheric Physics,
Erice, Italy.   F. Verniani (Editor).
CNR Institute of Atmospheric Physics and University of Bologna. Editrice Compositori, Bologna, Italy. 31 + 461 pp.

Price $ 20.00 (1971). – Review in Journ. British Interplanet.
Soc., Vol. 25, 444 (1972).

003.169   **Das Leben der Sterne.**   A. J. Meadows.
Translated from English by P. Wellmann.
Verlag Chemie GmbH, Weinheim (Germany). 144 pp. Price
DM 18.80 (1972).

003.170   **Bibliography of stellar radial velocities.**
H. A. Abt, E. S. Biggs.
Kitt Peak National Observatory, Tucson, Arizona. 509 pp.
Price $ 9.00 (1972).

003.171   **The emerging universe. Essays on contemporary
astronomy.**   W. C. Saslaw, K. C. Jacobs (Editors).
University Press of Virginia, Charlottesville, Virginia. 188 pp.
Price $ 6.95 (1972).

004 History of Astronomy, Chronology

004.001 **Kepler and natural sciences. 400th birthday.**
A. I. Eremeeva.
Zemlya i Vselennaya, No. 1, p. 34 - 40 (1972). In Russian.

004.002 **"A dream" or on the astronomy of the moon – the posthumous work of Kepler.** Yu. A. Belyj.
Zemlya i Vselennaya, No. 1, p. 40 - 44 (1972). In Russian.

004.003 **Aus der Geschichte der Erforschung des Planetensystems. Vulkan – der Planet zwischen Sonne und Merkur.**
SuW, Vol. 11, 4 - 5 (1972).

004.004 **Dunkle Sternbilder.** T. Schmidt-Kaler.
SuW, Vol. 11, 10 - 11 (1972).

004.005 **Hevelius and his star catalogue.** J. Ashbrook.
Sky Telescope, Vol. 43, 223 - 224 (1972).

004.006 **How did Kepler discover his first two laws?**
C. Wilson.
Sci. American, Vol. 226, No. 3, p. 92 - 96, 99 - 106 (1972).

004.007 **Quel jour était-ce?** A. Hamon.
L'Astronomie, 86ᵉ année, p. 96 - 99 (1972).

004.008 **Edwin Hubble and a relativistic, expanding model of the universe.** N. S. Hetherington.
Astron. Soc. Pacific, Leaflet No. 509, 8 pp. (1971).

004.009 **Kepler's dream.** A. V. Douglas.
Journ. Roy. Astron. Soc. Canada, Vol. 66, 59 - 61 (1972).

004.010 **The origin of the lunar craters: An eighteenth-century view.** R. W. Home.
Journ. History Astron., Vol. 3, 1 - 10 (1972).

004.011 **The Carnac alignments.** A. Thom, A. S. Thom.
Journ. History Astron., Vol. 3, 11 - 26 (1972).

004.012 **Precession and trepidation in Indian astronomy before A.D. 1200.** D. Pingree.
Journ. History Astron., Vol. 3, 27 - 35 (1972).

004.013 **Aristotelian planetary theory in the renaissance: Giovanni Battista Amico's homocentric spheres.**
N. Swerdlow.
Journ. History Astron., Vol. 3, 36 - 48 (1972).

004.014 **The nebular hypothesis of Isaac Orr.**
R. L. Numbers.
Journ. History Astron., Vol. 3, 49 - 51 (1972).

004.015 **Did Copernicus have an observatory?** J. Classen.
Priroda, No. 3.72, p. 92 - 94, with comments by
E. Rybka, p. 94 (1972). In Russian.

004.016 **Notes of the history of Dunsink Observatory. I. Henry Ussher at Dunsink, 1783 to 1790.**
P. A. Wayman.
Irish Astron. Journ., Vol. 10, 121 - 128 (1971).

004.017 **"In a mirror brightly": French attempts to build reflecting telescopes using platinum.**
S. L. Chapin.
Journ. History Astron., Vol. 3, 87 - 104 (1972).

004.018 **Remarks on the theoretical treatment of eclipses in antiquity.** A. Aaboe.
Journ. History Astron., Vol. 3, 105 - 118 (1972).

004.019 **The astronomical instruments of John Rowley in eighteenth-century Russia.** V. L. Chenakal.
Journ. History Astron., Vol. 3, 119 - 135 (1972).

004.020 **The original formulation of the Titius-Bode law.**
S. L. Jaki.
Journ. History Astron., Vol. 3, 136 - 138 (1972).

004.021 **Astronomy in Japan.** W. Hartner.
Journ. History Astron., Vol. 3, 139 - 145 (1972).
Essay review.

004.022 **The Titius-Bode law: A strange bicentenary.**
S. L. Jaki.
Sky Telescope, Vol. 43, 280 - 281 (1972).

004.023 **More about sundials – old and new.** H. Egger.
Sky Telescope, Vol. 43, 288 - 289 (1972).

004.024 **The astrolabe: Its history, design and use.**
P. Simon.
Journ. Astron. Soc. Victoria, Vol. 24, 66 - 78 (1971). – The second Philipp Simon lecture.

004.025 **The constellation 'Crucis Australis'.** H. L. Hunt.
Journ. British Astron. Ass., Vol. 82, 99 - 105 (1972).

004.026 **Our heritage in Canadian astronomy.**
J. E. Kennedy.
Journ. Roy. Astron. Soc. Canada, Vol. 66, 83 - 98 (1972). –
An address delivered at the General Assembly of the R.A.S.C.
held at McMaster University, Hamilton, Ontario, on May 22, 1971.

004.027 **De geschiedenis van de sterrenkunde (1).**
G. W. E. Beekman.
Hemel en Dampkring, Vol. 70, 155 - 158 (1972).

004.028 **Die Universitäts-Sternwarte in Wien – Gründung und Baugeschichte (1874 - 1878).** P. Müller.
SuW, Vol. 11, 155 - 159 (1972).

004.029 **Empirical demonstration of precession and changes of the obliquity of the ecliptic in the works of ash-Shirazi.** M. Shermatov.
Uch. zap. Dushanbin. gos. ped. in-t, Vol. 81, 84 - 90 (1971).
In Russian. – Abstr. in Referativ. Zhurn. 51. Astron., 6.51.2 (1972).

004.030 **Ash-Shirazi's comments on the star catalogue of as-Sufi.** M. Shermatov.
Uch. zap. Dushanbin. gos. ped. in-t, Vol. 81, 73 - 83 (1971).
In Russian. – Abstr. in Referativ. Zhurn. 51. Astron., 6.51.3 (1972).

004.031 **Chronologic table and calendar used by scientists during the XVth - XVIIIth centuries.**
T. Shodiev, G. Sobirov.
Uch. zap. Dushanbin. gos. ped. in-t, Vol. 81, 30 - 72 (1971).

In Russian. – Abstr. in Referativ. Zhurn. 51. Astron. 6.51.4 (1972).

004.032 **Ulugh Begh's works in the papers of the members of the St. Petersbourg Academy of Sciences (XVIIIth century).** N. I. Nevskaya.
Dokl. AN UzSSR, 1971, No. 12, p. 5 - 7. In Russian. – Abstr. in Referativ. Zhurn. 51. Astron., 6.51.7 (1972).

004.033 **Georgian translation of Ulugh Begh's "Zidzh".** R. R. Orbeli.
Dokl. AN UzSSR, 1971, No. 12, p. 3 - 4. In Russian. – Abstr. in Referativ. Zhurn. 51. Astron., 6.51.8 (1972).

004.034 **Zur Geschichte des Entwicklungsgedankens in der Astronomie.** D. B. Herrmann.
Astron. in der Schule, 9. Jahrgang, p. 33 - 37 (1972).

004.035 **Historical chronicle.**
Urania Kraków, Vol. 43, 27 - 28, 59 - 61, 123 - 124 (1972). In Polish.

004.036 **Johannes Kepler and his time.** P. M. Ðurković.
Vasiona, Vol. 20, 8 - 11 (1972). In Serbo-Croatian.

004.037 **De geschiedenis van de sterrenkunde (2).** G. W. E. Beekman.
Hemel en Dampkring, Vol. 70, 181 - 184 (1972).

004.038 **Studies on Senmyō-Reki or Hsüan-ming-li (4): Records of solar eclipses in Japanese documents.** M. Utida.
Tokyo Astron. Obs., Report No. 59, Vol. 15, 693 - 709 (1972). In Japanese.

004.039 **An approach to the history of early astronomy.** R. Palter.
Studies History Philos. Sci., Vol. 1, 93 - 133 (1970). – Abstr. in Zentralblatt Math. Grenzgebiete, Vol. 224, No. 01010 (1972).

004.040 **La sphéropée, ou la mécanique au service de la découverte du monde.** G. Aujac.
Rev. Histoire Sci. Appl., Vol. 23, 93 - 107 (1970). – Abstr. in Zentralblatt Math. Grenzgebiete, Vol. 224, No. 01013 (1972).

004.041 **On the oriental sources of the Regiomontanus' trigonometrical treatise.** N. G. Hairetdinova.
Arch. Internat. Histoire Sci., Vol. 23, 61 - 66 (1971). – Abstr. in Zentralblatt Math. Grenzgebiete, Vol. 224, No. 01017 (1972).

004.042 **Quellenstudien zur Geschichte der Astronomie.** D. Wattenberg.
Blick in das Weltall, Archenhold-Sternw. Berlin-Treptow, 1972, No. 1, p. 7 - 9 = Sonderdruck Archenhold-Sternw. Berlin-Treptow, No. 16 (1972).

004.043 **Drei Einblattdrucke aus dem 16. Jahrhundert.** D. Wattenberg.

Blick in das Weltall, Archenhold-Sternw. Berlin-Treptow, 1972, Nos. 7/8, p. 63 - 70.

004.044 **Tradition und Fortschritt in der Astronomie des Mittelalters.** W. Petri.
Veröff. Forschungsinst. Deutsch. Museum Geschichte Naturwiss., Technik, Ser. A, Kleine Mitt. No. 104, p. 633 - 644 (1972).

004.045 **Johannes Keplers Theorien der Planetenbewegung.** F. Schmeidler.
Naturwiss. Rundschau [Wiss. Verlagsges., Stuttgart], Vol. 24, 509 - 513 (1971).

004.046 **Weltharmonie oder Weltgesetz: Johannes Kepler.** D. Wattenberg.
Vorträge und Schriften, Archenhold-Sternw. Berlin-Treptow, No. 42, 60 pp. (1972).

004.047 **Astrophysik im 19. Jahrhundert.** D. B. Herrmann.
Wiss. Fortschritt, 21. Jahrgang, No. 10, p. 463 - 468 = Mitt. Archenhold-Sternw. Berlin-Treptow, No. 91 (1971).

004.048 **Investigation of Nasireddin Tusi's scientific heritage during the recent 20 years.**
G. D. Mamedbejli, N. A. Abdulkasumova.
Soobshch. Shemakhinsk. Astrofiz. Obs., vyp. (No.) 6, p. 44 - 50 (1971). In Russian.

004.049 **Zur Erforschung der leuchtenden Nachtwolken auf der Berliner Sternwarte in den Jahren 1885 - 1901.** W. Schröder.
Monatsber. Deutsch. Akad. Wiss. Berlin, Vol. 13, 625 - 627 (1971).

004.050 **History of sundials.** P. Příhoda.
Říše hvězd, Vol. 53, 150 - 153 (1972). In Czech.

Theses and disserations on the history of astronomy.
See Abstr. 002.027.

Astronomisch-chronologische Tafeln für Sonne, Mond und Planeten. See Abstr. 003.020.

Campanus of Novara and medieval planetary theory. See Abstr. 003.025.

The books of Autolykos: On a moving sphere and on risings and settings. See Abstr. 003.034.

Dom i środowisko rodzinne Mikołaja Kopernika. See Abstr. 003.057.

Time and space. Measuring instruments from the fifteenth to the nineteenth century. See Abstr. 003.061.

Strejftog gennem himmelmekanikkens historie. See Abstr. 042.062.

005 Biography

005.001 **Yu. A. Mirkalov – mechanician and designing engineer of astronomical instruments.**
O. A. Mel'nikov, V. S. Popov, N. Yu. Lukina.
Zemlya i Vselennaya, No. 1, p. 66 - 69 (1972). In Russian.

005.002 **À la mémoire de Camille Flammarion.**
A. Duplay.
L'Astronomie, 86ᵉ année, p. 91 - 95 (1972).

005.003 **Personal profile: Dr. J. P. Vinti.**
Spaceflight, Vol. 14, 50 - 52 (1972).

005.004 **Thomas Romney Robinson (1792–1882), director of Armagh Observatory (1823–1882).**
D. Crowe.
Irish Astron Journ., Vol. 10, 93 - 101 (1971).

005.005 **Astronomers and inventors from the north-east of England.** A. Brown.
Irish Astron. Journ., Vol. 10, 102 - 108 (1971).

005.006 **Giovanni Giacomo Marinoni, the mathematician, topographer and astronomer of Udine. – Life,**
studies and works of this famous Italian. I. Candiloro.
L'Universo, Anno 52, p. 427 - 438 (1972). In Italian.

005.007 **Lewis Swift and the Lowe Observatory.**
J. Ashbrook.
Sky Telescope, Vol. 43, 364 - 366 (1972).

005.008 **Johannes Kepler 1571 - 1630.** J. L. Perdrix.
Journ. Astron. Soc. Victoria, Vol. 24, 86 - 91
(1971).

005.009 **Thomas Clausen als Astronom.**
K.-R. Biermann.
Janus, Vol. 57, 299 - 305 (1970). – Abstr. in Zentralblatt
Math. Grenzgebiete, Vol. 223, No. 01013 (1972).

005.010 **Willem de Sitter 1872 - 1934.** W. H. McCrea.
Journ. British Astron. Ass., Vol. 82, 178 - 181
(1972).

005.011 **Jean Kepler, son œuvre scientifique et astronomique.**
B. Morando.
L'Astronomie, 86ᵉ année, p. 209 - 222 (1972).

005.012 **Johannes Kepler, 1571–1630.**

T. J. Drewnowska.
Urania Kraków, Vol. 43, 2 - 3 (1972). In Polish.

005.013 **Tableau chronologique des principales œuvres de Johann Kepler. Etudes le concernant.** R. Taton.
L'Astronomie, 86ᵉ année, p. 288 - 303 (1972).

005.014 **Vom Musiker zum Astronomen. Zum 150. Todestag Wilhelm Herschels am 25. August 1972.**
G. Buttmann.
SuW, Vol. 11, 183 - 186 (1972).

005.015 **A visit to uncle Charlie. A. C. Gifford, M.A., F.R.A.S. (1861 - 1948).** G. A. Eiby.
Southern Stars, Vol. 24, 109 - 113 (1972).

005.016 **Kepler–der Revolutionär astronomischen Denkens.**
R. Buser.
Orion Schaffhausen, 30. Jahrgang, p. 107 - 108 (1972).
Report on a lecture by A. Schürer.

005.017 **Friedrich Wilhelm Bessel. Zum 125. Todestag des Astronomen.** D. Wattenberg.
Wiss. Fortschritt, 21. Jahrgang, No. 3, 3 pp. = Mitt. Archenhold-Sternw. Berlin-Treptow, No. 89 (1971).

005.018 **Nicolaus Copernicus (6).** S. R. Brzostkiewicz.
Urania Krakow, Vol. 43, 162 - 170 (1972). In Polish.

005.019 **Edmund Halley (1656 - 1742).** I. Chromek.
Kozmos, Vol. 3, 26 - 27 (1972). In Slovak.

005.020 **Augustin Seydler (1849 - 1891).** V. Guth.
Říše hvězd, Vol. 53, 44 - 45 (1972). In Czech.

005.021 **Augustin Seydler (1849 - 1891).** J. Široký.
Kozmos, Vol. 3, 59 (1972). In Czech.

Johannes Kepler, 1571–1630. See Abstr. 003.024.

Nowe materiały do działalności publicznej Mikołaja Kopernika z lat 1512 - 1537. See Abstr. 003.026.

Galileo reappraised. See Abstr. 003.056.

Johannes Hevelius and his catalog of stars.
See Abstr. 003.136.

006 Personal Notes

C. G. Abbot, 100th birthday.
Sky Telescope, Vol. 43, 352 (1972).

H. Alfvén received the Lomonosov Gold Medal.
Phys. Blätter, 28. Jahrgang, p. 143 (1972).

E. M. Burbidge, Director of the Royal Greenwich Observatory.
Mercury, (Journ. Astron. Soc. Pacific), Vol. 1, No. 1, p. 8 (1972).

E. M. Burbidge, Director of the Royal Greenwich Observatory at Herstmonceux.
Sky Telscope, Vol. 44, 17 (1972).

G. Field, Director of the Harvard College Observatory in July 1973.
Mercury, (Journ. Astron. Soc. Pacific), Vol. 1, No. 1, p. 8 (1972).

L. Goldberg, assumed the responsibilities of director of the Kitt Peak National Observatory.
Mercury, (Journ. Astron. Soc. Pacific), Vol. 1, No. 1, p. 7 (1972).

F. J. Heyden, director of the solar division of the Manila Observatory.
Mercury, (Journ. Astron. Soc. Pacific), Vol. 1, No. 2, p. 13 (1972).

J. Hoppe, 65. birthday.
Astron. in der Schule, 9. Jahrgang, p. 61 (1972).

R. P. Kraft, acting director of the Lick Observatory.
Mercury, (Journ. Astron. Soc. Pacific), Vol. 1, No. 2, p. 13 (1972).

P. Ledoux received the Eddington Medal of the Royal Astronomical Society.

Nature, Vol. 235, 186 (1972).

B. Mason received the Leonard Medal.
Meteoritics, Vol. 7, 75 (1972).

W. W. Morgan, was awarded an honorary Doctor's degree by the University of Córdoba.
Inform. Bull. Southern Hemisph., No. 19, p. 47 (1971).

D. E. Osterbrock, director of Lick Observatory.
Mercury, (Journ. Astron. Soc. Pacific), Vol. 1, No. 3, p. 9 (1972).

D. E. Osterbrock, Director of the Lick Observatory in California.
Sky Telescope, Vol. 44, 17 (1972).

M. Ryle, Astronomer Royal.
Nature, Vol. 237, 477 (1972).

B. Šternberk, 75. birthday.
Říše hvězd, Vol. 53, 12 - 13 (1972).   In Czech.

B. Strömgren, was awarded an honorary Doctor's degree by the University of Córdoba.
Inform. Bull. Southern Hemisph., No. 19, p. 49 (1971).

F. Zwicky received the Gold Medal of the Royal Astronomical Society.
Nature, Vol. 235, 186 (1972).

F. Zwicky received the Gold Medal of the Royal Astronomical Society.   E. Wiedemann.
Orion Schaffhausen, 30. Jahrgang, p. 32 (1972).

F. Zwicky received the Gold Medal of the Royal Astronomical Society.
Sky Telescope, Vol. 43, 289 (1972).

007 Obituaries

L. Arbey died 1972 March 27.
L'Astronomie, 86ᵉ année, p. 222 (1972).

N. P. Barabashov, 1894 March 30 –1971, April 20.
Astron. Zhurn. Akad. Nauk SSSR, Vol. 49, 227 - 229 (1972).
In Russian. – English translation in Soviet Astron. AJ, Vol. 16,
No. 1.

W. M. Baxter, 1896 Nov. 30 - 1971 Dec. 9.
M. R. Whippey.
Journ. British Astron. Ass., Vol. 82, 211 - 212 (1972).

L. Brown, 1903 March 10 - 1971 Oct. 26.
J. L. White.
Journ. British Astron. Ass., Vol. 82, 134 - 135 (1972).

P. de Bruyn, 1910 - 1971 Nov. 20.
J. J. B. van Eijk van Voorthuijsen.
Hemel en Dampkring, Vol. 70, 22 (1972).

J. D. Buddhue died 1971 December 21.
Meteoritics, Vol. 7, 75 (1972).

J. Camus, 1893 - 1971.   A. Hamon.
L'Astronomie, 86ᵉ année, p. 148 - 149 (1972).

F. Le Coultre, 1891 - 1971 December 31.
L'Astronomie, 86ᵉ année, p. 260 (1972).

H. E. Crull died 1972 April 25.
Sky Telescope, Vol. 43, 352 (1972).

V. G. Fesenkov, 1889 January 1 – 1972 March 12.
Astron. Zhurn. Akad. Nauk SSSR, Vol. 49, 678 - 682 (1972).
In Russian. English translation in Soviet Astron. AJ, Vol. 16,
No. 3.

V. G. Fesenkov, 1889 January 13 – 1972 March 12.
Sky Telescope, Vol. 43, 284 (1972).

V. G. Fesenkov, 1889 Jan. 13 - 1972 March 12.
Zemlya i Vselennaya, 1972, No. 3, p. 46 - 47. In Russian.

W. A. Heiskanen, 1895 July 23 - 1971 October 23.
Bull. Géod., Nouvelle Sér., Année 1972, No. 103, p. 3.

W. A. Heiskanen died on October 23, 1971.
Science, Vol. 175, 975 (1972).

C. Hoffmeister, 1892 - 1968.   R. Kippenhahn.
IAU Colloquium No. 15, (see 012.006), p. 315 - 316 (1972).

F. Kadavý died 1972 May 6.   O. Hlad.
Říše hvězd, Vol. 53, 135 - 136 (1972).   In Czech.

H. C. Lagerwey, 1912 April 29 - 1972 February 10.
A. W. J. Cousins, P. A. T. Wild.
Monthly Notes Astron. Soc. Southern Africa, Vol. 31, 32
(1972).

C. Lombardi, 1900 October 4 - 1971 June 30.
F. Zagar.
Mem. Soc. Astron. Italiana, Nuova Ser., Vol. 43, 223 (1972).

C. Luplau Janssen, 1889 - 1971
L. Tartois.
L'Astronomie, 86ᵉ année, p. 196 (1972).

C. Luplau Janssen, 1889 - 1971.
Sky Telescope, Vol. 44, 17 (1972).

R. K. Marshall died 1972 May 27.
Sky Telescope, Vol. 44, 3, 27 (1972).

R. S. McLaughlin, 1871 – 1972 January 6.
J. E. Kennedy.
Journ. Roy. Astron. Soc. Canada, Vol. 66, 125 - 126 (1972).

M. G. Pereira de Barros, 1908 - 1971 January 31.
J. Osorio.
Quarterly Journ. Roy. Astron. Soc., Vol. 13, 109 (1972).

A. Rybarski, 1889–1972.   L. Zajdler.
Urania Kraków, Vol. 43, 187 - 188 (1972). In Polish.

R. Shibahara died 1971.
J. Yoshida.
Celestial mechanics. Symposium Tokyo, (see 012.020), p. 2 - 3
(1972). In Japanese.

J. Q. Stewart died 1972 March 19.
Sky Telescope, Vol. 43, 284 (1972).

E. O. Tancock, 1886 Jan. 20 - 1971 Nov. 3.
E. A. Beet.
Journ. British Astron. Ass., Vol. 82, 213 (1972).

Y. Väisälä died 1971 July 21.
Å. Wallenquist.
Astron. Tidssk., Årg. 5, p. 100 (1972).

Y. Väisälä, 1891 September 6 - 1971 July 21.
Bull. Géod., Nouvelle Sér., Année 1972, No. 103, p. 4.

V. V. Vitkevich, 1917 July 2 – 1972.
Astron. Zhurn. Akad. Nauk SSSR, Vol. 49, 683 (1972). In
Russian. English translation in Soviet Astron. AJ, Vol. 16,
No. 3.

C. B. Watts, 1889 October 7 - 1971 July 17.
F. P. Scott.
Quarterly Journ. Roy. Astron. Soc., Vol. 13, 110 - 112 (1972).

L. R. Whitby, 1913 Oct. 5 - 1972 Feb. 6.
J. L. Perdrix, W. G. H. Tregear.
Journ. Astron. Soc. Victoria, Vol. 25, 22 (1972).

008 Observatories, Institutes

Reports, communications and publications of observatories and astronomical institutes are recorded in this section; included are numbered series of reprints. Whenever possible, the numbers of the abstracts referring to the publications are given. Observatories and institutes are listed in alphabetical order of their towns. In some cases observatory publications do not give the name of the town; the following list which gives names and towns of some institutions may serve as an aid in such cases.

Aarne Karjalainen Observatory	Oulu, Finland
Algonquin Radio Observatory	Lake Traverse, Ontario, Canada
Allegheny Observatory	Pittsburgh, Pennsylvania
Archenhold-Sternwarte	Berlin-Treptow, Germany
Arthur J. Dyer Observatory	Nashville, Tennessee
Astronomical Latitude Station, Polish Academy of Sciences	Borowiec, Poland
Bosscha Observatory	Lembang, Indonesia
Boyden Observatory	Bloemfontein, South Africa
Bureau International de l'Heure	Paris, France
Cajigal Observatory	Caracas, Venezuela
California Institute of Technology	Pasadena, California
Cape of Good Hope	Cape Town, South Africa
Carter Observatory	Wellington, New Zealand
Catalina Station	Tucson, Arizona
Cavendish Laboratory	Cambridge, England
Ceskoslovenská Akademie Ved Astronomický Ustav	Praha, Czechoslovakia
Chamberlin Observatory, University of Denver	Denver, Colorado
Commonwealth Observatory	Canberra, Australia
Corralitos Observatory	Las Cruces, New Mexico
David Dunlap Observatory, University of Toronto	Richmond Hill, Ontario
Dearborn Observatory	Evanston, Illinois
Department of Astronomy and Observatory, Univ. California	Los Angeles, California
Department of Astronomy, University of Texas	Austin, Texas
Division Radiophysics, C.S.I.R.O. University Grounds	Sydney, N.S.W., Australia
Dominion Astrophysical Observatory	Victoria, British Columbia
Dominion Observatory	Ottawa, Ontario
Dominion Radio Astrophysical Observatory	Penticton, British Columbia
Dudley Observatory	Albany, New York
Dunsink Observatory	Dublin, Ireland
Engelhardt Observatory	Kazan, R.S.F.S.R.
European Southern Observatory	Hamburg, Federal German Republic
Five College Observatories	Amherst, Massachusetts
Florida State University Radio Observatory	Tallahassee, Florida
Flower and Cook Observatories, University of Pennsylvania	Philadelphia, Pennsylvania
Fraunhofer Institut	Freiburg, Federal German Republic
Georgetown Observatory	Washington, D.C.
Goddard Space Flight Center	Greenbelt, Maryland
Goethe Link Observatory, University of Indiana	Bloomington, Indiana
Hale Observatories	Pasadena, California
Harvard College Observatory	Cambridge, Massachusetts

Harvard Radio Astronomy Station	Cambridge, Massachusetts
Haystack Observatory	Westford, Massachusetts
Heinrich-Hertz-Institut	Berlin, Germany
High Altitude Observatory, University of Colorado	Boulder, Colorado
Institute for Astronomy, University of Hawaii	Honolulu, Hawaii
Institute for Theoretical Astronomy (Institut Teoreticheskoj Astronomii)	Leningrad, R.S.F.S.R.
Institute of Theoretical Astrophysics, Blindern	Oslo, Norway
Inter-American Observatory	Cerro-Tololo, (La Serena), Chile
International Latitude Observatory	Mizusawa, Japan
Joint Institute for Laboratory Astrophysics (JILA)	Boulder, Colorado
Kandilli Observatory	Istanbul, Turkey
Kansas University Observatory	Lawrence, Kansas
Kapteyn Astronomical Laboratory	Groningen, Netherlands
Karl-Schwarzschild-Observatorium	Tautenburg, German Democratic Republic
Kenneth Mees Observatory	Rochester, New York
Kwasan Observatory	Kyoto, Japan
Lamont-Hussey Observatory	Bloemfontein, South Africa
Leander McCormick Observatory University of Virginia	Charlottesville, Virginia
Lee Observatory	Beirut, Lebanon
Leopold-Figl-Observatorium	Wien, Austria
Leuschner Observatory	Berkeley, California
Lick Observatory	Santa Cruz, (Mount Hamilton), California
Lindheimer Astronomical Research Center	Evanston, Illinois
Lockheed Solar Observatory	Saugus, California
Lohrmann-Observatorium für Geodätische Astronomie	Dresden, German Democratic Republic
Louisiana State University Observatory	Baton Rouge, Louisiana
Lowell Observatory	Flagstaff, Arizona
Lunar and Planetary Laboratory	Tucson, Arizona
Max-Planck-Institut für Astronomie	Heidelberg, Federal German Republic
Max-Planck-Institut für Phyik und Astrophysik	München, Federal German Republic
Max-Planck-Institut für Radioastronomie	Bonn, Federal German Republic
McDonald Observatory	Fort Davis, Texas
McMath Hulbert Observatory	Pontiac, Michigan
Michigan State University Observatory	East Lansing, Michigan
Molonglo Radio Observatory, University of Sydney	Sydney, New South Wales
Mount Cuba Observatory	Wilmington, Delaware
Mount John Observatory	Lake Tekapo, New Zealand
Mount Palomar Observatory	Pasadena, California
Mount Wilson Observatory	Pasadena, California
Mullard Radio Astronomy Observatory	Cambridge, England
Narrabri Observatory, University of Sydney	Sydney, New South Wales

National Bureau of Standards **Washington**, D. C.
National Observatory,USA **Kitt Peak**, Arizona
National Radio Astronomy **Charlottesville**, Virginia
  Observatory **Green Bank**, West Virginia
  Tucson, Arizona
New Mexico State
  University Observatory **Las Cruces**, New Mexico
Nizamiah Observatory **Hyderabad**, India
Nuffield Radio Astronomy
  Laboratories, Jodrell Bank
  University of Manchester **Manchester**, England
Observatoire Royal de Belgique **Uccle**, Belgium
Observatorio de Cartuja **Granada**, Spain
Observatorio del Ebro **Tortosa**, Spain
Observatorio Fabra **Barcelona**, Spain
Observatory, University of
  Michigan **Ann Arbor**, Michigan
Ohio State University
  Radio Observatory **Columbus**, Ohio
Ole Roemer-Observatoriet **Aarhus**, Denmark
Owens Valley Radio **Pasadena**, California
  Observatory
Perkins Observatory, Ohio State
  and Wesleyan Universities **Delaware**, Ohio
Purple Mountain Observatory **Nanking**, China
Radcliffe Observatory **Pretoria**, South Africa
Remeis-Sternwarte **Bamberg**,
  Federal German Republic
Republic Observatory **Johannesburg**, South Africa
Rosemary Hill Observatory **Gainesville**, Florida
Royal Radar Establishment,
  Radio Astronomy Division **Malvern**, England

Sagamore Hill Radio Observatory **Bedford**, Massachusetts
Saint-Michel, l'Observatoire **Haute Provence**, France
San Fernando Observatory **El Segundo**, California
Smithsonian Astrophysical
  Observatory **Cambridge**, Massachusetts
Specola Astronomica Vaticana **Castel Gandolfo**, Italy
Specola di Padova **Asiago**, Italy
Sproul Observatory **Swarthmore**, Pennsylvania
Sternberg Observatory **Moscow**, R.S.F.S.R.
Steward Observatory,
  University of Arizona **Tucson**, Arizona
United States Naval Observatory **Washington**, D.C.
University of Florida,
  Radio Observatory **Gainesville**, Florida
University of Illinois Observatory **Urbana**, Illinois
University of Michigan
  Observatories **Ann Arbor**, Michigan
University of South Florida
  Observatory **Tampa**, Florida
Uttar Pradesh State Observatory **Naini Tal**, India
Van Vleck Observatory **Middletown**, Connecticut
Wallace Observatory **Cambridge**, Massachusetts
Warner and Swasey Observatory **Cleveland**, Ohio
Washburn Observatory **Madison**, Wisconsin
West Melton Observatory **Christchurch**, New Zealand
Yale University Observatory **New Haven**, Connecticut
Yerkes Observatory **Williams Bay**, Wisconsin
Zentralinstitut für Astrophysik,
  Sternwarte Babelsberg, (Fach-
  bereich Kosmische Physik) **Potsdam-Babelsberg**, German
  Democratic Republic

008.001 Albany

Dudley Observatory and Department of Astronomy and Space Science, State University of New York at Albany (SUNYA).—Observatory report.   J. L. Weinberg. Bull. American Astron. Soc., Vol. 4, 13 - 21 (1972). – This report covers activities for both institutions in the period January 1969 to September 1971 and activities not included in previous reports.

Dudley Observatory, *Albany, New York,* **Reports**, Nos. 4 (A. G. D. Philip, 07.012.024), 6 (C. Leinert, 07.034.119).

Dudley Observatory, *Albany, New York.* **Reprint** Nos. B35 (C. L. Hemenway, D. S. Hallgren, 05.105.049), B36 (C. L. Hemenway, D. S. Hallgren, A. T. Laudate, H. Patashnick, T. S. Renzema, O. K. Griffith, 06.105.058), B37 (D. S. Hallgren, C. L. Hemenway, 06.105.056), B38 (O. K. Griffith, T. S. Renzema, D. S. Hallgren, C. L. Hemenway, 06.105.057).

Dudley Observatory, *Albany, New York.* **Reprint** Nos. C17 (A. G. D. Philip, N. Sanduleak, 03.160.003), C18 (A. G. D. Philip, 03.160.004), C19 (A. G. D. Philip, 03.112. 003), C21 (J. L. Weinberg, 05.106.018), C23 (A. G. D. Philip, 04.113.019), C24 (A. G. D. Philip, J. S. Drilling, 04.114.127), C25 (J. S. Drilling, A. G. D. Philip, 04.114.128), C26 (J. M. Greenberg, M. S. Hanner, 04.132.019), C27 (M. S. Hanner, J. M. Greenberg, 04.132.020), C28 (A. G. D. Philip, 05.113. 031), C31 (J. M. Greenberg, 04.131.108), C32 (M. S. Hanner, 05.132.013), C33 (P. M. Millman, A. F. Cook, C. L. Hemenway, 05.104.043; C. L. Hemenway, A. Swider, C. Bowman, 05.104.042), C34 (A. G. D. Philip, L. E. Tifft, 06.113.007), C35 (J. M. Greenberg, R. Stoeckly, 05.131.048), C37 (J. M. Greenberg, G. A. Shah, 05.131.067), C39 (A. G. D. Philip, 07.154.007), C42 (J. M. Greenberg, R. T. Wang, L. Bangs, 05.131.052).

008.002 Alger

Université d'Alger. Annales de l'Observatoire Astronomique d'Alger, Tome 3, Fasc. 3 (A. Ghezloun, M. Benhocine, A. Marouf, J. Pham-Van, 07.041.044; A. Ghezloun, J. Pham-Van, Saglio, 07.098.026).

008.003 Alma Ata

Akademiya Nauk Kazakhskoj SSR. Trudy Astrofizicheskogo Instituta, *Alma-Ata*, Vol. 18 (G. Sh. Livshits, I. A. Fedulin, 07.082.113; A. I. Ivanov, G. Sh. Livshits, B. T. Tashenov, I. A. Fedulin, 07.082.114; A. I. Ivanov, G. Sh. Livshits, B. T. Tashenov, I. A. Fedulin, 07.082.115; V. N. Glushko, A. I. Ivanov, G. Sh. Livshits, B. T. Tashenov, I. A. Fedulin, 07.082.116; B. T. Tashenov, Eh. L. Tem, 07.082. 117; B. T. Tashenov, 07.082.118; G. Sh. Livshits, L. M. Musorina, Eh. L. Tem, 07.082.119; A. I. Ivanov, B. T. Tashenov, 07.082.120; B. T. Tashenov, 07.082.121; V. N. Glushko, B. T. Tashenov, Eh. L. Tem, 07.082.122; B. T. Tashenov, 07.082. 123; A. I. Ivanov, G. Sh. Livshits, Eh. L. Tem, 07.082.124; A. I. Ivanov, 07.082.125; A. I. Ivanov, B. T. Tashenov, I. A. Fedulin, 07.034.116; P. N. Bojko, V. E. Pavlov, Ya. A. Tejfel', 07.034.117; V. N. Glushko, G. Sh. Livshits, B. T. Tashenov, V. A. Molchanov, 07.082.126; T. P. Toropova, 07.082.127; T. P. Toropova, S. O. Obasheva, L. L. Solntseva, 07.082.128; K. M. Salamakhin, T. P. Toropova, 07.082.129; Yu. I. Rudnev, L. A. Sataeva-Egorova, 07.082.130; N. M. Ibraimov, 07.082. 131; G. A. Kirienko, T. P. Toropova, 07.082.132; G. A. Kirienko, 07.034.118; T. P. Toropova, K. M. Salamakhin, A. P. Ten, 07.082.133), 21 (V. M. Tereshchenko, A. V. Kharitonov, 07.114.132).

008.004 Ames

Erwin W. Fick Observatory, Iowa State University, Ames, Iowa.—Observatory report. W. I. Beavers.
Bull. American Astron. Soc., Vol. 4, 29 - 30 (1972).

008.005 Amherst

Five College Astronomy Department: Amherst College, Amherst, Massachusetts; Hampshire College, Amherst, Massachusetts; Mount Holyoke College, South Hadley, Massachusetts; Smith College, Northampton, Massachusetts; University of Massachusetts, Amherst, Massachusetts.—Observatory report. W. M. Irvine.
Bull. American Astron. Soc., Vol. 4, 30 - 33 (1972).

Department of Physics and Astronomy, University of Massachusetts, Amherst, Mass., Separate prints (R. L. Harkness, Jr., 02.062.023; G. R. Huguenin, J. H. Taylor, R. M. Hjellming, C. M. Wade, 06.141.240; J. H. Taylor, G. R. Huguenin, 06.141.022; A. Uesugi, W. M. Irvine, Y. Kawata, 05.091.028; W. M. Irvine, A. P. Lane, 06.100.011; D. J. van Blerkom, 06.097.036; E. R. Harrison, 06.151.011; G. R. Huguenin, R. N. Manchester, J. H. Taylor, 06.141.152; J. F. Appleby, W. M. Irvine, 06.101.001; T. Arny, 06.065.078; K. J. Gordon, 06.158.086; C. P. Gordon, 06.114.110; S.-Y. Ho, 07.034.115; E. R. Harrison, R. G. Lake, 07.151.010; D. van Blerkom, T. T. Arny, 07.133.001; J. H. Taylor, G. R. Huguenin, R. M. Hirsch, 07.142.011; W. M. Irvine, 07.008.005; R. N. Manchester, J. H. Taylor, 07.141.504; W. A. Dent, 07.141.029).

008.006 Ankara

Communication of the Department of Astronomy of Ankara University, Nos. 49 (C. Aydin, AJB 68, 104.05), 50 (S. İşlik, AJB 68, 104.55), 52 (C. Aydin, AJB 68, 104.118), 55 (N. Dŏgan, 07.075.019).

008.007 Ann Arbor

The University of Michigan, Department of Astronomy, Ann Arbor, Michigan.—Observatory report.
W. A. Hiltner.
Bull. American Astron. Soc., Vol. 4, 119 - 123 (1972).

008.008 Arcetri

Elenco dei lavori eseguiti dal personale dell'Osservatorio Astrofisico di Arcetri durante il 1970.
Boll. Geod. Sci. Affini, Anno 31, p. 97 - 99 (1972).

008.009 Arecibo

Arecibo Observatory today – I.
W. E. Shawcross.
Sky Telescope, Vol. 43, 214 - 217, 228 (1972).

Arecibo Observatory today – II.
W. E. Shawcross.
Sky Telescope, Vol. 43, 293 - 295 (1972).

008.010 Athen

Astronomical Institute, National Observatory of Athens. – Annual report 1970. D. Kotsakis.
Annual Rep. Astron. Inst. Greece 1970, p. 3 - 5 (1971).

Department of Astronomy, University of Athens. – Annual report 1970. D. Kotsakis.
Annual Rep. Astron. Inst. Greece 1970, p. 6 - 7 (1971).

Department of Astronomy, Technical University of Athens. – Annual report 1970. J. Argyrakos.
Annual Rep. Astron. Inst. Greece 1970, p. 10 - 11 (1971).

Research Center for Astronomy and Applied Mathematics, Academy of Athens. – Annual report 1970.
J. N. Xanthakis.
Annual Rep. Astron. Inst. Greece 1970, p. 11 - 12 (1971).

Contributions from the Research Center for Astronomy and Applied Mathematics, Academy of Athens, Series I (Astronomy), No. 22 (J. Xanthakis, 04.072.024), 23 (J. Xanthakis, 07.072.043), 24 (C. J. Macris, 07.073.071), 25 (J. Xanthakis, 07.073.072), 26 (C. E. Alissandrakis, C. J. Macris, 07.073.080), 27 (C. E. Alissandrakis, C. J. Macris, 06.073.059), 28 (S. Koutchmy, C. Macris, 06.073.075), 29 (C. E. Alissandrakis, C. J. Macris, 07.073.001).

008.011 Auckland

Auckland Observatory.
Inform. Bull. Southern Hemisph., No. 19, p. 25 (1971).
Current research report.

008.012 Bamberg

Dr. Remeis-Sternwarte Bamberg, Astronomisches Institut der Universität Erlangen-Nürnberg. – Report 1971.
W. Strohmeier.
Mitt. Astron. Ges., No. 31, p. 235 - 236 (1972).

Veröffentlichungen der Remeis-Sternwarte Bamberg, Astronomisches Institut der Universität Erlangen–Nürnberg, Vol. 9, No. 100 (07.012.006).

008.013 Baton Rouge

Louisiana State University Observatory, Baton Rouge, Louisiana.—Observatory report. A. U. Landolt.
Bull. American Astron. Soc., Vol. 4, 103 - 105 (1972).

Contributions of the Louisiana State University Observatory, Nos. 49 (H. E. Bond, A. U. Landolt, 06.126.012), 51 (P. Lee, C. L. Perry, 06.113.016), 52 (H. E. Bond, C. L. Perry, W. P. Bidelman, 06.112.014), 53 (P. Lee, 06.122.104), 54 (H. E. Bond, C. L. Perry, 06.153.022), 55 (A. U. Landolt, 06.118.005), 56 (D. L. Crawford, J. C. Golson, A. U. Landolt, 06.113.042), 57 (S. N. Rasband, 06.162.043).

008.014 Bedford

Sagamore Hill Radio Observatory, Air Force Cambridge Research Laboratories, Bedford, Massachusetts.

Observatory report.   J. P. Castelli.
Bull. American Astron. Soc., Vol. 4, 166 - 167 (1972).

008.015   Beirut

Lee Observatory, American University of Beirut, Lebanon. Monthly Bulletin, Astronomical Section, 1971 November – 1972 March (F. Bruin, H. Hourani, N. G. Bustati, 07.075.017).

008.016   Belo Horizonte

Instituto de Ciencias Exatas da Universidade Federal de Minas Gerais. (Institute of Exact Sciences of the Federal University of Minas Gerais).
Inform. Bull. Southern Hemisph., No. 19, p. 12 (1971).
Current research report.

008.017   Beograd

University of Beograd, Faculty of Sciences. Publications of the Department of Astronomy (Publications de la Chaire d'Astronomie), Nos. 1 (M. Vukićević-Karabin, 07.076.039; D. Djurović, V. Radogostić, 07.044.031; D. Djurović, V. Radogostić, 07.031.038), 3 (G. Teleki, B. Ševarlić, 07.031.039; M. Vukićević-Karabin, M. Dimitrijević, 07.077.055; J. Milogradov-Turin, 07.157.014; J. Lazović, 07.098.027).

008.018   Berkeley

Research Units and Academic Departments, University of California: Berkeley, Los Angeles, San Diego, and Santa Cruz. – I. Berkeley Campus. – Report 1970 – 1971.
Bull. American Astron. Soc., Vol. 4, 272 - 278 (1972).

008.019   Berlin

Lehrstuhl für Astrophysik der Technischen Universität. – Report 1971.   K. Hunger.
Mitt. Astron. Ges., No. 31, p. 237 - 238 (1972).

Heinrich-Hertz-Institut. Solare Beobachtungsergebnisse. Deutsche Akademie der Wissenschaften zu Berlin, Zentralinstitut für Solar-Terrestrische Physik, Berlin-Adlershof. HHI Solar Data, Vol. 22, 1971 November – December (E. A. Lauter, A. Böhme, F. W. Jäger, F. Fürstenberg, H. Künzel, D. Scholz, S. Böhm, 07.075.020); Vol. 23, 1972 January – February (C.-U. Wagner, A. Böhme, F. Fürstenberg, D. Scholz, S. Böhm, 07.075.021).

Heinrich-Hertz-Institut. Supplement Series of Solar Data. Deutsche Akademie der Wissenschaften zu Berlin, Zentralinstitut für Solar-Terrestrische Physik, Berlin-Adlershof. HHI Suppl. Ser. Solar Data, Vol. 2, No. 6 (H. Daene, V. V. Fomichev, 07.077.058).

008.020   Berlin-Treptow

75 Jahre Archenhold-Sternwarte, 1896 - 1971.

D. Wattenberg.
Vorträge und Schriften, Archenhold-Sternw. Berlin-Treptow, No. 41, 24 pp. (1971).

Archenhold-Sternwarte Berlin-Treptow. Sonderdruck No. 16 (D. Wattenberg, 07.004.042).

Mitteilungen der Archenhold-Sternwarte Berlin-Treptow, Nos. 89 (D. Wattenberg, 07.005.017), 90 (D. Wattenberg, 06.007.000), 91 (D. B. Herrmann, 07.004.047), 92 (D. Wattenberg, 06.005.018), 93 (D. Wattenberg, 06.009.022).

Veröffentlichungen der Archenhold-Sternwarte Berlin-Treptow, No. 5 (D. B. Herrmann, 07.002.019).

Vorträge und Schriften, Archenhold-Sternwarte, Berlin-Treptow, No. 39 (D. B. Herrmann, 07.013.009), 40 (E. Rothenberg, 07.047.023), 41 (D. Wattenberg, 07.008.020). 42 (D. Wattenberg, 07.004.046).

008.021   Besançon

Compte-rendu d'activité du Service Chronométrique du 1er avril 1970 au 31 mars 1971.   A. Remond, M. Vincent.
Ann. Françaises Chronométrie Micromécanique, Année 1971, p. 149 - 151.

Observatoire de Besançon: Palmarès du 69me concours chronométrique.
Ann. Françaises Chronométrie Micromécanique, Année 1971, p. 153 - 157.

008.022   Bloemfontein

Boyden Observatory.   A. H. Jarrett.
Inform. Bull. Southern Hemisph., No. 19, p. 28 (1971).
Current research report.

008.023   Bloomington

Goethe Link Observatory, Indiana University, Bloomington, Indiana. – Observatory report.
F. K. Edmondson.
Bull. American Astron. Soc., Vol. 4, 38 - 40 (1972).

008.024   Bochum

Astronomisches Institut. – Report 1971.
T. Schmidt-Kaler.
Mitt. Astron. Ges., No. 31, p. 238 - 243 (1972).

Bereich Extraterrestrische Physik in der Abteilung XII. – Report 1971.   R.-H. Giese.
Mitt. Astron. Ges., No. 31, p. 243 - 246 (1972).

008.025   Bologna

Programma di ricerca del Laboratorio Nazionale di Radioastronomia Bologna.   A. Braccesi.
Atti XIII Riunione Soc. Astron. Italiana, Trieste 1969, (see 012.010), p. 167 - 171 (1970).

Relazione sull'attività scientifica dell'Osservatorio Universitario di Bologna. G. Mannino.
Atti XIII Riunione Soc. Astron. Italiana, Trieste 1969, (see 012.010), p. 173 (1970).

Osservatorio Astronomico Universitario di Bologna. Notizie e Rassegne, Nos. 44 - 51 (F. S. Delli Santi, E. Nasi, 07.077.054).

Pubblicazioni dell'Osservatorio Astronomico, Universitario di Bologna, Vol. 10, Nos. 10 (V. Castellani, P. Giannone, A. Renzini, 04.154.012), 11 (V. Castellani, L. Puppi, A Renzini, 05.064.020), 12 (V. Castellani, P. Giannone, A. Renzini, 05.065.030), 13 (V. Castellani, P. Giannone, A. Renzini, 05.065.092), 14 (A. Renzini, 05.065.140), 15 (V. Castellani, P. Giannone, A. Renzini, 05.065.113), 16 (A. Renzini, 06.065.092), 17 (V. Castellani, P. Giannone, A. Renzini, 06.065.036), 18 (F. Bertola, F. Lucchin, E. Nasi, 06.158.130), 19 (P. L. Battistini, M. Fracassini, L. E. Pasinetti, 06.121.075).

008.026   Bonn

Astronomische Institute der Universität Bonn. I. Sternwarte mit Observatorium Hoher List; II. Radioastronomisches Institut; III. Institut für Astrophysik und Extraterrestrische Forschung. – Report 1971.
O. Hachenberg, W. Priester, H. Schmidt.
Mitt. Astron. Ges., No. 31, p. 246 - 262 (1972).

Max-Planck-Institut für Radioastronomie. – Report 1971.   O. Hachenberg, P. G. Mezger, R. Wielebinski.
Mitt. Astron. Ges., No. 31, p. 262 - 278 (1972).

Max-Planck-Institut für Radioastronomie, Bonn. Sonderdrucke, Nos. 10 (K. S. Stankevich, R. Wielebinski, W. E. Wilson, 04.066.032), 12 (R. Wielebinski, 04.033.011), 13 (T. L. Landecker, R. Wielebinski, 03.033.051), 14 (T. L. Landecker, R. Wielebinski, 04.157.012), 15 (R. Wielebinski, 05.033.003), 16 (N. J. Keen, 07.033.034), 17 (K. Rohlfs, 05.131.055), 18 (E. Fürst, 05.077.021), 19 (N. J. Keen, 05. 033.042), 20 (O. Hachenberg, 04.033.040), 21 (O. Hachenberg, P. G. Mezger, R. Wielebinski, 05.008.022), 22 (K. Rohlfs, 05.131.099), 23 (H. J. Wendker, 05.131.100), 25 (G. C. Hunt, 06.141.001), 26 (P. G. Mezger, 06.131.117), 27 (E. M. Berkhuijsen, C. G. T. Haslam, C. J. Salter, 06.155.009), 28 (E. M. Berkhuijsen, 07.157.004), 28a (E. M. Berkhuijsen, 06.157.002), 29 (C. G. Wynn-Williams, D. Downes, T. L. Wilson, 06.141. 120), 30 (K. Rohlfs, 07.131.150), 31 (R. Wielebinski, 07.033. 035), 32 (E. Churchwell, 06.132.031), 33 (W. Wassenberg, 06.077.032), 34 (M. Walmsley, M. Grewing, 06.131.096), 35 (K. Rohlfs, 07.155.022), 36 (K. Rohlfs, 07.131.013), 37 (F. F. Gardner, J. B. Whiteoak, 07.155.030), 38 (T. L. Wilson, W. J. Altenhoff, 07.141.024), 39 (R. Schwartz, 06.153.026), 40 (W. M. Goss, 07.141.117), 42 (J. B. Whiteoak, F. F. Gardner, 07.131.077), 44 (K. I. Kellermann, M. M. Davis, I. I. K. Pauliny-Toth, 06.141.189), 52 (J. B. Whiteoak, 07.158.123).

008.027   Borowiec

Polish Academy of Sciences, Astronomical Latitude Station, Borowiec, Circular, Nos. 119–120 (07.044.039).

008.028   Boulder

Joint Institute for Laboratory Astrophysics of the National Bureau of Standards and the University of Colorado, Boulder, Colorado.–Observatory report.   L. Oster.
Bull. American Astron. Soc., Vol. 4, 59 - 65 (1972).

008.029   Brno

Contributions of the Observatory and Planetarium in Brno, No. 13 (V. Znojil, 07.104.038).

008.030   Buenos Aires

Instituto de Astronomiá y Física del Espacio. (Institute of Astronomy and Space Physics).
Inform. Bull. Southern Hemisph., No. 19, p. 6 - 7 (1971). Current research report.

008.031   Byurakan

Byurakan Astrophysical Observatory, Armenia, USSR, Reprints, Nos. 67 (V. S. Oskanian, V. Yu. Terebizh, 05.122.116), 68 (V. V. Papoyan, D. M. Sedrakian, E. V. Chubarian, 05.126.046), 69 (Yu. L. Vartanian, A. V. Hovsepian, 05.126.047), 70 (E. A. Tyoonov, A. D. Chernin, 05.066.065), 71 (M. A. Arakelian, E. A. Dibay, V. F. Yesipov, B. E. Markarian, 06.158.033), 72 (A. T. Kalloghlian, 06.158.034), 73 (G. S. Hajian, Yu. L. Vartanian, 06.065.051), 74 (G. G. Arutyunian, D. M. Sedrakian, 06.065.052), 75 (V. S. Oskanian, V. Yu. Terebizh, 06.122.052), 76 (K. A. Grigorian, M. A. Eritsian, 06.122.053; Z. F. Seydov, T. A. Eminzade, 06.126.003), 77 (V. A. Ambartsumian, L. V. Mirzoyan, E. S. Parsamian, H. S. Chavushian, L. K. Erastova, 06.122.094), 78 (D. Chalonge, L. Divan, L. V. Mirzoyan, 06.114.081), 79 (E. Ye. Khachikian, D. W. Weedman, 06.158.092), 80 (M. A. Arakelian, 06.141.161), 81 (G. G. Arutyunian, D. M. Sedrakian, E. V. Chubarian, 06.126.013), 82 (E. S. Parsamian, 06.122.097).

Soobshcheniya Byurakanskoj Observatorii, vyp. (No.) 45 (G. A. Gurzadyan, 07.051.018; G. A. Gurzadyan, E. A. Harutyunyan, 07.051.019; V. I. Patsaev, J. Ohanesyan, G. A. Gurzadyan, 07.114.103; M. N. Krmoyan, A. Z. Zakharyan, Sh. M. Harutyunyan, 07.051.020; G. A. Gurzadyan, J. Ohanesyan, P. A. Yepremyan, 07.034.050; V. M. Uvarova, M. P. Shpolskiy, A. N. Oshurkova, J. Ohanesyan, N. V. Uvarova, 07.036.007).

008.032   Cambridge, Engl.

Astronomy at Cambridge: Reshuffle not according to Hoyle.   N. Wade.
Science, Vol. 176, 999 - 1000 (1972).

Departure of Professor Hoyle.
Nature, Vol. 236, 417 - 418 (1972).

Cambridge astronomy without Sir Fred.
Nature, Vol. 236, 419 (1972).

008.033   Cambridge, Mass.

Harvard College Observatory, Cambridge, Massachusetts.–Observatory report.   L. Goldberg.

Bull. American Astron. Soc., Vol. 4, 40 - 45 (1972).

Smithsonian Astrophysical Observatory, Cambridge, Massachusetts.—Observatory report.   F. L. Whipple.
Bull. American Astron. Soc., Vol. 4, 167 - 175 (1972).

Smithsonian Institution. Astrophysical Observatory. Research in Space Science. SAO Special Reports, Nos. 310 (S. I. Gaposhkin, 07.159.023), 337 (P. W. Hodge, 07.159.005), 338 (E. Chipman, 07.076.020), 340 (W. W. Hauck, Jr., 07.021. 001), 341 (J. L. Elliot, 07.142.031), 342 (E. M. Gaposchkin, 07.081.028).

Wallace Observatory, Massachusetts Institute of Technology, Cambridge, Massachusetts.—Observatory report. T. B. McCord.
Bull. American Astron. Soc., Vol. 4, 187 - 188 (1972).

008.034   Cape Town

Department of Astronomy, University of Cape Town. A. P. Fairall.
Inform. Bull. Southern Hemisph., No. 19, p. 28 (1971). Current research report.

Royal Observatory at the Cape Province of Good Hope.
Inform. Bull. Southern Hemisph., No. 19, p. 29 - 30 (1971). Current research report.

Royal Observatory Bulletins, (Joint Publications of the Royal Greenwich Observatory, Herstmonceux, Royal Observatory, Cape of Good Hope), Nos. 167 (A. L. T. Powell, 07.114.040), 168 (A. L. T. Powell, 07.064.015), 170 (R. Woolley, A. Savage, 07.122.029).

008.035   Carloforte

Annual report for the year 1971.   E. Proverbio.
Pubbl. Stazione Astron. Internazionale Latitudine, Carloforte - Cagliari, Nuova Ser., No. 25, 14 pp. (1972).

Programma di ricerca della Stazione di Carloforte e per l'organizzazione dell'Osservatorio Astronomico di Cagliari.   E. Proverbio.
Atti XIII Riunione Soc. Astron. Italiana, Trieste 1969, (see 012.010), p. 217 - 219 (1970).

Circolari della Stazione Astronomica Internazionale di Latitudine, Carloforte—Cagliari, Serie B (2), No. 2 (E. Proverbio, S. Uras, 07.045.032).

Pubblicazioni della Stazione Astronomica Internazionale di Latitudine, Carloforte—Cagliari, Nuova Serie, Nos. 10 (E. Proverbio, 07.045.033), 15 bis (E. Proverbio, 06.008.022), 16 (R. O. Vicente, 07.044.044), 17 (E. Proverbio, 05.082.083), 18 (E. Proverbio, F. Carta, F. Mazzoleni, 06.045.026), 19 (E. Proverbio, S. Uras, 06.045.029), 20 (E. Proverbio, S. Uras, 06.045.027), 23 (E. Proverbio, V. Quesada, 07.045.034), 25 (E. Proverbio, 07.008.035).

008.036   Castel Gandolfo

Specola Vaticana. Annual report 1971: Report of the Astronomical Observatory; Report of the Astrophysical Labo-

ratory.   P. J. Treanor, J. Junkes.
Printed in Vatican City, 14 pp. (1972).

Specola Vaticana.   M. McCarthy.
Atti XIII Riunione Soc. Astron. Italiana, Trieste 1969, (see 012.010), p. 215 (1970).

Ricerche Astronomiche. Specola Vaticana, Città del Vaticano, Vol. 8, No. 11 (G. V. Coyne, 07.131.148).

Vatican Observatory Publications, Specola Vaticana, Città del Vaticano, Vol. 1, Nos. 2 (O. Van De Vyver, 07.094. 232), 3 (F. C. Bertiau, 07.113.044).

008.037   Catania

Notizie sui programmi di ricerca dell'Osservatorio Astrofisico di Catania.   G. Godoli.
Atti XIII Riunione Soc. Astron. Italiana, Trieste 1969, (see 012.010), p. 175 - 189 (1970).

Report from the Catania Astrophysical Observatory (1971).   G. Godoli.
Mem. Soc. Astron. Italiana, Nuova Ser., Vol. 43, 179 - 184 (1972).

Osservatorio Astrofisico di Catania, Pubblicazione, No. 147 (G. Godoli, 07.008.037).

008.038   Cerro Tololo

Kitt Peak National Observatory, Tucson, Arizona, and Cerro Tololo Inter-American Observatory, La Serena, Chile.—Observatory reports.   V. M. Blanco, N. U. Mayall.
Bull. American Astron. Soc., Vol. 4, 66 - 97 (1972).

008.039   Charlottesville

Leander McCormick Observatory, University of Virginia, Charlottesville.—Observatory report. L. W. Fredrick.
Bull. American Astron. Soc., Vol. 4, 97 - 100 (1972).

Publications of the Leander McCormick Observatory of the University of Virginia, Vol. 16 (P. A. Ianna, 07.012.026).

National Radio Astronomy Observatory, Charlottesville, Virginia, Green Bank, West Virginia, and Tucson, Arizona.—Observatory reports.   D. S. Heeschen.
Bull. American Astron. Soc., Vol. 4, 140 - 149 (1972).

008.040   Cincinnati

Minor Planets Circulars (MPC), Nos. 3293 - 3352 (P. Herget, 07.098.029).

008.041   Cleveland

Warner and Swasey Observatory, Case Western Reserve University, Cleveland, Ohio.—Observatory report. W. P. Bidelman.

Bull. American Astron. Soc., Vol. 4, 188 - 193 (1972).

Warner and Swasey Observatory, Case Western Reserve University, Reprints, Nos. 186 (J. F. Dolan, 03.142.019), 193 (M. P. Fitzgerald, N. Houk, 03.114.034), 194 (S. B. Parsons, 03.064.013), 195 (J. F. Dolan, 03.142.013) 196 (S. Wyckoff, 04.114.075), 197 (M. P. Fitzgerald, 03.113.018), 198 (W. H. Wooden II, 03.113.038), 199 (S. W. McCuskey, 03.155.033), 200 (A. G. D. Philip, N. S. Sanduleak, 03.160.003), 201 (C. B. Stephenson, H. E. Ross, 03.114.104), 202 (W. J. Webster, Jr., W. J. Altenhoff, 03.132.012), 203 (S. B. Parsons, 05.064.024), 205 (W. J. Webster, Jr., W. J. Altenhoff, 04.141.061), 206 (W. B. Weaver, 04.152.003), 209 (S. B. Parsons, 05.115.005), 210 (S. B. Parsons, G. D. Bouw, 05.115.006), 211 (F. Stienon, 05.034.016), 212 (N. Sanduleak, 05.122.037), 214 (N. Sanduleak, 05.114.044), 215 (W. P. Bidelman, 05.122.040), 218 (C. B. Stephenson, 05.126.021).

008.042 Cluj

Ein Besuch der Universitäts-Sternwarte Cluj.
M. Bressler.
Sternenbote, 15. Jahrgang, p. 84 - 86 (1972).

008.043 College Park

Astronomy program, University of Maryland, College Park, Maryland.—Observatory report.   G. Westerhout.
Bull. American Astron. Soc., Vol. 4, 109 - 117 (1972).

008.044 Columbus

The Observatories of the Ohio State and Ohio Wesleyan Universities, Columbus and Delaware, Ohio.
Observatory reports.   A. Slettebak.
Bull. American Astron. Soc., Vol. 4, 152 - 156 (1972).

Ohio State University Radio Observatory, Columbus, Ohio.—Observatory report.   J. D. Kraus.
Bull. American Astron. Soc., Vol. 4, 157 (1972).

008.045 Córdoba

Observatorio Astronómico. (Astronomical Observatory, National University of Córdoba).
Inform. Bull. Southern Hemisph., No. 19, p. 7 (1971).
Current research report.

Boletin del Instituto de Matematica, Astronomía y Fisica, Universidad Nacional de Córdoba, Vol. 3, No. 2 (H. A. Dottori, 07.158.170; L. A. Milone, 07.152.013).

Observatorio Astronómico (*Universidad Nacional de Córdoba, Argentina*), **Tirada Aparte,** Nos. 186 (E. L. Agüero, 05.158.113), 188 (J. L. Sersic, 06.008.029), 191 (R. F. Sisteró, 06.121.005), 192 (R. F. Sisteró, 06.162.032), 195 (J. J. Clariá, 06.153.008), 196 (H. Dottori, G. Carranza, 06.155. 033), 197 (J. L. Sĕrsic, 07.158.097).

008.046 Delaware

The Observatories of the Ohio State and Ohio

Wesleyan Universities, Columbus and Delaware, Ohio.
Observatory reports.   A. Slettebak.
Bull. American Astron. Soc., Vol. 4, 152 - 156 (1972).

Contribution from the Perkins Observatory. The Ohio State University and Ohio Wesleyan University, Series II, Nos. 28 (R. F. Wing, 05.114.110), 29 (T. P. Roark, 06.122. 035), 30 (B. F. Peery, Jr., P. C. Keenan, I. R. Marenin, 06.114.076).

Contributions from the Perkins Observatory, Ohio State — Ohio Wesleyan Universities. Series I, Nos. 125 (W. S. Kovach, 06.133.011), 126 (S. J. Czyzak, L. H. Aller, J. B. Kaler, 06.133.009), 127 (P. C. Keenan, 06.114.003), 128 (G. W. Lockwood, R. F. Wing, 06.122.083), 129 (L. H. Aller, S. J. Czyzak, E. G. Buerger, P. Lee, 07.133.006), 130 (T. P. Roark, J. H. Baumert, N. M. White, 07.118.001), 131 (P. C. Keenan, 07.122.051), 132 (P. L. Byard, K. E. Kissell, 07.074.010).

008.047 Dresden

Mitteilungen des Lohrmann-Observatoriums der Technischen Universität Dresden, Nos. 20 (E. Maase, 05.034. 112), 21 (K.-G. Steinert, 07.032.052), 22 (K.-G. Steinert, 07.034.121), 23 (S. Wächter, 07.082.134), 24 (H. U. Sandig, 07.041.045).

Technische Universität Dresden, Lohrmann-Observatorium, Zirkular, Nos. 51 - 55 (07.045.035).

008.048 Dublin

Dunsink Observatory. — Report for the year ending 1971 March 31.   P. A. Wayman.
Quarterly Journ. Roy. Astron. Soc., Vol. 13, 90 - 92 (1972).

008.049 East Lansing

Michigan State University Observatory, East Lansing, Michigan. — Report 1970 — 1971.   A. P. Linnell.
Bull. American Astron. Soc., Vol. 4, 291 - 293 (1972).

008.050 Edinburgh

Communications from the Royal Observatory, Edinburgh, Nos. 108 (M. J. Smyth, G. M. W. Cork, J. Harris, T. Wallace, 05.114.090), 119 (B. N. G. Guthrie, 06.114.084), 120 (A. Kelly, 06.131.092), 121 (J. W. Campbell, 06.113.037), 124 (B. N. G. Guthrie, 07.114.041), 125 (R. D. Eberst, 07.081.015), 127 (T. J. Lee, L. Gowland, V. C. Reddish, 05.131.130).

008.051 El Segundo

The Aerospace Corporation, El Segundo, California: (I) Electronics Research Laboratory; (II) Space Physics Laboratory.—Observatory report.   G. A. Paulikas.
Bull. American Astron. Soc., Vol. 4, 1 - 5 (1972).
Research in astronomy at Aerospace is carried out by groups in the Electronics Research Laboratory and the Space Physics Laboratory, which operates the Aerospace Corpora-

tion's San Fernando Observatory and also conducts solar research utilizing satellites.

008.052 Evanston

Lindheimer Astronomical Research Center, and Dearborn Observatory, Evanston, Illinois; Corralitos Observatory, Las Cruces, New Mexico.—Observatory reports. J. A. Hynek.
Bull. American Astron. Soc., Vol. 4, 100 - 103 (1972).

008.053 Flagstaff

Lowell Observatory, Flagstaff, Arizona.—Observatory report. J. S. Hall.
Bull. American Astron. Soc., Vol. 4, 105 - 108 (1972).

Lovell Observatory Bulletin, *Flagstaff, Arizona,* No. 158 = Vol. 7, No. 21 (H. L. Giclas, R. Burnham, Jr., N. G. Thomas, 07.112.015).

008.054 Frankfurt

Astronomisches Institut der Universität. – Report 1971. W. Gleissberg.
Mitt. Astron. Ges., No. 31, p. 278 - 280 (1972).

008.055 Frascati

The Laboratorio di Astrofisica Spaziale of the National Research Council—Frascati (Rome). V. Castellani.
Mem. Soc. Astron. Italiana, Nuova Ser., Vol. 43, 185 - 193 (1972).

008.056 Freiburg

Fraunhofer-Institut mit den Observatorien Schauinsland und Anacapri. – Report 1971.
Mitt. Astron. Ges., No. 31, p. 281 - 285 (1972).

Fraunhofer Institut, Map of the Sun.
1972 January 1 – June 30 (07.075.012).

008.057 Gainesville

University of Florida Observatories, Gainesville, Florida: Rosemary Hill Observatory, F. B. Wood; University of Florida Radio Observatory, A. G. Smith.—Observatory reports.
Bull. American Astron. Soc., Vol. 4, 33 - 36 (1972).

Rosemary Hill Observatory, Department of Physics and Astronomy, University of Florida, Gainesville, Florida, Contributions, Nos. 18 (K.-Y. Chen, W. J. Rhein, 06.121.053), 19 (A. G. Smith, H. W. Schrader, W. W. Richardson, 06.036.003), 20 (F. B. Wood, 05.121.056), 23 (J. D. Rosendhal, M. S. Snowden, 06.159.009), 24 (R. M. Williamon, 06.121.040), 25 (T. F. Collins, 06.121.041), 26 (G. H. Fol-

som, A. G. Smith, R.˙L. Hackney, K. R. Hackney, R. J. Leacock, 06.141.188).

008.058 Gothenburg

Research Laboratory of Electronics and Onsala Space Observatory, Chalmers University of Technology, Gothenburg, Sweden. **Research Report,** Nos. 104 (G. P. R. Netzler, 07.035.008), 109 (T. Cato, J. Ellder, B. Höglund, O. E. H. Rydbeck, B. Rönnäng, A. Sume, 07.131.136).

008.059 Göttingen

Universitäts-Sternwarte Göttingen und Institut für Sonnenforschung Locarno-Orselina (Tessin). – Report 1971. W. Deinzer, R. Kippenhahn, E. H. Schröter, H. H. Voigt.
Mitt. Astron. Ges., No. 31, p. 285 - 292 (1972).

008.060 Graz

Astronomisches Institut (Universitäts-Sternwarte). Sonnenobservatorium Kanzelhöhe. – Report 1971. H. Haupt.
Mitt. Astron. Ges., No. 31, p. 292 - 295 (1972).

Mitteilungen der Universitätssternwarte Graz, No. 11 (H. J. Schober, 07.034.120).

008.061 Green Bank

National Radio Astronomy Observatory, Charlottesville, Virginia, Green Bank, West Virginia, and Tucson, Arizona.—Observatory reports. D. S. Heeschen.
Bull. American Astron. Soc., Vol. 4, 140 - 149 (1972).

National Radio Astronomy Observatory, *Green Bank,* **Reprints,** Series A, Nos. 219, (J. F. C. Wardle, 05.141.199), 220 (A. G. Willis, J. R. Dickel, 05.141.197), 221 (D. S. De Young, 06.141.116), 222 (W. J. Webster, Jr., W. J. Altenhoff, J. E. Wink, 06.132.033), 223 (F. F. Gardner, D. K. Milne, P. G. Mezger, T. L. Wilson, 04.141.015), 224 (C. E. Heiles, B. E. Turner, 05.131.075), 225 (J. F. C. Wardle, 05.158.057), 226 (R. H. Rubin, P. Palmer, 05.133.022), 227 (R. M. Hjellming, C. M. Wade, 06.141.109), 228 (J. Maslowski, 06.141.030), 229 (J. Maslowski, 07.141.016), 230 (D. Buhl, L. E. Snyder, 06.033.001), 231 (T. D. Kinman, E. K. Conklin, 06.141.173), 232 (J. H. Taylor, G. R. Huguenin, R. M. Hirsch, R. N. Manchester, 06.141.177), 233 (R. N. Manchester, 05.141.209), 234 (W. B. Burton, 07.131.012), 235 (J. Edrich, 07.033.024), 236 (R. J. Mattauch, 07.034.054), 237 (D. Buhl, 06.131.087), 238 (Y. Terzian, B. Balick, 07.133.002), 239 (F. J. Kerr, G. R. Knapp, 06.131.130), 240 (D. S. Heeschen, 07.008.061), 241 (A. H. Bridle, M. J. L. Kesteven, 06.141.223), 242 (M. M. Davis, 06.141.225).

National Radio Astronomy Observatory, *Green Bank,* **Reprints,** Series B, Nos. 271 (P. Thaddeus, R. W. Wilson, M. Kutner, A. A. Penzias, K. B. Jefferts, 06.132.015), 272 (R. W. Wilson, P. M. Solomon, A. A. Penzias, K. B. Jefferts, 06.114.075), 273 (R. H. Rubin, G. W. Swenson, Jr., R. C. Benson, H. L. Tigelaar, W. H. Flygare, 06.131.083), 274 (G. R. Huguenin, R. N. Manchester, J. H. Taylor, 06.141.152), 275 (K. J. Gordon, 06.158.086), 276 (R. N. Manchester, M. A.

Gordon, 06.131.099), 277 (B. E. Turner, 06.131.118), 278 (L. E. Snyder, D. Buhl, 06.131.123), 279 (B. Zuckerman, M. Morris, B. E. Turner, P. Palmer, 06.131.100), 280 (F. Biraud, 05.141.204), 281 (M. H. Cohen, W. Cannon, G. H. Purcell, D. B. Shaffer, J. J. Broderick, K. I. Kellermann, D. L. Jauncey, 06.141.202), 282 (C. M. Wade, R. M. Hjellming, K. I. Kellermann, J. F. C. Wardle, 06.158.098), 283 (K. I. Kellermann, M. M. Davis, I. I. K. Pauliny-Toth, 06.141.189), 284 (M. A. Gordon, M. S. Roberts, 06.131.106), 285 (B. E. Turner, R. H. Rubin, 06.131.109), 286 (M. Morris, B. Zuckerman, P. Palmer, B. E. Turner, 06.114.119), 287 (C. M. Wade, R. M. Hjellming, 06.142.077), 288 (B. E. Turner, C. Heiles, 06.131.108), 289 (E. Tademaru, 06.141.032), 290 (P. S. Berger, M. Simon, 07.074.001), 291 (B. E. Turner, 07.131.023), 292 (W. C. Saslaw, D. S. De Young, 06.151.049), 293 (M. S. Roberts, 07.158.066), 294 (R. N. Manchester, E. Tademaru, 06.141.241), 295 (R. N. Manchester, 07.141.508), 296 (G. R. Huguenin, J. H. Taylor, R. M. Hjellming, C. M. Wade, 06.141.240), 297 (J. H. Taylor, G. R. Huguenin, R. M. Hirsch, 07.142.011), 298 (M. R. Kundu, 06.077.044), 299 (J. J. Broderick, K. I. Kellermann, D. B. Shaffer, D. L. Jauncey, 07.141.027), 300 (I. I. K. Pauliny-Toth, K. I. Kellermann, M. M. Davis, 07.141.107), 301 (A. H. Bridle, M. M. Davis, 07.141.106), 302 (K. I. Kellermann, 07.158.088), 303 (G. K. Miley, G. H. MacDonald, 07.141.089), 304 (R. M. Hjellming, C. M. Wade, V. A. Hughes, A. Woodsworth, 06.142.090), 305 (M. R. Kundu, T. Velusamy, 06.141.145), 306 (N. Z. Scoville, P. M. Solomon, P. Thaddeus, 07.131.029), 307 (E. Tademaru, 07.022.013), 308 (T. Nakano, E. Tademaru, 07.131.063) 309 (S. Edelson, E. B. Mayfield, F. I. Shimabukuro, 05.077.045), 310 (D. Buhl, C. Ponnamperuma, 07.131.149), 311 (R. N. Manchester, W. L. Peters, 07.141.522), 312 (D. S. De Young, 07.161.007), 313 (C. M. Wade, R. M. Hjellming, 07.141.184), 314 (C. M. Wade, R. M. Hjellming, 07.141.185).

008.062 Greenwich

Royal Observatory Bulletins, (Joint Publications of the Royal Greenwich Observatory, Herstmonceux, Royal Observatory, Cape of Good Hope), Nos. 167 (A. L. T. Powell, 07.114.040), 168 (A. L. T. Powell, 07.064.015), 170 (R. Woolley, A. Savage, 07.122.029).

008.063 Groningen

Nederlandse Vereniging voor Weer- en Sterrenkunde. **Observations of Variable Stars. Report** (Kapteyn Astronomical Laboratory, Groningen – Netherlands), No. 21 (L. Plaut, H. Feijth, 07.123.048).

008.064 Hamburg

Hamburger Sternwarte (Hamburg-Bergedorf). – Report 1971. Mitt. Astron. Ges., No. 31, p. 296 - 300 (1972).

European Southern Observatory. Annual report 1970. A. Blaauw. Printed in the Federal Republic of Germany by Bergedorfer Buchdruckerei von E. Wagner, Hamburg-Bergedorf. 67 pp. (1972).

Deutsches Hydrographisches Institut (DHI). Report 1971. W. Horn. Mitt. Astron. Ges., No. 31, p. 295 (1972).

Deutsches Hydrographisches Institut, Hamburg. **Astronomische Zeit- und Breitenbestimmungen, Empfangszeiten von Zeitsignalen,** 1971 October - December (07.044.036).

008.065 Hannover

Astronomische Station des Instituts für Theoretische Geodäsie der Technischen Universität. – Report 1971. K. Pilowski. Mitt. Astron. Ges., No. 31, p. 300 - 306 (1972).

008.066 Haute Provence

Ein Besuch des Observatoriums Haute-Provence. B. Springschitz. Sternenbote, 15. Jahrgang, p. 82 - 84 (1972).

Publications de l'Observatoire de Haute Provence Vol. 10, Nos. 35 (E. Rebeirot, 03.153.012), 36 (N. Morguleff, M. P. Véron, 03.113.020), 37 (L. D. Kaplan, J. Connes, P. Connes, 02.097.017), 38 (G. Adam, J.-H. Bigay, A.-M. Delplace, M. Duval, R. Garnier, R. Herman, A. Peton, 02.114.074), 39 (R. Louise, 03.132.021), 40 (A. Terzan, 03.124.100), 41 (J. H. Bigay, R. Garnier, 03.113.021), 42 (M. F. Ingham, 02.082.059), 43 (A. Baranne, F. Spite, M. Spite, 03.114.021), 44 (Y. Andrillat, 03.114.030), 45 (Y. P. Georgelin, Y. M. Georgelin, 03.131.094), 46 (R. Louise, 03.131.096), 47 (M. Bloch, N. Morguleff, A. Terzan, 03.112.010), 48 (Y. P. Georgelin, Y. M. Georgelin, 03.155.055), 49 (J. M. Deharveng, A. Pellet, 04.158.007), 50 (Y. P. Georgelin, 04.131.014).

008.067 Heidelberg

Astronomisches Rechen-Institut. – Report 1971. W. Fricke. Mitt. Astron. Ges., No. 31, p. 306 - 312 (1972).

Astronomy and Astrophysics Abstracts, Vol. 6 (S. Böhme, W. Fricke, U. Güntzel-Lingner, F. Henn, D. Krahn, U. Scheffer, G. Zech, 07.002.017).

Landessternwarte und Max-Planck-Institut für Astronomie Heidelberg-Königstuhl. – Report 1971. H. Elsässer. Mitt. Astron. Ges., No. 31, p. 314 - 321 (1972).

Present state of the MPIA-project and instrumentation. H. Elsässer. Auxiliary instrumentation for large telescopes. Conference 1972, (see 012.018), p. 83 - 89 (1972).

Ein großer Tag für das Heidelberger Max-Planck-Institut für Astronomie. SuW, Vol. 11, 91 - 96 (1972). Bonn warnt vor Sternwartenbau in Südwestafrika; Auf dem "Gamsberg" wird die Sicht geprüft, (J. M. Raffelberg); Stellungnahme der MPG; Zum Guß des 4-m-Teleskop-Spiegelrohlings, (J. Petzold).

Max-Planck-Institut für Astronomie – Heidelberg-Königstuhl, Separate prints (H. Elsässer, 02.009.018; H. Elsässer, 04.082.174; T. Schmidt, 03.159.014; W. Böhm, D. Labs, 06.022.106; K. Bahner, J. Solf, 05.031.027; H. Elsässer, T. Neckel, K. Birkle, 05.009.016; W. Hofmann, U. Fahrbach, D. Lemke, K. Voelcker, 05.034.099, D. Lemke, U. Fahrbach,

W. Hofmann, K. Voelcker, 05.034.100; C. Thum, W. Hofmann,
D. Lemke, 05.034.101, K. Voelcker, W. Hofmann, H. Elsässer,
05.113.055; G. Klare, T. Neckel, G. Schnur, 05.131.013;
T. Schmidt, 05.132.020; C. H. McGruder, G. Schnur, 06.113.
005; J. Solf, 05.032.057; G. Klare, T. Neckel, G. Schnur,
07.131.011; T. Schmidt, 07.159.002).

Lehrstuhl für Theoretische Astrophysik. – Report
1971. B. Baschek, G. Traving.
Mitt. Astron. Ges., No. 31, p. 312 - 313 (1972).

008.068 Holmdel

**Bell Telephone Laboratories, Incorporated, Craw-
ford Hill Laboratory, Holmdel, New Jersey.**–Observatory report. A. A. Penzias.
Bull. American Astron. Soc., Vol. 4, 6 - 7 (1972).

008.069 Honolulu

**University of Hawaii, Institute for Astronomy,
Honolulu, Hawaii.**–Observatory report. J. T. Jefferies.
Bull. American Astron. Soc., Vol. 4, 45 - 50 (1972). – This report covers progress at the Institute for Astronomy during the
period July 1970 through June 1971.

008.070 Ioannina

Department of Astronomy, University of Ioannina.
– Annual report 1970. S. N. Svolopoulos.
Annual Rep. Astron. Inst. Greece 1970, p. 15 (1971).

008.071 Iowa City

The University of Iowa, Iowa City, Iowa.–Observa-
tory report. J. S. Neff.
Bull. American Astron. Soc., Vol. 4, 55 - 59 (1972).
Report 1 July 1970 – 1 Sept. 1971.

008.072 Istanbul

Publications of the Istanbul University Observatory,
Nos. 92 (S. Karaali, 07.153.029), 93 (F. Yilmaz, 07.113.047),
94 (S. Karaali, 07.075.008), 95 (F. Yilmaz, 07.075.009).

008.073 Kazan

**Izvestiya Astronomicheskoj Ehngel'gardtovskoj
Observatorii,** *Kazan',* No. 38 (A. A. Nefed'ev, 07.094.175;
S. G. Valeev, 07.094.176; A. S. Mamakov, 07.032.032; M. I.
Lavrov, 07.121.071).

008.074 Kiel

**Institut für Theoretische Physik und Sternwarte der
Universität.** – Report 1971. A. Unsöld.
Mitt. Astron. Ges., No. 31, p. 322 - 323 (1972).

008.075 Kitt Peak

**Kitt Peak National Observatory, Tucson, Arizona,
and Cerro Tololo Inter-American Observatory, La Serena,
Chile.**–Observatory reports. V. M. Blanco, N. U. Mayall.
Bull. American Astron. Soc., Vol. 4, 66 - 97 (1972).

008.076 Kraków

**Astronomical Observatory of the Jagellonian Univer-
sity.** K. Koziel.
Postępy Astron., Vol. 20, 161 - 164 (1972). In Polish.

008.077 Krim

Izvestiya Krymskoj Astrofizicheskoj Observatorii,
Akademiya Nauk SSSR, Tom (Vol.) 43 (N. N. Erjushev, M. V.
Tinin, L. I. Tsvetkov, 07.077.016; L. I. Yourovskaya, Y. F.
Yourovsky, 07.077.017; V. A. Efanov, I. G. Moiseev,
07.077.018; A. N. Viestavkin, V. A. Efanov, V. N. Listvin, I. G.
Moiseev, E. I. Popov, V. T. Potapov, 07.077.019; A. E. An-
drievski, E. E. Spangenberg, I. E. Valtz, A. G. Gorshkov, V. N.
Ivanov, V. K. Konnikova, V. N. Kurilchik, M. G. Larionov, I.
G. Moiseev, V. V. Nikitin, V. I. Portman, V. A. Soglasnov,
07.141.060; A. A. Stepanian, I. V. Pavlov, 07.142.039; R. E.
Gershberg, A. A. Korovjakovskaja, Yu. P. Korovjakovskij,
07.022.018; V. P. Grinin, 07.064.018; R. E. Gershberg,
07.122.039; E. A. Vitrichenko, 07.119.009; E. A. Vitrichenko,
07.119.010; T. M. Rachkovskaya, 07.121.012; I. I. Pronik, K.
K. Chuvaev, 07.158.046; V. V. Leushin, 07.114.051; N. V.
Steshenko, 07.073.026; N. V. Steshenko, 07.073.027; M. B.
Ogir, 07.073.028; M. B. Ogir, 07.073.029; S. I. Gopasyuk, T.
T. Tsap, 07.071.018; D. N. Rachkovsky, 07.063.007; V. V.
Prokofieva, S. I. Usliber, 07.093.011; A. A. Rusak, E. I. Terez,
07.034.010).

008.078 Kyoto

**Contributions from the Institute of Astrophysics
and Kwasan Observatory, University of Kyoto.** Nos. 182
(S. Mizuno, M. Nishida, 02.080.004), 184 (S. Miyamoto,
07.097.099), 185 (H. Kurokawa, S. Tominaga, J. Kubota,
I. Kawaguchi, 02.079.100), 187 (A. Uesugi, J. Tsujita, 02.063.
022), 188 (S. Ueno, R. E. Kalaba, H. H. Kagiwada, R. E. Bellman, 03.063.021), 189 (A. Uesugi, I. Fukuda, 04.116.005),
190 (R. E. Bellman, H. H. Kagiwada, R. E. Kalaba, S. Ueno,
03.091.012), 191 (J. Kubota, J. L. Leroy, 03.073.046), 192
(J. Gruschinske, S. Ueno, 04.063.016), 193 (I. Kawaguchi,
04.073.026), 194 (M. Matsumoto, 07.063.044), 195 (P. Chamaraux, M. Tadokoro, 05.161.002), 196 (T. Ishizawa, 05.064.
022), 197 (T. Ishizawa, 05.073.013), 198 (S. Ueno, 05.063.
025), 199 (K. Iwasaki, 06.093.001), 200 (I. Kawaguchi, N. Oda,
S. Mizuno, 07.073.006).

Department of Astronomy, University of Kyoto.
Reprints Nos. 36 (T. Kogure, AJB 68, 104.149), 37 (T. Kogure, 01.114.008), 38 (S. Kawai, T. Kogure, 02.031.020), 39
(T. Shimizu, Y. Baba, 02.151.032), 40 (T. Kogure, 01.064.
064), 41 (T. Kogure, N. Toya, 04.158.021).

008.079 Lake Tekapo

Mt. John University Observatory.

N. A. Doughty.
Inform. Bull. Southern Hemisph., No. 19, p. 1 - 3 (1971).

Mount John Observatory. (Department of Physics, University of Canterbury, Department of Astronomy, University of Pennsylvania, Department of Physics and Astronomy, University of Florida).
Inform. Bull. Southern Hemisph., No. 19, p. 25 - 26 (1971).
Current research report.

008.080   La Plata

Observatorio Astronómico. (Astronomical Observatory, National University of La Plata).
Inform. Bull. Southern Hemisph., No. 19, p. 7 - 8 (1971).
Current research report.

008.081   Las Cruces

Lindheimer Astronomical Research Center, and Dearborn Observatory, Evanston, Illinois; Corralitos Observatory, Las Cruces, New Mexico.—Observatory reports.
J. A. Hynek.
Bull. American Astron. Soc., Vol. 4, 100 - 103 (1972).

New Mexico State University, Department of Astronomy, Las Cruces, New Mexico.—Observatory report.
W. Reitmeyer.
Bull. American Astron. Soc., Vol. 4, 149 - 151 (1972).

008.082   Lausanne

Observatoire Universitaire de Lausanne, Communication Nos. 20 (B. Hauck, C. Nicollier, 07.113.027), 21 (E. Lindemann, B. Hauck, 07.113.043).

008.083   Lawrence

Kansas University Observatory, Lawrence, Kansas.
Observatory report.   P. A. Wehinger.
Bull. American Astron. Soc., Vol. 4, 65 - 66 (1972).

008.084   Lembang

Bosscha Observatory.
Inform. Bull. Southern Hemisph., No. 19, p. 22 - 24 (1971).
Current research report.

008.085   Leningrad

Byulletin' Instituta Teoreticheskoj Astronomii,
Akademiya Nauk SSSR, Vol. 13, No. 2 (Yu. V. Batrakov, L. L. Filenko, 07.052.006; D. V. Zagrebin, 07.081.010; V. T. Kondurar, T. K. Shinkarik, 07.042.020; V. I. Lapshina, 07.052.007; V. I. Voronenko, F. F. Kalihevich, 07.098.003; N. S. Chernykh, 07.098.004; L. I. Chernykh, 07.098.005).

008.086   Liège

Université de Liège (Belgique). Institut d'Astrophysique, Cointe-Sclessin, Collection in 4°, Nos. 202 (M. Gabriel, 03.065.065), 203 (R. Simon, 03.162.036), 204 (J. P. Macau, 07.034.110), 205 (J. Mawhin, 07.021.010), 206 (H. B. Bredohl, 03.036.006), 207 (H. Nussbaumer, J. P. Swings, 04.022.008), 208 (H. Bredohl, S. Gardier, J. Henrion, 07.034.111), 209 (F. Remy, 07.022.102), 210 (F. Remy, 07.022.103), 211 (N. Grevesse, 04.071.023), 212 (N. Grevesse, A. J. Sauval, 04.071.026), 213 (J.-C. Gérard, 07.084.045), 214 (G. Chanmugam, M. Gabriel, 05.065.018), 215 (I. Dubois, H. Leclercq, 07.022.104), 216 (M. L. Aizenman, J. Perdang, 05.065.066), 217 (P. Renson, 05.155.038), 218 (N. Grevesse, J. P. Swings, 05.072.043), 219 (G. Chanmugam, M. Gabriel, 06.065.007), 220 (G. Marette, 07.034.112), 221 (M. L. Aizenman, J. Perdang, 06.065.070), 222 (N. Grevesse, A. J. Sauval, 06.071.018), 223 (I. Dubois, H. Leclercq, 07.022.105), 224 (G. Chanmugam, M. Gabriel, 07.126.005), 225 (M. L. Aizenman, J. Perdang, J. R. Lesh, 07.065.025), 226 J. P. Swings, P. Swings, 07.114.024), 227 (M. Gabriel, G. Chanmugam, 07.065.023).

Université de Liège (Belgique). Institut d'Astrophysique, Cointe-Sclessin, Collection in 8°, Nos. 592 (G. Marette, 07.034.113), 593 (D. Macau-Hercot, 07.022.106), 594 (J. Mawhin, 07.021.011), 595 (I. Dubois, 07.022.107), 596 (R. Duysinx, A. Monfils, 03.084.034), 597 (J. P. Swings, P. Swings, 03.114.098), 598 (G. Marette, 07.034.114), 599 (R. Duysinx, 07.084.046), 600 (N. Grevesse, J. P. Swings, 04.071.001), 601 (R Simon, 04.162.007), 602 (R. Simon, 04.065.052), 603 (N. Grevesse, 05.071.027), 604 (J. Demaret, 07.141.186), 605 (H. Bredohl, F. Remy, 07.082.112), 606 (J. Mawhin, 07.021.012), 607 (R. Zander, 07.071.049), 608 (P. Renson, 07.021.013), 609 (N. Grevesse, A. J. Sauval, 05.071.022), 610 (P. Ledoux, 05.065.111), 611 (F. Dossin, 07.032.050), 612 (J. Manfroid, 06.061.051), 613 (P. Ledoux, 05.013.009), 614 (G. Chanmugam, M. Gabriel, 05.065.128), 615 (P. Swings, 06.061.062), 616 (A. Boury, M. Gabriel, 05.065.108), 617 (A. Lausberg, 07.066.080), 618 (J. M. Vreux, 07.084.047), 619 (F. Remy, 07.022.108), 620 (N. Grevesse, J. P. Swings, 07.071.001).

008.087   Lisbonne

Bulletin de l'Observatoire Astronomique de Lisbonne (Tapada), No. 18 (A. Baptista dos Santos, 07.045.031).

008.088   London, Canada

The Observatories of the University of Western Ontario, London, Canada.—Observatory report.
W. H. Wehlau.
Bull. American Astron. Soc., Vol. 4, 201 - 202 (1972).

008.089   Los Angeles

Research Units and Academic Departments, University of California: Berkeley, Los Angeles, San Diego, and Santa Cruz. – **II. Los Angeles Campus.** – Report 1970 – 1971.
Bull. American Astron. Soc., Vol. 4, 278 - 281 (1972).

008.090 Lund

Reports from the Observatory of Lund, Nos. 3
(C. Schalén, 07.155.078), 4 (G. Lyngå, G. Arinder,
07.034.107).

008.091 Madison

Washburn Observatory, University of Wisconsin,
Madison, Wisconsin.–Observatory report.
D. E. Osterbrock.
Bull. American Astron. Soc., Vol. 4, 193 - 198 (1972).

008.092 Manchester

University of Manchester. Nuffield Radio Astronomy Laboratories, Jodrell Bank. – Report for the year ending
1971 August 31. B. Lovell.
Quarterly Journ. Roy. Astron. Soc., Vol. 13, 93 - 103 (1972).

008.093 Menlo Park

Center for Radar Astronomy, Stanford University,
Stanford, California and Stanford Research Institute, Menlo
Park, California.–Observatory reports.
V. R. Eshleman, R. L. Leadabrand, A. M. Peterson.
Bull. American Astron. Soc., Vol. 4, 177 - 179 (1972).

008.094 Middletown

Van Vleck Observatory, Wesleyan University,
Middletown, Connecticut.–Observatory report.
A. R. Upgren.
Bull. American Astron. Soc., Vol. 4, 185 - 187 (1972).

008.095 Minneapolis

University of Minnesota, Minneapolis, Minnesota,
Separate prints (W. J. Luyten, A. E. La Bonte, 07.112.017;
W. J. Luyten, 07.117.036; W. J. Luyten, J. H. Anderson,
A. R. Sandage, 07.113.046; W. J. Luyten, 07.112.018).

008.096 Mizusawa

Annual report of geophysical observations made at
the International Latitude Observatory of Mizusawa for the
year 1969. T. Okuda.
Published by the International Latitude Observatory of Mizusawa, Japan. 2 + 49 pp. (1971).

Annual report of the meteorological observations
made at the International Latitude Observatory of Mizusawa
for the year 1969. T. Okuda.
Published by the International Latitude Observatory of Mizusawa, Japan. 4 + 30 pp. (1970).

Annual report of the meteorological observations
made at the International Latitude Observatory of Mizusawa
for the year 1970. T. Okuda.

Published by the International Latitude Observatory of Mizusawa, Japan. 4 + 30 pp. (1971).

Bulletins, Time Service of the Mizusawa Observatory
Vol. 14, No. 1 - 12, 1969 (S. Takagi, I. Okamoto, K. Yokoyama, T. Hara, G. Murakami, 07.044.045).

Monthly Notes of the International Polar Motion
Service, 1971 Nos. 11–12, 1972 Nos. 1–4 (07.045.036).

Proceedings of the International Latitude Observatory of Mizusawa, No. 11 (N. Kikuchi, 07.082.059; T. Goto,
07.032.013; E. Onodera, 07.082.060; G. Teleki, V. Milovanović, 07.045.014; K. Takahasi, 07.082.061; T. Goto,
07.044.013; E. Onodera, 07.082.062; K. Yokoyama, S. Sakai,
K. Iwadate, 07.041.017; K. Yokoyama, 07.044.014; H. Ishii,
07.045.015).

Publications of the International Latitude Observatory of Mizusawa, Vol. 7, No. 2 (S. Takagi, 07.041.016; R. O.
Vicente, S. Yumi, 07.045.009; S. Takagi, M. Ooe, S. Abe,
07.045.010; C. Sugawa, M. Ooe, 07.045.011; S. Takagi,
07.044.012; C. Kakuta, N. Kikuchi, 07.081.013).

008.097 Mons

Communications du Département d'Astrophysique
de la Faculté des Sciences de Mons. Mons Astrophysical
Papers, Nos. 18 (Y. Andrillat, L. Houziaux, 06.124.103), 23
(Y. Andrillat, L. Houziaux, 07.114.042), 24 (Y. Andrillat,
L. Houziaux, 07.114.043), 25 (L. Houziaux, 07.014.014),
26 (L. Houziaux, G. Houziaux, 07.041.046).

008.098 Moskva

Soobshcheniya Gosudarstvennogo Astronomiches-
kogo Instituta im. P. K. Shternberga. Izdatel'stvo Moskovskogo
Universiteta, Nos. 174 (N. P. Grushinski, L. P. Pellinen,
07.081.018; N. P. Grushinski, N. B. Sagina, T. P. Baskakova,
G. I. Torochkova, 07.081.019; N. P. Grushinski, N. B. Sagina,
07.081.020), 176 (E. D. Pavlovskaya, D. K. Karimova,
07.112.010; G. A. Starikova, 07.118.013; T. A. Uranova,
07.113.021; O. D. Dokuchaeva, 07.036.005).

Trudy Gosudarstvennogo Astronomicheskogo Instituta im. P. K. Shternberga. Izdatel'stvo Moskovskogo Universiteta, Vol. 42 (L. M. Hommik, 07.041.009; T. S. Meshkova,
07.041.010; A. G. Oborneva, 07.041.011; T. S. Meshkova,
A. G. Oborneva, 07.041.012; D. K. Karimova, E. D. Pavlovskaya, 07.112.008; L. P. Basurmanova-Gribko, 07.041.013;
D. N. Ponomarev, 07.041.014; A. A. Tochilina, 07.041.015;
N. P. Grushinsky, E. D. Korjakin, P. A. Stroev, G. E. Lasarev,
D. V. Sidorov, N. F. Virskaja, 07.081.012).

008.099 München

Universitäts-Sternwarte. – Report 1971.
P. Wellmann.
Mitt. Astron. Ges., No. 31, p. 350 - 354 (1972).

Max-Planck-Institut für Physik und Astrophysik,
Institut für Astrophysik und Institut für extraterrestrische
Physik. – Report 1971. L. Biermann, R. Lüst.
Mitt. Astron. Ges., No. 31, p. 323 - 350 (1972).

Max-Planck-Institut für Physik und Astrophysik, München, Separate prints (H. C. van de Hulst, A. Scheepmaker, B. N. Swanenburg, H. A. Mayer-Hasselwander, E. Pfeffermann, K. Pinkau, H. Rothermel, H. Schneider, W. Voges, J. Labeyrie, P. Keirle, J. Paul, G. Bellomo, G. Bignami, G. Boella, L. Scarsi, G. W. Hutchinson, A. J. Pearce, D. Ramsden, R. D. Wills, P. J. Wright, 06.061.021; H. A. Mayer-Hasselwander, K. Pinkau, K. H. Schenkl, W. Voges, H. J. Schneider, 06.034.047; H. A. Mayer-Haßelwander, E. Pfeffermann, K. Pinkau, H. Rothermel, M. Sommer, 07.142.132; E. Stingl, 07.022.111; L. Biermann, 07.107.011; L. Biermann, 07.102.022; L. Biermann, 07.131.153).

008.100 Münster

Astronomisches Institut der Universität. – Report 1971. H. Straßl.
Mitt. Astron. Ges., No. 31, p. 354 - 356 (1972).

Astronomisches Institut, Universität Münster. Sonderdruck No. 2 (M. P. Véron, P. Véron, A. Witzel, 07.158.042).

008.101 Napoli

Programmi di ricerca e loro sviluppo all'Osservatorio di Napoli. M. Rigutti.
Atti XIII Riunione Soc. Astron. Italiana, Trieste 1969, (see 012.010), p. 191 - 194 (1970).

008.102 Narrabri

Narrabri Observatory. R. Hanbury Brown.
Inform. Bull. Southern Hemisph., No. 19, p. 11 (1971). Current research report.

008.103 Nashville

Dyer Observatory, Vanderbilt University, Nashville, Tennessee.–Observatory report. A. M. Heiser.
Bull. American Astron. Soc., Vol. 4, 27 - 29 (1972). – Report 1970 September 15–1971 September 15.

The Arthur J. Dyer Observatory, Vanderbilt University, Nashville, Tennessee, Reprint, Series 1, Nos. 59 (D. W. Weedman, 05.158.100), 60 (M. G. Smith, D. W. Weedman, 06.159.008).

The Arthur J. Dyer Observatory, Vanderbilt University, Nashville, Tennessee, Reprint, Series 2, Nos. 2 (D. S. Hall, S. L. Weedman, 05.121.024), 3 (L. P. Lovell, D. S. Hall, 05.121.070), 4 (L. P. Lovell, D. S. Hall, 05.119.011), 5 (D. S. Hall, G. S. Hubbard, 06.121.054).

008.104 Neuchâtel

Rapport d'activité pour l'exercice 1971 et Rapport sur le Concours chronométrique 1971. J. Bonanomi.
Observatoire Cantonal de Neuchâtel, 25 pp. (1972).

Observatoire de Neuchâtel. Bulletin. Série B, 1971 June – December (07.044.042), Série D, 1971 May –

December (07.044.043).

008.105 New Haven

Yale University Observatory, New Haven, Connecticut.–Observatory report. P. Demarque.
Bull. American Astron. Soc., Vol. 4, 202 - 206 (1972).

008.106 Ottawa

Contributions Astrophysics Branch, National Research Council of Canada, Ottawa, Ontario, No. 12287 (W. J. Medd, 07.141.006).

Contributions from the Earth Physics Branch, Department of Energy, Mines and Resources, Ottawa, Ontario, Canada, No. 374 (P. H. Serson, K. Whitham, 06.034.081).

Publications of the Earth Physics Branch, Department of Energy, Mines and Resources, Ottawa, Canada, Vol. 42, No. 6 (E. G. Woolsey, 07.045.037); Vol. 43, Nos. 2 (G. J. van Beek, H. R. Reny, 07.084.285), 3 (G. J. van Beek, H. R. Reny. 07.084.286), 4 (G. J. van Beek, W. Piche, 07.084.287), 5 (G. J. van Beek, H. R. Reny, 07.084.288), 6 (G. J. van Beek, H. R. Reny, 07.084.289), 8 (P. H. Serson, F. Primdahl, 07.034.124).

008.107 Padova

Gruppo di Astrofisica Teorica dell'Università di Padova. N. Dallaporta.
Atti XIII Riunione Soc. Astron. Italiana, Trieste 1969, (see 012.010), p. 163 - 165 (1970).

Relazione programmatica sulle ricerche in corso presso l'Osservatorio di Padova-Asiago. R. Barbon.
Atti XIII Riunione Soc. Astron. Italiana, Trieste 1969, (see 012.010), p. 199 - 203 (1970).

008.108 Paris

Bureau International de l'Heure. Rapport annuel pour 1971. B. Guinot, M. Feissel, M. Granveaud.
Printing Office: Observatoire de Paris. 5 + A27 + B33 + C13 + D72 + E33 pp. (1972).
Contents: Méthodes de calcul; Tableaux et figures; Signaux horaires; Résidus des observations astronomiques; Annexe.

Activités du Bureau International de l'Heure (BIH), août 1967 – août 1971. B. Guinot.
Bull. Groupe Recherches Géod. Spatiale, Obs. Meudon, No. 2, p. 49 - 54 (1972). – Report prepared for the UGGI general assembly (Moscow, August 1971).

Bureau International de l'Heure, Circulaires B/C Nos. 189 - 195 (07.045.038).

Bureau International de l'Heure, Circulaires, D62 - D68 (07.044.049).

008.109 Pasadena

Owens Valley Radio Observatory, California Institute of Technology, Big Pine, California.—Observatory report. G. J. Stanley. Bull. American Astron. Soc., Vol. 4, 158 - 161 (1972).

008.110 Patras

Department of Astronomy, University of Patras. – Annual report 1970. B. Barbanis. Annual Rep. Astron. Inst. Greece 1970, p. 16 (1971).

008.111 Philadelphia

Flower and Cook Observatory, University of Pennsylvania, Philadelphia, Pennsylvania.—Observatory report. R. H. Koch. Bull. American Astron. Soc., Vol. 4, 36 - 38 (1972).

008.112 Pittsburgh

Allegheny Observatory, University of Pittsburgh, Pittsburgh, Pennsylvania.—Observatory report. J. Kiewiet de Jonge. Bull. American Astron. Soc., Vol. 4, 5 (1972).

008.113 Porto Alegre

Instituto de Astronomía da Universidade Federal do Rio Grande do Sul. (Astronomical Institute of the Federal University of Rio Grande do Sul). Inform. Bull. Southern Hemisph., No. 19, p. 12 (1971). Current research report.

008.114 Potsdam-Babelsberg

Deutsche Akademie der Wissenschaften zu Berlin. Zentralinstitut für Astrophysik, Sternwarte Babelsberg, Potsdam—Babelsberg, Mitteilungen. Neue Folge, Nos. 51 (H.-J. Treder, 07.066.081), 52 (H.-J. Treder, 07.066.082), 53 (U. Dyllong, 07.022.109), 54 (H.-J. Treder, 07.066.083), 55 (H.-J. Treder, 06.022.028), 56 (G. M. Richter, 06.141.197), 57 (K.-H. Schmidt, 06.141.198), 58 (U. Kasper, 06.066.063), 59 (K.-H. Schmidt, H. Oleak, 06.160.001), 60 (W. Bronkalla, 06.113.013), 61 (G. Dautcourt, K. Fritze, 06.065.022), 62 (P. Notni, H. Oleak, G. M. Richter, 07.141.020), 63 (H.-J. Treder, 06.004.035), 64 (U. Kasper, 06.066.062), 65 (K. Fritze, 07.065.018), 66 (H.-J. Treder, 07.066.086).

008.115 Praha

Académie Tchécoslovaque des Sciences, Institut Astronomique, Station de l'Heure à Praque, Série 5, No. 17 (L. Webrová, V. Ptáček, 07.044.040).

008.116 Princeton

Princeton University Observatory, Princeton, New Jersey.—Observatory report. L. Spitzer, Jr. Bull. American Astron. Soc., Vol. 4, 161 - 166 (1972).

008.117 Pulkovo

Trudy Glavnoj Astronomicheskoj Observatorii v Pulkove. Seriya 2, Vol. 78 (N. N. Pavlov, P. M. Afanasjeva, G. V. Staritsyn, 07.041.005; L. I. Medvedeva, A. A. Nemiro, 07.112.005; N. N. Pavlov, P. M. Afanasjeva, G. V. Staritsyn, 07.041.006; N. M. Bronnikova, 07.112.006; L. V. Zhukov, 07.154.012).

008.118 Quito

Boletin Astronomico del Observatorio Astronomico de Quito, Serie B, Nos. 3 (H. Dávila, 07.041.040; H. Dávila, S. Débarbat, J. Egred, L. Mena, A. Scheepmaker, 07.041.041), 4 (H. Dávila, 07.041.042; M. Alvaro, H. Dávila, S. Débarbat, J. Egred, L. Espín, A. Salvador, 07.041.043).

008.119 Richmond Hill

David Dunlap Observatory, University of Toronto, Richmond Hill, Ontario, Canada.—Observatory report. D. A. MacRae. Bull. American Astron. Soc., Vol. 4, 21 - 26 (1972). The following report of the work of the Observatory is for the period 1 July 1970 to 30 June 1971.

David Dunlap Observatory, Richmond Hill, Ontario. D. A. MacRae. Journ. Roy. Astron. Soc. Canada, Vol. 66, 75 - 77 (1972).

David Dunlap Observatory, University of Toronto. Report for the year ending 1971 June 30. D. A. MacRae. Quarterly Journ. Roy. Astron. Soc., Vol. 13, 71 - 87 (1972).

Communications from the David Dunlap Observatory, University of Toronto, Richmond Hill, Ontario, Canada, Nos. 303 (N. R. Walborn, 06.124.104), 304 (N. R. Walborn, 06.132.053), 305 (S. van Agt, C. Coutts, 06.125.022), 306 (G. G. Fahlman, S. P. S. Anand, 06.065.026), 307 (G. G. Fahlman, S. P. S. Anand, 06.065.027), 308 (R. C. Roeder, 06.141.123), 309 (S. van den Bergh, 07.158.007), 310 (R. C. Roeder, 07.141.022), 311 (S. van den Bergh, 06.158.102), 312 (S. van den Bergh, 07.158.065), 313 (E. R. Seaquist, 05.141.117).

008.120 Rio de Janeiro

Observatorio Nacional. (National Observatory). Inform. Bull. Southern Hemisph., No. 19, p. 12 - 13 (1971). Current research report.

Valongo Observatory, Federal University of Rio de Janeiro. Inform. Bull. Southern Hemisph., No. 19, p. 14 - 15 (1971). Current research report.

Contribuições do Observatório do Valongo, Universidade Federal do Rio de Janeiro, Série II, No. 13 (07.099.068).

Contribuições do Observatório do Valongo, Universidade Federal do Rio de Janeiro, Série III, No. 21 (07.096.021).

008.121 **Rochester**

C. E. Kenneth Mees Observatory, Rochester, New York.–Observatory report.   C. Sturch.
Bull. American Astron. Soc., Vol. 4, 117 - 119 (1972).

C. E. Kenneth Mees Observatory, University of Rochester, Rochester, N.Y. **Reprints**, Nos. 30 (C. Sturch, 07.008.000), 31 (H. M. van Horn, M. B. Richardson, C. J. Hansen, 07.065.027).

008.122 **Roma**

Programma dell'Osservatorio Astronomico di Roma.
M. A. Giannuzzi.
Atti XIII Riunione Soc. Astron. Italiana, Trieste 1969, (see 012.010), p. 195 - 198 (1970).

Monthly Bulletin. Osservatorio Astronomico di Roma, Nos. 167 - 168 (M. Cimino, M. Torelli, A. Cacciani, V. Croce, R. Flamini, U. Bartolini, 07.075.018).

008.123 **Rosario**

Observatorio Astronómico Municipal, Rosario – Argentina. Departamento de Fisica Solar, Contribuciones, Serie 1, No. 3 (V. Capolongo, J. A. Gutiérrez, O. F. Liesche, C. Sosa, R. Barbarroja, L. A. Mansilla, 07.075.015).

008.124 **San Diego**

Research Units and Academic Departments, University of California: Berkeley, Los Angeles, San Diego, and Santa Cruz. – III. San Diego Campus. – Report 1970–1971.
E. M. Burbidge.
Bull. American Astron. Soc., Vol. 4, 281 - 284 (1972).

008.125 **San Fernando**

Memoria de las actividades en 1971.
Separate print Inst. y Obs. de Marina, San Fernando (Cádiz). 13 pp. (1972).

008.126 **San Juan**

Observatorio "Felix Aguilar". (Observatory of the National University of Cuyo).
Inform. Bull. Southern Hemisph., No. 19, p. 8 - 9 (1971). Current research report.

008.127 **San Miguel**

Observatorio Nacional de Física Cosmica. (National Observatory of Cosmic Physics).
Inform. Bull. Southern Hemisph., No. 19, p. 9 - 10 (1971). Current research report.

Observatorio Nacional de Física Cósmica, San Miguel, Argentina. Boletín Meteorológico mensual, Vol. 25 (3) - (12), 1971 March - December (07.075.016).

008.128 **Santa Cruz**

Research Units and Academic Departments, University of California: Berkeley, Los Angeles, San Diego, and Santa Cruz. – **IV. Santa Cruz Campus:** A.) Board of Studies in Astronomy and Astrophysics, R. P. Kraft; B.) Lick Observatory, R. P. Kraft. – Report 1970 – 1971.
Bull. American Astron. Soc., Vol. 4, 284 - 290 (1972).

008.129 **Santiago**

Departamento de Astronomía, Universidad de Chile. (Astronomy Department, University of Chile).
Inform. Bull. Southern Hemisph., No. 19, p. 19 - 21 (1971).
A) National Astronomical Observatory, Cerro Calán; B) Cerro El Roble Astronomical Station; C) Maipú Radioastronomical Observatory. – Current research reports.

008.130 **Sao Paulo**

Centro de Radioastronomía e Astrofísica da Universidade Mackenzie. (Center of Radioastronomy and Astrophysics, Mackenzie University).
Inform. Bull. Southern Hemisph., No. 19, p. 15 - 17 (1971). Current research report.

Instituto Astronômico e Geofísico da Universidade de São Paulo. (Astronomical and Geophysical Institute).
Inform. Bull. Southern Hemisph., No. 19, p. 17 (1971). Current research report.

Observatorio Astronômico do Instituto Tecnologico de Aeronáutica. (I. T. A. Astronomical Observatory).
Inform. Bull. Southern Hemisph., No. 19, p. 17 - 18 (1971). Current research report.

008.131 **Seattle**

University of Washington, Astronomy Department, Seattle, Washington.–Observatory report.   P. W. Hodge.
Bull. American Astron. Soc., Vol. 4, 198 - 200 (1972).

008.132 **Sendai**

Sendai Astronomiaj Raportoj, No. 120 (K. Mimura, K. Suda, 06.065.108).

008.133 Shemakha

Soobshcheniya Shemakhinskoj Astrofizicheskoj Observatorii, Akademiya Nauk Azerbajdzhanskoj SSR, vyp. (No.) 6 (O. Kh. Gusejnov, 07.065.124; T. A. Guseva, 07.034.123; V. Ts. Gurovich, O. Kh. Gusejnov, 07.065.125; D. M. Kuli-Zade, Yu. A. Solonskij, 07.071.050; G. D. Mamedbejli, N. A. Abdulkasumova, 07.004.048; R. Sh. Yakh'yaev, 07.022.110; A. A. Aliev, 07.122.154; G. V. Akhundova, 07.032.053).

008.134 Sonneberg

Deutsche Akademie der Wissenschaften zu Berlin, Sternwarte Sonneberg. **Mitteilungen der Sternwarte zu Sonneberg**, No. 59 (W. Wenzel, J. Dorschner, C. Friedemann, 06.121.007; W. Thänert, 06.065.023; G. Jackisch, 06.141.029; G. A. Richter, 06.158.007; W. Götz, 06.153.023; L. Richter, N. B. Richter, W. Wenzel, 06.113.044, P. Ahnert, 04.096.008).

Deutsche Akademie der Wissenschaften zu Berlin, Zentralinstitut für Astrophysik. **Veröffentlichungen der Sternwarte in Sonneberg**, Vol. 8, No. 1 (L. Meinunger, 07.122.143).

Mitteilungen über Veränderliche Sterne, Sonneberg, Vol. 6, No. 2 (W. Wenzel, 07.121.081; W. Wenzel, W. Fürtig, 07.122.144; W. Wenzel, 07.123.045; L. Meinunger, 07.123.046; W. Götz, 07.122.145; H. Geßner, 07.123.047; W. Götz, W. Wenzel, 07.122.146).

008.135 Stanford

Center for Radar Astronomy, Stanford University, Stanford, California and Stanford Research Institute, Menlo Park, California.—Observatory reports. V. R. Eshleman, R. L. Leadabrand, A. M. Peterson. Bull. American Astron. Soc., Vol. 4, 177 - 179 (1972).

008.136 Strasbourg

Publications de l'Observatoire de Strasbourg, Vol. 2, Fasc. 1 (A. Florsch, 07.159.024), 2 (A. Schmitt, 07.031.036), 3 (07.046.031), 4 (A. Acker, 06.119.003), 5 (A. Schmitt, 07.054.020).

Centre de Données Stellaires. J. Jung. Centre de Données Stellaires, Inform. Bull. No. 3, (see 002.020) p. 1 (1972).
The Centre de Données Stellaires has been officially created at Strasbourg by the Institut National d'Astronomie et de Géophysique (INAG).

008.137 Swarthmore

Sproul Observatory, Swarthmore College, Swarthmore, Pennsylvania.—Observatory report. P. van de Kamp. Bull. American Astron. Soc., Vol. 4, 176 - 177 (1972).

Sproul Observatory, Swarthmore, Pennsylvania, **Reprints**, Nos. 193 (S. L. Lippincott, 03.011.002), 194 (P. van de Kamp, 03.008.125), 195 (W. D. Heintz, 04.117.007), 196 (P. van de Kamp, 05.008.122), 197 (B. H. Feierman, 05.118.003), 198 (W. D. Heintz, 05.118.005), 199 (P. van de Kamp, 05.126.017), 200 (P. van de Kamp, 07.117.046), 201 (W. D. Heintz, 05.118.019), 202 (S. L. Lippincott, 05.118.023), 203 (P. van de Kamp, 06.155.008), 204 (P. van de Kamp, 07.011.045), 205 (S. L. Lippincott, 07.041.051), 206 (W. D. Heintz, 07.111.009), 207 (P. van de Kamp, M. D. Worth, 06.118.027).

008.138 Sydney

Division of Radiophysics, C. S. I. R. O., Sydney (Epping, New South Wales, Australia), **Separate prints** (S. F. Smerd, G. A. Dulk, 06.077.024; W. M. Goss, J. L. Caswell, B. J. Robinson, 06.131.032; R. T. Hansen, C. J. Garcia, R. J.-M. Grognard, K. V. Sheridan, 06.074.046; R. T. Stewart, K. V. Sheridan, 06.077.026: J. C. Ribes, F. Biraud, 06.015.014; A. C. Riddle, K. V. Sheridan, 06.077.027; B. J. Robinson, J. L. Caswell, W. M. Goss, 06.131.078; H. C. Minnett, 06.032.010; B. M. Thomas, 06.033.081; K. V. Sheridan, D. J. McLean, 06.033.018; D. K. Milne, J. R. Dickel, 05.125.021; J. L. Caswell, 05.158.066; F. F. Gardner, J. C. Ribes, M. W. Sinclair, 06.141.185; L. H. Aller, D. K. Milne, 07.133.011; O. B. Slee, G. A. Dulk, 07.099.032; N. Fourikis, 06.033.071; J. L. Caswell, B. J. Robinson, H. R. Dickel, 06.122.041; C. S. Higgins, M. M. Komesaroff, O. B. Slee, 06.141.063; M. M. Komesaroff, J. G. Ables, P. A. Hamilton, 06.141.119; F. F. Gardner, J. C. Ribes, B. F. C. Cooper, 06.131.095; F. F. Gardner, J. C. Ribes, 06.131.094; M. M. Komesaroff, 06.141.175; R. W. Clarke, 07.141.132).

008.139 Tampa

University of South Florida Observatory, Tampa, Florida.—Observatory report. H. K. Eichhorn-von Wurmb. Bull. American Astron. Soc., Vol. 4, 175 - 176 (1972).

008.140 Tartu

Tartu Astronoomia Observatoorium, Teated, Nos. 35 (I. Pustylnik, 07.064.048; I. Pustylnik, 07.117.037; I. Pustylnik, 07.117.038), 36 (J. Einasto, L. Einasto, 07.151.099; J. Einasto, L. Einasto, 07.155.049; J. Einasto, U. Rümmel, 07.158.169), 37 (M. Jõeveer, 07.155.050; H. Eelsalu, 07.155.079), 38 (E. Saar, 07.162.054), 39 (A. Nikitin, T. Feklistova, 07.114.131).

008.141 Tashkent

Chronicle. Tsirk. Astron. Inst., *Tashkent*, No. 27 (374), p. 14 - 17 (1971). In Russian.

Tsirkulyar Astronomicheskogo Instituta, Akademiya Nauk Uzbekskoj SSR, Nos. 27 (374) (N. A. Omelina, 07.044.020; F. G. Mustaeva, 07.075.003; A. Sadikov, 07.071.033; F. G. Mustaeva, 07.072.029; 07.008.141), 28 (375) (N. A. Omelina, 07.044.020; F. G. Mustaeva, 07.075.003 A. Sadikov, 07.073.053; M. R. Ehshmatov, 07.096.010), 29 (376) (N. A. Omelina, 07.044.020; F. G. Mustaeva, 07.075.003; K. D. Avulov, 07.152.044; Eh. A. Sanakulov, 07.041.024), 30 (377) (N. A. Omelina, 07.044.020; F. G. Mustaeva, 07.075.003; G. M. Kaganovskij, 07.045.023; Eh. A. Sanakulov, 07.032.024), 31 (378) (N. A. Omelina, 07.044.020; F. G. Mustaeva,

07.075.003; G. G. Khodak, 07.041.025; G. M. Kaganovskij, 07.045.024), 32 (379) (N. A. Omelina, 07.044.020; F. G. Mustaeva, 07.075.003; G. M. Kaganovskij, 07.045.025; A. G. Rakhimov, A. Sadikov, L. I. Bashtova, 07.099.045).

008.142 Teramo

Relazione sull'attività dell'Osservatorio di Teramo. P. Tempesti.
Atti XIII Riunione Soc. Astron. Italiana, Trieste 1969, (see 012.010), p. 205 - 207 (1970).

008.143 Thessaloniki

Astronomical Department, University of Thessaloniki. — Annual report 1970.   G. Contopoulos.
Annual Rep. Astron. Inst. Greece 1970, p. 8 - 10 (1971).

Contributions from the Astronomical Department of the University of Thessaloniki, Nos. 59 (G. Contopoulos, 06.151.039), 60 (S. Persides, 07.066.091), 61 (S. Persides, 06.066.057).

Department of Geodetic Astronomy, University of Thessaloniki, Thessaloniki, Greece. — Annual report 1970. L. N. Mavridis.
Annual Rep. Astron. Inst. Greece 1970, p. 13 - 14 (1971).

Contributions from the Department of Geodetic Astronomy, University of Thessaloniki, Nos. 2 (L. N. Mavridis, 05.155.030), 3 (K. Bahner, L. N. Mavridis, 07.122.131), 4 (L. N. Mavridis, A. Tsioumis, 05.122.127), 5 (G. Asteriadís, L. N. Mavridis, 07.122.105).

Université de Thessaloniki, Annuaire de l'Institut Météorologique et Climatologique, 34 - 35 (G. C. Livadas, 07.082.135).

008.144 Tokyo

Contributions from the Department of Astronomy, University of Tokyo, Nos. 139 (S. Kato, 06.151.012), 140 (Y. Osaki, 06.065.061), 141 (H. Maehara, 06.122.061), 142 (T. Tsuji, 06.064.026), 143 (G.-i. Hori, 06.042.029), 144 (K. Kodaira, 06.158.074), 145 (H. Yoshimura, 06.072.004), 146 (M. Simoda, K. Tanikawa, 07.154.002; M. Simoda, 07.154.003), 147 (Y. Yamashita, 07.071.010), 148 (S. Kato, 07.151.007), 149 (T. Yoneyama, 07.065.010), 150 (Y. Nakada, D. Sugimoto, 07.065.012).

Time and Latitude Bulletins, Tokyo Astronomical Observatory, Vol. 45, Nos. 9 - 12 (07.044.046).

Tokyo Astronomical Bulletin, Tokyo Astronomical Observatory, Second Series, Nos. 216 (Y. Hatanaka, S. S. Rao, B. Lokanadham, 07.142.131), 217 (K. Tomita, H. Kosai, H. Shibasaki, 07.103.100), 218 (S. Isobe, N. Kawajiri, T. Oiima, N. Kawano H. Kurihara, 07.132.037).

Tokyo Astronomical Observatory, Reprints.
Nos. 400 (S. Isobe, 07.131.015), 401 (S. Isobe, 07.155.004), 402 (N. Sekiguchi, 07.045.002), 403 (S. Iijima, S. Okazaki, 07.044.001), 404 (H. Yasuda, S. Aoki, 07.045.003), 409 (A. Yamasaki, M. Kitamura, 07.121.031), 410 (H. Hirabaya-shi, T. Takahashi, 07.125.017), 411 (H. Yasuda, 07.041.018),

412 (B. Takase, 07.161.012), 413 (H. Tabara, D. Morris, N. Kawajiri, M. Konno, 07.141.158), 414 (K. Tanaka, E. Hiei, 07.073.067), 415 (K. Kodaira, K. Tanaka, 07.064.042), 416 (H. Kinoshita, 07.042.049).

University of Tokyo, Tokyo Astronomical Observatory, Report, No. 59, Vol. 15, No. 4 (K. Saito, S. Shinozawa, 07.079.109; T. Kanda, M. Nukariya, 07.098.024; M. Utida, 07.004.038; M. Nukariya, N. Kobayashi, 07.098.025; T. Kanda, 07.032.047; R. Fukaya, 07.032.048; H. Ishii, 07.032.049; S. Isobe, Y. Norimoto, 07.031.033; S. Isobe, N. Miyauchi, 07.041.038; H. Kinoshita, 07.042.052; E. Watanabe, M. Yutani, T. Noguchi, 07.034.105; Y. Shimizu, Y. Norimoto, 07.034.106; K. Fujiwara, T. Hara, T. Sakai, 07.035.009).

008.145 Torino

Attività dell'Osservatorio.   M. G. Fracastoro.
Annuario Torino 1972, (see 047.021), p. 27 - 35 (1971).

Report on the history, present and future activity of the Astronomical Observatory of Turn.   M. G. Fracastoro.
Mem. Soc. Astron. Italiana, Nuova Ser., Vol. 43, 173 - 178 (1972).

Contributi dell'Osservatorio Astronomico di Torino, (Pino Torinese), Nos. 56 (F. Rossati, 04.121.053), 57 (M. A. Vogliotti, V. Zappalà, 06.098.041), 58 (M. G. Fracastoro, 05.115.019), 59 (06.079.101).

Osservatorio Astronomico di Torino. Studio Mono-grafico, No. 7 (V. Zappalà, 07.032.051).

Osservatorio Astronomico di Torino, Pino Torinese. Time Service, Bulletin No. 1 (M. G. Fracastoro, 07.044.034).

Pubblicazioni Varie Fuori Serie dell'Osservatorio Astronomico di Torino (Pino Torinese), Nos. 44 (N. Missana, Review on AJB 67, 43.34), 45 (M. G. Fracastoro, M. A. Vo-gliotti, 04.098.029), 46 (M. G. Fracastoro, Review on 04.003.008), 47 (07.047.021).

008.146 Tortosa

Publicaciones del Observatorio del Ebro, Memoria No. 13 (H. G. Fournier, 07.084.284).

008.147 Trieste

Programma di sviluppo e di ricerca dell'Osservatorio di Trieste.   M. Hack.
Atti XIII Riunione Soc. Astron. Italiana, Trieste 1969, (see 012.010), p. 209 - 212 (1970).

008.148 Tübingen

Astronomisches Institut der Universität und Lehr-stuhl für Theoretische Astrophysik. — Report 1971. G. Elwert, J. Trümper, K. Walter.
Mitt. Astron. Ges., No. 31, p. 356 - 362 (1972).

Mitteilungen des Astronomischen Instituts der Universität Tübingen, Nos. 114 (F. Unz, 02.082.111), 124

(K. Walter, 05.121.066), 125 (H. Bräuninger, H. J. Einighammer, J. V. Feitzinger, H. H. Fink, D. H. Höhn, H. Koops, G. Krämer, U. Mayer, G. Möllenstedt, M. Mozer, 06.076.019), 126 (H. Urbarz, 07.077.057).

tory, Victoria, B. C., Vol. 14, Nos. 1 (E. H. Richardson, G. A. Brealey, R. Dancey, 07.034.065), 2 (J. B. Hutchings, J. Smolinski, J. Grygar, 07.124.104).

008.149 Tucson

National Radio Astronomy Observatory, Charlottesville, Virginia, Green Bank, West Virginia, and Tucson, Arizona.–Observatory reports. D. S. Heeschen.
Bull. American Astron. Soc., Vol. 4, 140 - 149 (1972).

Steward Observatory and the Department of Astronomy, University of Arizona, Tucson, Arizona.
Report 1970 – 1971. R. J. Weymann.
Bull. American Astron. Soc., Vol. 4, 293 - 303 (1972).

008.150 Uccle

Bulletin Astronomique. (Astronomisch Bulletin). Observatoire Royal de Belgique(Koninklijke Sterrenwacht van België), Vol. 7, Nos. 5 (G. Evrard, C. Gonze, A. Koeckelenbergh, 07.075.007), 6 (H. Debehogne, 07.098.014; H. Debehogne, 07.103.010; C. Gonze, R. Gonze, 07.077.052; J. Dommanget, O. Nys, 07.118.025; S. Arend, 07.118.026; R. R. de Freitas Mourão, 07.118.027).

Observatoire Royal de Belgique (Koninklijke Sterrenwacht van Belgie). Communications (Mededelingen), Série A, Nos. 14 (R. J. Dejaiffe, P. J. Melchior, 06.045.001), 18 (P. Cugnon, 05.131.085), 19 (E. W. Elst, 07.122.013).

Observatoire Royal de Belgique (Koninklijke Sterrenwacht van Belgie), Communications (Mededelingen), Série B, Nos. 61 (C. Gonze, R. Gonze, 07.077.022), 64 (J. Dommanget, 05.118.020; J. Dommanget, 05.118.021), 65 (J. Dommanget, 05.118.015), 66 (C. de Jager, L. Neven, 07.071.004), 67 (J. Dommanget, 07.094.083), 68 (R. Dejaiffe, 07.045.028), 70 (J. Dommanget, 07.117.039).

008.151 Urbana

University of Illinois Department of Astronomy, Urbana, Illinois.–Observatory report. G. W. Swenson, Jr.
Bull. American Astron. Soc., Vol. 4, 54 - 55 (1972).

008.152 Utrecht

Sterrekundig Instituut. Laboratorium voor Ruimteonderzoek, Utrecht. Reprint of the Meteor Section of the Netherlands Ass. for Astronomy and Meteorology, Nos. 50 (A. G. W. Maas, 05.103.111), 51 (B. Apeldoorn, E. J. Kaptein, 06.104.004).

008.153 Victoria

Dominion Astrophysical Observatory, Victoria, B.C. J. B. Hutchings.
Journ. Roy. Astron. Soc. Canada, Vol. 66, 128 (1972).

Publications of the Dominion Astrophysical Observa-

008.154 Warsaw

Warsaw University Observatory and Astronomical Institute, Polish Academy of Sciences, Reprint Nos. 323 (J. Smak, 07.117.045), 324 (M. Kubiak, 07.122.161).

Publikacje Działu Geodezji Wyższej i Astronomii, Geodezyjnej Zg. PAN, Nos. 17 (S. Domaradzki, S. Oszczak, 05.032.030), 20 (J. Łatka, 06.055.004).

008.155 Washington

U. S. Naval Observatory, Washington, D. C.
Observatory report. K. A. Strand.
Bull. American Astron. Soc., Vol. 4, 179 - 185 (1972).

Termination of the U.S. Naval Observatory solar program. A. E. Covington.
Journ. Roy. Astron. Soc. Canada, Vol. 66, 165 - 166 (1972).

Publications of the United States Naval Observatory, Washington, Second Series, Vol. 20, Part 4 (D. H. Ables, 07.158.051); Vol. 22, Part 2 (C. E. Worley, 07.118.029).

United States Naval Observatory. Washington, D. C., Circular Nos. 136 (B. L. Klock, D. K. Scott, 07.041.048), 137 (B. L. Morrison, V. Meiller, 07.099.063).

U. S. Naval Observatory, Washington, D. C. Time Service Publications, Series 4, Nos. 255 - 282; Series 7, Nos. 209 - 235; Series 11, No. 220; Series 14, Nos. 9 - 10 (07.044.050 - 07.044.053).

Georgetown Observatory Reprint, Nos. 38 (T. E. Margrave, Jr., AJB 68,54.70), 39 (T. E. Margrave, Jr., 01.121.017), 40 (K. Johnston, 04.121.036), 41 (J. E. O'Brien, 06.071.014).

Georgetown Observatory Reprint, Series 2, No. 34 (K. J. Johnston, 05.121.064).

National Aeronautics and Space Administration, Office of Space Science and Applications, Physics and Astronomy Programs, Washington, D. C.–Observatory report. N. G. Roman.
Bull. American Astron. Soc., Vol. 4, 123 - 136 (1972). – Concerning the reports of Ames Research Center; Goddard Space Flight Center; Jet Propulsion Laboratory; Langley Research Center; Manned Spacecraft Center; Marshall Space Flight Center.

National Bureau of Standards, Spectroscopy Section, Institute for Basic Standards, Washington, D. C.–Observatory report. J. L. Tech.
Bull. American Astron. Soc., Vol. 4, 136 - 139 (1972).

008.156 Wellington

Carter Observatory.
Inform. Bull. Southern Hemisph., No. 19, p. 27 (1971). Current research report.

008.157 Westford

Haystack Observatory, Northeast Radio Observatory Corporation, Westford, Massachusetts.—Observatory report. P. B. Sebring. Bull. American Astron. Soc., Vol. 4, 51 - 54 (1972). — This report covers the period 1 July 1970 through 30 June 1971.

008.158 Wien

Universitäts-Sternwarte mit Leopold-Figl-Observatorium für Astrophysik. — Report 1971. J. Meurers. Mitt. Astron. Ges., No. 31, p. 362 - 367 (1972).

Institut für Theoretische Astronomie. — Report 1971. K. Ferrari d'Occhieppo. Mitt. Astron. Ges., No. 31, p. 368 - 369 (1972).

008.159 Williams Bay

University of Chicago, Yerkes Observatory, Williams Bay, Wisconsin.—Observatory report.   C. R. O'Dell. Bull. American Astron. Soc., Vol. 4, 7 - 12 (1972).

008.160 Würzburg

Astronomisches Institut und Sternwarte. — Report 1971. H. Haffner. Mitt. Astron. Ges., No. 31, p. 369 - 371 (1972).

008.161 Zelenchukskaya

Chronicle. Astrofiz. Issled. Izv. Spets. Astrofiz. Obs., Vol. 3, 186 - 187 (1971). In Russian.

Astrofizicheskie Issledovaniya. Izvestiya Spetsial'noj Astrofizicheskoj Observatorii, Vol. 3 (B. K. Ionnisiani, 07.032.006; J. B. Vyatskin, A. S. Naishul, E. M. Neplokhov, 07.032.007; O. B. Vasilyev, N. F. Nelubin, 07.082.044; V. V. Leushin, 07.114.055; J. V. Glagolevsky, 07.116.004; J. V. Glagolevsky, N. M. Chunakova, 07.116.005; K. I. Kozlova, 07.114.056; A. A. Korovyakovskaya, J. P. Korovyakovsky, 07.122.046; M. N. Naugolnaya, 07.094.103; T. B. Pyatunina, 07.155.031; T. B. Pyatunina, 07.158.054; P. A. Fridman, 07.033.007; A. F. Dravskikh, S. G. Smolentsev, 07.033.008; V. M. Spitkovsky, 07.033.009; A. L. Alexandrov, V. M. Brylov, A. A. Stotsky, 07.033.010; N. F. Ryzhkov, V. A. Jakovlev, 07.033.011; J. K. Zverev, A. I. Kopylov, 07.033.012; J. K. Zverev, 07.033.013; V. E. Bakhchivandzhi, 07.031.010; Z. V. Dravskikh, 07.033.014).

008.162 Zürich

Tätigkeitsbericht der Eidgenössischen Sternwarte Zürich für das Jahr 1971.   M. Waldmeier. Zürich, 7 pp. (1972).

Astronomische Mitteilungen der Eidgenössischen Sternwarte Zürich, Nos. 306 (W. Stanek, 06.074.074), 310 (M. Waldmeier, 07.075.014).

Quarterly Bulletin on Solar Activity (Zürich), Nos. 173 - 174 (M. Waldmeier, R. Howard, R. Michard, G. Olivieri, M. Bernot, 07.075.010).

009 Notes on Observatories, Planetaria, and Exhibitions

009.001 **People's observatories and development of amateur astronomy in Yugoslavia.**   P. Djurković. Zemlya i Vselennaya, No. 1, p. 70 - 73 (1972). In Russian. Translated from Serbo-Croatian by M. N. Efremov.

009.002 **Observatory at Ridley Park, Pennsylvania.** N. Ignatuk, Jr. Sky Telescope, Vol. 43, 19 (1972).

009.003 **The Wise Observatory in Israel.** Sky Telescope, Vol. 43, 72 - 73 (1972).

009.004 **People's astronomical observatories and planetaria in our country (in Bulgaria).**   N. Nikolov. Mat. i fizika (NRB), Vol. 14, No. 4, p. 64 - 68 (1971). In Bulgarian. — Abstr. in Referativ. Zhurn. 51. Astron., 3.51.91 (1972).

009.005 **A new solar observatory on the Ottawa River.** V. Gaizauskas. Journ. Roy. Astron. Soc. Canada, Vol. 66, 69 (1972). — Abstr. Canadian Astron. Soc.

009.006 **Department of Natural Philosophy, University of Aberdeen.**—Report for the year ending 1971 September 30.   R. V. Jones.

Quarterly Journ. Roy. Astron. Soc., Vol. 13, 88 - 89 (1972).

009.007 Queen Mary College: Astrophysics Group (Physics Department). – Report for the year ending 1971 September 30. J. A. Bastin.
Quarterly Journ. Roy. Astron. Soc., Vol. 13, 104 - 106 (1972).

009.008 University College London: Theoretical Atomic Physics and Astrophysics (Department of Physics). Report 1970 January 1 to 1971 July 31. M. J. Seaton.
Quarterly Journ. Roy. Astron. Soc., Vol. 13, 106 - 108 (1972).

009.009 Instituto de Astronomía y Física del Espacio (IAFE). H. S. Ghielmetti.
Inform. Bull. Southern Hemisph., No. 19, p. 4 - 5 (1971).

009.010 Recent progress at the Strasenburgh Planetarium. T. Dickinson.
Journ. Roy. Astron. Soc. Canada, Vol. 66, 120 - 124 (1972).

009.011 Planetarium "Longines". P. Landman.
Hemel en Dampkring, Vol. 70, 101 - 103 (1972).

009.012 Geschichte und Tätigkeit des Astronomischen Instituts der Universität Bern. M. Schürer.
Orion Schaffhausen, 30. Jahrgang, p. 5 - 6 (1972).

009.013 Le nouvel observatoire de La Tour-de-Peilz. P. Bignens, R. Durussel, V. Fryder.
Orion Schaffhausen, 30. Jahrgang, p. 19 - 20 (1972).

009.014 Das Observatorium von Valasske Mezirici, CSSR. B. Malecek.
Orion Schaffhausen, 30. Jahrgang, p. 44 - 46 (1972).

009.015 Le stellarium de Nançy. B. Clouet.
L'Astronomie, 86ᵉ année, p. 198 - 200 (1972).

009.016 Een Amsterdams planetarium uit 1756. D. Koelbloed.
Hemel en Dampkring, Vol. 70, 143 - 150 (1972).

009.017 The upper atmosphere observatory. J. V. Evans.
Science, Vol. 176, 463 - 473 (1972).
American and Canadian scientists plan to build a major research center.

009.018 Eine Sternwarte für Kreuzlingen. E. Obreschkow.
Orion Schaffhausen, 30. Jahrgang, p. 110 - 111 (1972).

009.019 Via satellite i calcoli del più antico osservatorio astronomico del mondo.
Coelum, Vol. 40, 106 - 108 (1972).

009.020 Projet d'un centre d'études et de recherches géo-dynamiques et astronomiques en France (CERGA). J. Kovalevsky.

Ciel et Terre, Vol. 88, 196 - 204 (1972).

009.021 25 Jahre Wiener Volkssternwarte. W. Jaschek.
Sternenbote, 15. Jahrgang, p. 34 - 41 (1972).

009.022 Bericht über die Tätigkeit des Rates Westdeutscher Sternwarten im Jahr 1971. H. Elsässer.
Mitt. Astron. Ges., No. 31, p. 233 - 234 (1972).

009.023 Aus der Arbeit der „Jungen Astronomen" an der Volkssternwarte „Adolf Diesterweg" Radebeul. R. Kollar.
Astron. in der Schule, 9. Jahrgang, p. 65 - 67 (1972).

009.024 Country people observatory. I. M. Bezchastnov.
Zemlya i Vselennaya, 1972, No. 3, p. 76 - 77. In Russian.

009.025 Attività dell'Osservatorio "G. Horn d'Arturo". L. Baldinelli.
Atti XIII Riunione Soc. Astron. Italiana, Trieste 1969, (see 012.010), p. 213 - 214 (1970).

009.026 Coordinates of observatories.
IAU Circ., No. 2417 (1972).

009.027 University of Calgary Astrophysical Observatory opened. F. J. Howell.
Journ. Roy. Astron. Soc. Canada, Vol. 66, 161 - 164 (1972).

009.028 50 Jahre Volkssternwarte Stuttgart.
Edited by Verein Schwäbische Sternwarte e. V., Stuttgart. 64 pp. (1972).

009.029 Osservatorio Astronomico Nazionale: Rapporto N. 2. G. Righini (Editor).
Printed by Tipografia Antoniana, Padova. 113 pp. (1971).

009.030 Konstante narodne opservatorije. [Constants of the People's Observatory.] Z. Ivanović.
Vasiona, Vol. 20, 37 - 38 (1972).

009.031 Tätigkeitsbericht 1971 der Privatsternwarte Karlsruhe. W. Malsch.
Separate print: Erzbergerstr. 111c, Karlsruhe. 2 pp. (1972).

009.032 Scientific work at Czechoslovak public observatories. O. Obůrka.
Říše hvězd, Vol. 53, 29 - 33 (1972). In Czech.

009.033 Aus der Arbeit der Volkssternwarten.
Sterne, 48. Jahrgang, p. 122 - 124 (1972).
(1) Astronomiegeschichte an der Archenhold-Sternwarte im Jahre 1971 (*D. B. Herrmann*); Die Potsdamer Sternfreunde und ihre großes Fernrohr (*A. Zenkert*).

Burgenländische Landessternwarte.
Sternenbote, 15. Jahrgang, p. 18 - 19 (1972).

010 Societies, Associations, Organizations

010.001 American Association of Variable Star Observers
(AAVSO)

American Association of Variable Star Observers.
**Abstracts of papers presented at Rochester meeting, 1 May
1971.**   R. N. Mayall (Editor).
AAVSO Abstr. 1971, 9 pp. – Contents: Total solar eclipse
30 June 1973, *L. J. Boss;* T Tauri mystery stars, *L. J. Boss;*
The shield of Sobieski, *C. Borzelli;* New sequence for UY and
UZ CMa, *C. E. Scovil;* Computer-plotted light curves, *M. W.
Mayall;* Reawakening of the Nova Search Committee,
G. Diedrich; Recent work on charts, *C. B. Ford;* Total solar
eclipse July 1972, *C. H. Hossfield;* Classical cepheid program
for AAVSO, *T. A. Cragg.*

Occultation committee.   C. M. Good.
AAVSO Semi-annual Committee Rep., May 1971, p. 1.

Eclipsing Binary and RR Lyrae Committees.
M. E. Baldwin.
AAVSO Semi-annual Committee Rep., May 1971, p. 1 - 2.

Chart Committee.
AAVSO Semi-annual Committee Rep., May 1971, p. 2.

Nova Search Committee.   G. Diedrich.
AAVSO Semi-annual Committee Rep., May 1971, p. 3.

010.002 American Astronomical Society (AAS)

**The 136th meeting of the American Astronomical
Society, held 5–8 December 1971 at San Juan, Puerto Rico.
Abstracts of papers presented.**
Bull. American Astron. Soc., Vol. 3, 437 - 485 (1971).

**The 135th meeting of the American Astronomical
Society, held 24 through 27 August 1971 at Amherst, Massa-
chusetts. Addenda.**
Bull. American Astron. Soc., Vol. 3, 497 - 502 (1971).

**The 137th meeting of the American Astronomical
Society, held 8–12 April 1972 at Seattle, Washington. Ab-
stracts of papers presented.**
Bull. American Astron. Soc., Vol. 4, 207 - 245 (1972).

**Abstracts of papers presented at the meeting of the
APS/AAS Division of High Energy Astrophysics, held 1–3
December 1971 at San Juan, Puerto Rico.**
Bull. American Astron. Soc., Vol. 4, 255 -262 (1972).

**Abstracts of papers presented at the third regular
meeting of the AAS Division on Dynamical Astronomy,
held 2–3 March 1972 at the University of Maryland, College
Park, Maryland.**
Bull. American Astron. Soc., Vol. 4, 262 - 268 (1972).

**Late-paper abstracts from the 136th meeting of the
American Astronomical Society, held 5–8 December 1971
at San Juan, Puerto Rico.**
Bull. American Astron. Soc., Vol. 4, 268 - 271 (1972).

**The Cooperative Committee on the Teaching of
Science and Mathematics.**   B. F. Peery, Jr.
Bull. American Astron. Soc., Vol. 4, 303 - 304 (1972).

010.003 Association of Lunar and Planetary Observers
(ALPO)

The 1971 ALPO business meeting at Memphis.
Strolling Astronomer, Vol. 23, 127, 130 (1972).

Announcements.
Strolling Astronomer, Vol. 23, 147 - 148, 188 (1972).

010.004 Astronomical Society of Australia (ASA)

**Papers presented at a meeting held at the University
of Adelaide on 13, 14 and 15 December 1971.**
Proc. Astron. Soc. Australia, Vol. 2, 70 - 117 (1972).

010.005 Astronomical Society of Czechoslovakia

No publication received.

010.006 Astronomical Society of the Pacific (ASP)

Minutes of the meeting of the directors.
Mercury, (Journ. Astron. Soc. Pacific), Vol. 1, No. 2, p. 17 -
19 (1972).

010.007 Astronomical Society of Southern Africa (ASSA)

Notices.
Monthly Notes Astron. Soc. Southern Africa, Vol. 31, 1, 31,
43 - 44 (1972).

010.008 Astronomical Society of Victoria (ASV)

Annual report 1970.
G. A. Duncan, R. J. Lawrence, D. H. Walker, P. Simon,
C. W. Alexander, D. H. Whitehead, T. B. Tregaskis, J. B. Trai-
nor, B. A. J. Clark, L. R. Whitby, J. H. White, W. G. H. Tre-
gear, A. E. Coombs, B. S. Adcock.
Journ. Astron. Soc. Victoria, Vol. 24, 2 - 12 (1971). – Inclu-
ded are reports on the activities of different sections of the
Society.

Annual report 1971.
T. B. Tregaskis, R. J. Lawrence, D. H. Walker, J. L. Perdrix,
P. Simon, D. H. Whitehead, W. G. H. Tregear, A. E. Coombs,
J. H. White, B. S. Adcock, B. A. J. Clark.
Journ. Astron. Soc. Victoria, Vol. 25, 2 - 10 (1972). – Inclu-
ded are reports on the activities of different sections of the
Society.

Society notes.
Journ. Astron. Soc. Victoria, Vol. 24, 82 - 84, 99 - 100
(1971); Vol. 25, 11 - 12, 24 - 27 (1972).

010.009 Astronomical Society of Western Australia (ASWA)

**Reports of proceedings – 231st – 235th ordinary
meetings.**
Journ. Astron. Soc. Western Australia, Vol. 34 - 37, January –
May (1972).

010.010 Astronomische Gesellschaft (AG)

Versammlung der Astronomischen Gesellschaft im Rahmen der "Wissenschaftlichen Astronomischen Tagung Bonn 1971", 13. - 18. September 1971. Bericht über die Tagung. H. Schmidt.
Mitt. Astron Ges., No. 31, p. 7 - 8 (1972).

Mitgliederverzeichnis der Astronomischen Gesellschaft, (Stand 20. Mai 1972).
Mitt. Astron. Ges., No. 31, p. 373 - 379 (1972).

Mitteilungen der Astronomischen Gesellschaft.
No. 31 (edited by K. Schaifers).

010.011 Astronomisk Selskab Kobenhavn

No publication received.

010.012 British Astronomical Association (BAA)

Solar Section.
British Astron. Ass., Circ. No. 539 (1972).

Variable Star Section.
British Astron. Ass., Circ. No. 539 (1972).

VSS Binocular Sub-group.
British Astron. Ass., Circ. No. 539 (1972).

Notices.
Journ. British Astron. Ass., Vol. 82, 82 - 84, 161 - 164, 243 - 246 (1972).

Meetings of the Association.
Journ. British Astron. Ass., Vol. 82, 85 - 94, 165 - 177, 247 - 256 (1972).

Lunar Section. P. A. Ringsdore.
Journ. British Astron. Ass., Vol. 82, 284 - 286 (1972).

Solar Section. W. M. Baxter.
Journ. British Astron. Ass., Vol. 82, 138 (1972).

Variable Star Section.
Journ. British Astron. Ass., Vol. 82, 138, 298 - 299 (1972).

Meteor Section.
Journ. British Astron. Ass., Vol. 82, 214 - 219, 287 - 298 (1972).
Historical Section.
Journ. British Astron. Ass., Vol. 82, 300 (1972).

New members elected.
Journ. British Astron. Ass., Vol. 82, 150 - 159, 236 - 239, 314 - 317 (1972).

010.013 British Interplanetary Society (BIS)

Society news.
Spaceflight, Vol. 14, 34 - 36, 74 - 75, 157 - 160, 275 (1972).

27th annual general meeting.
Spaceflight, Vol. 14, 151 - 152 (1972).

The report of the Council for the year ended 31 December 1971.
Spaceflight, Vol. 14, 153 - 156 (1972).

010.014 Committee on Space Research (COSPAR)

No publication received.

010.015 European Space Research Organization (ESRO)

No publication received.

010.016 International Astronautical Federation (IAF)

Dvadesetdrugi medunarodni astronautički kongres u Brislu — Belgija. [22nd international astronautical congress in Brussels — Belgium]. A. Stojanović.
Vasiona, Vol. 20, 12 - 14 (1972).

010.017 International Astronomical Union (IAU)

International Astronomical Union, Information Bulletin, Nos. 27, 28, [printed by D. Reidel, Dordrecht — Holland], 35 + 26 pp. (1972).   C. de Jager.
Contents: General assemblies; Executive committee; Commissions, IAU symposia and colloquia; Other scientific meetings; IAU publications; Other publications; Other international organizations; Membership.

Das IAU - Colloquium über Veränderliche.
W. Quester.
BAV Rundbrief, 21. Jahrgang, p. 3 - 5 (1972).

I. A. U. Symposium No. 49: 'Wolf-Rayet and hot temperature stars'.
Inform. Bull. Southern Hemisph., No. 19, p. 38 (1971).
Instituto de Astronomía y Física del Espacio, Buenos Aires, Argentina, August 9 - 14, 1971.

I. A. U. Symposium No. 50: 'Spectral classification and multicolor photometry'.
Inform. Bull. Southern Hemisph., No. 19, p. 39 - 43 (1971).
Córdoba Observatory, October 19 - 23, 1971.

010.018 Meteoritical Society

No publication received.

010.019 Nederlandse Vereniging voor Weer- en Sterrenkunde

Verenigingsnieuws.
Hemel en Dampkring, Vol. 70, 21, 89, 119 - 125, 160 - 161, 190 (1972).

Jongerenwerkgroep.
Hemel en Dampkring, Vol. 70, 29 - 31, 63 - 69, 98 - 106, 203 - 205 (1972).

Werkgroep Meteoren.
Hemel en Dampkring, Vol. 70, 25 - 28 (1972).
De campagne-begroting 1972 voor fotografische en visuele waarnemers (*B. Apeldoorn*), Een fraaie meteooropname (*B. Apeldoorn*), Herberekening van een meteoor (*E. J. Kaptein*).

Werkgroep Veranderlijke Sterren V.V.S.
F. van Loo.
Hemel en Dampkring, Vol. 70, 94 (1972).

010.020 **Polskie Towarzystwo Astronomiczne (PTA)**

Resolutions of the plenary meeting of the Polish Astronomical Society (8th September, 1971).   A. Żytkow.
Postępy Astron., Vol. 20, 179 - 183 (1972). In Polish.

Report of the executive council of the Polish Astronomical Society on activities in the period 1969–1971.
J. Smak, W. Zonn.
Postępy Astron., Vol. 20, 183 - 185 (1972). In Polish.

15th meeting of the Polish Astronomical Society.
K. Ziołkowski.
Urania Kraków, Vol. 43, 89 - 91 (1972). In Polish.

010.021 **Polskie Towarzystwo Miłośników Astronomii (PTMA)**

PTMA Chronicle.
Urania Kraków, Vol. 43, 28 - 29, 78 - 86, 177 - 181 (1972).
In Polish.

50 years of our Society.   L. Zajdler.
Urania Kraków, Vol. 43, 34 - 45 (1972). In Polish.

010.022 **Royal Astronomical Society (RAS)**

Meetings of the Society.
Observatory, Vol. 92, 1 - 8, 25 - 33, 34 - 40 (1972).

Meetings of the Society.
Quarterly Journ. Roy. Astron. Soc., Vol. 13, 3 - 9, 117 - 123, 127 - 129 (1972).

Royal Astronomical Society meeting on the solar wind.   H. Rishbeth, P. C. Kendall.
Quarterly Journ. Roy. Astron. Soc., Vol. 13, 67 - 69 (1972).

Anniversary meeting of 1972 February 11.
Quarterly Journ. Roy. Astron. Soc., Vol. 13, 124 - 126 (1972).

010.023 **Royal Astronomical Society of Canada (RAS Canada)**

The Royal Astronomical Society of Canada 1968: Minutes of the Annual Meeting, May 20, 1972. M. Fidler.
Journ. Roy. Astron. Soc. Canada, Vol. 66, 171 - 172 (1972).

General assembly at Vancouver, May 19 - 22, 1972.
M. (Fidler) Litchinsky.
Journ. Roy. Astron. Soc. Canada, Vol. 66, L13 - L15 (1972).

Annual report 1971.
Suppl. Journ. Roy. Astron. Soc. Canada, April 1972, 24 pp.
Included are the reports of different sections of the Society.

010.024 **Royal Astronomical Society of New Zealand (RAS New Zealand)**

Royal Astronomical Society of New Zealand.
Variable Star Section.
Inform. Bull. Southern Hemisph., No. 19, p. 26 - 27 (1971).
Current research report.

Taking stock.   J. B. Mackie.
Southern Stars, Vol. 24, 81 - 93 (1972). – Presidential address

delivered at the annual general meeting held in Christchurch, 1971 November 27.
Recent developments; The role of the R.A.S.N.Z.; The growth of professional astronomy; Alternative types of organisation; Essential lines of action.

Annual report of council for the year ended 1971, September 30.   J. B. Mackie, D. J. Cameron.
Southern Stars, Vol. 24, 94 - 99 (1972).

Report of Variable Star Section.
F. M. Bateson.
Southern Stars, Vol. 24, 100 -107 (1972).

Lunar and Planetary Section.   P. A. Read.
Southern Stars, Vol. 24, 116 - 117 (1972).

010.025 **Schweizerische Astronomische Gesellschaft (SAG)**

Aus der SAG und den Sektionen.
Orion Schaffhausen, 30. Jahrgang, p. 22 - 24, 110 (1972).

Berichte des Präsidenten und des Generalsekretärs der SAG, erstattet an der Generalversammlung der Gesellschaft, 5./6. Mai 1972 in Zürich.   W. Studer, H. Rohr.
Orion Schaffhausen, 30. Jahrgang, p. 116 - 121 (1972).

010.026 **Sociedad Astronómica de México**

Actividades de la Sociedad.
El Universo, No. 93, Vol. 24, 83 - 86 (1970).

010.027 **Società Astronomica Italiana (SAI)**

La XIIIᵃ Riunione della Società Astronomica Italiana, Trieste, 25–28 settembre 1969.
Atti XIII Riunione Soc. Astron. Italiana, Trieste 1969, (see 012.010), p. 3 - 4 (1970).

Relazione del presidente della SAIt. per l'anno 1968- 69.   G. Righini.
Atti XIII Riunione Soc. Astron. Italiana, Trieste 1969, (see 012.010), p. 7 - 14 (1970).

Verbale dell'assemblea generale della Società Astronomica Italiana 28 settembre 1969.
Atti XIII Riunione Soc. Astron. Italiana, Trieste 1969, (see 012.010), p. 15 - 22 (1970).

Verbale di scrutinio delle schede di votazione per la elezione del consiglio direttivo della Società Astronomica Italiana per il biennio ottobre 1969 – settembre 1971.
Atti XIII Riunione Soc. Astron. Italiana, Trieste 1969, (see 012.010), p. 23 - 24 (1970).

Elenco dei membri della Società Astronomica Italiana. (31 dicembre 1969).
Atti XIII Riunione Soc. Astron. Italiana, Trieste 1969, (see 012.010), p. 221 - 237 (1970).

010.028 **Société Astronomique de France (SAF)**

Les séances de la Société.   B. Clouet.
L'Astronomie, 86ᵉ année, p. 152 - 158, 200, 244 - 247, 311 - 317 (1972).

Nouvelles du Groupe d'Alsace.
P. Lacroute, M. Piriou.
L'Astronomie, 86ᵉ année, p. 201 - 202 (1972).

010.029   Société Astronomique "R. Bosković"

No publication received.

010.030   Société Chronométrique de France

Liste des membres de la Société Chronométrique de France.
Ann. Françaises Chronométrie Micromécanique, Année 1971, p. 165 - 171.

010.031   Société Belge d'Astronomie, de Météorologie et de Physique du Globe

Séance mensuelle.
Ciel et Terre, Vol. 88, 144 - 145 (1972).

Réunions mensuelles.
Ciel et Terre, Vol. 88, 146 - 149, 229 - 230, 231 - 232 (1972).

010.032   Svenska Astronomiska Sällskapet

No publication received.

010.033   VAGO (Astronomical-Geodetical Society of the USSR)

No publication received.

010.034   Vereniging voor Steerenkunde, Belgie

No publication received.

010.035   Argentine Astronomical Association

No publication received.

010.036   Canadian Astronomical Society

The meeting of the National Research Council Associate Committee on Astronomy at Toronto November 11–13, 1971.   G. A. H. Walker.
Journ. Roy. Astron. Soc. Canada, Vol. 66, 63 - 64 (1972).

First regular meeting of the Canadian Astronomical Society at the University of Toronto, Nov. 11–13, 1971.
P. M. Millman.
Journ. Roy. Astron. Soc. Canada, Vol. 66, 65 (1972).

010.037   Nachrichten der Vereinigung der Sternfreunde e.V.
SuW, Vol. 11, 20 - 21, 52 - 53, 81 - 83, 112 - 113, 142 - 143, 170 - 172, 200 - 202 (1972).

010.038   Société Royale d'Astronomie d'Anvers. Cinquantedeuxième rapport 1971.   J. Storms.
La Prévoyance, Antwerpen. 36 pp. (1972). In French and Flemish.

010.039   The Nantucket Maria Mitchell Association.
Seventieth annual report for the year ending December 31, 1971.
The Nantucket Maria Mitchell Association, Vestal Street, Nantucket, Mass., 60 pp. (1972). – Included is the annual report of the director of Maria Mitchell Observatories, by *D. Hoffleit*.

011 Reports on Colloquia, Congresses, Meetings, Symposia, and Expeditions

011.001 **25 years seminar on the evolution of the earth.**
S. V. Kozlovskaya.
Zemlya i Vselennaya, No. 1, p. 33 (1972). In Russian.

011.002 **In honour of Kepler.** (Symposium in Leningrad, 1971, Aug. 21). Z. K. Sokolovskaya.
Zemlya i Vselennaya, No. 1, p. 45 - 48 (1972). In Russian.

011.003 **International geochemical congress in Moscow.**
V. P. Volkov.
Zemlya i Vselennaya, No. 1, p. 48 - 49 (1972). In Russian.
1971, July 20–25.

011.004 **Conference on the problem of the Tungusic meteorite.** E. M. Kolesnikov.
Zemlya i Vselennaya, No. 1, p. 54 - 55 (1972). In Russian. – Novosibirsk, 1971 April.

011.005 **Meeting review: The upper atmosphere of Venus.**
C. W. Snyder.
Icarus, Vol. 15, 555 - 557 (1971/72).

011.006 **Symposium on the geology and geochemistry of the oldest sedimentary-volcanic series on earth: The Swaziland sequence.** B. Nagy.
Icarus, Vol. 15, 558 - 559 (1971/72).

011.007 **Long-baseline interferometry.**
A. E. E. Rogers, P. Morrison.
Science, Vol. 175, 218, 220 (1972). – Brookline, Mass., 1971 April 13–14.

011.008 **International seminar on particle acceleration in cosmic space.** L. I. Miroshnichenko.
Geomagn. Aeronom., Vol. 12, 170 (1972). In Russian.

011.009 **IAU-Colloquium über Veränderliche in Bamberg.**
W. Quester.
SuW, Vol. 11, 15 (1972). – Bamberg, 1971 August 31 – September 3.

011.010 **European amateurs meet in West Germany.**
G. D. Roth.
Sky Telescope, Vol. 43, 18 (1972).

011.011 **The third lunar science conference – I.**
T. L. Page.
Sky Telescope, Vol. 43, 145 - 150 (1972). – Houston, 1972, January 10–13.

011.012 **The third lunar science conference – II.**
T. L. Page.
Sky Telescope, Vol. 43, 218 - 222 (1972).

011.013 **Cosmic rays and astrophysics. Seminar in Moscow.**
V. S. Berezinskij.
Vestn. AN SSSR, 1971, No. 10, p. 71 - 74. In Russian.

011.014 **First Soviet-American conference on communication with extra-terrestrial intelligence (CETI).**
Spaceflight, Vol. 14, 18 - 19 (1972). – Byurakan, 1971 Sept. 5 - 11.

011.015 **Kolloquium "Geodätische Astronomie" des Lohrmann-Observatoriums der TU Dresden.**

H.-U. Sandig.
Sterne, 48. Jahrgang, p. 30 - 31 (1972).

011.016 **XIII. Internationaler Kongreß für Geschichte der Wissenschaften.** D. B. Herrmann.
Sterne, 48. Jahrgang, p. 31 - 34 (1972).

011.017 **Scientific session of the Department of General Physics and Astronomy and of the Department of Nuclear Physics of the USSR Academy of Sciences, 26 - 27 May 1971.**
Uspekhi fiz. nauk, Vol. 105, 775 (1971). In Russian. – Abstr. in Referativ. Zhurn. 51. Astron., 4.51.27 (1972).

011.018 **Scientific session of the Department of General Physics and Astronomy of the USSR Academy of Sciences, 23 - 24 June 1971.**
Uspekhi fiz. nauk, Vol. 105, 782 (1971). In Russian.

011.019 **Joint scientific session of the Department of General Physics and Astronomy of the USSR Academy of Sciences together with the Department of Physico-technical and Mathematical Sciences of the Uzbekian SSR Academy of Sciences, 14 - 16 April 1971.**
Uspekhi fiz. nauk, Vol. 105, 743 - 758 (1971). In Russian.

011.020 **Joint scientific session of the Department of General Physics and Astronomy of the USSR Academy of Sciences together with the Department of Physico-technical and Chemical Sciences of the Turkmenian SSR Academy of Sciences, 20 - 22 April 1971.**
Uspekhi fiz. nauk, Vol. 105, 758 (1971). In Russian.

011.021 **Sternfreunde-Colloquium in Regensburg.**
W. Sandner.
Sterne, 48. Jahrgang, p. 52 - 53 (1972).

011.022 **CETI-71.** L. M. Gindilis.
Zemlya i Vselennaya, 1972, No. 2, p. 49 - 53. In Russian. – Byurakan, 1971 Sept. International conference on the connection with extraterrestrial civilizations.

011.023 **3rd international planetarium directors conference, Vienna 1969.** R. Mucke, H. Mucke.
Owner, Editor and Publisher: Astronomical Bureau, Vienna (Austria). 1 + 66 pp. (1972). In German and English.

011.024 **Seventh international symposium on remote sensing of environment, May 17 - 21, 1971.**
G. J. Zissis.
Icarus, Vol. 16, 401 - 403 (1972).

011.025 **First Soviet-American conference on communication with extraterrestrial intelligence (CETI).**
Icarus, Vol. 16, 412 - 414 (1972).

011.026 **ESO – CTIO – University of Chile – Catholic University meeting.** R. Havlen.
Inform. Bull. Southern Hemisph., No. 19, p. 34 - 38 (1971). E. S. O., Santiago, Chile, June 2, 1971.

011.027 **Protokoll der 117. Sitzung der Schweiz. Geodätischen Kommission vom 18. Juni 1971 an der Eidg.**
Techn. Hochschule in Zürich mit Auszügen aus den Berichten über die Tätigkeit im Jahre 1970.

Société Helvétique des Sciences Naturelles, (Schweiz. Natur-forschende Gesellschaft). Spross+Co., Kloten. 90 pp. (1972).

011.028 **Report on the Cambridge meeting of the A.A.S. Working Group on Photographic Materials in Astronomy—Part I.**
AAS Photo-Bull. 1972, No. 1, p. 17 - 21. — The individual contributions see abstracts 034.040, 113.020.

011.029 **Conferinţa naţională de astronomie cu ocazia a 50 de ani de existenţă a Observatorului Astronomic al Universităţii din Cluj.** C. Cristescu.
Stud. Cerc. Astron., Vol. 17, 161 - 162 (1972).

011.030 **The Minnaert Memorial Conference on education in, and the history of, modern astronomy.**
Mercury, (Journ. Astron. Soc. Pacific), Vol. 1, No. 1, p. 9 (1972).

011.031 **A Sonoran adventure: The University of Arizona's eclipse expedition of September, 1923.**
G. G. Sykes.
Mercury, (Journ. Astron. Soc. Pacific), Vol. 1, No. 1, p. 12 - 17 (1972).

011.032 **Polish-Czechoslovakian conference on the subject of dynamics of minor planets, comets, and meteors.**
K. Ziołkowski.
Urania Kraków, Vol. 43, 121 - 123 (1972). In Polish. — 1971 October 25—29.

011.033 **Report on the XXIInd International Astronautical Congress, Brussels, Belgium, 20—25 September 1971.**
Astronaut. Acta, Vol. 17, 279 - 287 (1972).

011.034 **Physics of solar prominences. Report on an inter-national colloquium held at the German Solar Observatory, Anacapri, September 29 to October 1, 1971.**
A. Bruzek, M. Kuperus.
Solar Physics, Vol. 24, 3 - 17 = Mitt. Fraunhofer Inst., *Freiburg*, No. 111 (1972).

011.035 **CETI-71.** L. M. Gindilis.
Zemlya i Vselennaya, 1972, No. 3, p. 48 - 55. In Russian. — Byurakan, 1971 Sept. International conference on the connection with extraterrestrial civilizations.

011.036 **Aviation and cosmonautics — past and presence.**
Yu. S. Voronkov.
Zemlya i Vselennaya, 1972, No. 3, p. 55 - 57. In Russian. Report of the XIII. international congress on the history of science, Moscow, 1971 August.

011.037 **Apollo 15 investigator's symposium.** The Lunar Science Institute, Houston, Texas.
The Moon, Vol. 4, 505 (1972). — 1971 November 30 - December 2.

011.038 **The early history of the earth and moon.** AAAS Symposium, Philadelphia.
The Moon, Vol. 4, 506 (1972). — 1971 December 28.

011.039 **Fifth National Australian Convention of Amateur Astronomers (NACAA).** J. L. Perdrix.
Journ. Astron. Soc. Victoria, Vol. 25, 14 - 17 (1972). — Melbourne, 1972 April.

011.040 **Symposium on 'The new astronomy'.** P. A. Wayman.
Monthly Notes Astron. Soc. Southern Africa, Vol. 31, 45 - 52 (1972).

011.041 **Meeting of the astronomical society "Ruđer Bošković".**
Vasiona, Vol. 20, 46 - 47 (1972). In Serbo-Croatian.

011.042 **Astronomical conferences and meetings.**
Astrophys. Letters, Vol. 11, 45 - 46, 100 - 101, 163 - 165 (1972).

011.043 **IUAA-kongressen i Malmö.** P. Linde.
Astron. Tidssk., Årg. 5, p. 93 - 94 (1972).

011.044 **The sixth international conference on general relativity and gravitation. Copenhagen, 5 - 9 July, 1971.**
V. B. Braginskij.
Uspekhi fiz. nauk, Vol. 106, 566 - 568 (1972). In Russian. Abstr. in Referativ. Zhurn. 51. Astron., 7.51.728 (1972).

011.045 **Introductory remarks to the fourth astrometric conference.** P. van de Kamp.
Proc. fourth astrometric conference, (see 012.026), p. 9 - 14 = Sproul Obs. Repr. No. 204 (1971).

011.046 **1972 Japanese comet conference.**
Yamamoto Circ., No. 1748 (1972). In Japanese.

011.047 **The solar wind conference, March 21 - 26, 1971.**
C. T. Russell.
Earth Extraterr. Sci., Vol. 1, 219 - 228 (1971).

011.048 **Report on the fifth Texas Symposium on relativistic astrophysics.** P. Hajicek.
General Relativity Gravitation, Vol. 2, 173 - 181 (1971).

012 Proceedings of Colloquia, Congresses, Meetings, and Symposia

012.001 **The International Symposium on the 1970 solar eclipse**, with a preface by J. Houtgast.
Solar Physics, Vol. 21, 259 - 495 (1971). – The individual contributions are included in their corresponding subject categories – see abstracts 047.001, 071.003, 073.002, 073.003, 074.004 – 074.021; 076.002, 076.003, 077.004 – 077.006; 079.101, 102.002.

012.002 **The eclipse of 7 March 1970.**
A symposium held in Seattle, U.S.A., 1971 June 18–21. The aeronomic and ionospheric papers are collected and prepared for publication by M. H. Rees, H. Rishbeth.
Journ. Atmosph. Terr. Phys., Vol. 34, 559 - 744 (1972). – The individual papers are included in their corresponding subject categories – see abstracts 079.101, 079.104, 079.105.

012.003 **Advanced Electronic Systems for Astronomy – 1971 symposium**, with an introduction by L. B. Robinson, E. J. Wampler.
Publ. Astron. Soc. Pacific, Vol. 84, 74 - 224 (1972). – The individual contributions are included in their corresponding subject categories – see abstracts 031.012 - 031.016, 032.008, 032.009, 033.015, 034.015 - 034.029, 051.007, 054.009.

012.004 **Gravitational N-body problem.** Proceedings of IAU Colloquium No. 10, held in Cambridge, England, August 12 - 15, 1970.   M. Lecar (Editor).
Astrophysics and Space Science Library, Vol. 31. D. Reidel Publishing Company, Dordrecht–Holland. 11 + 441 pp. Price Dfl. 115.00 (1972). – The individual contributions are included in their corresponding subject categories – see abstracts 042.024, 062.016 - 062.017, 117.012, 151.026 - 151.062.

012.005 **External galaxies and quasi-stellar objects.** International Astronomical Union, Symposium No. 44, held in Uppsala, Sweden, 10–14 August 1970.
D. S. Evans (Editor), assisted by D. Wills, B. J. Wills.
D. Reidel Publishing Company, Dordrecht–Holland. 17 + 549 pp. Price Dfl. 100.00, $ 33.00 respectively (1972). – The individual contributions are included in their corresponding subject categories – see abstracts 066.035, 125.016, 141.082, 141.084 - 141.112, 158.065 - 158.102, 159.010, 160.016 - 160.019, 161.008, 162.026 - 162.031.

012.006 **New directions and new frontiers in variable star research.** International Astronomical Union, Colloquium No. 15 (5th colloquium on variable stars). Combined colloquium of the Commissions 27 and 42, Bamberg, 1971, 31st August–3rd September.
Veröff. Remeis-Sternw. Bamberg – Astron. Inst. Univ. Erlangen–Nürnberg, Vol. 9, No. 100, 326 pp. (1972). – The individual contributions are included in their corresponding subject categories – see abstracts 034.037, 065.087 - 065.088, 114.076 - 114.082, 116.009 - 116.010, 117.015 - 117.020, 118.012, 119.016, 120.002 - 120.003, 121.034 - 121.038, 122.072 - 122.087, 123.007 - 123.008, 124.005, 124.101, 126.018, 131.087 - 131.091, 141.539, 142.060 - 142.061, 155.045.

012.007 **Coordinate systems in astronomy.**
V. P. Shcheglov (Editor).
Akademiya Nauk Uzbekskoj SSR. Astronomicheskij Institut. Izdatel'stvo "FAN" uzbekskoj SSR, Tashkent. 95 pp. Price 50 Kop. (1971). In Russian. – Conference at the International Latitude Station Ulugbek in Kitab 1970, Oct. 14 - 16. – The individual papers are included in their corresponding subject categories – see abstracts 032.025, 032.026, 041.026 - 041.030, 045.026, 045.027, 046.018.

012.008 **Extra collection of papers contributed to the IAU Symposium No. 48: "Rotation of the earth".**
Edited by S. Yumi, with a preface by T. Okuda.
Published for the Local Organizing Committee with the auspices of the International Latitude Observatory of Mizusawa, Mizusawa. 5 + 179 pp. (1971). – The individual contributions are included in their corresponding subject categories – see abstracts 021.005 - 021.006, 032.017 - 032.018, 041.019 - 041.023, 042.027, 044.016 - 044.018, 044.023, 045.017 - 045.022, 046.017, 081.016 - 081.017.

012.009 **Papers presented at a meeting held at the University of Adelaide on 13, 14 and 15 December 1971.**
Proc. Astron. Soc. Australia, Vol. 2, 70 - 117 (1972). – The individual contributions are included in their corresponding subject categories – see abstracts 033.022, 033.023, 034.053, 061.038, 061.039, 065.104, 066.056, 077.033 - 077.037, 080.024, 080.025, 104.034, 106.013 - 106.015, 122.130, 126.021, 133.020, 134.006, 134.007, 142.101, 143.057, 143.058, 159.015.

012.010 **Atti della XIII Riunione della Società Astronomica Italiana,** Trieste, 25 - 28 settembre 1969, with words of welcome by M. Hack.
Printed by Tipografia Baccini & Chiappi, Firenze, 239 pp. (1970). – The individual contributions are included in their corresponding subject categories – see abstracts 009.025, 065.112, 072.035, 077.047, 080.028, 085.006 - 085.008, 091.020, 113.033, 114.108, 141.159, 141.555, 158.157.

012.011 **Physics of the solar corona.** Proceedings of NATO Advanced Study Institute on physics of the solar corona held at Cavouri-Vouliagmeni, Athens, Greece 6 - 17 September 1970.
Edited by C. J. Macris, with a summary by L. Goldberg.
D. Reidel Publishing Company, Dordrecht–Holland.
Astrophys. and Space Science Library, Vol. 27, 12 + 345 pp. Price Dfl. 80.00 (1971). – The individual contributions are included in their corresponding subject categories – see abstracts 073.069 - 073.076, 074.078 - 074.087, 076.033, 076.034, 077.048, 077.049.

012.012 **The sun: Part I of solar-terrestrial physics/1970,** comprising the proceedings of the international symposium on solar-terrestrial physics held in Leningrad, U.S.S.R. 12–19 May 1970.
C. de Jager (Editor).
Astrophys. Space Sci. Library, Vol. 29, Part I, 2 + 181 pp. (1972). – The individual papers are included in their corresponding subject categories – see abstracts 072.037 - 072.039, 073.077, 073.078, 074.088, 074.089, 076.035, 077.050, 078.023, 078.024, 080.029, 080.030.

012.013 **The interplanetary medium: Part II of solar-terrestrial physics/1970,** comprising the proceedings of the international symposium on solar-terrestrial physics held in Leningrad, U.S.S.R. 12–19 May 1970.
E. R. Dyer, J. G. Roederer, A. J. Hundhausen (Editors).
Astrophys. Space Sci. Library, Vol. 29, Part II, 2 + 205 pp. (1972). – The individual papers are included in their corresponding subject categories – see abstracts 074.090, 074.091, 078.025, 106.019 - 106.021, 143.061, 143.062.

012.014 **The magnetosphere: Part III of solar-terrestrial physics/1970,** comprising the proceedings of the inter-

national symposium on solar-terrestrial physics held in Leningrad, U.S.S.R. 12–19 May 1970.
E. R. Dyer, J. G. Roederer (Editors).
Astrophys. Space Sci. Library, Vol. 29, Part III, 2 + 317 pp. (1972). – The individual papers are included in their corresponding subject categories – see abstracts 084.043, 084.270 - 084.277, 084.408, 084.409.

012.015 **The upper atmosphere: Part IV of solar-terrestrial physics/1970**, comprising the proceedings of the international symposium on solar-terrestrial physics held in Leningrad, U.S.S.R. 12–19 May 1970.
S. A. Bowhill (Editor).
Astrophys. Space Sci. Library, Vol. 29, Part IV, 2 + 211 pp. (1972). – The individual papers are included in their corresponding subject categories – see abstracts 082.096, 082.097, 083.072 - 083.079, 084.278.

012.016 **Solar-terrestrial physics/1970.** Proceedings of the international symposium on solar-terrestrial physics, held in Leningrad, U.S.S.R. 12–19 May 1970.
E. R. Dyer (General Editor).
Astrophys. and Space Sci. Library, Vol. 29, Parts I–IV.
D. Reidel Publishing Company, Dordrecht – Holland. 8 + 183 + 207 + 319 + 213 pp. Price Dfl. 215.00 (1972). – For the single parts – see abstracts 012.012 - 012.015.

012.017 **Selected topics in molecular physics.** Proceedings of the International Symposium at Ludwigsburg (Germany) from October 12–14, 1970.
E. Clementi (Editor).
Verlag Chemie, Weinheim/Bergstr. (Germany). 212 pp. Price DM 68.00 (1972). – Included is a paper by L. Biermann (see 064.044).

012.018 **Proceedings of ESO/CERN conference on auxiliary instrumentation for large telescopes**, Geneva, May 2–5, 1972.
Edited by S. Laustsen, A. Reiz, with opening words by A. Blaauw, and a summary by B. Strömgren.
Printed in Switzerland. ESO – Telescope Project Division CERN, Geneva. 14 + 525 pp. (1972). – The individual contributions are included in their corresponding subject categories – see abstracts 008.000, 013.008, 031.031, 031.032, 032.040 - 032.045, 034.066 - 034.103, 114.123.

012.019 **Galactic astronomy, Vol. 1, 2.** Proceedings of the Summer Institute for Astronomy and Astrophysics, held at the State University of New York at Stony Brook – Summer 1968. H.-Y. Chiu, A. Muriel (Editors).
Gordon and Breach Science Publishers, New York – London – Paris. Vol. 1: 9+334 pp.; Vol., 2: 9+300 pp. Price £ 14.60 (1970). – The individual contributions are included in their corresponding subject categories – see abstracts (1) Vol. 1: 122.137, 131.143, 151.090, 155.074, 155.075, 157.012; (2) Vol. 2: 061.047, 131.144, 151.091 - 151.094, 155.076, 155.077.

012.020 **Proceedings of a symposium on "Celestial mechanics"**, held in Tokyo, Japan, January 27–28, 1972.
Edited by Y. Kozai, H. Kinoshita, with a preface by Y. Kozai.
Tokyo Astron. Obs., 4 + 111 pp. (1972). – The individual contributions are included in their corresponding subject categories – see abstracts 021.009, 042.053 - 042.060, 102.019, 151.095 - 151.098.

012.021 **Atti della Iª riunione del "Gruppo del seeing OAN".** Frascati, 25 giugno 1971.
Oss. Astron. Nazionale, Rapp. No. 2, p. 29 - 74 (1971), with a report of the meeting by A.Righini, F. Bertola, p. 33 - 36. – The individual contributions: Considerazioni sulla scelta della

località per l'OAN, (*A. Mammano*), p. 37 - 45; Esame comparativo delle tracce di Walker, (*G. di Tullio*), p. 46 - 50; Remarks regarding the Vatican and British interest in the site testing project, (*P. J. Treanor*), p. 51 - 52; Rapporto sulla stazione di St. Barthélemy (Valle d'Aosta), (*F. Scaltriti*), p. 53 - 55; Rapporto su Campo Catino e su alcune zone del Lazio, (*R. Viotti*), p. 56 - 58; Rapporto preliminare sulla stazione di Monte Sacro, (*B. Caccin, L. Milano*), p. 59 - 64; Rapporto sulla stazione di S. Venere e considerazioni su altri possibili "sites" limitrofi, (*S. Catalano, S. Cristaldi*), p. 65 - 72.

012.022 **Atti della IIª riunione del "Gruppo del seeing OAN".** Arcetri, 28 luglio 1971.
Oss. Astron. Nazionale, Rapp. No. 2, p. 75 - 87, with a report of the meeting by A. Righini, F. Bertola, p. 77 - 81 and with a contribution on testing Pizzo Antenna, Monte Ferro, Pizzo Carbonara and Monte Mufara by S. Cristaldi, p. 82 - 87 (1971).

012.023 **Atti della IIIª riunione del "Gruppo del seeing OAN".** Bologna, 7 ottobre 1971.
Oss. Astron. Nazionale, Rapp. No. 2, p. 89 - 111 (1971), with a report of the meeting by A. Righini, F. Bertola, p. 91 - 93. The individual contributions: Indagine sulle condizioni meteorologiche a Pescopagano e Piano Battaglia, (*A. Cassatella, R. Viotti*), p. 94 - 95; Relazione conoscitiva sulla zona di Pescopagano, (*B. Caccin, L. Milano*), p. 96 - 103; Preventivo di spesa per la stazione di Pescopagano, (*R. Barletti, G. Ceppatelli, F. Mazzucconi*), p. 104 - 105; Condizioni meteorologiche di alcune località proposte per la scelta del sito OAN, (*A. Mammano*), p. 106 - 107; Proposta britannica di lavoro in Italia, (*B. McInnes*), p. 108 - 109; Sopraluogo nelle zone di Monte Cammarata e Monte S. Salvatore, (*S. Cristaldi*), p. 110 - 111.

012.024 **The evolution of population II stars.** The proceedings of a conference held at the State University of New York at Stony Brook, December, 1970.
A. G. D. Philip (Editor).
Dudley Obs., *Albany, New York,* Rep. No. 4, 6 + 216 pp. (1972). – The individual papers are included in their corresponding subject categories – see abstracts 065.119 - 065.123, 115.017, 122.147 - 122.149, 154.027 - 154.033, 158.171, 159.025.

012.025 **The Galaxy and the distance scale.** Proceedings of a conference in honour of Professor Sir Richard Woolley, Astronomer Royal, at Herstmonceux Castle on 1971 August 17–20.
Edited by D. Lynden-Bell, B. D. Yallop, with introductory remarks by J. Carroll.
Quarterly Journ. Roy. Astron. Soc., Vol. 13, 130 - 302 (1972). – The individual contributions are included in their corresponding subject categories – see abstracts 111.001 - 111.003, 114.133, 115.018, 117.040, 121.082, 122.150 - 122.153, 131.151, 141.188, 151.103, 155.080 - 155.085, 159.026, 162.055.

012.026 **Proceedings of the fourth astrometric conference,** held at the University of Virginia, Charlottesville, 1969 October 8 - 10. P. A. Ianna (Editor), with a summary by D. Hoffleit.
Publ. Leander McCormick Obs., Univ. Virginia, *Charlottesville,* Vol. 16, 305 pp. (1971). – The individual contributions are included in their corresponding subject categories – see abstracts 011.045, 041.051 - 041.053, 111.004 - 111.011.

012.027 **Physical studies of minor planets.** IAU Colloquium No. 12, held in Tucson, Arizona, March 6–10, 1971.
Edited by T. Gehrels, with a foreword by H. E. Newell, and an introduction by T. Gehrels, J. R. Gill, J. W. Haughey.
National Aeronautics and Space Administration, Scientific and

Technical Information Office, Washington, D.C., NASA SP-267. [For sale by the Superintendent of Documents, U.S. Government Printing Office, Washington, D.C.], 27 + 687 pp. Price $ 3.00 (1971/72). – The individual papers are included in their corresponding subject categories – see abstracts 042.064, 051.032 - 051.036, 052.034 - 052.036, 053.022, 053.023, 097.106, 098.030 - 098.076, 102.024, 102.025, 105.065 - 105.067, 106.027, 106.028, 107.013 - 107.015.

012.028 **From plasma to planet.** Proceedings of the 21st Nobel symposium, Saltsjöbaden, Sweden, 1971 September 6–10.
Almqvist & Wiksell, Stockholm, Sweden. 389 pp. (1972). The individual contributions are included in their corresponding subject categories – see abstracts 051.031, 051.041 - 051.043, 052.033, 062.041, 062.042, 080.039, 082.140, 084.290, 091.030, 098.077 - 098.079, 102.026, 102.027, 103.129, 104.044, 104.045, 105.069 - 105.071, 107.017 - 107.020, 117.048, 131.157.

012.029 **Relativity and gravitation.** International seminar held in Haifa, July 1969.
C. G. Kuper, A. Peres (Editors).
Gordon and Breach Science Publishers, New York – London – Paris. 11 + 324 pp. Price £ 10.25 (1971). – The individual contributions within the subject scope of Astronomy and Astrophysics Abstracts are included in their corresponding categories – see abstracts 022.117, 066.131 - 066.140, 162.077 - 162.081.

012.030 **The significance of space research for fundamental physics. Proceedings of an ESRO colloquium held at Interlaken, Switzerland, on 4 September 1971.**
E.S.R.O., S. P. (*France*), No. 52, 175 pp. (1971). – The individual contributions are included in their corresponding subject categories – see abstracts 051.046, 066.145 - 066.147, 155.089, 162.086, 162.087.

013 Reports on Astronomy in various Countries and Particular Fields, International Cooperation

013.001 **Astronomy in Mid-Atlantic.** J. E. Packer.
Sky Telescope, Vol. 43, 74, 96 (1972).

013.002 **Successes of spectroscopy and astrophysics.** Z. L. Ponizovskij.
Priroda, No. 1.72, p. 108 - 109 (1972). In Russian.

013.003 **Canonical satellite theory based on independent variables different from time.** (Report to ESRO). G. Scheifele, E. Stiefel.
Prepared under ESOC–Contract No. 219/70/AR by G. Scheifele, E. Stiefel, Swiss Federal Institute of Technology (ETH), Zürich, Switzerland, for European Space Research Organization. Juris Druck, Zürich. 162 pp. (1972). – The individual contributions are included in their corresponding subject categories – see abstracts 052.008 - 052.013.

013.004 **Realizări recente și tendințe în cunoașterea cosmosului** (Recent progress and trends in space exploration). I. D. Ilie.
Terra, Vol. 3 (23), No. 4, p. 32 - 46 (1971).

013.005 **Vistas in astronomical development in Uzbekistan.** V. P. Shcheglov.
Uspekhi fiz. nauk, Vol. 105, 749 - 752 (1971). In Russian.
Abstr. in Referativ. Zhurn. 51. Astron., 6.51.28 (1972).

013.006 **Space physics in Poland: Present state and postulates for future.** S. Grzędzielski, A. Wernik.
Postępy Astron., Vol. 20, 171 - 178 (1972). In Polish.

013.007 **Impulse für die astronomische Forschung.** W. Fricke.
Mitt. Astron. Ges., No. 31, p. 9 - 13 (1972). – Ansprache des Vorsitzenden der Astronomischen Gesellschaft, gehalten zur Eröffnung der Wissenschaftlichen Astronomischen Tagung verbunden mit der 52. ordentlichen Mitgliederversammlung der Astronomischen Gesellschaft in Bonn 1971.

013.008 **Review: Some problems in extragalactic astronomy.** E. M. Burbidge.
Auxiliary instrumentation for large telescopes. Conference 1972, (see 012.018), p. 9 - 19 (1972).

013.009 **Wissensexplosion in der Astronomie.** D. B. Herrmann.
Vorträge und Schriften, Archenhold-Sternw. Berlin-Treptow, No. 39, 23 pp. (1971).

013.010 **German Democratic Republic. National Report:** IAGA, presented at the XV general assembly of the International Association of Geomagnetism and Aeronomy, Moscow 1971.
Nationalkomitee Geod. Geophys., Deutsche Demokrat. Republik, Deutsche Akad. Wiss. Berlin. 44 pp. (1971).

013.011 **Advances in astronomy in the year 1971.** J. Grygar.
Říše hvězd, Vol. 53, 25 - 29, 46 - 55 (1972). In Czech.

014 Teaching in Astronomy

014.001 Some remarks on the contents of stellar astrophysical questions in the program of lectures on general astronomy at pedagogical colleges. M. T. Emel'yanenko.
Uch. zap. Kujbyshev. gos. ped. in-ta, 1971, vyp. (No.) 95, p. 64 - 70. In Russian. − Abstr. in Referativ. Zhurn. 51. Astron., 3.51.40 (1972).

014.002 Practical exercises in the field of astronomy (Handbook for teachers at secondary schools).
D. G. Makedonski, M. M. Gogoshev.
Nar. prosv., Sofiya. 107 pp. Price 0.44 Lv. (1971).
In Bulgarian.

014.003 Geocentric and heliocentric parallax demonstrations.
M. J. Pryor.
Journ. Roy. Astron. Soc. Canada, Vol. 66, 53 - 55 (1972).

014.004 Seminar für Didaktik der Astronomie in Dortmund.
A. Kunert.
SuW, Vol. 11, 25 (1972).

014.005 On revitalizing astronomy education.
R. Berendzen.
Mercury, (Journ. Astron. Soc. Pacific), Vol. 1, No. 3, p. 2 - 3 (1972).

014.006 On the role of astronomy within the scope of Soviet high school education and pedagogics.
V. N. Astashenko.
Nauch. trudy Tashkent un-t, 1971, vyp. (No.) 413, p. 13 - 14. In Russian. − Abstr. in Referativ. Zhurn. 51. Astron., 6.51.33 (1972).

014.007 Astrophysical problems at the high school course of astronomy. G. I. Malakhova.
Uch. zap. Yaroslav. gos. ped. in-t, 1971, vyp. (No.) 81, p. 84 - 120. In Russian. − Abstr. in Referativ. Zhurn. 51. Astron., 6.51.34 (1972).

014.008 Über Ziele, Aufgaben und einige Probleme der Arbeitsgemeinschaften nach Rahmenprogrammen in den Klassen 9 und 10. M. Schukowski, P. Klein.
Astron. in der Schule, 9. Jahrgang, p. 4 - 9 (1972).

014.009 Physikalische Experimente im Astronomieunterricht.
H. Albert, W. Gebhardt.
Astron. in der Schule, 9. Jahrgang, p. 9 - 14 (1972).

014.010 Über die außerunterrichtliche und außerschulische Arbeit auf dem Gebiet der Astronomie in der sowjetischen Oberschule. J. P. Lewitan.
Astron. in der Schule, 9. Jahrgang, p. 37 - 40 (1972).

014.011 Zum Problem der Festigung des astronomischen Lehrstoffs. K. Lindner.
Astron. in der Schule, 9. Jahrgang, p. 58 - 61 (1972).

014.012 Einige Bemerkungen zum Verhältnis Lehrplan − Rahmenprogramm für die Arbeitsgemeinschaft „Astronomie". J. Stier.
Astron. in der Schule, 9. Jahrgang, p. 62 - 65 (1972).

014.013 Anregungen für die Arbeit mit dem Tageslichtprojektor „Polylux". B. Schmidt.
Astron. in der Schule, 9. Jahrgang, p. 67 - 69 (1972).

014.014 Un essai d'individualisation de l'enseignement de notions élémentaires d'astronomie. L. Houziaux.
Univ. Mons, Dép. Astrophys., Commun. No. 25, 1 pp. (1972).

014.015 Measurement of distant celestial bodies.
B. M. Ševarlić.
Vasiona, Vol. 20, 40 - 46 (1972). In Serbo-Croatian.

014.016 Astronomically determined dates and alignments.
L. Winkler.
American Journ. Phys., Vol. 40, 126 - 132 (1972).
This is a tutorial article concerned with problems related to dates and geometrical alignments determined from astronomical considerations.

014.017 Astronomy and aids for teachers. M. B. Stewart.
Mercury, (Journ. Astron. Soc. Pacific), Vol. 1, No. 3, p. 12 - 13 (1972). − Tracking the moon with cross-staff and computer.

015 Miscellanea

015.001 Signals of extraterrestrial civilizations – of what
kind can they be?   Yu. P. Kuznetsov.
Zemlya i Vselennaya, No. 1, p. 30 - 33 (1972). In Russian.

015.002 Chemical basis of extraterrestrial life.
C. Ponnamperuma.
Bull. American Astron. Soc., Vol. 3, 458 (1971). – Abstr. AAS.

015.003 Stages in the chemical origin of life.   J. Keosian.
Bull. American Astron. Soc., Vol. 3, 458 - 459
(1971). – Abstr. AAS.

015.004 Some deep-sky photographs.   K. Rihm.
Sky Telescope, Vol. 43, 22 - 24 (1972).

015.005 Cours d'astronomie de la S.A.F.: 3. La lune.
G. Oudenot.
L'Astronomie, 86ᵉ année, p. 15 - 31 (1972).

015.006 Cours d'astronomie de la S.A.F.: 4. Le soleil.
M. Dumont.
L'Astronomie, 86ᵉ année, p. 119 - 136 (1972).

015.007 Astrogenic environments: The effect of stellar spec-
tral classes on the evolutionary pace of life.
K. A. Ehricke.
Spaceflight, Vol. 14, 2 - 14 (1972).

015.008 Signals from other worlds.   B. Belitsky.
Spaceflight, Vol. 14, 17 - 18 (1972).

015.009 On the imperturbability of elevator operators. LVII.
S. Candlestickmaker, communicated by J. Sykes.
Quarterly Journ. Roy. Astron. Soc., Vol. 13, 63 - 66 (1972).

015.010 ¿Está sola la humanidad en el universo ?
V. Ambartsumian.
El Universo, No. 93, Vol. 24, 66 - 69 (1970).

015.011 On the search for extra-solar intelligence.
R. P. Haviland.
Spaceflight, Vol. 14, 217 - 219, 223 (1972).

015.012 New power sources.   J. A. Strother.
IEEE Spectrum, Vol. 9, No. 5, p. 12 (1972).

015.013 Cours d'astronomie de la S. A. F. 5. Le système
solaire: Étude générale.   P. Oudenot.
L'Astronomie, 86ᵉ année, p. 175 - 188 (1972).

015.014 The shape of the analemma.   B. M. Oliver.
Sky Telescope, Vol. 44, 20 - 22 (1972).

015.015 Cours d'astronomie de la S.A.F. 6. Le système
solaire: Étude physique.   M. Dumont.
L'Astronomie, 86ᵉ année, p. 261 - 287 (1972).

015.016 Sterilization probability for small particles entering
the Martian atmosphere.   R. C. Corlett.
Astronaut. Acta, Vol. 17, 229 - 238 (1972).
  Spacecraft orbiting Mars may eject particles consisting of
or containing viable spores of terrestrial origin. Some such particles may pass through the Martian atmosphere and arrive at
the planet's surface. This possibility must be taken into account in provision for planetary quarantine prior to completion of the search for indigenous life.

015.017 On an interesting regularity of cosmic objects and
cosmic phenomena.   M. Kopecký.
Kozmos, Vol. 3, 39 - 40 (1972).   In Czech.

015.018 Astronomy and astrophysics and scientific progress.
V. Vanýsek.
Říše hvězd, Vol. 53, 1 - 4 (1972).   In Czech.

  Interstellar molecules and the origin of life.
See Abstr. 131.149.

Applied Mathematics, Physics

021 Mathematics, Computing, Machine Programs

021.001 **Foundations for estimation by the method of least squares.** W. W. Hauck, Jr.
Smithsonian Astrophys. Obs., *Cambridge, Mass.*, Special Rep. No. 340, 3 + 92 pp. (1971).
This paper discusses least-squares estimation from the point of view of a statistician. Much of the emphasis will be on problems encountered in application and, more specifically, on questions involving assumptions — what assumptions are needed, when are they needed, what happens if they are not valid, and if they are invalid, how can we detect that fact.

021.002 **Frequency distribution of astronomical objects and phenomena according to their age and lifetime.**
M. Kopecký.
Bull. Astron. Inst. Czechoslovakia, Vol. 23, 51 - 54 (1972).
Relations between the frequency distribution functions of phenomena and objects are derived according to their lifetime, age, and certain other factors, and certain possibilities of applying these relations to astronomical and astrophysical research are discussed.

021.003 **Upper bound on compression ratio for run-length encoding.** K. G. Gray, R. S. Simpson.
Proc. IEEE, Vol. 60, 148 (1972).
Based upon a first-order Markov model of video data, an upper bound on compression ratio is found for run-length encoding. The bound is compared with that of the Markov source.

021.004 **The ultimate computer.** W. H. Ware.
IEEE Spectrum, Vol. 9, No. 3, p. 84 - 91 (1972).
The digital processing of Mariner 6 and 7 television signals exemplifies typical large-computer usage, without which it would be virtually impossible to obtain accurate pictures of the Martian surface. The processing required includes rectification of video system noises, spacecraft video encoding nonlinearities, spatial variations in vidicon sensitivity, system geometric distortions, and system resolution limitations.

021.005 **The selection of an optimum smoothing parameter.** V. I. Sakharov, O. B. Vasilyev.
Extra collection of papers to IAU Symposium No. 48, (see 012.008), p. 113 - 122 (1971).

021.006 **The separation of discrete lines of the frequency spectrum of a compound process.**
N. P. Godisov, O. B. Vasilyev.
Extra collection of papers to IAU Symposium No. 48, (see 012.008), p. 123 - 137 (1971).

021.007 **A comparison for high order Runge-Kutta methods.** D. G. Bettis, W. H. Moore.
Bull. American Astron. Soc., Vol. 4, 262 (1972). – Abstr. AAS.

021.008 **A precompiler for the formula manipulation system TRIGMAN.** W. H. Jefferys.
Bull. American Astron. Soc., Vol. 4, 264 (1972). – Abstr. AAS.

021.009 **On the parametric stability of symplectic mappings.** R. Nagai.
Celestial mechanics. Symposium Tokyo, (see 012.020), p. 72 - 77 (1972). In Japanese.

021.010 **Une généralisation de théorèmes de J. A. Marlin.** J. Mawhin.
International Journ. Non-Linear Mechanics, Vol. 5, 335 - 339 = Univ. Liège, Inst. d'Astrophys., Coll. 4°, No. 205 (1970).
This paper is an existence study of periodic solutions in some non-linear differential systems of order $2n$ with symmetries.

021.011 **Solution périodiques de systèmes différentiels fortement non linéaires n'appartenant pas nécessairement à la classe D au sens de Levinson.** J. Mawhin.
Colloque Equations Différent. Non Linéaires, Louvain 1970, p. 43 - 55 = Univ. Liège, Inst. d'Astrophys., Coll. 8°, No. 594 (1970).

021.012 **Existence of periodic solutions for higher-order differential systems that are not of class D.**
J. Mawhin.
Journ. Differential Equations, Vol. 8, 523 - 530 = Univ. Liège, Inst. d'Astrophys., Coll. 8°, No. 606 (1970).

021.013 **Vector analysis formulae.** P. Renson.
Bull. Soc. Roy. Sci. Liège, Vol. 39, 590 - 594 = Univ. Liège, Inst. d'Astrophys., Coll. 8°, No. 608 (1970).

An A. P. L. computer program for predictions and heliocentric corrections of the observable minima of eclipsing binaries. See Abstr. 121.086.

022 Physical Papers Related to Astronomy and Astrophysics

022.001 The broadening of radio recombination lines by electron collisions. G. Peach.
Astrophys. Letters, Vol. 10, 129 - 130 (1972).

The broadening by electron collisions, of radio recombination lines in hydrogen, has been calculated using the impact theory of Baranger. The degeneracy of the energy levels $n\ell$, $\ell = 0, 1, 2...$ with respect to ℓ is explicitly taken into account and it is proved that the net effect of collisions which do not cause a change of principal quantum number n is negligible. This is in agreement with the assumptions of earlier workers.

022.002 Spectral lines in the Be I isoelectronic sequence. H. Nussbaumer.
Astron. Astrophys., Vol. 16, 77 - 80 (1972).

Atomic data for C III, N IV, O V, Ne VII, Na VIII, Mg IX, Si XI, S XII, Ca XV and Fe XXIII are discussed. Transition probabilities were calculated and examples of excitation cross sections calculated with the impact parameter method are given.

022.003 The interpretation of total line intensities from optically thin gases. I. A general method.
J. T. Jefferies, F. Q. Orrall, J. B. Zirker.
Solar Physics, Vol. 22, 307 - 316 (1972).

We describe a general method for inferring, from the line emission of an optically thin medium, the physical state of the gas along the column in the line of sight which is sampled by the observations. We show that, given sufficient observational data, it is in principle possible to determine the distribution function μ (n, T), describing the partitioning of the matter in the gas over the density and temperature, and the chemical composition. The method is devised for application to the astronomical case, especially for studies of the solar corona, the chromosphere-corona transition region, planetary nebulae and other optically thin sources.

022.004 On the integral of bipolar momenta.
V. V. Radzievsky.
Astron. Zhurn. Akad. Nauk SSSR, Vol. 49, 153 - 156 (1972). In Russian. English translation in Soviet Astron. AJ, Vol. 16, No. 1.

It is shown that the integral of bipolar momenta cannot be obtained as a combination of the angular momentum and energy integrals only.

022.005 The influence of combinative scattering on the distribution of molecules according to vibrational levels.
A. P. Sarychev.
Astron. Zhurn. Akad. Nauk SSSR, Vol. 49, 214 - 216 (1972). In Russian. English translation in Soviet Astron. AJ, Vol. 16, No. 1. – Short note.

022.006 Vacuum-ultraviolet spectrum of neutral boron (B I). D. Goorvitch, F. P. J. Valero.
Astrophys. Journ., Vol. 171, 643 - 645 (1972).

022.007 Overlap of argon ion spectra with satellites to the Lyman-alpha lines of carbon and boron.
N. W. Jalufka, J. Cooper.
Astrophys. Journ., Vol. 171, 647 - 649 (1972).

A theta pinch has been employed to investigate satellites to the H-like and He-like ion resonance lines of boron and carbon. Identification of the spectra shows an overlap of the Ar IX, Ar XI, and Ar XII lines with the predicted satellite wavelengths, and it is concluded that many of the previously reported satellite lines are due to argon impurities.

022.008 Excitation of Mg⁺ by electron collisions.
M. Blaha.
Astron. Astrophys., Vol. 16, 437 - 442 (1972).

The Mg II resonance doublet at $\lambda 2800$ represents a suitable diagnostic tool for the study of physical conditions in the solar chromosphere. The interpretation of intensities and line profiles requires the knowledge of all electron collision cross sections, which can affect directly or indirectly the population of upper levels of the doublet. For this purpose collision cross sections for transitions between levels $3s$, $4s$, $5s$, $6s$, $3p$, $4p$, $5p$, $3d$, $4d$, $5d$ have been calculated in the unitarized Coulomb-Born approximation without exchange.

022.009 Lifetime measurements of optical levels of Ni I.
J. Marek.
Astron. Astrophys., Vol. 17, 83 - 87 (1972).   In German.

Using the phase shift method, the mean lifetimes of some Ni I levels have been determined. Relative transition probabilities of all lines depopulating the investigated levels have been measured by a number of authors. Thus, the absolute transition probabilities of the strongest transitions have been derived. Some of the values can be compared with the results of absorption experiments by Lawrence, Link and King (1965).

022.010 The absorption spectrum of Tl I in the vacuum ultraviolet. J. P. Connerade.
Astrophys. Journ., Vol. 172, 213 - 227 (1972).

The absorption spectrum of Tl I vapor has been observed in the vacuum ultraviolet and a large number of new transitions has been found. An analysis and a comparison with recent work on similar spectra are given.

022.011 Fe I oscillator strengths determined from anomalous dispersion of shock-heated gases.
M. C. E. Huber, W. H. Parkinson.
Astrophys. Journ., Vol. 172, 229 - 247 (1972).

This study demonstrates the effectiveness of the hook method for quantitative spectroscopy on shock-heated gases and points up its usefulness for gas diagnostics on optically thick lines. If one assumes no change in atmospheric models, our f-value scale would suggest a solar abundance at the lower limit of the recently published values: $\log (A_{Fe}/A_H) + 12 = 7.4$.

022.012 Fokker-Planck equations for charged-particle transport in random fields. J. R. Jokipii.
Astrophys. Journ., Vol. 172, 319 - 326 (1972).

The Fokker-Planck equations for charged-particle dynamics are rederived, extending somewhat the elegant discussion of Hasselmann and Wibberenz. It is shown that the usual results are obtained and the conclusions in many cases are correct over a very broad range in energy. In particular, the rate for pitch-angle scattering (and hence $\kappa_\parallel$) may be accurately given down to energies much lower than previously thought.

022.013 Cyclo-synchrotron radiation at small angles.
E. Tademaru.
Astrophys. Journ., Vol. 172, 327 - 330 (1972).

Cyclo-synchrotron emission at small angles θ to the magnetic field has a spectral form in frequency quite different from the usual synchrotron spectrum. Thus care must be taken in interpreting the spectral index of sources with special geometries.

022.014 Polynomial approximation of partition functions for rare-earth elements.

M. F. Aller, C. H. M. Everett.
Astrophys. Journ., Vol. 172, 447 - 450 (1972).

We have computed temperature-dependent partition functions for a number of neutral, singly ionized, and doubly ionized rare-earth elements, using the atomic data presently available.

022.015 Electron excitation for the $2s \to nf$ transitions in lithium-like ions.   D. Petrini.
Astron. Astrophys., Vol. 17, 410 - 412 (1972).

Coulomb-Born I cross-sections have been evaluated for the $2s \to nl$ ($n = 4, 5, 6; l = 0, 1, 2, 3$) transitions in the lithium sequence.

022.016 Excitation by proton impact of fine structure transitions in positive ions of p^2 configuration.
F. Masnou-Seeuws, R. McCarroll.
Astron. Astrophys., Vol. 17, 441 - 444 (1972).

Proton induced transitions among the fine structure states of positive ions are investigated, using semi-classical theory. Results are presented for the $^3P_0 - ^3P_1$, $^3P_0 - ^3P_2$ and $^3P_1 - ^3P_2$ transitions in N II, O III, Si IX and Fe XIII. Also given are cross sections for the $^2P_{1/2} - ^2P_{3/2}$ transition in Fe XIV.

022.017 The absorption spectrum of S I in the vacuum ultraviolet.   G. Tondello.
Astrophys. Journ., Vol. 172, 771 - 783 (1972).

An analysis of the absorption spectrum of S I in the region $\lambda\lambda 1830 - 900$ resulted in the classification of more than 100 new lines, of which some are strongly autoionized. Parameters are given for the autoionized lines and for the continuous photoionization cross-section of S I.

022.018 The collisional excitation and ionization rates of the hydrogen atoms.
R. E. Gershberg, A. A. Korovjakovskaja, Yu. P. Korovjakovskij.
Izv. Krymskoj Astrofiz. Obs., Vol. 43, 49 - 51 (1971). In Russian.

The collisional excitation and ionization rates of hydrogen atoms calculated in accordance to classical theory and by quantum mechanical formulae are compared. It is reasonable to use the classical formulae in the most cases of astrophysical calculations.

022.019 Photoelectron spectra and partial photoionization cross sections for NO, N_2O, CO, CO_2 and NH_3.
J. L. Bahr, A. J. Blake, J. H. Carver, J. L. Gardner, V. Kumar.
Journ. Quant. Spectrosc. Radiat. Transfer, Vol. 12, 59 - 73 (1972).

Photoelectron spectra have been obtained for NO, N_2O, CO, CO_2 and NH_3. The observations cover the wavelength range 584 to 890 Å. Partial photoionization cross sections have been measured for all these gases for processes involving transitions to the ground states and the various excited electronic states of the residual ions; in the case of ammonia, cross sections for dissociative photoionization have also been determined.

022.020 Lifetime and quenching rates for the H_2 continuum.   R. T. Thompson, R. G. Fowler.
Journ. Quant. Spectrosc. Radiat. Transfer, Vol. 12, 117 - 121 (1972).

When measured down to low pressures, the radiation processes of the H_2 continuum have proved to be unexpectedly complex, involving at least two states.

022.021 Franck−Condon factors, r-centroids and oscillator strength of BaO ($A^1\Sigma - X^1\Sigma$).
T. Wentink, Jr., R. J. Spindler, Jr.
Journ. Quant. Spectrosc. Radiat. Transfer, Vol. 12, 129 - 148 (1972).

The spectroscopy and potential well constants of the BaO molecular band system $A - X$ have been reviewed and are reported. Morse and Rydberg−Klein−Rees (RKR) potentials were computed and used to calculate matrices of Franck−Condon factors and r-centroids. These matrices are reported for $v' = 0$ through 10 and $v'' = 0$ through 23 for the RKR case. Absolute intensity constants, level lifetimes, f-values, and the electronic transition moment are presented and discussed.

022.022 Determination of Van der Waals broadening at temperatures of astrophysical interest.
J. M. Evans, Jr., J. Cooper.
Journ. Quant. Spectrosc. Radiat. Transfer, Vol. 12, 259 - 265 (1972).

This experiment analyzes the widths of shock-excited emission lines at temperatures of about 5000°K. Line widths due to the broadening of argon were measured for Si I λ4102 and λ5948 and for Cs I λ4593. The latter line was also measured as broadened by neon and the shift of Si I λ4102 was measured for broadening by argon. These data are compared with other experimental data to determine the temperature dependence of the broadening.

022.023 Comparison of experimental and theoretical pressure-broadened atomic line shapes.
A. K. Atakan, H. C. Jacobson.
Journ. Quant. Spectrosc. Radiat. Transfer, Vol. 12, 289 - 298 (1972).

Line-structure calculations of the longer wavelength resonance line of cesium pressurized by argon are compared with experimental values over a range of perturber relative densities from 0.5 to 61. The main features of the line shapes are reproduced by the general pressures method using Lennard−Jones potentials.

022.024 Radiative transition probabilities and recombination coefficients of the ion C IV.   E. M. Leibowitz.
Journ. Quant. Spectrosc. Radiat. Transfer, Vol. 12, 299 - 306 (1972).

Bound-bound and bound-free radiative transition probabilities, as well as radiative recombination coefficients of the ion C IV, are computed with a semi-empirical polarization potential method. The non-hydrogenic probabilities and coefficients are given for all bound states of the ion up to the principal quantum number $n = 7$.

022.025 Relative intensity calculations for nitrous oxide.   L. D. G. Young.
Journ. Quant. Spectrosc. Radiat. Transfer, Vol. 12, 307 - 322 (1972).

A tabulation of calculated rotational line intensities, relative to the integrated intensity of a vibration-rotation band, is given for $\Sigma - \Sigma$, $\Pi - \Sigma$, $\Sigma - \Pi$, $\Pi - \Pi$, and $\Delta - \Pi$ transitions of $^{14}N^{16}O_2$. These calculations were made for a temperature of 250°K (typical for the earth's atmosphere) and for 300°K (representative of laboratory conditions). A summary of band-intensity measurements is also given.

022.026 Photoionization cross-sections for atoms and ions of aluminum, silicon, and argon.
R. D. Chapman, R. J. W. Henry.
Astrophys. Journ., Vol. 173, 243 - 245 (1972).

Photoionization cross-sections for all levels belonging to configurations $1s^2 2s^2 2p^6 3s^2 3p^q$, $q > 0$, of atoms and ions of aluminum, silicon, and argon have been calculated using Hartree-Fock bound-electron wave functions and close-coupling approximation free-electron wave functions. The results are presented in the form of a computationally convenient interpolation formula.

022.027 Simulation des Intensitätsverlaufs im Raman-Spek-

trum von Sauerstoff unter Berücksichtigung der Spinaufspaltung.
K. Altmann, G. Strey, J. G. Hochenbleicher, J. Brandmüller.
Zeitschr. Naturforschung, Vol. 27a, 56 - 64 (1972).

022.028 **Micrometeoroid simulation studies on metal targets.**
H. Dietzel, G. Neukum, P. Rauser.
Journ. Geophys. Res., Vol. 77, 1375 - 1395 (1972).

Experiments on impact craters and impact ionization of dust particles have been performed using iron microparticles from a 2-Mv Van de Graaff accelerator. As a result of these studies insight into some physical processes of hypervelocity impact has been gained.

022.029 **The effect of secondary autoionization on dielectronic recombination.** M. Blaha.
Astrophys. Letters, Vol. 10, 179 - 182 (1972).

During the process of dielectronic recombination, the energy of a recombined ion after a resonance capture and subsequent stabilization may still be higher than the first ionization limit. If that happens, a secondary autoionization can take place and reduce the recombination coefficient. The importance of this effect is considered for iron ions Fe^{+9} to Fe^{+13}.

022.030 **Étude en laboratoire des structures hyperfines des raies du multiplet e^6D-y^6P du manganèse I. Comparaison avec les structures de ce multiplet détectées dans le spectre solaire.** P. Luc, S. Gerstenkorn.
Astron. Astrophys., Vol. 18, 209 - 214 (1972).

Three lines belonging to the e^6D-y^6P multiplet of manganese I exhibit hyperfine structures measurable in the solar spectrum between 8650 Å and 8750 Å. The lines of the multiplet e^6D-y^6P were recorded, in emission, using two different high-resolution apparatus. This study enables us to confirm the astronomical measurements and to determine the hyperfine constants of the terms e^6D and y^6P.

022.031 **Calculations of the energy dependence of the angular distribution of photoelectrons from atomic oxygen.** D. J. Kennedy, S. T. Manson.
Planet. Space Sci., Vol. 20, 621 - 624 (1972).

The angular distribution of photoelectrons ejected from the ground configuration of atomic oxygen is calculated and is found to differ markedly from previous predictions.

022.032 **CO quantum yield in the photolysis of CO_2 at $\lambda 1470$ and $\lambda\lambda 1500-1670$.** E. C. Y. Inn.
Journ. Geophys. Res., Vol. 77, 1991 - 1993 (1972). – Letter.

022.033 **A new method for the direct measurement of spectral line strengths and widths.** R. E. Meredith.
Journ. Quant. Spectrosc. Radiat. Transfer, Vol. 12, 455 - 484 (1972).

The most important sources of error incurred in the measurements of spectral-line parameters arise from uncertainty in the determination of the 100 per cent transmittance and in the distortion of the line profile by the spectrometer. These errors have been investigated numerically by passing an idealized spectrometer slit function over several assumed line profiles. In this way, families of correction curves have been constructed. from which spectral-line strengths, widths, and peak absorption coefficients may be determined from apparent values measured directly from the chart recorder. The effect of the form of the slit function has been investigated by using triangular, Gauss, Cauchy and combination Gauss-Cauchy slit functions. The effect of uncertainties in the line shape has been investigated by using Doppler, Lorentz, and Voigt line shapes.

022.034 **Radiative lifetimes and absolute oscillator strengths for the SiO $A^1\Pi-X^1\Sigma^+$ transition.**
W. H. Smith, H. S. Liszt.

Journ. Quant. Spectrosc. Radiat. Transfer, Vol. 12, 505 - 509 (1972).

022.035 **On the Stark broadening of Lyman-β.**
M. E. Bacon.
Journ. Quant. Spectrosc. Radiat. Transfer, Vol. 12, 519 - 523 (1972).

022.036 **Total radiative intensity calculations for 100% H_2 and 87% $H_2-13\%$ He.** G. H. Stickford, Jr.
Journ. Quant. Spectrosc. Radiat. Transfer, Vol. 12, 525 - 529 (1972).

022.037 **Collisional relaxation and rotational intensity distributions in spectra of aeronomic interest.**
D. C. Nicholls, W. F. J. Evans, E. J. Llewellyn.
Journ. Quant. Spectrosc. Radiat. Transfer, Vol. 12, 549 - 558 (1972).

The weak-interaction model, which may be used to describe the rotational relaxation of a system of rotating oscillators, has been extended to include the effects of radiative losses. It is suggested that the model may be extended to include the relaxation of systems which exhibit a multi-quantum transition at a single kinetic collision. Effective emitting populations are calculated for hydroxyl for various collision frequencies.

022.038 **The study of dimeric molecules in ammonia vapour by submillimeter wave spectroscopy.**
G. G. Gimmestad, G. W. F. Pardoe, H. A. Gebbie.
Journ. Quant. Spectrosc. Radiat. Transfer, Vol. 12, 559 - 567 (1972).

022.039 **Spontaneous radiative dissociation in molecular hydrogen.** T. L. Stephens, A. Dalgarno.
Journ. Quant. Spectrosc. Radiat. Transfer, Vol. 12, 569 - 586 (1972).

The spontaneous radiative dissociations of the discrete vibrational levels of the $B^1\Sigma_u^+$ electronic states of H_2, HD and D_2 of the $C^1\Pi_u$ electronic state of H_2 into the vibrational continuum of the ground $X^1\Sigma_g^+$ state are calculated as a function of the emission wavelength. The vibrational radiative lifetimes are tabulated as are the fractions of radiative decays that lead to dissociation. The effects of centrifugal distortion are discussed briefly. An appendix describes a sum rule used to check the numerical accuracy of the calculations.

022.040 **Realistic Franck-Condon factors and related integrals for diatomic molecules–II. The O_2 Herzberg I system.** W. R. Jarmain.
Journ. Quant. Spectrosc. Radiat. Transfer, Vol. 12, 603 - 617 (1972).

022.041 **Absolute oscillator strength measurements of the $(v''= 0, v'= 0-3)$ bands of the $(A^2\Sigma-X^2\Pi)$ γ-system of nitric oxide.**
A. J. D. Farmer, V. Hasson, R. W. Nicholls.
Journ. Quant. Spectrosc. Radiat. Transfer, Vol. 12, 627 - 633 (1972).

022.042 **Absolute oscillator strength estimates for some bands of the β-system of nitric oxide.**
A. J. D. Farmer, V. Hasson, R. W. Nicholls.
Journ. Quant. Spectrosc. Radiat. Transfer, Vol. 12, 635 - 638 (1972).

022.043 **Franck-Condon factors and r-centroids for halogen molecules–II. The $B^3\Pi(0_u^+)-X^1\Sigma_g^+$ system of $^{79}Br^{81}Br$.** J. A. Coxon.
Journ. Quant. Spectrosc. Radiat. Transfer, Vol. 12, 639 - 650 (1972).

022.044 Application of atomic absorption spectroscopy to physical investigations. B. V. L'vov.
Journ. Quant. Spectrosc. Radiat. Transfer, Vol. 12, 651 - 681 (1972). In Russian.

Some physical applications of atomic absorption spectroscopy for determining the damping parameter a, oscillator strength values f and temperature T have been critically discussed.

022.045 Intensity and half width measurements of the $(00°2-00°0)$ band of N_2O. J. S. Margolis.
Journ. Quant. Spectrosc. Radiat. Transfer, Vol. 12, 751 - 757 (1972).

022.046 On the time-evolution operator in the semiclassical theory of Stark-broadening of hydrogen lines. H. Pfennig.
Journ. Quant. Spectrosc. Radiat. Transfer, Vol. 12, 821 - 837 (1972).

In the dipole-only case, the time-evolution-operator describing how the quantum-state of a hydrogen atom in a plasma will change when a perturbing particle passes can be exactly expressed in terms of elementary functions in a simple way.

022.047 Line shape parameters for HCl and HF in a CO_2 atmosphere.
P. Varanasi, S. K. Sarangi, G. D. T. Tejwani.
Journ. Quant. Spectrosc. Radiat. Transfer, Vol. 12, 857 - 872 (1972).

022.048 Band intensity and line half-width measurements in N_2O near 4.5 μ. J. E. Lowder.
Journ. Quant. Spectrosc. Radiat. Transfer, Vol. 12, 873 - 880 (1972).

022.049 On the adiabatic approach in the theory of Stark broadening of hydrogen lines.
V. I. Kogan, V. S. Lisitsa.
Journ. Quant. Spectrosc. Radiat. Transfer, Vol. 12, 881 - 892 (1972). In Russian.

022.050 Electric quadrupole transitions in Fe XVI, Co XVII and Ni XVIII.
C. E. Tull, M. Jackson, R. P. McEachran, M. Cohen.
Journ. Quant. Spectrosc. Radiat. Transfer, Vol. 12, 893 - 900 (1972).

Theoretical multiplet-strengths for electric quadrupole transitions between ^2S, $^2P^0$, ^2D and $^2F^0$ levels of the ions Fe XVI, Co XVII and Ni XVIII have been calculated using Hartree−Fock orbital wave-functions of frozen-core type. Relativistic corrections to the energy levels have been calculated by means of first-order perturbation theory, and yield results in excellent agreement with the recent measurements of Feldman et al.

022.051 RKR Franck−Condon factors for blue and ultraviolet transitions of some molecules of astrophysical interest and some comments on the interstellar abundance of CH, CH$^+$ and SiH$^+$. H. S. Liszt, W. H. Smith.
Journ. Quant. Spectrosc. Radiat. Transfer, Vol. 12, 947 - 958 (1972).

RKR Franck−Condon factors for thirteen of the blue and ultaviolet transitions of AlF, AlO, BH, BD, CH, CD, CH$^+$, SiO and SiH$^+$ have been calculated. The interstellar abundances of CH, CH$^+$ and SiH$^+$ are discussed with regard to recent laboratory measurements, our Franck−Condon factors, and observations of the sun and the interstellar medium.

022.052 On the Stark profile of the Ly-α-line in the transition region from shock to quasi-statistical broadening. V. S. Lisitsa, G. V. Sholin.

Journ. Quant. Spectrosc. Radiat. Transfer, Vol. 12, 985 - 988 (1972). In Russian.

022.053 The $^7Li(p, \alpha)^4He$ cross-section at low energies. F. C. Barker.
Astrophys. Journ., Vol. 173, 477 - 480 (1972).

Values of the S-factor for the $^7Li(p, \alpha)^4He$ reaction at stellar energies are obtained by extrapolation, using an R-matrix fit to experimental data at higher energies that takes account of levels in the compound nucleus ^8Be.

022.054 Analytical expressions for excitation and ionization cross-sections and rate coefficients of hydrogen-like ions by electron and proton impact. A. Jacobs.
Journ. Quant. Spectrosc. Radiat. Transfer, Vol. 12, 243 - 257 (1972).

Excitation and ionization cross-sections in the Bethe− Born approximation have been computed for a model of the hydrogen-like ion consisting of seven discrete energy levels and a continuum. We propose an interpolation formula connecting the high energy behaviour (Bethe−Born) of the cross-sections and their values at threshold (Coulomb−Born I). We compare this formula with other −frequently used− formulae. Excitation and ionization rate coefficients in a Maxwellian plasma have been computed.

022.055 Intensity measurements and transition probabilities for the C_2 Swan band system.
L. Danylewych, R. W. Nicholls.
Journ. Roy. Astron. Soc. Canada, Vol. 66, 73 - 74 (1972). Abstr. Canadian Astron. Soc.

022.056 Oscillator strength measurements on the (0,0) band of the NO Gamma system from dispersion measurements. V. Hasson, A. J. D. Farmer, R. W. Nicholls.
Journ. Roy. Astron. Soc. Canada, Vol. 66, 74 (1972). − Abstr. Canadian Astron. Soc.

022.057 On the energy release modelling a meteorite impact. M. M. Rusakov, M. A. Lebedev.
Kosm. Issled., Vol. 10, 128 - 129 (1972). In Russian. − Brief information.

022.058 Number of free electrons in condensed substance depending on its density. A. M. Resikian.
Astrofizika, Vol. 7, 655 - 662 (1971). In Russian.
English translation in Astrophysics, Vol. 7, No. 4.

By means of the Thomas-Fermi statistical model the number of free electrons appearing in the substance under high pressure (i.e. high densities) is found. These calculations are suitable for nonrelativistic densities only. Thus they are suited for part of white dwarfs.

022.059 On the interpretation of radio recombination line observations. M. Brocklehurst, M. J. Seaton.
Monthly Notices Roy. Astron. Soc., Vol. 157, 179 - 210 (1972).

A substantial amount of data from atomic physics is required for the interpretation of radio recombination line observations. A critical review of the available data is presented. The equation of transfer for the line radiation is discussed and a convenient linearized form is obtained. The use of constant density models is discussed. A spherically symmetric model is constructed for the Orion nebula. The electron density N_e is tabulated as a function of the distance r from the centre. From the success achieved with this model it is concluded that the basic theory used for the interpretation of the recombination lines is correct.

022.060 Low energy cross sections for transitions between highly excited states of atoms.

I. C. Percival, M. J. Seaton.
Astrophys. Letters, Vol. 11, 31 - 33 (1972).

It is shown that the conditions required for the validity of binary encounter theory are not satisfied at the low energies which occur in regions of neutral hydrogen.

022.061 The true potential energy curves, Morse r-centroids and Franck–Condon factors for the bands of the $E^2\Sigma^+ - X^2\Sigma^+$ system of CN. T. V. R. Rao, S. V. J. Lakshman.
Journ. Quant. Spectrosc. Radiat. Transfer, Vol. 12, 1063 - 1066 (1972).

022.062 Oscillator strengths for Sc III.
B. Warner.
Observatory, Vol. 92, 50 (1972). – Note.

022.063 A convergent theory of spectral line broadening in the impact approximation.
R. J. Dyne, B. J. O'Mara.
Astron. Astrophys., Vol. 18, 363 - 372 (1972).

The results obtained in the two-level approximation using the modified Vainshtein-Sobel'man procedure are compared with the original method of Vainshtein and Sobel'man and with an equivalent analysis based on the theory of Griem et al. These theories are then applied to two representative spectral lines with the inclusion of one or more perturbing levels in the calculations. The modified method of Vainshtein and Sobel'man is also applied to the electron impact broadening of a number of lines of neutral helium and the results compared with experiment.

022.064 A universal function for ionization of atoms, ions and molecules by structureless charged particles of arbitrary mass and charge. O. Bely, P. Faucher.
Astron. Astrophys., Vol. 18, 487 - 492 (1972).

Using classical impulse approximation and the experimental results, a semi-empirical expression is obtained for the ionization cross-section of an arbitrary target by structureless particles of given mass and charge. The method is discussed and compared with various experiments. The range of its validity is given.

022.065 On the He^+ triplet line intensities.
E. Daltabuit, D. P. Cox.
Astrophys. Journ., Vol. 173, 727 - 729 (1972).

It is shown that theory and observations of helium-triplet line intensities agree within the respective uncertainties, implying that the population of the $2\,^3S$ metastable level is now predictable within about 30 percent.

022.066 Charge-impact excitation of the highly excited hydrogen atoms in the dipole form of the semiclassical impact-parameter and Born approximations. K. Omidvar.
Astrophys. Journ., Vol. 173, 731 - 736 (1972).

Expressions for the excitation cross-section have been derived in the dipole approximation of the semi-classical impact parameter and the Born approximations, making use of a formula given by Menzel for the asymptotic expansion of the oscillator strength of the hydrogen-like atoms.

022.067 The square of the radial integral for calculating the dipole oscillator strengths for transitions in hydrogen atoms in their 3(s, p, or d) and 4(s, p, d, or f) states.
G. S. Khandelwal, E. E. Fitchard.
Astrophys. Journ., Vol. 173, 737 - 740 (1972).

Explicit expressions for the square of the radial integral associated with the dipole oscillator strengths for transitions in hydrogen atoms in their 3(s, p, or d) and 4(s, p, d, or f) states are obtained from Gordon's general formula. Numerical values of the square of the radial integrals are tabulated for $10 \leq n \leq 60$. The radial integrals are asymptotically expanded to order $1/n^9$ to facilitate calculations for large n.

022.068 The broadening of the sodium D-lines.
E. L. Lewis, L. F. McNamara, H. H. Michels.
Solar Physics, Vol. 23, 287 - 288 (1972). – Research note.

022.069 Tables for constructing logarithmic profiles of spectral lines. I. D. Piunov.
Solnechnye Dannye 1972 Byull., No. 1, p. 101 - 107 (1972). In Russian.

022.070 Remeasurement of the rest frequency of the 36-centimeter radio line of methanol. H. E. Radford.
Astrophys. Journ., Vol. 174, 207 - 208 (1972).

A new laboratory measurement under high-resolution conditions yields 834.267 ± 0.002 MHz as the rest frequency for the $1_{11} - 1_{10}$ methanol emission from the galactic center. Revised Doppler velocities for Sgr A and Sgr B2 are 39 ± 5 and 54 ± 9 km s^{-1} respectively.

022.071 The lithium-like spectra of K XVII through Mn XXIII in the extreme-ultraviolet region.
S. Goldsmith, U. Feldman, L. Oren (Katz), L. Cohen.
Astrophys. Journ., Vol. 174, 209 - 214 (1972).

Identifications and classifications of spectral lines in the lithium-like spectra of the elements K through Mn is presented for the first time. The extrapolation procedures used to calculate these spectra are presented in detail. The excellent agreement between the predicted and measured lines establish the long-range extrapolation.

022.072 Some transitions of Ne VIII and Ar VIII between levels of principal quantum numbers 3, 4, and 5.
M. Druetta, R. U. Datla, H.-J. Kunze.
Astrophys. Journ., Vol. 174, 215 - 217 (1972).

Transitions of Ne VIII and Ar VIII between levels of principal quantum numbers 3, 4, and 5 have been identified in a high-temperature plasma produced in a theta-pinch discharge. The transition probabilities for these lines have been computed by using a method of A. Burgess.

022.073 Semiempirical calculations for electron-impact excitation of the hydrogen atom.
M. M. Felden, M. A. Felden.
Astrophys. Journ., Vol. 174, 219 - 226 (1972).

Cross-sections for electron-impact excitation have been calculated for the hydrogen atom using the wave function $\psi_0(r_1, r_2) = \phi_0(r_1)\, g(r_1, r_2)$.

022.074 Approximations for collisional and radiative transition rates in atomic hydrogen. L. C. Johnson.
Astrophys. Journ., Vol. 174, 227 - 236 (1972).

A simple, accurate, and complete set of approximate formulae is presented for evaluating collisional and radiative transition rates for bound-bound and bound-free transitions in atomic hydrogen. The approximations are chosen in a form which facilitates computer programming and solution of nonequilibrium excitation problems of interest in plasma physics and astrophysics.

022.075 Matter in superstrong magnetic field.
B. B. Kadomtsev.
Priroda, No. 5.72, p. 7 - 12 (1972). In Russian.

022.076 Physical constants – are they constant?
Ya. M. Kramarovsky, V. P. Chechev.
Priroda, No. 5.72, p. 46 - 51 (1972). In Russian.

022.077 Transition probabilities for the vibration-rotation bands of silicon monoxide.
J. Hedelund, D. L. Lambert.

Astrophys. Letters, Vol. 11, 71 - 75 (1972).

022.078 Optical pumping. W. Happer.
Rev. Modern Phys., Vol. 44, 169 - 249 (1972).
Optical pumping of ground-state and metastable atoms and ions is reviewed. We present a critical survey of the literature on pumping mechanisms, light propagation, relaxation mechanisms, spin exchange, and experimental details on the various atomic species which have been successfully pumped.

022.079 Radiation from accelerated point dipoles in circular and Keplerian orbits.
J. J. Monaghan, C. M. Shapcott.
Australian Journ. Phys., Vol. 25, 197 - 206 (1972).
The frequency spectrum of electromagnetic radiation from accelerated magnetic dipoles is computed for circular and Keplerian motion. For circular motion three orientations of the dipole are considered. For Keplerian motion the orientation is taken as perpendicular to the orbit. Expressions that are valid for velocities arbitrarily small and arbitrarily close to the velocity of light are derived. The mean power radiated is determined numerically.

022.080 Algorithms for continuous opacities of light elements. R. W. Avery.
Bull. American Astron. Soc., Vol. 4, 212 (1972). – Abstr. AAS.

022.081 Oscillator strength measurements for Ti I and Ti II.
S. J. Wolnik, R. O. Berthel, G. W. Wares.
Bull. American Astron. Soc., Vol. 4, 212 (1972). – Abstr. AAS.

022.082 The statistics of the radiation from astronomical masers.
N. J. Evans II, R. E. Hills, O. E. H. Rydbeck, E. Kollberg.
Bull. American Astron. Soc., Vol. 4, 225 - 226 (1972). – Abstr. AAS.

022.083 Laboratory measurements of the H_2 Lyman band absorptions. G. R. Carruthers, K. L. Bromberg.
Bull. American Astron. Soc., Vol. 4, 269 - 270 (1972). Abstr. AAS.

022.084 Radiative lifetimes for ultraviolet transitions in C, N and O ions by the pulse decay method.
J. V. Mallow, J. Burns.
Journ. Quant. Spectrosc. Radiat. Transfer, Vol. 12, 1081 - 1087 (1972).

022.085 A low-lying resonance in the spectrum of H⁻ II. The $2^1 P$ state. W. Van Rensbergen.
Journ. Quant. Spectrosc. Radiat. Transfer, Vol. 12, 1105 - 1113 (1972).
This study was started as an attempt to find support for identification of the diffuse interstellar line at 6180 Å and the interstellar triplet (4430, 4760, 4890 Å) by a transition in the H⁻ ion, as was first suggested by Rudkjøbing.

022.086 Franck—Condon factors and R-centroids for the $B^2\Sigma \to A^2\Pi$ band system of CN. L. Schoonveld.
Journ. Quant. Spectrosc. Radiat. Transfer, Vol. 12, 1139 - 1145 (1972).

022.087 ^{146}Sm: A chronometer for p-process nucleosynthesis. J. Audouze, D. N. Schramm.
Nature, Vol. 237, 447 - 449 (1972).

022.088 Precision measurement of relative oscillator strengths – I. Fundamental technique: A first application to Mn I. D. E. Blackwell, B. S. Collins.
Monthly Notices Roy. Astron. Soc., Vol. 157, 255 - 271 (1972).
A discussion of the importance of accurate measurements of atomic oscillator strengths in astrophysics is followed by an analysis of the accuracy that is potentially available together with a critical survey of the most useful techniques.

022.089 ^{176}Lu and s-process nucleosynthesis.
J. Audouze, W. A. Fowler, D. N. Schramm.
Nature, Phys. Sci., Vol. 238, 8 - 11 (1972).

022.090 Experimental investigation of the ^{16}O + ^{16}O total reaction cross-section at astrophysical energies.
H. Spinka, H. Winkler.
Astrophys. Journ., Vol. 174, 455 - 461 (1972).
Results are presented for the ^{16}O + ^{16}O total reaction cross-section measured between E_{cm} = 6.8 and 11.85 MeV, near the region of astrophysical interest. The total reaction cross-section was found to be a smooth function of the ^{16}O + ^{16}O energy, in contrast to the behavior of ^{12}C + ^{12}C and ^{12}C + ^{16}O.

022.091 Laboratory measurement of the 2-centimeter, $2_{11}-2_{12}$, transition of normal formaldehyde and its carbon-13 and oxygen-18 species.
K. D. Tucker, G. R. Tomasevich, P. Thaddeus.
Astrophys. Journ., Vol. 174, 463 - 466 (1972).
The hyperfine components of the $2_{11}-2_{12}$ transitions of H_2CO, $H_2^{13}CO$, and $H_2C^{18}O$ have been measured to an accuracy of about 100 Hz.

022.092 Laboratory measurement of the millimeter-wavelength spectrum of formaldehyde. R. B. Nerf, Jr.
Astrophys. Journ., Vol. 174, 467 - 468 (1972).
Frequencies of millimeter-wavelength transitions of $H_2^{12}CO$ and $H_2^{13}CO$ of astronomical interest have been measured to a precision generally exceeding one part in 10^6.

022.093 Microwave spectrum of isocyanic acid, HNCO.
W. H. Hocking, M. C. L. Gerry, G. Winnewisser.
Astrophys. Journ., (*Letters*), Vol. 174, L93 - L96 (1972).
We report laboratory measurements on rotational transitions of isocyanic acid, HNCO, which are of potential interest in radio astronomy.

022.094 Accurate ground-term combinations in Ne I.
V. Kaufman, L. Minnhagen.
Journ. Optical Soc. America, Vol. 62, 92 - 95 (1972).

022.095 Photometric comparison between two calculable vacuum-ultraviolet standard radiation sources: synchrotron radiation and plasma-blackbody radiation.
D. Stuck, B. Wende.
Journ. Optical Soc. America, Vol. 62, 96 - 100 (1972).
Two important principles exist for the calibration of the vuv spectral radiance of secondary-standard lamps by means of calculable standard radiation sources: (1) use of blackbody radiation emitted by a plasma of known temperature and (2) use of electron-synchrotron radiation. This study demonstrates that in the long-wavelength vuv at $\lambda = 165$ nm, significant errors do not occur in the spectral-radiance calibration with either synchrotron radiation and plasma-blackbody radiation. Furthermore, this study shows that it is possible to do photometry in the vuv by use of secondary-standard gas discharge lamps in a way comparable to the routine technique with tungsten-strip lamps in the visible.

022.096 Carbon-like spectra of Sc XVI, Ti XVII, and V XVIII in the range 16—22 Å.
S. Goldsmith, U. Feldman, A. Crooker, L. Cohen.
Journ. Optical Soc. America, Vol. 62, 260 - 264 (1972).

022.097   **Spectra of Rb II, Sr III, Y IV, Zr V, Nb VI, and Mo VII in the vacuum ultraviolet.**
J. Reader, G. L. Epstein, J. O. Ekberg.
Journ. Optical Soc. America, Vol. 62, 273 - 284 (1972).

022.098   **Broadening of infrared absorption lines at reduced temperatures: Carbon dioxide.**
L. D. Tubbs, D. Williams.
Journ. Optical Soc. America, Vol. 62, 284 - 289 (1972).
An evacuated high-resolution Czerny–Turner spectrograph, which is described in this paper, has been used to determine the strengths S and self-broadening parameters γ^0 for lines in the R branch of the ν_3 fundamental of $^{12}C^{16}O_2$ at 298 and at 207 K.

022.099   **Broadening of infrared absorption lines at reduced temperatures, II. Carbon monoxide in an atmosphere of carbon dioxide.**   L. D. Tubbs, D. Williams.
Journ. Optical Soc. America, Vol. 62, 423 - 427 (1972).
The strengths of the rotational lines in the R branch of the CO fundamental have been determined at temperatures of 298, 202, and 132 K by means of a high-resolution spectrograph. Parameters describing the self-broadening and carbon dioxide broadening of CO lines have been determined at 298 and 202 K. The results are compared with other recent experimental and theoretical studies.

022.100   **Laboratory detection of microwave spectrum for HNO.**   S. Saito, K. Takagi.
Astrophys. Journ., (Letters), Vol. 175, L47 - L48 (1972).

022.101   **The mass difference of the muon and the electron.**
F. Hoyle, J. V. Narlikar.
Nature, Vol. 238, 86 - 87 (1972).
The fact that the muon and the electron have different masses can be understood in terms of a Machian theory of inertia. The masses of particles are not fixed and immutable but depend on the detailed particle composition of the universe.

022.102   **Étude spectroscopique d'une décharge à cathode de carbone fonctionnant sous basse pression dans la zone de transition de sa courbe caractéristique courant-tension.**
F. Remy.
Comptes Rendus Acad. Sci. Paris, Sér. B, Vol. 270, 1653 - 1656 = Univ. Liège, Inst. d'Astrophys., Coll. 4°, No. 209 (1970).

022.103   **Quelques remarques relatives aux processus d'excitation du radical C_3 dans une décharge à cathode de carbone sous faible pression d'hydrogène.**   F. Remy.
Comptes Rendus Acad. Sci. Paris, Sér. B, Vol. 271, 17 - 20 = Univ. Liège, Inst. d'Astrophys., Coll. 4°, No. 210 (1970).

022.104   **New electronic transition of the propargyl radical.**
I. Dubois, H. Leclercq.
Canadian Journ. Phys., Vol. 49, 174 - 176 = Univ. Liège, Inst. d'Astrophys.,Coll. 4°, No. 215 (1971).
A new system of diffuse bands in the spectral region 2000–3000 Å has been found after a flash discharge in allene, propylene, or acetylene. Those bands are assigned to a new transition of the propargyl radical $CH_2 - C \equiv CH$.

022.105   **Absorption spectrum of Si_2 in the visible and near-ultraviolet region.**   I. Dubois, H. Leclercq.
Canadian Journ. Phys., Vol. 49, 3053 - 3054 = Univ. Liège, Inst. d'Astrophys.,Coll. 4°, No. 223 (1971).
A very strong absorption spectrum of Si_2 has been observed between 4800 and 3000 Å, yielding new data on the first excited electronic states of this molecule.

022.106   **Application de l'effet de peau à la détermination de la densité électronique dans les décharges pulsées excitées par microondes.**   D. Macau-Hercot.
Bull. Soc. Roy. Sci. Liège, Vol. 38, 700 - 707 = Univ. Liège, Inst. d'Astrophys., Coll. 8°, No. 593 (1969).

022.107   **Le spectre d'absorption de la molécule TeO_2 dans la région de 5000 à 2000 Å.**   I. Dubois.
Bull. Soc. Roy. Sci. Liège, Vol. 39, 63 - 70 = Univ. Liège, Inst. d'Astrophys., Coll. 8°, No. 595 (1970).
The absorption spectra of the TeO_2 molecule, recorded by R. Migeotte in 1941, have been reinvestigated in the spectral region 5000–2000 Å on the basis of our results concerning SO_2 and SeO_2.

022.108   **Time resolved spectroscopy of a pulsed discharge through water vapor: Observations of emissions from the $C^2 \Sigma^+$ state of OH.**   F. Remy.
Spectroscopy Letters, Vol. 4, 319 - 327 = Univ. Liège, Inst. d'Astrophys., Coll. 8°, No. 619 (1971).

022.109   **Zur Statistik neutraler Fermionen mit anomalem magnetischen Moment im äußeren Magnetfeld.**
U. Dyllong.
Ann. Physik, 7. Ser., Vol. 27, 182 - 188 = Zentralinst. Astrophys., Sternw. Babelsberg, Mitt. New Ser., No. 53 (1971).
In this paper we calculate the equation of state of a relativistic uncharged Fermi particle gas in a magnetic field, using the exact solutions of the Dirac-Pauli equation for an uncharged Fermi particle with an anomalous magnetic moment. We find that the stress-energy tensor is anisotropic. The anisotropy is introduced by the anisotropy of energy states.

022.110   **Once more on the neutrino annihilation of a lepton-antilepton pair.**   R. Sh. Yakh'yaev.
Soobshch. Shemakhinsk. Astrofiz. Obs., vyp. (No.) 6, p. 51 - 58 (1971). In Russian.

022.111   **The ionization of boron and the isoelectronic ions carbon II, nitrogen III, and oxygen IV by electron impact.**   E. Stingl.
Journ. Phys. B, Atomic Molecular Phys., Vol. 5, 1160 - 1174 = Separate print Inst. Phys. Astrophys. München (1972).
Electron impact ionization of boron, carbon II, nitrogen III, and oxygen IV, in their ground state configuration $1s^2 2s^2 2p$, is treated, where allowance is made for inner shell ionization.

022.112   **Tired light and the 'missing mass' problem.**
W. Yourgrau, J. F. Woodward.
Acta Phys. Acad. Sci. Hungaricae, Vol. 30, 323 - 329 (1971).
It is demonstrated that the tired light hypothesis is a viable solution to the discrepancy between number mass and dynamical mass for clusters of galaxies; it is compatible with observational results. Proof of the finite photon mass would support the tired light hypothesis. An experiment is suggested to test the hypothesis.

022.113   **Microwave spectrum, ground state structure and dipole moment of thioformaldehyde.**
D. R. Johnson, F. X. Powell, W. H. Kirchhoff.
Journ. Molecular Spectroscopy, Vol. 39, 136 - 145 (1971).
JDM

022.114   **Millimeter and submillimeter wave rotational spectrum and centrifugal distortion efffects of HDO.**
F. C. de Lucia, R. L. Cook, P. Helminger, W. Gordy.
Journ. Chem. Phys. (USA), Vol. 55, 5334 - 5339 (1971).–JDM

022.115   **Photoionization and absorption spectrum of formaldehyde in the vacuum ultraviolet.**
J. E. Mentall, E. P. Gentieu, M. Krauss, D. Neumann.

Journ. Chem. Phys. (*USA*), Vol. 55, 5471 - 5479 (1971).–*JDM*

022.116 **Densities of generalized oscillator strengths for Ar with the inclusion of correlations.**
M. Ya. Amusia, S. I. Sheftel, N. A. Cherepkov, L. V. Cherny-sheva.
Phys. Letters A (*Netherlands*), Vol. 40A, 5 - 6 (1972).

022.117 **A class of exact solutions for the motion of a particle in a monopole-prolate quadrupole field.**
A. Armenti, Jr., P. Havas.
Relativity and gravitation, Haifa 1969, (see 012.029), p. 1 - 15 (1971).

022.118 **Solar neutrinos and the ^{37}Cl neutrino absorption experiment.** P. Bandyopadhyay.

Journ. Phys. A, General Phys., Vol. 5, L19 - L23 (1972).

022.119 **Réalisation d'un corps noir à haute température par chauffage haute fréquence.**
C. T. Hua, R. Peyturaux.
Nouv. Rev. Optique Appliquée, Vol. 3, No. 1, p. 31 - 36 (1972).

Errata

022.901 **Erratum: "Radiative recombination coefficients for complex ions"** [Astrophys. Journ., Vol. 168, 313 - 316 (1971)]. C. B. Tarter.
Astrophys. Journ., Vol. 172, 251 (1972).

Instruments and Astronomical Techniques

031 Optics, Methods of Observation and Reduction

031.001 **Effect of an error in plate reduction.**
H. Debehogne.
Astron. Astrophys., Suppl. Ser., Vol. 5, 185 - 201 (1972).
In French.
In order to improve the reduction methods of photographic plates we have to know the effect of an error attributed to a reference star 1) at the point where this error is applied; 2) on the whole plate; for 1) several numbers N of stars of reference; 2) various methods of reduction; 3) all the points on the plate where the error can be applied successively. A systematic investigation has been performed when the natural distribution of real stars is replaced by theoretical stars (or fictitious stars) forming a regular network on the plate.

031.002 **Methods of radar astrometric observations.**
Y. N. Alexandrov, B. I. Kuznetsov, G. M. Petrov,
O. N. Rzhiga.
Astron. Zhurn. Akad. Nauk SSSR, Vol. 49, 175 - 185 (1972).
In Russian. English translation in Soviet Astron. AJ, Vol. 16, No. 1.
Measurement methods of distances to planets and their radial velocities, which have been devised during 1965–1969 at the Institute of Radioelectronics of the USSR Academy of Sciences, are described. Main principles of radar construction are given. Estimates of measurement errors are listed for radar observations of Venus.

031.003 **Television microphotometry.**
G. S. Tsarevsky, A. Ya. Khesin, A. V. Nikonenko,
B. A. Yanson, I. K. Iljin.
Astron. Zhurn. Akad. Nauk SSSR, Vol. 49, 204 - 209 (1972).
In Russian. English translation in Soviet Astron. AJ, Vol. 16, No. 1.
A television unit for the analysis of photographic images is worked out. This unit fixes equidensities and carries out the automatic measurement of their areas at fixed levels of optical density. Measurements of calibrated photographs of stars have shown that with the help of such a method characteristic curves, linear in a wide interval of stellar magnitudes, are obtained.

031.004 **Direct reduction of astronomical X-ray data.**
J. F. Dolan.
Bull. American Astron. Soc., Vol. 3, 457 (1971). – Abstr. AAS.

031.005 **Two new data reduction techniques for the intensity modulation method.**
F. D. Rosenberg, J. F. Villamediana, L. W. Fredrick.
Bull. American Astron. Soc., Vol. 3, 464 - 465 (1971). – Abstr. AAS.

031.006 **Two adapters for projection photography.**
R. C. Price, T. D. Ross.
Sky Telescope, Vol. 43, 123 - 126 (1972).

031.007 **Hartmann test of aspherical mirrors.**
D. Malacara.
Applied Optics, Vol. 11, 99 - 101 (1972).
Astronomical aspherical mirrors may be tested in the optical shop with a Hartmann null test. A null test is ob-
tained by placing small wedges over each hole of the Hartmann screen. The wedges have an angle between the two faces such that the spherical aberration with the object and the image at the center of curvature is just compensated.

031.008 **Lightweight center-mounted 152-cm $F/2.5$ Cer-Vit mirror.** W. E. Carter.
Applied Optics, Vol. 11, 467 - 468 (1972). – Letter.

031.009 **Stellar photoelectric photometry for amateurs.**
E. P. Majden.
Journ. Roy. Astron. Soc. Canada, Vol. 66, L 1 - L 4 (1972).

031.010 **On employment of electronic-ionic technology in astronomical instrumentation.**
V. E. Bakhchivandzhi.
Astrofiz. Issled. Izv. Spets. Astrofiz. Obs., Vol. 3, 180 - 182 (1971). In Russian.

031.011 **Optimum photomultiplier photon counting photometry.** G. Sedmak.
Astron. Astrophys., Vol. 18, 232 - 241 (1972).
This paper deals with a general theoretical analysis of the conditions for the optimum use of photomultipliers in photon counting photometry. Some experimental results, which confirm the theory, are shown. Moreover a method is given, which allows the selection of the optimum working point of the photometer by means of measurements by the photometer itself, in most of the practical cases.

031.012 **CAMAC concepts.** D. A. Mack.
Publ. Astron. Soc. Pacific, Vol. 84, 167 - 175 (1972).
Paper presented at the symposium "Advanced Electronic Systems for Astronomy", 1971, Santa Cruz.

031.013 **An inexpensive astronomical control computer system.** J. Hämeen-Anttila.
Publ. Astron. Soc. Pacific, Vol. 84, 185 - 189 (1972). – Paper presented at the symposium "Advanced Electronic Systems for Astronomy", 1971, Santa Cruz.

031.014 **The Hale Observatories' computer system.**
E. W. Dennison.
Publ. Astron. Soc. Pacific, Vol. 84, 190 - 193 (1972). – Paper presented at the symposium "Advanced Electronic Systems for Astronomy", 1971, Santa Cruz.

031.015 **Experiences with the Cerro Tololo data-acquisition system.** B. M. Lasker.
Publ. Astron. Soc. Pacific, Vol. 84, 207 - 211 (1972). – Paper presented at the symposium "Advanced Electronic Systems for Astronomy", 1971, Santa Cruz.

031.016 **CAMAC multicrate systems.** I. G. van Breda.
Publ. Astron. Soc. Pacific, Vol. 84, 212 - 216 (1972). – Paper presented at the symposium "Advanced Electronic Systems for Astronomy", 1971, Santa Cruz.

031.017 **Präzisions-Beugungsgitter aus Jena.** P. Kröplin.

Jenaer Rundschau, (Jena Rev.), 17. Jahrgang, p. 68 - 71 (1972).

031.018 Unified multiplex projector.
Yu. N. Lipskij, V. I. Chikmachev.
Kosm. Issled., Vol. 10, 133 - 136 (1972). In Russian. – Brief information.

031.019 Precise stellar positions using GALAXY-machine measures of a Schmidt plate. R. J. Dodd.
Astron. Journ., Vol. 77, 306 - 311 (1972).
   A method of transforming Cartesian coordinates to equatorial coordinates using 20 plate constants, and a correction procedure involving the use of a weighted inverse squared error function are described. A plate of the Pleiades cluster taken with a 40/60/150-cm Schmidt telescope was measured by the GALAXY machine ; the results are reported.

031.020 Finding and guiding on dark skies. H. E. Dall.
Journ. British Astron. Ass., Vol. 82, 202 - 203 (1972).

031.021 La photographie des planètes. R. Bucaille.
L'Astronomie, 86ᵉ année, p. 161 - 174 (1972).

031.022 Höchste Auflösung für Teleskope. J. Rösch.
Umschau, 72. Jahrgang, p. 361 (1972).
   The observation of the sun and the planets as well as galaxies, clusters and stars require a high resolving power of the telescopes. Some future perspectives of improved resolution are given.

031.023 Analysis of control methods for optical surfaces of large diameter. N. L. Lazareva.
Trudy Mosk. vyssh. tekhn. uch-shcha im. N. Eh. Baumana, 1971, No. 143, p. 184 - 192. In Russian. – Abstr. in Referativ. Zhurn. 51. Astron., 6.51.91 (1972).

031.024 On the interference error arising while graduating an antenna by means of the artificial moon method.
K. S. Stankevich, V. P. Ivanov.
Izv. vyssh. ucheb. zavedenij. Radiofizika, Vol. 14, 1787 - 1790 (1971). In Russian. – Abstr. in Referativ. Zhurn. 51. Astron., 6.51.278 (1972).

031.025 Automation in angle measurements using the pulse (digital) theodolite designed in Poland.
H. Z. Kowalski.
Geodezja Kartografia, Vol. 21, 111 - 121 (1972). In Polish.

031.026 Moderne Stellar- und Nebel-Photographie.
E. Alt, G. Klaus.
Orion Schaffhausen, 30. Jahrgang, p. 83 - 88 (1972).

031.027 Nomogramm für die Sternfeld-Photographie.
H. Sigg.
Orion Schaffhausen, 30. Jahrgang, p. 98 - 99 (1972).

031.028 Zur Wahl des optischen Systems für ein Großteleskop. K. Bahner.
Mitt. Astron. Ges., No. 31, p. 118 - 119 (1972).

031.029 X-ray and γ-ray imaging with multiple-pinhole cameras using a posteriori image synthesis.
G. Groh, G. S. Hayat, G. W. Stroke.
Applied Optics, Vol. 11, 931 - 933 (1972).
   An image having considerably increased SNR compared with a single pinhole camera image of starlike sources, may be synthesized from the multiplicity of images recorded at X-ray and γ-ray wavelengths with the multiple-pinhole camera proposed by R. H. Dicke, using the method of extended-source

Fourier-transform holography.

031.030 Reflectance of evaporated rhenium and tungsten films in the vacuum ultraviolet from 300 to 2000 Å.
J. T. Cox, G. Hass, J. B. Ramsey, W. R. Hunter.
Journ. Optical Soc. America, Vol. 62, 781 - 785 (1972).

031.031 Photoelectric determination of stellar radial velocities. R. F. Griffin.
Auxiliary instrumentation for large telescopes. Conference 1972, (see 012.018), p. 171 - 173 (1972).

031.032 The use of CAMAC for telescope instrumentation.
C. L. Stephens, I. G. van Breda.
Auxiliary instrumentation for large telescopes. Conference 1972, (see 012.018), p. 499 - 514 (1972).

031.033 Image splitting prism for photographic photometry.
S. Isobe, Y. Norimoto.
Tokyo Astron. Obs., Report No. 59, Vol. 15, 760 - 765 (1972). In Japanese.

031.034 Direct electronography of astronomical objects.
T. Hawarden.
Monthly Notes Astron. Soc. Southern Africa, Vol. 31, 49 (1972). – Abstract.

031.035 Equidensieten. M. Drummen.
Hemel en Dampkring, Vol. 70, 195 - 197 (1972).

031.036 Formulae of photographic reduction and the astronomical refraction. A. Schmitt.
Publ. Obs. Astron. Strasbourg, Vol. 2, Fasc. 2, 2 pp. (1972).

031.037 Tips für die Astropraxis.
SuW, Vol. 11, 16 - 19, 47 - 51, 78 - 80, 107 - 111, 141, 166 - 168, 195 - 198 (1972).
Langbrennweitige Stellarphotographie, (*K. Rihm*), p. 16 - 17; Das Protuberanzenfernrohr als Hochleistungsinstrument VII, (*G. Nemec*), p. 17 - 19; Das zweiäugige Sehen astronomischer Objekte, (*H. Wichmann*), p. 47 - 50; Das Protuberanzenfernrohr als Hochleistungsinstrument VIII, (*G. Nemec*), p. 50 - 51; Sucher mit Komfort? (*G. Walterspiel*), p. 78 - 79; Das teure Loch im Protuberanzenfernrohr (*H. Baderschneider*), p. 79 - 80; Eine billige Stundennachführung ohne Schneckenrad, (*O. Nögel*), p. 80; Einfache Netzgeräte, (*J. Biel*), p. 107; Ein Zenit- und Sonnenprisma, (*W. Rummel*), p. 107 - 108; Ein einfaches Gerät zur Zentrierung optischer Systeme, (*O. Nögel*), p. 108; Ein bequemer Sucher mit größtem Gesichtsfeld, (*O. Nögel*), p. 108 - 109; Das Protuberanzenfernrohr als Hochleistungsinstrument IX (Schluß), (*G. Nemec*), p. 109 - 111; Über die Prüfung von Fernrohrobjektiven (*K. Teicher*), p. 141; Die elektronische Bildanalyse (*J. Biel*), p. 166 - 167; Ein Binokulartubus für lichtstarke Spiegelteleskope (*C. Albrecht*), p. 167 - 168; Das zweiäugige Sehen bei astronomischen Objekten (*R. Brandt*), p. 168; Zeiss Mini-Fernglas nun auch für Brillenträger, p. 168; Astrospektrographie mit Spiegelteleskopen, (*C. Albrecht*), p. 195; Ein Sonnenprisma, (*F. Fleig*, mit einer Bemerkung von *D. Schmadel*), p. 196; Versuche mit Äquidensiten an Jupiter (*C. Kowalec*), p. 196 - 197; Auswertung von Planetenaufnahmen mit dem Äquidensitenverfahren (*B. Wedel*), p. 197 - 198.

031.038 Les erreurs d'inclinaison et d'azimut aux instruments des passages. D. Djurović, V. Radogostić.
Publ. Dep. Astron. Univ. Beograd, No. 1, p. 13 - 20 (1969).
   Dans le present article nous exposons un peu plus en détail le problème de la détermination de l'inclinaison et de l'azimut de l'axe horizontale.

031.039 On the determination of anomalous refraction out

of astrometrical measurements in the zenith zone.
G. Teleki, B. Ševarlić.
Publ. Dep. Astron. Univ. Beograd, No. 3, p. 5 - 16 (1971).

031.040 **Amateur astronomers construct a telescope mirror.**
N. P. Barabashov.
Vasiona, Vol. 20, 17 - 21, 29 - 35 (1972). In Serbo-Croatian.
(Translated from Russian by D. Pekez).

031.041 **Untersuchungen zur Argelandermethode.**
U. Bastian.
BAV Rundbrief, 21. Jahrgang, p. 19 - 22 (1972).

031.042 **Laser process sharpens images from telescope.**
Photogr. Applied Sci. Techn. & Med., Vol. 6, No. 4,
p. 28 (1971). — See Phys. Abstr., Vol. 75, No. 20714 (1972).

031.043 **Méthodes physiques de l'astronomie.**
A. Lallemand.
Annu. Coll. Française, Vol. 70, 101 - 109 (1970/71). — See

Bull. Signal., Vol. 33, Section 120, No. 2654 (1972).

031.044 **High angular resolution astrophysical observations
from space.** C. de Jager.
Earth Extraterr. Sci., Vol. 1, 243 - 250 (1971).

031.045 **Comparaison de positions de réseaux optiques par
méthodes opto-électroniques.** D. Bernfeld.
Thesis, Univ. Louis Pasteur, Strasbourg. Centre Documentation, C. N. R. S., (1971-06-25), 87 pp. (1971). — See Bull.
Signal., Vol. 33, Section 120, No. 6970 (1972).

The photoelectric meridian circle of Bergedorf/Perth.
See Abstr. 032.031.

Beitrag zur Weiterentwicklung des Zirkumzenitals.
See Abstr. 032.052.

**Complex program for reducing positional observa-
tions of lunar objects on computers.**
See Abstr. 094.176.

032 Astronomical Instruments

032.001 **Supernova search by a computer controlled telescope – specifications and status.**
S. A. Colgate, E. P. Moore.
Bull. American Astron. Soc., Vol. 3, 450 (1971). – Abstr. AAS.

032.002 **A 4 $^1/_2$-inch unobstructed Maksutov telescope.**
A. E. Crowe, Jr.
Sky Telescope, Vol. 43, 47 - 52 (1972).

032.003 **An 18 $^1/_2$-inch telescope in Memphis.**
D. H. Bratton.
Sky Telescope, Vol. 43, 114 - 116 (1972).

032.004 **Small optical telescopes on the moon.**   E. H. Wells.
Spaceflight, Vol. 14, 90 - 94 (1972).

032.005 **Design parameters of paraboloid-hyperboloid telescopes for X-ray astronomy.**
L. P. VanSpeybroeck, R. C. Chase.
Applied Optics, Vol. 11, 440 - 445 (1972).
We have evaluated the principal optical characteristics of paraboloid-hyperboloid X-ray telescopes by a ray-tracing procedure; we find that our results for resolution, focal plane curvature, and finite source distance effects may be approximated in terms of the design parameters by simple empirical formulas.

032.006 **The six-meter telescope. I. Basic data.**
B. K. Ioannisiani.
Astrofiz. Issled. Izv. Spets. Astrofiz. Obs., Vol. 3, 3 - 19 (1971). In Russian.

032.007 **A scale model of the large telescope on altazimuth mounting.**
J. B. Vyatskin, A. S. Naishul, E. M. Neplokhov.
Astrofiz. Issled. Izv. Spets. Astrofiz. Obs., Vol. 3, 20 - 25 (1971). In Russian.

032.008 **The large space telescope instrumentation.**
A. B. Underhill.
Publ. Astron. Soc. Pacific, Vol. 84, 84 - 90 (1972). – Paper presented at the symposium "Advanced Electronic Systems for Astronomy", 1971, Santa Cruz.

032.009 **The MIT automated astrophysical observatory.**
T. B. McCord, G. Snellen, S. Paavola.
Publ. Astron. Soc. Pacific, Vol. 84, 220 - 224 = Contr. Planet. Astron. Lab., Dep. Earth Planet. Sci., Mass. Inst. Technology, No. 43 (1972). – Paper presented at the symposium "Advanced Electronic Systems for Astronomy", 1971, Santa Cruz.

032.010 **Present status of the Canadian 24-inch telescope on Las Campanas.**   D. A. MacRae.
Journ. Roy. Astron. Soc. Canada, Vol. 66, 69 (1972). – Abstr. Canadian Astron. Soc.

032.011 **A 7.6-meter (300-inch) equivalent optical telescope array.**   E. H. Richardson, G. J. Odgers.
Journ. Roy. Astron. Soc. Canada, Vol. 66, 69 - 70 (1972). – Abstr. Canadian Astron. Soc.

032.012 **Ripristino dello strumento dei passaggi Bamberg N. 15000 e prime osservazioni.**
G. Cocito, N. Missana, C. Moranzino.
Mem. Soc. Astron. Italiana, Nuova Ser., Vol. 43, 69 - 81 (1972).

032.013 **On the temperature field around the visual zenith telescope.**   T. Goto.
Proc. International Latitude Obs. Mizusawa, No. 11, p. 38 - 46 (1971). In Japanese.

032.014 **X-ray telescope on the moon.**   I. P. Tindo.
Zemlya i Vselennaya, 1972, No. 2, p. 10 - 13. In Russian.

032.015 **An Italian 54-inch reflector of unusual design.**
G. de Mottoni.
Sky Telescope, Vol. 43, 296 - 297 (1972).

032.016 **An Australian 11 $^1/_2$-inch Wright telescope.**
D. H. Whitehead.
Sky Telescope, Vol. 43, 320 - 326 (1972).

032.017 **On the importance and possibilities of increasing accuracy by visual zenith telescopes.**
V. S. Milovanović.
Extra collection of papers to IAU Symposium No. 48, (see 012.008), p. 25 - 29 (1971).

032.018 **Use of "circumzenithal" for astronomical measurements of high precision.**   K.-G. Steinert.
Extra collection of papers to IAU Symposium No. 48, (see 012.008), p. 31 - 34 (1971).

032.019 **The tower telescope of the Norwegian solar observatory.**   B. R. Pettersen.
Journ. British Astron. Ass., Vol. 82, 112 - 115 (1972).

032.020 **Constructing a beginner's telescope.**   C. J. R. Lord.
Journ. British Astron. Ass., Vol. 82, 116 - 121 (1972).

032.021 **Metrication and the telescope.**   D. Clarke.
Journ. British Astron. Ass., Vol. 82, 188 - 191 (1972).

032.022 **A simple Cassegrain-Newtonian telescope.**
E. J. Hysom.
Journ. British Astron. Ass., Vol. 82, 204 - 205 (1972).

032.023 **A 7.6-meter (300-inch) equivalent optical telescope array.**   E. H. Richardson, G. J. Odgers.
Journ. Roy. Astron. Soc. Canada, Vol. 66, 99 - 108 = Contr. Dominion Astrophys. Obs., *Victoria*, No. 173 (1972).
The equivalent of a 300-inch telescope operating at the coudé focus could be achieved by means of an array of twenty-five 60-inch telescopes having a common coudé focal point.

032.024 **Investigation of the pivots of the Bamberg transit instrument No. 85951.**   Eh. A. Sanakulov.
Tsirk. Astron. Inst., *Tashkent*, No. 30 (377), p. 22 - 28 (1971). In Russian.

032.025 **Photographic zenith tube.**   A. A. Mikhajlov.
Coordinate systems in astronomy, (see 012.007), p. 67 - 68 (1971). In Russian.

032.026 **Experience with and prospects for the work of the Moscow photographic zenith tube.**
D. N. Ponomarev.
Coordinate systems in astronomy, (see 012.007), p. 69 - 79 (1971). In Russian.

032.027 **Behaviour of the Bucharest meridian circle for declination at the zenith zone.**
T. Ionescu, M. Tudor, E. Toma.
Stud. Cerc. Astron., Vol. 17, 7 - 8 (1972). In Romanian.

032.028 **On the flexure of a meridian telescope.**
I. Rusu.
Stud. Cerc. Astron., Vol. 17, 9 - 18 (1972). In Romanian.

032.029 **Isochromatic mirror-lens telescope systems with spherical optics.** P. P. Argunov.
Astron. vestn., Vol. 6, 52 - 61 (1972). In Russian. – Abstr. in Referativ. Zhurn. 51. Astron., 6.51.86 (1972).

032.030 **Method for making large aspherical mirrors.**
T. D. Savostin.
Trudy Mosk. vyssh. tekhn. uch-shcha im. N. Eh. Baumana, 1971, No. 143, p. 153 - 156. In Russian. – Abstr. in Referativ. Zhurn. 51. Astron., 6.51.90 (1972).

032.031 **The photoelectric meridian circle of Bergedorf/Perth.**
E. Høg.
Astron. Astrophys., Vol. 19, 27 - 40 (1972).
The new meridian techniques introduced in Bergedorf have the aim of improving systematic and accidental errors, of increasing the number of observations per night, and of making the data ready for reduction by an off-line computer. The methods for reducing the observations of stars, nadir and meridian marks are given in detail.

032.032 **Investigation of the heliometer of the Astronomical Engelhardt-Observatory.** A. S. Mamakov.
Izv. Astron. Ehngel'gardt. Obs., *Kazan,* No. 38, p. 97 - 110 (1970). In Russian.

032.033 **Astronomical observation systems.** P. B. Fellgett.
Journ. British Astron. Ass., Vol. 82, 257 - 273 (1972).

032.034 **Air currents above mirrors.** E. J. Hysom.
Journ. British Astron. Ass., Vol. 82, 274 - 278 (1972).

032.035 **Das Maksutov-Cassegrain-Teleskop als Amateur-Instrument.** E. Wiedemann.
Orion Schaffhausen, 30. Jahrgang, p. 88 - 91 (1972).

032.036 **Bemerkungen zur Maksutov-Kamera.**
E. Wiedemann.
Orion Schaffhausen, 30. Jahrgang, p. 91 - 93 (1972).

032.037 **Construction d'un télescope à champ riche.**
R. Durussel.
Orion Schaffhausen, 30. Jahrgang, p. 112 - 113 (1972).

032.038 **Stratoscope II, een kijker aan een ballon.**
F. P. Israel.
Hemel en Dampkring, Vol. 70, 186 - 188 (1972).

032.039 **De rol van Schmidt-telescopen in de sterrenkunde.**
J. Denoyelle.
Hemel en Dampkring, Vol. 70, 188 - 190 (1972).

032.040 **Instrumentation for the KPNO and CTIO 4 meter reflectors.** A. A. Hoag.
Auxiliary instrumentation for large telescopes. Conference 1972, (see 012.018), p. 39 - 53 (1972).

032.041 **Instrumentation and some principal programs of the McDonald Observatory 2.75-meter (107-inch) reflector.** H. J. Smith.

Auxiliary instrumentation for large telescopes. Conference 1972, (see 012.018), p. 55 - 64 (1972).

032.042 **Basic instrumentation for the Anglo-Australian telescope.** S. C. B. Gascoigne, H. Wehner.
Auxiliary instrumentation for large telescopes. Conference 1972, (see 012.018), p. 65 - 73 (1972).

032.043 **Instrumentation planning for the ESO 3.6 m telescope.** A. Reiz.
Auxiliary instrumentation for large telescopes. Conference 1972, (see 012.018), p. 75 - 81 (1972).

032.044 **Projets d'instrumentation pour le telescope de 3.60 m de l'I.N.A.G.** R. Cayrel.
Auxiliary instrumentation for large telescopes. Conference 1972, (see 012.018), p. 91 - 93 (1972).

032.045 **The coudé of the 1.2 meter telescope at Victoria.**
E. H. Richardson.
Auxiliary instrumentation for large telescopes. Conference 1972, (see 012.018), p. 285 - 290 (1972).

032.046 **The Gartly-Black telescope.** J. L. Perdrix.
Journ. Astron. Soc. Victoria, Vol. 25, 20 - 21 (1972).

032.047 **Reflectivity of the 188-cm reflector of the Okayama Observatory.** T. Kanda.
Tokyo Astron. Obs., Report No. 59, Vol. 15, 733 - 736 (1972). In Japanese.

032.048 **On the instrumental constants of meridian circle.**
R. Fukaya.
Tokyo Astron. Obs., Report No. 59, Vol. 15, 737 - 754 (1972). In Japanese.

032.049 **On the azimuth error and the level error of meridian circle.** H. Ishii.
Tokyo Astron. Obs., Report No. 59, Vol. 15, 755 - 759 (1972). In Japanese.

032.050 **Notes sur le «Grand Schmidt» Liège – O. H. P. – 1.**
F. Dossin.
Bull. Soc. Roy. Sci. Liège, Vol. 40, 66 - 67 = Univ. Liège, Inst. d'Astrophys., Coll. 8°, No. 611 (1971).

032.051 **Caratteristiche ottiche di alcuni tipi di telescopi astronomici.** V. Zappalà.
Oss. Astron. Torino, Studio Monografico, No. 7, 67 pp. (1971).

032.052 **Beitrag zur Weiterentwicklung des Zirkumzenitals.**
K.-G. Steinert.
Mitt. Lohrmann-Obs. Techn. Univ. Dresden, No. 21, 79 pp. (1970). – Reprinted from Geod. Geophys. Veröff., Deutsche Akad. Wiss. Berlin, Ser. 3, No. 20 (1970).

032.053 **Investigation of the optics of the AST-452 telescope of the Shemakha Astrophysical Observatory.**
G. V. Akhundova.
Soobshch. Shemakhinsk. Astrofiz. Obs., vyp. (No.) 6, p. 69 - 77 (1971). In Russian.

032.054 **Telescopes that see red.** P. Connes.
Sci. Progr. Découverte, No. 3439, p. 39 - 47 (1971). In French.

032.055 **Considerations about telescopes for infra-red astronomy.** F. Melchiorri.
Nuovo Cimento B, Ser. 11, Vol. 8B, 167 - 182 (1972).
The choice of a telescope which may be used for ground-based infra-red observations is discussed. The efficiency of

these instruments for the study of localized sources and for sky surveys is discussed.

Observations de latitude et tubes zénithaux visuels.
See Abstr. 045.019.

033   Radio Telescopes and Equipment

033.001   **The Parkes survey of 21-centimeter absorption in discrete-source spectra. I. The Parkes hydrogen-line interferometer.**   V. Radhakrishnan, J. W. Brooks, W. M. Goss, J. D. Murray, U. J. Schwarz.
Astrophys. Journ., Suppl. Ser., No. 203, Vol. 24, 1 - 14 (1972).

A multichannel 21-cm-line interferometer is now in regular operation at the Parkes Observatory. We describe the instrument which incorporates certain novel features. Among the more interesting of the early observations made with this instrument are the detection of extragalactic H I absorption in the Large Magellanic Cloud and evidence of small-scale structure in the galactic H I concentrations in the direction of Orion A.

033.002   **A portable detector for the automatic gain-stabilization of radiometers.**   M. J. Yerbury.
Bull. American Astron. Soc., Vol. 3, 438 (1971). – Abstr. AAS.

033.003   **Capabilities of almucantar radio telescopes.**   P. D. Usher.
Bull. American Astron. Soc., Vol. 3, 446 (1971). – Abstr. AAS.

033.004   **The Arecibo upgrading project.**   F. D. Drake.
Bull. American Astron. Soc., Vol. 3, 466 (1971). Abstr. AAS.

033.005   **Adjustment of the large Pulkovo radio telescope by the autocollimation method.**
N. Khodzhamukhammedov, A. A. Stotskij.
Izv. AN TurkmSSR. Ser. fiz.-tekhn., khim. i geol. n., 1971, No. 4, p. 33 - 41. In Russian. – Abstr. in Referativ. Zhurn. 51. Astron., 3.51.127 (1972).

033.006   **Polarization observations using a varying profile antenna.**
N. A. Esepkina, V. Yu. Petrun'kin, N. S. Soboleva, A. V. Rejner.
Izv. vyssh. uchebn. zavedenij. Radiofizika, Vol. 14, 1149 - 1159 (1971). In Russian. – Abstr. in Referativ. Zhurn. 51. Astron., 3.51.131 (1972).

033.007   **A correction of the radio astronomical image distorted by atmospheric turbulence.**   P. A. Fridman.
Astrofiz. Issled. Izv. Spets. Astrofiz. Obs., Vol. 3, 135 - 141 (1971). In Russian.

An expression is derived for the root-mean-square deviation of a radio-astronomical image from the true brightness-temperature distribution over an object. The possibility of the radio-image correction by way of optimal filtering by the criterion of minimum of the mean standard error is discussed.

033.008   **On a method of spectral analysis in radio astronomy.**
A. F. Dravskikh, S. G. Smolentsev.
Astrofiz. Issled. Izv. Spets. Astrofiz. Obs., Vol. 3, 142 - 153 (1971). In Russian.

A scheme for construction of a multichannel radio-astronomical spectrograph is suggested without a modulator. This makes it possible to simplify the waveguide circuit and to improve the fluctuation sensitivity. An expression is presented describing the spectrum of the received radio emission with allowance for disturbing factors, and a technique of spectral measurements is described.

033.009   **The effect of a periscopic illumination system upon the noise temperature of a variable profile antenna.**
V. M. Spitkovsky.
Astrofiz. Issled. Izv. Spets. Astrofiz. Obs., Vol. 3, 154 - 158 (1971). In Russian.

A technique for calculating the noise temperature component of a variable profile antenna due to the system reflector is presented for one of the possible laws of the field amplitude distribution over the secondary mirror aperture. It is shown how the periscopic component of the noise temperature depends on the reflecting properties of the underlying surface behind the main mirror. A comparison of our calculations with experimental data at 3.2 cm supports the suitability of the accepted technique for calculating the periscopic component of the noise temperature of a variable profile antenna in the centimeter-wavelength range.

033.010   **A six-channel phase comparator of electrical lengths.**
A. L. Alexandrov, V. M. Brylov, A. A. Stotsky.
Astrofiz. Issled. Izv. Spets. Astrofiz. Obs., Vol. 3, 159 - 164 (1971). In Russian.

A six-channel phase radio range-finder system (a phase comparator) in the centimeter-wavelength range is described intended for study on the spatial and time characteristics of phase fluctuations of radio waves. It can also be used for geodetic and adjustment works when tuning large antennas. The comparator provides the accuracy of comparing electrical lengths of traces and of measuring length fluctuations of the order of 0.1 mm at lengths of several hundred meters.

033.011   **A multichannel recording device for a radio spectrograph.**   N. F. Ryzhkov, V. A. Jakovlev.
Astrofiz. Issled. Izv. Spets. Astrofiz. Obs., Vol. 3, 165 - 169 (1971). In Russian.

033.012   **An investigation of the reflecting surface of the secondary mirror of the large Pulkovo radio telescope.**   J. K. Zverev, A. I. Kopylov.
Astrofiz. Issled. Izv. Spets. Astrofiz. Obs., Vol. 3, 170 - 175 (1971). In Russian.

A technique is described for investigating the topography of the reflecting surface of the secondary mirror of the large Pulkovo radio telescope. The main results of such an investigation are given.

033.013 Orienting of the polar axis of a radio telescope.
J. K. Zverev.
Astrofiz. Issled. Izv. Spets. Astrofiz. Obs., Vol. 3, 176 - 179
(1971). In Russian.
Experience is described in setting the polar axis of a radio
telescope having a paraboloid mirror of 3 m in diameter at the
angle equal to the latitude of the site. Recommendations for
orienting the polar axis in azimuth are given.

**033.014 An electronic commutator for a solar radio spectro-
graph.** Z. V. Dravskikh.
Astrofiz. Issled. Izv. Spets. Astrofiz. Obs., Vol. 3, 183 - 185
(1971). In Russian.
A circuit of an electronic commutator for the twelve-
channel solar radio spectrograph is briefly described.

**033.015 A data acquisition and control system for a 46-me-
ter radio telescope.**
D. J. Bradt, L. A. Higgs, S. G. Jones, T. G. O'Neill, J. L. Wolfe.
Publ. Astron. Soc. Pacific, Vol. 84, 194 - 202 (1972). – Paper
presented at the symposium "Advanced Electronic Systems
for Astronomy", 1971, Santa Cruz.

**033.016 Das 100 m-Teleskop, ein neues Instrument für die
Radioastronomie.** B.-H. Grahl.
Sterne, 48. Jahrgang, p. 65 - 75 (1972).

**033.017 A general mapping program for the ARO radio tele-
scope.** L. A. Higgs.
Journ. Roy. Astron. Soc. Canada, Vol. 66, 66 (1972). – Abstr.
Canadian Astron. Soc.

**033.018 Radio interferometer with independent reception at
the frequency of 86 MHz.** V. A. Alekseev,
M. A. Antonets, V. V. Vitkevich, E. D. Gatélyuk, P. S. Zhivora,
V. D. Krotikov, A. E. Kryukov, V. S. Troitskij, A. I. Chikin,
V. A. Shemagin, M. V. Yankavtsev, B. P. Fateev.
Izv. vyssh. ucheb. zavedenij. Radiofizika, Vol. 14, 1303 -
1314 (1971). In Russian. – Abstr. in Referativ. Zhurn. 51.
Astron., 4.51.126 (1972).

**033.019 Baseline determination for 21-cm hydrogen line
astronomy.** G. T. Wrixon, C. Heiles.
Astron. Astrophys., Vol. 18, 444 - 449 (1972).
A brief review is given of the problems associated with
baseline determination in 21-cm line work. The 20-foot horn
reflector antenna at Crawford Hill and its associated receiving
equipment are described. Measurements of 11 regions of very
low hydrogen emission made with the Crawford Hill receiving
system are presented.

**033.020 Astrometric applications of very-long-baseline inter-
ferometry.** I. I. Shapiro.
Bull. American Astron. Soc., Vol. 4, 266 (1972). – Abstr. AAS.

**033.021 The confusion error on the flux estimate of a point
source.** W. R. Burns.
Astron. Astrophys., Vol. 19, 41 - 44 (1972).
The estimation of the flux of a point source is considered.
The error, due to confusion, on the estimate is given and the
relation of this error to the shape of the antenna power pattern. Expressions are given for the variance on three quantities,
the confusion noise on the record, the flux estimate assuming
a known baseline, and the baseline estimate. Some examples
are given and their implications discussed.

033.022 Long coherence intercontinental interferometry.
J. S. Gubbay.
Proc. Astron. Soc. Australia, Vol. 2, 114 - 115 (1972).

**033.023 Preliminary results obtained with an aperture syn-
thesis telescope at 30 MHz.**
E. A. Finlay, B. B. Jones.
Proc. Astron. Soc. Australia, Vol. 2, 115 - 117 (1972).

**033.024 Low-noise parametric amplifiers tunable over one
full octave.** J. Edrich.
IEEE Journ. Solid-State Circuits, Vol. SC-7, No. 1, p. 32 - 37 =
National Radio Astron. Obs., *Green Bank*, Repr. Ser. A,
No. 235 (1972).

033.025 Das Spektrum der Radiostrahlung.
R. Wielebinski, W. E. Wilson.
Mitt. Astron. Ges. No. 31, p. 94 - 99 (1972).

**033.026 Probleme der Justierung und elastischen Verformung
beim 100-m-Teleskop.** B. H. Grahl.
Mitt. Astron. Ges., No. 31, p. 101 (1972). – Abstract.

**033.027 Prozeßsteuerung des 100-Meter-Teleskops – Astro-
nomisches Konzept.** P. Stumpff.
Mitt. Astron. Ges., No. 31, p. 101 - 107 (1972).

**033.028 Prozeßsteuerung des 100-m-Teleskops – Digitale
Regelung.** H. G. Girnstein, W. Voss.
Mitt. Astron. Ges., No. 31, p. 108 - 111 (1972).

**033.029 Prozeßsteuerung des 100-m-Teleskops – Kommuni-
kation des Beobachters mit dem rechnergesteuerten
Teleskop.** J. Schraml.
Mitt. Astron. Ges., No. 31, p. 112 - 117 (1972).

**033.030 Eine Entwicklungsstudie für ein 65-m-Radioteleskop
für den mm-Wellenbereich.** H. Eschenauer.
Mitt. Astron. Ges., No. 31, p. 209 - 218 (1972).

033.031 Two gravity-wave detectors: A comparison.
S. N. Rasband, P. B. Pipes, W. O. Hamilton,
S. P. Boughn.
Phys. Rev. Letters, Vol. 28, 253 - 255 (1972).

**033.032 Ein Satelliten-Radioteleskop mit 1500-m-Spiegel für
die Langwellen-Radioastronomie.** H. Urbarz.
SuW, Vol. 11, 191 (1972).

**033.033 Long-baseline radio interferometry with indepen-
dent frequency standards.** W. K. Klemperer.
Proc. IEEE, Vol. 60, 602 - 609 (1972).
The present stage of developement of long-baseline interferometry using independent atomic frequency standards is
reviewed. The technique is being applied to precision geodesy
and astrometry and shows great promise as a means for the
intercomparison of national time scales at the 1-ns error level.
This paper discusses the development of radio interferometry
and, in particular, the theory and practice of interferometry
over a very wide (effective) bandwidth. The superb phase stability of the hydrogen maser frequency standard in the range
of 100 s to several hours makes it the ideal local oscillator for
long-baseline interferometry.

**033.034 Avalanche diodes as transfer noise standards for
microwave radiometers.** N. J. Keen.
Radio Electronic Engineer, Vol. 41, 133 - 136 = Max-Planck-
Inst. Radioastronomie Bonn, Sonderdruck No. 16 (1971).
Transfer calibrations of radiometers are performed by
injecting noise via a directional coupler. The avalanche diode
appears to have all of the physical attributes of the plasma
noise tube, except for a reasonably predictable absolute noise
level. Its saving in space, weight, cost and power consumption
render it an attractive proposition for transfer noise calibrations. Results of measurements at three frequencies between

1 GHz and 3 GHz are given.

033.035 On some errors of geodetic adjustment of large radio telescopes. Ya. V. Naumov.
Trudy TsNII geod., aehros"emki i kartogr., 1972, vyp. (No.) 169, p. 62 - 70. In Russian. – Abstr. in Referativ. Zhurn. 51. Astron., 7.51.83 (1972).

033.036 Interferometric measurement at 1415 MHz of radiation pattern of paraboloidal antenna at Dwingeloo Radio Observatory. A. P. Hartsuijker, J. W. M. Baars, S. Drenth, L. Gelato-Volders.
IEEE Trans. Antennas Propagation, Vol. AP-20, 166 - 176 (1972).

The complete radiation pattern of a 25-m diameter radio telescope has been measured at 21-cm wavelength to a level of 60 dB below the main beam response. Strong cosmic radio sources with known flux density were used as a signal source. The measurement employed an interferometer consisting of the radio telescope and a 7.5-m diameter reference antenna. The paper discussed the theoretical basis of the technique. It describes the layout of the interferometer, the electronic equipment, and the methods of observation and reduction of the data. Some 19 000 points in the pattern were measured. The pattern is displayed by contour plots of equal level.

033.037 Technik der Radioastronomie. R. Wielebinski.
Kleinheubacher Berichte, Vol. 14, 115 - 122 = Max-Planck-Inst. Radioastronomie Bonn, Sonderdruck No. 31 (1971).

This review paper describes the development of the techniques in radio astronomy.

033.038 Pointing calibration of a high-resolution millimeter-wave antenna by star observations. M. Burak.
U. S. Air Force Cambridge Res. Lab. Air Force Systems Command Microwave Phys. Lab. Phys. Sci. Res. Paper No. 450, AFCRL-71-0220, 61 pp. (1971).

An extensive pointing calibration program for the drive system and mount of the AFCRL 29-ft millimetre-wave antenna is described, the star tracking program, which includes a method of star selection, precession correction, and refraction correction, is described.-*ACM*

033.039 Microwave solar radiometer at 2800 MHz. K. Narayanan, R. V. Bhonsle.
Proc. Indian Acad. Sci., Ser. A, Vol. 74, 142 - 151 (1971).

A Dicke-type microwave radiometer has been developed for daily measurement of solar flux at 2800 MHz. The antenna system, consists of a 5 foot parabolic dish with horn feed, is equatorially mounted and is capable of tracking the sun for about 8 hours each day.

033.040 A transportable 16-GHz solar telescope for atmospheric transmissivity measurements. K. C. O'Brien, R. C. Heidt, G. J. Owens.
Radio Sci. (*USA*), Vol. 7, 215 - 221 (1972). – See Phys. Abstr., Vol. 75, No. 33772 (1972).

033.041 The use of a very fast routing algorithm for printed circuit board design. K. H. Hosking.
Marconi Rev., No. 182, Vol. 34, 207 - 226 (1971). – *ACM*

033.042 Spherical waves in antenna problems. P. J. Wood.
Marconi Rev., No. 182, Vol. 34, 149 - 172 (1971).

Two expansions analogous to the classical Taylor and Laurent series are discussed, with the spherical wave coefficients determined directly in terms of specified surface current sheets.-*ACM*.

033.043 Attenuation in corrugated rectangular waveguide. R. Baldwin, P. A. McInnes.
Electronics Letters, No. 26, Vol. 7, 770 - 772 (1971).

Attenuation characteristics of corrugated rectangular waveguides are discussed. As with circular corrugated waveguide, the attenuation is found to be significantly lower than that in smooth walled waveguide. – *DNC*

033.044 Miniature refrigerators for electronic devices. A. Daniels, F. K. Du Pré.
Philips Techn. Rev., Vol. 32, 49 - 56 (1971).

The article describes miniature low power stirling-cycle refrigerators. They are discussed with reference to infrared detectors where temperature down to 25 deg K may be required. – *DNC*

033.045 A new high-performance phase-sensitive detector. J. B. Grimbleby, D. W. Harding.
Journ. Phys. E, Sci. Instruments, Vol. 4, 941 - 944 (1971).

033.046 Design of normal-mode helix antennas. R. J. F. Guertler.
Proc. Instn. Radio Electronics Engineers Australia, Vol. 33, 23 - 24 (1972). – *NF*

033.047 Modified log periodic dipole antenna with plane reflector. M. D. Singh, S. P. Kosta.
International Journ. Electronics, Vol. 31, 565 - 572 (1971).

A theoretical analysis of the modified log-periodic dipole antenna (MLPD) with a plane reflector at its apex is presented.

033.048 Automatic control system for the solar patrol radio telescopes. A. Staniforth.
Bull. Radio Electr. Engineering Div. NRC Canada, Vol. 21, No. 2, p. 1 - 7 (1971).

A preliminary report on a computer controlled system which will operate the solar patrol radiometers (2700 MHz) automatically. When the system is fully operational the antennas will track the sun and the received signals will be calibrated three times daily. – *NF*

033.049 Evaluation of the Haystack antenna and radome. M. L. Meeks, J. Ruze.
IEEE Trans. Antennas Propagation, Vol. AP-19, 723 - 728 (1971).

Evaluation of the radome-enclosed Haystack antenna as a radio-astronomical instrument. Advantages of the radome enclosure include high pointing precision under all wind conditions, (pointing errors being 7 sec of arc in azimuth and elevation). – *MWS*

033.050 The development of an interferometer for millimeter wavelengths. R. Hills, W. Hoffman, M. A. Janssen, D. D. Thornton, S. Silver, W. J. Welch.
California Univ. Space Sci. Lab., Ser. 12, No. 64, 91 pp. (1971). – Annual report covering period July 1, 1970 through June 30, 1971.-*JWB*

033.051 The sun as a test source for boresight calibration of microwave antennas. W. Graf, R. N. Bracewell, J. H. Deuter, J. S. Rutherford.
IEEE Trans. Antennas Propagation, Vol. AP-19, 606 - 612 (1971). – *BMT*

033.052 Computer-aided design of parametric amplifiers. M. Maeda, A. Sumioka.
IEEE Trans. Mircrowave Theory Techn., Vol. MTT-19, 916 - 921 (1971). – *BMT*

033.053 Radiation patterns of circular apertures with structural shadows. C. J. E. Phillips, H. C. Minnett.

Proc. Instn. Electr. Engineers, Vol. 118, 1214 (1971). *— BMT*

033.054 A digital encoder of angular position using synchros and a digital computer.
A. J. Shimmins, D. J. Cole.
Proc. Instn. Radio Electronics Engineers Australia, Vol. 32, No. 4, p. 162 - 163 (1971).
 A description of the method used for the measurement of the pointing direction of the Parkes 210-ft radio telescope.

033.055 Metrology and radio performance of reflector of Chilbolton aerial. R. H. Slater.
Proc. Instn. Electr. Engineers, Vol. 118, 1691 - 1697 (1971).
 Methods for measuring the profile of the 25-m Chilbolton U. K. paraboloidal reflector are described. Reflector profiles, and measured radiation patterns at a wavelength of 3 cm are included. *— BMT*

033.056 The ring-loaded corrugated waveguide.
 Y. Takeichi, T. Hashimoto, F. Takeda.
IEEE Trans. Microwave Theory Techn., Vol. MTT-19, 947 - 951 (1971). *— BMT*

033.057 Digitized antenna measurements. J. Dijk, C. Kramer, E. J. Maanders, A. C. A. van der Vorst.
IEEE Trans. Microwave Theory and Techn., Vol. MTT—20, 48 - 51 (1972).
 A low-cost automated measuring method provides gain and phase of antenna feed systems in digital form. Output data may then be used as input data for computing secondary patterns of arbitrary reflectors. Stepping motors allow cheaper and easier operations. *— ACM*

033.058 Computation of the fast Fourier transform from data stored in external auxiliary memory for any general radix $\gamma = 2^n$, $n \geqslant 1$. M. Drubin.
IEEE Trans. Computers, Vol. C—20, 1552 - 1563 (1971).

033.059 Errors in S_{11} measurements due to the residual standing-wave ratio of the measuring equipment.
R. V. Garver, D. E. Bergfried, S. J. Raff, B. O. Weinschel.
IEEE Trans. Microwave Theory and Techn., Vol. MTT—20, 61 - 69 (1972). *— ACM*

033.060 A discrete point approach to the measurement of radiated power of planar apertures. A. C. Kak.
IEEE Trans. Microwave Theory and Techn., Vol. MTT—20, 74 - 81 (1972). *— ACM*

033.061 Maximum gain, mutual coupling and pattern control in array antennas.
J. F. McIlvenna, C. J. Drane, Jr.
Radio Electronic Engineer, Vol. 41, 569 - 572 (1971). *— ACM*

033.062 Applications of time-domain metrology to the automation of broad-band microwave measurements.
A. M. Nicolson, C. L. Bennett, Jr., D. Lamensdorf, L. Susman.
IEEE Trans Microwave Theory and Techn., Vol. MTT—20, 3 - 9 (1972). *— ACM*

033.063 Holographic approach to radiation pattern measurement. 1. General theory. R. H. T. Bates.

International Journ. Engineering Sci., Vol. 9, 1107 - 1121 (1971).
 The author develops a theory for the reconstruction of unknown source distributions from "Fraunhofer Holograms".

033.064 Holographic approach to radiation pattern measurement. 2. Experimental vertification.
P. J. Napier, R. H. T. Bates.
International Journ. Engineering Sci., Vol. 9, 1193 - 1208 (1971).
 Experiments performed with acoustic radiations to verify the theory developed in part I of the paper are discussed. Several methods of providing the reference source are presented. A possible application to source mapping in radio astronomy is suggested. *— DNC*

033.065 Amplificateurs paramétriques pour un interféromètre en radioastronomie.
R. Davies, R. E. Pearson.
Techn. Philips (*France*), No. 4, p. 21 - 34 (1971). — See Bull. Signal., Vol. 33, Section 120, No. 6976 (1972).

033.066 Conical-reflector antennas. A. C. Ludwig.
 IEEE Trans. Antennas Propagation, Vol. AP—20, 146 - 152 (1972). *— BMT*

033.067 Dielectric-loaded horn antenna. T. Satoh.
 IEEE Trans. Antennas Propagation, Vol. AP—20, 199 - 201 (1972). *— BMT*

033.068 Operating noise-temperature calibrations of low-noise receiving systems. C. T. Stelzried.
Microwave Journ., Vol. 14, 41 - 46, 48 (1971).
 Use is made of an ambient termination to calibrate the maser receivers used for *JPL/NASA* deep space communications systems. *— BMT*

033.069 Gain measurements of standard electromagnetic horns in the K and K_a bands.
G. T. Wrixon, W. J. Welch.
IEEE Trans. Antennas Propagation, Vol. AP—20, 136 - 142 (1972). *— BMT*

033.070 Les horloges atomiques dans l'interférométrie radioastronomique à base intercontinentale.
C. Audoin, P. Grivet.
Rev. Phys. Appliquée, Vol. 6, 247 - 254 (1971). — See Bull. Signal., Vol. 33, Section 120, No. 117 (1972).

033.071 Conception et réalisation d'un interféromètre en ondes décamétriques. M. Amaudric du Chaffaut.
Thesis Univ. Paris, Centre Documentation C.N.R.S. (1972-01-25), 112 pp. (1971). — See Bull. Signal., Vol. 33, Section 120, No. 4908 (1972).

Les radiotélescopes. See Abstr. 003.028.

Radio telescopes. See Abstr. 003.042.

Two new data reduction techniques for the intensity modulation method. See Abstr. 031.005.

034 Astronomical Accessories

034.001 **A nebular, two-etalon Fabry-Perot monochromator.**
J. Meaburn.
Astron. Astrophys., Vol. 17, 106 - 112 (1972).
The design, construction and use of a nebular two-etalon, propane scanned, photoelectric Fabry-Perot monochromator is described.

034.002 **Re-evaluation of the energy scale of the Michigan OSO III ion chamber soft X-ray photometer.**
R. G. Teske, P. W. Hodge, S. P. Worden.
Bull. American Astron. Soc., Vol. 3, 439 (1971). – Abstr. AAS.

034.003 **Surface absorptions recorded by the 1969 Mariner infrared spectrometer.**
K. C. Herr, P. B. Forney, G. C. Pimentel.
Bull. American Astron. Soc., Vol. 3, 466 - 467 (1971). – Abstr. AAS.

034.004 **Nebular photometry with an echelle spectrograph.**
T. J. Bohuski.
Bull. American Astron. Soc., Vol. 3, 481 (1971). – Abstr. AAS.

034.005 **Van meubelplaat tot astrokometencamera.**
B. Apeldoorn.
Hemel en Dampkring, Vol. 70, 56 - 59 (1972).

034.006 **Preliminary results from the X-ray experiment SL904.** D. J. Adams, A. F. Janes.
Journ. British Interplanet. Soc., Vol. 25, 201 - 208 (1972).
This paper describes the cosmic X-ray experiment made by Leicester University on Skylark 904 launched on 19 November 1970. Preliminary results are reported in the second half of the paper.

034.007 **The use of magnetometers on balloon-borne X-ray astronomy experiments.**
R. M. Thomas, P. J. N. Davison.
Planet. Space Sci., Vol. 20, 331 - 351 (1972).
The use of magnetometers for azimuth determination of balloon-borne X-ray astronomy payloads is examined, with special emphasis placed on the experiments of the University of Adelaide Cosmic Ray Group.

034.008 **Utilization of the electronic camera for the photometric measurement of close visual binaries.**
R. Despiau, P. Laques.
Astron. Astrophys., Vol. 18, 16 - 25 (1972). In French.
The electronic camera being a receiver well adapted to photometry, we have studied a method of measurement of the electronographic plates of double stars. We essentially measure the difference of magnitude (Δm) from the images registered on the plate. The measurement consists of exploring the images with a microdensitometer by using a 20 micron square hole, at the level of the plate.

034.009 **The AFCRL lunar laser ranging experiment in Arizona.** M. S. Hunt.
Sky Telescope, Vol. 43, 86 - 90 (1972).

034.010 **Study of the distribution of relative sensitivity over the photocathodes of photomultipliers.**
A. A. Rusak, E. I. Terez.
Izv. Krymskoj Astrofiz. Obs., Vol. 43, 206 - 212 (1971). In Russian.
Zonal characteristics of photomultipliers.

034.011 **Rotating wedge filter photometer for high altitude sounding rocket application.**
C. Holm, B. N. Maehlum, B. T. Narheim.
Applied Optics, Vol. 11, 421 - 427 (1972).
A scanning photometer is described, utilizing a rotating wedge interference filter as the wavelength scanning element around 6300 Å. A detailed description of the filter production is given, emphasizing the procedure for in situ wavelength control during fabrication. Subsequently, the complete photometer is briefly described, and some results from its applications on an auroral sounding rocket flight are presented.

034.012 **A rapid scanning radial velocity spectrometer.**
J. R. Stilborn, J. M. Fletcher, F. D. A. Hartwick.
Journ. Roy. Astron. Soc. Canada, Vol. 66, 49 - 52 = Contr. Dominion Astrophys. Obs., Victoria, No. 162 (1972).
A brief progress report on the construction and preliminary testing of a rapid scanning radial velocity spectrometer is given.

034.013 **An evaluation of cameras for photographing stars.**
R. Berry.
Journ. Roy. Astron. Soc. Canada, Vol. 66, L5 - L8 (1972).

034.014 **Astronomy: TV cameras are replacing photographic plates.** W. D. Metz.
Science, Vol. 175, 1448 - 1449 (1972).

034.015 **Electronic systems for the new multichannel spectrometer at Sacramento Peak.**
R. W. Hobbs, G. D. Harris, G. Epstein.
Publ. Astron. Soc. Pacific, Vol. 84, 74 - 83 (1972). – Paper presented at the symposium "Advanced Electronic Systems for Astronomy", 1971, Santa Cruz.

034.016 **The III-V photocathode: A major detector development.** W. E. Spicer, R. L. Bell.
Publ. Astron. Soc. Pacific, Vol. 84, 110 - 122 (1972). – Paper presented at the symposium "Advanced Electronic Systems for Astronomy", 1971, Santa Cruz.

034.017 **A calibration model for a stellar photometer using a SEC vidicon.** W. A. Deutschman.
Publ. Astron. Soc. Pacific, Vol. 84, 123 - 126 (1972). – Paper presented at the symposium "Advanced Electronic Systems for Astronomy", 1971, Santa Cruz.

034.018 **A digital image recorder.** H. G. Sachs.
Publ. Astron. Soc. Pacific, Vol. 84, 127 - 132 (1972). – Paper presented at the symposium "Advanced Electronic Systems for Astronomy", 1971, Santa Cruz.

034.019 **A silicon vidicon photometer.**
J. A. Westphal, T. B. McCord.
Publ. Astron. Soc. Pacific, Vol. 84, 133 = Contr. Planet. Astron. Lab., Dep. Earth Planet. Sci., Mass. Inst. Technology, No. 42 = Contr. Div. Geol. Planet. Sci., California Inst. Technology, No. 2073 (1972). – Paper presented at the symposium "Advanced Electronic Systems for Astronomy", 1971, Santa Cruz.

034.020 **The multianode photomultiplier.**
C. E. Catchpole, C. B. Johnson.
Publ. Astron. Soc. Pacific, Vol. 84, 134 - 136 (1972). – Paper presented at the symposium "Advanced Electronic Systems for Astronomy", 1971, Santa Cruz.

034.021 **Two-dimensional area scanning with image dissectors.** W. G. Tifft.
Publ. Astron. Soc. Pacific, Vol. 84, 137 - 144 (1972). – Paper presented at the symposium "Advanced Electronic Systems for Astronomy", 1971, Santa Cruz.

034.022 **An image-tube scanner for the Wisconsin echelle spectrograph.**
J. F. McNall, D. E. Michalski, T. L. Miedaner.
Publ. Astron. Soc. Pacific, Vol. 84, 145 - 148 (1972). – Paper presented at the symposium "Advanced Electronic Systems for Astronomy", 1971, Santa Cruz.

034.023 **Multichannel area photometry.** R. E. Nather.
Publ. Astron. Soc. Pacific, Vol. 84, 149 - 153 (1972). – Paper presented at the symposium "Advanced Electronic Systems for Astronomy", 1971, Santa Cruz.

034.024 **Single photoelectron excitation of phosphors.**
S. R. Smith, J. L. Lowrance.
Publ. Astron. Soc. Pacific, Vol. 84, 154 - 160 (1972). – Paper presented at the symposium "Advanced Electronic Systems for Astronomy", 1971, Santa Cruz.

034.025 **The Lick Observatory image-dissector scanner.**
L. B. Robinson, E. J. Wampler.
Publ. Astron. Soc. Pacific, Vol. 84, 161 - 166 = Lick Obs. Bull., No. 620 (1972). – Paper presented at the symposium "Advanced Electronic Systems for Astronomy", 1971, Santa Cruz.

034.026 **The automation of the Wisconsin echelle spectrograph.**
J. F. McNall, D. E. Michalski, T. L. Miedaner.
Publ. Astron. Soc. Pacific, Vol. 84, 176 - 181 (1972). – Paper presented at the symposium "Advanced Electronic Systems for Astronomy", 1971, Santa Cruz.

034.027 **A note on the calculation of photon loss in an image-tube scanner.** J. F. McNall.
Publ. Astron. Soc. Pacific, Vol. 84, 182 - 184 (1972). – Paper presented at the symposium "Advanced Electronic Systems for Astronomy", 1971, Santa Cruz.

034.028 **The computer-controlled spectrometers at McDonald Observatory.** D. C. Wells.
Publ. Astron. Soc. Pacific, Vol. 84, 203 - 206 (1972). – Paper presented at the symposium "Advanced Electronic Systems for Astronomy", 1971, Santa Cruz.

034.029 **Increasing the time resolution of a multichannel scaler.** E. C. Silverberg, C. A. Steggerda.
Publ. Astron. Soc. Pacific, Vol. 84, 217 - 219 (1972). – Paper presented at the symposium "Advanced Electronic Systems for Astronomy", 1971, Santa Cruz.

034.030 **A compact grating spectroheliograph for the MgII resonance lines.** B. Bates, M. W. McDowell.
Solar Physics, Vol. 23, 26 - 29 (1972).
A grating spectroheliograph with a spectral bandpass of 0.2 Å is described for operation in a rocket or balloon-borne payload. The same instrument can be made stigmatic for observations in the CaII resonance lines.

034.031 **Taratura di due filtri internazionali tipo Lyot-Öhman.**
R. Baldini, T. Grisendi, F. Mazzucconi, A. Righini.
Mem. Soc. Astron. Italiana, Nuova Ser., Vol. 43, 105 - 116 (1972).
After an introduction on the theory of Lyot-Öhman interferential filters, we discuss the results obtained in tuning the Arcetri Astrophysical Observatory filters, Zeiss type A and Halle n. 14.

034.032 **Descrizione del nuovo fotometro UBV dell' Osservatorio Astronomico di Bologna.** A. Piccioni.
Mem. Soc. Astron. Italiana, Nuova Ser., Vol. 43, 159 - 160 (1972). – Letter.

034.033 **The instrumental profile of the Herstmonceux 30-inch coudé spectrograph.**
R. A. E. Fosbury, C. F. W. Harmer.
Observatory, Vol. 92, 54 - 60 (1972). – Note.

034.034 **An automatic fast digital-photoelectric photometer with polarimeter.** N. Visvanathan.
Publ. Astron. Soc. Pacific, Vol. 84, 248 - 253 (1972).
The design of an automatic photoelectric photometer and polarimeter, along with a fast digital-recording system, is described.

034.035 **The vacuum method of making corrector plates.**
R. E. Cox.
Sky Telescope, Vol. 43, 388 - 393 (1972).

034.036 **Comparison of some methods for determining the values of a revolution of a position contact micrometer of an astronomical universal instrument.**
A. V. Gozhij.
Geod., kartogr. i aehrofotos"emka, Mezhved. resp. nauch.-tekhn. sb., 1971, vyp. (No.) 13, p. 23 - 27. In Russian.
Abstr. in Referativ. Zhurn. 52. Geod. Aehros"emka, 5.52.137 (1972).

034.037 **Applications of the Dollfus polarization modulator in stellar polarimetry.** J. Tinbergen.
IAU Colloquium No. 15, (see 012.006), p. 36 - 40 (1972).

034.038 **The design of a spot sensitometer for astronomical use.** C.-C. Wu, C. M. Anderson, J. D. Rosendhal, L. W. Marty, D. W. Bucholtz.
AAS Photo-Bull. 1972, No. 1, p. 9 - 12.
Several aspects of the design of a portable spot sensitometer used for the calibration of astronomical spectrograms are discussed.

034.039 **A method of constructing large composite gelatin filters.** J. J. Schreur.
AAS Photo-Bull. 1972, No. 1, p. 15 - 16.

034.040 **A sensitometer box for solar corona photometry.**
V. Pizzo, J. T. Gosling.
AAS Photo-Bull. 1972, No. 1, p. 19 - 21.

034.041 **Générateurs transistorisés pour l'entraînement électrique des équatoriaux.**
G. Florsch, J. Ballèvre.
L'Astronomie, 86ᵉ année, p. 189 - 195 (1972).

034.042 **The photoelectric photometer of the Cluj Astronomical Observatory. Determination of the extinction coefficients and of the system constants.**
I. Todoran, V. Ureche, V. Pop, D. Chiş.
Stud. Cerc. Astron., Vol. 17, 33 - 43 (1972). In Romanian.

034.043 **Observing with a digicon.** E. A. Beaver.
Bull. American Astron. Soc., Vol. 4, 215 (1972).
Abstr. AAS.

034.044 **The use of magnetometers and asymmetric antenna patterns in flight vehicle orientation analysis.**
T. E. Graedel, A. N. Delfico, R. Pringle, Jr.

Bull. American Astron. Soc., Vol. 4, 218 - 219 (1972). – Abstr. AAS.

034.045 X-ray observation of Vir X-1 from 1 to 60 keV with the MIT instrument on OSO-7.
G. Clark, G. Sprott, W. Lewin, H. Bradt, H. Schnopper.
Bull. American Astron. Soc., Vol. 4, 258 (1972). – Abstr. AAS.

034.046 Cosmic-ray induced radioactivity in scintillation detectors. G. J. Fishman, T. A. Parnell.
Bull. American Astron. Soc., Vol. 4, 259 (1972). – Abstr. AAS.

034.047 Cameras on the moon with Apollos 15 and 16.
T. Page.
Mercury, (Journ. Astron. Soc. Pacific), Vol. 1, No. 2, p. 4 - 11 (1972).

034.048 An inexpensive canvas dome for an 8-inch telescope.
S. S. Vogt.
Sky Telescope, Vol. 44, 31 - 32 (1972).

034.049 A novel drive on a 20-inch telescope.
R. Tuthill.
Sky Telescope, Vol. 44, 47 - 51 (1972).

034.050 Energy calibration of the "Orion" spectrograph.
G. A. Gurzadyan, J. Ohanesyan, P. A. Yepremyan.
Soobshch. Byurakan. Obs., No. 45, p. 36 - 41 (1972).
In Russian.

034.051 Langzeit-Astrophotographie mit Offset-Guiding für Amateure. M. Lammerer.
Orion Schaffhausen, 30. Jahrgang, p. 93 - 97 (1972).

034.052 Nuovo metodo e strumentazione relativa per il rilevamento del tempo di occultazione lunare.
F. Rosset.
Coelum, Vol. 40, 100 - 105 (1972).

034.053 The design of an infra-red interferometer.
K. Harwood.
Proc. Astron. Soc. Australia, Vol. 2, 113 - 114 (1972).

034.054 A simple vacuum system substrate heater.
R. J. Mattauch.
Rev. Sci. Instruments, Vol. 43, 148 - 149 = National Radio Astron. Obs., *Green Bank*, Repr. Ser. A, No. 236 (1972).

034.055 Bau und Erprobung eines 9-Kanal-Photometers.
G. V. Schultz, W. Wiemer.
Mitt. Astron. Ges., No. 31, p. 174 - 176 (1972).

034.056 Infrarotphotometer für $\lambda \geqslant 8\,\mu$m. H. Hefele.
Mitt. Astron. Ges., No. 31, p. 176 - 177 (1972).

034.057 Method of calculating the lag of a phase-photoelectric device of the time service. M. I. Malyshev.
Astron. Zhurn. Akad. Nauk SSSR, Vol. 49, 624 - 629 (1972).
In Russian. English translation in Soviet Astron. AJ, Vol. 16, No. 3.

The lag of a phase-photoelectric device depends on the arrival signal shape and on the pass band of the narrow-band amplifier. Formulae for the calculation of the lag of signal with rectangular and trapeziform complex amplitudes were obtained. They are analogous to Pavlov's formulae for the lag of signals in the RC-filter of a direct current amplifier.

034.058 Ein Fabry-Perot-Interferometer zur Messung von Radialgeschwindigkeiten an H II-Regionen.
H.-H. Hippelein.
Inaugural-Diss. Naturwiss. Gesamtfakultät der Ruprecht-Karl-Univ. Heidelberg. 4 + 54 pp. (1972).

034.059 Five spectrograph camera designs. C. G. Wynne.
Monthly Notices Roy. Astron. Soc., Vol. 157, 403 - 418 (1972).

The cameras described are wide aperture two-mirror systems, suitable for use with image tube receptors, and with better aberration correction than can be achieved with Maksutov cameras. The first two cameras described, of focal length 25 cm and relative aperture f/1.67, have air-spaced mirrors, in one case with two and the other with three lens elements. The remaining three cameras, 19 cm f/1.25 systems of semi-solid construction, consist of two with spherical surfaced lenticular correctors, with a Schmidt–Cassegrain design for comparison.

034.060 Computer simulation of an air-borne gamma-ray spectrometer.
R. B. Clark, J. S. Duval, Jr., J. A. S. Adams.
Journ. Geophys. Res., Vol. 77, 3021 - 3031 (1972).

A computer model is presented that simulates the data from a flight of an air-borne gamma-ray spectrometer over a surface containing gamma-ray sources of arbitrary horizontal dimension and coordinates. The mathematical formalism is given, together with a discussion of the assumptions or approximations that are made. Representative computational results are shown for simulated flights over circular sources and across linear sources.

034.061 Far infrared filters for a rocket-borne radiometer.
H. V. Romero, J. Gursky, A. G. Blair.
Applied Optics, Vol. 11, 873 - 880 (1972).

The paper describes the theory, fabrication, and laboratory transmission measurements of prototype grid filters investigated in a study prior to the construction of flight filters. Characteristics of the final flight filters are also presented.

034.062 Alignment of Michelson stellar interferometer.
R. H. Miller.
Applied Optics, Vol. 11, 955 - 956 (1972).

034.063 URSIES: an ultravariable resolution single interferometer echelle scanner.
A. A. Wyller, T. Fay.
Applied Optics, Vol. 11, 1152 - 1162 (1972).

A Fabry-Perot interferometer in a Ramsay mount is used in tandem with an echelle Hilger monochromator with pinholes instead of slits. The instrument, URSIES, is enclosed within a pressure chamber filled with Freon. Photoelectric pulse counting techniques and pressure scanning are used to record the spectrum. Measurements demonstrating the advantages of the design are discussed.

034.064 750-mm ruling engine producing large gratings and echelles.
G. R. Harrison, S. W. Thompson, H. Kazukonis, J. R. Connell.
Journ. Optical Soc. America, Vol. 62, 751 - 756 (1972).

034.065 An efficient coudé spectrograph system.
E. H. Richardson, G. A. Brealey, R. Dancey.
Publ. Dominion Astrophys. Obs., Victoria, Vol. 14, (No. 1), 1 - 15 (1971).

The amount of light that enters the coudé spectrograph of the 48-inch telescope at Victoria has been increased by about four times by changes made to the telescope. First, the amount has been more than doubled by the use of image slicers. Then an increase of at least 40% over the light reflected by aluminum-coated mirrors has been obtained by the use of high-reflectance coatings on the mirrors that reflect light to the coudé focus.

034.066 Luminosité et efficacité des spectrographes à fente.

C. Fehrenbach.
Auxiliary instrumentation for large telescopes. Conference 1972, (see 012.018), p. 99 - 109 (1972).

034.067   Intermediate dispersion Cassegrain spectrograph for the AAT.   D. R. Palmer, J. W. Gietzen.
Auxiliary instrumentation for large telescopes. Conference 1972, (see 012.018), p. 111 - 117 (1972).

034.068   Échelle spectrographs.   D. J. Schroeder.
Auxiliary instrumentation for large telescopes. Conference 1972, (see 012.018), p. 119 - 130 (1972).

034.069   Échelle grating spectrographs for the Cassegrain focus of the AAT.   R. C. M. Learner.
Auxiliary instrumentation for large telescopes. Conference 1972, (see 012.018), p. 131 - 140 (1972).

034.070   Spectrographic cameras of the Maksutov and related types.   C. G. Wynne.
Auxiliary instrumentation for large telescopes. Conference 1972, (see 012.018), p. 141 - 148 (1972).

034.071   A new Cassegrain grating spectrograph. Part 1: Optical design.   R. N. Wilson, A. Opitz.
Auxiliary instrumentation for large telescopes. Conference 1972, (see 012.018), p. 149 - 155 (1972).

034.072   A new Cassegrain grating spectrograph. Part 2: Mechanical design and operation.   R. Schlegelmilch.
Auxiliary instrumentation for large telescopes. Conference 1972, (see 012.018), p. 157 - 163 (1972).

034.073   A computer-controlled digital spectrum scanner for La Silla.   H. J. Wood.
Auxiliary instrumentation for large telescopes. Conference 1972, (see 012.018), p. 165 - 168 (1972).

034.074   Some features of the Leiden radial velocity instrument.   T. Walraven, J. H. Walraven.
Auxiliary instrumentation for large telescopes. Conference 1972, (see 012.018), p. 175 - 183 (1972).

034.075   Photoelectric determination of radial velocities. L. Karsten.
Auxiliary instrumentation for large telescopes. Conference 1972, (see 012.018), p. 185 - 192 (1972).

034.076   Diffraction gratings for large telescopes. E. G. Loewen.
Auxiliary instrumentation for large telescopes. Conference 1972, (see 012.018), p. 193 - 202 (1972).

034.077   Holo-gratings and spectrograph design. A. Labeyrie, J. Flamand, G. Pieuchard.
Auxiliary instrumentation for large telescopes. Conference 1972, (see 012.018), p. 203 - 208 (1972).

034.078   Stellar spectroscopy with holographic gratings. G. Schmahl, D. Rudolph.
Auxiliary instrumentation for large telescopes. Conference 1972, (see 012.018), p. 209 - 215 (1972).

034.079   Mosaique de réseaux.   A. Bayle, J. Espiard.
Auxiliary instrumentation for large telescopes. Conference 1972, (see 012.018), p. 217 - 222 (1972).

034.080   Équipement spectrographique du foyer coudé du télescope de 3.60 mètres. Etude de faisabilité d'un spectrographe universel.   A. Baranne.
Auxiliary instrumentation for large telescopes. Conference

1972, (see 012.018), p. 227 - 239 (1972).

034.081   Le spectrographe coudé "Echel.E.C. 152" (Montage à pupille blanche pour caméra électronique).
A. Baranne, M. Duchesne.
Auxiliary instrumentation for large telescopes. Conference 1972, (see 012.018), p. 241 - 245 (1972).

034.082   Design study of the coudé spectrographs for the 2.2-m telescopes of the MPI for Astronomy.
K. Bahner, J. Solf.
Auxiliary instrumentation for large telescopes. Conference 1972, (see 012.018), p. 247 - 257 (1972).

034.083   The coudé spectrograph and echelle scanner of the 2.7 m telescope at McDonald Observatory.
R. G. Tull.
Auxiliary instrumentation for large telescopes. Conference 1972, (see 012.018), p. 259 - 274 (1972).

034.084   Image slicers for image-tube spectrographs. E. H. Richardson.
Auxiliary instrumentation for large telescopes. Conference 1972, (see 012.018), p. 275 - 284 (1972).

034.085   A crossed field fast scanning image dissector. C. D. Johnson, V. A. Stanley.
Auxiliary instrumentation for large telescopes. Conference 1972, (see 012.018), p. 291 - 294 (1972).

034.086   Performance of the UCL image photon counting system.   A. Boksenberg.
Auxiliary instrumentation for large telescopes. Conference 1972, (see 012.018), p. 295 - 316 (1972).

034.087   An integrating television system for astronomy. D. C. Morton.
Auxiliary instrumentation for large telescopes. Conference 1972, (see 012.018), p. 317 - 331 (1972).

034.088   A photoelectron-counting spectrophotometer. R. W. Airey, F. C. Delori, J. D. McGee, B. L. Morgan.
Auxiliary instrumentation for large telescopes. Conference 1972, (see 012.018), p. 333 - 343 (1972).

034.089   Les appareils de haute résolution au foyer des grands télescopes.   P. Bouchareine.
Auxiliary instrumentation for large telescopes. Conference 1972, (see 012.018), p. 349 - 356 (1972).

034.090   High resolution Michelson interferometry in the visible.   J. Ring, C. L. Stephens, R. C. Wayte.
Auxiliary instrumentation for large telescopes. Conference 1972, (see 012.018), p. 357 - 365 (1972).

034.091   A high precision Fourier spectrometer for the visible. J. Brault.
Auxiliary instrumentation for large telescopes. Conference 1972, (see 012.018), p. 367 - 373 (1972).

034.092   Spectrographes à fente longue (reducteurs focaux - spectrographes Perot-Fabry - spectrographes à réseau).   G. Monnet.
Auxiliary instrumentation for large telescopes. Conference 1972, (see 012.018), p. 375 - 387 (1972).

034.093   Speckle interferometer for 0".02 stellar resolution. A. Labeyrie.
Auxiliary instrumentation for large telescopes. Conference 1972, (see 012.018), p. 389 - 393 (1972).

034.094 **The application of electronography in astronomical photometry.** M. F. Walker.
Auxiliary instrumentation for large telescopes. Conference 1972, (see 012.018), p. 399 - 419 (1972).

034.095 **Photométrie d'astres trés faibles par electronographie.** G. Wlérick, avec la collaboration technique de D. Michet.
Auxiliary instrumentation for large telescopes. Conference 1972, (see 012.018), p. 421 - 431 (1972).

034.096 **Electronographic image tubes for stellar field photometry.** D. McMullan.
Auxiliary instrumentation for large telescopes. Conference 1972, (see 012.018), p. 433 - 443 (1972).

034.097 **Les spectrographes et l'électronographie.**
M. Combes, P. Felenbok, J. P. Picat, B. Fort, R. Cayrel.
Auxiliary instrumentation for large telescopes. Conference 1972, (see 012.018), p. 445 - 453 (1972).

034.098 **Photographic attachment types 24 and 30.**
C. Kühne.
Auxiliary instrumentation for large telescopes. Conference 1972, (see 012.018), p. 457 - 462 (1972).

034.099 **Achromatic polarization modulators for multichannel polarimeters.** J. Tinbergen.
Auxiliary instrumentation for large telescopes. Conference 1972, (see 012.018), p. 463 - 471 (1972).

034.100 **The image tube nebular spectrograph of the Asiago Observatory.** F. Bertola.
Auxiliary instrumentation for large telescopes. Conference 1972, (see 012.018), p. 473 - 475 (1972).

034.101 **Realization of an automatic guiding system.**
J.-L. Poncet, P. Bartholdi.
Auxiliary instrumentation for large telescopes. Conference 1972, (see 012.018), p. 479 - 483 (1972).

034.102 **Observer's facilities at the Cassegrain focus.**
J. C. Farrell.
Auxiliary instrumentation for large telescopes. Conference 1972, (see 012.018), p. 485 - 491 (1972).

034.103 **A proposed standard mechanical interface for Cassegrain acquisition-guider heads.**
I. G. van Breda, P. W. Hill.
Auxiliary instrumentation for large telescopes. Conference 1972, (see 012.018), p. 493 - 498 (1972).

034.104 **Das neue Göttinger Photometer—ein prozeßrechnergesteuertes Mikrophotometer.** H. Wöhl.
SuW, Vol. 11, 187 - 188 (1972).

034.105 **The new high-resolution scanning equipment attached to the Coudé spectrograph of the 188-cm reflector.** E. Watanabe, M. Yutani, T. Noguchi.
Tokyo Astron. Obs., Report No. 59, Vol. 15, 813 - 819 (1972). In Japanese.

034.106 **Synchronous three-color photometer with a sampling circuit.** Y. Shimizu, Y. Norimoto.
Tokyo Astron. Obs., Report No. 59, Vol. 15, 820 - 824 (1972). In Japanese.

034.107 **The computer controlled micro-photometer at the Lund Observatory.** G. Lyngå, G. Arinder.
Rep. Obs. Lund, No. 4, 15 pp. (1972).

034.108 **A photoheliograph.**
G. S. Minasjants, S. O. Obashev.
Solnechnye Dannye 1972 Byull., No. 3, p. 98 - 100 (1972). In Russian.
A photoheliograph has been designed and mounted at the High-Altitude Solar Station (Alma-Ata peak). The photoheliograph circuit with instrumental distortion factors reduced to minimum permits to take photographs of the photosphere with resolution of $0\overset{''}{.}8$.

034.109 **A program automatic device for the AFR-2 chromospheric telescope.** V. S. Degtjarev.
Solnechnye Dannye 1972 Byull., No. 3, p. 100 - 108 (1972). In Russian.

034.110 **Use of far-ultraviolet photomultipliers as gamma-counters.** J. P. Macau.
Nuclear Instruments and Methods, Vol. 82, 317 - 318 = Univ. Liège, Inst. d'Astrophys., Coll. 4°, No. 204 (1970).

034.111 **Photometric analyser for television cameras.**
H. Bredohl, S. Gardier, J. Henrion.
Journ. Phys. E, Sci. Instruments, Vol. 3, 681 - 684 = Univ. Liège, Inst. d'Astrophys., Coll. 4°, No. 208 (1970).
A new, fairly inexpensive method of extracting information from a composite video signal is presented.

034.112 **Quantitative evaluation of unwanted doubly diffracted radiation in Ebert-Fastie spectrometers.**
G. Marette.
Optics Commun., Vol. 4, No. 1, p. 33 - 34 = Univ. Liège, Inst. d'Astrophys.,Coll. 4°, No. 220 (1971).
The theoretical predictions for double dispersion and related intensities in Ebert-Fastie spectrometers are illustrated by means of a FORTRAN program written for a quantitative treatment.

034.113 **Lumière parasite dans les spectromètres Czerny-Turner à faisceaux croisés.** G. Marette.
Bull. Soc. Roy. Sci., Liège, Vol. 38, 687 - 694 = Univ. Liège, Inst. d'Astrophys., Coll. 8°, No. 592 (1969).
Wavelength distribution of stray light due to double dispersions has been calculated and observed for a crossed beam Czerny-Turner grating monochromator and an attempt made to give the ratio in order of magnitude of the stray light to first order signals.

034.114 **Lumière parasite dans les spectromètres Ebert-Fastie à simple passage.** G. Marette.
Bull. Cl. Sci. Acad. Roy. Belgique. 5e Sér., Vol. 56, 310 - 321 = Univ. Liège, Inst. d'Astrophys., Coll. 8°, No. 598 (1970).
The wavelength distribution and relative intensity of stray radiation due to double dispersions have been calculated for single-pass in-plane Ebert-Fastie grating monochromators in order to correct previous spectroscopic results.

034.115 **New calibration method for prism infrared spectrometers.** S.-Y. Ho.
Applied Optics, Vol. 10, 1584 - 1586 = Separate print Dep. Phys. Astron., Univ. Mass., Amherst (1971).

034.116 **Spectroelectrophotometer for atmospheric-optical measurements in the near infrared spectral region.**
A. I. Ivanov, B. T. Tashenov, I. A. Fedulin.
Trudy Astrofiz. Inst., *Alma-Ata,* Vol. 18, 63 - 66 (1971). In Russian.

034.117 **Spectropolarimeter of the daytime sky for the ultraviolet spectral region.**
P. N. Bojko, V. E. Pavlov, Ya. A. Tejfel'.

Trudy Astrofiz. Inst., *Alma-Ata*, Vol. 18, 67 - 69 (1971). In Russian.

034.118 **Determination of the instrumental profile of the DFS-13 diffraction spectrograph.** G. A. Kirienko. Trudy Astrofiz. Inst., *Alma-Ata*, Vol. 18, 137 - 140 (1971). In Russian.

034.119 **Stray light suppression in sunshields and optical systems: Measurements on three space experiments.** C. Leinert. Dudley Obs., *Albany, New York*, Rep. No. 6, 6 + 45 pp. (1971).

The stray light suppression of three photometric space experiments was determined by laboratory measurements. The instruments involved are the Pioneer F/G Imaging Photopolarimeter, the Skylab zodiacal light photometer and a rocket photometer for observations of noctilucent clouds. The method and the results of the measurements are described in detail and the implications for the operation of the experiments are discussed.

034.120 **Die Registrierung der Zeit mit einem Magnetbandgerät.** H. J. Schober. Internationale Elektron. Rundschau, Jahrgang 26, p. 125 - 126 = Mitt. Univ.-Sternw. Graz, No. 11 (1972).

034.121 **Mikrometrische Registrierung der Almukantaratdurchgänge von Sternen mit dem Zirkumzenital.** K.-G. Steinert. Wiss. Zeitschr. Techn. Univ. Dresden, Vol. 19, 109 - 112 = Mitt. Lohrmann-Obs. Techn. Univ. Dresden, No. 22 (1970).

034.122 **Protuberanskikkerten.** P. Darnell. Astron. Tidssk., Årg. 5, p. 88 - 92 (1972).

034.123 **Investigation of the camera with elastic film of the Panajotov system at the Shemakha Astrophysical Observatory of the Academy of Sciences of the Azerbajdzhan SSR.** T. A. Guseva. Soobshch. Shemakhinsk. Astrofiz. Obs., vyp. (No.) 6, p. 9 - 16 (1971). In Russian.

034.124 **Bibliography of magnetometers.** P. H. Serson, F. Primdahl. Publ. Earth Phys. Branch, Ottawa, Vol. 43, 501 - 506 (1972).

034.125 **Superconducting magnetic spectrometer for cosmic ray nuclei.** L. H. Smith, A. Buffington, M. A. Wahlig, P. Dauber. Rev. Sci. Instruments, Vol. 43, 1 - 11 (1972).

Describes the design, calibration, and operation of a magnetic spectrometer for particle astronomy.

034.126 **SAS-B digitized spark chamber gamma ray telescope.** S. M. Derdeyn, C. H. Ehrmann, C. E. Fichtel, D. A. Kniffen, R. W. Ross. Nuclear Instrum. & Meth. (*Netherlands*), Vol. 98, 557 - 566 (1972). – See Phys. Abstr., Vol. 75, No. 33721 (1972).

034.127 **Two-dimensional silicon vidicon astronomical photometer.** T. B. McCord, J. A. Westphal. Applied Optics, Vol. 11, 522 - 526 (1972).

Describes an integrating two-dimensional silicon diode array vidicon photometer which is exceptionally well suited for use with telescopes.

034.128 **SHIRX-a high-resolution rocket-borne X-ray spectrometer.** S. Singer, W. P. Aiello, J. A. Bergey, F. J. Edeskuty. K. D. Williamson.

IEEE Trans. Nuclear Sci., Vol. NS−19, 626 - 631 (1972).

A rocket-borne X-ray spectrometer is described which used cooled lithium-drifted semiconductor detectors to measure the differential X-ray energy spectrum of Sco XR-1 in the range 0.6−20 keV.

034.129 **Design of the solar X-ray instrument for OSO-H.** T. M. Harrington, J. O. Maloy, D. L. McKenzie, L. E. Peterson. IEEE Trans. Nuclear Sci., Vol. NS−19, 596 - 605 (1972). See Phys. Abstr., Vol. 75, No. 30609 (1972).

034.130 **A gamma ray monitor for the OSO-7 spacecraft.** P. R. Higbie, E. L. Chupp, D. J. Forrest, I. U. Gleske. IEEE Trans. Nuclear Sci., Vol. NS−19, 606 - 612 (1972). See Phys. Abstr., Vol. 75, No. 30610 (1972).

034.131 **A high-resolution, multiple particle spectrometer for the measurement of solar particle events.** J. B. Reagan, J. C. Bakke, J. R. Kilner, J. D. Matthews, W. L. Imhof. IEEE Trans. Nuclear Sci., Vol. NS−19, 554 - 561 (1972).

The spectrometer consists of a multi-element, silicon detector stack mounted within a plastic scintillator guard counter to form a particle telescope.

034.132 **An all-sky X-ray monitor for the UK-5 satellite.** U. D. Desai, S. S. Holt. IEEE Trans. Nuclear Sci., Vol. NS−19, 592 - 595 (1972).

An X-ray imaging system using a pin-hole camera and a position sensitive proportional counter at the image plane has been designed for an all-sky X-ray monitor for UK-5 satellite.

034.133 **Rocket-borne celestial X-ray proportional counter system.** G. A. Burginyon, C. M. Cornell, R. W. Hill, A. Tantuono. Nuclear Instruments & Meth. (*Netherlands*), Vol. 96, 461 - 468 (1971). – See Phys. Ber., Vol. 51, No. 3−4606 (1972).

034.134 **Pointing error correction for millimeter wave spectroheliograms.** M. J. Neary. U.S. Air Force Cambridge Res. Lab. Air Force Systems Command Microwave Phys. Lab. Phys. Sci. Res. Papers No. 452. AFCRL−71−0236, 22 pp. (1971).

Description of the derivation of solar spectroheliograms at 8.6 mm, for solar forecasting purposes. − *MWS*

034.135 **Lunar orbital mass spectrometer.** J. H. Hoffmann. International Journ. Mass Spectromet. and Ion Phys. (*Netherlands*), Vol. 8, 403 - 416 (1972). – See Phys. Abstr., Vol. 75, No. 41174 (1972).

034.136 **Stellar interferometer.** P. L. Kebabian. Quarterly Progr. Rep. Res. Lab. Electronics, Mass. Inst. Technology (*USA*), No. 103, p. 20 - 34 (1971).

Describes the construction of the signal processor now nearing completion.

034.137 **Filters for solar observing.** L. A. Jones. Journ. Astron. Soc. Western Australia, Vol. 36, March, p. 3 - 6 (1972).

034.138 **A large area gas Čerenkov telescope for high energy gamma-ray astronomy.** P. Albats, S. E. Ball, J. P. Delvaille, K. I. Greisen, D. G. Koch, B. McBreen, G. G. Fazio, D. R. Hearn, H. F. Helmken. Nuclear Instruments and Methods (*Netherlands*), Vol. 95, 189 - 194 (1971). – See Bull. Signal., Vol. 33, Section 120, No. 478 (1972).

034.139 A balloon borne instrument for the study of cosmic X-rays.
P. C. Agrawal, V. S. Iyengar, M. A. Kalgaonkar, A. P. Kamat, P. K. Kunte, R. K. Manchanda, K. V. Srinivasan.
Nuclear Instruments and Methods (*Netherlands*), Vol. 95, 29 - 37 (1971). – See Bull. Signal., Vol. 33, Section 120, No. 479 (1972).

034.140 The technique of mass determination of slow cosmic ray particles using a large aperture range telescope.   F. Ashton, H. J. Edwards, G. N. Kelly.
Nuclear Instruments and Methods (*Netherlands*), Vol. 95, 109 - 118 (1971). – See Bull. Signal., Vol. 33, Section 120, No. 480 (1972).

034.141 Large-area pulse ionization chamber for measurement of extremely heavy cosmic rays.
J. W. Epstein, J. I. Fernandez, M. H. Israel, J. Klarmann, R. A. Mewaldt, W. R. Binns.
Nuclear Instruments and Methods (*Netherlands*), Vol. 95, 77 - 85 (1971). – See Bull. Signal., Vol. 33, Section 120, No. 481 (1972)

034.142 A detector for determining the direction and energy of fast neutrons in cosmic radiation.
S. Biswas, N. Durgaprasad, P. J. Kajarekar.
Nuclear Instruments and Methods (*Netherlands*), Vol. 95, 69 - 76 (1971). – See Bull. Signal., Vol. 33, Section 120, No. 482 (1972).

034.143 Alpha-particle densitometer with variable response.   W. Brandt, M. D. d'Agostino, A. J. Favale.
Nuclear Technol. (*USA*), Vol. 11, 99 - 104 (1971). – See Bull. Signal., Vol. 33, Section 120, No. 1315 (1972).

034.144 Aberrations of spectrographs with the correction of the Schmidt camera effected in the collimator.
J. M. Simon, L. R. de Novarini, R. Platzeck.
Optica Acta, Vol. 18, 829 - 837 (1971).

034.145 Ein Bildverstärker mit SEC-Vidicon für astronomische Anwendungen. II.   J. D. Schumann.
Messtechnik (Zeitschr. Instrumentenkunde), Vol. 79, 189 - 195 (1971).

034.146 A digital multichannel photometer.
E. A. Beaver, C. E. McIlwain.
Rev. Sci. Instruments, Vol. 42, 1321 - 1324 (1971).

034.147 On the precision of momentum measurements with a magnetic spectrometer.   A. Frohlich.
Nuclear Instruments & Methods, Vol. 97, 315 - 318 (1971). – See Bull. Signal., Vol. 33, Section 120, No. 6245 (1972).

034.148 Développement d'un nouveau modèle de caméra électronique Lallemand à focalisation électrostatique.   J. Guerin.
Thesis, Univ. Paris. 84 pp. (1971). – See Bull. Signal., Vol. 33, Section 120, No. 6975 (1972).

Optimum photomultiplier photon counting photometry.   See Abstr. 031.011.

The large space telescope instrumentation.
See Abstr. 032.008.

Advanced X-ray observatories.
See Abstr. 051.007.

The SAS-D ultraviolet astronomical satellite.
See Abstr. 054.009.

The metal-to-hydrogen ratio in F1–F5 stars, as determined by a model-atmosphere analysis of photoelectric observations of a group of weak metal lines.
See Abstr. 064.043.

Transportable lunar-ranging system.
See Abstr. 094.099.

035 Clocks and Frequency Standards

035.001 Spectre de phase et stabilité à très court terme des oscillateurs.
J. J. Gagnepain, G. Marianneau, M. Olivier.
Ann. Françaises Chronométrie Micromécanique, Année 1971, p. 43 - 47. – Communication présentée au Congrès de Chronométrie Franco-Allemand–Constance 1970.

035.002 Amélioration de la stabilité à court terme des oscillateurs à quartz.   M. Olivier, G. Marianneau.
Ann. Françaises Chronométrie Micromécanique, Année 1971, p. 115 - 120. – Communication présentée au Congrès de Chronométrie Franco-Allemand, Grenoble 1971.

035.003 Invertor circuits and motors for clock drives.
R. A. Knox.
Journ. British Astron. Ass., Vol. 82, 182 - 187 (1972).

035.004 Eine richtiggehende Sonnenuhr.   H. Schlüter.
Orion Schaffhausen, 30. Jahrgang, p. 51 - 53 (1972).

035.005 Method and apparatus for comparison of chronometers with second radio signals of exact time on field conditions.   E. M. Vinnikov, S. S. Tovchigrechko.
Geod., kartogr. i aehrofotos"emka. Mezhved. resp. nauch.-tekhn. sb., 1971, vyp. (No.) 13, p. 15 - 22. In Russian.
Abstr. in Referativ. Zhurn. 51. Astron., 6.51.229 (1972).

035.006 Hunting sundials.   R. N. Mayall.
Sky Telescope, Vol. 44, 14 - 15 (1972).

035.007 Eine analemmatische Doppelsonnenuhr. – Wer führt sie praktisch aus?   F. Heger.
Sternebote, 15. Jahrgang, p. 2 - 10 (1972).

035.008 A system for automatic time synchronization based on the timing capability of the Loran-C navigational network.   G. P. R. Netzler.
Onsala Space Res. Obs., Res. Lab. Electronics, Chalmers Univ. Techn., Gothenburg, Sweden, Res. Rep. No. 104, 3 + 42 pp.

(1971).

This report describes some new concepts in dealing with the circuitry for Loran-C automatic timing systems. Thus the conventional analogue techniques associated with phase shifting networks have been replaced by an incremental digital phase shift device. The format identification and decoding equipment together form a system which takes into account the information of every Loran-C pulse. For VLBI purposes and transcontinental use, the accuracy of this system will be better than 1μ s, when post corrections, supplied by the US Naval Observatory, are taken into account.

035.009 Precise clock comparison by Loran-C signals.
K. Fujiwara, T. Hara, T. Sakai.
Tokyo Astron. Obs., Report No. 59, Vol. 15, 825 - 837 (1972). In Japanese.

035.010 Path delay, its variations, and some implications for the field use of precise frequency standards.
G. M. R. Winkler.
Proc. IEEE, Vol. 60, 522 - 529 (1972).

In order to assess the practical utility of high-performance frequency standards in field applications, the effect of the propagation medium must be taken into account. After a brief summary of noise processes, a discussion of the performance of current frequency standards is given, together with an estimate of the relative cost, reliability, and necessary system support for each type.

035.011 Precise time and frequency dissemination via the Loran-C system. C. E. Potts, B. Wieder.
Proc. IEEE, Vol. 60, 530 - 539 (1972).

035.012 VLF timing: Conventional and modern techniques including Omega. E. R. Swanson, C. P. Kugel.
Proc. IEEE, Vol. 60, 540 - 551 (1972).

035.013 Time synchronization via lunar radar. W. H. Higa.
Proc. IEEE, Vol. 60, 552 - 557 (1972).

The advent of round-trip radar measurements has permitted the determination of the ranges to the nearby planets with greater precision than was previously possible. When the distances to the planets are known with high precision, the propagation delay for electromagnetic waves reflected by the planets may be calculated and used to synchronize remotely located clocks. Details basic to the operation of a lunar radar indicate a capability for clock synchronization to $\pm20\mu$s. One of the design goals for this system was to achieve a simple semiautomatic receiver for remotely located tracking stations.

035.014 The role of time/frequency in Navy Navigation Satellites. R. L. Easton.
Proc. IEEE, Vol. 60, 557 - 563 (1972).

035.015 The role of time-frequency in satellite position determination systems. E. Ehrlich.
Proc. IEEE, Vol. 60, 564 - 571 (1972).

035.016 Time transfer using near-synchronous reception of optical pulsar signals. D. W. Allan.
Proc. IEEE, Vol. 60, 625 - 627 (1972).

035.017 Frequency standard work in Egypt.
A. Loutfy el Sayed.
Proc. IEEE, Vol. 60, 627 - 628 (1972).

035.018 Ruggedized rubidium frequency standard.
M. E. Frerking.
Proc. IEEE, Vol. 60, 628 - 629 (1972).

035.019 A time code for the Omega world-wide navigation system. L. Fey.
Proc. IEEE, Vol. 60, 630 (1972).

035.020 Worldwide synchronization using the TRANSIT satellite system. L. M. Laidet.
Proc. IEEE, Vol. 60, 630 - 632 (1972).

035.021 Satellite time synchronization of a NASA network.
S. C. Laios.
Proc. IEEE, Vol. 60, 632 - 633 (1972).

035.022 Dual transponder time synchronization at C band using ATS-3. W. E. Mazur, Jr.
Proc. IEEE, Vol. 60, 633 - 634 (1972).

035.023 Results of active line-1 TV timing. D. A. Howe.
Proc. IEEE, Vol. 60, 634 - 637 (1972).

Les cadrans solaires anciens d'Alsace.
See Abstr. 003.113.

Long-baseline radio interferometry with independent frequency standards. See Abstr. 033.033.

Standard time and frequency generation.
See Abstr. 044.029.

Progress and feasibility for a unified standard for frequency, time, and length. See Abstr. 044.038

036 Photographic Auxiliaries

036.001 **Spectral sensitization and special photographic materials for scientific use.** J. Spence.
Applied Optics, Vol. 11, 4 - 12 (1972).

A brief account is given of the work in the Kodak Research Laboratories on spectral sensitization that contributed to the manufacture of a diverse array of photographic plates and films for scientific use. For the most part it is concerned with plates supplied for spectroscopy and astronomy from 1930 to 1950, years of growth in these branches of science and years when plates with different spectral sensitizations appeared in increasing number.

036.002 **Comparison of three films for use in electronographic cameras.** G. R. Carruthers.
AAS Photo-Bull. 1972, No. 1, p. 3 - 5.

036.003 **Large-scale negatives from solar patrol films.** P. S. McIntosh.
AAS Photo-Bull. 1972, No. 1, p. 12 - 14.

036.004 **Water-hypersensitization of Kodak special plates, type 098-02.** H. Spinrad, J. Wilder.
AAS Photo-Bull. 1972, No. 1, p. 14 - 15.

036.005 **Study of the influence of several packing factors during low temperature storage of Kas NII Tech-photoproect astronomical films.** O. D. Dokuchaeva.
Soobshch. Gos. Astron. Inst. Shternberga, No. 176, p. 24 - 34 (1972). In Russian.

036.006 **Photographie astronomique.** A. Hamon.
L'Astronomie, 86e année, p. 197 (1972).

036.007 **An investigation of the photographic characteristics of the film UFSH-4 used in the "Orion" equipment.**
V. M. Uvarova, M. P. Shpolskiy, A. N. Oshurkova, J. Ohanesyan, N. V. Uvarova.
Soobshch. Byurakan. Obs., No. 45, p. 42 - 47 (1972). In Russian.

036.008 **Polaroid film in undergraduate astronomical photography.**
K. A. Innanen, C. R. Purton, R. Berry, W. G. Weller.
Journ. Roy. Astron. Soc. Canada, Vol. 66, 157 - 160 (1972).

We report on several Polaroid emulsions that have proved satisfactory for undergraduate astronomical photography where the normal darkroom procedure may be impractical.

036.009 **Entraînement automatique d'un porte-châssis pour la photographie en astronomie.** G. Jeansaume.
Electron. Microelectron. Industrielles (France), No. 147, p. 53 - 54 (1971). − See Bull. Signal., Vol. 33, Section 120, No. 4912 (1972).

Moderne Stellar- und Nebel-Photographie.
See Abstr. 031.026.

Positional Astronomy. Celestial Mechanics

041 Positional Astronomy, Star Catalogues and Atlases

041.001 On the use of observations of the moon for the improvement of zero-points of star catalogues.
V. A. Fomin.
Astron. Zhurn. Akad. Nauk SSSR, Vol. 49, 186 - 190 (1972). In Russian. English translation in Soviet Astron. AJ, Vol. 16, No. 1.

Bessel's classical method of the improvement of the zero-points of star catalogues is modified for the reduction of lunar observations. The corrections to the zero-points of the FK4 catalogue and to some elements of the lunar orbit are obtained from Washington 1925–1968 meridian observations of the moon.

041.002 Systematic errors in the AGK 3 in the region of the Pleiades.   R. S. Harrington.
Bull. American Astron. Soc., Vol. 3, 464 (1971). – Abstr. AAS.

041.003 Blockadjustment methods in photographic astrometry.   C. de Vegt, H. Ebner.
Astron. Astrophys., Vol. 17, 276 - 285 (1972).

The present paper intends to summarize briefly some of the widely spread results concerning theoretical accuracy investigations with special emphasis on astrometric applications. Furthermore, suitable reduction models and the application to existing catalogue material, including estimates of computing effort, will be discussed.

041.004 Peculiarities in reducing the astrographic catalogue by means of AGK2–AGK3.
P. Lacroute, A. Valbousquet.
Astron. Astrophys., Vol. 17, 296 - 300 (1972). In French.

We discuss the properties of the AGK2–AGK3 catalogue in terms of truly random errors, and local systematic errors arising from random errors in the plate constants. The theoretically correct reduction procedure cannot be applied. But we describe simplified methods which, in practice, yield nearly equivalent results.

041.005 On the General Catalogue of the USSR Time Service.
N. N. Pavlov, P. M. Afanasjeva, G. V. Staritsyn.
Trudy Glav. Astron. Obs. Pulkovo, Ser. 2, Vol. 78, 4 - 45 (1971). In Russian.

To compile the General Catalogue 20 catalogues obtained at nine Time Services of the USSR were used. Altogether 185371 star observations were used, 80522 of them being made at Pulkovo. The General Catalogue is based on the reference catalogue compiled from the latest three Pulkovo catalogues of a high precision. The other 17 catalogues were compared with the reference catalogue and reduced to its system.

041.006 The General Catalogue of the USSR Time Service.
N. N. Pavlov, P. M. Afanasjeva, G. V. Staritsyn.
Trudy Glav. Astron. Obs. Pulkovo, Ser. 2, Vol. 78, 59 - 98 (1971). In Russian.

The General Catalogue of the USSR Time Service contains the right ascensions and proper motions of 687 basic and 120 additional stars on the epoch and equinox 1950.0 and 1975.0.

041.007 On the calculation of an ephemeris for physical observations of the sun from planetary surfaces.
V. K. Abalakin, N. N. Petrova, T. A. Polozhentseva, B. M. Rubashev.
Solnechnye Dannye 1971 Byull., No. 12, p. 88 - 92 (1972). In Russian.

The formulas for calculating a physical ephemeris of the sun destined to observations from solar system planets are given and the program for computers is described.

041.008 La Carte du Ciel.   P. Sémirot.
Vistas in astronomy, Vol. 13, (see 003.001), 153 - 154 (1972).

041.009 Catalogue d'ascensions droites de 589 étoiles de la liste FKSZ (1953 - 1958) ramenées au système FK4.
L. M. Hommik.
Trudy Gos. Astron. Inst. Shternberga, Vol. 42, 3 - 18 (1972). In Russian.

Ce catalogue est le résultat d'une révision complète des observations obtenues au moyen du cercle méridien de Repsold dans l'espace de temps 1953 - 1958. Il est strictement ramené au système FK4. L'erreur quadratique moyenne d'une observation est égale à $\pm 0\overset{s}{.}019$ sec δ.

041.010 Observations d'ascensions droites de 1323 étoiles KSZ dans la zone de déclinaisons $(+60° - +90°)$.
T. S. Meshkova.
Trudy Gos. Astron. Inst. Shternberga, Vol. 42, 19 - 21 (1972). In Russian.

L'article donne une description générale de la méthode d'observation et de réduction des ascensions droites des étoiles KSZ (zone 60°–90°), rapportées au système d'ensemble PFKSZ et FK3R. Les observations ont été faites les années 1957–1964 à l'aide du cercle méridien de Moscou. L'erreur quadratique moyenne d'une observation d'ascension droite est $\pm 0\overset{s}{.}019$ sec δ.

041.011 Observations de déclinaisons des étoiles KSZ $(+60° - +90°)$ au cercle méridien Repsold de Moscou.
A. G. Oborneva.
Trudy Gos. Astron. Inst. Shternberga, Vol. 42, 22 - 31 (1972). In Russian.

Description de la methode d'observation et de reduction des déclinaisons d'étoiles KSZ (zone 60°–90°) rapportées au système d'ensemble PFKSZ et FK3R. Les observations ont été faites pendant les années 1959 - 1969 à l'aide du cercle méridien Repsold de Moscou. L'erreur quadratique moyenne d'une observation de déclinaison est $\pm 0''.35$.

041.012 Catalogue d'ascensions droites et de déclinaisons des étoiles KSZ comprises dans la zone $(+60° - +90°)$.
T. S. Meshkova, A. G. Oborneva.
Trudy Gos. Astron. Inst. Shternberga, Vol. 42, 32 - 62 (1972). In Russian.

Le catalogue d'ascensions droites et de déclinaisons des étoiles KSZ (zone 60°–90°) est composé d'après les observations faites à l'aide du cercle méridien Repsold de l'observatoire de Moscou au cours des années 1957 - 1969. Le catalogue est rapporté au système d'ensemble PFKSZ et FK3R, à l'équinoxe 1950.0 et à l'époque des observations. Compléments du catalogue: Ascensions droites des étoiles supplémentaires KSZ situées aux environs des nébuleuses extragalactiques; Catalogue des corrections individuelles pour les coordonnées des étoiles fondamentales des catalogues PFKSZ et FK3R.

041.013 Catalogue of astrometric standards for determination of the value of a revolution of the micrometer screw of ILS zenith-telescopes. L. P. Basurmanova-Gribko. Trudy Gos. Astron. Inst. Shternberga, Vol. 42, 80 - 98 (1972). In Russian.

041.014 Catalogue of the declinations of Moscow PZT stars. D. N. Ponomarev. Trudy Gos. Astron. Inst. Shternberga, Vol. 42, 99 - 107 (1972). In Russian.
The PZT observations obtained at Moscow Observatory in the period September 1962 to May 1965 were used in order to determine corrections to the declinations of the AGK2 system. A catalogue of the declinations for 266 Moscow PZT stars is given.

041.015 A catalogue of the right ascensions of 248 stars observed with the PZT of the Moscow Observatory during 1963 - 1965. A. A. Tochilina. Trudy Gos. Astron. Inst. Shternberga, Vol. 42, 108 - 114 (1972). In Russian.
A catalogue is given of the right ascensions of 248 stars observed with Moscow PZT 1963 - 1965. The observational program and the method of reducing as well as the comparison of the obtained catalogue with various catalogues are described.

041.016 The new star list for the Mizusawa PZT observation. S. Takagi. Publ. International Latitude Obs. Mizusawa, Vol. 7, 97 - 107 (1970).

041.017 The signal generating system of astrolabe observations using a magnetic switch. K. Yokoyama, S. Sakai, K. Iwadate. Proc. International Latitude Obs. Mizusawa, No. 11, p. 109 - 114 (1971). In Japanese.

041.018 Confirmation of Δa_a in the FK4 system. H. Yasuda. Publ. Astron. Soc. Japan, Vol. 24, 247 - 260 = Tokyo Astron. Obs. Repr., No. 411 (1972).
A method is proposed for deriving a fundamental right ascension system from differential observations with meridian circles. The existence of the deviation Δa_a in the FK4 system as derived from minor planet observations with meridian circles is confirmed independently by applying this method to the observations made at Tokyo for the SRS program.

041.019 Catalogue of I. L. S. and I. P. M. S. stars. P. Melchior, R. Dejaiffe. Extra collection of papers to IAU Symposium No. 48, (see 012.008), p. 47 - 66 (1971).

041.020 Global comparison between meridian and latitude positions of the I.L.S. stars. R. Dejaiffe, P. Melchior. Extra collection of papers to IAU Symposium No. 48, (see 012.008), p. 67 - 96 (1971).

041.021 PZT star systems as compared with the SAO Catalog. S. Iijima, Y. Niimi. Extra collection of papers to IAU Symposium No. 48, (see 012.008), p. 97 (1971). − Abstract. − The full paper is published in the Ann. Tokyo Astron. Obs., Second Ser., Vol. 12, 241 - 262 (1971), (see 06.041.007).

041.022 Status of PZT catalogs in current use. H. Yasuda. Extra collection of papers to IAU Symposium No. 48, (see 012.008), p. 99 (1971). − Abstract. − The full paper is published in Ann. Tokyo Astron. Obs., Second Ser., Vol. 13, 80 - 92 (1971), (see 06.041.027).

041.023 Northern PZT stars observation program. H. Yasuda, I. Kamijo. Extra collection of papers to IAU Symposium No. 48, (see 012.008), p. 101 (1971). − Abstract. − The full paper is published in Ann. Tokyo Astron. Obs., Second Ser., Vol. 13, 1 - 79 (1971), (see 06.041.026).

041.024 On the influence of temperature variation on the determination of right ascensions of stars. Eh. A. Sanakulov. Tsirk. Astron. Inst., *Tashkent,* No. 29 (376), p. 20 - 23 (1971). In Russian.

041.025 Daytime observations of right ascensions of the sun and major planets with the Tashkent meridian circle in 1968 - 1969. G. G. Khodak. Tsirk. Astron. Inst., *Tashkent,* No. 31 (378), p. 12 - 18 (1971). In Russian.

041.026 On some questions of constructing an inertial coordinate system with the methods of astronomy. V. K. Abalakin. Coordinate systems in astronomy, (see 012.007), p. 26 - 34 (1971). In Russian.

041.027 On principles of constructing coordinate systems applied in astronomy. E. P. Fedorov. Coordinate systems in astronomy, (see 012.007), p. 35 - 57 (1971). In Russian.

041.028 Inertial coordinate system and some tasks of astronomy. V. V. Podobed. Coordinate systems in astronomy, (see 012.007), p. 58 - 64 (1971). In Russian.

041.029 On the reduction of astronomical observations to a fixed pole. A. A. Izotov. Coordinate systems in astronomy, (see 012.007), p. 65 - 66 (1971). In Russian.

041.030 Resolution of the USSR conference on the problem of constructing an inertial coordinate system and of organizing observations with zenith tubes. Coordinate systems in astronomy, (see 012.007), p. 91 - 92 (1971). In Russian.

041.031 Amelioration of right ascensions for 238 FKSZ stars from the −20° to +24° declination zone. E. Toma, M. Tudor. Stud. Cerc. Astron., Vol. 17, 3 - 6 (1972). In Romanian.

041.032 Some current developments in astrometry. C. A. Murray. Mitt. Astron. Ges., No. 31, p. 23 - 38 (1972). − Review article.

041.033 First supplement to the list of catalogues available at the Centre de Données Stellaires. J. Jung. Centre de Données Stellaires, Inform. Bull. No. 3, (see 002.020) p. 2 - 5 (1972).

041.034 The general catalogue of stellar identifications. A) Progress report − B) Statistics pertaining to the overlap between catalogues included in the C.S.I. J. Jung, F. Ochsenbein.

Centre de Données Stellaires, Inform. Bull. No. 3, (see 002. 020), p. 6 - 9 (1972).

041.035 List of errors found in the Yale Zone Catalogues, the GC, and the Cape Photographic Catalogues.
M. Bischoff.
Centre de Données Stellaires, Inform. Bull. No. 3, (see 002. 020), p. 10 - 12 (1972).

041.036 List of catalogues in machine-readable form available at the Royal Greenwich Observatory. C. A. Murray.
Centre de Données Stellaires, Inform. Bull. No. 3, (see 002. 020), p. 13 - 14 (1972).

041.037 On the problem of the identification of cluster stars.
J. C. Mermilliod.
Centre de Données Stellaires, Inform. Bull. No. 3, (see 002. 020), p. 19 - 21 (1972).

041.038 Errors of the observations of Southern Reference Star Program. S. Isobe, N. Miyauchi.
Tokyo Astron. Obs., Report No. 59, Vol. 15, 766 - 788 (1972). In Japanese.

041.039 Basic data of the fundamental stars. A. M. Sinzi.
Reprinted from Researches in Hydrography and Oceanography – Commemoration Publication of the Centenary of the Hydrographic Department of Japan, Tokyo, p. 271 - 353 (1972).
The present list has been prepared to facilitate for investigating various characters of the stars of FK4 themselves. The list contains (I) spectral type and luminosity class in the MK system for 1337 stars, (II) magnitude and colours in the UBV system for 1407 stars, (III) radial velocity for 1445 stars, and (IV) parallax for 1304 stars. Kinds of double stars, types of variable stars, and memberships to a star stream, cluster, association, and group as well as high velocity stars are also remarked.

041.040 Resultados de las observaciones efectuadas con el astrolabio Danjon en el Observatorio de Quito, 1°. de enero - 31 de diciembre 1969. H. Dávila.
Bol. Astron. Obs. Astron. Quito, Ser. B, No. 3, p. 1 - 14 (1971)

041.041 Observaciones de Jupiter efectuadas con el astrolabio A.Danjon del Observatorio de Quito, durante el año 1964. H. Dávila, S. Débarbat, J. Egred, L. Mena, A. Scheepmaker.
Bol. Astron. Obs. Astron. Quito, Ser. B, No. 3, p. 15 - 16 (1971).

041.042 Resultados de las observaciones efectuadas con el astrolabio Danjon en el Observatorio de Quito, 1°. de enero - 31 de diciembre 1970. H. Dávila.
Bol. Astron. Obs. Astron. Quito, Ser. B, No. 4, p. 1 - 9 (1972).

041.043 Observaciones de Saturno con el astrolabio del Observatorio de Quito. M. Alvaro, H. Dávila, S. Débarbat, J. Egred, L. Espín, A. Salvador.
Bol. Astron. Obs. Astron. Quito, Ser. B, No. 4, p. 10 - 13 (1972).

041.044 Résultats des observations faites à Alger avec l'astrolabe impersonnel A. Danjon OPL 8. Temps et latitude 1970.
A. Ghezloun, M. Benhocine, A. Marouf, J. Pham-Van.
Ann. Obs. Astron. Alger, Vol. 3, Fasc. 3, p. 3 - 21 (1972).

041.045 The inertial system and its importance for geodesy.
H. U. Sandig.
Acta Geod., Geophys., Montanist. Acad. Sci. Hung., Vol. 5,

389 - 395 = Mitt. Lohrmann-Obs. Techn.Univ. Dresden, No. 24 (1970). In German.
A definition of a suitable inertial system is given and three stages of approach of its realization, viz. the fundamental catalogue, the catalogue of faint stars, and joining to the galaxies is discussed. This joining in its total form is of primary importance first of all for geodesy while astronomy is in the first place interested in the changes in direction.

041.046 Bright stars charts in ecliptic coordinates with magnitudes at 2000 Angströms. L. Houziaux, G. Houziaux.
Univ. Mons, Dép. Astrophys., Commun. No. 26, 4 pp. + 9 charts (1972).
We present here a complete chart of the sky in 1970.0 ecliptic coordinates for objects bright enough in the ultraviolet. Computed magnitudes refer to 2000 Å. Star positions, magnitudes, star selection criteria, spectral types, and brightnesses are given. A few non-stellar objects have been located on the charts.

041.047 El astrolabio impersonal del Observatorio de Marina de San Fernando, estación astronómica permanente de precisión. A. Orte.
Bol. Inform. Service Geogr. Ejér., 1970, No. 12, p. 7 - 10.

041.048 Observations of the sun, moon, and planets. Six-inch transit circle results. B. L. Klock, D. K. Scott.
United States Naval Obs., *Washington, D. C.*, Circ. No. 136, 20 pp. (1972).
This circular contains positions of the sun, moon, and planets observed with the six-inch transit circle between 8 August, 1969 and 31 December, 1970. These results are provisional. Definitive results will be published later in a volume of Publications of the U. S. Naval Observatory.

041.049 On the accuracy of determining systematic errors of a fundamental catalogue and of the orbital elements of minor planets from common observations of them.
V. N. Bojko.
Uch. zap. Dal'nevost. un-ta, Vol. 51, 220 - 228 (1970). In Russian. – Abstr. in Referativ. Zhurn. 51. Astron., 7.51.121 (1972).

041.050 The effect of duration and intensity of minor planet observations upon the accuracy of determination of their orbits, the orbit of the earth and the zero points of a fundamental system of star positions. V. N. Bojko.
Uch. zap. Dal'nevost. un-ta, Vol. 51, 229 - 236 (1970). In Russian. – Abstr. in Referativ. Zhurn. 51. Astron., 7.51.122 (1972).

041.051 Astrometric accuracy of photographs taken with the Sproul 24-inch refractor: (1) Positions. (2) Parallaxes. S. L. Lippincott.
Proc. fourth astrometric conference, (see 012.026), p. 59 - 129, with a discussion, p. 131 - 147 = Sproul Obs. Repr. No. 205 (1971).

041.052 Astrometry: First aid for the spectroscopist.
W. P. Bidelman.
Proc. fourth astrometric conference, (see 012.026), p. 231 - 241, with a discussion p. 243 - 252 (1971).

041.053 The astrometric reliability of the Thaw Photographic refractor of the Allegheny Observatory.
K. W. Kamper, Jr.
Proc. fourth astrometric conference, (see 012.026), p. 285 - 290 (1971).

041.054 Tables of definitive plate constants for the zones

Potsdam, Hyderabad, Uccle, Oxford of the Astrographic Catalogue (Carte du Ciel). A. Günther, H. Kox. Astron. Astrophys., Suppl. Ser., Vol. 6, 201 - 247 (1972).

For the zones between +32° and +40° declination of the Astrographic Catalogue the tables give plate centres and linear constants that together with the also given formulae of systematic corrections to the printed coordinates permit derivation of spherical coordinates in the FK 4-system for the equinox 1950.0 retaining the epochs of the original plates.

041.055 Positions de grosses planètes et de la lune observées à l' équatorial photographique de 0, 33 m. G. Soulié, Dupouy, Teulet, Broqua, Dulou, Ralite. Astron. Astrophys., Suppl. Ser., Vol. 6, 311 - 326 (1972).

In order to improve our knowledge of the astronomical constants, great interest must be paid to accurate positions of major planets and their satellites. Likewise, accurate astrometry of the moon on stellar background is essential to determine the Ephemeris Time. The negative were obtained on Kodak II Ao plates with our 13-inch photographic refractor. Each plate is measured twice on an Asco-Zeiss.

Effect of an error in plate reduction. See Abstr. 031.001.

Precise stellar positions using GALAXY-machine measures of a Schmidt plate. See Abstr. 031.019.

Astrometric applications of very-long-baseline interferometry. See Abstr. 033.020.

A possible explanation of the discrepancy between polar motions in time and in latitude observations by PZT. See Abstr. 045.003.

The Ottawa PZT observations—1956-70, their comparison with BIH values by graphical, spectral and fourier analyses. See Abstr. 045.037.

Photographic positions of comet Ikeya-Seki (1967 n) with a summary of the Leyden reduction procedure. See Abstr. 103.101.

Rigorous determination of astrometric quantities, especially parallaxes, by photography. See Abstr. 111.008.

On the proper motions of stars from the General Catalogue of the USSR Time Service. See Abstr. 112.005.

Catalogue of relative proper motions of 4211 stars in the region of Taurus. See Abstr. 112.006.

The positions and proper motions of 127 stars in the region of WX Centauri. See Abstr. 112.011.

Contribution of Danjon's astrolabe to the study of stellar proper motions. See Abstr. 112.014.

042 Celestial Mechanics

042.001 Limitations on outer planet mass determinations from their mutual perturbations.
P. K. Seidelmann.
Celestial Mechanics, Vol. 5, 3 - 7 (1972).

The limitations on determining the masses of the outer planets from their mutual perturbations are investigated based on the magnitudes of the periodic perturbations given in general theories and the accuracy of the observational data.

042.002 A second-order solution of the ideal resonance problem by Lie series. A. H. Jupp.
Celestial Mechanics, Vol. 5, 8 - 26 (1972).

A second-order libration solution of the ideal resonance problem is constructed using a Lie-series perturbation technique.

042.003 Two-body problem with slowly decreasing mass.
F. Verhulst.
Celestial Mechanics, Vol. 5, 27 - 36 (1972).

Evolution of the orbital elements of a two-body system with slowly decreasing mass according to Jeans' mode is described by a non-linear, non-autonomous system of differential equations. By comparing the analytical results with the numerical calculations of Hadjidemetriou we explain the rapid rotation of the line of apsides which occurs if the initial value of e is nearly-circular.

042.004 Note concerning a conjecture by A. Wintner.
J. Waldvogel.
Celestial Mechanics, Vol. 5, 37 - 40 (1972).

The conjecture mentioned in the title is concerned with the existence of non-flat central configurations of N point masses; it was put forward in A. Wintner's 'Analytical Foundations of Celestial Mechanics' (1941), § 360 (iii), p. 279.

042.005 Some fundamental elements of universe mechanics.
Y. Y. Neiman.
Celestial Mechanics, Vol. 5, 55 - 66 (1972).

An essential part in the mechanics under study is taking into consideration the effect of motions of the universe objects upon that of an individual one surrounded by them including those infinitely far from it. Only macro-objects of the universe are meant here.

042.006 Some remarks on the generalized many-body problem. G. N. Duboshin.
Celestial Mechanics, Vol. 5, 67 - 79 (1972).

This paper deals with the generalized problem of motion of a system of a finite number of bodies. We suppose here that every point of the system acts on another one with a force (attractive or repulsive) directed along the straight line connecting these two points, and proportional to the product of their masses and a certain function of time, mutual distance and its derivatives of the first and second order (Duboshin, 1970). The results obtained may be applied for the investigation of motion in some isolated stellar systems, where the laws of mechanics may be different from the laws in our solar system.

042.007 Sur l'application des transformations de Lie aux problèmes résonnants de la mécanique céleste.
S. Ferraz-Mello, R. V. Martins.
Comptes Rendus Acad. Sci. Paris, Sér. A, Vol. 272, 521 - 524 (1972).

Analyse de la méthode des séries de Lie et des hypothèses implicites qui ne sont pas satisfaites lorsqu'il existe une résonance paramétrique. L'usage des développements par rapport à la racine carrée du petit paramètre est justifiée.

042.008 Le problème principal de la lune et le problème restreint des trois corps. J. Henrard.
Comptes Rendus Acad. Sci. Paris, Sér. A, Vol. 274, 1015 - 1017 (1972).

Le problème principal de la lune tel qu'il est défini par Brown n'est rien d'autre qu'un problème restreint plan des trois corps où le moyen mouvement des deux corps principaux est légèrement modifié. La reconnaissance de ce fait est propre à faciliter la vérification numérique des théories analytiques du mouvement de la lune.

042.009 La détermination de distribution zonale des vitesses angulaires dans le phénomène de rotation anomale des corps sphériques cosmiques. K. Stiegler.
Comptes Rendus Acad. Sci. Paris, Sér. A, Vol. 274, 1026 - 1028 (1972).

On poursuit ici les considérations commencées dans la note précédente (1971) en représentant une méthode générale de détermination de distribution zonale des vitesses angulaires d'un corps sphérique cosmique à condition que sa vitesse angulaire équatoriale, et sa vitesse angulaire au voisinage de l'équateur soient connues. La méthode donnée est appliquée à Jupiter et à Saturne.

042.010 Jacobian integral as a classificational and evolutionary parameter of interplanetary bodies.
L'. Kresák.
Bull. Astron. Inst. Czechoslovakia, Vol. 23, 1 - 34 (1972).

Different simplified approximations of the Jacobian integral in the restricted three-body problem, with the sun and Jupiter as the two massive bodies, are computed for smaller members of the solar system. The variability of this parameter, due to the imperfections of the model and nongravitational effects is evaluated. The selection effects on the statistics of Jacobian integrals are discussed, and the relevant orbital characteristics are listed for 30 samples of interplanetary bodies covering all types of objects known at present. Particular attention is paid to the possible transitional phases between comets and asteroids, the problem of extinct comets, the significance of resonances, the dividing line between cometary and asteroidal orbits, and the comet-meteor and asteroid-meteor relationships.

042.011 On the periodic solutions of slowly spinning gravity gradient system. V. J. Modi, J. E. Neilson.
Celestial Mechanics, Vol. 5, 126 - 143 (1972).

The paper emphasizes the importance of periodic solutions in the dynamic stability study of an axi-symmetric satellite in presence of gravity gradient torques. Initial conditions for periodic solutions are presented over a range of system parameters for motion in circular and elliptic orbits. The variational stability of periodic solutions is examined using an extension of the Floquet theory to a fourth order system.

042.012 Regularization in the ideal resonance problem.
B. Garfinkel.
Celestial Mechanics, Vol. 5, 189 - 203 (1972).

The ideal resonance problem, defined by the Hamiltonian $F = B(y) + 2\epsilon A(y) \sin^2 x$, $\epsilon \ll 1$, has been solved in Garfinkel *et al.* (1971). There the solution has been *regularized* by means of a special function ϕ_j, introduced into the new Hamiltonian F', under the tacit assumption that A and B'' are of order unity. This assumption is replaced here by the weaker assumption of *normality*. With the modified ϕ_j, and with the assumption of normality, the solution is regularized for all values of B', B'', and A. A regularized first-order algorithm is constructed here as an illustration and a check.

042.013   Les mouvements rectilignes dans le problème des trois corps lorsque la constante des forces vives est nulle.   M. Irigoyen, F. Nahon.
Astron. Astrophys., Vol. 17, 286 - 295 (1972).

Nous donnons la forme simplifiée d'une équation différentielle du premier ordre en coordonnées polaires, et nous faisons l'étude qualitative des solutions au moyen d'une inégalité déduite du théorème du viriel.

042.014   An experiment to measure the earth's orbital velocity.   D. Hoff.
Sky Telescope, Vol. 43, 9 - 10 (1972).

042.015   A new method for analytical solution of the two-body problem.   M. M. Dagaev, V. V. Radzievskij.
Astron. vestn., Vol. 5, 245 - 247 (1971). In Russian. – Abstr. in Referativ. Zhurn. 51. Astron., 3.51.147 (1972).

042.016   Two problems concerning the gravitational potential of planets.   A. N. Bogayevsky.
Trudy Mosk. in-ta radiotekhn., ehlektron. i avtomatiki, 1971, vyp. (No.) 52, p. 12 - 26. In Russian. – Abstr. in Referativ. Zhurn. 51. Astron., 3.51.175 (1972).

042.017   A revolution around the smallest primary of a restricted four-body problem.   V. Matas.
Bull. Astron. Inst. Czechoslovakia, Vol. 23, 113 - 116 (1972).

An analytical theory describing a close motion of an infinitesimal mass around the smallest material point from the three finite ones, during one period, is outlined.

042.018   Studies on planetary satellites. Satellite capture in the three-body elliptical problem.
J. M. Bailey.
Astron. Journ., Vol. 77, 177 - 182 (1972).

Satellite capture is examined for planets in elliptical orbits. Expressions are developed for the Jacobian constant of a satellite in a coordinate system rotating with the primaries about the center of mass. These are transformed to a coordinate system centered upon and moving with the planet, and enable the orbital elements of a captured satellite to be derived in terms only of the mass ratio and the eccentricity of the planetary orbit. Capture may only take place at perihelion or aphelion. It is shown that the masses of planets and the eccentricities of the orbits are such that all planets may effect the capture of satellites when at perihelion.

042.019   The stability of fast rotations of an axisymmetric satellite in a gravitational field.
V. V. Beletskii, A. P. Torzhevskii.
Dokl. Akad. Nauk SSSR, Ser. Mat. Fiz., Vol. 203, 50 - 53 (1972). In Russian.

042.020   Libration points in the restricted generalized problem of three bodies.
V. T. Kondurar, T. K. Shinkarik.
Byull. Inst. Teoret. Astron., *Leningrad,* Vol. 13, 102 - 110 (1972). In Russian.

The issue of the existence of libration points in the restricted problem about translational-rotational motion of a dynamically symmetric satellite under the gravitation of two spherical bodies is considered. It is proved that there are some solutions of the problem different from the classic ones to which the motion of bodies correspond, when their mass centers form an isosceles triangle or lie on a straight line passing through the mass centers of the given bodies.

042.021   Recent advances of celestial mechanics in the Soviet Union.   Y. Hagihara.
Vistas in astronomy, Vol. 13, (see 003.001), 15 - 47 (1972).

042.022   Recherche et étude des trajectoires planes du centre d'inertie d'un corps de révolution placé dans un champ newtonien.   A. Klat, M. Pascal.
Comptes Rendus Acad. Sci. Paris, Sér. A, Vol. 274, 1381 - 1384 (1972).

Considérant le problème du mouvement d'un corps de révolution placé dans un champ newtonien de centre fixe O, on recherche tous les mouvements pour lesquels le centre d'inertie du corps décrit une trajectoire plane dans un plan passant par O. Une étude plus détaillée des cas de planéité de la trajectoire du centre d'inertie est faite lorsque le corps possède, en outre, un plan de symétrie perpendiculaire à l'axe de révolution.

042.023   Soluzioni multiple nella determinazione di orbite circolari.   L. Buffoni, A. Manara.
Mem. Soc. Astron. Italiana, Nuova Ser., Vol. 43, 17 - 25 (1972).

The classical formula to determine circular orbits gives sometimes doubtful solutions. We prove how sometimes it is possible to get more than one solution (2 or 3) of which only one corresponds to the true one. Besides we study the boundary regions where the problem may present 2 or 3 solutions.

042.024   A multi-particle regularisation technique.
D. C. Heggie.
Gravitational N-body problem. IAU Colloquium No. 10, (see 012.004), p. 148 - 152 (1972) = 06.042.046.

042.025   Expansion of the eccentric anomaly in terms of Levi–Civita's variable.
É. A. Borisov, G. V. Morozova.
Trudy Mosk. in-ta inzh. geod., aehrofotos"emki i kartogr., 1971, vyp. (No.) 58, p. 23 - 25. In Russian. – Abstr. in Referativ. Zhurn. 51. Astron., 5.51.103 (1972).

042.026   A remark on the Jacobian integral of the restricted problem of three bodies.   H. Fiedler.
Journ. London Math. Soc., Ser. 2, Vol. 3, 437 - 438 (1971). – Abstr. in Zentralblatt Math. Grenzgebiete, Vol. 221, No. 70002 (1972).

042.027   Finite two body problem.   G.-i. Hori.
Extra collection of papers to IAU Symposium No. 48, (see 012.008), p. 141 - 148 (1971).

042.028   The significance of Newton's laws of motion.
M. W. Ovenden.
Journ. British Astron. Ass., Vol. 82, 106 - 111 (1972).

042.029   Étude de la stabilité de certains équilibres relatifs d'un satellite symétrique.   M. Pascal.
Comptes Rendus Acad. Sci. Paris, Sér. A, Vol. 274, 1590 - 1593 (1972).

042.030   Axiomatization of the invariant motion of two bodies.   L. Sofonea, N. Ionescu-Pallas.
Stud. Cerc. Astron., Vol. 17, 53 - 69 (1972). In Romanian.

042.031   Normality condition in the ideal resonance problem.   B. Garfinkel.
Bull. American Astron. Soc., Vol. 4, 262 - 263 (1972). Abstr. AAS.

042.032   Periodic librations in the restricted problem of three bodies.   G. E. O. Giacaglia.
Bull. American Astron. Soc., Vol. 4, 263 (1972). – Abstr. AAS.

042.033 Spinor regularization of conservative central fields.
G. E. O. Giacaglia, V. J. Nuotio.
Bull. American Astron. Soc., Vol. 4, 263 (1972). – Abstr. AAS.

042.034 Numerical accuracy of the symplectic check for Hamiltonian systems. C. F. Peters.
Bull. American Astron. Soc., Vol. 4, 264 - 265 (1972).
Abstr. AAS.

042.035 Generalization of the elements of Delaunay.
G. Scheifele.
Bull. American Astron. Soc., Vol. 4, 266 (1972). – Abstr. AAS.

042.036 An exact closed-form solution of Kepler's equation.
C. E. Siewert, E. E. Burniston.
Bull. American Astron. Soc., Vol. 4, 266 (1972). – Abstr. AAS.

042.037 On the stability of the solutions of the general problem of three bodies. E. M. Standish, Jr.
Bull. American Astron. Soc., Vol. 4, 267 (1972). – Abstr. AAS.

042.038 On the approximate calculation of the main part of perturbations in the intrinsic restricted three-body problem. Yu. V. Plakhov, A. M. Safiulin.
Izv. vyssh. ucheb. zavedenij. Geod. i aehrofotos"emka, 1971, No. 4, p. 33 - 39. In Russian. – Abstr. in Referativ. Zhurn. 51. Astron., 6.51.142; 62. Issled. kosm. prostranstva, 6.62. 257 (1972).

042.039 The effects of viscous friction on axial rotation of celestial bodies. Z. Kopal.
Astrophys. Space Sci., Vol. 16, 3 - 51 (1972).
The aim of the present paper will be to formulate explicitly the differential equations which govern three-dimensional rotation of deformable self-gravitating bodies of arbitrary structure, and consisting of fluid material whose viscosity is an arbitrary function of spatial coordinates; with special respect to a description of the effects of viscous friction exhibited in binary systems which consist of a close pair of such configurations. A numerical application of the results to the earth–moon system is made.

042.040 On the equinoctial orbit elements.
R. A. Broucke, P. J. Cefola.
Celestial Mechanics, Vol. 5, 303 - 310 (1972).
This paper investigates the equinoctial orbit elements for the two-body problem, showing that the associated matrices are free from singularities for zero eccentricities and zero and ninety degree inclinations. The matrix of the partial derivatives of the position and velocity vectors with respect to the orbit elements is given explicitly, together with the matrix of inverse partial derivatives. The application of the equinoctial orbit elements to general and special perturbations is discussed.

042.041 Transformations to extend the range of application of power series solutions of differential equations of motion. J. M. A. Danby.
Celestial Mechanics, Vol. 5, 311 - 316 (1972).
Let the solution of a differential equation, expanded in powers of the independent variable t, have radius of convergence T. Let τ, where $t = t(\tau)$, be a new independent variable, and let the corresponding power series in τ have radius of convergence S. Then $t(S)$ will not in general be equal to T. If $t(S) > T$, then the series in powers of τ may have advantages over those in powers of t. In this note the practical applications of some transformations are investigated.

042.042 The collision singularity in a perturbed N-body problem. H. J. Sperling.
Celestial Mechanics, Vol. 5, 396 - 406 (1972).
Consider the perturbed N-body problem and assume that at the instant t^* all N bodies collide, while the perturbing forces remain bounded. It is shown that the motion prior to the instant of collision is essentially the same as in the unperturbed N-body problem i.e., the same asymptotic relations are valid in both cases as $t \to t^*$.

042.043 The stability of an area-preserving mapping.
B. Z. Jenkins, J. H. Bartlett.
Celestial Mechanics, Vol. 5, 407 - 427 (1972).
The mapping $T(x, y) = (x', y')$: $x' = x + a(y - y^3)$, $y' = y - a(x' - x'^3)$ was studied in order to determine methods of readily compartmentalizing the plane into regions of stable and unstable behavior under many applications of T, without resorting to costly and frequently inaccurate methods requiring computation of thousands of maps.

042.044 Les lois de Kepler et la mécanique céleste moderne.
J. Kovalevsky.
L'Astronomie, 86e année, p. 304 - 310 (1972).

042.045 The systems of canonical variables of Adel Soudan.
J. Meffroy.
Mitt. Astron. Ges., No. 31, p. 183 (1972). – Abstract.

042.046 The decomposition of the force function of two homogeneous spheroids whose symmetry planes do not coincide. V. V. Vidiakin.
Astron. Zhurn. Akad. Nauk SSSR, Vol. 49, 641 - 646 (1972). In Russian. English translation in Soviet Astron. AJ, Vol. 16, No. 3.

042.047 Regularization of conservative central fields.
G. E. O. Giacaglia.
Publ. Astron. Soc. Japan, Vol. 24, 381 - 389 (1972).
It is shown that regularization in two dimensions for potentials of the type $V = k/r^n$, as a generalization of Levi-Civita's transformation, is only possible for the cases $n = 1$ and $n = 3$. The case $n = 3$ is extended to 3-dimensional space and gives a generalized K-S transformation.

042.048 Improved criteria for hyperbolic-elliptic motion in the general three-body problem. J. Yoshida.
Publ. Astron. Soc. Japan, Vol. 24, 391 - 408 (1972).
Two criteria for the hyperbolic-elliptic motion in the general three-body problem are shown improving on Khil'mi (1951) and Merman (1952). A criterion obtained as a corollary of the former includes criteria by Merman (1954), Tevzadze (1962), and Standish (1971), and is found to be effective for the Pythagorean problem.

042.049 Stationary motions of a triaxial body and their stabilities. H. Kinoshita.
Publ. Astron. Soc. Japan, Vol. 24, 409 - 417 = Tokyo Astron. Obs. Repr., No. 416 (1972).
Stationary motions are investigated for the system of two rigid bodies, one spherical and one triaxial, which experience no other forces than the mutual gravitational attraction, and their stabilities (both secular and ordinary) are discussed.

042.050 Further periodic solutions of the three-dimensional restricted problem. II.
W. H. Jefferys, E. M. Standish, Jr.
Astron. Journ., Vol. 77, 394 - 400 (1972).
Discrepancies between a previous paper on the subject by the authors and work by Kozai are investigated. More accurate figures are presented for 23 values of the mean motion, giving combinations of eccentricity and inclination which can be continued to periodic three-dimensional orbits.

042.051 Variational equations of Hamiltonian systems with two degrees of freedom. J. D. Hadjidemetriou.

Journ. Mécan. Paris, Vol. 10, 433 - 448 (1971).

In this paper, the properties of the solution of the first order variational equations of a Hamiltonian system with two degrees of freedom, corresponding to a symmetric periodic orbit of the Hamiltonian system, are studied.

042.052   **Theory of non-linear oscillations for non-conservative systems.**   H. Kinoshita.
Tokyo Astron. Obs., Report No. 59, Vol. 15, 789 - 812 (1972). In Japanese.

042.053   **Circular motion with non-uniform velocity.**   G. Hori.
Celestial mechanics. Symposium Tokyo, (see 012.020), p. 4 - 6 (1972). In Japanese.

042.054   **Sur le choc et la régularisation dans le problème des trois corps.**   T. Inoue.
Celestial mechanics. Symposium Tokyo, (see 012.020), p. 21 - 25 (1972). In Japanese.

042.055   **On criterions for hyperbolic-elliptic motion in three-body problem.**   J. Yoshida.
Celestial mechanics. Symposium Tokyo, (see 012.020), p. 26 - 34 (1972). In Japanese.

042.056   **Numerical integration of general three body problem.**   M. Yuasa.
Celestial mechanics. Symposium Tokyo, (see 012.020), p. 35 - 39 (1972). In Japanese.

042.057   **Ergodicity and scattering of orbits in the negative curvature zone of classical dynamical systems.**
Y. Aizawa.
Celestial mechanics. Symposium Tokyo, (see 012.020), p. 40 - 49 (1972). In Japanese.

042.058   **Finite two body problem.**   H. Kinoshita.
Celestial mechanics. Symposium Tokyo, (see 012.020), p. 57 - 71 (1972). In Japanese.

042.059   **Évaluation de l'intervalle du temps dans lequel la théorie des perturbations reste valable.**
T. Inoue.
Celestial mechanics. Symposium Tokyo, (see 012.020), p. 78 - 86 (1972). In Japanese.

042.060   **Celestial mechanics at present and in future.**   Y. Kozai.
Celestial mechanics. Symposium Tokyo, (see 012.020), p. 108 - 111 (1972). In Japanese.

042.061   **Über die Störungen der von Kepler verwendeten Marsbeobachtungen.**   F. Schmeidler.
Separate print from Kepler Festschrift 1971, [Naturwiss. Verein, Regensburg], p. 141 - 158.

042.062   **Strejftog gennem himmelmekanikkens historie.**   H. Q. Rasmusen.
Astron. Tidssk., Årg. 5, p. 53 - 66 (1972).

042.063   **Differential coefficients of rectangular coordinate variations using universal elements.**
B. T. Khukhunajshvili.
Vestn. Leningr. un-ta, 1972, No. 1, p. 150 - 154. In Russian. Abstr. in Referativ. Zhurn. 51. Astron., 7.51.97; 62. Issled. kosm. prostranstva, 7.62.276 (1972).

042.064   **Motion of small particles in the solar system.**   H. Alfvén.
IAU Colloquium No. 12, (see 012.027), p. 315 - 317 (1971/72).

042.065   **Computational aspects of asymptotic matching in the restricted three-body problem.**   L. M. Perko.
AIAA Journ., Vol. 10, 540 - 542 (1972).

042.066   **Numerical integration in series of the equations of the restricted three-body problem.**   A. Kranjc.
Rendiconti Ist. Lombardo A, Vol. 105, 96 - 104 (1971). In Italian.

042.067   **Periodic orbits of collision in the plane circular problem of four bodies.**   K. B. Bhatnagar.
Indian Journ. Pure and Applied Math., Vol. 2, 583 - 596 (1971). — See Phys. Abstr., Vol. 75, No. 41038 (1972).

042.068   **Two-to-one resonances near the equilateral libration points.**   A. H. Nayfeh.
AIAA Journ., Vol. 9, 23 - 27 (1971).

042.069   **Stability in incompressible systems.**   D. G. Saari.
Indiana Univ. Math. Journ. (*USA*), Vol. 20, 975 - 981 (1971). — See Bull. Signal., Vol. 33, Section 120, No. 16 (1972).

042.070   **Applications and computer implementation of Hansen's planetary theory.**
R. Broucke, E. Chapman.
Journ. Astronaut. Sci., Vol. 18, 368 - 382 (1971).

042.071   **Periodic solution of plane librational motion of a satellite.**   F. C. Liu.
AIAA Journ., Vol. 9, 1240 - 1244 (1971).

Celestial mechanics. Vol. 2.   See Abstr. 003.063.

Les fondements de la mécanique céleste.
See Abstr. 003.129.

Canonical satellite theory based on independent variables different from time.   See Abstr. 013.003.

On the parametric stability of symplectic mappings.
See Abstr. 021.009.

The perturbed motion of asteroids in proximity.
See Abstr. 098.027.

Dynamics of comets.   See Abstr. 102.019.

Errata

042.901   **Errata: 'On the generalized restricted problem of three bodies'** [Celestial Mechanics, Vol. 4, 423 - 444 (1971)].   G. N. Duboshin.
Celestial Mechanics, Vol. 5, 258 (1972).

042.902   **Errata: 'Numerical experiments on the N-body problem'** [Astrophys. Space Sci., Vol. 14, 20 - 34 (1971)].   S. J. Aarseth.
Astrophys. Space Sci., Vol. 15, 348 (1972).

042.903   **Errata: 'A multi-particle regularization technique'** [Astrophys. Space Sci., Vol. 14, 35 - 39 (1971)].
D. C. Heggie.
Astrophys. Space Sci., Vol. 15, 348 (1972).

042.904   **Errata: 'Direct integration methods of the N-body problem'** [Astrophys. Space Sci., Vol. 14, 118 - 132 (1971)].   S. J. Aarseth.
Astrophys. Space Sci., Vol. 15, 348 (1972).

043   Astronomical Constants

043.001   **Astronomical constants from analysis of inner-
planet radar data.**   M. E. Ash, D. B. Campbell,
R. B. Dyce, R. P. Ingalls, G. H. Pettengill, I. I. Shapiro.
Bull. American Astron. Soc., Vol. 3, 474 (1971). – Abstr.
AAS.

043.002   **Precessional corrections, apex and galactic rotation
derived from proper motions by a maximum likeli-
hood method.**   M. O. Mennessier.
Astron. Astrophys., Vol. 17, 220 - 225 (1972). In French.
   A maximum likelihood method is explained and tested
by numerical experiments, and the accuracy of the estimates
is determined. Several determinations are made from FK 4 and
GC proper motions, the stars of each sample being selected according to spectral types. The numerical experiments show
that the value of the sun's velocity is sensitive to the adopted
luminosity function, whereas the coordinates of the apex are
not. The results from the early type stars and K stars are the
most reliable. The estimates show that Oort's galactic constant
A should be greater than its conventional value; it should be
between 16 and 17 km s^{-1} kpc^{-1}. The galactic rotation near
the sun is $\omega \cong 30$ km s^{-1} kpc^{-1}.

043.003   **Die Masse des Erde-Mond-Systems.**
C. A. van den Bosch.

SuW, Vol. 11, 125 - 127 (1972).

043.004   **A determination of the motion of the ecliptic.**
R. E. Laubscher.
Bull. American Astron. Soc., Vol. 4, 264 (1972). – Abstr. AAS.

043.005   **Astronomical and geodynamical constants.**
J. W. Siry.
Bull. American Astron. Soc., Vol. 4, 266 - 267 (1972).
Abstr. AAS.

043.006   **Determination of the gravitational constant.**
E. Groten, S. von Thyssen-Bornemisza.
Zeitschr. Vermessungswesen (*Germany*), Vol. 97, 105 - 106
(1972). In German.

   **The earth's acceleration as deduced from al-Biruni's
solar data.**   See Abstr. 044.025.

   **Possible variation of the gravitational constant over
the elements.**   See Abstr. 066.070.

   **Astronomical evidence concerning non-gravitational
forces in the earth–moon system.**   See Abstr. 094.199.

044   Time, Rotation of the Earth

044.001   **Short period terms in the rate of rotation and in the
polar motion of the earth.**
S. Iijima, S. Okazaki.
Publ. Astron. Soc. Japan, Vol. 24, 109 - 125 = Tokyo Astron.
Obs. Repr. No. 403 (1972).
   Short period terms from 0.3 to 6.0 year periods in the
UT1 and in the polar motion are resolved on periodograms
calculated from respective data covering the same period of
13.5 years from 1955.5 to 1969.0.

044.002   **Estimation of the earth's rotational motion using
satellite observations.**
B. D. Tapley, B. E. Schutz.
Bull. American Astron. Soc., Vol. 3, 467 (1971). – Abstr. AAS.

044.003   **A step in time. Changes in standard frequency and
time signal broadcasts.**   A. R. Chi, H. S. Fosque.
IEEE Spectrum, Vol. 9, No. 1, p. 82 - 86 (1972).
   On January 1, 1972 an improved Coordinated Universal
Time (UTC) system has been adopted by the International Radio Consultative Committee. The new UTC system eliminates
the frequency offset of 300 parts in 10^{10} between the old UTC
and Atomic Time, thus making the broadcast time interval,
the UTC second, constant and defined by the resonant frequency of cesium atoms.

044.004   **Poles of rotation.**   C. G. A. Harrison.
   Earth Planet. Sci. Letters, Vol. 14, 31 - 38 (1972).
   Owing to the incompatibility of explaining all relative

movements between pairs of crustal plates using poles of rotation fixed relative to one of the plates in each pair, it must be
expected that there will be movements in the locations of the
poles of rotation. Because of these small movements, it is un-
desirable to calculate average poles of rotation by looking at
the movement of an individual point. The average pole of rotation in such a case could be in error by 20° or more. To establish poles of rotation, it is necessary to consider the movement
of at least two points.

044.005   **A reply to "Remarks on uncertainties in poles of
rotation in continental fitting" by E. C. Bullard and
D. P. McKenzie [Earth Planet. Sci. Letters, Vol. 11, 263 - 264
(1971)].**   M. Al-Chalabi.
Earth Planet. Sci. Letters, Vol. 14, 73 - 74 (1972).

044.006   **Time data and navigation.**   H. M. Smith.
   Journ. Navigation, *London*, Vol. 25, 13 - 31 (1972).
   In 1972 an atomic time standard based on the fundamental properties of the caesium atom is being substituted for
Greenwich Universal Time, based on the rotation of the earth
on its axis, as the ultimate control for international time signals. The autor outlines the current proposals for time signals,
their correction to Universal Time and the implications for the
navigator.

044.007   **The new system of coordinated Universal Time.**
D. H. Sadler.
Journ. Navigation, *London*, Vol. 25, 32 - 42 (1972).

The purpose of this article is twofold: firstly to give readers a general account of the new time-system that will be introduced on 1 January 1972, and secondly to give greater detail in respect of the corrections that a relatively few navigators may need to apply.

044.008 Irrégularités de la mesure du temps déduites de 13,5 années d'observations astronomiques.
S. Débarbat.
Ann. Françaises Chronométrie Micromécanique, Année 1971, p. 37 - 39. — Communication présentée au Congrès de Chronométrie Franco-Allemand — Constance 1970.

044.009 The description of Foucault's pendulum.
W. B. Somerville.
Quarterly Journ. Roy. Astron. Soc., Vol. 13, 40 - 62 (1972).

044.010 Mesure des conditions de réception des signaux horaires à basse fréquence.
J. de Prins, B. Colinet.
Ann. Françaises Chronométrie Micromécanique, Année 1971, p. 125 - 130.

044.011 La rotation de la Terre.   B. Guinot.
Vistas in astronomy, Vol. 13, (see 003.001), 49 - 50 (1972).

044.012 Numerical process of the analysis of the rotation of the earth. 1. Fundamental equations of the rotation of the earth.   S. Takagi.
Publ. International Latitude Obs. Mizusawa, Vol. 7, 149 - 172 (1970).

044.013 On the effect of the anomalous refraction on time observations at Mizusawa (I).   T. Goto.
Proc. International Latitude Obs. Mizusawa, No. 11, p. 95 - 99 (1971). In Japanese.

044.014 Note on erroneous coefficients of semi-annual solar nutation terms.   K. Yokoyama.
Proc. International Latitude Obs. Mizusawa, No. 11, p. 115 - 117 (1971). In Japanese.

Wako (1970) showed that the origin of annual z-term can be attributed to erroneous coefficients of the semi-annual solar nutation term caused by the effect of the liquid core of the earth. Granted that it is true, time determination must be affected in the same way.

044.015 New system of co-ordinated universal time.
Journ. Astron. Soc. Victoria, Vol. 24, 91 - 92 (1971).

044.016 Irrigation projects and the earth's rotation.
Yu. V. Batrakov.
Extra collection of papers to IAU Symposium No. 48, (see 012.008), p. 19 - 23 (1971).

044.017 Experiments of detecting the rotational velocity of the earth by means of the ring laser rotation sensor.
S. Takagi.
Extra collection of papers to IAU Symposium No. 48, (see 012.008), p. 41 - 42 (1971).

044.018 Remarks on the calculation of definitive universal time.   G. P. Pil'nik.
Extra collection of papers to IAU Symposium No. 48, (see 012.008), p. 43 - 45 (1971).

044.019 'Time-telling' techniques.
J. L. Jespersen, L. Fey.
IEEE Spectrum, Vol. 9, No. 5, 51 - 58 (1972).

044.020 Determination of time (TU1).   N. A. Omelina.
Tsirk. Astron. Inst., *Tashkent*, Nos. 27 (374) - 32 (379) (1971/72). In Russian. — 1970 November - 1971 October.

044.021 Séries Pugliano observées pendant 1969–1971 à l'Observatoire Astronomique de Bucarest.
M. Stavinschi.
Stud. Cerc. Astron., Vol. 17, 23 - 32 (1972). In Romanian.

044.022 Chronometry: Absolute time?
Nature, Vol. 237, 423 (1972).

044.023 Some remarks on seasonal variation in the rotation of the earth.
G. Hemmleb, H. Krüger, M. Meinig.
Extra collection of papers to IAU Symposium No. 48, (see 012.008), p. 15 - 17 (1971).

044.024 Change in the system of radio time signals.
L. Zajdler.
Urania Kraków, Vol. 43, 70 - 76 (1972). In Polish.

044.025 The earth's acceleration as deduced from al-Biruni's solar data.   R. R. Newton.
Mem. Roy. Astron. Soc., Vol. 76, 99 - 128 (1972).
In the process of finding the position of Ghazni relative to Baghdad, al-Biruni used about 65 observations of the sun, mostly measurements of meridian altitudes made at various places. The dates of the observations range from 829 to 1019. They yield an acceleration of the earth's spin of -26.5 ± 5.8 parts in 10^9 per century, and they yield a rate of change of the obliquity of -47.9 ± 2.0 sec per century. Al-Biruni also left a few miscellaneous observations of eclipses. These, taken in conjunction with the solar data, yield a lunar acceleration of -46.4 ± 6.0 sec per century per century.

044.026 The measurement of time.   J. H. Reid.
Journ. Roy. Astron. Soc. Canada, Vol. 66, 135 - 148 (1972).

044.027 International time and frequency coordination.
H. M. Smith.
Proc. IEEE, Vol. 60, 479 - 487 (1972).
The significant advances in the development of international coordination in time determination and dissemination are briefly reviewed.

044.028 The foundation of time and frequency in various countries.   J. T. Henderson.
Proc. IEEE, Vol. 60, 487 - 493 (1972).
The purpose of this paper is to stimulate discussion of applications of time and frequency methods and of the standards on which these methods rest. The emphasis is placed on interesting those who have not had occasion to follow recent developments in the subject.

044.029 Standard time and frequency generation.
P. Kartaschoff, J. A. Barnes.
Proc. IEEE, Vol. 60, 493 - 501 (1972).

044.030 Characterization and concepts of time–frequency dissemination.
J. L. Jespersen, B. E. Blair, L. E. Gatterer.
Proc. IEEE, Vol. 60, 502 - 521 (1972).
We point out fundamental techniques of time and frequency dissemination and describe similarities between systems for time dissemination and navigation.

044.031 Application de la méthode des paires à la détermination astronomique de l'heure.

D. Djurović, V. Radogostić.
Publ. Dep. Astron. Univ. Beograd, No. 1, p. 8 - 12 (1969).

044.032 The introduction of a new system of co-ordinated universal time. J. D. Laing.
Monthly Notes Astron. Soc. Southern Africa, Vol. 31, 53 - 55 (1972).

044.033 Ergebnisse im internationalen Zeit- und Frequenzvergleich mit Längstwellen.
G. Becker, B. Fischer.
Messtechnik (Zeitschr. Instrumentenkunde), 79. Jahrgang, p. 277 - 283 (1971).

044.034 Time service. M. G. Fracastoro.
Oss. Astron. Torino (Pino Torinese) Bull. No. 1, 3 pp. (1972). – The aim of this Bulletin is to publish and give rapid diffusion of the results of the time determinations carried out at Torino. The results for 1972 January - April are given.

044.035 Hur vår tidsenhet definieras. N. Hansson.
Astron. Tidssk., Årg. 5, p. 67 - 73 (1972).

044.036 Astronomische Zeit- und Breitenbestimmungen, Empfangszeiten von Zeitsignalen.
Edited by Deutsches Hydrographisches Institut, Hamburg. 1971 Oktober - Dezember (1972).

044.037 Sekundensprung in UTC am 30. Juni 1972.
Communicated by Deutsches Hydrographisches Institut, Hamburg. 1 pp. (1972).

044.038 Progress and feasibility for a unified standard for frequency, time, and length.
D. Halford, H. Hellwig, J. S. Wells.
Proc. IEEE, Vol. 60, 623 - 625 (1972).

044.039 Time and latitude service.
Polish Acad. Sci., Astron. Latitude Station, Borowiec, Circ. Nos. 119–120 (1971/1972). – 1971 July – December.

044.040 Détermination astronomique de l'heure et heures demi-définitives de réception des signaux horaires.
L. Webrová, V. Ptáček.
Acad. Tchécoslov. Sci., Inst. Astron., Station de l'Heure, Prague, Sér. 5, No. 17 (1971). – 1971 September – October.

044.041 Astronomische Zeit- und Breitenbestimmungen. Empfangszeiten von Zeitsignalen.
Deutsche Akad. Wiss. Berlin, Zentralinst. Phys. Erde, Bereich II (Geod., Gravimetrie), Potsdam, Abt. Astron., Jahrgang 1971, Nos. 1–4. – 1971 January – August.

044.042 Détermination astronomique de l'heure et de la latitude.
Obs. Neuchâtel, Bull. (B), 1971 June – December (1972).

044.043 L'heure astronomique définitive de l'Observatoire de Neuchâtel.
Obs. Neuchâtel, Bull. (D), 1971 May – December (1972).

044.044 La rotation de la terre: Théorie et observations.
R. O. Vicente.
Pubbl. Stazione Astron. Internazionale Latitudine, Carloforte-Cagliari, Nuova Ser., No. 16, 9 pp. (1971).

044.045 Time Service of the Mizusawa Observatory.
Bulletins, Vol. 14, No. 1 - 12, 1969.
S. Takagi, I. Okamoto, K. Yokoyama, T. Hara, G. Murakami.

Edited by the International Latitude Observatory of Mizusawa, Mizusawa-Shi, Iwate-Ken, Japan. 4 + 58 pp. (1971).
This Bulletin contains the results of Time Service and the astronomical observations made at the Mizusawa Observatory during the period beginning with January 1969 to December 1969.

044.046 Time and Latitude Bulletins, Tokyo Astronomical Observatory.
Tokyo Astron. Obs. Mitaka, Tokyo, Japan. Vol. 45, Nos. 9 - 12, p. 55 - 81 (1971/72). – 1971 September – December: Coordinates of instantaneous pole and corrections for UT2; Times of emission of radio time signals on UT2; Times of arrival of GBR and Iwôjima Loran-C signals; Astronomical observations made with the PZT.

044.047 Corrections to Czechoslovak time signals.
V. Ptáček.
Říše hvězd, Vol. 53, 20, 34, 61, 77, 95, 117, 141, 158 (1972). 1971 Oct. - 1972 May.

044.048 Ephemeris time, lunar orbital elements and FK4 equinox corrections from the observations of the moon by the method of equal altitudes. J. Vondrák.
Studia, Vol. 16, 197 - 200 (1972).

044.049 Universal time and coordinates of the pole; Emission time of time signals; Coordinated time; Independant local atomic time scales; Informations.
Bureau International de l'Heure, (BIH), Paris, Circ. D62 - D68 (1972). – 1971 November – 1972 May.

044.050 Daily phase values.
U. S. Naval Obs., Washington, D. C., Time Service Publ., Ser. 4, Nos. 255 - 282 (1972). – 1971 Dec. 22 – 1972 June 28.

044.051 Preliminary times and coordinates of the pole.
U. S. Naval Obs., Washington, D. C., Time Service Publ., Ser. 7, Nos. 209 - 235 (1972). – 1971 Dec. 30 – 1972 June 29.

044.052 Time Service Bulletin.
U. S. Naval Obs., Washington, D. C., Time Service Publ., Ser. 11, No. 220 (1972). – Times of emission, UTC (USNO) – UTC (Transmitter); Time scales; Observations; Polar coordinates, seasonal and polar (longitude) variation corrections. – 1970 Dec. 29 - 1972 Jan. 3.

044.053 Time Service Announcement. G. M. R. Winkler.
U. S. Naval Obs., Washington, D. C., Time Service Publ., Ser. 14, Nos. 9 - 10 (1972). – Transmission of DUT1; UTC time step.

044.054 The metrology of time. C. Egidi, S. Leschiutta.
Nuovo Cimento Rivista, Ser. 2, Vol. 1, 496 - 518 (1971). In Italian.
A detailed review of the present state of theory and practice of precise time measurement, is presented. The main physical standards and their characteristics are discussed. Methods for comparing standards are summarized and typical numerical results are given.

044.055 Study of the rotation of the earth.
J. de Prins, R. Gillard.
Journ. Interdiscipl. Cycle Res. (*Netherlands*), Vol. 2, 325 - 329 (1971). – See Bull. Signal., Vol. 33, Section 120, No. 1187 (1972).

044.056 NBS frequency and time broadcast services. Radio stations WWV, WWVH, WWVB, and WWVL.

National Bureau Standards (*USA*), Special Publ., No. 236, 14 pp. (1971). – See Bull. Signal., Vol. 33, Section 120, No. 2523 (1972).

Medieval chronicles and the rotation of the earth. See Abstr. 003.102.

A time code for the Omega world-wide navigation system. See Abstr. 035.019.

Résultats des observations faites à Alger avec l'astrolabe impersonnel A. Danjon OPL 8. Temps et latitude 1970.

See Abstr. 041.044.

Timing and geodesy. See Abstr. 046.027.

Lithospheric momenta and the deceleration of the earth. See Abstr. 081.030.

Annual and yearly variations of the atmospheric pressure and the earth's rotation. See Abstr. 082.059.

Paleontological evidence on the earth's rotational history since Early Precambrian. See Abstr. 094.201.

045 Latitude Determination, Polar Motion

045.001 On the International Parallel.
M. G. Kaganovskij, O. S. Tursunov.
Zemlya i Vselennaya, No. 1, p. 50 - 53 (1972). In Russian.

045.002 On some properties of the excitation and damping of the polar motion. N. Sekiguchi.
Publ. Astron. Soc. Japan, Vol. 24, 99 - 108 = Tokyo Astron. Obs. Repr. No. 402 (1972).
Values of the Chandlerian period and the damping coefficient of the polar motion are calculated for ten year intervals from 1900.0 to 1970.0 and discussed in detail.

045.003 A possible explanation of the discrepancy between polar motions in time and in latitude observations by PZT. H. Yasuda, S. Aoki.
Publ. Astron. Soc. Japan, Vol. 24, 127 - 133 = Tokyo Astron. Obs. Repr. No. 404 (1972).
It is suggested that catalog errors and the annual part of polar motion cannot be completely separated from each other on the basis of only PZT observations. Most of the existing discrepancy between the polar motions in time and in latitude observations in fact can be attributed to the errors in the current PZT star catalogs; this is obtained from a comparison of the positions between AGK3R and PZT catalogs.

045.004 Compiling of working ephemerides of Talcott pairs for the tropical zone (for latitudes from 0° to 15°).
Tran duy Thoan.
Astron. Zhurn. Akad. Nauk SSSR, Vol. 49, 218 - 221 (1972). In Russian. English translation in Soviet Astron. AJ, Vol. 16, No. 1.
The present ephemerides of Talcott pairs are described and the lack of such ephemerides for latitudes from 0° to 15°, in which Vietnam and other countries of Asia, Africa, and Central America are situated, is pointed out. Compiling a catalogue of 3352 southern stars for the epoch 1980.0, the choice of Talcott pairs and the calculation of working ephemerides for latitudes from 0° to 15° are described.

045.005 Study of latitude observations performed at the stations of Mizusawa, Kitab, Gaithersburg, Ukiah during the period 1935.0–1947.0.
E. Fichera, A. Pugliano.
Astron. Astrophys., Vol. 18, 1 - 9 (1972).
The difficulties which have been met in the computations involved in the ILS, tenth volume, are reviewed and final latitudes of the international stations at Mizusawa, Kitab, Gaithersburg, Ukiah are given using only completely observed series and the FK4 system for declinations and proper motions. These results are more homogeneous than those obtained reducing every observed pair in the GC-system with intergroup corrections $\Delta\delta$, $\Delta\mu$.

045.006 Analysis of latitude observations made on the Danjon astrolabe at the Paris Observatory between 1956.5 and 1970.8. F. C. Chollet, S. Débarbat.
Astron. Astrophys., Vol. 18, 133 - 142 (1972). In French.
The results given here concern an analysis of latitude observations made in Paris Observatory between 1956.5 and 1970.8. Before any analysis, we calculate once more the results of these observations with the new astronomical constants used since 1968.0 and the FK4 catalogue. We get periodograms by Fourier analysis of individual observations. Using the results, we have made a representation of latitude variations with 10 periodical terms and one non-periodical term, the mean value of the latitude.

045.007 Rotational inertia of continents: A proposed link between polar wandering and plate tectonics.
M.F. Kane.
Science, Vol. 175, 1355 - 1357 (1972).
A mechanism is proposed whereby displacement between continents and the earth's pole of rotation (polar wandering) gives rise to latitudinal transport of continental plates (continental drift) because of their relatively greater rotational inertia. When extended to short-term polar wobble, the hypothesis predicts an energy change nearly equivalent to the seismic energy rate.

045.008 Le mouvement du pôle instantané; la variation des latitudes et des longitudes. A. Stoyko.
Vistas in astronomy, Vol. 13, (see 003.001), 51 - 134 (1972).

045.009 Revised values (1941–1962) of the coordinates of the pole referred to the CIO [*Conventional International Origin*]. R. O. Vicente, S. Yumi.
Publ. International Latitude Obs. Mizusawa, Vol. 7, 109 - 112 (1970).

045.010 Some problems and a new program in the determination of the micrometer constant.
S. Takagi, M. Ooe, S. Abe.

Publ. International Latitude Obs. Mizusawa, Vol. 7, 113 - 121 (1970).

045.011 On the nearly diurnal nutation term derived from the ILS z term. C. Sugawa, M. Ooe.
Publ. International Latitude Obs. Mizusawa, Vol. 7, 123 - 147 (1970).

045.012 The motion of the earth's poles.
A. A. Mikhajlov.
Uspekhi fiz. nauk, Vol. 105, 776 - 777 (1971). In Russian.
Abstr. in Referativ. Zhurn. 52. Geod. Aehros"emka, 4.52.169 (1972).

045.013 Notes on progress in geophysics. Palaeomagnetic directions and pole positions – XII.
M. W. McElhinny.
Geophys. Journ. Roy. Astron. Soc., Vol. 27, 237 - 257 (1972).
Pole numbers 12/1 to 12/180.

045.014 Proposition for the cooperation between latitude services in Belgrade and Jozefosław.
G. Teleki, V. Milovanović.
Proc. International Latitude Obs. Mizusawa, No. 11, p. 52 - 56 (1971). In Japanese and English.

045.015 On the relation between the ILS closing error and the diurnal variation of latitude. H. Ishii.
Proc. International Latitude Obs. Mizusawa, No. 11, p. 118 - 148 (1971). In Japanese.

045.016 Motion of the earth's poles. A. A. Mikhajlov.
Uspekhi fiz. nauk, Vol. 105, 776 - 777 (1971). In Russian. – Abstr. in Referativ. Zhurn. 51. Astron., 5.51.161 (1972).

045.017 Critical analysis of reduction methods of latitude observations for the period 1900.0–1906.0.
E. Fichera, L. Milano, A. Pugliano.
Extra collection of papers to IAU Symposium No. 48, (see 012.008), p. 1 - 11 (1971).

045.018 The free polar motion in 1916–1933.
N. P. J. O'Hora, S. F. McWilliam.
Extra collection of papers to IAU Symposium No. 48, (see 012.008), p. 13 - 14 (1971).

045.019 Observations de latitude et tubes zénithaux visuels.
Y. Väisälä.
Extra collection of papers to IAU Symposium No. 48, (see 012.008), p. 35 - 40 (1971).

045.020 On the Chandler and annual ellipses in the polar motion as obtained from every 12 year period.
S. Iijima.
Extra collection of papers to IAU Symposium No. 48, (see 012.008), p. 103 - 111 (1971).

045.021 On the nearly diurnal nutation term derived from the ILS z term. C. Sugawa, M. Ooe.
Extra collection of papers to IAU Symposium No. 48, (see 012.008), p. 139 - 140 (1971).

045.022 On the determination of the coordinates x and y from the time determinations and the meridian transit instruments. V. S. Milovanović, V. J. Radogostić.
Extra collection of papers to IAU Symposium No. 48, (see 012.008), p. 175 - 179 (1971).

045.023 Observational program for the determination of the latitude of Tashkent. G. M. Kaganovskij.

Tsirk. Astron. Inst., *Tashkent,* No. 30 (377), p. 12 - 21 (1971). In Russian.

045.024 On the accuracy of the Tashkent latitude series 1895 - 1896. G. M. Kaganovskij.
Tsirk. Astron. Inst., *Tashkent,* No. 31 (378), p. 19 - 22 (1971). In Russian.

045.025 Results of an investigation of Talcott levels of the Bamberg transit instrument No. 100306.
G. M. Kaganovskij.
Tsirk. Astron. Inst., *Tashkent,* No. 32 (379), p. 12 - 15 (1971/72). In Russian.

045.026 History of latitude investigations in Central Asia till 1954. V. P. Shcheglov.
Coordinate systems in astronomy, (see 012.007), p. 80 - 86 (1971). In Russian.

045.027 Work of the International Latitude Station Ulugbek in Kitab from 1954 to the present time.
A. M. Kalmykov.
Coordinate systems in astronomy, (see 012.007), p. 87 - 90 (1971). In Russian.

045.028 La détermination astronomique des mouvements du pôle par les procédés en usage dans les stations internationales de latitude. R. Dejaiffe.
Ciel et Terre, Vol. 88, 157 - 195 (1972).

045.029 Pole position for 1971 based on Doppler satellite observations. R. J. Anderle.
U. S. Naval Weapons Lab., Dahlgren, Virginia, NWL Techn. Rep. TR–2734, 4 + 9 + A 10 + B 9 + C 51 + D 7 pp. (1972).
The instantaneous position of the earth's spin axis with respect to the crust has been computed daily since 1969 on the basis of Doppler satellite observations. Improved sets of observing station coordinates and gravity field coefficients obtained in 1970 were used throughout 1971 yielding pole positions with a standard error of about 0.5 meters for the mean of five days of data. However, for some time periods, a bias of about 0.5 meters also exists between the Doppler pole positions and positions reported by the Bureau International de L'Heure (BIH) based on astronomical data. Spot checks of latitude residuals and pole positions based on recomputed satellite orbits for 1968 and 1969 yielded results consistent with recent data, indicating significant compensation of station coordinate errors by changes in orbit constants.

045.030 The Chandlerian wobble from 1900 to 1970.
B. Guinot.
Astron. Astrophys., Vol. 19, 207 - 214 (1972).
The properties of the Chandlerian wobble were obtained from the study of individual series of latitudes measured in 20 stations.

045.031 Résultats des observations faites pendant la période 1969.5 à 1970.6 par photographie d'étoiles zénithales à l'instrument des passages établi dans le I. vertical. Détermination de la latitude. A. Baptista dos Santos.
Bull. Obs. Astron. Lisbonne (Tapada), No. 18, 4 pp. (1971).

045.032 Results of latitude observations for the year 1970 with VZT. E. Proverbio, S. Uras.
Circ. Stazione Astron. Internazionale Latitudine, Carloforte – Cagliari, Ser. B(2), No. 2, 200 pp. (1972).

045.033 Local term and declination corrections for zenith star program and latitude polar variations.
E. Proverbio.
Rend. Seminario Fac. Sci. Univ. Cagliari, Vol. 41, 223 - 248 =

Pubbl. Stazione Astron. Internazionale Latitudine, Carloforte-Cagliari, Nuova Ser., No. 10 (1971).

The latitude observations carried out in the period 1961.7–1969.0 with a transit instrument using a program of zenithal stars have been analysed using a reduction method which differs from the chain method.

045.034 **Long term components in polar motion.**
E. Proverbio, V. Quesada.
Ann. Geofis., Vol. 25, 37 - 54 = Pubbl. Stazione Astron. Internazionale Latitudine, Carloforte - Cagliari, Nuova Ser., No. 23 (1972).

In this work, the latitudes observed in the period 1900.0 - 1969.0 reduced to a single homogeneous system are analyzed separately for the five stations of the ILS. The analysis of this quantity points out the existence of two different long term components in polar motion of about 16 and 26 years. One possible interpretation of these phenomena is based on "beat" phenomena between the nutation of the mantle and the inner core of the earth.

045.035 **Breitenbestimmungen.**
Techn. Univ. Dresden, Lohrmann-Obs. Zirk. Nos. 51 - 55 (1971/1972). – 1971 May – 1972 February.

045.036 **Monthly Notes of the International Polar Motion Service.**
IPMS Monthly Notes, International Latitude Obs. Mizusawa (Japan). 1971 Nos. 11–12, p. 85 - 100; 1972 Nos. 1–4, p. 1 - 36 (1972). – Announces the values of latitudes observed at the collaborating stations during 1971 November until 1972 April.

045.037 **The Ottawa PZT observations–1956-70, their comparison with BIH values by graphical, spectral and fourier analyses.** E. G. Woolsey.
Publ. Earth Phys. Branch, Ottawa, Vol. 42, 183 - 214 (1972).

Results of measurements of latitude and time made with the Ottawa Photographic Zenith Tube (PZT) from 1956 to 1970 are given. The observations, on magnetic tape, are available to recognized scientific organizations at nominal cost. A star catalogue is published to contribute to the knowledge of stellar positions.

045.038 **Coordonnées du pôle instantané rapportées à l'origine conventionnelle internationale et corrections**

de longitude TU1 – TU0, à 0h TU.
Bureau International de l'Heure, (BIH), Paris, Circ. B/C, Nos. 189 - 195 (1972). – Valeurs interpolées et extrapolées.

The separation of discrete lines of the frequency spectrum of a compound process.   See Abstr. 021.006.

Beitrag zur Weiterentwicklung des Zirkumzenitals. See Abstr. 032.052.

PZT star systems as compared with the SAO Catalog.   See Abstr. 041.021.

Résultats des observations faites à Alger avec l'astrolabe impersonnel A. Danjon OPL 8. Temps et latitude 1970. See Abstr. 041.044.

Short period terms in the rate of rotation and in the polar motion of the earth.   See Abstr. 044.001.

Irrigation projects and the earth's rotation. See Abstr. 044.016.

Astronomische Zeit- und Breitenbestimmungen, Empfangszeiten von Zeitsignalen.   See Abstr. 044.036.

Time and latitude service.   See Abstr. 044.039.

Astronomische Zeit- und Breitenbestimmungen. Empfangszeiten von Zeitsignalen.   See Abstr. 044.041.

Détermination astronomique de l'heure et de la latitude.   See Abstr. 044.042.

Universal time and coordinates of the pole; Emission time of time signals; Coordinated time; Independant local atomic time scales; Informations.   See Abstr. 044.049.

Static deformation of a multilayered sphere by internal sources.   See Abstr. 081.011.

On the spectral analysis of the Chandler wobble from meteorological elements in the upper atmosphere. See Abstr. 082.061.

046 Geodetic Astronomy, Navigation

046.001 **Longitude difference determination between astronomic points Ia (University Astronomical Observatory) and Im (VNIIFTRI Siberian Department) in Irkutsk in 1969.**   L. N. Nadeev.
Astron. Zhurn. Akad. Nauk SSSR, Vol. 49, 216 - 218 (1972). In Russian. English translation in Soviet Astron. AJ, Vol. 16, No. 1. – Short note.

046.002 **Three-dimensional geodesy for terrestrial network adjustment.**   D. M. J. Fubara.
Journ. Geophys. Res., Vol. 77, 796 - 807 (1972).

A rigorous practical application of three-dimensional geodesy to the adjustment of a terrestrial network was investigated to determine (1) the precision obtainable from a given set

of measurements and data quality; (2) the roles, the types, and the number of geodetic observations required; and (3) the implications of the type of geodetic coordinate system involved and stability problems in the matrix systems for the weighted least squares solutions.

046.003 **Precision geodesy via radio interferometry: First results.**   H. F. Hinteregger, R. Ergas, C. A. Knight, D. S. Robertson, I. I. Shapiro, A. R. Whitney, A. E. E. Rogers, T. A. Clark.
Bull. American Astron. Soc., Vol. 3, 467 (1971). – Abstr. AAS.

046.004 **Determining the azimuth from observations of star groups near their elongations.**   M. A. Vajsov.

Izv. vyssh. uchebn. zavedenij. Geod. i aehrofotos"emka, 1970, No. 4, p. 53 - 59. In Russian. – Abstr. in Referativ. Zhurn. 51. Astron., 3.51.206; 52. Geod. Aehros"emka, 3.52.166 (1972).

046.005 **Le système géodésique de référence 1967 et le calcul des anomalies de la pesanteur.** J. J. Levallois.
Bull. Géod., Nouvelle Sér., Année 1972, No. 103, p. 85 - 91.

046.006 **Sir William Thomson and the intercept method.** C. H. Cotter.
Journ. Navigation, *London*, Vol. 25, 91 - 98 (1972).

046.007 **Eine zeichnerische Probiermethode zur geographischen Ortsbestimmung.** E. Lehmann.
Sterne, 48. Jahrgang, p. 44 - 51 (1972).

046.008 **International experiment of satellite geodesy.** S. K. Tatevjan.
Nauchn. Informatsii, vyp. (No.) 18, p. 108 - 112 (1970). In Russian.

046.009 **On the joint determination of length and direction of the chord from artificial earth satellite observations.** O. S. Razumov.
Izv. vyssh. uchebn. zavedenij. Geod. aehrofotos"emka, 1971, No. 3, p. 43 - 47. In Russian. – Abstr. in Referativ. Zhurn. 62. Issled. kosm. prostranstva, 4.62.271 (1972).

046.010 **On the influence of the errors of latitude and star coordinates on the results of observations by Zinger's method.** P. A. Petroshkyavichyus.
Izv. vyssh. uchebn. zavedenij. Geod. i aehrofotos"emka, 1971, No. 3, p. 73 - 77. In Russian. – Abstr. in Referativ. Zhurn. 51. Astron., 4.51.207; 52. Geod. Aehros"emka, 4.52.157 (1972).

046.011 **Analysis of instrumental errors of longitude determinations.** I. I. Krasnorylov.
Izv. vyssh. uchebn. zavedenij. Geod. i aehrofotos"emka, 1970, No. 6, p. 74 - 77. In Russian. – Abstr. in Referativ. Zhurn. 51. Astron., 4.51.208; 52. Geod. Aehros"emka, 4.52.161 (1972).

046.012 **On the estimates of the refraction error in Zinger's method.** V. B. Egorov.
Izv. vyssh. uchebn. zavedenij. Geod. i aehrofotos"emka, 1970, No. 6, p. 78 - 80. In Russian. – Abstr. in Referativ. Zhurn. 51. Astron., 4.51.209; 52. Geod. Aehros"emka, 4.52.158 (1972).

046.013 **On the determination of the astronomical azimuth by the method of equal hour angles.**
A. I. Piotrovskaya.
Geod. i kartografiya, 1971, No. 10, p. 24 - 31. In Russian. Abstr. in Referativ. Zhurn. 51. Astron., 4.51.210; 52. Geod. Aehros"emka, 4.52.162 (1972).

046.014 **Determination of the geodetical azimuth by multiple observations of bright stars near the meridian.**
V. G. L'vov.
Geod. i kartografiya, 1971, No. 11, p. 13 - 20. In Russian. Abstr. in Referativ. Zhurn. 51. Astron., 4.51.211; 52. Geod. Aehros"emka, 4.52.163 (1972).

046.015 **Determination of the azimuth by measurements of zenith distances near the prime vertical.**
V. G. Vasil'ev.
Izv. vyssh. uchebn. zavedenij. Geod. i aehrofotos"emka, 1970, No. 6, p. 72 - 73. In Russian. – Abstr. in Referativ. Zhurn. 51. Astron., 4.51.212; 52. Geod. Aehros"emka, 4.52.164 (1972).

046.016 **On the simultaneous determination of latitude and longitude by the equal altitude method.**

L. V. Cheskidova.
Trudy Mosk. in-ta inzh. geod., aehrofotos"emki i kartogr., 1971, vyp. (No.) 58, p. 91 - 95. In Russian. – Abstr. in Referativ. Zhurn. 51. Astron., 5.51.163; 52. Geod. Aehros": emka, 5.52.133 (1972).

046.017 **Practical techniques for the establishment of a world geodetic system from gravity data.** R. S. Mather.
Extra collection of papers to IAU Symposium No. 48, (see 012.008), p. 155 - 171 (1971).

046.018 **Establishment of a geocentric system of coordinate axes connected with the earth.** I. D. Zhongolovich.
Coordinate systems in astronomy, (see 012.007), p. 8 - 25 (1971). In Russian.

046.019 **Space trilateration by means of artificial satellites with only laser measurements.**
M. Caputo, G. Folloni, L. Pieri, M. Unguendoli.
Boll. Geod. Sci. Affini, Anno 31, p. 57 - 70 (1972). In Italian.
The problems connected with the practical realization of a space trilateration only with distance measurements to artificial satellites are examined.

046.020 **Le réseau WESTA, élaboration et analyse préliminaire des matériaux d'observations.**
L. Minowska, K. Minowski.
Geodezja Kartografia, Vol. 21, 63 - 78 (1972). In Polish.
L'article présente la première étape de l'élaboration expérimentale du réseau est-européen de triangulation satellitaire (WESTA).

046.021 **The spatial transformation in the satellite geodesy.**
A. Czarnecki.
Geodezja Kartografia, Vol. 21, 79 - 90 (1972). In Polish.
The author investigates some cases of three-dimensional transformation of large geodetic networks. The transformation may be applied where joining continental networks, by satellite method, is considered.

046.022 **Réduction sur la surface d'une sphère – par moyen de distances zénithales (ou d'angles verticaux) – d'une distance determinée à l'aide d'un télémètre électromagnétique.** W. Dąbrowski.
Geodezja Kartografia, Vol. 21, 105 - 109 (1972). In Polish.
Dans l'article présent l'auteur propose un procès de réduction, dans lequel la distance zénithale ou l'angle vertical est – au lieu de la différence des hauteurs – l'élément de la réduction. Les mesures doivent être fait à l'aide d'un théodolite qui permet en même temps la mesure électromagnétique des distances.

046.023 **Geodetic latitude and altitude from geocentric coordinates.** B. C. Getchell.
Celestial Mechanics, Vol. 5, 300 - 302 (1972).
This paper contains simplified formulas for the geodetic latitude and for the height of a body whose position is given in terms of geocentric equatorial coordinates. The results are applicable to the problem of finding the sub-vehicle point and height of a spaceship or missile.

046.024 **Ortsbestimmung mit künstlichen Erdsatelliten.** G. Gerstbach.
Sternenbote, 15. Jahrgang, p. 66 - 74 (1972).

046.025 **Ortung und Navigation mit Hilfe von künstlichen Erdsatelliten.** K.-W. Schrick.
Mitt. Astron. Ges., No. 31, p. 183 (1972). – Abstract.

046.026 **Astronomisch-geodätische Beobachtungen auf den**

Stationen Onstmettingen und Emmingen (1964),
Bochum and Velbert (1965). K.-W. Schrick, H. Walter.
Deutsche Geod. Kommission, Bayer. Akad. Wiss., Reihe B,
Heft No. 145 = Mitt. Inst. Angew. Geod., Frankfurt, No. 99,
12 + 28 pp. (1971).

046.027 **Timing and geodesy.** W. E. Carter.
Proc. IEEE, Vol. 60, 610 - 613 (1972).

046.028 **Utilisation des observations laser en géodésie géo-**
métrique. M. Bivas.
Bull. Groupe Recherches Géod. Spatiale, Obs. Meudon, No. 3,
p. 1 - 32 (1972).

046.029 **Résultats récents en géodésie géométrique.**
M. J. L. Pieplu, A. Cazenave.
Bull. Groupe Recherches Géod. Spatiale, Obs. Meudon, No. 3,
p. 33 - 38 (1972).

046.030 **Réduction des données laser françaises d'ISAGEX.**
G. Brachet, C. Brossier.
Bull. Groupe Recherches Géod. Spatiale, Obs. Meudon, No. 4,
4 + 90 pp. (1972).

046.031 **Programme géodésique de l'Europe de l'Ouest.**
Station: Strasbourg 05002. Tableau des erreurs
moyennes quadratiques internes.
Publ. Obs. Astron. Strasbourg, Vol. 2, Fasc. 3, 2 pp. (1972).

046.032 **Preparation of working ephemerides for determining**
the latitude in the Antarctic. F. D. Zablotskij.
Izv. vyssh. uchebn. zavedenij. Geod. i aehrofotos"emka, 1971,
No. 4, p. 47 - 53. In Russian. – Abstr. in Referativ. Zhurn. 52.
Geod. Aehros"emka, 7.52.110 (1972).

046.033 **Field tables for computing hour angle and azimuth**
of the Pole star for 1970 - 1980 (for latitudes from
20 to 65°). V. L. Kagan.
Nauka, Moskva, 52 pp. Price 15 Kop. (1971). In Russian.
Review in Referativ. Zhurn. 52. Geod. Aehros"emka, 7.52.112
(1972).

046.034 **Über den hypothetischen Charakter terrestrisch-**
dreidimensionaler Triangulationsergebnisse.
H. Wolf.
Mitt. Inst. Theor. Geod., Univ. Bonn, No. 5, 2 + 9 pp. (1972).

Global comparison between meridian and latitude
positions of the I.L.S. stars. See Abstr. 041.020.

The inertial system and its importance for geodesy.
See Abstr. 041.045.

Time data and navigation. See Abstr. 044.006.

Space geodesy and dynamics of the earth and the
moon. Recent research; Programs in France.
See Abstr. 094.184.

047 Ephemerides, Almanacs, Calendars

047.001 **The use of solar eclipse timings to compare the**
reference systems of Newcomb's tables of the sun
and of the Improved Lunar Ephemeris.
R. L. Duncombe, R. F. Haupt, J. S. Duncombe.
Solar Physics, Vol. 21, 260 - 262 (1971).
Photoelectric, photographic, and visual observations of
second and third contacts of the total solar eclipses of 12 November, 1966 and 7 March, 1970 are analyzed. The observations indicate a difference in the reference system of Newcomb's tables of the sun as compared to the reference system
of the Improved Lunar Ephemeris (J = 2) of $0.^{s}911 \pm 0.^{s}034$
m.e.

047.002 **Astronomical Yearbook of the USSR for the year**
1974.
V. K. Abalakin (Editor).
Institut Teoreticheskoj Astronomii Akademii Nauk SSSR.
Izdatel'stvo "Nauka", Leningradskoe Otdelenie, Leningrad.
719 pp. Price 7 Rbl. 55 Kop. (1972). In Russian.

047.003 **Events of 1972 in the graphic time table.**
Sky Telescope, Vol. 43, 33 - 35 (1972).
This graphic time table is published annually by the Mary-
land Academy of Sciences, Baltimore.

047.004 **The Handbook of the British Astronomical Associa-**
tion 1972.
Prepared by the Computing Section of the Association under
the supervision of C. Dinwoodie.

Office of the Association: Burlington House, Piccadilly, London. 81 pp. Price 11s., $ 1.60 respectively (1971).

047.005 **The Observer's Handbook 1972.**
J. R. Percy (Editor).
Royal Astronomical Society of Canada, Toronto, Ontario,
Canada. 101 pp. Price $ 2.00 (1971).

047.006 **Almanac for Geodetic Engineers 1972.**
Prepared under the supervision of S. V. Inciong.
Republic of the Philippines – Department of Commerce and
Industry – Weather Bureau, Quezon City. 11 + 23 pp. (1971).

047.007 **Tables of Sunrise, Sunset, Twilight, Moonrise and**
Moonset 1972.
Prepared under the supervision of S. V. Inciong.
Republic of the Philippines – Department of Commerce and
Industry – Weather Bureau, Quezon City. 10 + 33 pp. (1971).

047.008 **The Air Almanac 1972, May - August.**
Her Majesty's Stationery Office, London; United
States Naval Observatory, Washington. 2 + 248 + A84 + F4 pp.
Price £ 1.75 (1971).

047.009 **Annuario della Specola Cidnea per l'anno bisestile**
1972. Compiled by A. Valetti.
Published by Civica Specola Astronomica Cidnea, Municipio
di Brescia (Italia). 32 pp. (1972).

047.010 **Connaissance des Temps** ou des mouvements célestes pour l'an 1973 à l'usage des astronomes et des navigateurs.
Publié par le Bureau des Longitudes. Gauthier-Villars éditeur, Paris. 42 + 645 pp. Price F 220 (1972).

047.011 **Astronomical Yearbook 1972.**
Published by the Astronomical Society of Victoria, Melbourne. 36 pp. Price 40c., 60c. respectively (1971).

047.012 **Das Himmelsjahr 1972.** Sonne, Mond und Sterne im Jahre 1972.
Compiled by M. Gerstenberger.
Kosmos-Verlag, Franckh'sche Verlagshandlung, Stuttgart. 110 pp. Price DM 6.80 (1971).

047.013 **Kalender für Sternfreunde 1972.** Kleines astronomisches Jahrbuch.
P. Ahnert (Editor).
Johann Ambrosius Barth, Leipzig. 216 pp. Price DM 4.80 (1971).

047.014 **Philippine Astronomical Handbook 1972.**
Prepared under the supervision of S. V. Inciong.
Republic of the Philippines – Department of Commerce and Industry – Weather Bureau, Quezon City. 12 + 51 pp. (1971).

047.015 **Astronomische Grundlagen für den Kalender 1974.**
Edited by Astronomisches Rechen-Institut in Heidelberg. Verlag G. Braun GmbH., Karlsruhe. 84 pp. Price DM 28.00 (1972).

047.016 **The Handbook of the British Astronomical Association 1972.**
Prepared by the Computing Section of the Association under the supervision of C. Dinwoodie.
Office of the Association: Burlington House, Piccadilly, London. 84 pp. Price £ 0.40, £ 0.55 respectively (1971).

047.017 **Elementi astronomici per il calendario dell'anno 1972 calcolati all'Osservatorio Astrofisico di Arcetri-**
Firenze. G. Righini.
Boll. Geod. Sci. Affini, Anno 31, p. 81 - 95 (1972).

047.018 **Kalendarzyk astronomiczny na 1972 rok.**
G. Sitarski.
Appendix to Urania Kraków, Vol. 43, No. 1, 12 pp. (1972).

047.019 **Phénomènes astronomiques rares.** J. Meeus.
Ciel et Terre, Vol. 88, 205 - 211 (1972).

047.020 **Éphémérides Nautiques pour l'an 1973.** Ouvrage publié par le Bureau des Longitudes spécialement à l'usage des marins.
Gauthier-Villars Éditeur, Paris. 479 pp. (1972).

047.021 **Annuario 1972.**
Edited by Osservatorio Astronomico di Torino, [printed by Scuola Salesiana del Libro, Catania]. Pubbl. Varie Fuori Ser., Oss. Astron. Torino, No. 47, 67 pp. (1971).

047.022 **The American Ephemeris and Nautical Almanac for the year 1973.**
Issued by Nautical Almanac Office, United States Naval Observatory, Washington; Her Majesty's Nautical Almanac Office, Royal Greenwich Observatory, London, U.S. Government Printing Office, Washington. 8 + 584 pp. Price $ 6.25 (1971).

047.023 **Blick in die Sternenwelt 1972.** Astronomischer Kalender der Archenhold-Sternwarte. E. Rothenberg.
Vorträge und Schriften, Archenhold-Sternw. Berlin-Treptow, No. 40, 48 pp. (1971).

047.024 **The Air Almanac 1972, September – December.**
Her Majesty's Stationery Office, London; United States Naval Observatory, Washington. 246 + A84 + F4 pp. Price £ 1.75 (1972).

047.025 **Almanaque Nautico, 1973.**
Published by Instituto y Observatorio de Marina, San Fernando (Cádiz). Printed in Spain by Imprenta del Observatorio de Marina, San Fernando. 4 + 412 + 30* pp. (1972).

Space Research

051 Extraterrestric Research, Spaceflight Related to Astronomy

051.001 **Introduction to the Viking symposium.**
T. Gold.
Icarus, Vol. 16, II - III (1972).

051.002 **The Soviet space program: Effort said to surpass peak U.S. level.** R. Gillette.
Science, Vol. 175, 731 - 736 (1972).

051.003 **L'avenir du programme spatial américain ou l'exploration de l'espace au service de l'aménagement de la terre.** R. Dejaiffe.
Ciel et Terre, Vol. 88, 73 - 104 (1972).

051.004 **Interstellar beacons.** J. W. Macvey.
Spaceflight, Vol. 14, 14 - 16, 25 (1972).

051.005 **Space astronomy.** J. S. Griffith.
Spaceflight, Vol. 14, 139 - 142 (1972).

051.006 **Trojan communication systems.** J. Strong.
Spaceflight, Vol. 14, 143 - 144 (1972).

051.007 **Advanced X-ray observatories.** H. Gursky.
Publ. Astron. Soc. Pacific, Vol. 84, 99 - 109 (1972).
Paper presented at the symposium "Advanced Electronic Systems for Astronomy", 1971, Santa Cruz.

051.008 **NASA plans for 1972.** R. N. Watts, Jr.
Sky Telescope, Vol. 43, 92 - 93 (1972).

051.009 **Soviet space activity.** R. N. Watts, Jr.
Sky Telescope, Vol. 43, 357 (1972).

051.010 **Some preliminary results of the Soviet solar stratospheric observatory during the third flight.**
V. A. Krat.
Uspekhi fiz. nauk, Vol. 105, p. 769 - 770 (1971). In Russian.
Abstr. in Referativ. Zhurn. 51. Astron., 5.51.346 (1972).

051.011 **The complex rocket experiment "sun–atmosphere".**
G. S. Ivanov-Kholodnyj, T. V. Kazachevskaya.
Vestn. AN SSSR, 1971, No. 11, p. 50 - 56. In Russian.
Abstr. in Referativ. Zhurn. 51. Astron., 5.51.461; 62. Issled. kosm. prostranstva, 5.62.284 (1972).

051.012 **Prospects of extraatmospheric astronomical research.**
V. G. Kurt.
Priroda, No. 5.72, p. 13 - 19 (1972). In Russian.

051.013 **Improved OSO for 1 arc sec observations of the quiet sun.**
S. P. Maran, J. M. Thole, K. I. Kissin, R. J. Thomas.
Bull. American Astron. Soc., Vol. 4, 209 (1972). – Abstr. AAS.

051.014 **Twenty-five years of rocket astronomy.**
H. Friedman.
Bull. American Astron. Soc., Vol. 4, 255 (1972). – Abstr. AAS.

051.015 **Kooperation und Integration der sozialistischen Staaten in der Raumforschung.** H. Hoffmann.
Astron. in der Schule, 9. Jahrgang, p. 28 - 33 (1972).

051.016 **American plan of space exploration in the seventies.**
K. Ziołkowski.
Urania Kraków, Vol. 43, 112 - 115 (1972). In Polish.

051.017 **Radiations in outer space.** F. Zwicky.
Astronaut. Acta, Vol. 17, 129 - 135 (1972).
In the present study we present a brief review of the various aspects of the many types of radiations that spacecraft and astronauts will encounter on their travels.

051.018 **On the mode of operation of an orbital astronomical observatory manned by a cosmonaut.**
G. A. Gurzadyan.
Soobshch. Byurakan. Obs., No. 45, p. 5 - 11 (1972).
In Russian.
The paper deals with the problem of high-precision astronomical observations with the help of an astrophysical observatory installed in a multipurpose orbital station and manned by a cosmonaut who is not a professional astronomer.

051.019 **The Orbital Astrophysical Observatory "Orion".**
G. A. Gurzadyan, E. A. Harutyunyan.
Soobshch. Byurakan. Obs., No. 45, p. 12 - 19 (1972).
In Russian.
A description is given of the construction, optical and kinematic parts of the Astrophysical Observatory "Orion", installed in the orbital station "Salyut".

051.020 **The control system of the automatic operation of the "Orion" equipment.**
M. N. Krmoyan, A. Z. Zakharyan, Sh. M. Harutyunyan.
Soobshch. Byurakan. Obs., No. 45, p. 27 - 35 (1972).
In Russian.

051.021 **De suksesvolle lancering van ESRO's Thor-Delta-1A satelliet.** H. F. van Beek, C. de Jager, H. Lamers.
Hemel en Dampkring, Vol. 70, 171 - 178 (1972).

051.022 **Mond- und Planetenforschung im Spiegel des Weltraumfluges.** S. Marx.
Sterne, 48. Jahrgang, p. 95 - 103 (1972).

051.023 **The sun, moon and planets as observed from space vehicles.** D. H. Menzel.
Monthly Notes Astron. Soc. Southern Africa, Vol. 31, 51 - 52 (1972). – Abstract.

051.024 **The study of the moon, planets and stars using the instruments of space research.** I. Almar.
Fiz. Szemle, Vol. 21, 248 - 251 (1971). In Hungarian.
The results of the American and Soviet space programmes for the study of the moon, Mars and Venus from 1959 are reviewed. The programmes include radar and laser studies conducted from earth.

051.025 **Natural nuclear radiation environments for the**

Grand Tour missions. J. W. Haffner.
IEEE Trans. Nuclear Sci., Vol. NS-18, 443 - 453 (1971). – See Phys. Abstr., Vol. 75, No. 23113 (1972).

051.026 **SOLRAD 10(C).**
Naval Res. Rev., (*USA*), Vol. 24, No. 7, p. 1 - 10 (1971).
SOLRAD 10(C) is the third in a series of solar radiation spacecraft launched in a joint NRL/NASA program. Solar activity can be used to predict periods of probability of solar flares. This will be of great importance to communications, meteorology and manned space flights.

051.027 **An experimental study on the visual detection of stars in a spacecraft environment.** R. P. Heinisch.
Journ. Spacecraft & Rockets, (*USA*), Vol. 8, 852 - 858 (1971).

051.028 **Trajectory requirements for comet rendezvous.**
A. L. Friedlander, J. C. Niehoff, J. I. Waters.
Journ. Spacecraft & Rockets, (*USA*), Vol. 8, 858 - 866 (1971). – See Phys. Abstr., Vol. 75, No. 26864 (1972).

051.029 **Water and oxygen particles in space from Apollo missions.**
R. D. Sharma, M. L. Kratage, A. C. Buffalano.
Journ. Spacecraft & Rockets (*USA*), Vol. 8, 1230 - 1232 (1971). – See Phys. Abstr., Vol. 75, No. 30600 (1972).

051.030 **Design of the Particles Experiment Subsystem of the Apollo lunar subsatellite.**
J. J. Baum, R. W. Harman, R. G. Maronde, T. A. Hornbuckle.
IEEE Trans. Nuclear Sci., Vol. NS–19, 632 - 639 (1972).
The Particles Experiment Subsystem of the Apollo 15 Particles and Fields Subsatellite is now providing electron and proton energy spectra and flux information from lunar orbit. The experiment employs two silicon detector telescopes and a set of spherical plate electrostatic analyzers. Design of the overall instrument package is discussed and some of the data obtained presented.

051.031 **Potential contributions of the United States Space Program to exploration of the solar system.**
H. E. Newell, D. H. Herman, P. Tarver.
From plasma to planet. 21st Nobel symposium 1971, (see 012.028), p. 285 - 314 (1972).

051.032 **Exploration in the solar system with electric spacecraft.** E. Stuhlinger.
IAU Colloquium No. 12, (see 012.027), p. 489 - 501 (1971/72).

051.033 **Manned mission to an asteroid.** H. Hall.
IAU Colloquium No. 12, (see 012.027), p. 539 - 541 (1971/72).

051.034 **Design and science instrumentation of an unmanned vehicle for sample return from the asteroid Eros.**
H. F. Meissinger, E. W. Greenstadt.
IAU Colloquium No. 12, (see 012.027), p. 543 - 560 (1971/72).

051.035 **Potentials of asteroid space missions.**
A. Bratenahl.
IAU Colloquium No. 12, (see 012.027), p. 561 - 566 (1971/72).

051.036 **Manmade objects – a source of confusion to asteroid hunters?** K. Aksnes.
IAU Colloquium No. 12, (see 012.027), p. 649 - 652 (1971/72).

051.037 **Astronautics in the year 1971.** J. Bouška.
Říše hvězd, Vol. 53, 121 - 128 (1972). In Czech.
Progress report.

051.038 **Astronautica.**
Coelum, Vol. 40, 22 - 26, 66 - 68, 113 - 117 (1972).

051.039 **Space report.**
Spaceflight, Vol. 14, 28 - 29, 58 - 61, 94 - 95, 117, 128 - 131, 133 (1972).
X-ray source identified, p. 29; Animals around the moon, p. 59 - 60; Major Soviet 'Sky Watch' project, p. 60 - 61; Solar radio emissions in stereo, p. 128; Solar flare optically recorded, p. 128; OSO observes solar activity, p. 129; Artificial gravity experiments, p. 129 - 130; Is there life on Mars?, p. 130.

051.040 **Space report.**
Spaceflight, Vol. 14, 173 - 178, 182, 184, 224 - 226, 252 - 261, 264, 270 (1972).
Mars probes still active, p. 173; 'Double umbra' photo's, p. 175; Vistas of Mars, p. 176 - 178; Plans for Venus, p. 224; Space map of Mars, p. 224; Vistas of Mars, p. 225; On course for Jupiter, p. 226; Astronomy on the moon, p. 252 - 253; Skylab solar studies, p. 255; Grand tour substitution, p. 258; Venera 8, p. 259; New light on Mars, p. 259 - 260; Pioneers to Venus, p. 260.

051.041 **Remarks on techniques for future studies.**
T. Gehrels.
From plasma to planet. 21st Nobel symposium 1971, (see 012.028), p. 377 - 380 (1972).

051.042 **Planning of space experiments.** H. E. Newell.
From plasma to planet. 21st Nobel symposium 1971, (see 012.028), p. 381 (1972).

051.043 **Studies of other planetary systems with space techniques.** Z. Kopal.
From plasma to planet. 21st Nobel symposium 1971, (see 012.028), p. 381 - 384 (1972).

051.044 **Missions spatiales vers les planètes. IV. Physique des surfaces planétaires.**
ESRO, S.P. No. 56, 4 + 39 pp. (1970). – See Bull. Signal., Vol. 33, Section 120, No. 1491 (1972).

051.045 **ESRO study programme for a space experiment on gravitation theories.** G. M. Israël.
ESRO Sci. Memorandum, No. 78, 5 + 9 pp. (1971). – See Bull. Signal., Vol. 33, Section 120, No. 2678 (1972).

051.046 **Space research and observational cosmology.**
D. W. Sciama.
E. S. R. O., S. P., No. 52, (see 012.030), p. 5 - 8 (1971).

051.047 **The moon's first astronomical observatory.**
H. J. P. Arnold.
British Journ. Photography, Vol. 119, 310 - 311, 313 (1972).
The equipment consists of a combined far-ultra-violet camera/spectrograph, and the primary purpose of the experiment will be to measure the amount and excitation of hydrogen in nearby and distant regions of the universe.

Space travel to feature grand tour of outer planets.
IEEE Spectrum, Vol. 9, No. 1, p. 108 - 109 (1972).

Outer planets: From Jupiter to Saturn.
Nature, Vol. 236, 8 (1972).

Where next with Apollo?
Nature, Vol. 236, 418 (1972).

Energy calibration of the "Orion" spectrograph.
See Abstr. 034.050.

Stray light suppression in sunshields and optical systems: Measurements on three space experiments. See Abstr. 034.119.

See Abstr. 036.007.

More about Apollo 16.   See Abstr. 094.177.

An investigation of the photographic characteristics of the film UFSH-4 used in the "Orion" equipment.

Investigation of solar system evolution by automatic vehicles on the moon.   See Abstr. 107.020.

052 Astrodynamics and Navigation of Space Vehicles

052.001 On the use of the Hill variables in artificial satellite theory: Brouwer's theory.   K. Aksnes.
Astron. Astrophys., Vol. 17, 70 - 75 (1972).

By reformulating Brouwer's first-order satellite theory in terms of the Hill variables $(\dot{r}, G, H, r, u, h)$, it is demonstrated that these variables possess several advantages over the Delaunay variables (L, G, H, l, g, h). An algorithm for the computation of position and velocity is presented which is significantly simpler and faster in use than Brouwer's algorithm. By means of numerical integration of the equations of motion, it is found that the algorithm is capable of predicting the position of a close earth satellite to ±60 meters over at least six days.

052.002 Optimum aim point biasing in case of a planetary quarantine constraint.
G. S. Gedeon, V. N. Dvornychenko.
Celestial Mechanics, Vol. 5, 144 - 156 (1972).

NASA has initiated a planetary quarantine program, allowing a certain probability of impact for each mission. This paper discusses how planetary quarantine affects midcourse guidance. In the first part the concept of a quarantine contour is introduced, in the second part it is shown how to find the optimum biased aim points on this contour in order to minimize midcourse requirements.

052.003 Interaction between attitude libration and orbital motion of a rigid body in a near Keplerian orbit of low eccentricity.   S. N. Mohan, J. V. Breakwell, B. O. Lange.
Celestial Mechanics, Vol. 5, 157 - 173 (1972).

Interaction between orbital motion and attitude libration dynamics of an arbitrary rigid body moving in a central Newtonian field is considered to second order. An averaged solution to the nonlinear inplane-pitch equations is presented. The results show that the near-resonant motion is characterized by a periodic interchange of energy between the attitude and orbital motion.

052.004 An intermediate matching technique for solving two point boundary value problems using the perturbation method.   W. E. Williamson.
Celestial Mechanics, Vol. 5, 174 - 188 (1972).

The perturbation method, a numerical method for solving two point boundary value problems (TPBVP), is modified to attempt to improve inherent instability and sensitivity problems associated with the method. The desired solution to the TPBVP is divided into two time intervals. The equations for both time segments can be integrated simultaneously. To show the effectiveness of the method, two impulse trajectories which minimize the total velocity increment required to transfer a spacecraft from an earth orbit into a lunar orbit are calculated.

052.005 Theory of an experiment in an orbiting space laboratory to determine the gravitational constant.
J. P. Vinti.
Celestial Mechanics, Vol. 5, 204 - 254 (1972).

The paper analyzes an experiment in an orbiting laboratory to determine the gravitational constant G. A massive sphere, according to a suggestion of L. S. Wilk, is to have three tunnels drilled through it along mutually perpendicular diameters. The sphere either floats in the orbiting laboratory or is tethered to the spacecraft. Each tunnel contains a small test object, which is held on the tunnel's axis by means of a suspension system and held at rest relative to the sphere by slowly rotating the latter by means of inertia reaction wheels, governed by a servomechanism. Fundamentally, one balances the gravitational forces on the test objects by centrifugal force, determines the latter by measuring the components of angular velocity, and calculates G from the resulting balance. In Appendix B another version of the experiment is devised.

052.006 Motion of an earth satellite under the influence of tesseral harmonics.
Yu. V. Batrakov, L. L. Filenko.
Byull. Inst. Teoret. Astron., *Leningrad*, Vol. 13, 73 - 91(1972). In Russian.

Analytical expressions have been obtained for the perturbations of the first order of the orbital elements of an earth satellite due to the tesseral harmonics of the earth potential with $k, l \leqslant 4$. The coefficients of the trigonometric terms of these expressions have been given as polynomials in powers of sinus and cosinus of orbital inclination and as power series in eccentricity with an accuracy up to e^4.

052.007 The programming of literal expansions in the theory of the motion of earth satellites in case of perturbations due to zonal harmonics of arbitrary order.
V. I. Lapshina.
Byull. Inst. Teoret. Astron., *Leningrad*, Vol. 13, 111 - 124 (1972). In Russian.

The modern methods for determining literal expressions in the problems of celestical mechanics by a computer are given. Formal features of literal expressions programs are pointed out. The program of analytical computations resulting in explicit form of perturbations of the orbital elements due to zonal harmonics of the gravitational potential of the earth according to the theory of Yu. V. Batrakov (1971) is described in detail.

052.008 Canonical satellite theory. General discussion and conclusions.   E. Stiefel.
Canonical satellite theory, (see 013.003), p. 9 - 23 (1972).

052.009 Generalized Delaunay elements with special regard

to the true anomaly as the independent variable (TR-theory).   G. Scheifele.
Canonical satellite theory, (see 013.003), p. 25 - 90 (1972).

052.010   **Refinements of the TR-theory and numerical results.**   R. Kunz.
Canonical satellite theory, (see 013.003), p. 91 - 125 (1972).

052.011   **Discussion of small eccentricities or inclinations in the TR-theory.**   N. Sigrist.
Canonical satellite theory, (see 013.003), p. 127 - 137 (1972).

052.012   **Numerical treatment of additional perturbations based on a TR-orbit.**   G. Scheifele, N. Sigrist.
Canonical satellite theory, (see 013.003), p. 139 - 150 (1972).

052.013   **Short description of the theory in polar-KS-variables (PKS-theory).**   U. Kirchgraber.
Canonical satellite theory, (see 013.003), p. 151 - 160 (1972).

052.014   **Computing of satellite coordinates for arbitrary inclinations and eccentricities smaller than the Laplacian limit.**   V. V. Terent'ev.
Vestn. Leningr. un-ta, 1971, No. 19, p. 140 - 146. In Russian. Abstr. in Referativ. Zhurn. 51. Astron., 5.51.131 (1972).

052.015   **Optimal trajectories between earth and Mars in their true planetary orbits.**
J. P. Gravier, C. Marchal, R. D. Culp.
Journ. Optimization Theory Appl. Vol. 9, 120 - 136 (1972).

052.016   **Interstellar navigation.**   J. R. Wertz.
Spaceflight, Vol. 14, 206 - 216 (1972).

052.017   **On the possibility of using the gravitational field of Jupiter for the flight to a given distance from the sun and for leaving the ecliptic plane.**
N. G. Khavenson, P. E. Ehl'yasberg.
Kosm. Issled., Vol. 10, 159 - 166 (1972). In Russian.

052.018   **On some methods to increase the accuracy of computation of interplanetary trajectories.**
V. P. Kuraev, A. A. Panchukov.
Kosm. Issled., Vol. 10, 167 - 171 (1972). In Russian.

052.019   **Analytical estimates of the accuracy of determination of the parameters of motion of artificial earth satellites in reducing astronomical measurements with correlated errors.**   L. F. Porfir'ev, V. V. Smirnov.
Kosm. Issled., Vol. 10, 172 - 180 (1972). In Russian.

052.020   **On the determination of characteristics of the optimum orbit of an artificial satellite of a planet analyzing the flight earth – planet – earth.**   L. B. Livanov.
Kosm. Issled., Vol. 10, 295 - 296 (1972). In Russian. – Brief information.

052.021   **Interplanetary trajectories by quasilinearization with Chebyshev series.**   T. Feagin.
Bull. American Astron. Soc., Vol. 4, 262 (1972). – Abstr. AAS.

052.022   **A numerical artificial satellite theory.**
P. E. Nacozy.
Bull. American Astron. Soc., Vol. 4, 264 (1972). – Abstr. AAS.

052.023   **A satellite paradox.**   A. Drożyner.
Urania Kraków, Vol. 43, 76 - 78 (1972). In Polish.

052.024   **Optimal deterministic guidance for bounded-thrust spacecrafts.**   A. L. Kornhauser, P. M. Lion.
Celestial Mechanics, Vol. 5, 261 - 281 (1972).

The minimum-propellant deterministic guidance law for bounded-thrust, constant jet-exhaust velocity, spacecrafts is developed using the neighboring extremal theory. The solution is applied to a three-burn earth-Mars transfer.

052.025   **Analytical guidance in the neighborhood of optimal multi-impulse trajectories.**
A. L. Kornhauser, P. M. Lion.
Celestial Mechanics, Vol. 5, 282 - 299 (1972).
A first-order minimum propellant guidance law is developed for multi-impulse trajectories in an inverse-square gravitational field. A second-order variational analysis is used to formulate the guidance problem as an accessory minimum problem. It is shown that for multi-impulse transfers it is, in general, non-optimal to add impulses. All corrections to the trajectory should be made by a combination of small changes in timing, magnitude and direction of the nominal impulses.

052.026   **A perturbation technique applied to an optimal re-entry control problem.**
K. R. Bucher II, B. L. Pierson.
Astronaut. Acta, Vol. 17, 239 - 244 (1972).

052.027   **Theoretical determination of the aerodynamic forces on satellites.**   S. Nocilla.
Astronaut. Acta, Vol. 17, 245 - 258 (1972).
In the present paper a brief critical survey of theoretical methods is given. Results are reported about lift and drag on cones, cylinders and spheres. Comparisons with other theoretical results are also reported.

052.028   **Encounter in space.**   V. B. Sokolov.
Zemlya i Vselennaya, 1972, No. 3, p. 14 - 19.
In Russian.

052.029   **Stationary satellites.**   V. I. Levantovskij.
Zemlya i Vselennaya, 1972, No. 3, p. 34 - 39.
In Russian.

052.030   **Secular inequalities in the motion of artificial earth satellites.**   E. P. Aksenov, I. P. Prokhorova.
Astron. Zhurn. Akad. Nauk SSSR, Vol. 49, 630 - 640 (1972). In Russian. English translation in Soviet Astron. AJ, Vol. 16, No. 3.
Secular perturbations of elements of artificial satellites caused by non-sphericity of the earth are investigated. The dependence of secular perturbations on the semi-major axis, eccentricity, inclination and order of spherical harmonics of the earth's potential is investigated.

052.031   **Termes à courtes périodes du second ordre dans la théorie d'un satellite artificiel.**   P. Bretagnon.
Bull. Groupe Recherches Géod. Spatiale, Obs. Meudon, No. 2, p. 1 - 33 (1972).

052.032   **Comparison théorie analytique—intégration numérique.**   X. Berger.
Bull. Groupe Recherches Géod. Spatiale, Obs. Meudon, No. 2, 35 - 47 (1972).

052.033   **A computer-generated motion picture showing a rendezvous mission with comet Encke.**
H. E. Newell.
From plasma to planet. 21st Nobel symposium 1971, (see 012.028), p. 115 - 116 (1972).

052.034   **Asteroid rendezvous missions.**
D. F. Bender, R. D. Bourke.
IAU Colloquium No. 12, (see 012.027), p. 503 - 511 (1971/72).

052.035   **Sample-return missions to the asteroid Eros.**

A. C. Mascy, J. Niehoff.
IAU Colloquium No. 12, (see 012.027), p. 513 - 526 (1971/72).

052.036 Multiple asteroid flyby missions.
D. R. Brooks, W. F. Hampshire II.
IAU Colloquium No. 12, (see 012.027), p. 527 - 537 (1971/72).

052.037 Satellite orbit computations using gravity anomalies.
R. H. Rapp.
Studia, Vol. 16, 1 - 9 (1972).

052.038 Solution of the low-altitude satellite equations.
Tyn Myint-U.
Ann. New York Acad. Sci., Vol. 172, 679 - 719 (1971). – See
Bull. Signal., Vol. 33, Section 120, No. 4876 (1972).

052.039 Approximations of interplanetary trajectories by Chebyshev series. P. E. Nacozy, T. Feagin.
AIAA Journ., Vol. 10, 243 - 244 (1972).
A method to obtain rapidly convergent series approximations of trajectories with close planetary encounters is presented. The expansions are obtained as finite Chebyshev polynomial series with numerical coefficients.

On the periodic solutions of slowly spinning gravity gradient system. See Abstr. 042.011.

Recherche et étude des trajectoires planes du centre d'inertie d'un corps de révolution placé dans un champ new-tonien. See Abstr. 042.022.

Computational aspects of asymptotic matching in the restricted three-body problem. See Abstr. 042.065.

Geodetic latitude and altitude from geocentric coordinates. See Abstr. 046.023.

Space geodesy and dynamics of the earth and the moon. Recent research; Programs in France. See Abstr. 094.184.

053 Lunar and Planetary Probes and Satellites

053.001 A message from earth.
C. Sagan, L. Salzman Sagan, F. Drake.
Science, Vol. 175, 881 - 884 (1972).

053.002 Soviet exploration of Mars. R. N. Watts, Jr.
Sky Telescope, Vol. 43, 91 - 92 (1972).

053.003 Mariner star photographs. R. N. Watts, Jr.
Sky Telescope, Vol. 43, 92 (1972).

053.004 Plans for Apollo 16. R. Hillenbrand.
Sky Telescope, Vol. 43, 151 - 154 (1972).

053.005 Pioneers to Jupiter. D. Baker.
Spaceflight, Vol. 14, 111 - 117 (1972).

053.006 Objects on the moon.
Compiled by G. Falworth.
Spaceflight, Vol. 14, 145 - 150 (1972).

053.007 Possible intersection of a geomagnetic trace at the distance of 3000 earth radii during the flight of the automatic interplanetary station Mars 3.
O. L. Vaisberg, A. V. Bogdanov, N. F. Borodin, V. G. Kova-
lenko, A. A. Zertsalov, B. V. Polenov, S. A. Romanov, V. I.
Subbotin.
Dokl. Akad. Nauk SSSR, Ser. Mat. Fiz., Vol. 203, 309 - 310
(1972). In Russian.

053.008 Lunokhod 1 on the lunar surface.
K. P. Florenskij.
Zemlya i Vselennaya, 1972, No. 2, p. 6 - 8. In Russian.

053.009 Laser experiment on Lunokhod 1. Zh. Yusson.
Zemlya i Vselennaya, 1972, No. 2, p. 13 - 15. In
Russian. – Translated from French by M. D. Zakolupina.

053.010 Pioneer 10 mission to Jupiter. R. N. Watts, Jr.
Sky Telescope, Vol. 43, 299 (1972).

053.011 Soviet Venus probe. R. N. Watts, Jr.
Sky Telescope, Vol. 43, 302 (1972).

053.012 Exploration of the moon and planets by space probes. B. A. Volynskij, G. I. Malakhova.
Uch. zap. Yaroslav. gos. ped. in-t, 1971, vyp. (No.) 81, p. 5 -
28. In Russian. – Abstr. in Referativ. Zhurn. 51. Astron.,
5.51.251; 62. Issled. kosm. prostranstva, 5.62.216 (1972).

053.013 Heat flow and convection demonstration experi-ments aboard Apollo 14.
P. G. Grodzka, T. C. Bannister.
Science, Vol. 176, 506 - 508 (1972).
A group of experiments was conducted by Apollo 14
astronaut Stuart A. Roosa during the lunar flyback on 7 February 1971 to obtain information on heat flow and convection
in gases and liquids in an environment of less than 10^{-6} g
gravity. Flow observations and thermal data are discussed.

053.014 Mission to Descartes – 1. D. Baker.
Spaceflight, Vol. 14, 246 - 251 (1972).

053.015 Pioneer 10 on course. R. N. Watts, Jr.
Sky Telescope, Vol. 44, 13 (1972).

053.016 Analytical estimate of the descent distance of a spaceship for hyperbolic return trajectories.
N. M. Ivanov, V. G. Sobolevskij.
Kosm. Issled., Vol. 10, 326 - 337 (1972). In Russian.

053.017 Artificial satellites of Mars. I. Smirnov.
Nauka i zhizn', 1972, No. 3, p. 19. In Russian.
Abstr. in Referativ. Zhurn. 51. Astron.,7.51.62 (1972).

053.018 Pioneer 10 on flight to Jupiter. D. Knežević.
Vasiona, Vol. 20, 11 - 12 (1972). In Serbo-Croatian.

053.019 Pretposlednje putovanje na Mesec brodom "Apolo-16". [The last but one travel to the moon with Apol-lo 16.] D. Knežević.

Vasiona, Vol. 20, 25 - 27 (1972).

053.020 **The Pioneer mission to Jupiter.**
National Aeronautics and Space Administration.
NASA SP-268, 46 pp. (1971). — *IDP*

053.021 **A method of orbit determination using overlapping television pictures.**   J. K. Campbell.
Journ. Spacecraft & Rockets, (*USA*), Vol. 8, 867 - 872 (1971).
See Phys. Abstr., Vol. 75, No. 26845 (1972).

053.022 **Description of Pioneer F and G asteroid belt penetration experiment.**
W. H. Kinard, R. L. O'Neal.

IAU Colloquium No. 12, (see 012.027), p. 607 - 615 (1971/72).

053.023 **Precision of ephemerides for space missions.**
B. G. Marsden.
IAU Colloquium No. 12, (see 012.027), p. 639 - 642 (1971/72).

The Viking missions to Mars.
See Abstr. 097.008.

Imaging experiment: The Viking Mars Orbiter.
See Abstr. 097.009.

Radio science experiments: The Viking Mars Orbiter and Lander.   See Abstr. 097.012.

054   Artificial Earth Satellites

054.001 **The change in satellite orbital inclination caused by a rotating atmosphere with day-to-night density variation.**   D. G. King-Hele, D. M. C. Walker.
Celestial Mechanics, Vol. 5, 41 - 54 (1972).

The theory specifying the change Δi in a satellite's orbital inclination due to atmospheric rotation, in terms of the decrease in orbital period ΔT, has been extended to an atmosphere with simusoidal variation of density between day and night.

054.002 **Analytical study of the earth's shadowing effects on satellite orbits.**   S. Ferraz-Mello.
Celestial Mechanics, Vol. 5, 80 - 101 (1972).

A new mathematical model is proposed for the study of the effects of the direct solar radiation pressure on the orbit of an artificial earth satellite. The model leads to the non-existence of pure secular perturbations owing to the direct solar radiation pressure on the metric elements: semi-major axis, eccentricity and inclination. Numerical examples built with an approximation for the shadow function show that the secular inequalities on the angle variables — longitude, perigee and node — are very small.

054.003 **Ephemeris of a highly eccentric orbit: Explorer 28.**
B. E. Lowrey.
Celestial Mechanics, Vol. 5, 107 - 125 (1972).

An ephemeris has been obtained for Explorer 28 (IMP 3) which agrees well with 2 years of radio observations and with SAO observations a year later. This ephemeris is generated over the 3 year lifetime by a numerical integration method utilizing a set of initial conditions at launch and without requiring further differential correction. This ephemeris may be used as a standard for computing highly eccentric orbits in the earth-moon system.

054.004 **On the calculation of a satellite's coordinates with arbitrary inclinations and arbitrary eccentricities smaller than the Laplacian limit.**   V. V. Terent'ev.
Vestn. Leningr. un-ta, 1971, No. 19, p. 140 - 146. In Russian.
Abstr. in Referativ. Zhurn. 62. Issled. kosm. prostranstva, 3.62.262 (1972).

054.005 **On some special cases of evolution of orbits of light earth satellites.**   E. N. Polyakhova.
Vestn. Leningr. un-ta, 1971, No. 19, p. 133 - 139. In Russian.
Abstr. in Referativ. Zhurn. 62. Issled. kosm. prostranstva, 3.62.263 (1972).

054.006 **The stability of stationary motions of a satellite with a gyroscope in a central Newtonian field.**
R. S. Sulikashvili.
Izv. AN SSSR. Mekh. tverd. tela, 1971, No. 6, p. 3 - 6. In Russian. — Abstr. in Referativ. Zhurn. 62. Issled. kosm. prostranstva, 3.62.268 (1972).

054.007 **On the stability of stationary motions of a four-gyroscopic vertical mounted at a satellite in a Newtonian central field.**   V. I. Popov.
Izv. AN SSSR. Mekh. tverd. tela, 1971, No. 6, p. 25 - 28. In Russian. — Abstr. in Referativ. Zhurn. 62. Issled. kosm. prostranstva, 3.62.269 (1972).

054.008 **The use of basic satellite observations for improvement of orbital parameters for short time intervals.**
T. W. Kasimenko.
Nauchn. Informatsii, vyp. (No.) 18, p. 37 - 53 (1970).
In Russian.

A method of improving several orbital elements for a short time interval is given. The rectangular geocentric coordinates of a satellite obtained from simultaneous observations are used for improving orbital elements (mainly n) for a twenty-four hours interval.

054.009 **The SAS-D ultraviolet astronomical satellite.**
A. B. Underhill.
Publ. Astron. Soc. Pacific, Vol. 84, 91 - 98 (1972). — Paper presented at the symposium "Advanced Electronic Systems for Astronomy", 1971, Santa Cruz.

054.010 **On some particular cases of evolution of orbits of light earth satellites.**   E. N. Polyakhova.
Vestn. Leningr. un-ta, 1971, No. 19, p. 133 - 139. In Russian.
Abstr. in Referativ. Zhurn. 51. Astron., 4.51.164 (1972).

054.011 **Determination of the orientation of artificial earth satellites from magnetometric measurements.**
I. G. Khatskevich.
Kosm. Issled., Vol. 10, 3 - 13 (1972). In Russian.

054.012 **Orbital station "Salyut".**   K. P. Feoktistov.
Zemlya i Vselennaya, 1972, No. 2, p. 2 - 4. In Russian.

054.013 **Successful ESRO launch.**   R. N. Watts, Jr.
Sky Telescope, Vol. 43, 299 - 300 (1972).

054.014 Probabilistic theory and statistical distribution of earth satellites. N. G. Dennis.
Journ. British Interplanet. Soc., Vol. 25, 333 - 376 (1972).
Orbit motion is expressed in terms of probability densities of a number of variables. Densities and distributions, relating to both probability and expectation, are derived for altitude and latitude of a single satellite and a large population of satellites.

054.015 On a method for approximate computation of the motion of a stationary artificial earth satellite.
M. A. Vashkov'yak.
Kosm. Issled., Vol. 10, 147 - 158 (1972). In Russian.

054.016 Determination of the angular position of a non-oriented astronomical artificial earth satellite.
Eh. N. Kaurov, V. I. Prokhorenko.
Kosm. Issled., Vol. 10, 217 - 222 (1972). In Russian.

054.017 Precise ephemerides of artificial satellites using electronic computers.
M. Cîrşmaru, H. Alexandrescu.
Stud. Cerc. Astron., Vol. 17, 19 - 22 (1972). In Romanian.

054.018 The influence of shadow effects on the motion of artificial earth satellites. E. N. Polyakhova.
Vestn. Leningr. un-ta, 1972, No. 1, p. 138 - 144. In Russian. Abstr. in Referativ. Zhurn. 62. Issled. kosm. prostranstva, 6.62.259 (1972).

054.019 Table of earth satellites, Volume 2, Parts 1, 2 & 3: 1969, 1970 & 1971.
J. A. Pilkington, D. G. King-Hele, H. Hiller.
Published by Royal Aircraft Establishment, Farnborough, Hants, England. 5 + 116 pp. (1972).

054.020 Principle of satellite position determination by look angles from two stations using the great circle of simultaneousness and determination of the relative systematic error in time between the two stations. A. Schmitt.
Publ. Obs. Astron. Strasbourg, Vol. 2, Fasc. 5, 3 pp. (1972).

054.021 The influence of shadow effects on the motion of artificial earth satellites. E. N. Polyakhova.
Vestn. Leningr. un-ta, 1972, No. 1, p. 138 - 144. In Russian. Abstr. in Referativ. Zhurn. 51. Astron., 7.51.100 (1972).

054.022 Kunstmanen. J. Meeus.
Hemel en Dampkring, Vol. 70, 14 - 16, 158 - 160 (1972). – 1971 May – December.

054.023 Satellite digest.
Compiled by G. Falworth.
Spaceflight, Vol. 14, 30 - 31, 62 - 63, 96 - 98, 132 - 133, 179 - 180, 227 - 230, 262 - 264 (1972). – A monthly listing of all known artificial satellites and spacecraft, 1971 July – 1972 January.

A guide to earth satellites. See Abstr. 003.051.

055 Observations of Earth Satellites, Lunar and Planetary Probes

055.001 Resultaten van kunstmaanwaarnemingen.
J. Meeus.
Hemel en Dampkring, Vol. 70, 24 - 25 (1972).

055.002 Poljot 1 – 1963 43A (satellite). Equatorial coordinates (January–August 1969).
Rezul'taty Nablyud. Sovet. Iskusstv. Sputnikov Zemli, No. 131, 99 pp. (1970). In Russian.

055.003 Poljot 1 – 1963 43A (satellite). Equatorial coordinates (September–December 1969).
Rezul'taty Nablyud. Sovet. Iskusstv. Sputnikov Zemli, No. 132, 44 pp. (1970). In Russian.

055.004 Poljot 1 – 1963 43A (satellite). Horizontal coordinates (January–August 1969).
Rezul'taty Nablyud. Sovet. Iskusstv. Sputnikov Zemli, No. 133, 80 pp. (1970). In Russian.

055.005 Poljot 1 – 1963 43A (satellite). Horizontal coordinates (September–December 1969).
Rezul'taty Nablyud. Sovet. Iskusstv. Sputnikov Zemli, No. 134, 53 pp. (1970). In Russian.

055.006 Photographic tracking of artificial satellites.
A. G. Massevitch, A. M. Losinsky.
Nauchn. Informatsii, vyp. (No.) 18, p. 3 - 36 (1970). In Russian.

A review of the development of photographic tracking techniques in different countries during the recent 5–6 years is given. Main results of scientific investigations based on observations with Baker-Nunn cameras are considered. New large Soviet, English, French, German cameras are described. Important international projects on photographic tracking for satellite geodesy are reviewed. Main directions of research in this domain are discussed.

055.007 A determination of the time moments of topocentric directions for bright satellites on AFU-75 photographs. Ja. C. Balodis.
Nauchn. Informatsii, vyp. (No.) 18, p. 76 - 92, appendix, 18 pp. (1970). In Russian.
Formulae for determination of the time moments on AFU-75 photographs for bright satellites using electronic computers are developed. The final errors due to uncertain parameters of the shutter are given. The author proposes formulae for time determination in the case of a shutter moving non-uniformly.

055.008 A programme for determining the coordinates of tracking stations using the method of short arc orbit extrapolation. N. A. Sorokin, S. K. Tatevjan.
Nauchn. Informatsii, vyp. (No.) 18, p. 93 - 100 (1970). In Russian.
The description of a programme operated at the Astronomical Council for determinating the coordinates of satellite

tracking stations is given. A method of orbit extrapolation for short time intervals is used.

055.009 **The international programme of photographic tracking of the satellite Pageos Europe – Africa.**
N. N. Kovalenko.
Nauchn. Informatsii, vyp. (No.) 18, p. 101 - 107 (1970).
In Russian.

055.010 **Photographische Beobachtung künstlicher Erdsatelliten ohne Hilfe registrierender Zeiteinrichtungen.**
Teil II. R. Rajchl.
Bull. Astron. Inst. Czechoslovakia, Vol. 23, 181 - 194 (1972).

055.011 **Visual observations of artificial earth satellites in Finland, 1971 January – 1971 December.**
Directed and collected by P. Järvi.
Published by the University of Helsinki, 8 + 137 pp. (1972).

This volume contains visual satellite observations made at the Jokioinen meteorological observatory during 1971. At the beginning of 1971 the zero point of azimuth was changed so that it is expressed from zero at N, through 90 at E, 180 at S and 270 at W. The zero points of azimuth and the angle of elevation were checked every night on which satellite observations were made by recording the directions of four mires before and after the satellite observations. The reading accuracy of the observations is $0^s.1$ in time and $0°.1$ in position.

Theoretical Astrophysics

061 General Theoretical Problems of Astrophysics, Gravitational Instability, Neutrino Astronomy, X Ray- and Gamma Ray-Astronomy, Frequency and Origin of Elements etc.

061.001 Results and problems in the investigation of the synchrotron instability.
V. V. Zheleznyakov, E. V. Suvorov.
Astrophys. Space Sci., Vol. 15, 1 - 23, 24 - 43 (1972). In Russian and English.

A brief resumé of the results of the linear theory of the synchrotron instability is presented, obtained using the quantum method with the Einstein coefficients and the classical method with the kinetic equation. The expression for the growth-rate of synchrotron radiation in a system of relativistic electrons with an anisotropic momentum distribution is also found and investigated. Such a distribution might be realised in certain sources of cosmic radio emission (for example in pulsars). Certain problems are noted which have not yet been properly solved in studies of the synchrotron instability (the instability of a highly rarified plasma, the quasi linear theory of the synchrotron instability, and so on).

061.002 Instabilité gravitationnelle, analyse linéaire.
J.-P. Petit.
Comptes Rendus Acad. Sci. Paris, Sér. B, Vol. 274, 510 - 512 (1972).

L'équation de l'instabilité gravitationnelle, fournie par J. Jeans en 1902 se trouve réétablie par le truchement de la théorie cinétique.

061.003 Instabilité gravitationnelle, analyse non linéaire.
J.-P. Petit.
Comptes Rendus Acad. Sci. Paris, Sér. B, Vol. 274, 574 - 576 (1972).

Une analyse non linéaire de l'instabilité gravitationnelle est donnée où l'on suppose que la distribution de vitesse est localement maxwellienne. On utilise les concepts de la théorie cinétique des gaz et le formalisme de Chapman et Cowling.

061.004 X-ray astronomy: Observations of new phenomena.
W. D. Metz.
Science, Vol. 175, 397 - 399 (1972).

061.005 Synchrotron radiation from ultra-anisotropic particle distributions. R. I. Epstein.
Bull. American Astron. Soc., Vol. 3, 498 - 499 (1971). – Abstr. AAS.

061.006 The crystallization of neutronic matter.
J. W. Clark, N.-C. Chao.
Nature, Phys. Sci., Vol. 236, 37 - 39 (1972).

Solidification pressure of neutron matter has been estimated on the basis of the quantum mechanical principle of corresponding states, using density-dependent characteristic reduction parameters derived from a realistic two-nucleon interaction.

061.007 A variational principle for wave propagation in random media. I. Lerche.
Astrophys. Space Sci., Vol. 15, 206 - 213 (1972).

We set up a variational integral appropriate for discussing the 'eigenvalues' of the *exact* probability equation describing wave propagation in a turbulent medium. We illustrate the power, and accuracy, of the variational approach by several illustrative examples.

061.008 Equilibrium thermodynamics of nonideal reacting gases. H. C. Graboske, Jr.
Astrophys. Journ., Vol. 172, 689 - 698 (1972).

An analysis is made of the equilibrium thermodynamic properties of ionizing and dissociating gas mixtures in a system where many-body interactions are significant. A statistical-mechanical procedure is outlined which, for an appropriate many-body potential, includes a complete set of perturbed eigenstates, a nondivergent bound-state sum, and continuous and differentiable thermodynamic properties. Several theoretical models are applied to a solar-composition mixture to evaluate the region of validity of ideal-gas theory and the importance of various interaction effects.

061.009 Monotonously decreasing functions as the fundamental type of a frequency distribution function of cosmic objects and phenomena with respect to their importance. M. Kopecký.
Bull. Astron. Inst. Czechoslovakia, Vol. 23, 107 - 113 (1972).

The paper shows that a considerable part of cosmic objects and phenomena have the property that their frequency distribution with respect to their importance decreases monotonously. The discussion deals with some problems which occur in connection with this regularity. The main aim of the paper is to subject the presented problem to further criticism.

061.010 Selected properties of matter at high density and temperature. Equation of state and rates of beta processes and associated neutrino losses.
Z. Barkat, J.-R. Buchler, J. C. Wheeler.
Astrophys. Journ., Vol. 173, 183 - 194 (1972),

Current data on relevant nuclear parameters and rates of β-processes for a large number of nuclei are used to compute the key properties of matter which is assumed to be in nuclear statistical equilibrium at high density and temperature. We discuss the qualitative behavior of our results and the possible errors associated with them.

061.011 Variation of single particle mid-infrared emission spectrum with particle size.
G. R. Hunt, L. M. Logan.
Applied Optics, Vol. 11, 142 - 147 (1972).

The emission spectra of single particles of inorganic solids as a function of particle size have been recorded from $6\,\mu$ to $11.8\,\mu$. The data provide source functions necessary for determining the emission behavior of particulate samples in which temperature gradients exist, such as on the lunar surface. The data are of particular interest for interpreting the spectral behavior of circumstellar silicate particles.

061.012 Double-quantum photon-plasmon transition
$2S_{1/2} - 1S_{1/2}$.
S. A. Kaplan, E. B. Kleiman, I. M. Oringel.
Astron. Zhurn. Akad. Nauk SSSR, Vol. 49, 294 - 297 (1972).
In Russian. English translation in Soviet Astron. AJ, Vol. 16,
No. 2.
   The probabilities of double-quantum induced photon-
plasmon transition $2S_{1/2} - 1S_{1/2}$ for hydrogen atoms are calculated. Conditions under which the processes are more effective
than usual photon-photon emission are investigated. The role
of the processes in nebulae and in stellar chromospheres is discussed.

061.013 Neutron rich nuclei in Fermi gas.
   Yu. L. Vartanian, N. K. Ovakimova.
Astron. Zhurn. Akad. Nauk SSSR, Vol. 49, 306 - 315 (1972).
In Russian. English translation in Soviet Astron. AJ, Vol. 16,
No. 2.
   The equation of state at subnuclear densities when the
matter consists of atomic nuclei embedded in a degenerate
electron-neutron gas is considered.

**061.014 Non-linear theory of gravitational instability in the
expanding universe. III.** K. Tomita.
Progr. Theor. Phys., *Japan,* Vol. 47, 416 - 443 (1972).
   Quantitative analyses for the motion of inhomogeneities
are carried out by the use of hydrodynamical equations derived previously. The lower limits for initial perturbations
leading to galaxy formation are evaluated, and also a possibility for the formation of comparatively small galaxies is indicated.

**061.015 Magnetic properties of a degenerate electron gas and
implications for metals, white dwarfs, and neutron
stars.** R. F. O'Connell, K. M. Roussel.
Astron. Astrophys., Vol. 18, 198 - 200 (1972).
   In previous work we have considered the absolute stability of the Landau Orbital Ferromagnetic states for both a nonrelativistic and a relativistic electron gas at $T = 0$. We now extend this work to include the effects of a) non-zero temperatures, b) Coulomb interactions, c) effective mass values other
than unity.

061.016 Sternmaterie. I. Zustandsformen der Materie.
   F. Stöckmann.
Phys. Blätter, 28. Jahrgang, p. 153 - 160 (1972).

**061.017 Simplified method for measuring mean opacities
and effective absorption coefficients.** R. W. Patch.
Journ. Quant. Spectrosc. Radiat. Transfer, Vol. 12, 969 - 972
(1972).
   A method is given for measuring mean opacities and effective absorption coefficients by using wide bandwidths and
a number of path lengths.

061.018 Nonlinear condensations in self-gravitating media.
   J.-L. Tassoul, G. Pellieux.
Journ. Roy. Astron. Soc. Canada, Vol. 66, 67 (1972). – Abstr.
Canadian Astron. Soc.

061.019 The $^{12}C + ^{12}C$ thermonuclear reaction rates.
   G. Michaud.
Journ. Roy. Astron. Soc. Canada, Vol. 66, 68 (1972). – Abstr.
Canadian Astron. Soc.

**061.020 Ion-ion correlation effects on electron-nucleus neu-
trino bremsstrahlung.** G. De Zotti.
Mem. Soc. Astron. Italiana, Nuova Ser., Vol. 43, 89 - 93
(1972).

An analytical approximation formula for the neutrino
energy-loss rate due to electron-nucleus neutrino bremsstrahlung is given. The neutrino bremsstrahlung emissivity is
recalculated taking into accont the ion-ion correlation in a one-
component plasma.

**061.021 Convectively unstable layer in a periodic gravitation-
al field.** L. N. Ivanov.
Vestn. Leningr. un-ta, 1971, No. 19, p. 125 - 132. In Russian.
Abstr. in Referativ. Zhurn. 51. Astron., 4.51.244 (1972).

**061.022 Creation of particles and vacuum polarization in an
anisotropic gravitational field.**
Ya. B. Zel'dovich, A. A. Starobinskij.
Zhurn. ehksperim. i teor. Fiz., Vol. 61, 2161 - 2175 (1971).
In Russian. – Abstr. in Referativ. Zhurn. 51. Astron., 4.51.747
(1972).

061.023 Origin of strong magnetic fields.
   V. Canuto, S. Kumar, H. J. Lee.
Nature, Phys. Sci., Vol. 235, 9 - 10 (1972).
   The physical origins of the previously proposed LOFER
mechanism are further investigated. LOFER mechanism is
still a phenomenological approach since no detailed microscopic interaction has been proposed by the authors. It is neverthe-
less a possible thermodynamic state of matter. We pointed out
that its main feature is of being metastable or quasi-stable.
This is not a defect, as incorrectly pointed out by O'Connell,
since in astrophysics there are abundant example of quasi-
stable states, whose lifetime for decay into lower states is so
long to make them stable under practical circumstances.

**061.024 α-effect dynamos, by the Bullard-Gellman formal-
ism.** P. H. Roberts, M. Stix.
Astron. Astrophys., Vol. 18, 453 - 466 (1972).
   The Bullard-Gellman method of obtaining homogeneous
dynamos of spherical form is generalized to solve the mean
field induction equation. In the same way in which Bullard
and Gellman expanded velocity and magnetic field in spherical
harmonics, we expand the α-effect scalar field and compute
the corresponding interactions. We used this formalism to
examine several models of the solar dynamo. We also considered how our results are altered when magnetic diffusivity is
assumed to depend on distance from the solar center. We also
examined the effect of meridional shear. We studied the effect
of a meridional circulation by means of time-latitude plots of
the mean toroidal magnetic field.

061.025 Weak s-process irradiations.
   J. G. Peters, W. A. Fowler, D. D. Clayton.
Astrophys. Journ., Vol. 173, 637 - 648 = Contr. Louisiana
State Univ. Obs., *Baton Rouge,* No. 61 (1972).
   We calculate the overabundances of rare neutron-rich
species produced when seed nuclei between silicon and nickel
are exposed to small neutron irradiations in order to ascertain
whether those species may owe their natural abundances to
such a cause. Particular attention is given to ^{54}Cr and ^{58}Fe,
two relatively prominent neutron-rich species that have been
bypassed by calculation of explosive nucleo-synthesis.

**061.026 Re-calculation of efficiency factors for radiation
pressure.** C. Friedemann, R.-H. Giese.
Astrophys. Space Sci., Vol. 15, 401 - 403 (1972).
   Corrected values of the efficiency factor of light pressure
are presented for spherical particles of graphite and metallic
materials.

**061.027 Symmetric fission of superheavy nuclei and over-
abundance of rare earth elements.**
T. Ohnishi, H. Okamoto.
Nature, Phys. Sci., Vol. 236, 27 - 29 (1972).

The abundances of superheavies in the region of $260 \lesssim A$ were calculated by the r-process nucleosynthesis model. Abundance curves of fission fragments for four nuclides of A = 260, 270, 290 and 325 were also deduced by the statistical model of fission. It has become clear that the superheavy of A = 325 makes symmetric spontaneous fission to yield an abundance curve having a peak at around A = 160, and this fission fragments curve can explain well the overabundances of the r-nuclides in the rare earth region.

061.028 Basic methods of astrophysical research.
B. A. Volynskij.
Uch. zap. Yaroslav. gos. ped. in-t, 1971, vyp. (No.) 81, p. 55 - 86. In Russian. − Abstr. in Referativ. Zhurn. 51. Astron. 5.51.54 (1972).

061.029 Rates for Urca neutrino processes.
G. Beaudet, E. E. Salpeter, M. L. Silvestro.
Astrophys. Journ., Vol. 174, 79 - 90 (1972).
Simple fitting formulae are presented for all types of allowed β-transitions (direct and $e^{\pm}$-capture) as functions of density ρ and temperature T. For a number of the most common stable even-even nuclei under equilibrium conditions, nuclear abundances and the overall rates of Urca energy loss are calculated as a function of ρ and T.

061.030 Exchange contribution to the thermodynamic potential of a partially degenerate semirelativistic electron gas. A. Kovetz, D. Q. Lamb, H. M. Van Horn.
Astrophys. Journ., Vol. 174, 109 - 120 (1972).
We have obtained asymptotic formulae for the exchange contribution to the thermodynamic potential of a partially degenerate, semirelativistic electron gas in the limits of weak degeneracy, strong degeneracy, and nonrelativistic degeneracy. Our results are of wider interest than merely for this one astrophysical application, however, and we have therefore summarized our findings in this paper.

061.031 On the acceleration of particles by magnetic dipole radiation. F. Pacini, M. Salvati.
Bull. American Astron. Soc., Vol. 4, 222 (1972). − Abstr. AAS.

061.032 Centered difference equations and numerical instability. E. Simpson.
Bull. American Astron. Soc., Vol. 4, 240 (1972). − Abstr. AAS.

061.033 The nucleosynthesis of the light elements (A < 12). H. Reeves.
Bull. American Astron. Soc., Vol. 4, 256 - 257 (1972). − Abstr. AAS.

061.034 Gamma-ray astronomy with large scintillation detectors. J. D. Kurfess.
Bull. American Astron. Soc., Vol. 4, 260 (1972). − Abstr. AAS.

061.035 Phase transition in hadron matter.
Y. C. Leung, C. G. Wang.
Bull. American Astron. Soc., Vol. 4, 260 (1972). − Abstr. AAS.

061.036 An exploratory XUV astronomy survey from rocket altitudes. G. R. Riegler, G. P. Garmire.
Bull. American Astron. Soc., Vol. 4, 271 (1972). − Abstr. AAS.

061.037 On the possibility of applying the X- and gamma -ray astronomy to solving the problems of cosmochemistry and nuclear astrophysics. B. Kuchowicz.
Postępy Astron., Vol. 20, 147 - 150 (1972). In Polish.
The study of linear X-ray spectra may enable us to detect the superheavy elements (from the hypothetical island of stability near the atomic number $Z = 114$) in at least some of the stars in which one may hope to find them. There are certain

arguments in favour of the presence of these elements in the peculiar A stars. It is worth studying the X and gamma spectra of the Ap stars, of supernova remnants, and of the surroundings of pulsars.

061.038 The effect of rotation on thermal convection.
J. O. Murphy, R. Van der Borght.
Proc. Astron. Soc. Australia, Vol. 2, 92 - 93 (1972).

061.039 Effect of a magnetic field on finite amplitude convection.
J. O. Murphy, E. A. Spiegel, R. Van der Borght.
Proc. Astron. Soc. Australia, Vol. 2, 93 - 94 (1972).

061.040 Near resonant charge transfer processes at thermal energies. R. L. Brown.
Astrophys. Space Sci., Vol. 16, 274 - 283 (1972).
A useful procedure is described for the evaluation of near resonant charge transfer cross-sections at thermal energies. The rate coefficient is calculated for the reaction $O^+ + H \rightleftharpoons H^+ + O$ as a function of temperature between 10 and 10^4 K; the results are compared to those obtained using the 'orbiting approximation'. A brief application of these rates to the interstellar medium is made where it is found that substantial differences emerge between cases where the gas is heated by cosmic rays or by X-rays.

061.041 Dynamics of self-gravitating magnetized cylinders.
G. Bhowmik, S. P. Talwar.
Publ. Astron. Soc. Japan, Vol. 24, 365 - 373 (1972).
The problem of finite amplitude radial oscillations in a uniformly rotating gravitating cylinder carrying a volume distribution of electric current is investigated. The radial motions caused by the introduction of a uniform rotating or a current system in an otherwise static configuration are also studied.

061.042 Spatial inhomogeneities of nucleosynthesis.
H. Reeves.
Astron. Astrophys., Vol. 19, 215 - 223 (1972).
Data on the expansion of supernova remnants is used to study the mixing time scale of newly made material to the galactic gas. The distribution of regions where star formation and star explosion takes place, together with the dynamics of the mixing processes suggest the existence of spatial inhomogeneities in the element concentrations of the galactic gas. These inhomogeneities may be reflected in the abundances of fossil radioactivities in the solar system, and perhaps also in the stellar Z abundances.

061.043 Are neutrinos stable particles?
J. N. Bahcall, N. Cabibbo, A. Yahil.
Phys. Rev. Letters, Vol. 28, 316 - 318 (1972).
It is pointed out that neutrinos with a finite mass could be unstable. We discuss the consequences of this possibility for solar-neutrino experiments.

061.044 Neutrinos of nonzero rest mass.
S. Pakvasa, K. Tennakone.
Phys. Rev. Letters, Vol. 28, 1415 - 1418 (1972).
Some implications of neutrinos having nonzero rest masses and having finite lifetimes are considered.

061.045 Nucleosynthesis of rare nuclei from seed nuclei in explosive carbon burning.
W. M. Howard, W. D. Arnett, D. D. Clayton, S. E. Woosley.
Astrophys. Journ., Vol. 175, 201 - 216 (1972).
We demonstrate that a population I concentration of primordial heavy seed nuclei, when present in the carbon- and oxygen-rich core of the presupernova star and exposed to the temperature and free nucleon densities of explosive carbon burning, is efficiently transmuted into the rare species ^{36}S, ^{40}K,

^{40}Ar, ^{43}Ca, ^{46}Ca, ^{48}Ca, ^{45}Sc, ^{47}Ti, ^{49}Ti, ^{50}Ti, ^{50}V, ^{62}Ni, ^{64}Ni, ^{68}Zn, ^{70}Zn, and ^{76}Ge in approximately their solar-system abundance ratios. If the temperature is high enough, the nuclei ^{65}Cu, ^{67}Zn, ^{71}Ga, ^{73}Ge, and perhaps ^{75}As can also be synthesized.

061.046 Mean lifetimes and equilibrium abundances in the fast CN cycle. G. R. Caughlan, W. A. Fowler.
Nature, Phys. Sci., Vol. 238, 23 - 25 (1972).

We present here an analysis of the mean lifetimes and equilibrium abundances for the fast CN cycle and show that production of small nitrogen to carbon ratios is possible, in contrast to the equilibrium production in the ordinary CN cycle, but that associated with such production are high ratios of ^{13}C/^{12}C and of ^{15}N/^{14}N. The final N/C, ^{13}C/^{12}C, and ^{15}N/^{14}N ratios actually depend on the conditions under which cessation of hydrogen burning occurs in astrophysical circumstances.

061.047 Evolution of elements in the galaxies. A. G. W. Cameron.
Galactic astronomy, Vol. 2, (see 012.019), 171 - 200 (1970).

061.048 Exobiology of porphyrins. G. W. Hodgson.
Theory and experiment in exobiology, Vol. 1, (see 003.011), 83 - 103 (1971).

061.049 Source models with electron diffusion. L. Gratton.
Astrophys. Space Sci., Vol. 16, 81 - 100 (1972).

The variation of the energy distribution of relativistic electrons continuously generated by a central source and diffusing spherically in the surrounding space will be considered. Although the solution will be given in a fairly general form, only a greatly simplified model will be discussed in detail. It is remarkable that, notwithstanding its simplicity, this model seems to account quite satisfactorily for the general features of the continuous emission of the Crab nebula from the radio to the X-ray spectral region.

061.050 Superfluidity and superconductivity in astrophysics. V. L. Ginzburg.
Physica, Vol. 55, 207 - 212 (1971).

The problem of superfluidity and superconductivity in the universe is briefly summarized.

061.051 Problems and achievements of nuclear astrophysics. I. Nuclidic abundances and cosmic synthesis.
B. Kuchowicz.
Postępy Fiz., Vol. 22, 495 - 509 (1971). In Polish.

A short survey of problems considered in nuclear astrophysics is given. The question of the abundance of chemical elements and their isotopes which forms the basis for studies on the origin of the elements is outlined. Various theories of cosmic nucleosynthesis are presented, and a historical outline of the development of nuclear astrophysics is given.

061.052 Recent developments in astrophysics and cosmology. S. Hayakawa.
Butsuri (Japan), Vol. 26, 182 - 191 (1971).

The following six are the main problems; quasars, X-ray astronomy, discovery of OH etc., 3K blackbody radiation, infrared astronomy and pulsars.

061.053 Application of IVB-theory to a neutrino process of astrophysical significance. M. Abak.
Fortschr. Phys., Vol. 19, 529 - 557 (1971).

061.054 The magnetic susceptibility of dense neutron matter. V. R. Pandharipande, V. K. Garde, J. K. Srivastava.
Phys. Letters B (Netherlands), Vol. 38B, 485 - 486 (1972).

The paramagnetic susceptibility of dense neutron matter is calculated for the Reid soft-core potential with the lowest order variational method.

061.055 Angular-momentum changes in astrophysical processes due to neutrino emission. K. Tennakone.
Nuovo Cimento Lettere, Ser. 2, Vol. 3, 583 - 584 (1972).

It is shown that a star can gain or lose angular momentum by emitting a large flux of neutrinos in the presence of a very strong magnetic field. Neutrons in a star would tend to get aligned with their spins parallel or antiparallel to the magnetic field. Possible effects of this process on neutron stars, pulsars and collapsing stars are discussed.

061.056 Can supernovae explain the r-process elemental abundances? T. Ohnishi.
Techn. Rep. Inst. Atomic Energy Kyoto Univ. (Japan), No. 154, p. 1 - 18 (1971).

The possibility of the formation of heavy neutron rich nuclides in the universe is investigated by comparing the r-process time scale with the exploding or the cooling time scales of various types of stars. It becomes clear from the study of time scales that the r-process nuclides except the mass region of $70 \leqslant A \leqslant 120$ can not be produced in the expanding shells of type-I supernovae, but the heavier nuclides than $A \simeq 120$ can be easily synthesized in the cooling shells of neutron stars or in the expanding shells of supermassive stars.

061.057 Nuclear laboratory astrophysics. T. De Graaf.
Nederl. Tijdschr. Natuurk., Vol. 38, No. 7, p. 107 - 117 (1972). In Dutch.

This is a survey article discussing a few instances where nuclear physics has a bearing on astrophysical processes. Also discussed is the connection between laboratory measurements and astrophysical processes.

061.058 The astrophysical importance of heavy leptons. T. de Graaf.
Nuovo Cimento Lettere, Ser. 2, Vol. 2, 979 - 984 (1971).

061.059 Astrophysical importance of the reaction ^{16}O (p, a) ^{13}N. S. W. Woosley, W. D. Arnett, D. D. Clayton.
Phys. Letters B, (Netherlands), Vol. 38 B, 196 - 198 (1972).

061.060 A detection method for gamma-ray astronomy researches up to 0.4 MeV.
D. Brini, G. di Cocco.
Nuclear Instruments and Methods (Netherlands), Vol. 95, 181 - 188 (1971). – See Bull. Signal., Vol. 33, Section 120, No. 483 (1972).

061.061 Astrophysique théorique. J.-C. Pecker.
Annu. Coll. Française, Vol. 70, 93 - 100 (1970/71). See Bull. Signal., Vol. 33, Section 120, No. 2613 (1972).

061.062 Modification of the Coulomb law in a strong magnetic field. Yu. Loskutov, V. Skobelev.
Phys. Letters A, (Netherlands), Vol. 36A, 405 - 406 (1971).

X-ray and γ-ray imaging with multiple-pinhole cameras using a posteriori image synthesis. See Abstr. 031.029.

Advanced X-ray observatories.
See Abstr. 051.007.

Sternmaterie. II. Zustandsformen der Materie in stabilen Sternen. See Abstr. 065.086.

Solar rare gases and the abundances of the elements.
See Abstr. 105.024.

Cosmic and solar system abundances of deuterium and helium-3. See Abstr. 107.004.

062 Magneto-Hydrodynamics, Plasma

062.001 **Determination of magnetohydrodynamic shock normals.** B. Abraham-Shrauner.
Journ. Geophys. Res., Vol. 77, 736 - 739 (1972). – Brief report.

062.002 **Thermal instability with velocity shear.** S. P. Talwar, M. Singla.
Astrophys. Journ., Vol. 172, 155 - 163 (1972).

The instability arising due to a tangential discontinuity in velocities in a dilute plasma subject to heating and cooling mechanisms is investigated by using the normal-mode technique. The modifications introduced in the criteria of thermal instability in a static fluid due to the velocity shear are discussed, taking the ambient magnetic field aligned or transverse to flow in each half-space. It is suggested that this instability may play an important role in producing condensations in cosmic fluids.

062.003 **Radiative recombination of hydrogenic ions in high-temperature plasmas.**
P. Gratreau, B. V. Robouch.
Astrophys. Journ., Vol. 172, 201 - 203 (1972).

The rate of recombination and the rate of the mean kinetic energy loss of recombining electrons are calculated numerically and compared with the values reported by Seaton. The results show that Seaton's estimate of the uncertainty of his results is pessimistic.

062.004 **Ratio of line intensities in helium-like ions as a density indicator.**
G. R. Blumenthal, G. W. F. Drake, W. H. Tucker.
Astrophys. Journ., Vol. 172, 205 - 212 (1972).

The ratio R of the forbidden- to intercombination-line intensities in helium-like ions has been evaluated as a function of temperature. The effects of cascades to the $n = 2$ levels following recombination and collisional excitation of higher levels have been considered. The use of R as a density indicator in solar active regions and flares is discussed. We also discuss how in thermal cosmic X-ray sources this method may give information on the electron density in the source or on the ultraviolet flux in the vicinity of the hot plasma.

062.005 **Kinetic instability of a limited electron beam in a plasma.** T. D. Dolan.
Geomagn. Aeronom., Vol. 12, 18 - 23 (1972). In Russian.

062.006 **Electromagnetic fluctuations in a magnetized plasma in the presence of outer electromagnetic radiation.** A. K. Yukhimuk.
Geomagn. Aeronom., Vol. 12, 116 - 119 (1972). In Russian. Brief information.

062.007 **Circular polarization of synchrotron radiation in the presence of hydromagnetic waves.**
E. R. Seaquist.
Astrophys. Space Sci., Vol. 15, 284 - 292 (1972).

It is shown that relativistic electrons in the presence of circularly polarized hydromagnetic waves emit synchrotron radiation which is partially circularly polarized. The relation between the degree of polarization of the radiation and the energy density and wavelength of the waves is derived, and the factors determining the sense of polarization are discussed.

062.008 **Continuum radiative transfer in a hot plasma, with application to Scorpius X-1.**
J. E. Felten, M. J. Rees.

Astron. Astrophys., Vol. 17, 226 - 242 (1972).

Some interesting features of a spectrum emergent from a plasma cloud can be understood from simple physical arguments; in particular, the expected shape of the infrared, visible, ultraviolet, and X-ray continuum can be derived from known solutions in terms of Chandrasekhar's H-, X-, and Y-functions. We discuss the data on Sco X-1 and apply the theoretical results to obtain parameters for simple models of the source. Then we discuss the extent to which the results can be generalized to asymmetric and inhomogeneous source models.

062.009 **Hydrogen line ratios as electron temperature indicators in nonequilibrium plasmas.** C. Park.
Journ. Quant. Spectrosc. Radiat. Transfer, Vol. 12, 323 - 370 (1972).

The ratio of intensities of hydrogen Balmer lines $H_\alpha - H_\beta$ is proposed as an indicator of electron temperature in a nonequilibrium plasma. For such a plasma, the intensity ratios of the $H_\alpha - H_\beta$ lines are computed. The results are tabulated in the form of a conversion table between the measured excitation temperature and the true electron temperature. The ranges of applicability of the conversion table are also computed and are presented in separate tables. An example is shown in which particle densities are consistent with the Saha equilibrium condition at the apparent excitation temperature even though the plasma is in nonequilibrium at a different true electron temperature.

062.010 **The laws of reflection and refraction of incompressible magnetohydrodynamic waves at a fluid-solid interface.** D. D. Skiles.
Phys. Earth Planet. Interiors, Vol. 5, 90 - 98 (1972).

The early work of Ferraro (1954) and Roberts (1955) on the reflection and refraction of plane, periodic MHD waves at the plane surface S separating two incompressible, perfectly conducting fluids in the presence of a constant and uniform magnetic field is extended to the case of an infinitely conducting fluid and a finitely conducting solid.

062.011 **On the transmission of the energy in an incompressible magnetohydrodynamic wave into a conducting solid.** D. D. Skiles.
Phys. Earth Planet. Interiors, Vol. 5, 99 - 109 (1972).

The present paper continues the foregoing cited one and examines amplitude relations and energy reflection and transmission coefficients.

062.012 **Pressure broadening of multiply ionized carbon lines.** P. Bogen.
Zeitschr. Naturforschung, Vol. 27a, 210 - 214 (1972).

In a helium plasma with carbon impurity, the pressure broadened profiles of seven C III and two C IV lines have been measured. The measurements of the profiles are described and an interpretation of the line broadening data is presented.

062.013 **Time development of an axisymmetric shock. Quasi-stationary velocity approximation.**
Y. Hashimoto.
Progr. Theor. Phys., *Japan,* Vol. 47, 83 - 104 (1972).

Propagation formulae for an axisymmetric shock wave in a continuous, inhomogeneous medium are derived from hydrodynamic considerations on the state of the flow immediately behind the shock. Quasi-stationary velocity of isentropic flow in a steady gravity field is approximately assumed. The method is applied to an explosion of a highly rotating oblate star.

062.014 Premiers résultats d'une étude du rayonnement continu d'un plasma d'argon dans les domaines proche ultraviolet, visible et en haute pression.
C. Goldbach, G. Nollez.
Comptes Rendus Acad. Sci. Paris, Sér. B, Vol. 274, 970 - 973 (1972).

062.015 Absorption of gravitational energy by a conducting fluid in the presence of magnetic fields.
F. P. Esposito.
Astrophys. Journ., Vol. 173, 423 - 430 (1972).
The equations which determine the response of hydromagnetic systems to incident gravitational radiation are derived in the linearized approximation to general relativity. These equations are solved for a uniform incompressible fluid occupying a half-space region in the presence of a uniform magnetic field normal to the free surface of the fluid.

062.016 Computer simulation of plasmas.   J. M. Dawson.
Gravitational N-body problem. IAU Colloquium No. 10, (see 012.004), p. 315 - 336 (1972) = 06.062.026.

062.017 Enhancement of relaxation processes by collective effects.   R. M. Kulsrud.
Gravitational N-body problem. IAU Colloquium No. 10, (see 012.004), p. 337 - 346 (1972) = 06.062.027.

062.018 Nonlinear interaction of plasma waves in a cold magnetized plasma.   Y.-C. Chin.
Planet. Space Sci., Vol. 20, 711 - 720 (1972).
The general coupling equation for the wave-wave scattering in a cold magnetized plasma is derived. Specifically, the transformation of electrostatic waves is studied in connection with solar type-III radiation. The relevant numerical analysis is also presented.

062.019 Root mean square fluctuation of a weak magnetic field in an infinite medium of homogeneous stationary turbulence.   B.-C. Low.
Astrophys. Journ., Vol. 173, 549 - 555 (1972).
This paper considers the generation of magnetic field by statistically homogeneous, stationary velocity turbulence. The generation of rms magnetic fluctuation is explicitly demonstrated in the limit of short turbulence correlation time.

062.020 The equilibrium and stability of uniformly rotating gaseous systems in hydromagnetics. I: Mathematical technique.   S. P. S. Anand.
Astrophys. Space Sci., Vol. 15, 415 - 425 (1972).
The theory for investigating the equilibrium and stability of a uniformly rotating gaseous system with a prevalent magnetic field is developed by using the virial tensor approach. We have obtained the nine modes of oscillations, grouped into the transverse shear, toroidal and pulsation modes. From this analysis we have also found the conditions under which the sequence of a uniformly rotating axially symmetric configuration in the presence of a magnetic field should have a point of bifurcation. Applications to the cosmogonic fission problem, the study of the pulsation of rotating magnetic stars and some radio astrophysical problems are briefly discussed.

062.021 Diffusion of particles across a magnetic field under the action of large-scale hydromagnetic turbulence.
I. N. Toptygin.
Geomagn. Aeronom., Vol. 12, 329 - 331 (1972). In Russian. Brief information.

062.022 About three-dimensional hydromagnetic disturbances generated by a magnetic dipole in an anisotropic plasma.   L. L. Vanyan, A. S. Lipatov.
Geomagn. Aeronom., Vol. 12, 355 - 357 (1972). In Russian.

Brief information.

062.023 Plasma radiation from collisionless MHD shock waves. II. Plasma-wave generation and transformation processes.   D. F. Smith.
Astrophys. Journ., Vol. 174, 121 - 134 (1972).
The distribution of plasma waves which results when beams of electrons produced in a collisionless magnetohydrodynamic shock wave enter into the upstream plasma is examined. Nonlinear processes are analyzed to determine the angular distribution of plasma waves. Processes for the transformation of plasma waves into radiation are considered.

062.024 Gyrosynchrotron radiation and its transfer in a magnetoactive plasma.   K. Sakurai.
Astrophys. Journ., Vol. 174, 135 - 149 (1972).
Gyrosynchrotron radiation fields from mildly relativistic electrons in a magnetoactive plasma are asymptotically calculated by using the Green tensor and the Fourier transformation. Taking into account these fields, the emissivities and the absorption coefficients from an arbitrary distribution of electrons are calculated in order to discuss the intensity, spectrum, and polarization of gyrosynchrotron radiation. A consideration is given on the problem of radiative transfer in relation to the Stokes parameters.

062.025 Strong electromagnetic waves in overdense plasmas.
C. Max, F. Perkins.
Bull. American Astron. Soc., Vol. 4, 260 (1972). – Abstr. AAS.

062.026 A study of a high-pressure thermal argon plasma as a high-radiance standard.
C. Goldbach, G. Nollez, R. Peyturaux.
Journ. Quant. Spectrosc. Radiat. Transfer, Vol. 12, 1089 - 1104 (1972).
This work represents a first step in the realization of a plasma source emitting blackbody radiation. A theoretical evaluation of the conditions to be fulfilled in order to obtain strictly planckian radiation from a pure argon plasma has been made. Spectroscopic diagnostics of the plasma have been performed by using various methods and using the hypothesis that L. T. E. obtains.

062.027 Ultrarelativistic cosmic plasma.   Yu. P. Ochelkov, O. F. Prilutskij, I. L. Rozental', I. B. Shukalov.
Izv. AN SSSR. Ser. fiz., Vol. 35, 2453 - 2457 (1971). In Russian. – Abstr. in Referativ. Zhurn. 51. Astron., 6.51. 263 (1972).

062.028 On the theory of magneto-conjugate transfer of a cool plasma into the outer ionosphere of lower latitudes.   Yu. S. Sitnov, M. N. Fatkullin.
Kosm. Issled., Vol. 10, 459 - 461 (1972). In Russian. – Brief information.

062.029 Self-similar solution of nonstationary hydrodynamical accretion.   G. S. Bisnovaty-Kogan, Ja. B. Zeldovich, D. K. Nadezhin.
Astron. Zhurn. Akad. Nauk SSSR, Vol. 49, 483 - 488 (1972). In Russian. English translation in Soviet Astron. AJ, Vol. 16, No. 3.
A self-similar solution is found which describes falling of gas in a constant, uniform gravitational field on a hard wall. The solution can be used for describing the initial stage of matter accretion on a neutron star in supernova remnants.

062.030 The influence of electron scattering on the spectrum of emission from semi-opaque plasma.
N. I. Shakura.
Astron. Zhurn. Akad. Nauk SSSR, Vol. 49, 652 - 654 (1972). In Russian. English translation in Soviet Astron. AJ, Vol. 16,

No. 3.

The flux of emission from a homogeneous semi-opaque cloud in the presence of multiple electron scattering without change of frequency is considered. Distinctive breaks permit to determine the parameters of the emitting cloud.

062.031 An asymmetrically rotating fluid disc with applications. M. L. White.
Astrophys. Space Sci., Vol. 16, 295 - 310 (1972).

The rotation of a compressible inviscid fluid disc of (1) slowly varying density or (2) nonuniform density (cold gas approximation) or (3) nonuniform density (hot, but tenuous) is considered. Perturbation methods for solving the basic equation for conservation of vorticity are used. The possible role of the jet streams and the steady state long vortex waves in the formation and evolution of the solar system is also discussed. Comparisons are made with the von Weizsäcker (1944) and Chandrasekhar (1946) model of turbulent eddies in the solar nebula and with the particle (asteroidal) jet streams of Alfvén and Arrhenius (1970).

062.032 Frictional effects with neutrals and the gravitational instability of a plasma. P. K. Bhatia.
Publ. Astron. Soc. Japan, Vol. 24, 375 - 379 (1972).

The gravitational instability of a plasma has been studied taking into account the frictional effects with neutrals. The effects of finite conductivity and Hall currents have also been included. It is found that Jeans' criterion remains unchanged in the presence of finite conductivity, Hall currents, and frictional effects with neutrals.

062.033 A Lagrangian derivation of the action-conservation theorem for density waves. R. L. Dewar.
Astrophys. Journ., Vol. 174, 301 - 307 (1972).

A simple, physical derivation of the continuity equation for wave action is obtained from Hamilton's principle and an averaged Lagrangian density. As an example we briefly discuss electrostatic plasma waves, but the primary purpose of the paper is to derive the form of the wave action density for galactic spiral waves in a manner that is both simpler and more easily generalized than that of Shu.

062.034 On the resolution of a paradox in kinematic turbulence theory. I. Lerche.
Astrophys. Journ., Vol. 174, 309 - 319 (1972).

Using the thermal conduction of heat in an infinite, incompressible fluid possessing a turbulent velocity field as an illustration, we show that the existence of growing modes of a system is by no means a guarantee that the system at hand is unstable. We show that for our problem such growing modes are mathematically correct but physically absurd, giving rise, therefore, to a paradox.

062.035 Nonstationary behavior of collisionless shocks. D. L. Morse, W. W. Destler, P. L. Auer.
Phys. Rev. Letters, Vol. 28, 13 - 16 (1972).

Laboratory measurements indicate that a collisionless shock wave formed in a plasma-wind-tunnel device is nonstationary on the time scale of the ion gyroperiod. This behavior suggests a possible interpretation of recent data on magnetic field structure in the vicinity of earth's bow shock. Comparison is made between the laboratory measurements and earlier computer simulations in which ion gyromotion was included.

062.036 Topological dissipation and the small-scale fields in turbulent gases. E. N. Parker.
Astrophys. Journ., Vol. 174, 499 - 510 (1972).

It is shown that a large-scale magnetic field possesses a hydrostatic equilibrium only if the pattern of small-scale variations is uniform along the large-scale field. Thus equilibrium obtains only if the variations in the field consist of simple twisting of the lines, with the twists extending uniformly the full length of the field. Any more complicated topology is without equilibrium. The result is rapid dissipation and field-line merging. It follows from this general theorem that line merging has important consequences in turbulent fields. The effect has important astrophysical implications. It explains the absence of strong small-scale fields in the solar photosphere and in interstellar space in spite of the vigorous turbulence.

062.037 The self-consistent test-particle approach to relativistic kinetic theory. W. B. Thompson.
General relativity, (see 003.017), p. 243 - 254 (1972).

062.038 On some exact solutions in magnetohydrodynamics with astrophysical applications. C. Sozou.
Journ. Fluid. Mech., Vol. 51, 33 - 38 (1972).

Some exact solutions of the steady magnetohydrodynamic equations for a perfectly conducting inviscid self-gravitating incompressible fluid are discussed.

062.039 Plasma phenomena in astrophysics. C. G. Falthammar.
10th international conference on phenomena in ionized gases, Oxford, England, [Donald Parsons, Oxford], p. 1 - 35 (1971).

062.040 Electromagnetic field and internal motions of a spherical fluid model of a celestial body. E. Schmutzer.
Experimen. Techn. Phys., (*Germany*), Vol. 20, 95 - 105 (1972). In German.

062.041 Chemical effects in plasma condensation. G. Arrhenius.
From plasma to planet. 21st Nobel symposium 1971, (see 012.028), p. 117 - 132 (1972).

062.042 The critical velocity of gas-plasma interaction and its possible heterogonic relevance. J. C. Sherman.
From plasma to planet. 21st Nobel symposium 1971, (see 012.028), p. 315 - 341 (1972).

062.043 One-dimensional simulation of relativistic streaming instabilities. C. F. McKee.
Phys. Fluids, Vol. 14, 2164 - 2176 (1971).

062.044 Etude d'un plasma d'argon haute-pression comme étalon de rayonnement. C. Goldbach, Nollez, Peytureaux.
Bull. Inform. Bureau National Métrol., Vol. 2, No. 5, p. 3 - 6 (1971). – See Bull. Signal., Vol. 33, Section 120, No. 4886 (1972).

062.045 Self-consistent kinetic equations and the evolution of a relativistic plasma in an ambient magnetic field. M. A. Lee.
Plasma Phys., Vol. 13, 1079 - 1098 (1971).

062.046 On some exact solutions in magnetohydrodynamics with astrophysical applications. C. Sozou.
Journ. Fluid Mechanics, Vol., 51, 33 - 38 (1972).

062.047 Applications de la théorie cinétique des gaz à la physique des plasmas et à la dynamique des galaxies. J. P. Petit.
Thesis, Sci. Phys., Univ. Provence. Centre Documentation, C. N. R. S. (1972-02-01), 199 pp. (1972). – See Bull. Signal., Vol. 33, Section 120, No. 7092 (1972).

Hydromagnetic stability of a stratified atmosphere flowing over a liquid. See Abstr. 064.023.

063 Radiative Transfer

063.001 **The moment method in relativistic radiative transfer.**
J. L. Anderson, E. A. Spiegel.
Astrophys. Journ., Vol. 171, 127 - 138 (1972).
The moment method of Grad is applied to the problem of radiative transfer in a medium with relativistic differential motions. If a mean absorption coefficient is used, the method readily leads to a closed system of equations. The first approximation gives the relativistic analog of the classical Eddington approximation. In the limit of small photon mean free path, the Eddington approximation does not reproduce Thomas's radiative-viscosity terms which were derived from the exact transfer equation. To recover Thomas's results it is necessary to go to the second approximation.

063.002 **Sur une nouvelle méthode pour l'étude de la formation des raies spectrales, quand la fonction source est indépendante de la fréquence.** E. Simonneau.
Comptes Rendus Acad. Sci. Paris, Sér. B, Vol. 274, 85 - 88 (1972).
On présente une méthode pour résoudre l'équation de transfert dans une atmosphère plane, pour un atome à deux niveaux d'énergie. On trouve des propriétés communes à d'autres méthodes et certains avantages.

063.003 **An approximation to the solution of the radiative transfer equation in line frequencies.** V. V. Ivanov.
Astron. Zhurn. Akad. Nauk SSSR, Vol. 49, 115 - 120 (1972). In Russian. English translation in Soviet Astron. AJ, Vol. 16, No. 1.
A simple approximate expression is given for the resolvent function of the integral equation of transfer of line radiation in a plane layer. The approximation is based on information on the asymptotic properties of the solutions of the transfer equation.

063.004 **Time dependent radiative transfer. Damping of a temperature fluctuation. I. Near L.T.E. grey approximation.** P. Delache, C. Froeschlé.
Astron. Astrophys., Vol. 16, 348 - 355 (1972).
The time of flight of the photon is taken into account to establish the equations describing the evolution of a small perturbation applied to a grey and homogeneous medium near L.T.E. It is shown that the solution decays exponentially and thus we generalize a previous result obtained by Spiegel (1957). It appears that the analytic solution of this problem is possible only for a particular class of perturbations. However in using Eddington's approximation the same problem is solved for more general initial conditions. In particular, we obtain three types of homogeneous perturbations, which behave in very different ways and which are easy to interpret physically.

063.005 **Time dependent radiative transfer. Damping of a temperature fluctuation. II. The two-level atom case. Applications to the interstellar medium.** F. Le Guet.
Astron. Astrophys., Vol. 16, 356 - 360 (1972).
We study the radiative damping of harmonic temperature fluctuations by means of monochromatic radiation in an infinite homogeneous gas in L.T.E. In order to find out the time of capture, we solve simultaneously the transfer equation and the equations of statistical equilibrium. Preliminary results show that in interstellar clouds the time of capture can be much longer than the time of flight.

063.006 **Charged particle propagation in a scattering medium with constant and variable transport run when magnetic focusing is taken into account.**

S. N. Vernov, E. V. Gorchakov, G. A. Timofeev.
Geomagn. Aeronom., Vol. 12, 3 - 9 (1972). In Russian.

063.007 **On the theory of radiative transfer in a magnetic field.** D. N. Rachkovsky.
Izv. Krymskoj Astrofiz. Obs., Vol. 43, 190 - 200 (1971). In Russian.
The formation of absorption lines in a faint magnetic field is considered. Taking into account the phase correlation of the radiation emitted from different sublevels, the scattering matrix is deduced.

063.008 **Probability distributions for photon exit.**
G. D. Finn.
Journ. Quant. Spectrosc. Radiat. Transfer, Vol. 12, 35 - 58 (1972).
A linear integral equation is formulated for the probability that a photon in a spectral line formed at a specified depth in a scattering atmosphere eventually escapes at a particular frequency in the spectral line. Numerical solutions are obtained for an isothermal atmosphere and their physical meaning discussed under a variety of conditions.

063.009 **Statistical functions in radiative transfer.**
G. D. Finn.
Journ. Quant. Spectrosc. Radiat. Transfer, Vol. 12, 149 - 167 (1972).
Integral relations are obtained for the mean and mean square number of scatterings undergone by a photon formed at a specified depth in an isothermal atmosphere and destined to escape from the atmosphere at a particular frequency in a spectral line, there being complete redistribution of frequency in the scattering process. Theoretical estimates of these functions are compared with numerical solutions for cases in which the frequency profile of the absorption coefficient has Doppler and Voigt forms.

063.010 **Application of the mean absorption coefficient in nonisothermal infrared radiating gases.**
K.-Y. Chien.
Journ. Quant. Spectrosc. Radiat. Transfer, Vol. 12, 379 - 386 (1972).
A nongrey treatment of the nonisothermal, infrared radiative-transfer problem is formulated in terms of a rationally determined mean absorption coefficient, α_a. The relation between α_a and the equivalent isothermal total-band absorptance is established. A simplified procedure for using a two-term Gaussian-quadrature formula to evaluate the integrals in the general formulation is shown to be adequate. Results compare well with other theories and with experiments.

063.011 **Equations of the transfer of electrons and photons of large energies in magnetic fields.**
Yu. P. Ochelkov, I. L. Rozental, I. B. Shukalov.
Astron. Zhurn. Akad. Nauk SSSR, Vol. 49, 298 - 305 (1972). In Russian. English translation in Soviet Astron. AJ, Vol. 16, No. 2.
The transit of a beam of relativistic electrons of large density in strong magnetic fields is considered. This case is characterized by a collective interaction of electrons and photons.

063.012 **Approximate solution to the radiative equilibrium between concentric spheres.**
B. F. Armaly, A. N. Saad.
Journ. Quant. Spectrosc. Radiat. Transfer, Vol. 12, 511 - 518 (1972).
An approximate method was developed for the study of

the radiative energy transfer through a gray gas layer enclosed between black concentric spheres which are maintained at uniform but different temperatures. The approximate solutions have been found to agree very well (6 per cent maximum deviation) with the existing exact numerical solution throughout the entire range of optical thickness.

063.013 On the half-range orthogonality theorem appropriate to the scattering of polarized light.
C. E. Siewert.
Journ. Quant. Spectrosc. Radiat. Transfer, Vol. 12, 683 - 694 (1972).

A half-range orthogonality theorem relevant to the normal modes of a two-vector equation of transfer is proved. All appropriate normalization integrals are evaluated so that required expansion coefficients may be expressed concisely in terms of inner products.

063.014 An initial value method for the Ambarzumian integral equation. J. Buell, R. Kalaba, A. Fymat.
Journ. Quant. Spectrosc. Radiat. Transfer, Vol. 12, 769 - 776 (1972).

It is shown that a general class of nonlinear integral equations may be transformed into a Cauchy system. That this leads to an effective numerical scheme is demonstrated by solving the Ambarzumian integral equation. The new method does not involve successive approximations or series expansions.

063.015 A proof of the relation between reflectivity and emissivity in an isothermal scattering and absorptive atmosphere. J. L. Linsky.
Journ. Quant. Spectrosc. Radiat. Transfer, Vol. 12, 777 - 781 (1972).

A simple relation between directional emissivity and directional hemispherical reflectivity is proven for an isothermal coherent scattering and absorptive atmosphere and also for the case of complete redistribution in a line.

063.016 An explicit closed-form result for the discrete eigenvalue in studies of polarized light.
C. E. Siewert, E. E. Burniston.
Astrophys. Journ., Vol. 173, 405 - 406 (1972).

An exact closed-form result for the discrete eigenvalue pertinent to the combination of Rayleigh and isotropic scattering of polarized light is given.

063.017 Scattering by rough cylindrical particles.
G. A. Shah, M. S. Vardya.
Nature, Phys. Sci., Vol. 235, 115 - 116 (1972).

A model of a rough cylinder consisting of a central cylinder surrounded symmetrically by smaller cylinders has been considered and compared with the experimental results of Greenberg et al. (Nature, Phys. Sci., Vol. 230, 110 - 112 (1971)). It seems that the extinction and polarization can differ significantly, though not drastically, from that of the equivalent smooth particles in some region(s) of the spectrum.

063.018 Noncoherent scattering I. Isotropic scattering.
N. B. Yengibarian.
Astrofizika, Vol. 7, 573 - 586 (1971). In Russian.
English translation in Astrophysics, Vol. 7, No. 4.

The linear problem of noncoherent scattering in a plane-parallel finite layer is considered. Some concrete calculations in the case of Doppler line broadening have been carried out.

063.019 Line formation in a magnetic field. III. Formation of a Zeeman-triplet with unsplitted upper level.
Estimate of the influence of the magnetic field. H. Domke.
Astrofizika, Vol. 7, 587 - 604 (1971). In Russian.
English translation in Astrophysics, Vol. 7, No. 4.

The exact solution of the problem of formation of a Zeeman-triplet in an isothermic atmosphere without absorption in the continuous spectrum is considered. The dependence of the H-functions on the magnetic field is studied for Doppler and Lorentz frequency profiles of transitions. Two methods of approximate solution for problems of line formation in a magnetic field are proposed: a) finite field approximation, b) zero field approximation.

063.020 The radiative transfer problem in freely expanding gaseous clouds and its application to barium cloud experiments. S. W. Drapatz.
Planet. Space Sci., Vol. 20, 663 - 682 (1972).

The radiative transfer in freely expanding gaseous clouds, which are illuminated by external sources, is investigated on the basis of a point cloud model (spherical, collisionless flow from a point explosion). The essential quantity derived is the ratio of level populations of the atomic states. The calculations are used to determine the yield of vapourized material in artificial barium clouds and to describe the mechanism of photoionization.

063.021 The scattering of line radiation—III. The source function. J. D. Argyros.
Journ. Quant. Spectrosc. Radiat. Transfer, Vol. 12, 1001 - 1022 (1972).

After discussing the problems encountered in any attempt to interpret frequency-dependent source functions in terms of the population numbers, a method is developed for doing this to a certain class of frequency-dependent source functions. Solutions in the form of source functions are presented and discussed for a semi-infinite uniform atmosphere for five different forms of frequency redistribution during the scattering event. The different population distributions through the atmosphere, that can be deduced, are compared for the different cases of frequency redistribution used. Because of the nature of the results obtained, the feasibility of setting up iterative schemes to obtain frequency-dependent source functions for more complex problems is also considered.

063.022 Radiative transfer with partially coherent scattering.
M. J. Hemsch, J. H. Ferziger.
Journ. Quant. Spectrosc. Radiat. Transfer, Vol. 12, 1029 - 1046 (1972).

The Boltzmann equation for a gas in a plane-parallel atmosphere which scatters radiation partially coherently and partially by total redistribution is treated. The eigenfunctions are found and full-range completeness is proved. A method for solving half-range problems is also given, and numerical results for an isothermal semi-infinite atmosphere are presented.

063.023 Compton and inverse Compton scattering.
G. C. Pomraning.
Journ. Quant. Spectrosc. Radiat. Transfer, Vol. 12, 1047 - 1061 (1972).

A general expression is derived which gives the effects of motion of the scattering centers on an arbitrary scattering law for photons. Particular attention is paid to the case of photon scattering from a free electron gas in a relativistic Maxwellian velocity distribution. Exact numerical results for the scattering kernel in this case are presented.

063.024 Analytical solutions of Jefferies' equations for the grand source functions. H. Shimooda.
Publ. Astron. Soc. Japan, Vol. 24, 287 - 290 (1972).

Jefferies' (1968) simultaneous integral equations for the grand source functions for three-level transfer problems are analytically solved by the kernel approximation method. Furthermore a pair of line source functions is derived from the solutions thus obtained, and briefly discussed.

**063.025 A general formulation of the transfer equation.
II. Line formation with general redistribution.**
C. J. Cannon.
Australian Journ., Phys., Vol. 25, 177 - 195 (1972).

The multidimensional equation of transfer for spectral line radiation under a general redistribution law is studied. It is shown that the equation may be rewritten as a system of equations of the "Feautrier" form. It is also shown that the inclusion of a multidimensional differential macroscopic velocity field does not alter the functional form of the equations obtained.

**063.026 Interpolation functions, accuracy, and stability in
multidimensional transfer problems.**   H. P. Jones.
Bull. American Astron. Soc., Vol. 4, 211 - 212 (1972). – Abstr. AAS.

**063.027 A probabilistic formulation of the non-coherent
scattering problem.**   R. G. Athay.
Bull. American Astron. Soc., Vol. 4, 212 (1972). – Abstr. AAS.

**063.028 Transfer of resonant line radiation in differentially
expanding atmospheres.**
L. Caroff, P. D. Noerdlinger, J. D. Scargle.
Bull. American Astron. Soc., Vol. 4, 270 (1972). – Abstr. AAS.

**063.029 The induced light pressure under astrophysical con-
ditions.**
E. V. Levich, R. A. Sunyaev, Ya. B. Zeldovich.
Astron. Astrophys., Vol. 19, 135 - 139 (1972).

The radiation of compact sources with high brightness temperature in the low frequency region creates great force acting on free electrons due to induced Thomson scattering. This induced radiation pressure exceeds the well known pressure due to spontaneous scattering. The effect is of importance for plasma dynamics in the vicinity of pulsars, quasars, galactic nuclei and maser type sources.

**063.030 Cherenkov and transient radiation of uniformly
moving charge in random inhomogeneous medium.**
V. V. Tamoykin.
Astrophys. Space Sci., Vol. 16, 120 - 129 (1972).

Radiation of a uniformly moving charged particle in a medium with weak random inhomogeneities of dielectric permittivity isconsidered. The limiting cases of small- and large-scale inhomogeneities are studied. Peculiarities of energy losses near the threshold of Cherenkov radiation are clarified. The conclusions obtained in our paper appear to be useful for interpretating the experiments of observations of light generated by the cosmic rays penetrating in the earth's atmosphere.

**063.031 On combined operations method for transfer pro-
blems in homogeneous, cylindrical media.**
T. H. Kho, K. K. Sen.
Astrophys. Space Sci., Vol. 16, 151 - 166 (1972).

The problem of diffuse reflection by a homogeneous, isotropically scattering, infinite cylindrical medium has been considered. For a medium having cylindrical distribution of source in addition to the incident flux at the outer surface, the integro-differential equation for the emergent intensity has been established.

**063.032 On the determination of the radiation field in an op-
tically dense medium with strongly anisotropic
scattering.**   Yu. L. Biryukov, L. G. Titarchuk.
Kosm. Issled., Vol. 10, 400 - 410 (1972). In Russian.

**063.033 Stochastic magnetic field influence on the charac-
teristics of stellar absorption lines.**

A. Z. Dolginov, G. G. Pavlov.
Astron. Zhurn. Akad. Nauk SSSR, Vol. 49, 555 - 567 (1972).
In Russian. English translation in Soviet Astron. AJ, Vol. 16, No. 3.

Shapes of absorption spectral lines in the presence of a stochastic magnetic field are determined. The Stokes parameters are calculated for radiation transferring an optically thin layer of matter if the field a) has definite direction but random value, or b) is statistically isotropic. In both cases broadening of the σ-components of Zeeman triplet occurs.

**063.034 Invariant imbedding and Chandrasekhar's planetary
problem of radiative transfer.**
R. Bellman, S. Ueno.
Astrophys. Space Sci., Vol. 16, 241 - 248 (1972).

In connection with Chandrasekhar's planetary problem of radiative transfer the total scattering and the diffuse transmission functions have been discussed by several authors. With the aid of the Bellman-Krein formula for the resolvent kernel of the auxiliary equation governing the source function, we show how the invariant imbedding equations governing the diffuse scattering and transmission functions can readily be obtained.

**063.035 The escape of resonance-line radiation from extreme-
ly opaque media.**   T. F. Adams.
Astrophys. Journ., Vol. 174, 439 - 448 (1972).

We consider the escape of resonance-line radiation from a low-density medium in which photon destruction processes are negligible. Numerical solutions are presented for media with large enough linecenter optical depths τ_0 that the effects of the damping wings dominate. The numerical results show that the mean number of scatterings $\langle N \rangle \propto \tau_0$ in this limit, rather than $\langle N \rangle \propto \tau_0^2$ as predicted by Osterbrock. Further, the results show that $\langle N \rangle$ is independent of the value of the damping constant a.

**063.036 Energy straggling and radiation reaction for magne-
tic bremsstrahlung.**   C. S. Shen, D. White.
Phys. Rev. Letters, Vol. 28, 455 - 459 (1972).

When the average energy of the photon emitted by synchrotron radiation becomes appreciable compared to the energy of the particle, the particle will undergo straggling in its energy loss. The energy distributions of particles and the emitted photons are calculated using the method of quantum electrodynamics. The results are presented together with effects due to classical radiative reaction for experimental test. The significance of energy straggling in astrophysics is discussed briefly.

**063.037 Laboratory simulation of diffuse reflectivity from a
cloudy planetary atmosphere.**
J. S. Margolis, D. J. McCleese, G. E. Hunt.
Applied Optics, Vol. 11, 1212 - 1216 (1972).

For the first time measurements in the multiple scattering regime of the diffuse reflectivity as a function of single scattering albedo have been made in a geometry that may be simulated by a plane parallel atmosphere of large optical depth. A comparison between the measurements and a theoretical computation of the diffuse reflectivity is presented.

**063.038 Influence of beam shape on the phase fluctuations
of an electromagnetic field propagated through a
turbulent medium.**
A. Consortini, G. Fidanzati, A. Mariani, L. Ronchi.
Applied Optics, Vol. 11, 1229 - 1233 (1972).

The phase fluctuations of a beam propagated in a turbulent medium are shown to depend in a nonnegligible way on the shape of the beam. Some examples are presented and discussed on the basis of the von Karman model of the turbulence.

063.039 Rational approximation for the Voigt line profile.
K. G. Harstad.
Journ. Optical Soc. America, Vol. 62, 827 - 828 (1972). —
Letter.

063.040 Astrophysical masers. I. Source size and saturation.
P. Goldreich, D. A. Keeley.
Astrophys. Journ., Vol. 174, 517 - 525 = Contr. Lick Obs.,
No. 350 (1972).

The relation between the apparent and actual sizes of
maser sources is important in determining their internal physical conditions. This relation is investigated for models of homogeneous maser clouds having spherical and tube-shape geometries. The treatment of maser radiation uses rate equations
for the level populations and the ordinary equation of transfer.
Approximate analytic calculations are presented for all degrees
of saturation.

**063.041 On existence and uniqueness theorems concerning
the *H*-matrix of radiative transfer.**
C. E. Siewert, E. E. Burniston.
Astrophys. Journ., Vol. 174, 629 - 641 (1972).

A solution to the system of singular-integral equations
and the linear constraint which define mathematically the
H-matrix relevant to the scattering of polarized light is shown
to exist and to be unique. A discussion of a class of canonical
solutions to the matrix Riemann problem is given, and the
H-matrix is then expressed in terms of a convenient canonical
solution. Cauchy's theorem is then used to develop the non-
linear integral equation convenient, when used with the linear
constraint, for computing the *H*-matrix, and the required
existence and uniqueness theorem is proved. For the special
case of conservative Rayleigh scattering, an explicit analytical
result for the appropriate canonical matrix is given.

**063.042 Anisotropic nonconservative scattering in a semi-in-
finite medium.** L. Wang.
Astrophys. Journ., Vol. 174, 671 - 678 (1972).

Approximate formulae of plane albedo and spherical albedo of a semi-infinite medium are obtained by the application of the exponential kernel approximation. These formulae
are shown to be accurate for all values of single-scattering albedos, for strongly elongated scattering phase functions, and
for large angles of incidence.

**063.043 Redistribution of resonance radiation. I. The effect
of collisions.**
A. Omont, E. W. Smith, J. Cooper.
Astrophys. Journ., Vol. 175, 185 - 199 (1972).

The techniques of modern line-broadening theory are
used to investigate the scattering of polarized radiation in the
rest frame of an atom undergoing collisions. The formulation explicitly includes both elastic and inelastic (quenching) collisions. When the lower state has zero width, a form for the
redistribution function similar to that of Zanstra is obtained,
but with the redistribution in the neighborhood of the resonance line being caused solely by elastic collisions. In the limit
of no collisions, but with both levels of finite lifetime, the result of Weisskopf and Woolley is obtained. The effect of level-
degeneracy is also explicitly included.

063.044 A new derivation of a decomposition formula.
M. Matsumoto.
Journ. Math. Phys. Sci., Vol. 3, 393 - 397 = Contr. Inst. Astro-
phys., Kwasan Obs., Univ. Kyoto, No. 194 (1969).

Starting with the functional equations in the internal radiation field provided by us we derive a decomposition formula which expresses the internal intensity in the diffuse radiation
field in terms of Chandrasekhar's *X*- and *Y*-functions, the
source function, and two additional functions for which the
integro-differential equations can be solved with their initial
values.

**Continuum radiative transfer in a hot plasma, with
application to Scorpius X-1.** See Abstr. 062.008.

**Gyrosynchrotron radiation and its transfer in a
magnetoactive plasma.** See Abstr. 062.024.

**Monte Carlo treatment of Lyman-alpha radiation in
a plane-parallel atmosphere.** See Abstr. 082.108.

064 Stellar Atmospheres, Stellar Envelopes

064.001 On the excitation of C III in Wolf-Rayet envelopes.
J. I. Castor, H. Nussbaumer.
Monthly Notices Roy. Astron. Soc., Vol. 155, 293 - 304 (1972).

The model of a spherically expanding envelope is employed to find equivalent widths of C III lines in the rocket ultra-violet. The calculations include electron—ion collisions, and absorptions and emissions associated with the stellar and diffuse radiation fields. Effects of optical depth are taken into account by calculating photon escape probabilities.

064.002 Rotationally extended stellar envelopes—IV. The detached shell of HD 187399.
J. B. Hutchings, P. G. Laskarides.
Monthly Notices Roy. Astron. Soc., Vol. 155, 357 - 371 (1972).

Spectrographic observations were obtained at 6.5 Å mm⁻¹ of the shell star HD 187399 before, during and after an outburst, in June and July 1970. Velocity and line profile measurements are presented. A model for the stellar envelope is deduced from these data and calculations of stellar and shell line profiles.

064.003 Macroturbulence in O star atmospheres?
M. Scholz.
Astrophys. Letters, Vol. 10, 137 - 139 (1972).

Observations of lines in O type spectra suggest that macroturbulence rather than rotation is the major line broadening agent in many O star atmospheres.

064.004 Excitation temperatures in late G- and K-giant atmospheres. D. Koelbloed.
Astron. Astrophys., Vol. 16, 230 - 236 (1972).

Excitation temperatures for Fe I and Ti I for late G- and K-type giants are obtained, using curves of growth. These values are compared with model atmosphere predictions. Whereas there is a satisfactory agreement for Ti I, the COG-results for Fe I give higher temperatures than the LTE-models considered.

064.005 A fine analysis of two early B-type supergiants, HD 96248 and HD 164402, relative to the main sequence star τ Scorpii. P. L. Dufton.
Astron. Astrophys., Vol. 16, 301 - 314 (1972).

Model atmosphere analyses are presented for two supergiants HD 96248 (B1Iab) and HD 164402 (B0Ib) relative to the main sequence star τ Scorpii (B0V). Atmospheric parameters are T_{eff} = 26000 °K, log g = 3.3, and ξ_t = 10 km/s for HD 96248 and (33000, 4.1, 10) and (34000, 4.8, 5) for HD 164402 and τ Sco.

064.006 Non-LTE model atmospheres. VII. The hydrogen and helium spectra of the O stars.
L. H. Auer, D. Mihalas.
Astrophys. Journ., Suppl. Ser., No. 205, Vol. 24, 193 - 246 (1972). – Abstr. in Astrophys. Journ., Vol. 171, 411 - 412 (1972).

An extensive series of non-LTE calculations of the H, He I, and He II spectra of O stars has been carried out by using relatively complete atomic models, allowing for several levels and lines simultaneously. Results are presented for continuum fluxes and for equivalent widths and profiles of the lines. Detailed comparisons are made with the spectra of eleven O and B0 stars.

064.007 Analysis of the metal-poor red giant star HD 122563.
W. Wolffram.
Astron. Astrophys., Vol. 17, 17 - 33 (1972). In German.

The extremely metal-poor K giant HD 122563 is analysed using non-grey model atmospheres. Flux constancy is obtained following Lucy's iteration procedure. Rayleigh scattering is explicitly taken into account in the calculation of the opacity and the source function. The Rayleigh scattering affects the temperature stratification as well as the continuous and line spectra. These effects are studied also using a simple Milne-Eddington approximation. Finally the model is corrected for line blanketing and convection.

064.008 Theoretical effect of various broadening parameters on ultraviolet line profiles. E. Peytremann.
Astron. Astrophys., Vol. 17, 76 - 82 (1972).

We study the relative importance of various damping parameters in ultraviolet lines under conditions of normal stellar atmospheres ($10000°$K $\leq T_{eff} \leq 30000°$K, log g = 4).

064.009 Can the collapse of an iron core trigger an explosion of the envelope?
J. C. Wheeler, Z. Barkat, G. Rakavy.
Bull. American Astron. Soc., Vol. 3, 452 (1971). – Abstr. AAS.

064.010 Line-blanketed model stellar atmospheres.
J. W. Fowler.
Bull. American Astron. Soc., Vol. 3, 455 (1971). – Abstr. AAS.

064.011 Interpretation of photometric observations with blanketed model atmospheres. E. Peytremann.
Bull. American Astron. Soc., Vol. 3, 455 (1971). – Abstr. AAS.

064.012 Dynamic instabilities in stellar envelopes.
G. S. Kutter, W. M. Sparks.
Bull. American Astron. Soc., Vol. 3, 484 (1971). – Abstr. AAS.

064.013 Rapid differential rotation in radiative stellar envelopes. M. J. Clement.
Bull. American Astron. Soc., Vol. 3, 484 - 485 (1971). Abstr. AAS.

064.014 Mass loss in intermediate supergiants.
D. M. Peterson.
Bull. American Astron. Soc., Vol. 3, 485 (1971). – Abstr. AAS.

064.015 Spectral lines for curve-of-growth analysis of late-type stars, I. A. L. T. Powell.
Roy. Obs. Bull., [Roy. Greenwich Obs., Herstmonceux], No. 168, p. 317 - 349 (1971).

Spectral line constants, values of solar equivalent widths and their desaturated values for the bifurcated and unbifurcated case are given.

064.016 The $UBVr$ colors of supergiants.
E. Böhm-Vitense.
Astron. Astrophys., Vol. 17, 335 - 353 (1972).

The atmospheres of supergiants are studied in the range $4000° \leq T_{eff} \leq 9000°$. The influence of radiation pressure and spherical effects are discussed. The observed colors of supergiants are obtained with radiative equilibrium and hydrostatic models. A calibration of $B-V$ in terms of T_{eff} is given. The T_{eff} obtained here for G and K supergiants are about 10% lower than adopted so far.

064.017 Radiative relaxation of sound waves in an optically thin isothermal atmosphere. P. Souffrin.

Astron. Astrophys., Vol. 17, 458 - 467 (1972).

The radiative damping of oscillatory modes in an optically thin isothermal atmosphere is investigated within the approximation of Newton's law of cooling. Special attention is paid to vertical modes sustained by harmonic boundary conditions and to the discussion of the validity of the adiabatic and isothermal limiting cases. The influence of relaxation on the response of an infinite model to an applied body-force is studied in the one-dimensional case. Application to the solar conditions indicates that the dissipation in the chromosphere has little effect on the dynamics of the chromospheric oscillation.

064.018 **The reflection effect in eruptive stars.**
V. P. Grinin.
Izv. Krymskoj Astrofiz. Obs., Vol. 43, 52 - 65 (1971).
In Russian.

The problem of the non-stationary radiation of a semi-infinite homogeneous medium illuminated by a point source, whose power depends on time, has been solved. The values which allow to calculate the energy reflected by the medium at any luminosity change law of the source have been obtained. As an example, the case when the luminosity change corresponds to the irreversible recombination of an optically thin layer of a hydrogen gas at constant temperature is considered. The colour change of total radiation is considered also.

064.019 **Radiative opacity due to the red system of CN.**
H. R. Johnson, I. R. Marenin, S. D. Price.
Journ. Quant. Spectrosc. Radiat. Transfer, Vol. 12, 189 - 205 (1972).

Recent observational and theoretical work has indicated that CN may play a major role in forming the spectrum and in fixing the thermal structure in a wide variety of cool stellar atmospheres. As the first step in investigating that role, we have calculated straight mean absorption coefficients over intervals of 100 cm^{-1} for the red system $(A^2\Pi - X^2\Sigma)$ of CN for the entire spectrum ($0-25,000$ cm^{-1}) by summing the integrated absorption coefficient for every line within the chosen interval. Over 100,000 lines from all bands with $-3 \le \Delta v \le 12$ and $0 \le v'' \le 18$ are included. The accuracy of this smoothing in the opacity is investigated. Complete calculations are made for four temperatures (1000, 2000, 3500 and 5000 K), and the resulting mean absorption coefficients ($\bar{\alpha}$) are fitted by the equation $\log \bar{\alpha} = A_1 + A_2\theta + A_3\theta^2$, where θ is the reciprocal temperature ($\theta = 5040/T$). The coefficients A_1, A_2 and A_3 are given for each wavenumber interval for both $C^{12}N^{14}$ and $C^{13}N^{14}$.

064.020 **A note on circulation currents in differentially rotating stellar envelopes.** D. Brand.
Monthly Notices Roy. Astron. Soc., Vol. 156, 325 - 335 (1972).

A simple perturbation method for calculating the circulation velocities in radiative stellar envelopes for any cylindrical rotation law is presented, and applied to four such laws.

064.021 **Convective envelopes and radial pulsation of massive red supergiants.** R. Stothers.
Astron. Astrophys., Vol. 18, 325 - 328 (1972).

A grid of convective envelope models covering a wide range of astrophysical parameters (including mass loss) has been constructed for M-type supergiants with the help of theoretical evolutionary tracks, mixing-length theory, and the inclusion of hydrogen and helium ionization zones. Linear adiabatic pulsation theory has been used to obtain the first four normal modes of radial pulsation for all the models.

064.022 **Theory of stellar atmospheres.** K. H. Böhm.
Vistas in astronomy, Vol. 13, (see 003.001), 165 - 167 (1972).

064.023 **Hydromagnetic stability of a stratified atmosphere flowing over a liquid.** S. P. Talwar.
Astrophys. Journ., Vol. 173, 407 - 421 (1972).

The purpose of this paper is to investigate the stability of a configuration consisting of a stratified plasma atmosphere flowing over an infinitely conducting liquid subject to a downward gravity. Such situations, it may be pointed out, often arise in stellar atmospheres. We consider that the infinitely deep liquid is permeated with a uniform magnetic field (normal to gravity) while the stratified isothermal atmosphere carries a magnetic field (aligned to the field in the liquid) whose magnitude varies with height in such a manner as to render the Alfvén velocity constant for the entire atmosphere.

064.024 **Circulation in radiative stellar envelopes.**
M. J. Clement.
Journ. Roy. Astron. Soc. Canada, Vol. 66, 67 - 68 (1972).
Abstr. Canadian Astron. Soc.

064.025 **The theoretical behavior of NaI 5890 in a solar type star.** D. F. Gray.
Journ. Roy. Astron. Soc. Canada, Vol. 66, 70 - 71 (1972).
Abstr. Canadian Astron. Soc.

064.026 **Selected model stellar atmospheres.**
E. J. Mausser, R. J. Doyle.
Journ. Roy. Astron. Soc. Canada, Vol. 66, 73 (1972). – Abstr.
Canadian Astron. Soc.

064.027 **Convection in the envelopes of main sequence A–F stars.** E. Ergma.
Astrofizika, Vol. 7, 605 - 609 (1971). In Russian.
English translation in Astrophysics, Vol. 7, No. 4.

For the convective envelopes of late A and earlier F stars it is shown that various assumptions for the mixing length lead to different values of the convective velocities.

064.028 **On the problem of stability of adiabatic envelopes of stars.**
G. S. Bisnovaty-Kogan, Ya. B. Zeldovich, N. I. Shakura.
Astrofizika, Vol. 7, 617 - 624 (1971). In Russian.
English translation in Astrophysics, Vol. 7, No. 4.

The stability of adiabatic envelopes with adiabatic index γ in the external field of the nucleus is investigated. For the case of hard core the condition of stability is $\gamma > 1$. The condition of stability is obtained also for the case of compressible core.

064.029 **A modification of the Avrett-Krook temperature-correction procedure.** A. H. Karp.
Astrophys. Journ., Vol. 173, 649 - 652 (1972).

A variation of the Avrett-Krook temperature-correction procedure is presented which improves the convergence properties in the surface layers.

064.030 **Limb darkening for B-type main sequence stars in the infrared.** J. Grygar.
Bull. Astron. Inst. Czechoslovakia, Vol. 23, 175 - 177 (1972).

Limb darkening laws and the respective linear coefficients u' are tabulated for seven model stellar atmospheres and four wavelengths in the infrared region (14588 A - 58353 A). The models, constructed by Underhill, were identified with main-sequence stars of spectral types B3 to B9. A survey of available infrared limb darkening calculations is also given.

064.031 **Thermodynamics of the gray atmosphere. IV. Entropy transfer and production.** R. Wildt.
Astrophys. Journ., Vol. 174, 69 - 77 (1972).

In strict radiative equilibrium, i.e., if heat transport by conduction and convection is negligible, the local rate of entropy production must equal the divergence of the net flux of radiant entropy. Verification that this divergence is positive for LTE, is straightforward even in the nongray case. Entropy production throughout the interior can, in the absence of mass ejection, be disposed of only by the flux of radiant entropy escaping at the stellar surface, to which there is an (unattainable) upper limit depending solely on the effective temperature, namely, the entropy loss from the surface of an isothermal blackbody radiating at the same temperature.

064.032 The effect of stochastic magnetic fields on absorption lines. A. Z. Dolginov, G. G. Pavlov.
Astrophys. Letters, Vol. 11, 63 - 69 (1972).

The passage of light through an optically thin gas layer in a stochastic magnetic field is considered. Formulas connecting the profiles of four Stokes parameters of the incident and transmitted light are obtained. It is shown that the non-polarized component in the sunspot umbra located in the disk centre can be explained by the available stochastic magnetic field.

064.033 Analysis of normal and strong-lined K-type stars.
V. Oinas.
Bull. American Astron. Soc., Vol. 4, 211 (1972). – Abstr. AAS.

064.034 Multiple solutions for cepheid envelopes.
D. Lauterborn.
Bull. American Astron. Soc., Vol. 4, 217 (1972). – Abstr. AAS.

064.035 Expanding optically thin model atmospheres for early type stars. J. P. Cassinelli, J. I. Castor.
Bull. American Astron. Soc., Vol. 4, 230 (1972). – Abstr. AAS.

064.036 Stellar wind in cool carbon stars.
D. S. Balamore, L. B. Lucy.
Bull. American Astron. Soc., Vol. 4, 234 (1972). – Abstr. AAS.

064.037 Is Alpha Orionis reddened? T. D. Faÿ.
Bull. American Astron. Soc., Vol. 4, 234 (1972). Abstr. AAS.

064.038 Helium D_3 in stellar chromospheres.
J. M. Pasachoff, E. C. Lepler.
Bull. American Astron. Soc., Vol. 4, 235 (1972). – Abstr. AAS.

064.039 Microturbulence in G supergiant atmospheres.
W. Buscombe, J. Neatrour, E. Albert.
Bull. American Astron. Soc., Vol. 4, 237 (1972). – Abstr. AAS.

064.040 A relation between thermal diffusion and secular time scales in simple stellar envelopes.
C. J. Hansen.
Astron. Astrophys., Vol. 19, 71 - 75 (1972).

Some simple calculations are described which indicate explicitly the relation between thermal diffusion times of photons in stellar envelopes and secular time scales.

064.041 Diffusion processes in stars. E. Basińska.
Postępy Astron., Vol. 20, 119 - 137 (1972).
In Polish.

In the first section the physical background of the diffusion phenomenon is given. In subsequent sections a review of diffusion in different spectral type stars is presented with the emphasis on solar conditions.

064.042 Oxygen abundances of three population II horizontal-branch stars. K. Kodaira, K. Tanaka.
Publ. Astron. Soc. Japan, Vol. 24, 355 - 364 = Tokyo Astron. Obs. Repr., No. 415 (1972).

The oxygen abundances of three population II horizontal-branch stars, HD 86986, HD 109995, and HD 161817, are studied based on spectrograms taken at the Palomar Observatory ($\lambda\lambda$ 6100–8500 Å, 27 Å mm^{-1}).

064.043 The metal-to-hydrogen ratio in F1–F5 stars, as determined by a model-atmosphere analysis of photoelectric observations of a group of weak metal lines.
B. Gustafsson, P. E. Nissen.
Astron. Astrophys., Vol. 19, 261 - 282 (1972).

Photoelectric observations of the strength of a group of weak metal lines in the spectra of 74 F1–F5 stars, including eight Hyades and six Pleiades stars, were carried out with an echelle spectrometer at the Observatoire de Haute Provence. They were analysed as a function of the fundamental atmospheric parameters with the aid of hydrogen-line-blanketed model atmospheres, which were computed by a new and very effective method of the Feautrier type.

064.044 Small molecules in astrophysics: in stellar atmospheres, in comets, and in interstellar space.
L. Biermann.
Selected topics in molecular physics, (see 012.017), p. 17 - 18 (1972).

064.045 CaOH, a new triatomic molecule in stellar atmospheres. P. Pesch.
Astrophys. Journ., (Letters), Vol. 174, L155 - L156 (1972).

A strong absorption band at ~$\lambda\lambda$5500–5560 which occurs in spectra of late-type M-dwarf stars but not in those of late-type M giants is tentatively attributed to the triatomic molecule CaOH.

064.046 Meridian circulation with rapid differential rotation in radiative stellar envelopes. M. J. Clement.
Astrophys. Journ., Vol. 175, 135 - 145 (1972).

It is assumed that the turbulent flow found by Smith at the outer boundary of radiative envelopes leads to a conservative rotation law for which the stellar surface is a streamline. Under these conditions, the currents in low-density regions are relatively slow and steady with the possible exception of critical rotation. For all but the fastest rotators, the differential rotation eliminates the outer circulation zone which exists in the presence of rigid rotation.

064.047 Transfer of polarized radiation in a stellar atmosphere. G. W. Collins II.
Astrophys. Journ., Vol. 175, 147 - 156 (1972).

We present a method for solution of the equation of transfer for elliptically polarized radiation in an illuminated stellar atmosphere. Although the method is completely general, it is specifically applied to the gray-atmosphere problem in order to demonstrate its accuracy and stability.

064.048 Radiative transfer in atmospheres of Algol-type binaries. I. I. Pustylnik.
Tartu Astron. Obs. Teated, No. 35, p. 3 - 49 (1971).

The principles for constructing model atmospheres for the components of Algol-type close binaries are developed. The explicit expressions are derived for the distributions of integral radiative flux, surface temperature and gravitational acceleration over the surface of a non-spherical star identified with its Roche lobe. A special consideration is given to the effects of departures from the state of LTE both in the continuum and spectral lines in case of atmospheres of the secondary components. For the actual case of two non-spherical stars formulae for irradiative flux are derived. The problem of mutual eclipses is shortly discussed.

064.049 Helium in hot stars.
A. I. Poland.
Observatory, Vol. 92, 17 - 18 (1972). – Letter.

064.050 Helium in hot stars, still a problem.
A. B. Underhill.
Observatory, Vol. 92, 18 - 19 (1972). – Letter.

064.051 Molecules in atmospheres of cool giant stars.
D. Alexander, J. Collins, T. D. Fay, H. R. Johnson.
26th symposium on molecular structure and spectroscopy,
Columbus, Ohio, p. 34 - 35 (1971). – See Phys. Abstr., Vol.
75, No. 16996 (1972).

064.052 Stellar winds and breezes.
P. H. Roberts, A. M. Soward.
Proc. Roy. Soc. London, Ser. A, Vol. 328, 185 - 215 (1972).
Steady stellar winds are generally divided into two classes: (I) the winds proper, for which the energy flux per unit solid angle, E_∞, is non zero, and (II) the breezes, for which $E_\infty = 0$. The winds are examined in the limit $E_\infty \to 0$, and the relation with the breezes is studied.

064.053 On a non-linear differential equation of stellar atmospheres. S. K. Majumdar.
Indian Journ. Pure and Applied Math., Vol. 2, 762 - 768
(1971).
A static state of the stellar atmosphere incorporating radiation and heat conduction terms in the energy equation is derived. With certain simplifying assumptions it is shown that the temperature in such a static atmosphere is governed by a third-order non-linear ordinary differential equation.

064.054 Spectrum formation without the hypothesis of LTE physical principles. R. N. Thomas.
JILA Rep. 1970, No. 106, 27 pp. – See Phys. Ber., Vol. 51,
No. 4 - 4457 (1972).

064.055 Surface fluxes for model atmospheres for the central stars of planetary nebulae.
D. G. Hummer, D. Mihalas.
JILA Rep. 1970, No. 101, 27 pp.

064.056 Radiative transfer in spherical shell atmospheres with radial symmetry.
S. Veno, H. Kagiwada, R. Kalaba.
Journ. Math. Phys., *New York*, Vol. 12, 1279 - 1286 (1971).

064.057 Contribution à l'étude de la supermétallicité dans les atmosphères stellaires. M. J. Vaziaga.
Thesis, Univ. Paris. 26 pp. (1971). – See Bull. Signal., Vol. 33,
Section 120, No. 7066 (1972).

064.058 Étude de la déficience en métaux dans l'atmosphère d'une étoile F: 15 Pegasi. M. Thomas.
Thesis, 3ᵉ Cycle, Spéc. Astrophys., Univ. Paris. 32 pp. (1971).
See Bull. Signal., Vol. 33, Section 120, No. 7067 (1972).

Nuclear reactions in stellar surfaces and their relations with stellar evolution.   See Abstr. 003.110.

Basic physics of stellar atmospheres.
See Abstr. 003.122.

Sur une nouvelle méthode pour l'étude de la formation des raies spectrales, quand la fonction source est indépendante de la fréquence.   See Abstr. 063.002.

Convection in stars.   See Abstr. 065.061.

Relativistic effects on envelope formation during the collapse of a rotating star.   See Abstr. 066.032.

Revised *gf*-scale and solar curves-of-growth.
See Abstr. 071.010.

The effect of radiative equilibrium on the photospheric angular velocity.   See Abstr. 071.012.

Observations of Zeta Puppis.   See Abstr. 114.006.

Equivalent width data for several Am stars in clusters and comparison standards.   See Abstr. 114.008.

Contribution to the determination of the effective temperature of A-type stars.   See Abstr. 114.009.

A study of HD 4180 as compared to Be stars with lightly extended envelopes.   See Abstr. 114.050.

The atmospheres of the F-type supergiants. I. Calibration of the luminosity-sensitive O I λ 7774 line.
See Abstr. 114.052.

The atmospheres of the F-type supergiants. II. The line and continuous spectra.   See Abstr. 114.053.

An investigation of the atmospheres of metallic-line stars. I. A quantitative analysis of the ζ Lyr A, κ Ari, μ Aqr, 14 Psc and γ Equ atmospheres by the curve-of-growth method.
See Abstr. 114.056.

The ultraviolet flux envelopes of B type stars.
See Abstr. 114.092.

Mass outflow from hot stars.   See Abstr. 114.102.

Spettroscopia galattica in Italia nel periodo ottobre '68–ottobre '69.   See Abstr. 114.108.

A model atmosphere analysis of the Ap star HR 465.
See Abstr. 114.111.

Diffusion and the ³He abundance in 3 Cen A.
See Abstr. 114.117.

Stellar rotation and a simulation of the intensity interferometer.   See Abstr. 115.013.

Low-dispersion luminosity criteria in A and F type stars.   See Abstr. 115.019.

Light variations in magnetic stars.
See Abstr. 116.001.

Analysis of the atmospheres of magnetic stars by the curve-of-growth method. II. 10 Aql. and ι CrB.
See Abstr. 116.004.

Analysis of the atmospheres of magnetic stars by the curve-of-growth method. III. Standard stars and constructing of a temperature scale.   See Abstr. 116.005.

On the atmosphere of Epsilon Aurigae.
See Abstr. 121.015.

The spectra and element abundances of cool white dwarfs.   See Abstr. 126.008.

Hydrodynamic studies of thermonuclear runaways in helium rich white dwarfs with hydrogen rich envelopes.
See Abstr. 126.010.

Errata

064.901 Errata: "Meridional circulation in rotating stellar atmospheres–II. The effect of variable opacity"
[Monthly Notices Roy. Astron. Soc., Vol. 154, 293 - 300
(1971)].   D. Brand, R. C. Smith.
Monthly Notices Roy. Astron. Soc., Vol. 155, 383 (1972).

065 Stellar Structure, Stellar Evolution, Stellar Nucleosynthesis

065.001 **The p-process in explosive nucleosynthesis.**
J. W. Truran, A. G. W. Cameron.
Astrophys. Journ., Vol. 171, 89 - 92 (1972).

The limiting conditions consistent with p-process synthesis in supernova envelopes are inferred from calculations of the appropriate rates of proton capture and neutron photodisintegration.

065.002 **The pulsations of polytropic masses in rapid, uniform rotation.**
N. R. Lebovitz, G. W. Russell.
Astrophys. Journ., Vol. 171, 103 - 105 (1972).

The variational principle for the oscillations of gaseous masses is used to find the effect of a uniform rotation, which is not assumed slow, on the fundamental pulsation frequency of a polytrope of index $n = 3$. The results are compared with perturbation calculations based on the assumption that the rotation is slow.

065.003 **Nuclear forces, compressibility of neutron matter and the maximum mass of neutron stars.**
J. W. Clark, H. Heintzmann, W. Hillebrandt, M. Grewing.
Astrophys. Letters, Vol. 10, 21 - 25 (1972).

It is shown that if the published values of the MIT neutron-matter calculations are adopted at densities above 3×10^{14} g cm^{-3}, neutron-star models of about $0.8\,M_\odot$ can be obtained. If a solid-state model at high densities is adopted, even soft nuclear forces will lead to a maximum mass of $1.7\,M_\odot$ for stable neutron stars.

065.004 **Star formation by condensation, I.**
G. P. Horedt.
Astrophys. Space Sci., Vol. 15, 129 - 136 (1972).

We describe two possibilities for the formation of a star by transition of molecular hydrogen from gaseous to the liquid phase in the gravitational field of a spherical mass. Model A : Contraction through condensation of a polytropic respectively isothermal gas sphere. Model B: Condensation in the isothermal atmosphere which surrounds a core.

065.005 **Surface composition of magnetic neutron stars.**
L. C. Rosen, A. G. W. Cameron.
Astrophys. Space Sci., Vol. 15, 137 - 152 (1972).

The relative abundances of seven constituent nuclei, He^4, C^{12}, O^{16}, Ne^{20}, Mg^{24}, Si^{28} and Fe^{56}, are calculated as a function of time for neutron star atmospheres within which exist magnetic fields of the order of 10^{13} G. The opacity, equation of state of the electrons, and cooling rate of the magnetic star are discussed. Computations are performed both for a constant mass atmosphere and for an atmosphere in which mass is being ejected.

065.006 **Accretion of matter by condensed objects.**
F. C. Michel.
Astrophys. Space Sci., Vol. 15, 153 - 160 (1972).

The equations of motion for steady-state spherical symmetric flow of matter into or out of a condensed object (e.g. neutron stars, 'black holes', etc.) are displayed and solved for simple polytropic gases. The two fluid (electrons and ions separately) approach is also examined.

065.007 **The thermodynamics of white dwarf matter. II.**
G. Shaviv, A. Kovetz.
Astron. Astrophys., Vol. 16, 72 - 76 (1972).

An equation of state is constructed for a dense plasma in which the electrons are strongly degenerate. This equation of state is used together with the one obtained in Paper I to de-termine the conditions for a phase transition between liquid and solid. Some numerical examples for this are given.

065.008 **Surface temperatures and the curves of growth for Population I and Population II stars.**
E. Böhm-Vitense.
Astron. Astrophys., Vol. 16, 81 - 84 (1972).

The dependence of the curve of growth on metal abundance is studied by comparing the solar curve of growth with that of a star with the same effective temperature and gravity but a metal content reduced by a factor 100 and also with that of a giant with reduced metal abundance.

065.009 **Loops in the H-R diagram during core helium burning and the effect of rotation.**
E. Meyer-Hofmeister.
Astron. Astrophys., Vol. 16, 282 - 285 (1972).

Evolutionary tracks were computed for non-rotating and for rotating stars of 5, 6 and 9 solar masses to study the influence of rotation on the loops in the H-R diagram during core helium burning.

065.010 **On the fragmentation of a contracting hydrogen cloud in an expanding universe.** T. Yoneyama.
Publ. Astron. Soc. Japan, Vol. 24, 87 - 98 (1972).

The gravitational contraction and the fragmentation of a pure hydrogen gas cloud in an expanding universe are studied. A cloud contracts adiabatically at the initial phase, but at a later phase, hydrogen molecules are formed in the cloud and they act as an effective cooling source. As the contraction proceeds, fragmentation takes place in the cloud and massive stars with masses greater than $60\,M_\odot$ will be formed.

065.011 **On the anharmonic pulsations of a massive star in the helium-burning phase.** C. Mohan.
Publ. Astron. Soc. Japan, Vol. 24, 135 - 139 (1972).

The effects of including fifth and sixth modes in the equations of anharmonic pulsations of a 15.6 solar mass star in the helium-burning phase of its evolution are studied. The results indicate that much higher modes are not active in most of the cepheid-type pulsating stars.

065.012 **Critical core mass for carbon-detonation supernovae.**
Y. Nakada, D. Sugimoto.
Publ. Astron. Soc. Japan, Vol. 24, 141 - 144 (1972).

A lower limit of the core mass, for which carbon detonation can be generated, is obtained semi-analytically. A contradiction between complete disruption of stars by carbon detonation and a pulsar remnant still remains.

065.013 **Semiconvection in the core-helium-burning phase of stellar evolution.** J. W. Robertson, D. J. Faulkner.
Astrophys. Journ., Vol. 171, 309 - 315 (1972).

A treatment of the expansion by convective overshooting of the helium-burning convective cores in stars over a wide mass range is described. A semiconvective zone forms outside the core, and this zone has been followed through its evolution. In some circumstances there is a fully convective region embedded in this zone. To illustrate the semiconvection treatment, the core-helium-burning phases of $5\,M_\odot$ and $1.5\,M_\odot$, Population I stars have been computed to near exhaustion of helium in the core.

065.014 **Equilibrium, pulsational, and secular stability properties of the lower carbon-burning main sequence.**
J. T. Mariska, C. J. Hansen.
Astrophys. Journ., Vol. 171, 317 - 321 (1972).

Results are presented for stellar models composed of equal amounts of ^{12}C and ^{16}O in complete equilibrium within the mass range $0.82 \leq M/M_\odot \leq 2.31$. The character of the static models is analogous to the lower helium main sequence in that they are double-valued with respect to mass whereas the central density is monotonic along the sequence. All the models are pulsationally unstable, but only one, near the minimum mass, is secularly unstable.

065.015 Rotation and lithium abundance in solar-type stars.
R. H. Dicke.
Astrophys. Journ., Vol. 171, 331 - 362 (1972).

It is concluded that the Goldreich-Schubert-Fricke instability cannot extend much deeper than the onset of fast burning of ^7Li. By contrast, the assumption that a mild thermally driven turbulence is terminated at the boundary of a rapidly rotating core is used to give a quantitative account of the depletion of ^7Li in the sun and solar-type stars. Lithium-6 should be strongly depleted, but ^9Be not at all. The requirements that for the sun both the rotation and the ^7Li abundance agree with the observations give for each possible choice of core radius a value for the solar-wind torque and a lower bound for the angular velocity of the core.

065.016 Neutron star magnetism.
H. Heintzmann, M. Grewing.
Zeitschr. Physik, Vol. 250, 254 - 262 (1972).

It is shown that 1. an appreciable change of magnetic moment of a neutron star cannot occur via ohmic dissipation; 2. Pulsars provide evidence for large internal magnetic fields in main sequence stars. If pulsars are born from stars with masses exceeding $3M_\odot$ the internal field must be of the order of $10^3 - 10^4$ Gauss while if they are derived from less massive stars 10^2 Gauss are sufficient to give rise to a magnetic moment of $\sim 10^{30}$ Gauss cm^3.

065.017 Zur Dynamotheorie magnetischer Sterne: Der "symmetrische Rotator" als Alternative zum "schiefen Rotator". F. Krause.
Astron. Nachr., Vol. 293, 187 - 193 (1971/72).

The excitation of non-axisymmetric magnetic fields by dynamo action of stars is discussed by investigating models described earlier by Steenbeck and Krause (1969).

065.018 Über den Aufbau von Sternen der unteren Hauptreihe. K. Fritze.
Astron. Nachr., Vol. 293, 203 - 210 (1971/72).

Models of 0.6, 0.8 and 1.0 $M_\odot$ are computed for two chemical compositions ($X = 0.602$, $Z = 0.044$ and $X = 0.900$, $Z = 0.001$) representing the extreme populations I and II, and several values of the mixinglength-parameter α. Particular attention is paid to the treatment of ionisation equilibrium in the outer convection zone. The lowering of ionisation potential by pressure ionisation as well as the influence of metals on the electron pressure is taken into account. Theoretical Hertzsprung-Russell- and mass-luminosity-diagrams are compared with observations.

065.019 Evolutionary aspects of the cepheid stage.
K. J. Fricke, P. A. Strittmatter.
Monthly Notices Roy. Astron. Soc., Vol. 156, 129 - 150 (1972).

The structural and evolutionary conditions which allow high mass stars ($M > 4 M_\odot$) to become cepheid variables are examined. The important internal parameters for determining the location of stars in H-R diagram are isolated from computed sequences of static models. The interplay of the various processes which take place during actual evolution is concisely expressed in a semi-empirical relation between effective temperature and core mass (or helium content in the core) for the static situation. A qualitative interpretation of the computed results is offered. The existence of two branches of chemically identical solutions of the stellar structure equations—one stable, the other unstable—is demonstrated.

065.020 Advanced evolution of population II stars. III. Some uncertainties in horizontal-branch models.
P. Demarque, J. G. Mengel.
Astrophys. Journ., Vol. 171, 583 - 591 (1972).

Evolution from the zero-age horizontal branch is investigated for stars with the chemical composition parameters $(X, Z) = (0.75, 10^{-3})$ and masses 0.50, 0.56, and 0.60 $M_\odot$. For the 0.60 $M_\odot$ models, the effects of semiconvection and of the choice of interpolation between opacity tables are studied. Astronomical implications are briefly discussed.

065.021 The evolution of growing stellar cores up to carbon ignition: From whence the pulsars?
Z. Barkat, J. C. Wheeler, J.-R. Buchler.
Astrophys. Journ., Vol. 171, 651 - 662 (1972).

It has been argued that stars in the main-sequence mass range $4 \lesssim M/M_\odot \lesssim 10$ will explode and leave a collapsed remnant, as implied by empirical evidence, if the central density at carbon ignition (ρ_{ign}) is greater than some limit (ρ_{crit}). Herein the factors influencing ρ_{ign} are examined in detail and a comparison between ρ_{ign} and ρ_{crit} and their respective uncertainties are given. The proposed scenario for pulsar formation is within the realm of possibility, but current uncertainties forestall any firm conclusion.

065.022 New information on the rate of the carbon-burning reaction. M. Mazarakis, W. Stephens.
Astrophys. Journ., (Letters), Vol. 171, L97 - L99 (1972).

New measurements of the cross-section for the $^{12}C + ^{12}C$ carbon-burning nuclear reaction are reported, extending down to a center of mass energy of 2.45 MeV, equivalent to a thermal temperature of 10^9 °K. Reaction rates are tabulated for various burning temperatures.

065.023 Effects of nuclear reactions on the stability of degenerate stars. M. Gabriel, G. Chanmugam.
Astron. Astrophys., Vol. 16, 369 - 373 (1972).

The effects of changes in chemical composition on the stability of degenerate stars toward adiabatic radial perturbations are considered.

065.024 Stars with central helium burning and the occurrence of loops in the H-R diagram. III. Horizontal branch stars.
D. Lauterborn, S. Refsdal, R. Stabell.
Astron. Astrophys., Vol. 17, 113 - 127 (1972).

The structure and evolution of horizontal branch stars is discussed by considering separately the effects of changes in the mass, radius and luminosity of the He-core, and the X-profile above the core. It is discussed why the type of evolutionary track mainly depends on the relative luminosity (L_c/L) of the He-core.

065.025 The evolution of a 5 $M_\odot$ star.
M. L. Aizenman, J. Perdang, J. R. Lesh.
Astron. Astrophys., Vol. 17, 139 - 141 (1972).

A 5 $M_\odot$ star having an initial chemical composition of $X = 0.72$, $Y = 0.25$, and $Z = 0.03$ has been evolved to the stage where hydrogen burning in a thick shell takes place. This model is compared with other models having the same mass but different chemical compositions.

065.026 Thermohaline convection in stellar interiors.
R. K. Ulrich.
Astrophys. Journ., Vol. 172, 165 - 177 (1972).

The important conclusions of this study are: Thermally limited modes dominate mixing by thermohaline convection in stellar interiors; An inverted gradient of mean molecular weight is too short-lived to play an important role in the β Cephei pulsations; Mixing between the central ^{4}He region and the ^{12}C shell after a degenerate ^{4}He shell flash takes longer than 10^6 years. The last conclusion may have interesting consequences since it suggests that a central helium core may persist until near the end of the life of a low-mass star. In this case the course of evolution could be significantly different from what one would expect with a pure ^{12}C core.

065.027 **Radial pulsations of pre-white-dwarf stars. I. Linear quasi-adiabatic analysis.**
H. M. Van Horn, M. B. Richardson, C. J. Hansen.
Astrophys. Journ., Vol. 172, 181 - 199 (1972).
Numerical calculations of the periods, eigenfunctions, and stability integrals in the quasi-adiabatic limit have been carried out for the lowest radial pulsation modes of homogeneous, pre-white-dwarf stellar models. Detailed numerical results are given, and the effects of neutrino emission, differences in stellar mass, and structural changes accompanying the evolution are discussed.

065.028 **Carbon stars and the CNO bi-cycle.**
R. I. Thompson.
Astrophys. Journ., Vol. 172, 391 - 393 (1972).
It is shown that in the context of present stellar evolutionary theory, carbon stars cannot be the result of mixing CNO bi-cycle products to the surface. This is because insufficient material is processed in the stellar interior to affect the ^{12}C/^{16}O ratio significantly at the surface of the star. A 9 $M_\odot$ star from Iben's work is discussed as an example. Four speculative possibilities are listed for the origin of carbon stars.

065.029 **The rate of the ^{14}N(α, γ)^{18}F reaction.**
R. G. Couch, H. Spinka, T. A. Tombrello, T. A. Weaver.
Astrophys. Journ., Vol. 172, 395 - 401 (1972).
New experimental and theoretical results are presented. A new resonance has been found at E_a = 0.56 MeV with $(2J + 1)\Gamma_a\Gamma_\gamma/\Gamma = 2.8 \pm 0.5 \times 10^{-4}$ eV (lab). The reaction rate is calculated, and the significance of this rate is discussed. It is found, that in most stars α-capture by ^{14}N takes place after the initiation of helium burning by the $3\alpha \rightarrow {}^{12}$C reaction.

065.030 **Oscillatory thermal instabilities at the onset of helium shell burning.** R. Härm, M. Schwarzschild.
Astrophys. Journ., Vol. 172, 403 - 417 (1972).
The onset of thermal instability has been investigated in detail for globular-cluster stars in the evolutionary phase during which the helium-burning shell is just forming. The investigation shows that the main unstable phase ("second-red-giant phase") is preceded by a short and inconsequential preliminary unstable phase during which the instability is weak and of an oscillatory character. The oscillatory instability gives way to a simple exponential instability which causes the repetitive helium-shell flashes described in earlier investigations.

065.031 **Polarimetric observations and models of late-type stars.** S. J. Shawl.
Bull. American Astron. Soc., Vol. 3, 442 (1971). – Abstr. AAS.

065.032 **The effect of beta-processes on the post-detonation evolution of dense stellar cores.** S. W. Bruenn.
Bull. American Astron. Soc., Vol. 3, 452 (1971). – Abstr. AAS.

065.033 **Is the interior of a neutron star a quantum crystal?**
V. Canuto.
Bull. American Astron. Soc., Vol. 3, 452 (1971). – Abstr. AAS.

065.034 **Microquakes and macroquakes in neutron stars.**
D. Pines, J. Shaham.

Bull. American Astron. Soc., Vol. 3, 452 - 453 (1971). – Abstr. AAS.

065.035 **Resonant pulsations.** N. R. Simon, V. K. Sastri.
Bull. American Astron. Soc., Vol. 3, 479 (1971).
Abstr. AAS.

065.036 **Effects of semi-convection on the horizontal branch.**
P. Demarque, A. Sweigart.
Bull. American Astron. Soc., Vol. 3, 479 (1971). – Abstr. AAS.

065.037 **The evolutionary status of the blue halo stars.**
E. B. Newell.
Bull. American Astron. Soc., Vol. 3, 479 (1971). – Abstr. AAS.

065.038 **The effects of autoionization on the opacity of stellar mixtures.** A. L. Merts, N. H. Magee.
Bull. American Astron. Soc., Vol. 3, 483 (1971). – Abstr. AAS.

065.039 **New Los Alamos opacities.**
A. N. Cox, N. H. Magee, A. L. Merts.
Bull. American Astron. Soc., Vol. 3, 483 (1971). – Abstr. AAS.

065.040 **The effect of rare earths on the opacity of Ap stars.**
W. F. Huebner, G. D. Koontz, M. F. Argo.
Bull. American Astron. Soc., Vol. 3, 483 (1971). – Abstr. AAS.

065.041 **Entropy production and the stellar upper mass limit.** W. Rose, R. Smith, B. Carney.
Bull. American Astron. Soc., Vol. 3, 483 (1971). – Abstr. AAS.

065.042 **Theoretical evolution of stars of very low metal content.** R. L. Wagner.
Bull. American Astron. Soc., Vol. 3, 501 (1971). – Abstr. AAS.

065.043 **A discussion on the dense baryon matter.**
Y. C. Leung, C. G. Wang.
Bull. American Astron. Soc., Vol. 3, 502 (1971). – Abstr. AAS.

065.044 **The evolution of 1.40 $M_\odot$ pure He star.**
G. Shaviv, N. Vidal.
Astrophys. Space Sci., Vol. 15, 195 - 205 (1972).
The evolution of 1.40 $M_\odot$ pure He star is calculated from the stage of the ignition in the center up to the very advanced stage of evolution where mass ejection by the very luminous He shell could occur. It is found that C^{12} does not ignite by a modest margin. Subsequent evolution and relation to the central stars of planetary nebulae is discussed.

065.045 **On stellar activity cycles.**
B. R. Durney, J. O. Stenflo.
Astrophys. Space Sci., Vol. 15, 307 - 312 (1972).
The relation between the average magnetic field B, the angular velocity Ω, and the period P of stellar activity cycles is studied. For the calculations we have used Leighton's (1969) model for the solar cycle with the additional assumption that the differential rotation and the cyclonic turbulence are both proportional to Ω.

065.046 **Star formation in the Galaxy.**
B. W. Pendred, I. P. Williams.
Astrophys. Space Sci., Vol. 15, 334 - 339 (1972).
A previous theory of the authors regarding the planetary

system is generalized in an attempt to include star formation. It is found that the theory predicts the correct mass and radius for stellar clusters and also the general shape of the Galaxy.

065.047 Observational studies relating to star formation. III.
A. F. Aveni, J. H. Hunter.
Astron. Journ., Vol. 77, 17 - 23, 105 - 110 (1972).

Data are presented and the total mass content estimated for a number of early type clusters associated with reflection nebulosities noted by van den Bergh and Racine.

065.048 Secular stability. III. The secular spectrum of stars of 2.5, 5, 10, and 15 solar masses.
M. L. Aizenman, J. Perdang.
Astron. Astrophys., Vol. 17, 190 - 192, with a correction in Vol. 19, 315 (1972).

We discuss the results obtained for stars of 2.5, 5, 10 and 15 $M_\odot$ whose initial composition is identical to that of a 1.13 $M_\odot$ star ($X = 0.75$, $Y = 0.22$, $Z = 0.03$).

065.049 On the generation of magnetic field in rotating relativistic objects.
G. S. Bisnovaty-Kogan, A. A. Ruzmaikin.
Astron. Astrophys., Vol. 17, 243 - 245 (1972).

It is shown that there is no magnetic field generation in a stationary rotating relativistic star if Ω = const in the case of a barotropic medium and in an isentropic star with $\Omega \neq$ const.

065.050 Note on the growth rate of convective modes in a self gravitating gas sphere.
P. Grisvard, P. Souffrin, M. Zerner.
Astron. Astrophys., Vol. 17, 309 - 311 (1972).

It is shown that within Cowling's approximation the growth rate of the eigen modes in a convectively unstable self gravitating gas sphere is a bounded monotonically increasing function of the degree of the harmonic.

065.051 The loss of angular momentum due to evolution of rapidly rotating early-type stars.
P. R. Gredley, E. F. Borra.
Astrophys. Journ., Vol. 172, 609 - 614 (1972).

Rigidly rotating stars of 5, 7, 9, and 10 $M_\odot$ have been evolved, with loss of angular momentum under the assumption of rotational ejection of mass, from the zero-age main sequence to hydrogen exhaustion. Rates of mass loss have been obtained in which the coupling between star and envelope was taken into account.

065.052 Detonation in stellar cores: The effect of nuclear statistical equilibrium.
J.-R. Buchler, J. C. Wheeler, Z. Barkat.
Astrophys. Journ., Vol. 172, 713 - 716 (1972).

In a previous paper the local conditions behind a Chapman-Jouguet detonation front and the self-consistency of the detonation were investigated for cases of astrophysical interest. The detonation was assumed to be driven by a single reaction. In this paper, rather than a single product, the reaction products are assumed to be the appropriate ensemble of nuclei in nuclear statistical equilibrium. The parameters of the detonation and the condition of self-consistency are reexamined with this assumption.

065.053 Hydrodynamic model calculations for supermassive stars I. The collapse of a nonrotating 0.75 × 10⁶ $M_\odot$ star. I. Appenzeller, K. Fricke.
Astron. Astrophys., Vol. 18, 10 - 15 (1972).

In order to test Fowler's models for QSOs and exploding galaxies we calculated the pre-main-sequence contraction and the relativistic dynamical collapse of a nonrotating 0.75 × 10⁶ $M_\odot$ supermassive star by solving numerically the (relativistic) hydrodynamic equations. In contrast to earlier

estimates we found that in such a star the rapid burning of hydrogen cannot halt the collapse before the Schwarzschild radius is reached.

065.054 Fragen des Gravitationskollaps. M. Reinhardt.
Naturwissenschaften, 59. Jahrgang, p. 7 - 12 (1972).

After some remarks on the final states of stellar evolution (white dwarfs, neutron stars), the gravitational collapse of a spherical object is discussed. The issue of anticollapse is critically reviewed. The collapse of a rotating object, and the possibilities of extracting energy from black holes are treated. Some ways to detect collapsed objects are discussed.

065.055 The effect of composition gradients on a convective core. R. J. Boyle.
Astrophys. Journ., Vol. 173, 103 - 108 (1972).

The effects of composition gradients due to helium burning during the core flash in a low-mass Population II star on the fundamental laminar mode of convection are investigated. The results imply that the expected composition gradients are insufficient to inhibit convection even in the center of the core.

065.056 Studies in stellar evolution. X. Hydrostatic adjustment. L. G. Henyey, R. K. Ulrich.
Astrophys. Journ., Vol. 173, 109 - 120 (1972).

We analyze in detail the process of hydrostatic adjustment first discussed by Schwarzschild and Härm. This detailed analysis leads to a better understanding of the mechanism by which a star becomes a red giant. As numerical examples we treat a perfect-gas polytrope of index 3 and a red giant of mass 2 $M_\odot$ just prior to helium-core ignition.

065.057 Nucleosynthesis in silicon burning.
G. Michaud, W. A. Fowler.
Astrophys. Journ., Vol. 173, 157 - 181 (1972).

A study has been made of silicon burning in stellar explosions. A variety of initial compositions has been considered, and an approximate method for summing the contribution of different zones in the star has been developed. It is emphasized that summing over zones in which the initial conditions vary results in realistic agreement with natural abundances for nuclei with $28 \leq A \leq 59$. The effects of some neglected aspects of the freezing of the participating nuclear reactions are discussed.

065.058 The neutron-proton ratio in stars exploding from dense, high-temperature states.
D. N. Schramm, Z. Barkat.
Astrophys. Journ., Vol. 173, 195 - 204 (1972).

Detailed calculations are presented showing the dependence of the freeze-out neutron-proton ratio (NPR) on relevant parameters for matter exploding from a hot, dense configuration. It is found that the freeze-out NPR value is extremely sensitive to the expansion rate as well as to the maximum temperature and density attained by the ejected matter. Application of these results to the study of nucleosynthesis in supernova models is discussed.

065.059 Nonstationary hydrodynamical accretion on a neutron star.
Ja. B. Zeldovich, L. N. Ivanova, D. K. Nadezhin.
Astron. Zhurn. Akad. Nauk SSSR, Vol. 49, 253 - 264 (1972). In Russian. English translation in Soviet Astron. AJ, Vol. 16, No. 2.

Nonstationary accretion of an optically thick gaseous cloud on a neutron star is considered. The accretion of a small cloud with mass 10^{-5} $M_\odot$ is investigated numerically taking into account radiative conductivity and neutrino emission.

065.060 About nuclear energy sources in superdense celestial bodies. R. M. Avakian, G. S. Sahakian.

Astron. Zhurn. Akad. Nauk SSSR, Vol. 49, 316 - 323 (1972). In Russian. English translation in Soviet Astron. AJ, Vol. 16, No. 2.

In the interiors of white dwarfs and in the envelopes of barionic stars there exists enough energy to keep them in a hot state during some milliards of years. In the case of white dwarfs this energy is sufficient to explain their observed luminosity.

065.061 **Convection in stars.**
E. V. Ergma, A. G. Massevitch.
Astron. Nachr., Vol. 293, 145 - 154 (1971/72).

Difficulties of the mixing length theory when applied to convective envelopes of main sequence stars, red giants and white dwarfs are discussed. It is shown that not only the value of the mixing length, but also other parameters of the theory influence strongly the obtained solution. The necessity of improvement of the theory is stressed.

065.062 **Stellar evolution toward pre-supernova stage. I. Carbon and oxygen stars of 5 $M_\odot$, 10 $M_\odot$ and 30 $M_\odot$.**
S. Ikeuchi, K. Nakazawa, T. Murai, R. Hōshi, C. Hayashi.
Progr. Theor. Phys., *Japan*, Vol. 46, 1713 - 1737 (1971).

The evolution of carbon-oxygen stars of 5 $M_\odot$, 10 $M_\odot$ and 30 $M_\odot$ is computed from a pre-carbon-burning contraction stage to a pre-supernova stage through the phases of central burning of C, Ne, O, Mg, Si and Ni. These stars are regarded as representing the cores inside the helium shell-source of ordinary evolved stars. The computation is made for the two alternative cases, with and without neutrino production by electron-neutrino interaction.

065.063 **Theory of superdense stars.**
T. Kodama, M. Yamada.
Progr. Theor. Phys., *Japan*, Vol. 47, 444 - 459 (1972).

A theory of cold static superdense stars composed of two or more components is presented. With the help of the variational principle and the equations of general relativity, simultaneous equations for stellar structure are derived. These equations are solved numerically for a stellar model in which protons and neutrons are treated as independent components by neglecting the β-decay.

065.064 **Uncertainties in the observable properties of lower main sequence stellar models.** D. L. Moss.
Monthly Notices Roy. Astron. Soc., Vol. 156, 291 - 296 (1972).

The uncertainties in the properties of lower main sequence stars arising from different methods of treating the non-adiabatic parts of the convection zones are investigated numerically.

065.065 **Composition changes during stellar evolution.**
P. P. Eggleton.
Monthly Notices Roy. Astron. Soc., Vol. 156, 361 - 376 (1972).

This paper presents an algorithm by means of which composition changes in stars due to nuclear reactions, convection and semiconvection can be relatively simply calculated. The algorithm is based on a simple physical picture of convective mixing as a diffusion of convective cells throughout a region of a star. We attempt to show that the method tends, in a limiting sense, to give results which are identical with the 'standard' treatment of mixing by convection or semiconvection, for a wide range of possible physical models of convective mixing.

065.066 **Theoretical evolution of 15 $M_\odot$ and 11 $M_\odot$ stars from the main sequence to the He-exhaustion phase.**
G. Barbaro, C. Chiosi, L. Nobili.
Astron. Astrophys., Vol. 18, 186 - 197 (1972).

The evolution of 15 $M_\odot$ and 11 $M_\odot$ stars of extreme Population I ($X = 0.602$, $Z = 0.044$) from the main sequence to the helium-exhaustion phase is studied. Both stars have been analyzed by taking as the criterion for neutrality in the intermediate convective regions the Schwarzschild-Härm condition.

065.067 **Spallation processes in stellar surfaces: Anomalous helium ratios.** S. Vauclair, H. Reeves.
Astron. Astrophys., Vol. 18, 215 - 223 (1972).

The evidence regarding spallation processes in stellar surfaces is reviewed. The spallation origin of the $^3He/^4He$ ratio in 3 Centauri A is studied in view of its implications on various physical parameters.

065.068 **Secular stability of a 1.1 $M_\odot$ star during the gravitational contraction and the main-sequence phase.**
M. Gabriel.
Astron. Astrophys., Vol. 18, 242 - 252 (1972).

The meaning of the secular stability problem is discussed especially for quasi-static models. The problem is applied to a 1.1 $M_\odot$ star during the gravitational contraction and the main-sequence phases.

065.069 **Galactic evolution. I. Differential effects of radiation pressure around stars of different spectral types on neutral atoms and dust grains.** J.-C. Pecker.
Astron. Astrophys., Vol. 18, 253 - 266 (1972). In French.

The ratios, and actual values, for forces exerted both by radiation pressure and gravitation on atoms of H I, He I, C I, Fe I, Eu I, and on dust grains of various sizes are computed, at a large distance of the star, for stars of all spectral types (assuming that the medium is essentially neutral, and that atoms are in the ground state) and for a schematical galaxy.

065.070 **Final evolution of a low-mass star. II.**
W. K. Rose, R. L. Smith.
Astrophys. Journ., Vol. 173, 385 - 391 (1972).

Calculations are presented that describe the physical properties of the static envelopes of a luminous red giant and an advanced thermal-relaxation oscillation. The thermal-relaxation oscillation produces a sufficiently high luminosity at the base of the convective envelope that hydrostatic equilibrium becomes impossible.

065.071 **Hydrostatic oxygen burning in stars. I. Oxygen stars.**
W. D. Arnett.
Astrophys. Journ., Vol. 173, 393 - 400 (1972).

The general properties of oxygen stars ($M/M_\odot$= 2, 4, 8, and 16), up to and during core oxygen burning, are presented. The reasons for these properties are discussed quantitatively. A "single-zone" approximation for typical conditions during hydrostatic oxygen burning is developed which involves balanced power from $^{16}O + ^{16}O$ energy generation and from neutrino energy loss. Some implications for nucleosynthesis theory are discussed.

065.072 **On the nature of the horizontal branch. II. Extremely blue halo stars: A theoretical viewpoint.**
J. Faulkner.
Astrophys. Journ., Vol. 173, 401 - 404 = Contr. Lick Obs., No. 354 (1972).

It is shown that Greenstein's extremely blue sequence of halo subdwarfs may be entirely consistent with the theory of the horizontal branch in globular clusters.

065.073 **URCA neutrino processes in stellar evolution.**
G. Beaudet, E. E. Salpeter, M. L. Silvestro.
Journ. Roy. Astron. Soc. Canada, Vol. 66, 68 (1972). – Abstr. Canadian Astron. Soc.

065.074 **Stability and radial pulsations of rotating neutron**

stars.
Yu. L. Vartanian, A. V. Hovsepian, G. S. Hajian.
Astrofizika, Vol. 7, 625 - 641 (1971). In Russian.
English translation in Astrophysics, Vol. 7, No. 4.

Radial pulsations of rotating cold neutron stars that are near the state of stability loss are studied by the energetic method. The effects of general relativity are taken into account. The integral parameters and frequency of radial pulsations for different equilibrium configurations are calculated.

065.075 **The internal dynamics of the oblique rotator.**
L. Mestel, H. S. Takhar.
Monthly Notices Roy. Astron. Soc., Vol. 156, 419 - 436 (1972).

The field of motions within a magnetic oblique rotator is constructed by a double perturbation technique, assuming strict adiabaticity. The dissipation of energy through radiative conduction and turbulent viscosity is estimated roughly, yielding time scales for rotation of the angle between the magnetic and rotation axes.

065.076 **The collapse of a rotating cloud.**   R. B. Larson.
Monthly Notices Roy. Astron. Soc., Vol. 156, 437 - 458 (1972).

Numerical calculations have been made for the early stages of collapse of an axisymmetric cloud, both with and without rotation. The implications of the results for our understanding of star formation and fragmentation are discussed.

065.077 **The evolution of spherical protostars with masses 0.25 M$_\odot$ to 10 M$_\odot$.**   R. B. Larson.
Monthly Notices Roy. Astron. Soc., Vol. 157, 121 - 145 (1972).

The calculations reported in a previous paper for the collapse of a spherical protostar have been improved and extended to a wider range of masses with calculations for masses of 0.25, 0.5, 1.0, 1.5, 2.0, 3.0, 5.0 and 10 solar masses. The model calculations have been compared with most of the observations thought to relate to star formation or newly-formed stars.

065.078 **Evolution and secular stability of a 0.8 M$_\odot$ carbon star.**   A. Noels.
Astron. Astrophys., Vol. 18, 350 - 362 (1972).

The evolution of a 0.8 M$_\odot$ star composed initially of 50 per cent of C^{12} and 50 per cent O^{16} has been computed by the Henyey method from the early phase of gravitational contraction to the final cooling toward the white dwarf stage. Neutrino losses were not taken into account. The secular stability has then been studied numerically throughout the evolution.

065.079 **Non-radial oscillations and vibrational stability of a 0.5 M$_\odot$ star.**   H. Robe, P. Ledoux, A. Noels.
Astron. Astrophys., Vol. 18, 424 - 427 (1972).

The non-radial adiabatic oscillations and the vibrational stability of a 0.5 M$_\odot$ star, composed of a radiative core and an extended convective envelope are studied. It is found that this star remains vibrationally stable.

065.080 **Thermonuclear detonation and beta-induced reimplosion of dense stellar cores.**   S. W. Bruenn.
Astrophys. Journ., Suppl. Ser., No. 207, Vol. 24, 283 - 318 (1972).

Detonation self-consistency tests are performed for matter consisting of the products of helium burning and the products of carbon burning in the density and temperature ranges $5 \times 10^6 \leq \rho \leq 3 \times 10^{10}$ g cm^{-3}, and $10^{8\,\circ} \leq T \leq 5 \times 10^{8\,\circ}$ K. Two effects not included in prior investigations are included here. The effect of β-processes on the postdetonation evolution of dense stellar cores is studied by detailed numerical hydro-

dynamic calculations. These were performed for a total of 14 different models with initial densities in the range $6 \times 10^9 \leq \rho \leq 3 \times 10^{10}$ g cm^{-3}, and compositions consisting of helium- or carbon-burning products. The implications of the results for nucleosynthesis and the initial collapse to neutron densities are discussed.

065.081 **Envelope opacities and the cepheid loops at 5 M$_\odot$.**
J. W. Robertson.
Astrophys. Journ., Vol. 173, 631 - 636 (1972).

Theoretical evolutionary tracks for the stage of core helium burning have previously been unable to explain the detailed distribution in the H-R diagram of the red giants in NGC 1866. Models calculated with improved envelope opacities develop extremely deep convective envelopes at the tip of the red-giant branch. The convective envelope modifies the hydrogen distribution in the region near the hydrogen-burning shell, which leads to core-helium-burning evolution in better agreement with the observations.

065.082 **The evolution of massive stars with mass loss.**
G. S. Bisnovatyi-Kogan, D. K. Nadyozhin.
Astrophys. Space Sci., Vol. 15, 353 - 374 (1972).

The evolution of massive stars (30M$_\odot$) is investigated in the phases of hydrogen and helium burning, taking into account the mass-loss due to light pressure in optically thick media. The evolution of a 9 M$_\odot$ star is calculated. The formation of infrared stars and Wolf-Rayet stars is discussed.

065.083 **Self-accretion of matter, red subluminous stars and early evolution of low-mass stars.**
V. Castellani, N. Panagia.
Astrophys. Space Sci., Vol. 15, 462 - 466 (1972).

At low luminosities, accretion of the gas surrounding low-mass stars can make the stars to follow a peculiar evolutionary course which can account for certain types of red subluminous stars. The efficiency of the accretion mechanism can also account for some peculiarities in the spectra of the stars of low mass.

065.084 **Formation of neutron star spots and its connection with pulsars. I.**   M. Fujimoto, T. Murai.
Publ. Astron. Soc. Japan, Vol. 24, 269 - 279 (1972).

The active surface of a neutron star is investigated under the assumption that a strong magnetic dipole is situated at the center with its axis parallel to the rotation axis.

065.085 **Out of what do the stars form?**   Yu. N. Efremov.
Priroda (NRB), Vol. 20, No. 6, p. 10 - 17 (1971).
In Bulgarian.

065.086 **Sternmaterie.II. Zustandsformen der Materie in stabilen Sternen.**   F. Stöckmann.
Phys. Blätter, 28. Jahrgang, p. 212 - 219 (1972).

065.087 **Models for R Coronae Borealis stars.**
P. Biermann, R. Kippenhahn.
IAU Colloquium No. 15, (see 012.006), p. 54 - 59 (1972).

065.088 **Time scales of stars during the crossing of the cepheid strip.**
D. Lauterborn, S. Refsdal, M. L. Roth.
IAU Colloquium No. 15, (see 012.006), p. 96 (1972).

065.089 **Carbon ignition in degenerate stellar cores.**
B. Paczyński.
Astrophys. Letters, Vol. 11, 53 - 55 (1972).

URCA processes within highly degenerate convective carbon-burning cores prevent carbon detonation in stars with masses below 8 M$_\odot$. Two types of supernovae and two kinds of pulsars are likely to be produced as a result of collapse following carbon exhaustion.

065.090 On the theory of stability of a star with a toroidal magnetic field. Yu. V. Vandakurov.
Astron. Zhurn. Akad. Nauk SSSR, Vol. 49, 324 - 333 (1972). In Russian. English translation in Soviet Astron. AJ, Vol. 16, No. 2.
Criteria for stability of adiabatic low-frequency non-radial oscillations of a gravitating configuration with an internal toroidal magnetic field are studied.

065.091 Post-main sequence semiconvection. S. R. Sreenivasan, K. E. Ziebarth.
Bull. American Astron. Soc., Vol. 4, 240 - 241 (1972). – Abstr. AAS.

065.092 Thermonuclear reaction rates at high temperatures. W. A. Fowler, D. N. Schramm, B. A. Zimmerman.
Bull. American Astron. Soc., Vol. 4, 259 (1972). – Abstr. AAS.

065.093 An apparent stellar mass gap near $3\,^1/_2\,M_\odot$. E. J. Devinney.
Bull. American Astron. Soc., Vol. 4, 269 (1972). – Abstr. AAS.

065.094 On the rotation of stellar models in Newtonian theory.
E. M. Skhtoryan, E. V. Chubaryan, V. V. Papoyan.
Dokl. AN ArmSSR, Vol. 53, 84 - 88 (1971). In Russian. Abstr. in Referativ. Zhurn. 51. Astron., 6.51.572 (1972).

065.095 Vogts og Russells sætning om stjernemodellers entydighed. J. O. Petersen.
Astron. Tidssk., Årg. 5, p. 37 - 40 (1972).

065.096 Sternbild Orion – Stätte der Sternentstehung. W. Büttner.
Astron. in der Schule, 9. Jahrgang, p. 16 - 20 (1972).

065.097 Secular stability of the lower hydrogen-burning main sequence. R. F. Stellingwerf, J. P. Cox.
Astron. Astrophys., Vol. 19, 8 - 12 (1972).
The secular stability of the lower hydrogen-burning main sequence and its continuation – the high-density branch – is investigated using polytropic models which are modified to include the effects of ionization and photospheric boundary conditions. A form of "Jeans criterion" is derived for this case.

065.098 Short period variable stars. IX. Rotation and mixing in the outer layers of A stars. Turbulent mixing due to the meridional circulation velocity field. A. Baglin.
Astron. Astrophys., Vol. 19, 45 - 50 (1972).
It is shown that mixing (or non mixing) due to the meridional circulation velocity field associated with fast (or slow) rotation can explain most of the key observational properties of A stars (Breger, 1970). At present, it seems that the data are all consistent with the following scheme: slow rotators have abundance anomalies and are vibrationally stable, fast rotators are variable. This paper fixes the critical velocity at approximately 50 km/s.

065.099 Pulsational instability of stars during deuterium burning. E. Toma.
Astron. Astrophys., Vol. 19, 76 - 81 (1972).
Pre-main-sequence evolutionary tracks for extreme Population I stars of 0.2, 0.6, 1.0 and 2.0 $M_\odot$ were calculated in order to study the effect of deuterium burning on pulsational stability. The vibrational instability manifests itself through the fundamental mode of radial oscillation and the pulsation amplitude will be multiplied in these stars by factors of $10^8 - 10^{15}$ at the end of the deuterium burning phase. An at-

tempt is made to relate this result with T-Tauri-type stars in the young cluster NGC 2264.

065.100 Influence of the excited states of target nuclei in the vicinity of the iron peak on stellar reaction rates. M. Arnould.
Astron. Astrophys., Vol. 19, 92 - 98 (1972).
With the aid of the statistical theory, we examine quantitatively the influence on the stellar reaction rates of the existence of rather heavy target nuclei in their various excited states. From some specific examples, we show that the usual approximations commonly used to take such an effect into account sometimes fail to a great extent.

065.101 Evolution of a population I 8 $M_\odot$ star during central helium burning and the influence of the computational technique. A. Noels, M. Gabriel.
Astron. Astrophys., Vol. 19, 140 - 143 (1972).
In order to test the sensitivity of the evolution during the central helium burning phase, to the computational technique, the evolution of a 8 $M_\odot$ population I star is followed, using two schemes of difference for the discretisation of the equation of energy conservation. It is found that the tracks in the H-R diagram are identical except near the "blue end" of the loops where the maximum values of the effective temperature differ only by 2 %.

065.102 Secular stability of pure helium stars. C. J. Hansen, J. P. Cox, M. A. Herz.
Astron. Astrophys., Vol. 19, 144 - 154 (1972).
Pure helium stellar models in complete equilibrium are analyzed for secular instabilities. With only one exception, all of the models are stable in the "fundamental" mode. The one exception is a model near the minimum mass of the sequence. Higher modes are also investigated, and are found to be stable in all cases.

065.103 Equilibrium configurations (6). T. Kwast.
Urania Kraków, Vol. 43, 17 - 20 (1972). In Polish.

065.104 Thermal pulses in helium shell-burning stars. D. J. Faulkner, P. R. Wood.
Proc. Astron. Soc. Australia, Vol. 2, 105 - 106 (1972).

065.105 Sternmaterie. III. Das "schwarze Loch" und das expandierende Weltall. F. Stöckmann.
Phys. Blätter, 28. Jahrgang, p. 269 - 272 (1972).

065.106 Protostars. I. Appenzeller.
Mitt. Astron. Ges., No. 31, p. 39 - 47 (1972).
Review article.

065.107 Properties of neutron stars. M. Grewing, H. Heintzmann.
Mitt. Astron. Ges., No. 31, p. 87 - 90 (1972).

065.108 R-Coronae-Borealis-Sternmodelle. P. Biermann, R. Kippenhahn.
Mitt. Astron. Ges., No. 31, p. 124 (1972). – Abstract.

065.109 Model calculations of RR-Lyrae stars. K. von Sengbusch.
Mitt. Astron. Ges., No. 31, p. 124 - 125 (1972). – Abstract.

065.110 Hydrodynamic model calculations for dynamically unstable supermassive stars.
I. Appenzeller, K. Fricke.
Mitt. Astron. Ges., No. 31, p. 126 - 129 (1972).

065.111 On the problem of the rotation law of polytropic configurations. S. I. Blinnikov.

Astron. Zhurn. Akad. Nauk SSSR, Vol. 49, 654 - 658 (1972). In Russian. English translation in Soviet Astron. AJ, Vol. 16, No. 3.

An example of the exact solution of the structure equations for the polytrope n = 1 demonstrates the possibility of nonuniform rotation of polytropic stellar models, in contradiction to the result of a paper by Porfir'ev (1959).

065.112 **Evoluzione stellare.**   P. Giannone.
Atti XIII Riunione Soc. Astron. Italiana, Trieste 1969, (see 012.010), p. 113 - 123 (1970).

065.113 **Mixed models for blue horizontal branch stars.**
J. O. Petersen.
Astron. Astrophys., Vol. 19, 197 - 199 (1972).

The evolution of population II models assumed to be mixed by the helium flash is investigated. It is shown that such models can describe the bluest stars of the horizontal branches of globular clusters without the assumption of mass-loss prior to the horizontal branch stage, but not the main group of blue horizontal branch stars.

065.114 **The effects of recent opacity corrections on a main-sequence stellar model of 2.25 $M_\odot$.**   E. Novotny.
Astrophys. Journ., Vol. 174, 425 - 434 (1972).

A comparison is made between models calculated with the opacities of Carson, Mayers, and Stibbs and the electron-scattering corrections of Watson with models based on the opacities of Cox, Stewart, and Eilers. The models have a mass of 2.25 $M_\odot$ and a chemical composition with $X = 0.596$ and $Z = 0.02$.

065.115 **The influence of local conditions in the interstellar medium upon star formation.**   J. H. Hunter.
Monthly Notices Roy. Astron. Soc., Vol. 157, 17P - 19P (1972).

The gravitational stability of a spherical cloud surrounding a stellar core in the interstellar medium is discussed utilizing a modified form of the virial theorem. It is argued that the masses of the circumstellar shells surrounding most infra-red objects should not exceed $\sim 10^{-2} M_\odot$.

065.116 **Electron gas in superstrong magnetic fields: Wigner transition.**   J. I. Kaplan, M. L. Glasser.
Phys. Rev. Letters, Vol. 28, 1077 - 1079 (1972).

It is suggested that in extreme magnetic fields ($H > 10^{12}$G) a high-density electron gas (in a uniform positive background) undergoes a transition to an ordered structure. It is proposed that the ordered structure is a two-dimensional hexagonal lattice of "charged rods". The lattice spacing is evaluated for densities of astrophysical interest relevant to white dwarfs and pulsars.

065.117 **Magnetically distorted polytropes: Structure and radial oscillations.**   S. K. Trehan, M. S. Uberoi.
Astrophys. Journ., Vol. 175, 161 - 169 (1972).

The equilibrium structure of a gaseous polytrope with a toroidal and a poloidal magnetic field is examined under the assumption that the magnetic energy is small compared with the gravitational energy of the configuration. The radial oscillations of the configuration are examined by means of a variational principle. The frequencies of oscillation are tabulated for various values of γ.

065.118 **Stellar nucleosynthesis; preview and prospect.**
P. R. Warren.
Monthly Notes Astron. Soc. Southern Africa, Vol. 31, 46 (1972). − Abstract.

065.119 **Horizontal-branch and post horizontal-branch evolution.**   I. Iben, Jr.
The evolution of population II stars, (see 012.024), p. 1 - 26 (1972).

065.120 **Some uncertainties on models for horizontal-branch stars.**   P. Demarque, J. G. Mengel.
The evolution of population II stars, (see 012.024), p. 27 (1972). − Abstract.

065.121 **Remarks on Greenstein's "puzzling" horizontal-branch extension.**   J. Faulkner.
The evolution of population II stars, (see 012.024), p. 45 - 49 = Contr. Lick Obs., No. 353 (1972).

065.122 **Mass ejection from red giants and white dwarfs.**
W. K. Rose.
The evolution of population II stars, (see 012.024), p. 57 - 68 (1972).

065.123 **Composition changes in stellar evolution.**
P. Eggleton.
The evolution of population II stars, (see 012.024), p. 203 - 208 (1972).

065.124 **Collapse of a hot helium star.**   O. Kh. Gusejnov.
Soobshch. Shemakhinsk. Astrofiz. Obs., vyp. (No.) 6, p. 3 - 8 (1971). In Russian.

065.125 **Dispersion and pulsations of an initially compressed superdense configuration.**
V. Ts. Gurovich, O. Kh. Gusejnov.
Soobshch. Shemakhinsk. Astrofiz. Obs., vyp. (No.) 6, p. 17 - 24 (1971). In Russian.

065.126 **Note on the structure of the star with 0.28 solar masses in the phase towards white dwarfs.**
Y. Tanaka, T. Ishizuka.
Bull. Fac. Education Ibaraki Univ., No. 20, 205 - 209 (1970).

The possibility of a hydrogen envelope remaining in the phase of white dwarfs is discussed. The model sequence for the star with 0.28 solar masses is then discussed by the method of the $U - V$ curves. It is shown that in the H−R diagram, the star cools down towards a white dwarf with the hydrogen envelope of a few per cent of total mass.

065.127 **Problems and achievements of nuclear astrophysics. II. Nucleosynthesis of chemical elements as the result of energy production in stars.**   B. Kuchowicz.
Postępy Fiz., Vol. 22, 601 - 622 (1971). In Polish.

The problem of stellar evolution is viewed on from the standpoint of energy sources and nucleosynthesis of chemical elements. The major nuclear burning stages in stellar evolution are considered: hydrogen burning (especially the $p-p$ cycle), helium burning and advanced burning stages, finally the question of nuclear statistical equilibrium in stellar matter and its relation to the e process.

065.128 **Proton superfluidity in neutron-star matter.**
N.-C. Chao, J. W. Clark, C.-H. Yang.
Nuclear Phys. A, Vol. 179, 320 - 332 (1972).

The authors present a microscopic investigation of isotropic proton superfluidity in neutron-star matter, in the framework of the method of correlated basis functions.

065.129 **Superfluidity in neutron matter.**   E. Krotscheck.
Zeitschr. Phys., Vol. 251, 135 - 140 (1972).

The question of superfluidity in neutron matter is investigated in the framework of BCS-Bogoljubov-theory. Solving the gap-equation for a semirealistic hard-core and a solf-core potential, a rapidly converging numerical method is developed. The results are applied to neutron star models.

065.130 **First and zero sound in neutron matter.**   J. Nitsch.

Zeitschr. Phys., Vol. 251, 141 - 151 (1972).

In the framework of the Landau theory transport properties of neutron matter are investigated. An important problem is the propagation of first and zero sound and their damping.

065.131 Magnetic properties of neutron matter. J. Pfarr.
Zeitschr. Phys., Vol. 251, 152 - 158 (1972).

The energy of interacting neutron matter with spin polarization is calculated in Hartree-Fock approximation by the method of unitary transformations for a hard core and a soft core potential.

065.132 Neutron-matter equations of state.
M. Miller, C. W. Woo, J. W. Clark, W. J. Ter Louw.
Nuclear Phys. A (*Netherlands*), Vol. A184, 1 - 12 (1972).

The equation of state of dense matter is essential to the theoretical prediction of the structure of neutron stars on the global as well as local scale. The authors present two parallel microscopic calculations of the equation of state of a uniform extended system of neutrons interacting by the Reid potential.

065.133 Investigations in the physics of neutron stars.
C. E. Rhoades, Jr.
Thesis, Princeton Univ., Princeton, N.J. [Available from Univ. Microfilms, Ann Arbor, Mich., U.S.A. Order No. 72-2741], 196 pp. (1972).

This dissertation is concerned primarily with the following aspects of the problem of neutron stars: The effect of a new equation of state valid at densities above 10 times nuclear densities on equilibrium configurations of neutron stars and their stability. The effects of the periodic crystal potential in neutron star crusts upon both electron and neutron transport properties. The use of a variational principal to obtain an equation of state for intermediate (that is near nuclear) densities which maximizes the critical mass of a neutron star.

065.134 Superfluidity in neutron stars.
G. Chanmugam, R. F. O'Connell, A. K. Rajagopal.
Phys. Letters A, (*Netherlands*), Vol. 39A, 285 - 286 (1972).

Arguments are presented to show that the BCS theory of superfluidity in its original form may not be applicable to neutron star matter over a wide range of density.

065.135 High-density nucleon localization.
R. L. Coldwell.
Phys. Rev. D, Particles and Fields, Vol. 5, 1273 - 1285 (1972).

065.136 Entwicklung und Innenleben der Sterne. Erkenntnisse astronomischer Forschung. A. Weigert.
Universitas [Wiss. Verlagsgesellschaft, Stuttgart (Germany)], Vol. 27, 51 - 60 (1972).

065.137 Extreme Materiezustände. D. Kirschnitz.
Ideen exakt. Wiss. [Deutsche Verlagsanstalt, Stuttgart], 1972, No. 2, p. 91 - 99. – Concerning neutron stars.

065.138 Properties of the neutron gas and application to neutron stars. J.-R. Buchler, L. Ingber.
Nuclear Phys. A, (*Netherlands*), Vol. A170, 1 - 11 (1971).

065.139 Vortices in anisotropic superfluid neutron star matter near $T = T_c$. R. W. Richardson.
Phys. Letters A, (*Netherlands*), Vol. 35A, 342 - 344 (1971).
See Bull. Signal., Vol. 33, Section 120, No. 1254 (1972).

065.140 Stellar opacity. T. R. Carson.
Progr. High Temperature Phys. Chem. (*GB*), Vol. 4, 99 - 137 (1971). – See Bull. Signal., Vol. 33, Section 120, No. 2605 (1972).

065.141 Nucleosynthesis by charged-particle reactions.

C. A. Barnes.
Advances Nuclear Phys. (*USA*), Vol. 4, 133 - 204 (1971).

065.142 Nucleosynthesis and neutron-capture cross sections.
B. J. Allen, J. H. Gibbons, R. L. Macklin.
Advances Nuclear Phys. (*USA*), Vol. 4, 205 - 259 (1971).

065.143 Neutron star matter.
G. Baym, H. A. Bethe, C. J. Pethick.
Nuclear Phys. A, (*Netherlands*), Vol. A175, 225 - 271 (1971).
See Bull. Signal., Vol. 33, Section 120, No. 3726 (1972).

065.144 Stellar evolution and variable stars.
A. J. Penny, A. L. T. Powell.
Earth Extraterr. Sci., Vol. 1, 229 - 241 (1971).

065.145 Variational solutions of some nonlinear free boundary problems. J. F. G. Auchmutty, R. Beals.
Arch. Rational Mechanics and Analysis (*Germany*), Vol. 43, 255 - 271 (1971). – See Bull. Signal., Vol. 33, Section 120, No. 6948 (1972).

065.146 Critère de Jeans et formation d'étoiles dans le modèle d'étoile supermassive matière-antimatière d'Omnès. T. Montmerle.
Thesis, 3e Cycle, Spéc. Astrophys., Univ. Paris. 140 pp. (1971).
See Bull. Signal., Vol. 33, Section 120, No. 6950 (1972).

Nuclear reactions in stellar surfaces and their relations with stellar evolution. See Abstr. 003.110.

Theory of gravitation and the evolution of stars. See Abstr. 003.145.

Equilibrium thermodynamics of nonideal reacting gases. See Abstr. 061.008.

Selected properties of matter at high density and temperature. Equation of state and rates of beta processes and associated neutrino losses. See Abstr. 061.010

Magnetic properties of a degenerate electron gas and implications for metals, white dwarfs, and neutron stars. See Abstr. 061.015.

Angular-momentum changes in astrophysical processes due to neutrino emission. See Abstr. 061.055.

Astrophysical importance of the reaction $^{16}(p, a)$ ^{13}N. See Abstr. 061.059.

Time development of an axisymmetric shock. Quasistationary velocity approximation. See Abstr. 062.013.

Convective envelopes and radial pulsation of massive red supergiants. See Abstr. 064.021.

A limiting case of relativistic equilibrium. See Abstr. 066.089.

Spherical gravitational collapse with escaping neutrinos. See Abstr. 066.116.

General relativistic fluid spheres. V. Non-charged static spheres of perfect fluid in isotropic coordinates. See Abstr. 066.126.

Spheres with varying density in general relativity. See Abstr. 066.130.

The rotating Einstein-Rosen bridge.

See Abstr. 066.132.

Calculation of stellar structure. II. Determination of the helium abundance of the sun by the theoretical prediction of line and continuum radiation from solar-model photospheres.　　See Abstr. 071.052.

Time scales for Ca II emission decay, rotational braking, and lithium depletion.　　See Abstr. 114.020.

Rotation of evolving A and F stars.
See Abstr. 116.008.

Viscous evolutions of rapidly rotating stars.
See Abstr. 116.013.

Models for contact binaries.　　See Abstr. 117.002.

Theoretical pulsation constants and cepheid masses.
See Abstr. 122.002.

Rußende Sterne.　　See Abstr. 122.031.

On an apparent discrepancy between pulsation and evolution masses for cepheids.　　See Abstr. 122.043.

Pulsating helium stars.　　See Abstr. 122.065.

Theoretical aspects of population II variables.
See Abstr. 122.148.

Quasi-radial pulsations of rotating white dwarfs and neutron stars in Newton's theory of gravitation.
See Abstr. 126.016.

On the nature of the Monoceros supernova remnant.
See Abstr. 125.005.

Decoupling of magnetic fields in dense clouds with angular momentum.　　See Abstr. 131.063.

Galactic shocks in an interstellar medium with two stable phases.　　See Abstr. 131.080.

Phase transition in the interstellar medium.
See Abstr. 131.152.

Microquakes and macroquakes in neutron stars.
See Abstr. 141.519.

Stelle di neutroni e pulsars.　　See Abstr. 141.555.

Pulsar speedups related to metastability of the superfluid neutron-star core.　　See Abstr. 141.557.

Dissipative processes in neutron-star crusts and the production of blackbody X-ray sources.
See Abstr. 142.037.

A neutron star in Centaurus X-3.
See Abstr. 142.063.

The binary star model for compact galactic X-ray sources: Its place in stellar evolution.
See Abstr. 142.078.

On the acceleration of charged particles to cosmic ray energies.　　See Abstr. 143.059.

Polytropic and isothermal plane-symmetric configurations.　　See Abstr. 151.010.

A numerical experiment in the problem of accretion.
See Abstr. 151.086.

Shock formation and star formation in galactic spirals.　　See Abstr. 151.092.

066 Relativistic Astrophysics (without Cosmology), Background Radiation, Gravitation Theory

066.001 Electromagnetic fields produced by relativistic rotating disks.
E. T. Scharlemann, R. V. Wagoner.
Astrophys. Journ., Vol. 171, 107 - 125 (1972).

The quasi-stationary gravitational collapse of a uniformly rotating, current-carrying relativistic thin disk is investigated. By neglecting the effects of the electromagnetic field on the gravitational field, a general solution for the electromagnetic field is found which has the same basic structure as that found by Bardeen and Wagoner for the gravitational field. Numerical results are presented for the specific case of a perfectly conducting disk having no net charge, within which the magnetic field is uniform in the Newtonian limit.

066.002 Quelques univers magnétohydrodynamiques du type de Gödel. M. Bray.
Comptes Rendus Acad. Sci. Paris, Sér. A, Vol. 274, 809 - 812 (1972).

066.003 Sur quelques univers magnétohydrodynamiques du type de Gödel. M. Bray.
Comptes Rendus Acad. Sci. Paris, Sér. A, Vol. 274, 874 - 876 (1972).

066.004 Sur une interprétation possible du déplacement vers le rouge des raies spectrales dans le spectre des objets astronomiques. J.-C. Pecker, A. P. Roberts, J.-P. Vigier.
Comptes Rendus Acad. Sci. Paris, Sér. B, Vol. 274, 765 - 768 (1972).

L'origine cosmologique du déplacement vers le rouge étant désormais sérieusement remise en question, les auteurs suggèrent une interprétation qui fait appel aux interactions inélastiques entre photons doués d'une masse propre non nulle, et proposent une expérience susceptible de confirmer ou d'infirmer cette interprétation.

066.005 Intensification of gravitational radiation by a massive rotator. J. K. Lawrence.
Astrophys. Journ., Vol. 171, 483 - 484 (1972).

It is hypothesized that a massive, rotating, oblate object exists at the center of our Galaxy, aligned with its plane. Under favorable circumstances, gravitational waves emitted from the interior of the object would be strongly focused by its gravitational field into the galactic plane. An average intensification at the earth of about an order of magnitude would result.

066.006 Relativity parameter at high degeneracy.
R. Mitalas.
Astrophys. Journ., Vol. 172, 179 - 180 (1972).

It is emphasized that the relativity parameter under conditions of high degeneracy is not $\beta = KT/mc^2$ but $\beta' = \beta\psi$, where ψ is the degeneracy parameter.

066.007 Measurement of the gravitational deflection of radio waves. D. S. Robertson, C. A. Knight, A. E. E. Rogers, I. I. Shapiro, A. R. Whitney, T. A. Clark, G. E. Marandino, N. R. Vandenberg, R. M. Goldstein.
Bull. American Astron. Soc., Vol. 3, 474 - 475 (1971).
Abstr. AAS.

066.008 Tachyons and gravitational Čerenkov radiation.
A. S. Lapedes, K. C. Jacobs.
Nature, Phys. Sci., Vol. 235, 6 - 7 (1972).

Tachyons—particles moving faster than c—should generate a cone of gravitational radiation, analogous to the Čerenkov emission of electromagnetic radiation. This article examines such an effect and some of the consequences.

066.009 Exact cosmological solutions in Brans and Dicke's scalar-tensor theory, II. H. Dehnen, O. Obregón.
Astrophys. Space Sci., Vol. 15, 326 - 333 (1972).

In addition to our previous paper (Dehnen and Obregón, 1971) the exact cosmological solutions of Brans and Dicke's scalar-tensor theory allowing a power law between the gravitational constant κ and the radius of curvature R of the universe are sought in case that — in contrast to our previous paper — the initial condition $R(t = 0) = 0$ is avoided.

066.010 Comments on the "vibrations" of a black hole.
C. J. Goebel.
Astrophys. Journ., (Letters), Vol. 172, L95 - L96 (1972).

It is shown that the "vibrations of a black hole" of Press are gravitational waves in spiral orbits close to the well-known unstable circular orbit at $r = 3M$. The corresponding "vibrations" of a spinning black hole are discussed.

066.011 Gibt es kosmische Energie-Quellen? P. Jordan.
Phys. Blätter, 28. Jahrgang, p. 113 - 118 (1972).

066.012 The problem of n bodies with variable gravitational constant and some dynamical features of large-scale cosmic systems. T. B. Omarov.
Astron. Zhurn. Akad. Nauk SSSR, Vol. 49, 441 - 446 (1972). In Russian. English translation in Soviet Astron. AJ, Vol. 16, No. 2.

066.013 Substitution law and identity for long-range potentials. K. Hiida, H. Okamura.
Progr. Theor. Phys., Japan, Vol. 46, 1885 - 1904 (1971).

We discuss the relations between two-body and many-body static potentials for long-range interactions. There is a simple substitution law to get the two-body potential in the $2n$-th order from the $(n + 1)$-body potential obtained from tree diagrams for $(n + 1)$-body scatterings. When the interaction Lagrangian density is proportional to mass in the static limit, an identity holds. The reason why the substitution law exists and the identity holds is discussed in detail.

066.014 Gravitational waves of second order in empty space.
T. Tokuoka.
Progr. Theor. Phys., Japan, Vol. 47, 693 - 702 (1972).

The gravitational waves of second order in empty space with any gravitational field are investigated theoretically by the method of characteristics. Three kinds of waves, transverse, longitudinal and shear, are derived formally. It is proved that real waves are transverse, and longitudinal and shear waves can be canceled out by appropriate coordinate transformations.

066.015 Gravitational waves of second order for scalar-tensor field. T. Tokuoka.
Progr. Theor. Phys., Japan, Vol. 47, 703 - 707 (1972).

Gravitational waves of second order for Brans and Dicke's scalar-tensor field theory with any field strength are investigated by the method of characteristics. Only the transverse wave is real and it has three independent components.

066.016 On the gravitational collapse in Brans-Dicke theory of gravity. T. Matsuda.
Progr. Theor. Phys., Japan, Vol. 47, 738 - 740 (1972).

066.017 **Black holes and gravitational theory.**
R. Penrose.
Nature, Vol. 236, 377 - 380 (1972).
There have been several recent claims to have detected black holes. In this article the author describes the theoretical basis of black holes and some of the recent thinking on the subject.

066.018 **Numerical study of fluid flow in a Kerr space.**
J. R. Wilson.
Astrophys. Journ., Vol. 173, 431 - 438 (1972).
The purpose of the present calculation is to understand how gaseous material falling into an already formed black hole behaves—in particular, whether shock waves could be formed or material ejected.

066.019 **A constraint on astrophysical sources of gravity waves.** R. B. Partridge, G. T. Wrixon.
Astrophys. Journ., (*Letters*), Vol. 173, L75 - L78 (1972).
A refined search has been conducted for pulses of microwave emission arriving at the times Weber reports pulses of gravitational radiation. Two radiometers 100 km apart tracked both the galactic center and the Crab nebula. The low upper limit set on the microwave flux imposes an interesting constraint on possible models for the source of the events observed by Weber.

066.020 **Radio pulses from the direction of the galactic centre.** V. A. Hughes, D. Retallack.
Journ. Roy. Astron. Soc. Canada, Vol. 66, 66 (1972). – Abstr. Canadian Astron. Soc.

066.021 **On the detection of black holes.**
C. Leibovitz, D. P. Hube.
Journ. Roy. Astron. Soc. Canada, Vol. 66, 73 (1972). – Abstr. Canadian Astron. Soc.

066.022 **Study of the hydrodynamic analogy for some effects of general relativity.** A. P. Vovchenko.
Ukr. fiz. zhurn., Vol. 16, 2036 - 2042 (1971). In Russian.
Abstr. in Referativ. Zhurn. 51. Astron., 4.51.797 (1972).

066.023 **On the three-dimensional variational formalism for an interacting conservative compact medium.**
A. A. Sokol'skij, A. V. Minkevich.
Vestsi AN BSSR. Ser. fiz.-mat. n.; Izv. AN BSSR. Ser. fiz.-mat. n., 1971, No. 6, p. 76 - 81. In Russian. – Abstr. in Referativ. Zhurn. 51. Astron., 4.51.798 (1972).

066.024 **Rigid reference systems in the ergosphere of Kerr space.** I. S. Syaglo.
Vestsi AN BSSR. Ser. fiz.-mat. n.; Izv. AN BSSR. Ser. fiz.-mat. n., 1971, No. 6, p. 82 - 87. In Russian. – Abstr. in Referativ. Zhurn. 51. Astron., 4.51.803 (1972).

066.025 **Hypothesis concerning the gravitational radiation of a neutron star with high-speed rotation.**
A. I. Tsygan.
Pis'ma v ZhEhTF, Vol. 14, 465 - 468 (1971). In Russian.
Abstr. in Referativ. Zhurn. 51. Astron., 4.51.813 (1972).

066.026 **Radiation of a relativistic charge in the field of a gravitational wave.** F. A. Dimanshtejn.
Ukr. fiz. zhurn., Vol. 16, 1874 - 1884 (1971). In Ukrainian; Ukr. fiz. zhurn., Vol. 16, 1877 - 1887 (1971). In Russian.
Abstr. in Referativ. Zhurn. 51. Astron., 4.51.815 (1972).

066.027 **The problem of gravitational absorption.**
E. Groten.
Geophys. Journ. Roy. Astron. Soc., Vol. 27, 447 - 448 (1972). Letter.

066.028 **Eppur Si Muove.** D. W. Sciama.
Comments Astrophys. Space Phys., Vol. 4, 35 - 39 (1972).
A recent attempt (see P. S. Henry, Nature, Vol. 231, 516-518 (1971)) to measure the velocity of the earth through the microwave background prompts me to write a Comment surveying the implications of such measurements.

066.029 **About the radiation of electromagnetic and gravitational waves by the sources moving with velocity greater than velocity of light in a vacuum.** V. L. Ginzburg.
Comments Astrophys. Space Phys., Vol. 4, 41 - 46 (1972). – Review article.

066.030 **The crisis about the origin of irreversibility and time anisotropy.** B. Gal-Or.
Science, Vol. 176, 11 - 17 (1972).
New interrelationships advocated by the astrophysical school of thermodynamics.

066.031 **Self-scattering of a smooth toroidal gravitational pulse wave.** L. Marder.
Nature, Vol. 235, 379 - 380 (1972).
Brief details are given of the wave zone in a new class of exact solutions of the vacuum field equations in general relativity. The solutions represent axisymmetric toroidal pulse waves emanating from a circular ring, and in the case of a sufficiently smooth pulse the solution remains valid (without the development of a singularity) throughout the self-collision which occurs on the symmetry axis. No significant focusing occurs. One special case is time-symmetric and non-singular, even on the ring, providing an instance where the distant gravitational field of a bounded region of radiation may be studied. (Fuller details of this solution are given in Proc. Roy. Soc. London, Ser. A, Vol. 327, 123 - 130 (1972)).

066.032 **Relativistic effects on envelope formation during the collapse of a rotating star.**
P. R. Amnuel, O. H. Guseinov, F. K. Kasumov.
Astrofizika, Vol. 7, 651 - 654 (1971). In Russian.
English translation in Astrophysics, Vol. 7, No. 4.
The collapse of a rotating star is considered. Rotational instability leads to separation of an envelope of small mass.

066.033 **Introducing the black hole.**
R. Ruffini, J. Wheeler.
Zemlya i Vselennaya, 1972, No. 2, p. 18 - 26. In Russian.
Translated from English by L. P. Grishchuk.

066.034 **Stars invisible for the world.** P. R. Amnuehl'.
Zemlya i Vselennaya, 1972, No. 2, p. 26 - 29. In Russian.

066.035 **Observations of large-scale anisotropy in the 3 K background radiation.** E. K. Conklin.
IAU Symposium No. 44, (see 012.005), p. 518 - 519 (1972). Abstract.

066.036 **Sur une interprétation possible du déplacement vers le rouge des raies spectrales dans le spectre des objets astronomiques. Suggestions en vue d'expériences directes.** J.-C. Pecker, A. P. Roberts, J.-P. Vigier.
Comptes Rendus Acad. Sci. Paris, Sér. B, Vol. 274, 1159 - 1162 (1972).
La mesure directe des sections de choc photon-photon et des propriétés de l'interaction semble possible; une analyse critique de certaines mesures publiées met en garde contre des erreurs possibles.

066.037 **Relativistic runaway from collapsed matter.**
L. A. Case.

Astrophys. Journ., Vol. 173, 665 - 669 (1972).

When an evolving distribution of matter radiates energy from a central dense core, orbiting satellite matter may be released (or run away) due to lessened gravitational attraction. Investigation of the criteria of release and effects of subreleasing changes is made in relativity, and comparison is made with the classical predictions.

066.038 The optical appearance of a star orbiting an extreme Kerr black hole.
C. T. Cunningham, J. M. Bardeen.
Astrophys. Journ., (Letters), Vol. 173, L137 - L142 (1972).

We present results of calculations which indicate that the modulation of the apparent luminosity of a star orbiting a massive black hole in the center of our Galaxy or a nearby galaxy would provide unambiguous evidence for a black hole.

066.039 Calculation of the perihelion advance of planets in a field approach to gravitation. V. Majerník.
Astrophys. Space Sci., Vol. 15, 375 - 382 (1972).

It is shown that, when taking into account the self-gravity of field energy of a gravitational field, one obtains a modified field equation for the intensity of gravitational field, the solution of which, when inserted in Kepler's problem, furnishes a formula for the perihelion precession of planets which (except of a fitting numerical factor) is identical with Einstein's.

066.040 Die Ablenkung des Lichtes an der Sonne und die Änderung seiner Geschwindigkeit und Wellenlänge.
B. Thüring.
Astrophys. Space Sci., Vol. 15, 467 - 478 (1972).

In this publication the deflection of light, decrease of the velocity of light, and displacement of spectral-lines are explained with the aid of a wavetheory. Hereby the assumption of a material medium carrying light-waves is sufficient, the density of which is below that of the empirical vacuum.

066.041 Orbital motion of rotating bodies and the stability of the latter from the viewpoint of relativistic gravitation. A. P. Ryabushko.
Izv. AN BSSR. Ser. fiz.-mat. n., 1971, No. 6, p. 66 - 75. In Russian. – Abstr. in Referativ. Zhurn. 51. Astron., 5.51.800 (1972).

066.042 On the curvature of the physical space and the difference between gravitational and inertial fields.
R. F. Polishchuk.
Zhurn. ehksperim. i. teor. fiz., Vol. 62, 5 - 13 (1972). In Russian. – Abstr. in Referativ. Zhurn. 51. Astron., 5.51.806 (1972).

066.043 Gravitational waves. A. Polnarev.
Nauka i zhizn', 1972, No. 1, p. 65 - 71. In Russian. Abstr. in Referativ. Zhurn. 51. Astron., 5.51.818 (1972).

066.044 Equivalence of inertial and gravitational masses.
V. B. Braginskij, V. I. Panov.
Uspekhi fiz. nauk, Vol. 105, 779 - 780 (1971). In Russian. Abstr. in Referativ. Zhurn. 51. Astron., 5.51.824 (1972).

066.045 The quadratic Lagrangians in general relativity.
V. N. Folomeshkin.
Commun. math. Phys., Vol. 22, 115 - 120 (1971). – Abstr. in Zentralblatt Math. Grenzgebiete, Vol. 221, No. 70005 (1972).

066.046 Black holes. R. Penrose.
Sci. American, Vol. 226, No. 5, p. 38 - 46 (1972).

066.047 "Black holes" in the universe.
Ya. B. Zeldovich, I. D. Novikov.
Priroda, No. 4.72, p. 28 - 31 (1972). In Russian.

066.048 Thomas precession and the relativistic disk.
D. P. Whitmire.
Nature, Phys. Sci., Vol. 235, 175 - 176 (1972).

It is shown that an arbitrary finite section of a rapidly spinning disk does not undergo a Thomas rotation as previously assumed, but instead experiences "Thomas shear stresses". Only the individual "rigid" units of the disk rotate with the Thomas precession rate.

066.049 Solutions of the relativistic two-body problem.
I. Classical mechanics. J. L. Cook.
Australian Journ. Phys., Vol. 25, 117 - 139 (1972).

The electromagnetic and gravitational point source interactions are derived. Relativistic corrections are applied to the problem of planetary motion, a model for the relativistic Coulomb interaction is explored, and the relativistic harmonic oscillator is evaluated.

066.050 Influence of the gravitational fields of collapsed bodies on electromagnetic waves. P. C. Peters.
Bull. American Astron. Soc., Vol. 4, 222 (1972). – Abstr. AAS.

066.051 Pointless limits to gravitational radiation.
J. Faulkner.
Bull. American Astron. Soc., Vol. 4, 222 (1972). – Abstr. AAS.

066.052 Small scale anisotropy in the cosmic background radiation at $\lambda = 3.3$ mm.
P. E. Boynton, R. B. Partridge.
Bull. American Astron. Soc., Vol. 4, 243 (1972). – Abstr. AAS.

066.053 Spherically symmetric similarity solutions of the Einstein field equations for a perfect fluid.
M. Cahill, A. H. Taub.
Bull. American Astron. Soc., Vol. 4, 244 (1972). – Abstr. AAS.

066.054 Photography of relativistically moving objects.
A. W. Guess.
Bull. American Astron. Soc., Vol. 4, 244 (1972). – Abstr. AAS.

066.055 Gravitational radiation (3–7). B. Kuchowicz.
Urania Kraków, Vol. 43, 7 - 17, 45 - 50, 106 - 112, 136 - 139, 170 - 176 (1972). In Polish.

066.056 The development of a black hole – The Oppenheimer-Snyder dust cloud. P. Szekeres.
Proc. Astron. Soc. Australia, Vol. 2, 110 - 111 (1972).

066.057 Self-focusing and self-trapping of gravitational waves.
S. O. Olsen.
Nature, Phys. Sci., Vol. 238, 12 - 13 (1972).

The large amount of gravitational energy emitted from the galactic centre which Weber's observations would indicate is apparently at variance with established astrophysics. Here I try to reconcile the two by using the concept of self-focusing and self-trapping of radiation.

066.058 Can the lumpy distribution of galaxies be detected by X-ray observations? A. S. Webster.
Nature, Vol. 238, 20 - 24 (1972).

Published data from X-ray telescopes are not adequate to reveal the fluctuations in the surface brightness of the background radiation caused by clustering or superclustering of the sources giving rise to the background.

066.059 Kollaps einer rotierenden Staubwolke.
P. Biermann, R. Kippenhahn, W. Tscharnuter.
Mitt. Astron. Ges., No. 31, p. 124 (1972). – Abstract.

066.060 Exact solutions (particular and general) of the Einstein equations for space filled with matter.

B. Kuchowicz.
Postępy Astron., Vol. 20, 151 - 154 (1972). In Polish.

The most practically important matter distribution for which several exact solutions of the Einstein equations have been obtained, is a spherically symmetric distribution of a perfect fluid. Using the isotropic coordinates it was possible to derive a particularly simple form of the general formal solution in the static case.

066.061 Equation of state and adiabatic stability for general relativistic spheres. B. Kuchowicz.
Postępy Astron., Vol. 20, 155 - 158 (1972). In Polish.

Matter inside relativistic spheres has to fulfill certain physical conditions (e.g. positive definiteness of density and pressure, stability for static spheres etc.). The general formal solution for static spheres in isotropic coordinates enables us to present all these conditions in the simple form of differential inequalities containing two functions u and v which depend on the radial variable r.

066.062 The gravitational redshift: a three-body effect. A. Peres.
American Journ. Phys., Vol. 40, 398 - 400 (1972).

The nonlinearity of Einstein's gravitational equations produces three-body forces in the equations of motion. One of their consequences is that the motion of an electron around a nucleus (or a satellite around a planet) is slowed down by the existence of a distant massive body. This is the familiar gravitational redshift, which can be derived without invoking the equivalence principle.

066.063 Stand und Aussichten der Gravoastronomie. P. Kafka, F. Meyer.
Mitt. Astron. Ges., No. 31, p. 129 - 132 (1972).

066.064 On the Ptolemaic – Copernican problem. R. F. Polishchuk.
Astron. Zhurn. Akad. Nauk SSSR, Vol. 49, 669 - 671 (1972). In Russian. English translation in Soviet Astron. AJ, Vol. 16, No. 3.

066.065 Relativistic beaming. W. H. McCrea.
Monthly Notices Roy. Astron. Soc., Vol. 157, 359 - 365 (1972).

Working in terms of photon-fluxes rather than other radiational parameters, a short self-contained treatment is given of the observable properties of a revolving light-source recently discussed by F. G. Smith in connection with a theory of the radiation of pulsars.

066.066 Transcendence of the law of baryon-number conservation in black-hole physics. J. D. Bekenstein.
Phys. Rev. Letters, Vol. 28, 452 - 455 (1972).

The following result is stated: A black hole in its final state can be endowed with no exterior scalar, vector, or spin-2 meson fields. We conclude that a useful definition of baryon number cannot be given for such an object.

066.067 Gravitational radiation reaction. R. W. Lind, J. Messmer, E. T. Newman.
Phys. Rev. Letters, Vol. 28, 857 - 858 (1972).

We present the results of an exact calculation of the equations of motion (with gravitational radiation reaction terms) of a gravitating system subject to no external forces.

066.068 Can synchrotron gravitational radiation exist? M. Davis, R. Ruffini, J. Tiomno, F. Zerilli.
Phys. Rev. Letters, Vol. 28, 1352 - 1355 (1972).

A complete relativistic analysis for gravitational radiation emitted by a particle in circular orbit around a Schwarzschild black hole is presented in the Regge-Wheeler formalism. For completeness and contrast we also analyze the electromagnetic and scalar radiation emitted by a suitably charged particle. The three radiation spectra are drastically different. We stress some important consequences and astrophysical implications.

066.069 Cosmic background radiation at 1.32 mm. D. J. Hegyi, W. A. Traub, N. P. Carleton.
Phys. Rev. Letters, Vol. 28, 1541 - 1544 (1972).

The position of the $R(0)$, $R(1)$, and $R(2)$ lines of the 3874-Å band of interstellar CN in the spectrum of the star ζ Ophiuchi were scanned with a Fabry-Perot interferometer. At the expected position of the $R(2)$ line an absorption feature was found with an equivalent width of 0.077 ± 0.055 mÅ which implies a temperature of $2.9^{+3.4}_{-2.1}$ K for the cosmic background radiation at 1.32 mm. A conservative interpretation of the data is simply to give an upper limit (2σ) of 3.8 K to this temperature.

066.070 Possible variation of the gravitational constant over the elements. J. J. Gilvarry, P. M. Muller.
Phys. Rev. Letters, Vol. 28, 1665 - 1669 (1972).

We re-examine the theory and data of the Kreuzer experiment to measure the relative difference $\Delta\kappa/\kappa$ in the gravitational constant κ between two elements. Significant errors appear in the statistical analysis of Kreuzer which vitiate his conclusions.

066.071 Origin of the cosmic microwave background radiation in a chaotic universe. M. J. Rees.
Phys. Rev. Letters, Vol. 28, 1669 - 1671 (1972).

If the early universe possessed the maximum degree of irregularity compatible with its present large-scale uniformity and isotropy, dissipation of energy at redshifts $z \gtrsim 10^4$ would have generated a thermal microwave background with present temperature $\sim 3°$ K.

066.072 Nature of gravitational synchrotron radiation. D. M. Chitre, R. H. Price.
Phys. Rev. Letters, Vol. 29, 185 - 188 (1972).

Ultrarelativistic geodesic particle orbits in the Schwarzschild geometry produce radiation (GSR) with angular distributions like that of synchrotron radiation. The spectra of high-frequency scalar, electromagnetic, and gravitational GSR are compared to each other and to the spectrum of synchrotron radiation. The differences among the spectra are explained in terms of the shape of the effective potential for GSR and the inapplicability of geometric optics.

066.073 Rotating shells in general relativity. P. S. Florides, R. Wingate.
Bull. Soc. Math. Grèce, Nouvelle Sér., Vol. 11, No. 1, p. 172 - 230 (1971).

The method of successive approximations developed by the first author and J. L. Synge is applied to calculate the gravitational field of a spherical shell of finite thickness rotating steadily about one of its diameters. The fields of a thin rotating shell, of a rotating sphere and of a stationary sphere are obtained as special cases.

066.074 Time evolution of a rotating black hole immersed in a static scalar field. W. H. Press.
Astrophys. Journ., Vol. 175, 243 - 252 (1972).

We compute the time evolution of a Kerr (rotating) black hole which is immersed in a perturbing scalar field, uniform at large distances from the hole. The perturbing field produces a torque on the hole which (i) is perpendicular to the field lines; (ii) causes the perpendicular component of the hole's angular momentum to decrease exponentially with time, bringing the hole's total angular momentum J into eventual alignment with the field; and (iii) accomplishes this alignment by converting rotational energy of the black hole into irreducible mass. We conjecture extensions of these results to black holes perturbed by external electromagnetic or gravitational fields.

066.075 Radiation pressure on a test particle in general relativity. P. D. Noerdlinger.
Nature, Vol. 237, 30 - 31 (1972).
In contrast with Newtonian theory, where gravity and radiation pressure from a point source have the same radial dependence, we find that in general relativity the radiation pressure on a particle varies more rapidly with radial distance than the gravity force. This makes stable equilibria possible. Applications to QSO's are suggested, although for gas clouds collective effects or variations in opacity may dominate the effect found here.

066.076 Limits to the sub-millimetre isotropic background. J. E. Beckman, P. A. R. Ade, J. S. Huizinga, E. I. Robson, D. G. Vickers, J. E. Harries.
Nature, Vol. 237, 154 - 157 (1972).
A high resolution emission spectrum of the stratosphere in the wavenumber range $9\,cm^{-1}$ to $25\,cm^{-1}$ (400 m to 1.1 mm), obtained from an aircraft flown at 12.1 km, is presented. The combination of helium-cooled InSb detector with phase-modulated Michelson interferometer led to excellent signal to noise ratio and a resolution of $0.067\,cm^{-1}$. Photometric analysis of the many O_3 lines placed the continuum to better than $2^1/_2\%$, ruling out the possibility of radiation with flux greater than $10^{-9}\,watt\,cm^{-2}\,sr^{-1}$ in a broad spectral feature (of half width $3\,cm^{-1}$), greater than $4\times10^{-10}\,watt\,cm^{-2}\,sr^{-1}$ in integrated line radiation and $2\times10^{-11}\,watt\,cm^{-2}\,sr^{-1}$ in any single line, originating from outside the atmosphere. This result negates those of observers who reported large departures of the microwave background from a 2.7 K continuum in the sub-millimetre region.

066.077 Non-velocity redshifts and photon-photon interactions. J. C. Pecker, A. P. Roberts, J. P. Vigier.
Nature, Vol. 237, 227 - 229 (1972).
The authors assume photons have a non zero rest mass $m_\gamma \leqslant 10^{-48}g$ and introduce inelastic photon-photon scattering which: a) have a strong forward peak; b) imply a constant fractional energy loss $\delta\nu/\nu$; c) do not perturb quantum electrodynamics. This introduces a redshift for photons crossing a strong radiation field (for black body one gets $Z = -AT^3L$ with $A\sim 2.2\times10^{-29}$) which: 1) explains anomalous quasar and galaxy chain redshifts (Z being a source effect); 2) accounts for the Hubble shift as 3° K background effect; 3) interprets anomalous solar redshifts. This can be tested in laboratory experiments.

066.078 Theory and detection of gravitational waves. D. de Jongh.
Monthly Notes Astron. Soc. Southern Africa, Vol. 31, 49 (1972). — Abstract.

066.079 Black holes. J. Kovalesky.
Southern Stars, Vol. 24, 118 - 119 (1972).

066.080 On the inertial effects induced by a shell of finite thickness. A. Lausberg.
Bull. Cl. Sci. Acad. Roy. Belgique, 5ᵉ Sér., Vol. 57, 125 - 153 = Univ. Liège, Inst. d'Astrophys., Coll. 8°, No. 617 (1971).
The equations of motion of general relativity for a test-particle in a rotating shell are expressed in terms of $G\omega$, $G\omega^2$ and $G^2\omega^2$, where G is Newton's gravitational constant and ω the angular velocity of the shell relative to the "fixed stars". The Coriolis field (of order $G\omega$) is calculated in the case of a shell of finite extension. It is shown that the terms of order $G\omega^2$ can be interpreted as Newtonian gravitational effects.

066.081 Remarks on "gravity and velocity of light". H. J. Treder.

Ann. Physik, 7. Ser., Vol. 27, 125 = Zentralinst. Astrophys., Sternw. Babelsberg, Mitt. New Ser., No. 51 (1971).

066.082 Zur Konstanz der Teilchenzahlen in der allgemeinen Relativitätstheorie. H.-J. Treder.
Ann. Physik, 7. Ser., Vol. 27, 47 - 56 = Zentralinst. Astrophys., Sternw. Babelsberg, Mitt. New Ser., No. 52 (1971).
This paper suggests that the "spontaneous creation of matter in strong gravitation fields" — which is asserted in some papers — may be a result of an inconsistent approximation for the simultaneous systems of Einstein field equations and of matter field equations.

066.083 Über die Schwere des Lichtes in der Gravitationstheorie. II. H.-J. Treder.
Ann. Physik, 7. Ser., Vol. 27, 177 - 181 = Zentralinst. Astrophys., Sternw. Babelsberg, Mitt. New Ser., No. 54 (1971).
If the vacuum velocity v of light is a function of the gravitation field only (Einstein-effect), then v is independent from the light frequency ν. This statement results from the Maxwell equations only and is independent from the special form of gravitation field theory. The frequency independence of the velocity of light is in a good arrangement with the astrophysical facts.

066.084 The two-body problem in the theory of gravitation without potential. H. Oja.
Ann. Acad. Sci. Fennicae, A.VI, No. 377, p. 5 - 40 (1971)
The theory of gravitation without potential is based on eight axioms. The force law deduced from the axioms contains in the form treated here, eight structure constants three of which have been determined in connection with the one-body problem. The value of one more constant will be obtained from our results concerning the two-body problem. The theory does not assume constancy of masses. In the two-body problem the masses are found to change periodically. The perturbations caused by this change are of the same type as those due to the other terms in the equation of relative motion, i. e. advance of periastron, acceleration of the centre of mass and dilatation of the sidereal period.

066.085 A correlation between "gravitational signals" in Weber's experiments and the earth's magnetic activity. R. A. Adamyants, A. D. Alekseev, N. I. Kolosnitsyn.
Pis'ma v ZhEhTF, Vol. 15, 277 - 279 (1972). In Russian.
Abstr. in Referativ. Zhurn. 51. Astron., 7.51.734 (1972).

066.086 Zur Erhaltung der Teilchenzahlen in der allgemeinen Relativitätstheorie. H.-J. Treder.
Ann. Physik, 7. Ser., Vol. 28, 94 - 96 = Zentralinst. Astrophys., Sternw. Babelsberg, Mitt. New Ser., No. 66 (1972).

066.087 Einstein's path from special to general relativity. C. Lanczos.
General relativity, (see 003.017), p. 5 - 19 (1972).

066.088 Rotating bodies in general relativity. P. S. Florides.
General relativity, (see 003.017), p. 167 - 181 (1972).

066.089 A limiting case of relativistic equilibrium. S. Chandrasekhar.
General relativity, (see 003.017), p. 185 - 199 (1972).

066.090 The relativistic Boltzmann equation. W. Israel.
General relativity, (see 003.017), p. 201 - 241 (1972).

066.091 A new approximation method for wave theories. S. Persides.
Journ. Math. Phys., Vol. 12, 2355 - 2361 = Contr. Astron. Dep. Univ. Thessaloniki, No. 60 (1971).

The scalar field due to a bounded source and obeying the wave equation is analyzed. As a result of this, a set of rules is derived for solving a class of "wave theories," including electrodynamics and general relativity, by expanding the field in a power series of c^{-1} in null-spherical coordinates. The method is applied for the Maxwell equations to give all the well-known results without the use of Fourier analysis and Bessel functions.

066.092 Gravitational field and quantum theory.
Y. Iwasaki.
Butsuri (*Japan*), Vol. 26, 915 - 924 (1971).   In Japanese.
See Phys. Abstr., Vol. 75, No. 11256 (1972).

066.093 On the 'derivation' of Einstein's field equations.
S. Chandrasekhar.
American Journ. Phys., Vol. 40, 224 - 234 (1972).
An attempt is made to clarify the physical and the mathematical reasonings that underlie Einstein's laws of gravitation.

066.094 Waves from nowhere or from somewhere else.
C. Vilain.
Sci. Progr. Découverte, No. 3438, p. 4 - 10 (1971). In French.
The author reviews the various mechanisms which have been proposed to explain the origin of gravitational waves: the collapse of a star, the formation of neutron stars, the existence of black holes, the clustering of neutron stars at the center of the Galaxy.

066.095 Inertial reference frames in Einstein's theory of gravitation.   J. Audretsch.
International Journ. Theor. Phys., Vol. 4, No. 1, p. 1 - 9 (1972).
A physical definition of the inertial reference frame (IRF) is given, and the properties of solutions of the Einstein equation (with cosmological constant Λ), which admit an IRF are investigated.

066.096 Reversible transformations of a charged black hole.
D. Christodoulou, R. Ruffini.
Phys. Rev. D, Particles and Fields, Vol. 4, 3552 - 3555 (1971).
A formula is derived for the mass of a black hole as a function of its 'irreducible mass', its angular momentum, and its charge. It is shown that 50% of the mass of an extreme charged black hole can be converted into energy as contrasted with 29% for an extreme rotating black hole.

066.097 Colliding plane gravitational waves.   P. Szekeres.
Journ. Math. Phys., *New York,* Vol. 13, 286 - 294 (1972).
The equations governing the collision of two plane gravitational waves are derived. The general exact solution representing this situation when both waves are linearly polarized are found, and some special solutions of possible physical interest are discussed in detail.

066.098 The dispersion and attenuation of light by a weak random gravitational-wavefield background.
R. Bourret.
Nuovo Cimento B, Ser. 11, Vol. 7B, 195 - 219 (1972).
The effect of a random gravitational-wave metric upon the mean value of an electromagnetic wave is calculated and discussed.

066.099 Absence of gravitational scintillation.
B. Bertotti, D. Trevese.
Nuovo Cimento B, Ser. 11, Vol. 7B, 240 - 246 (1972).
The effect of gravitational waves on electromagnetic propagation was studied in the physical optics approximation.

066.100 Vacuum-field solutions in the Brans-Dicke theory.
J. O'Hanlon, B. O. J. Tupper.

Nuovo Cimento B, Ser. 11, Vol. 7B, 305 - 312 (1972).
The field equations of the Brans-Dicke theory are solved for a vacuum with the aid of a space-time metric of Friedmann type. Nonstatic solutions are found showing that, in general, a Birkoff theorem does not exist for the Brans-Dicke theory. Solutions are also found which may be interpreted as being contrary to Mach's principle.

066.101 A note on energy extremal properties for rotating stars in general relativity.   J. Katz.
Journ. Phys. A, General Phys., Vol. 5, 781 - 785 (1972).
Using Komar's localized energy and spin conserved vectors associated with the Killing fields of a stationary axisymmetric universe, the author derives a covariant identity which describes in terms of variations of fields and matter functions a perfect fluid in general relativity, exhibiting in a clear and straight-forward way a variety of known extremal theorems, or analogous ones, for energy or entropy.

066.102 Behaviour of fast particles in the Schwarzschild field.   M. Carmeli.
Nuovo Cimento Lettere, Ser. 2, Vol. 3, 379 - 383 (1972).
The author shows that the phenomenon that photons 'slow down' in the gravitational field of a fixed moon M also holds for very fast test particles when they move in a gravitational field.

066.103 Nonmeasurability of the lepton number of a black hole.   C. Teitelboim.
Nuovo Cimento Lettere, Ser. 2, Vol. 3, 397 - 400 (1972).
The author concludes that even though the neutrino field does not decouple from the source in the limit $a \to 2m$, it is still not possible to determine the lepton number of an astrophysical black hole by means of weakly interacting neutrinos.

066.104 The nature of time and its measurement.
S. J. Prokhovnik.
Journ. Proc. Roy. Soc. New South Wales, (*Australia*), Vol. 103, 123 - 126 (1970). – See Phys. Abstr., Vol. 75, No. 23252 (1972).

066.105 On the criteria of space homogeneity in general relativity theory.   L. Grishchuk.
Bull. Acad. Pol. Sci., Sci. Math., Astron., Phys., Vol. 19, 1129 - 1133 (1971).
At present two criteria of the space homogeneity are widely used in general relativity. It is shown that each of them can be obtained as a generalization of the definitions of homogeneity in the flat space. An example of metric is given which emphasizes the main difference between those criteria.

066.106 Black holes in general relativity.   S. W. Hawking.
Commun. Math. Phys., Vol. 25, 152 - 166 (1972).
See Phys. Abstr., Vol. 75, No. 23256 (1972).

066.107 Black holes in the Brans-Dicke theory of gravitation.
S. W. Hawking.
Commun. Math. Phys., Vol. 25, 167 - 171 (1972). – See Phys. Abstr., Vol. 75, No. 23257 (1972).

066.108 Relativistic motion in two space-like dimensions.
N. W. Taylor, L. McL. Marsh.
Journ. Proc. Roy. Soc. New South Wales, (*Australia*), Vol. 103, 119 - 122 (1970).
In relativity theory a set of co-ordinate transformations suitable for investigating conditions experienced directly by an accelerated observer can be found. These transformations simplify the discussion of topics such a uniform rotation and the Thomas precession.

066.109 A method for generating new solutions of Einstein's equation. II. R. Geroch.
Journ. Math. Phys., *New York*, Vol. 13, 394 - 404 (1972).
See Phys. Abstr., Vol. 75, No. 23259 (1972).

066.110 On uniqueness of the Kerr-Newman black holes. R. Wald.
Journ. Math. Phys., *New York*, Vol. 13, 490 - 499 (1972).
See Phys. Abstr., Vol. 75, No. 23260 (1972).

066.111 Electromagnetic field of a particle moving in a spherically symmetric black-hole background.
R. Ruffini, J. Tiomno, C. V. Vishveshwara.
Nuovo Cimento Lettere, Ser. 2, Vol. 3, 211 - 215 (1972).
The theoretical basis for the analysis of the electromagnetic radiation emitted by a charge moving in the gravitational field of spherically symmetric black holes is presented.

066.112 Spherically symmetric spacetimes in general relativity: Invariant formulation of Einstein's equations.
T. G. Clark.
Nuovo Cimento Lettere, Ser. 2, Vol. 3, 317 - 319 (1972).

066.113 Differential conditions for physically meaningful fluid spheres in general relativity. B. Kuchowicz.
Phys. Letters A, (*Netherlands*), Vol. 38A, 369 - 370 (1972).
Such physical conditions as positive definiteness of matter density and pressure, adiabatic stability etc. are presented in the form of inequalities relating certain functions which appear in the general formal solution; the latter solution has been derived previously by the author in isotropic coordinates for spherically symmetric matter distributions.

066.114 Light tracks near a dense charged star.
R. Burman.
Journ. Proc. Roy. Soc. New South Wales, (*Australia*), Vol. 103, 87 - 90 (1970).
Integrals of the Einstein-Maxwell equations for static spherically symmetric situations in regions outside matter, with one of the functions in the metric left arbitrary, are used to investigate light tracks in the presence of a charged body. Conditions for the existence of circular photon orbits are obtained and non-circular paths are discussed.

066.115 Testing general relativity: progress, problems, and prospects. I. I. Shapiro.
Proc. AIP Conference, No. 2, p. 286 - 301 (1971).
Discusses the testing of general relativity by the retardation of radar signals, deflection of radio waves, relativistic perihelion advance of Mercury, and the time variation of the gravitational constant. All these tests tend to confirm the general theory, within the experimental error.

066.116 Spherical gravitational collapse with escaping neutrinos. J. K. Rao.
Journ. Phys. A, General Phys., Vol. 5, 479 - 488 (1972).
The general relativity equations for the dynamics of spherically symmetric perfect fluid distributions with an outward neutrino flux are studied.

066.117 Bound geodesics in the Kerr metric. D. C. Wilkins.
Phys. Rev. D, Particles and Fields, Vol. 5, 814 - 822 (1972).
The bound geodesics (orbits) of a particle in the Kerr metric are examined.

066.118 An empirical test of Einstein's unified field theory. C. R. Johnson.
Nuovo Cimento B, Ser. 11, Vol. 8B, 391 - 398 (1972).
An outline of an analysis of Einstein's theory of the nonsymmetric field is presented showing that there are particles in the theory which interact in lowest approximation through the conventional classical electromagnetic interaction. Also associated with these particles is a weak nonconventional electromagnetic interaction important only over astronomical distances. The form of this nonconventional electromagnetic interaction is examined, and based on this examination a possible empirical test of Einstein's theory is described.

066.119 Cutting the Galaxy's losses. D. Sciama.
New Sci. & Sci. Journ., No. 783, Vol. 53, 373 - 374 (1972).

066.120 Mach's principle and a new gauge freedom in Brans-Dicke theory. J. O'Hanlon.
Journ. Phys. A, General Phys., Vol. 5, 803 - 811 (1972).
The relationship between Mach's principle and the Brans-Dicke scalar-tensor theory of gravitation is discussed.

066.121 An approximate solution for the static, spherically symmetric metric due to a point charged mass in Brans-Dicke theory. M. N. Mahanta, D. R. K. Reddy.
Journ. Math. Phys., *New York*, Vol. 13, 708 - 709 (1972).
An approximate solution of the field equations of Brans-Dicke theory is obtained for a static, spherically symmetric metric due to a charged point mass which can be considered to be an analog of the Reissner-Nordstrom metric in Einstein's theory.

066.122 Gravitational radiation: the theoretical aspect. S. W. Hawking.
Contemp. Phys. (*GB*), Vol. 13, 273 - 282 (1972).
The theory of relativity requires that there should be a gravitational analogue to electromagnetic radiation. It is impossible to generate measurable quantities in the laboratory but recent observations indicate that there may be a tremendous flux coming from the centre of our Galaxy. Collisions between gravitationally collapsed objects black holes seem to be the only mechanism which could produce this amount of radiation.

066.123 Gravitational radiation experiments. P. S. Aplin.
Contemp. Phys. (*GB*), Vol. 13, 283 - 293 (1972).
The experimental work on gravitational radiation from its prediction by Einstein as a part of his General Theory of Relativity is reviewed.

066.124 Investigations in gravitational collapse and the physics of black holes. D. L. Christodoulou.
Thesis, Princeton Univ., Princeton, N.J. [Available from Univ. Microfilms, Ann Arbor, Mich., U.S.A. Order No. 72-2696], 139 pp. (1972).
This thesis deals with the physics of gravitational collapse and its final states, the collapsed objects called black holes. A star, that is at the end point of thermonuclear evolution, will be unstable against gravitational collapse if its mass is larger than a certain critical mass. Will radiation be similarly unstable above a certain critical mass-energy and undergo gravitational collapse? This question is treated by considering an imploding and reexploding scalar pulse of radiation, and is answered in the affirmative.

066.125 A cosmological solution of Treder's gravitational theory. H. Oleak.
Ann. Physik, Vol. 28, 189 - 192 (1972). In German.
A solution of the field equations is given for the case of an universe filled with incoherent matter ($p = 0$). The observable universe behaves kinematically in a good approximation like the Milne-model but with flat subspaces t = const.

066.126 General relativistic fluid spheres. V. Non-charged static spheres of perfect fluid in isotropic coordi-

nates. B. Kuchowicz.
Acta Phys. Pol. B, Vol. B3, 209 - 229 (1972).

The search for new exact solutions of the gravitational field equations for space filled with matter is pursued in isotropic coordinates. These solutions describe static and spherically symmetric distributions of a non-charged, perfect fluid; they may be applied in a general relativistic treatment of stellar structure.

066.127 **Nonexistence of baryon number for static black holes.** J. D. Bekenstein.
Phys. Rev. D, Particles and Fields, Vol. 5, 1239 - 1246 (1972).

066.128 **Closed cosmological solutions to Einstein's field equations.** E. A. Rauscher.
Nuovo Cimento Lettere, Ser. 2, Vol. 3, 661 - 665 (1972).

066.129 **Gravitation theory and oscillating universe.** I. Goldman, N. Rosen.
Phys. Rev. D, Particles and Fields, Vol. 5, 1285 - 1287 (1972).

The field equations of a noncovariant theory of gravitation, based on the existence of a preferred frame of reference in the universe, are applied to the homogeneous isotropic cosmological model.

066.130 **Spheres with varying density in general relativity.** M. C. Durgapal, G. L. Gehlot.
Journ. Phys. A, General Phys., Vol. 4, 749 - 755 (1971).

066.131 **Applications of SU(2) technique in general relativity.** M. Carmeli.
Relativity and gravitation, Haifa 1969, (see 012.029), p. 69 - 76 (1971).

066.132 **The rotating Einstein-Rosen bridge.** J. M. Cohen.
Relativity and gravitation, Haifa 1969, (see 012.029), p. 87 - 97 (1971).

066.133 **Gravitational-scalar field coupling.** S. Deser, J. Higbie.
Relativity and gravitation, Haifa 1969, (see 012.029), p. 123 - 144 (1971).

066.134 **New experimental tests of relativity.** R. Fox, J. Shamir.
Relativity and gravitation, Haifa 1969, (see 012.029), p. 163 - 171 (1971).

066.135 **On the possibility of cosmological foundations for classical and relativistic thermodynamics.** B. Gal-Or.
Relativity and gravitation, Haifa 1969, (see 012.029), p. 173 - 176 (1971).

066.136 **Equations of motion in general relativity.** J. N. Goldberg.
Relativity and gravitation, Haifa 1969, (see 012.029), p. 189 - 193 (1971).

066.137 **Modification of the classical gravitational field equations due to a virtual quantized matter field.** L. Halpern.
Relativity and gravitation, Haifa 1969, (see 012.029), p. 195 - 197 (1971).

066.138 **Invariant evolution of gravitational field.** A. Peres.
Relativity and gravitation, Haifa 1969, (see 012.029), p. 269 - 273 (1971).

066.139 **Gravitational fields in matter.** P. Szekeres.

Relativity and gravitation, Haifa 1969, (see 012.029), p. 305 - 308 (1971).

066.140 **Gravitational radiation experiments.** J. Weber.
Relativity and gravitation, Haifa 1969, (see 012.029), p. 309 - 322 (1971).

066.141 **Gravitational radiation in the presence of a Schwarzschild black hole. A boundary value search.** M. Davis, R. Ruffini.
Nuovo Cimento Lettere, Ser. 2, Vol. 2, 1165 - 1168 (1971).

066.142 **On crucial tests of general relativity.** T. T. Arjun.
Chinese Journ. Phys. (*Taiwan*), Vol. 9, 29 - 30 (1971). – See Phys. Ber., Vol. 51, No. 4 - 281 (1972).

066.143 **General relativity and gravitational collapse.** R. U. Sexl.
Developments in high energy physics. Proc. Schladming 1970 [Acta Phys. Austriaca, Suppl. 7], p. 308 - 354 (1970).

066.144 **From Mendeléev's atom to the collapsing star.** J. A. Wheeler.
Trans. New York Acad. Sci., Vol. 33, 745 - 779 (1971). – See Bull. Signal., Vol. 33, Section 120, No. 4873 (1972).

066.145 **The experimental determination of the spacetime metric.** J. Blamont.
E.S.R.O., S. P., No. 52, (see 012.030), p. 9 - 15 (1971).

066.146 **Solar-system tests of gravitational theories.** I. I. Shapiro.
E.S.R.O., S. P., No. 52, (see 012.030), p. 17 - 21 (1971).

066.147 **Theoretical background and present status of the Stanford relativity-gyroscope experiment.** C. W. F. Everitt, W. M. Fairbank, L. I. Schiff.
E.S.R.O., S. P., No. 52, (see 012.030), p. 33 - 43 (1971).

066.148 **Multidirectional, multipolarization antennas for scalar and tensor gravitational radiation.** R. L. Forward.
General Relativity Gravitation, Vol. 2, 149 - 159 (1971).

Relativistische Astrophysik.
See Abstr. 003.016.

Theory of gravitation and the evolution of stars.
See Abstr. 003.145.

The mass difference of the muon and the electron.
See Abstr. 022.101.

A class of exact solutions for the motion of a particle in a monopole-prolate quadrupole field.
See Abstr. 022.117.

Two gravity-wave detectors: A comparison.
See Abstr. 033.031.

Absorption of gravitational energy by a conducting fluid in the presence of magnetic fields.
See Abstr. 062.015.

Ultrarelativistic cosmic plasma.
See Abstr. 062.027.

Hydrodynamic model calculations for supermassive stars. I. The collapse of a nonrotating $0.75 \times 10^6 \, M_\odot$ star.
See Abstr. 065.053.

Fragen des Gravitationskollaps.
See Abstr. 065.054.

Dispersion and pulsations of an initially compressed superdense configuration.   See Abstr. 065.125.

New measurement of the solar gravitational red shift.   See Abstr. 071.047.

Mercury's perihelion advance: Determination by

radar.   See Abstr. 092.010.

Relativistic Poincaré limitation for the speed of rotation of a star.   See Abstr. 116.015.

No evidence for black holes in eccentric binary systems.   See Abstr. 117.027.

Upper limit on the X-ray flux associated with gravitational radiation.   See Abstr. 142.109.

Sun

071 Solar Photosphere, Spectrum

071.001 The solar abundance of silicon from forbidden lines of Si I. N. Grevesse, J. P. Swings.
Astrophys. Journ., Vol. 171, 179 - 184 (1972).
The low-excitation (E.P. = 0.78 eV) [Si I] line at 10,991.41 Å is blended by a telluric water-vapor line when observed at the center of the solar disk, but can be detected on Doppler-shifted spectra obtained at the west limb of the sun. A description of our observations at $\mu = \cos \theta = 1.0$ and at $\mu = 0.2, 0.13$, and 0.11 is given.

071.002 Fluctuations of temperature and density in the photosphere. B. Schmieder.
Astron. Astrophys., Vol. 16, 44 - 52 (1972). In French.
The first part of this paper deals with the theoretical calculations of the intensities of two lines of Fe I (5196.067 Å and 5195.482 Å), as well as these of a point in the wing of the line b_1 of Mg I and of the neighbouring continuum in a perturbed atmosphere. In the second part, the study of the fluctuations of the measured relative intensities allows us to determine the r.m.s. of the temperature and density fluctuations at two altitudes. We try to evaluate the contribution of respectively the oscillatory phenomenon and the granulation. In the third part, we introduce the fluctuations associated to the granulation in a solar atmosphere model. In this way, we obtain perturbed models which give variations of the relative brightness fluctuations across the solar disk.

071.003 Spectral lines from photosphere to chromosphere, observed during the March 1970 eclipse; A first comparison with theory. J. Houtgast, O. Namba, R. J. Rutten, J. W. Wijbenga.
Solar Physics, Vol. 21, 281 - 285 (1971).
A brief report is presented on observations of line profile variations in the transition region from the photosphere to the chromosphere, based on high-resolution eclipse spectrograms.

071.004 On the applicability of Goldberg and Unno's method to the determination of microturbulent velocities in an atmosphere with convection.
C. de Jager, L. Neven.
Solar Physics, Vol. 22, 49 - 52 (1972).
The method of Goldberg and Unno for the determination of microturbulent velocities in a stellar atmosphere is only applicable if there are no macroturbulent or convective motions. If such motions occur, as in the solar photosphere, the derived results are false and may lead to misinterpretations such as an increase of the microturbulent velocity with depth or anisotropic microturbulence.

071.005 Remarks on the convergency of photospheric model conceptions and the solar quasi continuum.
D. Labs, H. Neckel.
Solar Physics, Vol. 22, 64 - 69 (1972).
A comparison is made of the observed intensity in the solar continuous spectrum with those predicted by some models. Arguments are given that the bend-off observed for $\lambda < 0.6\ \mu$ is a real phenomenon, and due to a veiled line haze.

071.006 Some properties of velocity fields in the solar photosphere. IV. Long periods, five minute oscillations, and the supergranulation at lower layers.
F.-L. Deubner.
Solar Physics, Vol. 22, 263 - 275 (1972).
Photoelectric measurements of photospheric velocity fields have been carried out with the Sacramento Peak Doppler Zeeman Analyzer. The most frequent 'wave-numbers' in the spectra of 5-min oscillations (as well as of the low frequency field) agree with those derived from a model assuming statistically independent oscillators of 10″ to 20″ diameter. These two velocity fields are anti-correlated spatially. Different explanations are suggested.

071.007 Spectral analyses of solar photospheric fluctuations. I. Power, coherence and phase spectra calculated by fast-Fourier-transform techniques.
F. N. Edmonds, Jr., C. J. Webb.
Solar Physics, Vol. 22, 276 - 296 (1972).
Alternative methods for smoothing raw spectra, direct averaging and indirect truncation of the correlation function, are compared and indirect smoothing is compared with spectra calculated by mean-lagged-product methods. Besides providing the raw spectrum, fast-Fourier-transform techniques easily allow computing a series of spectra with varying amounts of smoothing.

071.008 Turbulent velocity in undisturbed and active photosphere. O. G. Badalyan, M. A. Livshits.
Solar Physics, Vol. 22, 297 - 306 (1972).
Profiles of weak Fraunhofer lines have been recorded photoelectrically in both the faculae and the undisturbed photosphere. For the undisturbed photosphere a turbulent velocity of about 2.6 km/s was found with no appreciable increase with the depth. In facular filaments all lines formed above $\tau = 0.1$ showed smaller turbulence velocities, the velocity differences being in the range between 0.2 and 0.6 km/s.

071.009 The mean photospheric magnetic field from solar magnetograms: Comparisons with the interplanetary magnetic field. P. H. Scherrer, J. M. Wilcox, R. Howard.
Solar Physics, Vol. 22, 418 - 424 (1972).
Large-scale averages of daily solar magnetograms have been compared by cross-correlation with the interplanetary magnetic sector pattern during a $2^1/_2$ yr interval. A significant correlation was found at a lag of about $4^1/_2$ days, with the amplitude of the correlation depending on the area included in the magnetogram averages.

071.010 Revised gf-scale and solar curves-of-growth. Y. Yamashita.
Publ. Astron. Soc. Japan, Vol. 24, 49 - 60 (1972).
In order to check the validity of the revised gf-scale and to establish a standard in differential analyses of stellar spectra, the solar curve-of-growth for Fe I is studied with the use of new gf-values. Similar analyses are made for Fe II, Cr I, and Ni I with the use of gf-values in the revised scale.

071.011 Observations of the variation of temperature with latitude in the upper solar photosphere.
R. C. Altrock, R. C. Canfield.

Astrophys. Journ., (*Letters*), Vol. 171, L71 - L74 (1972).

We made photoelectric meridional and equatorial limb-darkening scans during the period 15–22 June 1971. At this time the temperature relative to the equatorial temperature was enhanced by 8°K ± 2.5°K at active-region latitudes and 5°K ± 2°K at latitude 50°N.

071.012 The effect of radiative equilibrium on the photospheric angular velocity. B. R. Durney.
Astrophys. Journ., Vol. 172, 479 - 484 (1972).

The photospheric angular velocity $[\omega = \omega(r)]$ is evaluated in the range of optical depths $0.05 < \tau < \tau_c \sim 0.8$ under the assumption of radiative equilibrium and vanishing von Zeipel currents. The angular velocity decreases outward, and significant pole-equator temperature differences develop. The von Zeipel currents are estimated for the case of uniform rotation.

071.013 The solar abundance of silver.
J. E. Ross, L. H. Aller.
Bull. American Astron. Soc., Vol. 3, 438 - 439 (1971). – Abstr. AAS.

071.014 The dynamics of a toroidal magnetic ring and their implications in solar phenomena.
C. G. Lilliequist, M. D. Altschuler, Y. Nakagawa.
Bull. American Astron. Soc., Vol. 3, 442 (1971). – Abstr. AAS.

071.015 Thin solar convection zone and sunspots.
D. J. Mullan.
Nature, Phys. Sci., Vol. 235, 58 - 59 (1972).

071.016 On the solar curve of growth of iron. R. Foy.
Astron. Astrophys., Vol. 18, 26 - 38 (1972).

We constructed a solar curve of growth of iron using new oscillator strengths now available in literature. This curve of growth shows that its damping part is unquestionably split into several branches, each of them corresponding to a different value of the damping constant. Different correlations between damping and other parameters – as parity of multiplets, excitation potentials of the levels of the transitions, quantum numbers J and L – are discussed.

071.017 On a hypothesis on the solar red shift.
A. G. Gasanalizade.
Solnechnye Dannye 1971 Byull., No. 12, p. 101 - 103 (1972). In Russian.

071.018 The field of velocities and brightness in the solar atmosphere. S. I. Gopasyuk, T. T. Tsap.
Izv. Krymskoj Astrofiz. Obs., Vol. 43, 174 - 189 (1971). In Russian.

The field of velocities and brightness in active and quiet regions of the sun has been studied. The recordings of radial velocities and brightness were made in the Fe I 5250 Å, Ca I 6103 Å, Hβ, Hα, K_3 Ca II lines. It is shown that a good correlation exists between the brightness at different levels in the solar atmosphere. In the places of highest brightness of plage knots neutral lines of photospheric and chromospheric radial velocities coincide as a rule. But the motions close to these neutral lines may have the same or different directions in the photosphere and chromosphere. A correlation between the plage knot brightness and the gradient of radial velocity is found.

071.019 The solar abundance of gold.
J. E. Ross, L. H. Aller.
Solar Physics, Vol. 23, 13 - 17 (1972).

From 13 scans obtained with a double-band pass spectrograph at the Snow telescope at Mount Wilson, interpreted by the method of spectral synthesis, the abundance of gold turns

out to be log $[N(Au)/N(H)] + 12 = 0.70$, assuming log $gf = -0.57$.

071.020 Measurements of the limb darkening in the forbidden MgI line at 4571.1 Å.
O. R. White, R. C. Altrock, J. W. Brault, C. D. Slaughter.
Solar Physics, Vol. 23, 18 - 25 (1972).

We report high resolution measurements of the center-to-limb variation of the MgI line at 4571.1 Å. Comparison of our measurements with the Harvard-Smithsonian Reference Atmosphere show that the line center radiation originates in the temperature minimum region from 330 to 550 km above the point where $\tau_{continuum} = 1$. Observations near the limb confirm that the temperature minimum is ~ 4200 K.

071.021 Intensity fluctuations in Fraunhofer lines.
F. Q. Orrall.
Solar Physics, Vol. 23, 30 - 46 (1972).

An expression is derived for the fluctuation $\psi(t)$ in emergent intensity caused by a perturbation in temperature $\theta(z, t)$ in the sun's atmosphere. It is found that $\psi_\lambda(t) \sim M_\lambda \theta(z_\lambda, t) + N_\lambda p(z_\lambda, t)$ where M_λ and N_λ depend on the structure of the atmosphere. We compute M, N, and contribution functions for several values of μ and $\Delta\lambda$ in the inner wings of the K line (λ3933 CaII).

071.022 Spectral analyses of solar photospheric fluctuations. II. Profile fluctuations in the wings of the λ5183.6 MgI b_1 line. F. N. Edmonds, Jr.
Solar Physics, Vol. 23, 47 - 57 (1972).

Residual intensity fluctuation measurements within the wings of the λ5183.6 MgI b_1 line, obtained from two, high-resolution, high-dispersion, Sacramento Peak Observatory spectrograms, have been subtracted from intensity fluctuations in the adjacent continuum in order to isolate fluctuations associated exclusively with line formation. The useable spectral range for studying these line-formation fluctuations is restricted to wavelengths between 1040 and 7170 km. Power and cross-power (coherence and phase) spectra proved to be valuable diagnostic tools in isolating line-formation fluctuations.

071.023 Table of solar diatomic molecular lines spectral range 4900–6441 Å. P. Sotirovski.
Astron. Astrophys. Suppl, Vol. 6, 85 - 115 (1972). – Abstr. in Astron. Astrophys., Vol. 18, 494 (1972).

The present table of diatomic molecular lines observed in sunspot spectra (λλ 4900 – 6440 Å) contains about 6000 lines; about 10 per cent of them are masked by atomic lines; whilst others are blended with solar molecular or telluric lines. A total of seven molecules have been identified in the three sunspots. For each identified line the rotation branch, quantum number and the vibration band are indicated. We measured the equivalent widths for all lines with profiles as free from blend as possible, the effect of scattering from the photosphere being taken into account. For other lines estimates are given.

071.024 Observations of photospheric pole-equator temperature differences.
R. C. Altrock, R. C. Canfield.
Solar Physics, Vol. 23, 257 - 264 (1972).

Using photoelectric methods we have repeated Plaskett's (1970) measurements of pole-equator temperature differences. We average many limb-darkening scans to reduce statistical errors. We then analyze the differences between the average polar and equatorial scans. Our data yield a pole-equator temperature difference of 1.5K ± 0.6K.

071.025 Empirical NLTE analyses of solar spectral lines. I. A

method and some applications to earlier analyses.
J. W. Wijbenga, C. Zwaan.
Solar Physics, Vol. 23, 265 - 286 (1972).

A method is suggested for empirical NLTE analyses of solar spectral lines. Special depth-dependent departure coefficients are introduced and the formulas are given for further application. From test calculations it is shown that the separation of the departure coefficients for the upper and the lower level from each other and from the uncertainties in several input parameters is greatly facilitated when spectral lines are analyzed on the disk, around the limb, as well as in the flash spectrum. We discuss the advantages of translating the departure coefficients into a set of temperatures. Some published LTE and NLTE analyses are compared in terms of these temperatures.

071.026 Time behavior of Ca II K_{2v} spectral features in nonmagnetic regions of the solar disk.
S. Y. Liu, N. R. Sheeley, Jr., E. v. P. Smith.
Solar Physics, Vol. 23, 289 - 291 (1972).

The results presented here were obtained from time sequences of K_{2v} spectroheliograms taken at the rate of 10 s/frame during excellent seeing conditions at the Kitt Peak National Observatory.

071.027 The solar abundance of manganese.
D. E. Blackwell, B. S. Collins, A. D. Petford.
Solar Physics, Vol. 23, 292 - 293 (1972). — Research note.

071.028 Observations of short period oscillations in two dimensions. J. Harvey, R. Howard.
Solar Physics, Vol. 23, 300 - 303 (1972).

Observations of the photospheric velocity field at the disk center with a cadence of five frames per second strongly support the idea that short period oscillations arise from a combination of image motion and horizontal gradients of the line of sight velocity field. Any genuine solar short period oscillations are effectively masked by these false short period oscillations.

071.029 On the power spectrum of the photospheric resonance oscillations. F.-L. Deubner.
Solar Physics, Vol. 23, 304 - 306 = Mitt. Fraunhofer Inst., *Freiburg,* No. 110 (1972). — Research note.

071.030 Sur la détermination du rapport d'intensité des raies infrarouges de l'ion Fe XIII.
G. Ratier, J.-P. Rozelot.
Solar Physics, Vol. 23, 394 - 405 (1972).

The ratio of the two infrared lines 10747 Å and 10798 Å of Fe XIII is studied from a theoretical and observational point of view. The results and the discussion of the two determinations are summarized into two tables. Arising from a measured ratio of the two lines, a variation of the electron density as a function of dilution factor is given, taking into account the proton impact mechanism in the resolution of excitation equilibrium.

071.031 Measuring solar photospheric magnetic fields.
W. C. Livingston.
Sky Telescope, Vol. 43, 344 - 349 (1972).

071.032 Filamentary structure of solar magnetic fields, according to the data of the third flight of the Soviet stratospheric observatory. V. A. Krat.
Vestn. AN SSSR, 1971, No. 12, p. 28 - 33. In Russian.— Abstr. in Referativ. Zhurn. 51. Astron., 5.51.357 (1972).

071.033 On the possible reason for the variation of the source function with the heliographic latitude.
A. Sadikov.

Tsirk. Astron. Inst., *Tashkent,* No. 27 (374), p. 11 - 12 (1971). In Russian.

071.034 Solar Fe abundance deduced from solar particle measurements made with nuclear emulsions.
C. E. Fichtel, D. L. Bertsch, C. J. Pellerin, D. V. Reames.
Bull. American Astron. Soc., Vol. 4, 259 (1972). – Abstr. AAS.

071.035 Solar photospheric abundances of seven heavier elements by spectrum synthesis. R. P. Boyle.
Bull. American Astron. Soc., Vol. 4, 268 (1972). – Abstr. AAS.

071.036 Observations of the variation of temperature with latitude in the upper solar photosphere.
R. C. Altrock, R. C. Canfield.
Bull. American Astron. Soc., Vol. 4, 268 (1972). – Abstr. AAS.

071.037 A first order analysis of variations of the limb darkening and the shapes for solar Fraunhofer lines.
R. G. Athay, B. W. Lites, O. R. White, J. W. Brault.
Solar Physics, Vol. 24, 18 - 27 (1972).

New center-to-limb measurements in Fe I lines show changes in both the line profiles and the limb darkening curves that appear to be characteristic of many other solar lines. Here we seek the constraints placed on the atmospheric model by these effects.

071.038 On the choice of boundary conditions for integration of transfer equations. J. M. Katz.
Solar Physics, Vol. 24, 28 - 31 (1972).

This paper describes a method which makes it possible to choose the boundary conditions for the construction of the transfer equation solution.

071.039 Some new Dy II identifications in the solar spectrum.
J. C. Howard.
Solar Physics, Vol. 24, 32 - 42 (1972).

Extension of the identification of singly ionized dysprosium lines in the solar chromosphere and disk has been made from a recently published laboratory analysis of the first and second spectrum.

071.040 Space and time variations of the solar Na D line profiles. C. D. Slaughter, A. M. Wilson.
Solar Physics, Vol. 24, 43 - 58 (1972).

Preliminary results are presented of observations of the solar Na D lines obtained with high space and time resolution $(2.4'' \times 2.4'')$, (6 s).

071.041 On the temperature of the helium emission regions in the solar atmosphere. R. A. Gulyaev.
Solar Physics, Vol. 24, 72 - 78 (1972).

On the basis of data on the He I Lyman continuum, obtained by Dupree and Reeves from OSO−4, the electron temperature of undisturbed helium regions is determined: $T_e = 12500$ K.

071.042 The velocity fields in active regions.
R. Howard.
Solar Physics, Vol. 24, 123 - 128 (1972).

From line-shift observations in two spectrum lines it is determined that the downward motions observed in plages may represent a real downward transport of material, not an apparent downward flow due to brightness or ionization differences in a multistream velocity model.

071.043 Granulare Geschwindigkeiten in der Sonnenatmosphäre. A. Nesis, W. Mattig.
Mitt. Astron. Ges., No. 31, p. 150 (1972). — Abstract.

071.044 Magnetooptische Effekte in Fraunhoferlinien mit

Zeeman-Aufspaltung. A. Wittmann.
Mitt. Astron. Ges., No. 31, p. 150 - 153 (1972).

071.045 **New measurements of the polarization of photospheric light near the solar limb.** J. L. Leroy.
Astron. Astrophys., Vol. 19, 287 - 292 (1972). In French.

The measurements were performed, at the Pic-du-Midi
Observatory, during the period 1970–1971, for 6 wavelength
intervals. A comparison with previous results (Dollfus, 1960)
and theoretical computations (Débarbat, 1970) is given.

071.046 **The continuum near the H and K lines of Ca⁺ in the
solar spectrum.** S. Matsushima, K. Kawabata.
Astrophys. Letters, Vol. 11, 103 - 106 (1972).

A method is proposed for the determination of the true
continuum for the spectral region where no continuum is visible. The method is based on the use of the characteristic weak-
ening of absorption lines due to blending with a strong line and
thereby eliminates various uncertainties involved in theoretical
methods attempted earlier. The continuum determined by this
method for the H and K lines of Ca⁺ is about 10 to 15 per cent
higher than the newest empirical determination, and is in good
agreement with higher values suggested by earlier workers.

071.047 **New measurement of the solar gravitational red
shift.** J. L. Snider.
Phys. Rev. Letters, Vol. 28, 853 - 856 (1972).

The red shift of the solar potassium absorption line at
7699 Å has been measured by means of an atomic-beam reso-
nance-scattering technique. The shift at the center of the solar
disk, corrected for earth's motion, is $(\Delta\lambda)_{exp} = (1.01 \pm 0.06)$
$(\Delta\lambda)_{theor}$.

071.048 **New lines in the region from λ 6335 Å to λ 6356 Å
of the solar spectrum.**
E. K. Kokhan, V. A. Krat, N. I. Pechinskaya.
Solnechnye Dannye 1972 Byull., No. 3, p. 92 - 97 (1972).
In Russian.

Eighteen new telluric and several solar lines were detected
on photoelectric records made with the Pulkovo spectrophotometer with narrow intermediate slit.

071.049 **Observations, par ballon stratosphérique, du spectre
solaire à 1,85 microns avec un pouvoir de résolution
de 135.000.** R. Zander.
Bull. Cl. Sci. Acad. Roy. Belgique, 5 ᵉ Sér., Vol. 56, 729 - 739 =
Univ. Liège, Inst. d'Astrophys., Coll. 8°, No. 607 (1970).

071.050 **On the difference of Fraunhofer lines in the spectrum at the pole and equator of the solar disc.**
D. M. Kuli-Zade, Yu. A. Solonskij.
Soobshch. Shemakhinsk. Astrofiz. Obs., vyp. (No.) 6, p. 25 -
43 (1971). In Russian.

071.051 **The classification of Ni XI to XVII and Co X to
XVII emission lines from $3s^2$ $3p^n$-3s $3p^{n+1}$ and
$3p^n$-$3p^{n-1}$ 3d transitions and the identification of nickel solar
lines.** B. C. Fawcett, R. W. Hayes.
Journ. Phys. B, Atomic and Molecular Phys., Vol. 5, 366 - 370
(1972).

071.052 **Calculation of stellar structure. II. Determination of
the helium abundance of the sun by the theoretical**

prediction of line and continuum radiation from solar-model
photospheres. C. A. Rouse.
Progr. High Temperature Phys. Chem. (*GB*), Vol. 4, 139 - 191
(1971). – See Bull. Signal., Vol. 33, Section 120, No. 2604
(1972).

**Fe I oscillator strengths determined from anomalous
dispersion of shock-heated gases.** See Abstr. 022.011.

**Pointing error correction for millimeter wave spec-
troheliograms.** See Abstr. 034.134.

Diffusion processes in stars. See Abstr. 064.041.

**Über die Schwere des Lichtes in der Gravitations-
theorie. II.** See Abstr. 066.083.

**The identification of $^{13}C^{16}O$ in the infrared sunspot
spectrum and the determination of the solar $^{12}C/^{13}C$
abundance ratio.** See Abstr. 072.008.

**The behaviour of CO at temperatures close to
5000°K.** See Abstr. 072.034.

**A comparison between the helium 10830 Å and the
hydrogen Hα chromospheres.** See Abstr. 073.005.

**Magnetic coupling of the active chromosphere to
the solar interior.** See Abstr. 073.036.

**The relations between chromospheric features and
photospheric magnetic fields.** See Abstr. 073.059.

**Correlation of photospheric magnetic field strength
with coronal brightness on 7 March, 1970.**
See Abstr. 074.021.

**Analyse des renforcements coronaux à travers quel-
ques acquisitions spectroscopiques récentes des émissions monochromatiques du fer ionisé (X à XV).** See Abstr. 074.022.

**Differential rotation and the structure and energy
content of coronal magnetic fields.** See Abstr. 074.027.

**The blocking effect of lines on solar UV continuum
between 2000 Å and 3000 Å.** See Abstr. 076.011.

**The interpretation of XUV rocket measurements of
intensity ratios of solar spectral lines of the lithiumlike ions
O VI, Ne VIII, and Mg X.** See Abstr. 076.026.

**A comparison of time profiles of type IV bursts at
1.5 meters with the total flux of Hα emission.**
See Abstr. 077.017.

Formation of the solar Hα profile.
See Abstr. 080.005.

Can astrophysical abundances be taken seriously?
See Abstr. 114.011.

**A search for C_2^- in spectra of HD 201626 and the
sun.** See Abstr. 114.073.

072 Sunspots, Faculae, Solar Activity

072.001 Solar dynamo theory and the models of Babcock and Leighton. J. H. Piddington.
Solar Physics, Vol. 22, 3 - 19 (1972).

The dynamo theory of the solar cycle as developed by Parker and others, and the observational models of Babcock and Leighton have been examined, with the conclusion that the dynamo theory is not applicable to the sun and that the models fail. That means that a shallow, reversing field is unacceptable, and that a deeply penetrating field is required. Reference is made to an alternative theory of the solar cycle based on a deep magnetic field.

072.002 The crossover and magneto-optical effects in sunspot spectra. V. M. Grigorjev, J. M. Katz.
Solar Physics, Vol. 22, 119 - 128 (1972).

Two peculiarities of the magnetic splitting of lines in sunspot spectra have been investigated: In a rather small region of the penumbra, near the umbra-penumbra boundary, the π-component is absent in the spectra of one circular polarization while both σ-components are present. In the spectra of the opposite circular polarization, the σ-components are absent but the π-component is present. The second peculiarity consists of the anomalous splitting of the π-component of Zeemann triplets which are of the same and opposite signs in comparison with splitting of the σ-components.

072.003 On the fine structure of the Evershed effect. M. Mamadazimov.
Solar Physics, Vol. 22, 129 - 136 (1972).

The fine structure of the Evershed effect was studied using spectrograms obtained on 3 July 1969 at the Pulkovo Observatory. The results of the study of Fe I and Ca I lines show that the outward motion in the penumbra is concentrated only in dark filaments. The velocity variation along a few dark filaments shows that maximum velocity is at a distance $0.8\,R_s$ from the center of the sunspot. The mean velocity in the interfilamentary elements is of the same order as that in the photosphere directly adjoining the penumbra. The mean velocity at the outer boundaries of the dark filaments (Ti I) is 1.5–2.0 km/s.

072.004 Magnetic fields in umbral atmospheres under 'similarity' configuration. H. S. Yun.
Solar Physics, Vol. 22, 137 - 139 (1972). – Research note.

072.005 Alfvén waves in umbral flux tubes. P. R. Wilson.
Solar Physics, Vol. 22, 434 - 442 (1972).

The hypothesis requires that Alfvén waves travel along the closed flux tubes linking the umbra either with the umbra of another spot or with the surrounding faculae and passing through regions of variable field strength and density. It is shown that, for a very simplified model, standing waves are possible in a symmetrical field configuration. It is further shown that mechanical dissipation of these waves in local regions of the flux tube may contribute to the heating of faculae.

072.006 Photometry of a sunspot at $\lambda = 3.75\,\mu$. P. Coupiac, S. Koutchmy.
Astron. Astrophys., Vol. 16, 272 - 275 (1972). In French.

A sunspot with an umbra of 16″ was observed in the N° 11145 MacMath plage during Feb. 1971. Good photometric scans were obtained by spatial cancellation techniques using a modulation frequency of 330 Hz and amplitudes < 4″ with a 0.34 μ band pass filter and InSb detector at a wavelength of 3.75 μ.

072.007 On the mechanism of formation of sunspots. Yu. B. Ponomarenko.
Astron. Zhurn. Akad. Nauk SSSR, Vol. 49, 148 - 152 (1972). In Russian. English translation in Soviet Astron. AJ, Vol. 16, No. 1.

In the mechanism under discussion the sunspots are the result of concentration of the magnetic flux of an active region.

072.008 The identification of $^{13}C^{16}O$ in the infrared sunspot spectrum and the determination of the solar $^{12}C/^{13}C$ abundance ratio. D. N. B. Hall, R. W. Noyes, T. R. Ayres.
Astrophys. Journ., Vol. 171, 615 - 620 (1972).

The presence of the first-overtone vibration-rotation bands of $^{13}C^{16}O$ in the infrared sunspot spectrum has been established on the basis of wavenumber and relative intensity consistency of 30 weak lines. Nine particularly clean lines have been used to obtain a solar $^{12}C/^{13}C$ abundance ratio of 90 with a probable error of 15 percent.

072.009 Sunspot theory based on magneto-fluid-mechanic turbulent damping. P. S. Lykoudis.
Bull. American Astron. Soc., Vol. 3, 461 (1971). – Abstr. AAS.

072.010 Horizontal magnetic stresses generated by differential rotation and their influence upon the poleward migration of solar magnetic features.
G. W. Pneuman, M. A. Raadu.
Bull. American Astron. Soc., Vol. 3, 462 (1971). – Abstr. AAS.

072.011 The number of spotgroups formed and their average lifetime in solar activity cycle No 19.
M. Kopecký.
Bull. Astron. Inst. Czechoslovakia, Vol. 23, 55 - 60 (1972).

The paper complements the values of the number of formed sunspot groups and their average lifetimes already published earlier. It is shown that the very high value of the Wolf number in cycle No 19 is a product of superimposing the 80-year period of the mean spotgroup's lifetime and the very long secular variation of the number of new formed spotgroups.

072.012 Line profiles in sunspots. F. Kneer.
Astron. Astrophys., Vol. 18, 39 - 46 = Mitt. Fraunhofer Inst., *Freiburg*, No. 108 (1972). In German.

Photoelectric measurements of the brightness ratio spot/photosphere at the wavelengths $\lambda 5131$ Å and 6222 Å as well as the profiles of three magnetically unaffected Fe I lines ($\lambda 5123.7$, 5434.5 and 5576.1 Å) obtained from high-resolution photographic spectra are given. A hydrostatic model for the umbra is adjusted to the observations, which is in a good agreement with the model given by Stellmacher and Wiehr (1970).

072.013 Profiles of magnetically split lines in sunspots. F. Kneer.
Astron. Astrophys., Vol. 18, 47 - 50 = Mitt. Fraunhofer Inst., *Freiburg*, No. 109 (1972). In German.

The gradient of the magnetic field dH/dh in sunspot umbrae is determined from profiles of magnetically split lines. For this purpose 17 line profiles in the range $\lambda\lambda 6200$–6300 Å are selected from spectra of the main spot of the group 5454 (Rome-number) taken on June 12 and 13, 1969. The magnetic field strength is determined for each line by fitting calculated intensity profiles to the measured ones.

072.014 On the 22-year solar cycles. N. P. Chirkov.
Solnechnye Dannye 1971 Byull., No. 11, p. 75 - 79 (1972). In Russian.

The relation between the length of the 22-year solar cycle and the sum of maximum Wolf numbers in it is considered. Some new regularities for this cycle have been detected, being of interest for solar activity forecasts.

072.015 Statistical analysis and forecast of the Zürich series of Wolf numbers with the regular part excepted.
K. A. Kandaurova.
Solnechnye Dannye 1971 Byull., No. 11, p. 80 - 89 (1972). In Russian.

The Zürich series of Wolf numbers with the regular part excepted within the interval of 2—5 years complies with the normal law of distribution and is stationary in a limited sense.

072.016 Analysis and calculations of a magnetic field in the solar atmosphere. II. The magnetic field of bipolar spots. O. S. Korolev.
Solnechnye Dannye 1971 Byull., No. 11, p. 97 - 104 (1972). In Russian.

The calculations of the magnetic field of bipolar spots in the solar atmosphere have been made by the normalized formulas of dipole. For bipolar groups of sunspots the parameter d/b is given, which determines the probability of flare activity.

072.017 Analysis and calculation of the magnetic field of spots in the solar atmosphere. III.
O. S. Korolev.
Solnechnye Dannye 1971 Byull., No. 12, p. 77 - 84 (1972). In Russian.

Supposing that some flares arise in the region where the longitudinal field is equal to zero ($H_z = 0$), one can determine the heights of the beginning of flares according to the measurements on the solar disk, as well as according to the location of the longitudinal $H_z = 0$ field component.

072.018 The (0,0) band of the γ-system of TiO in the umbral spectrum: The isotopic abundances of Ti.
D. L. Lambert, E. A. Mallia.
Monthly Notices Roy. Astron. Soc., Vol. 156, 337 - 348 (1972).

The primary purpose of the present paper is to report on the solar isotopic abundance ratios of titanium. The observed equivalent widths of the ^{48}TiO lines, in conjunction with a model atmosphere, enables the oscillator strength of the transition to be deduced. This result is compared with a recent laboratory determination. Finally, analysis of the line profiles provides an estimate of the microturbulent velocity field in the umbra.

072.019 Photoelectric line profiles in umbral spectra.
T. D. Fay, A. A. Wyller, H. S. Yun.
Solar Physics, Vol. 23, 58 - 77 (1972).

Umbral line profiles of Hα, Na D_2 and HeI 10830 have been observed photoelectrically with a pressure scanning spectrometer. Intermittent pulse counting techniques are applied with integration times as short as 0.8 s which permits selection of moments of good seeing and guiding even in poor climates.

072.020 Note on the characteristics of sunspot groups which produce solar proton flares. K. Sakurai.
Solar Physics, Vol. 23, 142 - 145 (1972).

It is shown that the formation of magnetic field gradients is associated with the rotating motion of sunspot groups. Hence, the sunspot groups which show a reversed polarity distribution are very effective for the production of solar proton flares.

072.021 Interferometry applied to visible solar features.
J. W. Harvey.
Nature, Phys. Sci., Vol. 235, 90 - 91 (1972).

It is shown that wave-front division interferometry can be applied to the study of small solar features in the visible spectrum. The first results demonstrate the existence of structures in sunspots with characteristic dimensions as small as 100 km.

072.022 Some observations relating to umbral dots.
E. A. Mallia, A. D. Petford.
Monthly Notices Roy. Astron. Soc., Vol. 157, 73 - 84 (1972).

Observations of Fraunhofer lines of singly-ionized metals in the spectrum of a sunspot are used to derive information about inhomogeneities in umbrae.

072.023 Detection of blends in the vicinity of Zeeman lines.
A. Wittmann.
Solar Physics, Vol. 23, 294 - 299 (1972).

High resolution spectrograms taken in polarized light have revealed the presence of significant blending within the profiles of some important Zeeman multiplets of a large umbra. Wavelength and equivalent width of each depictable blend have been derived from the corrected spectrograms and some preliminary identifications have been made.

072.024 Faculae and east-west asymmetry of sunspot area.
C. Sawyer, M. W. Haurwitz.
Solar Physics, Vol. 23, 429 - 437 (1972).

Asymmetry of sunspot area with respect to the central meridian is found to depend strongly on the location of the spot group in its chromospheric facula or plage. Qualitatively, the observed asymmetries can be explained by supposing that the apparent area of the spot is decreased by overlying bright facula, especially west of central meridian where the spot (in the usual preceding position) is viewed through the relatively bright and extensive follower part of the plage.

072.025 Real-time masking of solar photographs with photochromic glass. J. W. Harvey.
AAS Photo-Bull. 1972, No. 1, p. 5 - 6.

072.026 Variations of magnetic fields of sunspots at two levels in connection with the development of active regions. A. N. Koval, N. N. Stepanjan.
Solnechnye Dannye 1972 Byull., No. 1, p. 83 - 91 (1972). In Russian.

Variations of the magnetic field in sunspots at two levels in the photosphere in connection with the development of an active region (appearance and evolution of sunspots, moustaches, surges and flares) were studied. Visual and photographic observations of the magnetic field in two lines (λ 6103 Ca I — upper level — and λ 6302 Fe I — lower level) were used.

072.027 Photographic measurements of the magnetic fields of sunspots influenced by the contrast of the solar image. N. N. Stepanjan.
Solnechnye Dannye 1972 Byull., No. 1, p. 91 - 94 (1972). In Russian.

Graphic dependences between the solar image contrast and measured intensities of the magnetic fields for sunspots with different diameters have been determined from observations.

072.028 Photographische Sonnenbeobachtungen eines Amateurs. G. Klaus.
Orion Schaffhausen, 30. Jahrgang, p. 47 - 50 (1972).

072.029 Distribution of the mean areas of spot groups according to the heliographic latitudes in the 11-year cycle of solar activity. F. G. Mustaeva.
Tsirk. Astron. Inst., *Tashkent*, No. 27 (374), p. 13 - 14 (1971). In Russian.

072.030 The active longitudes of sunspots and solar flares

in the 20th cycle. G. Mariş.
Stud. Cerc. Astron., Vol. 17, 71 - 77 (1972). In Romanian.

The active longitudes of different indices for sunspots and solar flares are investigated by the isoline method for the period July 1, 1964 – December 31, 1969. The problems connected with north–south asymmetry of solar activity and of the concentration rate in the active longitudes are also considered.

072.031 Variations of the winter anomaly in the solar activity cycle. E. K. Vasilieva.
Geomagn. Aeronom., Vol. 12, 553 - 554 (1972). In Russian.
Brief information.

072.032 Linienprofile in Sonnenflecken. F. Kneer.
Mitt. Astron. Ges., No. 31, p. 149 - 150 (1972).
Abstract.

072.033 On heating and decay of sunspots.
Yu. B. Ponomarenko.
Astron. Zhurn. Akad. Nauk SSSR, Vol. 49, 568 - 578 (1972). In Russian. English translation in Soviet Astron. AJ, Vol. 16, No. 3.

It is shown that the suppression of the supergranular motion in a sunspot gives rise to heating of its invisible lower part. The energy of heating is equal to the "missing" radiative energy of the spot. It is supposed that heating of the spot leads to renewal of the supergranular motion in it. This motion results in decay of the spot and daily oscillations of its magnetic field.

072.034 The behaviour of CO at temperatures close to 5000°K. V. V. Polonsky.
Astron. Zhurn. Akad. Nauk SSSR, Vol. 49, 663 - 666 (1972). In Russian. English translation in Soviet Astron. AJ, Vol. 16, No. 3.

The spectrum of faculae and the photosphere of the sun is investigated. On the basis of spectrograms obtained of 25 lines of the first overtone spectrum of the CO molecule in the region $\lambda\lambda\, 2.3 - 2.4\,\mu$ the equivalent widths of these lines are determined. It is established that on conditions of low-temperature plasma, the CO molecule is very sensible to small temperature variations.

072.035 Relazione sull'attività solare. G. Godoli.
Atti XIII Riunione Soc. Astron. Italiana, Trieste 1969, (see 012.010), p. 61 - 83 (1970).

072.036 A working model for sunspot umbrae.
G. Stellmacher, E. Wiehr.
Astron. Astrophys., Vol. 19, 293 - 297 (1972).

The three "best umbra models" by Zwaan, Hénoux and Stellmacher-Wiehr have been tested by the much stronger criterion of the center-to-limb variation of the non-split line Fe λ 5434.5 and the Na D-lines.

072.037 Local magnetic fields on the sun. A. B. Severny.
Astrophys. Space Sci. Library, Vol. 29, Part I, (see 012.012), p. 38 - 48 (1972).

072.038 Properties of solar active regions. A. Bruzek.
Astrophys. Space Sci. Library, Vol. 29, Part I, (see 012.012), p. 49 - 60 = Mitt. Fraunhofer Inst., *Freiburg*, No. 101 (1972).

072.039 Time variations in solar activity.
H. W. Dodson, E. R. Hedeman.
Astrophys. Space Sci. Library, Vol. 29, Part I, (see 012.012), p. 151 - 172 (1972).

072.040 On the motion in a complex sunspot group.
H. I. Abdussamatov, M. N. Stoyanova.

Solnechnye Dannye 1972 Byull., No. 2, p. 88 - 93 (1972). In Russian.

The structure of the motions in a complex sunspot group is investigated on the basis of a series of spectra obtained by the method of escalation. The structure of the observed phenomenon is explained by spatial cross direction of penumbra filaments in the spots with complicated contours of umbra, or it may be a special type of a surge.

072.041 On the problem of the influence of galactic factors upon solar activity.
G. Y. Vassilyeva, D. A. Kuznetsov, A. A. Shpitalnaya.
Solnechnye Dannye 1972 Byull., No. 2, p. 99 - 106 (1972). In Russian.

A hypothesis on the electromagnetic nature of interaction between the sun and the planets in the presence of the external galactic field is suggested on the basis of an analysis of regularities in the position of Jupiter on the orbit during the maxima of solar activity, the orientation of the solar system in space being taken into account.

072.042 On the fragmentation of sunspots. A. A. Solovjev.
Solnechnye Dannye 1972 Byull., No. 3, p. 109 - 114 (1972). In Russian.

A critical analysis of the "dropping model" of a sunspot suggested by V. Kassinsky is given. It is shown that only certain external forces can be responsible for splitting a magnetic rope into separate thin ropes. The case when the rising rope splitted under the influence of the lifting force gradient, developing due to inhomogeneity of the atmosphere, is considered.

072.043 On a relation between the indices of solar activity in the photosphere and the corona.
J. Xanthakis.
Praktika Acad. Athens, Vol. 44, 153 - 185 = Contr. Res. Center Astron., Applied Math., Acad. Athens, Ser. 1 (Astron.), No. 23 (1970).

072.044 Acceleration of charged particles in the neutral sheet. T. Mukai.
Journ. Geomagn. Geoelectr. (*Japan*), Vol. 23, 225 - 233 (1971).

The plasma which flows into the contact region of two sunspots of opposite polarity (which is named 'the neutral sheet') induces an electric field. This induced electric field can accelerate a part of the charged particles. It is shown that the charged particle can reach an energy of a few Mev within a short time less than one second.

072.045 Solar activity in the year 1971. L. Schmied.
Říše hvězd, Vol. 53, 134 - 135 (1972). In Czech.

072.046 Solar activity prediction.
R. J. Slutz, T. B. Gray, M. L. West, F. G. Stewart, M. Leftin.
National Aeronautics and Space Administration. NASA Contractor Rep. NASA CR-1939, 112 pp. (1971).

Formulae for predicting sunspot variation and relating this to 10.7 cm flux.-*IDP*

α-effect dynamos, by the Bullard-Gellman formalism. See Abstr. 061.024.

The effect of stochastic magnetic fields on absorption lines. See Abstr. 064.032.

On stellar activity cycles.
See Abstr. 065.045.

Thin solar convection zone and sunspots.

See Abstr. 071.015.

Table of solar diatomic molecular lines spectral range 4900 - 6441 Å.   See Abstr. 071.023.

Fine structure of solar magnetic fields. See Abstr. 073.004.

Suppression of the kink instablility for magnetic flux ropes in the chromosphere.   See Abstr. 073.010.

A connection of Ca and Hα flocculi and sunspot groups conjugate with them.   See Abstr. 073.032.

Comments on filament-disintegration and its relation to other aspects of solar activity.   See Abstr. 073.046.

On the development of the sources of the S-component of solar radio emission at 9.0 cm. See Abstr. 077.021.

On a relationship between solar activity and solar cosmic rays.   See Abstr. 078.021.

Evolution of solar magnetic fields over an 11-year period.   See Abstr. 080.019.

Large-scale magnetic fields and activity patterns on the sun.   See Abstr. 080.030.

Force-free magnetic-field structures and their role in solar activity.   See Abstr. 080.032.

073 Solar Chromosphere, Flares, Prominences

073.001 Isophotometry of the chromospheric bright and dark mottles on the solar disk.
C. E. Alissandrakis, C. J. Macris.
Astrophys. Letters, Vol. 10, 59 - 60 (1972).

An isophote map of bright and dark mottles at the center of Hα is given. The relative positions of the mottles are discussed.

073.002 High resolution observations of the chromosphere at mm and cm wavelengths. M. Simon.
Solar Physics, Vol. 21, 297 - 304 (1971).

High resolution observations of the sun at 3.3 mm, 3.5 mm, 1.35 cm, and 1.95 cm which were obtained during the eclipses 11 September 1969 and 7 March 1970 are described. The observations indicate that the region emitting at these wavelengths is very irregular with typical length scales smaller than one half minute of arc.

073.003 Localization of the source of flare X-ray emission during the eclipse of 7 March, 1970.
R. W. Kreplin, R. G. Taylor.
Solar Physics, Vol. 21, 452 - 459 (1971).

An occultation of X-ray emission from a solar flare was observed by an NRL instrument aboard the OSO-5 satellite. The X-ray flare was observed to be confined within a region 136000 km in one dimension. The measurements indicate the existence of a denser core 54000 km wide in the direction of advance of the moon's limb. Comparison of these results with X-ray photographs of flare regions are made and a model for the developement of the soft X-ray flare is proposed.

073.004 Fine structure of solar magnetic fields. H. Zirin.
Solar Physics, Vol. 22, 34 - 48 (1972).

The deduction of magnetic fields from chromospheric structure is extended to active regions and transverse fields. In this paper we summarize relationships deduced from new high resolution Hα pictures and then test them by comparing the predicted magnetic field with a high resolution magnetogram. We then apply the method to understanding the structure and history of several active regions. The Hα pictures enable us to evaluate the direction and strength of transverse fields and to trace some of these lines of force and infer others.

073.005 A comparison between the helium 10830 Å and the hydrogen Hα chromospheres.
R. G. Giovanelli, D. N. B. Hall, J. W. Harvey.
Solar Physics, Vol. 22, 53 - 63 (1972).

Spectroheliograms obtained simultaneously in He 10830 Å and Hα show in the network a very close agreement in position of dark Hα mottles and of bright Hα plage remnants with 10830 Å absorption. The typical intensity in 10830 Å, corrected for overlapping lines, is $I \approx 0.91$ of the continuum. Some parts of the network do not appear in 10830 Å. This line is much weaker over supergranule centres ($I \approx 0.98$), though near active regions dark Hα fibrils coincide with faint 10830 Å fibrils ($I \approx 0.93 - 0.98$).

073.006 Observation of a smoke ring on October 30, 1970.
I. Kawaguchi, N. Oda, S. Mizuno.
Solar Physics, Vol. 22, 140 - 141 (1972). – Research note.

073.007 Heating of the solar flare plasma by high energy electrons. C.-C. Cheng.
Solar Physics, Vol. 22, 178 - 188 (1972).

It is shown that for large flares the heating is enough to produce a thermal plasma of a temperature up to a few times of $10^7 K$ rapidly in the initial phase of the flares. Thus thermal bremsstrahlung in addition to non-thermal bremsstrahlung should be considered for the X-ray emission of solar flares in the initial phase.

073.008 Inhomogeneous structure of the solar chromosphere from Lyman-continuum data.
J. E. Vernazza, R. W. Noyes.
Solar Physics, Vol. 22, 358 - 374 (1972).

We describe a new model of the chromosphere based on Lyman-continuum observations by Harvard spectrometers aboard the satellites OSO 4 and OSO 6. The model assumes (a) that a random distribution of optically thick inhomogeneities overlies a plane-parallel homogeneous atmosphere, and (b) that the Lyman continuum in the chromosphere is optically thick and the only significant opacity source between 600 and 912 Å. The temperature, gas pressure, electron pressure, particle densities, and the hydrogen ground-state departure coefficient are calculated as a function of height in the chromosphere.

073.009 Multi-component models for the formation of the chromospheric Ca II K line. II. The effect of velocity fields. L. E. Cram.
Solar Physics, Vol. 22, 375 - 386 (1972).

It is shown that the inclusion of quite reasonable velocity fields permits the reproduction of not only the high resolution profiles of the small scale emission features but also the qualitative centre-limb behaviour of the spatially averaged profiles.

073.010 Suppression of the kink instability for magnetic flux ropes in the chromosphere. M. A. Raadu.
Solar Physics, Vol. 22, 425 - 433 (1972).

Energy storage in chromospheric flux ropes is discussed, in the context of solar flares. The structure is represented by a cylindrically symmetric magnetic field of finite length. The field is assumed to be approximately force-free. It is shown that for a moderate degree of twisting the fields are stable to a kink perturbation. Thus energy can be stored in cylindrical fields prior to release in a solar flare.

073.011 A perturbation analysis of radiative-conductive coupling. R. D. Cess.
Astron. Astrophys., Vol. 16, 327 - 328 (1972).

It is illustrated that the recent analyses of Frisch (1970, 1971), concerning radiative-conductive coupling in the solar chromosphere, may be interpreted within the framework of a singular perturbation.

073.012 On estimates of the nonradiative energy input to the solar chromosphere from the H⁻ emission.
F. Praderie, R. N. Thomas.
Astrophys. Journ., Vol. 172, 485 - 490 (1972).

The inclusion of non-LTE effects in the computation of mechanical heating in a gray atmosphere changes previous LTE estimates by a factor of $4b_{H^-}$.

073.013 Preliminary results of identifications in the XUV spectrum of a solar flare.
J. D. Purcell, R. Tousey, K. G. Widing.
Bull. American Astron. Soc., Vol. 3, 448 (1971). – Abstr. AAS.

073.014 Extreme ultraviolet spectroheliograms of a solar flare.
J. D. Purcell, R. Tousey.
Bull. American Astron. Soc., Vol. 3, 448 (1971). – Abstr. AAS.

073.015 Solar flares in EUV observed from OSO-5.

W. A. Rense, P. T. Kelly.
Bull. American Astron. Soc., Vol. 3, 448 - 449 (1971). – Abstr AAS.

073.016 **Iron-line emission during solar flares.**
G. A. Doschek, J. F. Meekins, R. D. Cowan.
Bull. American Astron. Soc., Vol. 3, 461 (1971). – Abstr. AAS.

073.017 **The physical conditions in solar flare plasmas.**
G. R. Blumenthal, W. H. Tucker, A. T. Wood.
Bull. American Astron. Soc., Vol. 3, 461 (1971). – Abstr. AAS.

073.018 **Evidence for magnetic field reconnection in flares.**
H. Zirin.
Bull. American Astron. Soc., Vol. 3, 461 (1971). – Abstr. AAS.

073.019 **A new slant on discrepancies given by weak shock theory in the solar chromosphere.** S. D. Jordan.
Bull. American Astron. Soc., Vol. 3, 462 (1971). – Abstr. AAS.

073.020 **Structure and energy balance of the chromosphere-corona transition region.** R. L. Moore, P. C. W. Fung.
Bull. American Astron. Soc., Vol. 3, 501 (1971). – Abstr. AAS.

073.021 **Dynamics of the solar prominences. I. Kinematics of the solar prominence of 10 September 1956.**
P. Paľuš.
Bull. Astron. Inst. Czechoslovakia, Vol. 23, 60 - 68 (1972).
A special type of solar prominence is studied, namely the AS type, according to the Menzel-Evans classification. The motion of the knot shows a spiral pattern. The analysis of the data showed that a hyperbolic spiral on a cone surface is the best approximation to the observed motion.

073.022 **Flare-associated solar wind disturbances and type II and IVm radio bursts.** Š. Pintér.
Bull. Astron. Inst. Czechoslovakia, Vol. 23, 69 - 75 (1972).
The paper deals with the relations between flare-associated solar wind disturbances and type II and IVm radio bursts. The flare-associated solar wind disturbances were observed by means of different satellites between 1965 and 1969. The observed characteristics of the solar wind are the proton density and the flow velocity. The kinetic energy of the flow was computed from these values for the individual flare-associated disturbances of the solar wind. The relations between the interplanetary shock wave and the II and IVm type bursts are explained by means of a flare mechanism, suggested by Křivský (1968).

073.023 **The effect of resonance-line transfer on hydrogen ionization.** H. A. Beebe, R. W. Milkey.
Astrophys. Journ., (Letters), Vol. 172, L111 - L114 (1972).
Non-LTE calculations have been carried out on a model of a hydrogen atom in the solar chromosphere. In evaluating the assumption of detailed balance in Lyman-α, we have found that radiative transfer with a Voigt profile alters the hydrogen ionization equilibrium above a Lyman-continuum optical depth of about 10^4.

073.024 **Supergranulation-driven Alfvén waves in the solar chromosphere and related phenomena.**
J. V. Hollweg.
Cosmic Electrodynamics, Vol. 2, 423 - 444 (1972).
It has recently been recognized that Alfvén waves frequently dominate the microstructure of the solar wind at the orbit of the earth. We seek a solar source for these waves, and consider here their excitation by the supergranular motions. The interaction of Alfvén wave motions associated with adjacent supergranules is discussed qualitatively. Finally, we advance a theory for spicule formation, in which spicules form as a result of matter being squeezed upward, out of the compression region between adjacent supergranules.

073.025 **Zeeman splitting in some bright prominences.**
E. Wiehr.
Astron. Astrophys., Vol. 18, 79 - 81 (1972).
The Zeeman splitting of the Mg b_2, Na D_2 and Ca^+ 8542 lines in several bright prominences has been determined photographically with $\lambda/4$ plate and Wollaston prism.

073.026 **The location of the continuum emission regions of solar flares.** N. V. Steshenko.
Izv. Krymskoj Astrofiz. Obs., Vol. 43, 130 - 151 (1971). In Russian.
The location of the continuum emission in the spectra of 47 flares observed with the tower solar telescope of the Crimean Astrophysical Observatory is measured. It is found that in most cases the continuum emission of flares appears in other regions of the flare than emission of lines.

073.027 **A comparison of physical conditions in proton and non-proton flares.** N. V. Steshenko.
Izv. Krymskoj Astrofiz. Obs., Vol. 43, 152 - 156 (1971). In Russian.
The available data on physical conditions in 16 proton and 54 non-proton flares are compared. Neither of the main characteristics of proton flares is found to have essential difference in comparison with those of non-proton flares.

073.028 **On the motions of two details in chromospheric flare knots.** M. B. Ogir.
Izv. Krymskoj Astrofiz. Obs., Vol. 43, 157 - 164 (1971). In Russian.
Using spectrograms and Hα-films, the motions of two details of a flare branch in three chromospheric flares are considered. It is shown that the radial velocities of both details of a single flare knot often differ in size and direction. It is pointed out that the complex picture of motions observed in flares, may be due to relative motions of both details and macroscopic inner motions in each of them.

073.029 **The connection of the motions of chromospheric flares and surges with the magnetic field.**
M. B. Ogir.
Izv. Krymskoj Astrofiz. Obs., Vol. 43, 165 - 173 (1971). In Russian.
The motions of flare sprays and loops relative to the transverse magnetic field are discussed. The existence of close conformity between the direction of this motion and that of the transverse component of the magnetic field is shown.

073.030 **Spectrophotometric study of the chromospheric flare on July 8, 1966.** M. M. Musaev.
Izv. AN AzSSR. Ser. fiz.-tekhn. i mat. n., 1971, No. 1, p. 73 - 78. In Russian. – Abstr. in Referativ. Zhurn. 51. Astron., 3.51.489 (1972).

073.031 **A dark band near the solar limb studied with the 53-cm Lyot coronograph.**
V. I. Makarov, Y. V. Platov.
Solnechnye Dannye 1971 Byull., No. 12, p. 72 - 76 (1972). In Russian.
The structure of a dark band near the solar limb has been investigated using filtergrams and spectrograms in the Hα line.

073.032 **A connection of Ca and Hα flocculi and sunspot groups conjugate with them.**
T. M. Schchenikova.
Solnechnye Dannye 1971 Byull., No. 12, p. 84 - 87 (1972). In Russian.

The growth of flocculi areas in Hα and K$_{232}$ Ca II is shown to be synchronous, however it lags behind the evolution of the sunspot group.

073.033 Development and spatial structure of proton flares near the limb and coronal phenomena. IV. Proton flare on November 2, 1969 and its active region.
V. Bumba, L. Křivský, J. Sýkora.
Bull. Astron. Inst.Czechoslovakia, Vol. 23, 85 - 93, 146a (1972).

The proton flare on November 2nd, 1969, generated at about 10 30 UT approximately 8–10° behind the west limb (N 14°) has been investigated. On the basis of examination of flares of the same character and using other results of the author we can deduce certain mechanisms of the acceleration of particles in the complex process of the flare.

073.034 Development and spatial structure of proton flares near the limb and coronal phenomena. V. Emissions and effects of the proton flare and active processes from November 2, 1969.
L. Křivský, A. Tlamicha, J. Halenka, J. Laštovička, P. Tříska, Š. Pintér, J. Ilenčik.
Bull. Astron. Inst. Czechoslovakia, Vol. 23, 94 - 104, 146b (1972).

From the character of the X-ray burst measured on satellites the maximum flux in particular channels and the total radiated energy has been found, and the course of electron temperature and emission has been calculated. The course and character of the radio activity and of ionospheric effects is described. The diffusion of the particles of sub-cosmic radiation has been found. The geomagnetic and ionospheric situation after the proton flare is described.

073.035 Plasma heating by energetic electrons and non-thermal X-ray emission in solar flares.
S. I. Syrovatskii, O. P. Shmeleva.
Astron. Zhurn. Akad. Nauk SSSR, Vol. 49, 334 - 347 (1972). In Russian. English translation in Soviet Astron. AJ, Vol. 16, No. 2.

Power transferred to a plasma by a beam of energetic particles due to collision losses has been estimated as a function of layer depth for power law energy distributions of particles in the beam. The evaluation of the total energy contributed by energetic electrons into a plasma for some known flares has been revised.

073.036 Magnetic coupling of the active chromosphere to the solar interior. P. Foukal.
Astrophys. Journ., Vol. 173, 439 - 444 (1972).

Evidence is summarized to show that the configuration of field lines which governs the appearance of Hα fine structure in active regions is set mainly by motions in the subphotosphere. It is shown that Hα fine structure is directly coupled to a layer probably more than 5000 km below the photosphere, and little distortion of the strong fields is expected in the intervening layers. The shorter rotation period of active regions observed by Howard and others (compared to the photospheric gas) is interpreted as a result of this direct coupling of the strong field to a more rapidly rotating solar interior.

073.037 Structure of the chromosphere-corona transition region. R. L. Moore, P. C. W. Fung.
Solar Physics, Vol. 23, 78 - 102 (1972).

The structure and energy balance of the chromosphere-corona transition region is investigated by means of a static, planar model which is compared with the results of XUV-resonance-line observations. The model explains the observational finding of Noyes et al. (1970) that the number density and the downward heat flux both increase by the same factor from quiet regions to active regions. The implications of these results are discussed with regard to spicules.

073.038 Remark on rotational motions in flares and prominences. Y. Öhman.
Solar Physics, Vol. 23, 134 - 141 (1972).

From material secured with the McMath Solar Telescope of the Kitt Peak National Observatory rotational (orbital) motion has been found in a prominence ejected from a rotating flare. A period of rotation of 32 min has been derived from the study of a periodic asymmetry of the Al I 3961 absorption line.

073.039 A physical mechanism for the production of solar flares. H. K. Sen, M. L. White.
Solar Physics, Vol. 23, 146 - 154 (1972).

The weakly ionized photospheric layer in the sunspot environment satisfies certain dynamo inequalities resulting in photospheric Hall current systems. The corresponding Joule dissipation is associated with the surrounding plage area. For critical values of the 'driving' or convective winds (speeds $\geqslant$ 1 km/s), two stream instability results. The computed energy is of the order of that found in solar flares.

073.040 2700 MHz observations and comments for the flare of October 24, 1969. A. E. Covington.
Journ. Roy. Astron. Soc. Canada, Vol. 66, 74 (1972). – Abstr. Canadian Astron. Soc.

073.041 Metallic abundances in the solar chromosphere. K. Nakayama.
Publ. Astron. Soc. Japan, Vol. 24, 177 - 199 (1972).

Abundances and excitation temperatures of metals in the lower chromosphere are determined as functions of height using the 1962 eclipse data of the joint expedition of HAO and SPO, without the assumptions of local thermodynamic equilibrium and a constant source function. Lower term densities are determined from the self-absorption of flash lines, and estimates are made for the abundances of iron, titanium, chromium, scandium, strontium, barium, and zirconium.

073.042 A bright arch of April 24, 1971.
J. Kubota, T. Tamenaga, Y. Funakoshi, T. Kureizumi.
Publ. Astron. Soc. Japan, Vol. 24, 281 - 286 (1972).

A brief description is given of a bright arch prominence that appeared outside the west limb on April 24, 1971.

073.043 Macroscopic motions in prominences. I: The prominence of 26th March, 1971.
M. E. Machado, H. Grossi Gallegos.
Solar Physics, Vol. 23, 340 - 345 (1972).

Macroscopic velocity fields have been studied in a solar prominence. The spectra and monochromatic images were analyzed, and the existence of a contracting motion, possibly due to a pinch effect, is discussed. A helical shape of the prominence is proposed.

073.044 The internal motion of quiescent prominences. O. Engvold.
Solar Physics, Vol. 23, 346 - 352 (1972).

A study has been made of fine structure wavelength shift in the K line spectra from quiescent prominences. A persistent small scale motion is found in the prominence main body.

073.045 Analysis of two active prominences. M. E. Machado.
Solar Physics, Vol. 23, 353 - 359 (1972).

We discuss the observations of two eruptive prominences, and the formation of condensations during the phenomena. The density and intensity variations of the condensations are analyzed spectroscopically in one of the events. Some hypothesis about the magnetic field configuration have been used in order to explain the observational data.

073.046 **Comments on filament-disintegration and its relation to other aspects of solar activity.**
H. W. Dodson, E. R. Hedeman, M. R. de Miceli.
Solar Physics, Vol. 23, 360 - 368 (1972).

Studies of 'disparitions brusques' in solar cycles 19 and 20 (to 1969) indicate that such events occur frequently. Approximately 30 % of all large filaments in these cycles disintegrated in the course of their transit across the solar disk. Relationships between a disintegrating filament on July 10–11, 1959, a prior major flare, a newly formed spot, and concomitant growth of Hα plage are presented.

073.047 **A classification of magnetic field configurations associated with solar flares.** P. A. Sturrock.
Solar Physics, Vol. 23, 438 - 443 (1972).

On the assumption that solar flares are due to instabilities which occur in current sheets in the sun's atmosphere, one may classify magnetic-field configurations associated with flares into two types. One is characterized by 'closed', the other by 'open' current sheets. Flares associated with open current sheets can produce type III radio bursts and high-energy-particle events, but flares associated with closed current sheets cannot.

073.048 **Properties of white light flares. I: Association with Hα flares and sudden frequency deviations.**
P. S. McIntosh, R. F. Donnelly.
Solar Physics, Vol. 23, 444 - 456 (1972).

Solar flares that emit light throughout the optical continuum, white light flares, are seldom observed. This paper summarizes properties of white light flares, including their relationship to EUV emission as deduced from sudden frequency deviations.

073.049 **Location of the electron acceleration region in solar flares.** S. R. Kane, R. P. Lin.
Solar Physics, Vol. 23, 457 - 466 (1972).

Observations of impulsive solar flare X-rays $\gtrsim$ 10 keV by the OGO-5 satellite and the measurements of energetic solar electrons made with the Explorer-35 and Explorer-41 (IMP-5) satellites during the period March 1968–September 1969 have been analyzed in order to determine the ion density in the X-ray source region as well as the location of the electron acceleration region in the solar atmosphere.

073.050 **Solar flares and solar wind helium enrichments: July 1965–July 1967.**
J. Hirshberg, S. J. Bame, D. E. Robbins.
Solar Physics, Vol. 23, 467 - 486 (1972).

We have studied the 43 spectra with He/H $\geqslant$ 15 % that were observed among 10300 spectra collected by Vela 3 between July 1965–July 1967. The 43 spectra were distributed among 16 distinct periods of helium enhancement, 12 of which (containing 75 % of the spectra) were associated with solar flares. Six new flare-enhancement events are discussed in this paper. It is concluded that the association of helium enhancements with major flares is real, non-random and very strong.

073.051 **Some peculiarities of development of active regions with proton flares.** V. P. Kuleshova.
Geomagn. Aeronom., Vol. 12, 328 - 329 (1972). In Russian. Brief information.

073.052 **Sonneneruptionen.** E. Obreschkow.
Orion Schaffhausen, 30. Jahrgang, p. 39 - 44 (1972).

073.053 **Investigation of the H and K Ca II lines in the solar chromosphere.** A. Sadikov.
Tsirk. Astron. Inst., *Tashkent*, No. 28 (375), p. 11 - 19 (1971). In Russian.

073.054 **A "ducky" prominence spectrum.**
W. C. Livingston.
Mercury, (Journ. Astron. Soc. Pacific), Vol. 1, No. 3, p. 4 - 6 (1972).

073.055 **Solar flare time development: Three phases.**
S. I. Syrovatskii.
Comments Astrophys. Space Phys., Vol. 4, 65 - 70 (1972).

The development of a solar flare in time is not gradual. Instead, it is interrupted by the periods of rapid, explosion-like changes in many flare characteristics. We discuss the three stages or phases in flare development: initial, explosive, and decay phases.

073.056 **The chromosphere in continuum emission observed at the total solar eclipse on 7 March 1970.**
M. Makita.
Solar Physics, Vol. 24, 59 - 71 (1972).

Direct images of the sun were photographed in continuum emission centered at 6900 Å by the jumping film method near the second contact of the Mexico eclipse on 7 March 1970. The intensity distribution of the solar outer layers obtained shows a steep decrease by a factor of 0.9 in logarithmic units around 2500 km. Spicules observed at 3500 km are explained by log n_e = 11.25 and $T_e \approx$ 6000 K. Discussions are made in relation to the other observations and some chromosphere models.

073.057 **Hα mottles.** C. Sawyer.
Solar Physics, Vol. 24, 79 - 86 (1972).

Mottles located in the dark lanes that form the Hα chromospheric network are longer lived than those that lie within network cells. The two groups of mottles are similar in size and shape. Bright mottles are found to be slightly larger than dark ones, with a different distribution of lifetimes, but no difference in shape or in number of mottles.

073.058 **Intensity oscillation in Hα-fine structure.**
A. Bhatnagar, K. Tanaka.
Solar Physics, Vol. 24, 87 - 97 (1972).

Using a new technique of directly measuring the intensity variation from the 16 mm time-lapse filtergram movies taken in the blue wing and in the line center of Hα, we found periodic intensity oscillations in the center of Hα-supergranulation network, in rosette centers and in plage granules. Oscillatory intensity fluctuations have been also observed in the sunspot umbra.

073.059 **The relations between chromospheric features and photospheric magnetic fields.** E. N. Frazier.
Solar Physics, Vol. 24, 98 - 112 (1972).

High resolution photographic magnetograms are compared with Hα filtergrams (both on - and off - band) for a wide variety of solar features. It is verified that Hα filaments overlie neutral lines or bands and that Hα plages always occur at magnetic field clumps. However, the brightness of Hα plages bear no relation to magnetic field strength or polarity, and the direction of the magnetic field with respect to threads and filaments remains obscure. Basic questions about the usefulness and final research goal of filtergrams and magnetograms are raised.

073.060 **EUV observations of the chromospheric network.**
E. M. Reeves, W. H. Parkinson.
Solar Physics, Vol. 24, 113 - 117 (1972).

Extreme ultraviolet observations of a quiet region of the sun on August 18, 1969, with the Harvard spectroheliometer on OSO 6 indicate that the chromospheric network can be observed in lines of the chromosphere and transition region with almost identical structure. At coronal heights, the network changes but some residual structure can still be discerned in Mg X and perhaps Si XII.

073.061 On the relation between filaments (prominences) and Hα loops. A. Bruzek.
Solar Physics, Vol. 24, 118 - 122 = Mitt. Fraunhofer Inst., *Freiburg*, No. 112 (1972).

The relation between occurrence of Hα loops and filaments is discussed on the occasion of the observation of a new type of transient loops during a flare associated filament activation. Considering all known types of loop systems crossing neutral lines it is concluded that concurrent existence of stable filaments and Hα loops is incompatible.

073.062 On polarimetry in solar active regions. V. The magnetic field immediately before and after a flare.
E. Wiehr.
Solar Physics, Vol. 24, 129 - 132 (1972).

High resolved magnetograms were obtained 3 hrs before and 1 hr after a 1b flare, respectively, the only bright flare reported for that active region. Careful comparison between both magnetograms shows that the line-of-sight component of the active region magnetic field remains constant.

073.063 Polarization of solar active regions at 9.5 mm wavelength. M. R. Kundu, T. P. McCullough.
Solar Physics, Vol. 24, 133 - 141 (1972).

A study of the circular polarization structure of solar active regions has been made from data obtained at 9.5 mm wavelength, using the 85 ft reflector and polarimeter at the Naval Research Laboratory Maryland Point Observatory. All important active regions observed at 9.5 mm are bipolar in nature, the degree of polarization is about the same for both right and left circular components and it ranges up to about 4%.

073.064 Spectral analysis of highly inhomogeneous chromospheric flares. Z. Svestka.
Solar Physics, Vol. 24, 154 - 168 (1972).

Analysis of a hydrogen flare spectrum is carried out, assuming that the flare radiation is diluted due to a highly inhomogeneous space distribution of the flare elements in the chromosphere. It is shown that one obtains the correct physical parameters in the flare elements, irrespective of the extent of dilution, if all the elements are optically thin in the spectral regions considered. However, if this is not true for all the elements, the parameters deduced are in error, which increases with the extent of dilution.

073.065 Models of the chromosphere corona transition region. P. K. Raju.
Mitt. Astron. Ges., No. 31, p. 154 (1972). − Abstract.

073.066 Investigation of the chromosphere in the D_3He line during the eclipse on September 22, 1968.
I. L. Belkina, N. P. Dyatel.
Astron. Zhurn. Akad. Nauk SSSR, Vol. 49, 588 - 594 (1972). In Russian. English translation in Soviet Astron. AJ, Vol. 16, No. 3.

Results of observations of the D_3He line for active and quiet regions of the sun during the total eclipse of September 22, 1968 are given. The height distribution of the surface brightness of the lower and mean chromosphere is obtained. An additional maximum of the surface brightness of the chromosphere is discovered at the height near 1500 km in some measured regions.

073.067 The chromospheric continuum observed at the total solar eclipse of 12 November 1966 and a model of the lower chromosphere. K. Tanaka, E. Hiei.
Publ. Astron. Soc. Japan, Vol. 24, 323 - 341 = Tokyo Astron.

Obs. Repr., No. 414 (1972).

We have obtained the Balmer continuum and the continuum at 4000 Å in the chromosphere by observing the total solar eclipse of 12 November 1966. Existing chromospheric models are compared with comments. By using this eclipse data and the continuum at 6900 Å observed by Makita (1971) at the total solar eclipse of 7 March 1970, we have constructed a model chromosphere, which can explain the eclipse observations and also the observations of UV and mm radiation.

073.068 Intensity ratios of He I and H lines in a prominence and the chromosphere.
J. Kubota, T. Tamenaga, K. Yoshikawa.
Publ. Astron. Soc. Japan, Vol. 24, 343 - 354 (1972).

The intensity ratios $I(D_3)/I(H\beta)$ and $I(D_3)/I(6678)$ in a prominence observed at the 1970 eclipse are studied by comparing monochromatic images in these lights. Based on Ishizawa's (1972) calculation of the hydrogen model and the new calculation on He I with seven discrete levels and the continuum, it is shown that the increase of the intensity ratio $I(D_3)/I(H\beta)$ can be interpreted as being caused by the decrease of the total number density of prominence matter in the faint part. Using the data at the 1962 eclipse (Dunn et al. 1968), the intensity ratios $I(D_3)/I(H\beta)$ and $I(D_3)/I(6678)$ in the chromosphere are briefly discussed in comparison with those in quiescent prominences.

073.069 The chromosphere−corona transition region. R. G. Athay.
Astrophys. Space Sci. Library, Vol. 27, (see 012.011), 36 - 65 (1971).

073.070 Observational effects of flare-associated waves. S. F. Smith, K. L. Harvey.
Astrophys. Space Sci. Library, Vol. 27, (see 012.011), 156 - 167 (1971).

073.071 The surges. C. J. Macris.
Astrophys. Space Sci. Library, Vol. 27, (see 012.011), 168 - 178 (1971).

073.072 Relations between the areas index and different phenomena in the chromosphere, the corona and the interplanetary space. J. Xanthakis.
Astrophys. Space Sci. Library, Vol. 27, (see 012.011), 179 - 191 (1971).

073.073 Models of the quiet and active solar atmosphere from Harvard OSO data. R. W. Noyes.
Astrophys. Space Sci. Library, Vol. 27, (see 012.011), 192 - 218 (1971).

073.074 The determination of chromospheric-coronal structure from solar XUV observations.
C. Jordan, R. Wilson.
Astrophys. Space Sci. Library, Vol. 27, (see 012.011), 219 - 236 (1971).

073.075 Note on a recent identification on the solar flare 1.9 Å line feature. K. J. H. Phillips.
Astrophys. Space Sci. Library, Vol. 27, (see 012.011), 254 - 256 (1971).

073.076 Models of the solar transition region chromosphere−corona. G. Noci.
Astrophys. Space Sci. Library, Vol. 27, (see 012.011), 308 - 316 (1971).

073.077 Particle acceleration and plasma ejection from the sun. S. I. Syrovatsky.

Astrophys. Space Sci. Library, Vol. 29, Part I, (see 012.012), p. 119 - 133 (1972).

073.078 Methods for the forecasting of solar flares.
A. B. Severny, N. V. Steshenko.
Astrophys. Space Sci. Library, Vol. 29, Part I, (see 012.012), p. 173 - 181 (1972).

073.079 Stationäre Filamente, ihre koronale Umgebung sowie ihre Beziehung zum solaren Magnetfeld.
W. Stanek.
Abhandlung zur Erlangung der Würde eines Doktors der Naturwissenschaften der Eidgenössischen Technischen Hochschule Zürich.
Diss. No. 4795. Juris Druck + Verlag, Zürich. 3 + 88 pp. Price DM 16.00 (1971).

After calibration of the intensity scale used at Arosa by comparison with the measurements in absolute units of Pic du Midi Observatory the intensity contours over prominences are evaluated and discussed. Model calculations for the distribution of the electrons in the surroundings of prominences up to 2 solar radii were made, based on photographs of solar streamers in integral light. A theoretical model based on the assumption to have not only radial components of the magnetic field in the streamer but also tangential ones leads to a theoretical density distribution in the streamer parallel to the solar surface in good agreement with the observed values. Based on maps of the solar magnetic field and on the intensity distribution of the coronal line 6374 A, so-called "streets of prominences" can be postulated.

073.080 On the relationships between bright mottles and spicules of the solar chromosphere.
C. E. Alissandrakis, C. J. Macris.
Praktika Acad. Athens, Vol. 46, 107 - 115 = Contr. Res. Center Astron., Applied Math., Acad. Athens, Ser. 1 (Astron.), No. 26 (1971).

Isophotometry of Hα photographs of the solar limb reveals that in most cases bright mottles appear at spicule roots. Discussion and comparison with related limb and disk observations follow.

073.081 Chromospheric flares as sources of directed corpuscular streams. A. T. Nesmyanovich, N. G. Ul'yanich.
Astron. vestn., Vol. 6, 3 - 6 (1972). In Russian. – Abstr. in Referativ.Zhurn. 51. Astron., 7,51.383 (1972).

073.082 Solar spicules and chromospheric heating.
R. N. Thomas.
Topics in modern physics, [Colorado Associated University Press, Boulder, Colo.], p. 331 - 339 (1971).

073.083 Étude expérimentale et théorique de la région de transition chromosphère-couronne solaire par l'utilisation des longueurs d'onde radioélectriques. P. Lantos.
Thesis, Univ. Paris. Centre Documentation, C. N. R. S. (1971-11-26), 30 pp. (1971). – See Bull. Signal., Vol. 33, Section 120, No. 3786 (1972).

073.084 Contributions à l'étude de la chromosphère solaire. Influence de la dissipation sur la remontée de température chromosphérique et couplage entre la conduction thermique et le transfert radiatif. H. Frisch.
Thesis, Sci. Phys., Univ. Paris. Centre Documentation, C. N. R. S., (1972-04-12), 40 pp. (1971). – See Bull. Signal., Vol. 33, Section 120, No. 6992 (1972).

Physics of solar prominences. Report on an international colloquium held at the German Solar Observatory, Anacapri, September 29 to October 1, 1971. See Abstr. 011.034.

Ratio of line intensities in helium-like ions as a density indicator. See Abstr. 062.004.

Radiative relaxation of sound waves in an optically thin isothermal atmosphere. See Abstr. 064.017.

Spectral lines from photosphere to chromosphere, observed during the March 1970 eclipse; A first comparison with theory. See Abstr. 071.003.

The field of velocities and brightness in the solar atmosphere. See Abstr. 071.018.

Measurements of the limb darkening in the forbidden MgI line at 4571.1 Å. See Abstr. 071.020.

Some new DyII identifications in the solar spectrum. See Abstr. 071.039.

Note on the characteristics of sunspot groups which produce solar proton flares. See Abstr. 072.020.

Photographische Sonnenbeobachtungen eines Amateurs. See Abstr. 072.028.

The active longitudes of sunspots and solar flares in the 20th cycle. See Abstr. 072.030.

Local magnetic fields on the sun. See Abstr. 072.037.

Properties of solar active regions. See Abstr. 072.038.

On the coronal lines in the chromosphere at the 1970 eclipse. See Abstr. 074.007.

Spectrophotometry of the corona and a quiescent prominence based on observations of the total solar eclipse of 7 March, 1970 in Mexico. See Abstr. 074.009.

Some newly discovered coronal emission lines from high altitude infrared observations of the 7 March, 1970, solar eclipse. See Abstr. 074.011.

A model for the chromosphere-corona transition region based on radio observations and on hydrodynamical conservation equations. See Abstr. 074.026.

Coronal active regions and flare-associated events. See Abstr. 074.086.

The 1–55 Å X-ray emission from an active limb prominence. See Abstr. 076.021.

A search for high energy gamma-rays from solar active regions. See Abstr. 076.023.

Solar flares in the extreme ultraviolet. I. The observations. See Abstr. 076.030.

Solar flares in the extreme ultraviolet. II. Comparisons with other observations. See Abstr. 076.031.

X-ray spectroscopy of solar active regions and flares. See Abstr. 076.033.

Struktur solarer Aktivzonen im mm-Wellengebiet. See Abstr. 077.042.

High energy electrons detected during solar flares.

See Abstr. 078.012.

Evidence for solar particle production above ~75 GeV.   See Abstr. 078.022.

Formation of the solar Hα profile.
See Abstr. 080.005.

Ionospheric effects of solar flares—I. The statistical relationship between X-ray flares and SID's.
See Abstr. 083.020.

Ionospheric effects of solar flares—II. The flare spectrum below 10 Å deduced from satellite observations.

See Abstr. 083.021.

Ionospheric effects of solar flares—III. The quantitative relationship of flare X-rays to SID's.
See Abstr. 083.022.

Ionospheric effects of solar flares—IV. Electron density profiles deduced from measurements of SCNA's and VLF phase and amplitude.   See Abstr. 083.023.

Ionospheric effects of solar flares—V. The flare event of 30 January 1968 and its implications.
See Abstr. 083.024.

074   Solar Corona, Solar Wind

074.001   **Solar radio recombination lines.**
P. S. Berger, M. Simon.
Astrophys. Journ., Vol. 171, 191 - 200 (1972).
We have carried out a search at 85–92 GHz for solar recombination lines, particularly those in which dielectronic recombination is expected to overpopulate the high-n states. We observed at the frequencies of hydrogen and helium, and all ions up to $Z = 15$. We have used the dielectronic recombination rates of Shore to revise the theoretical estimates, and find the revised estimates to be consistent with our observations. We develop criteria for selecting detectable ions, and estimate the strengths of their lines in the radio and far-infrared regions.

074.002   **The termination of the solar wind.**
T. R. McDonough, N. M. Brice.
Icarus, Vol. 15, 505 - 510 (1971/72).
Analytic solutions for the solar wind flux and velocity as a function of distance have been obtained, yielding estimates of the slowing down and termination distance as a function of the various input parameters (solar wind flux, velocity, solar Lyman-continuum flux, interstellar hydrogen density and velocity, and magnetic field strength).

074.003   **Diffuse external reinforcements in the solar corona from the 1970 March 7 solar eclipse plates.**
S. Koutchmy.
Astron. Astrophys., Vol. 16, 103 - 107 (1972). In French.
Observations of diffuse external reinforcements (R.E.D.) during 1970 March 7 total eclipse are reported. The reality of the R.E.D. is demonstrated by the agreement between the results obtained using different methods and by different authors. The morphology and brightness distribution are studied.

074.004   **The coronal and interplanetary magnetic fields at the time of the solar eclipse of 7 March, 1970.**
J. O. Stenflo.
Solar Physics, Vol. 21, 263 - 271 (1971).
The field-line maps are compared with eclipse photographs showing coronal structures out to about $12\,r_\odot$. The projected field lines as well as the observed streamers appear straight. The calculations show that the angular velocity of the coronal plasma decreases rapidly with distance. The relation between magnetic fields and density enhancements is discussed. The calculations show a sector structure of the interplanetary field, which agrees well with spacecraft observations.

074.005   **Coronal electron density maps for 7 March, 1970, derived from MgX λ625 spectroheliograms.**
G. L. Withbroe, A. K. Dupree, L. Goldberg, M. C. E. Huber, R. W. Noyes, W. H. Parkinson, E. M. Reeves.
Solar Physics, Vol. 21, 272 - 280 (1971).
We have analyzed daily MgX λ625 spectroheliograms for a 28-day period centered on 7 March, 1970. These data are used to construct maps of the variation across the solar disk of the electron density at the base of the corona. The correspondence of high and low density regions with regions of enhanced and reduced emission in white light and MgX pictures made during or near the time of the eclipse are described.

074.006   **Fine structure in the inner corona observed at the 1970 eclipse.**
T. Tsubaki, H. Kurokawa, M. Kanno.
Solar Physics, Vol. 21, 305 - 313 (1971).
On the slit spectrogram obtained at the 1970 eclipse in Mexico, the intensities of four coronal lines (Ni XIII λ5116, Fe XIV λ5303, Fe X λ6374, and Ni XV λ6702) and the continuum were measured as a function of distance along the slit. It is found that there exist a lot of fine fluctuations both in the lines and in the continuum intensities superposed on a large scale formation.

074.007   **On the coronal lines in the chromosphere at the 1970 eclipse.**   M. Kanno, T. Tsubaki, H. Kurokawa.
Solar Physics, Vol. 21, 258, 314 - 324 (1971).
The integrated intensities of Fe XIV λ5303, Fe X λ6374, and the continuum were measured on the spectrograms as a function of height above the sun's limb. It was found that a large amount of emission in the coronal lines originates in the interspicular regions of the chromosphere. Analysis of the data yielded that the interspicular regions consist of coronal material of $T_e = 1.6 \times 10^6 - 1.2 \times 10^6$ and $\log N_e = 8.5–9.5$, and that a decrease in T_e and an increase in N_e occur with decreasing height.

074.008   **The physical conditions in inner corona derived from spectral data of the solar eclipse on 7 March, 1970.**   E. A. Gurtovenko, K. V. Alikayeva.
Solar Physics, Vol. 21, 325 - 331 (1971).
From the analysis of the line halfwidths it is concluded that the emission of coronal lines of various groups arises in different volumes of the corona. The lines λ 5303 and λ4231 are analyzed in detail and by the use of the equations of excitation and ionization balance. The distribution of electron temperature T_e and density n_e vs height in the solar corona has been found.

074.009   **Spectrophotometry of the corona and a quiescent prominence based on observations of the total solar eclipse of 7 March, 1970 in Mexico.**
G. M. Nikolsky, R. A. Gulyaev, K. I. Nikolskaya.
Solar Physics, Vol. 21, 332 - 350 (1971).
Slit spectrograms of a quiescent prominence and the inner corona ($h \leqslant 2.5$ arc min) in the range λλ3400–7000 Å were obtained. The electron density in the prominence $n_e = (7 \pm 3) \times 10^{10}\,\text{cm}^{-3}$ was deduced. The kinetic temperature T_κ and the non-thermal velocities v_t are $T_\kappa \approx 10000$ K and $v_t \approx 6$ km/s. In the coronal spectrum 24 coronal lines were found. 13 of these lines were identified. The line-of-sight and non-thermal velocities are $v_r \leqslant 10$ km/s and $v_t \approx 25$ km/s. The coronal lines originate in at least three types of regions with different temperatures. The emission measure as a function of the ionization temperature was determined. The abundances of V, Cr, Mn, Co were estimated. The degree of inhomogeneity in the corona was estimated: $\overline{n_e^2}/(\bar{n}_e)^2 \approx 3 - 10$.

074.010   **Observations of the infrared Fe XIII lines in the solar corona of 12 November, 1966.**
P. L. Byard, K. E. Kissell.
Solar Physics, Vol. 21, 351 - 359 (1971).
These are the first eclipse observations of the infrared Fe XIII lines which indicate that proton collisions are important in the excitation of the coronal lines. The results are compared with the recent theoretical calculations of Chevalier and Lambert. Our observations are consistent with an electron density of 4×10^8 in the inner corona. The observed equivalent widths of the lines lie in the range 10 to 30 Å for the 10747 line and 5 to 12 Å for the 10798 line. The ratio of these equivalent widths is found to vary from 2.3 in the inner corona to 6 at a point 1.36 solar radii from the center of the sun.

074.011   **Some newly discovered coronal emission lines from**

high altitude infrared observations of the 7 March, 1970, solar eclipse.
K. H. Olsen, C. R. Anderson, J. N. Stewart.
Solar Physics, Vol. 21, 360 - 371 (1971).

A survey of the infrared coronal spectrum between 1 μ and 3 μ was made. In addition to well known chromospheric lines of HI and HeI, nine additional lines were seen. Evidence is presented for the tentative assignment of these lines to forbidden transitions in highly ionized atoms of magnesium, aluminum, silicon, sulphur, and chromium.

074.012 The identification of new forbidden coronal lines in the solar EUV spectrum. C. Jordan.
Solar Physics, Vol. 21, 381 - 391 (1971).

Identifications are proposed for 20 of the 28 coronal lines observed in the spectra obtained during a rocket flight into the path of the 7 March, 1970 solar eclipse. Most of the lines identified are from forbidden transitions between levels in the ground $2p^n$ and $3p^n$ configurations in high ions of magnesium, silicon, sulphur, iron, and nickel. The temperature range represented is from 6.9×10^5 K to 2.5×10^6 K.

074.013 Measurements on the Lyman alpha corona.
A. H. Gabriel.
Solar Physics, Vol. 21, 392 - 400 (1971).

During the solar eclipse on 7 March, 1970, measurements have been made of the brightness of Lyman α from the corona, at heights between 5×10^4 and 5×10^5 km above the limb. The emission is shown to occur primarily through the resonance scattering of chromospheric Lyman α from the residual neutral hydrogen in the corona.

074.014 The relation between the white light and XUV coronas on 7 March, 1970.
R. Tousey, M. J. Koomen.
Solar Physics, Vol. 21, 401 - 407 (1971).

The white-light corona from 3–9 R_s and the XUV (170–500 Å) corona are compared with the X-ray corona, the H Ly-α corona, the Fe XIV 5303 Å corona, and total eclipse photographs in white-light and infrared.

074.015 Rocket-coronagraph photometry of the 7 March, 1970 corona from 3 to 8.5 R_s.
J. D. Bohlin, M. J. Koomen, R. Tousey.
Solar Physics, Vol. 21, 408 - 417 (1971).

On the basis of our analyses we conclude that the quiet equatorial and polar K + F corona of the 1970 eclipse is nearly identical to the model advanced by Blackwell. Our streamer model is not particularly unique, but does demonstrate how much variation can result in derived densities due to the choice of the geometrical model.

074.016 A polarization-color effect in the K-corona.
D. E. Billings, Young Oh.
Solar Physics, Vol. 21, 418 - 424 (1971).

An observation of the corona during the 7 March, 1970 eclipse through a Wollaston prism-red and blue filter combination was carried out for the purpose of confirming the prediction that the component of the K-corona with electric vector radial to the sun is redder than the tangential component. An analysis of a portion of the resulting data seems to confirm this prediction.

074.017 Results of polarization observations of the outer corona from a jet aircraft. C. F. Keller.
Solar Physics, Vol. 21, 425 - 429 (1971).

White-light photographs of the solar corona were taken during the March 1970 eclipse at an altitude of 36380 ft. The corona was recorded to distances beyond 12 $R_\odot$. A preliminary study of per cent polarization as a function of position with respect to the solar disk for a set of 1.0 s exposures shows an inversion in per cent polarization in the region 6 to 8 $R_\odot$ – polarization decreasing outward to the region and increasing again beyond it. This inversion is most apparent along the major streamers.

074.018 Photometric intensity and polarization measurements of the solar corona. D. S. McDougal.
Solar Physics, Vol. 21, 430 - 438 (1971).

During the 7 March, 1970, total eclipse of the sun, simultaneous measurements were made of the two orthogonal components of coronal light in the B and R bands of the UBVRI system. For the first time, simultaneous multicolor intensity, degree, and angle of polarization profiles are computed from photoelectric measurements. Comparison of the variations of the measurements for each spiral scan yield a detailed picture of the intensity and polarization features in the K corona.

074.019 The polarization of coronal emission lines.
J. M. Beckers, W. J. Wagner.
Solar Physics, Vol. 21, 439 - 447 (1971).

During the 7 March, 1970 solar eclipse we attempted to measure both the amount and direction of linear polarization of all emission lines between 3400 and 9000 Å in the inner corona ($1.034 \leqslant r/r_0 \leqslant 1.085$). Only the green and red coronal lines have been analyzed in detail. Neither of these lines shows polarization exceeding the probable error of 1.0% for λ5303 and 1.8% for λ6374. None of the other 17 coronal lines show any obvious (>5%) polarization.

074.020 Interferometric studies of spectral lines in the solar corona.
J. G. Hirschberg, A. Wouters, L. Hazelton, Jr.
Solar Physics, Vol. 21, 448 - 451 (1971).

A photographic Fabry-Perot interferometer was used to measure the breadth and wavelength of the Fe XIV spectral line at 5303 Å in the solar corona, during the eclipse of 7 March, 1970, in Mexico.

074.021 Correlation of photospheric magnetic field strength with coronal brightness on 7 March, 1970.
G. C. J. Suffolk, S. M. Smith.
Solar Physics, Vol. 21, 481 (1971). – Abstract.

074.022 Analyse des renforcements coronaux à travers quelques acquisitions spectroscopiques récentes des émissions monochromatiques du fer ionisé (X à XV).
J. P. Rozelot.
Solar Physics, Vol. 22, 88 - 113 (1972).

Theoretical computations of the coronal spectrum have been performed and compared with observations carried out at the Pic du Midi Observatory. A possible explanation can be given of some anomalies found in the behaviour Fe X, indicator of young active centers. A strong correlation between the intensity of Fe XI and Fe XIV indicates high temperature regions. Fe XIII is on the contrary a sensitive indicator of strong electronic density regions. The case of Fe XII is discussed apart. The non-uniformity of temperature and density is studied along the line of sight. It is shown that a gaussian distribution of the density, together with a bi-squared variation of the temperature fits best with our observations.

074.023 Note on the helium-like ion line emission in solar plasmas. R. Mewe.
Solar Physics, Vol. 22, 114 - 118 (1972).

An analysis is presented of the rate coefficients occurring in the Gabriel-Jordan theory on the relative intensities of the forbidden, intercombination, and resonance lines of helium-like ions in a steady-state plasma. Simple expressions are given to show the dependence on atomic number and electron temperature. The influence of proton collisions on the excitation $2^3S \rightarrow 2^3P$ is estimated and deviations from the theory under non-equilibrium conditions are briefly discussed.

074.024 **The interpretation of total line intensities from optically thin gases. II. The coronal forbidden lines.**
J. T. Jefferies, F. Q. Orrall, J. B. Zirker.
Solar Physics, Vol. 22, 317 - 326 (1972).

We discuss the application of a general diagnostic procedure, developed in part I of this series, to the inference of the physical state of coronal condensations from a knowledge of their forbidden line emission.

074.025 **The interpretation of total line intensities from optically thin gases. III. Application to coronal forbidden line spectra.**
J. T. Jefferies, F. Q. Orrall, J. B. Zirker.
Solar Physics, Vol. 22, 327 - 343 (1972).

The diagnostic method developed in the two preceding papers of this series is applied to coronal forbidden line intensity data obtained at eclipses in 1952, 1961, 1965, 1966, and 1970. The application of the method is limited by the nature of the data but allows a first inference of the relationship between electron density and temperature in the condensations observed at these eclipses, and of the distribution of the electrons within the temperature range samples by the observations – effectively 10^6 to 2.4×10^6 K. We determine the relative abundance of nickel to iron in the corona.

074.026 **A model for the chromosphere-corona transition region based on radio observations and on hydrodynamical conservation equations.** P. Lantos.
Solar Physics, Vol. 22, 387 - 401 (1972).

6- and 11-cm maps obtained at cycle maximum are used in combination with existing measurements of the central brightness temperatures at centimetric wavelengths to obtain an equatorial, quiet solar atmosphere model between 10 000 K and 300 000 K in the interspicular regions. This model introduces a large ascending velocity in these regions. Some of the consequences of the model concerning the heating and replenishment of the corona are discussed. An explanation of sudden disappearance of quiescent prominences is suggested.

074.027 **Differential rotation and the structure and energy content of coronal magnetic fields.**
M. A. Raadu.
Solar Physics, Vol. 22, 443 - 449 (1972).

It is argued that differential rotation of the photospheric magnetic fields will induce currents in the corona. The work done against surface magnetic stresses will increase the energy content of the coronal magnetic field. The electrical conductivities are high and the foot points of field lines move with the differential rotation. The force-free field equations are solved with this constraint to obtain a minimum estimate of the energy increase for a quadrupole field. During a solar rotation the magnetic energy increases by 25%.

074.028 **The observation of an eruptive phenomenon in the solar corona.** Ts. S. Khetsuriani, E. I. Tetruashvili.
Astron. Zhurn. Akad. Nauk SSSR, Vol. 49, 213 - 214 (1972).
In Russian. English translation in Soviet Astron. AJ, Vol. 16, No. 1.

The emergence and disappearance of a coronal phenomenon is observed in red and green coronal lines, having radial velocities of 15 km/sec and 12 km/sec respectively. The phenomenon is identified with a surge observed in Hα filtograms.

074.029 **On the domains of existence of the three types of supersonic solutions of the inviscid solar-wind equations.** B. R. Durney, N. Werner.
Astrophys. Journ., Vol. 171, 609 - 613 (1972).

The approximate energy equation for radial distances larger than the critical point is solved with "critical point"

boundary conditions. This clarifies the domains of existence of the Parker, Whang and Chang, and $r^{-4/3}$ supersonic solutions of the solar-wind equations.

074.030 **Transfer characteristics of solar radiation in a scattering corona.**
C. Caroubalos, M. Aubier, Y. Leblanc, J. L. Steinberg.
Astron. Astrophys., Vol. 16, 374 - 378 (1972).

The purpose of this paper is to estimate the effects of coronal inhomogeneities in electron density on the propagation of the radio radiation from a white spectrum point source observed from different directions. We present the results of a study including the effects of coronal scattering.

074.031 **Abundances in the solar corona.**
K. S. de Boer, H. Olthof, S. R. Pottasch.
Astron. Astrophys., Vol. 16, 417 - 430 (1972).

On the basis of the coronal measurements of the 30 May 1965 eclipse, and using improved atomic parameters, we discuss the coronal abundance determination. Two new line identifications are suggested and several existing suggestions for possible identifications are discussed on the basis of predicted intensities and abundances. Finally an extensive list is given of lines which are possibly observable, together with predicted intensity and equivalent width.

074.032 **Observation of the green line with an electronic camera during the eclipse of March 7, 1970.**
B. Fort, J. P. Picat, M. Combes, P. Felenbok.
Astron. Astrophys., Vol. 17, 55 - 59 (1972).

We have studied the west part of the solar corona using an electrostatic camera derived from the Lallemand type. Results given here are the brightness measured in the 5303 Å line and in the neighbouring continuum up to $\rho = 1.7$ solar radii from the center of the sun. The electron density and mean temperature derived from the ionization degree are given for a real height h in the condensation and above the surface of the sun from $0.2\,R_\odot$ to $0.9\,R_\odot$.

074.033 **Collisionless solar wind in the spiral magnetic field.**
W. M. Chen, C. S. Lai, H. E. Lin, W. C. Lin.
Journ. Geophys. Res., Vol. 77, 1 - 11 (1972).

The two-fluid model for the solar wind of Hollweg (1970) is reconsidered with the inclusion of the spiral structure of the interplanetary magnetic field. In the present model, the protons are assumed to become collisionless beyond 0.1 AU from the sun, whereas the electrons are treated hydrodynamically and the electron temperature is supposed to obey the polytropic law.

074.034 **Solar-wind speed variations 1964–1967: An autocorrelation analysis.** J. T. Gosling, S. J. Bame.
Journ. Geophys. Res., Vol. 77, 12 - 26 (1972).

An autocorrelation analysis has been performed on the Vela 2 and 3 solar-wind speed data obtained from July 1964 through December 1967 with two goals: to establish the degree to which speed structures recur from one solar rotation to the next, and to examine the duration and origin of 'persistence' in the solar-wind speed. The results are described in detail.

074.035 **Structure and orientations of solar-wind interaction fronts: Pioneer 6.** G. L. Siscoe.
Journ. Geophys. Res., Vol. 77, 27 - 34 (1972).

An analysis is given of five stream-stream interaction events observed in the Pioneer 6 plasma and field data. That the time profiles of all the parameters are consistent among the events and with previous descriptions given in the literature substantiates the notion of a common interaction type. This study is largely restricted to an exposition of certain synoptic aspects of the stream-stream interaction. The only dynamics concerns the origin of the zonal deflections of the flow.

074.036 Solar-wind structure determined by corotating coronal inhomogeneities. 2. Arbitrary perturbations.
G. L. Siscoe, L. T. Finley.
Journ. Geophys. Res., Vol. 77, 35 - 45 (1972).

The problem of the solar-wind structure resulting from long-lived inhomogeneities in the solar corona has been extended within the framework of a linearized hydrodynamic approach to allow for arbitrary perturbations in the plasma parameters, namely, velocity, density, and temperature. A Parker model for the zero-order flow speed was used.

074.037 Excitation and propagation of an upstreaming electromagnetic wave in the solar wind. A. Hasegawa.
Journ. Geophys. Res., Vol. 77, 84 - 90 (1972).

The slow electron cyclotron mode at the low-frequency and long-wavelength limit is shown to generate a left-hand-polarized wave with a frequency near the ion cyclotron frequency at the bow shock and to propagate the wave upstream in the solar wind. The sense of polarization and the frequency of the generated wave is consistent with the recent observations of the circularly polarized electromagnetic waves in the solar wind made by Russell et al. (1971).

074.038 Three-dimensional solar wind. S. T. Suess.
Journ. Geophys. Res., Vol. 77, 567 - 574 (1972).

A restricted three-dimensional model of the solar wind is studied by a perturbation technique with spherically symmetric boundary conditions. The results give the velocities and fields at 1 AU with an internal accuracy of about 1 % and can be used to examine the accuracy and applicability of the results of previous calculations. Although only a polytropic gas is specifically treated, a simple consequence of the mathematics allows the azimuthal velocity to be calculated with great accuracy by using a conductive heat equation.

074.039 Theory of discrete wave packets in the solar wind.
C. S. Wu.
Journ. Geophys. Res., Vol. 77, 575 - 587 (1972).

The present communication is motivated by the OGO 5 data. We suggest a possible plasma process that may be responsible for the observed wave packets. Qualitative comparison between our theory and the observational results seem to show plausibility. However, in an attempt to draw more confirmative conclusions, we still need more observational data.

074.040 The sun's changing white light corona, viewed from OSO-7.
M. J. Koomen, G. E. Brueckner, R. Tousey.
Bull. American Astron. Soc., Vol. 3, 440 (1971). –Abstr. AAS.

074.041 Electron density from 1.5 – 10 R.. inferred from observations of the solar corona.
R. Calbert, D. B. Beard.
Bull. American Astron. Soc., Vol. 3, 447 (1971). – Abstr. AAS.

074.042 The minimum size and minimum solar approach of interplanetary dust.
D. B. Beard, R. Calbert.
Bull. American Astron. Soc., Vol. 3, 482 - 483 (1971).
Abstr. AAS.

074.043 Effect of differential refraction and differential extinction on coronal observations. V. Rušin.
Bull. Astron. Inst. Czechoslovakia, Vol. 23, 75 - 82 (1972).

As mentioned by Sýkora (1971), among many other errors in observing the intensities of coronal lines, there are also differential refraction and differential extinction. The analysis of coronal observations made at the observatories of Pic du

Midi and Lomnický Štit does not prove the effects mentioned above uniquely. It was found that the errors due to the observational method and photometry were larger.

074.044 Heterogeneous solar wind.
L. P. Smirnova, V. P. Shabansky.
Geomagn. Aeronom., Vol. 12, 10 - 17 (1972). In Russian.

074.045 On the significance of the existence of classes for visible coronal lines.
F. Magnant, B. Fort, D. R. Flower.
Astron. Astrophys., Vol. 17, 403 - 409 (1972).

We show that there exists an almost linear relationship between the logarithm of ionization potential of a given ion and the logarithm of the temperature at which the ionization function of this ion reaches its maximum. This enables ions to be classified according to their ionization potential. This explains the existence of empirically defined classes for visible coronal lines.

074.046 Velocity of the solar wind as determined from interplanetary scintillations.
J. R. Jokipii, L. C. Lee.
Astrophys. Journ., Vol. 172, 729 - 737 (1972).

The "method of smooth perturbations" is used to derive the equations for interplanetary scintillation in a (statistically) spherically symmetric solar wind with constant wind velocity. The expressions, valid in the limit of small scintillation index, are used to discuss the relation between the radial wind velocity and the observed projected velocity of the scintillation pattern. Application to various observations is discussed.

074.047 Heating of the solar wind ions.
N. D'Angelo, V. O. Jensen.
Cosmic Electrodynamics, Vol. 2, 396 - 398 (1972).

The suggestion is advanced that the high ion temperatures in the solar wind, observed near the earth's orbit, may be associated with the presence of an influx of neutral hydrogen from the boundary of the heliosphere.

074.048 Electron-temperature asymmetry and the structure of the solar wind. M. Schulz, A. Eviatar.
Cosmic Electrodynamics, Vol. 2, 402 - 422 (1972).

The observational fact that, in the frame of the solar wind, outgoing thermal electrons have a somewhat larger mean energy than those incoming corresponds to an outward 'heat flow' that can be simulated in the extreme by a purely exospheric model. In the present work, micro-instabilities of the interplanetary plasma are studied within the framework of an underlying electron-proton solar exosphere.

074.049 Solar wind for a magnetized plasma with tensor plasma pressure. M. Tan, B. Abraham-Shrauner.
Cosmic Electrodynamics, Vol. 3, 71 - 80 (1972).

A solar wind model for a magnetized solar wind is presented using one-fluid hydromagnetic equations with generalized polytrope equations of state for the two tensor components of the plasma pressure. Fluid and magnetic field variables are calculated at the earth using certain boundary conditions at the sun.

074.050 Dynamics of the azimuthally dependent solar wind.
T. Matsuda, T. Sakurai.
Cosmic Electrodynamics, Vol. 3, 97 - 115 (1972).

The dynamics of the steady, rotating, azimuthally dependent solar wind under the influence of a magnetic field is considered. Numerical calculations based on the newly proposed hypervelocity approximation, which includes the convective terms due to the azimuthal component of the solar wind velocity are performed. The typical sawtoothed shape of the azimuthal distributions of radial velocity and spiked configurations of density and magnetic field near the orbit of the earth are obtained on the basis of purely sinusoidal distributions at $30\,R_\odot$.

074.051 Radial dependence of electron temperature in two-fluid models of the solar wind. E. J. Weber.
Cosmic Electrodynamics, Vol. 3, 116 - 118 (1972).
Research note.

074.052 On the propagation of shock waves in the solar corona causing radio bursts of type II.
V. V. Fomichev.
Astron. Zhurn. Akad. Nauk SSSR, Vol. 49, 348 - 354 (1972).
In Russian. English translation in Soviet Astron. AJ, Vol. 16, No. 2.
The problem of the character of changing parameters of shock waves in the case of their distribution in the solar corona from the data of 9 radio bursts of type II is considered. It is noted that the variety in the behavior of shock waves in the solar corona is apparently reflected in the variety of the behavior of shock waves in the interplanetary space.

074.053 Solar wind expansion beyond the heliosphere.
V. P. Bhatnagar, H. J. Fahr.
Planet. Space Sci., Vol. 20, 445 - 460 (1972).
A model of solar wind expansion beyond the heliosphere is given on the basis of the assumption of a continuous, stationary out-flow. The distributions of the hot solar wind protons, and the constituents resulting from charge-exchange collisions of these protons with the cool interstellar atomic hydrogen, like hot neutral atomic hydrogen and cold protons, have been investigated for the region beyond the heliosphere. The effect of the relative motion between cool interstellar atomic hydrogen and the solar wind protons in the interstellar space has been included.

074.054 Explorer 35 observations of solar-wind electron density, temperature, and anisotropy.
G. P. Serbu.
Journ. Geophys. Res., Vol. 77, 1703 - 1712 (1972).
Measurement of the electron integral spectrum yields electron temperatures ranging from 1 to 4×10^5 °K and having an average value of 1.82×10^5 °K, electron densities having an average value of 4.6 cm^{-3}, and electron-temperature anisotropies ranging from 1 to 1.4.

074.055 Coronal emission line polarization. L. L. House.
Solar Physics, Vol. 23, 103 - 119 (1972).
A discussion of a program for the computation of coronal emission line polarization is presented. Sample results are presented for the numerical computation of the angle of maximum polarization and the degree of maximum polarization to be expected from idealized magnetic field configurations such as radial and dipole. A computation is included for a realistic field configuration predicted to exist at the time of the 1966 eclipse. It is the purpose of the sample calculations to demonstrate how the measurement of emission polarization measurements can be interpreted in terms of the direction of coronal magnetic fields.

074.056 Neutralization and stabilization of particle streams in the corona and type III radio bursts.
D. F. Smith.
Solar Physics, Vol. 23, 191 - 203 (1972).
The processes by which streams of charged particles become charge and current neutralized in the corona are investigated. It is shown that a large amplitude plasma wave, which is related to precursor phenomenon in type III bursts and possibly plasma radiation from type IV bursts, will be excited at the head of the stream. The energy extracted from the stream to produce this plasma wave is computed and used to

set conservative upper limits on the densities of possible excitors for type III bursts. Since electron streams cannot produce their own stabilizing ion-acoustic waves, other mechanisms for producing ion-acoustic waves in the corona are examined. Another stabilization mechanism due to velocity inhomogeneity is investigated.

074.057 Critical point regularity conditions and asymptotic solutions to the time stationary, linearized, inhomogeneous solar wind flow problem.
G. L. Siscoe, R. L. Carovillano.
Solar Physics, Vol. 23, 211 - 222 (1972).
The assumption of the existence of steady state, linearized, inhomogeneous solar wind flow imposes a relation between the perturbation variables such that the solutions are regular at the critical points. The critical point regularity conditions are derived for the case of hydromagnetic flow with no rotation, for which there are two critical points, and the case of hydrodynamic flow with rotation and one critical point. Analytic asymptotic solutions are also derived to complete the description given previously of the intermediate range flow obtained by numerical integrations. The pressure perturbation for expected solar wind conditions is found to increase with distance and the velocity perturbations to decrease.

074.058 On neutral sheets in the solar wind.
G. W. Pneuman.
Solar Physics, Vol. 23, 223 - 237 (1972).
The structure and dynamics of neutral sheets in the solar wind is examined. The internal magnetic topology of the sheet is argued to be that of thin magnetic tongues greatly distended outward by the expansion inside the sheet. The energy release in the form of Joulean dissipation inside the sheet is estimated. It is concluded that ohmic heating in current sheets is not a significant source of energy for the overall solar wind expansion, however, the local energy release through this mechanism is found to be large.

074.059 A two-fluid solar wind model with anisotropic proton temperature. E. Leer, W. I. Axford.
Solar Physics, Vol. 23, 238 - 250 (1972).
A two-fluid model of the solar wind with anisotropic proton temperature and allowing for extended coronal proton-heating is considered for the case of a purely radial and of a spiral magnetic field. Reasonable values are obtained for the flow-velocity, number density and electron-temperature near the orbit of the earth.

074.060 On the identification on rotational discontinuities in the solar wind. K. G. Ivanov.
Kosm. Issled., Vol. 10, 131 - 133 (1972). In Russian. – Brief information.

074.061 On the abundance of calcium in the solar corona.
K. S. de Boer, S. R. Pottasch.
Solar Physics, Vol. 23, 406 - 409 (1972). – Research note.

074.062 On determining the electron density distribution of the solar corona from K-coronameter data.
M. D. Altschuler, R. M. Perry.
Solar Physics, Vol. 23, 410 - 428 (1972).
The electron density distribution of the inner solar corona ($r \leqslant 2\,R_\odot$) as a function of latitude, longitude, and radial distance is determined from K-coronameter polarization-brightness (pB) data. A Legendre polynomial is assumed for the electron density distribution, and the coefficients of the polynomial are determined by a least-mean-square regression analysis of several days of pB-data. The method is particularly useful in determining the longitudinal extent of coronal streamers and enhancements and in resolving coronal features whose projections on the plane of the sky overlap.

074.063 Study of the solar wind according to the data on the 11-year cosmic ray modulation.
A. N. Charakhch'yan, T. N. Charakhch'yan.
Izv. AN SSSR. Ser. fiz., Vol. 35, 2488 - 2491 (1971). In Russian. – Abstr. in Referativ. Zhurn. 51. Astron., 5.51.391 (1972).

074.064 Dissipation mechanisms in a pair of solar-wind discontinuities.
T. W. J. Unti, G. Atkinson, C.-S. Wu, M. Neugebauer.
Journ. Geophys. Res., Vol. 77, 2250 - 2263 (1972).

A pair of sharp closely spaced discontinuities in the solar wind was recorded by the high time resolution instruments aboard Ogo 5 on March 14, 1968. There is plasma turbulence within the double structure, and there appear to be small-amplitude hydromagnetic waves radiating from the discontinuities. The generation of the plasma turbulence is discussed in terms of magnetic drift waves.

074.065 Polarization of the solar corona.
W. A. Feibelman.
Journ. Roy. Astron. Soc. Canada, Vol. 66, 109 - 110 (1972).

074.066 Magnetic energy flow in the solar wind.
J. L. Modisette.
Astrophys. Journ., Vol. 174, 151 - 152 (1972).

The effect on magnetic energy flow of rotation of the sun is shown to result in most of the magnetic energy being transported by magnetic shear stress near the sun.

074.067 Electron density irregularity spectrum in the solar wind.
D. C. Backer, R. V. E. Lovelace, G. A. Zeissig.
Bull. American Astron. Soc., Vol. 4, 209 (1972). – Abstr. AAS.

074.068 On the relationship between the solar coronal green line intensity and the limiting primary rigidity of the solar diurnal anisotropy of cosmic rays.
H. S. Ahluwalia.
Bull. American Astron. Soc., Vol. 4, 257 (1972). – Abstr. AAS.

074.069 About cyclicity and geometry of the solar wind.
N. P. Chirkov.
Geomagn. Aeronom., Vol. 12, 331 - 332 (1972). In Russian. Brief information.

074.070 Turbulence of solar plasma.
S. A. Kaplan, S. B. Pikel'ner, V. N. Tsytovich.
Comments Astrophys. Space Phys., Vol. 4, 71 - 78 (1972).

The objective of the present paper is to present an up-to-date general picture of plasma turbulence in the upper layers of the solar atmosphere.

074.071 Observations of the region of interaction between the solar wind plasma and Mars.
O. L. Vajsberg, A. V. Bogdanov, N. F. Borodin, E. M. Vasil'ev, A. V. D'yachkov, A. A. Zertsalov, B. V. Polenov, S. A. Romanov.
Kosm. Issled., Vol. 10, 462 - 463 (1972). In Russian. – Brief information.

074.072 The solar wind H and He⁺ content.
C. T. Gregory.
Planet. Space Sci., Vol. 20, 841 - 847 (1972).

Charge exchange collisions between interplanetary neutral H atoms and solar wind protons may lead to fluxes of neutral H atoms and He^+ ions in the solar wind. Photoionization of interplanetary helium atoms may also contribute to the He^+ flux. The expected fluxes of He^+ ions and neutral H atoms in the solar wind are computed. A simple model is used to compute the intensity of resonantly backscattered solar He II ($\lambda 304$ Å) and Lyman α radiation.

074.073 A model for drift pair and hook burst emission from the solar corona. W. K. Yip.
Solar Physics, Vol. 24, 197 - 209 (1972).

Combination scattering of the Cerenkov plasma waves generated by a fast electron beam on the electron density fluctuations in a magnetoactive plasma is assumed to be the cause of the emission of the drift pair (or the hook burst) from the solar corona. The features of the combination emission are studied numerically with parameters appropriate to the solar corona condition. It is found that the major properties of the drift pair and the hook burst can be accounted for.

074.074 Hydrogen and helium velocities in the solar wind.
K. W. Ogilvie, H. J. Zwally.
Solar Physics, Vol. 24, 236 - 242 (1972).

This paper concerns an examination of the bulk speeds of helium and hydrogen in the solar wind.

074.075 Der Anregungsmechanismus der Koronalinie 5303 Å.
M. Waldmeier.
Mitt. Astron. Ges., No. 31, p. 153 - 154 (1972).

074.076 Faraday rotation of linearly polarized radio waves from the Crab nebula by the solar corona.
Y. Sofue, K. Kawabata, N. Kawajiri, N. Kawano.
Publ. Astron. Soc. Japan, Vol. 24, 309 - 321 (1972).

Position angles of linearly polarized radio waves from the Crab nebula were measured at wavelength $\lambda=7.2$ cm, in the middle of June 1971, at which time the source was occulted by the outer solar corona. The observed variation in the rotation measure suggests that there exist local structures in the magnetic field and electron density of the corona.

074.077 Study of the solar wind using the power spectrum of interplanetary scintillation of radio sources.
G. Bourgois.
Astron. Astrophys., Vol. 19, 200 - 206 (1972).

We give a discussion of the Fourier and Bessel transforms of the time autocorrelation function of the interplanetary scintillations. We discuss the conditions of appearance of extrema and show that their observation would allow in certain conditions to determine the size of the blobs and the velocity of the solar wind.

074.078 Introduction to research on the solar corona.
J. W. Evans.
Astrophys. Space Sci. Library, Vol. 27, (see 012.011), 1 - 12 (1971).

074.079 Atomic processes in the solar corona. G. Noci.
Astrophys. Space Sci. Library, Vol. 27, (see 012.011), 13 - 28 (1971).

074.080 Magnetohydrodynamics and plasma physics of the solar corona. F. Meyer.
Astrophys. Space Sci. Library, Vol. 27, (see 012.011), 29 - 35 (1971).

074.081 Coronal magnetic fields. G. Newkirk, Jr.
Astrophys. Space Sci. Library, Vol. 27, (see 012.011), 66 - 87 (1971).

074.082 The extended coronal magnetic field.
J. M. Wilcox.
Astrophys. Space Sci. Library, Vol. 27, (see 012.011), 88 - 96 (1971).

074.083 Investigations on coronal monochromatic emissions

in the optical range.   A. Dollfus.
Astrophys. Space Sci. Library, Vol. 27, (see 012.011), 97 - 113 (1971).

074.084   Coronal events observed in 5303 Å.   R. B. Dunn.
Astrophys. Space Sci. Library, Vol. 27, (see 012.011), 114 - 129 (1971).

074.085   The solar corona in the eleven-year cycle.
M. Waldmeier.
Astrophys. Space Sci. Library, Vol. 27, (see 012.011), 130 - 139 (1971).

074.086   Coronal active regions and flare-associated events.
J. B. Zirker.
Astrophys. Space Sci. Library, Vol. 27, (see 012.011), 140 - 155 (1971).

074.087   Studies of the outer corona through space radio astronomy.   M. D. Papagiannis.
Astrophys. Space Sci. Library, Vol. 27, (see 012.011), 317 - 332 (1971).

074.088   Structure and dynamics of the solar corona.
M. Kuperus.
Astrophys. Space Sci. Library, Vol. 29, Part I, (see 012.012), p. 9 - 20 (1972).

074.089   Particle diffusion in the solar corona.   H. Elliot.
Astrophys. Space Sci. Library, Vol. 29, Part I, (see 012.012), p. 134 - 150 (1972).

074.090   Composition and dynamics of the solar wind plasma.
A. J. Hundhausen.
Astrophys. Space Sci. Library, Vol. 29, Part II, (see 012.013), p. 1 - 31 (1972).

074.091   Interaction of the solar wind with the moon.
N. F. Ness.
Astrophys. Space Sci. Library, Vol. 29, Part II, (see 012.013), p. 159 - 205 (1972).

074.092   Twenty-seven-day recurrences in the solar-wind speed: Mariner 2.
J. T. Gosling, V. Pizzo, M. Neugebauer, C. W. Snyder.
Journ. Geophys. Res., Vol. 77, 2744 - 2751 (1972).
   The analysis gives a correlation at lags near 27 days of the order of 0.4, similar to that previously obtained from the Vela 2 and 3 analysis. High and low speeds appear to be about equally responsible for the correlation observed. The details and results of this analysis are presented.

074.093   Solar wind plasma.   T. Toichi.
Butsuri (Japan), Vol. 26, 584 - 586 (1971).
In Japanese. – See Phys. Abstr., Vol. 75, No. 17059 (1972).

074.094   Helium and heavy ions in the solar wind.
V. Formisano, G. Moreno.
Nuovo Cimento Rivista, Ser. 2, Vol. 1, 365 - 422 (1971).
   A picture of the behaviour of helium and heavier ions through the expanding solar corona and in the solar wind is presented. The relative abundance of these elements observed in the solar wind is composed with solar abundances as known from direct observations and from theoretical predictions. The solar atmosphere composition is discussed and theoretical models of the expanding corona reviewed.

074.095   Thermal properties of the solar wind.
E. Amata, V. Formisano.
Nuovo Cimento Lettere, Ser. 2, Vol. 4, 23 - 26 (1972).
   Data obtained by the ESRO satellite Heos-1 for the ener-gy distribution of the positive component of the solar wind have been used for studying some features of the solar-wind thermal properties.

074.096   On the activity of solar coronal condensations discussed after long-enduring microwave events.
P. Kaufmann.
Rev. Brasil. Fis. (São Paulo), Vol. 1, 289 - 295 (1971). – See Phys. Abstr., Vol. 75, No. 41234 (1972).

074.097   The determination of coronal polarization.
D. H. Menzel.
Topics in modern physics [Colorado Associated University Press, Boulder, Colo.], p. 245 - 250 (1971).

074.098   Interaction of solar wind with the moon and possibly other planetary bodies.   C. P. Wang.
AIAA Journ., Vol. 9, 1148 - 1153 (1971).

074.099   Accélération du vent solaire par gradient de pression d'ondes d'Alfvén.   J. Alazraki.
Thesis, 3e Cycle, Spéc. Astrophys., Univ. Paris. 27 pp. (1971). See Bull. Signal., Vol. 33, Section 120, No. 7031 (1972).

Mysteriöse Ionen im Sonnenwind.
Umschau, 72. Jahrgang, p. 366, 368 (1972).

Royal Astronomical Society meeting on the solar wind.   See Abstr. 010.022.

The interpretation of total line intensities from optically thin gases. I. A general method.
See Abstr. 022.003.

Electric quadrupole transitions in Fe XVI, Co XVII and Ni XVIII.   See Abstr. 022.050.

Determination of magnetohydrodynamic shock normals.   See Abstr. 062.001.

Properties of solar active regions.
See Abstr. 072.038.

On a relation between the indices of solar activity in the photosphere and the corona.
See Abstr. 072.043.

Structure and energy balance of the chromosphere-corona transition region.   See Abstr. 073.020.

Structure of the chromosphere-corona transition region.   See Abstr. 073.037.

Solar flares and solar wind helium enrichments: July 1965–July 1967.   See Abstr. 073.050.

Models of the chromosphere corona transition region.   See Abstr. 073.065.

The chromosphere–corona transition region.
See Abstr. 073.069.

The surges.   See Abstr. 073.071.

Relations between the areas index and different phenomena in the chromosphere, the corona and the interplanetary space.   See Abstr. 073.072.

Models of the quiet and active solar atmosphere from Harvard OSO data.   See Abstr. 073.073.

The determination of chromospheric-coronal structure from solar XUV observations.   See Abstr. 073.074.

Models of the solar transition region chromosphere—corona.   See Abstr. 073.076.

Stationäre Filamente, ihre koronale Umgebung sowie ihre Beziehung zum solaren Magnetfeld.
See Abstr. 073.079.

Étude expérimentale et théorique de la région de transition chromosphère-couronne solaire par l'utilisation des longueurs d'onde radioélectriques.   See Abstr. 073.083.

Observed heights of EUV lines formed in the transition zone and corona.   See Abstr. 076.006.

Recent investigation about solar X-ray emitted by the solar corona by means of Solrad satellites.
See Abstr. 076.034.

Generation of bremsstrahlung X-rays from solar corona.   See Abstr. 076.045.

Coronal scattering of radiobursts at hectometer and kilometer wavelengths.   See Abstr. 077.026.

Pulsating modulations and peculiar absorptions of type IV emissions from the solar corona.
See Abstr. 077.053.

Transport of cosmic rays in the solar corona.
See Abstr. 078.011.

On the directional dependence of the emission of acoustic noise by convective turbulence in a gravitational atmosphere.   See Abstr. 080.004.

The polarization of skylight near the sun and its influence on polarimetric K corona observations.
See Abstr. 082.013.

Unipolar interaction of Mercury with the solar wind; the steady state bow shock problem.   See Abstr. 092.005.

Comet-like interaction of Venus with the solar wind, I.   See Abstr. 093.009.

Lunar fossil magnetism and perturbations of the solar wind.   See Abstr. 094.032.

Measurements of lunar magnetic field interaction with the solar wind.   See Abstr. 094.104.

Possible magnetic interaction of asteroids with the solar wind.   See Abstr. 098.069.

'Cometary' suggestion.   See Abstr. 102.027.

Heavy ions from interplanetary dust.
See Abstr. 106.006.

Shock waves and magnetic field configuration in interplanetary space.   See Abstr. 106.008.

Flare induced shocks and corotating streams in the interplanetary medium.   See Abstr. 106.013.

Scattering and scintillations of discrete radio sources as a measure of the interplanetary plasma irregularities.
See Abstr. 106.020.

Cosmic and solar system abundances of deuterium and helium-3.   See Abstr. 107.004.

Backscatter of solar resonance radiation – I.
See Abstr. 131.118.

Coronal broadening of the Crab nebula 1969-71. Observations.   See Abstr. 134.006.

Coronal broadening of the Crab nebula 1969-71. Interpretation.   See Abstr. 134.007.

Observations with the Haystack-Goldstone interferometer of phase scintillations due to the solar corona.
See Abstr. 141.036.

Simultaneous, multifrequency observations of interplanetary scintillations.   See Abstr. 141.049.

Galactic cosmic ray modulation by the interplanetary medium (including the problem of the outer boundary).
See Abstr. 143.061.

Errata

074.901   Addendum and erratum: 'The solution of one-fluid equations with modified thermal conductivity for the solar wind' [Cosmic Electrodynamics, Vol. 1, 205 - 217 (1970)].   S. Cuperman, A. Harten.
Cosmic Electrodynamics, Vol. 3, 119 - 120 (1972).

075 Solar Patrol

075.001 **Solar prominences in 1970.** M. K. V. Bappu.
Quarterly Journ. Roy. Astron. Soc., Vol. 13, 70 (1972).

075.002 **Curves of the solar radio radiation from observations of the Observatory of the Department of Astronomy at the Kiev University in Lesnikakh.**
Kometn. Tsirk., *Kiev,* Nos. 127, 129 - 131 (1972). In Russian. 1972 January – April.

075.003 **Solar activity.** F. G. Mustaeva.
Tsirk. Astron. Inst., *Tashkent,* Nos. 27 (374) - 32 (379) (1971/72). In Russian. – 1970 November - 1971 October.

075.004 **Nombres relatifs de Wolf pour l'année 1971.** M. Waldmeier.
L'Astronomie, 86ᵉ année, p. 241 (1972).

075.005 **Definitive Sonnenflecken-Relativzahlen für 1971.** R. A. Naef.
Orion Schaffhausen, 30. Jahrgang, p. 104 (1972).

075.006 **Definitive Sonnenflecken-Relativzahlen für 1971.** M. Waldmeier.
Sterne, 48. Jahrgang, p. 112 (1972).

075.007 **Observations photosphériques et chromosphériques solaires faites à Uccle en 1970.**
G. Evrard, C. Gonze, A. Koeckelenbergh.
Bull. Astron. Obs. Roy. Belgique, Vol. 7, 201 - 271 (1971).
Observations photosphériques: 1.) Observations visuelles à la lunette Merz-Grubb: Nombre relatif de Wolf journalier à Uccle, position et classification des groupes de taches; 2.) Observations photographiques à l'héliographe Zeiss; 3.) Les rotations solaires, l'évolution des groupes de taches et tableaux récapitulatifs. Observations chromosphériques: 1.) Observations au filtre de Lyot-Hα; 2.) Observations au filtre de Halle-K.

075.008 **Observations of the solar chromosphere in 1964–66.** S. Karaali.
Publ. Istanbul Univ. Obs., No. 94, 11 pp. (1971).

075.009 **The sunspot observations made in 1965.** F. Yilmaz.
Publ. Istanbul Univ. Obs., No. 95, 6 pp. (1971).

075.010 **Sunspots** (sunspot relative numbers and sunspot-areas); **Synoptic charts of solar magnetic fields** (Mount Wilson Observatory); **Eruptions chromosphériques brillantes; Intensité de la couronne solaire; Solar radio emission.**
M. Waldmeier, R. Howard, R. Michard, G. Olivieri, M. Bernot.
Quarterly Bull. Solar Activity (published by Eidgen. Sternw. Zürich), Nos. 173 - 174, p. 1 - 71 (1972). – Observations for the co-operating observatories for 1971 January – June are given.

075.011 **Geomagnetic and solar data.**
J. V. Lincoln (Editor).
Journ. Geophys. Res., Vol. 77, 278, 788, 1347, 2005, 2413 - 2416, 3003 (1972). – 1971 September - 1972 February.

075.012 **Map of the sun.**
Edited by Fraunhofer Institut, Freiburg.
1972 January 1 – June 30.

075.013 **Observation of the sun in 1971.** V. Čelebonović.
Vasiona, Vol. 20, 38 - 40 (1972). In Serbo-Croatian.

075.014 **Sunspot relative numbers for 1971.** M. Waldmeier.
Astron. Mitt. Sternw. Zürich, No. 310, 9 pp. (1972).

075.015 **Actividad solar julio – diciembre 1971.**
V. Capolongo, J. A. Gutiérrez, O. F. Liesche, C. Sosa, R. Barbarroja, L. A. Mansilla.
Obs. Astron. Municipal, Rosario – Argentina, Dep. Fis. Solar, Contr. Ser. 1, No. 3, 36 pp. (1972).

075.016 **Boletín Meteorológico mensual.**
Edited by Obs. Nacional de Física Cósmica, San Miguel, Argentina. Vol. 25 (3) - (12), 1971 March - December (1971/72).

075.017 **Solar photospheric observations.**
F. Bruin, H. Hourani, N. G. Bustati.
Lee Obs., American Univ. Beirut, Monthly Bull., Astron. Section, 1971 November – 1972 March (1972).
Sunspot relative numbers; Heliographic mean position and classification of the sunspot groups; Number of facular zones.

075.018 **Solar phenomena.** M. Cimino, M. Torelli, A. Cacciani, V. Croce, R. Flamini, U. Bartolini.
Oss. Astron. Roma, Monthly Bull. Nos. 167 - 168 (1972).
1971 November – December: Daily total areas of sunspot-groups; Heliographic position, classification and area of sunspot-groups; Longitudinal sunspot magnetic fields; Hours of K-line cinematographic patrol; Hours of Hα cinematographic patrol; S.C.N.A. and S.E.A.; Explanation.

075.019 **Observations des taches solaires en 1970.**
N. Doğan.
Commun. Fac. Sci. Univ. Ankara, Sér. A, Vol. 20 A, 87 - 123 = Commun. Dep. Astron. Ankara Univ. No. 55 (1971).

075.020 **Solare Beobachtungsergebnisse (Solar Data).**
E. A. Lauter, A. Böhme, F. W. Jäger, F. Fürstenberg, H. Künzel, D. Scholz, S. Böhm.
Zentralinst. für Solar-Terrestrische Physik (Heinrich-Hertz-Inst.), Deutsche Akad. Wiss. Berlin, HHI Solar Data, Vol. 22, November – December (1971). – Solar radio emission; Sunspot magnetic data.

075.021 **Solare Beobachtungsergebnisse (Solar Data).**
C.-U. Wagner, A. Böhme, F. Fürstenberg, D. Scholz, S. Böhm.
Zentralinst. für Solar-Terrestrische Physik (Heinrich-Hertz-Inst.), Deutsche Akad. Wiss. Berlin, HHI Solar Data, Vol. 23, January – February (1972). – Solar radio emission.

075.022 **Solar photosphere charts.** L. Schmied.
Říše hvězd, Vol. 53, 35, 57 (1972). – Rotations Nos. 1572 - 1577.

075.023 **Daily maps of the sun and geophysical graphs.**
Solnechnye Dannye 1971 Byull., No. 11, p. 2 - 66; No. 12, p. 2 - 71; 1972 Byull., No. 1, p. 3 - 77; No. 2, p. 2 - 87; No. 3, p. 2 - 91 (1972). In Russian.

075.024 **Magnetic fields of sunspots.**
Prilozhenie k Byulletenyu "Solnechnye Dannye", 1971, Nos. 11 - 12; 1972, Nos. 1 - 3. In Russian.

075.025 **Sunspot numbers.**
Sky Telescope, Vol. 43, 48, 129, 197, 267, 329, 398; Vol. 44, 15 (1972). – 1971 November - 1972 May.

075.026 **Fenomeni solari.**
F. Mazzucconi, S. Delli Santi, M. L. Sturiale, A. Abrami.
Coelum, Vol. 40, 29 - 34, 75 - 81, 124 - 128 (1972). – 1971 September – 1972 February.

075.027 **Osservatorio Magnetico de l'Aquila. Bollettino magnetico.**
Coelum, Vol. 40, 35, 82, 129 (1972). – 1971 August – 1972 January.

075.028 **Centro Universitario Fenomeni fluttuanti – Firenze. Test P.**
Coelum, Vol. 40, 36, 83, 130 (1972). – 1971 September – 1972 February.

075.029 **Zonnevlekkengetallen.**
Hemel en Dampkring, Vol. 70, 107, 141, 203 (1972). – 1971 September – 1972 April.

075.030 **L'activité solaire.**   M.-J. Martres.
L'Astronomie, 86e année, p. 44 - 46, 102 - 103, 158 - 159, 204 - 205, 253 - 255, 322 - 323 (1972). – Rotations 1575 - 1581.

075.031 **Solar and solar system activity.**
R. J. J. Langton, J. R. Smith.
Journ. British Astron. Ass., Vol. 82, 136 - 138, 220 - 222, 301 - 308 (1972). – 1971 September – 1972 February.

075.032 **Indices of geomagnetic activity.**
Journ. Atmosph. Terr. Phys., Vol. 34, 173, 349, 555, 745, 953, 1161, 1305 (1972). – 1971 September – 1972 March.

075.033 **Provisional sunspot-numbers.**
Yamamoto Circ., Nos. 1747, 1749, 1750, 1753, 1754, 1755 (1972). – 1971 December – 1972 May.

075.034 **Predictions for the smoothed monthly sunspot-numbers.**
Yamamoto Circ., Nos. 1747, 1749, 1750, 1753 (1972). 1972 January – September.

076 Solar UV, X Rays, Gamma Radiation

076.001 Missing solar ultraviolet opacity and diatomic molecules. S. P. Tarafdar, M. S. Vardya.
Astrophys. Journ., Vol. 171, 185 - 190 (1972).

The band absorption coefficient of some of the molecules abundant in the solar photosphere has been compared with the total absorption coefficient of metals and hydrogen. It is found that the band absorptions considered are possibly large enough to account for the missing opacity in the solar ultraviolet spectrum, except in a small interval between 2500 and 3000 Å.

076.002 Eclipse observations in the rocket ultraviolet.
T. L. J. Jones, W. H. Parkinson, R. J. Speer, C. Yang.
Solar Physics, Vol. 21, 372 - 380 (1971).

At 7 March, 1970, an aerobee 150 was launched into the umbra, carrying two intensity calibrated EUV Wadsworth grating spectrographs. The recorded spectra lie in the wavelength range 850 - 2150 Å. Lyman-α limb spectroheliograms reveal complex activity extending high into the corona in contrast to the modest level of activity observed in Hα.

076.003 XUV image of the sun from eclipse observations of the ionospheric E-region. R. T. Marriott, D. E. S. John, R. M. Thorne, S. V. Venkateswaran.
Solar Physics, Vol. 21, 483 - 494 (1971).

Ionospheric E-region observations during the total solar eclipse of 7 March, 1970 are used for reconstructing the distribution of the ionizing XUV radiation over the solar disk. The derived solar image compares reasonably well with the EUV and X-ray pictures of the sun obtained from rockets.

076.004 Results from OSO-IV: The long term behavior of X-ray emitting regions.
A. Krieger, F. Paolini, G. S. Vaiana, D. Webb.
Solar Physics, Vol. 22, 150 - 177 (1972).

A grazing incidence X-ray telescope on board the OSO-IV spacecraft obtained images of the sun in the 2.5 to 12 Å waveband nearly continuously from 27 October 1967 to 12 May 1968. Variations in the absence of flares of as much as a factor of 10 in the X-ray output of individual regions were observed, with typical durations ranging from several hours to several days. The X-ray time variations are related to observations at optical and radio wavelengths. The results are interpreted.

076.005 Note on the energy scale of the Michigan OSO III ion chamber.
R. G. Teske, P. E. Hodge, S. P. Worden.
Solar Physics, Vol. 22, 235 - 239 (1972).

The energy scale of the Michigan OSO III soft X-ray ion chamber has been assessed by using realistic theoretical X-ray spectra.

076.006 Observed heights of EUV lines formed in the transition zone and corona.
G. W. Simon, R. W. Noyes.
Solar Physics, Vol. 22, 450 - 458 (1972).

The heights of formation of a number of extreme ultraviolet lines in active regions have been measured from OSO-IV spectroheliograms. We find heights for He I, He II, C III, N III, O IV, O VI, Ne VIII, Mg X, Si XII, Fe XV and Fe XVI that are in approximate agreement with models based on analysis of EUV emission intensities. The height of C II is anomalously high.

076.007 Calculated solar X-radiation from 1 to 60 Å.
R. Mewe.
Solar Physics, Vol. 22, 459 - 491, with a correction Vol. 23, 508 (1972).

The fluxes of about 230 spectral lines in the range 1−60 Å from coronal ions of C, N. O, Ne, Na, Mg, Al, Si, S, Ar, K, Ca, Ti, Cr, Mn, Fe, and Ni are computed for a range of electron temperature from 10^5 to 10^9 K. The relative ion abundances are derived from Jordan's ionization equilibrium calculations. The continuum emission is derived from computations of Landini and Fossi with a correction for the free-free emission.

076.008 The normalization of solar X-ray data from many experiments. C. D. Wende.
Solar Physics, Vol. 22, 492 - 502 (1972).

A conversion factor is used to convert Geiger tube count rates or ion chamber currents into units of the incident X-ray energy flux in a specified passband. A method is described which varies the passband to optimize these conversion factors such that they are relatively independent of the spectrum of the incident photons.

076.009 The relation of X-ray radiation of the coronal condensation with the evolution of an active region.
I. A. Zhitnik, M. A. Livshits.
Astron. Zhurn. Akad. Nauk SSSR, Vol. 49, 137 - 147 (1972). In Russian. English translation in Soviet Astron. AJ, Vol. 16, No. 1.

An intercomparison of the radiation rigidity of separate local sources on the sun has been made through X-ray heliograms obtained from Cosmos-166 during relatively low solar activity. The flux in the spectrum ranges 8−14 and 2−8 Å has been measured for 550 similar maxima on the heliograms. A relative difference of the radiation rigidity for coronal condensations located over spot-groups and flocculi without spots has been discovered.

076.010 Compton backscattering of solar X-ray emission.
F. F. Tomblin.
Astrophys. Journ., Vol. 171, 377 - 389 (1972).

Compton backscattering of solar flare emission, $\lambda < 2$ Å, appears to have several interesting effects when X-ray spectra are considered in detail. Line emission generates secondary spectra shifted up to ~ 0.08 Å, thereby broadening the long-wavelength wing. The iron lines at 1.9 Å are discussed in detail, and the observed spectra obtained between 1.88 and 1.92 Å are shown to be significantly affected by Compton backscattering for flares occurring on the solar disk.

076.011 The blocking effect of lines on solar UV continuum between 2000 Å and 3000 Å.
D. Sacotte, R. M. Bonnet.
Astron. Astrophys., Vol. 17, 60 - 69 (1972).

A synthetic solar spectrum using 2400 lines is computed for the near UV region of 2000 Å to 3000 Å and is compared with our observations. We show that the evaluation of the damping constants is highly open to criticism, and leads us to attribute the "disturbed" component to wide metallic lines. We suggest that the linear component might be attributed to a great numer of weak lines and we attempt to evaluate their number.

076.012 The structure of coronal X-ray features.
A. S. Krieger, A. F. Timothy, G. S. Vaiana.
Bull. American Astron. Soc., Vol. 3, 439 (1971). − Abstr. AAS.

076.013 The sun's changing XUV corona, as viewed from OSO-7. D. J. Michels, R. Tousey.

Bull. American Astron. Soc., Vol. 3, 439 - 440 (1971). – Abstr. AAS.

076.014 **Si I lines in the XUV solar spectrum.**
A. S. Milman.
Bull. American Astron. Soc., Vol. 3, 448 (1971). – Abstr. AAS.

076.015 **Relationship between impulsive solar flare X-rays and type III solar radio bursts.** S. R. Kane.
Bull. American Astron. Soc., Vol. 3, 449 (1971). – Abstr. AAS.

076.016 **X-ray (E> 10 keV), Hα and microwave emission during the impulsive phase of flares.** J. Vorpahl.
Bull. American Astron. Soc., Vol. 3, 449 (1971). – Abstr. AAS.

076.017 **Comments on the decay phase of impulsive solar X-ray bursts.** R. W. Milkey.
Bull. American Astron. Soc., Vol. 3, 460 (1971). – Abstr. AAS

076.018 **On the interpretation of the relative intensities of the solar XUV lines of ions in the lithium isoelectronic sequence.** D. R. Flower.
Astron. Astrophys., Vol. 17, 201 - 206 (1972).

Observations of solar XUV emission lines of ions in the lithium isoelectronic sequence have been re-examined in the light of recent calculations of electron excitation cross-sections for the relevant transitions. The temperatures deduced from the observations are not in satisfactory agreement with the temperatures at which the observed lines are most efficiently produced according to currently accepted theories of the ionization equilibrium.

076.019 **Time variation of X-ray flux around solar proton events.** T. Galanová, Š. Pintér, L. Křivský.
Bull. Astron. Inst.Czechoslovakia, Vol. 23, 105 - 106 (1972).

We investigate the daily changes of the solar X-ray flux in a spectral region 1–8 and 8–20 Å several days before and after a proton flare and they are dealt with using the method of superposition of zero-epochs.

076.020 **Analysis of solar ultraviolet lines.** E. Chipman.
SAO, *Cambridge, Mass.,* Special Rep. No. 338, 7 + 176 + A10 pp. (1971).

We have made a detailed study of the formation of the strongest ultraviolet emission lines of Mg II, O I, C II, and C III in the solar atmosphere. We solve the equations of statistical equilibrium and radiative transfer for each ion, using a general computer program that is capable of solving non-LTE line-formation problems for arbitrary atmospheric and atomic models. We interpret the results in terms of the structure of the solar atmosphere, and also investigate the structure of the solar atmosphere in the range 20,000 to 100,000 K and the effects of microturbulence on the formation of lines. We solve some approximate analytic line-formation problems relevant to the more exact solutions derived later. In the Appendix we attempt to make the best possible fit to the Ca II K line center-to-limb profiles with a one-component atmosphere, with an assumed source function and microturbulent velocity.

076.021 **The 1–55 Å X-ray emission from an active limb prominence.** A. C. Brinkman, M. L. Shaw.
Solar Physics, Vol. 23, 120 - 133 (1972).

The solar X-ray emission spectrum between 1 and 55 Å has been studied for one solar event, using the two X-ray experiments on-board of the ESRO 2 satellite. By assuming the radiation processes in the source region to be thermal, temperatures and emission measures against time for the source region have been derived using the most recent models for free-free continuum and line emission due to thermal plasmas. The temperatures derived have been used to calculate cooling

rates due to radiation losses. An estimate of the total injected energy is made.

076.022 **Soft X-ray and microwave observations of hot regions in solar flares.** H. S. Hudson, K. Ohki.
Solar Physics, Vol. 23, 155 - 168 (1972).

Hot regions in solar flares produce X-radiation and microwaves by thermal processes. Recent X-ray data make it possible to specify the temperature and emission measure of the soft X-ray source, by using, for instance, a combination of the 1–8 Å and the 0.5–3 Å broad-band photometers. The temperatures and emission measures thus derived satisfactorily explain the radio fluxes.

076.023 **A search for high energy gamma-rays from solar active regions.** R. K. Sood.
Solar Physics, Vol. 23, 183 - 190 (1972).

The Elliot model for solar flares predicts weak γ-ray emission from the flare region prior to large flares. A search has been made for such γ-radiation of energy >50 MeV. The experiment was performed using balloon-borne detectors flown from an equatorial station during the 1967/1968 solar maximum. A number of small flares were observed, but no associated γ-rays were detected.

076.024 **Il problema dei bursts X non termici.**
R. Pallavicini.
Mem. Soc. Astron. Italiana, Nuova Ser., Vol. 43, 195 - 219 (1972).

Some solar X-ray bursts are presumably due to an interaction of non-Maxwellian electrons with the surrounding plasma. In this paper some problems concerning these events are discussed and the most important both theoretical and experimental results obtained in the latest years are reported.

076.025 **The solar spectrum 2100–3200 Å.**
A. L. Broadfoot.
Astrophys. Journ., Vol. 173, 681 - 689 (1972).

The spectral irradiance of the sun between the wavelengths of 2100 and 3200 Å was measured with a spectrometer from a stabilized Aerobee rocket. Previous spectra are in generally good agreement with these new measurements. A significant difference was found in the 2500–2600 Å region.

076.026 **The interpretation of XUV rocket measurements of intensity ratios of solar spectral lines of the lithium-like ions O VI, Ne VIII, and Mg X.**
L. Heroux, M. Cohen, M. Malinovsky.
Solar Physics, Vol. 23, 369 - 393 (1972).

Experimental ratios were compared with the ratios calculated by using specific theoretical values of the ionization equilibrium in which dielectronic recombination was included in the processes establishing ionization balance. A reliable measurement of the electron temperature in the lower corona was obtained from the experimental ratios for Mg X.

076.027 **Intensity measurements of the solar Lα-radiation with non-optical methods aboard Vertical 1.**
L. Martini, N. M. Shyutte, K. I. Gringauz, B. Shtark.
Kosm. Issled., Vol. 10, 255 - 260 (1972). In Russian.

076.028 **Satellite lines in the solar X-ray spectrum of helium-like magnesium.** J. H. Parkinson.
Nature, Phys. Sci., Vol. 236, 68 - 70 (1972).

New observations are reported of the $1s^2.2l–1s2p.2l$ lines lying close to the resonance line of helium-like magnesium in the solar X-ray spectrum. It is confirmed that the dominant mechanism of forming these lines is dielectronic recombination. The intensity of the $1s^22s\ ^2S–(1s2p\ ^1P)2s$ line relative to the resonance line is sensitive to temperature, and it is demonstrated that observations of these lines lead to a powerful new method of determining temperatures in active regions.

076.029 Spectral development of a solar X-ray burst observed on OSO-7.
D. L. McKenzie, D. W. Datlowe, L. E. Peterson.
Bull. American Astron. Soc., Vol. 4, 209 (1972). – Abstr. AAS.

076.030 Solar flares in the extreme ultraviolet. I. The observations. A. T. Wood, Jr., R. W. Noyes, A. K.
Dupree, M. C. E. Huber, W. H. Parkinson, E. M. Reeves, G. L. Withbroe.
Solar Physics, Vol. 24, 169 - 179 (1972).

Solar-flare observations in the extreme ultraviolet (300–1350 Å) are reported. Some 269 flares observed by the Harvard College Observatory (HCO) experiment on OSO 4 and 211 flares observed by the HCO experiment on OSO 6 have been analyzed.

076.031 Solar flares in the extreme ultraviolet. II. Comparisons with other observations.
A. T. Wood, Jr., R. W. Noyes.
Solar Physics, Vol. 24, 180 - 196 (1972).

Extreme-ultraviolet (300–1350 Å) observations of nearly 500 solar flares from the satellites OSO 4 and OSO 6 have been compared with data in X-ray and radio wavelengths.

076.032 Theoretical studies of the flux and energy spectrum of gamma radiation from the sun. C.-C. Cheng.
Space Sci. Rev., Vol. 13, 3 - 123 (1972).

The purpose of this work is to study the various γ-ray-production mechanisms in solar flares and to calculate the flux, the spectrum, and the decay curves of γ radiation. Using the continuity equation and taking into account the energy losses for solar-flare-accelerated particles, we obtain the time-dependent particle distribution and thus the time behavior of the resulting γ rays. The important processes for producing γ rays in solar flares are found to be nonthermal electron bremsstrahlung, decay of neutral π mesons, positron annihilation, neutron capture, and decay of excited nuclei. The results are applied to several known solar flares.

076.033 X-ray spectroscopy of solar active regions and flares.
W. M. Neupert.
Astrophys. Space Sci. Library, Vol. 27, (see 012.011), 237 - 253 (1971).

076.034 Recent investigations about solar X-rays emitted by the solar corona by means of Solrad satellites.
M. Landini, B. C. Fossi.
Astrophys. Space Sci. Library, Vol. 27, (see 012.011), 257 - 266 (1971).

076.035 Evidence that solar X-ray emission is of purely thermal origin (also observation of far UV flash during 28 August 1966 proton flare). T. A. Chubb.
Astrophys. Space Sci. Library, Vol. 29, Part I, (see 012.012), p. 99 - 118 (1972).

076.036 Evidence for the 300-second oscillation from OSO-7 extreme-ultraviolet observations.
R. D. Chapman, S. D. Jordan, W. M. Neupert, R. J. Thomas.
Astrophys. Journ., (*Letters*), Vol. 174, L97 - L99 (1972).

Evidence is presented for a 300-second oscillation in the intensity of solar extreme-ultraviolet emission lines of He II, Mg VIII, and Mg IX as observed by OSO-7.

076.037 Extreme ultraviolet solar images televised in-flight with a rocket-borne SEC vidicon system.
R. Tousey, I. Limansky.

Applied Optics, Vol. 11, 1025 - 1031 (1972).

A TV image of the entire sun while an importance 2N solar flare was in progress was recorded in the extreme ultraviolet radiation band 171–630 Å and transmitted to ground from an Aerobee-150 rocket on 4 November 1969 using *S*-band telemetry. A number of images were recorded with integration times between $^1/_{30}$ sec and 2 sec. Reconstruction of pictures was enhanced by combining several to reduce the noise.

076.038 Solar ultra-violet studies from a balloon platform.
M. McDowell.
Monthly Notes Astron. Soc. Southern Africa, Vol. 31, 47 (1972). – Abstract.

076.039 Flux of solar XUV radiation inferred from ionospheric variations during the eclipse of 20th May 1966. M. Vukićević-Karabin.
Publ. Dep. Astron. Univ. Beograd, No. 1, p. 3 - 7 (1969).

In this paper the incident solar flux was estimated by measuring electron density and ion production in the ionosphere during the solar eclipse of 20th May, 1966 over Belgrade. The obtained results are compatible with the rocket data.

076.040 Fast increase of solar X-ray intensity of December 10, 1970. G. E. Kocharov, Yu. E. Charikov,
A. A. Kharchenko, G. V. Gusev, A. V. Baskakov.
Pis'ma v ZhEhTF, Vol. 15, 153 - 156 (1972). In Russian.
Abstr. in Referativ. Zhurn. 51. Astron., 7.51.365 (1972).

076.041 Sudden frequency deviations, solar extreme ultraviolet bursts, and solar radio bursts.
D. W. Richards.
U.S. Air Force Cambridge Res. Lab. Air Force Systems Command Ionosph. Phys. Lab., Environmental Res. Papers No. 363. AFCRL–71–0392, 23 pp. (1971).

Includes comparison between radio and EUV bursts – no striking common feature. – *IDP*

076.042 On the X-ray emission of the active centres on the solar disc.
J. Mergentaler, Z. Kordylewski, M. Hlond.
Bull. Acad. Pol. Sci., Sci. Math., Astron., Phys., Vol. 19, 1065 - 1067 (1971).

In order to investigate the distribution of X-ray radiation on the solar disc a battery of pinhole cameras was mounted on the geophysical rocket Vertical I. The photographs of 8 distinct active regions were obtained, connected with groups of sunspots.

076.043 Solar electron temperatures and X-ray flare activity.
D. M. Horan, R. W. Kreplin.
AIAA Journ., Vol. 9, 1634 - 1636 (1971).

Discusses a technique for obtaining an X-ray source region's electron temperature and emission measure as functions of time. Using this technique an investigation of the relation between flare activity and electron temperature is made.

076.044 Characteristics of the energetic solar X-ray burst of 11 February 1970 observed at balloon altitude.
M. Kodama, M. Kusunose, K. Ogura.
Rep. Ionosph. & Space Res. Japan, Vol. 25, 285 - 300 (1971).
See Phys. Abstr., Vol. 75, No. 33764 (1972).

076.045 Generation of bremsstrahlung X-rays from solar corona. R. N. Singh, R. Y. Prasad.
Indian Journ. Pure & Applied Phys., Vol. 9, 833 - 836 (1971).
See Phys. Abstr., Vol. 75, No. 37420 (1972).

076.046 Flare-time temperature in soft X-ray sources.
S. D. Deshpande, J. N. Tandon.

Astrophys. Journ., Vol. 175, 253 - 259 (1972).

Temperatures in flare-time soft X-ray sources have been obtained for selected events by using measurements in 0.5−3 Å and 1−8 Å bands from OGO-4/SR-9 satellites. Higher temperature is assumed for emission below 3 Å, and two temperatures for 0.5−3 Å and 3−8 Å spectral ranges.

Localization of the source of flare X-ray emission during the eclipse of 7 March, 1970.   See Abstr. 073.003.

Location of the electron acceleration region in solar flares.   See Abstr. 073.049.

The relation between the white light and XUV coronas on 7 March, 1970.   See Abstr. 074.014.

The correlation of type III radio bursts and 4 keV X-ray events.   See Abstr. 077.013.

Radio and X-ray emissions associated with the chro- mospheric flares of November 15 and 16, 1970. See Abstr. 077.031.

Relativistic electrons from the sun observed by IMP-4.   See Abstr. 078.003.

Measurements of solar Lyman-α radiation during the eclipse of 7 March 1970.   See Abstr. 079.101.

Rocket observations of solar X-rays during the eclipse of 7 March 1970.   See Abstr. 079.101.

Rocket observations of solar UV radiation during the eclipse of 7 March 1970.   See Abstr. 079.101.

D-region extraionization and solar X-ray flux derived from SPA and SES measurements.   See Abstr. 083.028.

Backscatter of solar resonance radiation − I. See Abstr. 131.118.

077 Solar Radio Radiation

077.001 Decay of the magnetic field in a region generating a solar microwave burst. D. Basu, E. Scalise, Jr.
Astrophys. Letters, Vol. 10, 13 - 15 (1972).
The decay of the magnetic field in the region of a solar radio burst has been evaluated from the time profile of the degree of circular polarization of a burst at 7 GHz.

077.002 High-resolution observations of type III solar radio bursts. Ø. Elgarøy, E. Lyngstad.
Astron. Astrophys., Vol. 16, 1 - 12 (1972).
Solar radio bursts of spectral type III have been observed using high-resolution radio spectrographs operating in the frequency ranges 310–340 MHz and 145–175 MHz. The results are interpreted in the light of current ideas about type III bursts and the structure of the solar corona.

077.003 Solar radio spike bursts.
G. L. Tarnstrom, K. W. Philip.
Astron. Astrophys., Vol. 16, 21 - 27 (1972).
We review spike burst literature to obtain a working definition for spike bursts, then proceed to summarize the bursts thus qualifying as spike bursts in the observations of the high-time resolution radiospectrograph of the Geophysical Institute. Possible spike burst mechanisms will be briefly discussed.

077.004 Observations of the 7 March, 1970 total solar eclipse at wavelengths of 3.2 and 8.3 mm.
J. P. Hagen, P. N. Swanson, R. W. Haas, F. L. Wefer, R. W. Vogt.
Solar Physics, Vol. 21, 286 - 296 (1971).
Excellent eclipse curves were obtained at 3 mm. The 8 mm brightness distribution is subject to some uncertainty, but tends to show limb brightening. The 3 mm brightness distribution shows a well defined complex limb brightening within about 1 arc min of the optical limb. The maximum brightening is approximately 30% above the average disc temperature.

077.005 Eclipse of radio emission on 7 March, 1970 at 10 cm wavelength from the active region associated with McMath plage 10618.
E. B. Mayfield, G. A. Chapman, R. M. Straka.
Solar Physics, Vol. 21, 460 - 468 (1971).
The Sagamore Hill data indicate the region was about 3.8′ and contributed about 0.21 of the total radiation from the disk. The SFO data gave about 5.4′ for the size of the southern half of the region and showed that about 0.20 of the total radiation came from there.

077.006 Spectral radio observations of a solar eclipse.
R. M. Straka.
Solar Physics, Vol. 21, 469 - 480 (1971).
Measurements were made of the 7 March, 1970 solar eclipse in Hamilton, Mass., on the wavelengths of 0.86, 1.95, 3.4, 6.0, 11.1, 21.2, 49.5, and 122.5 cm. Source flux spectra for the intense sources show gyro-resonance spectral peaking. An estimated flux spectrum of the undisturbed radio sun for 7 March, 1970 is given and compared to the spectrum for the solar minimum of 1964.

077.007 Quiet-sun center-limb observations at 6 and 11 cm during cycle maximum.
J. C. Ceballos, P. Lantos.
Solar Physics, Vol. 22, 142 - 146 (1972).
Simultaneous 6 and 11 cm maps of the quiet sun have been obtained with the Nançay radiotelescope during the 1968–69 cycle maximum.

077.008 Pencil beam observation of a large microwave outburst at 94.8 GHz. J. R. Cogdell.
Solar Physics, Vol. 22, 147 - 149 (1972).
On 27 March 1969 at approximately 1326 UT an impulsive outburst of the sun was fortuitously observed with the 16-ft radiotelescope of the University of Texas Millimeter Wave Observatory, Fort Davis, Texas.

077.009 Total flux and polarization measurements of the S-component of the sun at mm-wavelengths.
G. Feix.
Astron. Astrophys., Vol. 16, 268 - 271 (1972).
Observations at 81 GHz and 250 GHz of the slowly varying component have been performed. The polarization degree at 81 GHz was measured. and compared with that at 36 GHz. The optical depths of the radio source at mm-wavelengths are given by evaluating the QL-propagation of mm-waves on the assumption of the magnetoionic theory.

077.010 Characteristics of 3.3-mm bursts.
F. I. Shimabukuro.
Bull. American Astron. Soc., Vol. 3, 449 (1971). – Abstr. AAS.

077.011 Polarization of solar active regions at 9 millimeter wavelength. M. R. Kundu, T. P. McCullough.
Bull. American Astron. Soc., Vol. 3, 449 - 450 (1971). – Abstr. AAS.

077.012 High time resolution (0.01 sec) spectral observations of solar radio bursts at decimetric wavelengths.
B. L. Gotwols.
Bull. American Astron. Soc., Vol. 3, 450 (1971). – Abstr. AAS.

077.013 The correlation of type III radio bursts and 4 keV X-ray events. S. W. Kahler.
Bull. American Astron. Soc., Vol. 3, 450 (1971). – Abstr. AAS.

077.014 Solar radio bursts out to 1 A.U.
J. Fainberg, R. G. Stone.
Bull. American Astron. Soc., Vol. 3, 482 (1971). – Abstr. AAS.

077.015 Spike burst – type III burst associations.
G. L. Tarnstrom, K. W. Philip.
Astron. Astrophys., Vol. 17, 267 - 275 (1972).
High-time-resolution spectral observations of meter wavelength solar radio type III burst events reveal occasional spike burst-type III burst associations. Also observed are type III bursts of $\approx$ 0.1 s duration. A plasma hypothesis interpretation of spike bursts requires burst exciters of spatial extent less than $\approx 10^3 - 10^4$ km, stream density $\approx 1 - 10$ electrons/cm^3, and coronal temperature $\approx 10^6$ °K.

077.016 The observations of circular polarization of solar radio emission at 3.15 cm.
N. N. Erjushev, M. V. Tinin, L. I. Tsvetkov.
Izv. Krymskoj Astrofiz. Obs., Vol. 43, 3 - 16 (1971).
In Russian.
The observational results of circular polarization of local sources and bursts at 3.15 cm are given. The observations were made with the 22-meter radio telescope at the Crimean Observatory during June–July, 1968.

077.017 A comparison of time profiles of type IV bursts at 1.5 meters with the total flux of Hα emission.
L. I. Yourovskaya, Y. F. Yourovsky.
Izv. Krymskoj Astrofiz. Obs., Vol. 43, 17 - 20 (1971).

In Russian.

The time profiles of six type IV bursts are considered. The curves of total flux of Hα emission have been obtained for corresponding flares. In some cases the profiles of type IV bursts at 1.5 meters coincide with those of the total flux of Hα emission.

077.018 **Observations of the solar radio emission at 8, 13 and 16 mm.** V. A. Efanov, I. G. Moiseev.
Izv. Krymskoj Astrofiz. Obs., Vol. 43, 21 - 25 (1971).
In Russian.

The observational results of the solar radioemission within the range 8 to 16 mm are given. The observations were made with the 22-meter radio telescope at the Crimean Observatory. Brightness temperatures of the undisturbed sun $(7950 \pm 500, 9700 \pm 600$ and $11750 \pm 700°$K) at 8.1, 13.2 and 16.6 mm respectively and the frequency spectrum of six local sources have been determined.

077.019 **Radioastronomical observations in the range 0.9–1.5 mm with the 22-meter radiotelescope and the receiver made of n-InSb.** A. N. Viestavkin, V. A. Efanov, V. N. Listvin, I. G. Moiseev, E. I. Popov, V. T. Potapov.
Izv. Krymskoj Astrofiz. Obs., Vol. 43, 26 - 29 (1971).
In Russian.

The local sources of radioemission of the sun were found.

077.020 **Dynamic theory of solar radio bursts of type III.** V. V. Zajtsev, N. A. Mityakov, V. O. Rapoport.
N.-i. radiofiz. in-t. Preprint No. 11. Gor'kij, 1971. 28 pp.
In Russian. – Abstr. in Referativ. Zhurn. 51. Astron., 3.51.500 (1972).

077.021 **On the development of the sources of the S-component of solar radio emission at 9.0 cm.**
Sh. B. Akhmedov.
Solnechnye Dannye 1971 Byull., No. 11, p. 67 - 74 (1972).
In Russian.

It is shown that the radio source flux increases at the appearance of sunspot groups with an area of Sp $\geq 60-100$ millionths of the visible solar hemisphere in plages; in this case the effective size of the sources connected with these regions decreases. At the decay of sunspot groups the radio flux falls while the size of the source rises. The development and decay of sunspots with the area of Sp $< 60-100$ millionths of the visible solar hemisphere has no influence on the parameters of the sources.

077.022 **Activité radioélectrique solaire en 1969.**
C. Gonze, R. Gonze.
Ciel et Terre, Vol. 88, 37 - 56 (1972).

077.023 **On the correlation of the polarization of type III solar radio bursts at 23.5 and 30 MHz.**
A. Krüger, V. V. Fomichev, I. M. Chertok.
Astron. Zhurn. Akad. Nauk SSSR, Vol. 49, 355 - 359 (1972).
In Russian. English translation in Soviet Astron. AJ, Vol. 16, No. 2.

A comparison of the mean polarization degree of type III radio bursts at 23.5 and 30 MHz is carried out by observational data during 1965–1968.

077.024 **On the temperature and emission measure of thermal radio bursts.** F. I. Shimabukuro.
Solar Physics, Vol. 23, 169 - 177 (1972).

Once the area of a thermal burst region has been determined, it is possible to obtain the temperature and emission measure of the burst by examination of the flux spectrum. Such determinations have been made for three events.

077.025 **Possible long-period oscillations in solar radio emission at microwaves.** P. Kaufmann.
Solar Physics, Vol. 23, 178 - 182 (1972).

Long-enduring bursts in the sun at microwaves can occur in a succession, and rare examples of 41-min periodic structure are shown. The experimental difficulties for identifying such phenomena are discussed; their possible association with oscillations in quiescent prominences is suggested.

077.026 **Coronal scattering of radiobursts at hectometer and kilometer wavelengths.** J.-L. Steinberg.
Astron. Astrophys., Vol. 18, 382 - 389 (1972).

We use a Monte-Carlo ray-tracing technique to study the scattering of low-frequency bursts, to derive the angular size of their scattered image and the variation of their apparent intensity with their heliographic longitude. These results are compared with the few available observations.

077.027 **Emission at frequencies much higher than the plasma frequency and type III solar radio bursts.**
D. F. Smith.
Astron. Astrophys., Vol. 18, 403 - 407 (1972).

The possibility that radiation at frequencies much higher than the plasma frequency plays a role in type III burst sources at low frequencies is examined. It is concluded that most of the radiation of type III bursts at low frequencies is near the fundamental of the plasma frequency by considering the observed or apparent directivity of the radiation. Hence the plasma hypothesis is valid all the way to the orbit of the earth.

077.028 **The solar radio emission service in the USSR.**
M. S. Durasova, O. I. Yudin.
Solnechnye Dannye 1972 Byull., No. 1, p. 78 - 82 (1972).
In Russian.

The principal data on the Soviet stations of the solar radio emission service and their equipment are given. The methods of calibration of receivers used at different stations are described.

077.029 **The frequency drift and time splitting of decameter solar radio bursts.** C. V. Sastry.
Astrophys. Letters, Vol. 11, 47 - 51 (1972).

The frequency drift and time splitting of solar radio bursts are observed with a multi-channel radiometer and a polarimeter at frequencies around 25 MHz. It is found that a majority of noise storm bursts have drift rates between +1.0 and –1.0 MHz/sec. Two types of unusual bursts with frequency drift are described. The two components of a double burst are found to be polarized to the same degree and in the same sense.

077.030 **Solar non-thermal radio emissions in the absence of flares.**
M. J. Martres, M. Pick, I. Soru-Escaut, F. Axisa.
Nature, Phys. Sci., Vol. 236, 25 - 27 (1972).

Observations showing that metric type III bursts can also occur in the absence of flares and in association with a chromospheric transient motion along a magnetic inversion line at the photospheric level are presented. This study is performed by using radio data from the 169 MHz Nançay radio heliograph which provided the East-West position of the bursts and spectrographic data from different observatories. Optical data were obtained at Meudon Observatory and Sines Station by using an Hα telescope fitted with a 0.75 Å band pass Lyot filter. Three frames are obtained at the Hα center and ±0.75 Å every minute, allowing us to detect radial velocities up to 30 km/s. Thus the new association proposed between type III bursts and the chromospheric feature are based on both time and position coincidence. Such observations may provide meaningful information about the acceleration of fast particles in absence of flares.

077.031 Radio and X-ray emissions associated with the chromospheric flares of November 15 and 16, 1970.
G. Mariş.
Stud. Cerc. Astron., Vol. 17, 79 - 84 (1972). In Romanian.
The radio bursts and X-ray emission associated with the chromospheric flares of November 15 and 16, 1970 are presented. An explanation of these radio and X-ray emissions is proposed, in agreement with the comprehensive model of T. Takakura and K. Kai.

077.032 Position observations of simultaneous continuum and type III bursts at decametric wavelengths.
R. J. Fitzenreiter, J. Fainberg.
Bull. American Astron. Soc., Vol. 4, 209 (1972). – Abstr. AAS.

077.033 A model of the evolution of the flux density and polarization in isolated moving type IV bursts.
E. J. Schmahl.
Proc. Astron. Soc. Australia, Vol. 2, 95 - 98 (1972).

077.034 The effect of scattering on solar radio sources at 80 MHz. A. C. Riddle.
Proc. Astron. Soc. Australia, Vol. 2, 98 - 100 (1972).

077.035 Relative positions of fundamental and second harmonic type III bursts. R. T. Stewart.
Proc. Astron. Soc. Australia, Vol. 2, 100 - 101 (1972).

077.036 The coronal site of a type III burst as a source of interplanetary electrons. I. D. Palmer, R. P. Lin.
Proc. Astron. Soc. Australia, Vol. 2, 101 - 103 (1972).

077.037 Solar radio observations of the proton event of 1971 January 24.
I. D. Palmer, S. F. Smerd, A. C. Riddle.
Proc. Astron. Soc. Australia, Vol. 2, 103 - 105 (1972).

077.038 Slowly varying component spectrum of the solar radio emission at millimetre wavelengths.
V. A. Efanov, A. G. Kislyakov, I. G. Moiseev.
Solar Physics, Vol. 24, 142 - 153 (1972).
This paper deals with the observed data on the solar S-component sources at millimetre wavelengths. The observations were made in 1968 and 1969 using the 22-m radio telescope of the Crimean Astrophysical Observatory at six wavelengths: 2, 4, 6, 8, 13 and 17 mm.

077.039 Frequency separation in structure of solar continuum radio bursts. H. Rosenberg, G. Tarnstrom.
Solar Physics, Vol. 24, 210 - 214 (1972).
A plot of frequency separation in fine structure in solar continuum radio bursts against emission frequency indicates that the frequency structure cannot represent local proton plasma frequency modulation. However, the observations are consistent with the interpretation of the frequency structure as harmonics of the local electron cyclotron frequency and lead to reasonable estimates of the ratio between magnetic and kinetic pressures in stable coronal magnetic field configurations producing continuum radio sources.

077.040 Results of observation of spectra and polarization of meter solar radio emission with high time resolution: May–June, 1969. G. P. Chernov, I. M. Chertok, V. V. Fomichev, A. K. Markeev.
Solar Physics, Vol. 24, 215 - 232 (1972).
Simultaneous observations of spectra and polarization of two noise storms with high time resolution have been performed in IZMIRAN during the periods: May 17–23 and June

7–13, 1969. The results of the analysis show that for different noise storms type I bursts and chains of type I bursts possess different spectral and polarization characteristics and different tendencies in variation of these characteristics from day to day. In addition, 112 type III bursts with weak or moderate polarization were observed.

077.041 On the temporal distribution of type IV burst-active centres over the solar cycle.
A. Krüger, B. Trinkkeller.
Solar Physics, Vol. 24, 233 - 235 (1972). – Research note.

077.042 Struktur solarer Aktivzonen im mm-Wellengebiet.
G. Feix.
Mitt. Astron. Ges., No. 31, p. 155 - 158 (1972).

077.043 Über die Abklingphase von Mikrowellenbursts.
E. Fürst.
Mitt. Astron. Ges., No. 31, p. 158 - 163 (1972).

077.044 Homologe komplexe Radiobursts vom Typ IV - II in der Sonnenatmosphäre. H. Urbarz.
Mitt. Astron. Ges., No. 31, p. 163 - 164 (1972).

077.045 Some characteristics of type IV radio burst groups.
S. T. Akinjan.
Astron. Zhurn. Akad. Nauk SSSR, Vol. 49, 579 - 587 (1972).
In Russian. English translation in Soviet Astron. AJ, Vol. 16, No. 3.
For the components of three main groups of type IV radio bursts the following characteristics are obtained on the basis of solar observations at fixed frequencies: 1) the time delay of the burst start at different frequencies within the limits of one component, 2) the time delay between components belonging to the same group, 3) the time delay of a type IV burst comparatively to the corresponding one of type II. The conditions of phenomena development in groups are established from these characteristics.

077.046 The experience of the correlation analysis of the continuum of radio emission of noise storms.
A. A. Gnezdilov.
Astron. Zhurn. Akad. Nauk SSSR, Vol. 49, 666 - 669 (1972).
In Russian. English translation in Soviet Astron. AJ, Vol. 16, No. 3.
For two noise storms at frequencies of 111, 202, 224, 287 MHz and two others only at 202 MHz, the results of a correlation spectral analysis of radio emission fluctuations are given.

077.047 Relazione sulla radioastronomia solare e planetaria.
A. Abrami.
Atti XIII Riunione Soc. Astron. Italiana, Trieste 1969, (see 012.010), p. 29 - 41 (1970).

077.048 Radio emission of the quiet sun.
M. Felli, G. Tofani.
Astrophys. Space Sci. Library, Vol. 27, (see 012.011), 267 - 286 (1971).

077.049 Solar bursts at decameter and hectometer wavelengths. M. R. Kundu.
Astrophys. Space Sci. Library, Vol. 27, (see 012.011), 287 - 307 (1971).

077.050 Solar radiobursts. A. Boischot.
Astrophys. Space Sci. Library, Vol. 29, Part I, (see 012.012), p. 87 - 98 (1972).

077.051 Plasma radiation from collisionless MHD shock waves. III. Type II solar radio bursts. D. F. Smith.

Astrophys. Journ., Vol. 174, 643 - 658 (1972).

The theory developed in Papers I and II of this series is applied to type II solar radio bursts. Observations are reviewed, and basic requirements for a theory are established. The case of a shock at the 80-MHz plasma level with a velocity of 1000 km s^{-1}, an Alfvén Mach number of 2.5, and an angle between the plane of the shock and the ambient magnetic field of 3°42' is examined in detail. A possible explanation for the difference in apparent positions of the sources of fundamental and second-harmonic radiation close to the limb is presented. The connection between type II and type IV solar radio bursts is briefly discussed.

077.052 **Observations radioélectriques solaires faites sur 600 MHz en 1970 au Laboratoire de Radioastronomie de Humain-Rochefort.** C. Gonze, R. Gonze.
Bull. Astron. Obs. Roy. Belgique, Vol. 7, 281 - 300 (1972).

077.053 **Pulsating modulations and peculiar absorptions of type IV emissions from the solar corona.**
A. Abrami.
Nature, Phys. Sci., Vol. 238, 25 - 28 (1972).

Between 1967 and the present, we have observed twenty-six pulsating events at a frequency of 237 MHz, with high temporal resolution (10 ms of max time response) allowing accurate recording of the time profile of the pulsation cycle and of other transient details of the emissions. We have summarized the most evident single-frequency features of these peculiar emission phases that are usually associated with the maximum flux enhancements during type IV radio events.

077.054 **327 MHz solar radio observations.**
F. S. Delli Santi, E. Nasi.
Oss. Astron. Univ. Bologna, Notizie Rassegne, Nos. 44 - 51 (1970/1971). – 1970 September – 1971 December.

077.055 **The criteria for homologous radio events associated with flares.** M. Vukićević-Karabin, M. Dimitrijević.
Publ. Dep. Astron. Univ. Beograd, No. 3, p. 17 - 24 (1971).

The purpose of this contribution is to derive a conclusion by analysing the homologous radio events as to what are the strong and weak points of Fokker's criteria. In addition, an attempt has been made to improve these criteria by introducing certain contributions.

077.056 **Curves of the solar radio radiation from observations of the Observatory of the Department of Astronomy at the Kiev University in Lesnikakh.**
Kometn. Tsirk., Kiev, Nos. 133, 134 (1972). In Russian.
1972 May – June.

077.057 **Das Verhalten von Schmalbandbursts vom Typ IV und der Vergleich mit der Theorie der Synchrotronstrahlung im Medium mit $n \leqslant 1$.** H. Urbarz.
Kleinheubacher Berichte, Vol. 14, 103 - 108 = Mitt. Astron. Inst., Univ. Tübingen, No. 126 (1971).

The features of 15 narrow band type IV-bursts recorded at Weissenau are described. The burst of this type recorded on May 26th, 1969 on film has also been obtained on magnetic tape in digital form. It was found that the corresponding flux density-frequency profiles show only small variations in shape while the maximum frequency decreases by 15 %. The theory of Ramaty and Lingenfelter predicting narrow band synchrotron emission in a medium with refractive index $n \leqslant 1$ is applied to determine the magnetic field and electron energy from the burst profiles.

077.058 **On a possible connection between stopped type III-bursts and U-bursts.** H. Daene, V. V. Fomichev.
Zentralinstitut für Solar-Terrestrische Physik (Heinrich-Hertz-Institut), Deutsche Akad. Wiss. Berlin, HHI Suppl. Ser. Solar

Data, Vol. 2, 241 - 252 (1971).

It is suggested by the shape of the distribution curve of type III-stop frequencies that there exist two subclasses of type III bursts guided by unipolar or bipolar magnetic fields. The latter subclass is evidently at least partly connected with the U-burst respectively I-burst phenomenon. The reasons responsible for the less development or absence of the descending branch are discussed and it is shown that by diverged field lines of the leading magnetic field a decrease of the radiation intensity can be explained without restraint.

077.059 **A U-type solar radio burst originating in the outer corona.** R. G. Stone, J. Fainberg.
Rep. NASA–TM–X–65484, National Aeronautics and Space Administration, Greenbelt, Maryland. [Available from NTIS, Springfield, Va.], 14 pp. (1971).

A discussion is presented of the very rare occurrence of a U-type burst observed between 5 and 0.7 MHz by RAE-1. A possible model is developed from the data.

077.060 **Components of the radio emission of the sun.**
J. Olmr.
Kozmos, Vol. 3, 8 - 10 (1972). In Czech.

077.061 **The solar radio burst activity index (I_b) and the burst incidence (B_i) for 1968-69.** D. Basu.
Rev. Brasil. Fis. (*São Paulo*), Vol. 1, 297 - 302 (1971). – See Phys. Abstr., Vol. 75, No. 41235 (1972).

The sun as a test source for boresight calibration of microwave antennas. See Abstr. 033.051.

High resolution observations of the chromosphere at mm and cm wavelengths. See Abstr. 073.002.

Flare-associated solar wind disturbances and type II and IVm radio bursts. See Abstr. 073.022.

A classification of magnetic field configurations associated with solar flares. See Abstr. 073.047.

On the propagation of shock waves in the solar corona causing radio bursts of type II. See Abstr. 074.052.

Neutralization and stabilization of particle streams in the corona and type III radio bursts. See Abstr. 074.056.

Studies of the outer corona through space radio astronomy. See Abstr. 074.087.

On the activity of solar coronal condensations discussed after long-enduring microwave events. See Abstr. 074.096.

Relationship between impulsive solar flare X-rays and type III solar radio bursts. See Abstr. 076.015.

X-ray (E > 10 keV), Hα and microwave emission during the impulsive phase of flares. See Abstr. 076.016.

Soft X-ray and microwave observations of hot regions in solar flares. See Abstr. 076.022.

Sudden frequency deviations, solar extreme ultraviolet bursts, and solar radio bursts. See Abstr. 076.041.

Correlation of solar radio bursts and sudden increases of the total electron content (SITEC) of the ionosphere. See Abstr. 083.066.

078 Solar Cosmic Radiation

078.001 Nuclear composition and energy spectra in the 1969 April 12 solar-particle event.
D. L. Bertsch, C. E. Fichtel, D. V. Reames.
Astrophys. Journ., Vol. 171, 169 - 177 (1972).

The charge composition for several of the multicharged nuclei and the energy spectra for hydrogen, helium, and medium ($6 \leq Z \leq 9$) nuclei were measured in the 1969 April 12 solar-particle event. By combining the results obtained here with previous work, improved estimates of the Ne/O and Mg/O values were obtained. Silicon and sulfur abundances relative to O were determined and 85 percent confidence upper limits for Ar and Ca relative to O were obtained.

078.002 Enrichment of very heavy nuclei in the composition of solar accelerated particles.
A. Mogro-Campero, J. A. Simpson.
Astrophys. Journ., (*Letters*), Vol. 171, L5 - L9 (1972).

The abundances of the nuclei C, N, O, Ne, Mg, Si, Ar, and Ca and the group Cr-Co relative to oxygen from seven solar energetic-particle events have been measured in the energy range $\sim 14-61$ MeV per nucleon with a solid-state detector telescope on the OGO-5 satellite, 1968–1971.

078.003 Relativistic electrons from the sun observed by IMP-4. G. M. Simnett.
Solar Physics, Vol. 22, 189 - 219 (1972).

Data are presented of 0.3–12 MeV electrons from the sun between May 24, 1967 and May 2, 1969. Correlations with contemporary proton intensity increases at energies above 1 MeV are studied. Categories of unusual events are defined and examples of each type are given. The differential electron energy spectrum (0.3–12 keV) from solar flares appears to be a constant of the flare process. Particle emission from solar flares contains a prompt component, which is injected into the interplanetary medium beyond the sun and which is responsible for the diffusion characteristics of solar particle events. Storage of electrons > 300 keV and protons > 1 MeV is essential to explain emission and propagation characteristics of solar particle events.

078.004 North-south anisotropies in the cosmic radiation.
M. A. Pomerantz, S. P. Duggal.
Journ. Geophys. Res., Vol. 77, 263 - 265 (1972). – Letter.

078.005 Rapid access of solar electrons to the polar caps.
J. P. Turtle, E. J. Oelbermann, Jr., J. B. Blake, L. J. Lanzerotti, A. L. Vampola, G. K. Yates.
Journ. Geophys. Res., Vol. 77, 730 - 735 (1972).

Simultaneous measurements of solar electrons and protons in interplanetary space and in the magnetotail were made during the onset of the November 2, 1969, solar-particle event. These particle measurements, when compared with continuous transpolar VLF measurements on three propagation paths, indicate that the solar electrons have access to the magnetotail and north polar cap with a time delay τ such that $0 \lesssim \tau < 1$ min.

078.006 Satellite measurements of the charge composition of solar cosmic rays.
B. J. Teegarden, T. T. von Rosenvinge, F. B. McDonald.
Bull. American Astron. Soc., Vol. 3, 439 (1971). – Abstr. AAS.

078.007 Proton measurements with the satellite Azur during the solar particle events of March 5 - 13, 1970.
E. Kirsch, J. W. Münch.
Planet. Space Sci., Vol. 20, 89 - 101 (1972).

078.008 Low-energy solar positrons. C. T. Gregory.
Journ. Geophys. Res., Vol. 77, 1316 - 1320 (1972).
Letter.

078.009 Enhanced abundances of low-energy heavy elements in solar cosmic rays.
L. J. Lanzerotti, C. G. Maclennan, T. E. Graedel.
Astrophys. Journ., (*Letters*), Vol. 173, L39 - L43 (1972).

Measurements of solar cosmic-ray O/He, Si/He, and Fe/He ratios during the onset phase of the 1971 January 25 flare event indicate a substantial (20–40 times) enhancement of the ratios above the solar photospheric abundances. It is suggested that preferential acceleration of the heavy elements within the solar-flare region is the most likely cause of the increased abundances.

078.010 Two-satellite observation of spatial and temporal particle flux variations over the polar caps.
V. Domingo, D. E. Page.
Journ. Geophys. Res., Vol. 77, 1971 - 1975 (1972). – Letter.

078.011 Transport of cosmic rays in the solar corona.
L. A. Fisk, K. H. Schatten.
Solar Physics, Vol. 23, 204 - 210 (1972).

A model is proposed to explain the transport of energetic protons in the solar corona. Comparison of predictions of the model with cosmic ray observations at ~ 1 AU provide some support for the model.

078.012 High energy electrons detected during solar flares.
C. Dilworth, D. Maccagni, F. Perotti, E. G. Tanzi, J. P. Mercier, A. Raviart, L. Treguer, M. Gros.
Solar Physics, Vol. 23, 487 - 500 (1972).

The S 79 experiment on board of the HEOS-A1 European Satellite has been designed to electrons detection whose kinetic energies should be equal or greater than 7.5 MeV. From December 1968 to July 1970, 11 events were observed. Their main characteristics are described in this article. Two different categories of events may be sorted out from these observations. The propagation conditions in the interplanetary space are now discussed to find out a possible interpretation.

078.013 About the possibility of determining the spectrum of solar cosmic rays and the geomagnetic cutoff rigidity by ionospheric data. L. I. Dorman, T. M. Krupitzkaya.
Geomagn. Aeronom., Vol. 12, 180 - 183 (1972). In Russian.

078.014 Generation of the nuclei of cosmic rays of solar origin. S. S. Konyakhina, L. V. Kurnosova, V. I. Logachëv, L. A. Razorënov, V. G. Sinitsyna, M. I. Fradkin.
Izv. AN SSSR. Ser. fiz., Vol. 35, 2446 - 2447 (1971). In Russian. – Abstr. in Referativ. Zhurn. 51. Astron., 5.51.389; 62. Issled. kosm. prostranstva, 5.62.247 (1972).

078.015 Some problems of the theory of cosmic ray modulation effects. M. B. Bagdasaryan, A. V. Belov, L. I. Dorman, B. A. Shakhov.
Izv. AN SSSR. Ser. fiz., Vol. 35, 2492 - 2497 (1971). In Russian. – Abstr. in Referativ. Zhurn. 51. Astron., 5.51.392 (1972).

078.016 On the propagation of solar protons in a medium with constant and variable transport path length at instantaneous injection near the sun.
S. N. Vernov, E. V. Gorchakov, G. A. Timofeev.
Izv. AN SSSR. Ser. fiz., Vol. 35, 2423 - 2427 (1971). In

Russian. – Abstr. in Referativ. Zhurn. 51. Astron., 5.51.444 (1972).

078.017 On the propagation of "localized" solar cosmic rays.
G. P. Lyubimov, N. N. Kontor, N. V. Pereslegina.
Izv. AN SSSR. Ser. fiz., Vol. 35, 2428 - 2433 (1971). In Russian. – Abstr. in Referativ. Zhurn. 51. Astron., 5.51.445 (1972).

078.018 Solar cosmic ray events in February – April 1969.
G. A. Bazilevskaya, V. V. Bayarevich, L. P. Borovkov, E. V. Vashenyuk, L. L. Lazutin, N. S. Svirzhevskij, A. V. Stozhkov.
Izv. AN SSSR. Ser. fiz., Vol. 35, 2530 - 2537 (1971). In Russian. – Abstr. in Referativ. Zhurn. 51. Astron., 5.51.465 (1972).

078.019 Solar cosmic ray events on 11 - 18 April 1969 and their effects on the lower ionosphere.
V. M. Driatskij, T. M. Krupitskaya, A. V. Shirochkov.
Izv. AN SSSR. Ser. fiz., Vol. 35, 2538 - 2542 (1971). In Russian. – Abstr. in Referativ. Zhurn. 51. Astron., 5.51.466 (1972).

078.020 Generation of nucleonic and electromagnetic components during solar flares.
L. E. Gajnova, V. I. Ivanov, K. Imazhanova, E. V. Kolomeets.
Izv. AN SSSR. Ser. fiz., Vol. 35, 2543 - 2546 (1971). In Russian. – Abstr. in Referativ. Zhurn. 51. Astron., 5.51.467 (1972).

078.021 On a relationship between solar activity and solar cosmic rays.
O. M. Kovrizhnykh, I. A. Savenko, G. I. Chukhrai.
Solnechnye Dannye 1972 Byull., No. 1, p. 107 - 114 (1972). In Russian.
About 200 events of increasing intensity of cosmic rays connected with solar flares have been considered. Variations of the number of proton flares with the 11-year cycle were studied. A possible relationship between proton flares and intensity of coronal lines has been detected.

078.022 Evidence for solar particle production above ~75 GeV. S. M. Schindler, P. D. Kearney.
Nature, Vol. 237, 503 - 505 (1972).
Observations of solar particles in the energy region above ~75 GeV provide good evidence for particle acceleration during the initial phase of a solar flare.

078.023 Permanent sources of particle emission from the sun. M. Pick.
Astrophys. Space Sci. Library, Vol. 29, Part I, (see 012.012), p. 61 - 71 (1972).

078.024 Solar discrete particle events. Z. Švestka.
Astrophys. Space Sci. Library, Vol. 29, Part I, (see 012.012), p. 72 - 86 (1972).

078.025 Energetic solar particles in the interplanetary medium. W. I. Axford.
Astrophys. Space Sci. Library, Vol. 29, Part II, (see 012.013), p. 110 - 134 (1972).

078.026 Polar-cap structures of solar protons observed during the passage of interplanetary discontinuities.
M. Scholer.
Journ. Geophys. Res., Vol. 77, 2762 - 2769 (1972).
Particle measurements from the low-altitude polar-orbiting satellite Azur and from the IMP 5 satellite inside the magnetosphere near the magnetopause and particle and magnetic-field measurements in interplanetary space from the Heos A1 satellite are presented for the solar-proton event of March 1970. The report is concerned with the structures of polar-cap intensity that are due to two sudden interplanetary flux increases during the March 1970 solar events.

The physics of cosmic X-ray, γ-ray, and particle sources. See Abstr. 003.015.

Chromospheric flares as sources of directed corpuscular streams. See Abstr. 073.081.

On the relationship between the solar coronal green line intensity and the limiting primary rigidity of the solar diurnal anisotropy of cosmic rays. See Abstr. 074.068.

Non-uniform entry of solar protons into the polar cap. See Abstr. 084.215.

Solar particles and the dayside limit of closed field lines. See Abstr. 084.219.

Interaction of solar and galactic cosmic-ray particles with the moon. See Abstr. 094.031.

Streaming of galactic cosmic rays in the interplanetary magnetic field. See Abstr. 143.005.

079   Solar Eclipses

079.001   **Die nächsten zwanzig in der BRD sichtbaren Sonnen-finsternisse.**   F. Dorst.
SuW, Vol. 11, 53 - 55 (1972).

079.002   **A suggested eclipse experiment.**
B. K. Dennison, G. L. Sego.
Journ. Roy. Astron. Soc. Canada, Vol. 66, L4 - L5 (1972).

079.003   **The problem of shadow band observations.**
A. T. Young.
Sky Telescope, Vol. 43, 291 - 292 (1972).

079.004   **Hints on photographing the eclipse.**
P. A. Leavens.
Sky Telescope, Vol. 43, 358 (1972).

079.005   **The shadow of the moon in the sky at a total solar eclipse: Part I.**   W. H. Glenn.
Strolling Astronomer, Vol. 23, 136 - 142 (1972).

079.006   **Solformørkelser.**   O. P. Sveen.
Astron. Tidssk., Årg. 5, p. 22 - 33 (1972).

079.007   **The shadow of the moon in the sky at a total solar eclipse: Part II.**   W. H. Glenn.
Strolling Astronomer, Vol. 23, 150 - 159 (1972).

Solar eclipses and the ionosphere.
See Abstr. 003.021.

Prediction and analysis of solar eclipse circumstances.   See Abstr. 003.141.

Studies on Senmyō-Reki or Hsüan-ming-li (4): Records of solar eclipses in Japanese documents.
See Abstr. 004.038.

The chromospheric continuum observed at the total solar eclipse of 12 November 1966 and a model of the lower atmosphere.   See Abstr. 073.067.

The interpretation of total line intensities from optically thin gases. III. Application to coronal forbidden line spectra.   See Abstr. 074.025.

079.100   **Solar eclipse, 1971 August 6**

Observation of the total solar eclipse on August 6, 1971.   V. A. Bronshtehn.
Zemlya i Vselennaya, No. 1, p. 74 - 76 (1972). In Russian.

079.101   **Solar eclipse, 1970 March 7**

Atmospheric electricity measurements at Waldorf, Maryland during the 7 March 1970 solar eclipse.
R. V. Anderson, H. Dolezalek.
Journ. Atmosph. Terr. Phys., Vol. 34, (see 012.002), 561 - 566 (1972).

Atmospheric electricity, turbulence and a pseudo-sunrise effect resulting from a solar eclipse.
R. V. Anderson.
Journ. Atmosph. Terr. Phys., Vol. 34, (see 012.002), 567 - 572 (1972).

The shadow band phenomenon.
J. J. Quann, C. J. Daly.
Journ. Atmosph. Terr. Phys., Vol. 34, (see 012.002), 577 - 583 (1972).
During the solar eclipse of 7 March 1970 shadow band data were gathered at the site established at Wallops Island, Virginia, in two ways: (1) Recording onto magnetic tape of the output from six collimated photocells whose spectral responses ranged from the ultraviolet to the infrared; (2) visual and mechanical measurement of the orientation and motion of the bands.

Brightness of forbidden OI lines and properties of shadow bands during the eclipse of 7 March 1970.
D. E. Kerr, G. G. Sivjee, W. McKinney, P. Takacs, W. G. Fastie.
Journ. Atmosph. Terr. Phys., Vol. 34, (see 012.002), 585 - 592 (1972).

Response of the neutral particle upper atmosphere to the solar eclipse of 7 March 1970.
J. J. Horvath, J. S. Theon.
Journ. Atmosph. Terr. Phys., Vol. 34, (see 012.002), 593 - 599 (1972).

Rocket observations of solar UV radiation during the eclipse of 7 March 1970.   L. G. Smith.
Journ. Atmosph. Terr. Phys., Vol. 34, (see 012.002), 601 - 611 (1972).
Photometers sensitive to narrow bands of radiation at Lyman-α (1216 Å) and at 2600 Å were included in the payloads of four Nike Apache rockets flown from Wallops Island before and during the eclipse. This paper presents some results of the solar UV radiation experiments and indicates where further analysis is planned.

Rocket observations of solar X-rays during the eclipse of 7 March 1970.
C. A. Accardo, L. G. Smith, G. A. Pintal.
Journ. Atmosph. Terr. Phys., Vol. 34, (see 012.002), 613 - 620 (1972).

Measurements of solar Lyman-α radiation during the eclipse of 7 March 1970.   P. H. G. Dickinson.
Journ. Atmosph. Terr. Phys., Vol. 34, (see 012.002), 621 - 625 (1972).

Ionization changes in the lower ionosphere during the solar eclipse of 7 March 1970.
J. S. Belrose, D. B. Ross, A. G. McNamara.
Journ. Atmosph. Terr. Phys., Vol. 34, (see 012.002), 627 - 640 (1972).

Electron loss coefficients for the D-region of the ionosphere from rocket measurements during the eclipses of March 1970 and November 1966.
E. A. Mechtly, C. F. Sechrist, Jr., L. G. Smith.
Journ. Atmosph. Terr. Phys., Vol. 34, (see 012.002), 641 - 646 (1972).

Ion composition measurements in the lower ionosphere during the November 1966 and March 1970 solar eclipses.
R. S. Narcisi, A. D. Bailey, L. E. Wlodyka, C. R. Philbrick.
Journ. Atmosph. Terr. Phys., Vol. 34, (see 012.002), 647 - 658 (1972).

The eclipsed lower ionosphere as investigated by natural very low frequency radio signals.
C. D. Reeve, M. J. Rycroft.
Journ. Atmosph. Terr. Phys., Vol. 34, (see 012.002), 667 - 672 (1972).

The electron heating rate and ion chemistry in the thermosphere above Wallops Island during the solar eclipse of 7 March 1970.   L. H. Brace, H. G. Mayr, M. W. Pharo III, L. R. Scott, N. W. Spencer, G. R. Carignan.
Journ. Atmosph. Terr. Phys., Vol. 34, (see 012.002), 673 - 688 (1972).

Rocket measurements of conjugate photoelectrons during the total solar eclipse of 7 March 1970 over Wallops Island.   E. J. Maier, B. C. N. Rao.
Journ. Atmosph. Terr. Phys., Vol. 34, (see 012.002), 689 - 694 (1972).

Ionospheric effects of two recent solar eclipses.
R. T. Marriott, D. E. St. John, R. M. Thorne, S. V. Venkateswaran, P. Mahadevan.
Journ. Atmosph. Terr. Phys., Vol. 34, (see 012.002), 695 - 712 (1972).

Neutral winds implied by electron content observations during the 7 March 1970 solar eclipse.
O. G. Almeida, H. Waldman, A. V. Da Rosa.
Journ. Atmosph. Terr. Phys., Vol. 34, (see 012.002), 713 - 717 (1972).

Ionospheric undulations during the solar eclipse of 7 March 1970.   P. R. Arendt.
Journ. Atmosph. Terr. Phys., Vol. 34, (see 012.002), 719 - 725 (1972).

Ionospheric HF Doppler dispersion during the eclipse of 7 March 1970 and TID analysis.   R. D. Sears.
Journ. Atmosph. Terr. Phys., Vol. 34, (see 012.002), 727 - 732 (1972).

Traveling ionospheric disturbances observed near the time of the solar eclipse of 7 March 1970.
G. M. Lerfald, R. B. Jurgens, J. F. Vesecky, T. W. Washburn.
Journ. Atmosph. Terr. Phys., Vol. 34, (see 012.002), 733 - 741 (1972).

Photography of the eclipse of 7 March, 1970 from two locations.   S. A. Korff, R. B. Mendell.
Solar Physics, Vol. 21, 482 (1971). − Abstract.

The International Symposium on the 1970 solar eclipse.   See Abstr. 012.001.

The chromosphere in continuum emission observed at the total solar eclipse on 7 March 1970.
See Abstr. 073.056.

Observation of the green line with an electronic camera during the eclipse of March 7, 1970.
See Abstr. 074.032.

Eclipse and noneclipse differential photoelectron flux.   See Abstr. 083.030.

079.102   Solar eclipse, 1955 June 20

Solar-eclipse effect on sporadic-E ionization.
See Abstr. 083.005.

079.103   Solar eclipse, 1972 July 10

Solar eclipse of July 10, 1972.
Sky Telescope, Vol. 43, 25 - 28 (1972).

The aurora borealis and the 1972 total solar eclipse.
R. S. Krassa, W. N. Hall.
Sky Telescope, Vol. 43, 192 - 194 (1972).

Canadian solar eclipse flight.   A. N. Cox.
Sky Telescope, Vol. 43, 290 (1972).

Stars and planets during totality on July 10th.
Sky Telescope, Vol. 43, 292 (1972).

Canadian eclipse rocket program.   J. Bird.
Sky Telescope, Vol. 43, 359 - 360 (1972).

Notes on the partial eclipse of July 10th.
Sky Telescope, Vol. 43, 360 - 361 (1972).

079.104   Solar eclipse, 1965 May 30

Polarimetry of the daytime sky during solar eclipses.
C. R. N. Rao, T. Takashima, J. G. Moore.
Journ. Atmosph. Terr. Phys., Vol. 34, (see 012.002), 573 - 576 (1972).
The degree of linear polarization of the residual illumination in the umbral region was measured in different colors at intermediate levels in the atmosphere (10−12 km) during the total solar eclipses of 30 May 1965 and 12 November 1966.

079.105   Solar eclipse, 1966 November 12

Eclipse rocket measurements of charged particle concentrations.   J. C. Ulwick.
Journ. Atmosph. Terr. Phys., Vol. 34, (see 012.002), 659 - 665 (1972).

Electron loss coefficients for the D-region of the ionosphere from rocket measurements during the eclipses of March 1970 and November 1966.   See Abstr. 079.101.

Ion composition measurements in the lower ionosphere during the November 1966 and March 1970 solar eclipses.   See Abstr. 079.101.

Polarimetry of the daytime sky during solar eclipses.
See Abstr. 079.104.

079.106   Solar eclipse, 1969 September 11

Ionospheric effects of two recent solar eclipses.
See Abstr. 079.101.

079.107   Solar eclipse, 1968 September 22

Results of observations during the partial solar eclipse on Sept. 22, 1968 at 3.1 cm.
A. F. Dravskikh, Z. V. Dravskikh, A. S. Kutuzov, S. G. Smolentsev.
Solnechnye Dannye 1971 Byull., No. 12, p. 106 - 107 (1972). In Russian.

079.108   Solar eclipse, 1971 February 25

Solformørkelsen 25. februar 1971.
T. Hansen, O. P. Sveen.
Astron. Tidssk., Årg. 5, p. 34 - 36 (1972).

079.109   Solar eclipse, 1887 August 19

Observations of the August 19, 1887, total solar e-
clipse in Japan— Second report.   K. Saito, S. Shinozawa.
Tokyo Astron. Obs., Report No. 59, Vol. 15, 617 - 675 (1972).
In Japanese.

079.110   Solar eclipse, 1966 May 20

Flux of solar XUV radiation inferred from iono-
spheric variations during the eclipse of 20th May 1966.
See Abstr. 076.039.

080 Solar Figure, Internal Constitution, Rotation, Miscellanea

080.001 **Solar neutrinos: Where are they?**
A. L. Hammond.
Science, Vol. 175, 505 (1972).

080.002 **On time variations of the solar differential rotation law and asymmetry of the global distribution of the solar activity.** H. Yoshimura.
Solar Physics, Vol. 22, 20 - 33 (1972).

The relation between the systematic time variations of the solar differential rotation at middle latitudes and the asymmetry of the solar activity is discussed in connection with the study of the maintenance of the solar differential rotation. If we adopt the working hypothesis of the solar equatorial acceleration maintained by the angular momentum transport due to the very large scale convection, the two phenomena are related through the concurrent presence of the neighboring modes with the presumed dominant mode of the very large scale convection.

080.003 **Semi-empirical solar line blanketing. II. The results of blanketing in solar model atmospheres.**
J. P. Mutschlecner, C. F. Keller.
Solar Physics, Vol. 22, 70 - 87 = Publ. Goethe Link Obs., Indiana Univ., *Bloomington,* No. 131 (1972).

We have attempted to evaluate the blanketing method in terms of the assumptions and approximations employed. The evaluation was carried out by variations in the assumptions and functions used and by comparison of blanketed models computed by this method with several previously published blanketed models. We examined the consequences of line blanketing in the solar atmosphere. We have considered the results of blanketing both in terms of the effect on the atmospheric structure and in terms of the effect on the predicted intensities and limb darkening. The details of these calculations for semi-empirical and theoretical solar atmospheres are given.

080.004 **On the directional dependence of the emission of acoustic noise by convective turbulence in a gravitational atmosphere.** M. Kuperus.
Solar Physics, Vol. 22, 257 - 262 (1972).

The directional dependence of the emission of sound waves in the solar atmosphere is studied. It is shown that quadrupole acoustic radiation generated in convective turbulence is strongly enhanced in the direction of the mean convective flow. The influence of the atmospheric cut off frequency on the transmitted acoustic spectrum is taken into account. It is suggested that low frequency atmospheric oscillations may modulate the flux of high frequency sound waves.

080.005 **Formation of the solar Hα profile.**
S. A. Schoolman.
Solar Physics, Vol. 22, 344 - 357 (1972).

A series of non-LTE radiative transfer solutions for Hα was computed using the integro-differential equation technique of Athay and Skumanich (1967). A model hydrogen atom consisting of three bound levels and a continuum was assumed. It was found that increasing the temperature of the chromosphere at the height of line formation decreases the central intensity of the line. The density structure of the atmosphere primarily affects the optical depth scale rather than the source function.

080.006 **On the filamentary nature of solar magnetic fields.**
R. Howard, J. O. Stenflo.
Solar Physics, Vol. 22, 402 - 417 (1972).

From our analysis of a large amount of magnetograph data obtained in the Fe I 5250 and 5233 Å lines during 26 dif-

ferent days, we draw the following tentative conclusions: (a) No more than 10% of the magnetic flux of the sun can exist in the form of weak interfilamentary fields. (b) The magnitude of this observed interfilamentary field is less than 3 G on the average. (c) The magnetic-field lines in a filament are confined to a small area in the photosphere, but spread out with height in the atmosphere. (d) The correction factor for recent Mount Wilson magnetograph measurements varies from about 1.8 near the center of the disk to a smaller value near the limb.

080.007 **Theoretical oscillations of the entire sun.**
C. L. Wolff.
Bull. American Astron. Soc., Vol. 3, 462 (1971). – Abstr. AAS.

080.008 **Die Rotation der Sonne.** K. D. Rakosch.
SuW, Vol. 11, 68 - 70 (1972).

080.009 **Forced diffusion of solar magnetic fields.**
G. W. Pneuman, M. A. Raadu.
Astrophys. Journ., Vol. 172, 739 - 748 (1972).

The observed latitude migration of solar magnetic features during the solar cycle is explained within the framework of a dynamic model employing subsurface horizontal forces. The calculated drift rates agree quite well with the latitude drifts observed for sunspots, filaments, unipolar magnetic regions, and coronal features.

080.010 **Oscillations in an isothermal atmosphere: The solar five-minute oscillations.** G. Worrall.
Astrophys. Journ., Vol. 172, 749 - 753 (1972).

An isothermal atmosphere, subject to pressure fluctuations at its base, responds very strongly at the frequency $\omega_g = (\gamma - 1)^{1/2} g/c$, regardless of the horizontal scale of the exciting fluctuations. Because of a misinterpretation of a result obtained by Lamb it has previously been supposed that this response occurs at the frequency $\omega_a = \gamma g/2c$.

080.011 **On the arising of magnetic ropes in the convective zone. III. Gas motions in a magnetic rope.**
A. A. Solovjev.
Solnechnye Dannye 1971 Byull., No. 11, p. 90 - 96 (1972). In Russian.

The character of gas motions in a magnetic rope is investigated. A condition for gas outflow from a rised part of the rope has been obtained. The Parker mechanism of field strengthening with height due to temperature differences outside and inside the magnetic tube leads to negative value of buoyancy force and cannot be realized in an arising rope (or tube).

080.012 **On the problem of oscillations of the solar radius.**
O. B. Vasilyev, B. M. Rubashev.
Solnechnye Dannye 1971 Byull., No. 12, p. 93 - 100 (1972). In Russian.

Italian observations 1876–1937 and Greenwich observations 1851–1937 of the solar radius were smoothed with the help of the Whittaker operator. In the smoothed series and in their frequency spectra common features appeared. The average curve was obtained for possible oscillations of the solar radius during the years 1876–1937.

080.013 **The sun as a magnetic rotator.** J. Tuominen.
Astrophys. Letters, Vol. 10, 175 - 177 (1972).

Counts of newly-formed sunspot groups indicate that the long-lasting magnetic fields of the sun as a whole, which have been observed by Severny, represent fields of the following components of sunspot groups and magnetic regions dispersed over large areas.

080.014 The use of similarity arguments for the study of the circulation of the sun. G. S. Golitsyn.
Astron. Zhurn. Akad. Nauk SSSR, Vol. 49, 360 - 366 (1972).
In Russian. English translation in Soviet Astron. AJ, Vol. 16, No. 2.

The similarity theory, recently proposed, for the general circulation of planetary atmospheres is modified in order to obtain estimates of some characteristics of the sun's circulation, i. e. of the differential rotation.

080.015 The general magnetic field of the sun and its changes with time. A. Severny.
Vistas in astronomy, Vol. 13, (see 003.001), 135 - 148 (1972).

080.016 On the dependence of the linear velocity of solar rotation on latitude and optical depth.
Y. A. Solonsky.
Solar Physics, Vol. 23, 3 - 12 (1972).

The law of solar axial rotation is investigated by the spectroscopic method from a great number of observations. The dependence of the solar rotation on the depth in the atmosphere was investigated. For this purpose the optical depths of the formation of the core of 800 spectral lines were determined. The main conclusion of this investigation is the existence of an anomaly in the rotational velocity of the sun at heliographic latitude 25°. Moreover the amplitude of the deflexion of the rotational velocity at the above-mentioned latitude varies with the optical depth.

080.017 Processes of energy transport by longitudinal waves and the problem of solar neutrinos.
J. E. Littleton, H. M. Van Horn, H. L. Helfer.
Astrophys. Journ., Vol. 173, 677 - 679 (1972).

Energy transport by longitudinal plasma waves in the sun is shown to contribute less than $\sim 5 \times 10^{-5}$ of the total energy flux and is thus insufficient to explain the solar-neutrino problem.

080.018 The evolution of the solar inner rotation by the Eddington-Sweet type circulation under the influence of the solar wind torque. T. Sakurai.
Publ. Astron. Soc. Japan, Vol. 24, 153 - 176 (1972).

The evolution of the axisymmetric rotation of the radiative interior of the sun is investigated. The basic equations are derived within the framework of the Eddington-Sweet theory of perturbation. The effect of the molecular weight gradient built up by the proton-proton reaction and the effect of the solar wind torque are taken into account. The initial state of the rotation is assumed to be rigid body rotation and to have an angular velocity about 20 to 65 times larger than the present surface value.

080.019 Evolution of solar magnetic fields over an 11-year period. J. O. Stenflo.
Solar Physics, Vol. 23, 307 - 339 (1972).

Digitized data on solar magnetic fields recorded at the Mount Wilson Observatory during the period August 1959—May 1970 have been used to study the large-scale evolution of the photospheric magnetic fields. The latitude distribution (butterfly diagram) of the magnetic field is compared with the distribution of sunspots, faculae, prominences and the intensity of the green-line corona. The evolution of the sector structure of the field is calculated. 36 synoptic charts, each representing an average of four solar rotations, illustrate the evolution of the magnetic field over the 11-year period.

080.020 On the relationship between possible oscillations of the solar radius and solar activity.
O. B. Vasilyev, B. M. Rubashev.

Solnechnye Dannye 1972 Byull., No. 1, p. 95 - 101 (1972). In Russian.

Average values of the solar radius and absolute values of its time derivative are compared with the Wolf numbers and their time integrals. Two close correlations are noticed.

080.021 On meridional circulation of solar magnetic field. K. L. McDonald.
Bull. American Astron. Soc., Vol. 4, 208 - 209 (1972). — Abstr. AAS.

080.022 Fe I ionization and excitation equilibrium in the solar atmosphere. B. W. Lites, R. G. Athay.
Bull. American Astron. Soc., Vol. 4, 212 - 213 (1972). — Abstr. AAS.

080.023 Description of solar structure and processes. E. G. Gibson.
Rev. Geophys. Space Phys., Vol. 10, 395 - 461 (1972).

A general introduction to solar structure and processes is presented. The sun is first viewed as a spherically symmetric steady-state system, and the energy generated in the core is traced as it flows outward. The various forms and manifestations of this energy flow and the resulting uniquely defined characteristics of different atmospheric layers are described. The sources of solar activity are assumed to be differential rotation and solar magnetic fields. The interaction of these sources to produce the observed solar cycle, active regions, the active-sun corona, and solar flares is discussed.

080.024 A model of the solar convection zone. B. E. Waters, R. Van der Borght.
Proc. Astron. Soc. Australia, Vol. 2, 92 (1972).

080.025 The temperature of the solar photosphere-chromosphere transition region.
J. H. Carver, B. H. Horton, G. W. A. Lockey.
Proc. Astron. Soc. Australia, Vol. 2, 94 - 95 (1972).

080.026 What cooks with solar neutrinos? W. A. Fowler.
Nature, Vol. 238, 24 - 26 (1972).

Two desperate explanations of the solar neutrino puzzle are proposed, one involving experimental nuclear physics and the other theoretical solar structure and evolution.

080.027 Magnetfelder in Kosmos. F. Krause.
Astron. in der Schule, 9. Jahrgang, p. 50 - 53 (1972).

080.028 Relazione sulla radiazione e struttura dell'atmosfera solare. M. Rigutti.
Atti XIII Riunione Soc. Astron. Italiana, Trieste 1969, (see 012.010), p. 45 - 59 (1970).

080.029 Solar energy sources. C. de Jager.
Astrophys. Space Sci. Library, Vol. 29, Part I, (see 012.012), p. 1 - 8 (1972).

080.030 Large-scale magnetic fields and activity patterns on the sun. V. Bumba.
Astrophys. Space Sci. Library, Vol. 29, Part I, (see 012.012), p. 21 - 37 (1972).

080.031 Circumsolar radiation and the International Pyrheliometric Scale. J. Geist, J. M. Kendall.
Applied Optics, Vol. 11, 1437 - 1439 (1972).

Values of solar irradiance at the surface of the earth measured by the absolute radiometers of Kendall and Willson are approximately 2% higher than values measured by Eppley-Angstrom pyrheliometers which have been calibrated against the International Pyrheliometric Scale of 1956. It has been suggested that the two types of instruments might collect different amounts of circumsolar radiation, and that this would explain the 2% discrepancy. The purpose of this note is to show that this explanation is very unlikely.

080.032 Force-free magnetic-field structures and their role in solar activity.   C. W. Barnes, P. A. Sturrock.
Astrophys. Journ., Vol. 174, 659 - 670 (1972).
Magnetic-field structures in solar active regions are expected to be substantially force-free. A method is proposed for calculating such structures by numerical methods. The method is applied to the study of the magnetic-field pattern associated with a sunspot of one polarity surrounded by a magnetic region of opposite polarity when the sunspot rotates with respect to the surrounding region.

080.033 Interaction contributions to the solar p-p reaction.   M. Gari, A. H. Huffman.
Astrophys. Journ., (Letters), Vol. 174, L151 - L153 (1972).
Interaction contributions (meson-exchange effects) to the solar p-p reaction are evaluated by using the low-energy theorem results. A correction to the cross-section factor of approximately +10 percent is found.

080.034 Comments on the destruction of ^{7}Be in the solar interior.   P. D. Parker.
Astrophys. Journ., Vol. 175, 261 - 264 (1972).
A comparison is made between the rate of the ^{7}Be$(p, \gamma)^8$B reaction and the rates of three other nuclear reactions which might have been important in destroying ^{7}Be in the solar interior. Alternate reactions for destroying ^{8}B are also examined.

080.035 On the structure of a magnetic rope over the photosphere.   A. A. Solovjev.
Solnechnye Dannye 1972 Byull., No. 2, p. 94 - 98 (1972). In Russian.
It is found that the arising condition obtained earlier retains for any isolated magnetic rope, apart from its inner structure. The question on the kind of force-free magnetic ropes on the solar surface is discussed.

080.036 On the dispersion of the velocity field in the solar atmosphere.   R. Kh. Salman-zade.
Vestn. Leningr. un-ta, 1972, No. 1, p. 145 - 149. In Russian. Abstr. in Referativ. Zhurn. 51. Astron., 7.51.313 (1972).

080.037 Solar neutrinos.   J. N. Bahcall.
Proc. AIP Conference, No. 2, p. 243 - 246 (1971).
Discusses the experiments currently undertaken to detect solar neutrinos. The apparent discrepancy between the calculated neutrino flux and that observed may be explained

by either the ^{8}Be reaction in the sun is somehow suppressed or the heavy element abundance on the surface is greater than the interior abundance.

080.038 The solar neutrino problem.   J. N. Bahcall.
Comments Nuclear and Particle Phys. (USA), Vol. 5, 59 - 64 (1972).
The author provides a starting point for those wishing to make an informed guess as to where the trouble regarding the discrepancy between astrophysical predictions and terrestrial experimental observations might lie.

080.039 On some plasma rotation phenomena on the sun.   T. Ohman.
From plasma to planet. 21st Nobel symposium 1971, (see 012.028), p. 343 - 347 (1972).

080.040 Elastic scattering of electrons by solar neutrinos and weak interaction theories.   P. R. Chaudhuri.
Journ. Phys. A, General Phys., Vol. 4, L109 - L111 (1971).

The effect of rotation on thermal convection.
See Abstr. 061.038.

Effect of a magnetic field on finite amplitude convection.   See Abstr. 061.039.
See Phys. Ber., Vol. 51, No. 3–4613 (1972).

Are neutrinos stable particles?
See Abstr. 061.043.

Neutrinos of nonzero rest mass.
See Abstr. 061.044.

Rotation and lithium abundance in solar-type stars.
See Abstr. 065.015.

On stellar activity cycles.
See Abstr. 065.045.

Intensity fluctuations in Fraunhofer lines.
See Abstr. 071.021.

On the choice of boundary conditions for integration of transfer equations.   See Abstr. 071.038.

Magnetic coupling of the active chromosphere to the solar interior.   See Abstr. 073.036.

Stationäre Filamente, ihre koronale Umgebung sowie ihre Beziehung zum solaren Magnetfeld.
See Abstr. 073.079.

Earth

081 Figure, Composition, and Gravity of the Earth

081.001 A search for life on earth at 100 meter resolution.
C. Sagan, D. Wallace.
Icarus, Vol. 15, 515 - 554 (1971/72).

A study of several thousand photos indicates that $\sim 10^{-2}$ of Gemini and Apollo photographs of the earth at 100 m resolution reveal signs of life − rectangular arrays due to human agricultural and urban territoriality, roads, canals, jet contrails, and industrial pollution. A curve is derived for the detectivity of contemporary life on earth, in a plot of ground resolution versus global coverage.

081.002 The geopotential in Cartesian coordinates.
P. Sconzo.
Bull. American Astron. Soc., Vol. 3, 500 (1971). − Abstr. AAS.

081.003 Interpretation of global gravity anomalies.
M. H. P. Bott.
Nature, Phys. Sci., Vol. 236, 23 - 24 (1972).

081.004 Analysis of paleomagnetic data and continental motion. D. Zidarov.
Geomagn. Aeronom., Vol. 12, 103 - 110 (1972). In Russian.

081.005 Détermination locale des anomalies de gravité et hauteur du géoïde à l'aide d'observations de satellites artificiels. M. Chapront-Touze.
Bull. Géod., Nouvelle Sér., Année 1972, No. 103, p. 47 - 62.

081.006 Statistical analysis of the gravity. L. P. Pellinen.
Izv. vyssh. uchebn. zavedenij. Geod. i aehrofotos"emka, 1970, No. 5, p. 43 - 50. In Russian. − Abstr. in Referativ. Zhurn. 52. Geod. Aehros"emka, 3.52.93 (1972).

081.007 Gosses Bluff impact structure, Australia.
D. J. Milton, B. C. Barlow, R. Brett, A. R. Brown, A. Y. Glikson, E. A. Manwaring, F. J. Moss, E. C. E. Sedmik, J. Van Son, G. A. Young.
Science, Vol. 175, 1199 - 1207 (1972).

Geological and geophysical techniques establish the origin of an analog of lunar craters.

081.008 La détermination du champ de la pesanteur.
B. Ducarme.
Ciel et Terre, Vol. 88, 1 - 24 (1972).

081.009 Gravimetrische Bestimmung von anomalen Dichtestrukturen für Lotkrümmungen und orthometrische Höhen. R.-D. Düppe.
Diss., Inst. Theor. Geodäsie, Rheinisch. Friedrich-Wilhelms-Univ., Bonn. 140 pp. (1972).

081.010 Sur la précision de la théorie du géoïde régularisé.
D. V. Zagrebin.
Byull. Inst. Teoret. Astron., *Leningrad,* Vol. 13, 92 - 101 (1972). In Russian.

On précise la formule généralisée de Stokes pour le cas de la surface de niveau de l'ellipsoïde; on corrige la valeur de la fonction, qui entre dans l'intégrale de type de Stokes. On applique la théorie précise obtenue pour le modèle de la terre, représentant le géoïde ellipsoïdale à trois axes inégaux. L'ondulation du géoïde pour le cas considéré a une erreur relative de l'ordre de 0.40 %.

081.011 Static deformation of a multilayered sphere by internal sources. H. R. Wason, S. J. Singh.
Geophys. Journ. Roy. Astron. Soc., Vol. 27, 1 - 14 (1972).

The Thomson−Haskell matrix device is used to solve the problem of the static deformation of a multilayered spherical earth model by buried sources. Explicit series expressions in terms of layer matrices are obtained for the displacements and stresses at any point in the medium for three sources. The singular case corresponding to the Legendre polynomial of the first degree has been discussed in detail.

081.012 Catalogue of the gravity stations in Antarctic.
N. P. Grushinsky, E. D. Korjakin, P. A. Stroev, G. E. Lasarev, D. V. Sidorov, N. F. Virskaja.
Trudy Gos. Astron. Inst. Shternberga, Vol. 42, 115 - 311 (1972). In Russian.

081.013 Variations of the geopotential of the P_2^2 mode for the period of decades. C. Kakuta, N. Kikuchi.
Publ. International Latitude Obs. Mizusawa, Vol. 7, 173 - 183 (1970).

081.014 Precession and the earth's magnetic field.
J. A. Jacobs, T. Chan, M. Frazer.
Nature, Phys. Sci., Vol. 236, 24 - 25 (1972).

The paper discusses precession as the cause of fluid motions in the earth's electrically conducting core which in turn give rise to the earth's magnetic field through a dynamo action. Two problems are discussed − the functional dependence of various parameters on the electrical conductivity σ of the core and the correct value of σ in the core. It is shown that different dynamo models yield very different functional dependences of the ohmic dissipation rate on σ. This is even more serious since the correct value of σ in the core is not known within orders of magnitude.

081.015 Earth satellites and the gravitational potential.
R. D. Eberst.
Nature, Phys. Sci., Vol. 235, 130 (1972).

Investigation of the 15th-order harmonics of the earth's gravitational potential can be made using satellites whose orbits are in resonance with these harmonics. Results are presented from a computer program which determines which satellites are of interest, and when they should be observed.

081.016 Sampling functions as an alternative to spherical harmonics. G. E. O. Giacaglia, C. A. Lundquist.
Extra collection of papers to IAU Symposium No. 48, (see 012.008), p. 149 - 153 (1971).

081.017 Observations of the tidal strains at Kinki District in Japan. I. Ozawa.
Extra collection of papers to IAU Symposium No. 48, (see 012.008), p. 173 - 174 (1971).

081.018 The Australian geoid.

N. P. Grushinski, L. P. Pellinen.
Soobshch. Gos. Astron. Inst. Shternberga, No. 174, p. 3 - 28 (1972). In Russian.

081.019   The gravity field of Australia and surrounding seas.
N. P. Grushinski, N. B. Sagina, T. P. Baskakova, G. I. Torochkova.
Soobshch. Gos. Astron. Inst. Shternberga, No. 174, p. 29 - 41 (1972). In Russian.

081.020   The main reductions of the gravity and some errors connected with them.
N. P. Grushinski, N. B. Sagina.
Soobshch. Gos. Astron. Inst. Shternberga, No. 174, p. 42 - 53 (1972). In Russian.

081.021   Depth of sources of gravity anomalies.
R. R. Allan.
Nature, Phys. Sci., Vol. 236, 22 - 23 (1972).

The hypotheses that the earth's gravity anomalies are due either to undulations in the depth of a density jump or to density variations uniformly distributed from a particular depth downwards, are tested against the geoid of Gaposchkin and Lambeck, assuming the variations are randomly distributed and uncorrelated over great distances. In either case the low degree anomalies, from $l = 2$ to $l = 6$ or 8, correspond to much greater effective depths (of order 1500 km) than the higher degree anomalies, which probably arise in the asthenosphere.

081.022   Formation of the earth's core.
D. L. Anderson, T. C. Hanks.
Nature, Vol. 237, 387 - 388 (1972).

The inhomogeneous accretion model is attractive because it helps to explain the chemical differences among the meteorites and among the terrestrial planets, including the moon. We propose that the inhomogeneous accretion hypothesis, when looked at in more detail, can explain early melting of the earth's core.

081.023   Forschungsbericht der Deutschen Forschungsgemeinschaft: "Unternehmen Erdmantel".
U. Vetter.
Umschau, 72. Jahrgang, p. 400 (1972).

081.024   Moving heaters as a model of continental drift.
J. A. Whitehead.
Phys. Earth Planet. Interiors, Vol. 5, 199 - 212 (1972).

This paper embraces convective behavior of a viscous fluid, differs from most others, however, in that not only is the effect of convection included but the effect of movement or convection of explicitly stated heat sources is also included.

081.025   L'isostasie et la forme du géoïde.   J. Lagrula.

Comptes Rendus Acad. Sci. Paris, Sér. B, Vol. 274, 1338 - 1342 (1972).

Plusieurs auteurs ont proposé de substituer à l'ellipsoïde de référence un sphéroïde défini, grâce aux satellites, par un développement en série de fonctions sphériques. Il semble indiqué, en géophysique, de limiter cette série à un très petit nombre de termes et, en outre, de doter le sphéroïde de référence de continents isostatiquement compensés.

081.026   Evaluation of 15th-order harmonics in the geopotential.   D. G. King-Hele.
Nature, Phys. Sci., Vol. 238, 13 (1972).

081.027   Un nouvel aspect concernant la détermination du champ gravifique extérieur de la terre.
R. Badesco, I. Diaconu.
Bull. Géod., Nouvelle Sér., Année 1972, No. 104, p. 211 - 219 (1972).

081.028   Empirical data and the variance–covariance matrix for the 1969 Smithsonian Standard Earth (II).
E. M. Gaposchkin.
SAO, *Cambridge, Mass.*, Special Report No. 342, 5 + 21 + A37 + B1 + C1 pp. (1972).

Empirical data used in the standard earth (II) are described. The variance-covariance matrix is available, and the format of this matrix on a standard magnetic tape is described.

081.029   Equations for 15th-order harmonics in the geopotential.   D. G. King-Hele, H. Hiller.
Nature, Phys. Sci., Vol. 235, 130 - 131 (1972).

081.030   Lithospheric momenta and the deceleration of the earth.   L. Knopoff, A. Leeds.
Nature, Vol. 237, 93 - 95 (1972).

We have recalculated the relative velocities of ten lithospheric plates, instead of the six plates considered by LePichon, using more recent information.

A reply to "Remarks on uncertainties in poles of rotation in continental fitting" by E. C. Bullard and D. P. McKenzie [Earth Planet. Sci. Letters, Vol. 11, 263 - 264 (1971)].   See Abstr. 044.005.

On inferring elastic properties of deep planetary interiors: Moon and Mars.   See Abstr. 094.105.

Space geodesy and dynamics of the earth and the moon. Recent research; Programs in France. See Abstr. 094.184.

Potassium distribution between metal and silicate and its bearing on the occurrence of potassium in the earth's core.   See Abstr. 105.033.

082 The Earth's Atmosphere including Refraction, Scintillation, Extinction, Airglow, Site Testing

082.001 **Atmospheric turbidity and surface temperature on the polar ice sheets.**
W. L. Hamilton, T. A. Seliga.
Nature, Vol. 235, 320 - 322 (1972).

082.002 **Nouvelle méthode pour calculer le chemin optique S' dans l'ionosphère et la troposphère. Application à la détermination de l'ordre, par rapport à la fréquence du terme de S' dû à la réfraction astronomique.**
L. Bertel, D. Geslin.
Comptes Rendus Acad. Sci. Paris, Sér. B, Vol. 274, 721 - 724 (1972).

082.003 **Spectral measurements of noctilucent clouds.**
B. Fogle, M. H. Rees.
Journ. Geophys. Res., Vol. 77, 720 - 725 (1972).

The first high-resolution, absolutely calibrated spectral intensity measurements of noctilucent clouds were made on the night of July 22-23, 1969, from Fort Nelson, British Columbia, with a scanning spectrophotometer. From these spectra we obtain absolute intensity data on NLC of different brightness and with different morphological forms, and estimates of NLC particle size and number density.

082.004 **A note on the density at high latitudes inferred from low-altitude satellite drag data.** B. K. Ching.
Journ. Geophys. Res., Vol. 77, 781 - 783 (1972).

The density data used in this study were obtained from the satellite OV1-15 in the period from July 11 to November 6, 1968. We show that a latitude-dependent discrepancy still exists between the data and the revised model (Jacchia, 1971).

082.005 **Tropical He I 10830-Å observations.**
A. B. Christensen, B. A. Tinsley, N. R. Teixeira, P. D. Angreji.
Journ. Geophys. Res., Vol. 77, 784 - 787 (1972). – Letter.

082.006 **The inference of atmospheric temperature profiles from ground-based measurements of microwave emission from atmospheric oxygen.**
G. F. Miner, D. D. Thornton, W. J. Welch.
Journ. Geophys. Res., Vol. 77, 975 - 991 (1972).

082.007 **Infrared absorption by atmospheric aerosol substances.** F. E. Volz.
Journ. Geophys. Res., Vol. 77, 1017 - 1031 (1972).

The main aim of our investigations was to obtain the typical absorption spectra and the imaginary part of the refractive index from 2.5 to 15 μm for different components (water solubles, dust, soot, and benzene extract) of dry atmospheric aerosol substances, pressed into KBr pellets.

082.008 **Measurement of stratospheric water vapor.**
J. Cooney.
Journ. Geophys. Res., Vol. 77, 1078 - 1080 (1972). – Brief report.

082.009 **The spectrum of the airglow.** M. F. Ingham.
Sci. American, Vol. 226, No. 1, p. 78 - 85 (1972).

082.010 **Measurement of the temperature of the neutral atmosphere by means of the device RIM-901 aboard Cosmos 320.**
N. V. Djorjio, E. M. Nikolaenko, L. G. Oldekop, E. E. Khavin.
Geomagn. Aeronom., Vol. 12, 77 - 82 (1972). In Russian.

082.011 **On the O(^{1}D) distribution in the upper atmosphere from 60 to 140 km.** M. N. Vlasov.
Geomagn. Aeronom., Vol. 12, 144 - 146 (1972). In Russian. Brief information.

082.012 **Backscattering by turbulent irregularities: A new analytical description.** A. D. Wheelon.
Proc. IEEE, Vol. 60, 252 - 265 (1972).

The purpose of this paper is to present a new analytical description of backscattering which can be used to connect experimental results with models of the turbulent atmosphere.

082.013 **The polarization of skylight near the sun and its influence on polarimetric K corona observations.**
J. L. Leroy, R. Muller, P. Poulain.
Astron. Astrophys., Vol. 17, 301 - 308 (1972).

We first recall the main features of the polarization of skylight near the sun according to the presently available information on the subject. We then give some results obtained at the Pic-du-Midi Observatory as a part of a four-year program of observations of the K corona. In conclusion, we summarize the different sources of inaccuracy in the observations. Our general conclusion is that the K corona polarimetric measurements are not safe beyond 8' from the solar limb.

082.014 **Parallel study of 6300 Å airglow emission and ionospheric scintillations.**
H. Mullaney, M. D. Papagiannis, J. F. Noxon.
Planet. Space Sci., Vol. 20, 41 - 46 (1972).

During periods of enhanced ionospheric scintillations, fluctuations in the intensity of the 6300 Å OI airglow line are considerably higher than during periods of low scintillation activity. On the basis of our present data, it is not possible to choose between particle precipitation and travelling compressional waves as the cause of the ionospheric irregularities and the associated red line fluctuations.

082.015 **Applicability of a diffusion model to lateral transport in the terrestrial and lunar exospheres.**
R. R. Hodges, Jr.
Planet. Space Sci., Vol. 20, 103 - 115 (1972).

Kinetic theory is used to determine a series expansion of the vertical flux of particles in an exosphere in terms of time and space derivatives of particle concentration, exobase velocity, and temperature. For sufficiently large scale variations of these parameters in time and space, the series can be truncated to a form that is similar to a diffusion equation. It is shown that the approximation of exospheric lateral flow as a diffusion process is applicable to global scale problems concerning terrestrial helium and heavier gases, and lunar gases heavier than helium.

082.016 **The diffusive motion of an initially spherically symmetric cloud of ionization in the earth's upper atmosphere.** W. M. Pickering.
Planet. Space Sci., Vol. 20, 149 - 156 (1972).

The diffusive motion of an initially symmetric ion cloud in the presence of a magnetic field (the geomagnetic field) is investigated. The electric field is assumed to be simply the space-charge electric field and possible effects of neutral air motion are ignored.

082.017 **Quenching of NO $A^2\Sigma^+$ by $O_2 X^3\Sigma_g^-$.**
L. A. Melton, W. Klemperer.
Planet. Space Sci., Vol. 20, 157 - 160 (1972).

082.018 **Extreme ultra-violet absorption cross-sections in the earth's upper atmosphere.**
D. E. Knight, R. Uribe, B. E. Woodgate.
Planet. Space Sci., Vol. 20, 161 - 164 (1972).

Relative absorption cross-sections between 180 Å and 304 Å, and between 584 Å and 304 Å are obtained for atomic oxygen in the upper atmosphere, by observing the attenuation of solar radiation using a satellite instrument.

082.019 **Resonance radiation in artificial strontium clouds.**
H. G. Horak, D. M. Kerr, M. S. Tierney.
Planet. Space Sci., Vol. 20, 165 - 182 (1972).

We have considered the problem of the resonance scattering of sunlight by a spherically symmetric, optically thick gas cloud. The results of the calculations are compared with photometric data obtained by the Los Alamos Scientific Laboratory (LASL) from the barium-strontium release of the DOGWOOD event in the ARPA-sponsored SECEDE I series (Puerto Rico, May 1968).

082.020 **Ozone determinations by lunar rocket photometry.**
J. H. Carver, B. H. Horton, R. S. O'Brien, B. Rofe.
Planet. Space Sci., Vol. 20, 217 - 223 (1972).

Atmospheric ozone number densities have been determined over the altitude range 30−75 km by measuring the absorption of lunar u.v. radiation in a number of wavelength bands between 2400 Å and 2900 Å. We used lunar photometry from a rocket and the object of the present work was to obtain further measurements of atmospheric ozone by means of this technique.

082.021 **Temporal variations in night-time hydroxyl rotational temperature.**
G. G. Sivjee, K. A. Dick, P. D. Feldman.
Planet. Space Sci., Vol. 20, 261 - 269 (1972).

Airborne and groundbased OH measurements showing short-term temporal variations in OH rotational temperature are presented. The variations are interpreted on the basis of a dynamic model of the OH emitting region; they are also employed to clarify the apparent contradiction in latitude dependence of OH intensity reported in the literature.

082.022 **The diurnal phase anomaly in the upper atmospheric density and temperature.**
C. H. Cummack, P. H. Butler.
Planet. Space Sci., Vol. 20, 289 - 292 (1972). – Research note.

082.023 **Zenith angle dependence of the geocoronal Lyman-alpha glow.**
F. Paresce, S. Kumar, S. Bowyer.
Planet. Space Sci., Vol. 20, 297 - 299 (1972). – Research note.

082.024 **On the effect of small-scale wind shears on radio meteor echoes.** J. Jones.
Planet. Space Sci., Vol. 20, 301 - 312 (1972).

The author presents a theory of how small scale wind shears, that is, those wind shears with vertical wavelengths less than about 6 km, affect the reflection of radio waves from meteor trains. Comparison of the theory with observational data indicates that in the meteor region the root-mean-square turbulent velocity associated with these small scale wind shears is < 1.6 m sec^{-1}.

082.025 **Bremsstrahlung in the atmosphere.**
M. J. Berger, S. M. Seltzer.
Journ. Atmosph. Terr. Phys., Vol. 34, 85 - 108 (1972).

Calculations are described pertaining to the emission of bremsstrahlung by electrons in the upper atmosphere and the penetration of this radiation to atmospheric depths of 3−10 g cm^{-2} where it can be measured by balloon-borne detectors. The basic calculations were done for incident monoenergetic electron beams with energies between 20 keV and 2 MeV. The results from these calculations were then synthesized to obtain corresponding results for incident electron beams with exponential or power-exponential energy spectra.

082.026 **Methods for determining the angle of refraction through a mean atmosphere.** J. E. Bennett.
Journ. Atmosph. Terr. Phys., Vol. 34, 149 - 157 (1972).

A comparison of the optical angle of refraction through the earth's atmosphere is presented using actual mean atmospheric data up to an altitude of 30 km versus mean data at the surface level. The results show: (1) the angle of refraction is linearly dependent on the refractivity at the earth's surface; (2) the iteration method for calculating the angle and a previous method based upon data at the surface level produces angle of refraction values to within a few tenths of an arc second of each other at a wavelength of 6328 Å.

082.027 **Determination of total atmospheric water vapour content from solar observations at millimetre wavelengths.** T. O. Aro.
Journ. Atmosph. Terr. Phys., Vol. 34, 305 - 314 (1972).

082.028 **EUV resonance radiation from helium atoms and ions in the geocorona.** R. R. Meier, C. S. Weller.
Journ. Geophys. Res., Vol. 77, 1190 - 1204 (1972).

We examine the mechanism of resonance scattering of solar 584-Å and 304-Å photons through the spherical atmosphere to assess the degree of contribution to the observed signals and to determine the distribution of helium atoms and ions responsible for the emissions from the geocorona.

082.029 **Nitric oxide in the D region.**
D. F. Strobel.
Journ. Geophys. Res., Vol. 77, 1337 - 1339 (1972). – Letter.

082.030 **Radiation, evaporation and the maintenance of turbulence under stable conditions in the lower atmosphere.** W. Brutsaert.
Boundary-Layer Meteorology, Vol. 2, 309 - 325 (1972).

A method for calculating the critical Richardson number is developed to apply to the lower atmosphere through the introduction of some different physical assumptions and of the effect of vertical water vapor flux and through use of experimental turbulence data that have become available in the meantime. In addition, an analysis is presented of the rate of destruction of the variance of the temperature fluctuations as a result of radiation by considering the radiative flux divergence for a stratified atmosphere and by using a simple functional relationship to represent empirical emissivity data. The results are compared with some recent data.

082.031 **Atmospheric waves observed in the planetary boundary layer using an acoustic sounder and a microbarograph array.**
W. H. Hooke, J. M. Young, D. W. Beran.
Boundary-Layer Meteorology, Vol. 2, 371 - 380 (1972).

Acoustic sounder and microbarograph records of atmospheric waves propagating in the planetary boundary layer over Table Mountain, Colorado, are presented and compared. The two observing techniques are complementary in that the array provides wave amplitude, horizontal phase speed, direction, and wavelength, while the sounder provides a detailed picture of temporal changes in the structure of the lowermost kilometer or so of the earth's atmosphere.

082.032 **Staubmessungen in leuchtenden Nachtwolken.**
H. Fechtig.
Umschau, 72. Jahrgang, p. 61 - 62 (1972).

Experiments for collection of dust particles between 70 and 110 km altitude indicate, that a great deal of the dust particles in noctilucent clouds may be micrometeorites coated with ice and/or CO_2.

082.033 **Concerning the tasks of optical studies of the upper atmosphere and interplanetary space.**
V. G. Fesenkov.
Astron. vestn., Vol. 5, 217 - 221 (1971). In Russian. – Abstr. in Referativ. Zhurn. 51. Astron., 3.51.390 (1972).

082.034 **Equipment for measurements of thermal fluctuations in the air during the daytime.**
A. I. Khlystov, N. I. Kozhevnikov.
Solnechnye Dannye 1971 Byull., No. 11, p. 104 - 110 (1972). In Russian.
Registrations of the thermal fluctuations have been made by means of the EhPP–09 electronic potentiometer. Charts for thermal fluctuations in the air inside the solar telescope pavilion are given.

082.035 **Photographie d'une partie du massif du Canigou se détachant sur le disque du soleil couchant par un groupe de membres de la Société Scientifique Flammarion de Marseille.** G. Guigay.
L'Astronomie, 86e année, p. 73 - 85 (1972).

082.036 **Component of the seeing conditions near the earth's surface during the daylight.** N. I. Kozhevnikov.
Solnechnye Dannye 1971 Byull., No. 12, p. 104 - 105 (1972). In Russian.
The results of investigations of the instability of a point-source image due to the effect of terrestrial atmospheric layers, extending to 5–8 meters from the earth's surface are given.

082.037 **Radiometer for remote sounding of the upper atmosphere.** F. W. Taylor, J. T. Houghton, G. D. Peskett, C. D. Rodgers, E. J. Williamson.
Applied Optics, Vol. 11, 135 - 141 (1972).
The pressure-modulated CO_2 radiometer is a new kind of instrument capable of making temperature soundings in the 40–80-km region of the earth's atmosphere. It is intended to be mounted on a polar-orbiting satellite, where it will give global coverage of the upper atmosphere in a region that is not well understood at present. Tests with a laboratory prototype and a balloon-borne instrument show that the device, if mounted outside the atmosphere, could detect changes of around 1 K in the temperature at 65-km altitude.

082.038 **Model for the visible-light scattering properties of clouds.** R. T. Hall, R. D. Rawcliffe.
Applied Optics, Vol. 11, 468 - 469 (1972). – Letter.

082.039 **Astronomers look at lightning.** L. E. Salanave.
Astron. Soc. Pacific, Leaflet No. 510, 8 pp. (1971).

082.040 **Observing conditions on Mount Kobau.**
E. Brosterhus, E. Pfannenschmidt, F. Younger.
Journ. Roy. Astron. Soc. Canada, Vol. 66, 1 - 48 = Contr. Dominion Astrophys. Obs., Victoria, No. 167 (1972).
An evaluation of the Mount Kobau Observatory Site in British Columbia is presented from data accumulated over the period 1967–1970. The site's characteristics and the effects of meteorological conditions on the seeing as well as the instruments and methods used in this systematic study are discussed.

082.041 **Sunset in the mirage zone.** I. A. Galperin.
Priroda, No. 3.72, p. 88 - 90 (1972). In Russian.

082.042 **On the meteorologist I. A. Galperin's observation.**
A. D. Zamorsky.
Priroda, No. 3.72, p. 90 - 92 (1972). In Russian.

082.043 **On the analysis of fluctuations of the parameters of the earth's upper atmosphere. II.**
V. E. Chertoprud.
Nauchn. Informatsii, vyp. (No.) 18, p. 54 - 75 (1970). In Russian.
The development of methods for estimating the parameters of the upper atmosphere begun in paper I is being continued with the help of a correlation analysis of temperature fluctuations of the atmosphere.

082.044 **A statistical study on astroclimatic and meteorological characteristics and an estimate of the possibility to predict seeing conditions on the BTA site.**
O. B. Vasilyev, N. F. Nelubin.
Astrofiz. Issled. Izv. Spets. Astrofiz. Obs., Vol. 3, 26 - 35 (1971). In Russian.

082.045 **Density variations in the exosphere from June 1968 to December 1970.** G. E. Cook.
Planet. Space Sci., Vol. 20, 473 - 482 (1972).
Two satellites, Calsphere (1964-63C) and Dodecapole (1965-16G), have been used to determine weekly mean values of atmospheric density at heights near 1070 and 900 km respectively, between June 1968 and December 1970.

082.046 **Diurnal variation of atomic hydrogen in the thermosphere.** L. Wallace, D. F. Strobel.
Planet. Space Sci., Vol. 20, 521 - 531 (1972).
The diurnal variation of atomic hydrogen concentrations in the thermosphere is calculated from the time-dependent continuity and diffusion equations for a range of exospheric temperatures using Jacchia's formulation for the thermal and density structure of the background atmosphere.

082.047 **Resonance scattering of the first positive system of N_2 in the dayglow.** P. D. Feldman.
Planet. Space Sci., Vol. 20, 549 - 555 (1972).
Rocket measurements of the differential photoelectron flux and the volume emission rate of the $N_2(0, 0)$ second positive band are used to calculate the daytime concentration of metastable $A^3\Sigma_u^+$ molecules, and consequently the resonance scattering contribution to the N_2 first positive emission in the dayglow.

082.048 **Oxygen and hydrogen ion densities above Millstone Hill.** R. W. Schunk, J. C. G. Walker.
Planet. Space Sci., Vol. 20, 581 - 589 (1972).
From observations at Millstone Hill we have derived the rate of loss of O^+ in the charge-exchange reaction with hydrogen, the rate of photoionization of atomic oxygen, and the rate of loss of O^+ in reactions with N_2 and O_2. In combination with laboratory and theoretical results, these rates can be interpreted to yield number densities of the corresponding neutral species.

082.049 **Excitation rates of the vibrational levels of hydroxyl and nightglow intensities.**
W. F. J. Evans, E. J. Llewellyn.
Planet. Space Sci., Vol. 20, 624 - 627 (1972). – Research note.

082.050 **Atmospheric density variations at 140 kilometers deduced from precise satellite radar tracking data.**
L. L. DeVries, L. Schusterman, R. W. Bruce.
Journ. Geophys. Res., Vol. 77, 1905 - 1913 (1972).
The technique of evaluating density values from precise radar-tracking data of satellites in the altitude region 130 to 140 km is discussed. Ten days of high-resolution density data deduced from orbital decay of each of three satellites are presented.

082.051 **On the semiannual change in exospheric tempera-**

ture.   J. E. Titheridge.
Journ. Geophys. Res., Vol. 77, 1978 - 1981 (1972). – Letter.

082.052   **Condizioni meteorologiche e turbolenza ottica osservate in alcune stazioni OAN.**
R. Barbon, G. di Tullio Vanzani.
Mem. Soc. Astron. Italiana, Nuova Ser., Vol. 43, 33 - 54 (1972).

082.053   **Considerazioni su alcuni risultati della campagna JOSO 71 a Izaña, Tenerife.**
G. Ceppatelli, A. Righini.
Mem. Soc. Astron. Italiana, Nuova Ser., Vol. 43, 123 - 134 (1972).
In this paper we discuss some results obtained by Hansen *et al.* at Tenerife during the JOSO 71 site testing program.

082.054   **On anomalies of astronomical refraction near the horizon.**   V. V. Kirichuk.
Izv. vyssh. uchebn. zavedenij. Geod. i aehrofotos"emka, 1971, No. 3, p. 67 - 71. In Russian. – Abstr. in Referativ. Zhurn. 52. Geod. Aehros"emka, 4.52.159 (1972).

082.055   **Studies in astrophysics and physics of the upper atmosphere in the Turkmenian Soviet Republic.**
Kh. Gul'medov, A. P. Savrukhin.
Uspekhi fiz. nauk, Vol. 105, 764 (1971). In Russian.

082.056   **Dayglow of Mg I and Mg II.**   M. Gadsden.
Journ. Geophys. Res., Vol. 77, 1330 - 1331 (1972).
Letter.

082.057   **Visual observations of the night-time, twilight and day-time horizons of the earth from guided spaceships.**
A. A. Buznikov, K. Ya. Kondrat'ev, A. I. Lazarev, M. M. Miroshnikov, A. G. Nikolaev, V. I. Sevast'yanov, O. I. Smoktij, E. V. Khrunov.
Kosm. Issled., Vol. 10, 100 - 112 (1972). In Russian.

082.058   **Upper atmosphere heating near the auroral zones.**
R. R. Allan.
Nature, Vol. 235, 100 - 102 (1972).
Orbit analysis of the 12 h sub-synchronous communications satellite Molniya 1K (1968-85A) implies high and variable air densities near 400 km height in the southern auroral zones in October–November 1969. The highest values occur near 18h local time, and exceed the model equatorial densities by at least a factor 6. The heat source might be protons with energy typical of the quiet solar wind, possibly entering directly.

082.059   **Annual and yearly variations of the atmospheric pressure and the earth's rotation.**   N. Kikuchi.
Proc. International Latitude Obs. Mizusawa, No. 11, p. 11 - 28 (1971). In Japanese.
The world-wide distribution of the atmospheric pressure is analysed by spherical harmonics in terms of Ocean Function to study extensively the effect of the atmospheric pressure on the rotation of the earth.

082.060   **Effect of water vapour on star images.**
E. Onodera.
Proc. International Latitude Obs. Mizusawa, No. 11, p. 47 - 51 (1971). In Japanese.
A theoretical study is tried to explain the observational result that steadiness of star images increases proportionally to the amount of water vapour in the atmosphere.

082.061   **On the spectral analysis of the Chandler wobble from meteorological elements in the upper atmosphere.**   K. Takahasi.

Proc. International Latitude Obs. Mizusawa, No. 11, p. 57 - 80 (1971). In Japanese.
A study of finding spectra of the Chandler wobble in various meteorological elements of the upper atmosphere has been performed by using aerological data. The results of the spectral analysis show that the Chandler motion of the atmosphere is of different feature between the stratosphere and the troposphere, and depends on the distribution of ocean and land.

082.062   **On the profile of vertical velocity of the atmosphere measured at Mizusawa.**   E. Onodera.
Proc. International Latitude Obs. Mizusawa, No. 11, p. 100 - 108 (1971). In Japanese.
An attempt to derive seasonal variations of the diurnal change in the profile of the vertical velocity in the atmosphere has been made for studying the structure of the atmosphere at Mizusawa.

082.063   **Circumscribed halos.**
R. G. Greenler, A. J. Mallmann.
Science, Vol. 176, 128 - 131 (1972).
Circumscribed halos which appear around the sun are simulated by a computer treatment of a simple model.

082.064   **Superorkane in der Hochatmosphäre.**   M. Römer.
Umschau, 72. Jahrgang, p. 296 (1972).
Winds with velocities of up to 1.5 km/s have been observed at altitudes above 140 km in the upper atmosphere in the course of a severe geomagnetic storm.

082.065   **On the annual and semiannual variations of the thermospheric density.**
H. Volland, C. Wulf-Mathies, W. Priester.
Journ. Atmosph. Terr. Phys., Vol. 34, 1053 - 1063 (1972).
The semiannual density variation is decomposed into Fourier terms to find out the physical nature of this effect. In order to represent the observational data of the 'semiannual' effect it is sufficient to take into account the first two harmonics—the annual and the semiannual component. From the satellite drag observations we deduced different height profiles for the two Fourier components which suggests that both components originate from different energy sources.

082.066   **Molecular oxygen concentrations in the equatorial mesosphere.**
B. H. Subbaraya, S. Prakash, P. N. Pareek.
Journ. Atmosph. Terr. Phys., Vol. 34, 1141 - 1144 (1972).
A nitric oxide filled ion chamber with lithium fluoride window was used to measure the absorption profile of solar Lyman alpha radiation in the earth's atmosphere on a rocket flight from Thumba during relatively undisturbed conditions on the sun. The results were used to derive molecular oxygen concentrations in the 70–95 km region.

082.067   **Lunar influence on atmospheric ozone.**
G. M. Shah.
Nature, Vol. 237, 275 (1972).

082.068   **On the imitation of star images.**
A. I. Birulin, O. K. Yurkin.
Izv. vyssh. uchebn. zavedenij. Geod. i aehrofotos"emka, 1970, No. 6, p. 129 - 134. In Russian. – Abstr. in Referativ. Zhurn. 51. Astron., 5.51.99 (1972).

082.069   **Diffusion model for the phase delay between thermospheric density and temperature.**
H. G. Mayr, H. Volland.
Journ. Geophys. Res., Vol. 77, 2359 - 2367 (1972).
Considering a two-dimensional time-dependent model in which the thermosphere dynamics is excited by the UV heat

input within the thermosphere, it is shown that the wind-induced variations in the diurnal component of atomic oxygen dominate over its temperature-induced variations up to 200 km. The assumption of diffusive equilibrium is therefore in general not valid for O within the lower thermosphere.

082.070   **Latitudinal diurnal variation of some atmospheric parameters determined by a simultaneous analysis** of incoherent-scatter and satellite-drag data.
D. Alcaydé, P. Bauer, C. Jaeck, J. L. Falin.
Journ. Geophys. Res., Vol. 77, 2368 - 2376 (1972).

Oxygen concentrations obtained from satellite-drag and incoherent-scatter observations over the period of February 6–10, 1969, are presented for comparison. A good agreement is shown to exist between both types of data, which are also compatible with the Jacchia (1970) model.

082.071   **On the solar-cyclic intensity variation of surplus radiation at small latitudes.**   V. S. Tsaplin.
Kosm. Issled., Vol. 10, 301 - 302 (1972). In Russian. – Brief information.

082.072   **La spedizione JOSO all'isola di Lampione nel settembre 1971.**   G. Ceppatelli, A. Righini.
Coelum, Vol. 40, 43 - 54 (1972).

082.073   **Amplitude variations in artificially stimulated v.l.f. emissions.**   M. O. Diesendorf.
Nature, Phys. Sci., Vol. 236, 56 - 58 (1972).

A study was made of amplitudes of VLF emissions triggered artificially from 100 watt 10.2 kHz Omega pulses transmitted from Forest Port, N.Y. and received, after following a magnetospheric trajectory, at Eights station Antarctica (L≃4). Deep beating between falling or rising tones and the main pulses was observed, and irregularly spaced "blobs" of high amplitude ($100-200 \mu$ V/m) and lengths 2–20 msec were found within the pulses. The amplitude fluctuations should be included in theories of VLF emissions involving the trapping of gyroresonant particles.

082.074   **Atmospheric extinction coefficients.**
W. Blitzstein, G. Spear.
Bull. American Astron. Soc., Vol. 4, 210 (1972). – Abstr. AAS.

082.075   **Terrestrial refraction and anomalies of astronomical refraction on large zenith distances (z = 88, 89, 90°).**
V. V. Kirichuk.
Geod., kartogr. i aehrofotos"emka. Mezhved. resp. nauch.-tekhn. sb., 1971, vyp. (No.), 13, p. 36 - 40. In Russian.

082.076   **Terrestrial atmospheric composition from stellar occultations.**
P. B. Hays, R. G. Roble, A. N. Shah.
Science, Vol. 176, 793 - 794 (1972).

The stellar ultraviolet photometers aboard the OAO-2 satellite have measured the intensity changes of several stars during occultation of the star by the earth's atmosphere. From the occultation data the nighttime vertical number density profiles of molecular oxygen at altitudes from 120 to 200 kilometers and of ozone at altitudes from 60 to 100 kilometers have been obtained.

082.077   **Turbidity of the atmosphere: Source of its background variation with the season.**
H. W. Ellsaesser, R. F. Pueschel, H. T. Ellis.
Science, Vol. 176, 814 - 815 (1972).

082.078   **Upper atmosphere zonal winds.**
D. G. King-Hele.
Nature, Vol. 237, 451 - 452 (1972).

082.079   **About defining the parameters of the upper atmosphere by measuring the coefficient of ambipolar diffusion by the method of radiolocation of meteor traces.**
Yu. I. Portnyagin.
Geomagn. Aeronom., Vol. 12, 464 - 467 (1972). In Russian.

082.080   **Investigation of the motion of artificial ionized clouds in the upper atmosphere.**
L. A. Katasev, V. F. Chepura.
Geomagn. Aeronom., Vol. 12, 473 - 476 (1972). In Russian.

082.081   **Oscillation-excited nitrogen in the upper atmosphere.**   M. N. Vlasov.
Geomagn. Aeronom., Vol. 12, 477 - 480 (1972). In Russian.

082.082   **Time and spatial distribution of intensity of surplus radiation in the equatorial region.**
V. S. Tzaplin, L. V. Zubareva.
Geomagn. Aeronom., Vol. 12, 536 - 537 (1972). In Russian. Brief information.

082.083   **Night variations of intensity of 6300 Å emission on quiet geomagnetic conditions.**   Yu. L. Tiutze.
Geomagn. Aeronom., Vol. 12, 561 - 564 (1972). In Russian. Brief information.

082.084   **Atmospheric seeing.**   V. Barocas.
Journ. British Astron. Ass., Vol. 82, 279 - 281 (1972).

082.085   **Fluctuations of water vapour content in the troposphere as derived from interferometric observations** of celestial radio sources.   R. Hinder.
Journ. Atmosph. Terr. Phys., Vol. 34, 1171 - 1186 (1972).

Observations of extra-terrestrial radio sources with ground-based microwave interferometers provide evidence upon irregularities of refractive index which occur in the troposphere. The present paper describes observations carried out with the one-mile telescope at the Mullard Radio Astronomy Observatory, Cambridge and discusses the nature of the irregularities which give rise to the variations of the refractive index.

082.086   **Measurements of molecular oxygen in the thermosphere.**   A. E. Parker, K. H. Stewart.
Journ. Atmosph. Terr. Phys., Vol. 34, 1223 - 1232 (1972).

Results are given from 363 observations of molecular oxygen densities at heights near 180 km made by the Ariel 3 satellite between May and November 1967.

082.087   **Prognosis for the density of the upper atmosphere for the time of existence of artificial earth satellites.**
V. F. Kameko, V. M. Kovtunenko, Eh. P. Yaskevich, Yu. T. Reznichenko, V. P. Al'bokha.
Kosm. Issled., Vol. 10, 450 - 453 (1972). In Russian. – Brief information.

082.088   **Experimental evidences for a transient ion layer formation in connection with sudden ionospheric disturbances in the height range 20–50 km.**
G. Rose, H. U. Widdel, A. Azcarraga, L. Sanchez.
Planet. Space Sci., Vol. 20, 871 - 876 (1972).

Results of four rocket flights in which the height profile of charged carrier concentration was measured suggest that a transient ion layer is formed below 50 km with a maximum in stratospheric height levels (18–35 km) during or shortly after S.I.D. events. Furthermore, it was shown experimentally that the current measured with rocket-borne Langmuir probes is predominantly caused by free electrons in heights over 63 km.

082.089   **On the possibility of a simultaneous measurement of wind speed, wind direction, air density and air temperatures at heights which correspond to the upper D-region (max. 95 km) with chaff cloud sensors.**
G. Rose, H. U. Widdel.
Planet. Space Sci., Vol. 20, 877 - 889 (1972).

It is shown that, in addition to wind measurements, air density and air temperatures can be measured in the upper atmosphere up to about 95 km with chaff sensors provided that the chaff elements are properly dimensioned. Results of experiments are presented and a comparison between two different methods of evaluation is made.

082.090   **The orbit of Cosmos 307 rocket and its use in atmospheric research.**   H. Hiller.
Planet. Space Sci., Vol. 20, 891 - 898 (1972).

About 1500 observations from 46 observing stations were used to determine the orbit of Cosmos 307 rocket (1969-94 B) at 25 epochs spread throughout its nine-month orbital life. The values of orbital inclination obtained were adjusted to account for various perturbations and then used to find the mean rotational speed of the earth's atmosphere. Values of density scale height of the upper atmosphere for 1969-70 were calculated from the change in perigee distance.

082.091   **Photochemistry of the lower troposphere.**
H. Levy II.
Planet. Space Sci., Vol. 20, 919 - 935 (1972).

A steady state model of the chemistry in the lower troposphere is developed in which the number densities of major atmospheric species, N_2, O_2, H_2O, CH_4, CO, O_3, N_2O, NH_3 and $NO + NO_2$ are held constant at values appropriate for a summer latitude of 34°. Hourly daytime concentrations are then calculated for HO_2, OH, CH_3O_2, CH_3O, $H_2C=O$, NO, NO_2, NO_3, N_2O_5, HNO_3 and HNO_2 along with a representative evening value.

082.092   **Große Schwankungen in der äußeren Atmosphäre.**
L. G. Jacchia.
Umschau, 72. Jahrgang, p. 394 - 395 (1972).

082.093   **On the size distribution of turbulent elements in the earth's atmosphere.**   A. Greve.
Solar Physics, Vol. 24, 243 - 246 (1972).

Solar scintillation measurements made by Wessely and Mitchell are used to derive the size distribution of turbulent elements in the earth's atmosphere. Assuming a stationary state of the earth's atmosphere, the production rate of the turbulent elements is calculated.

082.094   **Fortgang des Projektes JOSO (Joint Organization for Solar Observation).**   K. O. Kiepenheuer.
Mitt. Astron. Ges., No. 31, p. 147 - 149 (1972).

082.095   **For amateur astronomers − on the investigation of astroclimate.**   S. B. Novikov.
Zemlya i Vselennaya, 1972, No. 3, p. 73 - 75. In Russian.

082.096   **Transport processes in the thermosphere.**
F. S. Johnson.
Astrophys. Space Sci. Library, Vol. 29, Part IV, (see 012.015), p. 53 - 67 (1972).

082.097   **The exosphere and geocorona.**   P. Mange.
Astrophys. Space Sci. Library, Vol. 29, Part IV, (see 012.015), p. 68 - 86 (1972).

082.098   **Lyman-alpha measurements of neutral hydrogen in the outer geocorona and in interplanetary space.**
G. E. Thomas, R. C. Bohlin.
Journ. Geophys. Res., Vol. 77, 2752 - 2761 (1972).

We describe over 2 years of Lyman-α data taken during that portion of the orbit of OGO 5, when the spacecraft was above the interfering effects of the Van Allen radiation belts. These data pertain to the outer region of the earth's hydrogen geocorona and to the neutral hydrogen component of the interplanetary medium.

082.099   **Photodissociation rates of molecular oxygen in the mesosphere and lower thermosphere.**
R. D. Hudson, S. H. Mahle.
Journ. Geophys. Res., Vol. 77, 2902 - 2914 (1972).

082.100   **Absorption measurements of upper-atmospheric sodium at Boca Raton, Florida, 1967−1971.**
C. R. Burnett, W. E. Lammer, W. T. Novak, V. L. Sides.
Journ. Geophys. Res., Vol. 77, 2934 - 2941 (1972).

082.101   **Lunar semidiurnal variation in O I(5577 Å) nightglow.**   J. M. Forbes, M. A. Geller.
Journ. Geophys. Res., Vol. 77, 2942 - 2947 (1972).

The effect of lunar tidal dynamics on the O I(5577 Å) airglow is studied for Chapman's (1931) reaction, Barth's (1964) mechanism, and the F-region contribution to the green-line intensity. It is concluded that the dynamics of the lunar tide are sufficient to produce a lunar variation in green-line airglow that is similar to the observed variation.

082.102   **Measurements of the absorption of solar radiation by O_2 and O_3 in the 2150-Å region.**
G. C. Tisone.
Journ. Geophys. Res., Vol. 77, 2971 - 2974 (1972).

Simultaneous measurements of absorption of the solar radiation in the wavelength regions of 2150 and 2500 Å as a function of altitude between 53 and 68 km have been made. These measurements have been used to obtain O_3 densities and to obtain an absorption cross section for O_2 at 2150 Å.

082.103   **Simple model for the calculation of the flux of solar radiation through the atmosphere.**   R. Shapiro.
Applied Optics, Vol. 11, 760 - 764 (1972).

The flux of solar radiation through a model atmosphere composed of n homogeneous layers and a ground surface is represented by $2(n + 1)$ linear simultaneous equations. The system is solved explicitly for the albedo and for the radiation reaching the ground as a function of the vertically incident radiation and known or assumed reflection and transmission coefficients for each of the n layers and ground surface. Sample calculations are made for the case of $n = 7$ for a wide variety of cloud states in order to illustrate the properties of the system.

082.104   **Intensity−half-width products for seven lines in the 6.3-μm water-vapor band.**
B. Fridovich, J. R. Kinard.
Journ. Optical Soc. America, Vol. 62, 542 - 544 (1972).

Intensity−half-width products have been measured at intermediate resolution using pure-water-vapor samples at pressures from 0.5 to 3.5 torr and a path length of 9355 cm. Intensities, derived from the results using theoretical values of the half-width, have been compared with those reported by other investigators.

082.105   **Irradiance fluctuations in optical transmission through the atmosphere.**
D. J. Torrieri, L. S. Taylor.
Journ. Optical Soc. America, Vol. 62, 145 - 147 (1972). − Letter.

082.106   **Irradiance fluctuations in optical transmission through the atmosphere.**   R. S. Lawrence.
Journ. Optical Soc. America, Vol. 62, 701 (1972). − Letter.

082.107 **On the night errors in astronomical determinations. Refraction anomalies or variable systematic instrumental errors?** V. S. Milovanović, H. Pannwitz.
Bull. Géod., Nouvelle Sér., Année 1972, No. 104, p. 185 - 193 (1972).

082.108 **Monte Carlo treatment of Lyman-alpha radiation in a plane-parallel atmosphere.**
S. B. Modali, J. C. Brandt, S. O. Kastner.
Astrophys. Journ., Vol. 175, 265 - 274 (1972).

A Monte Carlo technique involving Stokes vectors is used to obtain the state of polarization and intensity of solar Lα photons as they diffuse through a plane-parallel homogeneous model of earth's hydrogen envelope. Fine structure of Lα and Doppler redistribution of frequencies are taken into account.

082.109 **Leuchtende Nachtwolken.** B. Haurwitz.
SuW, Vol. 11, 180 - 182 (1972).

082.110 **Eine Serie schöner Halo-Erscheinungen.** W. Sandner.
SuW, Vol. 11, 193 - 194 (1972).

082.111 **Dust in the upper atmosphere.**
B. R. Clemesha, Y. Nakamura.
Nature, Vol. 237, 328 - 329 (1972).

A large influx of dust into the upper atmosphere was observed at São José dos Campos (23° S, 46° W) during October 1972. The observations were made by laser radar. Dust was first observed at a height of 47 km in a layer a few km thick on October 19, but the initial influx could have occurred up to one month prior to this date. Assuming one micron radius particles the peak concentration was calculated to be about 500 m^{-3}. Measurements made after October 19 showed dust from 35 km to at least 80 km. In January 1972 dust was still being observed, although the concentration was considerably decreased.

082.112 **Direct recording of isophotes in artificial clouds by means of television techniques.**
H. Bredohl, F. Remy.
Eldo-Cecles/ESRO-CERS, Sci. Techn. Rev., Vol. 2, 357 - 362= Univ. Liège, Inst. d'Astrophys., Coll. 8°, No. 605 (1970).

A method is described of using a photometric analyser in conjunction with a TV screen to monitor the evolution of an isophote in an artificial cloud diffusing into the atmosphere. A brief account is given of the use of the technique during the launching by ESRO of four cloud-release experiments in Sardinia.

082.113 **Formula for the brightness of the daytime sky.**
G. Sh. Livshits, I. A. Fedulin.
Trudy Astrofiz. Inst., *Alma-Ata*, Vol. 18, 3 - 8 (1971).
In Russian.

082.114 **Brightness and polarization of the sky in the almucantar of the sun in the near infrared spectral region.**
A. I. Ivanov, G. Sh. Livshits, B. T. Tashenov, I. A. Fedulin.
Trudy Astrofiz. Inst., *Alma-Ata*, Vol. 18, 9 - 13 (1971).
In Russian.

082.115 **The scattering indicatrix of light in the atmosphere in the near infrared spectral region.** A. I. Ivanov,
G. Sh. Livshits, B. T. Tashenov, I. A. Fedulin.
Trudy Astrofiz. Inst., *Alma-Ata*, Vol. 18, 14 - 17 (1971).
In Russian.

082.116 **The interconnection of the absolute brightness indicatrix of the daytime sky with the optical thickness of the atmosphere.** V. N. Glushko, A. I. Ivanov, G. Sh. Livshits, B. T. Tashenov, I. A. Fedulin.

Trudy Astrofiz. Inst., *Alma-Ata*, Vol. 18, 18 - 21 (1971).
In Russian.

082.117 **Optical characteristics of the daytime clear sky in the visual spectral region and atmospheric aerosol.**
B. T. Tashenov, Eh. L. Tem.
Trudy Astrofiz. Inst., *Alma-Ata*, Vol. 18, 22 - 28 (1971).
In Russian.

082.118 **Solar aureole and atmospheric aerosol.**
B. T. Tashenov.
Trudy Astrofiz. Inst., *Alma-Ata*, Vol. 18, 29 - 37 (1971).
In Russian.

082.119 **Spectral measurements of the brightness of the sky.**
G. Sh. Livshits, L. M. Musorina, Eh. L. Tem.
Trudy Astrofiz. Inst., *Alma-Ata*, Vol. 18, 38 - 42 (1971).
In Russian.

082.120 **Spectral brightness of the sky at small angular distances from the sun and the spectrum of aerosol.**
A. I. Ivanov, B. T. Tashenov.
Trudy Astrofiz. Inst., *Alma-Ata*, Vol. 18, 43 - 46 (1971).
In Russian.

082.121 **On the mean optical parameters of the atmospheric aerosol in summer and winter periods.**
B. T. Tashenov.
Trudy Astrofiz. Inst., *Alma-Ata*, Vol. 18, 47 - 51 (1971).
In Russian.

082.122 **Solar aureole and the spectrum of aerosol on steppe conditions of Kazakhstan.**
V. N. Glushko, B. T. Tashenov, Eh. L. Tem.
Trudy Astrofiz. Inst., *Alma-Ata*, Vol. 18, 52 - 53 (1971).
In Russian.

082.123 **On the connection of the number of large and small particles in the spectrum of atmospheric aerosol.**
B. T. Tashenov.
Trudy Astrofiz. Inst., *Alma-Ata*, Vol. 18, 54 - 55 (1971).
In Russian.

082.124 **Seasonal variations of the albedo of an area from polarization measurements of sky light.**
A. I. Ivanov, G. Sh. Livshits, Eh. L. Tem.
Trudy Astrofiz. Inst., *Alma-Ata*, Vol. 18, 56 - 58 (1971).
In Russian.

082.125 **On a possibility of polarimetric control of the optical stability of the atmosphere.** A. I. Ivanov.
Trudy Astrofiz. Inst., *Alma-Ata*, Vol. 18, 59 - 62 (1971).
In Russian.

082.126 **Investigation on the absorption of light in aerosol particles.** V. N. Glushko, G. Sh. Livshits,
B. T. Tashenov, V. A. Molchanov.
Trudy Astrofiz. Inst., *Alma-Ata*, Vol. 18, 70 - 73 (1971).
In Russian.

082.127 **On the spectral transparency and scattering indicatrixes of some turbid media.** T. P. Toropova.
Trudy Astrofiz. Inst., *Alma-Ata*, Vol. 18, 74 - 88 (1971).
In Russian.

082.128 **Optical properties of polydisperse media with different distribution of particles according to size.**
T. P. Toropova, S. O. Obasheva, L. L. Solntseva.
Trudy Astrofiz. Inst., *Alma-Ata*, Vol. 18, 89 - 110 (1971).
In Russian.

082.129 **On measurements of the transparency of the circumterrestrial layer of the atmosphere at various wavelengths.** K. M. Salamakhin, T. P. Toropova.
Trudy Astrofiz. Inst., *Alma-Ata*, Vol. 18, 111 - 114 (1971). In Russian.

082.130 **Use of the Fabry-Perot étalon for investigation of the absorption spectra of the atmosphere.**
Yu. I. Rudnev, L. A. Sataeva-Egorova.
Trudy Astrofiz. Inst., *Alma-Ata*, Vol. 18, 115 - 120 (1971). In Russian.

082.131 **The spectral course of transparency and scattering indicatrixes of city haze.** N. M. Ibraimov.
Trudy Astrofiz. Inst., *Alma-Ata*, Vol. 18, 121 - 127 (1971). In Russian.

082.132 **On seasonal variations of telluric lines of oxygen and water vapour.** G. A. Kirienko, T. P. Toropova.
Trudy Astrofiz. Inst., *Alma-Ata*, Vol. 18, 128 - 136 (1971). In Russian.

082.133 **Scattering indicatrixes of light in the circumterrestrial layer of the atmosphere in the region of large angular distances.**
T. P. Toropova, K. M. Salamakhin, A. P. Ten.
Trudy Astrofiz. Inst., *Alma-Ata*, Vol. 18, 141 - 147 (1971). In Russian.

082.134 **Die Turbulenzverhältnisse auf dem Collm, dem Geisingberg, dem Kahleberg und dem Valtenberg und Beurteilung der Eignung dieser Berge für astronomische Beobachtungen.** S. Wächter.
Wiss. Zeitschr. Techn. Univ. Dresden, Vol. 19, 113 - 116 = Mitt. Lohrmann-Obs. Techn. Univ. Dresden, No. 23 (1970).

082.135 **Observations météorologiques de Thessaloniki, 1966–1967.**
Published by G. C. Livadas.
Univ. Thessaloniki, Annuaire Inst. Météorologique et Climatologique, 34 - 35, 30 + 30 pp. (1972).

082.136 **Criteri per la scelta della località ove erigere l'Osservatorio Astronomico Nazionale.**
M. G. Fracastoro.
Oss. Astron. Nazionale, Rapp. No. 2, p. 7 - 17 (1971).

082.137 **Relazione sui criteri e lo stato dei lavori per la ricerca del sito.** M. G. Fracastoro.
Oss. Astron. Nazionale, Rapp. No. 2, p. 18 - 28 (1971).

082.138 **Excess radiation at heights from 200 to 500 km.**
S. S. Konyakhina, L. V. Kurnosova, V. I. Logachev, L. A. Razorenov, V. G. Sinitsina, M. I. Fradkin.
Izv. AN SSSR. Ser. fiz., Vol. 35, 2472 - 2475 (1971). In Russian. Abstr. in Referativ. Zhurn. 62. Issled. kosm. prostranstva, 7.62.245 (1972).

082.139 **Upper atmosphere emission spectrum at millimetre wavelengths.**
I. G. Nolt, J. V. Radostitz, R. J. Donnelly.
Nature, Vol. 236, 444 - 445 (1972).
Our experiment was carried out aboard the NASA Convair 990 aircraft during the Mars opposition flight of August 8 and 9, 1971, near the Hawaiian Islands. Observations were made from an altitude of ~37,000 feet at approximately local midnight.

082.140 **Some suggested research activities.**
H. S. W. Massey.
From plasma to planet. 21st Nobel symposium 1971, (see 012.028), p. 368 - 373 (1972).

Atti della I^a riunione del "Gruppo del seeing OAN".
See Abstr. 012.021.

Atti della II^a riunione del "Gruppo del seeing OAN".
See Abstr. 012.022.

Atti della III^a riunione del "Gruppo del seeing OAN".
See Abstr. 012.023.

Calculations of the energy dependence of the angular distribution of photoelectrons from atomic oxygen.
See Abstr. 022.031.

On the effect of the anomalous refraction on time observations at Mizusawa (I). See Abstr. 044.013.

6300 Å quantum efficiency of the recombination mechanism in the night-time F layer.
See Abstr. 083.015.

Seasonal changes in planetary circulations.
See Abstr. 091.009.

Laser radar observations of dust from comet Bennett. See Abstr. 103.100.

Atomic and molecular reactions in space.
See Abstr. 131.157.

083 Ionosphere

083.001 **Diurnal and seasonal model of the F_1 layer at medium to high latitudes.**
D. G. Torr, M. R. Torr, D. P. Laurie.
Journ. Geophys. Res., Vol. 77, 203 - 211 (1972).

083.002 **An electron temperature gradient instability and its possible application to the ionosphere.**
D. M. Cunnold.
Journ. Geophys. Res., Vol. 77, 224 - 233 (1972).

083.003 **Thermalization of colliding ion streams beyond the plasmapause.** M. Schulz, H. C. Koons.
Journ. Geophys. Res., Vol. 77, 248 - 254 (1972). — Brief report.

083.004 **Two-body mutual neutralization rates of $O_2^+ + O^-$, $NO^+ + O^-$, and $Na^+ + O^-$ obtained with merged beams.** J. T. Moseley, W. Aberth, J. R. Peterson.
Journ. Geophys. Res., Vol. 77, 255 - 259 (1972). — Brief report.

083.005 **Solar-eclipse effect on sporadic-E ionization.**
R. N. Datta.
Journ. Geophys. Res., Vol. 77, 260 - 262 (1972). — Brief report.

083.006 **Radio-auroral aspect sensitivity and the two-stream instability.** D. R. Moorcroft.
Journ. Geophys. Res., Vol. 77, 765 - 768 (1972). — Letter.

083.007 **Ionospheric shock front from Apollo 15 launching.**
P. R. Arendt.
Nature, Phys. Sci., Vol. 236, 8 - 10 (1972).

083.008 **Absorption in the nightly lower ionosphere as an indicator of ionization processes under normal conditions and in the presence of disturbances. II.** G. Nestorov.
Geomagn. Aeronom., Vol. 12, 33 - 43 (1972). In Russian.

083.009 **Exponential model of the ionospheric D region.**
G. Nestorov.
Geomagn. Aeronom., Vol. 12, 44 - 53 (1972). In Russian.

083.010 **Mean profiles of electron concentration of the D layer and their dependence on the zenith angle of the sun.** V. Zinger, I. Bremer.
Geomagn. Aeronom., Vol. 12, 54 - 59 (1972). In Russian.

083.011 **Spatio-temporal appearance of the seasonal anomaly in the outer ionosphere.** N. M. Boenkova.
Geomagn. Aeronom., Vol. 12, 119 - 121 (1972). In Russian. Brief information.

083.012 **The influence of solar activity on the geometrical parameters of the F1 layer.**
I. A. Freizon, B. S. Shapiro.
Geomagn. Aeronom., Vol. 12, 122 - 124 (1972). In Russian. Brief information.

083.013 **Latitudinal variations of disturbance of the ionospheric F region in the subauroral zone.**
A. P. Mamrukov.
Geomagn. Aeronom., Vol. 12, 129 - 131 (1972). In Russian. Brief information.

083.014 **Structure of the E_s layer from spatially spread sounding.**
O. Ovezgeldiev, K. N. Vasiliev, G. V. Michailova.
Geomagn. Aeronom., Vol. 12, 131 - 133 (1972). In Russian. Brief information.

083.015 **6300 Å quantum efficiency of the recombination mechanism in the night-time F layer.**
W. E. Brown, W. R. Steiger.
Planet. Space Sci., Vol. 20, 11 - 24 (1972).
Simultaneous airglow and electron content measurements made at Hawaii are used to infer the number of 6300 and 6364 Å quanta produced per electron lost in the night-time F layer of the ionosphere.

083.016 **A search for an ionospheric dynamo current effect of the galactic X-ray ionization.** W. H. Campbell.
Planet. Space Sci., Vol. 20, 61 - 71 (1972).
An ionization contribution from galactic X-ray stars may contribute to a measurable night-time dynamo current modulation which could cause a consistent and measurable sidereal time variation of the geomagnetic field at the earth's surface. The results of the study indicate a diurnal, sidereal time variation in H of about 0.7 to 1.3 γ with opposite phase in the two hemispheres and a maximum value near noon in the north.

083.017 **Forbush decreases in the flux of galactic cosmic rays and associated VLF night-time propagation phenomena.** S. Ananthakrishnan, B. Hackradt.
Planet. Space Sci., Vol. 20, 81 - 87 (1972).
In this paper we present the effects produced by three well known Forbush decrease events on the phase and amplitude of VLF signals received at São Paulo, Brazil, from several distant transmitters. The effect of each event is discussed individually and reasons are advanced to account for the fact that even comparatively low latitude paths crossing the anomaly can at times be affected by decreases in the cosmic ray flux.

083.018 **Low energy electron precipitation and the ionospheric F-region in and north of the auroral zone.**
D. S. Evans, T. Jacobsen, B. N. Maehlum, G. Skovli, T. Wedde.
Planet. Space Sci., Vol. 20, 233 - 251 (1972).

083.019 **Effects of X-ray stars on VLF signal phase.**
J. Svennesson, F. Reder, J. Crouchley.
Journ. Atmosph. Terr. Phys., Vol. 34, 49 - 72 (1972).
Stellar X-ray effects from Sco XR-1, Cen XR-2 and Cen XR-4 have been found on several of the numerous VLF phase recordings collected on a world-wide basis since 1965. Only marginal VLF phase anomalies could be associated with Tau XR-1, and none with the Sco XR-1 flares of 3 June 1967. One of the conclusions is that continuous VLF phase tracking at frequencies above 20 kHz can provide some valuable astrophysical information complementing space probe data.

083.020 **Ionospheric effects of solar flares—I. The statistical relationship between X-ray flares and SID's.**
S. D. Deshpande, C. V. Subrahmanyam, A. P. Mitra.
Journ. Atmosph. Terr. Phys., Vol. 34, 211 - 227 (1972).
The Sudden Ionospheric Disturbances (SID's) produced in the earth's ionosphere by enhanced solar radiation during solar flares are examined in the light of recent information on flare X-ray emission. It is found that the occurrence probability of SID's with X-ray flares increases as the emission extends towards the shorter end of X-ray spectrum. An examination of time profiles of X-rays, radio bursts and SID effects is made for a number of events.

083.021 Ionospheric effects of solar flares—II. The flare spectrum below 10 Å deduced from satellite observations. S. D. Deshpande, A. P. Mitra.
Journ. Atmosph. Terr. Phys., Vol. 34, 229 - 242 (1972).

The spectral distribution and its time development for X-rays below 10 Å during solar flares are deduced from the measurements of integrated flux levels by satellite-borne detectors in the 0.5–3 Å and 1–8 Å bands. An exponential distribution and a power law distribution are considered, as well as a composite spectrum with two different electron temperatures in the bands 0.5–3 Å and 3–8 Å.

083.022 Ionospheric effects of solar flares—III. The quantitative relationship of flare X-rays to SID's.
S. D. Deshpande, A. P. Mitra.
Journ. Atmosph. Terr. Phys., Vol. 34, 243 - 253 (1972).

This paper describes the effects of changes in the spectral distribution of solar X-rays below 10 Å, on the nature and magnitude of Sudden Cosmic Noise Absorption (SCNA) and compares 50 SCNA's with X-ray flares for which spectral distributions could be deduced.

083.023 Ionospheric effects of solar flares—IV. Electron density profiles deduced from measurements of SCNA's and VLF phase and amplitude.
S. D. Deshpande, A. P. Mitra.
Journ. Atmosph. Terr. Phys., Vol. 34, 255 - 266 (1972).

A method is presented by which electron density profiles in the D-region of the ionosphere may be derived during solar flares from: (1) SCNA measurements at two frequencies; (2) simultaneous measurements of SCNA's and SPA's at VLF. A parabolic profile is assumed.

083.024 Ionospheric effects of solar flares—V. The flare event of 30 January 1968 and its implications.
S. D. Deshpande, S. Ganguly, V. C. Jain, A. P. Mitra.
Journ. Atmosph. Terr. Phys., Vol. 34, 267 - 281 (1972).

From riometer and pulse absorption measurements on three frequencies for the large X-ray flare event that occurred on 30 January 1968, electron density profiles are derived for the entire course of the flare.

083.025 Electron content and other related measurements for a low latitude station obtained at sunspot maximum using a geostationary satellite.
G. O. Walker, S. D. Ting.
Journ. Atmosph. Terr. Phys., Vol. 34, 283 - 294 (1972).

083.026 Comparison of simultaneous satellite measurements of auroral particle precipitation with bottomside ionosonde measurements of the electron density in the F-region. T. Turunen, L. Liszka.
Journ. Atmosph. Terr. Phys., Vol. 34, 365 - 372 (1972).

083.027 A classification of night-time electron-density profiles. P. Knight.
Journ. Atmosph. Terr. Phys., Vol. 34, 401 - 410 (1972).

083.028 D-region extraionization and solar X-ray flux derived from SPA and SES measurements.
R. Barletti, G. L. Tagliaferri.
Journ. Atmosph. Terr. Phys., Vol. 34, 543 - 547 (1972). Short paper.

083.029 Thermal positive ions in the outer ionosphere and magnetosphere from OGO 1.
M. Ahmed, R. C. Sagalyn.
Journ. Geophys. Res., Vol. 77, 1205 - 1220 (1972).

083.030 Eclipse and noneclipse differential photoelectron flux. W. C. Knudsen, G. W. Sharp.

Journ. Geophys. Res., Vol. 77, 1221 - 1232 (1972).

Differential photoelectron flux in the energy range 3 to 50 electron volts has been measured in the lower ionosphere both during the March 7, 1970, solar eclipse and during a period 24 hours earlier. The results of the present study are of significance for comparison with theory and other measurements in the lower-altitude region of the ionosphere.

083.031 Mechanism of solar activity effect on the upper atmosphere's motions. G. M. Teptin.
Solnechnye Dannye 1971 Byull., No. 11, p. 110 - 114 (1972). In Russian.

The correlations between solar activity and ionospheric motions at some observation points for many years are analyzed. The radio solar flux at $\lambda = 10.7$ cm is taken as an indicator of solar activity. The maximum coherence of these two processes is observed for the periods of 4, 6, 8 days.

083.032 Field-strength measurements taken in Cluj during the partial solar eclipse of February 25, 1971. (Preliminary report). T. I. László, G. Lazar.
Stud. Univ. Babeş-Bolyai, Ser. Phys., Anul 17, Fasc. 1, p. 87 - 90 (1972).

083.033 Relationship between E_S and the earth's magnetic field at middle latitudes. G. L. Goodwin.
Australian Journ. Phys., Vol. 25, 97 - 101 (1972). – Short communication.

083.034 World-wide electron density changes and associated thermospheric winds during an ionospheric storm.
T. Obayashi.
Planet. Space Sci., Vol. 20, 511 - 520 (1972).

Global changes in the ionospheric electron density associated with a geomagnetic storm of June 15–21, 1965 are studied, using data of thirty ground-based ionospheric sounders and of Alouette I satellite. $\Delta N_m/N_m$ at the $F2$ peak level and $\Delta N_s/N_s$ at the satellite level are mapped on L-value vs. storm-time diagrams. A theory is proposed to explain a characteristic of ionospheric storms in terms of thermospheric winds generated across the plasmapause.

083.035 Atomic nitrogen ions in the F-region.
G. J. Bailey, R. J. Moffett.
Planet. Space Sci., Vol. 20, 616 - 621 (1972). – Research note.

083.036 Topside ionosphere disturbance effects during different phases of two successive magnetic storms in September, 1963. M. N. Fatkullin.
Planet. Space Sci., Vol. 20, 627 - 636 (1972).

Using the data obtained by means of the Alouette-1 satellite, the distribution of electron density in the region of the $F2$-layer maximum and topside ionosphere during different phases of two successive magnetic storms on September 13 and 16, 1963 have been studied.

083.037 Twilight and nighttime ionospheric temperatures from oxygen λ6300 and λ5577 spectral-line profiles. W. A. Feibelman, R. D. Hake, Jr., D. P. Sipler, M. A. Biondi.
Journ. Geophys. Res., Vol. 77, 1869 - 1877 (1972).

Fabry-Perot interferometer measurements of atomic-oxygen λ6300- and λ5577-line profiles from twilight and nightglow are used to determine the neutral temperatures in the F_2 and E regions of the earth's ionosphere.

083.038 Registration of low-frequency radiation and of signals aboard Intercosmos 3.
L. V. Vernova, Ya. Vojta, F. Irzhichek, V. V. Korobovkin, V. I. Lebedeva, Ya. I. Likhter, Ya. P. Sobolev, L. A. Timofeeva, P. Triska.

Kosm. Issled., Vol. 10, 82 - 91 (1972). In Russian.

083.039　Electron-ion and ion-ion reaction rate coefficients at low altitudes during a PCA event.　T. R. Larsen, M. Jespersen, J. Murdin, T. S. Bowling, H. F. Van Beek, G. A. Stevens.
Journ. Atmosph. Terr. Phys., Vol. 34, 787 - 794 (1972).

083.040　Ionospheric effects of solar flares—VI. Changes in D-region ion chemistry during solar flares.
A. P. Mitra, J. N. Rowe.
Journ. Atmosph. Terr. Phys., Vol. 34, 795 - 806 (1972).

We examine the nature and magnitude of the decrease of the effective recombination coefficient during solar flares as a function of flare flux, develop model recombination coefficient profiles for different flare conditions, and then offer a theory of how these decreases can occur.

083.041　Distortion of the E-region by the Sq current system.　E. C. Butcher.
Journ. Atmosph. Terr. Phys., Vol. 34, 903 - 909 (1972).

The amount of distortion of the E-region electron density at the magnetic equator caused by the Sq current system has been computed using measured electron density and current density profiles. The results obtained are used to explain other measurements on the E-region affected by Sq.

083.042　Electron density measurements in the ionosphere using a radio propagation method.
W. J. G. Beynon, A. D. Maude, G. Ratcliff.
Planet. Space Sci., Vol. 20, 809 - 819 (1972).

A rocket C.W. radio propagation experiment has been used to measure electron density profiles and the results compared with values calculated from ionograms.

083.043　A study of ion-pair production rates and electron number densities in the ionospheric D-region.
K. Folkestad, E. V. Thrane, B. Landmark.
Journ. Atmosph. Terr. Phys., Vol. 34, 963 - 968 (1972).

The objective of this report is to explore how the quantities $q/(1+\lambda)$ and N_e^2 are related at 74 and 80 km. Ground-based and rocket measurements are used in the analysis.

083.044　The ion chemistry and thermal balance of the E- and lower F-regions of the daytime ionosphere: An experimental study.
A. F. Timothy, J. G. Timothy, A. P. Willmore, J. H. Wager.
Journ. Atmosph. Terr. Phys., Vol. 34, 969 - 1035 (1972).

The processes of heating and ionization in the E- and F-regions of the daytime ionosphere are discussed in depth. Construction of a sounding-rocket payload, consisting of an assembly of extreme ultraviolet (EUV) spectrometers and plasma probes designed for simultaneous measurement of related ionospheric parameters, is described, and results obtained from a flight of this payload on a sun-pointing Skylark sounding rocket at Woomera, S. Australia (31°S, 137°E) on 3 April 1969 are presented in detail.

083.045　Equatorial sporadic E and plasma instabilities.
R. G. Rastogi.
Nature, Phys. Sci., Vol. 237, 73 - 75 (1972).

083.046　Heating of the ionosphere in a transverse electric field.　E. E. Tzedilina.
Geomagn. Aeronom., Vol. 12, 201 - 207 (1972). In Russian.

083.047　Direct investigations of the distribution of the ionospheric reflection field.
A. G. Vologdin, S. F. Mirkotan, S. M. Saveliev.
Geomagn. Aeronom., Vol. 12, 226 - 229 (1972). In Russian.

083.048　Planetary distribution of the electron concentration in the outer ionosphere at various phases of the magnetic storm on December 17−21, 1962.
M. N. Fatkullin, E. S. Zayarnaya, L. F. Mamonova.
Geomagn. Aeronom., Vol. 12, 230 - 236 (1972). In Russian.

083.049　Heat influx to the electron gas at heights of $h > 180$ km.　G. L. Gdalevitch, N. M. Shyutte.
Geomagn. Aeronom., Vol. 12, 266 - 272 (1972). In Russian.

083.050　The seasonal anomaly at latitudes below the electron concentration maximum of the F2 layer depending on latitude and time.
N. M. Boenkova, N. V. Mednikova.
Geomagn. Aeronom., Vol. 12, 335 - 337 (1972). In Russian. Brief information.

083.051　Ion and electron kinetics in the disturbed ionosphere at heights of 100–200 km.
B. P. Bykov, S. I. Kozlov.
Geomagn. Aeronom., Vol. 12, 340 - 342 (1972). In Russian. Brief information.

083.052　Measurements of horizontal drifts in the E and F regions at Millstone Hill.　J. V. Evans.
Journ. Geophys. Res., Vol. 77, 2341 - 2352 (1972).

At Millstone Hill an oblique-incidence incoherent-scatter radar system has been employed to observe the drift of the ionospheric plasma in the E and F regions in directions across the earth's magnetic field.

083.053　Barium cloud growth in a highly conducting medium.　A. Simon, A. M. Sleeper.
Journ. Geophys. Res., Vol. 77, 2353 - 2358 (1972).

An equation of motion is obtained for the spatial and temporal growth of an artificial barium cloud in the ionosphere. The resulting diffusion equation for the ion density is solved by numerical means.

083.054　Magnetic apex coordinates: A magnetic coordinate system for the ionospheric F_2 layer.
T. E. VanZandt, W. L. Clark, J. M. Warnock.
Journ. Geophys. Res., Vol. 77, 2406 - 2411 (1972). − Letter.

083.055　Polarization investigation of the anomalous absorption of cosmic radio radiation in the ionosphere at mean latitudes.　A. V. Antonov.
Kosm. Issled., Vol. 10, 223 - 227 (1972). In Russian.

083.056　Determining the frequency of electron collisions in the ionospheric E-region from data of rocket and terrestrial measurements during the flight of Vertical 1.
K. Bishoff, K. I. Gringauz, G. Klyajn, I. A. Knorin, G. Kober, V. A. Rudakov.
Kosm. Issled., Vol. 10, 228 - 238 (1972). In Russian.

083.057　Investigation of non-stationary processes in the ionosphere and cosmic space using quantum stabilizers of frequency.
G. S. Ivanov-Kholodny, A. A. Korchak, L. A. Lobachevsky, I. P. Stakhanov, B. M. Chikhachev.
Geomagn. Aeronom., Vol. 12, 398 - 406 (1972). In Russian.

083.058　The structure of ionospheric irregularities according to simultaneous observations at two magneto-ionic components.　V. D. Gusev, A. A. Kashin, N. B. Lan.
Geomagn. Aeronom., Vol. 12, 427 - 430 (1972). In Russian.

083.059　The ionospheric disturbances on March 6−10, 1970.
N. P. Ben'kova, G. V. Bukin, V. M. Driatzky, I. A.

Dzhulin.
Geomagn. Aeronom., Vol. 12, 431 - 439 (1972). In Russian.

083.060 **A possibility to estimate the energetic particle flux in the ionospheric D-region at sunrise and on daytime conditions.**   G. Nestorov.
Geomagn. Aeronom., Vol. 12, 440 - 445 (1972). In Russian.

083.061 **Dependence of sporadic ionization in the E layer of the ionosphere of high latitudes on magnetic activity.**   A. S. Besprozvannaya, T. I. Tschuka.
Geomagn. Aeronom., Vol. 12, 453 - 457 (1972). In Russian.

083.062 **About peculiarities of the spectrum of VLF-noises when the resonator earth − ionosphere is excited by cosmic sources.**   A. P. Nikolaenko.
Geomagn. Aeronom., Vol. 12, 458 - 463 (1972). In Russian.

083.063 **Measurement of the integral parameters of the night ionosphere by observations of the signals from ASE Intercosmos 2.**
G. K. Solodovnikov, V. M. Migunov, G. V. Vladimirova, Yu. F. Vorontzov, A. R. Yagovkin.
Geomagn. Aeronom., Vol. 12, 541 - 543 (1972). In Russian.
Brief information.

083.064 **About non-stationary distribution of electron-ion gas in the ionosphere.**
A. G. Khantadze, G. M. Khocholava, I. I. Mebagishvili.
Geomagn. Aeronom., Vol. 12, 543 - 545 (1972). In Russian.
Brief information.

083.065 **Real height variations of the ionospheric $F2$-layer above some pairs of geomagnetically conjugate stations.**   E. F. Petelski.
Journ. Atmosph. Terr. Phys., Vol. 34, 1163 - 1170 (1972).

083.066 **Correlation of solar radio bursts and sudden increases of the total electron content (SITEC) of the ionosphere.**   D. A. Matsoukas, M. D. Papagiannis, J. Aarons, J. A. Klobuchar.
Journ. Atmosph. Terr. Phys., Vol. 34, 1275 - 1283 (1972).
   We have tried to study the sudden increases of the total electron content (SITEC) of the ionosphere using as a guide continuous solar radio flux data in the centimeter and decimeter range. An effort was also made to correlate radio burst intensities and spectra with SITEC events in order to identify the types of radio bursts which are associated with SITEC's and the types that are not.

083.067 **Inhomogeneities of the ion concentration in the ionosphere at a height of 600 km.**
V. N. Ponomarev.
Kosm. Issled., Vol. 10, 368 - 375 (1972). In Russian.

083.068 **Experimental evidence of an electron temperature enhancement in the wake of an ionospheric satellite.**
U. Samir, G. L. Wrenn.
Planet. Space Sci., Vol. 20, 899 - 904 (1972).
   A study of electron temperature measurements made by a Langmuir probe mounted on the skin of the Explorer 31 satellite indicates that electron temperature in the very near wake exceeds that of the ambient electron gas. The results are compared with observations from the Gemini/Agena wake experiment and are found to be in reasonable agreement.

083.069 **Observations of D-region modifications at low and very low frequencies.**
T. B. Jones, K. Davies, B. Wieder.
Nature, Vol. 238, 33 - 34 (1972).

083.070 **Theoretical models of ionospheric storms.**
N. Matuura.
Space Sci. Rev. Vol. 13, 124 - 189 (1972).
   This review will briefly describe the morphology of ionospheric storms in the F-region and topside ionosphere and attempt to discuss some processes yielding the disturbance effects on the electron density of the F-region and some theoretical models of ionospheric storms.

083.071 **Quasi-linear interaction of whistler-mode waves and nonthermal electrons.**   P. C. W. Fung.
Astrophys. Space Sci., Vol. 16, 249 - 273 (1972).
   We attempt to analyze the quasi-linear cyclotron instability (under the weak turbulence regime) for whistler-mode waves due to pitch angle anisotropy of nonthermal electrons. The time evolution of the growth rate and the induced waves spectrum for a loss cone type of nonthermal electrons is analyzed numerically. The diffusion of particles in pitch angles due to quasilinear cyclotron instability is illustrated. Some predictions of the theory is given and suggestions for further research are presented.

083.072 **Interpretation of ionospheric effects of solar flares.**   A. P. Mitra.
Astrophys. Space Sci. Library, Vol. 29, Part IV, (see 012.015), p. 1 - 26 (1972).

083.073 **Behaviour of ionized components in the chemistry and aeronomy of D and E regions.**   A. D. Danilov.
Astrophys. Space Sci. Library, Vol. 29, Part IV, (see 012.015), p. 27 - 40 (1972).

083.074 **The chemistry of neutral species in the D and E regions.**   S. A. Bowhill.
Astrophys. Space Sci. Library, Vol. 29, Part IV, (see 012.015), p. 41 - 52 (1972).

083.075 **Electric fields and their effects in the ionosphere.** (Summary of observations).   G. Haerendel.
Astrophys. Space Sci. Library, Vol. 29, Part IV, (see 012.015), p. 87 - 116 (1972).

083.076 **Electric fields and their effects in the ionosphere.** (A theoretical treatment).
E. N. Bramley, M. I. Pudovkin.
Astrophys. Space Sci. Library, Vol. 29, Part IV, (see 012.015), p. 117 - 141 (1972).

083.077 **The structure of the plasmasphere on the basis of direct measurements.**   K. I. Gringauz.
Astrophys. Space Sci. Library, Vol. 29, Part IV, (see 012.015), p. 142 - 164 (1972).

083.078 **The structure of the plasmasphere on the basis of indirect measurements.**   R. A. Helliwell.
Astrophys. Space Sci. Library, Vol. 29, Part IV, (see 012.015), p. 165 - 175 (1972).

083.079 **Theoretical model of F-region storms.**
T. Obayashi, N. Matuura.
Astrophys. Space Sci. Library, Vol. 29, Part IV, (see 012.015), p. 199 - 211 (1972).

083.080 **Photodetachment of electrons from major negative ions in the lower D region.**
L. Thomas, P. M. Gondhalekar, M. R. Bowman.
Nature, Vol. 238, 89 - 90 (1972).
   The photodetachment of electrons from negative ions by solar radiations has been used to explain increases in electron concentrations which occur prior to ground sunrise.

On the theory of magneto-conjugate transfer for a cool plasma into the outer ionosphere of lower latitudes. See Abstr. 062.028.

Flux of solar XUV radiation inferred from ionospheric variations during the eclipse of 20th May 1966. See Abstr. 076.039.

Solar cosmic ray events on 11 - 18 April 1969 and their effects on the lower ionosphere. See Abstr. 078.019.

Parallel study of 6300 Å airglow emission and ionospheric scintillations. See Abstr. 082.014.

Theory on the precipitation of magnetospheric electrons at the time of a sudden commencement. See Abstr. 084.206.

Summary of discussions at symposium on magnetosphere—ionosphere interactions: A review. See Abstr. 084.239.

On the interaction between the magnetosphere and the ionosphere. See Abstr. 084.278.

084   Aurorae, Geomagnetic Field, Radiation Belts

Aurorae

084.001   Observations of magnetic-field-aligned auroral-electron precipitation.   B. A. Whalen, I. B. McDiarmid.
Journ. Geophys. Res., Vol. 77, 191 - 202 (1972).

This report concerns the energetic-particle results from a Black Brant VB (AKD-VB-27) sounding rocket that was instrumented to measure electron and ion precipitation, electric and magnetic fields, electron densities, and auroral optical emissions.

084.002   Midday auroras and magnetospheric substorms.
S.-I. Akasofu.
Journ. Geophys. Res., Vol. 77, 244 - 247 (1972).

The equatorward motion of the midday auroras was observed during substorms, and its significance is discussed.

084.003   Pitch-angle distributions of 100- to 300-kev protons measured by the ESRO IB satellite.
R. Amundsen, F. Søraas, H. R. Lindalen, K. Aarsnes.
Journ. Geophys. Res., Vol. 77, 556 - 566 (1972).

The pitch-angle distribution of 100- to 300-kev protons in the early morning/late afternoon local magnetic-time sector has been studied by using data from the early lifetime of the ESRO IB (Boreas) satellite. The pitch-angle distribution is described as a function of invariant latitude for different geomagnetic activity levels.

084.004   Systematics in auroral energy spectra.
R. H. Eather, S. B. Mende.
Journ. Geophys. Res., Vol. 77, 660 - 673 (1972).

This paper presents statistical analyses of auroral spectral data obtained on NASA Airborne Auroral Expeditions during January–March 1968 and November–December 1969. These optical measurements are interpreted in terms of particle precipitation parameters.

084.005   Vertical gradients in the theory of radio aurora.
D. R. Moorcroft.
Journ. Geophys. Res., Vol. 77, 769 - 772 (1972). – Letter.

084.006   Balloon observations of simultaneous auroral X-ray and visible bursts.
D. J. Hofmann, R. A. Greene.
Journ. Geophys. Res., Vol. 77, 776 - 780 (1972). – Letter.

084.007   The intensity ratio I(5577)/I(4278) and the effective lifetime of O(^1S) atoms in pulsating aurora.
A. Brekke, K. Henriksen.
Planet. Space Sci., Vol. 20, 53 - 60 (1972).

The intensity ratio I(5577)/I(4278) in pulsating aurora and the effective lifetime of the O(^1S) atoms has been calculated for a large number of cases in Tromsø during the winter 1969/70. The purpose of this paper is to make a more thorough search for any systematic variations both in the intensity ratio and the effective lifetime.

084.008   The frequency of pulsating aurora and its relationship to other characteristic parameters.
A. Brekke.
Planet. Space Sci., Vol. 20, 285 - 289 (1972). – Research note.

084.009   An infrasonic pressure disturbance study of two polar substorms.   R. E. Johnson.
Planet. Space Sci., Vol. 20, 313 - 329 (1972).

Infrasonic pressure disturbances generated during the growth and decay of two polar substorms of 31 October 1968 are studied utilizing pressure records from four stations spanning continental North America. Correlation with all-sky camera photographs shows that a reversal in direction of motion of the luminous aurora may follow poleward expansions which then generate infrasonic pressure disturbances into the lower atmosphere.

084.010   Observations of fast auroral waves.
J. S. Boyd, T. N. Davis, N. B. Brown, T. J. Hallinan, D. D. Wallis.
Planet. Space Sci., Vol. 20, 437 - 440 (1972). – Research note.

084.011   Behavior of aurora at a midlatitude station.
B. S. Dandekar.
Journ. Atmosph. Terr. Phys., Vol. 34, 127 - 138 (1972).

The 6-yr study of aurorae from a midlatitude station, Sacramento Peak, New Mexico shows that these aurorae exhibit a good correlation with solar activity as regards recurrence and phase of the solar cycle.

084.012   On observations of auroral-zone X-rays with energies up to and greater than 200 keV.
V. Jentsch, G. Kremser.
Journ. Atmosph. Terr. Phys., Vol. 34, 499 - 511 (1972).

084.013   Secondary electron excitation in the aurora.
J. C. Gérard.
Journ. Atmosph. Terr. Phys., Vol. 34, 531 - 535 (1972).
Short paper.

084.014   The morphology of displays of pulsating auroras.
G. R. Cresswell.
Journ. Atmosph. Terr. Phys., Vol. 34, 549 - 554 (1972).

In the present study the overall aim was to fit the characteristics of displays of pulsating auroras into the framework of the auroral substorm. For a 2-yr period all nights in which numerous careful visual observations had been made in association with pulsation photometer and/or television recordings at College were analysed – a total of 33 displays of pulsating auroras on 21 nights.

084.015   Correlated satellite measurements of low-energy electron precipitation and ground-based observations of a visible auroral arc.
K. L. Ackerson, L. A. Frank.
Journ. Geophys. Res., Vol. 77, 1128 - 1136, with a correction, p. 3002 (1972).

We report comprehensive observations of low-energy electron and proton intensities with the polar-orbiting low-altitude satellite Injun 5 over a visual auroral arc as recorded simultaneously by an all-sky camera at Fort Churchill, Manitoba, Canada.

084.016   Observed correlations between auroral and VLF emissions.   S. R. Mosier, D. A. Gurnett.
Journ. Geophys. Res., Vol. 77, 1137 - 1145 (1972).

We present the results of a program of simultaneous observations of VLF radio noise made by using the NASA and University of Iowa Injun 5 satellite and all-sky observations of aurora at the Auroral Observatory in Fort Churchill, Manitoba, Canada.

084.017   Further low-energy auroral-ion composition measurements.   B. A. Whalen, I. B. McDiarmid.
Journ. Geophys. Res., Vol. 77, 1306 - 1310 (1972). – Brief report.

084.018 **Auroral infrasonic waves and poleward expansions of auroral substorms at Inuvik, N. W. T., Canada.**
C. R. Wilson.
Geophys. Journ. Roy. Astron. Soc., Vol. 26, 179 - 181 (1971).

084.019 **Observations of acoustic aurora in the 1−16 Hz range.** R. W. Procunier.
Geophys. Journ. Roy. Astron. Soc., Vol. 26, 183 - 189 (1971).

084.020 **Auroral activity during 1970.**
J. Paton.
Observatory, Vol. 92, 21 - 23 (1972).

084.021 **The upper atmosphere as a regulator of subauroral red arcs.**
S. Chandra, E. J. Maier, P. Stubbe.
Planet. Space Sci., Vol. 20, 461 - 472 (1972).

The mechanisms for producing a subauroral red arc (SARARC) are studied by solving a system of basic ionospheric and atmospheric equations. It is shown that many of the observed features of a SARARC can be explained within the framework of the two processes generally responsible for the ionospheric behavior during a magnetic storm: these are (1) energy conduction from the magnetosphere to the ionosphere and (2) the changes in neutral composition of the lower atmosphere caused by the increase in turbulent mixing.

084.022 **Time-dependent behaviour of a stable auroral red arc excited by an electric field.**
R. G. Roble, R. E. Dickinson.
Planet. Space Sci., Vol. 20, 591 - 605 (1972).

The present paper develops a model for red arcs produced by electric fields by including the response of the neutral atmosphere to the resulting ion drift and by considering recent observations of electron densities within the red arc and electron temperatures above the arc.

084.023 **Observations of SAR [stable auroral red] arcs from OV1–10.** S. R. LaValle, D. D. Elliott.
Journ. Geophys. Res., Vol. 77, 1802 - 1809 (1972).

Three midlatitude red arcs were measured by the night airglow photometer on board OV1–10 (1966-111B) during the winter of 1966–1967. These observations are shown to be consistent with the theory that SAR arcs are generated at the plasmapause as a consequence of the turbulent dissipation of ring current energy.

084.024 **Auroral spectrum between 1200 and 4000 Angstroms.**
W. E. Sharp, M. H. Rees.
Journ. Geophys. Res., Vol. 77, 1810 - 1819 (1972).

Spectroscopic observations of an auroral event were made simultaneously by airborne and satellite-borne scanning spectrometers, in the wavelength region between 1200 and 4000 A. Photon emission rates of several vibrational bands were recorded. Model calculations of the emission rates of the observed features are found to be in reasonable agreement with the measurements.

084.025 **Relative motion of auroral conjugate points during substorms.**
H. C. Stenbaek-Nielsen, T. N. Davis, N. W. Glass.
Journ. Geophys. Res., Vol. 77, 1844 - 1858 (1972).

We present detailed observations of latitudinal and longitudinal displacements of auroras relative to their conjugates, the bulk of the data coming from the conjugate flight of March 26, 1968. We select this night because a cleanly defined substorm occurs during the observing period.

084.026 **Discussion of paper by G. Atkinson, 'Auroral arcs: Result of the interaction of a dynamic magnetosphere with the ionosphere'** [Journ. Geophys. Res., Vol. 75,

4746 - 4755 (1970)]. T. Sato, T. Ogawa.
Journ. Geophys. Res., Vol. 77, 1994 - 1995, with a reply by G. Atkinson, p. 1996 - 1997 (1972).

084.027 **A theory of discrete radio aurora.** F. B. Knox.
Journ. Atmosph. Terr. Phys., Vol. 34, 747 - 764 (1972).

Electric fields and ionization density gradients in the edges of optical auroral forms are such that the drift-gradient (cross-field) instability may easily occur, and give rise to radio aurora observable at metre wavelengths. The predictions of theory are compared with observations of type B_2 and B_3 (short and long discrete) radio aurora, and show good quantitative agreement in order of magnitude.

084.028 **The effect of the ring current and polar electrojet on the oval of proton aurorae.**
L. S. Evlashin, Yu. L. Truttse, Ya. I. Feldstein.
Journ. Atmosph. Terr. Phys., Vol. 34, 859 - 866 (1972).

The present paper makes an attempt to separate the effect of the magnetospheric ring current and the polar ionospheric electrojet on the position of the southern boundary of proton aurorae and to find a correlation between proton-glow and the intensity of the ring current in the magnetosphere.

084.029 **Ion-acoustic waves and aspect sensitivity in radio aurora.** J. Hofstee, P. A. Forsyth.
Journ. Atmosph. Terr. Phys., Vol. 34, 893 - 902 (1972).

The present experiment reveals the presence of ion-acoustic waves propagating in directions which diverge significantly (approx 6°) from the preferred plane. It is suggested that the existence of a large horizontal sheet current could account for the observations by distorting the magnetic field more or less uniformly over a large volume of the ionosphere.

084.030 **Mid-latitude VLF emissions and the mechanism of dayside auroral particle precipitation.**
J. K. Hargreaves, K. Bullough.
Planet. Space Sci., Vol. 20, 803 - 807 (1972).

Simultaneous observations of VLF noise on the satellite Ariel 3 and of daytime auroral radio absorption at a high-latitude station are interpreted to show a linear relationship between the flux of auroral particles and the intensity of mid-latitude VLF emissions. The results are consistent with particle precipitation by a process of weak pitch-angle diffusion.

084.031 **Photographs of the front of the expanding auroral bulge during an auroral substorm.** S.-I. Akasofu.
Planet. Space Sci., Vol. 20, 821 - 822 (1972) − Research note.

084.032 **The dynamics of auroral particle invasion during substorms.** E. M. Zhulina, V. M. Driatzky, T. I. Tshuka, I. A. Zhulin, F. Kambou, A. Saint-Marc.
Geomagn. Aeronom., Vol. 12, 243 - 251 (1972). In Russian.

084.033 **The aurorae of the polar cap and the DCF-field intensity.** I. A. Alexandrova, S. A. Zaitzeva.
Geomagn. Aeronom., Vol. 12, 350 - 351 (1972). In Russian. Brief information.

084.034 **Dayside auroral-oval plasma density and conductivity enhancements due to magnetosheath electron precipitation.** C. F. Kennel, M. H. Rees.
Journ. Geophys. Res., Vol. 77, 2294 - 2302 (1972).

084.035 **Midday auroras at the south pole during magnetospheric substorms.** S.-I. Akasofu.
Journ. Geophys. Res., Vol. 77, 2303 - 2308 (1972).

It is shown that there is a close relationship between the range of the equatorward shift of the midday aurora and the magnitude of a negative bay at an auroral-zone station in the midnight sector. This relationship can be taken to be a clear indication of a close association of magnetospheric substorms with the transfer of the magnetic flux from the front of the magnetosphere to the magnetotail.

084.036 **Polar aurora over Kiev.** N. I. Dzyubenko.
Kometn. Tsirk., *Kiev*, No. 126 (1972). In Russian.

084.037 **Dynamics of auroral absorption.**
V. M. Driatzky, O. I. Shumilov, A. I. Frank-Kamenetzky.
Geomagn. Aeronom., Vol. 12, 446 - 452 (1972). In Russian.

084.038 **X-ray bremsstrahlung in the stratosphere and the auroral activity on January 21, and February 3, 1969.** A. M. Novikov, V. D. Sokolov, Yu. G. Shafer, Yu. A. Nadubovitch, V. F. Lutenko, O. V. Kirienko.
Geomagn. Aeronom., Vol. 12, 468 - 472 (1972). In Russian.

084.039 **Short-time bursts of integral radiation from aurorae.**
L. P. Kuzakova.
Geomagn. Aeronom., Vol. 12, 560 - 561 (1972). In Russian. Brief information.

084.040 **Regular oscillations of auroral intensity and of the geomagnetic field.**
O. M. Raspopov, V. K. Roldugin.
Geomagn. Aeronom., Vol. 12, 566 - 568 (1972). In Russian. Brief information.

084.041 **About the possibility of acceleration of auroral electrons by an electric field of pulsations.**
V. M. Barsukov, L. L. Vanyan, M. I. Pudovkin.
Geomagn. Aeronom., Vol. 12, 568 - 569 (1972). In Russian. Brief information.

084.042 **N_2 positive and N_2^+ band systems and the energy spectra of auroral electrons.**
D. E. Shemansky, T. M. Donahue, E. C. Zipf, Jr.
Planet. Space Sci., Vol. 20, 905 - 917 (1972).

The relative emission rates of the auroral N_2 positive and N_2^+ band systems can be used to limit the permissible range of differential electron fluxes in auroras, due to remarkable differences in electron excitation functions for the two kinds of systems. Use of recently measured electron cross sections and many observational data from ground based and rocket studies shows that the results are consistent with spectra equivalent to a power law $E^{-1.4}$ for primaries and secondaries combined.

084.043 **Auroras and associated phenomena.**
Y. I. Feldstein.
Astrophys. Space Sci. Library, Vol. 29, Part III, (see 012.014), p. 152 - 191 (1972).

084.044 **Altitudes of the $\lambda 6300$-Å, $\lambda 5577$-Å, and $\lambda 4278$-Å emissions of the stable auroral red arcs of March 8–9. 1970.**
L. L. Smith, R. J. Hoch, R. W. Owen, G. Hernandez, E. Marovich.
Journ. Geophys. Res., Vol. 77, 2987 - 2996 (1972). – Letter.

084.045 **Metastable oxygen ions distribution and related optical emission in the aurora.** J.-C. Gérard.
Ann. Géophys. , Vol. 26, 777 - 781 = Univ. Liège, Inst. d'Astrophys., Coll. 4°, No. 213 (1970).

084.046 **Présence du système γ de NO dans le spectre des aurores polaires.** R. Duysinx.
Bull. Soc. Roy. Sci. Liège, Vol. 39, 157 - 167 = Univ. Liège, Inst. d'Astrophys., Coll. 8°, No. 599 (1970).

084.047 **Evidence d'une excitation préférentielle de N_2^+ (λ 3914 Å) dans une aurore.** J. M. Vreux.
Bull. Cl. Sci. Acad. Roy. Belgique, 5e Sér., Vol. 57, 655 - 668 = Univ. Liège, Inst. d'Astrophys., Coll. 8°, No. 618 (1971).
Intensities of 5577 Å [OI] and 3914 Å (N_2^+) emissions have been recorded during two rocket flights. The results are compared with recent observations of ultra-violet emissions.

Geomagnetic Field

084.201 Fluctuating magnetic fields in the magnetosphere. I. ELF and VLF fluctuations.
C. T. Russell, R. L. McPherron, P. J. Coleman, Jr.
Space Sci. Rev., Vol. 12, 810 - 856 (1972).

The study of extremely-low-frequency and very-low-frequency waves in space has been intensively pursued in the past decade. In order to fully understand the state of our present knowledge of these waves, we will first examine the instrumentation that has been flown on spacecraft. Following this, we will discuss the various whistler phenomena: their characteristics, their formation and their uses in probing the magnetosphere. Emissions have been studied principally in two separate altitude ranges: near the ionosphere from about 200 to 2000 km altitude; and at high altitudes in the magnetosphere near the magnetic equator.

084.202 Maximum entropy power spectrum of long period geomagnetic reversals. T. Ulrych.
Nature, Vol. 235, 218 - 219 (1972).

084.203 Effectiveness of cyclonic convection for producing the geomagnetic field. E. H. Levy.
Astrophys. Journ., Vol. 171, 621 - 633 (1972).

The dynamo equations for the production of astrophysical magnetic fields are solved for a spherical shell of shear and two axisymmetric rings of cyclonic convective cells. The results of the calculations show that, for the fluid dynamics expected in the earth's core, dipole-like fields can be maintained with any locations of the rings of cyclones, although convective cells at mid-latitudes are most efficient. Quadrupole-like fields are maintained only by high-latitude cyclones and generally require more vigorous fluid motions.

084.204 Kinematic reversal schemes for the geomagnetic dipole. E. H. Levy.
Astrophys. Journ., Vol. 171, 635 - 642 (1972).

Fluctuations in the distribution of cyclonic convective cells, in the earth's core, can reverse the sign of the geomagnetic field. Two kinematic reversal schemes are discussed. In the first scheme, a field maintained by cyclones concentrated at low latitude is reversed by a burst of cyclones at high latitude. Conversely, in the second scheme, a field maintained predominantly by cyclones in high latitudes is reversed by a fluctuation consisting of a burst of cyclonic convection at low latitude. It appears that a fluctuation in the distribution of cyclonic cells over latitude can cause a geomagnetic reversal.

084.205 Internal structure of the geomagnetic neutral sheet.
K. Schindler, N. F. Ness.
Journ. Geophys. Res., Vol. 77, 91 - 100 (1972).

084.206 Theory on the precipitation of magnetospheric electrons at the time of a sudden commencement.
G. E. Perona.
Journ. Geophys. Res., Vol. 77, 101 - 111 (1972).

084.207 Changes in the distribution of low-energy trapped protons associated with the April 17, 1965, magnetic storm. A. L. Burns, S. M. Krimigis.
Journ. Geophys. Res., Vol. 77, 112 - 130 (1972).

084.208 Irregular structure of thermal ion plasma near the plasmapause observed from OGO 3 and Pc 1 measurements. H. Kikuchi, H. A. Taylor, Jr.
Journ. Geophys. Res., Vol. 77, 131 - 142 (1972).

084.209 OGO 5 magnetic-field data near the earth's bow shock: A correlation with theory.

J. K. Guha, D. L. Judge, J. H. Marburger.
Journ. Geophys. Res., Vol. 77, 604 - 610 (1972).

Magnetic-field data obtained in the earth's bow-shock region with a high-resolution triaxial fluxgate magnetometer aboard the OGO 5 satellite have been correlated with a theory of Tidman and Northrop (1968).

084.210 IMP 5 magnetic-field measurements in the high-latitude outer magnetosphere near the noon meridian.
D. H. Fairfield, N. F. Ness.
Journ. Geophys. Res., Vol. 77, 611 - 623 (1972).

IMP 5 magnetic-field measurements at geomagnetic latitudes up to 75° and at distances beyond 6 R_E reveal the permanent existence of a broad depressed field region centered on the polar or dayside cusp. Field strengths at 7 R_E on cusp field lines that connect to the earth are typically only 50%–70% of that of an undistorted dipole field.

084.211 Dynamics of the magnetosphere and geomagnetic pulsations of Pc5 type.
O. M. Raspopov, L. T. Afanasieva, B. V. Kiselev, G. A. Loginov.
Geomagn. Aeronom., Vol. 12, 150 - 153 (1972). In Russian. Brief information.

084.212 On the Kelvin-Helmholtz instability of the earth's magnetopause. R. S. B. Ong, N. Roderick.
Planet. Space Sci., Vol. 20, 1 - 10 (1972).

The effect of the finite thickness of the shear layer on the Kelvin-Helmholtz instability of the earth's magnetopause boundary is investigated. The thickness of the layer stabilizes the boundary with respect to short wavelength perturbations, which were previously found to be unstable in the zero thickness analysis. Compressibility effects further stabilize the layer.

084.213 A model of a magnetic field in the geomagnetosphere. I. I. Alekseev, V. P. Shabansky.
Planet. Space Sci., Vol. 20, 117 - 133 (1972).

A new magnetospheric field model is proposed. The magnetospheric surface is considered to be a tangential discontinuity and is approximated by a paraboloid of revolution. The magnetic field within the magnetosphere has been obtained as a solution of the magnetostatics problem. The model makes it possible to obtain the magnetospheric field for an arbitrary inclination of the dipole axis to the earth-sun line and to include correctly the returning currents streaming over the magnetopause and caused by the neutral sheet in the magnetotail.

084.214 The annual variation of the earth-magnetic activity according to the character figures Ci.
J. W. Münch.
Planet. Space Sci., Vol. 20, 225 - 231 (1972).

084.215 Non-uniform entry of solar protons into the polar cap. M. Scholer, B. Häusler, D. Hovestadt.
Planet. Space Sci., Vol. 20, 271 - 283 (1972).

Observations of the temporal and spatial behavior of energetic solar protons are presented for the solar proton event that began on March 6, 1970. Differences in the fluxes of low energy protons of up to a factor of 8 are observed between the central polar cap and the low latitude region of the polar cap. The results are compared with interplanetary proton flux and magnetic field measurements as obtained by the HEOS A1 satellite.

084.216 Semiannual and annual modulation of the magnetic field. B. N. Bhargava.
Planet. Space Sci., Vol. 20, 423 - 427 (1972).

084.217 Absence of the plasma sheet at lunar distance du-

ring geomagnetically quiet times.
W. J. Burke, D. L. Reasoner.
Planet. Space Sci., Vol. 20, 429 - 436 (1972).

084.218 Auroral-zone electron precipitation events observed before and at the onset of negative magnetic bays.
T. Pytte, H. Trefall.
Journ. Atmosph. Terr. Phys., Vol. 34, 315 - 337 (1972).

084.219 Solar particles and the dayside limit of closed field lines.
I. B. McDiarmid, J. R. Burrows, M. D. Wilson.
Journ. Geophys. Res., Vol. 77, 1103 - 1108 (1972).
Low-altitude dayside profiles of solar-electron latitudes are examined as a function of magnetic activity. The solar-electron measurements are compared both with outer-zone electron measurements and with solar-proton measurements in an attempt to determine the dayside limit of closed geomagnetic-field lines.

084.220 Large-scale coherence and high velocities of the earth's bow shock on February 12, 1969.
E. W. Greenstadt, P. C. Hedgecock, C. T. Russell.
Journ. Geophys. Res., Vol. 77, 1116 - 1122 (1972).

084.221 Distributions of electron plasma oscillations upstream from the earth's bow shock.
R. W. Fredricks, F. L. Scarf, I. M. Green.
Journ. Geophys. Res., Vol. 77, 1300 - 1305 (1972).
We present data from the 14.5- and 30-kHz plasma-wave detector channels aboard OGO 5 for the period December 2, 1968, to April 8, 1969, to demonstrate the relatively isotropic occurrence of electron plasma oscillations upstream from the bow shock.

084.222 The earth's bow shock in an oblique interplanetary field. W.-W. Shen.
Cosmic Electrodynamics, Vol. 2, 381 - 395 (1972).
The pressure, magnetic field, temperature, particle density, and stream velocity throughout the magnetosheath have been calculated in the plane containing the interplanetary field and solar wind velocity vector for various orientations of the interplanetary magnetic field and various assumed ratios of specific heats of the compressed solar wind. Jump conditions at the bow shock gave initial conditions in the shocked plasma from which the appropriate hydromagnetic equations were integrated numerically back to the surface of the magnetosphere.

084.223 Solar current penetration into the magnetosphere.
R. Parthasarathy.
Cosmic Electrodynamics, Vol. 2, 399 - 401 (1972).
The plausibility and consequences of a current flow extending from the sun and into the interplanetary medium is considered speculatively. Order of magnitude calculations are made of the several other observable perturbations in the magnetosphere.

084.224 Intrusion of fast particles into the magnetosphere (model experiments).
I. M. Podgorny, E. M. Dubinin.
Cosmic Electrodynamics, Vol. 2, 445 - 452, 453 - 459 (1972). In Russian and English.
Precipitation of fast particles in the magnetosphere has been investigated using simulation experiments. The model magnetosphere was created by a plasma stream interaction with a dipole magnetic field. We have also found the penetration of fast particles on night-side in the region which separates the magnetic tail lines and lines of the closed magnetosphere.

084.225 Geomagnetic tail oscillations.
A. I. Ershkovich, A. A. Nusinov.
Cosmic Electrodynamics, Vol. 2, 460 - 470, 471 - 480 (1972). In Russian and English.
A model of the earth's magnetic tail is considered as a plasma cylinder, immersed into the solar wind plasma. A dispersion equation describing oscillations is obtained. Oscillation amplitudes excited due to variations in solar wind velocity are estimated. The Kelvin-Helmholtz instability (arising when the solar wind velocity is high enough) is investigated.

084.226 Structure of the geomagnetic tail and substorm activity. A. Hruška.
Cosmic Electrodynamics, Vol. 2, 481 - 490 (1972).
The analysis of the IMP-3 magnetic data shows that the rate of change of the magnetic field in the solar direction varies with position in the tail. The thickness of the plasma sheet changes with substorm activity and the magnetic field strength decreases from the tail center towards its flanks. The correlation between time derivatives of the magnetic field and time variation of the AE-index is discussed.

084.227 Two-dimensional equilibrium solution of the plasma sheet and its application to the problem of the tail magnetosphere. T. Toichi.
Cosmic Electrodynamics, Vol. 3, 81 - 96 (1972).
Two-dimensional self-consistent solutions are obtained which are appropriate to describe the structure of the plasma sheet in the magnetosphere. These are compared with the satellite magnetic field and plasma observations.

084.228 Polar magnetic substorms.
G. Rostoker.
Rev. Geophys. Space Phys., Vol. 10, 157 - 211 (1972).
A brief review of the historical development of some of the concepts of substorm generation, and the relative merits of these concepts, are discussed against the background of the large number of observational data now available. The over-all picture presented suggests that substorms are a drastic mechanism by which energy stored in the nightside magnetosphere is dissipated.

084.229 Solar energetic particles and the configuration of the magnetosphere. L. J. Lanzerotti.
Rev. Geophys. Space Phys., Vol. 10, 379 - 393 (1972).
Since the original study of PCA riometer measurements, solar energetic particles have been increasingly employed as test particles in the magnetosphere. The characteristics of the access of these particles to the magnetosphere are studied to determine if substantial connection exists between the earth's dipole field and the solar-generated interplanetary field.

084.230 Mathematical models of the earth's magnetic field.
M. Kono.
Phys. Earth Planet. Interiors, Vol. 5, 140 - 150 (1972).
Palaeomagnetic data for the last 10^7 y show that the geomagnetic field had been almost dipolar, that the reversals of polarity took place in a manner described by the Bernoulli trials, and that the distribution of palaeointensities is of normal type both for normal and reversed polarities. Properties of mathematical models which satisfy these observational facts are discussed.

084.231 Horizontal polarization in array studies of anomalous geomagnetic variations.
D. J. Bennett, F. E. M. Lilley.
Nature, Phys. Sci., Vol. 237, 8 - 9 (1972).

084.232 Effects on the plasmasphere of a time-varying convection electric field. A. J. Chen, R. A. Wolf.

Planet. Space Sci., Vol. 20, 483 - 509 (1972).

The time evolution of the plasmasphere has been investigated theoretically, using simple computational models. The magnetic field is assumed to be dipolar and time-independent, but the convection electric field is allowed to vary in time. For purposes of comparison, various spatial distributions of the magnetospheric electric field are considered. Plasmasphere flux tubes are assumed to be filled by diffusion of plasma upwards from the dayside ionosphere.

084.233 **Observation of nonuniform structure of the earth's bow shock correlated with interplanetary field orientation.**   E. W. Greenstadt.
Journ. Geophys. Res., Vol. 77, 1729 - 1738 (1972).

Explorer 33 and 35 magnetometers, on the western and eastern flanks of the earth's bow shock, respectively, observed the boundary concurrently between 0130 and 0430 UT, October 30, 1968. Contrasting shock structures were recorded. The observations support a model of the shock in which perpendicular and oblique collisionless structures coexist and form a nonuniform magnetosheath outer boundary.

084.234 **Average plasma-sheet configuration near 60 earth radii.**   C.-I. Meng, J. D. Mihalov.
Journ. Geophys. Res., Vol. 77, 1739 - 1755 (1972).

Nearly three years of Ames magnetometer data from Explorer 35 are used to establish plasma-sheet configurations in the geomagnetic tail at the lunar orbit. The plasma-sheet configuration at the lunar orbit is found to be similar to the distribution at $18 R_E$ observed by Vela satellites. The seasonal dependence and variations with geomagnetic activity are also presented.

084.235 **Radial diffusion of ionized helium and protons: A probe for magnetospheric dynamics.**
J. M. Cornwall.
Journ. Geophys. Res., Vol. 77, 1756 - 1770 (1972).

The comparison of radial diffusion of magnetospheric helium ions and protons is a sensitive test of the diffusion process assumed (magnetic or electrostatic fluctuations) and thus valuable as a probe of magnetospheric dynamics for protons, electrons, and heavy ions, as well as for the helium ions themselves. In this paper radial profiles of ionized helium flux and protons are calculated, and certain physical effects ignored in previous treatments are taken into account.

084.236 **The ring current in the magnetosphere and the polar magnetic substorms.**
O. A. Troshichev, Ya. I. Feldstein.
Journ. Atmosph. Terr. Phys., Vol. 34, 845 - 858 (1972).

Measurements of charged particles made in the ATS-1, OGO-3, and Explorer-34 satellites and simultaneous measurements of the magnetic field at the ground are used to investigate the temporal-spatial characteristics of some magnetic storms.

084.237 **The long-term statistical forecast of geomagnetic pulsations of type Pc 1 activity.**
E. T. Matveyeva, V. A. Troitskaya, A. V. Gul'elmi.
Planet. Space Sci., Vol. 20, 637 - 638 (1972).

084.238 **Thermal plasma densities determined from digital sonagrams of hydromagnetic whistlers.**
Y. Higuchi, J. L. Kisabeth, J. A. Jacobs.
Planet. Space Sci., Vol. 20, 707 - 710 (1972).

084.239 **Summary of discussions at symposium on magnetosphere—ionosphere interactions: A review.**
K. Folkestad, E. V. Thrane, J. O. Thomas.
Journ. Atmosph. Terr. Phys., Vol. 34, 955 - 962 (1972).

Several theoretical and experimental aspects of the physics of the D-region, the upper polar ionosphere, the magnetosphere and wave particle interactions were treated at a Study Institute arranged at Dalseter, Norway, in April 1971. Some main features of the broad spectrum of topics considered at the symposium are given in this paper.

084.240 **A three-layer model of a small-scale dynamo taking into account the horizontal component of the magnetic field.**   L. M. Alexeeva, Yu. S. Vardanyan.
Geomagn. Aeronom., Vol. 12, 208 - 217 (1972). In Russian.

084.241 **About the equations of potential electric fields in the magnetosphere and ionosphere of the earth.**
A. L. Krylov, V. P. Tsherbakov.
Geomagn. Aeronom., Vol. 12, 218 - 225 (1972). In Russian.

084.242 **About the form and character of the variation of autocorrelation functions of magnetic and gravitational anomalies.**   V. N. Lugovenko, S. A. Serkerov.
Geomagn. Aeronom., Vol. 12, 306 - 310 (1972). In Russian.

084.243 **The dipolar system of coordinates and some of its peculiarities.**
M. N. Fatkullin, Yu. S. Sitnov.
Geomagn. Aeronom., Vol. 12, 333 - 335 (1972). In Russian. Brief information.

084.244 **Dynamics of the inner boundary of the plasma layer in the magnetosphere's tail during substorms.**
Ya. I. Feldshtein.
Geomagn. Aeronom., Vol. 12, 359 - 362 (1972). In Russian. Brief information.

084.245 **The quasi-stationary pinch effect and the structure of the magnetosphere.**
O. M. Raspopov, V. K. Koshelevsky.
Geomagn. Aeronom., Vol. 12, 362 - 364 (1972). In Russian. Brief information.

084.246 **Study of waves in the earth's bow shock.**
R. E. Holzer, T. G. Northrop, J. V. Olson, C. T. Russell.
Journ. Geophys. Res., Vol. 77, 2264 - 2273 (1972).

The perturbation vectors of waves upstream and downstream from the region of maximum compression in the bow shock were examined on Ogo 5 under particularly steady solar-wind conditions.

084.247 **Plasma waves in the dayside polar cusp. I. Magnetospheric observations.**
F. L. Scarf, R. W. Fredricks, I. M. Green, C. T. Russell.
Journ. Geophys. Res., Vol. 77, 2274 - 2293 (1972).

We present a general survey of the Ogo 5 plasma-wave measurements for the dayside polar-cusp encounters of November 1, 1968, and a detailed analysis of the observations at the low-altitude ($r \simeq 3-5 R_E$) cusp boundaries. The survey section contains an over-all discussion of the ULF magnetic-field wave levels and the VLF electric-field amplitude ranges measured from perigee out to $9 R_E$ on November 1, 1968.

084.248 **Hydromagnetic oscillations of the earth's core.**
R. Hide, K. Stewartson.
Rev. Geophys. Space Phys., Vol. 10, 579 - 598 (1972).

The present paper reviews theoretical work on free hydromagnetic oscillations of a bounded rotating fluid in the context of the geomagnetic problem.

084.249 **Geomagnetic field distortions and their effects on radiation belt particles.**   J. G. Roederer.
Rev. Geophys. Space Phys., Vol. 10, 599 - 630 (1972). Review article.

084.250   On the theory of the equatorial electrojet.
A. V. Gurevitch, A. L. Krylov, V. P. Tsherbakov.
Geomagn. Aeronom., Vol. 12, 413 - 420 (1972). In Russian.

084.251   Electron and proton acceleration in the outer layers of the magnetosphere during polar substorms.
S. N. Kuznetzov, E. N. Sosnovetz, L. V. Tverskaya, A. M. Teltzov, O. V. Khorosheva.
Geomagn. Aeronom., Vol. 12, 481 - 484 (1972). In Russian.

084.252   About the development of the pinch effect in the earth's magnetosphere.
O. M. Raspopov, V. K. Koshelevsky.
Geomagn. Aeronom., Vol. 12, 485 - 487 (1972). In Russian.

084.253   About the latitudinal distribution of the energy of geomagnetic disturbances.
A. Best, L. L. Vanyan, V. G. Dubrovsky, S. A. Kramarenko, G.-R. Lemann.
Geomagn. Aeronom., Vol. 12, 488 - 491 (1972). In Russian.

084.254   About excitation of magnetospheric resonators.
L. L. Vanyan, M. B. Gokhberg.
Geomagn. Aeronom., Vol. 12, 492 - 497 (1972). In Russian.

084.255   Secular variation of the geomagnetic field of the epoch 1965–1970 by data from observatories and satellites.   Sh. Sh. Dolginov, M. P. Ivchenko, V. P. Orlov, A. N. Pushkov, L. O. Tyurmina, T. N. Cherevko.
Geomagn. Aeronom., Vol. 12, 503 - 512 (1972). In Russian.

084.256   About the character of the drift of the main eccentric geomagnetic dipole.
I. M. Pudovkin, G. E. Valueva.
Geomagn. Aeronom., Vol. 12, 513 - 518 (1972). In Russian.

084.257   On the possibility of defining the secular variation of the geomagnetic field components from the distribution of the variations of the module of the total vector.
A. N. Pushkov.
Geomagn. Aeronom., Vol. 12, 519 - 523 (1972). In Russian.

084.258   Spherical analyses of the main geomagnetic field, 1550–1800.   S. I. Braginsky.
Geomagn. Aeronom., Vol. 12, 524 - 529 (1972). In Russian.

084.259   Penetration of hydromagnetic waves into the polar cap from the magnetosphere's tail.
O. V. Bolshakova, A. V. Gulielmi.
Geomagn. Aeronom., Vol. 12, 569 - 571 (1972). In Russian.
Brief information.

084.260   Calculation of the angular distribution of particles in the magnetosphere.
U. Kasimov, V. P. Shabansky.
Geomagn. Aeronom., Vol. 12, 571 - 573 (1972). In Russian.
Brief information.

084.261   About the location of the geomagnetic pole in the past.   M. Z. Nodia, Z. A. Chelidze.
Geomagn. Aeronom., Vol. 12, 581 - 582 (1972). In Russian.
Brief information.

084.262   The charts of isoclinic lines for the past epochs and prognosis for the year 2000.   E. N. Tarkhov.
Geomagn. Aeronom., Vol. 12, 586 - 587 (1972). In Russian.
Brief information.

084.263   Lunar semidiurnal variations of the geomagnetic field determined from the 2.5-min data scalings.
S. Matsushita, W. H. Campbell.

Journ. Atmosph. Terr. Phys., Vol. 34, 1187 - 1200 (1972).

084.264   Some problems of electromagnetic induction in the equatorial electrojet region – I. Magneto-telluric relations.   R. Hutton.
Geophys. Journ. Roy. Astron. Soc., Vol. 28, 267 - 284 (1972).
General theoretical expressions have been obtained for time variations of the magnetic and electric fields in the vicinity of the magnetic dip equator. From these, magneto-telluric ratios have been determined for a number of earth models, including the general case of n stratified layers. Model apparent resistivity curves have been computed for the same earth models as used by other research workers and the differences between them are discussed. Some of the problems which have to be considered in the practical application of the magneto-telluric expressions are noted and attention is drawn to the rapid change in the value of the apparent resistivity with electrojet width for variations of period greater than several hours.

084.265   Palaeomagnetic differences between normal and reversed field sources, and the problem of far-sided and right-handed pole positions.   R. L. Wilson.
Geophys. Journ. Roy. Astron. Soc., Vol. 28, 295 - 304 (1972).
New data strengthen the case for a world-wide eastward declination of the geomagnetic field during normal regimes, and westward (west of south) declination during reversed regimes. The question of how such fields could be maintained is discussed.

084.266   Indices of geomagnetic pulsations.
V. A. Troitskaya, A. V. Gul'elmi, O. V. Bolshakova, E. T. Matveyeva, R. V. Schepetnov.
Planet. Space Sci., Vol. 20, 849 - 858 (1972).
A system of spectral and energetic indices based on pulsations is proposed. The principle of the proposed system is to ascribe to each index a clear physical meaning. This principle is illustrated on the example of the R index, characterising the degree of contraction of the magnetosphere by the solar wind. The perspective of presenting the results of diagnostics in the form of indices is discussed. It is stressed that only the practice of long-term exploitation of an index can answer the question of its utility.

084.267   Consequences of an isotropic static plasma sheet in models of the geomagnetic tail.   G. L. Siscoe.
Planet. Space Sci., Vol. 20, 937 - 953 (1972).
The equation of momentum balance and magnetic flux conservation are given for a static tail model with an isotropic plasma sheet. The possibility of magnetic field leakage into the solar wind and across the neutral sheet is allowed. Numerical integrations for a wide variety of adjustable model parameters are presented that give the dependence on distance from earth of all tail parameters (field strength inside and outside of the plasma sheet, plasma pressure, plasma sheet area, tail radius, and normal field component to the neutral sheet).

084.268   Palaeosecular variation of the geomagnetic field in the Aleutian Islands, Alaska.
D. K. Bingham, D. B. Stone.
Geophys. Journ. Roy. Astron. Soc., Vol. 28, 317 - 335 (1972).

084.269   Secular variation in the Pacific Ocean region.
D. K. Bingham, D. B. Stone.
Geophys. Journ. Roy. Astron. Soc., Vol. 28, 337 - 355 (1972).

084.270   The configuration of the geomagnetic field.
D. H. Fairfield.
Astrophys. Space Sci. Library, Vol. 29, Part III, (see 012.014), p. 1 - 24 (1972).

084.271   Particle motions in the earth's magnetosphere.

V. P. Shabansky.
Astrophys. Space Sci. Library, Vol. 29, Part III, (see 012.014), p. 25 - 65 (1972).

084.272 Sources, losses, and transport of magnetospherically trapped particles. D. J. Williams.
Astrophys. Space Sci. Library, Vol. 29, Part III, (see 012.014), p. 66 - 130 (1972).

084.273 Magnetospheric substorms: A model.
S.-I. Akasofu.
Astrophys. Space Sci. Library, Vol. 29, Part III, (see 012.014), p. 131 - 151 (1972).

084.274 Magnetospheric plasma. V. M. Vasyliunas.
Astrophys. Space Sci. Library, Vol. 29, Part III, (see 012.014), p. 192 - 211 (1972).

084.275 The role of the geomagnetic tail in substorms.
A. J. Dessler.
Astrophys. Space Sci. Library, Vol. 29, Part III, (see 012.014), p. 212 - 218 (1972).

084.276 Instabilities in the magnetosphere (theoretical treatment). J. W. Dungey.
Astrophys. Space Sci. Library, Vol. 29, Part III, (see 012.014), p. 219 - 235 (1972).

084.277 Gyroresonant wave-particle interactions.
R. Gendrin.
Astrophys. Space Sci. Library, Vol. 29, Part III, (see 012.014), p. 236 - 269 (1972).

084.278 On the interaction between the magnetosphere and the ionosphere. B. Hultqvist.
Astrophys. Space Sci. Library, Vol. 29, Part IV, (see 012.015), p. 176 - 198 (1972).

084.279 High-latitude low-energy electron fluxes and variation of the magnetospheric structure with the direction of the interplanetary magnetic field and with the geomagnetic activity. A. Hruška, J. R. Burrows, I. B. McDiarmid.
Journ. Geophys. Res., Vol. 77, 2770 - 2779 (1972).

084.280 Computation of the lunar daily variation and error analysis. H. F. Petersons.
Journ. Geophys. Res., Vol. 77, 2948 - 2956 (1972).
The Chapman-Miller method is modified so as to decrease the number of terms that have to be neglected and the number of computations required for the determination of the lunar daily variation. A method for estimating confidence limits is presented. Four years (1952–1955) of hourly values of horizontal intensity at Toolangi geomagnetic observatory have been analyzed with the modified method.

084.281 Geomagnetic observations at the Kanozan Geodetic Observatory.
Bull. Geograph. Survey Inst., *Tokyo*, Vol. 16, Part 2, p. 19 - 45 (1971).
This is a table of geomagnetism observed for the period from Jan. 1969 to Dec. 1969 at the Kanōzan Geodetic Observatory, and a continuation of previous reports.

084.282 Geomagnetic observations at the Mizusawa Geodetic Observatory.
Bull. Geograph. Survey Inst., *Tokyo*, Vol. 16, Part 2, p. 47 - 71 (1971).
In this report the outline of the geomagnetic observations, and hourly mean values of the total force, horizontal component and declination, conducted by a digital processing system are illustrated (Apr. - Dec. 1969).

084.283 The second order magnetic survey of Japan (6).
Bull. Geograph. Survey Inst., *Tokyo*, Vol. 17, Part 1, p. 12 - 35 (1971).
This report contains the results of the survey from 1964 to 1968 in the western half of Japan, and is a continuation of the previous report which presented the results obtained in the Kantō and Tōhoku districts.

084.284 Ensayo histórico sobre los conocimientos magneto-telúricos.
H. G. Fournier. Translated from French by M. B. de O'Neill.
Publ. Obs. Ebro, Memoria No. 13, 16 + 320 pp. (1969).

084.285 Record of observations at Baker Lake Magnetic Observatory 1969. G. J. van Beek, H. R. Reny.
Publ. Earth Phys. Branch, Ottawa, Vol. 43, 75 - 148 (1972).
Mean hourly value for each hour of the day, and the mean daily value for each day of the month for 1969, for north component of horizontal intensity (X), east component of horizontal intensity (Y), and vertical intensity (Z); Summary by month, season and year of mean hourly values of X, Y and Z for all days and for the international quiet and disturbed days for 1969; Hourly ranges in 10-gamma units in X and Y for 1969. A set of three-component standard-run Ruska variometers record the X (north), Y (east), and Z (vertical) components of the earth's magnetic field. The time scale of the Ruska magnetograms is 20 mm/hr.

084.286 Record of observations at Great Whale River Magnetic Observatory 1969.
G. J. van Beek, H. R. Reny.
Publ. Earth Phys. Branch, Ottawa, Vol. 43, 149 - 222 (1972).
Mean hourly value for each hour of the day, and the mean daily value for each day of the month for 1969, for horizontal intensity (H), declination (D) and vertical intensity (Z); Summary by month, season and year of mean hourly values of H, D, Z for all days and for the international quiet and disturbed days for 1969; Hourly ranges in 10-gamma units in H and D for 1969. A standard Ruska three-component variograph records declination (D), horizontal intensity (H) and vertical intensity (Z). Time scale is 20 mm/hr.

084.287 Record of observations at Mould Bay Magnetic Observatory 1969. G. J. van Beek, W. Piche.
Publ. Earth Phys. Branch, Ottawa, Vol. 43, 223 - 298 (1972).

084.288 Record of observations at Resolute Bay Magnetic Observatory 1969. G. J. van Beek, H. R. Reny.
Publ. Earth Phys. Branch, Ottawa, Vol. 43, 299 - 372 (1972).

084.289 Record of observations at Fort Churchill Magnetic Observatory 1969. G. J. van Beek, H. R. Reny.
Publ. Earth Phys. Branch, Ottawa, Vol. 43, 373 - 446 (1972).

084.290 Fossil magnetic fields. S. K. Runcorn.
From plasma to planet. 21st Nobel symposium 1971, (see 012.028), p. 373 - 377 (1972).

Nonstationary behavior of collisionless shocks.
See Abstr. 062.035.

The radiative transfer problem in freely expanding gaseous clouds and its application to barium cloud experiments. See Abstr. 063.020.

Polar-cap structures of solar protons observed during the passage of interplanetary discontinuities.
See Abstr. 078.026.

Precession and the earth's magnetic field.
See Abstr. 081.014.

Superorkane in der Hochatmosphäre.
See Abstr. 082.064.

Thermal positive ions in the outer ionosphere and magnetosphere from OGO 1.   See Abstr. 083.029.

Relationship between E_S and the earth's magnetic field at middle latitudes.   See Abstr. 083.033.

Magnetic apex coordinates: A magnetic coordinate system for the ionospheric F_2 layer.   See Abstr. 083.054.

Interplanetary sector structure at solar maximum. See Abstr. 106.004.

Discontinuities and shock waves in the interplanetary medium and their interaction with the magnetosphere. See Abstr. 106.021.

Radiation Belts

084.401 Response of trapped particles to a collapsing dipole moment.   H. H. Heckman, P. J. Lindstrom.
Journ. Geophys. Res., Vol. 77, 740 - 743 (1972). – Brief report.

084.402 Secular magnetic variation and the inner proton belt.   M. Schulz, G. A. Paulikas.
Journ. Geophys. Res., Vol. 77, 744 - 747 (1972). – Brief report.

084.403 The composition of radiation registered below the region of radiation belts at mean latitudes.
V. S. Tzaplin, P. I. Shavrin, O. I. Savun.
Geomagn. Aeronom., Vol. 12, 177 - 179 (1972). In Russian.

084.404 Formation of the belt of DR-currents.
S. A. Zaitzeva, A. Glazhevska.
Geomagn. Aeronom., Vol. 12, 296 - 300 (1972). In Russian.

084.405 Differential energy spectrum of geomagnetically trapped protons with the ESRO 2 satellite.
P. Valot.
Journ. Geophys. Res., Vol. 77, 2309 - 2318 (1972).
The proton energy spectrum in the inner Van Allen zone was investigated in the range 30 to 300 MeV with a solid state dE/dx telescope aboard the ESRO 2 satellite and was tentatively interpreted with respect to recent theories.

084.406 Particle-flux limits in the synchronous orbit.
F. S. Mozer.
Journ. Geophys. Res., Vol. 77, 2401 - 2403 (1972). – Letter.

084.407 Observations of low-energy protons in July–August 1970 aboard Molniya 1.
S. N. Vernov, V. P. Borodulin, M. I. Panasyuk, I. A. Rubinshtejn, I. A. Savenko, Eh. N. Sosnovets.
Kosm. Issled., Vol. 10, 376 - 381 (1972). In Russian.

084.408 Diffusion and acceleration in the earth's radiation belts.   C.-G. Fälthammar.
Astrophys. Space Sci. Library, Vol. 29, Part III, (see 012.014), p. 270 - 296 (1972).

084.409 Electric fields in the magnetosphere and the origin of trapped radiation.   B. A. Tverskoy.
Astrophys. Space Sci. Library, Vol. 29, Part III, (see 012.014), p. 297 - 317 (1972).

084.410 Geomagnetically trapped carbon, nitrogen, and oxygen nuclei.   A. Mogro-Campero.
Journ. Geophys. Res., Vol. 77, 2799 - 2818 (1972).

084.411 Observation of energetic particles at very low altitudes near the geomagnetic equator.
D. Hovestadt, B. Häusler, M. Scholer.
Phys. Rev. Letters, Vol. 28, 1340 - 1344 (1972).
A considerable flux on the order of 7 particles/cm^{2}sec sr (probably protons) in the energy range above 500 keV is observed in the vicinity of the magnetic equator. The particles show a narrow distribution around 90° pitch angle. The particles are probably injected at several hundred kilometers altitude by a yet unknown mechanism.

084.412 A new source for the large-scale electric fields in the magnetosphere.   C. S. Wang, J. S. Kim.
Nature, Vol. 238, 91 - 92 (1972).
We show, using model calculations, that a decaying ring current is capable of producing the large scale electric fields in the magnetosphere.

084.413 Charged particles trapped in the earth's magnetic field.   D. J. Williams.
Advances Geophys., (USA), Vol. 15, 137 - 218 (1971). – See Bull. Signal., Vol. 33, Section 120, No. 5138 (1972).

084.414 A theory of the Van Allen belt.   G. Pizzella.
Nuovo Cimento B, Ser. 11, Vol. 8B, 35 - 52(1972).

Quasi-linear interaction of whistler-mode waves and nonthermal electrons.   See Abstr. 083.071.

Errata

084.901 Correction: 'Multiple crossings of the earth's bow shock at large geocentric distances' [Journ. Geophys. Res., Vol. 76, 5970 - 5977 (1971)].
B. Bavassano, F. Mariani, U. Villante, N. F. Ness.
Journ. Geophys. Res., Vol. 77, 2004 (1972). – Concerning 06.084.236.

085 Solar-Terrestrial Relations

085.001 **Observation de sursauts de rayonnement liés à l'activité solaire dans la basse atmosphère.**
J. Blamont, J.-P. Pommereau.
Comptes Rendus Acad. Sci. Paris, Sér. B, Vol. 274, 203 - 206 (1972).

085.002 **On the geomagnetically active flares of September 1963.**   M. C. Ballario.
Mem. Soc. Astron. Italiana, Nuova Ser., Vol. 43, 55 - 64 (1972).
The solar and geomagnetic phenomena recorded during the intervals Sept. 9 - Oct. 5 and June 9 - July 3, 1963, are examined.

085.003 **Intensity variations of surplus radiation at heights between 200 and 350 km.**
V. S. Tsaplin, P. I. Shavrin, L. V. Zubareva.
Kosm. Issled., Vol. 10, 136 - 140 (1972). In Russian. – Brief information.

085.004 **About solar activity display in long-term variations of the baric field of the northern hemisphere of the earth.**   B. A. Sleptzov-Shevlevitch.
Geomagn. Aeronom., Vol. 12, 326 - 327 (1972). In Russian. Brief information.

085.005 **Comparative characteristics of the radiation conditions at the earth's surface during the periods of the quiet and disturbed sun.**   N. I. Gojsa, V. I. Grishko.
Trudy Ukr. n.-i. gidrometeorol. in-ta, 1971, vyp. (No.) 120, p. 67 - 74. In Russian. – Abstr. in Referativ. Zhurn. 51. Astron., 5.51.482 (1972).

085.006 **Intervento dopo la relazione del Prof. Abrami.**
G. Piccardi.
Atti XIII Riunione Soc. Astron. Italiana, Trieste 1969, (see 012.010), p. 43 (1970).

085.007 **L'attività solare vista dai non astronomi.**
G. Piccardi.

Atti XIII Riunione Soc. Astron. Italiana, Trieste 1969, (see 012.010), p. 85 - 86 (1970).

085.008 **Considerazioni generali, programma di lavoro ed una proposta per lo studio dei fenomeni fluttuanti.**
G. Piccardi.
Atti XIII Riunione Soc. Astron. Italiana, Trieste 1969, (see 012.010), p. 159 - 161 (1970).

085.009 **Aspekte der solar-terrestrischen Forschung.**
F. W. Jäger.
Sterne, 48. Jahrgang, p. 87 - 95 (1972).

085.010 **Probabili cause e conseguenze delle oscillazioni climatiche.**   A. Murri.
L'Universo, Anno 52, p. 617 - 626 (1972).

085.011 **Solar activity and variability of meteorological conditions.**   A. A. Dmitriev.
Solnechnye Dannye 1972 Byull., No. 2, p. 107 - 115 (1972). In Russian.
The level of solar activity (Wolf numbers) has been compared with the character of variations of the atmospheric conditions (air temperature, wind velocity, relative humidity, precipitation). It is shown that the extremum values of meteorological elements are not consistent with those of Wolf numbers. An attempt is made to interpret this fact as the result of nonlinear connection between Wolf numbers and the solar constant.

085.012 **Dependence of the $\cos \chi$ index on solar and magnetic activities.**   A. S. M. Rao, B. R. Rao.
Indian Journ. Pure and Applied Phys., Vol. 9, 763 - 764 (1971). See Phys. Ber., Vol. 51, No. 5 - 4272 (1972).

085.013 **22 year solar cycle and chemical tests.**
G. Piccardi.
Geofis. Meteorol. (*Italy*), Vol. 20, 104 - 105 (1971). – See Bull. Signal., Vol. 33, Section 120, No. 7001 (1972).

Planetary System

091 Physics of the Planetary System (Planetary Atmospheres, Figure, Interior, Magnetic Fields, Rotation, etc.)

091.001 **News of the atmospheres of planets.**
G. S. Golitsyn.
Zemlya i Vselennaya, No. 1, p. 18 - 19 (1972). In Russian.

091.002 **On the planetary distance law.**
V. L. Polyachenko, A. M. Fridman.
Astron. Zhurn. Akad. Nauk SSSR, Vol. 49, 157 - 164 (1972). In Russian. English translation in Soviet Astron. AJ, Vol. 16, No. 1.
The stability of the protoplanetary cloud model of the solar system as a flat gaseous-dusty disk is investigated.

091.003 **Frequencies of occultations of stars by planets, satellites, and asteroids.** B. O'Leary.
Science, Vol. 175, 1108 - 1112 (1972).
Calculations show that several occultations of stars by the large satellites of the outer planets, Pluto, and the large asteroids could be observed each decade with existing equipment at earth-based telescopes.

091.004 **The collision-induced absorption of hydrogen and helium.** E. L. Wright.
Bull. American Astron. Soc., Vol. 3, 482 (1971). – Abstr. AAS.

091.005 **Ultraviolet photometry from the Orbiting Astronomical Observatory. III. Observations of Venus, Mars, Jupiter, and Saturn longward of 2000 Å.**
L. Wallace, J. J. Caldwell, B. D. Savage.
Astrophys. Journ., Vol. 172, 755 - 769 (1972).
A number of high-quality spectra of Mars, Jupiter, the combined disk and rings of Saturn, as well as lower-quality spectra of Venus, have been obtained with an objective-grating spectrometer of the Wisconsin Experiment Package of OAO-2. An intercomparison of the Mars, Jupiter, and Saturn spectra has failed to disclose any discrete planetary absorptions. Continuous albedos have been derived for the four planets by using OAO-2 observations of G-type stars for $\lambda > 2700$ Å and a new photoelectric solar spectrum by A. L. Broadfoot for $\lambda < 2700$ Å.

091.006 **Formation of spectral lines in planetary atmospheres – I. Theory for cloudy atmospheres; application to Venus.** G. E. Hunt.
Journ. Quant. Spectrosc. Radiat. Transfer, Vol. 12, 387 - 404 (1972).
The theory of the formation of spectral lines in a cloudy planetary atmosphere is studied in detail. The theory must be developed using an inhomogeneous (gravitational) model of a planetary atmosphere, accurately incorporating all the physical processes of radiative transfer. Such a model of the lower Venus atmosphere, consistent with our present knowledge, is constructed.

091.007 **Progress in planetary studies.** D. Ya. Martynov.
Vestn. AN SSSR, 1971, No. 10, p. 3 - 11. In Russian. – Abstr. in Referativ. Zhurn. 51. Astron., 3.51.309 (1972).

091.008 **On the theory of planetary spectra.**
V. V. Sobolev.
Astron. Zhurn. Akad. Nauk SSSR, Vol. 49, 397 - 405 (1972). In Russian. English translation in Soviet Astron. AJ, Vol. 16, No. 2.
The profiles and the equivalent widths of absorption lines are determined for two cases: 1) an atmosphere of small optical thickness limited by a reflecting surface, 2) a semi-infinite atmosphere with small true absorption. An arbitrary scattering indicatrix is assumed.

091.009 **Seasonal changes in planetary circulations.**
R. E. Newell, C. Boyer.
Planet. Space Sci., Vol. 20, 607 - 612 (1972).
Analogies are drawn between the atmospheric general circulation of the earth, Jupiter and Saturn as part of a search for seasonal variations. Such variations can give an indication of the factors governing the temperature of planetary atmospheres.

091.010 **On a formula of mass distribution of some planets.**
A. Kitov.
Godishn. Vissh. tekhn. uchebni zaved. Fiz., Vol. 5, sb. 2, p. 115 - 123 (1968, 1970). In Bulgarian. – Abstr. in Referativ. Zhurn. 51. Astron., 4.51.155 (1972).

091.011 **On the distribution of masses and kinetic moments of some planets.** A. Kitov.
Godishn. Vissh. tekhn. uchebni zaved. Fiz., Vol. 6, No. 1, p. 95 - 106 (1969, 1971). In Bulgarian. – Abstr. in Referativ. Zhurn. 51. Astron., 4.51.156 (1972).

091.012 **On the effect of departures from spherical symmetry upon the bistatic-radar occultation experiment.** P. T. McCormick, R. C. Whitten.
Planet. Space Sci., Vol. 20, 822 - 825 (1972). – Research note.

091.013 **Formation of spectral lines in planetary atmospheres–III. The use of analytic scattering diagrams in computations of synthetic spectra for cloudy atmospheres.**
G. E. Hunt.
Journ. Quant. Spectrosc. Radiat. Transfer, Vol. 12, 1023 - 1028 (1972).
We present results of synthetic spectra computed from realistic models of a cloudy planetary atmosphere, and demonstrate that the spectroscopic features of line profiles and equivalent widths computed for cloud models whose scattering properties are represented by the Heyney–Greenstein function are identical to those features when the Mie theory is used to compute the cloud properties.

091.014 **De schijnbare diameters van de planeetmanen.**
J. Meeus.
Hemel en Dampkring, Vol. 70, 88 - 89 (1972).

091.015 **Our solar system.** A. Bonev.
Aviatsiya i kosmonavtika (NRB), Vol. 13, No. 9, p. 8 - 10 (1971). In Bulgarian. – Abstr. in Referativ. Zhurn.

51. Astron., 6.51.75 (1972).

091.016 New orbital elements for moon and planets.
C. Oesterwinter, C. J. Cohen.
Celestial Mechanics, Vol. 5, 317 - 395 (1972).
The results of a simultaneous solution for the orbital elements of moon and planets are given and their derivation is discussed. A modern Cowell integrator is used for orbit computations, and least-squares fits are made to some 40000 optical observations taken since 1913. The model includes relativistic terms, the leading zonal harmonics of earth and moon, the precession of the lunar equator, and the tidal couple between earth and moon. The solution presented here is believed to be the only simultaneous improvement of the orbits of moon and planets. The planetary orbital elements compete with efforts of similar scope and accuracy at the Massachusetts Institute of Technology and the Jet Propulsion Laboratory.

091.017 Allgemeine Untersuchung der Positionen maximalen Glanzes bei inneren Planeten. B. Stanek.
Orion Schaffhausen, 30. Jahrgang, p. 81 - 82 (1972).

091.018 Karakteristieken van planeten en hun manen.
W. de Rop.
Hemel en Dampkring, Vol. 70, 200 - 201 (1972).

091.019 About the total spectrum of pulsation frequency of a polytropic atmosphere. R. S. Oganesian.
Astron. Zhurn. Akad. Nauk SSSR, Vol. 49, 672 - 674 (1972).
In Russian. English translation in Soviet Astron. AJ, Vol. 16, No. 3.

091.020 Lo studio dei pianeti da terra e dallo spazio.
G. de Mottoni.
Atti XIII Riunione Soc. Astron. Italiana, Trieste 1969, (see 012.010), p. 147 - 156 (1970).

091.021 Secular variations of the first order for the four major planets. Comparison with Le Verrier and Gaillot.
J. Chapront, J. L. Simon.
Astron. Astrophys., Vol. 19, 231 - 234 (1972). In French.
The authors present the results they have obtained for the secular variations of the osculating elements for the major planets of the solar system, Jupiter, Saturn, Uranus and Neptune. Three sets of results are given.

091.022 Characteristic γ- and X-radiation in the planetary system. D. Kuroczkin.
Postępy Astron., Vol. 20, 79 - 83 (1972). In Polish.

091.023 The tenth planet? T. Clarke.
Journ. Roy. Astron. Soc. Canada, Vol. 66, L17 - L18 (1972).

091.024 Planetary exploration–exobiology. R. S. Young.
Theory and experiment in exobiology, Vol. 1, (see 003.011), 143 - 160 (1971).

091.025 Satelliti retrogradi del sistema solare. V. Banfi.
Annuario *Torino* 1972, (see 047.021), p. 37 - 45 (1971).

091.026 Monochromatic brightness coefficients of giant planets. L. A. Bugaenko.
Astron. vestn., Vol. 6, 19 - 21 (1972). In Russian. – Abstr. in Referativ. Zhurn. 51. Astron., 7.51.214 (1972).

091.027 Aeronomy of CO_2 atmospheres. D. M. Hunten.
Comments Astrophys. Space Phys., Vol. 4, 1 - 5 (1972).
Two major questions: Why is there so little O, O_2, and CO in an atmosphere of CO_2, which is strongly dissociated by solar radiation?; How can relatively low temperatures in the thermosphere and exosphere be reconciled with the large electron densities, which imply strong solar heating; or is the heating efficiency in CO_2 much less than expected? are discussed.

091.028 On the exospheric temperature of hydrogen-dominated planetary atmospheres. S. H. Gross.
Journ. Atmosph. Sci., Vol. 29, 214 - 218 (1972).
All the planets may have been surrounded originally by atmospheres dominated by hydrogen or one of its compounds. Calculations of exospheric temperatures for the smaller planets preclude loss of hydrogen by the selective process of thermal escape since exceedingly high values are found for these planets. Instead, a general outflow or 'planetary wind' is expected to deplete rapidly the thermosphere of all constituents. Low values are obtained for the major planets and their retention of hydrogen is understandable. Expressions are given for the exospheric temperature of a two-component diffusive equilibrium model.

091.029 Exploration of planetary surfaces by mass spectrometry. K. Biemann.
Journ. Vacuum Sci. and Technology (*USA*) Vol. 9, 456 (1972).
See Phys. Abstr., Vol. 75, No. 41191 (1972).

091.030 On the existence of a resonance-captured 'quasi-satellite' of the earth. L. Danielsson, W.-H. Ip.
From plasma to planet. 21st Nobel symposium 1971, (see 012.028), p. 353 - 358 (1972).

091.031 Collision-induced spectra of hydrogen in the first and second overtone regions with applications to planetary atmospheres. A. R. W. McKellar, H. L. Welsh.
Proc. Roy. Soc. London, Ser. A, Vol. 322, 421 - 434 (1971).

091.032 Phase equilibria in fluid mixtures at high pressures: the neon-argon system. W. B. Streett, J. L. E. Hill.
Journ. Chem. Phys. (*USA*), Vol. 54, 5088 - 5094 (1971).
See Bull. Signal., Vol. 33, Section 120, No. 263 (1972).

091.033 On the effects of topography on planetary atmospheric circulation. S. L. Blumsack.
Journ. Atmosph. Sci., Vol. 28, 1134 - 1143 (1971).

091.034 Comments on the infrared absorption line formation in cloudy planetary atmospheres. N. Fukuta.
Journ. Atmosph. Sci., Vol. 28, 1511 - 1513 (1971).

091.035 Calculation of the temperature dependence for absorption in CO_2 in the 1750 - 1200 Å region.
P. S. Julienne, D. Neumann, M. Krauss.
Journ. Atmosph. Sci., Vol. 28, 833 - 837 (1971).

091.036 The photolysis of CO_2 at wavelengths exceeding 1740 Å. E. C. Y. Inn, J. M. Heimerl.
Journ. Atmosph. Sci., Vol. 28, 838 - 841 (1971).

091.037 Circular polarization of sunlight reflected by clouds.
J. E. Hansen.
Journ. Atmosph. Sci., Vol. 28, 1515 - 1516 (1971).

Moons and planets: An introduction to planetary science. See Abstr. 003.064.

Physique et dynamique planétaires.
See Abstr. 003.094.

Aus der Geschichte der Erforschung des Planetensystems. Vulkan – der Planet zwischen Sonne und Merkur.

See Abstr. 004.003.

The Titius-Bode law: A strange bicentenary.
See Abstr. 004.022.

Cours d'astronomie de la S.A.F. 6. Le système solaire: Etude physique. See Abstr. 015.015.

Photoelectron spectra and partial photoionization cross sections for NO, N_2O, CO, CO_2 and NH_3.
See Abstr. 022.019.

CO quantum yield in the photolysis of CO_2 at $\lambda 1470$ and $\lambda\lambda 1500-1670$. See Abstr. 022.032.

Broadening of infrared absorption lines at reduced temperatures: Carbon dioxide. See Abstr. 022.098.

Positions de grosses planètes et de la lune observées à l' équatorial photographique de 0, 33 m.
See Abstr. 041.055.

Limitations on outer planet mass determinations from their mutual perturbations. See Abstr. 042.001.

Missions spatiales vers les planètes. IV. Physique des surfaces planétaires. See Abstr. 051.044.

Laboratory simulation of diffuse reflectivity from a cloudy planetary atmosphere. See Abstr. 063.037.

Anisotropic nonconservative scattering in a semi-infinite medium. See Abstr. 063.042.

Interaction of solar wind with the moon and possibly other planetary bodies. See Abstr. 074.098.

Relazione sulla radioastronomia solare e planetaria.
See Abstr. 077.047.

New data on the moon and planets.
See Abstr. 094.073.

Some dynamic aspects of the Jovian and other atmospheres. See Abstr. 099.065.

092 Mercury

092.001 **Measuring of position angles and contact moments during the passage of Mercury across the disk of the sun on May 9, 1970.** N. A. Nesterko, V. E. Solov'ev.
Astron. vestn., Vol. 5, 257 - 260 (1971). In Russian. – Abstr. in Referativ. Zhurn. 51. Astron., 3.51.226 (1972).

092.002 **Merkurbeobachtung bei geringem Winkelabstand vom Sonnenrand.** F. Dorst.
SuW, Vol. 11, 102 (1972).

092.003 **Radar measurements of Mercury: Topography and scattering characteristics at 3.8 cm.**
R. P. Ingalls, L. P. Rainville.
Astron. Journ., Vol. 77, 185 - 190 (1972).
During the April 1971 inferior conjunction, the planet Mercury was observed with the Haystack radar operating at a wavelength of 3.8 cm. Sufficient sensitivity was available to measure scattering characteristics and topography over part of the equatorial region. The reflections corresponded typically to mean surface slopes of about 10° and although these showed some measurable variation with longitude, the reflections were far less variable than on Mars or Venus. Topographic variations of the order of a kilometer were observed, but the planet appears to have less variation in this respect than either Venus or Mars.

092.004 **There is no evidence for ice caps on Mercury.** G. E. Hunt.
Observatory, Vol. 92, 16 (1972). – Letter.

092.005 **Unipolar interaction of Mercury with the solar wind; the steady state bow shock problem.**
D. S. Colburn, C. P. Sonett, K. Schwartz.
Earth Planet. Sci. Letters, Vol. 14, 325 - 337 (1972).
The steady state electromagnetic interaction of the solar wind with the planet Mercury is computed for a spectrum of electrical conductivity functions using the assumption that no atmosphere or planetary magnetic field prohibits the direct interaction. The form of the induction is described by the unipolar effect and corresponds to the zero frequency limit of a transverse magnetic (TM) mode. Calculations are included to determine the effective surface temperature of the planet. These calculations include the apparent motion of the sun in the Hermean sky.

092.006 **Observation de Mercure au voisinage de la phase nulle.** G. Ratier.
Icarus, Vol. 16, 318 - 320 (1972).
The planet Mercury has been observed near superior conjunction with the aid of a Lyot coronagraph. The observed magnitude is in accord with the value obtained by extrapolation from the formula of Danjon.

092.007 **An attempt to measure the diameter of Mercury by Hertzsprung's method.**
J. Rösch, H. Camichel, F. Chauveau, M. Hugon, G. Ratier.
Icarus, Vol. 16, 321 - 327 (1972).
During the transit of Mercury on 9 May 1970, the photoelectric Hertzsprung method was used to measure the diameter of the planet. A careful examination of the various causes of errors, in combination with previous measurements, leads to a value $\geqslant 4920$ km, about 1% larger than the radar value. Sources of systematic errors in a variety of optical methods are discussed.

092.008 **Merkur, Venus und Spika im Oktober 1971.** F. Dorst.
SuW, Vol. 11, 169 (1972).

092.009 **Transit of Mercury across the disk of the sun on May 9, 1970 according to observations at Poltava.**

S. A. Onishchenko, B. F. Sincheskul.
Astron. vestn., Vol. 6, 62 - 63 (1972). In Russian. – Abstr. in
Referativ. Zhurn. 51. Astron. 6.51.207 (1972).

092.010 **Mercury's perihelion advance: Determination by
radar.** I. I. Shapiro, G. H. Pettengill, M. E. Ash,
R. P. Ingalls, D. B. Campbell, R. B. Dyce.
Phys. Rev. Letters, Vol. 28, 1594 - 1597 (1972).
Measurements of echo delays of radar signals transmitted
from earth to Mercury have yielded an accurate value for the
advance of the latter's perihelion position. Given that the sun's
gravitational quadrupole moment is negligible, the result in
terms of the Eddington–Robertson parameters is $(2+2\gamma-\beta)/3 \simeq$
1.005 ± 0.007, where $\gamma=\beta=1$ in general relativity, and where
0.007 represents the statistical standard error. Inclusion of the
probable contribution of systematic erros raises the uncertainty to about 0.02.

092.011 **Mercury in 1966 and 1967.** R. G. Hodgson.

Strolling Astronomer, Vol. 23, 170 - 174 (1972).

092.012 **Dråpeforma til Merkur og Venus ved passasjar.**
J. Bønes.
Astron. Tidssk., Årg. 5, p. 74 - 78 (1972).

092.013 **Beregning av Merkurs dråpeform på solskiven.**
R. Brahde.
Astron. Tidssk., Årg. 5, p. 79 - 82 (1972).

092.014 **Merkur.** J. Chodak.
Ideen Exakt. Wiss.[Deutsche Verlagsanstalt, Stutt-
gart], 1972, No. 1, p. 55 - 60.

Venus and Mercury. See Abstr. 003.103.

**Continuum brightness temperature of Mars and
Mercury at 21 cm wavelength.** See Abstr. 097.045.

093 Venus

093.001 **On the occurrence of ferrous chloride in the clouds
of Venus.** D. P. Cruikshank, A. B. Thomson.
Icarus, Vol. 15, 497 - 503, with a reply by G. P. Kuiper, p. 504
(1971/72).
We examine the observational basis for Kuiper's (1969)
conclusion that the upper cloud layers of Venus are composed
of dihydrated ferrous chloride particles. We find that the evidence for dihydrated ferrous chloride in the atmosphere of
Venus is inconclusive.

093.002 **Venus: Topography revealed by radar data.**
D. B. Campbell, R. B. Dyce, R. P. Ingalls, G. H.
Pettengill, I. I. Shapiro.
Science, Vol. 175, 514 - 516 (1972).
Surface height variations over the entire equatorial region
on Venus have been estimated from extended series of meas-
urements of interplanetary radar echo delays. The mean equa-
torial radius was found to be 6050.0 ± 0.5 kilometers.

093.003 **Venus clouds: A dirty hydrochloric acid model.**
B. Hapke.
Science, Vol. 175, 748 - 751 (1972).
The spectral and polarization data for Venus are consistent with micrometer-sized aerosol cloud particles of hydrochloric acid with soluble and insoluble iron compounds, whose
source could be volcanic or crustal dust.

093.004 **Composition of the Venus cloud tops in light of
recent spectroscopic data.** J. S. Lewis.
Astrophys. Journ., (Letters), Vol. 171, L75 - L79 = Contr.
Planet. Astron. Lab., Dep. Earth Planet. Sci., Mass. Inst.
Techn., Cambridge, No. 46 (1972).
The observed chemical composition of the atmosphere
near the Venus cloud tops, the refractive index of cloud particles as deduced from polarization data, and the $3-4\mu$ reflectivity of the clouds are all consistent with and strongly support
the possibility that the topmost clouds of Venus are composed
of concentrated aqueous HCl solutions.

093.005 **Venus: Topography revealed by radar data.**
R. B. Dyce, D. B. Campbell, R. P. Ingalls, G. H.
Pettengill, I. I. Shapiro.
Bull. American Astron. Soc., Vol. 3, 466 (1971). – Abstr. AAS.

093.006 **Venus: Measurements of microwave brightness
temperatures and interpretations of the radio and
radar spectra.** W. W. Warnock, J. R. Dickel.
Bull. American Astron. Soc., Vol. 3, 475 (1971). – Abstr.
AAS.

093.007 **Retrograde rotation of the upper atmosphere of
Venus.** J. Caldwell.
Bull. American Astron. Soc., Vol. 3, 501 (1971). – Abstr.
AAS.

093.008 **The topography of a swath around the equator of
the planet Venus from the wavelength dependence
of the radar cross section.**
A. E. E. Rogers, R. P. Ingalls, L. P. Rainville.
Astron. Journ., Vol. 77, 100 - 103 (1972).
The topography of the swath around the equator traced
out by the subradar point has been estimated from the changes
of cross section with wavelength and longitude. The topography has a maximum altitude variation of 6 km.

093.009 **Comet-like interaction of Venus with the solar
wind, I.** M. K. Wallis.
Cosmic Electrodynamics, Vol. 3, 45 - 59 (1972).
Clear differences of the plasma flow past Venus from
that outside the magnetosphere – in particular, absence of an
upstream bow shock and of an extended ionospheric tail –
characterize the Venus-solar wind interaction. Using data of
the Venus probes, the interaction is interpreted here from the
point of view of ion creation in an extended Venusian atmosphere, similar to the outer coma of comets. The new ions both
decelerate the solar wind and themselves gain large energies.

093.010 Observing Venus near the sun.
A. T. Young, L. G. Young.
Sky Telescope, Vol. 43, 140 - 144 (1972).

093.011 Observations of Venus with a TV device.
V. V. Prokofieva, S. I. Usliber.
Izv. Krymskoj Astrofiz. Obs., Vol. 43, 201 - 205 (1971).
In Russian.
Photographs of Venus in ultraviolet and red light have been taken with a high sensitive TV device at the Crimean Astrophysical Observatory. No markings of Venus were observed in red light.

093.012 Composition of the upper clouds of Venus.
D. G. Rea.
Rev. Geophys. Space Phys., Vol. 10, 369 - 378 (1972).
An analysis of the Mariner 5 occultation data has led to improved temperature and pressure profiles. When these are combined with transit data, it is concluded that there is an optically thin cloud layer with a top at 81-km altitude. The inclusion of temperatures derived from the near-infrared CO_2 bands leads to the postulate of a second cloud deck with a top at 61-km altitude. Additional important constraints on cloud models are imposed by the measured abundances of HCl and H_2O, by the polarization data, and by the reflection and emission spectra.

093.013 Formation of spectral lines in a planetary atmosphere – II. Spectroscopic evidence for the structure of the visible Venus clouds. G. E. Hunt.
Journ. Quant. Spectrosc. Radiat. Transfer, Vol. 12, 405 - 419 (1972).
We demonstrate that there is spectroscopic evidence for the structure of the visible Venus cloud layers. It is proved that Venus cannot have a single dense cloud layer, but must have two scattering layers; a thin aerosol layer situated in the lower stratosphere, overlying a dense cloud deck.

093.014 Phasenbeobachtungen der Venus vor und nach der unteren Konjunktion 1970 November 10.
P. Ahnert.
Sterne, 48. Jahrgang, p. 55 - 57 (1972).

093.015 Spectrum of Venus with a double-cloud model.
J. W. Chamberlain, G. R. Smith.
Astrophys. Journ., Vol. 173, 469 - 475 (1972).
We investigate the hypothesis that the CO_2 absorption lines in Venus's spectrum could be formed between an upper thin cloud and a lower thick one. The principal conclusion is that the present data do not allow a unique interpretation of the cloud–gas mixing in Venus's atmosphere.

093.016 Venus im August und September 1971.
F. Dorst.
SuW, Vol. 11, 24 (1972).

093.017 Phase dependence of the 2.7 cm wavelength radiation of Venus. T. P. McCullough.
Icarus, Vol. 16, 310 - 313 (1972).
Observations of Venus made during the period July 1968 to July 1970 at 2.7 cm wavelength show no evidence for variation of brightness temperature over a complete phase cycle greater than $\pm 9° K$ measurement uncertainty.

093.018 Search for a Venus halo effect during 1970.
D. Ward, B. O'Leary.
Icarus, Vol. 16, 314 - 317 (1972).
Photometric observations of Venus during 1970 inferior conjunction, in contrast to observations made during the 1969 inferior conjunction, show no evidence of a Venus halo effect at 158° phase angle.

093.019 Theoretical interpretation of the Venus 1.05-micron CO_2 band and the Venus 0.8189-micron H_2O line.
J. L. Regas, L. P. Giver, R. W. Boese, J. H. Miller.
Astrophys. Journ., Vol. 173, 711 - 725 (1972).
A new analysis of the 1.05-μ CO_2 band and the 0.8189-μ H_2O line in the Venus spectrum has been undertaken, using the synthetic-spectrum technique. We constructed our synthetic spectra with a model which takes into account both isotropic scattering and the inhomogeneity in the Venus atmosphere. Also we used the Potter-Hansen correction factor to correct for anisotropic scattering, so that our synthetic spectra are the first which contain all the essential physics of line formation. We applied objective statistical techniques to judge the quality of the match of the theory and the observations.

093.020 Venus section report: The eastern (evening) apparition of 1970. D. P. Cruikshank.
Strolling Astronomer, Vol. 23, 142 - 146 (1972).

093.021 Ultraviolet observations of Venus in 1969. Comments on the cusp or polar caps and the 4-day atmospheric period of rotation. J. Hiscott.
Journ. British Astron. Ass., Vol. 82, 198 - 199 (1972).

093.022 Wind velocity and turbulence in the Venus atmosphere from data of Doppler measurements of the velocity of Venera 4, Venera 5, Venera 6.
V. V. Kerzhanovich.
Kosm. Issled., Vol. 10, 261 - 273 (1972). In Russian.

093.023 On the determination of the parameters of the Venus atmosphere below the level of critical refraction from radio obscuring measurements.
D. S. Lukin, Yu. G. Spiridonov, V. A. Shkol'nikov, S. I. Fominykh.
Kosm. Issled., Vol. 10, 274 - 278 (1972). In Russian.

093.024 The atmosphere of Venus. J. Michel.
Urania Kraków, Vol. 43, 66 - 70 (1972). In Polish.

093.025 Wind velocity and some characteristics of the surface of Venus obtained with Venera 7.
V. V. Kerzhanovich, M. K. Rozhdestvenskij, B. N. Andreev, V. M. Gotlib, V. P. Lysov, Yu. N. Shnygin.
Kosm. Issled., Vol. 10, 390 - 399 (1972). In Russian.

093.026 Some optical properties of the Venus atmosphere and possibilities of interpreting photometric and polarization measurements. E. M. Fejgel'son.
Kosm. Issled., Vol. 10, 411 - 422 (1972). In Russian.

093.027 Venus-Dichotomie im April 1972. J. Alean.
Orion Schaffhausen, 30. Jahrgang, p. 100 - 101 (1972).

093.028 Venus cloud cover. J. T. Bartlett, G. E. Hunt.
Nature, Phys. Sci., Vol. 238, 11 - 12 (1972).
We show that there need be no cloud free region on Venus. The continuous cloud cover on Venus can be readily understood by making certain simple assumptions.

093.029 Temperaturberechnung der Venusatmosphäre bis 80 km Höhe aufgrund solarer und thermischer Strahlungsströme sowie konvektiver und turbulenter Wärmetransporte. E. Roeckner.
Mitt. Max-Planck-Inst. Aeronomie, Lindau (Harz), No. 46, 54 pp. To be obtained by Springer-Verlag, Berlin–Heidelberg–New York. Price DM 20.00 (1972).
The purpose of the paper is to investigate whether the temperature of the Venus surface can be interpreted by means of a greenhouse model with high solar radiation absorption of

the atmosphere and the clouds. An attempt is made to calculate a stationary temperature distribution of a $(CO_2 + H_2O)$ Venus atmosphere up to 80 km height as asymptotic solution of an initial value problem, taking into account the solar radiation, the thermal emission of the atmosphere, as well as convective and turbulent heat transports.

093.030 Ion clusters and the Venus ultraviolet haze layer.
        A. C. Aikin.
Nature, Phys. Sci., Vol. 235, 10 - 11 (1972).

The daytime ionosphere of Venus is observed between 100 and 500 km altitude and has a peak electron concentration of 5×10^5 cm^{-3} at 140 km. Below 200 km CO_2^+ is thought to be the principal ion unless oxygen is present. I suggest that at altitudes less than 130 km $CO_2^+ \cdot CO_2$ is an important ionic constituent of the Venus ionosphere. Below 100 km ion clustering processes combine with the low temperature at the mesopause to form coagulates giving rise to the ultraviolet haze layer which has frequently been observed.

093.031 A Venus seismic experiment for the late 1970's.
        J. S. Derr.
Bull. Seismol. Soc. America, Vol. 61, 1731 - 1737 (1971).

093.032 Numerical experiments on the general circulation of the Venus atmosphere.
D. V. Chalikov, A. S. Monin, V. G. Turikov, S. S. Zilitinkevich.
Tellus, Vol. 23, 483 - 488 (1971).

A number of numerical experiments on Venusian atmospheric circulation with a time-dependent two-layer primitive equations model was made.

093.033 The evolution of water vapor in the atmosphere of Venus. L. L. Smith, S. H. Gross.
Journ. Atmosph. Sci., Vol. 29, 173 - 178 (1972).

The atmosphere of Venus appears to be deficient in water vapor by a factor of about 10^4 compared with the total amount of water on earth. The feasibility of loss of water vapor from the Venus atmosphere is examined, assuming H_2O as the sole initial constituent. A steady-state model is constructed.

093.034 Enhanced microwave absorption in the lower atmosphere of Venus. K. R. Richter.
Radio Sci., Vol. 7, 443 - 447 (1972).

The enhanced microwave attenuation in the altitude range between 35 and 52 km is explained by the assumption of an atmospheric layer consisting of water vapor and water droplets. Independent measurements lead to a reasonable altitude distribution of the water droplets and the water vapor content. The results show good agreement with assumptions which have to be made to explain the brightness temperature of the planet Venus.

093.035 Infrared and microwave transmission in the atmosphere of Venus. R. K. L. Poon, D. H. Staelin.
Quarterly Progr. Rep. Res. Lab. Electronics, Mass. Inst. Technology (USA), No. 103, p. 34 - 42 (1971).

093.036 An experimental investigation of the spectral characteristics of the Venusian atmosphere at high temperatures. A. A. Kon'kov, S. G. Kulagin.
High Temperature, [Consultants Bureau, New York, N. Y.], Vol. 9, 453 - 459 (1971). – Translated from Russian.

093.037 Venus: Surface temperature variations. J. S. Lewis.
Journ. Atmosph. Sci., Vol. 28, 1084 - 1086 (1971).

093.038 Photochemistry of HCl and other minor constituents in the atmosphere of Venus. R. G. Prinn.
Journ. Atmosph. Sci., Vol. 28, 1058 - 1068 (1971).

093.039 A numerical study of the atmospheric circulation on Venus. T. Sasamori.
Journ. Atmosph. Sci., Vol. 28, 1045 - 1057 (1971).

093.040 Circulation and thermal structure of the Venusian thermosphere. R. E. Dickinson.
Journ. Atmosph. Sci., Vol. 28, 885 - 894 (1971).

Venus and Mercury.   See Abstr. 003.103.

Mariner–Venus 1967: Final project report.
See Abstr. 003.148.

Meeting review: The upper atmosphere of Venus.
See Abstr. 011.005.

Methods of radar astrometric observations.
See Abstr. 031.002.

Formation of spectral lines in planetary atmospheres – I. Theory for cloudy atmospheres; application to Venus.   See Abstr. 091.006.

Aeronomy of CO_2 atmospheres.
See Abstr. 091.027.

Dråpeforma til Merkur og Venus ved passasjar.
See Abstr. 092.012.

Comparative estimate of the surface roughness of the moon and Venus.   See Abstr. 094.162.

The surfaces of Mars and Venus.
See Abstr. 097.041.

Radiative transfer in the atmosphere of Mars and that of Venus above the cloud deck.   See Abstr. 097.063.

Das Wettergeschehen auf Mars, Venus und Jupiter.
See Abstr. 097.067.

Results and interpretation of the S-band occultation experiments on Mars and Venus.   See Abstr. 097.112.

The electron distributions in the Mars and Venus upper atmospheres.   See Abstr. 097.114.

Aeronomy of CO_2 atmospheres: A review.
See Abstr. 097.115.

Observations of Jupiter, Venus, and source 3C 273 at 2 and 8 mm wavelengths.   See Abstr. 099.071.

345-micron ground-based observations of M17, M82, and Venus.   See Abstr. 113.002.

094 Moon

094.001 **Interaction between lunar ground and solid bodies.**
A. A. Silin, V. V. Shvarev.
Zemlya i Vselennaya, No. 1, p. 27 - 29 (1972). In Russian.

094.002 **Geological provinces of the near side of the moon.**
J. F. McCauley, D. E. Wilhelms.
Icarus, Vol. 15, 363 - 367 (1971/72).
Systematic geologic mapping of the near side of the moon has provided the basis for defining and delineating the major geological provinces of the near side. From the nature of the provinces and their distribution patterns a general historical sequence evolves. Five main surface-shaping periods are recognized.

094.003 **Two former faces of the moon.**
D. E. Wilhelms, D. E. Davis.
Icarus, Vol. 15, 368 - 372 (1971/72).
Systematic geologic mapping of the lunar near side has resulted in the assignment of relative ages to most visible features. As a derivative of this work, geologic and artistic interpretations have been combined to produce reconstructions of the moon's appearance at two significant points in its history.

094.004 **The meteorite flux at the lunar surface.**
W. Wiesel.
Icarus, Vol. 15, 373 - 383 (1971/72).
The relative meteorite flux values at the lunar surface as a function of position and earth-moon distance are of prime importance in applications of crater counting. A general method for obtaining these values from an assumed distribution function far from the earth is proposed, and solved numerically in the case where the moon's mass is neglected. Some implications of these results for lunar history are explored.

094.005 **Moon: Possible nature of the body that produced the Imbrian Basin, from the composition of Apollo 14 samples.**
R. Ganapathy, J. C. Laul, J. W. Morgan, E. Anders.
Science, Vol. 175, 55 - 59 (1972).
Soils from the Apollo 14 site contain nearly three times as much meteoritic material as soils from the Apollo 11, Apollo 12, and Luna 16 sites. Part of this material consists of the ubiquitous micrometeorite component, of primitive (carbonaceous-chondrite-like) composition. This material seems to be debris of the Cyprus-sized planetesimal that produced the Imbrian Basin. It resembles the initial composition of the earth as postulated by the two-component model. Apparently the Imbrian planetesimal was an earth satellite swept up by the moon during tidal recession or capture, or an asteroid deflected by Mars into terrestrial space.

094.006 **Chondrules in Apollo 14 samples: Implications for the origin of chondritic meteorites.**
E. A. King, Jr., M. F. Carman, J. C. Butler.
Science, Vol. 175, 59 - 60 (1972).
Chondrules have been observed in several breccia samples returned by the Apollo 14 mission. These lunar chondrules are believed to have formed during a large impact event, perhaps the one that formed the Imbrian Basin. This suggests that some meteoritic chondrules are also formed by impact processes.

094.007 **Lunar gravity via Apollo 14 Doppler radio tracking.**
W. L. Sjogren, P. Gottlieb, P. M. Muller, W. Wollenhaupt.
Science, Vol. 175, 165 - 168 (1972).
Gravity measurements at high resolution were obtained over a 100-kilometer band from +70° to −70° of longitude during the orbits of low periapsis altitude (approximately 16 kilometers). The line-of-sight accelerations are plotted on Aeronautical Chart and Information Center mercator charts (scale 1 : 1,000,000) as contours at 10-milligal intervals. Direct correlations between gravity variations and surface features are easily determined.

094.008 **Chemical composition of sawdust from lunar rock 12013 and comparison of a Java tektite with the rock.** D. L. Showalter, H. Wakita, R. H. Smith, R. A. Schmitt, D. E. Gillum, W. D. Ehmann.
Science, Vol. 175, 170 - 172 (1972).
Abundances of 11 major and minor elements and 11 trace elements have been determined by instrumental neutron activation analysis of two Apollo 12013 rock fragments, a sample of rock 12013,17 sawdust, and a Java tektite (J2). Although the abundances of major elements in tektite J2 are similar to those of rock 12013, comparison of the minor and trace elements shows that no fragment or sawdust of rock 12013 that has been analyzed to date is chemically similar to tektite glass.

094.009 **The Apollo 15 lunar samples: A preliminary description.**
Apollo 15 Preliminary Examination Team: P. W. Gast, W. C. Phinney, M. B. Duke, L. T. Silver, N. J. Hubbard, G. H. Heiken, P. Butler, D. S. McKay, J. L. Warner, D. A. Morrison, F. Horz, J. Head, G. E. Lofgren, W. I. Ridley, A. M. Reid, H. Wilshire, J. F. Lindsay, W. D. Carrier, P. Jakes, M. N. Bass, P. R. Brett, E. D. Jackson, J. M. Rhodes, B. M. Bansal, J. E. Wainwright, K. A. Parker, K. V. Rodgers, J. E. Keith, R. S. Clark, E. Schonfeld, L. Bennett, M. Robbins, W. Portenier, D. D. Bogard, W. R. Hart, W. C. Hirsch, R. B. Wilkin, E. K. Gibson, C. B. Moore, C. F. Lewis.
Science, Vol. 175, 363 - 375 (1972).
The morphology, mineralogy, petrology, and chemistry of the Apollo 15 samples are described.

094.010 **Geologic setting of the Apollo 15 samples.**
Apollo Lunar Geology Investigation Team: G. A. Swann, N. G. Bailey, R. M. Batson, V. L. Freeman, M. H. Hait, H. E. Holt, K. B. Larson, V. S. Reed, G. G. Schaber, R. L. Sutton, E. W. Wolfe, K. A. Howard, H. G. Wilshire, J. W. Head, J. B. Irwin, D. R. Scott, W. R. Muehlberger, L. T. Silver, J. J. Rennilson.
Science, Vol. 175, 407 - 415 (1972).

094.011 **Chemistry, geochronology, and petrogenesis of lunar sample 15555.**
B. W. Chappell, W. Compston, D. H. Green, N. G. Ware.
Science, Vol. 175, 415 - 416 (1972).
Lunar sample 15555 is a mare type basalt generally similar in chemical composition to the Apollo 12 basalts. Sample 15555 is older than any Apollo 12 basalt but younger than Apollo 14 basalts analyzed thus far.

094.012 **Argon-40−Argon-39 dating of Apollo sample 15555.**
E. C. Alexander, Jr., P. K. Davis, R. S. Lewis.
Science, Vol. 175, 417 - 419 (1972).
An age of $3.33 \pm 0.05 \times 10^9$ years was obtained for Apollo 15 sample 15555 by argon-40−argon-39 dating.

094.013 **Rubidium-strontium and potassium-argon age of lunar sample 15555.**
V. Rama Murthy, N. M. Evensen, Bor-ming Jahn, M. R. Coscio, Jr., J. C. Dragon, R. O. Pepin.

Science, Vol. 175, 419 - 421 (1972).

The lunar mare basalt 15555 from the edge of Hadley Rille has been dated at 3.3×10^9 years by both rubidium-strontium and potassium-argon techniques.

094.014   Rare gas record in the largest Apollo 15 rock.
         K. Marti, B. D. Lightner.
Science, Vol. 175, 421 - 422 (1972).

The spallation krypton data from rock chip 15555,23 indicate a well-shielded location during most of the time during which the rock was exposed to cosmic rays. A krypton-krypton exposure age of $81^{+17}_{-7} \times 10^6$ years is calculated, and the gas retention ages are estimated.

094.015   Gas-retention and cosmic-ray exposure ages of lunar rock 15555.
F. A. Podosek, J. C. Huneke, G. J. Wasserburg.
Science, Vol. 175, 423 - 425 (1972).

The last lava flow in the Hadley Rille area of Mare Imbrium, as inferred from an argon-40–argon-39 experiment on a plagioclase separate from the lunar basalt 15555, occurred $3.31 \pm 0.03 \times 10^9$ years ago. A cosmic-ray exposure age of $90 + 10 \times 10^6$ years is determined from the ratio of spallogenic argon-38 to calcium.

094.016   Geochemistry of Apollo 15 basalt 15555 and soil 15531.
C. C. Schnetzler, J. A. Philpotts, D. F. Nava, S. Schuhmann, H. H. Thomas.
Science, Vol. 175, 426 - 428 (1972).

Major and trace element concentrations have been determined by atomic absorption spectrophotometry, colorimetry, and isotope dilution in Apollo 15 mare basalt 15555 from the Hadley Rille area; trace element concentrations have also been determined in plagioclase and pyroxene separates from basalt 15555 and in soil 15531 from the same area.

094.017   Age of a lunar anorthosite.
L. Husain, O. A. Schaeffer. J. F. Sutter.
Science, Vol. 175, 428 - 430 (1972).

The crystallization age of an Apollo 15 anorthosite rock, 15415,9, returned from the lunar highlands has been measured to be $(4.09 \pm 0.19) \times 10^9$ years.

094.018   Mineralogic and petrologic study of lunar anorthosite slide 15415,18.
R. B. Hargraves, L. S. Hollister.
Science, Vol. 175, 430 - 432 (1972).

094.019   Lunar anorthosite 15415: Texture, mineralogy, and metamorphic history.   O. B. James.
Science, Vol. 175, 432 - 436 (1972).

094.020   Apollo 15 geochemical X-ray fluorescence experiment: Preliminary report.
I. Adler, J. Trombka, J. Gerard, P. Lowman, R. Schmadebeck, H. Blodget, E. Eller, L. Yin, R. Lamothe, P. Gorenstein, P. Bjorkholm.
Science, Vol. 175, 436 - 440 (1972).

094.021   Primordial radioelements and cosmogenic radionuclides in lunar samples from Apollo 15.
G. D. O'Kelley, J. S. Eldridge, E. Schonfeld, K. J. Northcutt.
Science, Vol. 175, 440 - 443 (1972).

094.022   Crystallization studies of lunar igneous rocks: Crystal structure of synthetic armalcolite.
M. D. Lind, R. M. Housley.
Science, Vol. 175, 521 - 523 (1972).

094.023   Ultrathin amorphous coatings on lunar dust grains.
         J. P. Bibring, J. P. Duraud, L. Durrieu, C. Jouret, M. Maurette, R. Meunier.
Science, Vol. 175, 753 - 755 (1972).

Ultrathin amorphous coatings have been observed by high-voltage electron microscopy on micrometer-sized dust grains from the Apollo 11, Apollo 12, Apollo 14, and Luna 16 missions. Calibration experiments show that these coatings result from an "ancient" implantation of solar wind ions in the grains.

094.024   Absolute dating techniques.
         G. J. Wasserburg, J. C. Huneke, F. A. Podosek, L. Husain, O. A. Schaeffer, J. F. Sutter.
Science, Vol. 175, 795 (1972).

094.025   The experience of the preparation of a spectral zone map of an area of the lunar surface.
Yu. N. Lipsky, V. V. Shevchenko.
Astron. Zhurn. Akad. Nauk SSSR, Vol. 49, 170 - 174 (1972). In Russian. English translation in Soviet Astron. AJ, Vol. 16, No. 1.

On the basis of results of spectral zone photography in the range with effective wavelength 380 mμ and 640 mμ, a contour map of the "reddish coefficient" of Mare Imbrium was prepared. A correlation between boundary of colour areas and some relief details is found.

094.026   On the problem of values of coefficients C_{20}, C_{22} of the expansion of the gravitational field of the moon.
Sh. T. Khabibullin, Yu. A. Chikanov.
Astron. Zhurn. Akad. Nauk SSSR, Vol. 49, 222 - 223 (1972). In Russian. English translation in Soviet Astron. AJ, Vol. 16, No. 1.

The most probable values of Cassini's equator inclination to the ecliptic and the parameter g' determining the moon's radial density distribution are proposed as a criterion for selection of the gravitational field coefficients C_{20}, C_{22}.

094.027   High-frequency electromagnetic response of the moon.   G. Schubert, K. Schwartz.
Journ. Geophys. Res., Vol. 77, 76 - 83 (1972).

It is shown that the contribution of higher harmonics to the lunar transfer functions for the tangential components of the surface magnetic field is significant at frequencies greater than 0.01 Hz (Apollo 12 and Explorer 35 magnetometer data extend to frequencies as high as 0.04 Hz). We outline a theory leading to the definitions of the tangential magnetic-field transfer functions and discuss their relationship to the low-frequency approximations. Transfer functions are then presented for several lunar electrical conductivity models and are qualitatively compared with the experimental data.

094.028   Energetic ion bursts on the nightside of the moon.
         J. W. Freeman, Jr.
Journ. Geophys. Res., Vol. 77, 239 - 243 (1972).

Recurrent bursts of positive ions are a persistent feature of the data from the Apollo 12 and Apollo 14 Alsep suprathermal ion detectors. Most remarkable is the occasional occurrence of these events when the Alsep is deep in the lunar night. One such series of bursts seen during December 1969 starting approximately 4.7 days before sunrise at the Apollo 12 Alsep site is reported and analyzed.

094.029   Technique for rapid determination of relative ages of lunar areas from orbital photography.
L. A. Soderblom, L. A. Lebofsky.
Journ. Geophys. Res., Vol. 77, 279 - 296 = Contr. Planet. Astron. Lab., Dep. Earth Planet. Sci., Mass. Inst. Technology,

Cambridge, No. 20 (1972).

A technique for determining relative ages of regions of the lunar surface using orbital photography was developed from a model of small-impact erosion. The erosion model relates the shape of a crater to the integrated flux of debris that has impacted the surface since that crater was fresh. The shape of the most modified crater of a particular diameter is thereby related to the relative age of the surface. Application of the technique for many lunar areas leads to relative ages which are given in a table. A discussion of the errors is included.

094.030 Calculation of the lunar photon albedo from galactic and solar proton bombardment.
T. W. Armstrong.
Journ. Geophys. Res., Vol. 77, 524 - 536 (1972).

The lunar photon albedo due to cosmogenic and primordial photon sources has been calculated. The individual photon leakage spectra from prompt photons produced by galactic cosmic ray (GCR) and solar cosmic ray (SCR) induced nuclear interactions, from the decay of GCR- and SCR-induced radionuclides, and from the decay of naturally occurring radionuclides are given. An approximate estimate of the leakage from the photon-electron cascade initiated by the decay of neutral pions is also given.

094.031 Interaction of solar and galactic cosmic-ray particles with the moon. R. C. Reedy, J. R. Arnold.
Journ. Geophys. Res., Vol. 77, 537 - 555 (1972).

The rates of formation of radionuclides as a function of depth in the moon are calculated for bombardments by galactic-cosmic-ray particles and by solar protons. The fluxes and spectra of galactic-cosmic-ray particles and of solar protons as a function of depth in the moon are first determined semi-empirically. For galactic cosmic rays, the model emphasizes the production of secondary particles and the attenuation of particles by nuclear interactions. Solar proton calculations cover a range of observed spectral parameters. The calculated production rates are given for a range of depths in the moon and are compared with experimental results.

094.032 Lunar fossil magnetism and perturbations of the solar wind. C. P. Sonett, J. D. Mihalov.
Journ. Geophys. Res., Vol. 77, 588 - 603 (1972).

Perturbations of the solar wind downstream of the moon and lying outside of the rare faction wave that defines the diamagnetic cavity are used to define possible source regions comprised of intrinsically magnetized areas of the moon. A map of the moon is constructed showing that a model in which the sources are exposed to the grazing solar wind during the lunation yields a selenographically invariant set of regions strongly favoring the lunar highlands over the maria. An alternative model with the source due to electromagnetic induction is explored.

094.033 Discussion of paper by J. A. Morrison and P. R. Norton, 'The heat capacity and thermal conductivity of Apollo 11 lunar rocks 10017 and 10046 at liquid helium temperatures' [Journ. Geophys. Res., Vol. 75, 6553 - 6557 (1970)]. R. M. Housley.
Journ. Geophys. Res., Vol. 77, 965 - 966, with a reply by J. A. Morrison, P. R. Norton, p. 967 - 968 (1972).

094.034 Optical constants for terrestrial analogs of lunar minerals from 0.33 to 2.3 μm.
W. G. Egan, T. W. Hilgeman.
Bull. American Astron. Soc., Vol. 3, 500 (1971). – Abstr. AAS.

094.035 Lunar Research: No agreement on evolutionary models. A. L. Hammond.

Science, Vol. 175, 868 - 870 (1972).

094.036 New zirconium-rich minerals from Apollo 14 and 15 lunar rocks.
A. Peckett, R. Phillips, G. M. Brown.
Nature, Vol. 236, 215 - 217 (1972).

Two more members of the suggested Zr−Ti−Fe mineral group of lunar oxides have been discovered. They have similar optical and chemical properties to the two such compounds previously detected, but all four show considerable differences in their detailed chemistry.

094.037 Das aschgraue Mondlicht. U. Köster.
SuW, Vol. 11, 56 - 57 (1972).

094.038 Rare gas studies of the galactic cosmic ray irradiation history of lunar rocks.
J. C. Huneke, F. A. Podosek, D. S. Burnett, G. J. Wasserburg.
Geochim. Cosmochim. Acta, Vol. 36, 269 - 301 (1972).

Eight analyses of all five noble gases in whole rock samples and mineral separates from lunar rocks 10017, 10044 and 10069, in conjunction with available literature data, permit qualitative conclusions concerning average irradiation depths and enable internally consistent exposure ages for Tranquillity Base rocks to be calculated. Surface production rates for Xe^{126} and Ne^{21} yielded exposure ages for fourteen Tranquillity Base rocks which are in reasonable agreement for most rocks. Five low−K rocks have exposure ages around 100 m.y.

094.039 Measurement on the lunar surface of impact-produced plasma clouds.
D. L. Reasoner, B. J. O'Brien.
Journ. Geophys. Res., Vol. 77, 1292 - 1299 (1972).

Simultaneous enhancements of low-energy ions and negative-particle fluxes due to the impact of the Apollo 14 lunar module were observed by the lunar-based charged-particle lunar-environment experiment. It is argued that the observed charged particles could not have energized at the instant of impact but rather that the impact produced expanding gas clouds and that constituents of these clouds were ionized and accelerated by some continuously active acceleration mechanism.

094.040 Total emittance of lunar fines.
R. C. Birkebak, A. Abdulkadir.
Journ. Geophys. Res., Vol. 77, 1340 - 1341 (1972).

The total emittance of lunar fines is needed for the calculation of lunar-surface temperatures. We present total emittance as a function of temperature with the density large enough to eliminate its effect on the total emittance.

094.041 More about the visibility of the lunar crescent. J. Ashbrook.
Sky Telescope, Vol. 43, 95 - 96 (1972).

094.042 The geologic setting of the Luna 16 landing site. J. F. McCauley, D. H. Scott.
Earth Planet. Sci. Letters, Vol. 13, 225 - 232 (1972).

The purposes of this paper are: (1) to provide a context for discussion of detailed studies of the Luna 16 sample by describing its regional setting, (2) to compare the setting of Luna 16 with those of Apollos 11 and 12, from which petrologic and geochemical data are now abundantly available, and (3) to discuss the possible source of non-basaltic materials in the Luna 16 and Apollo 11 and 12 samples suites, particularly those of the regolith.

094.043 Microprobe studies of three Luna 16 basalt fragments.
R. A. F. Grieve, G. A. McKay, D. F. Weill.
Earth Planet. Sci. Letters, Vol. 13, 233 - 242 (1972).

Three small fragments of Luna 16 basalt (NASA sample designation G38) have been analyzed with an electron microprobe. The samples indicate a magmatic history (rapid crystallization from reduced melts) similar to that inferred from studies of Apollo 11 and 12 basalts.

094.044 Lithic fragments, glasses and chondrules from Luna 16 fines. K. Keil, G. Kurat, M. Prinz, J. A. Green. Earth Planet. Sci. Letters, Vol. 13, 243 - 256 (1972).

Emphasis in the present study was placed on the measurement of the bulk compositions of 18 igneous and 7 microbreccia lithic fragments, 25 glasses, and 6 chondrules as a means of determining possible genetic relationships between these materials.

094.045 Petrology of a portion of the Mare Fecunditatis regolith. P. Jakeš, J. Warner, W. I. Ridley, A. M. Reid, R. S. Harmon, R. Brett, R. W. Brown. Earth Planet. Sci. Letters, Vol. 13, 257 - 271 (1972).

1300 microprobe analyses of glasses, pyroxenes, feldspars, oxides, olivines, troilite and metal in two 0.025 g samples of the Luna 16 return were made in order to characterize the Mare Fecunditatis regolith. The results are reported in many tables and diagrams.

094.046 Silicate melt inclusions and glasses in lunar soil fragments from the Luna 16 core sample. E. Roedder, P. W. Weiblen. Earth Planet. Sci. Letters, Vol. 13, 272 - 285 (1972).

2099 fragments were studied microscopically, and electron microprobe analyses were made of 39 selected areas, from a few square mm of polished surface, through 75- to 425-μm fragments of lunar soil from two samples of the Luna 16 core. The results are compared with those of Apollo 11 and 12.

094.047 Luna 16: Relative proportions and petrologic significance of particles in the soil from Mare Fecunditatis. J. B. Reid, Jr., G. J. Taylor, U. B. Marvin, J. A. Wood. Earth Planet. Sci. Letters, Vol. 13, 286 - 298 (1972).

We have classified some 2380 lithic and vitreous particles from two levels of the Luna 16 core-tube sample. We report the relative proportions of particle types from each of the two layers and discuss some petrological and mineralogical aspects of the lithic and vitreous components of the Luna 16 regolith.

094.048 Petrology of basaltic and monomineralic soil fragments from the Sea of Fertility. A. E. Bence, W. Holzwarth, J. J. Papike. Earth Planet. Sci. Letters, Vol. 13, 299 - 311 (1972).

Basaltic and monomineralic fragments from the 150–425 μm size fractions of the Luna 16 core obtained from the Sea of Fertility were studied by optical petrographic, electron microprobe, and single-crystal X-ray diffraction techniques.

094.049 Luna 16 sample G36: another crystalline product of an extremely mafic magma. L. S. Hollister, C. G. Kulick. Earth Planet. Sci. Letters, Vol. 13, 312 - 315 (1972).

Luna 16 sample G36 is a microbasalt containing skeletal olivine, plagioclase, ilmenite, and interstitial pyroxene. It apparently resulted from very rapid crystallization of a highly fractionated, totally liquid mafic magma. Although different in many details, G36 is generally similar to the ferromagnesian-rich Apollo 11 and 12 basalts. In this respect, it emphasizes the continuing problem of identifying a process on the moon which generated highly mafic magmas.

094.050 Shock metamorphic effects in the Luna 16 soil sample from Mare Fecunditatis. B. M. French. Earth Planet. Sci. Letters, Vol. 13, 316 - 322 (1972).

Shock metamorphic effects characteristic of meteorite impact and virtually identical to those observed in Apollo samples are common in fragments of the Luna 16 soil sample from Mare Fecunditatis. Two types of shock effects are described.

094.051 Mineral and bulk compositions of three fragments from Luna 16. I. M Steele, J. V. Smith. Earth Planet. Sci. Letters, Vol. 13, 323 - 327 (1972).

Electron microprobe analyses of fragment G37 show that it is a mare basalt with high aluminum and low magnesium content. Detailed analyses are given of olivine, pyroxene, ilmenite, ulvöspinel and plagioclase. Fragment G46 is a cinder with bulk composition fairly similar to G37. Fragment G51 is entirely composed of plagioclase, An_{93-96}, with low contents of minor elements.

094.052 Luna 16: An opaque mineral study and a systematic examination of compositional variations of spinels from Mare Fecunditatis. S. E. Haggerty. Earth Planet. Sci. Letters, Vol. 13, 328 - 352 (1972).

This study reports on the opaque mineralogy of the Luna 16 soil sample, with an emphasis on a detailed survey of the compositional variations of the $Fe-Ti-Cr-Al-Mg$ spinels. Comparisons are made between the Luna 16 samples and published and unpublished data on the Apollo samples.

094.053 Mineralogy, petrology, and chemistry of a Luna 16 basaltic fragment, sample B-1. A. L. Albee, A. A. Chodos, A. J. Gancarz, E. L. Haines, D. A. Papanastassiou, L. Ray, F. Tera, G. J. Wasserburg, T. Wen. Earth Planet. Sci. Letters, Vol. 13, 353 - 367 (1972).

Luna 16 sample B-1 was the largest fragment (62 mg) obtained in the sample exchange with the USSR. It is a fine-grained ophitic basalt but is distinguished from the Apollo samples by containing a single pyroxene, predominantly pigeonitic, an ilmenite content (7%) intermediate to that of the Apollo 11 and 12 samples, and subequal amounts of pyroxene (50%) and plagioclase (40%).

094.054 Rb-Sr age of a Luna 16 basalt and the model age of lunar soils. D. A. Papanastassiou, G. J. Wasserburg. Earth Planet. Sci. Letters, Vol. 13, 368 - 374 (1972).

A Rb-Sr internal isochron on a 62 mg basaltic boulder from the Luna 16 mission yields an age of 3.42 ± 0.18 AE. Lunar soils from Apollo 11, 12, 14, 15 and Luna 16 all lie very close to a 4.6 AE isochron. This appears to be a universal characteristic of lunar soils.

094.055 Gas retention and cosmic-ray exposure ages of a basalt fragment from Mare Fecunditatis. J. C. Huneke, F. A. Podosek, G. J. Wasserburg. Earth Planet. Sci. Letters, Vol. 13, 375 - 383 (1972).

An $^{40}Ar-^{39}Ar$ gas retention age and an $^{38}Ar-^{37}Ar$ cosmic ray exposure age have been determined on a total rock sample of the basalt fragment B-1 returned from Mare Fecunditatis by the Luna 16 mission. This sample defines a reasonably good high-temperature plateau at 3.45 ± 0.04 AE. This is presumed to date a period of igneous activity in Mare Fecunditatis.

094.056 Neutron capture on Gd and Sm in the Luna 16, G-2 soil. G. P. Russ III. Earth Planet. Sci. Letters, Vol. 13, 384 - 386 (1972).

The Gd and Sm isotopic compositions have been measured in the Luna 16, G-2 soil. This sample has the largest low energy neutron fluence yet observed in a lunar sample. The ratio of the number of neutrons captured per atom by ^{149}Sm to ^{157}Gd is 0.76 which is distinct from the value of 0.86 observed at the Apollo 11, 12 and 14 sites.

094.057 **Rare gas studies in Luna-16-G-7 fines by stepwise heating technique. A low fission solar wind Xe.**
W. A. Kaiser.
Earth Planet. Sci. Letters, Vol. 13, 387 - 399 (1972).

He, Ne, Ar, Kr, and Xe were examined in a dust sample of Luna-16 in twelve temperature steps with especially small intervals in the low temperature range. The gas concentrations, as well as their relative abundances, are in general agreement with values reported by Vinogradov for Luna-16 and values found in Apollo 11 fines.

094.058 **Inert gases in twelve particles and one "dust" sample from Luna 16.**
D. Heymann, A. Yaniv, S. Lakatos.
Earth Planet. Sci. Letters, Vol. 13, 400 - 406 (1972).

The inert gases were measured mass-spectrometrically in 12 fragments and 1 "dust" sample from Luna 16. The fragments were classified petrologically by microscopic inspection. The major petrologic types were breccias and basalts. The former were much richer in trapped gases than the latter. The $^4He/^{20}Ne$ ratio of the breccias was systematically smaller than that of the basalts. Exposure ages of four fragments are several hundred million years.

094.059 **The particle track record of the Sea of Plenty.**
G. M. Comstock, R. L. Fleischer, H. R. Hart, Jr.
Earth Planet. Sci. Letters, Vol. 13, 407 - 409 (1972).

We have measured the particle track densities in 36 grains taken from two levels of the soil column returned from the Sea of Plenty by Luna 16. We conclude that all of the Luna 16 soil has been irradiated very close to the surface, and that the regolith is both unusually thin at the Luna 16 site and extremely old ($\sim$ 3 by and more).

094.060 **Extreme radiation damage in soil from Mare Fecunditatis.** P. P. Phakey, P. B. Price.
Earth Planet. Sci. Letters, Vol. 13, 410 - 418 (1972).

High-voltage electron microscopy has been used to compare radiation effects in micron-size soil grains from the Luna 16 site and the four Apollo landing sites. Radiation damage by heavy solar particles is strikingly greater in the Luna 16 sample than in the other four samples.

094.061 **Fossil track and thermoluminescence studies of Luna 16 material.** R. Walker, D. Zimmerman.
Earth Planet. Sci. Letters, Vol. 13, 419 - 422 (1972).

Track densities in feldspar crystals from two samples range from 2.5×10^8/cm^2 to $> 2 \times 10^9$/cm^2. No significant difference is found between the two positions. The track densities are similar to those observed in heavily irradiated samples of Apollo 11, 12 and 14. Thermoluminescence is similar to the Apollo 12 core results at the same depths.

094.062 **Chemical features of the Luna 16 regolith sample.** N. J. Hubbard, L. E. Nyquist, J. M. Rhodes, B. M. Bansal, H. Wiesmann, S. E. Church.
Earth Planet. Sci. Letters, Vol. 13, 423 - 428 (1972).

The Luna 16 regolith sample differs from Apollo 11, 12 and 14 regolith and basalt samples by having smaller negative Eu and Sr anomalies and nearly chondritic Eu/Sm and Eu/Sr ratios although the overall rare earth elements Ba, Sr and U concentrations are 25 to 45 times chondrites.

094.063 **Luna 16: Some Li, K, Rb, Sr, Ba, rare-earth, Zr, and Hf concentrations.**
J. A. Philpotts, C. C. Schnetzler, M. L. Bottino, S. Schuhmann, H. H. Thomas.
Earth Planet. Sci. Letters, Vol. 13, 429 - 435 (1972).

Concentrations of Li, K, Rb, Sr, Na, rare-earths, Zr and Hf have been determined for some Luna 16 core materials by mass-spectrometric isotope-dilution. The results are shown in two tables.

094.064 **Determination of 29 elements in Luna 16 soil by non-destructive activation analysis.**
D. Y. Jerome, J.-C. Philippot, E. Brichet.
Earth Planet. Sci. Letters, Vol. 13, 436 - 440 (1972).

Abundances of 29 elements, including 11 rare earth elements were obtained in Luna 16 soil via instrumental neutron activation analysis. The results are compared with those of Apollo 11, 12, and 14.

094.065 **Rare earths and other trace elements in Luna 16 soil.** P. A. Helmke, L. A. Haskin.
Earth Planet. Sci. Letters, Vol. 13, 441 - 443 (1972).

We have analyzed for rare earths (REE), Ba, Co, Cr, Ga, Hf, K, and Sc in four small samples of material brought to earth by the Luna 16 mission. Modifications of our earlier procedures for neutron activation analysis were used.

094.066 **Bulk and rare earth abundances in the Luna-16 soil levels A and D.**
D. E. Gillum, W. D. Ehmann, H. Wakita, R. A. Schmitt.
Earth Planet. Sci. Letters, Vol. 13, 444 - 449 (1972).

Abundances of the major, minor and trace elements (SiO_2, O, TiO_2, Al_2O_3, FeO, CaO, Na_2O, K_2O, Cr_2O_3, MnO, Sc, V, Co, La, Sm, Eu, Yb and Lu) have been determined via sequential instrumental neutron activation analysis in two Luna-16 soils, levels A ($\sim$ 7 cm) and D ($\sim$ 30 cm depth).

094.067 **Meteoritic and non-meteoritic trace elements in Luna 16 samples.**
J. C. Laul, R. Ganapathy, J. W. Morgan, E. Anders.
Earth Planet. Sci. Letters, Vol. 13, 450 - 454 (1972).

Two Luna 16 soils have been analyzed for Ag, Au, Bi, Br, Cd, Co, Cs, Cu, Ga, Ge, In, Ur, Ni, Rb, Re, Sb, Se, Te, Tl, and Zn. A meteoritic component similar to that in Apollo 11 and 12 soils seems to be present, corresponding to $\sim$ 1.5 to 2% Cl chondrites or equivalent. It probably consists largely of micrometeorites.

094.068 **Oxygen isotope composition of the Luna 16 soil.**
R. N. Clayton.
Earth Planet. Sci. Letters, Vol. 13, 455 - 456 (1972).

094.069 **U—Th—Pb analyses of soil from the Sea of Fertility.**
F. Tera, G. J. Wasserburg.
Earth Planet. Sci. Letters, Vol. 13, 457 - 466 (1972).

In this paper we report analyses of lead, uranium and thorium in soil samples A and G from the Luna 16 mission and for comparison the results on soil samples 14163 and 14259 from the Apollo 14 mission.

094.070 **Effects of vaporization and condensation on Apollo 11 glass spherules: Implications for cooling rates.**
G. Kurat, K. Keil.
Earth Planet. Sci. Letters, Vol. 14, 7 - 13 (1972).

Fourteen of 40 glass spherules present in a section of the Apollo 11 microbreccia 10019,22 were found to exhibit steep concentration gradients at their rims: the oxides of the relatively volatile elements Na, K, and P increase by, respectively, factors of up to 67, 16, and > 54, at the spherule rims in comparison to the homogeneous centers. It is suggested that the concentration gradients observed are diffusion gradients and are the result of the impact event which produced the glasses.

094.071 **Crystallization of plagioclase in lunar basalts and its significance.** A. E. Ringwood, D. H. Green.
Earth Planet. Sci. Letters, Vol. 14, 14 - 18 (1972).

A new series of crystallization experiments upon an average high K, Rb type Apollo 11 basalt has been carried out in order to check the suggestion that earlier results by several groups of investigators were invalid because of failure to achieve equilibrium. The new experiments confirm the earlier conclusion that plagioclase is not a liquidus phase in this magma, and indeed, does not separate until more than 30 % of the magma has crystallized as olivine, pyroxene and ore minerals.

094.072 **Radioactive heat sources in the lunar interior.**
J. F. Hays.
Phys. Earth Planet. Interiors, Vol. 5, 77 - 84 (1972).
Chemical data on Apollo samples show that the moon is depleted relative to chondrites in volatile elements, including potassium, and possibly enriched relative to chondrites in refractory elements, including uranium and thorium. Additional thermal models have therefore been investigated in order to set upper limits on lunar radioactivity consistent with the proposed temperature distribution.

094.073 **New data on the moon and planets.**
Aviatsiya i kosmonavtika (NRB), Vol. 13, No. 6, p. 6 - 8 (1971). In Bulgarian.

094.074 **^{40}Ar-^{39}Ar age and cosmic ray irradiation history of the Apollo 15 anorthosite, 15415.**
G. Turner.
Earth Planet. Sci. Letters, Vol. 14, 169 - 175 (1972).
The ^{40}Ar-^{39}Ar method of age determination has been applied to the Apollo 15 anorthosite sample, 15415. Release of argon from the most retentive crystal sites indicates an age of (4.05 ± 0.15) aeons. The cosmic ray exposure age of 15415 is 112 my.

094.075 **Apollo 14 inverted pigeonites: Possible samples of lunar plutonic rocks.** J. J. Papike, A. E. Bence.
Earth Planet. Sci. Letters, Vol. 14, 176 - 182 (1972).
"Inverted pigeonites" have been found in Apollo 14 samples 14082 and 14083 and have been studied by a combination of optical, electron probe and single-crystal X-ray diffraction techniques. We interpret these "inverted pigeonites" to be samples of plutonic rocks that have been blasted out of the Imbrium Basin.

094.076 **Chromian spinels from Apollo 14 rocks.**
I. M. Steele.
Earth Planet. Sci. Letters, Vol. 14, 190 - 194 (1972).
Electron microprobe analysis of 13 pink, isotropic, high-relief grains from Apollo 14 clastic rock 14063,14 and a lithic fragment from the 1−2mm fines, 14002,7, identifies them as spinel minerals dominated by the spinel component $MgAl_2O_4$ associated with a moderate content of chromite and hercynite.

094.077 **U, Th, Pb and REE abundances and ^{207}Pb/^{206}Pb ages of individual minerals in returned lunar material by ion microprobe mass analysis.**
C. A. Andersen, J. R. Hinthorne.
Earth Planet. Sci. Letters, Vol. 14, 195 - 200 (1972).
The results reported here indicate that bulk isotopic and chemical analyses are significantly affected by small populations of accessory minerals that contain high concentrations of the elements under study. These minerals should be taken into consideration in discussions of lunar age and petrogenesis.

094.078 **Apollo 14 active seismic experiment.**
J. S. Watkins, R. L. Kovach.
Science, Vol. 175, 1244 - 1245 (1972).
Explosion seismic refraction data indicate that the lunar near-surface rocks at the Apollo 14 site consist of a regolith 8.5 meters thick and characterized by a compressional wave velocity of 104 meters per second. The regolith is underlain by a layer with a compressional wave velocity of 299 meters per

second. The thickness of this layer, which we interpret to be the Fra Mauro Formation, is between 16 and 76 meters. The layer immediately beneath this has a velocity greater than 370 meters per second. We found no evidence of permafrost.

094.079 **Geochemical evidence for the origin of the moon.**
J. A. O'Keefe.
Naturwissenschaften, 59. Jahrgang, p. 45 - 52 (1972).
It is supposed that the moon separated from the earth after the formation of the core. This was followed by an episode of heating from the outside, possibly due to tidal interaction, and formation of a large common atmosphere for earth and moon, which ablated away much of the original lunar mass. Since this time, the heated zone of the moon has gradually migrated toward the center.

094.080 **Wasser am Mond?** H. Hintenberger.
Phys. Blätter, 28. Jahrgang, p. 119 - 121 (1972).

094.081 **Selten sichtbare Mondlandschaften.**
C. Albrecht.
SuW, Vol. 11, 102 - 103 (1972).

094.082 **Observations photographiques de la lune à l'Observatoire de Besançon.** G. Hilaire.
Ciel et Terre, Vol. 88, 114 - 118 (1972).

094.083 **Observation de positions de la lune relativement au champ stellaire environnant.** J. Dommanget.
Ciel et Terre, Vol. 88, 119 - 123 (1972).

094.084 **History of the moon.** S. W. Richardson.
Spaceflight, Vol. 14, 82 - 85 (1972).

094.085 **Exploration at Hadley-Rille.** R. J. Fryer.
Spaceflight, Vol. 14, 86 - 89, 98 (1972).

094.086 **Lunar spectral types.**
T. B. McCord, M. P. Charette, T. V. Johnson, L. A. Lebofsky, C. Pieters, J. B. Adams.
Journ. Geophys. Res., Vol. 77, 1349 - 1359 (1972).
The spectral reflectance properties $(0.3−1.1 \mu)$ of a number of lunar mare, upland, and bright crater areas were observed with the use of ground-based telescopes. These new data are discussed in view of earlier studies in an attempt to provide a basis for more detailed interpretation. The spectral reflectivity curves $(0.3−1.1 \mu)$ for all lunar areas studied consist of a positive sloping continuum with a superimposed symmetric absorption band centered at 0.95μ. Upland, mare, and bright crater materials can be identified by their spectral curves.

094.087 **Investigation of a possible solar-wind darkening of the lunar surface by photoelectron spectroscopy.**
L. I Yin, S. Ghose, I. Adler.
Journ. Geophys. Res., Vol. 77, 1360 - 1367 (1972).
Detailed sequential studies by photoelectron spectroscopy of the change in valence states of iron in Fe_2O_3, Fe_3O_4, FeO, and Fe foil samples show that chemical reduction does take place on the sample surface when argon ion bombardment occurs. Our primary purpose in using photoelectron spectroscopy is therefore to determine whether ion bombardment actually causes reduction and the consequent darkening of samples containing iron and/or whether the darkening observed in the laboratory is due to a buildup of surface contaminants, such as hydrocarbons.

094.088 **Descartes region: Evidence for Copernican-age volcanism.** J. W. Head III, A. F. H. Goetz.
Journ. Geophys. Res., Vol. 77, 1368 - 1374 (1972).
A model that suggests that the high-albedo central region

of the Descartes Formation was formed by Copernican-age volcanism was developed from Orbiter photography, Apollo 12 multispectral photography, earth-based spectrophotometry, and thermal IR and radar data. On the basis of these data, the bright unit is thought to be a young pyroclastic deposit mantling older volcanic units of the Descartes Formation.

094.089 Relative deformations of selenodetic nets of coordinates.
Sh. T. Habibullin, Yu. A. Chikanov, V. S. Kisliuk.
The Moon, Vol. 3, 371 - 385 (1972).

The investigations concerning the comparison of catalogues and the elucidation of their systematic differences are of very recent date. Various methods of interpretation of the systematic differences between catalogues have been proposed. Without an attempt to encompass the whole problem in what follows, we shall describe one method for comparative study of catalogues based on the theory of the deformation of continuous media.

094.090 Results from the Apollo lunar seismic experiment.
G. V. Latham.
The Moon, Vol. 3, 386 - 387 (1972). – Research note.

094.091 Recent results on the mass, gravitational field and moments of inertia of the moon.
W. H. Michael, Jr., W. T. Blackshear.
The Moon, Vol. 3, 388 - 402 (1972).

Doppler tracking data from the Lunar Orbiter series of spacecraft have been used. The value obtained for the mass of the moon, $GM = 4902.84$ km^3 s^{-2}, is in good agreement with previous results. Acceleration contour plots, derived from the gravitational coefficients, show correlations with surface features on the near side of the moon. Based on the most recent gravitational field data, the current estimate for the polar moment of inertia is given.

094.092 KREEP minerals and the origin of Luny Rock – 1.
A. Peckett.
The Moon, Vol. 3, 403 - 407 (1972).

Luny Rock – 1, dated at 4.44×10^9 yr by Albee and Chodos (1970), is one of a number of norite fragments rich in potassium, rare-earth elements and phosphorus (KREEP) found in the Apollo 11 and 12 soils.

094.093 The system of lunar craters, revised. C. A. Wood.
The Moon, Vol. 3, 408 - 411 (1972).

A new catalogue of lunar crater diameters, depths and morphologies based on Orbiter photography is described.

094.094 Lunar rilles and Hawaiian volcanic features: Possible analogues. D. P. Cruikshank, C. A. Wood.
The Moon, Vol. 3, 412 - 447 (1972).

In this paper we consider the origin of rilles on the lunar maria, both sinuous and those having straight line segments, from the point of view of lava tubes formed in surface lava flows, and also in terms of collapses along active fissures. Terrestrial examples serve as the basis of comparison with lunar topography shown on Orbiter photographs of the moon.

094.095 Laser measurements of earth-moon distances.
J. Rösch.
The Moon, Vol. 3, 448 - 455 (1972).

094.096 The lunar disturbance effect. B. W. Hapke.
The Moon, Vol. 3, 456 - 460 (1972).

Apollo photographs indicate that the lunar disturbance effect is due to changes in the photometric function of the lunar soil caused by rearrangement of the soil particles, rather than to any physical or chemical differences between the uppermost layer and the underlying materials.

094.097 Model for radon diffusion through the lunar regolith. L. J. Friesen, D. Heymann.
The Moon, Vol. 3, 461 - 471 (1972).

A model for radon diffusion through the lunar regolith is proposed in which the atom migrates by random walk. Because of the greatly different mean lifetimes against radioactive decay of ^{219}Rn, ^{220}Rn, and ^{222}Rn, the regolith acts as a powerful 'filter' for these species. Results of calculations of radon escape from a 4 m thick regolith are given.

094.098 On the modelling of possible processes of formation of lunar regolith glassy particles.
M. D. Nusinov, K. P. Florenskii, A. V. Kuznetsov, A. I. Kosolapov, Iu. B. Cherniak, L. I. Ivanov, V. A. Janushkevich, L. V. Obukhov, V. V. Vysochkin.
Dokl. Akad. Nauk SSSR, Ser. Mat. Fiz., Vol. 202, 811 - 814 (1972). In Russian.

094.099 Transportable lunar-ranging system.
C. G. Lehr, M. R. Pearlman, J. A. Monjes, W. F. Hagen.
Applied Optics, Vol. 11, 300 - 304 (1972).

Ruby lasers in large, fixed installations are being used by groups in France, Japan, the Soviet Union, and the U.S. for measuring distances to retroreflectors on the moon. This paper describes a transportable transmitting unit that can be installed at any astronomical observatory where a large telescope is available to detect the received signal.

094.100 Laboratory simulation of lunar craters.
H. Fechtig, D. E. Gault, G. Neukum, E. Schneider.
Naturwissenschaften, 59. Jahrgang, p. 151 - 157 (1972).

Techniques are described for accelerating μg-, mg- and g-sized particles to meteoroid velocities (up to 50 km/sec). Typical results are illustrated and discussed. Conclusions on meteoroid fluxes, erosion processes and the mechanics of cratering are summarized.

094.101 Determination of the mutual orientation of nine selenodetic catalogues by Euler's angles.
V. A. Nikonov, T. P. Skobeleva.
Astron. Zhurn. Akad. Nauk SSSR, Vol. 49, 436 - 440 (1972). In Russian. English translation in Soviet Astron. AJ, Vol. 16, No. 2.

The mutual positions of the centres of nine selenodetic systems were determined and these systems were reduced to a common centre – to the centre of figure of the visible hemisphere of the moon. Further, the mutual orientation of coordinate axes was determined by Euler's angles from the discrepancies of longitudes and latitudes of common points. The equated positions of the centre of figure in the system of nine selenodetic catalogues and the results of determination of elements of the mutual orientation of coordinate axes for 21 pairs of catalogues are presented.

094.102 TOR-1 examining the lunar ground.
E. A. Dukhovskoi, A. A. Silin, V. V. Shvarev.
Priroda, No. 1.72, p. 83 - 86 (1972). In Russian.

094.103 Some results of a lunar surface study in the infrared range. M. N. Naugolnaya.
Astrofiz. Issled. Izv. Spets. Astrofiz. Obs., Vol. 3, 116 - 123 (1971). In Russian.

On the basis of our observational data the angles of incidence and reflection of light rays for 101 features of the lunar surface are calculated. Variation curves of «color indices» of these features are plotted as a function of the angle of incidence of light rays. A comparison is made between «blue» and infrared photographs of the moon obtained by using contrast printing technique, and differences between them are explained.

094.104 **Measurements of lunar magnetic field interaction with the solar wind.**
P. Dyal, C. W. Parkin, C. W. Snyder, D. R. Clay.
Nature, Vol. 236, 381 - 385 (1972).

This article reports a study of the compression of the remanent lunar magnetic field by the solar wind, based on measurements of remanent magnetic fields at four Apollo landing sites and of the solar wind at two of these sites. Measurements at the Apollo 12 site show, to first order, a compression of the remanent field in direct proportion to the solar wind pressure.

094.105 **On inferring elastic properties of deep planetary interiors: Moon and Mars.**
G. H. A. Cole, D. Parkinson.
Planet. Space Sci., Vol. 20, 557 - 569 (1972).

The planet is represented as a spherical body with a crust enclosing an interior in hydrostatic equilibrium. The interior is composed of a mantle enclosing an iron core, the radius of which is regarded as a parameter of the theory. It is found on the present model that the core radius for Mars could be rather larger than one third the total Martian radius; for the moon, a core is unlikely but present observational uncertainties would be compatible with a core of radius one tenth the total lunar radius. The present model is also applied to the earth as a test, and provides data in satisfactory agreement with that already available by alternative refined methods.

094.106 **Density and flux distributions of neutral gases in the lunar atmosphere.** T. T. J. Yeh, G. K. Chang.
Journ. Geophys. Res., Vol. 77, 1720 - 1728 (1972).

Using the kinetic theory of gases, we have calculated numerically the density and flux distributions in the lunar atmosphere resulting from variations in the assumed surface gas temperature and density.

094.107 **Differentiation and volcanism in the lunar highlands: Photogeologic evidence and Apollo 16 implications.** N. J. Trask, J. F. McCauley.
Earth Planet. Sci. Letters, Vol. 14, 201 - 206 (1972).

The purpose of this note is to describe the deposits of probable volcanic origin on the highlands as seen on Orbiter photographs, to present evidence bearing on their age relative to the lunar maria and other major features of the lunar crust, to present limited photogeologic evidence bearing on their composition, and to discuss the implications of the existence extensive of highland volcanic rocks.

094.108 **Rubidium-strontium isotope characteristics of lunar soils.** R. A. Cliff, C. Lee-Hu, G. W. Wetherill.
Journ. Geophys. Res., Vol. 77, 2007 - 2013 (1972).

K, Rb, and Sr concentrations and Sr isotopic composition are reported for hand-picked fractions from the 175- to 1000-μm Apollo 12 fines samples 12032 and 12070. In addition, two <1-mm bulk fines samples, 14163,160 and 14259,21, from Apollo 14 were analyzed.

094.109 **Lunar conductivity models from the Apollo 12 magnetometer experiment.** W. R. Sill.
The Moon, Vol. 4, 3 - 17 (1972). – Commun. presented at the Conference on Lunar Geophysics 1971, Houston, Texas.

094.110 **Post-sunset cooling behavior of the lunar surface.**
W. W. Mendell, F. J. Low.
The Moon, Vol. 4, 18 - 27 (1972). – Commun. presented at the Conference on Lunar Geophysics 1971, Houston, Texas.

094.111 **Moments of inertia of the lunar globe, and their bearing on chemical differentiation of its outer layers.** Z. Kopal.
The Moon, Vol. 4, 28 - 34 (1972). – Commun. presented at

the Conference on Lunar Geophysics 1971, Houston, Texas.

094.112 **Electrical conductivity of olivine and the lunar temperature profile.** R. M. Housley, F. J. Morin.
The Moon, Vol. 4, 35 - 38 (1972). – Commun. presented at the Conference on Lunar Geophysics 1971, Houston, Texas.

094.113 **Comments on the moon's magnetism.**
R. Hide.
The Moon, Vol. 4, 39 (1972). – Commun. presented at the Conference on Lunar Geophysics 1971, Houston, Texas.

094.114 **The lunar seismogram.** A. F. Gangi.
The Moon, Vol. 4, 40 - 48 (1972). – Commun. presented at the Conference on Lunar Geophysics 1971, Houston, Texas.

094.115 **Thermoremanent magnetization (TRM) of lunar samples.** J. R. Dunn, M. Fuller.
The Moon, Vol. 4, 49 - 62 (1972). – Commun. presented at the Conference on Lunar Geophysics 1971, Houston, Texas.

094.116 **Lunar properties from transient and steady magnetic field measurements.** P. Dyal, C. W. Parkin.
The Moon, Vol. 4, 63 - 87 (1972). – Commun. presented at the Conference on Lunar Geophysics 1971, Houston, Texas.

094.117 **Thermal conductivity of Apollo 12 fines at intermediate density.** C. J. Cremers.
The Moon, Vol. 4, 88 - 92 (1972). – Commun. presented at the Conference on Lunar Geophysics 1971, Houston, Texas.

094.118 **Polarized absorption spectra of single crystals of lunar pyroxenes and olivines.**
R. G. Burns, F. E. Huggins, R. M. Abu-Eid.
The Moon, Vol. 4, 93 - 102 (1972). – Commun. presented at the Conference on Lunar Geophysics 1971, Houston, Texas.

094.119 **A proposed lunar orbiting gravity gradiometer experiment.**
D. B. DeBra, J. C. Harrison, P. M. Muller.
The Moon, Vol. 4, 103 - 112 (1972). – Commun. presented at the Conference on Lunar Geophysics 1971, Houston, Texas.

094.120 **Lunar subsurface exploration with coherent radar.**
W. E. Brown, Jr.
The Moon, Vol. 4, 113 - 127 (1972). – Commun. presented at the Conference on Lunar Geophysics 1971, Houston, Texas.

094.121 **Apollo 12 thermal radiation properties.**
R. C. Birkebak.
The Moon, Vol. 4, 128 - 133 (1972). – Commun. presented at the Conference on Lunar Geophysics 1971, Houston, Texas.

094.122 **The effects of boundary condition asymmetries on the interplanetary magnetic field-moon interaction.**
A. C. Reisz, D. L. Paul, T. R. Madden.
The Moon, Vol. 4, 134 - 140 (1972). – Commun. presented at the Conference on Lunar Geophysics 1971, Houston, Texas.

094.123 **Trace ferric ion in lunar and meteoritic titanaugites.**
A. J. Cohen.
The Moon, Vol. 4, 141 - 154 (1972). – Commun. presented at the Conference on Lunar Geophysics 1971, Houston, Texas.

094.124 **Measurement of physical librations using laser retroreflectors.** J. D. Mulholland, E. C. Silverberg.
The Moon, Vol. 4, 155 - 159 (1972). – Commun. presented at the Conference on Lunar Geophysics 1971, Houston, Texas.

094.125 **Lunar rock magnetism.**

T. Nagata, R. M. Fisher, F. C. Schwerer.
The Moon, Vol. 4, 160 - 186 (1972). — Commun. presented at the Conference on Lunar Geophysics 1971, Houston, Texas.

094.126 **D. C. Electrical conductivity of lunar surface rocks.**
F. C. Schwerer, G. P. Huffman, R. M. Fisher, T. Nagata.
The Moon, Vol. 4, 187 - 189 (1972). — Commun. presented at the Conference on Lunar Geophysics 1971, Houston, Texas.

094.127 **Thermal evolution of the moon.**
M. N. Toksöz, S. C. Solomon, J. W. Minear, D. H. Johnston.
The Moon, Vol. 4, 190 - 213 (1972). — Commun. presented at the Conference on Lunar Geophysics 1971, Houston, Texas.

094.128 **Shock melting and vaporization of lunar rocks and minerals.** T. J. Ahrens, J. D. O'Keefe.
The Moon, Vol. 4, 214 - 249 (1972). — Commun. presented at the Conference on Lunar Geophysics 1971, Houston, Texas.

094.129 **Lunar theory.** J. S. Griffith.
Journ. Roy. Astron. Soc. Canada, Vol. 66, 68 - 69 (1972). — Abstr. Canadian Astron. Soc.

094.130 **On large-scale density fluctuations in the lunar interior.** V. N. Zharkov.
Izv. AN SSSR. Fiz. zemli, 1971, No. 11, p. 3 - 6. In Russian. Abstr. in Referativ. Zhurn. 51. Astron., 4.51.362 (1972).

094.131 **Investigation of the physical properties of the lunar ground and its terrestrial analogues.**
A. R. Golovkin, E. A. Dukhovskoj, R. G. Petrochenkov, V. V. Rzhevskij, A. A. Silin, V. V. Shvarev.
Kosm. Issled., Vol. 10, 120 - 124 (1972). In Russian.

094.132 **The position of the mass center of the moon from data of Lunar Orbiter 1 and of a photogrammetric analysis.** Yu. N. Lipskij, V. A. Nikonov, T. P. Skobeleva.
Kosm. Issled., Vol. 10, 141 - 142 (1972). In Russian. — Brief information.

094.133 **Lunar crater origin in the maria from analysis of Orbiter photographs.**
G. Fielder, R. J. Fryer, C. Titulaer, A. K. Herring, B. Wise.
Phil. Trans. Roy. Soc. London, Ser. A, Vol. 271, (No. 1215), 361 - 409 (1972).
One third of a million counts on 73127 craters in 22 test areas of lunabase (the dark areas of the moon, normally maria) have been made with the aim of diagnosing the origin of the craters that are predominantly between 100 and 2000 m in diameter by the use of a statistical method that is capable of measuring both the chaining and clustering of craters. Independent measurements of clustering have been made using photometry and using equi-areal counts to determine the number density of craters. Computer simulations of random fields of craters and of specially proportioned linear arrays and clusters of craters on random backgrounds have been used widely in the interpretation of the observations. By comparing the results of statistical tests on the real distributions of crater centres with those on the distributions created by computer it has been possible to set probable limits on the number of endogenic craters in each test area.

094.134 **Tides and the earth-moon system.** P. Goldreich.
Sci. American, Vol. 226, No. 4, p. 42 - 52 (1972).

094.135 **Lunar basalts.** H. C. Urey, K. Marti.
Science, Vol. 176, 117, 119 (1972). — Letter.

094.136 **Explanation of transient lunar phenomena based on**

lunar samples studies.
G. F. J. Garlick, G. A. Steigmann, W. E. Lamb.
Nature, Vol. 235, 39 - 40 (1972).
Transient surface brightenings reported in sunlit lunar areas are ascribed to marked rises in albedo when dust flows or is disturbed. Experimental studies of lunar dust samples in a 'fluidiser' vibrator system show large increases in reflectivity to occur when intergrain cohesion is destroyed. The effect is a maximum for viewing angle of 20° to the surface. No spectral change accompanies the rise in albedo. A possible cause of temporary rises in zodiacal light after solar disturbances, namely a change in light scattering, is also suggested.

094.137 **Chromian pleonaste and aluminous picotite in two Apollo 14 microbreccias 14306 and 14055.**
H. I. Drever, R. Johnston, F. G. F. Gibb.
Nature, Phys. Sci., Vol. 235, 30 - 31 (1972).
Tiny grains of pink spinel, observed in two partially shock-melted feldspathic microbreccias, are described and analyzed by electron microprobe. Their chemical composition, being anomalously rich in Al and Mg, is contrasted with that of spinels from mare basalts. It is suggested that the source of these microbreccias, forming part of the Imbrium ejecta blanket in the Fra Mauro region, was a layered ultrabasic intrusion within the Imbrium basin.

094.138 **Simulation study of lunar carbon chemistry.**
C. T. Pillinger, P. H. Cadogan, G. Eglinton, J. R. Maxwell, B. J. Mays, W. A. Grant, M. J. Nobes.
Nature, Phys. Sci., Vol. 235, 108 - 109 (1972).
The possible role of solar wind hydrogen and carbon in the synthesis of lunar hydrocarbons and "carbides" has been demonstrated by laboratory simulations involving irradiation of a number of targets including the lunar fines themselves with $^{13}C^+$ and D_2^+ at energies similar to those in the solar wind. Low resolution mass spectrometry of C_1 compounds released by acid dissolution (HCl) of the targets showed that hydrocarbons were generated either completely ($^{13}CD_4$) or partly ($^{12}CD_4$) from the irradiating species. The presence of material reacting as carbide in metal targets was indicated by the release of $^{13}CH_4$.

094.139 **Lunar ground from data of Lunokhod 1.**
V. V. Gromov, A. K. Leonovich, V. V. Shvarev.
Zemlya i Vselennaya, 1972, No. 2, p. 8 - 10. In Russian.

094.140 **Apollo 15 investigates the moon.**
D. Yu. Gol'dovskij.
Zemlya i Vselennaya, 1972, No. 2, p. 15 - 17. In Russian.

094.141 **U-Th-Pb systematics in three Apollo 14 basalts and the problem of initial Pb in lunar rocks.**
F. Tera, G. J. Wasserburg.
Earth Planet. Sci. Letters, Vol. 14, 281 - 304 (1972).
The isotopic composition of Pb and the elemental concentration of U, Th and Pb were measured on 'total' rock samples 14053, 14073 and 14310 and on mineral separates of 14310 and 14053. The first Pb-U isochrons obtained for lunar basalts indicate a reasonable solution to the previous discrepancy between the different methods of 'absolute' age determination. The resulting U-Pb isochron ages are compatible with the Rb-Sr and K-Ar ages on the same rocks.

094.142 **Temperature-dependent Mg, Fe distribution in a lunar olivine.** D. Virgo, S. S. Hafner.
Earth Planet. Sci. Letters, Vol. 14, 305 - 312 (1972).
The nuclear hyperfine doublets of ^{57}Fe at the two nonequivalent octahedrally coordinated positions $M1$ and $M2$ in an olivine $Mg_{0.34}Fe_{0.66}SiO_4$ from lunar basalt 12018 have been analyzed in the temperature range between 250 and 450°C.

094.143 Lunar zirkelite: A uranium-bearing phase.
F. D. Busche, M. Prinz, K. Keil, G. Kurat.
Earth Planet. Sci. Letters, Vol. 14, 313 - 321 (1972).
Zirkelite (simplified $CaZrTiO_5$) containing 7–17 wt% oxides of trivalent elements (largely yttrium and the rare earths) and minor amounts of U, Th, and Pb is described from an Apollo 12 feldspathic peridotite (12036,9) and two KREEP-type norite lithic fragments separated from Apollo 14 loose fines (14163,39 and 14257,3).

094.144 Alpha spectrometry of a surface exposed lunar rock.
T. Grjebine, G. Lambert, J. C. Le Roulley.
Earth Planet. Sci. Letters, Vol. 14, 322 - 324 (1972).
Surface alpha radioactivity of a moon rock was measured in order to find the activity of ^{210}Po which could be designated as a descendant of radon outgassed from the lunar soil.

094.145 Thermal history and early magmatism in the moon.
J. A. Wood.
Icarus, Vol. 16, 229 - 240 (1972).
Lunar heat-flow calculations are carried out for a model moon in which (a) near-surface initial temperatures are very high (as the occurrence of a surface anorthositic layer seems to require), and (b) heat-generating radionuclides are transported upward when melting occurs.

094.146 Suprathermal ions near the moon.
J. W. Freeman, Jr., M. A. Fenner, H. K. Hills, R. A. Lindeman, R. Medrano, J. Meister.
Icarus, Vol. 16, 328 - 338 (1972).
This paper reports some preliminary results from the Suprathermal Ion Detectors deployed on the lunar surface by the Apollo 12 and 14 astronauts. Salient features of these results include: the possible observation of sporadic venting of gas from the lunar surface; evidence for a prompt ionization and acceleration mechanism operating in the lunar exosphere; and a preliminary measurement yielding approximately 1 month for the e-folding decay time for the heavier components of the exhaust gases from the Apollo lunar landing systems.

094.147 Comparative analyses of observations of lunar transient phenomena. W. S. Cameron.
Icarus, Vol. 16, 339 - 387 (1972).
From the author's collection of more than 900 reports of lunar transient phenomena covering the period 1540 - 1970, 771 positive plus 112 negative observations (several times more than any previously published analyses) with sufficient ancillary data were analyzed for five hypotheses of causes.

094.148 Lava filled craters and the thermal history of the lunar surface. C. S. Beals.
Nature, Vol. 237, 226 - 227 (1972).
To provide a semi-quantitative test of the hypothesis of two periods of high surface temperature separated by a period of cooling and thickening crust, 8 lava filled craters or lava covered areas were examined using the number of small craters 2 km diameter and over, per $10^4 km^2$, as an indication of relative age.

094.149 Micrometeorite craters on lunar glass particles: The relationship between radial fracture zones and spall zones. B. P. Glass.
Meteoritics, Vol. 7, 47 - 49 (1972).

094.150 A first report on Apollo 16. R. N. Watts, Jr.
Sky Telescope, Vol. 43, 353 - 357 (1972).

094.151 Formaciones volcánicas en la tierra y en la luna.

W. Sandner.
El Universo, No. 93, Vol. 24, 79 - 82 (1970).

094.152 Luna incognita: The last frontier? J. E. Westfall.
Strolling Astronomer, Vol. 23, 118 - 122 (1972).

094.153 Lunar notes. J. E. Westfall.
Strolling Astronomer, Vol. 23, 130 - 136 (1972).
Additions to the A.L.P.O. Lunar Photograph Library: Amateur and Apollo-11, -12, and -13 photographs; Luna incognita observing schedule, 1972.

094.154 Whitlockite and apatite from lunar rock 14310 and from Ödegården, Norway.
W. L. Griffin, R. Åmli, K. S. Heier.
Earth Planet. Sci. Letters, Vol. 15, 53 - 58 (1972).
Microprobe analyses show that whitlockite from lunar rocks is enriched in REE relative to the associated apatite, whereas a terrestrial whitlockite is severely depleted in REE relative to the associated apatite.

094.155 An ESCA study of lunar and terrestrial materials.
W. T. Huntress, Jr., L. Wilson.
Earth Planet. Sci. Letters, Vol. 15, 59 - 64 (1972).
The ESCA technique is used to obtain rapid, nondestructive elemental analysis of selected lunar samples, and the chemical shift of the Fe(2p) line in lunar material is found characteristic of iron in the Fe^{2+} state. A difference in binding energy of approximately 0.5 eV is observed between the 0(1s) levels of the terrestrial minerals fayalite and quartz, and effects due to surface oxidation and adsorption are also observed in terrestrial materials.

094.156 Structural studies of rim augite and core pigeonite from lunar rock 12052. H. Takeda.
Earth Planet. Sci. Letters, Vol. 15, 65 - 71 (1972).
Euhedral clinopyroxenes in a vug from Apollo 12 rock 12052 are composed of honey-yellow pigeonite cores overgrown, with approximately (100) in common, by dark-brown augite, which itself exsolves very small amouts of pigeonite on (001). The crystal structures, cation distributions, and anisotropic 'thermal' ellipsoids of this pyroxene pair have been refined by least-squares techniques using X-ray intensities measured with a single-crystal diffractometer.

094.157 Transient lunar phenomena – a possible explanation? R. A. Jahn.
Journ. British. Astron. Ass., Vol. 82, 122 - 125 (1972).

094.158 Third Lunar Science Conference.
A. Burlingame, D. Burnett, B. Doe, D. Gault, L. Haskin, H. Schnoes, D. Heymann, W. Melson, J. Papike, R. Tilling, N. Toksoz, J. Wood.
Science, Vol. 176, 975 - 981 (1972). – Houston, Texas, 1972 January 10–13.
Primal igneous activity in the outer layers of the moon generated a feldspathic crust 40 kilometers thick.

094.159 Lunar crust: Structure and composition.
M. N. Toksöz, F. Press, K. Anderson, A. Dainty, G. Latham, M. Ewing, J. Dorman, D. Lammlein, G. Sutton, F. Duennebier, Y. Nakamura.
Science, Vol. 176, 1012 - 1016 (1972).
Lunar seismic data from artificial impacts recorded at three Apollo seismometers are interpreted to determine the structure of the moon's interior to a depth of about 100 kilometers.

094.160 De Apollo 15 op de maan gefotografeerd.
F. P. Israel.
Hemel en Dampkring, Vol. 70, 92 - 94 (1972).

094.161 **Het tekenen van de maan.** H. Nieuwenhuis.
Hemel en Dampkring, Vol. 70, 134 - 137 (1972).

094.162 **Comparative estimate of the surface roughness of the moon and Venus.** N. N. Krupenio.
Kosm. Issled., Vol. 10, 279 - 285 (1972). In Russian.

094.163 **Luna 20: Lunar soil to the earth.**
S. A. Nikitin.
Priroda, No. 4.72, p. 100 (1972). In Russian.

094.164 **Carbon chemistry of Apollo 14 size-fractionated fines.** P. T. Holland, B. R. Simoneit, P. C. Wszolek, W. H. McFadden, A. L. Burlingame.
Nature, Phys. Sci., Vol. 235, 106 - 108 (1972).

A sample of the 14240 SESC fines was sieved, yielding a size distribution very similar to that reported for the bottom-trench sample 14149. The acid dissolutions performed on the size fractions yielded CH_4, CD_4, ^{20}Ne and ^{36}Ar which directly correlated with the grain surface area, confirming a solar wind origin for these components. Similar evolution patterns of the CH_4, ^{20}Ne and ^{36}Ar during pyrolysis of these samples confirmed the common site and surface correlation of these gases.

094.165 **Distinct cooling histories of Apollo 15 basalts.**
N. Gray, S. S. Hafner, K. Schürmann, D. Virgo.
Nature, Phys. Sci., Vol. 236, 71 - 73, with a correction, Vol. 237, 16 (1972).

094.166 **Lunar glass: Interferometric evidence for low-temperature shock.** S. Tolansky.
Science, Vol. 176, 671 - 673 (1972).

Glass objects in the fines from the Apollo 11 and Apollo 12 missions are shown, by two-beam reflection interferometry, to have been subject to shock at temperatures below the melting or softening point of the glass. Possible causes for the glass fragmentation are discussed.

094.167 **Lunar laser ranging with decimeter accuracy.**
E. C. Silverberg, D. G. Currie.
Bull. American Astron. Soc., Vol. 4, 219 (1972). – Abstr. AAS.

094.168 **Preliminary filtering of lunar laser ranging experiment data at the University of Texas.**
P. J. Shelus.
Bull. American Astron. Soc., Vol. 4, 266 (1972). – Abstr. AAS.

094.169 **Making solutions from lunar laser ranging data.**
J. G. Williams, W. S. Sinclair, D. B. Holdridge, M. A. Slade.
Bull. American Astron. Soc., Vol. 4, 267 (1972). – Abstr. AAS.

094.170 **A source of observational bias in the correlation of lunar transient events with perigee.** M. T. Yates.
Journ. Geophys. Res., Vol. 77, 2613 - 2615 (1972). – Letter.

094.171 **Occurrence of diopside and Cr−Zr−armalcolite on the moon.** I. M. Steele, J. V. Smith.
Nature, Phys. Sci., Vol. 237, 105 - 106 (1972).

094.172 **A 4.3 aeon pre-Imbrium event.**
R. K. O'Nions, R. J. Pankhurst.
Nature, Vol. 237, 446 - 447 (1972).

We point out that published Rb-Sr data for Apollo 14 whole rock samples define an isochron age of 4.31 ± 0.04 aeon, which is interpreted as their time of igneous crystallization.

094.173 **Our solar system.** A. Bonev.
Aviatsiya i kosmonavtika (NRB), Vol. 13, No. 8, p. 4 - 6 (1972). In Bulgarian. – Abstr. in Referativ. Zhurn.

62.Issled. kosm. prostranstva, 6.62.199 (1972).

094.174 **First results of the Lunokhod 1 data.** A. Marks.
Urania Kraków, Vol. 43, 130 - 136 (1972). In Polish.

094.175 **The physcial libration of the moon.**
A. A. Nefed'ev.
Izv. Astron. Ehngel'gardt. Obs., *Kazan*, No. 38, p. 3 - 90 (1970). In Russian.

094.176 **Complex program for reducing positional observations of lunar objects on computers.**
S. G. Valeev.
Izv. Astron. Ehngel'gardt. Obs., *Kazan*, No. 38, p. 91 - 96 (1970). In Russian.

094.177 **More about Apollo 16.** R. N. Watts, Jr.
Sky Telescope, Vol. 44, 6 - 13 (1972).

094.178 **Infrared spectroscopy of regolith (Luna 16).**
M. V. Akhmanova, B. V. Dement'ev, M. N. Markov, M. M. Sushchinskij.
Kosm. Issled., Vol. 10, 423 - 432 (1972). In Russian.

094.179 **Extension of the lunar nomenclature for the back side of the moon.** K. B. Shingareva.
Kosm. Issled., Vol. 10, 433 - 438 (1972). In Russian.

094.180 **Electron paramagnetic resonance of radiation damage in a lunar rock.**
F.-D. Tsay, S. I. Chan, S. L. Manatt.
Nature, Phys. Sci., Vol. 237, 121 - 122 (1972).

We report here evidence for the presence of radiation induced electron paramagnetic resonance signals in one of the lunar samples (12021−42) we have examined.

094.181 **Weitere Ergebnisse des Apollo-Programms. Der Apollo 15-Flug.** E. Stuhlinger.
Phys. Blätter, 28. Jahrgang, p. 247 - 257 (1972).

094.182 **Plutonium-244 fission tracks: Evidence in a lunar rock 3.95 billion years old.**
I. D. Hutcheon, P. B. Price.
Science, Vol. 176, 909 - 911 (1972).

094.183 **Effect of fluidization on the polarization of reflected light from lunar dust layers.**
G. F. J. Garlick, G. A. Steigmann, W. E. Lamb.
Nature, Phys. Sci., Vol. 238, 13 - 14 (1972).

Following our discovery of the marked rise in albedo of lunar dust layers which accompanies the loss of intergrain cohesion when the layers are fluidized in a vibratory system, we have now studied the effect of such fluidization on the degree of polarization of light reflected from dust layers.

094.184 **Space geodesy and dynamics of the earth and the moon. Recent research; Programs in France.**
J. Kovalevsky.
Mitt. Astron. Ges., No. 31, p. 49 - 60 (1972). – Review article.

094.185 **A catalogue of selenographic coordinates of points of the libration zones and the reverse side of the moon.** Yu. N. Lipsky, V. I. Chikmachev, K. I. Dekhtyareva.
Astron. Zhurn. Akad. Nauk SSSR, Vol. 49, 613 - 620 (1972). In Russian. English translation in Soviet Astron. AJ, Vol. 16, No. 3.

A catalogue of 250 points obtained on the basis of the reduction of original moon photographs by Soviet and American spacecraft are given. The reduction is carried out by the method of rectification on a spherical screen.

094.186 **The hypsometric features of the visible side of the moon.** J. F. Rodionova.
Astron. Zhurn. Akad. Nauk SSSR, Vol. 49, 621 - 623 (1972).
In Russian. English translation in Soviet Astron. AJ, Vol. 16, No. 3.
    The medium altitudes of $10° \times 10°$ areas relative to the lunar center of mass by the method of constructing hypsographic curves are determined. Diagrams of distribution of altitudes over maria, continents and the visible side of the moon are given. The mean radius of the visible side relative to the center of mass is computed (R = 1736.5 km).

094.187 **A new uniform system of designation of objects on the surface of the moon.** V. A. Nikonov.
Astron. Zhurn. Akad. Nauk SSSR, Vol. 49, 674 - 677 (1972).
In Russian. English translation in Soviet Astron. AJ, Vol. 16, No. 3.
    The existent five-digit system of designations of objects (Arthur's system) is useless for the marginal zone and invisible hemisphere of the moon. A new six-digit system of designation is considered. The new system, based on spherical coordinates λ, φ is uniform for the whole surface of the moon.

094.188 **The motions of the moon in space.** A. Deprit.
Physics and astronomy of the moon, (see 003.005), p. 1 - 28 (1971).

094.189 **Librations of the lunar globe.** M. D. Moutsoulas.
Physics and astronomy of the moon, (see 003.005), p. 29 - 61 (1971).

094.190 **Dynamics of the earth–moon system.** G. W. Groves.
Physics and astronomy of the moon, (see 003.005), p. 63 - 99 (1971).

094.191 **Geometrical and dynamical properties of the moon.** C. L. Goudas.
Physics and astronomy of the moon, (see 003.005), p. 101 - 154 (1971).

094.192 **Optical properties of the lunar surface.** B. Hapke.
Physics and astronomy of the moon, (see 003.005), p. 155 - 211 (1971).

094.193 **Origin and history of the moon.** H. C. Urey, G. J. F. MacDonald.
Physics and astronomy of the moon, (see 003.005), p. 213 - 289 (1971).

094.194 **Microwave studies of thermal emission from the moon.** T. Hagfors.
Advances Astron. Astrophys., Vol. 8, (see 003.006), 1 - 28 (1971).

094.195 **Radar studies of the moon.** J. V. Evans, T. Hagfors.
Advances Astron. Astrophys., Vol. 8, (see 003.006), 29 - 105 (1971).

094.196 **Cratering and the moon's surface.** E. J. Öpik.
Advances Astron. Astrophys., Vol. 8, (see 003.006), 107 - 337 (1971).

094.197 **Infrared observation on the eclipsed moon.** R. W. Shorthill, J. M. Saari.
Advances Astron. Astrophys., Vol. 9, (see 003.007), 149 - 201 (1972).

094.198 **Infrared emission from the surface of the moon.** D. F. Winter.
Advances Astron. Astrophys., Vol. 9, (see 003.007), 203 - 243 (1972).

094.199 **Astronomical evidence concerning non-gravitational forces in the earth–moon system.** R. R. Newton.
Astrophys. Space Sci., Vol. 16, 179 - 200 (1972).
    Evidence indicates that the present acceleration of the moon is between -20 and -52 s of arc per century per century and that the present average acceleration of the earth is between -5 and -23 parts in 10^9 per century. Over the past 2000 yr, the average for the moon has been about -42s per century per century and for the earth has been about -28 parts in 10^9 per century; these values are probably correct within 10%. There are no satisfactory explanations of the accelerations. Existing theories of tidal friction are quite inadequate.

094.200 **The origin of the moon: Theories involving joint formation with the earth.** J. A. O'Keefe.
Astrophys. Space Sci., Vol. 16, 201 - 211 (1972).
    In the moon, there is clear chemical evidence that liquid iron was separated from the mass, but the moon has no detectable iron core. This points to some kind of joint origin, which put the metallic iron in the earth's core. If tektites come from the moon, then Darwin's suggestion is probably right; if they come from the earth, then the Öpik-Ringwood sediment ring may be the origin.

094.201 **Paleontological evidence on the earth's rotational history since Early Precambrian.** G. Pannella.
Astrophys. Space Sci., Vol. 16, 212 - 237 (1972).
    The daily growth layers arranged into seasonal and tidal patterns, present in calcified structures of many modern as well as fossil organisms, provide evidence on the length of lunar month and year in the geological past. This paper deals only with the evidence provided by fossils and presents the data gathered during seven years of investigations. The data available from literature are also discussed. The obvious conclusions are that the moon has been associated with earth since at least Early Precambrian times, that all theories implying a late capture of the moon should be revised and that the calculated secular changes in the earth's rotation rate cannot be accepted as representative for the all geological history.

094.202 **Compositional characteristics of olivines from Apollo 12 samples.** P. Butler, Jr.
Geochim. Cosmochim. Acta, Vol. 36, 773 - 785 (1972).
    This study examines the compositional populations of olivines from basalts with similar and apparently uncomplicated thermal histories in order to establish a baseline of inter- and intra-sample variations. These variations are then compared with the compositional variations of olivines from a sample that shows evidence of re-equilibration.

094.203 **Tidal triggering of moonquakes.** W. L. Hamilton, D. Lammlein, J. Dorman, G. Latham.
Science, Vol. 176, 1258 - 1259 (1972).

094.204 **Bemerkungen zur Morphologie der Mondoberfläche.** A. Priem.
Sterne, 48. Jahrgang, p. 103 - 111 (1972).

094.205 **Measurements of radon emanation from Apollo 11, 12, and 14 fines.** A. Yaniv, D. Heymann.
Earth Planet. Sci. Letters, Vol. 15, 95 - 100 (1972).

094.206 **Lunar pentlandite and sulfidization reactions in microbreccia 14315, 9.** P. Ramdohr.
Earth Planet. Sci. Letters, Vol. 15, 113 - 115 (1972).
    Detailed microscopic studies of the Apollo 14 breccia 14315, 9 indicate that sulfur and nickel migrations took place.

094.207 Lunar neutron stratigraphy.
G. P. Russ III, D. S. Burnett, G. J. Wasserburg.
Earth Planet. Sci. Letters, Vol. 15, 172 - 186 (1972).
Variations in the isotopic ratios $^{158}Gd/^{157}Gd$ and $^{150}Sm/^{149}Sm$ in the Apollo 15 deep drill stem show that the neutron fluence is a smoothly varying function of depth with a relatively symmetric peak at a depth of 190 g/cm^2.

094.208 Comments on "Lunar seismograms for LM and S-IVB impacts interpreted as modulation mirage"
by E. Strick. [Earth Planet. Sci. Letters, Vol. 13, 23 - 31 (1971/72)].   G. V. Latham, M. Ewing, F. Press, G. Sutton, J. Dorman, Y. Nakamura, N. Toksoz, D. Lammlein, F. Duennebier, with a reply by E. Strick.
Earth Planet. Sci. Letters, Vol. 15, 212 - 214 (1972).

094.209 Ferromagnetic and paramagnetic resonance spectra of lunar material: Apollo 12.
R. A. Weeks, J. L. Kolopus, S. Arafa.
The Moon, Vol. 4, 271 - 295 (1972).
Electron magnetic resonance spectra of specimens of two crystalline rocks (12021-55 and 12075-19) and of four specimens of fines < 1 mm (12001-16, 12030-16, 12033-50 and 12070-125) have been obtained as a function of spectrometer frequency (9 and 35 GHz), temperature (78 to 300 K), heat treatments (to 960°C), and mineral phases (plagioclase, olivine, pyroxferroite, glass, and basaltic fragments).

094.210 A new, earth-based radar technique for the measurement of lunar topography.   S. H. Zisk.
The Moon, Vol. 4, 296 - 306 (1972).
Radio interferometry is a new technique for the measurement of the surface topography of the moon. Elevation data may be obtained directly without regard for unambiguously-identified features, for any lunar surface element that yields a recognizable radar echo. A program has been undertaken at the Haystack Observatory for the topographic mapping of the major part of the lunar earthside hemisphere. Some results are presented for the Alphonsus-Arzachel region.

094.211 Lunar gravity traverse experiment.   M. Talwani.
The Moon, Vol. 4, 307 (1972).
Communication presented at the 'Conference on lunar geophysics', Houston, Texas, 1971 October 18 - 21.

094.212 Thermoluminescence of lunar samples: Measurement of temperature gradients in core material.
R. M. Walker, D. W. Zimmerman, J. Zimmerman.
The Moon, Vol. 4, 308 - 314 (1972).
Communication presented at the 'Conference on lunar geophysics', Houston, Texas, 1971 October 18 - 21.

094.213 Infrared and Raman spectra of lunar samples from Apollo 11, 12 and 14.   C. H. Perry, D. K. Agrawal, E. Anastassakis, R. P. Lowndes, A. Rastogi, N. E. Tornberg.
The Moon, Vol. 4, 315 - 336 (1972).
We report the room temperature infrared reflectance spectra of several lunar surface rocks in the form of polished slices or butt ends. The spectra were obtained over the frequency range 20–2000 cm^{-1} throughout the mid and far infrared (5–500 μ) region of the electromagnetic spectrum where the fundamental internal and lattice vibrational modes of all minerals and rocks occur. Some fines samples were examined as pressed pellets and their reflectivities compared with the bulk samples. Several terrestrial minerals and rocks were also investigated. Kramers-Kronig analyses of these reflectance spectra were undertaken and the dispersion of the dielectric response and the optical constants have been determined over this frequency range. The low frequency and high frequency (infrared) dielectric constants were also calculated from the reflectance

data. Raman light scattering measurements were made on all the samples supplied from the first three Apollo missions.

094.214 Comments on the figure of the moon based on preliminary results from laser altimetry.
W. R. Wollenhaupt, W. L. Sjogren.
The Moon, Vol. 4, 337 - 347 (1972).
Range measurements from the orbiting spacecraft to the lunar surface were made during the Apollo 15 mission using a laser altimeter. Analysis of measurements made during one complete lunar revolution indicates that the figure of the moon is very complex. The lunar far side appears to be considerably rougher than the near side in this plane. There appears to be a very large depression on the far side centered at approximately 180° longitude. These data provide some proof that there is a displacement between the center of figure and the center of mass of the moon.

094.215 Lunar color boundaries and their relationship to topographic features: A preliminary survey.
E. A. Whitaker.
The Moon, Vol. 4, 348 - 355 (1972).
Communication presented at the 'Conference on lunar geophysics', Houston, Texas, 1971 October 18 - 21.

094.216 Laboratory studies on seismic and electrical properties of the moon.   D. H. Chung.
The Moon, Vol. 4, 356 - 372 (1972).
A summary of all the laboratory characterization of the seismic and electrical properties of lunar samples is made; along with new measurements, data on both the laboratory seismic and electrical parameters are evaluated.

094.217 Moonquakes and lunar tectonism.
G. Latham, M. Ewing, J. Dorman, D. Lammlein, F. Press, N. Toksöz, G. Sutton, F. Duennebier, Y. Nakamura.
The Moon, Vol. 4, 373 - 382 (1972).
With the successful installation of a geophysical station at Hadley Rille, on July 31, 1971, on the Apollo 15 mission, and the continued operation of stations 12 and 14 the Apollo program for the first time achieved a network of seismic stations on the lunar surface. The major discoveries that have resulted to date from the analysis of seismic data from this network are summarized.

094.218 Evidence for objects of lunar mass in the early solar system.   H. C. Urey.
The Moon, Vol. 4, 383 - 389 (1972).
The problem of the accumulation of the moon is discussed on the assumption that the moon is a captured object. Evidence indicating that other massive objects were present at that time is presented. Also, it is pointed out that interior of the moon must contain normal solar proportions of the elements of intermediate volatility in the lunar interior, if the moon accumulated in a gas sphere.

094.219 The Apollo 15 lunar heat-flow measurement.
M. G. Langseth, Jr., S. P. Clark, Jr., J. L. Chute, Jr., S. J. Keihm, A. E. Wechsler.
The Moon, Vol. 4, 390 - 410 (1972).
The heat-flow experiment is one of the Apollo Lunar Surface Experiment Package (ALSEP) instruments that was emplaced on the lunar surface on Apollo 15. This experiment is designed to make temperature and thermal property measurements in the lunar subsurface so as to determine the rate of heat loss from the lunar interior through the surface. About 45 days ($1^1/_2$ lunations) of data has been analyzed in a preliminary way.

094.220 Apollo 15 gravity analysis from the S-band trans-

ponder experiment.
W. L. Sjogren, P. M. Muller, W. R. Wollenhaupt.
The Moon, Vol. 4, 411 - 418 (1972).
Communication presented at the 'Conference on lunar geophysics', Houston, Texas, 1971 October 18 - 21.

094.221 Satellite measurements of the moon's magnetic field: A preliminary report.
P. J. Coleman, Jr., G. Schubert, C. T. Russell, L. R. Sharp.
The Moon, Vol. 4, 419 - 429 (1972).
Communication presented at the 'Conference on lunar geophysics', Houston, Texas, 1971 October 18 - 21.

094.222 Q and structure.   N. Warren.
The Moon, Vol. 4, 430 - 441 (1972).
   A model is presented in which scattering occurs to a depth of about 10 km below the volcanic ridge in the Tonga-Fiji region. The case of the lunar maria is discussed in terms of the relation of measured seismic Q, scattering parameters, and direct ray propagation parameters. The results indicate that a fairly simple jointed bedrock model is compatible with observed lunar seismic data.

094.223 Infrared and radar maps of the lunar equatorial region.
R. W. Shorthill, T. W. Thompson, S. H. Zisk.
The Moon, Vol. 4, 442 - 446 (1972).
Communication presented at the 'Conference on lunar geophysics', Houston, Texas, 1971 October 18 - 21.

094.224 Thermophysical properties of lunar material returned by Apollo missions.
K.-I. Horai, N. Fujii.
The Moon, Vol. 4, 447 - 475 (1972).
   Data on thermophysical properties measured on lunar material returned by Apollo missions are reviewed. In particular, the effects of temperature and interstitial gaseous pressure on thermal conductivity and diffusivity have been studied.

094.225 Accretion process of the moon.
H. Mizutani, T. Matsui, H. Takeuchi.
The Moon, Vol. 4, 476 - 489 (1972).
   Recent geochemical and geophysical data suggest that the initial temperature of the moon was strongly peaked toward the lunar surface. To explain such an initial temperature distribution, a simple model of accretion process of the moon is presented. The model assumes that the moon was formed from the accumulation of solid particles or gases in an isolated, closed cloud. Numerical calculations are made for a wide range of the parameters particle concentration and particle velocity in the cloud.

094.226 Velocity structure and properties of the lunar crust.
M. N. Toksöz, F. Press, K. Anderson, A. Dainty, G. Latham, M. Ewing, J. Dorman, D. Lammlein, Y. Nakamura, G. Sutton, F. Duennebier.
The Moon, Vol. 4, 490 - 504 (1972).
   Lunar seismic data from three Apollo seismometers are interpreted to determine the structure of the moon's interior to a depth of about 100 km. The travel times and amplitudes of P arrivals from Saturn IVB and Lunar Module impacts are interpreted in terms of a compressional velocity profile.

094.227 Evidence for lunar vulcanism from lunar probes.
W. S. Cameron.
Strolling Astronomer, Vol. 23, 159 - 170 (1972).

094.228 ALPO selected areas program report: Kepler.
K. J. Delano.
Strolling Astronomer, Vol. 23, 174 - 179 (1972).

094.229 Lunar notes.   W. S. Cameron, K. J. Delano, H. D. Jamieson, R. C. Parish.
Strolling Astronomer, Vol. 23, 179 - 182 (1972). – The program for observations of lunar transient phenomena; The dark-haloed craters program; Progress report on a study of Messier and Pickering.

094.230 Moonblink im Aristarch: Erste Ergebnisse des Mondprogramms.   G. Küveler.
SuW, Vol. 11, 192 (1972).

094.231 Rotation of lunar dumbbell-shaped globules during formation.   M. J. Pugh.
Nature, Vol. 237, 158 - 159 (1972).
   It has been proposed that the elongated smooth globules found in lunar fines are produced by rotation of the molten glass during the formation process. In this paper measurements of surface curvatures of globules are used to support this theory, showing that rotation occurs about a minor axis of the figure.

094.232 Lunar maps of the XVIIth century.
O. Van De Vyver.
Vatican Obs. Publ., Vol. 1, No. 2, 15 + 32 pp. (1971).

094.233 Extreme declinations of the moon.
G. P. Können, J. Meeus.
Journ. British Astron. Ass., Vol. 82, 192 - 193 (1972).

094.234 Analysis of lunar samples for carbon compounds.
K. A. Kvenvolden.
Theory and experiment in exobiology, Vol. 1, (see 003.011), 105 - 121 (1971).

094.235 Recent exploration of the moon.   G. Fielder.
Geology and physics of the moon, (see 003.013), p. 1 - 13 (1971).

094.236 Lava flows and the origin of small craters in Mare Imbrium.   G. Fielder, J. Fielder.
Geology and physics of the moon, (see 003.013), p. 15 - 26 (1971) = 06.094.317.

094.237 Sinuous rilles.   J. B. Murray.
Geology and physics of the moon, (see 003.013), p. 27 - 39 (1971).

094.238 Centres of igneous activity in the maria.
J. E. Guest.
Geology and physics of the moon, (see 003.013), p. 41 - 53 (1971).

094.239 Multiphase eruptions associated with the craters Tycho and Aristarchus.   R. G. Strom, G. Fielder.
Geology and physics of the moon, (see 003.013), p. 55 - 92 (1971).

094.240 Geology of the farside crater Tsiolkovsky.
J. E. Guest.
Geology and physics of the moon, (see 003.013), p. 93 - 103 (1971).

094.241 Origins of lunar craters.   R. J. Fryer.
Geology and physics of the moon, (see 003.013), p. 105 - 114 (1971).

094.242 Photometric studies.   L. Wilson.
Geology and physics of the moon, (see 003.013), p. 115 - 123 (1971).

094.243 Polarimetric studies.   E. L. G. Bowell.

Geology and physics of the moon, (see 003.013), p. 125 - 133 (1971).

094.244 Chemical problems of the surface.
A. P. Brown, L. Wilson.
Geology and physics of the moon, (see 003.013), p. 135 - 142 (1971).

094.245 Thermal studies. J. A. Bastin, E. L. G. Bowell.
Geology and physics of the moon, (see 003.013), p. 143 - 155 (1971).

094.246 Concerning a transient phenomenon in the lunar crater Posidonius. V. P. Dzhapiashvili.
Astron. vestn., Vol. 6, 7 - 8 (1972). In Russian. – Abstr. in Referativ. Zhurn. 51. Astron., 7.51.230 (1972).

094.247 Tectonic map of the moon.
Compiled by V. V. Kozlov, E. D. Sulidi-Kondrat'ev. Yu. Ya. Kuznetsov (Scientific editor).
Ministry of Geology of the USSR. Scientific-research laboratory of the geology of foreign countries, NILZarubezhgeologiya, Moscow. Scale 1 : 15000000 (1969).

094.248 Tectonics of the moon. Explanatory description to the tectonic map of the moon on a scale of 1 : 7500000. V. V. Kozlov, E. D. Sulidi-Kondrat'ev.
Yu. Ya. Kuznetsov (Editor).
Ministerstvo Geologii SSSR. Nauchno-issledovatel'skaya laboratoriya geologii zarubezhnykh stran "NILZarubezhgeologiya", Moskva. 43 pp. (1969). In Russian.

094.249 On determining the moon's density function.
S. L. Levie, Jr.
Rep. NASA–CR–118892, Bellcomm. Inc., Washington, D.C., U.S.A. [Available from NTIS, Springfield, Va.], 12 pp. (1971).
A qualitative study is presented of the relations between the moon's density function, figure, external gravitational potential and physical libration constants.

094.250 Topographic measurements along the path of Lunokhod-1. B. N. Rodionov.
Fiz. Szemle, Vol. 21, 245 - 247 (1971). In Hungarian.
The topography of the lunar surface was studied up to the width of 150 metres in the vicinity of the Heraklidus mountains. The surface was photographed by four telephotometers and two directional cameras.

094.251 Density, pressure, and temperature effects on heat transfer in Apollo 11 fines. C. J. Cremers.
AIAA Journ., Vol. 9, 2180 - 2183 (1971).
The thermal conductivity of the Apollo 11 lunar fines is presented as a function of temperature for three values of bulk density corresponding to lightly disturbed, normal surface condition, and packed. The thermal conductivity is also presented as a function of ambient pressure at conditions of normal density and standard temperature.

094.252 A traverse gravimeter for the lunar surface.
G. Mamon.
IEEE Trans. Geosci. Electron., Vol. GE–10, No. 1, p. 64 - 72 (1972).
A semiautomatic, self-levelling lunar gravimeter has been designed for the purpose of measuring gravity at predetermined stops along the route of a Lunar Rover Vehicle to obtain a gravity profile.

094.253 On the moon with Apollo 15. G. Simmons.
Spaceworld (*USA*), Vol. H–10–94, p. 4 - 50 (1971).

094.254 Evidence for vulcanism on the moon. W. S. Caern.
Astron. & Space (*GB*), Vol. 1, 194 - 201 (1971).

094.255 Thermal conductivity and diffusivity of the Apollo 12 lunar fines at landing site conditions.
C. J. Cremers.
11th international conference on thermal conductivity, Albuquerque, New Mexico, p. 151 - 152 (1971).

094.256 Lunar surface temperatures at Tranquillity Base.
C. J. Cremers, R. C. Birkebak, J. E. White.
AIAA Journ., Vol. 9, 1899 - 1903 (1971).
The diurnal temperatures in the lunar soil in the vicinity of Tranquillity Base are calculated from the one-dimensional energy equation.

094.257 The moon – its origin and physiography.
S. Mukerjee.
Sci. & Culture (*India*), Vol. 37, 230 - 236 (1971).

094.258 The first stereoscopic pictures of the moon.
T. B. Greenslade, Jr.
American Journ. Phys., Vol. 40, 536 - 540 (1972). – See Phys. Abstr., Vol. 75, No. 30534 (1972).

094.259 New data on the shape and the surface of the moon as obtained from the transmitted pictures of Zond-6.
Ny. B. Rogyionov, J. V. Iszavnyina, Ju. F. Avgyejev, V. D. Blagov, A. Sz. Dorofejev, B. Sz. Dunajev, Ja. L. Ziman, V. V. Kiszeljov, V. A. Kraszikov, O. Ny. Lebegyev, A. B. Mihailovszkij, A. P. Tiscsenko, B. V. Nyepoklonov, V. K. Szamojlov, F. M. Truszkov, Ju. M. Cszesznokov, Ju. I. Fivenszkij.
Fiz. Szemle, Vol. 21, 267 - 274 (1971). In Hungarian. – See Phys. Abstr., Vol. 75, No. 30539 (1972).

094.260 Thickness of impact crater ejecta on the lunar surface. N. M. Short, M. L. Forman.
Modern Geol. (*GB*), Vol. 3, No. 2, p. 69 - 91 (1972). – See Phys. Abstr., Vol. 75, No. 30541 (1972).

094.261 The surface orientation of some Apollo 14 rocks.
F. Horz, D. A. Morrison, J. B. Hartung.
Modern Geol. (*GB*), Vol. 3, No. 2, p. 92 - 104 (1972). – See Phys. Abstr., Vol. 75, No. 30542 (1972).

094.262 Some indications as to the development of the earth and moon according to the thorium content of moon rock and terrestrial basalts. R. Lauterbach.
Wiss. Zeitschr. Karl-Marx-Univ. Leipzig, Ser. Math.-Nat., Vol. 20, 659 - 669 (1971). In German.
From consideration of the importance of the radioactivity of basalt rock in research into the interior of earth, a study is made of the thorium concentration in some basalts and a companion review study is made of moon rock. It is inferred that the earth and moon originated almost simultaneously from at least similar solar matter and both developed 'in the cold way'.

094.263 The morphology of free-growth crystal faces of lunar ilmenites. J. Jedwab.
Bull. Soc. Française Minéralogie Cristallographie, Vol. 94, 477 - 485 (1971). In French. – See Phys. Abstr., Vol. 75, No. 44905 (1972).

094.264 Directional behavior of thermal emission from a rough lunar surface. A. S. Adorjan.
Journ. Spacecraft & Rockets (*USA*), Vol. 9, 59 - 61 (1972).
Earth-based measurements of the infrared emission from the lunar surface reveal significant directionality effects. The

directionality of the emission can be explained qualitatively by the topographical roughness of the lunar surface. By integrating the radio-sites of both the craters and plain surfaces over a control area, and for given solar elevation and observation angles, the brightness temperatures can be determined.

094.265 **An experimental study of the seismic ringing of the moon.** M. R. Thapar.
Pure and Applied Geophys. (*Switzerland*), Vol. 94, No. 2, p. 15 - 32 (1972). – See Phys. Abstr., Vol. 75, No. 44926 (1972).

094.266 **Experimental investigation of the lunar seismic signals.** M. R. Thapar.
Pure and Applied Geophys. (*Switzerland*), Vol. 94, No. 2, p. 33 - 52 (1972). – See Phys. Abstr., Vol. 75, No. 44927 (1972).

094.267 **Geology of Fra Mauro.** V. Bouška.
Vesmír, Vol. 51, 111 - 112 (1972). In Czech.

094.268 **About the origin of some lunar features.**
M. Eliáš.
Říše hvězd, Vol. 53, 84 - 87 (1972). In Czech.

094.269 **Origin of lunar surface features.** J. Green.
Říše hvězd, Vol. 53, 68 - 73 (1972). In Czech.

094.270 **To the origin of the central peaks in the crater formations filled with the melt after impacts.**
L. Křivský.
Říše hvězd, Vol. 53, 9 - 12 (1972). In Czech.

094.271 **Thermal characteristics of the lunar surface layer.** C. J. Cremers, R. C. Birkebak, J. E. White.
International Journ. Heat & Mass Transfer (*GB*), Vol. 15, 1045 - 1055 (1972).
    The thermophysical properties of the fines from the Apollo 12 landing site have been determined as a function of their relevant parameters. These properties include the thermal conductivity, thermal diffusivity, directional reflectances and emittance.

094.272 **Vacuum measurements on the lunar surface.**
F. S. Johnson, J. M. Carroll, D. E. Evans.
Journ. Vacuum Sci. and Technology (*USA*), Vol. 9, 450 - 456 (1972). – See Phys. Abstr., Vol. 75, No. 41175 (1972).

094.273 **Plasmatheorie der Magnetfelder am Mond.**
F. Cap.
Anzeiger Österreich. Akad. Wiss., Math. - nat. Kl., 107. Jahrgang, p. 170 - 171 (1971).

094.274 **Temperature distribution in lunar rilles.**
A. S. Adorjan.
Journ. Spacecraft & Rockets, Vol. 8, 669 - 674 (1971).

094.275 **Subsurface temperature of the moon.**
W. W. Salisbury, D. L. Fernald.
Journ. Astronaut. Sci., Vol. 18, 236 - 243 (1971).

094.276 **Magmatic differentiation of the moon.** B. J. Levin.
Chem. der Erde (Germany), Vol. 30, 251 - 257 (1971).

094.277 **Search for magnetic monopoles in lunar material.** P. H. Eberhard, R. R. Ross, L. W. Alvarez.
Phys. Rev. D, Particles and Fields, Vol. 4, 3260 - 3272(1971).

Bringing back the lunar bacon.
Nature, Vol. 236, 3 (1972).

Lunar orbiter photographic atlas of the moon.
See Abstr. 003.029.

Beginnings of lunar soil science. Physico-mechanical properties of the lunar soil. See Abstr. 003.041.

Exploring the moon and the solar system.
See Abstr. 003.049.

Atlas and gazetteer of the near side of the moon.
See Abstr. 003.060.

Fundamentals of lunar astrometry.
See Abstr. 003.082.

Thermal characteristics of the moon. (Progress in Astrophysics and Aeronautics, Vol. 28).
See Abstr. 003.089.

Apollo 15: Preliminary science report.
See Abstr. 003.149.

The origin of the lunar craters: An eighteenth-century view. See Abstr. 004.010.

Small optical telescopes on the moon.
See Abstr. 032.004.

The AFCRL lunar laser ranging experiment in Arizona. See Abstr. 034.009.

Positions de grosses planètes et de la lune observées à l' équatorial photographique de 0, 33 m.
See Abstr. 041.055.

Le problème principal de la lune et le problème restreint des trois corps. See Abstr. 042.008.

The use of solar eclipse timings to compare the reference systems of Newcomb's tables of the sun and of the Improved Lunar Ephemeris. See Abstr. 047.001.

Variation of single particle mid-infrared emission spectrum with particle size. See Abstr. 061.011.

Interaction of the solar wind with the moon.
See Abstr. 074.091.

Interaction of solar wind with the moon and possibly other planetary bodies. See Abstr. 074.098.

Gosses Bluff impact structure, Australia.
See Abstr. 081.007.

Applicability of a diffusion model to lateral transport in the terrestrial and lunar exospheres.
See Abstr. 082.015.

New orbital elements for moon and planets.
See Abstr. 091.016.

Genetic implications of the shapes of Martian and lunar craters. See Abstr. 097.002.

On the origins of trapped helium, neon and argon isotopic variations in meteorites—I. Gas-rich meteorites, lunar soil and breccia. See Abstr. 105.012.

Impacts auf der Erde und auf dem Mond.
See Abstr. 105.052.

Magnetic-field anomalies in the lunar wake.
See Abstr. 106.007.

Evidence for objects of lunar mass in the early solar system and for capture as a general process for the origin of satellites. See Abstr. 107.010.

094.901 **Erratum:'Distinct cooling histories of Apollo 15 basalts'** [Nature, Phys. Sci., Vol. 236, 71 - 73 (1972)].

N. Gray, S. S. Hafner, K. Schürmann, D. Virgo.
Nature, Phys. Sci., Vol. 237, 16 (1972).

094.902 **Errata: 'Meteoritic and non-meteoritic trace elements in Luna 16 samples'** [Earth Planet. Sci. Letters, Vol. 13, 450 - 454 (1972)].
J. C. Laul, R. Ganapathy, J. W. Morgan, E. Anders.
Earth Planet. Sci. Letters, Vol. 14, 451 (1972).

095   Lunar Eclipses

095.001 **Observations of the total lunar eclipse on February 10, 1971 at 3.1 mm wavelength.** B. L. Ulich.
Icarus, Vol. 16, 304 - 309 (1972).
Observations were made of the total lunar eclipse on February 10, 1971 at a wavelength of 3.1 mm. Eclipse cooling curves were obtained for Copernicus, Mare Serenitatis, and a mountainous region. Cooling rates of about 6°K/hour were measured. The normalized solar insolation has also been calculated for each region.

095.002 **Beobachtung der totalen Mondfinsternis vom 6. August 1971 auf Korsika.** W. Hänig.
SuW, Vol. 11, 21 (1972).

095.003 **The lunar eclipse in January.**
Sky Telescope, Vol. 43, 60 - 61 (1972).

095.004 **January lunar eclipse synopsis.**
Sky Telescope, Vol. 43, 258 - 265 (1972).

095.005 **Photoelectric photometry of the penumbral lunar eclipse of September 25, 1969.**
J. Bouška, P. Mayer, A. Mrkos.
Bull. Astron. Inst. Czechoslovakia, Vol. 23, 139 - 143 (1972).
From photometric measurements the densities and isophotes of the penumbra in the spectral regions B and V were obtained. A very large flattening of the penumbral shadow was found.

095.006 **Die Mondfinsternis vom 6. August 1971.**
P. Ahnert.
Sterne, 48. Jahrgang, p. 57 (1972).

095.007 **Totale Mondfinsternis 1971 Aug. 6, beobachtet in Rožnava (ČSSR).** P. Ahnert.
Sterne, 48. Jahrgang, p. 58 (1972).

095.008 **How large was the earth's shadow?**
J. Ashbrook.
Sky Telescope, Vol. 43, 330 - 331 (1972).

095.009 **Photometry methods and the total lunar eclipse of January 30, 1972.** J. E. Westfall.
Strolling Astronomer, Vol. 23, 113 - 118 (1972).

095.010 **L'éclipse totale de lune du 6 août 1971, vue de Lausanne.** M. Roud.
Orion Schaffhausen, 30. Jahrgang, p. 21 (1972).

095.011 **Totale Mondfinsternis vom 6. August 1971.**
D. Böhme.
Orion Schaffhausen, 30. Jahrgang, p. 63 (1972).

095.012 **Lunar eclipses.** F. Link.
Advances Astron. Astrophys., Vol. 9, (see 003.007), 67 - 148 (1972).

095.013 **Rezultati promatranja potpune pomrčine Mjeseca 6. VIII 1971. [Results of an investigation on the total lunar eclipse on August 6, 1971.]**
S. Kulišić, K. Pavlovski.
Vasiona, Vol. 20, 27 - 29 (1972).

Infrared observation on the eclipsed moon.
See Abstr. 094.197.

096 Lunar Occultations

096.001 **Reduzierte Sternbedeckungen durch den Mond 1968 und 1969.** H. Krüger.
Astron. Nachr., Vol. 293, 237 (1971/72) = Mitt. Zentralinst. Phys. Erde, Potsdam, No. 144 (1972).

096.002 **Sternbedeckungen 1970.** P. Ahnert.
Astron. Nachr., Vol. 293, 239 (1971/72).

096.003 **De Meteoor: Sterren die in 1972 door de maan worden bedekt.** J. J. B. van Eijk van Voorthuijsen.
Hemel en Dampkring, Vol. 70, 21 - 22 (1972).

096.004 **Occultation highlights for the year 1972.** D. W. Dunham.
Sky Telescope, Vol. 43, 54 - 56 (1972).

096.005 **Sky and Telescope, 1972 occultation supplement. Predictions for United States and Canada.**
Prepared by H. M. Nautical Almanac Office, Royal Greenwich Observatory. Printed and distributed by Sky Publishing Corporation, Cambridge, Mass. 12 pp. (1971).

096.006 **La périodicité des occultations.** J. Meeus.
L'Astronomie, 86ᵉ année, p. 141 - 147 (1972).

096.007 **Grazing occultations over Canada during 1972.** L. V. Morrison.
Journ. Roy. Astron. Soc. Canada, Vol. 66, 56 - 58 (1972).

096.008 **The amateur scientist.** Conducted by C. L. Stong.
Sci. American, Vol. 226, No. 1, p. 108 - 111 (1972).
Groups are organized to observe the eclipse of stars by the moon.

096.009 **Rakende sterbedekkingen, juli - december 1972.** J. Meeus.
Hemel en Dampkring, Vol. 70, 96 - 97 (1972).

096.010 **Observations of occultations of stars by the moon in Tashkent in 1969 - 1970.** M. R. Ehshmatov.
Tsirk. Astron. Inst., *Tashkent*, No. 28 (375), p. 20 - 21 (1971). In Russian.

096.011 **Occultations rasantes en France, juillet - décembre 1972.** J. Meeus.
L'Astronomie, 86ᵉ année, p. 242 - 243 (1972).

096.012 **Fading at occultations.** P. Moore.
Journ. British Astron. Ass., Vol. 82, 282 - 283 (1972).

096.013 **Plejaden-Bedeckung am 29. Dezember 1971.** A. Doerr, U. Thein.
Orion Schaffhausen, 30. Jahrgang, p. 99 (1972).

096.014 **Grazing occultations.** J. L. Perdrix.
Journ. Astron. Soc. Victoria, Vol. 24, 13 - 16 (1971).

·096.015 **Two observations of the grazing occultation of Antares, 25 Sept. 1972.** L. M. G. Poole.
Monthly Notes Astron. Soc. Southern Africa, Vol. 31, 4 - 6 (1972).

096.016 **Further observations of occultations I. 1971 September 24 − ZC 2220 = ADS 9689 − Arthur's view.** J. Hers.
Monthly Notes Astron. Soc. Southern Africa, Vol. 31, 56 - 61 (1972).

096.017 **Further observations of grazing occultations II. 1971 October 25 − ZC 2771 − Dinokana.** J. Hers.
Monthly Notes Astron. Soc. Southern Africa, Vol. 31, 62 - 63 (1972).

096.018 **Further observations of grazing occultations III. 1972 February 6 − ZC 2084 − Bronkhorstspruit.** J. Hers.
Monthly Notes Astron. Soc. Southern Africa, Vol. 31, 64 - 65 (1972).

096.019 **Further observations of grazing occultations IV. 1972 February 12 − ZC 2811 − Johannesburg.** J. Hers.
Monthly Notes Astron. Soc. Southern Africa, Vol. 31, 66 - 67 (1972).

096.020 **Occultation of the Pleiades on March 19, 1972.** Z. Ivanović.
Vasiona, Vol. 20, 36 - 37 (1972). In Serbo-Croatian.

096.021 **Lunar occultations.**
Contr. Obs. Valongo, Univ. Federal Rio de Janeiro, Sér. III, No. 21 (1971). − 1971 April - June.

096.022 **Observations of lunar occultations.**
Yamamoto Circ., No. 1749, p. 3 - 4 (1972). In Japanese.

Nuovo metodo e strumentazione relativa per il rilevamento del tempo di occultazione lunare.
See Abstr. 034.052.

Infrared diameter of IRC+10216 determined from lunar occultations. See Abstr. 115.010.

Lunar occultation observations of 24 radio sources.
See Abstr. 141.132.

Identification of GX3 + 1 from lunar occultations.
See Abstr. 142.051.

097 Mars

097.001 The motions of the satellites of Mars.
  A. T. Sinclair.
Monthly Notices Roy. Astron. Soc. Vol. 155, 249 - 274 (1972).

Analytical expressions for the secular and periodic variations in the orbital elements of the satellites of Mars due to the combined action of the oblateness of Mars and the attraction of the sun are developed. Improved values of the arbitrary parameters of the theory are determined by an analysis of all available observations of the positions of the satellites during the period 1877–1969. It is concluded that the observational data are not sufficiently accurate to determine the secular accelerations, if any, of the mean longitudes of the satellites.

097.002 Genetic implications of the shapes of Martian and lunar craters. R. J. Pike.
Icarus, Vol. 15, 384 - 395 (1971/72).

The primary purpose of this paper is to report the shape of the logarithmic depth-diameter curve for large Martian craters and its similarity to lunar depth-diameter curves. The secondary objective is a generalized geologic explanation of the data.

097.003 Martian cratering II: Asteroid impact history.
  W. K. Hartmann.
Icarus, Vol. 15, 396 - 409 (1971/72).

This paper considers the extent to which Martian craters can be explained by considering asteroidal impact. Sections I, II, and III of this paper derive the diameter distribution of hypothetical asteroidal craters on Mars from recent Palomar-Leiden asteroid statistics and show that the observed Martian craters correspond to a bombardment by roughly 100 times the present number of Mars-crossing asteroids. Section IV discusses the early bombardment history of Mars, based on the capture theory of Öpik and probable orbital parameters of early planetesimals. Section V presents a comparison with results and terminology of other authors.

097.004 Martian cratering III: Theory of crater obliteration.
  W. K. Hartmann.
Icarus, Vol. 15, 410 - 428 (1971/72).

The purpose of this paper is to discuss a theoretical basis for interpretation of Martian craters, and particularly processes obliterating the smaller Martian craters. The mathematical form of the theory which is developed will be tested by application to the earth.

097.005 On the effects of large-scale temperature advection in the Martian atmosphere. S. L. Blumsack.
Icarus, Vol. 15, 429 - 442 (1971/72).

A theoretical model in which large-scale topography can generate steady winds is analyzed. The wind is determined by the radiative-equilibrium thermal structure, but gradient-level winds are affected to lowest order by the small temperature anomalies induced by large-scale temperature advection. The application of the model to the Martian general circulation is discussed and examples in which Martian parameters are used are presented and analyzed.

097.006 Mariner 7 ultraviolet spectrometer experiment: Photometric function and roughness of Mars' polar cap surface. K. Pang, C. W. Hord.
Icarus, Vol. 15, 443 - 453 (1971/72).

The Mariner 7 ultraviolet spectrometer observed the south polar cap of Mars. The observed phenomena make it possible to obtain a photometric function of the polar cap using the angular dependence of the data. Knowing the photometric function, we can deduce some information on the roughness of the polar cap surface.

097.007 The long winter model of Martian biology: A speculation. C. Sagan.
Icarus, Vol. 15, 511 - 514 (1971/72).

An estimated mean thickness ~ 1 km of frost in the Martian North Polar Cap summer remnant, if vaporized, would yield ~ 10^3 g cm^{-2} of atmosphere over the planet, higher global temperatures through the greenhouse effect, and a greatly increased likelihood of liquid water. Vaporization of such cap remnants may occur twice each equinoctial precession, and Martian organisms may now be in cryptobiotic repose awaiting the end of the long precessional winter.

097.008 The Viking missions to Mars.
  G. A. Soffen, A. T. Young.
Icarus, Vol. 16, 1 - 16 (1972).

The Viking Project will launch two unmanned spacecraft to Mars in 1975 for scientific exploration with special emphasis on the search for life. Each spacecraft will consist of an orbiter and a lander. The landing site will be selected after the spacecraft is in orbit. Twelfe investigations will be performed: three mapping experiments from the orbiter, one atmospheric experiment during the lander entry phase, seven analytical experiments on the surface of the planet, and one using the spacecraft radio and radar systems. The experiments on the surface will deal principally with biology, geology, and meteorology.

097.009 Imaging experiment: The Viking Mars Orbiter.
  M. H. Carr, W. A. Baum, G. A. Briggs, H. Masursky, D. W. Wise, D. R. Montgomery.
Icarus, Vol. 16, 17 - 33 (1972).

The general objectives of the Imaging Experiment on the Viking Orbiter are to aid the selection of Viking Lander sites, to map and monitor the chosen sites during lander operations, to aid in the selection of future landing sites, and to extend our knowledge of the planet.

097.010 The detection and mapping of water vapor in the Martian atmosphere.
C. B. Farmer, D. D. LaPorte.
Icarus, Vol. 16, 34 - 46 (1972).

The objectives of the water vapor investigations, to be carried out during the Viking missions, are described in the light of our current knowledge of Martian atmospheric conditions and the diurnal and seasonal variations of the water abundance. A discussion is given of the relative merits of the different experimental approaches which can be adopted for this purpose, followed by a summary of the detection limits achievable in the available spectral regions.

097.011 Infrared thermal mapping experiment: The Viking Mars Orbiter. H. H. Kieffer, G. Neugebauer, G. Munch, S. C. Chase, Jr., E. Miner.
Icarus, Vol. 16, 47 - 56 (1972).

The Mars infrared thermal mapper (IRTM) will be carried on the scan platform of the orbiter of the Viking 1975 mission. The IRTM is a multichannel radiometer with several detectors in each of six spectral regions. This instrument will measure the reflected solar radiation and surface thermal emission from the area viewed by the orbiter imaging system with nominal 5 km resolution.

097.012 Radio science experiments: The Viking Mars Orbiter and Lander.

W. H. Michael, Jr., D. L. Cain, G. Fjeldbo, G. S. Levy, J. G. Davies, M. D. Grossi, I. I. Shapiro, G. L. Tyler.
Icarus, Vol. 16, 57 - 73 (1972).

The objective of the radio science investigations is to extract the maximum scientific information from the data provided by the radio and radar systems on the Viking Orbiters and Landers. Unique features of the Viking missions include tracking of the landers on the surface of Mars, dual-frequency S- and X-band tracking data from the orbiters, lander-to-orbiter communications system data, and lander radar data.

097.013 **Entry science experiments for Viking 1975.**
A. O. Nier, W. B. Hanson, M. B. McElroy, A. Seiff, N. W. Spencer.
Icarus, Vol. 16, 74 - 91 (1972).

A review is given of our present knowledge of the Martian atmosphere with special emphasis on the results obtained by the Mariner 4, 6, and 7 fly-bys. A description is given of the neutral gas mass spectrometer and retarding potential analyzer experiments to be performed as the lander enters the upper atmosphere and the experiments planned for determining atmospheric structure as the lander approaches the surface of the planet.

097.014 **Imaging experiment: The Viking Lander.**
T. A. Mutch, A. B. Binder, F. O. Huck, E. C. Levinthal, E. C. Morris, C. Sagan, A. T. Young.
Icarus, Vol. 16, 92 - 110 (1972).

The purpose of this article is to give the reader some familiarity with the Viking Lander Imaging Experiment, the equipment, and the probable scientific yield.

097.015 **Mass spectrometric analysis of organic compounds, water and volatile constituents in the atmosphere and surface of Mars: The Viking Mars Lander.**
D. M. Anderson, K. Biemann, L. E. Orgel, J. Oro, T. Owen, G. P. Shulman, P. Toulmin III, H. C. Urey.
Icarus, Vol. 16, 111 - 138 (1972).

An experiment centering around a mass spectrometer is described, which is aimed at the identification of organic substances present in the top 10 cm of the surface of Mars and an analysis of the atmosphere for major and minor constituents as well as isotopic abundances. In addition, an indication of the abundance of water in the surface and some information concerning the mineralogy can be obtained by monitoring the gases produced upon heating the soil sample.

097.016 **Biological experiments: The Viking Mars Lander.**
H. P. Klein, J. Lederberg, A. Rich.
Icarus, Vol. 16, 139 - 146 (1972).

Biological interest in the exploration of Mars is briefly described as is the biological experiments package to be flown as part of the Viking 1975 lander payload.

097.017 **The carbon-assimilation experiment: The Viking Mars Lander.**
N. H. Horowitz, J. S. Hubbard, G. L. Hobby.
Icarus, Vol. 16, 147 - 152 (1972).

The carbon-assimilation experiment detects life in soils by measuring the incorporation of carbon from ^{14}CO and $^{14}CO_2$ into organic matter. It is based on the premise that Martian life, if it exists, is carbonaceous and exchanges carbon with the atmosphere, as do all terrestrial organisms.

097.018 **Detection of metabolically produced labeled gas: The Viking Mars Lander.** G. V. Levin.
Icarus, Vol. 16, 153 - 166 (1972).

A qualitative, nonspecific method will test for life on Mars in 1976 by supplying radioactive substrates to samples of the planetary surface material. If microorganisms are present, they may assimilate one or more of the simple labeled compounds and produce radioactive gas. The compounds have been selected on the basis of biological theory and terrestrial results. The Mars experiment is described and simulated return data are given.

097.019 **The Gas Exchange Experiment for life detection: The Viking Mars Lander.** V. I. Oyama.
Icarus, Vol. 16, 167 - 184 (1972).

The Gas Exchange Experiment of the Viking mission accepts a sample of Martian soil, incubates this soil with nutrient medium, and periodically samples the enclosed atmosphere over this soil for the gases H_2, N_2, O_2, CH_4, Kr, and CO_2. These gases are analyzed by an automated gas chromatograph, and the data are transmitted to earth.

097.020 **Light scattering experiment: The Viking Mars Lander.**
W. V. Vishniac, G. A. Welty.
Icarus, Vol. 16, 185 - 195 (1972).

Microbial growth can be detected with great sensitivity by light scattering. Even the low level of bacterial multiplication in distilled water can be reliably detected. It is proposed to observe the turbidity of an aqueous phase in contact with Martian soil for changes which can be interpreted as growth.

097.021 **Meteorology experiments: The Viking Mars Lander.**
S. L. Hess, R. M. Henry, J. Kuettner, C. B. Leovy, J. A. Ryan.
Icarus, Vol. 16, 196 - 204 (1972).

The purposes, procedures, and nature of the planned meteorology experiment of Viking, 1976 are described. The elements to be measured are pressure, temperature, wind speed, wind direction, and water vapor content of the atmosphere. The interactions with other Viking experiments are outlined and candidate sensors are described.

097.022 **Seismic investigations: The Viking Mars Lander.**
D. L. Anderson, R. L. Kovach, G. Latham, F. Press, M. N. Toksoz, G. Sutton.
Icarus, Vol. 16, 205 - 216 (1972).
The purpose of the seismic experiment is to provide preliminary information relevant to the structure, composition, and dynamics of the interior of Mars.

097.023 **Martian physical properties experiments: The Viking Mars Lander.**
R. W. Shorthill, R. E. Hutton, H. J. Moore, R. F. Scott.
Icarus, Vol. 16, 217 - 222 (1972).

The Viking Lander and its subsystems will be used in a manner similar to that used by the Surveyor program to define properties of the Martian "soil". Data for estimates of bearing strength, cohesion, angle of internal friction, porosity, grain size, adhesion, thermal inertia, dielectric constants, and homogeneity of the Martian surface materials will be collected.

097.024 **Magnetic properties investigation: The Viking Mars Lander.** R. B. Hargraves, N. Petersen.
Icarus, Vol. 16, 223 - 227 (1972).

In conjunction with the lander imaging system, a permanent magnet array will be used to detect the presence, color, relative abundance, and possibly the composition of magnetic particles in Martian surface material. The data obtained will be used as clues to the composition and degree of differentiation of the solid planet, and the extent of surface-atmosphere interaction.

097.025 **Mars as an active planet: The view from Mariner 9.**
A. L. Hammond.
Science, Vol. 175, 286 - 287 (1972).

097.026 **Mariner 9 science experiments: Preliminary results.**
R. H. Steinbacher, A. Kliore, J. Lorell, H. Hipsher,
C. A. Barth, H. Masursky, G. Münch, J. Pearl, B. Smith.
Science, Vol. 175, 293 - 294 (1972).

097.027 **Mariner 9 television reconnaissance of Mars and its satellites: Preliminary results.**
H. Masursky, R. M. Batson, J. F. McCauley, L. A. Soderblom,
R. L. Wildey, M. H. Carr, D. J. Milton, D. E. Wilhelms, B. A.
Smith, T. B. Kirby, J. C. Robinson, C. B. Leovy, G. A. Briggs,
T. C. Duxbury, C. H. Acton, Jr., B. C. Murray, J. A. Cutts,
R. P. Sharp, S. Smith, R. B. Leighton, C. Sagan, J. Veverka,
M. Noland, J. Lederberg, E. Levinthal, J. B. Pollack, J. T.
Moore, Jr., W. K. Hartmann, E. N. Shipley, G. de Vaucouleurs, M. E. Davies.
Science, Vol. 175, 294 - 305 (1972).

097.028 **Infrared spectroscopy experiment on the Mariner 9 mission: Preliminary results.**
R. A. Hanel, B. J. Conrath, W. A. Hovis, V. G. Kunde, P. D.
Lowman, J. C. Pearl, C. Prabhakara, B. Schlachman, G. V.
Levin.
Science, Vol. 175, 305 - 308 (1972).

097.029 **Infrared radiometry experiment on Mariner 9.**
S. C. Chase, Jr., H. Hatzenbeler, H. H. Kieffer, E.
Miner, G. Münch, G. Neugebauer.
Science, Vol. 175, 308 - 309 (1972).

The brightness temperatures at 10 and 20 micrometers measured by the Mariner 9 infrared radiometer differ substantially from those predicted by the radiometer results of Mariners 6 and 7. The results indicate a significant latitude-dependent contribution of the atmospheric dust to the observed thermal emission.

097.030 **Mariner 9 ultraviolet spectrometer experiment: Initial results.**
C. A. Barth, C. W. Hord, A. I. Stewart, A. L. Lane.
Science, Vol. 175, 309 - 312 (1972).

097.031 **Mariner 9 S-band Martian occultation experiment: Initial results on the atmosphere and topography of Mars.**
A. J. Kliore, D. L. Cain, G. Fjeldbo, B. L. Seidel, S. I. Rasool.
Science, Vol. 175, 313 - 317 (1972).

097.032 **Mariner 9 celestial mechanics experiment: Gravity field and pole direction of Mars.**
J. Lorell, G. H. Born, E. J. Christensen, J. F. Jordan, P. A.
Laing, W. L. Martin, W. L. Sjogren, I. I. Shapiro, R. D. Reasenberg, G. L. Slater.
Science, Vol. 175, 317 - 320 (1972).

097.033 **Martian dust storm: Its depth on 25 November 1971.** T. D. Parkinson, D. M. Hunten.
Science, Vol. 175, 323 (1972).

097.034 **Mars: An evolving atmosphere.** M. B. McElroy.
Science, Vol. 175, 443 - 445 (1972).

Photochemical reactions in the martian exosphere produce fast atoms of oxygen, carbon, and nitrogen and provide large escape fluxes of these elements. They appear to play a crucial role in the evolution of the martian atmosphere.

097.035 **Size frequency distribution of Martian craters and relative age of light and dark terrains.**
A. Woronow, E. A. King, Jr.
Science, Vol. 175, 755 - 757 (1972).

Light and dark terrains in and around Meridiani Sinus, mapped on the imagery from Mariner 6 and Mariner 7, were found to have significantly different cumulative size frequency distributions of craters.

097.036 **On the temperature conditions of the upper cover of Mars during the opposition 1971.**
V. D. Krotikov, O. B. Shchuko.
Astron. Zhurn. Akad. Nauk SSSR, Vol. 49, 165 - 169 (1972). In Russian. English translation in Soviet Astron. AJ, Vol. 16, No. 1.

The temperature conditions of the upper cover of Mars during the opposition 1971 are considered theoretically. Temperature maps of the illuminated part of the surface are obtained. Diurnal temperature variations at different areographic latitudes are presented as a harmonic series.

097.037 **Identification of carbon dioxide frost on the Martian polar caps.** H. P. Larson, U. Fink.
Astrophys. Journ., (*Letters*), Vol. 171, L91 - L95 (1972).

A spectrum of the Martian south polar cap shows 11 narrow features attributed to the composition of the polar cap. Through comparison with laboratory spectra all 11 have been identified as solid CO_2 absorptions.

097.038 **Mariner 6 and 7 ultraviolet spectrometer experiment: Implications of CO_2^+, CO, and O airglow.**
A. I. Stewart.
Journ. Geophys. Res., Vol. 77, 54 - 68 (1972).

The Mariner 6 and 7 ultraviolet spectrometer experiments observed intense emissions from CO, O, and CO_2^+ in the Martian airglow. Analysis shows that they are excited predominantly by the absorption of solar EUV photons by CO_2 and constitute a major energy-loss mechanism for the thermosphere. Models of the thermospheric temperature profile and the airglow layer that demonstrate the effects of neutral chemistry and ionospheric composition are developed.

097.039 **Production of CO ($a^3\Pi$) and other metastable fragments by electron impact dissociation of CO_2.**
W. C. Wells, W. L. Borst, E. C. Zipf.
Journ. Geophys. Res., Vol. 77, 69 - 75 (1972).

Recent observations of the intense CO Cameron band emission in the upper Martian atmosphere by Mariner 6 and 7 have generated a renewed interest in the dissociative excitation of CO_2 by both electron impact and photodissociation. To determine the role played by these processes, it is necessary to know the absolute collision and absorption cross sections. To this end we have studied the production of metastable fragments by electron impact on CO_2 in an attempt to arrive at an absolute cross section for the production of the CO ($a^3\Pi$) state.

097.040 **Internal constitution of Mars.** D. L. Anderson.
Journ. Geophys. Res., Vol. 77, 789 - 795 (1972).

The present astronomical data for Mars require that it have a dense core. The size of the core and its density can be traded off. By using the density of pure iron and the density of pure troilite (FeS), its radius can be considered to lie between 0.36 and 0.60 of the radius of the planet. The zero-pressure density of the mantle is confined to the range 3.54–3.49 g/cm^3. Mars contains 25-28 % iron, independent of assumptions about the over-all composition or distribution of the iron. Meteorite models for Mars that contain 25 wt % iron and 12 wt % core have been constructed.

097.041 **The surfaces of Mars and Venus.** C. Sagan.
Bull. American Astron. Soc., Vol. 3, 465 (1971).

097.042 **Mars: High-resolution radar topography.**
G. H. Pettengill, A. E. E. Rogers, I. I. Shapiro.
Bull. American Astron. Soc., Vol. 3, 466 (1971). – Abstr. AAS.

097.043 **Two-dimensional digital vidicon photometry of Mars.** T. B. McCord, J. A. Westphal.

Bull. American Astron. Soc., Vol. 3, 466 (1971). – Abstr. AAS.

097.044 Ephemeris of Mars for Mariner 9 based on radar range and optical data.
D. A. O'Handley, J. H. Lieske.
Bull. American Astron. Soc., Vol. 3, 467 (1971). – Abstr. AAS.

097.045 Continuum brightness temperature of Mars and Mercury at 21 cm wavelength.
F. H. Briggs, F. D. Drake.
Bull. American Astron. Soc., Vol. 3, 475 (1971). – Abstr. AAS.

097.046 Improved figure of Mars from Mariner radio occultation measurements. A. J. Kliore.
Bull. American Astron. Soc., Vol. 3, 498 (1971). – Abstr. AAS.

097.047 Mars opnieuw bezocht. F. P. Israel.
Hemel en Dampkring, Vol. 70, 3 - 7 (1972).

097.048 Mars door het oog van de Mariner 9. F. P. Israel.
Hemel en Dampkring, Vol. 70, 41 - 48 (1972).

097.049 De omlooptijd van een Mars-satelliet. J. Meeus.
Hemel en Dampkring, Vol. 70, 49 (1972).

097.050 Die Mars-Opposition 1971 in photographischer Sicht. M. Unkel.
SuW, Vol. 11, 23 (1972).

097.051 Mariner 6 und 7 – ein Rückblick. W. Carnuth.
SuW, Vol. 11, 41 - 42 (1972).

097.052 The Martian bow wave–Theory and observation.
J. R. Spreiter, A. W. Rizzi.
Planet. Space Sci., Vol. 20, 205 - 208 (1972).
The relationship between the trajectory of Mariner 4 in its fly-by of Mars and the calculated location of a proposed Martian bow arising from interaction of the solar wind and the ionosphere has been refined by inclusion of aberration effects of the planet's motion about the sun. The modified theory indicates that the bow wave was crossed twice during a 3-hr interval following the time of closest approach of Mariner 4 to Mars, instead of being missed slightly by the spacecraft as indicated in our previous paper.

097.053 Three spacecraft study the red planet.
R. N. Watts, Jr.
Sky Telescope, Vol. 43, 14 - 17 (1972).

097.054 Telescopic observations of Mars in 1971 – III.
G. de Vaucouleurs.
Sky Telescope, Vol. 43, 20 - 21 (1972).

097.055 Some Mariner 9 observations of Mars.
R. N. Watts, Jr.
Sky Telescope, Vol. 43, 208 - 213 (1972).

097.056 Der Mars ist ein staubiger Planet. H. Heuseler.
Umschau, 72. Jahrgang, p. 119 - 121 (1972).
The first results of the Mars orbiters Mariner 9 (USA), Mars 2 and 3 (USSR) about dust on the ground and in the atmosphere, vulcanism, the ice cap, the temperatures, and Phobos and Deimos are presented.

097.057 The great opposition of Mars in 1971. A. Bonov.
Mat. i fizika (NRB), Vol. 14, No. 4, p. 69 - 70 (1971). In Bulgarian.

097.058 On the nature of the Hellas area on Mars.

V. D. Davydov.
Astron. vestn., Vol. 5, 232 - 239 (1971). In Russian. – Abstr. in Referativ. Zhurn. 51. Astron., 3.51.333 (1972).

097.059 Mars from orbit – 1, 2.
Spaceflight, Vol. 14, 68 - 70, 118 - 120 (1972).

097.060 Minor planets and related objects. VIII. Deimos.
B. Zellner.
Astron. Journ., Vol. 77, 183 - 185 (1972).
Twelve polarization measurements of Deimos in blue light were obtained at phase angles between 5° and 31°. A well-developed negative branch, with minimum polarization approximately −1.5% near phase angle 11°, indicates that Deimos is not bare rock but has a particulate surface layer of dark dust or powder. The geometrical albedo of Deimos in blue light should be 0.07±0.01, and the radius 5.4±0.8 km.

097.061 On the origin of the polar caps of Mars.
V. I. Aleshin, O. B. Shchuko.
Astron. Zhurn. Akad. Nauk SSSR, Vol. 49, 406 - 412 (1972). In Russian. English translation in Soviet Astron. AJ, Vol. 16, No. 2.

097.062 Analysis of large-scale Martian topography variations–I. Data preparation from earth-based radar, earth-based CO_2 spectroscopy, and Mariners 6 and 7 CO_2 spectroscopy. R. A. Wells.
Geophys. Journ. Roy. Astron. Soc., Vol. 27, 101 - 133 (1972).
Primary problems associated with the preparation of a systematic distribution of surface heights from the four data sets involve the determination of a zero altitude level common to all sources and the reorganization of the data points onto a regular grid network. The results of this data combination process are the production of a surface height contour map and the preparation of a table of surface heights spaced on a $5° \times 5°$ grid interval. Only the contour map and difficulties associated with the data preparation are presented with this article.

097.063 Radiative transfer in the atmosphere of Mars and that of Venus above the cloud deck.
R. D. Cess, V. Ramanathan.
Journ. Quant. Spectrosc. Radiat. Transfer, Vol. 12, 933 - 945 (1972).
The dimensionless parameters which describe radiative transfer within a carbon dioxide atmosphere are shown to yield the same simplifications when applied to the atmospheres of Mars and Venus above the cloud deck. The radiative-transfer formulation correspondingly predicts a middle atmosphere temperature for Mars which is somewhat lower than previous theoretical estimates. Diurnal changes in the atmospheric temperature profile for Venus, which result from the time dependence of solar heating, are also considered. It is shown that there exists a region above the cloud deck, exceeding 20 km in height, within which atmospheric temperature is independent of time. In the middle atmosphere, assuming a four-day rotation for this portion of the atmosphere, a diurnal temperature variation of 33°K is predicted.

097.064 On possibilities of thermic sounding of the Martian atmosphere.
Yu. M. Timofeev, O. M. Pokrovskij, T. A. Dvorovik.
Kosm. Issled., Vol. 10, 92 - 99 (1972). In Russian.

097.065 Mars: The lineament systems.
A. B. Binder, D. W. McCarthy, Jr.
Science, Vol. 176, 279 - 281 (1972).
Analysis of the Mariner 4, Mariner 6, and Mariner 7 photographs shows that Mars has at least two distinct types of lineament systems. The most prominent is a well-developed

global-type system. The second consists of radial and concentric lineaments associated with the Hellas and south polar basins.

097.066 Relief-Darstellung der Marsoberfläche.
R. A. Wells.
Umschau, 72. Jahrgang, p. 293 - 295 (1972).

Altitude information of the grosser features of the surface has been derived from the two techniques: direct radar-ranging from the earth, and infrared CO_2 spectroscopy of the Martian atmosphere using both earth-based and onboard spacecraft spectrometers. The resulting topographic contour map shows the presence on Mars of block and basin features that are comparable in areal extent to the continents and ocean basins of the earth.

097.067 Das Wettergeschehen auf Mars, Venus und Jupiter.
A. Ghazi.
Umschau, 72. Jahrgang, p. 327 (1972).

The atmospheric circulation of Mars, Venus and Jupiter are explained. In principle this circulation is the result of the difference between the absorption of solar radiation as an energy source and the dissipation of kinetic energy.

097.068 Mariner 6 and 7 ultraviolet spectrometer experiment: Photometry and topography of Mars.
C. W. Hord.
Icarus, Vol. 16, 253 - 280 (1972).

The characteristic spectrum of nonpolar regions on Mars is discussed. An interpretation of the variation of the intensity at 3050 Å due to surface pressure variations is given. A comparison of the ultraviolet observations with pressures inferred by the infrared spectrometer is used to test this interpretation and to determine physical parameters of Mars which lead to the observed reflectance. Earth based pressures obtained by broadening of the 1μ band of CO_2 and radar topography are compared with the ultraviolet results.

097.069 Effects of surface roughness on the photometric properties of Mars. J. Veverka, L. Wasserman.
Icarus, Vol. 16, 281 - 290 (1972).
Starting with a macroscopically flat surface which at each point scatters according to the Minnaert law we investigate the photometric effects of increasing the large scale roughness of the Martian surface.

097.070 A new look at the Martian "violet haze" problem. I. Syrtis Major-Arabia, 1969.
P. B. Boyce, D. T. Thompson.
Icarus, Vol. 16, 291 - 303 (1972).

Photographic and photoelectric data obtained in 1969 have been combined to provide a relatively accurate and complete description of the so-called "blue clearing" phenomenon on Mars. The dark feature Syrtis Major and its surroundings are the subject of the present study.

097.071 Electron temperature in the Martian ionosphere.
M. Shimizu, O. Ashihara.
Publ. Astron. Soc. Japan, Vol. 24, 201 - 211 (1972).

Electron temperatures in the Martian ionosphere at the times of Mariner 4 and 6 missions are calculated in the continuous slowing down approximation by using new molecular data.

097.072 Beobachtungen während der Marsopposition 1971.
F. Kimberger.
SuW, Vol. 11, 133 - 136 (1972).

097.073 Mars' great storm of 1971.
C. F. Capen, L. J. Martin.
Sky Telescope, Vol. 43, 276 - 279 (1972).

097.074 Mariner 9 resumes work.   R. N. Watts, Jr.
Sky Telescope, Vol. 43, 300 - 302 (1972).

097.075 Observing Mars IV – The Martian central meridian.
C. F. Capen, V. W. Capen.
Strolling Astronomer, Vol. 23, 122 - 124 (1972).

097.076 The relative age of the transition zone between Hellas and the Martian cratered terrain.
C. H. Thorman, G. G. Goles.
Earth Planet. Sci. Letters, Vol. 15, 45 - 52 (1972).

Previous suggestions concerning the relationship of craters to scarps and ridges and the antiquity of Hellas relative to the cratered terrain seem questionable. An origin of Hellas by impact with later isostatic subsidence to account for the characteristics of the transition zone seems to be the best model, and may also account for the featureless floor of the basin.

097.077 Mapping Mars from Mariner pictures.   C. A. Cross.
Journ. British Astron. Ass., Vol. 82, 206 - 210 (1972).

097.078 De afplatting van de planeet Mars.   R. van Gent.
Hemel en Dampkring, Vol. 70, 100 - 101 (1972).

097.079 Amateur waarnemingen van Mars in 1971.
H. Nieuwenhuis.
Hemel en Dampkring, Vol. 70, 137 - 138 (1972).

097.080 Martian "peak".   I. K. Koval.
Priroda, No. 4.72, p. 2 - 9 (1972). In Russian.

097.081 Mars im Jahr 1971.   E. Wiedemann.
Orion Schaffhausen, 30. Jahrgang, p. 11 - 14 (1972).

097.082 Mars-Aufnahmen 1971 der Sternwarte Bochum.
T. Droste, P. Riepe, G. Weber, H.-U. Keller.
Orion Schaffhausen, 30. Jahrgang, p. 14 - 15 (1972).

097.083 Mariner 9 – Marsaufnahmen.
SuW, Vol. 11, 43 - 46 (1972). – Photo: USIS (U.S. Information Service).

097.084 Mid-infrared remote sensing of composition and its application to Mars.
G. R. Hunt, L. M. Logan, J. W. Salisbury.
Bull. American Astron. Soc., Vol. 4, 269 (1972). – Abstr. AAS.

097.085 Polarization of Deimos.   B. H. Zellner.
Bull. American Astron. Soc., Vol. 4, 269 (1972). Abstr. AAS.

097.086 Martian doublet craters.
V. R. Oberbeck, M. Aoyagi.
Journ. Geophys. Res., Vol. 77, 2419 - 2432 (1972).

A large number of Mars craters are nearly tangential to other craters. They occur in clusters or as isolated crater doublets. Clusters and doublets could be caused by meteoroid breakup resulting from stresses induced in the meteoroid by the gravitational field of Mars. Calculations are provided that show the required relationships between mass, entry angle, impact velocity, and tensile strengths of the meteoroid, and the separation distance between the centers of the resulting craters relative to crater sizes. It is concluded that, under certain conditions, doublets should be produced on Mars as a direct result of breakup of an impacting meteoroid.

097.087 Regional variations in degradation and density of Martian craters.   G. E. McGill, D. U. Wise.
Journ. Geophys. Res., Vol. 77, 2433 - 2441 (1972).

Martian craters visible on Mariner 6 and 7 imagery show a spectrum of topographic types from very fresh to highly degraded. A method of numerical scoring of rim, wall, and floor is proposed to yield a degradation number to classify each crater. Plots of degradation class versus density of large (diameter >16 km) craters are similar for all four regions studied, whereas similar plots for small (diameter 1–8 km) craters show marked differences between regions.

097.088 Dynamical characteristics of Phobos and Deimos. J. A. Burns.
Rev. Geophys. Space Phys., Vol. 10, 463 - 483 (1972).

The orbital properties of the two small Martian satellites, Phobos and Deimos, are discussed, as well as those dynamical constants of Mars that can be determined from the satellite orbits. It is proposed that the Martian satellites were born at the same time as Mars from equatorial dust clouds.

097.089 Mariner 9 – Marsaufnahmen.
SuW, Vol. 11, 152 - 154 (1972). – Photos USIS (U. S. Information Service).

097.090 Mars et ses satellites vus par Mariner 9. B. Morando.
L'Astronomie, 86e année, p. 231 - 241 (1972).

097.091 The spectrum of Mars between 8 and 13 microns. J. P. Verdet, Y. Zéau, J. Gay, T. Encrenaz, F. Sèvre.
Astron. Astrophys., Vol. 19, 159 - 163 (1972).

During June and July 1971, observations of Mars and the moon were performed at the Jungfraujoch Scientific Station, with a Michelson interferometer. We found a mean temperature of $230 \pm 15°K$ over the whole disk and 5 ± 2 mb as a partial pressure of CO_2 at the surface.

097.092 News of Mars. V. I. Moroz, L. V. Ksanfomaliti.
Zemlya i Vselennaya, 1972, No. 3, p. 3 - 9.
In Russian.

097.093 The Martian sky. V. V. Shevchenko.
Zemlya i Vselennaya, 1972, No. 3, p. 10 - 12.
In Russian.

097.094 The Martian star map. A. D. Marlenskij.
Zemlya i Vselennaya, 1972, No. 3, p. 12 - 13.
In Russian.

097.095 "Fire" on Mars. V. D. Davydov.
Zemlya i Vselennaya, 1972, No. 3, p. 58 - 63.
In Russian.

097.096 Martian topography from earth-based radar, earth-based CO_2, and Mariner 6 and 7 CO_2 measurements. R. A. Wells.
Astron. Zhurn. Akad. Nauk SSSR, Vol. 49, 607 - 612 (1972). In Russian. English translation in Soviet Astron. AJ, Vol. 16, No. 3.

All data of Martian topographical variations available at present have been combined and smoothed to the same spatial resolution limit. The total height range on Mars is found to be nearly 18 km corresponding to a CO_2 pressure range from 2 mb at 12 km to 10 mb at – 6 km.

097.097 Spectrophotometric studies of the photometric function, composition, and distribution of the surface materials of Mars. A. B. Binder, J. C. Jones.
Journ. Geophys. Res., Vol. 77, 3005 - 3020 (1972).

This paper presents the results of a program of spectrophotometric mapping of the surface of Mars and the results of a laboratory comparison program. The data derived from these programs have provided additional information on the

composition and physical state of surface materials and new information on the distribution and photometric function of the surface units.

097.098 Mars in focus. T. Dickinson.
Journ. Roy. Astron. Soc. Canada, Vol. 66, 168 - 170 (1972).

097.099 Meteorological observations of Mars during the 1969 opposition. S. Miyamoto.
Contr. Inst. Astrophys., Kwasan Obs., Univ. Kyoto, No. 184, 63 pp. (1970).

This is the seventh report of our series about the meteorological observation of Mars during the 1969 opposition. In this opposition, we secured records of meteorological phenomenae covering the Martian season from northern late summer to early winter, or from southern late winter to early summer.

097.100 Phobos og Deimos. T. Hansen.
Astron. Tidssk., Årg. 5, p. 83 - 84 (1972).

097.101 About the distribution function of the normal albedo of the surface of Mars.
Yu. V. Aleksandrov, D. F. Lupishko.
Astron. vestn., Vol. 6, 9 - 12 (1972). In Russian. – Abstr. in Referativ. Zhurn. 51. Astron., 7.51.201 (1972).

097.102 Perihelna opozicija Marsa 1971. godine. [Perihelion opposition of Mars in 1971.] K. Pavlovski.
Vasiona, Vol. 20, 1 - 7 (1972).

097.103 Electron collision frequency in Martian ionosphere. D. C. Agarwal.
Indian Journ. Phys., Vol. 44, 336 - 338 (1970).

Calculations of the F_2 region collision frequency of the Martian atmosphere are presented.

097.104 The present state of research of Mars. I. K. Koval.
Cesk. Cas. Fis. A, Vol. 21, 620 - 628 (1971).
In Czech.

The results of research of Mars accomplished with the aid of cosmic devices as well as by observations from the earth during the year 1971 are reviewed. The following topics are presented: chemical composition of the atmosphere, the total optical thickness of the atmosphere and atmospheric pressure, the shape of the surface and some properties of the surface layer.

097.105 Martian topography. A. B. Binder.
Modern Geol. (GB), Vol. 3, No. 2, p. 61 - 68 (1972).

New radar topographic data obtained between +2° and +12° latitude on Mars have been combined with earlier radar data obtained near +22° to construct a topographic map between the equator and +30°. The data show that the highest areas are deserts, the dark areas predominantly occur in low zones and on steep slopes, and the surface markings are related to the topography.

097.106 The Martian satellites. S. F. Singer.
IAU Colloquium No. 12, (see 012.027), p. 399 - 405 (1971/72).

097.107 Radio occultation measurements of the Mars atmosphere with Mariners 6 and 7.
J. S. Hogan, R. W. Stewart, S. I. Rasool.
Radio Sci., Vol. 7, 525 - 537 (1972).

An analysis of the Mariner 6 and 7 occultation data has been completed. Final profiles of temperature, pressure, and electron density have been obtained for the Mariner 6 and 7 entry and exit cases, and results are presented for both the lower atmosphere and the ionosphere.

097.108   The effect of dust on the temperature of the Martian atmosphere.   P. J. Gierasch, R. M. Goody.
Journ. Atmosph. Sci., Vol. 29, 400 - 402 (1972).

097.109   Mars since the dust settled.   E. Burgess.
New Scient. Sci. Journ., No. 784, Vol. 53, 420 - 423 (1972).
      Discusses the pictures of the Martian surface sent back by Mariner 9 since the dust storm cleared.

097.110   Die erste Landung auf dem Mars. Flug und Bedeutung der automatischen Station "Mars 3".
W. Lewski.
Universitas, [Wiss. Verlagsgesellschaft, Stuttgart (Germany)], Vol. 27, 275 - 280 (1972).

097.111   Neutral composition of the upper atmosphere of Mars as determined from the Mariner UV spectrometer experiments.   G. E. Thomas.
Journ. Atmosph. Sci., Vol. 28, 859 - 868 (1971).

097.112   Results and interpretation of the S-band occultation experiments on Mars and Venus.
S. I. Rasool, R. W. Stewart.
Journ. Atmosph. Sci., Vol. 28, 869 - 878 (1971).

097.113   Dissociation of CO_2 in the Martian atmosphere.
M. B. McElroy, J. C. McConnell.
Journ. Atmosph. Sci., Vol. 28, 879 - 884 (1971).

097.114   The electron distributions in the Mars and Venus upper atmospheres.   R. W. Stewart.
Journ. Atmosph. Sci., Vol. 28, 1069 - 1073 (1971).

097.115   Aeronomy of CO_2 atmospheres: A review.
T. M. Donahue.
Journ. Atmosph. Sci., Vol. 28, 895 - 900 (1971).

Mars: Lost in the dust.
Nature, Vol. 235, 67 (1972).

Vistas of Mars.
Spaceflight, Vol. 14, 265 - 267 (1972).

New experiment to investigate Mars.
Vasiona, Vol. 20, 24 (1972). In Serbo-Croatian.

The Mariner 6 and 7 pictures of Mars.
See Abstr. 003.044.

Mars − the mysterious planet.
See Abstr. 003.080.

Broadening of infrared absorption lines at reduced temperatures, II. Carbon monoxide in an atmosphere of carbon dioxide.   See Abstr. 022.099.

Observations of the region of interaction between the solar wind plasma and Mars.   See Abstr. 074.071.

On the effect of departures from spherical symmetry upon the bistatic-radar occultation experiment.
See Abstr. 091.012.

Aeronomy of CO_2 atmospheres.
See Abstr. 091.027.

The photolysis of CO_2 at wavelengths exceeding 1740 Å.   See Abstr. 091.036.

On inferring elastic properties of deep planetary interiors: Moon and Mars.   See Abstr. 094.105.

098 Minor Planets

098.001 **Photopolarimetric observations of the minor planet Flora.** J. Veverka.
Icarus, Vol. 15, 454 - 458 (1971/72).

Flora is an almost spherical asteroid. In order to clarify the question of its variability as well as to obtain the phase coefficients of Flora in several colors, photometric observations were made during the 1968–69 opposition using a standard UBV photometer on the Harvard 16-in. telescope. In addition, the polarization was measured using the Harvard 61-in. telescope and the Harvard two-channel Wollaston prism polarimeter.

098.002 **Photographische Kleinplaneten-Beobachtung durch Amateure.** H. Vehrenberg.
SuW, Vol. 11, 36 - 38 (1972).

098.003 **Observations of minor planets at the Nikolayev Observatory during 1968.**
V. I. Voronenko, F. F. Kalihevich.
Byull. Inst. Teoret. Astron., *Leningrad*, Vol. 13, 125 - 129 (1972). In Russian.

098.004 **Observations of minor planet 1620 Geographos at the Crimean Astrophysical Observatory.**
N. S. Chernykh.
Byull. Inst. Teoret. Astron., *Leningrad*, Vol. 13, 130 - 131 (1972). In Russian.

098.005 **Observations of minor planets made at the Crimean Astrophysical Observatory (17th report).**
L. I. Chernykh.
Byull. Inst. Teoret. Astron., *Leningrad*, Vol. 13, 132 - 136 (1972). In Russian.

098.006 **Some photometric parameters of the minor planet 2 Pallas.** R. Burchi.
Mem. Soc. Astron. Italiana, Nuova Ser., Vol. 43, 27 - 32 (1972).

The minor planet 2 Pallas has been observed photoelectrically during the opposition of 1970. A Fourier analysis of its light curves confirms the period of about $10^h 5$. Other photometric parameters, as the phase function, the phase coefficient, the absolute magnitude and the albedo, have been computed. Finally, the mean radius $\bar{r}$ has been computed.

098.007 **Fast moving object Kohoutek.**
Kometn. Tsirk., *Kiev*, No. 125 (1972). In Russian.

098.008 **Five fast moving asteroid objects of Gehrels.**
Kometn. Tsirk., *Kiev*, No. 129 (1972). In Russian.

098.009 **Precise positions of minor planets observed during 1967–1970.** C. Cristescu, V. I. Vlăsceanu,
I. Ghețu, G. Bocșa.
Stud. Cerc. Astron., Vol. 17, 85 - 159 (1972). In Romanian.

098.010 **A minor planet on an 'inner grand tour'.**
P. M. Janiczek, P. K. Seidelmann, R. L. Duncombe.
Bull. American Astron. Soc., Vol. 4, 263 (1972). – Abstr. AAS.

098.011 **Capture resonance of the asteroid 1685 Toro by the earth.** L. Danielsson, W.-H. Ip.
Science, Vol. 176, 906 - 907 (1972).

The asteroid 1685 Toro has its perihelion inside the earth's orbit and a period which is 8/5 of that of the earth. A calculation of Toro's orbit covering 200 years shows that this asteroid at present is captured in resonance with the earth. The capture is due to the gravitational interaction at close encounters between the bodies.

098.012 **Massenbestimmung von Planetoïden.** J. Schubart.
Mitt. Astron. Ges., No. 31, p. 182 - 183 (1972).

098.013 **Das photometrische Verhalten der Pallas.**
H. Haupt.
Mitt. Astron. Ges., No. 31, p. 208 - 209 (1972).

098.014 **Observations photographiques de petites planètes, effectuées à l'astrographe double de 40 cm au cours de l'année 1970.** H. Debehogne.
Bull. Astron. Obs. Roy. Belgique, Vol. 7, 274 - 279 (1972).

098.015 **Apparitions of two unusual minor planets (#1685 Toro and #433 Eros) in August, 1972.**
R. G. Hodgson.
Strolling Astronomer, Vol. 23, 149 - 150 (1972).

098.016 **1971 UA.**
C. Torres, J. Petit, H. Wroblewski.
IAU Circ., No. 2385 (1972).

098.017 **Five objects Gehrels.** T. Gehrels.
IAU Circ., No. 2391 (1972).

098.018 **1948 EA.** E. Roemer, Z. M. Pereyra.
IAU Circ., No. 2393 (1972).

098.019 **1972 FA, 1972 FC, 1972 FD, 1972 FE.**
H. Giclas, K. Aksnes.
IAU Circ., No. 2397 (1972). – Designation of Gehrels' objects No. 1, 3, 4, and 5 (see IAU Circ., No. 2391 (1972)).

098.020 **1972 FA and 1972 FE.** E. Roemer, C. T. Kowal.
IAU Circ., No. 2402 (1972).

098.021 **1971 UA.** C. Torres, J. Petit, H. Wroblewski,
B. G. Marsden.
IAU Circ., No. 2403 (1972).

098.022 **1221 Amor.**
C. Torres, J. Petit, H. Wroblewski.
IAU Circ., No. 2409 (1972).

098.023 **1948 EA.** C. D. Vesely.
IAU Circ., No. 2415 (1972).

098.024 **Simulative experiment on the light curve of asteroids.** T. Kanda, M. Nukariya.
Tokyo Astron. Obs., Report No. 59, Vol. 15, 690 - 692 (1972). In Japanese.

098.025 **Rotation of the minor planet (15) Eunomia.**
M. Nukariya, N. Kobayashi.
Tokyo Astron. Obs., Report No. 59, Vol. 15, 710 - 732 (1972). In Japanese.

098.026 **Petites planètes photographiées en 1970 sur plaques Kodak 103 AO.**
A. Ghezloun, J. Pham-Van, Saglio.
Ann. Obs. Astron. Alger, Vol. 3, Fasc. 3, p. 23 (1972).

098.027 **The perturbed motion of asteroids in proximity.**
J. Lazović.
Publ. Dep. Astron. Univ. Beograd, No. 3, p. 29 - 35 (1971).

Two asteroids are in proximity when they find themselves on their orbits in the position of their least distance. We consider the results of our examination of perturbed motion of one asteroid under the influence of another one at an interval that includes their proximity. We applied our calculations to the pair of quasicomplanar orbits of asteroids 589 Croatia and 1564 Srbija. For this pair we found that their shortest distance is 0.000498 AU.

098.028 On the accuracy of determining the orbital elements of minor planets from combined observations of them.   V. N. Bojko.
Vestn. Leningr. un-ta, 1972, No. 1, p. 133 - 137. In Russian.
Abstr. in Referativ. Zhurn. 51. Astron., 7.51.105 (1972).

098.029 Minor Planet Circulars, (MPC), Nos. 3293 - 3352 (1971).
Edited by Cincinnati Observatory, under the supervision of P. Herget.
    A repository of nearly all new data for numbered and unnumbered minor planets: Observations, elements and ephemerides, identifications, newly assigned numbers and names, occultations.

098.030 Astrometric observations.   E. Roemer.
IAU Colloquium No. 12, (see 012.027), p. 3 - 7 (1971/72).

098.031 The work at the Minor Planet Center.   P. Herget.
IAU Colloquium No. 12, (see 012.027), p. 9 - 12 (1971/72).

098.032 The use of asteroids for determinations of masses and other fundamental constants.   E. Rabe.
IAU Colloquium No. 12, (see 012.027), p. 13 - 23 (1971/72).

098.033 Diameter measurements of asteroids.   A. Dollfus.
IAU Colloquium No. 12, (see 012.027), p. 25 - 31 (1971/72).

098.034 Asteroid masses and densities.   J. Schubart.
IAU Colloquium No. 12, (see 012.027), p. 33 - 39 (1971/72).

098.035 The method of determining infrared diameters.   D. A. Allen.
IAU Colloquium No. 12, (see 012.027), p. 41 - 44 (1971/72).

098.036 Infrared observations of asteroids.   D. L. Matson.
IAU Colloquium No. 12, (see 012.027), p. 45 - 50 (1971/72).

098.037 A review of spectrophotometric studies of asteroids.   C. R. Chapman, T. V. Johnson, T. B. McCord.
IAU Colloquium No. 12, (see 012.027), p. 51 - 65 (1971/72).

098.038 Inferences from optical properties concerning the surface texture and composition of asteroids.   B. Hapke.
IAU Colloquium No. 12, (see 012.027), p. 67 - 77 (1971/72).

098.039 The physical meaning of phase coefficients.   J. Veverka.
IAU Colloquium No. 12, (see 012.027), p. 79 - 90 (1971/72).

098.040 Asteroid polarimetry: A progress report.   J. Veverka.
IAU Colloquium No. 12, (see 012.027), p. 91 - 94 (1971/72).

098.041 Physical studies of asteroids by polarization of the light.   A. Dollfus.

IAU Colloquium No. 12, (see 012.027), p. 95 - 116 (1971/72).

098.042 Photometric observations and reductions of lightcurves of asteroids.   R. C. Taylor.
IAU Colloquium No. 12, (see 012.027), p. 117 - 131 (1971/72).

098.043 Summary on orientations of rotation axes.   C. D. Vesely.
IAU Colloquium No. 12, (see 012.027), p. 133 - 140 (1971/72).

098.044 Lightcurve inversion and surface reflectivity.   A. A. Lacis, J. D. Fix.
IAU Colloquium No. 12, (see 012.027), p. 141 - 146 (1971/72).

098.045 Laboratory work on the shapes of asteroids.   J. L. Dunlap.
IAU Colloquium No. 12, (see 012.027), p. 147 - 154 (1971/72).

098.046 624 Hektor: a binary asteroid?   A. F. Cook.
IAU Colloquium No. 12, (see 012.027), p. 155 - 163 (1971/72).

098.047 Asteroid characteristics by radar.   R. M. Goldstein.
IAU Colloqtium No. 12, (see 012.027), p. 165 - 171 (1971/72).

098.048 Descriptive survey of families, Trojans, and jetstreams.   C. J. van Houten.
IAU Colloquium No. 12, (see 012.027), p. 173 - 175 (1971/72).

098.049 Proper elements, families, and belt boundaries.   J. G. Williams.
IAU Colloquium No. 12, (see 012.027), p. 177 - 181 (1971/72).

098.050 The Palomar-Leiden survey.   C. J. van Houten.
IAU Colloquium No. 12, (see 012.027), p. 183 - 186 (1971/72).

098.051 The distribution of asteroids in the direction perpendicular to the ecliptic plane.   T. Kiang.
IAU Colloquium No. 12, (see 012.027), p. 187 - 195 (1971/72).

098.052 Orbital selection effects in the Palomar-Leiden asteroid survey.   L. Kresák.
IAU Colloquium No. 12, (see 012.027), p. 197 - 210 (1971/72).

098.053 Asteroidal theories and experiments.   G. Arrhenius, H. Alfvén.
IAU Colloquium No. 12, (see 012.027), p. 213 - 223 (1971/72).

098.054 On the formation of the asteroids.   J. G. Hills.
IAU Colloquium No. 12, (see 012.027), p. 225 - 237 (1971/72).

098.055 Accumulation of chondrules on asteroids.   F. L. Whipple.
IAU Colloquium No. 12, (see 012.027), p. 251 - 256 (1971/72).

098.056 The alinement of asteroid rotation.   J. A. Burns.
IAU Colloquium No. 12, (see 012.027), p. 257 - 262 (1971/72).

098.057 Fragmentation and distribution of asteroids.   J. S. Dohnanyi.
IAU Colloquium No. 12, (see 012.027), p. 263 - 295 (1971/72).

098.058 Remarks on the size distribution of colliding and fragmenting particles.   L. W. Bandermann.
IAU Colloquium No. 12, (see 012.027), p. 297 - 303 (1971/72).

098.059 Internal constitution and mechanisms of asteroid

fragmentation.　A. Brecher.
IAU Colloquium No. 12, (see 012.027), p. 305 - 314 (1971/72).

098.060　A study of asteroid families and streams by computer techniques.　B. A. Lindblad, R. B. Southworth.
IAU Colloquium No. 12, (see 012.027), p. 337 - 352 (1971/72).

098.061　The profile of a jetstream.　L. Danielsson.
IAU Colloquium No. 12, (see 012.027), p. 353 - 362 (1971/72).

098.062　On the amount of dust in the asteroid belt.
F. L. Whipple.
IAU Colloquium No. 12 (see 012.027), p. 389 - 393 (1971/72).

098.063　Are meteors a tool for studying the asteroids? or vice versa?　R. E. McCrosky.
IAU Colloquium No. 12 (see 012.027), p. 395 - 397 (1971/72).

098.064　Trojans and comets of the Jupiter group.
E. Rabe.
IAU Colloquium No. 12, (see 012.027), p. 407 - 412 (1971/72).

098.065　Evolution of comets into asteroids?
B. G. Marsden.
IAU Colloquium No. 12, (see 012.027), p. 413 - 421 (1971/72).

098.066　A core-mantle model for cometary nuclei and asteroids of possible cometary origin.　Z. Sekanina.
IAU Colloquium No. 12, (see 012.027), p. 423 - 428 (1971/72).

098.067　Arguments for a mission to an asteroid.
H. Alfvén, G. Arrhenius.
IAU Colloquium No. 12, (see 012.027), p. 473 - 478 (1971/72).

098.068　Reasons for not having an early asteroid mission.
E. Anders.
IAU Colloquium No. 12, (see 012.027), p. 479 - 487 (1971/72).

098.069　Possible magnetic interaction of asteroids with the solar wind.　E. W. Greenstadt.
IAU Colloquium No. 12, (see 012.027), p. 567 - 575 (1971/72).

098.070　Feasibility of determining the mass of an asteroid from a spacecraft flyby.　J. D. Anderson.
IAU Colloquium No. 12, (see 012.027), p. 577 - 583 (1971/72).

098.071　Asteroid mass distribution measurement with gravity gradiometers.　R. L. Forward.
IAU Colloquium No. 12, (see 012.027), p. 585 - 593 (1971/72).

098.072　Estimate of particle densities and collision danger for spacecraft moving through the asteroid belt.
D. J. Kessler.
IAU Colloquium No. 12, (see 012.027), p. 595 - 605 (1971/72).

098.073　Asteroid detection from Pioneers F and G?
R. K. Soberman, S. L. Neste, A. F. Petty.
IAU Colloquium No. 12, (see 012.027), p. 617 - 631 (1971/72).

098.074　Observations in the asteroid belt with the imaging photopolarimeter of Pioneers F and G.
C. E. Kenknight.
IAU Colloquium No. 12, (see 012.027), p. 633 - 637 (1971/72).

098.075　Discovery and observation of close-approach asteroids.　E. Roemer.
IAU Colloquium No. 12, (see 012.027), p. 643 - 648 (1971/72).

098.076　Physical studies of minor planets: Future work.
T. Gehrels.
IAU Colloquium No. 12, (see 012.027), p. 653 - 659 (1971/72).

098.077　Physical parameters of asteroids and interrelations with comets.　T. Gehrels.
From plasma to planet. 21st Nobel symposium 1971, (see 012.028), p. 169 - 178 (1972).

098.078　On certain aerodynamic processes for asteroids and comets.　F. L. Whipple.
From plasma to planet. 21st Nobel symposium 1971, (see 012.028), p. 211 - 232 (1972).

098.079　The possibility of a trans-Saturnian belt of particulate matter.　B. A. Lindblad.
From plasma to planet. 21st Nobel symposium 1971, (see 012.028), p. 359 - 360 (1972).

098.080　Object Piksaev.　L. I. Chernykh, V. A. Piksaev.
Yamamoto Circ., No. 1747 (1972). In Japanese.

Twee nieuwe uitzonderlijke planetoiden.
Hemel en Dampkring, Vol. 70, 49 - 50 (1972).

Jacobian integral as a classificational and evolutionary parameter of interplanetary bodies.　See Abstr. 042.010.

Minor planets and related objects. VIII. Deimos.
See Abstr. 097.060.

Is water ice the major difference between comets and asteroids?　See Abstr. 102.024.

Structure of comets and the possible origin of faint asteroids.　See Abstr. 102.025.

Meteor and asteroid streams.
See Abstr. 104.045.

Interrelations of meteorites, asteroids, and comets.
See Abstr. 105.066.

Errata

098.901　Errata: 'Heldere kleine planeten en 1972'.
[Hemel en Dampkring, Vol. 69, 331 - 333 (1971)].
J. Meeus.
Hemel en Dampkring, Vol. 70, 52 (1972).

099 Jupiter

099.001 The atmospheres of Jupiter and Saturn.
V. G. Tejfel'.
Zemlya i Vselennaya, No. 1, p. 2 - 9 (1972). In Russian.

099.002 Decametric radio emission from Jupiter.
A. Cavaliere, A. Speranza.
Astrophys. Letters, Vol. 10, 9 - 12 (1972).
A new model for the control of the Jovian decametric radio emission by Io is proposed. The geometry of the model is based on the presence of two 'surface' sources on the planet, emitting independently of Io and modulated by the passage of Io. When Io sweeps past the sources, streams of particles associated with the lines of force passing through the satellite interact with the plasma waves in the primary sources, thus altering the direction, the time structure, the brightness, and the maximum frequency of the emission received at earth.

099.003 Vertical shear in the Jovian equatorial zone.
R. G. Layton.
Icarus, Vol. 15, 480 - 485 (1971/72).
The relative motions of features visible on photographs of Jupiter obtained with light of different colors may be interpreted as a vertical shear at visible cloud level in the Jovian atmosphere.

099.004 Narrow band characteristics of Jovian L-bursts.
C. H. Barrow.
Icarus, Vol. 15, 486 - 491 (1971/72).
Radio observations of Jupiter have been made during three apparitions using several receivers, each tuned to a slightly different frequently close to 18 MHz. Total power and left- and right-hand polarization components have been compared, for L-bursts, over various narrow frequency bands from 0 to 400 kHz and a list of characteristics is presented. Several interesting new features have been observed in addition to those listed by Douglas and Smith (1967) which are generally confirmed.

099.005 Absence of post-eclipse brightening of Io and Europa in 1970. F. W. Fallon, R. E. Murphy.
Icarus, Vol. 15, 492 - 496 (1971/72).
Observations of four eclipse reappearances of Io, three of Europa, and one of Ganymede were made from the Mauna Kea Observatory during the 1970 apparition. A single beam photometer was used on a 24-in. reflector for all of the observations. Wavelengths from $\lambda 4100$ to $\lambda 6500$ Å were used with aperture stops of 7 and 15 arc-sec. No post-eclipse brightening was observed on any of the satellites.

099.006 Determination of the mass of Jupiter using the motion of its ninth satellite and a Kalman-Bucy filter.
M. Garcia de Polavieja, C. Edelman.
Astron. Astrophys., Vol. 16, 66 - 71 (1972). – In French.
The aim of this work is the determination of the mass of Jupiter using the motion of its ninth satellite JIX. The discrete-time observations of this satellite are compared with the theoretical path by a sequential Kalman-Bucy filter. The equations of the satellite motion are integrated numerically by the Burlisch and Stoer method. The following result has been found for the reciprocal mass of Jupiter: 1047.352 ± 0.002 (p.e.).

099.007 Theoretical study of the Jovian spectrum between 6 and 14 microns. T. Encrenaz.
Astron. Astrophys., Vol. 16, 237 - 246 (1972).
The absorption coefficient in the Jovian atmosphere is calculated in the $6-14 \mu$ range as a function of frequency, using a line-by-line calculation of the contributions due to H_2, CH_4 and NH_3. In order to compute this absorption coefficient, the absolute intensities of the individual lines of the CH_4 and NH_3 bands are first estimated. The result is applied to the computation of the brightness temperature of Jupiter in the $6-14 \mu$ range, using different theoretical atmospheric models. Finally, the possible interpretation of experimental infra-red spectra at high resolution is briefly discussed.

099.008 Search for X rays from the planet Jupiter.
K. C. Hurley.
Journ. Geophys. Res., Vol. 77, 46 - 53 (1972).
Actively collimated balloon-borne scintillation counters employing a special phoswich anticoincidence technique were flown a total of 5 times from Palestine, Texas. Jupiter was observed for a total of 133 min, and an upper limit to the flux of X rays present at the observation time is 1.6×10^{-2} X rays/cm^2 sec in the energy range 30−100 kev. No X rays were being generated at the time of the observation.

099.009 Jupiter: Secular variations in its decimeter flux.
M. Klein, S. Gulkis, C. Stelzried, T. Sato.
Bull. American Astron. Soc., Vol. 3, 475 - 476 (1971).
Abstr. AAS.

099.010 Jupiter's decametric rotational period from flux density measurements. H. R. Miller, A. G. Smith.
Bull. American Astron. Soc., Vol. 3, 476 (1971). – Abstr. AAS.

099.011 Transhemispheric VLB interferometry of a Jovian decametric L burst.
M. A. Lynch, T. D. Carr, J. May.
Bull. American Astron. Soc., Vol. 3, 476 (1971). – Abstr. AAS.

099.012 Jovian ionosphere, Mark II.
S. S. Prasad, L. A. Capone.
Bull. American Astron. Soc., Vol. 3, 476 (1971). – Abstr. AAS.

099.013 The occultation of Beta Scorpii by Jupiter and Io. I. Jupiter.
W. B. Hubbard, R. E. Nather, D. S. Evans, R. G. Tull, D. C. Wells, G. W. van Citters, B. Warner, P. Vanden Bout.
Astron. Journ., Vol. 77, 41 - 59 (1972).
Three one-dimensional area-scanning photometers under computer control, equipped to observe in one color, were taken to observing stations in Australia, India, and Africa to observe the occultation of the Beta Scorpii system by Jupiter on 13 May 1971. Six high-quality light curves were obtained; three of the occultations of the brighter component Beta Sco A and three of Beta Sco C. Correlated departures of the light curve from a theoretical isothermal curve are reproduced in the three bright-star curves, and are thus not due to random density fluctuations in the Jovian atmosphere, but rather due to global stratification. Details of the stratification, which includes at least a number of warm layers, are examined by deconvolution of the light curves.

099.014 The occultation of Beta Scorpii by Jupiter and Io. II. Io. P. Bartholdi, F. Owen.
Astron. Journ., Vol. 77, 60 - 65 (1972).
Photoelectric observations of the occultation of β Sco C (β^2 Sco) by the Jovian satellite Io were made on 14 May 1971 from St. Thomas in the Virgin Islands. Accurate times of in-

gress and egress were determined, and the effects of Fresnel diffraction at the limb of Io were observed. In spite of the extreme sensitivity of such an observation to refraction, no effects due to the presence of an atmosphere were detected.

099.015 The occultation of Beta Scorpii by Jupiter and Io. III. Astrometry.
W. B. Hubbard, T. C. Van Flandern.
Astron. Journ., Vol. 77, 65 - 74 (1972).
The timings of the occultation events of 13–14 May 1971 are used to derive a value for the equatorial radius and oblateness of Jupiter at an atmospheric level corresponding to a refractivity of 10^{-9}. Two independent analyses are presented; discrepancies between the results give a measure of the sensitivity to initial assumptions. Analysis of the Io occultation data gives an additional determination of the declination of the center of Jupiter and permits a more precise determination of the Jovian oblateness than would have otherwise been possible.

099.016 Spectrophotometry of the 1.5-μm window of Jupiter. A. B. Binder.
Astron. Journ., Vol. 77, 93 - 99, 113 (1972).
High spatial resolution (2″3) infrared ($\lambda\lambda$ 1.4–1.7 μm) spectrophotometric data obtained for 86 areas on Jupiter were used to study variations in the absorptions of NH_3 and CH_4 over the Jovian belts, zones, and spots, and to study the limb darkening in the 1.5-μm region. Values of the limb-darkening coefficient of the Minnaert function are given for two representative wavelengths and for three different phase angles.

099.017 Beobachtungen während der Jupiteropposition 1970. C. Kowalec, R. Sopper.
SuW, Vol. 11, 72 - 75 (1972).

099.018 Lunar and Planetary Laboratory studies of Jupiter – I, II. G. P. Kuiper.
Sky Telescope, Vol. 43, 4 - 8, 75 - 81 (1972).

099.019 Optical properties and the structure of the Jovian atmosphere. IV. Results of a photoelectric spectrophotometry in the 6300–8100 Å range. V. G. Tejfel'.
Astron. vestn., Vol. 5, 222 - 231 (1971). In Russian. – Abstr. in Referativ. Zhurn. 51. Astron., 3.51.338 (1972).

099.020 Jupiter: Observation of deuterated methane in the atmosphere.
R. Beer, C. B. Farmer, R. H. Norton, J. V. Martonchik, T. G. Barnes.
Science, Vol. 175, 1360 - 1361 (1972).
A positive identification of singly deuterated methane has been made in the 4- to 5-micron spectrum of Jupiter.

099.021 L'occultation de β Scorpion par Jupiter le 13 mai 1971.
M. Combes, C. De Bergh, T. Encrenaz, D. Gautier, R. Laporte, J. Lecacheux, L. Vapillon, J.-P. Verdet, Y. Zeau.
L'Astronomie, 86ᵉ année, p. 1 - 14 (1972).

099.022 Passages de la terre devant le soleil vus de Jupiter.
J. Meeus.
L'Astronomie, 86ᵉ année, p. 47 - 49 (1972).

099.023 Jupiter en 1970 (opposition le 21 avril).
J. Lecacheux, A. Lecacheux.
L'Astronomie, 86ᵉ année, p. 57 - 72 (1972).

099.024 Long-baseline analysis of a Jovian decametric L burst. M. A. Lynch, T. D. Carr, J. May, W. F. Block, V. M. Robinson, N. F. Six.
Astrophys. Letters, Vol. 10, 153 - 158 (1972).
Three-station interferometer observations of a 150-msec

Jovian burst at 18 MHz yielded fringe visibilities of practically unity over baselines up to 462,000 λ. An approximately constant phase relationship between the burst modulation envelope and the 'carrier wave' beneath the envelope suggests that the envelope did not result from a sweeping beam or from solar wind interference, but instead kept the same shape it had on leaving the planet.

099.025 The high gravitational moments of Jupiter and Saturn.
V. N. Zharkov, V. P. Trubitsyn, A. B. Makalkin.
Astrophys. Letters, Vol. 10, 159 - 161 (1972).
The coefficients J_2, J_4 and J_6 of the external gravitational potential series of a rotating planet are computed for Jupiter and Saturn on the basis of available models. Estimates of two following coefficients (J_8 and J_{10}) are made. Logarithmic derivations of J_2, J_4 and J_6 with respect to the mean density of a planet are calculated for several models. These results are used to obtain corrections to the gravitational moments for the change of the mean radius.

099.026 Die Jupitermonde in den Oppositionsperioden 1969 und 1970. P. Ahnert.
Sterne, 48. Jahrgang, p. 34 - 42 (1972).

099.027 Der grosse rote Fleck (GRF) auf Jupiter 1971. P. Ahnert.
Sterne, 48. Jahrgang, p. 53 - 55 (1972).

099.028 EAS [extensive air showers] and the Jovian sporadic decametre radio emission.
V. P. Vasilyev, V. D. Volovik, I. I. Zalyubovsky.
Astron. Zhurn. Akad. Nauk SSSR, Vol. 49, 413 - 419 (1972). In Russian. English translation in Soviet Astron. AJ, Vol. 16, No. 2.
A new explanation for the nature of Jupiter's sporadic radio emission is proposed. This is in agreement with main observational data on decametre bursts.

099.029 On an adiabatic model of Jupiter.
V. P. Trubitsyn.
Astron. Zhurn. Akad. Nauk SSSR, Vol. 49, 420 - 426 (1972). In Russian. English translation in Soviet Astron. AJ, Vol. 16, No. 2.
The adiabatic lines of gaseous-liquid hydrogen and helium are considered. The distribution of the temperature is calculated by an adiabatic model of Jupiter. This model leads to temperatures of about $\sim 5 \times 10^4$ °K in the centre of the planet.

099.030 The flux density of Jupiter.
V. S. Artukh, A. D. Kuzmin, A. N. Makarov.
Astron. Zhurn. Akad. Nauk SSSR, Vol. 49, 456 - 458 (1972). In Russian. English translation in Soviet Astron. AJ, Vol. 16, No. 2.
The flux density of Jupiter was measured at the frequency of 102 MHz. Averaged over the visible disc the brightness temperature was found to be 3.8×10^5 °K.

099.031 Comments on the Jupiter–Beta Scorpii occultation of 13th May, 1971. A. H. Jarrett.
Irish Astron Journ., Vol. 10, 109 - 110 (1971).

099.032 80 MHz measurements of Jupiter's synchrotron emission. O. B. Slee, G. A. Dulk.
Australian Journ. Phys., Vol. 25, 103 - 105 (1972). – Short communication.

099.033 Inhomogeneous models of the atmosphere of Jupiter. L. Axel.
Astrophys. Journ., Vol. 173, 451 - 468 (1972).
Infrared observations and general theoretical considera-

tions lead to an inhomogeneous model of the Jovian atmosphere with two cloud layers: an upper semitransmitting cloud separated by a clear space from a lower cloud deck. Scattering takes place between the upper partially transmitting NH_3 cloud layer and the lower opaque cloud layer. In this model, the depth of the clear space between the clouds plays the same role as the scattering mean free path in homogeneous scattering models.

099.034 Variation of highest cutoff frequency of the Io-controlled Jovian decametric radiation.
C. Goertz, A. Haschick.
Nature, Phys. Sci., Vol. 235, 91 - 94 (1972).

The maximum cutoff frequency f_M of the decametric radiation shows a marked variation with Io λ III (1967). Radiation at 40 MHz is received only when Io is above $230°$ Jovian longitude. This variation is only observed for the Io-controlled radiation. The variation can be accounted for by a symmetrical variation of height above Jupiter's surface at which f_M is emitted. Predictions of the unipolar theory do not seem to agree with the observations.

099.035 Gravitational fields of Jupiter and Saturn.
V. N. Zharkov, V. P. Trubitsyn, A. B. Makalkin.
Dokl. Akad. Nauk SSSR, Ser. Mat. Fiz., Vol. 203, 1021 - 1023 (1972). In Russian.

099.036 Jupiter–giant in the solar system.
Ü.-I. Veltmann.
"Valgus", Tallinn. 60 pp. (1971).   In Estonian.

Review on observations and theories about the planet Jupiter. Under consideration are mechanical data about the planet and its orbit, belts and zones, disturbances, rotation period, temperature, atmosphere, radio radiation, existence of life, satellites and plans for Jupiter trips. Some original ideas of the author are also published. For example the decametric radio radiation is explained by the collective processes in the plasma. It is expected that life fields are existing on the planet Jupiter.

099.037 The hypothetical chemical and spectroscopic activity of germane in the atmosphere of Jupiter.
R. J. Corice, Jr., K. Fox.
Icarus, Vol. 16, 388 - 391 (1972).

It is suggested that in a reducing environment such as the Jovian atmosphere, GeH_4 may be spectroscopically active. Absorption features might be detectable in the 4.7μ window in the Jovian upper atmosphere.

099.038 The clouds of Jupiter: Observational characteristics.
T. Owen, J. A. Westphal.
Icarus, Vol. 16, 392 - 396 (1972).

Recent observations at 5μ show that the emitted radiation is strongly localized and that it exhibits temporal variations. It is not clear that meaningful temperatures can be derived from such measurements and this plus other observations discussed herein suggest that various models proposed for the clouds require reexamination.

099.039 The Jupiter greenhouse.
C. Sagan, G. Mullen.
Icarus, Vol. 16, 397 - 400 (1972).

A step function nongray approximation to infrared absorption by H_2, NH_3, and CH_4 on Jupiter implies a temperature at the lower clouds of 270 to $340°K$. Such clouds are conjectured to be aqueous. No contradiction is implied by reported lower rotational temperatures.

099.040 Temperatures of Titan and the Galilean satellites at 20 microns.
D. Morrison, D. P. Cruikshank, R. E. Murphy.
Astrophys. Journ., (Letters), Vol. 173, L143 - L146 (1972).

Broad-band $20-\mu$ observations made at Mauna Kea in 1971 give the following effective temperatures in degrees K: Titan, 93 ± 1; J I, 127 ± 3; J II, 119 ± 3; J III, 134 ± 4; J IV, 149 ± 5.

099.041 Observations of occultations of β Scorpii on 1971 May 13 and 14.   J. Dragesco.
Journ. British Astron. Ass., Vol. 82, 200 - 201 (1972).

099.042 Photoelectron emission from Io as the cause of enhancements of the Jovian decametric radiation.
F. S. Mozer, F. H. Bogott.
Astrophys. Journ., Vol. 174, 153 - 156 (1972).

A theoretical explanation for the dependence of the Jovian decametric radiation on the position of its satellite Io is advanced in terms of electrostatic instabilities created by injection into the hot natural plasma of cold photoelectrons from Io.

099.043 Exobiology, Jupiter and life.   P. M. Molton.
Spaceflight, Vol. 14, 220 - 223 (1972).

099.044 Jupiter: Présentation 1971.   S. Cortesi.
Orion Schaffhausen, 30. Jahrgang, p. 53 - 57 (1972).
Rapport No. 22 du «Groupement planétaire SAS».

099.045 Observations of the occultation of the star β Scorpii by Jupiter.
A. G. Rakhimov, A. Sadikov, L. I. Bashtova.
Tsirk. Astron. Inst., Tashkent, No. 32 (379), p. 16 (1971/72). In Russian.

099.046 Vertically inhomogeneous scattering models of Jupiter's atmosphere.   R. A. Stokes.
Bull. American Astron. Soc., Vol. 4, 219 (1972). – Abstr. AAS.

099.047 Jovian atmosphere absorption bands observed by means of Galilean satellite eclipse observations.
T. F. Greene, R. W. Shorthill.
Bull. American Astron. Soc., Vol. 4, 219 (1972). – Abstr. AAS.

099.048 Methods and approximations for the computation of transmission profiles in the ν_4 band of methane in the atmosphere of Jupiter.   F. W. Taylor.
Journ. Quant. Spectrosc. Radiat. Transfer, Vol. 12, 1151 - 1156 (1972).

This note discusses the validity of certain band models and scaling approximations for computing transmissions in the ν_4 band of methane along inhomogeneous paths in the atmosphere of Jupiter. It is shown that Goody's random band model approximates the results of a rigorous numerical line-by-line calculation of the transmission profile of a Jovian model atmosphere.

099.049 Occultations d'étoiles par Jupiter   J. Meeus.
L'Astronomie, 86e année, p. 223 - 229 (1972).

099.050 Opposition of the planet Jupiter.   Momchev.
Priroda (NRB), Vol. 20, No. 6, p. 58 - 59 (1971).
In Bulgarian. – Abstr. in Referativ. Zhurn. 51. Astron., 6.51.79 (1972).

099.051 On the formation of Jupiter.   S. S. Kumar.
Astrophys. Space Sci., Vol. 16, 52 - 54 (1972).
Arguments are presented to show that Jupiter could not have been formed as a star of mass 0.001 $M_\odot$ and that the observed excess energy is not due to gravitational contraction from an extended, gaseous state.

099.052 Strong beaming of Jupiter's non-Io-related radio

emission. G. M. Gruber, C. Way-Jones.
Nature, Phys. Sci., Vol. 237, 137 - 139 (1972).

Although Io's influence on decametric radiation has been known for some time, no effort has as yet been made to distinguish between Io-related and non-Io-related radiation when analysing the effect of declining probability of reception after opposition. Here we present such an analysis.

099.053 Does Io's ionosphere influence Jupiter's radio bursts?
D. L. Webster, A. Y. Alksne, R. C. Whitten.
Astrophys. Journ., Vol. 174, 685 - 696 (1972).

Goldreich and Lynden-Bell's theory of Jupiter's Io-correlated dekametric radiation sets a lower limit to Io's conductivity, high enough to carry the current associated with the radiated power. Dermott's analysis of conductivities of rocks and ice shows no such conductivity at Io's temperature. However, we show that if Io has even a small atmosphere it will have an ionosphere with adequate conductivity to meet the above criterion.

099.054 The plasma physics of the Jovian decameter radiation. M. L. Goldstein, A. Eviatar.
Astrophys. Journ., Vol. 175, 275 - 283 (1972).

We have assumed that the decameter radiation from Jupiter is produced near the local electron gyrofrequency and is amplified as it propagates out of the Jovian magnetosphere. Using the Vlasov-Maxwell equations, which describe the propagation of radiation in hot collisionless plasmas, we have derived the growth rate for radiation that propagates almost perpendicular to the direction of the magnetic field. When the electrons are described by a loss-cone distribution function, the growth rate is large enough to lead to a large amplification factor over a source region of 100–4000 km.

099.055 The flashing phenomena during the Jovian eclipse of β Scorpii. A. Fairall.
Monthly Notes Astron. Soc. Southern Africa, Vol. 31, 50 (1972). – Abstract.

099.056 The radioastronomy of Jupiter. J. A. Gledhill.
Monthly Notes Astron. Soc. Southern Africa, Vol. 31, 50 (1972). – Abstract.

099.057 Occultation of SAO 186800 by Jupiter III on 1972 June 7. G. E. Taylor.
IAU Circ., No. 2401 (1972).

099.058 Occultation of SAO 186800 by Jupiter III.
J. C. Bhattacharyya.
IAU Circ., No. 2412 (1972).

099.059 Émissions radioélectriques de la planète Jupiter (1).
R. Gonze.
Gaz. astron., *Antwerpen*, No. 3, p. 15 - 22 (1972).

099.060 Jupiter, the active planet. B. S. Adcock.
Journ. Astron. Soc. Victoria, Vol. 24, 79 - 81 (1971). – Lunar and Planetary Section report.

099.061 Chemistry and photochemistry of the atmosphere of Jupiter. J. S. Lewis, R. G. Prinn.
Theory and experiment in exobiology, Vol. 1, (see 003.011), 123 - 142 (1971).

099.062 Jupiters atmosfaere. K. Messell.
Astron. Tidssk., Årg. 5, p. 85 - 87 (1972).

099.063 Ephemeris of the radio longitude of the central meridian of Jupiter, System III (1957.0).
B. L. Morrison, V. Meiller.

United States Naval Obs., *Washington, D. C.,* Circ. No. 137, 5 pp. (1972).

099.064 Electro-polarimetric observations of the planet Jupiter. O. R. Bolkvadze.
AN GruzSSR, Vol. 65, 567 - 570 (1972). In Russian. – Abstr. in Referativ. Zhurn. 51. Astron., 7.51.210 (1972).

099.065 Some dynamic aspects of the Jovian and other atmospheres. V. P. Starr.
Tellus, Vol. 23, 489 - 499 (1971).

A fluid system comprised of a mass of gas isolated in space from outside mechanical influences is studied from the standpoint of its differential rotation.

099.066 Magnetospheres of Jupiter and Saturn.
J. W. Haffner.
AIAA Journ., Vol. 9, 2422 - 2427 (1971).

A series of calculations has been carried out to estimate the characteristics of the magnetospheres around the planets Jupiter and Saturn. Solar wind characteristics near these planets were deduced from those near the earth using standard hydrodynamic equations. Planetary magnetic fields were estimated from nonthermal rf emissions, which were assumed to be synchrotron radiation.

099.067 Upper atmospheric thermal structure of Jupiter with convective heat transfer.
W. E. McGovern, S. D. Burk.
Journ. Atmosph. Sci., Vol. 29, 179 - 189 (1972).

099.068 Satellites of Jupiter.
Contr. Obs. Valongo, Univ. Federal Rio de Janeiro, Sér. II, No. 13 (1971). – 1971 May - June.

099.069 Occultation of CoD −23° 14351 by Ganymede (Jupiter III) on 1972 June 7.
Yamamoto Circ., No. 1754 (1972). In Japanese.

099.070 Heating environment and protection during Jupiter entry. M. E. Tauber, R. M. Wakefied.
Journ. Spacecraft & Rockets, Vol. 8, 630 - 636 (1971).

099.071 Observations of Jupiter, Venus, and source 3C 273 at 2 and 8 mm wavelengths.
V. A. Yefanov, A. G. Kislyakov, I. G. Moiseyev, A. I. Naumov.
Journ. Astronaut. Sci., Vol. 18, 383 - 389 (1971).

099.072 Jovian ultraviolet reflectivity compared to absorption by solid ammonia.
R. C. Anderson, J. G. Pipes.
Journ. Atmosph. Sci., Vol. 28, 1086 - 1087 (1971).

099.073 Contribution à l'étude du rayonnement de Jupiter sur ondes décamétriques. L. Conseil.
Thesis, 3ᵉ Cycle, Spéc. Phys., Univ. Paris. 56 pp. (1971).
See Bull. Signal., Vol. 33, Section 120, No. 7021 (1972).

Cornell project to seek data on Jupiter's radiation.
IEEE Spectrum, Vol. 9, No. 4, p. 93 - 94 (1972).

Observaciones de Jupiter efectuadas con el astrolabio A. Danjon del Observatorio de Quito, durante el año 1964. See Abstr. 041.041.

The Pioneer mission to Jupiter.
See Abstr. 053.020.

Das Wettergeschehen auf Mars, Venus und Jupiter.
See Abstr. 097.067.

The Beta Scorpii system.   See Abstr. 117.003.

Symmetry of flashes during the Jovian occultation of β Scorpii.   See Abstr. 117.013.

099.901   **Erratum: 'The far infrared spectrum of Jupiter'**
[Astron. Astrophys., Vol. 11, 431 - 449 (1971)].
T. Encrenaz, D. Gauthier, L. Vapillon, J. P. Verdet.
Astron. Astrophys., Vol. 18, 167 (1972).

100   Saturn

100.001   **Saturn.**   M. S. Bobrov.
Zemlya i Vselennaya, No. 1, p. 10 - 17 (1972).   In Russian.

100.002   **Observations of Saturn at a wavelength of 49.5 cm.**
M. J. Yerbury, J. J. Condon, D. L. Jauncey.
Icarus, Vol. 15, 459 - 465 (1971/72).
Radio emission from the planet Saturn was detected and measured by an unusually efficient observing technique at a wavelength of 49.5 cm. The corresponding equivalent disk brightness temperature was hence determined to be $390 \pm 65°K$, providing further evidence for a mild enhancement in the emission at long wavelengths.

100.003   **Recent photographic measurements of Saturn.**
E. J. Reese.
Icarus, Vol. 15, 466 - 479 (1971/72).
Some recent measurements of Saturn, particularly latitudes of the faint atmospheric belts, are summarized.

100.004   **Saturn: Interpretation of its microwave spectrum.**
S. Gulkis, R. Poynter.
Bull. American Astron. Soc., Vol. 3, 481 (1971). – Abstr. AAS.

100.005   **On Saturn's rings thickness estimates obtained from the observations made in 1966.**
M. S. Bobrov.
Astron. Zhurn. Akad. Nauk SSSR, Vol. 49, 427 - 435 (1972).
In Russian. English translation in Soviet Astron. AJ, Vol. 16, No. 2.
Considered are the method and the results of the estimate of Saturn's rings thickness from photometric observations of the rings from the edge. Observational difficulties and means to overcome them are discussed. The photometric series obtained in 1966 by R. I. Kiladze (1969) and J. Focas and A. Dollfus (1969) are analysed.

100.006   **On the formation of Saturn's rings.**   K. Lumme.
Astrophys. Space Sci., Vol. 15, 404 - 414 (1972).
It is shown that the formation of Saturn's ring C can be explained by the action of solar radiation pressure on the small ring particles.

100.007   **Saturn central meridian ephemeris, 1972.**
J. E. Westfall.
Strolling Astronomer, Vol. 23, 126, 128 - 129 (1972).

100.008   **Iapetus and the glare of Saturn.**   R. G. Hodgson.
Strolling Astronomer, Vol. 23, 126 - 127 (1972).

100.009   **Die Rotation des Saturn.**   W. Sandner.
Orion Schaffhausen, 30. Jahrgang, p. 58 - 59 (1972).

100.010   **Saturn 1971.**   M. Unkel.
SuW, Vol. 11, 169 (1972).

100.011   **Photometric behaviour of Saturn's rings as a function of the Saturnocentric latitudes of the earth and the sun.**   K. A. Hämeen-Anttila, S. Pyykkö.
Astron. Astrophys., Vol. 19, 235 - 247 (1972).
The surface brightness of Saturn's rings A and B are determined for a constant phase $3°5$ as a function of the solar and terrestrial elevation angles. The light curve for ring A follows the Lommel-Seeliger law corrected for a minor effect produced by multiple scattering. The intensity of ring B increases with the elevation angles. A theoretical explanation for stratification and for the difference in albedos is given in terms of size differences.

100.012   **On the nature of Iapetus.**   B. Zellner.
Astrophys. Journ., (*Letters*), Vol. 174, L107 - L109 (1972).
Polarization observations of the leading and trailing hemispheres of Iapetus indicate an albedo difference consistent with the amplitude of the light curve.

100.013   **On the possible detection of H_2 in Titan's atmosphere.**   L. Trafton.
Astrophys. Journ., Vol. 175, 285 - 293 (1972).
We have detected weak absorption features in Titan's spectrum at the wavelengths of the $3-0\,S(1)$ H_2 quadrupole line and, with less certainty, the $S(0)$ line of the same band. We discuss the probable reality of these features in terms of a statistical analysis of the photometric data and present evidence that they are not scattered from Saturn by the sky or the telescope. We estimate the abundance implied if this absorption arises from H_2 and discuss briefly the implications this has for escape and outgassing rates.

100.014   **The bulk composition of Titan's atmosphere.**
L. Trafton.
Astrophys. Journ., Vol. 175, 295 - 306 (1972).
We show that the methane bands in Titan's spectrum are saturated and that, consequently, either the methane abundance is nearly an order of magnitude greater than is generally believed or else methane is a minor constituent in this atmosphere. Consideration of the physical constraints for Titan's atmosphere leads to a model which describes the bulk composition in terms of observable parameters. Intermediate-resolution photometric scans of both Saturn and Titan, including our scans of the Q-branch of Titan's $3\nu_3$ methane band, constrain these parameters in such a way that the model indicates the presence of another important atmospheric gas, namely, another bulk constituent or a significant thermal opacity.

100.015   **Evaluation of ammonia abundance in the subcloud** atmosphere of Saturn from radio astronomical measurements. A. D. Kuz'min, A. P. Naumov, T. V. Smirnova. Astron. vestn., Vol. 6, 13 - 18 (1972). In Russian. — Abstr. in Referativ. Zhurn. 51. Astron., 7.51.215 (1972).

Observaciones de Saturno con el astrolabio del Observatorio de Quito. See Abstr. 041.043.

The atmospheres of Jupiter and Saturn. See Abstr. 099.001.

The high gravitational moments of Jupiter and Saturn. See Abstr. 099.025.

Gravitational fields of Jupiter and Saturn. See Abstr. 099.035.

Temperatures of Titan and the Galilean satellites at 20 microns. See Abstr. 099.040.

Magnetospheres of Jupiter and Saturn. See Abstr. 099.066.

101   Uranus, Neptune, Pluto, Transplutonian Planet

101.001 The infrared spectral albedo of Uranus.
  A. B. Binder, D. W. McCarthy, Jr.
Astrophys. Journ., (Letters), Vol. 171, L1 - L3 (1972).
  New infrared ($\lambda\lambda 0.60$-$2.27\,\mu$) spectrophotometric data for Uranus are presented which show that CH_4 absorptions are important in controlling the infrared spectral albedo of the planet.

101.002 On the methane opacity for Uranus and Neptune.
  L. Trafton.
Astrophys. Journ., (Letters), Vol. 172, L117 - L120 (1972).
  The contribution of CH_4 to the thermal opacity in the atmospheres of Uranus and Neptune is negligible owing to the low CH_4 mixing ratio imposed by saturation in the radiative equilibrium regions of these atmospheres. Furthermore, that portion of the emergent flux sensitive to the He/H_2 ratio is not affected significantly by the pressure-induced absorption of CH_4–H_2 collisions.

101.003 Résonances dans le système Neptune-Pluton.
  J. Meeus.
L'Astronomie, 86ᵉ année, p. 33 - 36 (1972).

101.004 An analysis of the light curve of Pluto.
  A. A. Lacis, J. D. Fix.
Astrophys. Journ., Vol. 174, 449 - 453 (1972).
  The light curve of Pluto is analyzed in terms of a geomet-

rical model consisting of bright and dark areas which are assumed to exhibit either a diffuse or a geometrical type of reflectivity. However, on the basis of the presently available photometric data, the existence or absence of limb-darkened material cannot be established.

101.005 Interferometer observations of Uranus, Neptune, and Pluto at wavelengths of 11.1 and 3.7 centimeters.
W. J. Webster, Jr., A. C. Webster, G. T. Webster.
Astrophys. Journ., Vol. 174, 679 - 684 (1972).
  Visibility observations of Uranus at 8.085 GHz fit a uniform circular disk of diameter $3''.8 \pm 0''.2$. The disk temperature of Uranus and Neptune are, respectively $189° \pm 7°$K and $190° \pm 20°$K at 8.085 GHz, and $195° \pm 30°$K and $201° \pm 40°$K at 2.695 GHz. The 8.085-GHz observations yield an upper limit of $162°$K for the disk temperature of Pluto.

101.006 Uranus, the extraordinary planet behind an ordinary veil.
Sci. Dimension (Canada), Vol. 4, No. 1, p. 18 - 20 (1972). In English and French.
  Comparison of planetary and laboratory spectra shows that the Kuiper bands on the spectrum of Uranus are due to methane.

The effect of a trans-Plutonian planet on Halley's comet.   See Abstr. 102.010.

102   Comets

102.001   The origin of short-period comets.
E. Everhart.
Astrophys. Letters, Vol. 10, 131 - 135 (1972).

A numerical study of the evolution of comet orbits is described. This follows the orbits of tens of thousands of hypothetical comets as modified by Jupiter's perturbations for up to 2000 returns for each comet. It is shown that the short-period comets can evolve from original near-parabolic orbits of small inclination with a perihelion distance near the orbit of Jupiter. Simple considerations based on the Tisserand criterion support and explain these results.

102.002   Comet searches during four major eclipses.
H. C. Courten.
Solar Physics, Vol. 21, 495 (1971). – Abstract.

102.003   Stellar perturbations of orbits of long-period comets.   S. Yabushita.
Astron. Astrophys., Vol. 16, 395 - 403 (1972).

Stellar perturbations of extremely eccentric orbits are investigated. For distant encounters where the shortest distance, q_1 of a star to the sun is greater than the aphelion distance of a comet analytical expressions are obtained for the changes in binding energy and angular momentum of the comet in terms of orbital elements. It is found that although the energy perturbation is only a few per cent of the binding energy of the comet the variation of angular momentum is such that a change in perihelion distance of a few astronomical units can occur.

102.004   Planetary perturbation of orbits of long-period comets with large perihelion distances.
S. Yabushita.
Astron. Astrophys., Vol. 16, 471 - 477 (1972).

The change of the binding-energy of a long-period comet caused by planetary perturbation is calculated by assuming that the orbit is a parabola. It is shown that although the energy change is comparable with the cometary binding-energy, the perturbation of perihelion distance is so small that it can be regarded as remaining constant throughout the life history of the comet. In order to obtain statistical properties of the distribution of energy changes caused by random encounters with a planet, Monte Carlo calculation has been carried out.

102.005   Some aspects about detecting microwave lines in comets.   W. F. Huebner.
Bull. American Astron. Soc., Vol. 3, 500 - 501 (1971). – Abstr. AAS.

102.006   Does a continuous solid nucleus exist in comets?
R. A. Lyttleton.
Astrophys. Space Sci., Vol. 15, 175 - 184 (1972).

The implication of actual cometary observations for the physical nature of comets is briefly reviewed and brings out the complete conflict with observation of the ice-dust solid nucleus model put forward in recent years as representing the fundamental structure of comets.

102.007   Neutral hydrogen in cometary comas.
D. A. Mendis, T. E. Holzer, W. I. Axford.
Astrophys. Space Sci., Vol. 15, 313 - 325 (1972).

The strong $L\alpha$ radiation observed recently in comets Tago-Sato-Kosaka and Bennett can be explained in terms of the resonant scattering of solar $L\alpha$ radiation on neutral hydrogen formed by the photo-dissociation of H_2O which is vaporized from a nucleus having an ice core. Numerical results are computed in a typical case, and it is found that a temperature

of about 3000 K for the cometary atmosphere provides the best fit with observation.

102.008   On the eigenmodes of type I comet tails.
A. I. Ershkovich, A. A. Chernikov.
Astron. Zhurn. Akad. Nauk SSSR, Vol. 49, 458 - 461 (1972). In Russian. English translation in Soviet Astron. AJ, Vol. 16, No. 2.

It is shown that the quasiperiodic and the rotational motions observed in type I cometary tails can be interpreted as eigenmodes excited by fluctuations of solar wind parameters.

102.009   Cometary research.   P. Herget.
Vistas in astronomy, Vol. 13, (see 003.001), 149 - 151 (1972).

102.010   The effect of a trans-Plutonian planet on Halley's comet.   J. L. Brady.
Publ. Astron. Soc. Pacific, Vol. 84, 314 - 322 (1972).

An orbit and mass for a hypothetical trans-Plutonian planet is determined which reduces the residuals in the time of perihelion passage of Halley's comet at the seven apparitions from 1910 to 1456 by 93%. The effect of this hypothetical planet on the major planets is briefly discussed, and it is shown that the residuals of two similar periodic comets, Olbers and Pons-Brooks, are also improved.

102.011   On the connection between outbursts of comets and processes on the sun.   D. A. Andrienko.
Kometn. Tsirk., *Kiev*, No. 131 (1972). In Russian.

102.012   Physische Beobachtungen von Kometen. XVII.
M. Beyer.
Astron. Nachr., Vol. 293, 241 - 257 (1972).

In continuation of this series the results are given of systematic visual observations of 7 bright comets (1969 a, 1969 b, 1969 g, 1969 i, 1970 g, 1970 l, 1970 m) in the years 1969–70. As in preceding papers the photometric parameters H_0 and n are derived from the observed total magnitudes of the comet's head, and the correlation of magnitude with solar activity is examined.

102.013   Zur Bahnbestimmung von Kometen.
G. Schrutka-Rechtenstamm.
Sternenbote, 15. Jahrgang, p. 54 - 58 (1972).

102.014   Catalogue of cometary orbits.   B. G. Marsden.
Central Bureau Astron. Telegrams, IAU. Smithsonian Astrophys. Obs., Cambridge, Mass., 70 pp. Price $ 2.50 (1972).

102.015   Mehrfachstreuung in den Wasserstoffwolken von Kometen.   H. U. Keller.
Mitt. Astron. Ges., No. 31, p. 183 (1972). – Abstract.

102.016   On the thickness of cometary antitails.
K. Jockers.
Mitt. Astron. Ges., No. 31, p. 184 (1972). – Abstract.

102.017   Struktur der Kometen.   V. Vanýsek.
Endeavour, No. 113, Vol. 31, 60 - 66 (1972). Review article.

102.018   Origin of C_3 in comets.   L. J. Stief.
Nature, Vol. 237, 29 (1972).

Recent results in our laboratory on the photolysis of propyne ($CH_3 C \equiv CH$) at 1236 Å suggest that an important primary process is $C_3H_4 + h\nu \rightarrow C_3 + 2H_2$. It is therefore proposed that a potential source of C_3 in comets is the photodissociation of propyne by solar radiation, most probably in the region of the strong 1216 Å Lyman-α line. This tentative suggestion becomes more interesting in light of the recent discovery of propyne (methylacetylene) in interstellar space.

102.019 **Dynamics of comets.** S. Yabushita.
Celestial mechanics. Symposium Tokyo, (see 012.020), p. 87 - 101 (1972). In Japanese.

102.020 **Die Lyman α Strahlung der Wasserstoffatmosphären von Kometen. Ein Modell mit Mehrfachstreuung.**
H. U. Keller.
Inaugural-Diss. Nat.Fakultät, Ludwig-Maximilians-Univ., München.3 + 72 pp. (1971).

102.021 **L'origine delle comete.** S. Vaghi.
Annuario *Torino* 1972, (see 047.021), p. 47 - 62 (1971).

102.022 **Comets in the solar wind.** L. Biermann.
Cosmic Plasma Physics, [Plenum Publishing Corporation, New York, edited by K. Schindler], p. 123 - 135 = Separate print Inst. Phys. Astrophys. München (1972).
Two bright comets, Tago-Sato-Kosaka and Bennett, appeared in winter and spring of last year. They were the first ones to be observed by means of satellite-borne instruments in the UV down to approximately 900 Å. The most important discovery made during these observations was the existence of a huge atmosphere of atomic hydrogen visible in the resonance line Lyman α 1216 Å.

102.023 **Comets as cosmic measurement devices.**
O. Dobrovolskiy.
Soviet Sci. Rev. (*GB*), Vol. 3, No. 1, p. 49 - 55 (1972). – See Phys. Abstr., Vol. 75, No. 20685 (1972).

102.024 **Is water ice the major difference between comets and asteroids?** A. H. Delsemme.
IAU Colloquium No. 12, (see 012.027), p. 461 - 464 (1971/72).

102.025 **Structure of comets and the possible origin of faint asteroids.** V. Vanýsek.
IAU Colloquium No. 12, (see 012.027), p. 465 - 469 (1971/72).

102.026 **The structure and formation of comets.**
V. Vanysek.
From plasma to planet. 21st Nobel symposium 1971, (see 012.028), p. 233 - 259 (1972).

102.027 **'Cometary' suggestion.** D. Lal.
From plasma to planet. 21st Nobel symposium 1971, (see 012.028), p. 349 - 352 (1972).

102.028 **Comets and meteors.** P. L. Masters.
Journ. Astron. Soc. Western Australia, Vol. 36, April, p. 2 - 5 (1972).

Jacobian integral as a classificational and evolutionary parameter of interplanetary bodies. See Abstr. 042.010.

Trajectory requirements for comet rendezvous.
See Abstr. 051.028.

Small molecules in astrophysics: in stellar atmospheres, in comets, and in interstellar space.
See Abstr. 064.044.

A core-mantle model for cometary nuclei and asteroids of possible cometary origin. See Abstr. 098.066.

On certain aerodynamic processes for asteroids and comets. See Abstr. 098.078.

Small bodies in the solar system.
See Abstr. 104.037.

Cometary meteoroids. See Abstr. 104.044.

Interrelations of meteorites, asteroids, and comets.
See Abstr. 105.066.

Errata

102.901 **Erratum: 'Planetary perturbation of orbits of long-period comets with large perihelion distances'**
[Astron. Astrophys., Vol. 16, 471 - 477 (1972)].
S. Yabushita.
Astron. Astrophys., Vol. 18, 167 (1972).

103 Comets: Listed Objects

103.001 **Zur Genauigkeit visueller Helligkeitsschätzungen von Kometen.** V. Kasten.
SuW, Vol. 11, 24 (1972).

103.002 **Vorschau auf die periodischen Kometen der nächsten Jahre.** V. Kasten.
SuW, Vol. 11, 55 - 56 (1972).

103.003 **Die endgültige Bezeichnung der Kometen des Jahres 1970.**
SuW, Vol. 11, 77 (1972).

103.004 **Comets lost and found.**
Sky Telescope, Vol. 43, 155 (1972).

103.005 **Die Kometen des Jahres 1971.** R. Lukas.
SuW, Vol. 11, 86 (1972).

103.006 **Comet notes.**
Journ. Astron. Soc. Victoria, Vol. 24, 93 - 96 (1971).

103.007 **Definite designations of the comets of 1970.**
Kometn. Tsirk., *Kiev*, No. 128 (1972). In Russian.

103.008 **New catalogue of cometary orbits.**
Kometn. Tsirk., *Kiev*, No. 129 (1972). In Russian.

103.009 **Comet notes.** E. Roemer.
Mercury, (Journ. Astron. Soc. Pacific), Vol. 1, No. 1, p. 18 - 19; No. 3, p. 16 - 19 (1972).

103.010 **Observations photographiques de comètes, effectuées en 1970 à l'astrographe double de 40 cm.**
H. Debehogne.
Bull. Astron. Obs. Roy. Belgique, Vol. 7, 279 - 280 (1972). − Concerning the comets 1969g, 1969i, 1970g.

103.011 **Possible comet Antal.** M. Antal.
IAU Circ., No. 2410 (1972).

103.012 **Possible comet Antal.**
R. L. Waterfield, C. T. Kowal.
IAU Circ., No. 2412 (1972).

103.013 **Observations of comets.** L. Hartmann, D. Royce, J. Stein, A. Weitenbeck.
IAU Circ., No. 2415 (1972). − Concerning the comets 1971e, 1972a, 1972d.

103.014 **Comets in 1971.** A. C. Gilmore.
Southern Stars, Vol. 24, 114 - 115 (1972).

103.015 **Kometer 1971.** H. Q. Rasmusen.
Astron. Tidssk., Årg. 5, p. 95 (1972).

103.016 **Possible new comet Antal.**
Kometn. Tsirk., *Kiev*, No. 132 (1972). In Russian.

103.017 **Comets in the year 1971.** J. Bouška.
Vesmír, Vol. 51, 188 (1972). In Czech. − Progress report.

103.100 **Comet 1970 II Bennett**

  Direct infrared measurements of thermal radiation from the nucleus of comet Bennett. J. A. Myer.
Astrophys. Journ., (*Letters*), Vol. 175, L49 - L53 (1972).
  A technique of contrast radiometry has enabled us to make direct infrared measurements on the nucleus of comet 1969i. The intensity level of the observation of 9.0×10^{-15} watts cm^{-2} μ^{-1} is consistent with the measurements of two other groups of observers. A small but slightly hotter nucleus may explain the irregular thermal radiation continuum observed by one of the groups.

  Laser radar observations of dust from comet Bennett. F. Link.
Journ. Atmosph. Terr. Phys., Vol. 34, 343 (1972).
  Possibility of simultaneous influence of Eta-Aquarids. Other detections of the descending motion of meteoritic dust in the atmosphere.

  The nature of the tails and motion of condensations in the plasma tail of comet Bennett.
S. K. Vsekhsvyatskij, A. A. Demenko.
Kometn. Tsirk., *Kiev*, No. 126 (1972). In Russian.

  Spectral polarization of the radiation of the head of comet Bennett, 1969i. L. A. Bugaenko, O. I. Bugaenko, L. S. Galkin, V. P. Konopleva, A. V. Morozhenko.
Kometn. Tsirk., *Kiev*, No. 127 (1972). In Russian.

  Observations of comet Bennett, 1970 II (1969i).
A. G. Kirichenko, G. V. Moskaleva, V. P. Ryabov.
Kometn. Tsirk., *Kiev*, No. 128 (1972). In Russian.

  Photographs of comet Bennett 1969i.
K. Tomita, H. Kosai, H. Shibasaki.
Tokyo Astron. Bull., Second Ser., No. 217, p. 2529 - 2548 (1972).
  About 100 plates of comet Bennett 1969i were taken with the NIKON Schmidt camera (500/650/998.8 mm) by using a field flattener lens, at the Dodaira Station of the Tokyo Astronomical Observatory on 20 nights in the spring of 1970.

  Comets in the solar wind. See Abstr. 102.022.

103.101 **Comet 1968 I Ikeya-Seki**

  Photographic positions of comet Ikeya-Seki (1967 n) with a summary of the Leyden reduction procedure.
J. van Kuilenburg.
Astron. Astrophys., Vol. 18, 154 - 158 (1972).
  Photographic positions of comet Ikeya-Seki (1967 n) in March and April 1968 are presented. Their accuracy is somewhat better than 1″. A summary of the reduction procedure in use at the Leyden Observatory is given.

103.102 **Comet 1972b Grigg-Skjellerup**

  Comet P/Grigg-Skjellerup 1972 b.
J. B. Gibson, U. T. Gibson.
British Astron. Ass., Circ. No. 539 (1972).

  Une pluie d'étoiles filantes les 22−23 avril 1972?
J. Meeus.
Ciel et Terre, Vol. 88, 57 - 59 (1972).

Periodic comet Grigg-Skjellerup (1972b).
J. B. Gibson, U. T. Gibson.
IAU Circ., No. 2384 (1972).

Periodic comet Grigg-Skjellerup (1972b).
J. B. Gibson.
IAU Circ., No. 2385 (1972).

Periodic comet Grigg-Skjellerup (1972b).
B. Donn.
IAU Circ., No. 2386 (1972).

Ephemeris of comet Grigg-Skjellerup (1902 II = 1922 I).
Kometn. Tsirk., *Kiev*, No. 125 (1972). In Russian.

Recovery of comet Grigg-Skjellerup, 1972b.
Kometn. Tsirk., *Kiev*, No. 126 (1972). In Russian.

Short-periodic comet Grigg-Skjellerup, 1972b.
Kometn. Tsirk., *Kiev*, No. 127 (1972). In Russian.

Comet Grigg-Skjellerup, 1972b.
Kometn. Tsirk., *Kiev*, No. 128 (1972). In Russian.

Periodic comet Grigg-Skjellerup (1972 b).
Yamamoto Circ., No. 1748 (1972). In Japanese.

A possible meteor shower associated with comet P/Grigg-Skjellerup. See Abstr. 104.021.

Possible meteor shower associated with Comet P/Grigg-Skjellerup. See Abstr. 104.041.

103.103 Comet 1970o Wolf-Harrington

Periodic comet Wolf-Harrington (1970o).
A. Mrkos.
IAU Circ., No. 2381 (1972).

Observations of comets. T. Seki, N. Kojima.
IAU Circ., No. 2393 (1972).

Periodic comet Wolf-Harrington, 1970o.
Kometn. Tsirk., *Kiev*, No. 125 (1972). In Russian.

Periodic comet Wolf-Harrington, 1970o.
Kometn. Tsirk., *Kiev*, No. 127 (1972). In Russian.

Observations of comets at the Klet̆ Observatory.
Kometn. Tsirk., *Kiev*, No. 130 (1972). In Russian.

103.104 Comet 1971a Toba

Comet Toba (1971a). D. P. Elias.
IAU Circ., No. 2381 (1972).

Comet Toba (1971a).
C. Torres, J. Petit, H. Wroblewski, H. Debehogne.
IAU Circ., No. 2389 (1972).

Comet Toba, 1971a.
Kometn. Tsirk., *Kiev*, No. 125 (1972). In Russian.

Comet Toba (1971 a).
Yamamoto Circ., No. 1748 (1972). In Japanese.

103.105 Comet 1970 XV Abe

Comet Abe (1970 XV). H. K. Raudsaar.
IAU Circ., No. 2381 (1972).

Observations of comet Abe, 1970g, in Tartu.
Kometn. Tsirk., *Kiev*, No. 125 (1972). In Russian.

Photometric investigation of comet Abe, 1970g.
A. M. Bakharev.
Kometn. Tsirk., *Kiev*, No. 127 (1972). In Russian.

103.106 Comet 1969 IX Tago-Sato-Kosaka

The Lyman-alpha image of comet Tago-Sato-Kosaka (1969g). E. B. Jenkins, D. W. Wingert.
Astrophys. Journ., Vol. 174, 697 - 704 (1972).
   On 1970 January 25.105 UT an f/2 objective-grating spectrograph aboard an Aerobee rocket recorded a faint image of comet Tago-Sato-Kosaka (1969g) in $L\alpha$ emission. Some sharpening of the image and reduction of film grain noise was accomplished. After this processing the $L\alpha$ comet image appeared as a nearly circularly symmetric diffuse glow approximately 8×10^5 km in diameter. There was also the suggestion of a faintly visible core of emission whose dimensions are less than 10^5 km. The sublimation of ice in the nucleus of the comet and the subsequent dissociation of the free water molecules by sunlight could produce a large cloud of hydrogen atoms around the comet. Resonant re-radiation of solar $L\alpha$ emission by this cloud is probably responsible for most of the observed emission.

Variation in the spectrum of comet Tago-Sato-Kosaka, 1969g. K. I. Churyumov, F. I. Kravtsov.
Kometn. Tsirk., *Kiev*, No. 125 (1972). In Russian.

Geschwindigkeiten im Schweif des Kometen Tago-Sato-Kosaka 1969 IX. K. Jockers, Rh. Lüst.
Mitt. Astron. Ges., No. 31, p. 184 (1972).

Comets in the solar wind. See Abstr. 102.022.

103.107 Comet 1972a Tempel 1

Comet P/Tempel (1), 1972a.
E. Roemer, L. M. Vaughan.
British Astron. Ass., Circ. No. 539 (1972).

Comet P/Tempel (1) 1972a. G. Rutter.
British Astron. Ass., Circ. No. 541 (1972).

Periodic comet Tempel 1 (1966 VII = 1972a).
E. Roemer, L. M. Vaughn.
IAU Circ., No. 2383 (1972).

Periodic comet Tempel 1 (1972a).
A. Mrkos, R. Petrovic̆ová.
IAU Circ., No. 2391 (1972).

Periodic comet Tempel 1 (1972a).
A. Mrkos, R. Petrovic̆ová.
IAU Circ., No. 2399 (1972).

Periodic comet Tempel 1 (1972a).
V. L. Matchett, T. Seki, A. Mrkos, R. Petrovic̆ová.
IAU Circ., No. 2405 (1972).

Periodic comet Tempel 1 (1972a). T. Seki,
N. Kojima, G. H. Rutter, R. L. Waterfield, I. M. Purcell,
R. Petrovičová, A. Mrkos, M. Antal.
IAU Circ., No. 2410 (1972).

Ephemeris of comet Tempel 1.
Kometn. Tsirk., *Kiev*, No. 125 (1972). In Russian.

Recovery of comet Tempel 1, 1972a.
Kometn. Tsirk., *Kiev*, No. 126 (1972). In Russian.

Short-periodic comet Tempel 1, 1972a.
Kometn. Tsirk., *Kiev*, No. 127 (1972). In Russian.

Observations of comets at the Kleť Observatory.
Kometn. Tsirk., *Kiev*, No. 130 (1972). In Russian.

Observations of comet Tempel 1, 1972a.
Kometn. Tsirk., *Kiev*, No. 132 (1972). In Russian.

Observations of comet Tempel 1, 1972a and Giaco-
bini-Zinner, 1972d at the Kleť Observatory.
Kometn. Tsirk., *Kiev*, No. 133 (1972). In Russian.

P/Tempel 1 (1972a).
Yamamoto Circ., Nos. 1747, 1749 (1972). In Japanese.

103.108   **Comet 1971f Tsuchinshan 1**

Comet P/Tsuchinshan (1), 1971f.   E. Roemer.
British Astron. Ass., Circ. No. 538 (1972).

Periodic comet Tsuchinshan 1 (1971f).
E. Roemer, L. M. Vaughn.
IAU Circ., No. 2381 (1972).

Recovery of comet Tsuchinshan 1, 1971f.
Kometn. Tsirk., *Kiev*, No. 126 (1972).   In Russian.

Short-periodic comet Tsuchinshan 1, 1971f.
Kometn. Tsirk., *Kiev*, No. 127 (1972). In Russian.

P/Tsuchinshan 1 (1971f).
Yamamoto Circ., No. 1747 (1972). In Japanese.

103.109   **Comet 1972g Neujmin 3**

Comet P/Neujmin (3) 1972g.
E. Roemer, R. A. McCallister.
British Astron. Ass., Circ. No. 541 (1972).

Periodic comet Neujmin 3 (1972g).
E. Roemer, R. A. McCallister.
IAU Circ., No. 2400 (1972).

Ephemeris of comet Neujmin 3.
Kometn. Tsirk., *Kiev*, No. 126 (1972). In Russian.

Recovery of comet Neujmin 3, 1972g.
Kometn. Tsirk., *Kiev*, No. 131 (1972). In Russian.

Periodic comet Neujmin 3 (1972g).
Yamamoto Circ., No. 1753 (1972). In Japanese.

103.110   **Comet 1971e Shajn-Schaldach**

Comet P/Shajn-Schaldach 1971e.
R. L. Waterfield, M. Hendrie, R. South.
British Astron. Ass., Circ. No. 538 (1972).

Periodic comet Shajn-Schaldach (1971e).
A. C. Danks, M. G. Sause.
IAU Circ., No. 2383 (1972).

Observations of comets.   T. Seki, N. Kojima.
IAU Circ., No. 2393 (1972).

Periodic comet Shajn-Schaldach (1971e).
K. Tomita, H. Kosai.
IAU Circ., No. 2417 (1972).

Observations of comet Shajn-Schaldach, 1971e.
Kometn. Tsirk., *Kiev*, No. 127 (1972). In Russian.

103.111   **Comet 1972c Tempel 2**

Periodic comet Tempel (2) 1972c.
E. Roemer, J. Q. Latta.
British Astron. Ass., Circ. No. 540 (1972).

Periodic comet Tempel 2 (1972c).
E. Roemer, J. Q. Latta.
IAU Circ., No. 2386 (1972).

Recovery of comet Tempel 2, 1972c.
Kometn. Tsirk., *Kiev*, No. 128 (1972). In Russian.

Periodic comet Tempel 2 (1972c).
Yamamoto Circ., No. 1749 (1972). In Japanese.

103.112   **Comet 1969 II Gunn**

Periodic comet Gunn (1969 II).   B. G. Marsden.
IAU Circ., No. 2417 (1972).

New elements and motion of comet Gunn, 1969 II
for 200 years (1800 - 2000).
N. A. Belyaev, E. A. Reznikov.
Kometn. Tsirk., *Kiev*, No. 128 (1972). In Russian.

Ephemeris of comet Gunn, 1969 II.
Kometn. Tsirk., *Kiev*, No. 134 (1972). In Russian.

103.113   **Comet 1970 XII Kojima**

Periodic comet Kojima (1970 XII).   C. Torres.
IAU Circ., No. 2403 (1972).

Investigation of the motion of the short-periodic
comet Kojima, 1970 XII for 200 years (1800 - 2000).
N. A. Belyaev, E. A. Reznikov.
Kometn. Tsirk., *Kiev*, No. 128 (1972).   In Russian.

103.114   **Comet 1972e Gehrels**

New comet Gehrels 1972e.   T. Gehrels.
British Astron. Ass., Circ. No. 540 (1972).

Comet Gehrels (1972e).
T. Gehrels, H. L. Giclas.
IAU Circ., No. 2391 (1972).

Comet Gehrels (1972e).
T. Gehrels, C. D. Vesely, R. E. Sather, H. L. Giclas.
IAU Circ., No. 2393 (1972).

Comet Gehrels (1972e).   H. L. Giclas.
IAU Circ., No. 2399 (1972).

Comet Gehrels (1972e).   B. G. Marsden.
IAU Circ., No. 2403 (1972).

New comet Gehrels, 1972e.
Kometn. Tsirk., *Kiev,* No. 129 (1972). In Russian.

Comet Gehrels, 1972e.
Kometn. Tsirk., *Kiev,* No. 130 (1972). In Russian.

Comet Gehrels, 1972e.
Kometn. Tsirk., *Kiev,* No. 131 (1972). In Russian.

Comet Gehrels (1972 e).
Yamamoto Circ., No. 1754 (1972). In Japanese.

103.115   Comet 1972f Bradfield

New comet Bradfield 1972f.
W. A. Bradfield, M. P. Candy.
British Astron. Ass., Circ. No. 540 (1972).

Comet Bradfield 1972f.   M. P. Candy.
British Astron. Ass., Circ. No. 541 (1972).

Comet Bradfield (1972f).
W. A. Bradfield, T. B. Tregaskis.
IAU Circ., No. 2392 (1972).

Comet Bradfield (1972f).
R. R. D. Austin, M. V. Jones.
IAU Circ., No. 2394 (1972).

Comet Bradfield (1972f).   M. P. Candy, G. Punko,
J. B. Tregaskis, H. J. Edelman, R. C. Shinkfield.
IAU Circ., No. 2396 (1972).

Comet Bradfield (1972f).
Z. M. Pereyra, B. Oviedo, N. Keller.
IAU Circ., No. 2400 (1972).

Comet Bradfield (1972f).   T. B. Tregaskis,
W. J. H. Fisher, A. C. Gilmore, R. E. Millington, J. B. Gibson,
U. T. Gibson, C. Rogati, Zaparat, M. P. Candy.
IAU Circ., No. 2402 (1972).

Comet Bradfield (1972f).   Z. M. Pereyra,
B. Oviedo, C. E. E. Rogati, J. B. Gibson, U. T. Gibson.
IAU Circ., No. 2404 (1972).

Comet Bradfield (1972f).   Z. M. Pereyra, B.
Oviedo, R. Decostanzi, H. L. Giclas, V. L. Matchett.
IAU Circ., No. 2408 (1972).

Comet Bradfield (1972f).   J. A. Bruwer.
IAU Circ., No. 2409 (1972).

Comet Bradfield (1972f).
A. C. Gilmore, R. E. Millington.

IAU Circ., No. 2411 (1972).

Comet Bradfield (1972f).
Z. M. Pereyra, B. Oviedo, R. Decostanzi.
IAU Circ., No. 2420 (1972).

New comet Bradfield, 1972f.
Kometn. Tsirk., *Kiev,* No. 129 (1972). In Russian.

Southern comet Bradfield, 1972f.
Kometn. Tsirk., *Kiev,* No. 130 (1972). In Russian.

Southern comet Bradfield, 1972f.
Kometn. Tsirk., *Kiev,* No. 131 (1972). In Russian.

Comet Bradfield, 1972f.
Kometn. Tsirk., *Kiev,* No. 132 (1972). In Russian.

Comet Bradfield (1972 f).
Yamamoto Circ., Nos. 1753, 1754 (1972). In Japanese.

103.116   Comet 1972d Giacobini-Zinner

Periodic comet Giacobini-Zinner 1972d.
E. Roemer, R. A. McCallister.
British Astron. Ass., Circ. No. 540 (1972).

Periodic comet Giacobini-Zinner (1972d).
E. Roemer, R. A. McCallister.
IAU Circ., No. 2390 (1972).

Periodic comet Giacobini-Zinner (1972d).
R. E. McCrosky, C. Y. Shao, A. Posen, T. Seki, A. Mrkos,
J. E. Bortle.
IAU Circ., No. 2414 (1972).

Periodic comet Giacobini-Zinner (1972d).
A. Mrkos.
IAU Circ., No. 2419 (1972).

Recovery of comet Giacobini-Zinner, 1972d.
Kometn. Tsirk., *Kiev,* No. 129 (1972). In Russian.

Observations of comet Giacobini-Zinner, 1972d.
Kometn. Tsirk., *Kiev,* No. 132 (1972). In Russian.

**Observations of comet Tempel 1, 1972a and Giaco-
bini-Zinner, 1972d at the Kleť Observatory.**
Kometn. Tsirk., *Kiev,* No. 133 (1972). In Russian.

Comet Giacobini-Zinner, 1972d.
Kometn. Tsirk., *Kiev,* No. 134 (1972). In Russian.

P/Giacobini-Zinner (1972 d).
Yamamoto Circ., Nos. 1750, 1756 (1972). In Japanese.

103.117   Comet 1971d Tsuchinshan 2

Periodic comet Tsuchinshan 2 (1971d).
T. Seki.
IAU Circ., No. 2390 (1972).

Periodic comet Tsuchinshan 2 (1971d).
T. Seki, N. Kojima, C. Torres, J. Petit, H. Wroblewski.
IAU Circ., No. 2410 (1972).

Periodic comet Tsuchinshan 2, 1971d.
Kometn. Tsirk., *Kiev*, No. 131 (1972). In Russian.

Observations of comet Tsuchinshan 2, 1971d.
Kometn. Tsirk., *Kiev*, No. 132 (1972). In Russian.

103.118   Comet 1910 II Halley

The past orbit of Halley's comet.   T. Kiang.
Mem. Roy. Astron. Soc., Vol. 76, 27 - 66 (1972).
Perturbations of the orbit of Halley's comet over the last 28 revolutions are calculated and a parallel re-examination of Chinese records on this object made. For the calculation, the method used by Cowell and Crommelin in 1907/8 is refined in several respects. All available records, mainly Chinese ones, are re-examined with the aim of improving on the catalogue values of T. A set of remarkably precise data for the return of +837 are made use of for the first time, and the three recorded novae in the same year are shown to be not connected with Halley's comet.

The linking of four appearances of the Halley comet.
G. Sitarski.
Postępy Astron., Vol. 20, 167 - 169 (1972).   In Polish.

The effect of a trans-Plutonian planet on Halley's comet.   See Abstr. 102.010.

103.119   Comet 1963 III Alcock

Analysis of the spectra of the 1963 III comet (Alcock 1963b) taken before and after brightness increase (blast).   S. Grudzińska.
Postępy Astron., Vol. 20, 159 - 160 (1972).   In Polish.

103.120   Comet 1970 III Kohoutek

Comet Kohoutek (1970 III).
S. I. Gerasimenko, V. L. Afanas'ev.
IAU Circ., No. 2384 (1972).

103.121   Comet 1957 IV Schwassmann-Wachmann 1

Periodic comet Schwassmann-Wachmann 1.
IAU Circ., No. 2387 (1972).

Periodic comet Schwassmann-Wachmann 1.
Yamamoto Circ., No. 1749 (1972). In Japanese.

103.122   Comet 1970e Ashbrook-Jackson

Periodic comet Ashbrook-Jackson (1970e).
A. N. Dejch, V. D. Shkutov, N. A. Shkutova.
IAU Circ., No. 2393 (1972).

103.123   Comet 1969 VI Faye

Periodic comet Faye (1969VI).   G. Bocşa.
IAU Circ., No. 2401 (1972).

103.124   Comet 1965 VIII Ikeya-Seki

Photographic observations of comet Ikeya-Seki, 1965 VIII in the USSR (atlas).
AN UkrSSR. Glav. astron. observ. Naukova dumka, Kiev. 26 pp. Price 30 Kop. (1970). In Russian.

103.125   Comet 1971c Kearns-Kwee

Periodic comet Kearns-Kwee (1971c).
B. G. Marsden.
IAU Circ., No. 2405 (1972).

Ephemeris of periodic comet Kearns-Kwee, 1971c.
Kometn. Tsirk., *Kiev*, No. 132 (1972). In Russian.

103.126   Comet 1971b Holmes

Periodic comet Holmes (1971b).
IAU Circ., No. 2407 (1972).

Continuation of the ephemeris of comet Holmes, 1971b.
Kometn. Tsirk., *Kiev*, No. 133 (1972). In Russian.

103.127   Comet 1972h Sandage

Comet Sandage.
A. R. Sandage, D. Hoffleit, B. G. Marsden.
IAU Circ., Nos. 2413, 2414 (1972).

Comet Sandage (1972h).   A. R. Sandage, C. T. Kowal, K. D. Clardy, V. Ionescu-Vlăsceanu, R. L. Waterfield, I. M. Purcell, R. H. South, C. Day, B. Hatfield, B. G. Marsden.
IAU Circ., No. 2416 (1972).

Comet Sandage (1972h).   A. R. Sandage, C. T. Kowal, H. H. Lanning, K. D. Clardy, H. L. Giclas.
IAU Circ., No. 2418 (1972).

Comet Sandage (1972h).   B. Milet.
IAU Circ., No. 2420 (1972).

New comet Sandage, 1972h.
Kometn. Tsirk., *Kiev*, Nos. 132, 133, 134 (1972). In Russian.

Comet Sandage (1972 h).
Yamamoto Circ., No. 1756 (1972). In Japanese.

103.128   Comet 1965 V Reinmuth

Periodic comet Reinmuth 1.   G. Sitarski.
IAU Circ., No. 2419 (1972).

103.129   Comet 1970 XIV Whipple

Observation and feature variations of comet 1969e before and during the perihelion passage.   A. Mrkos.
From plasma to planet. 21st Nobel symposium 1971, (see 012.028), p. 261 - 272 (1972).

104 Meteors, Meteor Streams

104.001 A radar determination of the right ascension and declination of the Quadrantid meteor stream.
D. W. Hughes.
Monthly Notics Roy. Astron. Soc., Vol. 155, 395 - 402 (1972).
    Radio echo observations were made of the Quadrantid meteor shower during early January 1966. The mean range and influx rate of meteors having a radio magnitude greater than +7.5 was observed as a function of universal time. From this data the right ascension and declination (1950.0) were found to be 230.45 ± 0.18° and 47.0 ± 1.0° respectively.

104.002 The diurnal variation in the mass distribution of sporadic meteors.
D. W. Hughes, D. G. Stephenson.
Monthly Notices. Roy. Astron. Soc., Vol. 155, 403 - 413 (1972).
    The value of the exponent in the inverse power law describing the differential distribution of meteoroid mass was measured every half hour during 1969 September 9–14. Meteors were observed using a 25.5 MHz backscatter radar. The echoes from over 30000 underdense meteor trains were analyzed. The experimental data and their significance discussed. It is concluded that there is a genuine diurnal variation in the mass distribution index.

104.003 Radio-echo observations of the major night-time meteor streams–III. Quadrantids.
L. M. G. Poole, D. W. Hughes, T. R. Kaiser.
Monthly Notices Roy. Astron. Soc., Vol. 156, 223 - 241 (1972).
    The results of radio-echo observations of the Quadrantid meteor shower between 1964 and 1971 are presented. Radio magnitude distributions in the range +6.05^m to +7.55^m were determined as a function of solar longitude. Factors affecting the incident flux are discussed.

104.004 Variations in the rate of meteor head echoes with the diurnal motion of the radiant. A. Hajduk.
Bull. Astron. Inst. Czechoslovakia, Vol. 23, 35 - 39 (1972).
    The relative rate of meteor head echoes is derived from the Orionid observations in 1961–1965. The variation in the occurrence of head echoes is compared with that of the Perseid meteors obtained by McIntosh (1963). In both cases a very similar effect appears, from which an influence of the radiant position on the abundance of head echoes may be inferred. Some correlations between the rate of head echoes and other echo characteristics are found.

104.005 Meteor variation of cosmic ray intensity for type I and II meteor streams. S. A. Belsky.
Geomagn. Aeronom., Vol. 12, 114 - 116 (1972). In Russian. Brief information.

104.006 Das Perseidenmaximum 1971.
H. G. Schmidt, H. Rollenhagen, J. Faubel.
SuW, Vol. 11, 22 (1972).

104.007 The Leonids in November.
Sky Telescope, Vol. 43, 57 - 58 (1972).

104.008 The Leonids: Who saw them first?
D. J. Schove.
Sky Telescope, Vol. 43, 156 - 157 (1972).

104.009 The Quadrantid meteor stream. K. B. Hindley.
Sky Telescope, Vol. 43, 162 - 164 (1972).

104.010 Dutch meteor observations. B. C. J. Apeldoorn.
Sky Telescope, Vol. 43, 195 (1972).

104.011 Some results from the 1971 Geminid meteor shower.
Sky Telescope, Vol. 43, 196 - 197 (1972).

104.012 Determination of the moment of meteor flight by means of an optical shutter. I. T. Zotkin.
Astron. vestn., Vol. 5, 240 - 244 (1971). In Russian. – Abstr. in Referativ. Zhurn. 51. Astron., 3.51.391 (1972).

104.013 The number of meteors in the 1–7 magnitude range according to visual observations.
T. V. Bryzgalova, V. V. Martynenko.
Astron. vestn., Vol. 5, 248 - 251 (1971). In Russian. – Abstr. in Referativ. Zhurn. 51. Astron., 3.51.392 (1972).

104.014 Two-channel photoelectric observations of Perseids 1970 at Kuchino. D. L. Astavin-Razumin.
Astron. vestn., Vol. 5, 252 - 253 (1971). In Russian. – Abstr. in Referativ. Zhurn. 51. Astron., 3.51.393 (1972).

104.015 The radiative efficiency of potassium ions and atoms emitted from a meteor. S. H. Neff.
Astrophys. Journ., Vol. 173, 235 - 241 (1972).
    We have used an ion-beam technique to measure the emission cross-section for K I λ7665 for collisions of K^+ with N_2 at energies below 1000 eV. By observing the dependence of radiated intensity with target-gas pressure, we have also obtained an upper limit to the emission cross-section for collisions of neutral K atoms with N_2. Using these experimental results, we estimate the radiative efficiency for potassium atoms or ions emitted by a high-speed meteor to be ~5 photons per atom and ~0.5 photons per ion.

104.016 The detection of shower structure in the sporadic meteor background.
L. M. G. Poole, T. R. Kaiser.
Monthly Notices Roy. Astron. Soc., Vol. 156, 283 - 290 (1972).
    Inspections of range-time photographs of meteor echoes on an objective statistical basis have provided clear evidence for the existence of radiant structure within the background of sporadic meteors.

104.017 Results of determinations of meteor orbits using continuous radio emission. D. M. Smagin,
M. F. Lagutin, Kh. D. Gul'medov, A. Kh. Khanberdiev.
Izv. AN TurkmSSR. Ser. fiz-tekhn., khim. i geol. n., 1971, No. 4, p. 122 - 125. In Russian. – Abstr. in Referativ. Zhurn. 51. Astron., 4.51.394 (1972).

104.018 Investigating the number of meteors.
V. V. Martynenko.
Zemlya i Vselennaya, 1972, No. 2, p. 70 - 71. In Russian.

104.019 Nodal retrogression of the Quadrantid meteor stream. D. W. Hughes.
Observatory, Vol. 92, 41 - 43 (1972).
    By considering all available radar data the nodal retrogression rate for Quadrantid radio meteors has been found to be 0.41 ± 0.17 deg/century in comparison to a rate of 0.31 ± 0.04 deg/century for visual meteors. In 1968 Quadrantid radio meteors (mean mass 2 × 10^{-3} gm) maximised at a solar longitude of 282.525 ± 0.03 deg. in comparison to the visual meteors (mean mass 0.3 gm) which maximised at 282.53 ± 0.02 deg.

104.020 **Results of radar observations of the Quadrantid meteor shower in 1966 - 1969.**
Sh. O. Isamutdinov.
Izv. AN TadzhSSR. Otd. Fiz.-mat. i geol.-khim.n., 1971, No. 2, p. 11 - 20. In Russian. — Abstr. in Referativ. Zhurn. 51. Astron., 5.51.321 (1972).

104.021 **A possible meteor shower associated with comet P/Grigg-Skjellerup.** H. B. Ridley.
Journ. British Astron. Ass., Vol. 82, 95 - 98 (1972).

104.022 **Fireball of 1969 April 3.** H. Miles.
Journ. British Astron. Ass., Vol. 82, 126 - 129 (1972).

104.023 **Bolides of 1971 February 18 and March 2.**
H. Miles.
Journ. British Astron. Ass., Vol. 82, 194 - 197 (1972).

104.024 **Light-emission measurements of iron at simulated meteor conditions.** C. A. Boitnott, H. F. Savage.
Astrophys. Journ., Vol. 174, 201 - 206 (1972).
Ionization and spectral emission (3500–4200 Å) cross-sections have been measured for $Fe + N_2$, O_2 reactions between 350 and 2000 eV. Meteor luminous efficiencies from 3500 to 5400 Å were calculated from these measurements. These results are thought to be accurate within ±40 percent for meteor velocities greater than 35 km s $^{-1}$ and at conditions where free molecular flow prevails.

104.025 **Een meteorenregen op 22/23 April 1972??**
J. Meeus.
Hemel en Dampkring, Vol. 70, 96 (1972).

104.026 **De Lyriden: een fraaie voorjaarszwerm.**
B. Apeldoorn.
Hemel en Dampkring, Vol. 70, 96 (1972).

104.027 **Meer bijzonderheden over de bolide van 22 april 1971.** P. A. Koning.
Hemel en Dampkring, Vol. 70, 131 - 134 (1972).

104.028 **Beobachtung des Geminiden-Meteorstromes am 13. Dezember 1971.** R. Germann, R. A. Naef.
Orion Schaffhausen, 30. Jahrgang, p. 62 - 63 (1972).

104.029 **A Giacobinid meteor shower prediction for October, 1972.** D. K. Yeomans.
Bull. American Astron. Soc., Vol. 4, 268 (1972). — Abstr. AAS.

104.030 **Artificial meteor ablation studies: Iron oxides.**
M. B. Blanchard.
Journ. Geophys. Res., Vol. 77, 2442 - 2455 (1972).
Artificial meteor ablation was performed on natural minerals composed predominately of magnetite and hematite by using an arc-heated plasma stream of air. Analysis indicates that most of the ablated debris was composed of two or more minerals.

104.031 **Taurid meteor stream fireballs.**
K. B. Hindley.
Journ. British Astron. Ass., Vol. 82, 287 - 298 (1972).

104.032 **De Perseiden komen !!** B. Apeldoorn.
Hemel en Dampkring, Vol. 70, 191 - 192 (1972).

104.033 **Wenken bij het waarnemen en fotograferen van meteoren.** B. Apeldoorn.
Hemel en Dampkring, Vol. 70, 192 - 193 (1972).

104.034 **A southern hemisphere radio meteor orbit survey.**
G. Gartrell.

Proc. Astron. Soc. Australia, Vol. 2, 89 - 90 (1972).

104.035 **Giacobinid meteor shower: Prospect for 1972.**
B. A. McIntosh.
Journ. Roy. Astron. Soc. Canada, Vol. 66, 149 - 156 (1972).

104.036 **Effect of sunrise on the meteor region.**
D. W. Hughes, W. J. Baggaley.
Nature, Vol. 237, 224 - 226 (1972).
A marked increase in the number of observed radio meteor echoes with enhanced duration (> 2 sec) was found to occur during the 1972 Quadrantid return at 08.30 U. T. about 60 min after layer sunrise. At this time the solar zenith angle at the layer was 91°. A corresponding increase of short duration echoes was not observed. It is suggested that this sunrise effect is caused by (a) an increase in large-scale turbulence in the meteor region at sunrise and (b) a day-night change in the electron loss process in meteor trains.

104.037 **Small bodies in the solar system.** F. L. Whipple.
Monthly Notes Astron. Soc. Southern Africa, Vol. 31, 47 - 48 (1972). — Abstract.

104.038 **Instruction for meteor observations from more observing stations.** V. Znojil.
Contr. Obs. and Planetarium Brno, No. 13, 93 pp. (1971). In Czech.

104.039 **Deceleration of meteoric bodies in the visible part of the trajectory.** V. A. Vorob'eva, E. N. Kramer.
Astron. vestn., Vol. 6, 29 - 34 (1972). In Russian. — Abstr. in Referativ. Zhurn. 51. Astron., 7.51.274 (1972).

104.040 **Evaluation of the contribution of lunar particles to the meteoric sporadic background.** V. P. Orlov.
Astron. vestn., Vol. 6, 44 - 48 (1972). In Russian. — Abstr. in Referativ. Zhurn. 51. Astron., 7.51.276 (1972).

104.041 **Possible meteor shower associated with comet P/Grigg-Skjellerup.** H. B. Ridley.
British Astron. Ass., Circ. No. 538 (1972).

104.042 **The NASA LRC Faint Meteor Spectra Patrol.**
G. A. Harvey.
Rep. NASA–TN–D–6298, National Aeronautics and Space Administration, Langley Station, Va. [Available from NTIS, Springfield, Va.], 27 pp. (1971).
A description of the instrumentation facilities and patrol techniques used by the NASA LRC Faint Meteor Spectra Patrol is given. A list of the 319 meteor spectra obtained in the first 2 years of operation is included and four representative types of spectra obtained by the patrol are briefly described.

104.043 **Comparison of radio-meteor rate with abundance of sodium in the upper atmosphere.** M. Gadsden.
Ann. Géophys., Vol. 27, 401 - 406 (1971).
The recent publication of detailed, hourly, counts of radio meteors observed from Christchurch, New Zealand, makes possible a direct comparison of these data with the sodium abundance measurements made in 1964 at Lauder, New Zealand.

104.044 **Cometary meteoroids.** P. M. Millman.
From plasma to planet. 21st Nobel symposium 1971, (see 012.028), p. 157 - 168 (1972).

104.045 **Meteor and asteroid streams.** B. A. Lindblad.
From plasma to planet. 21st Nobel symposium 1971, (see 012.028), p. 195 - 210 (1972).

On the effect of small-scale wind shears on radio

meteor echoes.   See Abstr. 082.024.

Are meteors a tool for studying the asteroids? or vice versa?   See Abstr. 098.063.

Comets and meteors.   See Abstr. 102.028.

Une pluie d'étoiles filantes les 22–23 avril 1972? See Abstr. 103.102.

104.901 **Errata: Complete data on iron meteoroid** [Bull. Astron. Inst. Czechoslovakia, Vol. 17, 195 - 206 (1966)].   Z. Ceplecha.
Bull. Astron. Inst. Czechoslovakia, Vol. 23, 146 (1972).

104.902 **Errata: Fireballs and the physical theory of meteors** [Bull. Astron. Inst. Czechoslovakia, Vol. 21, 271 - 296 (1970)].   R. E. McCrosky, Z. Ceplecha.
Bull. Astron. Inst. Czechoslovakia, Vol. 23, 146 (1972).

105 Meteorites, Meteorite Craters

105.001 Occurrence and significance of formaldehyde in the Allende carbonaceous chondrite.
I. A. Breger, P. Zubovic, J. C. Chandler, R. S. Clarke, Jr.
Nature, Vol. 236, 155 - 158 (1972).

Meteorites may disperse formaldehyde to planets where, in the proper conditions, the compound could then serve as a precursor for carbohydrates.

105.002 Search for carbon 14 in tektites. R. S. Boeckl.
Journ. Geophys. Res., Vol. 77, 367 - 368 (1972).

An attempt was made to detect cosmic-ray-produced ^{14}C in australites and far-east tektites. No ^{14}C activity was found outside of statistical fluctuations. If a terrestrial age of 10^4 years is assumed for australites, a maximum cosmic-ray exposure time of 10^3 years is obtained for these tektites on the basis of a $2-\sigma$ statistical error.

105.003 Extinct superheavy element in meteorites: Attempted characterization.
E. Anders, J. W. Larimer.
Science, Vol. 175, 981 - 983 (1972).

If the unexplained fission xenon component in meteorites is due to a volatile superheavy element, this element must have had a heat of vaporization of 54 ± 3 kilocalories per mole and a normal boiling point of $2500° ± 400°K$. The prime candidates are elements 111 and 115, followed by 113, 114, 112, and 116.

105.004 Meteoritenfälle in Deutschland. H. Eisenlohr.
SuW, Vol. 11, 5 - 10 (1972).

105.005 Mikrospektraluntersuchungen des Meteoriten Pawel.
A. Petrakiew, G. Dimitrow, S. Beltschew, N. Nikolow.
Jenaer Rundschau, (Jena Rev.), 17. Jahrgang, p. 21 - 23 (1972).

105.006 Uranium content and radiogenic ages of hypersthene, bronzite, amphoterite and carbonaceous chondrites.
D. E. Fisher.
Geochim. Cosmochim. Acta, Vol. 36, 15 - 33 (1972).

U was measured by fission track analysis in 115 samples of hypersthene, bronzite, amphoterite and carbonaceous chondrites. On a weight basis the average values for the Cl carbonaceous and bronzite chondrites are similar to the "classic" value of 11 ppb, but the hypersthenes and amphoterites are ~50 per cent higher.

105.007 The effect of phosphorus on the formation of the Widmanstätten pattern in iron meteorites.
J. I. Goldstein, A. S. Doan, Jr.
Geochim. Cosmochim. Acta, Vol. 36, 51 - 69 (1972).

Most iron meteorites contain small to moderate amounts of elements other than Fe–Ni. One of the most important of these elements is phosphorus. It is the purpose of this paper to outline the effects of P on the development of the Widmanstätten pattern. We will base this discussion on our recent re-evaluation of the Fe–Ni–P equilibrium phase diagram (Doan and Goldstein, 1970) and the results of our laboratory cooling experiments using Fe–Ni–P alloys.

105.008 Thermodynamic models in cosmochemical systems.
P. R. Griffiths, C. W. Brown, E. R. Lippincott, M. O. Dayhoff.
Geochim. Cosmochim. Acta, Vol. 36, 109 - 128 (1972).

Generalized computer methods are developed for inferring details of the formation of cosmochemical systems. Compositions of ideal gas mixtures existing in equilibrium with multicomponent solid and liquid phases are calculated. A comparison of computed results with experimental data is made for the ternary system $MgO-FeO-SiO_2$. A model system containing the elements H, O, Si, Mg, S, C, Cl and F is investigated over a range of compositions involving the gas and ten solid phases, to show the power of the technique in dealing with complex gas-solid equilibria.

105.009 Oxygen isotope temperatures of "equilibrated" ordinary chondrites.
N. Onuma, R. N. Clayton, T. K. Mayeda.
Geochim. Cosmochim. Acta, Vol. 36, 157 - 168 (1972).

Measurements have been made of O^{18}/O^{16} ratios of co-existing minerals (plagioclase, pyroxene, olivine) from nine ordinary chondrites. The O^{18} content of a given mineral increases systematically from H through L to LL group. Isotope fractionations for mineral pairs fall on a concordancy line, suggesting that these minerals were crystallized in oxygen isotope equilibrium. The isotopic temperature is estimated to be 950 ± 100°C for seven of the type 5 and type 6 chondrites.

105.010 Origin of organic matter in early solar system – V. Further studies of meteoritic hydrocarbons and a discussion of their origin.
M. H. Studier, R. Hayatsu, E. Anders.
Geochim. Cosmochim. Acta, Vol. 36, 189 - 215 (1972).

In 1968 we proposed that the organic compounds in meteorites had formed in the solar nebula by Fischer-Tropsch-type reactions, i.e. catalytic combination of CO, H_2 and NH_3 on the surfaces of dust grains. The recent fall of two large carbonaceous chondrites, Allende and Murchison provided an opportunity to reexamine our earlier results. The Murchison meteorite contains aliphatic and aromatic hydrocarbons similar to those made in static Fischer-Tropsch-type syntheses. In the Allende meteorite, only methane, benzene, toluene and an aromatic polymer seem to be indigenous.

105.011 Chemical fractionations in meteorites – V. Volatile and siderophile elements in achondrites and ocean ridge basalts. J. C. Laul, R. R. Keays, R. Ganapathy, E. Anders, J. W. Morgan.
Geochim. Cosmochim. Acta, Vol. 36, 329 - 345 (1972).

Lunar and terrestrial basalts differ substantially in their contents of siderophile and volatile elements (Ganapathy et al., 1970; Anders et al., 1971). Presumably these differences reflect compositional differences among the planets, established during accretion from the solar nebula. It therefore seemed desirable to extend this comparison to meteoritic 'basalts'. Eighteen achondrites and 4 terrestrial basalts (3 ocean ridge, 1 continental) were analyzed by radiochemical neutron activation analysis for Ag, Au, Bi, Br, Cd, Co, Cs, Cu, Ga, In, Ir, Rb, Se, Tl and Zn. The results are shown in many tables.

105.012 On the origins of trapped helium, neon and argon isotopic variations in meteorites–I. Gas-rich meteorites, lunar soil and breccia. D. C. Black.
Geochim. Cosmochim. Acta, Vol. 36, 347 - 375 (1972).

This work is an attempt to synthesize the earlier discussions, as well as more recent studies of lunar material, and so delineate the origins of the various trapped compositions observed in gas-rich meteorites and carbonaceous chondrites. The compositions characteristically found in the former class of meteorites are discussed in this paper, while the compositions present in the latter class of meteorites are discussed in a second paper (see Abstr. 07.105.013).

105.013 On the origins of trapped helium, neon and argon isotopic variations in meteorites – II. Carbonaceous

meteorites. D. C. Black.
Geochim. Cosmochim. Acta, Vol. 36, 377 - 394 (1972).

This paper reports the results of rare gas analyses from stepwise heating of five carbonaceous chondrites. These data, in conjunction with corresponding isotopic patterns in gas-rich meteorites discussed in paper I (Black, 1972), are used to construct a unified working model for the trapped helium, neon and argon isotopic structures in carbonaceous chondrites.

105.014 Global deposition of extraterrestrial particles during October. J. Rosinski.
Journ. Atmosph. Terr. Phys., Vol. 34, 487 - 497 (1972).

Concentration peaks of magnetic spheroids were recorded simultaneously at different latitudes during October 1967 and 1969. The shapes of concentration curves in each hemisphere are very similar. Spheroids present in an exceptionally high concentration on 10 October 1969 in the southern hemisphere contained Mn, Cr, and Ni, a fact which supports the existence of an extraterrestrial source for those particles. There is a lack of association between the magnetic spheroid concentration peak and any well-defined meteor activity.

105.015 Glassy objects (impactite glasses?), a possible new evidence for meteoritic origin of the Lonar Crater, Maharashtra State, India. V. K. Nayak.
Earth Planet. Sci. Letters, Vol. 14, 1 - 6 (1972).

The glassy objects resembling impact glasses are recognized for the first time from the suspected Lonar Crater, India. Its physical, optical and chemical characteristics are furnished and discussed. The available data provide possible evidence to support an explosive meteorite impact origin of the Indian crater.

105.016 Chemical sub-groups amongst HL chondrites. T. S. McCarthy, L. H. Ahrens.
Earth Planet. Sci. Letters, Vol. 14, 97 - 102 (1972).

Recently, Van Schmus proposed a subdivision of the HL or Type III carbonaceous chondrites into two sub-groups – the Ornans and Vigarano sub-groups – on a basis of their chondrule morphology. Eight HL chondrites have been analyzed with a view to investigating in more detail the nature of the composition differences between the Vigarano and Ornans sub-groups.

105.017 Thermal release characteristics of spallogenic He, Ne, and Ar from the Carbo iron meteorite. L. E. Nyquist, J. C. Huneke, H. Funk, P. Signer.
Earth Planet. Sci. Letters, Vol. 14, 207 - 215 (1972).

We have investigated the diffusion of spallogenic noble gases from samples of the Carbo iron meteorite by means of an experimental procedure which, in addition to providing detailed information about the release pattern, eliminates ambiguities encountered in other diffusion studies.

105.018 Mass spectrometric analysis of gas inclusions in Muong Nong glass and Libyan Desert glass. E. Jessberger, W. Gentner.
Earth Planet. Sci. Letters, Vol. 14, 221 - 225 (1972).

Noble and non-noble gases in bubbles of Muong Nong- and Libyan Desert glass were released by vacuum crushing at room temperature and measured by high sensitivity mass spectrometry. The N_2 :Ar:Kr:Xe ratio as well as the rare gas isotope ratios were found to be atmospheric, indicating the terrestrial origin of these glasses.

105.019 On the mass distribution of fragments from Łowicz meteorite shower 1935. B. Lang.
Earth Planet. Sci. Letters, Vol. 14, 245 - 248 (1972).

The Łowicz meteorite shower fell on March 12, 1935, covering an area of 9 km². The histogram accounting for 96 specimens of this meteorite with masses ranging from 2 to 10^4 g, expressed in Krumbein phi-units, has been interpreted in terms of a lognormal frequency distribution.

105.020 Ein angeblicher Meteoritenkrater nahe der türkisch-iranischen Grenze. W. Sandner.
Sterne, 48. Jahrgang, p. 42 - 44 (1972).

105.021 Crater near the Ladoga Lake. E. V. Borovkova.
Priroda, No. 3.72, p. 119 - 120 (1972). In Russian.

105.022 Crater at Yenisei ridge. N. K. Mikhnyak.
Priroda, No. 3.72, p. 120 (1972). In Russian.

105.023 Crater near Moscow. A. K. Stanyukovich.
Priroda, No. 3.72, p. 120 - 121 (1972). In Russian.

105.024 Solar rare gases and the abundances of the elements. K. Marti, L. L. Wilkening, H. E. Suess.
Astrophys. Journ., Vol. 173, 445 - 450 (1972).

A distinct component of the rare gases present in gas-rich meteorites and in gas-rich lunar material has been directly implanted by solar-particle radiation. The elemental abundance ratios in this component agree with those predicted by interpolation in accordance with abundance rules, using analytical values for type I carbonaceous chondrites. The importance of this agreement in connection with astrophysical and cosmochemical problems is discussed.

105.025 On the astronomical interpretation of data on dust particles recorded by the spaceships "Pioneer". B. Yu. Levin, A. N. Simonenko.
Kosm. Issled., Vol. 10, 113 - 119 (1972). In Russian.

105.026 Some aspects of the geochemistry of Ru, Os, Ir and Pt in iron meteorites. J. H. Crocket.
Geochim. Cosmochim. Acta, Vol. 36, 517 - 535 (1972).

Neutron activation data for Ru, Os and Ir in 46 iron meteorites and Pt and Au in 38 iron meteorites are reported.

105.027 Origin of organic matter in early solar system—VI. Catalytic synthesis of nitriles, nitrogen bases and porphyrin-like pigments. R. Hayatsu, M. H. Studier, S. Matsuoka, E. Anders.
Geochim. Cosmochim. Acta, Vol. 36, 555 - 571 (1972).

A variety of nitrogen compounds have been synthesized by a static Fischer-Tropsch type reaction from CO, D_2 and ND_3, with Ni-Fe and Al_2O_3 catalysts. Such reactions may have been involved in the production of interstellar molecules, organic compounds in meteorites, and prebiotic organic matter on planets.

105.028 Radiation ages of different fragments of the Sikhote-Alin meteorite fall. E. M. Kolesnikov, A. K. Lavrukhina, A. V. Fisenko, L. K. Levsky.
Geochim. Cosmochim. Acta, Vol. 36, 573 - 576 (1972).

Seven specimens from five different fragments of the Sikhote-Alin iron meteorite have been analyzed for their content of ^{39}Ar, ^{38}Ar, ^{21}Ne, and ^{3}He. The ^{39}Ar–^{38}Ar–cosmic ray exposure ages calculated from these data are $(450 \pm 20) \times 10^6$ yr for four of the fragments, that of one fragment is only $(145 \pm 11) \times 10^6$ yr.

105.029 Are microtektites the result of cometary impacts with the earth? H. A. Kellner, S. Yabushita.
Nature, Vol. 235, 383, with a reply by S. A. Durrani (1972).

A comment is made on the hypothesis put forward by Durrani and Khan in Nature, Vol. 232, 320 - 323 (1971): The probability of cometary impact appears to be too small by a factor of 80 to allow an explanation of the frequency of geomagnetic reversal to which the microtectites' ages are related (about 0.71 and 0.88 m.y.).

105.030 Frost's rule and the fragments of the Allende meteorite. M. Kowalski, B. Lang.
Nature, Phys. Sci., Vol. 235, 149 - 150 (1972).

The consistence of spatial sorting of 60 fragments of the Allende meteorite, taken for statistical representative sample, with Frost's logarithmic rule (Meteoritics, Vol. 4, 217 - 232 (1969)) was examined. With least squares method two straight lines in semilog co-ordinates were fitted to a set of points as obtained from averaging the positions of stones, divided into groups by weight. This arrangement of fragments on the strewn field has been found to suggest a primary breakup of the meteoroid into two parts.

105.031 Terrestrial age of nineteen stony meteorites derived from their radiocarbon content. R. Boeckl.
Nature, Vol. 236, 25 - 26 (1972).

The contemporary C-14 content of 4 falls and 19 finds has been measured. With the assumption of a constant cosmic-ray intensity, terrestrial ages of the finds have been calculated. The average age of the finds was found to be 5,200 years, corresponding to a terrestrial "half-life" of 3,600 years for stony meteorites from the western USA.

105.032 Use of a CO_2 laser to prepare chondrule-like spherules from supercooled molten oxide and silicate droplets. L. S. Nelson, M. Blander, S. R. Skaggs, K. Keil.
Earth Planet. Sci. Letters, Vol. 14, 338 - 344 (1972).

Chondrule-like spherules were formed from individual freely falling subcooled droplets of alumina, enstatite, forsterite, enstatite-albite and forsterite-albite mixtures that had been melted with a focused continuous CO_2 laser beam. Their textures (rimmed, excentro-radial, barred, glassy) are strikingly similar to those of many meteoritic chondrules.

105.033 Potassium distribution between metal and silicate and its bearing on the occurrence of potassium in the earth's core. V. M. Oversby, A. E. Ringwood.
Earth Planet. Sci. Letters, Vol. 14, 345 - 347 (1972).

The hypothesis that a large fraction of the earth's potassium is present in an Fe-FeS core is discussed. Data on the distribution of potassium in meteorites, and experimental work on potassium distribution between silicate and metal-sulfide phases limit the amount of potassium in the core to a very small percentage of the total potassium in the earth.

105.034 Vagabond tektites. H. Faul, G. A. Wagner.
Earth Planet. Sci. Letters, Vol. 14, 357 - 359 (1972).

105.035 Irradiation ancienne et récente des aubrites. G. Poupeau, J. L. Berdot.
Earth Planet. Sci. Letters, Vol. 14, 381 - 396 (1972).

Three gas-rich aubrites (Bustee, Khor Temiki and Staroe Pesyanoe) have been studied and compared with five other non gas-rich aubrites (Aubres, Bishopville, Cumberland Falls, Norton County and Shallowater), by means of the track method.

105.036 Age d'exposition de la météorite Kiffa. J. Tobailem, C. Lalou.
Comptes Rendus Acad. Sci. Paris,Sér. B, Vol. 274, 1185 - 1187 (1972).

Les activités des radionuclides ^{48}V, ^{54}Mn, ^{22}Na, ^{60}Co, ^{26}Al et ^{40}K ont été mesurées par spectrométrie γ dans la météorite Kiffa (chute du 23 octobre 1970). L'âge d'exposition a été calculé à partir des teneurs en ^{26}Al et ^{22}Na; il a été trouvé être d'environ 2 millions d'années.

105.037 Remanent magnetization in meteorites. A. J. Meadows.
Nature, Vol. 237, 274 (1972).

105.038 The Timmersoi, Niger, hypersthene chondrite. D. G. W. Smith, J. D. Bell, T. Frisch.
Meteoritics, Vol. 7, 1 - 16 (1972).

The Timmersoi meteorite, a new type L5 hypersthene chondrite from the Niger Republic is described and microprobe analyses of its olivine, orthopyroxene, clinopyroxene, plagioclase, kamacite, taenite, troilite, chromite, whitlockite, chlorapatite and limonite presented.

105.039 On naming meteorites. V. F. Buchwald, J. T. Wasson.
Meteoritics, Vol. 7, 17 - 21 (1972).

To improve the labelling value of meteoritic names, new meteorites should be assigned geographical names which are not used for other meteorites, which are simple, and, if possible, of mnemonic value.

105.040 Calcium variation in olivines of the Murchison and Vigarano meteorites. R. Hutchison, R. F. Symes.
Meteoritics, Vol. 7, 23 - 29 (1972).

Olivine grains were analyzed by microprobe for Si, Ca, Mg, and Fe. More than one fractionation process is necessary to account for Ca variation superimposed on Mg: Fe fractionation observed in Murray, Murchison and Vigarano.

105.041 The Landes meteorite. T. E. Bunch, K. Keil, G. I. Huss.
Meteoritics, Vol. 7, 31 - 38 (1972).

Minerals and their compositions are very similar to those in Odessa-type silicate inclusions. The angular nature of the inclusions, recrystallization textures, and mineral compositions indicate a "xenolithic" origin for the inclusions.

105.042 Popigai basin — an explosion meteorite crater. V. L. Masaitis, M. V. Mikhailov, T. V. Selivanovskaya.
Meteoritics, Vol. 7, 39 - 46 (1972). — Translated from Dokl. Akad. Nauk SSSR, Vol. 197, No. 6 (1971) by T. K. Bodine.

105.043 Gravity reconnaissance at three Mauritanian craters of explosive origin. R. F. Fudali, W. A. Cassidy.
Meteoritics, Vol. 7, 51 - 70 (1972).

We have carried out reconnaissance gravity surveys across three Mauritanian craters: Aouelloul, an undoubted meteorite crater; Tenoumer, a probable meteorite crater with a unique array of concentric dikes on its outer rim flanks containing xenoliths of country rock showing abundant shock artifacts; and Temimichat Ghallaman, a crater of possible meteorite impact origin. All three have residual negative gravity anomalies associated with their interiors. In all cases the gravity values return to "normal" immediately outside their rims.

105.044 A possible recent meteorite impact site near the border between Zambia and Malawi. D. J. Mossman.
Meteoritics, Vol. 7, 71 - 74 (1972).

A crater, 80 m diameter by 6 m deep, was created in 1959 on the Nyika Plateau between Zambia and Malawi. It may have been the result of a meteorite.

105.045 Identificación de meteoritos. R. Compte Porta.
El Universo, No. 93, Vol. 24, 70 - 73 (1970).

105.046 Further evidence in support of the mixing model for howardite origin. T. S. McCarthy, L. H. Ahrens, A. J. Erlank.

Earth Planet. Sci. Letters, Vol. 15, 86 - 93 (1972).

Four diogenites, four howardites and seven eucrites have been analysed for major, minor and a number of trace elements by X-ray fluorescence spectrometry. Inter-element relationships between various elements in howardites, particularly refractory elements, are interpreted in the light of the recently proposed mixing model, whereby howardites are considered to be mixtures of eucritic and diogenitic material.

105.047 **Isotopic ratios of Gd, Sm and Eu in "Abee" enstatite chondrite.** A. J. Loveless, S. Yanagita, H. Mabuchi, M. Ozima, R. D. Russell.
Geochim. Cosmochim. Acta, Vol. 36, 685 - 698 (1972).

105.048 **Isotopic composition of lithium in meteorites.**
L. K. Levskij, A. N. Murin, V. G. Zaslavskij.
Geokhimiya, 1972, No. 2, p. 212 - 220. In Russian. — Abstr. in Referativ. Zhurn. 51. Astron., 6.51.429 (1972).

105.049 **Isotopic criterion of the extent of meteorite orbits.**
A. K. Lavrukhina, G. K. Ustinova.
Dokl. Akad. Nauk SSSR, Ser. Mat. Fiz., Vol. 204, 316 - 319 (1972). In Russian.

105.050 **Fragmentation of the Łowicz 1935 meteoroid.**
B. Lang.
Postępy Astron., Vol. 20, 141 - 145 (1972). In Polish.

The distribution of fragment masses from the Łowicz meteorite was examined. Lognormality of this distribution has been found. The mean value, standard deviation, skewness and peakedness (flatness) were estimated for fragment size distribution.

105.051 **Organic matter in meteorites.**
J. G. Lawless, C. E. Folsome, K. A. Kvenvolden.
Sci. American, Vol. 226, No. 6, p. 38 - 46 (1972).

The organic compounds of meteorites have been subjected to detailed analyses which indicate that they are not of biological origin.

105.052 **Impacts auf der Erde und auf dem Mond.**
A. Stilp.
Umschau, 72. Jahrgang, p. 432 - 433 (1972).

105.053 **The isotopic composition and elemental abundance of gallium in meteorites and in terrestrial samples.**
J. R. De Laeter.
Geochim. Cosmochim. Acta, Vol. 36, 735 - 743 (1972).

The isotopic composition of gallium in six iron meteorites and a terrestrial standard were measured using a solid source mass spectrometer. Isotopic abundances of meteoritic and terrestrial gallium agree to within 0.11 per cent. The concentration of gallium in 21 iron and 5 stone meteorites and in 13 standard rocks was determined.

105.054 **Gas retention chronology of Petersburg and other meteorites.** F. A. Podosek.
Geochim. Cosmochim. Acta, Vol. 36, 755 - 772 (1972).

This paper will be concerned with application of the neutron-activation isotope-correlation technique to rare gas radiochronology of meteorites, with xenon spallation corrections made by the correlation systematics. Argon and xenon data are presented for the eucrite Petersburg.

105.055 **Aouelloul glass: Aluminum 26 limit and some geochemical comparisons with Zli sandstone.**
P. J. Cressy, C. C. Schnetzler, B. M. French.
Journ. Geophys. Res., Vol. 77, 3043 - 3051 (1972).

Results of the analysis of Aouelloul crater glass for cosmogenic ^{26}Al place severe constraints on the possibility that the Aouelloul crater was formed by a glassy body that was

also the source of the Aouelloul glass. The existence of strong similarities between Aouelloul glass and Zli sandstone and the dissimilarities between Aouelloul glass and tektites or other impactite glasses strongly support terrestrial derivation of Aouelloul glass by fusion of the Zli sandstone during impact formation of the Aouelloul crater.

105.056 **^{129}I and ^{244}Pu abundances in white inclusions of the Allende meteorite.** F. A. Podosek, R. S. Lewis.
Earth Planet. Sci. Letters, Vol. 15, 101 - 109 (1972).

Data are presented for thermal-release xenon analysis of a neutron-irradiated sample of white inclusions from the carbonaceous chondrite Allende and of the neutron-irradiated carbonaceous chondrite Karoonda.

105.057 **Profile and depth of microcraters formed in glass.**
J. C. Mandeville.
Earth Planet. Sci. Letters, Vol. 15, 110 - 112 (1972).

105.058 **A specific comment on "An alternative model for the formation of iron meteorites", by M. R. Bloch,**
O. Müller. [Earth Planet. Sci. Letters, Vol. 12, 134 - 136 (1971)]. W. A. Kaiser, with a reply by M. R. Bloch, O. Müller.
Earth Planet. Sci. Letters, Vol. 15, 220, 221 - 222 (1972).

105.059 **Comments on paper by Dean R. Chapman, 'Australasian tektite geographic pattern, crater and ray of origin, and theory of tektite events** [Journ. Geophys. Res., Vol. 76, 6309 - 6338 (1971)]. R. B. Baldwin.
Journ. Geophys. Res., Vol. 77, 2616 - 2617, with a reply by D. R. Chapman, p. 2618 - 2619 (1972).

105.060 **On the capture of dust particles moving near the ecliptic by the earth.** N. I. Komarnitskaya.
Astron. vestn., Vol. 6, 35 - 43 (1972). In Russian. — Abstr. in Referativ. Zhurn. 51. Astron., 7.51.275 (1972).

105.061 **The mass distribution of individual specimens of the Sikhote-Alin meteorite shower.** V. I. Tsvetkov.
Astron. vestn., Vol. 6, 49 - 51 (1972). In Russian. — Abstr. in Referativ. Zhurn. 51. Astron., 7.51.286 (1972).

105.062 **Propagation of air waves from the Tunguska meteorite in connection with atmospheric inhomogeneity.**
V. A. Bronshtén.
Astron. vestn., Vol. 6, 22 - 28 (1972). In Russian. — Abstr. in Referativ. Zhurn. 51. Astron., 7.51.299 (1972).

105.063 **Evidence for amino-acids of extraterrestrial origin in the Orgueil meteorite.** J. G. Lawless, K. A. Kvenvolden, E. Peterson, C. Ponnamperuma, E. Jarosewich.
Nature, Vol. 236, 66 - 67 (1972).

105.064 **Meteoritic impacts.** V. Porubčan.
Kozmos, Vol. 3, 46 - 47, 51 (1972). In Slovak.

105.065 **The relationship of meteoritic parent body thermal histories and electromagnetic heating by a pre-main sequence T Tauri sun.** C. P. Sonett.
IAU Colloquium No. 12, (see 012.027), p. 239 - 245 (1971/72).

105.066 **Interrelations of meteorites, asteroids, and comets.**
E. Anders.
IAU Colloquium No. 12, (see 012.027), p. 429 - 446 (1971/72).

105.067 **Cometary versus asteroidal origin of chondritic meteorites.** G. W. Wetherill.
IAU Colloquium No. 12, (see 012.027), p. 447 - 460 (1971/72).

105.068 **Physical properties of meteorites.**

E. S. Gorshkov, E. G. Gus'kova, V. I. Pochtarev.
Modern Geol. (*GB*), Vol. 3, No. 2, p. 105 - 106 (1972). — See
Phys. Abstr., Vol. 75, No. 30598 (1972).

105.069 **Accretion processes leading to formation of meteorite parent bodies.** D. Lal.
From plasma to planet. 21st Nobel symposium 1971, (see
012.028), p. 49 - 64 (1972).

105.070 **Irradiation history of grain aggregates in ordinary
chondrites. Possible clues to the advanced stages of
accretion.** P. Pellas.
From plasma to planet. 21st Nobel symposium 1971, (see
012.028), p. 65 - 92 (1972).

105.071 **The meteoritic complex.** P. M. Millman.
From plasma to planet. 21st Nobel symposium
1971, (see 012.028), p. 367 (1972).

105.072 **Isotopic ratio of lithium in chondrite measured by
an ion probe mass spectrometer.**
H. Nishimura, J. Okano.
Japanese Journ. Applied Phys., Vol. 10, 1613 - 1622 (1971).
See Phys. Ber., Vol. 51, No. 3—4596 (1972).

105.073 **Tektites.** F. Ward.
Journ. Astron. Soc. Western Australia, Vol. 37,
May, p. 2 - 6 (1972).

105.074 **Artificial simulation of micrometeoroids.**
S. Auer.
AIAA Journ., Vol. 9, 516 - 518 (1971).

Micrometeoroid simulation studies on metal targets.
See Abstr. 022.028.

^{146}**Sm: A chronometer for** *p*-**process nucleosynthesis.** See Abstr. 022.087.

**Chondrules in Apollo 14 samples: Implications for
the origin of chondritic meteorites.** See Abstr. 094.006.

**Chemical composition of sawdust from lunar rock
12013 and comparison of a Java tektite with the rock.**
See Abstr. 094.008.

Laboratory simulation of lunar craters.
See Abstr. 094.100.

Trace ferric ion in lunar and meteoritic titanaugites.
See Abstr. 094.123.

Formaciones volcánicas en la tierra y en la luna.
See Abstr. 094.151.

Oxygen isotope cosmothermometer.
See Abstr. 107.002.

**Origin of planetary primordial rare gas: The possible
role of adsorption.** See Abstr. 107.003.

**Conditions in the early solar system as inferred from
meteorites.** See Abstr. 107.018.

Errata

105.901 **Erratum: 'Mineralogy and chemistry of the Kyle,
Texas, chondrite' [Meteoritics, Vol. 6, 71 - 79
(1971)].** R. R. Fodor, K. Keil, E. Jarosewich, G. I. Huss.
Meteoritics, Vol. 7, 75 (1972).

106 Interplanetary Matter, Interplanetary Magnetic Field, Zodiacal Light

106.001 **Density fluctuations in the interplanetary plasma: Agreement between space-probe and radio scattering observations.** W. M. Cronyn.
Astrophys. Journ., (Letters), Vol. 171, L101 - L105 (1972).

Measurements of interplanetary electron-density fluctuations, deduced from interplanetary scintillation spectra of compact radio sources, and measurements of interplanetary proton-density fluctuations made by space probes are in good agreement with a single, continuous power-law irregularity spectrum, index ~ -1.6 as measured by a space probe. The space-probe and radio scattering observations give complementary information about plasma density variations over distances of from 6 to 6×10^5 km.

106.002 **Effects of interplanetary magnetic field on the propagation of flare-generated interplanetary shock waves.** C. K. W. Tam, V. Yousefian.
Journ. Geophys. Res., Vol. 77, 234 - 238 (1972).

The effects of an interplanetary magnetic field on the propagation of flare-generated interplanetary shock waves are investigated with an approximate analytical method. It is found that the interplanetary magnetic field is relatively unimportant for strong shocks as far as the shock speed and transit time are concerned. It has more significant effects for weak shocks.

106.003 **Correlation of interplanetary-space B_z field fluctuations and trapped-particle redistribution.**
G. K. Parks, R. Pellat.
Journ. Geophys. Res.,Vol. 77, 266 - 269, with a correction, p. 2412 (1972). – Letter.

106.004 **Interplanetary sector structure at solar maximum.**
J. M. Wilcox, D. S. Colburn.
Journ. Geophys. Res., Vol. 77, 751 - 756 (1972). – Letter.

106.005 **A determination of the vaporization temperature of circumsolar dust at 4 $R_\odot$.** A. W. Peterson.
Bull. American Astron. Soc., Vol. 3, 500 (1971). – Abstr. AAS.

106.006 **Heavy ions from interplanetary dust.**
M. P. Nakada.
Journ. Geophys. Res., Vol. 77, 1713 - 1719 (1972).

Most atoms that are evaporated from interplanetary dust near the sun are quickly ionized and are probably carried away by the interplanetary magnetic field and the solar wind. The contribution of these heavy ions to the solar-wind flux has been estimated by using estimates of the mass required to maintain the zodiacal cloud. The contribution of the interplanetary glow to the background of twilight glow has been estimated.

106.007 **Magnetic-field anomalies in the lunar wake.**
Y. C. Whang, N. F. Ness.
Journ. Geophys. Res., Vol. 77, 1109 - 1115 (1972).

The interplanetary magnetic field is only slightly perturbed by the presence of the moon in the solar-wind flow. A statistical study of the umbral increases and penumbral decreases and increases was conducted with variation of the solar-wind plasma β value, the distance from the moon and the selenographic longitude of the limb regions of the lunar surface in the solar-wind flow.

106.008 **Shock waves and magnetic field configuration in** interplanetary space.
V. P. Korobeinikov, Yu. M. Nikolayev.
Cosmic Electrodynamics, Vol. 3, 3 - 24, 25 - 44 (1972).
In Russian and English.

We use the gas-dynamical approximation to calculate the formation and propagation of strong shock waves in interplanetary space after chromospheric flares. The dynamics of solar wind pressure after a flare may be described in terms of the blast wave model and coronal gas expansion in accordance with a power law piston model. The accurate analytical solution of the problem of the propagation of a strong shock wave in a medium, the density of which varies in inverse proportion to the square of the distance, has been obtained in the approximation of the point explosion model, including the velocity and initial pressure of the interplanetary plasma. A numerical solution of the piston problem has been obtained. The space vehicle observational data have been analyzed and interpreted physically. The theoretical results are compared to some results of experimental plasma measurements in interplanetary space.

106.009 **Interplanetary magnetic-field direction and high-latitude ionospheric currents.**
T. Stockflet Jørgensen, E. Friis-Christensen, J. Wilhjelm.
Journ. Geophys. Res., Vol. 77, 1976 - 1977 (1972). – Letter.

106.010 **Quiet-time electron increases: A measure of conditions in the outer solar system.**
L. A. Fisk, M. Van Hollebeke.
Journ. Geophys. Res., Vol. 77, 2232 - 2249 (1972).

One possible explanation for the increases in the intensity range of 3- to 12-MeV interplanetary electrons that McDonald et al. (1972) have labeled as 'quiet-time electron increases' is discussed. It is argued that the electrons in quiet-time increases are galactic in origin but that the observed increases are not the result of any variation in the modulation of these particles in the inner solar system. It is suggested instead that quiet-time increases may occur when more electrons than normal penetrate a modulating region that lies far beyond the orbit of the earth.

106.011 **Flux variations of charged particles of various energies according to measurements on space probes.**
S. N. Vernov, E. V. Gorchakov, P. P. Ignat'ev, N. G. Galach'ev.
Izv. AN SSSR. Ser. fiz., Vol. 35, 2418 - 2422 (1971).
In Russian. – Abstr. in Referativ. Zhurn. 51. Astron., 6.51. 498 (1972).

106.012 **About defining the speed of shock waves in the interplanetary space.**
A. E. Kuzmicheva, L. I. Dorman, N. S. Kaminer.
Geomagn. Aeronom., Vol. 12, 534 - 535 (1972). In Russian. Brief information.

106.013 **Flare induced shocks and corotating streams in the interplanetary medium.**
M. Wiseman, P. A. Dennison.
Proc. Astron. Soc. Australia, Vol. 2, 79 - 82 (1972).

106.014 **Bessellian spectral analysis of interplanetary scintillation.**
B. D. Ward, R. G. Blesing, P. A. Dennison.
Proc. Astron. Soc. Australia, Vol. 2, 82 - 84 (1972).

106.015 **Interplanetary scintillation studies at the Molonglo**

Radio Observatory. R. G. Milne.
Proc. Astron. Soc. Australia, Vol. 2, 88 - 89 (1972).

106.016 Zur Problematik von Zodiakallichtmodellen mit kugelförmigen Teilchen. R.-H. Giese.
Mitt. Astron. Ges., No. 31, p. 195 - 197 (1972).

106.017 Mikrowellen-Analogieversuche zur Lichtstreuung an kosmischen Staubpartikeln. R. Zerull.
Mitt. Astron. Ges., No. 31, p. 198 - 200 (1972).

106.018 Characteristics of interplanetary electron irregularities according to observations in 1967–1969.
V. V. Vitkevich, V. I. Vlasov.
Astron. Zhurn. Akad. Nauk SSSR, Vol. 49, 595 - 606 (1972).
In Russian. English translation in Soviet Astron. AJ, Vol. 16, No. 3.
The parameters of the electron irregularities in the interplanetary medium at distances of $0.35 - 1.2$ a.u. from the sun have been defined by scintillation observations using three spaced radio telescopes.

106.019 The configuration of the interplanetary magnetic field. L. Davis, Jr.
Astrophys. Space Sci. Library, Vol. 29, Part II, (see 012.013), p. 32 - 48 (1972).

106.020 Scattering and scintillations of discrete radio sources as a measure of the interplanetary plasma irregularities. V. V. Vitkevich.
Astrophys. Space Sci. Library, Vol. 29, Part II, (see 012.013), p. 49 - 66 (1972).

106.021 Discontinuities and shock waves in the interplanetary medium and their interaction with the magnetosphere. L. F. Burlaga.
Astrophys. Space Sci. Library, Vol. 29, Part II, (see 012.013), p. 135 - 158 (1972).

106.022 Observations of strong interplanetary scintillation at 74 Megahertz.
J. W. Armstrong, W. A. Coles, B. J. Rickett.
Journ. Geophys. Res., Vol. 77, 2739 - 2743 (1972).
Observations of the probability density for interplanetary scintillation are given. They are shown to follow a log-normal distribution better than the frequently used Rice-squared distribution. An explanation in terms of focusing by large-scale structure is given and shown to be consistent with spacecraft electron-density spectra.

106.023 Interplanetary magnetic-field variations and substorm activity. B. T. Tsurutani, C.-I. Meng.
Journ. Geophys. Res., Vol. 77, 2964 - 2970 (1972). – Brief report.

106.024 Observations of the zodiacal light from the ecliptic to the poles. J. G. Sparrow, E. P. Ney.
Astrophys. Journ., Vol. 174, 705 - 716 (1972).
The brightness and polarization of the zodiacal light have been measured from the satellite OSO-5, using two photometers of effective wavelengths 4180 and 6820 Å. Measurements have been made from the ecliptic to the poles. No temporal changes in zodiacal light have been found nor any significant differences in the intensities in the two hemispheres. The direction of polarization of the zodiacal light has been shown to be H-pass radial within the accuracy of the determination of $\pm 5°$.

106.025 Raketenexperiment zur Messung des Zodiakallichts bei Elongationen von 15° bis 30°. H. Link.
Inauguraldiss. Naturwiss. Gesamtfakultät, Ruprecht-Karl-Uni-

versität Heidelberg. 5 + 82 pp. (1972).

106.026 Study of interplanetary space by means of cosmic rays. A. Somogyi.
Fiz. Szemle, Vol. 21, 241 - 244 (1971). In Hungarian. – See Phys. Abstr., Vol. 75, No. 17043 (1972).

106.027 Spatial distribution of interplanetary dust.
R. G. Roosen.
IAU Colloquium No. 12 (see 012.027), p. 363 - 375 (1971/72).

106.028 Physical properties of the interplanetary dust.
M. S. Hanner.
IAU Colloquium No. 12, (see 012.027), p. 377 - 388 (1971/72).

The mean photospheric magnetic field from solar magnetograms: Comparisons with the interplanetary magnetic field. See Abstr. 071.009.

Relations between the areas index and different phenomena in the chromosphere, the corona and the interplanetary space. See Abstr. 073.072.

The coronal and interplanetary magnetic fields at the time of the solar eclipse of 7 March, 1970. See Abstr. 074.004.

Collisionless solar wind in the spiral magnetic field. See Abstr. 074.033.

The minimum size and minimum solar approach of interplanetary dust. See Abstr. 074.042.

Velocity of the solar wind as determined from interplanetary scintillations. See Abstr. 074.046.

The solar wind H and He$^+$ content. See Abstr. 074.072.

Study of the solar wind using the power spectrum of interplanetary scintillation of radio sources. See Abstr. 074.077.

Energetic solar particles in the interplanetary medium. See Abstr. 078.025.

Concerning the tasks of optical studies of the upper atmosphere and interplanetary space. See Abstr. 082.033.

Lyman-alpha measurements of neutral hydrogen in the outer geocorona and in interplanetary space. See Abstr. 082.098.

The earth's bow shock in an oblique interplanetary field. See Abstr. 084.222.

High-latitude low-energy electron fluxes and variation of the magnetospheric structure with the direction of the interplanetary magnetic field and with the geomagnetic activity. See Abstr. 084.279.

The effects of boundary condition asymmetries on the interplanetary magnetic field-moon interaction. See Abstr. 094.122.

Remarks on the size distribution of colliding and fragmenting particles. See Abstr. 098.058.

On the amount of dust in the asteroid belt. See Abstr. 098.062.

Interferometer visibility scintillation.
See Abstr. 141.125.

Interplanetary scintillation, interstellar scattering of two pulsars.   See Abstr. 141.506.

Streaming of galactic cosmic rays in the interplanetary magnetic field.   See Abstr. 143.005.

Low-energy cosmic rays in interplanetary space. See Abstr. 143.062.

107   Cosmogony of the Planetary System

107.001   **The formation of the hydrogen-rich planets.**
    J. G. Hills.
Bull. American Astron. Soc., Vol. 3, 481 - 482 (1971).
Abstr. AAS.

107.002   **Oxygen isotope cosmothermometer.**
    N. Onuma, R. N. Clayton, T. K. Mayeda.
Geochim. Cosmochim. Acta, Vol. 36, 169 - 188 (1972).
    Variations in oxygen isotopic abundances of meteoritic minerals, chondrules, whole meteorites, and planets are discussed in terms of a model involving isotopic exchange between primordial dust and a cooling solar nebular gas. From the temperature-dependence of the isotopic fractionation factors temperatures have been assigned to the processes of initial condensation, chondrule formation and planetary accretion. The ordinary chondrites, the earth and the moon have remarkably similar isotopic compositions, which are not readily accounted for in terms of mixtures of known materials of the primitive meteorites. Mean accretion temperatures of 450 - 470°K are estimated for members of this group.

107.003   **Origin of planetary primordial rare gas: The possible role of adsorption.**   F. P. Fanale, W. A. Cannon.
Geochim. Cosmochim. Acta, Vol. 36, 319 - 328 (1972).
    The degree of physical adsorption of Ne, Ar, Kr and Xe on pulverized samples of the Allende meteorite at 113°K has been measured. Results indicate that, at 113°K, a total nebular pressure of from 10^{-2} to 10^{-3} atm would be required to explain the Ar, Kr and Xe abundances in carbonaceous chondrites with an adsorption mechanism. The hypothesis that the presence of 'planetary primordial' Ar, Kr and Xe in carbonaceous chondrites is due to their adsorption from the gaseous portion of the pre-planetary cloud is offered as an alternative to the hypothesis that these gases were incorporated as the result of attainment of solubility equilibrium between dust and gas.

107.004   **Cosmic and solar system abundances of deuterium and helium-3.**   J. Geiss, H. Reeves.
Astron. Astrophys., Vol. 18, 126 - 132 (1972).
    From analysis of solar and solar wind abundances it is concluded that the D/H ratio in the protosolar gas was much smaller than it is found in ocean water or carbonaceous chondrites.

107.005   **On the origin of planets.**   A. G. Kitov.
    Godishn. Vissh. tekhn. uchebni zaved. Fiz., Vol. 5, sb. 1, p. 137 - 142 (1968, 1970). In Bulgarian. – Abstr. in Referativ. Zhurn. 51. Astron., 4.51.315 (1972).

107.006   **Condensation in the primitive solar nebula.**
    L. Grossman.
Geochim. Cosmochim. Acta, Vol. 36, 597 - 619 (1972).
    The purpose of the present study was to predict what

type of chemical fractionation trends would be produced during the equilibrium condensation of grains as the primitive solar nebula cooled. In order to accomplish this, all the optimum features of previous studies were employed.

107.007   **Low temperature condensation from the solar nebula.**   J. S. Lewis.
Icarus, Vol. 16, 241 - 252 (1972).
    Two models of the low temperature condensation accretion process for different extreme assumptions are investigated. In one extreme the time scale for accretion is taken to be long compared to the time scale for cooling of the nebula; in the opposite extreme, accretion is assumed to be rapid compared to the time scale for cooling. The results of these models are compared and their relevance to processes in the solar nebula is discussed.

107.008   **The transmission of mass and angular momentum from a satellite or planetary system to its primary.**
N. Aall Barricelli.
Astrophys. Space Sci., Vol. 15, 479 - 501 (1972).
    Main implications of basic properties detected in the satellite systems of Jupiter, Saturn and Uranus, and presented by the author in an earlier contribution (1971) are investigated. The similarity between the primary periods in the three systems, their apparent relation to the axial rotation periods of the three planets and other features suggesting that collisions with the planetary surfaces may have played a role in the evolution of the three satellite systems are interpreted by assuming that in each case a satellite of unusually large size was originally disintegrated at the Roche limit of its primary.

107.009   **Exploring the origin of the solar system by space missions to asteroids.**   H. Alfvén, G. Arrhenius.
Naturwissenschaften, 59. Jahrgang, p. 183 - 187 (1972).
    Theories about the origin and evolution of the solar system have directed interest to the study of asteroids which are likely to constitute, or be similar to, intermediate products in the formation of planets. Space missions to asteroids are now being seriously considered.

107.010   **Evidence for objects of lunar mass in the early solar system and for capture as a general process for the origin of satellites.**   H. C. Urey.
Astrophys. Space Sci., Vol. 16, 311 - 323 (1972).
    The problem of the accumulation of the moon is discussed on the assumption that the moon is a captured object. If it is such, it is highly improbable that it is the only object of this kind present in the early solar system. Evidence indicating that other massive objects were present at that time is presented. Also, it is pointed out that the interior of the moon must contain normal solar proportions of the elements of intermediate volatility in the lunar interior, if the moon accumulated in

a gas sphere.

107.011 New views on the origin of the planetary system.
L. Biermann.
Geol. Rundschau, (Stuttgart), Vol. 61, No. 1, p. 1 - 15 = Separate print Inst. Phys. Astrophys. München (1972). In German.

107.012 Implications of the inhomogeneous planetary accretion hypothesis. D. L. Anderson.
Comments Earth Sci. Geophys. (*GB*), Vol. 2, 93 - 98 (1971/72).

107.013 Preliminary results on formation of jetstreams by gravitational scattering. R. T. Giuli.
IAU Colloquium No. 12, (see 012.027), p. 247 - 250 (1971/72).

107.014 Jetstream formation through inelastic collisions.
D. C. Baxter, W. B. Thompson.
IAU Colloquium No. 12, (see 012.027), p. 319 - 326 (1971/72).

107.015 Collisional focusing of particles in space causing jetstreams. J. Trulsen.
IAU Colloquium No. 12, (see 012.027), p. 327 - 335 (1971/72).

107.016 Contribution to Kant and Laplace's nebular hypothesis. A. J. Rutgers.
Proc. Konnikl. Nederl. Akad. Wet., Ser. B, Vol. 75, 121 - 126 (1972).

It is shown, that the transversal thermal speed of the gasmolecules in a hot, slowly rotating sphere of atomic hydrogen gas prevents the angular momentum from moving towards the centre, when the shell looses the greater part of its molecules; hence the angular momentum remains concentrated in the outer shell, and the loss of molecules is compensated by an increase of the average transversal speed of the few remaining molecules; at last the outer shell or ring reaches the Huygens speed of revolution; it then is left behind, and transforms into a planet. The central sphere goes on shrinking, and by repetition of the process, will leave behind consecutively a number of such shells, which each will transform into a planet. The theory may also explain the observed difference in rotational speed between O, B and A stars on the one hand, and F, G... stars on the other.

107.017 A gasdynamical view on the motion, heating and accretion of solid bodies in the solar system.
H. Sato.
From plasma to planet. 21st Nobel symposium 1971, (see 012.028), p. 93 - 109 (1972).

107.018 Conditions in the early solar system as inferred from meteorites. E. Anders.
From plasma to planet. 21st Nobel symposium 1971, (see 012.028), p. 133 - 156 (1972).

107.019 Theory of jet streams. J. Trulsen.
From plasma to planet. 21st Nobel symposium 1971 (see 012.028), p. 179 - 194 (1972).

107.020 Investigation of solar system evolution by automatic vehicles on the moon. G. I. Petrov.
From plasma to planet. 21st Nobel symposium 1971, (see 012.028), p. 273 - 283 (1972).

107.021 Vulkanismus und Planetengeschichte.
S. Vsekhsvyatskij.
Ideen exakt. Wiss. [Deutsche Verlagsanstalt, Stuttgart], 1972, No. 2, p. 83 - 90.

On the planetary distance law.
See Abstr. 091.002.

Tides and the earth-moon system.
See Abstr. 094.134.

Evidence for objects of lunar mass in the early solar system. See Abstr. 094.218.

Thermodynamic models in cosmochemical systems.
See Abstr. 105.008.

Origin of organic matter in early solar system − V. Further studies of meteoritic hydrocarbons and a discussion of their origin. See Abstr. 105.010.

Solar rare gases and the abundances of the elements.
See Abstr. 105.024.

Stars

111 Stellar Parallaxes

111.001 **A theorist's view of the distance scale.**
D. Lynden-Bell.
Quarterly Journ. Roy. Astron. Soc., Vol. 13, 133 - 137
(1972). – Presented at the Woolley symposium (see 012.025).

111.002 **Errors in the trigonometric parallaxes.** W. Gliese.
Quarterly Journ. Roy. Astron. Soc., Vol. 13, 138 -
151 (1972). – Presented at the Woolley symposium (see
012.025).

111.003 **The statistical analysis of stellar kinematics.**
S. V. M. Clube.
Quarterly Journ. Roy. Astron. Soc., Vol. 13, 183 - 188
(1972). – Presented at the Woolley symposium (see 012.025).

111.004 **Trigonometric stellar parallaxes: Past, present, and**
future. K. A. Strand.
Proc. fourth astrometric conference, (see 012.026), p. 15 - 31,
with a discussion, p. 32 - 57 (1971).

111.005 **Parallax catalogues.** W. Gliese.
Proc. fourth astrometric conference, (see 012.026),
p. 149 - 160, with a discussion p. 161 - 196 (1971).

111.006 **Some present applications of statistical parallaxes.**
A. Blaauw.
Proc. fourth astrometric conference, (see 012.026), p. 197 -
202, with a discussion p. 203 - 229 (1971).

111.007 **Systematic errors of parallax telescopes.**
R. d'E. Atkinson.
Proc. fourth astrometric conference, (see 012.026), p. 259 -
266 (1971).

111.008 **Rigorous determination of astrometric quantities,**
especially parallaxes, by photography.
H. Eichhorn, W. H. Jefferys.
Proc. fourth astrometric conference, (see 012.026), p. 267 -
282 (1971).

111.009 **Dynamical parallaxes.** W. D. Heintz.

Proc. fourth astrometric conference, (see 012.026),
p. 283 - 284 = Sproul Obs. Repr. No. 206 (1971).

111.010 **The Van Vleck Observatory parallax program.**
A. R. Upgren.
Proc. fourth astrometric conference, (see 012.026), p. 291 -
295 (1971).

111.011 **Increase in accuracy of trigonometric parallaxes at**
the Yerkes Observatory. W. F. van Altena.
Proc. fourth astrometric conference, (see 012.026), p. 297 -
305 (1971).

Some current developments in astrometry.
See Abstr. 041.032.

Astrometric accuracy of photographs taken with the
Sproul 24-inch refractor. (1) Positions. (2) Parallaxes.
See Abstr. 041.051.

Astrometric study of four binary stars.
See Abstr. 118.008.

Parallax and orbital motion of the two nearby long-
period visual binaries Groombridge 34 and ADS 9090.
See Abstr. 118.009.

Parallaxes, mass ratios, and masses for the visual
binary stars BD + 75°403, ADS 9094, and ADS 9716.
See Abstr. 118.010.

Parallax and astrometric orbit of the spectroscopic
binary γ Geminorum from observations with the 20-inch re-
fractor of the Van Vleck Observatory.   See Abstr. 119.005.

A problem in distance-determination for Mira varia-
bles with an appendix on OB-star distances.
See Abstr. 122.050.

Kinematics and the distance scale.
See Abstr. 155.080.

112 Proper Motions, Radial Velocities, Space Motions

112.001 New kinematical data for bright southern OB stars.
J. R. Lesh.
Astron. Astrophys., Suppl. Ser., Vol. 5, 129 - 165 (1972).

New distance moduli and distances (based on the MK spectral types of Hiltner *et al.,* 1969, and *UBV* photometry from the Royal Observatory, Cape of Good Hope), new proper motions on the FK4 system, and newly evaluated and averaged radial velocities are presented for 440 bright O and early B stars south of declination −20°. These data have been prepared for a study of the velocity field of the southern portion of the Gould Belt.

112.002 Space motions of K3−M2 dwarfs in the solar neighborhood.   A. R. Upgren.
Astrophys. Journ., Vol. 172, 149 - 154 (1972).

Space motions have been determined from parallaxes, proper motions, and radial velocities of 62 dwarfs with $B - V$ color indices between +0.95 and +1.55. An analysis of the space motions of our K3−M2 dwarfs yields 18.0 km s^{-1} for the solar motion with apex at 19^{h}48^m(297°), +53° (1950). The *U-*, *V-*, and *W*-axes of the velocity ellipsoid are 28.7, 19.1, and 15.1 km s^{-1}, respectively and the vertex deviation is 16°. These results are compared with earlier analyses.

112.003 New proper motions in the McCormick regions.
P. A. Ianna, C. D. Garmany.
Bull. American Astron. Soc., Vol. 3, 464 (1971). − Abstr. AAS.

112.004 Attainable accuracy of reduction to absolute acceleration, determined by long-focus photographic astrometry, with illustration for Barnard's star.
P. van de Kamp.
Astron. Journ., Vol. 77, 88 - 92 (1972).

The accuracy of the reduction to absolute acceleration is limited by the internal dispersion in proper motion of the background stars; the resulting cosmic error is roughly proportional to the proper motion of the central star. This error cannot be reduced by increasing the observational accuracy. For Barnard's star, existing astrometric material does not permit any further improvement in accuracy for the absolute acceleration at this time.

112.005 On the proper motions of stars from the General Catalogue of the USSR Time Service.
L. I. Medvedeva, A. A. Nemiro.
Trudy Glav. Astron. Obs. Pulkovo, Ser. 2, Vol. 78, 46 - 58 (1971). In Russian.

Proper motions of 807 stars from the General Catalogue of the USSR Time Service (KCB) were derived with the use of right ascensions given in KCB, GC, FK3, FK4 and N30. The obtained proper motions are quite independent on the stellar proper motions given in fundamental catalogues. A comparison of the obtained proper motions with those of the GC, FK3, FK4 and N30 has shown that the new system of proper motions is close to the average value derived from the system of all four fundamental catalogues. The accidental errors of the values of the new proper motions will allow to use the KCB for determinations of time during 20−25 years from its mean epoch (1958).

112.006 Catalogue of relative proper motions of 4211 stars in the region of Taurus.   N. M. Bronnikova.
Trudy Glav. Astron. Obs. Pulkovo, Ser. 2, Vol. 78, 99 - 159 (1971). In Russian.

The proper motions of 4211 stars relative to the stars of 13^{m}2 in four areas of Taurus, bounding with the Parenago region No. 4, were determined using plates which had been tak-

en at Pulkovo with the normal astrograph, the difference of epochs being 57−60 years. Photographic magnitudes of all the measured stars were derived from the photographic magnitudes of 570 stars determined by A. D. Chuadze with the Schmidt telescope of the Abastumani Observatory. The proper motions obtained are compared with those determined in 1947 by A. N. Deutsch and V. V. Lavdovsky which used the same plates as first epochs. For the stars from the Yale and GC catalogues absolute proper motions were determined.

112.007 Radial velocities of 65 early-type stars.
H. A. Abt, S. G. Levy, T. L. Gandet.
Astron. Journ., Vol. 77, 138 - 143 (1972).

These stars are most of the 1061 OB stars and cepheids being observed for proper motions at the U. S. Naval Observatory and that previously lacked radial velocities. Although their space motions are not discussed here, these stars include three possible runaway stars and two newly discovered spectroscopic binaries with preliminary orbital elements. Half of the stars are variable in velocity, a fraction that is typical for OB stars.

112.008 Absolute proper motions of 407 Am and Ap stars.
D. K. Karimova, E. D. Pavlovskaya.
Trudy Gos. Astron. Inst. Shternberga, Vol. 42, 63 - 79 (1972). In Russian.

New proper motions of 117 stars and improved proper motions of 290 stars are given. The proper motions are in the FK4 system and for equinox 1950.0.

112.009 Reduced-proper-motion diagrams.   E. M. Jones.
Astrophys. Journ., Vol. 173, 671 - 676 (1972).

Photometric and astrometric observations of proper-motion objects have been used to construct analogs of H-R diagrams. The diagrams are used to predict the location of red degenerates.

112.010 Absolute proper motions of 111 O-stars.
E. D. Pavlovskaya, D. K. Karimova.
Soobshch. Gos. Astron. Inst. Shternberga, No. 176, p. 3 - 8 (1972). In Russian.

112.011 The positions and proper motions of 127 stars in the region of WX Centauri.
G. Gatewood, H. Eichhorn.
Monthly Notices Roy. Astron. Soc., Vol. 155, 415 - 424 (1972).

The positions and proper motions in the system of the FK4 of the star suggested by O. J. Eggen as the optical component of the X-ray source Cen X-2, and 126 other stars are given. WX Centauri does not seem to be an obvious member of a comoving group of stars. Its position in the proper motion plot does, however, suggest that it is a Population I object. Two relatively faint red stars are identified as possible members of the Scorpio−Centaurus Association.

112.012 Photoelectric radial velocities, Paper V. 69 southern HR stars.   R. F. Griffin.
Monthly Notices. Roy. Astron. Soc., Vol. 155, 449 - 461 (1972).

403 photoelectric observations have been made of the radial velocities of 69 stars lacking radial velocities in the Bright Star Catalogue; most of the stars are at declinations between 0° and −9°.

112.013 Kinematics of faint M stars near the north galactic pole, and the mass density in the solar neighbour-

hood.   C. A. Murray, N. Sanduleak.
Monthly Notices Roy. Astron. Soc., Vol. 157, 273 - 279
(1972).
 Proper motions of 21 faint M stars near the north galactic
pole are presented and discussed. It is shown that these are
probably all nearby dwarfs but that their transverse velocity
dispersion is likely to be only about 10 km s^{-1}. It is suggested
that these M stars belong to a numerous population, concentrated toward the galactic plane.

**112.014   Contribution of Danjon's astrolabe to the study of
           stellar proper motions.**   G. Billaud.
Astron. Astrophys., Vol. 19, 181 - 188 (1972). In French.
 More than 100 000 transits have been observed and the
results computed using the IAU system of astronomical constants. A study of the annual deviations of the star positions
and of the group corrections shows a general drift which is interpreted as $\Delta\mu$ and $\Delta\mu'$ errors in the proper motions of the
FK4 catalogue.

112.015   A Proper Motion Survey at the South Galactic Pole.
           H. L. Giclas, R. Burnham, Jr., N. G. Thomas.
Lowell Obs. Bull., *Flagstaff, Arizona*, No. 158, Vol. 7, (No. 21)
217 - 254 (1972).
 This Bulletin reports on the first results of the extension
of the Lowell Proper Motion Survey into the Southern Hemisphere. To this end, a special study, to a limit of motion of
0.″20/year, has been undertaken on the four regions G 266
through G 269. This publication summarizes the data for the
motion stars found between the magnitude limits of 8 and 17,
and with motions greater than 0.″19/year. Finder charts are
provided for each star listed. Also included is a supplementary
list of 135 white dwarf suspects and a list of 47 very red stars
with motions below our adopted program limit of 0.″20/year.

112.016   Eight stars of large proper motion.   W. J. Luyten.
           IAU Circ., No. 2386 (1972).

**112.017   Proper Motion Survey with the forty-eight inch
           Schmidt telescope. XXVIII. Faint proper motion
stars near the south galactic pole.**

W. J. Luyten, A. E. La Bonte.
Separate print Univ. Minnesota, Minneapolis, Minnesota.
4 pp. (1971).

**112.018   Proper motion survey with the forty-eight inch
           Schmidt telescope. XXX. Proper motions for 1 357
faint stars.**   W. J. Luyten.
Separate print Univ. Minnesota, Minneapolis, Minnesota. 28 pp.
(1972).

     Some current developments in astrometry.
See Abstr. 041.032.

     **First supplement to the list of catalogues available
at the Centre de Données Stellaires.**   See Abstr. 041.033.

     **The surface parameters and composition of some
high velocity A-stars.**   See Abstr. 114.066.

     Spectrum variations of the Be HD 217050 star.
See Abstr. 114.110.

     **Proper Motion Survey with the forty-eight inch
Schmidt telescope: XXIX. Double stars with common proper
motion.**   See Abstr. 117.036.

     **Catalogue of the photometric characteristics and
proper motions of stars in the neighbourhood of the globular
clusters M 3 and M 5. Proper motions and spce velocities of
the clusters.**   See Abstr. 154.012.

     **On the z-motion of some young clusters in the
Galaxy.**   See Abstr. 155.004.

     A test for relative motions of gas and young stars.
See Abstr. 155.043.

     **Spectrographic and photometric observations of
supergiants and foreground stars, in the direction of the Large
Magellanic Cloud.**   See Abstr. 159.022.

113 Stellar Magnitudes, Colors, Photometry

113.001 Four-color and Hβ photometry for the brighter A0 type stars.
D. L. Crawford, J. V. Barnes, J. Gibson, J. C. Golson, C. L. Perry, M. L. Crawford.
Astron. Astrophys., Suppl. Ser., Vol. 5, 109 - 128 (1972).
Intermediate and narrow-band photoelectric photometry is presented for 572 A0-type stars brighter than $m_v = 6^m.5$.

113.002 345-micron ground-based observations of M17, M82, and Venus.
R. R. Joyce, D. Y. Gezari, M. Simon.
Astrophys. Journ., (Letters), Vol. 171, L67 - L69 (1972).
Observations from the ground in the 345-μ atmospheric window show that the sources M17 and M82 have large infrared fluxes in this spectral region.

113.003 Effects of reddening on UBV colour transformations. II. A. Gutiérrez-Moreno, H. Moreno.
Astron. Astrophys., Vol. 17, 41 - 46 (1972).
The analysis of the effects of reddening on colour-systems transformations has been extended to larger reddenings, later type stars (F5), different interstellar absorption laws and different types of transformations.

113.004 Study of emission in narrow-band photometry in Be stars.
A. Peton, J. H. Bigay, R. Garnier, G. Paturel.
Astron. Astrophys., Vol. 17, 47 - 54 (1972).   In French.
Comparisons between B and Be stars having the same intrinsic index $(U-B)_0$, show that the Be stars do not follow the B sequence. Direct photoelectric measurements of Balmer decrements are compared with those obtained by conventional spectroscopy.

113.005 Brightness gradients for intrinsically variable celestial objects. F. I. Lukatskaya.
Astron. Astrophys., Vol. 17, 97 - 105 (1972).
Variations of photoelectric U, B, V magnitudes between maximum and minimum brightness of intrinsically variable celestial objects of different types are statistically proportional. B, V- and B, U-relations are linear regressions with angular coefficients ∇_V and ∇_U that have been called brightness gradients. On the (∇_U, ∇_V) plane variables are separated according to their types, thus forming a gradient diagram for variable stars.

113.006 Spectroscopic and photometric observations of M supergiants in Carina.
R. M. Humphreys, D. W. Strecker, E. P. Ney.
Astrophys. Journ., Vol. 172, 75 - 88 (1972).
Thirty southern-hemisphere M supergiants mostly in Carina have been studied spectroscopically in the blue and near-infrared and photometrically from 0.4 to 18 μ. The uncertainties in the determinations of interstellar extinction are discussed, and the spatial distribution of the M supergiants in the Carina arm is shown. The presence of the 11-μ excess attributed to silicate dust is a common feature. Three stars – VY CMa, VX Sgr, and HD 97671 – appear to be extreme examples of stars with large excesses over the entire long-wave region. It is suggested that these objects are surrounded by large amounts of particulate material over a great range of distances from the stars.

113.007 Narrow- and broad-band photometry of red stars. VII. Luminosities and temperatures for halo-population red stars of high luminosity. O. J. Eggen.
Astrophys. Journ., Vol. 172, 639 - 677 (1972).
(UBVRI) observations of red giants in halo-population clusters are discussed. The clusters include 47 Tucanae, NGC 362, ω Centauri, NGC 5897, M4, M5, M92, NGC 6752, M22, and M2, as well as the old disk-population cluster M67. Some 20 variables in these clusters have also been observed, including two new variables at the tip of the giant branch in 47 Tucanae. Variable 162 in ω Centauri may not be variable, and Nos. 6, 17, and 164 in that cluster are probably field variables.

113.008 A far-ultraviolet look at Orion.   R. C. Henry.
Sky Telescope, Vol. 43, 160 - 161 (1972).

113.009 Flächenphotometrie des Südhimmels mit einer Kugelspiegelkamera.
T. Schmidt-Kaler, W. Schlosser.
SuW, Vol. 11, 98 - 101 (1972).

113.010 UBV photographic photometry of stars in the region AR$_{1950}$: $17^h03^m - 17^h41^m$ Decl$_{1950}$: $-28.8°$ to $-33.4°$. III. The catalogue and identification maps of open star clusters: NGC 6405, NGC 6383, "NGC 6374", Av 2, NGC 6416 and H alpha emission regions: Gum 67 (Av 3), Gum 68 (Av 2). A. Antalová.
Bull. Astron. Inst. Czechoslovakia, Vol. 23, 126 - 139, 146c - d (1972).
Based on a UBV photometry of stars from the neighbourhood of the star clusters mentioned, the problem of assigning stars to star clusters was treated. Two-colour diagrams, identification maps and catalogues of the measured stars are given. The probable variable stars have been indicated following comparison of this photometry with the photometry of the stars carried out by other authors. The interstellar absorption was investigated in the regions mentioned as a function of distance.

113.011 Photometric data for 139 supergiants.
J. D. Fernie.
Astron. Journ., Vol. 77, 150 - 151 (1972).
Photometric data for 139 supergiants, originally determined on the Lick six-color system by Kron, are transformed to the Johnson UBVRI system. Approximate color excesses are given.

113.012 Aus der Entwicklung der Größenklassen-Definition im 19. Jahrhundert. I. Historische Variationen über $m - m_0 = 2.5 \times \log I_0/I$. D. B. Herrmann.
Sterne, 48. Jahrgang, p. 20 - 30 (1972).

113.013 An observation of variation of colour excesses of hot stars with time. R. M. Raznik.
Astron. Zhurn. Akad. Nauk SSSR, Vol. 49, 455 - 456 (1972). In Russian. English translation in Soviet Astron. AJ, Vol. 16, No. 2.
Systematic changes of colour excess $E_{(B-V)}$ of O9.5 spectral class stars of cluster NGC 6913 are noted.

113.014 Topical problems in stellar photometry.
V. B. Nikonov.
Vistas in astronomy, Vol. 13, (see 003.001), 155 - 159 (1972).

113.015 Interstellar absorption in the direction of Maffei 1 and 2. L. Kohoutek, U. Haug.
Nature, Phys. Sci., Vol. 236, 55 - 56 (1972).
Photographic photometry of more than 100 early type stars on Schmidt plates has been used to study the variation of absorption with distance in a field of 2 degrees in diameter. For distances greater than 1.5 kps the absorption is almost constant and close to 3^m.

113.016 Faint O–B2 stars in the Vela, Carina, Centaurus, and Crux sections of the southern Milky Way.
E. W. Miller.
Astron. Journ., Vol. 77, 216 - 229, 257 - 264 (1972).

Photographic *UBV* plates and photoelectric *UBV* standard sequences have been combined to search for faint blue stars in five regions of the southern Milky Way. The search has resulted in the identification of 39 new stars of spectral type B2 and earlier which are fainter than $V = 13$. A working model of the Carina spiral feature is presented which displays the distributions of neutral and ionized hydrogen, OB stars, cepheids, and cosmic dust.

113.017 Infrared observations of some southern IR–OH sources. A. R. Hyland, R. A. Hirst, G. Robinson, J. A. Thomas.
Astrophys. Letters, Vol. 11, 7 - 11 (1972).

Infrared photometry from 2.2 to 20μm is presented for the southern IR–OH sources V Mic, R. Hor and VX Sgr. On the basis of these results and previously published infrared photometry it is shown that the infrared colours of IR–OH sources correlate well with the velocity separation of the two OH emission peaks. It is further suggested that, in the case of Mira variables, there is an upper limit to the ratio of the OH flux to the integrated optical-infrared flux, independent of far infrared colour.

113.018 The effect of microturbulence on *UBV* colours.
J. B. Hearnshaw.
Observatory, Vol. 92, 43 - 47 (1972).

There are at least two stars which have been cited in the literature in recent years as having abnormal broad-band colours because of the effects of microturbulence on their spectra. It would therefore be of interest to be able theoretically to predict the colours of stars with abnormal microturbulence, and the purpose of the present paper is to describe an approximate method of doing this, without having to compute a complete synthetic spectrum.

113.019 Flächenphotometrie des Südhimmels mit einer Kugelspiegelkamera.
T. Schmidt-Kaler, W. Schlosser.
SuW, Vol. 11, 120 - 125 (1972).

113.020 Photometry of uncalibrated objective prism photographs. F. M. Stienon.
AAS Photo-Bull. 1972, No. 1, p. 17 - 18.

113.021 L'absorption de la lumière dans la région de la bifurcation de la Voie Lactée sur les données de la photométrie photographique. IV. Les magnitudes stellaires et les classes spectrales des étoiles dans une parcelle centrée ($\alpha = 18^h 52^m$, $\delta = + 13°00'$) 1900.0. T. A. Uranova.
Soobshch. Gos. Astron. Inst. Shternberga, No. 176, p. 15 - 23 (1972). In Russian.

On donne des listes avec les résultats de la détermination des classes spectrales et des magnitudes B et V pour les étoiles O–F 5 dans la parcelle No. 14 de la région No. 1 du Plan de P. P. Parenago. La courbe de l'absorption interstellaire pour cette parcelle est construite par les données de ce travail aussi par les données de la photométrie photoélectrique.

113.022 Intermediate-band and H-beta photometry of short-period variable stars. C. R. Chambliss.
Inform. Bull. Variable Stars, (I.A.U. Commission 27), Konkoly Obs., Budapest, No. 650 (1972).

113.023 Preliminary study of stars behind dense dust clouds.
J. G. Cohen.
Bull. American Astron. Soc., Vol. 4, 233 (1972). – Abstr. AAS.

113.024 Fotografisk fotometri. J. Andersen.
Astron. Tidssk., Årg. 5, p. 1 - 21 (1972).

113.025 Photometric and spectrophotometric investigations of the carbon star HD 59643.
J. Krempeć, S. Krawczyk.
Postępy Astron., Vol. 20, 139 - 140 (1972). In Polish.

113.026 Photoelectric *UBV* photometry in Cygnus and Cassiopeia. K. Bern, B. Virdefors.
Astron. Astrophys., Suppl. Ser., Vol. 6, 117 - 130 (1972).

Photoelectric measurements in the standard *UBV*-system are presented for 39 stars in Cygnus and 74 stars in Cassiopeia.

113.027 A–F supergiants in the Geneva Observatory photometric system. B. Hauck, C. Nicollier.
Obs. Univ. Lausanne, Commun. No. 20, 7 pp. (1972).

The photometric system of the Geneva Observatory permits a tridimensional representation of stars near the main sequence for spectral types between A0 and G5. It is interesting to test the validity of this representation for supergiant stars. The discussion is based upon some 40 stars, of spectral type later than B5.

113.028 Erste Resultate einer Vierfarbenphotometrie der Südlichen Milchstraße. W. Schlosser.
Mitt. Astron. Ges., No. 31, p. 165 (1972). – Abstract.

113.029 Lichtelektrische H$_a$-Photometrie heller Sterne auf der Südhalbkugel. J. Dachs, T. Schmidt-Kaler.
Mitt. Astron. Ges., No. 31, p. 166 (1972). – Abstract.

113.030 Integrated U–V color index of the southern Milky Way. C. Classen, J. Pfleiderer, F. E. Roach, L. L. Smith, R. W. Owen.
Mitt. Astron. Ges., No. 31, p. 166 - 167 (1972).

113.031 Flächenpolarimetrie des südlichen Gesamthimmels.
K. Wolf, J. Staude.
Mitt. Astron. Ges., No. 31, p. 167 (1972). – Abstract.

113.032 Studies of blue objects at high galactic latitudes. Faint blue objects in the field of BD + 29°2348 (the galactic north pole). C. Barbieri, L. Rosino.
Astrophys. Space Sci., Vol. 16, 324 - 335 (1972).

Continuing the survey for faint blue objects at high galactic latitudes, a field of 25 square degrees centered at the Coma Cluster of galaxies has been examined on two-color plates taken with the 90–67 cm Schmidt telescope of Asiago. 487 objects have been identified. A selected list of those with the largest negative color index is given in a table. Identification charts, notes and comments follow.

113.033 Report on progress in photometry and spectral classification in Italy 1966–1969. M. F. McCarthy.
Atti XIII Riunione Soc. Astron. Italiana, Trieste 1969, (see 012.010), p. 87 - 103 (1970).

113.034 A highly reddened star near NGC 6231.
G. H. Herbig.
Astrophys. Journ., (*Letters*), Vol. 174, L89 - L91 = Contr. Lick Obs., No. 334 (1972).

The very red ($B - V = +3.5$) twelfth-magnitude star discovered by Sanduleak and by Seggewiss near the edge of the young cluster NGC 6231 is a normal F5 Ia, with a radial velocity differing by 24 km s⁻¹ from the cluster mean. It is probably a background super-supergiant, and not an R CrB-type member of the cluster.

113.035 Simultaneous photometry and spectroscopy of the

Ap star 108 Aqr.   C. Megessier, R. Garnier.
Astrophys. Letters, Vol. 11, 113 - 116 (1972).
Spectroscopic and photometric observations of the Ap Si 4200 star 108 Aqr show profile shapes and intensity variations of the Ti II lines which seem to be in phase with the periodic light curves.

113.036 **On the nature of some faint infrared stars.**
T. A. Lee.
Astron. Journ., Vol. 77, 374 - 375 (1972).
Photometry extending from 0.7 to 3.4 μm is presented for a sample of Haro-Chavira infrared stars. Circumstellar radiation is not necessary to explain the observations. These objects appear to be ordinary reddened late-type M giants or Mira variables.

113.037 *UBV* and Hγ observations of early-type field stars.
J. H. Lutz, T. E. Lutz.
Astron. Journ., Vol. 77, 376 - 380 (1972).
Photometric *UBV* and Hγ data are presented for 157 early-type field stars. These stars all lie within 4° of the galactic plane, and many of them show considerable reddening.

113.038 **Aus der Entwicklung der Größenklassen-Definition im 19. Jahrhundert. II.**   D. B. Herrmann.
Sterne, 48. Jahrgang, p. 113 - 120 (1972).

113.039 **A new catalogue of stellar UBV photoelectric photometry.**   J. C. Mermilliod.
Centre de Données Stellaires, Inform. Bull, No. 3, (see 002.020), p. 15 - 16 (1972).

113.040 **U, B, V photometry of 500 southern stars.**
P. M. Corben, B. S. Carter, R. M. Banfield, G. M. Harvey.
Monthly Notes Astron. Soc. Southern Africa, Vol. 31, 7 - 22 (1972).

113.041 **Some photoelectric observations of the red variable HD 218348.**   M. Wisse, P. N. J. Wisse.
Monthly Notes Astron. Soc. Southern Africa, Vol. 31, 33 - 34 (1972).

113.042 **UBV photometry of some very bright stars.**
A. W. J. Cousins.
Monthly Notes Astron. Soc. Southern Africa, Vol. 31, 69 - 70 (1972).

113.043 **Comparaison des systèmes photométriques uvbyβ et de Genève.**   E. Lindemann, B. Hauck.
Bull. Soc. vaudoise Sci. nat., No. 338, Vol. 71, 201 - 210 = Obs. Univ. Lausanne, Commun. No. 21 (1972).
We compare the properties of the Strömgren and Geneva photometric systems. First for the temperature parameter, then the luminosity and blanketing parameters. For stars with spectral type between A 0 and G 5 the possibilities of both systems are equivalent.

113.044 **A field guide to the selection of bright stars for extinction measures.**   F. C. Bertiau.
Vatican Obs. Publ., Vol. 1, No. 3, 38 pp. (1971).
As an aid to astronomers working in site-testing expeditions with small portable telescopes, tables have been computed to facilitate the best choice of bright stars with UBV magnitudes, for a good determination of the extinction and its colour dependence. For various latitudes and for each hour of sidereal time, a blue, a yellow and a red star are listed for each of ten values of the airmass.

113.045 **Blue stars in the infra-red.**   D. A. Allen.
Astron. & Space (*GB*), Vol. 1, 232 - 235 (1971).

113.046 **A search for faint blue stars. XLVIII. A field centered at 9 : 32 +24. XLIX. A field centered at 12 : 34+30.**
W. J. Luyten, J. H. Anderson, A. R. Sandage.
Separate print Univ. Minnesota, Minneapolis, Minnesota. 10 pp. (1968).

113.047 **Ein extrem roter Stern in der Nähe des galaktischen Sternhaufens M 37.**   F. Yilmaz.
Publ. Istanbul Univ. Obs., No. 93, 2 pp. (1971).

The application of electronography in astronomical photometry.   See Abstr. 034.094.

Photométrie d'astres très faibles par electronographie.   See Abstr. 034.095.

The *UBV r* colors of supergiants.
See Abstr. 064.016.

The influence of lines of various metals on the solar colours.   See Abstr. 114.018.

Two new faint carbon stars at high galactic latitudes.   See Abstr. 114.059.

The temperature scale of F and G stars. II. Continuum photometry.   See Abstr. 114.121.

Zonal spectrophotometric standards. Investigation of the energy distribution in the spectra of 109 stars in absolute units.   See Abstr. 114.132.

The system for photometric three-dimensional classification of stars.   See Abstr. 115.007.

Contribution à l'étude des magnitudes absolues et des couleurs intrinséques des étoiles bleues de grande luminosité à partir de l'observation de systèmes multiples.
See Abstr. 117.049.

The classification of intrinsic variable stars. I. The red variables of type N.   See Abstr. 121.088.

High-speed photometry of cataclysmic variables.
See Abstr. 122.038.

Photoelectric photometry of selected galactic cepheids. I. Two-color observations of 6 cepheid variables.   See Abstr. 122.131.

uvby **photometry of white-dwarf stars.**
See Abstr. 126.011.

The interstellar reddening law in the ultraviolet deduced from filter photometry obtained by the OAO-II satellite.   See Abstr. 131.104.

Pekuliare A-Sterne in offenen Sternhaufen. I. Photoelektrische UBV-Photometrie und MK-Klassifikation von Sternen im Gebiet von NGC 7039.   See Abstr. 153.005.

Pekuliare A-Sterne in offenen Sternhaufen. II. Resultate der Suche nach pekuliaren A-Sternen im Gebiet des Haufens Tr 2 (Cr 29).   See Abstr. 153.006.

The determination of reddening and ultraviolet excess of clusters from *UBV* observations of red giants.
See Abstr. 153.020.

Catalogue of the photometric characteristics and proper motions of stars in the neighbourhood of the globular

clusters M 3 and M 5. Proper motions and space velocities of the clusters.   See Abstr. 154.012.

A search for UV-bright stars in 27 globular clusters. See Abstr. 154.017.

Photometry of blue horizontal-branch stars in globular clusters.   See Abstr. 154.029.

An empirical determination of the star density in the galactic halo with three-colour-photometric methods (IV, SA 71).   See Abstr. 155.060.

A study of interstellar reddening in a region in Ophiuchus.   See Abstr. 155.078.

Spectrographic and photometric observations of supergiants and foreground stars, in the direction of the Large Magellanic Cloud.   See Abstr. 159.022.

Errata

113.901   Erratum: 'Observations of OJ 287 between 0.36 and 3.4 μm'[Nature, Phys. Sci., Vol. 234, 71 - 73 (1971)].   H. M. Dyck, T. D. Kinman, G. W. Lockwood, A. U. Landolt.
Nature, Phys. Sci., Vol. 237, 48 (1972). – Concerning 06.113.045.

114 Stellar Spectra, Temperatures, Spectroscopy

114.001 On the origin of the Am phenomenon.
D. J. Stickland, J. A. J. Whelan.
Monthly Notices Roy. Astron. Soc., Vol. 155, 11P - 15P (1972).

New model stellar envelope calculations are combined with the element separation hypothesis to account for the appearance of the metallic-line stars, their position on the H-R diagram and their absence among rapidly rotating stars.

114.002 The spectrum of Omicron Virginis.
P. M. Williams.
Monthly Notices Roy. Astron. Soc., Vol. 155, 17P - 19P (1972).

From an analysis relative to ϵ Vir using 20 Å mm^{-1} spectrograms, o Vir is found to be generally metal-poor, but overabundant in heavy metals. Unlike the Ba II and CH stars, however, o Vir appears to be anomalously rich in oxygen.

114.003 An unusual absorption feature in the far-ultraviolet spectrum of early-type supergiants.
A. B. Underhill, D. S. Leckrone, D. K. West.
Astrophys. Journ., Vol. 171, 63 - 70 (1972).

The OAO-II satellite has been used to obtain far-ultraviolet scans of six early-type supergiants. The data reveal the presence of a distinct, broad absorption feature centered near 1720 Å.

114.004 The lithium isotope ratio in F and G field stars.
J. G. Cohen.
Astrophys. Journ., Vol. 171, 71 - 77 (1972).

A Fabry-Perot interferometer was used to obtain high-resolution profiles of the resonance line of Li I and λ6717.69 of Ca I in 14 bright F, G, and early K field stars. The observational data were fitted by computed theoretical line profiles to determine the rotational velocity, Li abundance, and Li isotope ratio for each star.

114.005 An abundance analysis of the photometric standard 29 Piscium. D. A. Klinglesmith III.
Astrophys. Journ., Vol. 171, 79 - 88 (1972).

A fine analysis of the photometric standard star 29 Psc has been performed by using flux-constant hydrogen-line-blanketed model atmospheres. The observable quantities — Balmer and Paschen slopes, Balmer discontinuity, and H and He I lines — were used to determine the final model parameters T_{eff}, $\log g$ and $N(\text{He})/N(\text{H})$. Lines from 14 elements besides H and He were identified.

114.006 Observations of Zeta Puppis. S. R. Heap.
Astrophys. Letters, Vol. 10, 49 - 53 (1972).

The visual line spectrum of ζ Puppis (O4f) cannot be matched by theoretical line profiles computed by Auer and Mihalas (1971) from their non-LTE models. The discrepancy between observation and prediction suggests that deviations from plane-parallel geometry and hydrostatic equilibrium must be accounted for in order for models to represent the atmospheric regions forming the observed lines. Relaxation of these classical assumptions may affect the value of effective temperature for this star derived from observations of its continuous spectrum.

114.007 Infrared excesses and forbidden emission lines in early-type stars. D. A. Allen, J. P. Swings.
Astrophys. Letters, Vol. 10, 83 - 87, with a correction Vol. 11, 99 (1972).

Low-excitation forbidden emission lines have been detected in the spectra of early-type stars found to have prominent infrared excesses. The relationship of these objects to Be stars and planetary nebulae is briefly discussed.

114.008 Equivalent width data for several Am stars in clusters and comparison standards. M. A. Smith.
Astron. Astrophys., Suppl. Ser., Vol. 5, 81 - 107 (1972).

The equivalent width data used in curve of growth analyses in another study (Smith 1971) are presented herein for sixteen Am stars and four standards. These data are compared with available equivalent width material from four other observatories.

114.009 Contribution to the determination of the effective temperature of A-type stars. M. O. Baylac.
Astron. Astrophys., Vol. 16, 85 - 94 (1972).

We demonstrate in this paper the possibility of determining effective temperatures of A stars of different effective gravities with an accuracy of about 0.02 in θ_e, using the $V - I$ index of Johnson for unreddened stars. A comparison is made between the theoretical and observational results; the influence of hydrogen and metallic line blocking, convection, rotation and interstellar and atmospheric absorption are taken into account.

114.010 A proposed test of the supermetallicity of Mu Leonis. G. Wallerstein.
Astron. Astrophys., Vol. 16, 153 - 154 (1972).

We suggest that measurement of the central depth of the Ca I line at λ6572 and the Mg I line at λ4571 should yield the ratio of boundary temperature to effective temperature for K giants and hence test the mechanism suggested by Strom, Strom and Carbon for the enhancement of strong lines in μ Leo and other super-metal-rich stars.

114.011 Can astrophysical abundances be taken seriously?
G. Worrall, A. M. Wilson.
Nature, Vol. 236, 15 - 18 (1972).

The determination of the abundances of elements rests on several assumptions, particularly the existence of microturbulence, for which there is no experimental or theoretical justification.

114.012 Étude des étoiles à raies métalliques par une méthode photométrique à bandes passantes étroites.
M. Gerbaldi.
Comptes Rendus Acad. Sci. Paris, Sér. B, Vol. 274, 669 - 672 (1972).

On montre que la mesure du gradient bleu, de la discontinuité de Balmer et de la raie Hβ permettent une séparation des étoiles à raies métalliques des étoiles normales.

114.013 Mariner 9 ultraviolet spectrometer experiment: Stellar observations.
C. F. Lillie, R. C. Bohlin, M. R. Molnar, C. A. Barth, A. L. Lane.
Science, Vol. 175, 321 - 322 (1972).

Photoelectric spectra have been obtained for a number of early-type stars in the 1100- to 2000-angstrom region with the Mariner 9 ultraviolet spectrometer.

114.014 Infrared observations of 1612 MHz IR/OH sources.
A. R. Hyland, E. E. Becklin, J. A. Frogel, G. Neugebauer.
Astron. Astrophys., Vol. 16, 204 - 219 (1972).

Infrared spectra and photometry are given for several infrared objects which are known to have 1612 MHz OH emission. These IR/OH sources are shown to be either M type Mira variables or M supergiants with photospheric temperatures around 2000°K. All those examined are oxygen rich and none are carbon stars which suggests a high abundance of OH in their photospheres.

114.015 **A spectral classification of close groups of stars in the case of very small dispersion of a prism camera.** K. T. Stojanova.
Astron. Zhurn. Akad. Nauk SSSR, Vol. 49, 210 - 212 (1972). In Russian. English translation in Soviet Astron. AJ, Vol. 16, No. 1.

A method of spectral classification of weak and close groups of stars with very low dispersion (of the order of 10000 Å/mm) is described. Main criterions of a rough classification of stars according to «image shape» of the spectrum are given.

114.016 **The spectra and colors of two blue horizontal-branch stars in Omega Centauri.** A. W. Rodgers.
Astrophys. Journ., Vol. 171, 257 - 258 (1972).

The spectra of two blue horizontal-branch stars in ω Cen are discussed and effective temperatures and gravities derived. They are shown to have significantly weak helium lines in their spectra.

114.017 **The strength of the O I λ7774 line in the brightest stars in the Magellanic Clouds.** P. S. Osmer.
Astrophys. Journ., Vol. 171, 393 - 396 (1972).

Image tube spectra indicate that the strength of the O I λ7774 line in the F-type supergiants in the Magellanic Clouds is consistent with the relation between line strength and luminosity shown by their galactic counterparts. However, the A-type supergiants in the Clouds have a weaker oxygen line than expected.

114.018 **The influence of lines of various metals on the solar colours.** R. A. Bell.
Monthly Notices Roy. Astron. Soc., Vol. 156, 13P - 17P (1972).

Synthetic spectra calculations are used to examine the influence of lines of different metals on the solar colours.

114.019 **Profiles of emission lines in Be stars.** S.-S. Huang.
Astrophys. Journ., Vol. 171, 549 - 564 (1972).

The broadening functions resulting from a gaseous ring in circular motion around a star according to Kepler's law have been studied. When the distribution in the gaseous ring has a circular symmetry in the equatorial plane, the broadening profile is related to the surface density distribution along the radius by an integral equation which can be solved analytically. Profiles corresponding to gaseous rings with a uniform density distribution but different widths are used to illustrate the general properties of the profile broadened by the gaseous ring in circular motion. When the distribution in the gaseous ring lacks circular symmetry, the profile becomes not only asymmetric with respect to the center of the line but also time dependent. The emission profile has also been studied for cases in which the circular motion does not follow Kepler's law. Finally, some unresolved problems concerning the formation of gaseous rings are briefly discussed.

114.020 **Time scales for Ca II emission decay, rotational braking, and lithium depletion.** A. Skumanich.
Astrophys. Journ., Vol. 171, 565 - 567 (1972).

A comparison of the Ca^+ emission luminosity—after correction for spectral-type effects—for the Pleiades, Ursa Major, and Hyades stars and the sun indicate an emission decay which varies as the inverse square root of the age. Further, the rotational decay curve is found to satisfy the same law. It is further suggested that lithium depletion follows the same law but only as far as the Hyades age, after which the depletion proceeds exponentially.

114.021 **Ultraviolet spectrophotometry of Arcturus from a rocket.** Y. Kondo.
Astrophys. Journ., Vol. 171, 605 - 608 (1972).

Spectrophotometric observations of Arcturus were obtained at the resolution of about 7 Å in the wavelength region 2700 - 4000 Å from an Aerobee rocket. The most significant feature in the ultraviolet shortward of the atmospheric cutoff is an emission feature at 2800 Å identified as the resonance line of the Mg II doublet.

114.022 **Spectral properties of luminous late-type stars.** G. F. Gahm, L. Hultqvist.
Astron. Astrophys., Vol. 16, 329 - 343 (1972).

A survey is given of some spectral characteristics observed from a material of 16 Å/mm plates covering the visual-red spectral region of 52 luminous stars of spectral classes K0 to M7. Of these stars 11 are of luminosity class III and 41 of class II or I, including 17 supergiants belonging to the I Per association. New spectral types are given for 13 stars. The D line profiles are examined, the $H\alpha$ line is discussed, and the Li abundances are determined; the Ba II lines are discussed and a few remarks on the spectral properties of BC Cyg and RW Cep are given.

114.023 **The spectrum of the supergiant ε Orionis (B0 Ia). II. Radial velocities curve of growth analysis.** H. J. Lamers.
Astron. Astrophys., Vol. 17, 34 - 40 (1972).

The visual spectrum of the supergiant ε Ori (B0 Ia, HD 37128) is discussed. Abundances and microturbulent velocities are derived by means of a simple curve of growth analysis.

114.024 **Is HD 45677 a variable star?** J. P. Swings, P. Swings.
Astron. Astrophys., Vol. 17, 142 - 145 (1972). In French.

Whereas HD 45677 has faded by approximately one magnitude since 1969, the amplitude of variation in photographic magnitude between 1899 and 1969 has always been smaller than 0.3 mag.

114.025 **Astronomical infrared spectroscopy with a Connes-type interferometer. III. Alpha Orionis, 2600–3450 cm^{-1}.**
R. Beer, R. B. Hutchison, R. H. Norton, D. L. Lambert.
Astrophys. Journ., Vol. 172, 89 - 115 (1972).

Recent spectra of α Ori in the 3–4 μ region show clear evidence of the $\Delta \nu = 1$ sequence of the rotation-vibration bands of OH. A detailed investigation of the rotational and vibrational populations suggests that the OH is close to being in LTE at an apparent temperature of 4100° ± 200°K. We deduce an OH abundance of 1.2×10^{20} molecules cm^{-2} and upper limits for H_2O and $H^{35}Cl$ of 8×10^{18} and 8×10^{17} molecules cm^{-2}, respectively. We further deduce that the rms turbulence velocity in the region of OH line formation is 11.5 ± 2 km s^{-1}.

114.026 **Unidentified bands in the near-infrared spectrum of S Lyrae.** S. Wyckoff, P. Wehinger.
Astrophys. Journ., Vol. 172, 117 - 119 (1972).

The near-infrared minimum-light spectrum of S Lyr has been found to exhibit the four unidentified bands previously observed in S-type spectra. Measured wavelengths of the band heads are given and possible molecular sources are discussed.

114.027 **Infrared objects in H II regions.** D. A. Allen.
Astrophys. Journ., (Letters), Vol. 172, L55 - L58 (1972).

This paper reports the discovery of 2-μ sources in NGC 2264, IC 2087, and M1-82, and of prominent infrared emission in Lk Hα 198 and V376 Cas.

114.028 Carbon stars with composite spectra. H. B. Richer. Astrophys. Journ., (*Letters*), Vol. 172, L63 - L65 (1972).

Observations of three carbon stars exhibiting composite spectra are presented. The visual absolute magnitudes of the carbon stars in these systems are derived. The results are : TU Tau, $M_v = -3.9$; $-26°2983$, $M_v = -2.4$; SZ Sgr, $M_v = -1.4$.

114.029 The temperature scale of F and G stars. E. G. Schmidt. Bull. American Astron. Soc., Vol. 3, 453 (1971). – Abstr. AAS.

114.030 Observations of the Mg II doublet from a balloon-borne ultraviolet stellar spectrometer. Y. Kondo, J. L. Modisette, R. T. Giuli. Bull. American Astron. Soc., Vol. 3, 453 - 454 (1971). – Abstr. AAS.

114.031 Classification of stars with carbon and nitrogen anomalies. N. R. Walborn. Bull. American Astron. Soc., Vol. 3, 454 (1971). – Abstr. AAS.

114.032 Ultraviolet emissions from Arcturus. H. W. Moos, G. J. Rottman. Bull. American Astron. Soc., Vol. 3, 454 (1971). – Abstr. AAS.

114.033 Line blanketing in Arcturus–Statistical descriptions of observations. F. N. Edmonds, Jr., T. E. Morgan. Bull. American Astron. Soc., Vol. 3, 454 - 455 (1971). – Abstr. AAS.

114.034 Observations of line profiles formed in differential expanding atmospheres of shell stars and QSO's. J. D. Scargle. Bull. American Astron. Soc., Vol. 3, 455 (1971). – Abstr. AAS.

114.035 Time scales for Ca II emission decay, rotational braking and lithium depletion. A. Skumanich. Bull. American Astron. Soc., Vol. 3, 455 (1971). – Abstr. AAS.

114.036 The spectrum of HDE 226868 = Cyg X-1(?). C. T. Bolton. Bull. American Astron. Soc., Vol. 3, 458 (1971). – Abstr. AAS.

114.037 OH microwave emission in infrared stars. M. M. Litvak, D. F. Dickinson. Bull. American Astron. Soc., Vol. 3, 460 (1971). – Abstr. AAS.

114.038 The nature of the Herbig Ae- and Be-type stars. K. M. Strom, S. E. Strom, J. Yost, L. Carrasco, G. Grasdalen. Bull. American Astron. Soc., Vol. 3, 478 - 479 (1971). Abstr. AAS.

114.039 A 1-micron search for young stellar objects. S. E. Strom, K. M. Strom, L. Carrasco, G. Grasdalen, S. Derenzo. Bull. American Astron. Soc., Vol. 3, 479 (1971). – Abstr. AAS.

114.040 The equivalent widths of spectral lines in twelve late F dwarfs. A. L. T. Powell. Roy. Obs. Bull., [Roy. Greenwich Obs., Herstmonceux], No. 167, p. 303 - 314 (1971).

This paper gives the equivalent widths and other relevant data for the spectral lines of twelve late F dwarfs. All the blue wavelength spectrograms were obtained on the 30-inch coudé at Herstmonceux. For one star (χ Her) plates in the red spectral region were obtained on the 100-inch Mount Wilson coudé. These data have been used as the basis of differential curve-of-growth analyses which have been published elsewhere.

114.041 The Os-Pt-Hg abundance peak in Ap stars and the problem of very heavy cosmic rays. B. N. G. Guthrie. Astrophys. Space Sci., Vol. 15, 214 - 228 (1972).

Relative abundances in the region $74 \leqslant Z \leqslant 83$ (W to Bi) are determined for 73 Dra, HR 4072, and some other Ap stars. The peak at $A \sim 201$ on Mn stars is discussed briefly.

114.042 Spectre continu dans le proche infrarouge de deux étoiles Be: HD 50138 et HD 51585. Y. Andrillat, L. Houziaux. Astrophys. Space Sci., Vol. 15, 236 - 239 (1972).

Continuous spectra of the Be objects HD 50138 and HD 51585 have been investigated between 1.6 and $1.2 \mu^{-1}$.

114.043 Spectres de quelques étoiles Be entre 7000 et 9600 Å. Y. Andrillat, L. Houziaux. Astrophys. Space Sci., Vol. 15, 240 - 244 (1972).

Kodak IM plates have been used in order to extend to 9600 Å the observed wavelength range for 23 O, B, Be Bp and A stars. Between 1.45 and $1.20 \mu^{-1}$, we have measured the parameter $d \log I_\lambda / d(1/\lambda)$.

114.044 Scanner observations of cool stars from 3400 to 11000 Å. T. Fay, R. K. Honeycutt. Astron. Journ., Vol. 77, 29 - 34 (1972).

Photoelectric scans of the M supergiant α Ori and the carbon stars 19 Psc, W Ori, and DS Peg were made at 20-Å resolution from 3400 to 6000 Å and at 40-Å resolution from 6000 to 11000 Å. The dominant spectral features are due to C_2, CN, and TiO; the variation of these features with spectral class is pointed out.

114.045 The spectral classification of the F stars of intermediate luminosity. W. W. Morgan, H. A. Abt. Astron. Journ., Vol. 77, 35 - 37, 111 (1972).

A general procedure in visual spectral classification is outlined which makes use of all significant spectral features in the available wavelength region. A grid of standard F-type stars of luminosity classes II-IV is given. This grid is on the MK system, but it defines a localized structure of higher internal precision than the latter. The revised (Morgan, Keenan, Abt) system is used for the determination of new types for 14 stars of the δ Scuti class.

114.046 The effect of rapid rotation on radiation from stars. IV. Weak absorption lines and spectral types. J. Hardorp, P. A. Strittmatter. Astron. Astrophys., Vol. 17, 161 - 164 (1972).

It is shown that for pole-on stars (a) rotation alone has a negligible effect on abundance determinations based on lines with equivalent widths in the range 25–150 mÅ; (b) rotation does not provide an explanation for the excessive strength of Si III lines in early B stars and (c) the relation between spectral type and color index $(b-y)$ is essentially unchanged by rotation.

114.047 Observed effects of gravity darkening in rotating B stars. J. Norris, M. Scholz. Astron. Astrophys., Vol. 17, 182 - 189 (1972).

In early B stars of a given Balmer discontinuity, the equivalent width of He I λ4471 is systematically larger in sharp-lined than in broad-lined stars. In late B stars, the difference between the projected rotational velocities derived from Mg II λ4481 on the one hand and from He I λ4471 on the other, increases towards later spectral types.

114.048 Characteristics of OH emission from infrared stars. W. J. Wilson, A. H. Barrett. Astron. Astrophys., Vol. 17, 385 - 402 (1972).

The characteristics of OH emission sources associated

with late M-type variable stars are discussed. A total of 456 IR stars and 18 other objects were surveyed. There are 29 OH emission sources associated with IR stars. The OH/IR stars (19 stars) with their OH emission strongest at 1612 MHz and concentrated in two narrow velocity ranges are discussed in this paper.

114.049 **An abundance analysis of the late type supergiant, Epsilon Pegasi.** P. R. Warren, D. W. Peat.
Astron. Astrophys., Vol. 17, 450 - 457 (1972).
The effective temperature, surface gravity, and metal abundances of the K-type supergiant are derived.

114.050 **A study of HD 4180 as compared to Be stars with lightly extended envelopes.** A. Peton.
Astron. Astrophys., Vol. 18, 106 - 110 (1972). In French.
The Hγ, Hδ, Hϵ equivalent widths variations of the star HD 4180 are characteristic of the shell. The emission decreases over a period of thirty years, until it disappears absolutely. Before it increases again, a very strong absorption by the external shell is observed, which affects the wings of the line profiles.

114.051 **Spectrophotometric investigation of Ap type stars. I. A two-dimensional quantitative spectral classification.** V. V. Leushin.
Izv. Krymskoj Astrofiz. Obs., Vol. 43, 113 - 129 (1971). In Russian.
56 Ap type stars and 28 normal stars have been analyzed on the basis of spectrograms with a dispersion of 14 Å/mm. An analysis of spectral classes and luminosities obtained from the spectral lines of helium, iron and hydrogen has been carried out.

114.052 **The atmospheres of the F-type supergiants. I. Calibration of the luminosity-sensitive O I λ 7774 line.** P. S. Osmer.
Astrophys. Journ., Suppl. Ser., No. 206, Vol. 24, 247 - 253 (1972).
The equivalent width of the O I λ7774 line has been measured with a narrow-band photoelectric technique in 59 stars of spectral types A0–G3 and luminosity classes Ia–V. By calibrating the line strength in 10 F-type supergiants of known luminosities, we find that it indicates their absolute magnitudes with an accuracy of 0.5 mag in the range $-4 > M_v > -9$. The calibration permits some of the most luminous stars known to be used as distance indicators.

114.053 **The atmospheres of the F-type supergiants. II. The line and continuous spectra.** P. S. Osmer.
Astrophys. Journ., Suppl. Ser., No. 206, Vol. 24, 255 - 282 (1972).
The present study has two general objectives: (1) to present observations of F-type supergiants of all luminosities; and (2) to apply standard model-atmosphere techniques to these stars. The purpose of the observations is to establish the empirical properties of the atmospheres as a function of luminosity and temperature and to provide data for comparison with theoretical predictions.

114.054 **A study of the spectral variable silicon Ap star 56 Ari.** I. A. Aslanov, V. L. Khokhlova.
Astron. Zhurn. Akad. Nauk SSSR, Vol. 49, 271 - 278 (1972). In Russian. English translation in Soviet Astron. AJ, Vol. 16, No. 2.
A study of the profiles of Si II, He I and hydrogen lines in the spectrum variable silicon Ap star 56 Ari was carried out using 30 spectrograms obtained in 1969 - 1970. The possibility of spectral variations with a period unequal to the rotational period is mentioned.

114.055 **A quantitative investigation of peculiarities in the spectra of Ap stars.** V. V. Leushin.
Astrofiz. Issled. Izv. Spets. Astrofiz. Obs., Vol. 3, 36 - 61 (1971). In Russian.
Quantitative characteristics of peculiarities in the spectra of Ap stars have been determined on the basis of their quantitative spectral classification. The characteristics are compared with the spectral class, magnetic field strength, Balmer discontinuity, and with each other. From these comparisons the following conclusions are drawn: (1) the peculiarity continuously increases in passing from normal stars to peculiar ones; (2) the «degree of peculiarity» correlates with the magnetic field strength and the Balmer discontinuity.

114.056 **An investigation of the atmospheres of metallic-line stars. I. A quantitative analysis of the ζ Lyr A, κ Ari, μ Aqr, 14 Psc and γ Equ atmospheres by the curve-of-growth method.** K. I. Kozlova.
Astrofiz. Issled. Izv. Spets. Astrofiz. Obs., Vol. 3, 83 - 100 (1971). In Russian.
A quantitative analysis of the atmospheres of four metallic-line stars, ζ Lyr A, κ Ari, μ Aqr, 14 Psc, and of an intermediate one, γ Equ, was carried out on the basis of 15 Å mm^{-1} spectrograms by using the curve-of-growth method. Microturbulent velocities, excitation and ionization temperatures of these stars are determined. Electron densities of the stellar atmospheres are determined using general methods.

114.057 **Fundamental spectroscopic data.** C. Moore-Sitterly.
Vistas in astronomy, Vol. 13, (see 003.001), 161 - 163 (1972).

114.058 **The interpretation of early-type spectra.** A. B. Underhill.
Vistas in astronomy, Vol. 13, (see 003.001), 169 - 206 (1972).
In this paper the present position concerning the meaning of spectral type among the O- and B-stars in terms of physical conditions in the stellar atmosphere and the abundances of the elements is reviewed. In Section 2 the question "what is a spectral type" is examined and it is noted that because absorption-line spectral types and photometrical spectral types are based on different criteria, the one is not necessarily always consistent with the other. In Section 3 the meaning of spectral type is examined from a theoretical viewpoint. First the simple physical concepts underlying present theories for constructing model atmospheres and predicting theoretical spectra are reviewed and then the results obtained from a series of model atmospheres representing, in principle, main-sequence stars of types B6 to O9 are compared with observed details in stellar spectra.

114.059 **Two new faint carbon stars at high galactic latitudes.** J. S. Drilling.
Publ. Astron. Soc. Pacific, Vol. 84, 35 - 36 = Contr. Louisiana State Univ. Obs., *Baton Rouge,* No. 59 (1972).

114.060 **The nature of the Herbig Ae- and Be-type stars associated with nebulosity.** S. E. Strom, K. M. Strom, J. Yost, L. Carrasco, G. Grasdalen.
Astrophys. Journ., Vol. 173, 353 - 366 (1972).
A quantitative spectrophotometric study of the Herbig Ae and Be stars associated with nebulosity demonstrates that the majority have surface gravities appropriate to pre-main-sequence stars while the remainder have surface gravities characteristic of the zero-age main sequence. The geometry of the circumstellar shells is discussed and the evidence seems to favor a nonspherical, possibly disklike shape. Finally, photographs taken with a 1-μ image-tube system reveal a large number of red stellar objects surrounding many of these stars. This suggests that the Herbig Ae and Be objects may be the brightest representatives of very recently formed stellar groups.

114.061 **BD+40°4124 and two nearby stars.** M. Cohen.
Astrophys. Journ., (*Letters*), Vol. 173, L61 - L64 (1972).

Multifilter infrared observations between 2 and 18 μ are presented for BD+40°4124 and two nearby emission-line stars which show that the 18-μ brightness is the inverse of the visual brightness.

114.062 **Two-dimensional spectral classification of O stars.** N. R. Walborn.
Journ. Roy. Astron. Soc. Canada, Vol. 66, 71 (1972). – Abstr. Canadian Astron. Soc.

114.063 **The spectrum and light variations in the peculiar A star HD 51418.**
A. F. Gulliver, D. A. MacRae, J. R. Percy, J. E. Winzer.
Journ. Roy. Astron. Soc. Canada, Vol. 66, 72 (1972). – Abstr. Canadian Astron. Soc.

114.064 **Remarks about X-rays from Eta Carinae.** K. Davidson, J. P. Ostriker.
Nature, Phys. Sci., Vol. 236, 46 - 48 (1972).

If a recent identification of a soft X-ray source with Eta Carinae is correct, this source may be explained as due to thermal excitation of a dense surrounding medium by the expanding shell of Eta Carinae. Some possible implications are mentioned.

114.065 **New Hα-emission stars around γ Cygni.** M. A. Kazarian, E. S. Parsamian.
Astrofizika, Vol. 7, 671 - 673 (1971). In Russian.
English translation in Astrophysics, Vol. 7, No. 4.

On the 40″ Schmidt telescope with 4° objective prism, 35 new Hα-emission stars have been found around γ Cyg.

114.066 **The surface parameters and composition of some high velocity A-stars.** A. W. Rodgers.
Monthly Notices Roy. Astron. Soc., Vol. 157, 171 - 178 (1972).

Continuum and line spectrophotometry of bright high velocity A-stars is reported and effective temperatures and surface gravities are derived. The surface abundance of Ca is found from the strength of the line Ca II K λ 3933 Å and compared with that in population I stars of the same temperature.

114.067 **Is 17 Leporis a shell star?** D. A. Allen, E. P. Ney.
Observatory, Vol. 92, 47 - 49 (1972).

In the 17 Leporis system we find the silicate feature despite the relatively luminous A-type star. Indeed, since M 1 giants do not normally exhibit detectable silicate emission, we must in this case attribute the existence of the shell to the presence of the early-type star. Neither star, on its own, would possess circumstellar dust, but in combination they do.

114.068 **An observational test of hydrogen line broadening theories.** P. L. Dufton.
Astron. Astrophys., Vol. 18, 335 - 340 (1972).

A method is described of deducing the gravity of an early-type star from the strength of the forbidden neutral helium line at 4045.16 Å. For two helium-rich stars, HD 96446 and HD 168785, gravities log g = 4.0 are found. These are shown to be significantly smaller, by approximately a factor of three, than the gravities deduced using the hydrogen lines and the ESW Stark broadening theory.

114.069 **Equivalent width of interstellar molecular lines. I. CO, SiO, H_2 and CS in the interstellar spectrum of ζ Ophiuchi.** K. S. K. Swamy, S. P. Tarafdar.
Astron. Astrophys., Vol. 18, 415 - 423 (1972).

An analysis of the equivalent widths of the band system ($A^1\Pi - X^1\Sigma^+$) of the CO molecule has been made using the observed equivalent widths of Smith and Stecher for the star ζ Ophiuchi. We have also calculated the expected equivalent widths of the Lyman band of H_2 and ($A^1\Pi - X^1\Sigma^+$) bands of SiO and CS for ζ Ophiuchi.

114.070 **Scanner abundance studies. IV. Observations of some strong-CN and miscellaneous evolved stars.**
B. J. Taylor, H. Spinrad, F. Schweizer.
Astrophys. Journ., Vol. 173, 619 - 630 (1972).

We present scanner observations of a number of strong-CN stars observed by Schmitt and also by McClure, together with observations of a number of other stars from various sources.

114.071 **Spectral classification of OB stars in both hemispheres and the absolute-magnitude calibration.**
N. R. Walborn.
Astron. Journ., Vol. 77, 312 - 318 (1972).

Some developments relating to the spectroscopic parallax system for the OB stars are described. New spectral-classification standards in the southern hemisphere are given and two-dimensional classifications for early O stars in both hemispheres are listed. An investigation of the absolute-magnitude calibration has been made. Some new OB supergiants with anomalous nitrogen or carbon spectra have been found.

114.072 **Spectrum variations in 56 Arietis. III. Lines of hydrogen, magnesium, and calcium.**
W. K. Bonsack.
Publ. Astron. Soc. Pacific, Vol. 84, 260 - 272 (1972).

Intensity and wavelength variations have been determined for the lines Hδ and Hγ, the MgII line at 4481 Å, and the K line of CaII. The equivalent widths of all of these features vary with a period of $0^d.728$ in synchronism with the variation of HeI found previously. However, the wavelength measurements do not exhibit variations similar to those found for HeI; at least in the case of the MgII line this result is difficult to reconcile with the rigid rotator model for this star.

114.073 **A search for C_2^- in spectra of HD 201626 and the sun.** T. D. Faÿ, H. R. Johnson.
Publ. Astron. Soc. Pacific, Vol. 84, 284 - 287 = Publ. Goethe Link Obs., Indiana Univ., *Bloomington*, No. 135 (1972).

Rotational lines of the (0, 0) and (1, 1) bands of the C_2^- ion are shown to be coincident with absorption features in the CH star HD 201626 (C1, 1). The identification of C_2^- must still be considered uncertain because the spectrum of HD 201626 is complex and many of the coincident lines are blended. A similar search for coincidences in the solar spectrum gave nagative results, which sets an upper limit on the solar ratio C_2^-/C_2 of $\approx 10^{-2}$.

114.074 **The spectrum of the periodic shell variable HR 2142, 1969–71.** G. J. Peters.
Publ. Astron. Soc. Pacific, Vol. 84, 334 - 354 (1972).

Extensive spectroscopic observations of HR 2142 made during 1969–71 are the data source for a discussion of this most unusual B-emission star. The spectrum as it appears in the star's short-termed, conspicuous shell state and in its intershell state is thoroughly described. The nine shell phases observed during the past three years are intercompared and variations seen in the duration and strength of the shell phase and in the ratio V/R of Hβ throughout a cycle are noted. The procedure adopted to determine the shell period $80^d.85$ is discussed.

114.075 **A magnesium feature in ultraviolet stellar spectra.** G. A. Gurzadyan.
Sky Telescope, Vol. 43, 350 (1972).

114.076 **Microwave emission from stars.**
P. R. Schwartz.
IAU Colloquium No. 15, (see 012.006), p. 44 - 50 (1972).

114.077 **Spectral changes in the Grygar variable HBV 475.**
A. Mammano, G. M. Righini.
IAU Colloquium No. 15, (see 012.006), p. 63 (1972).

114.078 **The visual and infrared spectrum 1968–71 of MHα 328-116 = V 1016 Cyg–Note III.**
F. Ciatti, A. Mammano, L. Rosino.
IAU Colloquium No. 15, (see 012.006), p. 64 - 73 (1972).

114.079 **V 1016 Cygni, spectral observations 1968–1971.**
M. P. Fitzgerald.
IAU Colloquium No. 15, (see 012.006), p. 73 - 84 (1972).

114.080 **Rapid Balmer line variability in Ap stars.**
H. J. Wood.
IAU Colloquium No. 15, (see 012.006), p. 176 - 177 (1972).

114.081 **The spectrum variable a Centauri (HD 125 823).**
D. A. Klinglesmith, P. L. Bernacca, H. Frey.
IAU Colloquium No. 15, (see 012.006), p. 205 - 216 (1972).

114.082 **Rapid spectral variations in Be stars.–New techniques and results.**
J. B. Hutchings, G. A. H. Walker, J. R. Auman.
IAU Colloquium No. 15, (see 012.006), p. 279 - 285 (1972).

114.083 **The coming shell phase of AX Monocerotis.**
M. Plavec, P. Harmanec.
Inform. Bull. Variable Stars, (I.A.U. Commission 27), Konkoly Obs., Budapest, No. 613 (1972).

114.084 **Photovisual and spectrophotometric observations of the carbon star HD 59643.**
S. Krawczyk, J. Krempeć.
Inform. Bull. Variable Stars, (I.A.U. Commission 27), Konkoly Obs., Budapest, No. 643, 3 pp. (1972).

114.085 **A new spectrum variable star in Ophiuchus.**
N. Sanduleak.
Inform. Bull. Variable Stars, (I.A.U. Commission 27), Konkoly Obs., Budapest, No. 663 (1972).

114.086 **Mariner 9 ultraviolet observations of Delta Per.**
M. R. Molnar.
Bull. American Astron. Soc., Vol. 4, 217 (1972). – Abstr. AAS.

114.087 **The abundances of interstellar carbon, oxygen and silicon from rocket observations of ultraviolet spectra**
E. B. Jenkins.
Bull. American Astron. Soc., Vol. 4, 226 (1972). – Abstr. AAS.

114.088 **A survey of carbon stars for microwave HCN emission.** P. R. Schwartz, W. J. Wilson, E. E. Epstein.
Bull. American Astron. Soc., Vol. 4, 226 - 227 (1972). – Abstr. AAS.

114.089 **Early Am stars as companions to two Hg stars.**
M. M. Dworetsky.
Bull. American Astron. Soc., Vol. 4, 229 (1972). – Abstr. AAS.

114.090 **Spectrophotometric luminosity criteria in O stars.**
R. S. Chaldu.
Bull. American Astron. Soc., Vol. 4, 229 (1972). – Abstr. AAS.

114.091 **Mariner 9 ultraviolet spectrometer experiment: Stellar observations.** C. F. Lillie.
Bull. American Astron. Soc., Vol. 4, 229 - 230 (1972). – Abstr. AAS.

114.092 **The ultraviolet flux envelopes of B type stars.**
A. B. Underhill.
Bull. American Astron. Soc., Vol. 4, 230 (1972). – Abstr. AAS.

114.093 **Water vapor microwave emission from infrared stars: New sources, and monitoring of time variations.**
K. P. Bechis, A. H. Barrett, D. F. Dickinson.
Bull. American Astron. Soc., Vol. 4, 232 (1972). – Abstr. AAS.

114.094 **Ten micron spectroscopy of circumstellar shells.**
R. H. Gammon, J. E. Gaustad, R. R. Treffers.
Bull. American Astron. Soc., Vol. 4, 234 (1972). – Abstr. AAS.

114.095 **On the variable inverse P Cygni profile in Theta Ori C.**
P. S. Conti.
Bull. American Astron. Soc., Vol. 4, 236 (1972). – Abstr. AAS.

114.096 **Line blanketing in Arcturus–Statistical descriptions of observations.** F. N. Edmonds, Jr., T. E. Morgan.
Bull. American Astron. Soc., Vol. 4, 236 - 237 (1972). – Abstr. AAS.

114.097 **Some properties of G stars near the north galactic pole.** C. R. Sturch, S. Sharpless.
Bull. American Astron. Soc., Vol. 4, 241 (1972). – Abstr. AAS.

114.098 **Mariner 9 observations of interstellar Lyman-alpha equivalent widths in OB stars.** R. C. Bohlin.
Bull. American Astron. Soc., Vol. 4, 242 (1972). – Abstr. AAS.

114.099 **The spectrum of Herbig Haro object No. 1.**
K. H. Böhm, J. Perry, R. Schwartz.
Bull. American Astron. Soc., Vol. 4, 243 (1972). – Abstr. AAS.

114.100 **Helium-Sterne.** K. Hunger.
SuW, Vol. 11, 160 - 163 (1972).

114.101 **Carbon stars.** N. Maron.
Urania Kraków, Vol. 43, 3 - 7 (1972). In Polish.

114.102 **Mass outflow from hot stars.** A. Zytkow.
Postępy Astron., Vol. 20, 105 - 118 (1972).
In Polish.
It is presented a short review of the observational data on hot stars exhibiting the P Cygni type lines in the far ultraviolet thus suggesting mass loss. The theoretical model by Lucy and Solomon of the mass outflow is described. The mass loss in this model is due to an outwardly directed acceleration acquired by the gas by the absorption of radiation in the ultraviolet resonance lines.

114.103 **Spectrograms of α Lyr and β Cen in the range of 2000–3800 Angströms.**
V. I. Patsaev, J. Ohanesyan, G. A. Gurzadyan.
Soobshch. Byurakan. Obs., No. 45, p. 20 - 26 (1972).
In Russian.
Six spectrograms of β Cen (June 18, 1971) and nine spectrograms of α Lyr (June 21, 1971) were obtained in the range of 2000–3800 Å. The spectrograms were obtained with the help of the Observatory "Orion" installed on the external surface of the "Salyut".

114.104 **Mehrdimensionale, quantitative Spektralklassifikation mittels Objektivprismenspektren.**
T. Schmidt–Kaler, G. Diaz, R. Rudolph.
Mitt. Astron. Ges., No. 31, p. 165 (1972). – Abstract.

114.105   Observations of six new helium-rich stars.
J. P. Kaufmann, K. Hunger.
Mitt. Astron. Ges., No. 31, p. 185 - 187 (1972).

114.106   H/He ratios and evolutionary status of WN stars.
L. F. Smith.
Mitt. Astron. Ges., No. 31, p. 187 (1972). – Abstract.

114.107   Spektrale Selbstumkehr, Index für Alter und Leucht-kraft der Sterne?   W. Strohmeier.
Mitt. Astron. Ges., No. 31, p. 188 - 190 (1972).

114.108   Spettroscopia galattica in Italia nel periodo ottobre '68–ottobre '69.   P. L. Bernacca.
Atti XIII Riunione Soc. Astron. Italiana, Trieste 1969, (see 012.010), p. 105 - 112 (1970).

114.109   The helium abundance in thirty-three main sequence B stars.   B. J. O'Mara, R. W. Simpson.
Astron. Astrophys., Vol. 19, 167 - 180 (1972).

In this paper we present helium abundance results for thirty-three main sequence B stars. The main purpose of this paper is to decide which neutral helium lines are unaffected by deviations from local thermodynamic equilibrium (LTE) and are thus suitable for use in an abundance analysis using the LTE assumption. The theoretical line profiles and curves of growth of O'Mara and Simpson (1971) are used to analyse the observations of Norris (1971), Leckrone (1971) and Watson (1971) in order to examine the importance of non LTE effects. A mean abundance of $N(He)/N(H) = 0.092 \pm 0.025$ is obtained for main sequence B stars, in good agreement with other determinations.

114.110   Spectrum variations of the Be HD 217050 star.
P. Granes.
Astron. Astrophys., Vol. 19, 224 - 230 (1972). In French.

The measurements of the radial velocity for lines produced by the envelope of HD 217050 during 11 years seem to indicate a time variation on the time-scale of 19 days. The variations of the $H\beta$ line profiles would be correlated to radial velocities variations.

114.111   A model atmosphere analysis of the Ap star HR 465.
M. F. Aller.
Astron. Astrophys., Vol. 19, 248 - 260 (1972).

A detailed model atmosphere analysis of the Ap spectrum variable HR 465 is presented for three well-spaced epochs. Overabundances of the order of 10^3 or more are found for the rare earths, Mo and Nb; of order 10^2 for Sr and Zr, and of order 10 for the iron-peak elements. Smaller overabundances are found for Al and Si. The results of the abundance analysis are discussed in terms of several current theories proposed to explain the Ap abundance anomalies.

114.112   Is there promethium in HR 465?
O. Havnes, E. P. J. van den Heuvel, with a comment by M. F. Aller, C. R. Cowley.
Astron. Astrophys., Vol. 19, 283 - 285, 286 (1972).

114.113   Second catalog of emission-line stars of the Orion population.   G. H. Herbig, N. Kameswara Rao.
Astrophys. Journ., Vol. 174, 401 - 423 = Contr. Lick Obs., No. 355 (1972).

The catalog contains 323 emission-line stars of the Orion population that have been observed with slit spectrographs. Accurate positions are given for most objects, together with UBV data where available, spectral classification, emission-line intensity class, type of star, light range and classification of the light curve, nebula with which the star is associated, and remark or indication when the star is nebulous or double. The catalog is based upon all published information available to

mid-1971, together with unpublished Lick data.

114.114   The far-ultraviolet emission spectrum of the K2 III star, Arcturus.   H. W. Moos, G. J. Rottman.
Astrophys. Journ., (Letters), Vol. 174, L73 - L77 (1972).

A moderate-resolution far-ultraviolet spectrum of the K2 IIIp star Arcturus (α Boo), obtained with a rocket-borne spectrometer, shows chromospheric emission features. Hydrogen Lα and O I λ1304 are clearly identified. The O I λ1304 stellar surface brightness is as great or greater than that of the sun. Other metal lines, including those of carbon, are weak compared to the O I line.

114.115   On the accretion of material by Theta Orionis C.
P. S. Conti.
Astrophys. Journ., (Letters), Vol. 174, L79 - L86 (1972).

Observations of an inverse P Cygni profile in the He II line λ4686 are reported and interpreted as infalling material. The profile of this line, and presumably the density of the infalling material, varies over a timescale of days.

114.116   Photoelectric measurements of the 6284 Å diffuse interstellar line.   P. Murdin.
Monthly Notices Roy. Astron. Soc., Vol. 157, 461 - 475 (1972).

A photoelectric index of the λ6284 diffuse interstellar line has been observed in 48 stars. It is shown that this line is probably exactly correlated with the line at λ4430. There are galactic variations in the ratio of λ6284 to interstellar reddening, correlated with similar variations of λ4430 to reddening. There is no clear evidence whether these variations are correlated with variations of the ratio of total-to-selective absorption.

114.117   Diffusion and the ³He abundance in 3 Cen A.
G. Michaud, S. Vauclair.
Astrophys. Letters, Vol. 11, 117 - 118 (1972).

It is shown that diffusion processes can account for the observed ³He abundance in 3 Cen A and similar stars if the zone at $T > 100,000$ K is stable, the upper zone being mixed somehow up to at least optical depth of 0.5.

114.118   The absolute magnitudes of the barium stars.
D. J. MacConnell, R. L. Frye, A. R. Upgren.
Astron. Journ., Vol. 77, 384 - 391, 403 (1972).

In the course of the Michigan Spectral Survey of the southern sky 205 new Ba stars have been discovered. Mean absolute visual magnitudes have been derived from the upsilon components of the proper motions reduced to the FK4 system. They are −0.4 and +1.0 for the certain and marginal groups separately, and +0.2 for a combined solution, assuming the motions reflect the standard solar motion and assuming a standard deviation in M_V of $\pm 0^m.8$. Both groups appear to possess kinematical properties resembling those of young stars. The mean luminosities of both are equivalent to normal giants; thus they are considerably fainter than normal stars with comparable Sr II-line intensities. It is estimated that Ba stars are about 1% as abundant as normal G and K giants in the vicinity of the sun.

114.119   Sur un spectre particulier de l'étoile symbiotique BF Cygni.   P. Merlin.
Comptes Rendus Acad. Sci. Paris, Sér. B, Vol. 275, 45 - 47 (1972).

Les spectres de BF Cygni de 1965 à 1970 révèlent que cette étoile peut durant quelques mois prendre un tout autre aspect que celui d'étoile symbiotique. Un essai d'interprétation est proposé.

114.120   The temperature scale of F and G stars. I. Hydrogen line profiles.   E. G. Schmidt.
Astrophys. Journ., Vol. 174, 595 - 603 (1972).

Hα profiles have been obtained for 28 cepheids and 67 nonvariable F, G, and K stars of all luminosity classes. These observed profiles are compared with theoretically calculated profiles to find effective temperatures. The temperatures are used to obtain color excesses for these stars. The absolute magnitudes of the G and K stars are obtained from the measured line-core widths following Kraft, Preston, and Wolff.

114.121 **The temperature scale of F and G stars. II. Continuum photometry.** E. G. Schmidt.
Astrophys. Journ., Vol. 174, 605 - 615 (1972).

Measurements of the slope of the Paschen continuum for 28 nonvariable F, G and early K stars and 20 cepheids are presented. From these data, line-blocking coefficients and temperatures are derived through comparison with model-atmosphere energy distributions. Color-temperature and spectrum-temperature relations are derived from this data and compared with previous results. From the data obtained in this paper and that of Paper I the location of the supergiants in the H-R diagram is considered. The hot edge of the instability strip from theoretical calculations is compared with the present temperature scale, and it is found that the hottest cepheids lie outside the theoretical instability strip even when a helium abundance of 45 percent is used.

114.122 **Studies of extremely young clusters. VI. Spectroscopic observations of the ultraviolet-excess stars in the Orion nebula cluster and NGC 2264.** M. F. Walker.
Astrophys. Journ., Vol. 175, 89 - 116 = Contr. Lick Obs., No. 361 (1972).

Spectroscopic observations of 25 ultraviolet-excess stars in the Orion nebula cluster and NGC 2264 are presented. These objects are found to be late-type gravitationally contracting stars with masses of $0.2-0.5\,M_\odot$, radii of $2-6\,R_\odot$, and absolute bolometric magnitudes of +2.5 to +5.8, surrounded by zones in which line and continuous emission originate.

114.123 **High resolution astronomical spectroscopy: its future.** G. Münch.
Auxiliary instrumentation for large telescopes. Conference 1972, (see 012.018), p. 21 - 33 (1972).

114.124 **Extremely red star.** C. T. Kowal.
IAU Circ., No. 2394 (1972).

114.125 **CoD $-42°14462$.**
J. E. Hesser, B. M. Lasker, P. S. Osmer.
IAU Circ., No. 2407 (1972).

114.126 **Peculiar red star.** H. Albers.
IAU Circ., No. 2412 (1972).

114.127 **On some phenomena in T Tau type stars.**
J. Sikorski.
Postępy Astron., Vol. 20, 37 - 50 (1972). In Polish.

Some peculiarities in the spectra of the T Tau type stars are reviewed. These may be caused by regions of atmospheric plasma being not in thermodynamical equilibrium. The intensity of phenomena caused by such a plasma depends on the evolutionary stage of the protostar.

114.128 **First ultraviolet stellar spectra from the orbiting stellar spectrophotometer S 59.**
R. Hoekstra, K. van der Hucht, T. Kamperman, H. Lamers, A. Hammerschlag, W. Werner.
Nature, Phys. Sci., Vol. 236, 121 - 122 (1972).

The stellar spectrophotometer S 59 on board the TD1A-satellite, launched March 12th 1972, is briefly described. Stellar spectra are scanned in three wavelength-regions, 90 Å wide and centred around 2110 Å, 2540 Å and 2820 Å, with a spectral resolution of about 1.8 Å. About 200 stars of spectral types

O, B and A with $m_V < 4.5$ mag will be observed at least 12 times in a period of half a year. The spectrum of β Aur (A 2 V) is presented, showing strong Mg II resonance lines and many other metallic lines. Part of the material will be available to guest observers.

114.129 **Classifications on red stars for statistical investigations.** A. G. Velghe.
Monthly Notes Astron. Soc. Southern Africa, Vol. 31, 45 - 46 (1972). – Abstract.

114.130 **Infrared astronomy.** I. Glass.
Monthly Notes Astron. Soc. Southern Africa, Vol. 31, 51 (1972). – Abstract.

114.131 **A study of the emission spectra of WR stars. I. On the structure of the spectra CIII, NIV, OV related to the upper excitation configurations $2pn, pd$.**
A Nikitin, T. Feklistova.
Izv. Akad. Nauk Ehstonskoj SSR, Vol. 21, (Fiz., Mat., 1972, No. 1), 31 - 40 = Tartu Astron. Obs. Teated, No. 39. In Russian.

The applicability of the one-configuration approximation and the LS-scheme for the ions of isoelectronic sequence CIII, NIV and OV is discussed. Estimates of the energy levels and wave-lengths of some lines corresponding to the transitions between the high-excitation levels obtained are given in a table.

114.132 **Zonal spectrophotometric standards. Investigation of the energy distribution in the spectra of 109 stars in absolute units.** V. M. Tereshchenko, A. V. Kharitonov.
Trudy Astrofiz. Inst., *Alma-Ata,* Vol. 21, 185 pp. (1972). In Russian.

114.133 **K-line luminosities and the calibration of MK classes for G and K giants.** B. E. J. Pagel.
Quarterly Journ. Roy. Astron. Soc., Vol. 13, 164 - 176 (1972). – Presented at the Woolley symposium (see 012.025).

114.134 **Peculiar red star.** R. M. Catchpole.
IAU Circ., No. 2420 (1972).

114.135 **Microturbulence and abundance anomalies in cluster metallic line A stars.** M. A. Smith.
Thesis, Univ. Arizona, Tucson. [Available from Univ. Microfilms, Ann Arbor, Mich., U.S.A. Order No. 71–15929], 241 pp. (1971). – See Phys. Abstr., Vol. 75, No. 13906 (1972).

114.136 **Ultraviolet emission from the stars.** B. Bates, C. D. McKeith.
Contemp. Phys. (*GB*), Vol. 13, 225 - 246 (1972).

Some scientific objectives of a programme in ultraviolet astronomy are considered, and a review made of the types of space vehicles and attitude control systems employed. A brief discussion is made of some recent results and their interpretation.

114.137 **Shell-stars.** P. Harmanec, P. Koubský.
Vesmir, Vol. 51, 113 - 114 (1972). In Czech.

Transition probabilities for the vibration-rotation bands of silicon monoxide. See Abstr. 022.077.

First supplement to the list of catalogues available at the Centre de Données Stellaires. See Abstr. 041.033.

Astrometry: First aid for the spectroscopist. See Abstr. 041.052.

Rotationally extended stellar envelopes–IV. The

detached shell of HD 187399.   See Abstr. 064.002.

A fine analysis of two early B-type supergiants, HD 96248 and HD 164402, relative to the main sequence star τ Scorpii.   See Abstr. 064.005.

Non-LTE model atmospheres. VII. The hydrogen and helium spectra of the O stars.   See Abstr. 064.006.

Analysis of the metal-poor red giant star HD 122563. See Abstr. 064.007.

Theoretical effect of various broadening parameters on ultraviolet line profiles.   See Abstr. 064.008.

Is Alpha Orionis reddened? See Abstr. 064.037.

Oxygen abundances of three population II horizontal-branch stars.   See Abstr. 064.042.

The metal-to-hydrogen ratio in F1—F5 stars, as determined by a model-atmosphere analysis of photoelectric observations of a group of weak metal lines. See Abstr. 064.043.

CaOH, a new triatomic molecule in stellar atmospheres.   See Abstr. 064.045.

Kinematics of faint M stars near the north galactic pole, and the mass density in the solar neighbourhood. See Abstr. 112.013.

Study of emission in narrow-band photometry in Be stars.   See Abstr. 113.004.

Spectroscopic and photometric observations of M supergiants in Carina.   See Abstr. 113.006.

Photometric data for 139 supergiants. See Abstr. 113.011.

The effect of microturbulence on UBV colours. See Abstr. 113.018.

Photometric and spectrophotometric investigations of the carbon star HD 59643.   See Abstr. 113.025.

A—F supergiants in the Geneva Observatory photometric system.   See Abstr. 113.027.

Report on progress in photometry and spectral classification in Italy 1966—1969.   See Abstr. 113.033.

Simultaneous photometry and spectroscopy of the Ap star 108 Aqr.   See Abstr. 113.035.

Rocket spectroscopy of Zeta Orionis. See Abstr. 118.006.

Orbit of the double-lined binary mercury star Chi Lupi.   See Abstr. 119.013.

On the nature of HD 161227. See Abstr. 119.014.

Étude spectroscopique de deux étoiles nouvelles VV Cephée: BD + 54°2698, BD + 63°3. See Abstr. 119.019.

The spectrum of the symbiotic star AG Pegasi. See Abstr. 122.070.

Calcium II K-line profiles in front of distant OB stars.   See Abstr. 131.054.

Interferometer positions of eighteen OH emission sources.   See Abstr. 131.055.

Abundances of interstellar CH and CH+ radicals. See Abstr. 131.071.

Structure of the OH/infrared object NML Cygnus. See Abstr. 131.073.

Polarization of infrared objects. See Abstr. 131.088.

Interstellar CN excitation at 2.64 mm. See Abstr. 131.133.

Spectral investigation of NGC 7635 and BD +60°2522.   See Abstr. 132.031.

Pekuliare A-Sterne in offenen Sternhaufen. I. Photoelektrische UBV-Photometrie und MK-Klassifikation von Sternen im Gebiet von NGC 7039.   See Abstr. 153.005.

Pekuliare A-Sterne in offenen Sternhaufen. II. Resultate der Suche nach pekuliaren A-Sternen im Gebiet des Haufens Tr 2 (Cr 29).   See Abstr. 153.006.

Lithium abundances of stars in open clusters. See Abstr. 153.008.

Spectrographic and photometric observations of supergiants and foreground stars, in the direction of the Large Magellanic Cloud.   See Abstr. 159.022.

115 Stellar Luminosities, Masses, Diameters, HR-Diagrams and Others

115.001 On the stellar luminosity function from the method of mean absolute magnitudes. J. F. Wanner.
Monthly Notices Roy. Astron. Soc., Vol. 155, 463 - 481 (1972).

The method of mean absolute magnitudes for the determination of the luminosity function is examined in detail and its reliability tested.

115.002 Ages and kinematics of the late F dwarfs in relation to their chemical composition. A. L. T. Powell.
Monthly Notices Roy. Astron. Soc., Vol. 155, 483 - 494 (1972).

The iron to hydrogen ratios, relative to the sun, of 93 late F field dwarfs within 25 pc of the sun have been determined from their ultra-violet excesses. The ages of these stars have been deduced from their positions in the magnitude-colour index diagram.

115.003 A comparison of the luminosities of Perseus-arm stars in the Hγ and MK systems. D. Crampton.
Astrophys. Journ., Vol. 171, 285 - 291 = Contr. Dominion Astrophys. Obs., Victoria, No. 169 (1972).

For the majority of these stars (B0–A3 supergiants) there are no significant differences in the luminosities determined from the two methods and random errors appear to be of comparable magnitude.

115.004 A new absolute calibration of Vega.
D. S. Hayes, D. W. Latham, S. H. Hayes.
Bull. American Astron. Soc., Vol. 3, 501 (1971). – Abstr. AAS.

115.005 A broad-band index for the effective temperatures of normal late type stars. J. P. Chaturvedi.
Bull. Astron. Inst. Czechoslovakia, Vol. 23, 48 - 50 (1972).

An index for stellar temperatures, based on B, V and R magnitudes, has been obtained.

115.006 Current problems on asymptotic branch stars.
V. Castellani, F. A. D'Antona.
Astrophys. Space Sci., Vol. 15, 340 - 347 (1972).

An effort is made to select, from the available observational data, information concerning the asymptotic branch stars. Particular care is devoted to derive luminosity functions for these stars in clusters representative of selected classes. Some theoretical predictions are briefly made. It is stressed that, from a theoretical point of view, the comparison between the one-shell H-burning evolutionary characteristics and the corresponding ones for the A.B. phase could be expected to provide a powerful means of investigation.

115.007 The system for photometric three-dimensional classification of stars.
V. Straižys, Z. Sviderskienė.
Astron. Astrophys., Vol. 17, 312 - 322 (1972).

The general scheme of three dimensional classification of any collection of stars, including the samples of different temperatures, luminosities, populations, interstellar reddenings and some kinds of peculiarities, is given. The errors of spectral class and M_V determinations are discussed.

115.008 Nonvariable supergiants in the Cepheid instability strip. E. G. Schmidt.
Astrophys. Journ., Vol. 172, 679 - 683 (1972).

The temperatures of 11 supergiants of spectral types F and G have been measured. The locations of these stars in the temperature-luminosity diagram are then found by using absolute magnitudes obtained from the literature. When these stars are compared with the location of the Cepheid instability strip, it is found that four of the stars lie more than 2 standard errors within the instability strip even though they are constant in light.

115.009 Speckle interferometry: Diffraction-limited measurements of nine stars with the 200-inch telescope.
D. Y. Gezari, A. Labeyrie, R. V. Stachnik.
Astrophys. Journ., (*Letters*), Vol. 173, L1 - L5 (1972).

A new method has enabled us to repeat most of the classical Michelson-Pease measurements of stellar diameters. Stellar images are photographed "coherently" with a special camera. Nine of the stars observed were resolved, showing angular dimensions as small as 0".016. Limb darkening is evidenced in α Ori, and a faint companion is found for β Cep.

115.010 Infrared diameter of IRC+10216 determined from lunar occultations. R. I. Toombs, E. E. Becklin, J. A. Frogel, S. K. Law, F. C. Porter, J. A. Westphal.
Astrophys. Journ., (*Letters*), Vol. 173, L71 - L74 (1972).

Lunar occultations of IRC+10216 have been used to determine the size of the regions emitting 2.2-, 3.5-, 4.8-, and 10-μ radiation. The results are interpreted as the first direct measurement of the physical size of a dust shell surrounding a late-type star.

115.011 The mass—luminosity relationship of the Hyades and other stars. J. B. Alexander.
Monthly Notices Roy. Astron. Soc., Vol. 157, 11P - 13P (1972). The conclusions of a previous paper are confirmed. Although the mass—luminosity relationship of the Hyades is poorly determined at present, it is probable that the unevolved stars in the cluster are somewhat undermassive when compared with typical unevolved field stars of the corresponding luminosity. For the stars in the general field, there is no evidence for two discrete mass—luminosity relationships.

115.012 Magnitud absoluta y luminosidad.
A. L. de la Barra.
El Universo, No. 93, Vol. 24, 74 - 78 (1970).

115.013 Stellar rotation and a simulation of the intensity interferometer. P. R. Jordahl.
Bull. American Astron. Soc., Vol. 4, 237 (1972). – Abstr. AAS.

115.014 Les étoiles supergéantes. C. Nicollier.
Orion Schaffhausen, 30. Jahrgang, p. 75 - 81 (1972).

115.015 The high-luminosity boundary of the β Cephei instability strip. J. R. Percy, K. Madore.
Astron. Journ., Vol. 77, 381 - 383 = Commun. David Dunlap Obs., Univ. Toronto, *Richmond Hill*, No. 330 (1972).

Twenty stars, slowly rotating, more luminous than $M_V = -5.0$, and lying slightly above the upper main sequence have been tested photometrically for short-period light variations of the type found in β Cephei stars. None of the stars show such variations. We conclude that β Cephei variability is found infrequently or not at all in stars more luminous than $M_V = -5.0$.

115.016 Variación luminosa de las estrellas de largo periodo.
L. Celis Santelices, with a preface by M. A. Cordona G.

Separate print Obs. Astrofis. "Manuel Foster", Santiago de Chile. 141 pp. (1970).

115.017 On the masses, luminosities, and compositions of horizontal-branch stars.
T. S. van Albada, N. Baker.
The evolution of population II stars, (see 012.024), p. 41 (1972). — Abstract.

115.018 The redetermination of the zero age main sequence.
T. L. Evans.
Quarterly Journ. Roy. Astron. Soc., Vol. 13, 177 - 182 (1972). — Presented at the Woolley symposium (see 012.025).

115.019 Low-dispersion luminosity criteria in A and F type stars. C. T. Bolton.
Thesis, Univ. Michigan, Ann Arbor. [Available from Univ. Microfilms, Ann Arbor, Mich., U.S.A. Order No. 71−15101], 48 pp. (1970). — See Phys. Abstr., Vol. 75, No. 13905 (1972).

Reduced-proper-motion diagrams.
See Abstr. 112.009.

Narrow- and broad-band photometry of red stars. VII. Luminosities and temperatures for halo-population red stars of high luminosity.   See Abstr. 113.007.

A−F supergiants in the Geneva Observatory photometric system.   See Abstr. 113.027.

Étude des étoiles à raies métalliques par une méthode photométrique à bandes passantes étroites.
See Abstr. 114.012.

The atmospheres of the F-type supergiants. I. Calibration of the luminosity-sensitive O I λ 7774 line.
See Abstr. 114.052.

The nature of the Herbig Ae- and Be-type stars associated with nebulosity.   See Abstr. 114.060.

The surface parameters and composition of some high velocity A-stars.   See Abstr. 114.066.

Spectral classification of OB stars in both hemispheres and the absolute-magnitude calibration.
See Abstr. 114.071.

Spectrophotometric luminosity criteria in O stars.
See Abstr. 114.090.

Spektrale Selbstumkehr, Index für Alter und Leuchtkraft der Sterne?   See Abstr. 114.107.

The absolute magnitudes of the barium stars.
See Abstr. 114.118.

K-line luminosities and the calibration of MK classes for G and K giants.   See Abstr. 114.133.

A physical interpretation of recent period-luminosity relations for RR Lyrae variables.   See Abstr. 122.017.

La classification spectrale des étoiles β Canis Majoris et leur position dans le diagramme HR.
See Abstr. 122.040.

The spectral classification of the β Cephei stars and their location in the theoretical Hertzsprung-Russell diagram.
See Abstr. 122.120.

Circumstellar shells in the young cluster NGC 2264. II. Infrared and further optical observations.
See Abstr. 153.003.

Gravities of blue horizontal-branch stars.
See Abstr. 154.007.

Gravities of blue horizontal-branch stars.
See Abstr. 154.010.

Pseudo color-magnitude diagrams for three southern hemispheric globular star clusters.   See Abstr. 154.018.

Comments on blue horizontal-branch stars.
See Abstr. 154.032.

116 Stellar Magnetic Field, Figure, Rotation

116.001 Light variations in magnetic stars. J. D. Trasco.
Astrophys. Journ., Vol. 171, 569 - 582 (1972).
A magnetic field exerting non-zero stresses is assumed to be present in a stellar atmosphere. The pressure density gradients between the magnetic and the nonmagnetic regions needed to maintain equilibrium are calculated. The effects of line absorption on the light flux are also considered.

116.002 Eleven-band photometry of the magnetic variable HD 125248. H. M. Maitzen, A. F. J. Moffat.
Astron. Astrophys., Vol. 16, 385 - 394 (1972). In German.
The continuum intensity distribution of the AOp-star HD 125248 has been obtained relative to that of the comparison star HD 124683 (B9 V) from photoelectric observations using the filters: UBV, $uvby$, Hβ-wide, Hα-wide and two intermediate band filters centred on λλ 5030 and 5240 Å.

116.003 A search for magnetic fields in four bright stars. J. R. Auman, O. G. Jensen, G. A. H. Walker.
Bull. American Astron. Soc., Vol. 3, 442 (1971). – Abstr. AAS.

116.004 Analysis of the atmospheres of magnetic stars by the curve-of-growth method. II. 10 Aql and ι CrB. J. V. Glagolevsky.
Astrofiz. Issled. Izv. Spets. Astrofiz. Obs., Vol. 3, 62 - 69 (1971). In Russian.
The atmospheres of the stars 10 Aql and ι CrB have been investigated by using the curve-of-growth method. Microturbulent velocities, electron densities, excitation temperatures, and ionization temperatures are determined.

116.005 Analysis of the atmospheres of magnetic stars by the curve-of-growth method. III. Standard stars and constructing of a temperature scale. J. V. Glagolevsky, N. M. Chunakova.
Astrofiz. Issled. Izv. Spets. Astrofiz. Obs., Vol. 3, 70 - 82 (1971). In Russian.
For a comparitive analysis of the properties of the atmospheres of magnetic and metallic-line stars investigated in previously published works the standard A stars γ Crv, γ Gem, η Vir, 2 Hya, and HD 25291 have been chosen. The parameters of their atmospheres: electron densities, excitation and ionization temperatures, and turbulent velocities are determined using traditional methods.

116.006 Enhanced accretion onto a rotating, gravitating, magnetic star. T. T. Chia, R. N. Henriksen.
Journ. Roy. Astron. Soc. Canada, Vol. 66, 67 (1972). – Abstr. Canadian Astron. Soc.

116.007 Coudé polarimeter measurements of weak magnetic fields in bright stars. E. F. Borra, J. D. Landstreet.
Journ. Roy. Astron. Soc. Canada, Vol. 66, 71 (1972). – Abstr. Canadian Astron. Soc.

116.008 Rotation of evolving A and F stars. I. J. Danziger, S. M. Faber.
Astron. Astrophys., Vol. 18, 428 - 443 (1972).
An analysis of stellar rotation among evolved stars of classes A and F is presented.

116.009 Some properties of magnetic variables. K. D. Rakoš.
IAU Colloquium No. 15, (see 012.006), p. 59 - 62 (1972).

116.010 Metal line-blanketing and opacity in the UV of α² CVn. M. R. Molnar.

IAU Colloquium No. 15, (see 012.006), p. 307 (1972). Abstract.

116.011 Zeeman effect observations at the Mauna Kea Observatory. W. K. Bonsack, S. C. Wolff.
Bull. American Astron. Soc., Vol. 4, 228 - 229 (1972). – Abstr. AAS.

116.012 The field and atmosphere of the magnetic A star Beta Cor. Bor. R. S. Freedman.
Bull. American Astron. Soc., Vol. 4, 229 (1972). – Abstr. AAS.

116.013 Viscous evolutions of rapidly rotating stars. R. H. Durisen.
Bull. American Astron. Soc., Vol. 4, 240 (1972). – Abstr. AAS.

116.014 An analysis of the magnetic field of 53 Camelopardalis and its implications for the decentered-dipole rotator model. J. Huchra.
Astrophys. Journ., Vol. 174, 435 - 438 (1972).
The mean surface field of 53 Cam has been estimated by measurement of resolved Zeeman splittings, where possible, and by use of a relation between observed line width and Zeeman pattern to estimate the field strength when the splittings are not resolved. A fit was then made to the decentered-dipole rotator model giving $a = 0.145$, $\beta = 80°$, $i = 50°$, $H_p = -28.4$ kilogauss.

116.015 Relativistic Poincaré limitation for the speed of rotation of a star. M. Abramowicz.
Postępy Astron., Vol. 20, 59 - 66 (1972). In Polish.

116.016 11-Farben-Photometrie des magnetischen Veränderlichen HD 125248. H. M. Maitzen.
Mitt. Astron. Ges., No. 31, p. 170 (1972). – Abstract.

Loops in the H-R diagram during core helium burning and the effect of rotation. See Abstr. 065.009.

Zur Dynamotheorie magnetischer Sterne: Der "symmetrische Rotator" als Alternative zum "schiefen Rotator". See Abstr. 065.017.

The internal dynamics of the oblique rotator. See Abstr. 065.075.

Magnetfelder im Kosmos. See Abstr. 080.027.

Contribution to Kant and Laplace's nebular hypothesis. See Abstr. 107.016.

The effect of rapid rotation on radiation from stars. IV. Weak absorption lines and spectral types. See Abstr. 114.046.

Observed effects of gravity darkening in rotating B stars. See Abstr. 114.047.

Errata

116.901 Erratum: Effective temperatures of some magnetic stars [Observatory, Vol. 91, 115 - 117 (1971)]. G. S. D. Babu.
Observatory, Vol. 92, 23 - 24 (1972).

117 Binary and Multiple Stars, Theory

117.001 Contact binary systems: Why the secondary might be hotter. J. A. J. Whelan.
Monthly Notices Roy. Astron. Soc., Vol. 156, 115 - 128 (1972).

Some aspects of the contact binary problem are considered. The case of energy transfer occurring in the superadiabatic part of the convective envelope is considered and compared with the methods of Lucy and Moss and Whelan. New results for ZAMS non-rotating model contact systems, which illustrate the effects of changes in several parameters, are presented. The importance, for the construction of systems with unequal components, of the depth of the convective envelope at the pp-chain, CN-cycle switchover mass is demonstrated. The results show good, but not sufficient, agreement with observations for W UMa systems.

117.002 Models for contact binaries.
P. Biermann, H.-C. Thomas.
Astron. Astrophys., Vol. 16, 60 - 65 (1972).

Zero age models for contact binaries were constructed using a method that allows for different adiabatic constants in the convective envelopes. The exchanged luminosity was determined by the condition that the component stars fill their respective Roche lobes and are thereby in contact. The models reproduce the main features of the period-color diagram. It is noteworthy that the theoretical models cover most of the range of periods observed for W UMa systems.

117.003 The Beta Scorpii system. T. C. van Flandern.
Bull. American Astron. Soc., Vol. 3, 443 (1971).
Abstr. AAS.

117.004 Ultra-short-period binaries, gravitational radiation and mass transfer. J. Faulkner.
Bull. American Astron. Soc., Vol. 3, 502 (1971). − Abstr. AAS.

117.005 Observational properties of models of detached close binaries. I. P. Giannone, M. A. Giannuzzi.
Astron. Astrophys., Vol. 18, 111 - 120 (1972).

Observational properties and the probability of discovery as eclipsing binaries have been studied for models of detached systems. Masses of the primaries were varied from $0.8\,M_\odot$ to $10\,M_\odot$ and mass ratios in the range $0.1-1$. Evolutionary stages of the component stars from the zero-age main sequence up to the red giant phase have been considered.

117.006 Binary systems after rapid mass loss.
Z. F. Seidov.
Bull. Astron. Inst. Czechoslovakia, Vol. 23, 123 - 125 (1972).

The distribution of semi-major axes and eccentricities of binary systems after rapid mass losses is discussed.

117.007 Time-dependent studies of dynamical instabilities in semidetached binary systems. G. T. Bath.
Astrophys. Journ., Vol. 173, 121 - 134 (1972).

Both linear and nonlinear time-dependent methods are used to investigate the stability of contact members of semidetached binary stars to dynamical perturbations.

117.008 Studies concerning the existence of runaway stars with collapsed companions. J. R. Gott, III.
Astrophys. Journ., Vol. 173, 227 - 234 (1972).

We have analyzed the runaway-star dynamics for the bound case. In the bound case, the first star of a binary system explodes and the collapsed residue of the explosion is captured in orbit around the second star. There are some indications that HD 59543 may be such a system. It consists of a B5 star and an unseen companion with a mass greater than $1.6\,M_\odot$. The unseen companion could possibly be a neutron star or a black hole.

117.009 Velocity effects in triple stellar systems.
V. Szebehely.
Astron. Journ., Vol. 77, 169 - 177 (1972).

Previous results concerning the evolution and eventual disintegration of triple stellar systems are extended to more general initial conditions. Specifically, the initial velocities used in this paper are different from zero. The initial value of the virial coefficient is 1 for some of the numerical experiments and less for others. The total energy of the system is negative in all cases investigated in this paper since the process of disintegration is trivial and predictable for positive energy. The results are reported and discussed.

117.010 Mass loss from binary components with convective outer layers. D. Lauterborn, A. Weigert.
Astron. Astrophys., Vol. 18, 294 - 300 (1972).

Shell source burning stars with deep outer convective layers are shown to expand as a reaction on rapid mass loss. Binary components of this type must start mass exchange on a time scale short compared with the thermal time scale. Using a simple analytical model for the envelope, the numerical results can be understood as due to the conditions for the photosphere, for the shell source, and for the entropy of convective layers.

117.011 Relative luminosity and surface brightness of the components of contact binary stars.
E. L. Robinson.
Publ. Astron. Soc. Pacific, Vol. 84, 51 - 55 (1972).

It is shown that the model of zero-age main-sequence contact binary stars in which a common convective envelope surrounds both stars incorrectly predicts the observed ratios of luminosity and mean surface brightness of the two components of W Ursae Majoris stars.

117.012 On the dissolution time of a class of binary systems.
C. Cruz-González, A. Poveda.
Gravitational N-body problem. IAU Colloquium No. 10, (see 012.004), p. 99 - 113 (1972) = 06.117.021.

117.013 Symmetry of flashes during the Jovian occultation of β Scorpii. A. P. Fairall.
Nature, Vol. 236, 342 (1972).

On 1971 May 13 when Jupiter occulted the bright star Beta Scorpii 1, the fading and brightening of the star were accompanied by a number of spectacular flashes. Cine film observations show that the disappearance flashes are time-symmetric with the reappearance ones. This implies global stratification in the Jovian atmosphere.

117.014 Double stars of equal components.
J. Stock, H. Wroblewski.
Astron. Astrophys., Vol. 18, 341 - 349 (1972).

A list of 271 double stars of nearly equal components with separations up to about one minute of arc is given. Their physical nature is demonstrated by their spectral type distribution which coincides with that of close double stars of equal components, but differs from that of general field stars.

117.015 Duplicity and its consequences among variable stars in general. G. Larsson-Leander.
IAU Colloquium No. 15, (see 012.006), p. 185 - 196 (1972).

117.016 **Eruptive binaries.** J. Smak.
IAU Colloquium No. 15, (see 012.006), p. 248 -
267 (1972).

117.017 **Models for contact binaries.**
P. Biermann, H.-C. Thomas.
IAU Colloquium No. 15, (see 012.006), p. 285 - 288 (1972).

117.018 **The evolution of a contact binary.**
J. Hazlehurst, E. Meyer-Hofmeister.
IAU Colloquium No. 15, (see 012.006), p. 289 (1972).

117.019 **Variability due to contact configurations.**
H. Mauder.
IAU Colloquium No. 15, (see 012.006), p. 290 - 291 (1972).

117.020 **Search for contracting close binaries.**
R. H. Koch.
IAU Colloquium No. 15, (see 012.006), p. 291 - 293 (1972).

117.021 **Suspected duplicity of the Ap star 53 Cam.**
R. Faraggiana.
Inform. Bull. Variable Stars, (I.A.U. Commission 27), Konkoly
Obs., Budapest, No. 618 (1972).

117.022 **The new companion of Theta Coronae Borealis.**
R. Burchi, P. Tempesti.
Inform. Bull. Variable Stars, (I.A.U. Commission 27), Konkoly
Obs., Budapest, No. 619, 3 pp. (1972).

117.023 **An old evolved binary in the galactic halo.**
A. Young, B. Nelson, R. Mielbrecht.
Bull. American Astron. Soc., Vol. 4, 210 - 211 (1972). – Abstr.
AAS.

117.024 **Stability criteria for triple stars.** R. S. Harrington.
Bull. American Astron. Soc., Vol. 4, 263 (1972).
Abstr. AAS.

117.025 **Eruptive binaries. Part I.** J. Smak.
Postępy Astron., Vol. 20, 91 - 103 (1972). In Polish.
This article contains a review of problems connected with
the physical properties of the components and the circumstellar matter and with the mechanisms of outbursts.

117.026 **Structure of close binaries. II: Theory of uniformly
rotating binaries.** M. D. T. Naylor, S. P. S. Anand.
Astrophys. Space Sci., Vol. 16, 137 - 150 (1972).
The stellar structure equations for uniformly rotating,
synchronous close binaries have been derived using a modified
double approximation technique which has been successful
in studies of rotating stars and close binary polytropes. Tables
are presented for different degrees of distortion, mass ratio,
and the two extremes of inclination which allow the apparent
luminosity and mean effective temperature to be determined
if the interior luminosity and separation of the components
are given.

117.027 **No evidence for black holes in eccentric binary
systems.** W. D. Evans, G. T. Bath.
Nature, Phys. Sci., Vol. 238, 7 - 8 (1972).
Following the suggestion by Gibbons and Hawking that
collapsed objects might be detected in eccentric binary systems
we have analysed additional systems in Batten's catalogue and
conclude that such objects cannot be found in this way at present. We find that the anomalous eccentricity distribution of
single line non-eclipsing systems found by them also occurs in
single line eclipsing systems, which are not thought to be able
to contain collapsed objects.

117.028 **Modelle für Kontaktsysteme von engen Doppel-**

sternen. P. Biermann, H.-C. Thomas.
Mitt. Astron. Ges., No. 31, p. 125 - 126 (1972).

117.029 **Sterntrupps und weite Mehrfachsysteme in Vulpe-
cula.** J. Hopmann.
Mitt. Astron. Ges., No. 31, p. 142 (1972). – Abstract.

117.030 **Sur la rotation des étoiles doubles serrées.**
J.-P. Zahn.
Comptes Rendus Acad. Sci. Paris, Sér. B, Vol. 274, 1443 -
1446 (1972).
On montre que la rotation lente des étoiles doubles
serrées à enveloppe radiative est due au freinage exercé par la
marée dynamique.

117.031 **The Roche model and its applications to close binary systems.** Z. Kopal.
Advances Astron. Astrophys., Vol. 9, (see 003.007), 1 - 65
(1972).

117.032 **Evolution of close binary systems with intermediate
initial mass ratios.** P. Giannone, M. A. Giannuzzi.
Astron. Astrophys., Vol. 19, 298 - 302 (1972).
The evolution of the original primaries of three close binary systems with total masses of $3.5\,M_\odot$ and initial separations of $10\,R_\odot$ has been computed through the phase of mass
exchange between the components. The initial values of the
mass ratio are varied in a wide range. The departure from thermal equilibrium of the models of the components losing matter is studied. The final remnants have masses around $0.32\,M_\odot$;
two of them become white dwarfs and one a helium star. The
results obtained for systems with the same total masses are discussed.

117.033 **Angular momentum loss from contact binary systems.** D. L. Moss.
Monthly Notices Roy. Astron. Soc., Vol. 157, 433 - 441 (1972).
The effect of angular momentum loss on zero-age main
sequence contact binary systems constructed according to the
Lucy model is investigated. It is found that the separation of
the mass centres decreases, the mass ratio decreases, and the
system adopts a dumb-bell configuration. If the system reaches
a configuration in which it fills its outer closed common equi-
potential surface it is unstable to mass loss with the local angular momentum through the outer Lagrange point.

117.034 **The shell star 88 Herculis: Pulsating envelope or
binary star?** P. Harmanec. P. Koubský, J. Krpata.
Astrophys. Letters, Vol. 11, 119 - 122 (1972).
The periodic variability of the radial velocity of the shell
star 88 Herculis was revealed with the semi-amplitude of about
10 km/sec. Possible consequences of this result are briefly discussed.

117.035 **On the fission theory of binary stars.**
N. R. Lebovitz.
Astrophys. Journ., Vol. 175, 171 - 183 (1972).
The fission theory is reformulated so as to be independent
of the action of viscosity. It is found that this results in evolution not along the Jacobi series, but along a similar series of ellipsoids, the lower self-adjoint series of Riemann. Certain ob-
jections that have been raised against the fission theory are reconsidered.

117.036 **Proper Motion Survey with the forty-eight inch
Schmidt telescope: XXIX. Double stars with common
proper motion.** W. J. Luyten.
Separate print Univ. Minnesota, Minneapolis, Minnesota.
11 pp. (1972).
In continuation of the lists given in Nos. XXI and XXV
of these publications data are herewith given for 501 further

double stars with common proper motion.

117.037 On the determination of the rotational velocities of non-spherical stars. I. Pustylnik.
Tartu Astron. Obs. Teated, No. 35, p. 51 - 62 (1971).

The problem of non-synchronous axial rotation in close double stars is rediscussed on the basis of the papers by Koch et al. (1965) and Stoeckley (1968). Some difficulties arising in connection with Stoeckley's interpretation of the observed line profiles of Altair are explained by peculiarities of the brightness distribution over the surface of a rotating star. The influence of the effects of non-sphericity of a rotating component on the values of rotational velocities obtained by Koch et al. is studied.

117.038 On a theory of figures of non-spherical stars. I. Pustylnik.
Tartu Astron. Obs. Teated, No. 35, p. 63 - 71 (1971).

Following Kopal's idea an approximate formula describing the figure of a non-spherical component of a close binary is derived in a form appropriate for computer calculations.

117.039 La distribution des distances de séparation des étoiles doubles et ses enseignements.
J. Dommanget.
Sciences, Vol. 2, (No. 3), 155 - 160 = Obs. Roy. Belgique, Commun. Sér. B, No. 70 (1971).

117.040 A preliminary investigation of the dependence of M_v (K) on [Fe/H].
O. C. Wilson, E. H. Olsen, P. Kjaergaard.
Quarterly Journ. Roy. Astron. Soc., Vol. 13, 161 - 163 (1972). — Presented at the Woolley symposium (see 012.025)

117.041 Double stars, indicators for the secular mass loss of a star. V. Čelebonović.
Vasiona, Vol. 20, 15 - 17 (1972). In Serbo-Croatian.

117.042 Duoblaj sistemoj kaj interna stelstrukturo. D. J. Martinov, translated by A. Heck.
Sci. Revuo Internacia Sci. As. Esperantista, (Beograd), Vol. 21, 45 - 50 (1970).

117.043 Pri la ekvacioj de la interna stelstrukturo. A. Heck.
Sci. Revuo Internacia Sci. As. Esperantista, (Beograd), Vol. 22, 189 - 192 (1971).

117.044 Evoluo kaj masintersanĉo en duoblaj steloj. A. Heck.
Sci. Revuo Internacia Sci. As. Esperantista, (Beograd), Vol. 22, 193 - 204 (1971).

117.045 Eruptive binaries. V. Period variations. J. Smak.
Acta Astron., Vol. 22, 1 - 9 = Warsaw Univ. Obs., Astron. Inst., Polish Acad. Sci., Repr. No. 323 (1972).

The observed rates of period variations are typically between d ln P/dt = ±10^{-10} and ±10^{-8} d^{-1}. In the three best studied cases of U Gem, RW Tri, and UX UMa the variations are of alternating character. It is shown that such variations can be due to the exchange between the orbital momentum and the rotational momentum of the disk. To account for the observed variations the masses of the disks must be larger than

10^{27} g and most likely larger than 10^{29} g.

117.046 Barnard's star. P. van de Kamp.
Sproul Obs. Repr., No. 200, 2 pp. (1971).

117.047 Other planetary systems. A. Hajduk.
Kozmos, Vol. 3, 5 - 6 (1972). In Slovak. — Review article.

117.048 Origin of the planetary systems. Astronomical evidence in other stars. Z. Kopal.
From plasma to planet. 21st Nobel symposium 1971, (see 012.028), p. 39 - 47 (1972).

117.049 Contribution à l'étude des magnitudes absolues et des couleurs intrinséques des étoiles bleues de grande luminosité à partir de l'observation de systèmes multiples. M.-L. Burnichon.
Thesis, Sci. Phys., Univ. Paris. Centre Documentation, C. N. R. S. (1972-03-10), 124 pp. (1972). — See Bull. Signal., Vol. 33, Section 120, No. 7058 (1972).

Studies of other planetary systems with space techniques. See Abstr. 051.043.

The collapse of a rotating cloud. See Abstr. 065.076.

An apparent stellar mass gap near 3 $^1/_2$ $M_\odot$. See Abstr. 065.093.

The occultation of Beta Scorpii by Jupiter and Io. I. Jupiter. See Abstr. 099.013.

The occultation of Beta Scorpii by Jupiter and Io. II. Io. See Abstr. 099.014.

The occultation of Beta Scorpii by Jupiter and Io. III. Astrometry. See Abstr. 099.015.

L'occultation de β Scorpion par Jupiter le 13 mai 1971. See Abstr. 099.021.

Observations of occultations of β Scorpii on 1971 May 13 and 14. See Abstr. 099.041.

Observations of the occultation of the star β Scorpii by Jupiter. See Abstr. 099.045.

The coming shell phase of AX Monocerotis. See Abstr. 114.083.

Evidence of mass loss from a close binary system: The Hyades member eclipsing binary BD + 16° 516. See Abstr. 121.011.

An old evolved binary in the galactic halo. See Abstr. 121.040.

EQ Pegasi: A double flare star. See Abstr. 122.003.

The effect of masses on the evolution of triple gravitational systems. See Abstr. 151.077.

118 Visual Binaries

118.001 **Near infrared photometry of the θ Coronae Borealis system.** T. P. Roark, J. H. Baumert, N. M. White.
Astrophys. Letters, Vol. 10, 55 - 57 (1972).

The integrated light output from the newly discovered binary system θ CrB was observed in the near infrared on 25 June 1970, 10 August and 17 August 1971. The observations indicate that the spectral type of the companion was between O5 and A5 at these times.

118.002 **New double stars (6th series) discovered at Nice with the 50 cm refractor.** P. Couteau.
Astron. Astrophys., Suppl. Ser., Vol. 5, 167 - 174 (1972). In French.

A list of 100 pairs discovered with the 50 cm refractor of the Nice Observatory is given.

118.003 **Orbit of the visual binary β 989 = ADS 15281.** P. Couteau, P. J. Morel.
Astron. Astrophys., Suppl. Ser., Vol. 5, 175 - 180 (1972). In French.

Improved orbital elements are given for the visual binary ADS 15281; $21^h42^m4 +25°25'$ (1950); 4,8–5,3 Sp F5.

118.004 **Orbits of two visual binaries.** P. Couteau.
Astron. Astrophys., Suppl. Ser., Vol. 5, 181 - 183 (1972). In French.

The orbital elements of two binaries ADS 16046 and ADS 16165 are given.

118.005 **Emploi de la transformation de Fourier pour la réduction des mesures d'étoiles doubles par balayage photoélectrique.** G. Coupinot, J. Rösch.
Comptes Rendus Acad. Sci. Paris, Sér. B, Vol. 274, 149 - 152 (1972).

118.006 **Rocket spectroscopy of Zeta Orionis.** A. M. Smith.
Astrophys. Journ., Vol. 172, 129 - 148 (1972).

A spectrum of ζ Ori extending from 922 to 1453 Å with approximately 0.8 Å resolution has been recorded at rocket altitudes. The mean radial velocities associated with seven P Cygni-type lines originating in the circumstellar envelope have been determined, including that of O VI ions as revealed by strong doublet lines at 1031.91 and 1037.61 Å. A large number of interstellar lines have been tentatively identified, but with the exception of the atomic hydrogen Lα line none can be analyzed quantitatively.

118.007 **A unique visual double star.** C. B. Stephenson, N. Sanduleak.
Astrophys. Journ., (*Letters*), Vol. 172, L115 (1972).

We have discovered the first known case of an apparent binary star consisting of a carbon star and an M star.

118.008 **Astrometric study of four binary stars.** W. D. Heintz.
Astron. Journ., Vol. 77, 160 - 165 (1972).

Orbits, parallaxes, and masses have been derived for the visual binaries ADS 8197, Wolf 424, 99 Her, and Kpr 112. The results are based on a total of more than 1000 Sproul astrometric plates measured by the author. The components of Wolf 424 appear to be the least massive visible stars currently known.

118.009 **Parallax and orbital motion of the two nearby long-period visual binaries Groombridge 34 and ADS 9090.** S. L. Lippincott.
Astron. Journ., Vol. 77, 165 - 168 (1972).

Measurements of photographs taken with the Sproul 24-inch refractor between 1937 and 1970 on Groombridge 34 A and B yield π_{abs} = +0".283±0".005 (p.e.). A provisional orbit was computed, including the Sproul B–A positions and those measured on plates from the McCormick Observatory in addition to all other available data. Measurements from plates taken with the Sproul refractor on ADS 9090 A and B over the interval 1938–1969 yield π_{abs} = +0".092±0".005.

118.010 **Parallaxes, mass ratios, and masses for the visual binary stars BD + 75°403, ADS 9094, and ADS 9716.** J. L. Hershey.
Astron. Journ., Vol. 77, 251 - 253 (1972).

Series of plates taken with the Sproul 24-inch refractor extending over more than three decades have been used to determine parallaxes and masses for three visual binaries. The parallaxes and mass ratios are quite accurate; the small parallaxes of these stars result in relatively large errors for the masses.

118.011 **Radial velocities of selected visual binaries.** C. D. Doucet, G. A. Bakos.
Journ. Roy. Astron. Soc. Canada, Vol. 66, 72 - 73 (1972). Abstr. Canadian Astron. Soc.

118.012 **Photometry of variables in close visual double stars.** O. G. Franz, R. L. Millis, N. M. White.
IAU Colloquium No. 15, (see 012.006), p. 230 - 234 (1972).

118.013 **Trajectories of the rectilinear relative motion of the components of visual double stars. IV.** G. A. Starikova.
Soobshch. Gos. Astron. Inst. Shternberga, No. 176, p. 9 - 14 (1972). In Russian.

Trajectories of the rectilinear relative motion of the components of ten visual double stars are given (ADS 2552, 3812, 4223, 4472, 6405, 7681, 7896, 8515, 10583, 16130). The possibility of orbital motions was investigated by means of Dommanget's criterion.

118.014 **Measurements of double stars made with the 38 cm equatorial refractor of the Paris Observatory.** P. Baize.
Astron. Astrophys., Suppl. Ser., Vol. 6, 147 - 175 (1972). In French.

Details of observations and annual means of 1813 measurements of double stars made with the 38 cm equatorial refractor of the Paris Observatory between 1966.7 and 1971.5 are given. For each pair remarks and the $O-C$ with the most reliable orbits are given. Orbital elements of the double stars ADS 6851, 8325, 9019, 11260, 12752, 16111, Kui 83, and Kui 99 have been inserted in the text.

118.015 **New double stars (7th series) discovered at Nice with the 50 cm refractor.** P. Couteau.
Astron. Astrophys., Suppl. Ser., Vol. 6, 177 - 184 (1972). In French.

We give a list of 100 double stars discovered at the 50 cm refractor.

118.016 **Measurements of double stars made at Nice.** P. Couteau.
Astron. Astrophys., Suppl. Ser., Vol. 6, 185 - 198 (1972). In French.

We give 400 measurements of 153 binaries made at the 74 cm refractor, and 160 measurements of 79 binaries made at the 50 cm refractor.

118.017 Orbites nouvelles. P. Muller.
Circ. Inform. (U.A.I. Commission des Étoiles Doubles), Obs. Meudon, No. 56 (1972).

118.018 Etoiles doubles découvertes à Nice, lunette de 50 cm. P. Couteau, P. Muller.
Circ. Inform. (U.A.I. Commission des Étoiles Doubles), Obs. Meudon, No. 56 (1972).

118.019 Orbites nouvelles. P. Muller.
Circ. Inform. (U.A.I. Commission des Etoiles Doubles), Obs. Meudon, No. 57 (1972).

118.020 Étoiles doubles nouvelles découvertes à Beograd, lunette de 65 cm. G. M. Popovic.
Circ. Inform. (U.A.I. Commission des Étoiles Doubles), Obs. Meudon, No. 57 (1972).

118.021 Étoiles doubles nouvelles découvertes à Nice, lunettes de 50 (n) et de 74 (N) cm.
P. Couteau, P. Muller.
Circ. Inform. (U.A.I. Commission des Étoiles Doubles), Obs. Meudon, No. 57 (1972).

118.022 Lichtelektrische Photometrie enger visueller Doppelsterne. K. D. Rakosch.
Mitt. Astron. Ges., No. 31, p. 170 - 173 (1972).

118.023 Photographic measures of double stars.
J. McK. Luck.
Mem. Roy. Astron. Soc., Vol. 76, 67 - 98 (1972).
  Separations in right ascension and declination of 111 double stars from 368 photographic plates taken on the Yale-Columbia 26-inch refractor, as well as seasonal means, together with separations and position angles, are given in a table. Median standard errors of $0\overset{''}{.}015$ in $\Delta\alpha$ and $0\overset{''}{.}013$ in $\Delta\delta$ were obtained.

118.024 Parallaxes and proper motions. VII. Double-star measures.

W. S. Mesrobian, T. D. Griess, J. C. Titter.
Astron. Journ., Vol. 77, 392 - 394 (1972).
  Parallaxes and proper motions are given for three known binaries and one suspected binary system based on photographs taken with the Van Vleck 20-inch refractor. New orbital information is given for the three known binaries, ADS 41, 48, and BD +31°3767. The suspected binary, Ross 413, shows no perturbation.

118.025 La résolution du système des équations de Thiele-Innes dans le calcul d'une orbite d'étoile double visuelle. J. Dommanget, O. Nys.
Bull. Astron. Obs. Roy. Belgique, Vol. 7, 301 - 304 (1972).

118.026 L'orbite circulaire d'une étoile double visuelle. S. Arend.
Bull. Astron. Obs. Roy. Belgique, Vol. 7, 305 - 310 (1972).

118.027 Orbite circulaire de l'étoile double visuelle ADS 9453 = β 239. R. R. de Freitas Mourão.
Bull. Astron. Obs. Roy. Belgique, Vol. 7, 311 - 313 (1972).

118.028 CoD −23°12133. H. Povenmire.
IAU Circ., No. 2380 (1972).

118.029 Micrometer measures of 1,343 double stars.
C. E. Worley.
Publ. United States Naval Obs., *Washington,* Second Ser., Vol. 22, Part II, 136 pp. (1971).

Utilization of the electronic camera for the photometric measurement of close visual binaries.
See Abstr. 034.008.

Dynamical parallaxes. See Abstr. 111.009.

Double stars of equal components.
See Abstr. 117.014.

On the identification of the eclipsing component of the visual binary BV 1481 Cet. See Abstr. 121.085.

X Ophiuchi, ein langperiodisch Veränderlicher des Mira-Typs und zugleich Doppelstern. See Abstr. 122.090.

119 Spectroscopic Binaries

119.001 **Spectroscopic study of o Per.** R. Foy.
Astron. Astrophys., Vol. 16, 108 - 114 (1972).
The spectroscopic binary o Per is analyzed in order to ascertain whether it is a β Cep star, and to determine, by means of an improvement of Petrie's method (1939), spectral types and absolute magnitudes of the two components; from that, masses, radii and absolute values of the orbital parameters are obtained. These data permit the discussion of the anomalous proper motion of o Per with respect to the association II Per, of which o Per has been considered as a member (Blaauw, 1952). o Per is found to be not a member of the association II Per.

119.002 **IC 4665, a cluster of binaries.**
H. A. Abt, C. T. Bolton, S. G. Levy.
Astrophys. Journ., Vol. 171, 259 - 266 (1972).
Radial velocities of the 19 brightest stars in this open cluster show that all but one are spectroscopic binaries. Orbital elements are derived for 13 systems, and one additional one is double-lined. It is suggested that the low mean rotational velocity for stars in this cluster is due to its reduction by tidal interactions in binary systems, although in most of the systems with periods greater than 6 days the rotational motion has not yet attained synchronization with the orbital motion.

119.003 **Spectroscopic binaries in the open cluster NGC 2516.**
H. A. Abt, S. G. Levy.
Astrophys. Journ., Vol. 172, 355 - 360 (1972).
A limited study of the radial velocities of the 16 brightest early-type stars in NGC 2516 yields approximate periods for six binaries. Tidal interaction in that number of binaries is insufficient to explain the low mean rotational velocity obtained for the cluster stars, but allowance for the additional Ap stars, in which decreased rotation by magnetic braking may have occurred, does account for the low rotational velocities.

119.004 **Seven new spectroscopic binaries in Cepheus.**
C. D. Garmany.
Astron. Journ., Vol. 77, 38 - 41 (1972).
Orbital elements are presented for seven new spectroscopic binaries in the association Cep OB 3. Their periods range from 1 to 100 days, and their semi-amplitudes range from 20 to 65 km/sec.

119.005 **Parallax and astrometric orbit of the spectroscopic binary γ Geminorum from observations with the 20-inch refractor of the Van Vleck Observatory.**
K. W. Kamper, Jr.
Astron. Journ., Vol. 77, 85 - 87 (1972).
Measurements of 130 parallax plates of γ Geminorum covering the time interval 1932–1958 confirm the $12\,^1/_2$-yr orbital period previously announced by Wagman, Daniel, and Crissman (1963). The parallax and photocentric semimajor axis are found to be $+0\overset{..}{.}031 \pm 0\overset{..}{.}003$ (m.e.) and $0\overset{..}{.}065 \pm 0\overset{..}{.}016$ (m.e.) from the Van Vleck material.

119.006 **De veranderlijke van de maand: Z Camelopardalis.**
H. Feijth.
Hemel en Dampkring, Vol. 70, 16 - 18 (1972).

119.007 **The absolute magnitudes and spectral types of the stars in the Gamma Velorum system.**
P. S. Conti, L. F. Smith.
Astrophys. Journ., Vol. 172, 623 - 630 (1972).
The ratio of the visual luminosities of the components of the binary system, γ^2 Vel, WC8 + O, is derived by comparison of the emission-line strengths in the spectrum with those of

HD 192103 (WC8). The assumption of similarity of the emission spectrum of γ^2 Vel to that of HD 192103 and the appearance of the absorption spectrum of the O9 I star are also discussed.

119.008 **A search for small-amplitude spectroscopic binaries among main-sequence F-type stars.**
K. S. Anderson, R. P. Kraft.
Astrophys. Journ., Vol. 172, 631 - 637 = Contr. Lick Obs., No. 347 (1972).
A search for small-amplitude spectroscopic binaries among sharp-line main-sequence stars between spectral types F3 and G1 was conducted principally with the coudé spectrograph of the Lick 120-inch reflector over a 2-year interval. Out of 45 previously supposed constant-velocity stars, seven certain and four possible stars of variable velocity were found.

119.009 **Improvement of the spectroscopic orbit of Y Cyg.**
E. A. Vitrichenko.
Izv. Krymskoj Astrofiz. Obs., Vol. 43, 71 - 75 (1971).
In Russian.
26 new radial velocity measurements have been obtained. These data and those of other authors were used for the improvement of the spectroscopic orbit by least squares. The change of γ-velocity is discussed in brief. Its variability is not confirmed.

119.010 **Double star HD 175514. III. Analysis of observations in 1968.** E. A. Vitrichenko.
Izv. Krymskoj Astrofiz. Obs., Vol. 43, 76 - 86 (1971).
In Russian.
The photoelectric light curves obtained in 1968 have been rectified and solved by least squares for a model of similar ellipsoids and uniform disks. New radial velocity measurements have been obtained. The spectroscopic orbit was improved. The mass ratio has been determined by the spectrophotometric method. Some peculiarities of the system are presented.

119.011 **Orbital elements of spectroscopic binary HD 95363.**
M. Imbert.
Astron. Astrophys., Vol. 18, 267 - 270 (1972). In French.
The orbital elements of the spectroscopic binary HD 95363 (m_{pg} 9.64, F7V) have been derived from 19 observations. The orbital parameters were computed, from measurements to the final elements, with four FORTRAN IV programs.

119.012 **Spectroscopic binaries with circular orbits.**
C. D. Scarfe.
Observatory, Vol. 92, 60 - 61 (1972). – Letter.

119.013 **Orbit of the double-lined binary mercury star Chi Lupi.** M. M. Dworetsky.
Publ. Astron. Soc. Pacific, Vol. 84, 254 - 259 (1972).
The orbital elements of χ Lupi were derived from radial velocity measures of ten high-dispersion coudé plates obtained in 1961–65 combined with measures of plates from the Lick Observatory's Chile expedition obtained in 1909–11. The orbit is circular, with $P = 15\overset{d}{.}2565$. The mass ratio is $M_A/M_B = 1.42$; $(M_A + M_B)\sin^3 i = 3.96\,M_\odot$. Both components are on or near the main sequence.

119.014 **On the nature of HD 161227.** M. A. Smith.
Publ. Astron. Soc. Pacific, Vol. 84, 281 - 283 = Lick Obs. Bull., No. 624 (1972).
Spectroscopic evidence is presented demonstrating that

HD 161227 is a double-lined binary. Its secondary's D-lines are visible in the visual but no lines of the secondary are visible in the blue spectral region. It is found that the primary is an F0m star located not far above the main sequence.

119.015 The mass function in spectroscopic binaries.
C. Jaschek, O. Ferrer.
Publ. Astron. Soc. Pacific, Vol. 84, 292 - 297 (1972).

The distribution of the mass function of spectroscopic binaries was examined for both giants and dwarfs. It was found that the distribution is very similar for both kinds of objects and that the distribution is rather insensitive to the distribution of the mass ratios.

119.016 A search for spectroscopic binaries among southern cepheids.   T. L. Evans.
IAU Colloquium No. 15, (see 012.006), p. 204 - 205 (1972).

119.017 A search for eclipses of HD 27149.
B. G. Jørgensen, E. H. Olsen.
Inform. Bull. Variable Stars, (I.A.U. Commission 27), Konkoly Obs., Budapest, No. 652 (1972).

119.018 The Hyades spectroscopic binary HD 27149.
A. H. Batten, G. Wallerstein.
Bull. American Astron. Soc., Vol. 4, 211 (1972). – Abstr. AAS.

119.019 Étude spectroscopique de deux étoiles nouvelles VV Cephée: BD + 54°2698, BD + 63°3.
M. Barbier.
Comptes Rendus Acad. Sci. Paris, Sér. B, Vol. 274, 1440–1442 (1972).

On donne les vitesses radiales des corps en émission et en absorption. Le calcium interstellaire situe BD + 63°3 à une distance minimale de 6100 pc. Sa magnitude absolue est ≤ à − 7.9.

119.020 The unresolved double-lined binary HR 7774.
M. A. Smith.
Astron. Astrophys., Vol. 19, 312 - 314 = Lick Obs. Bull., No. 626 (1972).

From variations in spectral line widths and depths, HR 7774 is shown to be an unresolved double-lined binary of period 61 days. A few characteristics of the system are discussed. Both components are probably metallic line A stars with true rotational velocities of the order of 120 km/s (very high values).

119.021 Binary nature of the B supergiant in the error box of the Vela X-ray source.
W. A. Hiltner, J. Werner, P. Osmer.
Astrophys. Journ., (Letters), Vol. 175, L19 - L22 (1972).

The B0.5 Ib star in the error box of 2U 0900–40 is a spectroscopic binary with a provisional period of 7.0 days and a semiamplitude of 26 km s^{-1}. The minimum mass of the secondary star is near 1.4 $M_\odot$.

119.022 U Geminorum – ein sensationeller Veränderlicher.
R. Lukas.
SuW, Vol. 11, 194 (1972).

119.023 Analysis of a variable spectroscopic double star.
D. Herbison-Evans, N. R. Lomb.
Computer Phys. Commun. (Netherlands), Vol. 2, 368 - 380 (1971). – See Bull. Signal., Vol. 33, Section 120, No. 3919 (1972).

Radial velocities of 65 early-type stars.
See Abstr. 112.007.

The shell star 88 Herculis: Pulsating envelope or binary star?   See Abstr. 117.034.

HD 21242: A new bright variable star.
See Abstr. 123.020.

Identification of Cygnus X-1 with HDE 226868.
See Abstr. 142.052.

Cygnus X-1 – a spectroscopic binary with a heavy companion?   See Abstr. 142.048.

Further radio observations of Cygnus X-1.
See Abstr. 142.069.

120   Variable Stars: Catalogues, Ephemerides, Miscellanea

120.001 Estrellas variables.   J. Rubí.
El Universo, No. 93, Vol. 24, 87 - 90 (1970).

120.002 Techniques and results of observations of rapid and ultrarapid variable stars.   B. Warner.
IAU Colloquium No. 15, (see 012.006), p. 144 - 159 (1972).

120.003 Variable star observations from outside the earth's atmosphere: Review and prospects.
Y. Kondo.
IAU Colloquium No. 15, (see 012.006), p. 298 - 307 (1972).

120.004 The Delta Scuti stars. An annotated catalogue and bibliography.   M. A. Seeds.
Inform. Bull. Variable Stars, (I.A.U. Commission 27), Konkoly Obs., Budapest, No. 625 (1972).

Announcement on the completion of the survey on Delta Scuti stars. The catalogue which can be requested from the author contains data for 58 known and 97 suspected Delta Scuti stars.

120.005 The main types of variable stars.   P. Flin.
Urania Kraków, Vol. 43, 98 - 106 (1972). In Polish.

120.006 Fotografische Veränderlichenbeobachtung.
K. Wälke.
BAV Rundbrief, 21. Jahrgang, p. 10 - 13 (1972).

Atlas of variable star identification charts.
See Abstr. 003.131.

First supplement to the third edition of the general catalogue of variable stars.   See Abstr. 003.154.

121 Eclipsing Variables

121.001 **A model for the totally eclipsing W Ursae Majoris system AW UMa.**
S. W. Mochnacki, N. A. Doughty.
Monthly Notices Roy. Astron. Soc., Vol. 156, 51 - 65 (1972).
　A method for the synthesis of theoretical light curves and line profiles for close binary systems is outlined, considering Roche geometry, non-linear limb-darkening and convective envelopes with irradiation and gravity-darkening. The method is applied to the totally eclipsing system AW UMa, for which the unusually low mass ratio of 0.079 is obtained.

121.002 **Bedeckungslichtwechsel und Aktivität des Nova-ähnlichen Veränderlichen V Sagittae.** M. Beyer.
Astron. Nachr., Vol. 293, 167 - 169 (1971/72).
　New elements for the eclipsing period of the presumable exnova V Sge, derived from 1513 visual observations obtained at Bergedorf between 1947 and 1970, are in good accordance with the results of G. H. Herbig et al. (1965). The light curves for the eclipsing variability as well as those for the eruptive activity are given.

121.003 **Models for five W Ursae Majoris systems.**
S. W. Mochnacki, N. A. Doughty.
Monthly Notices Roy. Astron. Soc., Vol. 156, 243 - 252 (1972).
　Assuming the Lucy model, the method of light curve synthesis has been applied to derive common-envelope models for the totally eclipsing W Ursae Majoris system: V 566 Ophiuchi, RR Centauri, RZ Tauri, FG Hydrae and DK Cygni. The applicability of the Lucy model is discussed.

121.004 **Investigations on the structure of W UMa stars.**
H. Mauder.
Astron. Astrophys., Vol. 17, 1 - 16 = Veröff. Remeis-Sternw. Bamberg, Astron. Inst. Univ. Erlangen-Nürnberg, Vol. 8, No. 92 (1972).
　Starting with the rectifiable model and the Roche model for W UMa stars it is shown that the photometric elements depend remarkably on the rectification procedure. Taking into account this effect the photometric elements for 34 W UMa systems are recalculated and a catalogue of preliminary elements is given. On the basis of these results, the model of Lucy (1968) for W UMa stars is tested.

121.005 **The O-type eclipsing binary, LY Aurigae.**
F. B. Wood.
Bull. American Astron. Soc., Vol. 3, 444 (1971). – Abstr. AAS.

121.006 **UBV ground based photometry and OAO UV photometry of the eclipsing binary star VV Orionis.**
H. L. Atkins, R. L. White.
Bull. American Astron. Soc., Vol. 3, 444 (1971). – Abstr. AAS.

121.007 **Rectification of synthetic light curves.**
E. F. Guinan, D. B. Wood.
Bull. American Astron. Soc., Vol. 3, 444 (1971). – Abstr. AAS.

121.008 **On the orbital eccentricity of an eclipsing binary system.** I. Todoran.
Astrophys. Space Sci., Vol. 15, 229 - 235 (1972).
　A new method is presented to precisely deduce the orbital eccentricity of an eclipsing binary from observed epochs of its light minima. Application to the system V526 Sagittarii gives $e = 0.2220 \pm 0.0016$.

121.009 **Sudden changes in the period of the eclipsing con-**tact binary 44 i Bootis.
J. Bergeat, M. Lunel, F. Sibille, F. van't Veer.
Astron. Astrophys., Vol. 17, 215 - 219 (1972).
　From photoelectric observations of 44 i Boo obtained during the years 1969–1971, we made 23 determinations of the times of conjunction from the observed minima. We concluded that this eclipsing system suffers periodic changes of the period. The interval between sudden period changes is of the order of 10 years.

121.010 **The binary system AU Puppis.** H. Bossen.
Astron. Astrophys., Vol. 17, 362 - 366, with a correction in Vol. 19, 315 (1972).
　Absolute orbital elements were calculated and they show that the system consists of a A0V star of 3.0 solar masses and of a A1.5V star of 2.1 solar masses. The system belongs to the detached ones and the components do not differ much from spherical shape.

121.011 **Evidence of mass loss from a close binary system: The Hyades member eclipsing binary BD + 16° 516.**
G. Vauclair.
Astron. Astrophys., Vol. 17, 437 - 440 (1972).
　The present evolved system being composed of a $\sim 0.8\,M_\odot$ K0V star on the main sequence and of a $\sim 0.7\,M_\odot$ hot white dwarf, it is argued that the evolution of such a system took place through both a mass exchange of $\sim 0.4\,M_\odot$ from the primary to the secondary and a mass loss of $\sim 0.8\,M_\odot$, probably through a planetary-nebula phase.

121.012 **Spectrophotometric study of the components of eclipsing variable systems.** T. M. Rachkovskaya.
Izv. Krymskoj Astrofiz. Obs., Vol. 43, 87 - 100 (1971). In Russian.
　Spectrograms of the eclipsing systems AH Cep, CW Cep, σ Aql, AG Per, U Oph and AR Aur were obtained. From these spectrograms the mean spectral types, the rotational velocities and absolute visual magnitudes have been determined for every component of the system.

121.013 **RI photoelectric photometry of Epsilon Coronae Austrinae.** C. A. Hernández.
Astron. Journ., Vol. 77, 152 - 154 (1972).
　Light and color curves of ϵ Cr A in RI are presented. These observations were simultaneously made with those by S. Tapia in UBV and whose results are already published.

121.014 **Photoelectric UBV photometry of TY Bootis.**
R. B. Carr.
Astron. Journ., Vol. 77, 155 - 159 (1972).
　An orbital solution for the short period W UMa-type eclipsing variable TY Boo has been found from 391 photoelectric UBV observations. The solution parameters are consistent with the photometric results which indicate main-sequence G3 and G7 components. Nonrectifiable distortions in the light curves limit the accuracy of the analysis.

121.015 **On the atmosphere of Epsilon Aurigae.**
D. J. Stickland, D. Branch.
Observatory, Vol. 92, 9 - 11 (1972).
　The nature of the unseen companion of the F-type supergiant ϵ Aurigae remains a matter of controversy. The possibility of an unusual chemical composition in the atmosphere of the F star has led us to study the spectrum of ϵ Aurigae, outside eclipse, relative to the normal composition supergiant α Persei (F 5 Ib).

121.016   **Observations of rapid blue variables — IV. WZ Sge.**
B. Warner, R. E. Nather.
Monthly Notices Roy. Astron. Soc., Vol. 156, 297 - 303 (1972).

We have observed the 82-min eclipsing binary WZ Sge to obtain light curves with 3-s time resolution. Our and other deductions from the light curve strongly support the model of WZ Sge recently suggested by Krzeminski and Smak.

121.017   **Observations of rapid blue variables — V. VV Puppis.**
B. Warner, R. E. Nather.
Monthly Notices Roy. Astron. Soc., Vol. 156, 305 - 313 (1972).

We report photoelectric photometry of the 100-min binary VV Pup with a time resolution of 3 s. The identification of eclipses and of other recurrent features of the light curves leads to a qualitative model for the system. In order to produce compatibility between our light curves and previously published spectra of VV Pup, it is necessary to re-interpret the spectra. The model for VV Pup is then similar to that already proposed for WZ Sge.

121.018   **The photometric behavior of close binary stars.**
R. H. Koch.
Publ. Astron. Soc. Pacific, Vol. 84, 5 - 24 (1972).

A recent catalog of eclipsing binary photometric solutions is used to assess present confidence in the eclipse analyses and their consistency with the interactions between the member stars and with the density distributions of stellar interior models. Selected systems are cited to illustrate the cautions necessary in interpreting photometric anomalies.

121.019   **Differential *UBV* photometry of LY Aurigae.**
D. S. Hall, A. M. Heiser.
Publ. Astron. Soc. Pacific, Vol. 84, 33 - 34 = Repr. Arthur J. Dyer Obs., Vanderbilt Univ., *Nashville, Tennessee,* Ser. 2, No. 6 (1972).

UBV observations (76) of the O9.5 III eclipsing binary LY Aur, obtained as part of the 1970-71 coordinated campaign, are presented.

121.020   **A photometric study of the eclipsing binary AS Cam.**
R. W. Hilditch.
Mem. Roy. Astron. Soc., Vol. 76, (Part 1), 1 - 26 (1972).

Three-colour photoelectric observations of this $3^{d}4$ period eclipsing binary are presented. Consistent geometric elements for the system are derived from each light curve independently. The eccentricity of the orbit and the longitude of periastron are also well determined. A small amount of third light is found to be present and evidence of some instability in the system is presented.

121.021   **Analysis of the Hα profile in spectra of VV Cephei.**
K. O. Wright.
Journ. Roy. Astron. Soc. Canada, Vol. 66, 70 (1972). — Abstr. Canadian Astron. Soc.

121.022   **Period changes in eclipsing variables II. The system VW Cephei.**   P. Tempesti, R. De Carlo.
Mem. Soc. Astron. Italiana, Nuova Ser., Vol. 43, 1 - 15 (1972).

Photoelectric observations of VW Cephei performed at the Teramo Observatory from 1967 to 1971 show light-curve patterns confirming the results of the previous investigators. An analysis of all the known times of minima supports the conclusion that the wide long-term variation of the O-C curve may be periodic with a period of 44 years and with an amplitude of $0^{d}.04$; this result makes very improbable the hypothesis of a third body long since advanced to explain the variations of the photometric period.

121.023   **Nature of the secondary component of Beta Lyrae.**
R. Stothers, L. B. Lucy.
Nature, Vol. 236, 218 - 219 (1972).

To explain the underluminosity of the disk-like secondary component of Beta Lyrae, it is here proposed that the disk encloses a massive "main-sequence" star in rapid, nonuniform rotation and that the system has attained its present state as a result of mass transfer from the primary.

121.024   **A numerical analysis of the variations in the light curve of W Ursae Majoris.**   P. V. Rigterink.
Astron. Journ., Vol. 77, 230 - 238 (1972).

Photoelectric observations of W Ursae Majoris in the *B* and *V* wavelength regions are presented. The light curves are compared with previous observations, and the changes in the light curve are noted. New orbital elements are derived for the system using the mean binary light curve. In addition, an analysis is made of the validity of the present and previous solutions.

121.025   **The light variation and orbital elements of RW Coronae Borealis.**   L. Binnendijk.
Astron. Journ., Vol. 77, 239 - 245 (1972).

A total of 639 observations in yellow light and a total of 639 observations in blue light of RW CrB are presented. The light curve is in good agreement with photoelectric observations made by Carr. The orbital elements agree satisfactorily with those derived by Pierce.

121.026   **The light variation of UZ Leonis.**
L. Binnendijk.
Astron. Journ., Vol. 77, 246 - 250 (1972).

A total of 537 observations in yellow light and a total of 533 observations in the blue light of UZ Leonis are presented. The light curve shows complete eclipses with the secondary minimum being the total eclipse. The components of the system are too distorted to give reliable elements according to the Russell model.

121.027   **A photometric study of the eclipsing binary AS Cam.**   R. W. Hilditch.
Monthly Notices Roy. Astron. Soc., Vol. 156, 471 (1972). — Abstract. — The full text of the above paper has appeared in Mem. Roy. Astron. Soc., Vol. 76, Part I (1972).

121.028   **Period changes in the W Ursae Majoris star XY Leo.**   U. K. Gehlich, J. Prölss, R. Wehmeyer.
Astron. Astrophys., Vol. 18, 477 - 480 (1972).

The period changes of XY Leo have been analyzed. An interpretation of these variations as a light time orbit implied that the third mass, if present, could be a white dwarf. Assuming sudden changes of the period ("jumps"), the observations could be represented by three periods.

121.029   **Analysis of the white-dwarf eclipsing binary BD+16°516.**   A. Young, B. Nelson.
Astrophys. Journ., Vol. 173, 653 - 664 (1972).

The present study consists of a more definitive analysis of this system based upon recent and more complete data.

121.030   **A new analysis of the eclipsing binary, U Pegasi.**   P. V. Rigterink.
Astron. Journ., Vol. 77, 319 - 329 (1972).

Photoelectric observations of U Pegasi in the *B* and *V* wavelength regions are presented. The light curves are compared with previous observations, and the changes in the light curve are noted. An analysis is made of the validity of the present and previous solutions.

121.031   **The close binary system TX Cancri.**
A. Yamasaki, M. Kitamura.
Publ. Astron. Soc. Japan, Vol. 24, 213 - 230 = Tokyo Astron.

Obs. Repr., No. 409 (1972).

TX Cnc is studied from *UBV* and narrow-band photo-electric light curves, Hβ photometry, and Cassegrain spectrograms. The mean *B*- and *V*-light curves are analyzed and the most plausible photometric elements are derived.

121.032 The limb darkening problem in eclipsing binaries.
J. Grygar, M. L. Cooper, I. Jurkevich.
Bull. Astron. Inst. Czechoslovakia, Vol. 23, 147 - 174 (1972).

During the past sixty years an increasing attention has been given to the problem of limb darkening in eclipsing binaries. Recent developments bearing on precise determinations of limb darkening as well as pertinent results are reviewed in the present survey.

121.033 Determination of the elements of synthetic close binaries. T. B. Horák.
Bull. Astron. Inst. Czechoslovakia, Vol. 23, 178 - 180 = Astron. Contr. Univ. South Florida, Tampa, No. 55 (1972).

Wilson's and Devinney's synthetic light curves of close binaries were used to determine the elements by means of the automatic iterative minimization method. The results are satisfactory if the eclipses are complete.

121.034 A highly flattened secondary star in the young e-clipsing binary BM Orionis. D. S. Hall.
IAU Colloquium No. 15, (see 012.006), p. 217 - 229 (1972).

121.035 A photoelectric investigation of the eclipsing binary V 78 in Omega Centauri (NGC 5139).
E. H. Geyer.
IAU Colloquium No. 15, (see 012.006), p. 235 - 237 (1972).

121.036 Variations of the period of AH Virginis.
G. A. Bakos.
IAU Colloquium No. 15, (see 012.006), p. 293 - 296 (1972).

121.037 The locating of hot spots on components of Algol systems. K. Walter.
IAU Colloquium No. 15, (see 012.006), p. 297 - 298 (1972).

121.038 OAO-2 observations of Beta Lyrae and a provisional interpretation.
Y. Kondo, G. E. McCluskey, Jr., T. E. Houck.
IAU Colloquium No. 15, (see 012.006), p. 308 - 313 (1972).

121.039 Recent observations of the minimum of Algol.
J. E. Isles.
Journ. British Astron. Ass., Vol. 82, 130 - 131 (1972).

121.040 An old evolved binary in the galactic halo.
A. Young, B. Nelson, R. Mielbrecht.
Astrophys. Journ., Vol. 174, 27 - 31 (1972).

The faint blue star HZ 22 is shown to be a very short period binary with properties which suggest that is is highly evolved. It is probably more than 1500 pc above the galactic plane, and is either a very old disk population I or a halo population II star.

121.041 Ergebnisse der Beobachtungen von Bedeckungsveränderlichen. R. Diethelm, J. Isles, K. Locher.
Orion Schaffhausen, 30. Jahrgang, p. 60 - 61 (1972).

121.042 Orbit of RW CrB not eccentric.
E. F. Milone, A. J. Wesselink.
Inform. Bull. Variable Stars, (I.A.U. Commission 27), Konkoly Obs., Budapest, No. 611, 5 pp. (1972).

121.043 On the spectral type of DI Pegasi. W. P. Bidelman.
Inform. Bull. Variable Stars, (I.A.U. Commission 27), Konkoly Obs., Budapest, No. 629 (1972).

121.044 Minima of eclipsing variables. H. Muthsam.
Inform. Bull. Variable Stars, (I.A.U. Commission 27), Konkoly Obs., Budapest, No. 631 (1972).

121.045 The period of EQ Tauri. B. S. Whitney.
Inform. Bull. Variable Stars, (I.A.U. Commission 27), Konkoly Obs., Budapest, No. 633 (1972).

121.046 Remark on the period of RW Coronae Borealis.
I. Todoran.
Inform. Bull. Variable Stars, (I.A.U. Commission 27), Konkoly Obs., Budapest, No. 636 (1972).

121.047 Minima of eclipsing variables. Z. Klimek.
Inform. Bull. Variable Stars, (I.A.U. Commission 27), Konkoly Obs., Budapest, No. 637 (1972).

121.048 HR 3327 - an eclipsing binary with eccentric orbit.
B. G. Jørgensen.
Inform. Bull. Variable Stars, (I.A.U. Commission 27), Konkoly Obs., Budapest, No. 641 (1972).

121.049 Photoelectric minima of eclipsing binaries.
E. Pohl, A. Kizilirmak.
Inform. Bull. Variable Stars, (I.A.U. Commission 27), Konkoly Obs., Budapest, No. 647, 3 pp. (1972).

121.050 UBV photometry of two eclipsing variables in the vicinity of M67. R. L. Millis.
Inform. Bull. Variable Stars, (I.A.U. Commission 27), Konkoly Obs., Budapest, No. 649 (1972).

121.051 Photoelectric observations of Zeta Aurigae near minimum 1971–72.
S. Milcheva, M. Winiarski.
Inform. Bull. Variable Stars, (I.A.U. Commission 27), Konkoly Obs., Budapest, No. 651 (1972).

121.052 The eclipsing binary BD + 16°516.
L. E. Anderson, M. A. Seeds.
Inform. Bull. Variable Stars, (I.A.U. Commission 27), Konkoly Obs., Budapest, No. 657 (1972).

121.053 On the binary system CV Ser. R. Burchi.
Inform. Bull. Variable Stars, (I.A.U. Commission 27), Konkoly Obs., Budapest, No. 658 (1972).

121.054 W UMa: Photoelectric minima and a new period variation. B. Cester, M. Pucillo.
Inform. Bull. Variable Stars, (I.A.U. Commission 27), Konkoly Obs., Budapest, No. 659 (1972).

121.055 Three new bright eclipsing binaries. W. Strohmeier.
Inform. Bull. Variable Stars, (I.A.U. Commission 27), Konkoly Obs., Budapest, No. 665, 3 pp. = Veröff. Re-meis-Sternw. Bamberg – Astron. Inst. Univ. Erlangen-Nürnberg, Vol. 10, No. 102 (1972).

121.056 Times of minima of eclipsing binaries. A. Meyer.
Inform. Bull. Variable Stars, (I.A.U. Commission 27), Konkoly Obs., Budapest, No. 668 (1972).

121.057 A note on HR6773 = HD165814. D. P. Hube.
Inform. Bull. Variable Stars, (I.A.U. Commission 27), Konkoly Obs., Budapest, No. 671 (1972).

121.058 Elements for IQ Persei. W. Bischof.
Inform. Bull. Variable Stars, (I.A.U. Commission 27), Konkoly Obs., Budapest, No. 673 = BAV–Mitt. No. 24 (1972).

121.059 Minima of R CMa. M. B. K. Sarma.
Inform. Bull. Variable Stars, (I.A.U. Commission 27), Konkoly Obs., Budapest, No. 675 (1972).

121.060 Photoelectric observations of the eclipsing variable Z Vulpeculae. H. Ogata, T. Outi.
Inform. Bull. Variable Stars, (I.A.U. Commission 27), Konkoly Obs., Budapest, No. 676 (1972).

121.061 A new short period eclipsing binary. D. Hoffleit.
Inform. Bull. Variable Stars, (I.A.U. Commission 27), Konkoly Obs., Budapest, No. 677 (1972).

121.062 HD 93206 (CSV 6797), an eclipsing system presenting observational problems.
W. S. G. Walker, B. F. Marino.
Inform. Bull. Variable Stars, (I.A.U. Commission 27), Konkoly Obs., Budapest, No. 681, 3 pp. (1972).

121.063 Radio flare on β Persei.
V. A. Hughes, A. Woodsworth.
Nature, Phys. Sci., Vol. 236, 42 - 43 (1972).

Observations were made on Algol at $\lambda = 2.8$ cm at the time of a radio flare, and were combined with the 3.7 and 11.1 cm observations by Hjellming, Wade and Webster. An attempt to separate the flare component from the quasi-steady component showed that the former had a decay rate of 2.5 hours at all three wavelengths and a spectrum which rises at the shorter wavelengths, similar to the spectrum of Cygnus X-1; the spectrum of the latter had a short wavelength cut-off which suggests an origin in a cloud of monoenergetic electrons, or more likely, in a cloud of electrons with a high energy cut-off.

121.064 Computer calculation of the preliminary elements of eclipsing binaries. H. Minţi.
Stud. Cerc. Astron., Vol. 17, 45 - 52 (1972).

An algorithm used for the determination of the preliminary elements of the eclipsing binaries is described on the basis of the χ method of Russell and Merrill. The method is applied to the rectified light curve of the eclipsing binary AB Andromedae.

121.065 Radio emission from binary star systems.
R. M. Hjellming, C. M. Wade.
Bull. American Astron. Soc., Vol. 4, 210 (1972). – Abstr. AAS.

121.066 W Ursae Majoris systems and the hotter secondary.
J. Whelan.
Bull. American Astron. Soc., Vol. 4, 210 (1972). – Abstr. AAS.

121.067 U, B, V photometry of the eclipsing binary V 505 Sagittarii. C. R. Chambliss.
Bull. American Astron. Soc., Vol. 4, 210 (1972). – Abstr. AAS.

121.068 A photometric and spectroscopic investigation of SV Centauri. J. B. Irwin, A. U. Landolt.
Bull. American Astron. Soc., Vol. 4, 269 (1972). – Abstr. AAS.

121.069 Eclipsing binary stars. P. Flin.
Urania Kraków, Vol. 43, 140 - 144 (1972). In Polish.

121.070 Photoelectric observations of six eclipsing binaries.
P. B. van der Wal, C. Nagel, H. R. Voordes, K. S. de Boer.
Astron. Astrophys., Suppl. Ser., Vol. 6, 131 - 145 (1972).

Observations of the eclipsing binaries: QS Aql, EG Cep, FZ Del, AK Her, LS Her and EW Lyr are reported. The suspected eclipsing binary AT Her was also measured and found to be of constant brightness. A special way of reduction for slowly varying extinction is explained.

121.071 Solution of brightness curves of eclipsing binaries with the computer "Nairi". M. I. Lavrov.
Izv. Astron. Ehngel'gardt. Obs., *Kazan*, No. 38, p. 111 - 116 (1970). In Russian.

121.072 Underluminosity and magnetic fields in β Lyrae.
R. Stothers.
Nature, Phys. Sci., Vol. 238, 5 - 7 (1972).

Models of the secondary component of β Lyrae involve an underluminous star embedded in a gaseous cloud, disk, or ring. Here I estimate quantitatively the amount of underluminosity of the secondary and show that magnetic fields do not provide a likely explanation for this underluminosity.

121.073 On the stratification of emission in the envelope of the eclipsing binary of Wolf–Rayet type V444 Cyg.
A. M. Cherepashchuk, A. V. Goncharsky, A. G. Yagola.
Astron. Zhurn. Akad. Nauk SSSR, Vol. 49, 533 - 543 (1972).
In Russian. English translation in Soviet Astron. AJ, Vol. 16, No. 3.

A restoration of the WR star's disc brightness distribution from curves of intensity variations of nine emissions obtained by Kuhi for the system V444 Cyg at the moment of the WR star eclipse by the O6 V component was made. It is shown that the WR envelope, disturbed by effects of the near O6 V component, for ions with $\chi > 60$ eV, has a considerable concentration of radiation to the WR core. In the WR envelope a straight stratification of ions by ionization states is outlined.

121.074 On eclipsing binaries light curves solutions by a generalized least squares method. V. M. Tabachnik.
Astron. Zhurn. Akad. Nauk SSSR, Vol. 49, 544 - 554 (1972).
In Russian. English translation in Soviet Astron. AJ, Vol. 16, No. 3.

New formulae for determination of elements of eclipsing binaries by the method of function minimization are deduced. The process of error determination of the calculated parameters is described.

121.075 Eclipsing-binary solutions by sequential optimization of the parameters.
B. Nelson, W. D. Davis.
Astrophys. Journ., Vol. 174, 617 - 628 (1972).

This is a method of solving the light curves of eclipsing binaries with small computers in reasonable amounts of time. Total, annular, and partial eclipses can be treated with equal ease in the case of spherical stars and in some cases of non-spherical stars. The program computes the area eclipsed at a given phase angle without the use of auxiliary functions and converges to a solution by minimizing the sum of $(O-C)^2$.

121.076 Neue Elemente für WW Cam. W. Braune.
BAV Rundbrief, 21. Jahrgang, p. 7 - 9 (1972).

121.077 β Persei. R. M. Hjellming, C. M. Wade, E. Webster, C. T. Bolton.
IAU Circ., No. 2388 (1972).

121.078 Eclipsing binaries in other galaxies. J. M. Kreiner.
Postępy Astron., Vol. 20, 51 - 58 (1972). In Polish.

Problems connected with investigations of eclipsing binaries in other galaxies as well as the main results of observations are presented. A description of S. Gaposhkin's method of determination of parallaxes of eclipsing binaries is given.

121.079 Photometric observations of W Crucis.
G. F. G. Knipe.
Monthly Notes Astron. Soc. Southern Africa, Vol. 31, 25 - 26 (1972).

121.080 A minimum of QS Aquilae. G. F. G. Knipe.

Monthly Notes Astron. Soc. Southern Africa, Vol. 31, 27 (1972).

121.081 UX Canum Venaticorum.   W. Wenzel.
MVS, *Sonneberg*, Vol. 6, 17 - 18 (1972).
The "faint blue star" UX CVn does not show any variations definitely larger than the normal photographic scattering on 500 Sonneberg patrol plates.

121.082 Parallaxes of eclipsing binaries.   D. J. K. O'Connell.
Quarterly Journ. Roy. Astron. Soc., Vol. 13, 222 - 225 (1972). – Presented at the Woolley symposium (see 012.025).

121.083 Lists of minima of eclipsing binaries.
Compiled by R. Diethelm, R. Germann, K. Locher, H. Peter.
BBSAG Bull. No. 1, p. 1 - 3; No. 2, p. 1 - 2; No. 3, p. 1 - 2; No. 4, p. 1 - 3 (1972). – 34th – 37th list of Swiss Astronomical Society's Eclipsing Variable Observers.

121.084 New elements for the eclipsing binary BU Vul.
R. Diethelm.
BBSAG Bull. No. 1, p. 3 - 4 (1972).

121.085 On the identification of the eclipsing component of the visual binary BV 1481 Cet.   K. Locher.
BBSAG Bull. No. 1, p. 4 (1972).

121.086 An A. P. L. computer program for predictions and heliocentric corrections of the observable minima of eclipsing binaries.   K. Locher.
BBSAG Bull. No. 2, p. 3 (1972).

121.087 New elements for the eclipsing binary AY Puppis.
R. Diethelm.
BBSAG Bull. No. 2, p. 4 (1972).

121.088 Probable change of period of BO Monocerotis.
R. Diethelm.
BBSAG Bull. No. 2, p. 4 - 5 (1972).

121.089 New period determination for V 505 Sagittarii.
R. Diethelm.
BBSAG Bull. No. 3, p. 3 (1972).

121.090 Elemente für IQ Per.   W. Bischof.
BAV Rundbrief, 21. Jahrgang, p. 17 - 19 (1972).

121.091 Radio variations of β Persei and β Lyrae.
R. M. Hjellming, C. M. Wade, E. Webster.
Nature, Phys. Sci., Vol. 236, 43 - 46 (1972).
This communication describes observations of these two stars with the National Radio Astronomy Observatory interferometer between January 11 and February 16, 1972, on simultaneous frequencies of 2,695 and 8,085 MHz at element spacings of 900, 1,800 and 2,700 m. During this period β Persei behaved in a manner spectacularly different from that observed previously: its flux density at 8,085 MHz varied between 0.01 and 0.34 f.u. with a typical time scale of a few hours. The β Lyrae system faded to a level below the detection limit of the instrument, thereby proving that it is also a variable radio source.

Radiative transfer in atmospheres of Algol-type binaries. I.   See Abstr. 064.048.

Is 17 Leporis a shell star?   See Abstr. 114.067.

Models for contact binaries.   See Abstr. 117.002.

The Roche model and its applications to close binary systems.   See Abstr. 117.031.

Eruptive binaries. V. Period variations.
See Abstr. 117.045.

Étude spectroscopique de deux étoiles nouvelles VV Cephée: BD + 54° 2698, BD + 63° 3.
See Abstr. 119.019.

A T Tauri-like star in the eclipsing binary RS Canum Venaticorum.   See Abstr. 122.071.

Simultaneous photometric and image-tube spectroscopic observations of short period binary systems and rapid intrinsic variables.   See Abstr. 122.086.

RW Arietis; an RR Lyrae variable star in an eclipsing system.   See Abstr. 122.087.

G61−29.   See Abstr. 126.025.

Radio stars Beta Persei and Beta Lyrae.
See Abstr. 141.184.

The limiting mass of Centaurus X-3.
See Abstr. 142.062.

Identification of Cen X-3.   See Abstr. 142.073.

LR Cen is not Cen X-3.
See Abstr. 142.105.

Centaurus X-3 and Roche limits of close binary systems.   See Abstr. 142.107.

On the optical identification of Centaurus X-3.
See Abstr. 142.115.

Discovery of a periodic pulsating binary X-ray source in Hercules from *Uhuru*.   See Abstr. 142.116.

Cen X-3.   See Abstr. 142.122.

Possible optical identification of X-ray source.
See Abstr. 142.123.

Optical counterpart of Cen X-3.
See Abstr. 142.134.

Disk and ring structure in the universe.
See Abstr. 158.044.

Errata

121.901 Addendum and Erratum: "Realization of accurate close-binary light curves: Application to MR Cygni"
[Astrophys. Journ., Vol. 166, 605 - 619 (1971)].
R. E. Wilson, E. J. Devinney.
Astrophys. Journ., Vol. 171, 413 (1972).

121.902 Erratum: '*UBV* photometry of the eclipsing binary Zeta Phoenicis' [Astron. Astrophys., Vol. 12, 286 - 296 (1971)].   J. Dachs.
Astron. Astrophys., Vol. 17, 154 (1972).

121.903 Erratum: 'Hyades membership of the white-dwarf eclipsing system BD + 16° 516 ' [Astrophys. Journ.,

(*Letters*), Vol. 166, L81 - L83 (1971)].
A. Young, R. W. Capps.
Astrophys. Journ., (*Letters*), Vol. 172, L35 (1972).

121.904 **Errata: "Monochromatic reflection effect of close binary stars"** [Publ. Astron. Soc. Pacific, Vol. 83, 449 - 458 (1971)]. K.-Y. Chen, W. J. Rhein.

Publ. Astron. Soc. Pacific, Vol. 84, 355 (1972).

121.905 **Correction: 'Minima of R CMa'** [Inform. Bull. Variable Stars, No. 675 (1972)]. M. B. K. Sarma.
Inform. Bull. Variable Stars, (IAU Commission 27), Konkoly Obs., Budapest, No. 692 (1972). – See 121.059.

122 Physical Variables, Flare Stars, Pulsation Theory

122.001 **Observations of three short period RR Lyrae variables.** M. J. Penston.
Monthly Notices Roy. Astron. Soc., Vol. 156, 103 - 113 (1972).

Observations of the RR Lyrae variables AE Boo, TV Lyn and RW Ari are presented. For all three variables, light curves in the UBV system are given. A period is derived for AE Boo for which no ephemeris was previously available and measures of the radial velocity of this variable are also given.

122.002 **Theoretical pulsation constants and cepheid masses.** J. P. Cox, D. S. King, R. F. Stellingwerf.
Astrophys. Journ., Vol. 171, 93 - 102 (1972).

Simple fitting formulae, based on extensive numerical calculations, are presented for the linear, non-adiabatic pulsation constants for the fundamental mode and first two harmonics of stellar models having radiative envelopes and He mass fraction $Y \approx 0.3$. The formulae are applied to a sample of classical cepheids to estimate their masses on the basis of radii reported in the literature, under various assumptions regarding the pulsation mode.

122.003 **EQ Pegasi: A double flare star.**
F. N. Owen, B. W. Bopp, T. J. Moffett, F. J. Lazor.
Astrophys. Letters, Vol. 10, 37 - 40 (1972).

Simultaneous high-speed and area-scanning photometry show that the bright component of EQ Peg is a flare star. Additional simultaneous spectroscopy and high-speed photometry confirm that the faint component also flares. This finding refutes the common assumption that the fainter component of a flaring double star is always the flare star.

122.004 **Short period variable stars IV: Periods of Delta Scuti stars.** J. C. Valtier.
Astron. Astrophys., Vol. 16, 38 - 43 (1972).

A method of analyzing short and non equi-spaced light curves is described and used to study all the published photometric data of Delta Scuti stars. A set of homogeneous results on periods is given, but of 34 stars studied, four are called "Delta Scuti" on very poor basis and nine exhibit more than 30 minutes difference between the different determinations of periods.

122.005 **The spectrum of HR 2902 in 1971.**
M. Jaschek, E. Brandi.
Astron. Astrophys., Vol. 16, 115 - 117 (1972).

The spectral features of HR 2902 in 1971 are described with special emphasis upon the far ultraviolet. A rapidly evolving shell is evident, which at this phase reproduces the spectral appearance of 1947—48 and agrees well with orbital predictions.

122.006 **Lichtkurve von TW Delphini aus visuellen Beobachtungen.** M. Beyer.
Astron. Nachr., Vol. 293, 171 - 173 (1971/72).

The results of 686 visual observations in 1947—1970 are given. Characteristics of the light curve, in particular sudden interchanges of primary and secondary minima, point to a near relation to RV Tauri-type stars. Only the irregularities of the period and the changing amplitude of the light variations classify TW Del as a SRb type variable.

122.007 **Zur Erklärung der Polarisation der Strahlung von μ Cephei.** C. Friedemann.
Astron. Nachr., Vol. 293, 179 - 186 (1971/72) = Mitt. Univ.-Sternw. Jena, No. 111 (1972).

Assuming a stellar magnetic dipole-field which aligns particles in the stellar atmosphere the observed polarization of the light of μ Cephei can be explained. The spatial orientation of the rotational axis, the angle between the magnetic axis and the rotational axis, and the period of rotation were derived from the observational values.

122.008 **The short period light and velocity variations in Alpha Virginis.**
R. R. Shobbrook, N. R. Lomb, D. Herbison-Evans.
Monthly Notices Roy. Astron. Soc., Vol. 156, 165 - 180 (1972).

The light variation of α Virginis (Spica) has been followed over a total of 26 nights in 1968, 1969 and 1970. On seven nights in 1970, three-colour observations were made. Both the 4-day variation with the orbital period and the 4-hr variation due to Beta Canis Majoris type pulsation are evident. New spectra have been obtained and the radial velocities discussed together with others published from 1890 to 1956. The 4-hr period is also found in part of this data but some of the earlier velocities indicate the presence of at least three more periodicities, including one of about 6 hr. The phase relationships between the light, colour and velocity variations during the 4-hr period are investigated and the temperature variation estimated.

122.009 **Two new broad-lined Beta Canis Majoris stars.**
R. R. Shobbrook, N. R. Lomb.
Monthly Notices Roy. Astron. Soc., Vol. 156, 181 - 188 (1972).

Light variations of small amplitude have been detected in the two bright southern B stars λ and κ Scorpii, indicating that they are Beta Canis Majoris variables. In a Fourier analysis of the brightness variations of λ Sco the following periods were found: a fundamental period of $0\overset{d}{.}2137015$, the first harmonic of this period and also a longer period of $10\overset{d}{.}15$. For κ Sco the Fourier analysis yielded a main variation with a period of $0\overset{d}{.}19987$, together with another short period of $0\overset{d}{.}20544$ and a longer period of $2\overset{d}{.}951$.

122.010 **The masses of cepheid variables.**
K. Fricke, R. S. Stobie, P. A. Strittmatter.
Astrophys. Journ., Vol. 171, 593 - 604 (1972).

The results of stellar pulsation and evolution computations are applied to a large sample of cepheid variables in order to derive independent estimates of their mass. That the masses derived from evolution theory are significantly greater than those obtained from pulsation theory confirms previous results. It is also shown that a significant discrepancy exists between independent pulsation mass determinations for cepheids with a secondary maximum in their light curves.

122.011 **Spectrophotometry of the Beta Cephei stars.**
R. D. Watson.
Astrophys. Journ., Suppl. Ser., No. 204, Vol. 24, 167 - 191 (1972). — Abstr. in Astrophys. Journ., Vol. 171, 411 (1972).

Scanner observations and hydrogen line profiles are presented for the majority of the classical β Cephei stars, some recently confirmed β Cephei stars, and some of Hill's β Cephei candidates. These data, together with data from other sources and a sample of nonvarying early B stars, are used to obtain stellar temperatures and gravities via the Morton model atmospheres. The existence of a β Cephei strip in the $(\theta_e, \log g)$-plane is demonstrated and its correspondence with the S-bend region of theoretical evolutionary tracks is noted. The normality of the β Cephei phenomenon for slow rotators is emphasized. For comparison, $(M_{bol}, \log T_e)$ values derived by using the Petrie luminosity calibration are presented. Masses and radii derived from the $(\theta_e, \log g)$ data are given.

122.012 **Photo-electric observations of the ultra short-period variable V567 Ophiuchi.**
A. G. de Bruyn.
Astron. Astrophys., Vol. 16, 478 - 481 (1972).

Photo-electric five-colour observations of V567 Oph show that there remains doubt about the period of this variable, as given by Hoffmeister (1941). Its relative spectral energy distribution appears to be the same as that of RS Gru, a stable variable with a period of $0^d 147$.

122.013 **Overtones in the lightcurve of CY Aquarii.**
E. W. Elst.
Astron. Astrophys., Vol. 17, 148 - 153 (1972).

A harmonic analysis of the observations of Zissell (1968) reveals in the lightcurve of CY Aquarii a set of overtones. New observations (1970) are accordingly interpreted.

122.014 **Forbidden emission lines of Mn I and Fe II in the spectra of long-period variables near minimum light.**
H. Nussbaumer, J. P. Swings.
Astrophys. Journ., Vol. 172, 121 - 128 (1972).

Departures from a statistical distribution of the metastable levels occur in the cases of neutral manganese and singly ionized iron, as seen from relative intensities of [Mn I] and [Fe II] lines in the spectra of several Mira variables near minimum light. The magnetic dipole and electric quadrupole transition probabilities are computed for the configurations $3d^5 4s^2$ and $3d^6 4s$ of Mn I.

122.015 **A photometric study of selected R Coronae Borealis variables.**
J. D. Fernie, V. Sherwood, D. L. DuPuy.
Astrophys. Journ., Vol. 172, 383 - 390 (1972).

UBV observations of 10 possible R CrB variables are presented. For eight of these only a few observations are available, and average values are presented. More extensive observations are available for ρ Cas and R CrB. The star ρ Cas was found to show low-amplitude quasi-periodic variability of the type previously observed by Brodskaya. Detailed observations at maximum light show that R CrB exhibits a sinusoidal variability of period about 44^d and amplitude about 0.15 mag. The light curve suggests RV Tauri-like behavior. The $B - V$ color curve seems irregular with an amplitude of about 0.04 mag, while the radial-velocity at maximum is also slightly variable by about 4 km s^{-1}.

122.016 **A W Virginis model using radiation transfer.**
C. G. Davis, Jr.
Astrophys. Journ., Vol. 172, 419 - 421 (1972).

Christy's model of W Vir is studied with the variable Eddington method of radiative transfer.

122.017 **A physical interpretation of recent period-luminosity relations for RR Lyrae variables.**
J. P. Cox, J. I. Castor, D. S. King.
Astrophys. Journ., Vol. 172, 423 - 434 (1972).

A simple physical interpretation is proposed of the period-luminosity relation for RR Lyrae variables which was found by Iben and Huchra on the basis of linearized pulsation calculations. The purpose of this paper is to suggest a possible physical interpretation which may apply to the Iben-Huchra transition line and may have some bearing on the Christy transition line. Although these two transition lines are no doubt closely related physically, there are certain differences which will be discussed. An interpretation is presented, and is tested by applying it to the Iben-Huchra transition line and to the Christy transition line.

122.018 **Spectroscopic and photometric changes in the peculiar infrared star VX Sagittarius.**
R. M. Humphreys, G. W. Lockwood.
Astrophys. Journ., (*Letters*), Vol. 172, L59 - L62 (1972).

Recent spectroscopic and photometric variations in the peculiar M supergiant-like star, VX Sgr, are reported. During the period of observation, its behavior resembled that of a Mira variable.

122.019 **The remarkable variations of CH Cygni.**
C. Y. Shao, W. Liller.
Bull. American Astron. Soc., Vol. 3, 443 (1971). – Abstr. AAS.

122.020 **Photoelectric observations of BL Lacertae.**
E. F. Milone.
Bull. American Astron. Soc., Vol. 3, 443 (1971). – Abstr. AAS.

122.021 **Infrared radiation from RV Tauri stars.**
R. D. Gehrz.
Bull. American Astron. Soc., Vol. 3, 454 (1971). – Abstr. AAS.

122.022 **A W Virginis model using radiation transfer.**
C. G. Davis, Jr.
Bull. American Astron. Soc., Vol. 3, 456 (1971). – Abstr. AAS.

122.023 **Recent observations of flare stars and pulsars.**
K. R. Lang.
Bull. American Astron. Soc., Vol. 3, 462 (1971). – Abstr. AAS.

122.024 **The relationship between the pulsation of long-period variables and the ejection of planetary nebulae.** W. M. Sparks, G. S. Kutter.
Bull. American Astron. Soc., Vol. 3, 484 (1971). – Abstr. AAS.

122.025 **Uncertainties in the theoretical determination of blue edges and estimates of Y and $L/L_\odot$ for RR Lyrae stars in globular clusters.** I. Iben, Jr., R. S. Tuggle.
Bull. American Astron. Soc., Vol. 3, 485 (1971). – Abstr. AAS.

122.026 **An analysis of the light variation of Mira variables.**
K. C. Leung.
Bull. American Astron. Soc., Vol. 3, 485 (1971). – Abstr. AAS.

122.027 **Pre-main sequence stars. I. Shell variability and pulsation in NGC 2264.** M. Breger.
Bull. American Astron. Soc., Vol. 3, 501 - 502 (1971). – Abstr. AAS.

122.028 **Kinematics of Mira variables.** T. G. Barnes III.
Bull. American Astron. Soc., Vol. 3, 502 (1971). – Abstr. AAS.

122.029 **Masses, radii and luminosities of RR Lyrae variable stars.** R. Woolley, A. Savage.
Roy. Obs. Bull., [Roy. Greenwich Obs., Herstmonceux], No. 170, p. 365 - 395 (1971).

The Baade-Wesselink method of determining radii of variable stars is examined. The idea is introduced that specific intensity in B and V is affected by gravity as well as by temperature, so that it is necessary to take into account the accelerations which occur at certain phases in the pulsation cycle. This means that the mass of the star enters as a parameter in the equations, and the Baade-Wesselink method is incomplete as a means of determining radius and hence absolute magnitude. It is necessary to appeal to an extra piece of information to arrive at values of both mass and radius. This may be an appeal to proper motions to establish mean absolute magnitudes, an appeal to a pulsation constant supposed known from theory, or even in one case a trigonometrical parallax. It is suggested that observations in *I* may supply the missing data. The galactic orbits are calculated, assuming an inverse square attracting field.

122.030 Maxima van mira-sterren.   G. W. E. Beekman.
Hemel en Dampkring, Vol. 70, 54 - 56 (1972).

122.031 Rußende Sterne.   R. Kippenhahn.
SuW, Vol. 11, 32 - 35 (1972). – R Corona Borealis stars.

122.032 Anmerkungen zum Lichtwechsel von T Cep.
E. Heiser.
SuW, Vol. 11, 75 - 77 (1972).

122.033 Fluorescent Fe I emission in long-period variables.
L. A. Willson.
Astron. Astrophys., Vol. 17, 354 - 361 (1972).

A statistical equilibrium analysis of the fluorescent emission of $\lambda\lambda 4202$, 4308 of Fe I in long period variables indicates that the coincidence mechanism first proposed by Thackeray (1937) is consistent with the observed intensities. In addition the analysis gives some indications of conditions present in the atmospheres of long period variables during the post-maximum phase.

122.034 Pulsation of models in the lower part of the cepheid instability strip and properties of AI Velorum and δ Scuti variables.   J. O. Petersen, H. E. Jørgensen.
Astron. Astrophys., Vol. 17, 367 - 377 (1972).

We have considered six chemical compositions ranging from extreme Pop II ($Z = 0.001$) to Pop I ($Z = 0.03$) and constructed models with masses from $0.5\,M_\odot$–$1.6\,M_\odot$. Comparing theoretically derived period ratios with observational data for AI Vel variables, it is shown that this group of stars can not belong to Pop I. For 53 δ Sct stars we have used *uvby*, β photometry to derive effective temperatures, luminosities, gravities and masses, assuming these stars to be Pop I stars.

122.035 A recent change in the circumstellar spectrum of U Monocerotis.   G. W. Preston.
Astrophys. Journ., (*Letters*), Vol. 172, L105 - L106 (1972).

Double circumstellar absorption components of the Ca II H- and K-lines, that were present in the spectrum of the RV Tauri star U Mon in 1963, have been replaced by strong single absorption features with a velocity intermediate between those of the previous lines.

122.036 o Velorum, a new Beta Canis Majoris star.
A. Van Hoof.
Astron. Astrophys., Vol. 18, 51 - 54 (1972).

Photoelectric and spectrographic observations reveal that o Vel is a variable star of the βCMa type with its brightness changing by 0^m03 and its RV by 9 km/s in a period of 3^h10^m. There is an "outflow" of H of 3 km/s with respect to Si II and Mg II and of 5 km/s with respect to He I.

122.037 Ap stars with variable period?   P. Renson.
Astron. Astrophys., Vol. 18, 159 - 162 (1972). In French.

The analysis of the data, especially for HD 32633 contradicts Rakos' suggestion of slow changes in *P* for periodic Ap stars. Thus HR 5597 is actually the only known Ap star with a variable period.

122.038 High-speed photometry of cataclysmic variables.
B. Warner, R. E. Nather.
Sky Telescope, Vol. 43, 82 - 85 (1972).

122.039 On the rotation of UV Cet type stars.
R. E. Gershberg.

Izv. Krymskoj Astrofiz. Obs., Vol. 43, 66 - 70 (1971). In Russian.

The upper limits of the rotational velocities of flare stars are evaluated by two methods – from photometric features of the stars and on the basis of the widths of the Hα emission line in the quiet chromosphere. Values have been derived for UV Cet, YZ CMi, EV Lac, AD Leo.

122.040 La classification spectrale des étoiles β Canis Majoris et leur position dans le diagramme HR.
J. Rountree-Lesh, M. Aizenman.
Comptes Rendus Acad. Sci. Paris, Sér. B, Vol. 274, 911 - 914 (1972).

D'après une étude spectrographique et photométrique, nous situons les étoiles variables du type β Canis Majoris dans le diagramme HR par rapport aux étoiles B «normales». Une comparaison avec des modèles évolutifs nous permet de faire une hypothèse sur le stade d'évolution des β *CMa*, qui serait postérieur à la «combustion» centrale de l'hydrogène.

122.041 Progressi nell' interpretazione e studio delle stelle variabili. VII. – Le cefeidi.   L. Rosino.
Coelum, Vol. 40, 3 - 16 (1972).

122.042 Les étoiles symbiotiques.   E. Stram.
L'Astronomie, 86e année, p. 37 - 44 (1972).

122.043 On an apparent discrepancy between pulsation and evolution masses for cepheids.
I. Iben, Jr., R. S. Tuggle.
Astrophys. Journ., Vol. 173, 135 - 155 (1972).

Results of new theoretical pulsation calculations in the linear nonadiabatic approximation are reported. Both Q-values and the location of blue edges in the H-R diagram are presented for masses and compositions that are of interest in connection with classical cepheids. It is shown that estimates of cepheid masses obtained by comparing observational data with theoretical Q-values can be made to agree with mass estimates obtained by comparing with results of theoretical evolution calculations.

122.044 HR 2957 – a cepheid variable of small amplitude.
R. S. Stobie.
Observatory, Vol. 92, 12 - 14 (1972).

To determine the nature of the variability of HR 2957, photoelectric observations were obtained at Mount Stromlo and at Siding Spring Observatories. These observations, together with the 1965 and 1967 astrographic observations, are listed in a table and discussed. It is concluded that all the evidence is consistent with interpreting HR 2957 as a population I cepheid of small amplitude.

122.045 Some values of ΔS for RR Lyrae stars.
R. B. Willis.
Observatory, Vol. 92, 14 - 15 (1972). – Note.

122.046 Theoretical light curves of luminous gas and their application to interprete flares of UV Cet type stars.
A. A. Korovyakovskaya, J. P. Korovyakovsky.
Astrofiz. Issled. Izv. Spets. Astrofiz. Obs., Vol. 3, 101 - 115 (1971). In Russian.

Theoretical light-curves of luminous gas in the *U, B, V* photometric regions are calculated and compared with observed light-curves of flares of UV Cet-type stars. The value of the parameter n_0, the density of luminous matter, is found. In 35 cases of 50 a reasonable agreement is obtained between the theoretical and the observed light-curves.

122.047 Three-colour photometry of DY Pegasi.
B. Warner, R. E. Nather.
Monthly Notices Roy. Astron. Soc., Vol. 156, 315 - 319

(1972).

Sequential three-colour observations of the 105-min dwarf cepheid DY Peg are described. These show that the colours vary smoothly with phase. The shape of the loop in the two-colour diagram is well defined. There is no evidence for any change of period in DY Peg.

122.048 Three-colour photometry of CY Aquarii.
R. E. Nather, B. Warner.
Monthly Notices Roy. Astron. Soc., Vol. 156, 321 - 323 (1972).

Sequential three-colour observations of the 88-minute dwarf cepheid CY Aqr are reported. These show smooth variations of colour with phase.

122.049 The light curve of RU Cam from 1969.8 to 1970.6 and the variation of its period.
P. Broglia, G. Guerrero.
Astron. Astrophys., Vol. 18, 201 - 208 (1972).

Two colour photoelectric light curves of RU Cam for the interval October 1969—August 1970 are presented. An analysis of the light curves confirms that an erratic component is superimposed on the periodic one. The substantially constant mean luminosity and colour of the variable support the hypothesis of Wallerstein on why the light pulsation ceased.

122.050 A problem in distance-determination for Mira variables with an appendix on OB-star distances.
M. W. Feast.
Vistas in astronomy, Vol. 13, (see 003.001), 207 - 221 (1972).

Difficulties in accepting the results on the kinematics of distant Mira variables obtained by Smak and Preston (1965) and in an earlier paper, are outlined. It is shown that proper allowance for statistical bias in the distance determinations together with an allowance for the different definition of maximum magnitude, as used in these two investigations and in the statistical parallax work, removes the difficulties.

122.051 Bands of the light molecules in Mira variables.
P. C. Keenan.
Vistas in astronomy, Vol. 13, (see 003.001), 223 - 240 (1972).

122.052 On the stability of the light-variations of RR Lyrae stars. V. Tsesevich.
Vistas in astronomy, Vol. 13, (see 003.001), 241 - 256 (1972).

122.053 A spectrophotometric survey of SS Cygni and RW Aurigae. E. K. Kharadse, R. A. Bartaya.
Vistas in astronomy, Vol. 13, (see 003.001), 257 - 263 (1972).

122.054 Photometry and polarimetry of V 1057 Cygni.
G. Rieke, T. Lee, G. Coyne.
Publ. Astron. Soc. Pacific, Vol. 84, 37 - 45 (1972).

We have obtained extensive photometric and polarimetric data over a six-month interval for V 1057 Cyg. A model characterized by two circumstellar shells which absorb and reradiate short wavelength stellar flux from an underlying A1-type star provides the best fit to the observational data.

122.055 Paschen beta emission in the spectrum of Omicron Ceti.
R. P. Kovar, A. E. Potter, N. S. Kovar, L. Trafton.
Publ. Astron. Soc. Pacific, Vol. 84, 46 - 50 (1972).

We report observations of Paschen beta emission in the spectrum of o Cet. These measurements are used with $H\gamma$ emission observations, taken independently in the same time period, to discuss the relative continuous opacity at $1.28\,\mu$ and $0.4\,\mu$, as well as the intensity variations in the hydrogen emission lines appearing in the spectra of long-period variables.

122.056 $UBV\beta$ photometry of EU Tauri. E. F. Guinan.

Publ. Astron. Soc. Pacific, Vol. 84, 56 - 60 (1972).

UBV and $H\beta$ observations of EU Tau are presented. On the basis of all the available evidence it is concluded that EU Tau is a classical cepheid with a period of $2^{d}1052$ instead of the W UMa variable as previously supposed. The photometric parameters of the star are determined.

122.057 Dust models for R Coronae Borealis and RY Sagittarii. K. S. Krishna Swamy.
Publ. Astron. Soc. Pacific, Vol. 84, 64 - 67 (1972).

The observed infrared emission from R CrB and RY Sgr can be explained as due to thermal emission from the shell of dust grains surrounding these stars. We have calculated the emission spectrum of the dust shell as a function of the visual magnitude and also the expected flux of the star as a function of the optical depth of the dust shell. The question of the condensation of graphite particles in these stars is briefly discussed.

122.058 A pulsating metallic-line star.
M. S. Bessell, O. J. Eggen.
Publ. Astron. Soc. Pacific, Vol. 84, 72 - 73 (1972).

HR 5491, a member of the Hyades group, has been found to be both an ultrashort-period cepheid ($0^{d}08$) and a classical, metallic-line star.

122.059 Photoelectric observations of CH Cygni in the interval 1967—70. B. Cester.
Mem. Soc. Astron. Italiana, Nuova Ser., Vol. 43, 83 - 88 (1972).

The behaviour of CH Cyg is studied, by means of photoelectric observations made at Trieste. The variations of the mean colour indices are described here till to the end of 1970. V magnitude seems to vary independently between 6.5 and 8 at long intervals (about 630 days). In 1970 an oscillation of about 0.3 mag in 100 days is visible.

122.060 X Persei an X-ray source?
L. L. E. Braes, G. K. Miley.
Nature, Vol. 235, 273 (1972).

It is suggested that the sixth magnitude irregular variable star X Per in the Per OB2 association is connected with the X-ray source 2ASE 0352+30. No radio emission brighter than 0.005 flux units has been detected from it at 1415 MHz with the Westerbork telescope.

122.061 Possible identification of X Persei with an X-ray source. S. van den Bergh.
Nature, Vol. 235, 273 - 274 (1972).

It is tentatively suggested that X-Persei coincides with a strong X-ray source position determined by the Uhuru satellite. Some support for this speculation is provided by Blaauw's suggestion that X-Persei is the stellar remnant of a former supernova.

122.062 Slow flare up in the Pleiades. E. S. Parsamian.
Astrofizika, Vol. 7, 547 - 555 (1971). In Russian.
English translation in Astrophysics, Vol. 7, No. 4.

During observations of the Pleiades on the 40″ Schmidt telescope of the Byurakan Observatory a slow flare up and two rapid flare ups of star No. 103 were revealed. Statistical data on the number of premaximal images of flare stars of different exposures in the Pleiades and in Orion are given. A comparison between slow flare stars in Orion and Pleiades has been made.

122.063 Fuors. V. A. Ambartsumian.
Astrofizika, Vol. 7, 557 - 572 (1971). In Russian.
English translation in Astrophysics, Vol. 7, No. 4.

The FU Orionis stars (fuors) have the peculiarity that during comparatively a short time they strongly increase their luminosity in the observable part of the spectrum. An explanation of this phenomenon is given, which is based on the assumption that before the increase of brightness directly around the star some energy sources exist which mostly radiate the high energy particles.

122.064   Radial-velocity measures of θ Herculis.
L. Shapiro.
Astron. Journ., Vol. 77, 215 (1972).
No radial-velocity variations are found for θ Her, indicating that it is not an eclipsing binary.

122.065   Pulsating helium stars.   V. Trimble.
Monthly Notices Roy. Astron. Soc., Vol. 156, 411 - 418 (1972).
Linear and non-linear pulsation computations have established the existence of an instability strip for helium stars. A $2 M_\odot$ model reproduces the period and velocity curve found by Feast for RY Sgr. The discrepancies between the observed and calculated light and temperature curves can be understood at least qualitatively.

122.066   The beat period of AX Velorum.
R. S. Stobie, T. Hawarden.
Monthly Notices Roy. Astron. Soc., Vol. 157, 157 - 165 (1972).
Analysis of the photoelectric observations of AX Vel has revealed two periodicities. The period ratio is similar to that found in previous analyses of beat period cepheids.

122.067   The beat period of Y Carinae.   R. S. Stobie.
Monthly Notices Roy. Astron. Soc., Vol. 157, 167 - 170 (1972).
Autocorrelation analysis of the photoelectric observations of Y Car has shown two periodicities. The period ratio is 0.7031.

122.068   Two new Beta Canis Majoris variables: ε Cen and δ Lup.   R. R. Shobbrook.
Monthly Notices Roy. Astron. Soc., Vol. 157, 5P - 9P (1972).
ε Cen and δ Lup have been discovered to be two additions to the rapidly growing list of broad-lined β CMa variables. The paper concludes with a discussion, concerning the photometric errors, of the reality of the deviations from the adopted periods in ε Cen, δ Lup and also in λ Sco and κ Sco.

122.069   HD 197481: A periodic dMe variable star.
C. A. O. Torres, S. Ferraz Mello, G. R. Quast.
Astrophys. Letters, Vol. 11, 13 - 14 (1972).
The quiescent light variations of the flare star HD 197481 are studied and the existence of enhanced ultraviolet activity at the light minimum is shown. The observations are analyzed and their connection with the dark spot model is considered.

122.070   The spectrum of the symbiotic star AG Pegasi 1970–71.   J. B. Hutchings, R. O. Redman.
Publ. Astron. Soc. Pacific, Vol. 84, 240 - 247 (1972).
Results are reported of velocity measurements and spectrophotometry of AG Pegasi made from spectrograms of medium and high dispersion. Phases in the periodic variation of the spectrum are identified and recent spectral changes are noted. A variable Balmer emission-line progression is found in all spectra. About 70 absorption lines of the M-type spectrum have been identified and some of them measured for velocity.

122.071   A T Tauri-like star in the eclipsing binary RS Canum Venaticorum.   D. S. Hall.
Publ. Astron. Soc. Pacific, Vol. 84, 323 - 333 = Repr. Arthur J. Dyer Obs., Vanderbilt Univ., *Nashville, Tennessee*, Ser. 2, No. 8 (1972).
In this paper I examine existing data on RS CVn, propose

a model which succeeds in accounting for all of the photometric complications and the period variation as well, and then go on to argue that RS CVn is most likely in pre-main-sequence contraction and that the cool component has many of the youthful properties of a T Tauri star.

122.072   Red variables with and without OH radio emission.
W. Strohmeier.
IAU Colloquium No. 15, (see 012.006), p. 51 - 53 (1972).

122.073   Photometry of R Coronae Borealis.   J. D. Fernie.
IAU Colloquium No. 15, (see 012.006), p. 54 (1972).

122.074   A photometric search for circumstellar shell variability and pre-main sequence Delta Scuti variables.
M. Breger.
IAU Colloquium No. 15, (see 012.006), p. 85 - 89 (1972).

122.075   Photometry of long period variables in the Magellanic Clouds.   C. J. Butler.
IAU Colloquium No. 15, (see 012.006), p. 90 - 95 (1972).

122.076   UV observations of Beta Cephei.
D. Fischel, W. M. Sparks.
IAU Colloquium No. 15, (see 012.006), p. 97 (1972).

122.077   Flare stars.   V. A. Ambartsumian, L. V. Mirzoyan.
IAU Colloquium No. 15, (see 012.006), p. 98 - 108 (1972).

122.078   A report on flare stars on the Sproul astrometric program.   S. L. Lippincott.
IAU Colloquium No. 15, (see 012.006), p. 109 - 113 (1972).

122.079   Radio observations of flare stars.   A. J. Wilson.
IAU Colloquium No. 15, (see 012.006), p. 114 - 115 (1972).

122.080   New flare stars in the Pleiades region.
L. Pigatto, L. Rosino.
IAU Colloquium No. 15, (see 012.006), p. 116 - 123 (1972).

122.081   Spectral investigations of UV Ceti-type flare stars and search for new variables of this type carried out at Crimea.   R. E. Gershberg, N. I. Shakhovskaya.
IAU Colloquium No. 15, (see 012.006), p. 126 - 137 (1972).

122.082   The red dwarf stars of the UV Ceti-type in the neighbourhood of the sun.
N. I. Shakhovskaya.
IAU Colloquium No. 15, (see 012.006), p. 138 - 143 (1972).

122.083   UBV observations of the ultrashort period variable HZ 29.   W. Krzemiński.
IAU Colloquium No. 15, (see 012.006), p. 178 - 181 (1972).

122.084   On Stothers' and Simon's binary-star hypothesis for Beta Canis Majoris star pulsation.
J. R. Percy, K. Madore.
IAU Colloquium No. 15, (see 012.006), p. 197 - 203 (1972).

122.085   The effect of binary motion on period changes in RR Lyrae stars.   C. M. Coutts.
IAU Colloquium No. 15, (see 012.006), p. 238 - 243 (1972).

122.086   Simultaneous photometric and image-tube spectroscopic observations of short period binary systems and rapid intrinsic variables.   M. F. Walker.
IAU Colloquium No. 15, (see 012.006), p. 243 - 247 = Contr. Lick Obs., No. 343 (1972).

122.087 RW Arietis; an RR Lyrae variable star in an eclipsing system. W. Z. Wisniewski.
IAU Colloquium No. 15, (see 012.006), p. 247 (1972). – Abstract.

122.088 The classification of intrinsic variable stars. I. The red variables of type N. O. J. Eggen.
Astrophys. Journ., Vol. 174, 45 - 55 (1972).

This is the first in a series of papers based on an examination of the classification system for intrinsic variables and concerns extensive (*UBVRI*) observations of the 35 N-type variables for which relatively accurate apparent motions are available. The N-type variables are found to be divided into all three population groups, (1) young disk (YD) objects such as the carbon stars in the Large Magellanic Cloud with $M_{bol} = -5$ mag, (2) old disk (OD) population stars of mass near 1.25 $M_\odot$ and $M_{bol} = -3$ mag, and (3) the halo population, which consists of CH stars with M_{bol} near –3.5 mag.

122.089 The conversion factor from radial to pulsational velocity and the radii of classical cepheids.
S. B. Parsons.
Astrophys. Journ., Vol. 174, 57 - 67 (1972).

The conversion factor from observed radial velocity to the pulsational velocity of a stellar atmosphere, commonly adopted as 24/17 = 1.41, is found to be about 1.31 from actual calculations of line profiles appropriate to classical cepheids. This 7 percent reduction still leaves the radii determined by Wesselink's method consistent for most cepheids with the period-luminosity-color relation of Sandage and Tammann.

122.090 X Ophiuchi, ein langperiodisch Veränderlicher des Mira-Typs und zugleich Doppelstern.
R. German.
Orion Schaffhausen, 30. Jahrgang, p. 16 - 17 (1972).

122.091 EV Lac and UV Cet.
K. Osawa, K. Ichimura, Y. Shimizu.
Inform. Bull. Variable Stars, (I.A.U. Commission 27), Konkoly Obs., Budapest, No. 608 (1972).

122.092 Photoelectric observations of the flare star UV Cet.
P. F. Chugainov, N. I. Shakhovskaya.
Inform. Bull. Variable Stars, (I.A.U. Commission 27), Konkoly Obs., Budapest, No. 615, 5 pp., with corrections, No. 632 (1972).

122.093 Photoelectric observations of the flare star EV Lac.
P. F. Chugainov, A. N. Kulapova, N. I. Shakhovskaya.
Inform. Bull. Variable Stars, (I.A.U. Commission 27), Konkoly Obs., Budapest, No. 616, 4 pp., with correction, No. 632 (1972).

122.094 Some recent observations of UV Ceti.
A. H. Jarrett, J. P. Eksteen.
Inform. Bull. Variable Stars, (I.A.U. Commission 27), Konkoly Obs., Budapest, No. 620, 4 pp. (1972).

122.095 A new flare star in Aquila.
B. Hidajat, M. U. Akyol.
Inform. Bull. Variable Stars, (I.A.U. Commission 27), Konkoly Obs., Budapest, No. 623 (1972).

122.096 A new flare star in Cygnus.
G. Haro, E. Chavira.
Inform. Bull. Variable Stars, (I.A.U. Commission 27), Konkoly Obs., Budapest, No. 624 (1972).

122.097 Photoelectric observations of AD Leo and EV Lac
during the 1971 international patrol intervals.
J. Arsenijević, A. Kubičela.
Inform. Bull. Variable Stars, (I.A.U. Commission 27), Konkoly Obs., Budapest, No. 627 (1972).

122.098 Photoelectric observations of YZ CMi.
P. F. Chugainov, N. I. Shakhovskaya.
Inform. Bull. Variable Stars, (I.A.U. Commission 27), Konkoly Obs., Budapest, No. 632, 4 pp. (1972).

122.099 YZ CMi. K. Osawa, Y. Shimizu, T. Okada, H. Koyano, K. Ichimura, E. Watanabe, M. Yutani, K. Okida.
Inform. Bull. Variable Stars, (I.A.U. Commission 27), Konkoly Obs., Budapest, No. 635 (1972).

122.100 Improved elements of 3 RR-Lyrae variables in Aquila. H. Busch, K. Häussler.
Inform. Bull. Variable Stars, (I.A.U. Commission 27), Konkoly Obs., Budapest, No. 638 (1972).

122.101 Photoelectric observations of AD Leo.
P. F. Chugainov, N. I. Shakhovskaya, K. Barlai, L. Szabados, B. Szeidl.
Inform. Bull. Variable Stars, (I.A.U. Commission 27), Konkoly Obs., Budapest, No. 640 (1972).

122.102 UBV observations of FU Ori, SU Tau and OJ 287.
A. U. Landolt.
Inform. Bull. Variable Stars, (I.A.U. Commission 27), Konkoly Obs., Budapest, No. 642 (1972).

122.103 Radii variations in RZ Lyrae with the Blashko-effect.
I. G. Kolesnik, Yu. S. Romanov.
Inform. Bull. Variable Stars, (I.A.U. Commission 27), Konkoly Obs., Budapest, No. 644, 4 pp. (1972).

122.104 Visual observations of AD Leonis. J. Bingham, A. Forno, J. Isles, W. Pennell.
Inform. Bull. Variable Stars, (I.A.U. Commission 27), Konkoly Obs., Budapest, No. 648 (1972).

122.105 Photoelectric observations of the flare star UV Cet.
G. Asteriadis, L. N. Mavridis.
Inform. Bull. Variable Stars, (I.A.U. Commission 27), Konkoly Obs., Budapest, No. 654, 4 pp. (1972).

122.106 Remarks on the observations of flare stars by Japanese astronomers. G. A. Gurzadyan.
Inform. Bull. Variable Stars, (I.A.U. Commission 27), Konkoly Obs., Budapest, No. 656, with an answer by K. Osawa, K. Ichimura, No. 691 (1972).

122.107 R CrB variable UW Cen. F. M. Bateson.
Inform. Bull. Variable Stars, (I.A.U. Commission 27), Konkoly Obs., Budapest, No. 661 (1972).

122.108 AD Leo. K. Osawa, Y. Shimizu, T. Okada, H. Koyano, K. Ichimura, E. Watanabe, M. Yutani, K. Okida.
Inform. Bull. Variable Stars, (I.A.U. Commission 27), Konkoly Obs., Budapest, No. 666 (1972).

122.109 Photoelectric observations of the flare star EV Lac.
M. E. Contadakis, L. N. Mavridis.
Inform. Bull. Variable Stars, (I.A.U. Commission 27), Konkoly Obs., Budapest, No. 669 (1972).

122.110 New flare stars in Orion. R. I. Kiladze.
Inform. Bull. Variable Stars, (I.A.U. Commission 27), Konkoly Obs., Budapest, No. 670 (1972).

122.111 Photoelectric observations of the flare star EV Lac.
D. Deming, J. C. Webber.
Inform. Bull. Variable Stars, (I.A.U. Commission 27), Konkoly Obs., Budapest, No. 672, 3 pp. (1972).

122.112 Photoelectric observations of V645 Cen.
B. F. Marino, W. S. G. Walker.
Inform. Bull. Variable Stars, (I.A.U. Commission 27), Konkoly Obs., Budapest, No. 678 (1972).

122.113 Visual observations of YZ CMi and V371 Ori.
F. M. Bateson.
Inform. Bull. Variable Stars, (I.A.U. Commission 27), Konkoly Obs., Budapest, No. 678 (1972).

122.114 Photoelectric observations of the flare star AD Leo during the 1972, February 9–22 international patrol. S. Cristaldi, M. Rodonò.
Inform. Bull. Variable Stars, (I.A.U. Commission 27), Konkoly Obs., Budapest, No. 682 (1972).

122.115 Soft X-ray background and flare stars.
G. Cavallo, H. Horstman.
Nature, Phys. Sci., Vol. 235, 110 (1972).
Flare stars might contribute in a substantial way to the soft X-ray background. While the number requirements are roughly satisfied, the time averaged soft X-ray luminosity must be between 10^{29} and 10^{30} erg/s. This requirement does not conflict with early estimates of the soft X-ray emission and of the flare frequency. The galactic latitude dependence of the background can be reproduced by appropriate choices of the hydrogen scale height, and of the hydrogen density.

122.116 Optical polarization of X Per.
B. Baud, J. Tinbergen.
Nature, Vol. 237, 29 - 30 (1972).
The suspected X-ray source X Per has a degree of circular polarization of less than 2×10^{-4}.

122.117 Is X Persei an X-ray source?
D. Crampton, J. B. Hutchings.
Nature, Vol. 237, 92 - 93 (1972).
Spectra of X Per (HD 24534) obtained early in 1972 show double emission peaks in the strong Balmer and He lines. Small variations were suspected in the longward (weaker) peaks The velocities and profiles are consistent with a rapidly rotating infalling circumstellar disk. Although X Per is not distinctly different from normal Be stars, it is similar to their X-ray source candidates.

122.118 The two W Virginis stars in M5.
C. M. Coutts, H. Sawyer Hogg.
Bull. American Astron. Soc., Vol. 4, 217 (1972). – Abstr. AAS.

122.119 An observational study of the RV Tauri stars.
D. L. DuPuy.
Bull. American Astron. Soc., Vol. 4, 217 - 218 (1972). – Abstr. AAS.

122.120 The spectral classification of the β Cephei stars and their location in the theoretical Hertzsprung-Russell diagram. J. R. Lesh, M. L. Aizenman.
Bull. American Astron. Soc., Vol. 4, 218 (1972). – Abstr. AAS.

122.121 Spectral variations of the β CMa star BW Vulpeculae.
B. A. Goldberg, G. A. H. Walker, J. R. Auman, G. J. Odgers.
Bull. American Astron. Soc., Vol. 4, 218 (1972). – Abstr. AAS.

122.122 Pulsating helium stars. V. L. Trimble.
Bull. American Astron. Soc., Vol. 4, 218 (1972).
Abstr. AAS.

122.123 The spectrum of RR Telescopii, a study in iron.
L. H. Aller, R. Polidan, E. Rhodes, G. W. Wares.
Bull. American Astron. Soc., Vol. 4, 236 (1972). – Abstr. AAS.

122.124 A rotational temperature of Mira at maximum light.
P. A. Wehinger, S. Wyckoff.
Bull. American Astron. Soc., Vol. 4, 237 (1972). – Abstr. AAS.

122.125 Line identifications in the near-infrared spectrum of Mira. P. Wehinger, S. Wyckoff.
Bull. American Astron. Soc., Vol. 4, 270 (1972). – Abstr. AAS.

122.126 X-ray emission from flare stars.
S. Kahler, S. Shulman,
Nature, Phys. Sci., Vol. 237, 101 - 102 (1972).
Grindlay and Gurzadyan have claimed that under good conditions a detectable X-ray flux due to flares of UV Ceti would be seen at the earth. We show that an empirical approach based on the strong similarities between solar and stellar flares can be used to calculate stellar X-ray fluxes. We then obtain an upper limit on the unresolved flare star contribution to the diffuse hard X-ray emission from the galactic disk, using a calculation similar to that of Edwards.

122.127 The U Gem type variable Z Cha. F. M. Bateson.
Inform. Bull. Variable Stars, (I.A.U. Commission 27), Konkoly Obs., Budapest, No. 678 (1972).

122.128 Photoelectric observations of the Mira variable S Carinae. P. N. J. Wisse, M. Wisse.
Astron. Astrophys., Vol. 19, 164 - 165 (1972). – Research note.

122.129 Progressi nell'interpretazione e studio delle stelle variabili. L. Rosino.
Coelum, Vol. 40, 91 - 99 (1972).
VIII - Le Cefeidi di popolazione II; IX - Le variabili di tipo RR Lyrae.

122.130 The variable polarization of T Cen. A. V. Peterson.
Proc. Astron. Soc. Australia, Vol. 2, 108 - 110 (1972).

122.131 Photoelectric photometry of selected galactic cepheids. I. Two-color observations of 6 cepheid variables. K. Bahner, L. N. Mavridis.
Ann. Fac. Tech., Univ. Thessaloniki, Vol. 5, 65 - 79 = Contr. Dep. Geod. Astron., Univ. Thessaloniki, No. 3 (1971).
In the present paper, a description is given of the methods of observation and reduction used during the measurement of 18 galactic cepheids followed by the results obtained for the 6 cepheids CD Cyg; X, Z, RR Lac; U Vul and TU Cas.

122.132 The RR Lyrae stars in Baade's field near NGC 6522.
F. D. A. Hartwick, J. E. Hesser, G. Hill.
Astrophys. Journ., Vol. 174, 573 - 582 = Contr. Dominion Astrophys. Obs., Victoria, No. 182 (1972).
The results of a reinvestigation of the RR Lyrae stars in Baade's field near NGC 6522 are presented.

122.133 Compact extragalactic nonthermal sources.
P. A. Strittmatter, K. Serkowski, R. Carswell, W. A. Stein, K. M. Merrill, E. M. Burbidge.
Astrophys. Journ., (Letters), Vol. 175, L7 - L13 (1972).
Results of photometry, polarimetry, spectroscopy, and direct photography at visual wavelengths of four additional objects of the BL Lacertae type are reported together with in-

frared observations to $\lambda = 10\,\mu$ of one of them (OJ 287). The data indicate that these objects are compact sources of non-thermal continuum radiation. They are likely to be extragalactic, and their possible relationship to QSOs is discussed.

122.134 Ein einzigartiger Veränderlicher: FG Sge.
K. Wälke.
BAV Rundbrief, 21, Jahrgang, p. 5 - 6 (1972).

122.135 BD −7°3007.
R. Frye, J. E. Hesser, B. M. Lasker.
IAU Circ., No. 2398 (1972).

122.136 R Coronae Borealis type stars.   J. Krełowski.
Postępy Astron., Vol. 20, 23 - 36 (1972). In Polish.
The paper gives a description of observational facts of R CrB type variables. Special attention is paid to spectroscopic, infrared and polarimetric observations. A brief review of hypotheses on the interpretation of the R CrB phenomenon is presented.

122.137 Cepheid variables in Large and Small Magellanic Clouds.   C. Payne-Gaposchkin.
Galactic astronomy, Vol. 1, (see 012.019), 245 - 329 (1970).

122.138 Short-period oscillations in Eta Carinae.
B. M. Lasker, J. E. Hesser.
Nature, Phys. Sci., Vol. 237, 73 (1972).
The optical variability of the Eta Carinae system in the range 0.028 to ≈ 400 seconds has been investigated; no short-period oscillations with amplitudes greater than 0.008 magnitudes, typically, have been found. The unknown dynamical response of the extended nebulosity complicates the direct interpretation of these results as a limit on a pulsar-like phenomenon in Eta Carinae.

122.139 Photoelectric observations of the Be star V771 Sagittarii.   P. N. J. Wisse, M. Wisse.
Monthly Notes Astron. Soc. Southern Africa, Vol. 31, 35 - 36 (1972).

122.140 Some recent observations of UV Ceti.
A. H. Jarrett, J. P. Eksteen.
Monthly Notes Astron. Soc. Southern Africa, Vol. 31, 37 - 41 (1972).

122.141 35 Crucis − an apology.   A. W. J. Cousins.
Monthly Notes Astron. Soc. Southern Africa, Vol. 31, 42 (1972). − Letter.

122.142 Composition data from variable stars.
P. A. Wayman.
Monthly Notes Astron. Soc. Southern Africa, Vol. 31, 48 (1972). − Abstract.

122.143 Untersuchungen an unregelmäßig veränderlichen Sternen hoher Effektivtemperatur.
L. Meinunger.
Veröff. Sternw. Sonneberg, Vol. 8, No. 1, 69 pp. (1971).
Properties of irregular variables with high effective temperature together with their positions as to the stellar evolution are investigated.

122.144 FG Sagittae 1971.   W. Wenzel, W. Fürtig.
MVS, *Sonneberg*, Vol. 6, 18 - 22 (1972).
Photoelectric observations of FG Sagittae, obtained 1971 in Sonneberg and in the branch station Shemakha (USSR) are reported. A detailed light-curve for 1971 July to October (containing roughly 70 night means) is given.

122.145 Über das Verhalten der Flare-Sterne in den Plejaden und in einigen anderen Assoziationen.   W. Götz.
MVS, *Sonneberg*, Vol. 6, 24 - 33 (1972).
The comparison of the results from flare stars in the Pleiades with those of other associations demonstrate some aspects of the general behaviour of the flare stars, which represent a group of evolutionary young stars with low masses.

122.146 Spektraltypen von Veränderlichen. Teil XVII.
W. Götz, W. Wenzel.
MVS, *Sonneberg*, Vol. 6, 35 - 36 (1972).

122.147 Observational aspects of RR Lyrae and W Virginis stars: Some conundrums of stellar populations and galactic distribution.   R. P. Kraft.
The evolution of population II stars, (see 012.024), p. 69 - 95 = Contr. Lick Obs., No. 352 (1972).

122.148 Theoretical aspects of population II variables.
J. P. Cox, D. S. King.
The evolution of population II stars, (see 012.024), p. 103 - 146 (1972).

122.149 Shock waves in population II variable stars.
J. I. Castor.
The evolution of population II stars, (see 012.024), p. 147 - 186 (1972).

122.150 Masses, radii and luminosities of RR Lyrae variable stars.   R. Woolley.
Quarterly Journ. Roy. Astron. Soc., Vol. 13, 189 - 190 (1972). − Presented at the Woolley symposium (see 012.025).

122.151 Red variables in globular clusters, in the galactic centre and in the solar neighbourhood.
M. W. Feast.
Quarterly Journ. Roy. Astron. Soc., Vol. 13, 191 - 201 (1972). − Presented at the Woolley symposium (see 012.025).

122.152 Classical cepheids: Cornerstone to extragalactic distances?   A. Sandage.
Quarterly Journ. Roy. Astron. Soc., Vol. 13, 202 - 221 (1972). − Presented at the Woolley symposium (see 012.025).

122.153 A survey for RR Lyrae stars at high galactic latitude.   T. D. Kinman.
Quarterly Journ. Roy. Astron. Soc., Vol. 13, 258 - 265 (1972). − Presented at the Woolley symposium (see 012.025).

122.154 Common properties of the spectrum of U Monocerotis.   A. A. Aliev.
Soobshch. Shemakhinsk. Astrofiz. Obs., vyp. (No.) 6, p. 59 - 68 (1971). In Russian.

122.155 Photoelectric observations of YZ CMi.
V. S. Oskanian, J. V. Khatchatrian.
Inform. Bull. Variable Stars, (IAU Commission 27), Konkoly Obs., Budapest, No. 684 (1972).

122.156 Photoelectric observations of AD Leo.
V. S. Oskanian.
Inform. Bull. Variable Stars, (IAU Commission 27), Konkoly Obs., Budapest, No. 685 (1972).

122.157 New flares in the Pleiades.
L. G. Balázs, L. Patkós.
Inform. Bull. Variable Stars, (IAU Commission 27), Konkoly Obs., Budapest, No. 688 (1972).

122.158 Multi-colour photometry of the UVn-type variable, NS Orionis, during an exceptionally long-lived outburst.   A. D. Andrews.

Inform. Bull. Variable Stars, (IAU Commission 27), Konkoly Obs., Budapest, No. 689, 4 pp. (1972).

122.159 Photometric observations of 13 Cepheid variables. W. Wamsteker.
Inform. Bull. Variable Stars, (IAU Commission 27), Konkoly Obs., Budapest, No. 690, 4 pp. (1972).

122.160 Radial velocity of RR Lyrae variable − RW Ari. H. A. Abt, W. Z. Wisniewski.
Inform. Bull. Variable Stars, (IAU Commission 27), Konkoly Obs., Budapest, No. 697 (1972).

122.161 Spectral and photoelectric observations of BW Vulpeculae. M. Kubiak.
Acta Astron., Vol. 22, 11 - 32 = Warsaw Univ. Obs., Astron. Inst., Polish Acad. Sci., Repr. No. 324 (1972).
Photoelectric observations of BW Vul in Stromgren *uvby* system reveal the existence of stillstand in colour, corresponding to the stillstands in brightness and radial velocities. The effective temperature varies by about 5000°K during the cycle. Measurements of 48 coudé spectrograms with the dispersion 9 A/mm, covering the whole cycle, reveal the pronounced changes of equivalent widths of the absorption lines.

122.162 Évolution et pulsation des étoiles variables à courte période de type δ Scuti. C. Chevalier.
Thesis, Univ. Paris. Centre Documentation, C. N. R. S. (1972-4-12), 39 pp. (1971). − See Bull. Signal., Vol. 33, Section 120, No. 7089 (1972).

122.163 Aus der Mirasternbeobachtung im Jahr 1971. E. Heiser.
BAV Rundbrief, 21. Jahrgang, p. 34 (1972).

122.164 EV Lacertae. R. J. Livesey.
British Astron. Ass., Circ. No. 538 (1972).

Multiple solutions for cepheid envelopes. See Abstr. 064.034.

The pulsations of polytropic masses in rapid, uniform rotation. See Abstr. 065.002.

Evolutionary aspects of the cepheid stage. See Abstr. 065.019.

Models for R Coronae Borealis stars. See Abstr. 065.087.

Short period variable stars. IX. Rotation and mixing in the outer layers of A stars. Turbulent mixing due to the meridional circulation velocity field. See Abstr. 065.098.

Model calculations of RR-Lyrae stars. See Abstr. 065.109.

Radial velocities of 65 early-type stars. See Abstr. 112.007.

Brightness gradients for intrinsically variable celestial objects. See Abstr. 113.005.

Spectroscopic and photometric observations of M supergiants in Carina. See Abstr. 113.006.

Infrared observations of some southern IR−OH sources. See Abstr. 113.017.

Intermediate-band and H-beta photometry of short-period variable stars. See Abstr. 113.022.

Infrared observations of 1612 MHz IR/OH sources. See Abstr. 114.014.

Characteristics of OH emission from infrared stars. See Abstr. 114.048.

Spettroscopia galattica in Italia nel periodo ottobre '68−ottobre '69. See Abstr. 114.108.

The temperature scale of F and G stars. I. Hydrogen line profiles. See Abstr. 114.120.

The temperature scale of F and G stars. II. Continuum photometry. See Abstr. 114.121.

On some phenomena in T Tau type stars. See Abstr. 114.127.

The high-luminosity boundary of the β Cephei instability strip. See Abstr. 115.015.

Duplicity and its consequences among variable stars in general. See Abstr. 117.015.

Eruptive binaries. See Abstr. 117.016.

The new companion of Theta Coronae Borealis. See Abstr. 117.022.

The shell star 88 Herculis: Pulsating envelope or binary star? See Abstr. 117.034.

A search for spectroscopic binaries among southern cepheids. See Abstr. 119.016.

Polarization of variable stars. See Abstr. 131.087.

The nebulosity surrounding the galactic cepheid RS Puppis. See Abstr. 132.003.

Diameter of PKS 1514−24 (AP Lib). See Abstr. 141.083.

The relation between the optical and centimetric polarized emission in BL Lac and other QSOs. See Abstr. 141.085.

21-cm absorption in BL Lac. See Abstr. 141.100.

Multiwavelength program for the active sources 3C 120, BL Lac and OJ 287. See Abstr. 141.190.

Pre-main-sequence stars. I. Light variability, shells, and pulsation in NGC 2264. See Abstr. 153.007.

RR Lyrae variables in NGC 5897. See Abstr. 154.014.

(U-B)-colours and U-magnitudes of 63 RR Lyrae variables of the globular cluster NGC 5139 (ω Cen). See Abstr. 154.023.

Distributions and masses of horizontal-branch stars in clusters. See Abstr. 154.028.

A more precise definition of structural peculiarities of the instability strip of classical cepheids from Gascoigne's photoelectric data in the Small Magellanic Cloud. See Abstr. 159.020.

The Large Magellanic Cloud: Its topography of 1830 variable stars.   See Abstr. 159.023.

The field RR Lyrae stars in the Large Magellanic Cloud.   See Abstr. 159.025.

Errata

122.901   **Erratum: 'Photometric observations of V1057 Cygni.'** [Astrophys. Journ., (*Letters*), Vol. 169, L117 - L118 (1971)].   E. E. Mendoza V. Astrophys. Journ., (*Letters*), Vol. 172, L77 (1972).

123 Variable Stars: Lists of Observations, Individual Observations

123.001 **Abgeleitete Maxima von Mirasternen.** R. Lukas.
SuW, Vol. 11, 77 (1972).

123.002 **A new red variable star in Camelopardalis.**
M. Kazarian, A. Terzan.
Astron. Astrophys., Vol. 17, 323 - 324 (1972).
From the infrared and red spectra, it seems to be a M0 star and to have $\overline{m_{pg}-m_r} \cong +2.1$.

123.003 **Observation d'étoiles variables de faible amplitude.**
A. Figer.
L'Astronomie, 86e année, p. 86 - 90 (1972).
Observations of W Boo and X Her are presented.

123.004 **OJ 287: An exceptionally active variable source.**
B. H. Andrew, G. A. Harvey, W. J. Medd.
Journ. Roy. Astron. Soc. Canada, Vol. 66, 65 (1972). – Abstr. Canadian Astron. Soc.

123.005 **Variable star notes.** M. W. Mayall.
Journ. Roy. Astron. Soc. Canada, Vol. 66, 79 - 82 (1972).

123.006 **Researches with the Schmidt telescope. V. New variable stars in the field of Eta Cygni.**
G. Pinto, G. Romano.
Mem. Soc. Astron. Italiana, Nuova Ser., Vol. 43, 135 - 155 (1972).
Thirty new variable stars in the field of Eta Cygni have been discovered. The results of the photographic observations of these stars are given. Nine already known variables have been also studied (EY, V 465, V 482, V 550, V 724, V 823, V 1008, V 1153 and V 1162 Cyg). A method for the determination of the elements of periodic variables through a Hewlett-Packard calculator is given in the Appendix.

123.007 **Some observations of the flare star YZ CMi.**
T. R. Bhatt, S. D. Sinvhal.
IAU Colloqium No. 15, (see 012.006), p. 124 (1972).

123.008 **Some information on the Catania flare star survey.**
S. Cristaldi.
IAU Colloquium No. 15, (see 012.006), p. 124 - 125 (1972).

123.009 **Variable star notes.** M. W. Mayall.
Journ. Roy. Astron. Soc. Canada, Vol. 66, 131 - 134 (1972).

123.010 **De veranderlijke van de maand: R Coronae Borealis.**
G. Comello.
Hemel en Dampkring, Vol. 70, 150 - 152 (1972).

123.011 **New elements for WW Cam.** W. Braune.
Inform. Bull. Variable Stars, (I.A.U. Commission 27), Konkoly Obs., Budapest, No. 609 = BAV – Mitt. No. 21 (1972).

123.012 **New bright southern variable stars.** W. Strohmeier.
Inform. Bull. Variable Stars, (I.A.U. Commission 27), Konkoly Obs., Budapest, No. 610, 4 pp. = Veröff. Remeis-Sternw. Bamberg – Astron. Inst. Univ. Erlangen-Nürnberg, Vol. 8, No. 99 (1972).

123.013 **Six variable stars in Sagittarius.** D. Hoffleit.
Inform. Bull. Variable Stars, (I.A.U. Commission 27), Konkoly Obs., Budapest, No. 617 (1972).

123.014 **V 462 Cygni.**
G. B. Baratta, A. Cassatella, R. Viotti.
Inform. Bull. Variable Stars, (I.A.U. Commission 27), Konkoly Obs., Budapest, No. 626 (1972).

123.015 **Photoelectric observations of 32 Cyg near minimum in 1971.** J. M. Kreiner, M. Winiarski.
Inform. Bull. Variable Stars, (I.A.U. Commission 27), Konkoly Obs., Budapest, No. 628 (1972).

123.016 **Tests of two suspected δ Scuti-variables.**
G. C. Kilambi.
Inform. Bull. Variable Stars, (I.A.U. Commission 27), Konkoly Obs., Budapest, No. 630 (1972).

123.017 **A new variable star near the nebula K3+50.**
G. Romano.
Inform. Bull. Variable Stars, (I.A.U. Commission 27), Konkoly Obs., Budapest, No. 634 (1972).

123.018 **Notes on four variables.** H. Busch, K. Häussler.
Inform. Bull. Variable Stars, (I.A.U. Commission 27), Konkoly Obs., Budapest, No. 639, 3 pp. (1972).

123.019 **New variable stars in Lyra.** G. Romano.
Inform. Bull. Variable Stars, (I.A.U. Commission 27), Konkoly Obs., Budapest, No. 645 (1972).

123.020 **HD 21242: A new bright variable star.**
R. E. Montle, D. S. Hall.
Inform. Bull. Variable Stars, (I.A.U. Commission 27), Konkoly Obs., Budapest, No. 646 (1972).

123.021 **A period of 60^d in the light fluctuations of FG Sge.**
J. Papoušek.
Inform. Bull. Variable Stars, (I.A.U. Commission 27), Konkoly Obs., Budapest, No. 646 (1972).

123.022 **New and suspected variable stars in VSF 193.**
D. Hoffleit.
Inform. Bull. Variable Stars, (I.A.U. Commission 27), Konkoly Obs., Budapest, No. 660, 40 pp. (1972).
Identification charts and tables are given for 275 variable or suspected variable stars in Sagittarius, together with a summary of the work thus far accomplished in VSF 193.

123.023 **Observations of variable stars in NGC 1261.**
C. Bartolini, F. Grilli, F. Morisi.
Inform. Bull. Variable Stars, (I.A.U. Commission 27), Konkoly Obs., Budapest, No. 662 (1972).

123.024 **New faint southern variable stars.** F. M. Sosna.
Inform. Bull. Variable Stars, (I.A.U. Commission 27), Konkoly Obs., Budapest, No. 664, 4 pp. = Veröff. Remeis-Sternw. Bamberg – Astron. Inst. Univ. Erlangen-Nürnberg, Vol. 10, No. 101 (1972).

123.025 **Welch's red variable star in Crux, an R Centauri like variable.** W. S. G. Walker, B. F. Marino.
Inform. Bull. Variable Stars, (I.A.U. Commission 27), Konkoly Obs., Budapest, No. 679 (1972).

123.026 **PI PsA.** M. Petit.
Inform. Bull. Variable Stars, (I.A.U. Commission 27), Konkoly Obs., Budapest, No. 680 (1972).

123.027 **New variable star in the open cluster NGC 7128.**

L. Kohoutek.
Inform. Bull. Variable Stars, (I.A.U. Commission 27), Konkoly Obs., Budapest, No. 683 (1972).

123.028 **Hovedfaser for langperiodisk variable stjerner.** O. Klinting.
Astron. Tidssk., Årg. 5, p. 42 - 44 (1972).

123.029 **Verhüllung 1972 von R Coronae Borealis.** K. Locher.
Orion Schaffhausen, 30. Jahrgang, p. 101 (1972).

123.030 **Rho Persei, August 1969 - Dezember 1970.** C. Göttig.
SuW, Vol. 11, 194 (1972).

123.031 **R. Coronae Borealis.** R. Hodgson, J. Bortle, J. Ashbrook, D. Milon, R. Sinnott, M. Seslar.
IAU Circ., No. 2390 (1972).

123.032 **R. Coronae Borealis.** E. H. Mayer, R. Lukas, D. Rosebrugh, E. E. Friton.
IAU Circ., No. 2391 (1972).

123.033 **Eruptive variable or intergalactic supernova.** P. Wild.
IAU Circ., No. 2392 (1972).

123.034 **R Coronae Borealis.** P. Moore, D. Rosebrugh, E. H. Mayer, C. S. Morris.
IAU Circ., No. 2394 (1972).

123.035 **R Coronae Borealis.** H. Dürbeck, E. Scovil.
IAU Circ., No. 2395 (1972).

123.036 **RY Sagittarii.** P. J. Andrews.
IAU Circ., No. 2399 (1972).

123.037 **R Coronae Borealis.** F.-M. Sosna, R. Lukas.
IAU Circ., No. 2401 (1972).

123.038 **R. Coronae Borealis.** J. M. Pasachoff.
IAU Circ., No. 2403 (1972).

123.039 **R. Coronae Borealis.** C. E. Scovil, P. Moore.
IAU Circ., No. 2405 (1972).

123.040 **Eruptive object.**
Goranskij, Tanskij, V. F. Esipov.
IAU Circ., No. 2408 (1972).

123.041 **R Coronae Borealis.** J. D. Fernie.
IAU Circ., No. 2408 (1972).

123.042 **Eruptive object in Ursa Major.** C. Bertaud.
IAU Circ., No. 2412 (1972).

123.043 **R Coronae Borealis.** H. Dürbeck.
IAU Circ., No. 2415 (1972).

123.044 **Notes on U Geminorum and Z Camelopardalis type variables during 1971.** M. W. Mayall.
Journ. Roy. Astron. Soc. Canada, Vol. 66, 179 - 182 (1972).

123.045 **SV Cephei.** W. Wenzel.
MVS, *Sonneberg,* Vol. 6, 22 (1972).

123.046 **Beobachtung von Veränderlichen auf Tautenburger M31-Platten.** L. Meinunger.
MVS, *Sonneberg,* Vol. 6, 23 (1972).
5 variable or suspected stars, observed by Sharov et al. and by Rosino, were rediscussed (SVS 1729, SVS 1732, object E, nova Sharov 7, Rosino 87).

123.047 **Bearbeitung von 58 Veränderlichen am Südhimmel.** (Feld β Apodis, Teil III). H. Geßner.
MVS, *Sonneberg,* Vol. 6, 34 - 35 (1972).

123.048 **Observations of variable stars, July - December 1971. Report No. 21.** L. Plaut, H. Feijth.
Kapteyn Astron. Lab., Groningen – Netherlands. 11 pp. (1972).
This report gives 3590 visual observations of 164 variable stars.

123.049 **Eruptive object in Ursa Maior.** H. Huth.
Inform. Bull. Variable Stars, (IAU Commission 27), Konkoly Obs., Budapest, No. 687 (1972).

123.050 **Eruptive object in Ursa Maior.** H. Huth.
Inform. Bull. Variable Stars, (IAU Commission 27), Konkoly Obs., Budapest, No. 692 (1972).

123.051 **Observations of AX Monocerotis.** F. B. Wood, R. H. Bloomer.
Inform. Bull. Variable Stars, (IAU Commission 27), Konkoly Obs., Budapest, No. 693, 3 pp. = Rosemary Hill Obs., Dep. Phys. Astron., Univ. Florida, Gainesville, Florida, Contr. No. 30 (1972).

123.052 **PI PsA.** M. Petit.
Inform. Bull. Variable Stars, (IAU Commission 27), Konkoly Obs., Budapest, No. 695 (1972).

123.053 **Rosino's object: evidence for a fourth outburst.** S. C. Lucchetti, P. D. Usher.
Inform. Bull. Variable Stars, (IAU Commission 27), Konkoly Obs., Budapest, No. 696, 4 pp. (1972).

123.054 **R CrB.** J. Koda.
Yamamoto Circ., Nos. 1750, 1755 (1972). In Japanese.

124 Novae

124.001 Formation of coronal lines in the spectra of novae. I.
V. G. Gorbatsky.
Astron. Zhurn. Akad. Nauk SSSR, Vol. 49, 42 - 53 (1972). In Russian. English translation in Soviet Astron. AJ, Vol. 16, No. 1.

The conditions of formation of coronal lines in spectra of novae are investigated. Due to the mass loss from the system some of novae and recurrent novae are embedded in extensive gaseous envelopes. Gas heated by shock radiates in coronal lines. The shock motion is calculated and variations of the ion and electron temperatures behind the shock are found.

124.002 On the estimate of frequency of novae in the Andromeda nebula and the Galaxy. A. S. Sharov.
Astron. Zhurn. Akad. Nauk SSSR, Vol. 49, 54 - 57 (1972). In Russian. English translation in Soviet Astron. AJ, Vol. 16, No. 1.

It is shown that the frequency of novae in the Andromeda nebula is 31 per year. Supposing that the distribution of novae in the Andromeda nebula and in the Galaxy is the same, the frequency of novae in our stellar system is about 260. This value can be considered as an argument in favour of the Galaxy being of Sb type.

124.003 Interpretation of radio variations of novae.
R. M. Hjellming, V. Herrero, C. M. Wade.
Bull. American Astron. Soc., Vol. 3, 443 (1971). – Abstr. AAS.

124.004 Nova hydrodynamics. W. K. Rose, R. L. Smith.
Astrophys. Journ., Vol. 172, 699 - 712 (1972).
Calculations are described that explore hydrodynamic processes predicted to arise during the nova outburst on the basis of a model proposal by Rose. Two mechanisms for rapid mass loss are studied: (1) a direct shock-wave ejection mechanism, and (2) rapid mass loss resulting from pulsational instability.

124.005 Minimum-light phenomena in nova-like variables.
G. S. Mumford.
IAU Colloquium No. 15, (see 012.006), p. 275 - 278 (1972).

124.006 Novae in M 31 discovered and observed at Asiago from 1963 to 1970. L. Rosino.
Inform. Bull. Variable Stars, (I.A.U. Commission 27), Konkoly Obs., Budapest, No. 622 (1972).

124.007 Observations of novae. F. M. Bateson.
Inform. Bull. Variable Stars, (I.A.U. Commission 27), Konkoly Obs., Budapest, No. 661 (1972). – Concerning the novae: Sct 1970, Ser 1970, Sgr 1969.

Bedeckungslichtwechsel und Aktivität des Nova-ähnlichen Veränderlichen V Sagittae. See Abstr. 121.002.

A review of stellar outbursts of the pre-telescopic epoch. See Abstr. 125.003.

Possible identification of X-ray source (IM Normae). See Abstr. 142.120.

124.100 Nova Mensae 1970b

Nova Mensae 1970b in the Large Magellanic Cloud.
R. J. Havlen, R. M. West, B. E. Westerlund.

Astron. Astrophys., Vol. 16, 404 - 407 (1972).
The present paper gives the data obtained of this nova at the European Southern Observatory (ESO) on Cerro La Silla in Chile.

124.101 Nova Delphini 1967

The metallic absorption lines of nova Delphini before the December 1967 maximum.
M. Friedjung, I. Malakpur.
Astron. Astrophys., Vol. 18, 310 - 317 (1972).
The profiles of metallic absorption lines of nova Delphini have been studied for three dates. The results give excitation temperatures, and suggest that electron scattering was important in the formation of the profiles.

Some physical properties of nova Delphini 1967 during its nebular stage. A. Sanyal.
Bull. American Astron. Soc., Vol. 4, 217 (1972). – Abstr. AAS.

A model of nova Delphini 1967 from its nebular spectrum. W. C. Seitter.
IAU Colloquium No. 15, (see 012.006), p. 268 - 274 (1972).

124.102 Nova Vulpeculae 1968 No. 1

Photometric observations of nova Vulpeculae 1968 No. 1. A. Terzan, M. Bally, A. Durand.
Astron. Astrophys., Vol. 18, 471 - 474 (1972). In French.
The light curve of the nova Vulpeculae 1968 No. 1 is drawn for the period from April 15, 1968 to December 12, 1970. Four new variable stars are revealed near the nova.

124.103 Nova Cephei 1971

The spectrum of nova Cephei 1971.
J. D. R. Bahng.
Bull. American Astron. Soc., Vol., 4, 236 (1972). – Abstr. AAS.

Évolution du spectre de nova Cephei 1971.
C. Fehrenbach, Y. Andrillat.
Comptes Rendus Acad. Sci. Paris, Sér. B, Vol. 274, 1179 - 1184 (1972).

124.104 Nova FH Serpentis 1970

Nova FH Serpentis 1970 – Zwischenbericht.
K. Locher.
Orion Schaffhausen, 30. Jahrgang, p. 17 - 19 (1972).

A spectroscopic study of Nova FH Serpentis 1970.
J. B. Hutchings, J. Smolinski, J. Grygar.
Publ. Dominion Astrophys. Obs., Victoria, Vol. 14, (No. 2), 17 - 43 (1972).
The results of detailed velocity measurements and of spectrophotometry of coudé spectrograms of Nova FH Serpentis 1970 are presented in tabular form. Several hundred spectral lines have been identified in the wavelength interval $\lambda\lambda$ 3870–

5900 on spectrograms obtained during the bright phases of the nova. Velocity curves and tracings of selected line-profiles are given. The distance of the nova is estimated to be 750 parsecs, and its absolute visual magnitude at maximum to be -6^m5.

124.105 Nova RS Ophiuchi 1898

A preliminary model for the shell ionisation of the nova RS Ophiuchi.   M. Grewing, H. J. Habing.
Mitt. Astron. Ges., No. 31, p. 190 - 192 (1972).

124.106 Nova Sagittarii 1969

Bateson's nova Sgr 1969.   P. R. Knight.
Inform. Bull. Variable Stars, (IAU Commission 27), Konkoly Obs., Budapest, No. 694 (1972).

125   Supernovae, Supernova Remnants

125.001 The possible line feature in the X-ray background.
B. M. Tinsley.
Astrophys. Letters, Vol. 10, 31 - 35 (1972).

This letter evaluates the effects of cosmological redshift and galactic evolution on Silk's (1971) theory of the X-ray background due to young pulsars in supernova remnants, including the possible line feature near 7 keV due to K-line photons from iron-group elements. For any plausible cosmological time scales, galactic ages, and pulsar parameters, the strength of the redshifted continuum radiation from the supernova shells is found to obscure the redshift-smeared line feature in the background spectrum, so that the feature is probably undetectable.

125.002 On the physical characteristics of the envelopes of type I supernovae during the first period of their expansion. II. The evolution of the spectra of type I supernovae after light-maximum.   E. R. Mustel.
Astron. Zhurn. Akad. Nauk SSSR, Vol. 49, 15 - 30 (1972). In Russian. English translation in Soviet Astron. AJ, Vol. 16, No. 1.

The first section contains a discussion on the presence of a velocity gradient and a temperature gradient in the envelopes of type I supernovae and on their combined influence on the spectrum of the star. The second section contains an analysis of the development of the spectra of supernovae with the use of the absorption lines of Fe II. The third section contains a discussion of the lines of Na I, Ca II, Sc II, Si II, S II. The "anomalies" in the spectrum of SN 1967, NGC 3389 are explained.

125.003 A review of stellar outbursts of the pre-telescopic epoch.   Yu. P. Pskovsky.
Astron. Zhurn. Akad. Nauk SSSR, Vol. 49, 31 - 41 (1972). In Russian. English translation in Soviet Astron. AJ, Vol. 16, No. 1.

A summary of 125 objects, classified as star outbursts, is given. Possible supernovae, recurrent novae (among those V 603 Aql, GK Per), and novae 393 - 1600, slow novae of 396 and 1431, as well as the nova of 1006 are considered.

125.004 On the masses of type II supernova remnants.
S. Sofia.
Astrophys. Journ., Vol. 172, 53 - 55 (1972).

It is shown that the mass ejected during a type II supernova explosion in which a pulsar is formed is larger than $5\,M_\odot$, and may be as large as $40\,M_\odot$.

125.005 On the nature of the Monoceros supernova remnant.
W. L. Gebel, S. N. Shore.
Astrophys. Journ., (Letters), Vol. 172, L9 - L12 (1972).

A dynamical expansion time of 3×10^5 years and an envelope of 100 km s^{-1} are derived for the filamentary Monoceros Loop supernova remnant. A thermal soft X-ray point source located at $\alpha(1950) = 6^h35^m5$, $\delta(1950) = +6°28'$ is predicted based on the hypothesis of a cooling neutron star. By analogy, we interpret the Cygnus Loop X-ray source as a neutron star.

125.006 Cooling effects on interstellar shock waves from supernovae.
W. C. Straka, P. Goldreich, W. L. W. Sargent.
Bull. American Astron. Soc., Vol. 3, 451 (1971). – Abstr. AAS.

125.007 High resolution polarimetric observations of supernova remnants.   J. R. Dickel, D. K. Milne,
J. G. Ables, A. R. Kerr, B. R. Hermann.
Bull. American Astron. Soc., Vol. 3, 451 (1971). – Abstr. AAS.

125.008 Brightness and polarization structure of four supernova remnants at 2.8 centimeter wavelength.
M. R. Kundu, T. Velusamy.
Bull. American Astron. Soc., Vol. 3, 499 (1971). – Abstr. AAS.

125.009 Soft X-rays from supernova remnants.
S. A. Ilovaisky, C. Ryter.
Astron. Astrophys., Vol. 18, 163 - 165 (1972).

Some of the results concerning the observability of soft X-ray emission from supernova remnants presented in a previous paper (Ilovaisky and Ryter, 1971) are reviewed in the light of several new facts. The maximum linear diameter up to which a given remnant is likely to be a soft X-ray source is discussed quantitatively and depends on remnant distance from the galactic plane. An up-dated list of probable soft X-ray sources is proposed.

125.010 Upper limits to the X-ray luminosities of five supernovae.
M. Ulmer, V. Grace, H. S. Hudson, D. A. Schwartz.
Astrophys. Journ., Vol. 173, 205 - 211 (1972).

We have examined data from the OSO-III X-ray telescope for evidence of X-ray emission from five optically detected extragalactic supernovae during the period 1967 March—1968

June. Upper limits to the X-ray emission in the range 7.7–113 keV near optical maximum fall in the range 10^{-8}–10^{-10} ergs $(cm^2 s)^{-1}$. Reasonable estimates of the distances to these supernovae lead to upper limits on the total energies of from 10^{50} to 10^{51} ergs.

125.011 On ionization zones left after supernovae explosions. G. S. Bisnovaty-Kogan.
Astron. Zhurn. Akad. Nauk SSSR, Vol. 49, 453 - 454 (1972). In Russian. English translation in Soviet Astron. AJ, Vol. 16, No. 2.
It is shown that the investigation of ionization zones left after supernovae explosions may give information about the mechanism of these explosions.

125.012 A study of galactic supernova remnants: I. Distances, radio luminosity function and galactic distribution. S. A. Ilovaisky, J. Lequeux.
Astron. Astrophys., Vol. 18, 169 - 185 (1972).
Recent radio-frequency absorption-line observations for supernova remnants (SNR) are used to derive kinematic distances for twenty sources. Using the improved distance scale and the data presented in various published SNR catalogues we derive distances for approximately one hundred SNR. The radio luminosity function for SNR is studied in detail through integral number-diameter diagrams. We have derived the galactic distribution of SNR in distance from the galactic center and in height from the galactic plane.

125.013 The Chinese guest star of A.D. 1054 and the Crab nebula. H. Peng-Yoke, F. W. Paar, P. W. Parsons.
Vistas in astronomy, Vol. 13, (see 003.001), 1 - 13 (1972).
In this article new and old original Chinese and Japanese sources are analyzed. The result proves that there is considerable doubt whether the object of A.D. 1054 and the Crab nebula are connected at all.

125.014 A high-sensitivity search for X-rays from supernova remnants in Aquila. D. A. Schwartz, R. D. Bleach, E. A. Boldt, S. S. Holt, P. J. Serlemitsos.
Astrophys. Journ., (Letters), Vol. 173, L51 - L56 (1972).
We have performed a high-sensitivity scan of the galactic plane from $l = 70°$ to $l = 30°$ to search for 2–20-keV X-rays from supernova remnants. We report here on the spectra of five X-ray sources detected between 44° and 31° longitude, of which only two might be associated with suggested supernova remnants. Upper limits are presented for the 19 possible supernova remnants scanned in this survey.

125.015 Relativistic shock propagation and search for electromagnetic pulses from supernovae. S. A. Colgate, C. R. McKee, B. Blevins.
Astrophys. Journ., (Letters), Vol. 173, L87 - L91 (1972).
An electromagnetic pulse from a supernova occurring within its own magnetic field was previously predicted. We have searched for such pulses and continued theoretical investigations of the relativistic shock strength in a supernova envelope.

125.016 Radio-emission from supernovae remnants in distant galaxies. P. Notni, H. Oleak, G.-M. Richter.
IAU Symposium No. 44, (see 012.005), p. 82 - 86 (1972).

125.017 Radio observations of two supernova remnants, HB 21 and IC 443, at 4170 MHz. H. Hirabayashi, T. Takahashi.
Publ. Astron. Soc. Japan, Vol. 24, 231 - 237 = Tokyo Astron. Obs. Repr., No. 410 (1972).
The extended supernova remnants, HB 21 and IC 443, were observed at 4170 MHz. Radio contour maps and inte-

grated flux densities are given. Making use of other continuum observations, spectra of the sources were examined and spectral indices were obtained.

125.018 Supernova 1955 in a peculiar galaxy. M. Lovas.
Inform. Bull. Variable Stars, (I.A.U. Commission 27), Konkoly Obs., Budapest, No. 612 (1972).

125.019 The supernova service of the Byurakan Astrophysical Observatory. R. G. Mnatsakanian.
Inform. Bull. Variable Stars, (I.A.U. Commission 27), Konkoly Obs., Budapest, No. 655 (1972).

125.020 Observations of soft X rays: Two supernova remnants in the constellation Lupus and the diffuse background. T. M. Palmieri, G. Burginyon, R. Hill, F. Seward.
Bull. American Astron. Soc., Vol. 4, 219 - 220 (1972). – Abstr. AAS.

125.021 Relativistic shock propagation and search for supernova electromagnetic pulses. S. A. Colgate, C. R. McKee, B. A. Blevins.
Bull. American Astron. Soc., Vol. 4, 258 (1972). – Abstr. AAS.

125.022 A high sensitivity search for X-rays from supernova remnants in Aquila. D. A. Schwartz, E. A. Boldt, S. S. Holt, P. J. Serlemitsos, R. D. Bleach.
Bull. American Astron. Soc., Vol. 4, 261 (1972). – Abstr. AAS.

125.023 Why supernovae flash up. G. S. Bisnovaty-Kogan.
Priroda, No. 6.72, p. 24 - 33 (1972). In Russian.

125.024 A model for type I supernovae. F. D. A. Hartwick.
Nature, Phys. Sci., Vol. 237, 137 (1972).
Because type I supernovae are found in elliptical galaxies which are old stellar systems we are faced with finding a mechanism which involves stars of mass $\lesssim 1 M_\odot$. I suggest that hydrogen burning caused by mass transfer from a star of 0.7 $M_\odot$ – 1 $M_\odot$ to a degenerate hydrogen star of mass $\sim 0.1 M_\odot$ may be responsible.

125.025 The effect of the interstellar gas on the continuum spectrum of 3C 391. M. A. Gordon.
Astrophys. Journ., Vol. 174, 361 - 363 (1972).
Toward low frequencies, the spectral flux density of the supernova remnant 3C 391 is known to decrease to 60 percent of its maximum at approximately 80 MHz. Observations of radio recombination lines from nearby regions believed to be free of discrete radio sources imply turnover frequencies $(\tau = 1)$ of 30–110 MHz for the diffuse interstellar gas, thereby suggesting that the low-frequency cutoff of the spectrum of 3C 391 is probably due simply to free-free absorption in the intervening interstellar medium rather than to a decrease in the emission intrinsic to the source.

125.026 The production of ionizing ultraviolet radiation by supernova ejecta. S. A. Colgate.
Astrophys. Journ., Vol. 174, 377 - 382 (1972).
The kinetic energy of supernova ejecta is the presumed source of the necessary heating. In the case of blackbody emission the constraints of high temperature and high efficiency result in an unrealistic requirement of the collision between ultrathin shells. In the limit of transparent emission, a significant fraction of the emitted photons are either too low in energy to ionize hydrogen or too high in energy either to ionize efficiently or to be absorbed.

125.027 Observations of the supernova remnant IC 443 at 1.42 GHz. I. E. Hill.
Monthly Notices Roy. Astron. Soc., Vol. 157, 419 - 431 (1972).

The supernova remnant IC 443 has been mapped at 1.42 GHz with a resolution of 1.05×2.71 arc, using the Half-Mile telescope at Cambridge. The map shows that the source has a thick shell extending over the eastern half of its circumference with no evidence of a shell on the western edge. There is remarkably detailed agreement of optical and radio features. Estimates of some physical parameters of the source are derived.

125.028 Acceleration of cosmic rays in supernova remnants.
R. M. Kulsrud, J. P. Ostriker, J. E. Gunn.
Phys. Rev. Letters, Vol. 28, 636 - 639 (1972).

We find that rotating neutron stars can produce the bulk of the galactic cosmic rays if we accept the electromagnetic theory of pulsars. Specifically, the simple integrations presented here show that particles accelerated within the supernova remnants have approximately the required total energy, energy distribution, and chemical composition.

125.029 Supernova or nova near NGC 3147.
R. Altizer.
IAU Circ., No. 2381 (1972).

125.030 Supernova or nova near NGC 3147.
J. R. Dunlap.
IAU Circ., No. 2383 (1972).

125.031 Supernova in anonymous galaxy. C. T. Kowal.
IAU Circ., No. 2385 (1972).

125.032 Supernovae. M. Lovas, D. Shaffer.
IAU Circ., No. 2409 (1972).

125.033 Supernovae: Evidence of type I features in a type II spectrum. A. P. Fairall.
Monthly Notes Astron. Soc. Southern Africa, Vol. 31, 23 - 24 (1972).

125.034 Supernova in anonymous galaxy. G. Romano.
Inform. Bull. Variable Stars, (IAU Commission 27), Konkoly Obs., Budapest, No. 695 (1972).

Supernova search by a computer controlled telescope – specifications and status. See Abstr. 032.001.

Can supernovae explain the r-process elemental abundances? See Abstr. 061.056.

Self-similar solution of nonstationary hydrodynamical accretion. See Abstr. 062.029.

The p-process in explosive nucleosynthesis.
See Abstr. 065.001.

The neutron-proton ratio in stars exploding from dense, high-temperature states. See Abstr. 065.058.

Stellar evolution toward pre-supernova stage. I. Carbon and oxygen stars of 5 $M_\odot$, 10 $M_\odot$ and 30 $M_\odot$.
See Abstr. 065.062.

Carbon ignition in degenerate stellar cores.
See Abstr. 065.089.

Eruptive variable or intergalactic supernova.
See Abstr. 123.033.

Cooling and recombination in interstellar material ionized by supernovae. See Abstr. 131.014.

Interstellar material ionized by supernovae.
See Abstr. 131.111.

Time-dependent models of the interstellar gas.
See Abstr. 131.129.

Spectrophotometric investigations of filamentary nebulae. See Abstr. 132.004.

Energy balance of the Crab nebula and the Vela X remnant. See Abstr. 134.005.

Low-frequency turnover in the spectrum of 3C 391.
See Abstr. 141.117.

Pulsars, supernova remnants, and the pulsar period-luminosity relation. See Abstr. 141.505.

A new pulsar–supernova association.
See Abstr. 141.527.

Una possibile causa di emissione gamma da parte di supernovae. See Abstr. 141.530.

The acceleration of cosmic rays in supernova remnants. See Abstr. 143.041.

125.100 Supernova in NGC 4214

The spectrum of the type I supernova of 1954 in NGC 4214. D. Branch.
Astron. Astrophys., Vol. 16, 247 - 251 (1972).

Intensity minima in the spectrum of SN 4214 are interpreted as blueshifted He I absorption lines, as originally suggested by McLaughlin. The light curve and spectral evolution during observed phases were not consistent with a simple model of an expanding photosphere in which lines and continuum were formed, nor was the apparent temperature high enough to account for excitation of He I lines.

125.101 Supernova in NGC 5457

Polarization observations of the supernova in M 101.
N. M. Shakhovskoy, Yu. S. Efimov.
Astron. Zhurn. Akad. Nauk SSSR, Vol. 49, 11 - 14 (1972). In Russian. English translation in Soviet Astron. AJ, Vol. 16, No. 1.

The results of multi-colour polarization observations of the type II supernova in M 101 obtained with the 2.6 m Shajn telescope in August–September 1970 are given.

First-epoch radio observations of supernova 1970g.
S. T. Gottesman, J. J. Broderick, R. L. Brown, B. Balick, P. Palmer.
Astrophys. Journ., Vol. 174, 383 - 388 (1972).

Radio observations of the supernova 1970g which occurred in the galaxy M 101 have been made at wavelengths of 11.1 and 3.7 cm over a 9-month interval and have revealed a pointlike source with an intensity $\sim 4 \times 10^{-29}$ W m^{-2} Hz^{-1}. The agreement in position of this source with the supernova is excellent.

125.102 Supernova in NGC 5055

On the magnitude of the supernova in M 63 on May 20, 1971. G. van Herk, A. A. Schoenmaker.

Astron. Astrophys., Vol. 17, 146 - 147 (1972).
A revised magnitude is given for the magnitude of the supernova in M 63 on the evening of May 20, 1971.

La supernova de 1971 dans NGC 5055 (Messier 63).
C. Bertaud, C. Pollas.
L'Astronomie, 86ᵉ année, p. 137 - 140 (1972).

125.103   Supernova in NGC 3811

Supernovae in the spiral galaxy NGC 3811.
Zemlya i Vselennaya, 1972, No. 2, p. 81. In Russian.

125.104   Supernova in NGC 4027

Supernova 1954 in the Sbp galaxy NGC 4027.
M. Lovas.
Inform. Bull. Variable Stars, (I.A.U. Commission 27), Konkoly Obs., Budapest, No. 653 (1972). — Remarks by F. Zwicky, No. 671 (1972).

125.105   Supernova in NGC 5236

Photometry of the supernova 1968 in M 83.
W. Wamsteker.
Astron. Astrophys., Vol. 19, 99 - 103 (1972).
The supernova, which seems to be of Zwicky's Type I, had an absolute magnitude $M_v = -17.7$, corrected for foreground absorption. The observations cover a period of 62 days. The colours are very close to those of a black body.

125.106   Supernova in NGC 493

Supernova in NGC 493.   L. Kohoutek.
IAU Circ., No. 2389 (1972).

125.107   Supernova in NGC 5253

Supernova in NGC 5253.   C. T. Kowal.
British Astron. Ass., Circ. No. 542 (1972).

Supernova in NGC 5253.   C. T. Kowal.
IAU Circ., No. 2405 (1972).

Supernova in NGC 5253.   G. H. Herbig,
C. B. Stephenson, W. Wisniewski, T. Lee, T. Wdowiak,
J. Michlovic, V. L. Matchett, E. H. Mayer,
IAU Circ., No. 2407 (1972).

Supernova in NGC 5253.
R. Barbon, F. Ciatti, A. L. Ardeberg, M. J. H. de Groot.
IAU Circ., No. 2411 (1972).

Supernova in NGC 5253.
Walker, Christie, Feasey, Freeth, K. Locher, J. E. Bortle.
IAU Circ., No. 2413 (1972).

Extragalactic Ca II absorption lines in the spectra of the supernova in NGC 5253.
R. F. Sisteró, M. F. Castore de Sisteró.
Inform. Bull. Variable Stars, (IAU Commission 27), Konkoly Obs., Budapest, No. 686, 3 pp. (1972).

Optical observations of the supernova in NGC 5253.
P. S. Osmer, J. E. Hesser, W. E. Kunkel, B. M. Lasker, M. F. McCarthy, A. U. Landolt.
Nature, Phys. Sci., Vol. 238, 21 - 22 (1972).
The type I supernova in NGC 5253 discovered recently by Kowal is of particular interest because of its brightness (m_v ~8.5) and because it is the second supernova to occur in this peculiar galaxy in the past 100 years. The galaxy is well placed for observation from Cerro Tololo, and we report here the initial results obtained from rather extensive studies.

Supernova in NGC 5253.
Yamamoto Circ., Nos. 1755, 1756 (1972). In Japanese.

126 Low-luminosity Stars, Subdwarfs, White Dwarfs

126.001 Observations of rapid blue variables — III. HL Tau-76. B. Warner, R. E. Nather.
Monthly Notices Roy. Astron. Soc., Vol. 156, 1 - 5 (1972).

We report new high speed photometric observations of the white dwarf HL Tau-76 which strongly suggest that the variability of this star is caused by an underlying driving mechanism with a period of 12.437 ± 0.003 min. Three-colour photometry of the outbursts in HL Tau-76 shows that there is an increase in temperature of ∼650°K during outburst.

126.002 The spectral dependence of circular polarization in Grw+70°8247.
J. R. P. Angel, J. D. Landstreet, J. B. Oke.
Astrophys. Journ., (*Letters*), Vol. 171, L11 - L15 (1972).

Sharp changes in circular polarization with wavelength are seen in new observations of Grw+70°8247 made with 80 Å resolution. Some of the structure is associated with the Minkowski bands. Continued broad-band measurements show that while the polarization below 6000 Å appears to remain constant, above this wavelength there have been significant changes in the past year.

126.003 Interpretation of the Minkowski bands in Grw+70°8247. J. R. P. Angel.
Astrophys. Journ., (*Letters*), Vol. 171, L17 - L21 (1972).

It is argued from the spectral structure of circular polarization in Grw+70°8247 that the absorption bands are at least in part molecular in origin.

126.004 Metal abundances and internal temperatures of DG white dwarfs. K. H. Böhm, T. C. Grenfell.
Astrophys. Letters, Vol. 10, 141 - 143 (1972).

The extremely low metal abundances which have recently been determined by Wegner (1971) for DF and DG white dwarfs have been used in calculating new convective envelopes for DG stars. We find a rather high degeneracy and temperatures of only 2.15×10^5 to 2.73×10^5 K at the bottom of the convection zone. It is argued that, if Wegner's abundances are correct, the central temperature cannot be much higher than 4×10^5 K, a value which is about a factor of 10 lower than the Debye temperature.

126.005 Magnetic fields of white dwarfs.
G. Chanmugam, M. Gabriel.
Astron. Astrophys., Vol. 16, 149 - 152 (1972).

The time scale of decay for the magnetic fields of white dwarfs has been determined for dipolar and toroidal field configurations, taking into account the non-uniform conductivity and effects due to cooling. It is shown that it is unlikely that field decay could be the cause of non-observability of magnetic fields in some white dwarfs. Alternative causes are discussed.

126.006 Spectrophotometry of the white dwarf van Maanen 2.
K. H. Böhm, R. Schwartz, P. Szkody, G. Wegner.
Astron. Astrophys., Vol. 16, 431 - 436 (1972).

The energy distribution in the 3500 Å to 6600 Å region of the spectrum of van Maanen 2 has been determined. We have used five spectra with dispersions of 51 Å/mm and 134 Å/mm. Line profiles and the structure of blends are visible in considerable detail.

126.007 Outer envelopes and cooling of white dwarfs.
D. Koester.
Astron. Astrophys., Vol. 16, 459 - 470 (1972).

The structure and energy transport of the nondegenerate envelopes of white dwarfs in the range 0.32 to 1.17 $M_\odot$ with effective temperatures from 6000 to 30000 °K is examined us-

ing newly evaluated opacities. Resulting relations between luminosities and central temperatures are used to calculate the cooling of white dwarfs. The density dependence of the specific heat of electrons and ions is fully taken into account. We discuss the relative importance of parameters such as mass, chemical composition of envelope and degenerate interior as well as different theories of cristallization on cooling and age of white dwarfs.

126.008 The spectra and element abundances of cool white dwarfs. G. Wegner.
Astrophys. Journ., Vol. 172, 451 - 478 (1972).

New spectroscopic observations of five white dwarfs were obtained at Kitt Peak with the 84-inch telescope. Three of these stars show sufficiently strong metal lines to carry out abundance determinations. Nongray model atmospheres are computed. The influence of convection is included and a temperature-correction procedure is described. Line profiles are computed by using the impact approximation with the (6, 12) Lennard-Jones potential to describe the interaction. The main result is that the atmospheres of the cool white dwarfs are found to be nearly pure He. The metal and hydrogen abundances are found to be of the order of 10^{-9} and 10^{-4} by number, respectively.

126.009 An interpretation of the 4135 band in the white dwarf Grw + 70°8247. J. R. P. Angel.
Bull. American Astron. Soc., Vol. 3, 442 - 443 (1971). – Abstr. AAS.

126.010 Hydrodynamic studies of thermonuclear runaways in helium rich white dwarfs with hydrogen rich envelopes.
S. G. Starrfield, W. M. Sparks, J. W. Truran, G. S. Kutter.
Bull. American Astron. Soc., Vol. 3, 484 (1971). – Abstr. AAS.

126.011 *uvby* photometry of white-dwarf stars.
J. A. Graham.
Astron. Journ., Vol. 77, 144 - 150 (1972).

New magnitudes and colors are presented for 36 white-dwarf stars from the Eggen–Greenstein lists. The photometry is on the intermediate-band *uvby* system devised by Strömgren. The new measurements are combined with the observations of 44 stars published previously to illustrate further some of the characteristics of the *uvby* system in this particular application. Attention is drawn to the $M_V - (b-y)$ diagram in which the white dwarfs appear to lie on a single sequence within the accuracy of the presently available absolute magnitudes.

126.012 The composition of the subdwarf Groombridge 1830. J. Tomkin.
Monthly Notices Roy. Astron. Soc., Vol. 156, 349 - 359 (1972).

A curve of growth analysis of the halo subdwarf Gmb 1830 relative to the sun shows that the metals are in general about 10 times less abundant than in the sun. Calcium and titanium are relatively overabundant and manganese is underabundant. Carbon may be slightly underabundant relative to the metals, and nitrogen certainly is by a factor of at least five.

126.013 *UBVr* photometry of white dwarfs and white-dwarf suspects. R. D. Schwartz.
Publ. Astron. Soc. Pacific, Vol. 84, 28 - 32 (1972).

Photoelectric *UBVr* photometry is presented for 33 white-dwarf suspects from the Lowell Observatory lists. On the basis of $(B-V)$ and $(U-B)$ color indices, eight probable white dwarfs are identified.

126.014 **The velocity, redshift, and pressure shift of van Maanen 2.** J. L. Greenstein.
Astrophys. Journ., Vol. 173, 377 - 384 (1972).

With probable Einstein shift, the expected Doppler shift of +200 km s^{-1} contrasts sharply with the newly measured velocity of Ca II, +39 ± 4 km s^{-1} from 15 spectra. The H-line shows a velocity +29 ± 7 km s^{-1} larger than the K-line. A tentative explanation suggested is extreme pressure broadening and the violet shift found in the laboratory for resonance lines of the alkalis with helium perturbers.

126.015 **Variable white dwarfs.** G. Chanmugam.
Nature, Phys. Sci., Vol. 236, 83 (1972).

An explanation of the periods of the recently discovered variable white dwarfs is given in terms of non-radial oscillations.

126.016 **Quasi-radial pulsations of rotating white dwarfs and neutron stars in Newton's theory of gravitation.**
V. V. Papoyan, D. M. Sedrakian, E. V. Chubarian.
Astrofizika, Vol. 7, 643 - 649 (1971) In Russian.
English translation in Astrophysics, Vol. 7, No. 4.

The frequencies of quasi-radial pulsations of rotating white dwarfs and neutron stars in the frame of Newton's theory of gravitation are calculated. The critical values of the central density in the sense of dynamical unstability are obtained.

126.017 **The chemical composition of the B-type subdwarf HD 4539.**
B. Baschek, W. L. W. Sargent, L. Searle.
Astrophys. Journ., Vol. 173, 611 - 618 (1972).

The spectrum of HD 4539 is compared to that of the standard main-sequence star ι Her and a differential model-atmosphere abundance analysis is reported.

126.018 **High-frequency stellar oscillations. The Cerro Tololo search for luminosity-variable white dwarfs.**
J. E. Hesser, B. M. Lasker.
IAU Colloquium No. 15, (see 012.006), p. 160 - 168 (1972).

126.019 **X-ray emission from white dwarfs.**
P. A. Strittmatter, K. Brecher, G. R. Burbidge.
Astrophys. Journ., Vol. 174, 91 - 99 (1972).

In this paper we suggest that (1) the contribution of the Galaxy to the soft X-ray background and (2) the periodic optical variations in certain white dwarfs may originate in "coronae" around white dwarfs. We shall accordingly examine each of these phenomena in the light of this proposal. An analysis of the energy requirements (and of other constraints) for the visible white dwarfs is made. The relationship between these postulated sources and the strong X-ray sources is examined.

126.020 **Thermal imbalance in white dwarfs.**
H. M. Van Horn, C. J. Hansen, J. P. Cox.
Bull. American Astron. Soc., Vol. 4, 240 (1972). – Abstr. AAS.

126.021 **CD – 42° 14462 – a rotating white dwarf (?).**
G. Wegner.
Proc. Astron. Soc. Australia, Vol. 2, 107 - 108 (1972).

126.022 **The low-luminosity star G51–15.**
C. C. Dahn, A. L. Behall, H. H. Guetter, J. B. Priser, R. S. Harrington, K. A. Strand, R. K. Riddle.
Astrophys. Journ., (*Letters*), Vol. 174, L87 (1972).

Photometric and astrometric observations of the high-proper-motion star G51–15 indicate that the star with M_v = 16.97 mag and $V - I$ = 4.34 mag lies on the low-luminosity extension of the main sequence and is one of the faintest stars known in intrinsic brightness.

126.023 **Polarized radiation from magnetic white dwarfs: Exact solution of Kemp's model.**
G. Chanmugam, R. F. O'Connell, A. K. Rajagopal.
Astrophys. Journ., Vol. 175, 157 - 159 (1972).

An exact quantum-mechanical calculation of Kemp's magnetoemission process, for magnetic white dwarfs at low temperatures, is presented and the implications discussed.

126.024 **The gravitational redshift of 40 Eridani B.**
J. L. Greenstein, V. Trimble.
Astrophys. Journ., (*Letters*), Vol. 175, L1 - L5 (1972).

Image-tube coudé spectra of the DA white dwarf 40 Eri B give an independent determination of the gravitational redshift as +27 km s^{-1}; prime-focus spectra give +16 km s^{-1}, or a mean +23 ± 5 km s^{-1}. The agreement with the theoretical mass-radius prediction is within the experimental errors of both.

126.025 **G61–29.** B. Warner.
IAU Circ., No. 2388 (1972).

Number of free electrons in condensed substance depending on its density. See Abstr. 022.058.

Magnetic properties of a degenerate electron gas and implications for metals, white dwarfs, and neutron stars. See Abstr. 061.015.

The thermodynamics of white dwarf matter. II. See Abstr. 065.007.

About nuclear energy sources in superdense celestial bodies. See Abstr. 065.060.

On the nature of the horizontal branch. II. Extremely blue halo stars: A theoretical viewpoint. See Abstr. 065.072.

Self-accretion of matter, red subluminous stars and early evolution of low-mass stars. See Abstr. 065.083.

Electron gas in superstrong magnetic fields: Wigner transition. See Abstr. 065.116.

Mass ejection from red giants and white dwarfs. See Abstr. 065.122.

A Proper Motion Survey at the South Galactic Pole. See Abstr. 112.015.

Interstellar Matter, Gaseous Nebulae, Planetary Nebulae

131 Interstellar Space, Interstellar Matter, Polarization of Starlight

131.001 **On the interaction of rotating interstellar grains with cosmic low frequency radiation.**
P. G. Martin.
Monthly Notices Roy. Astron. Soc., Vol. 155, 283 - 291 (1972).

The effect of the low frequency (MHz) interstellar radiation on the angular momentum distribution of interstellar grains with permanent dipole moments is examined. Some aspects of the origin of the low frequency radiation relevant to this problem are discussed.

131.002 **The Parkes survey of 21-centimeter absorption in discrete-source spectra. II. Galactic 21-centimeter observations in the direction of 35 extragalactic sources.**
V. Radhakrishnan, J. D. Murray, P. Lockhart, R. P. J. Whittle.
Astrophys. Journ., Suppl. Ser., No. 203, Vol. 24, 15 - 47 (1972).

A detailed comparison of emission and absorption spectra obtained in the direction of extragalactic sources permits an unambiguous separation of the contribution from a diffuse, high-temperature, optically thin component and that from denser and colder local concentrations of hydrogen.

131.003 **The Parkes survey of 21-centimeter absorption in discrete-source spectra. III. 21-centimeter absorption measurements on 41 galactic sources north of declination −48°.**
V. Radhakrishnan, W. M. Goss, J. D. Murray, J. W. Brooks.
Astrophys. Journ., Suppl. Ser., No. 203, Vol. 24, 49 - 121 (1972).

Forty-one galactic sources north of declination −48° have been observed for 21-cm absorption by using the single-dish and/or interferometer methods. The difficulties encountered by both methods on sources in the galactic plane are discussed. Detailed notes are presented for each source, together with the reference emission and absorption profiles and tables summarizing all the data.

131.004 **The Parkes survey of 21-centimeter absorption in discrete-source spectra. IV. 21-centimeter absorption measurements on low-latitude sources south of declination −46°.**
W. M. Goss, V. Radhakrishnan, J. W. Brooks, J. D. Murray.
Astrophys. Journ., Suppl. Ser., No. 203, Vol. 24, 123 - 159 (1972).

Twenty-one low-latitude sources in the range $280° < l < 340°$ have been observed for 21-cm absorption by using the Parkes 64-m telescope; twelve of these sources were also investigated with the 64-m/18-m interferometer. Distance estimates and detailed notes are presented for each source, together with profiles and tables summarizing all the data on the sources. The H I evidence on two of these sources is in support of the view that they are extragalactic.

131.005 **The Parkes survey of 21-centimeter absorption in discrete-source spectra. V. Note on the statistics of absorbing H I concentrations in the galactic disk.**
V. Radhakrishnan, W. M. Goss.
Astrophys. Journ., Suppl. Ser., No. 203, Vol. 24, 161 - 166 (1972).

A statistical analysis of the parameters of Gaussian components fitted to the absorption spectra presented in the preceding two papers leads to the conclusion that the hydrogen seen in absorption in the galactic disk has a mean value for N_H/T_s of 1.5×10^{19} atoms cm^{-2} °K^{-1} kpc^{-1}. A line of sight along the plane intersects on the average a minimum of 2.5 concentrations per kiloparsec.

131.006 **Interstellar sodium lines and the two-component model of galactic H I regions.**
L. M. Hobbs, B. Zuckerman.
Astrophys. Journ., Vol. 171, 17 - 20 (1972).

The intensities and widths of the interstellar sodium lines are considered as possible discriminants between two different two-component models recently proposed for galactic H I regions. Present observational evidence is insufficient to distinguish between these two possibilities.

131.007 **Photochemistry and lifetimes of interstellar molecules.**
L. J. Stief, B. Donn, S. Glicker, E. P. Gentieu, J. E. Mentall.
Astrophys. Journ., Vol. 171, 21 - 30 (1972).

This paper is concerned with a quantitative discussion of the photochemistry and lifetime against photodecomposition of five interstellar molecules: H_2CO, NH_3, H_2O, CH_4, and CO. For the first four molecules, primary photochemical decomposition processes yielding atomic and molecular hydrogen are considered in addition to photoionization.

131.008 **Search for interstellar furan and imidazole.**
R. L. Dezafra, P. Thaddeus, M. Kutner, N. Scoville, P. M. Solomon, H. Weaver, D. R. W. Williams.
Astrophys. Letters, Vol. 10, 1 - 3 (1972).

Results are reported of an unsuccessful 6-cm search for the hetrocyclic carbon ring molecules furan and imidazole. Upper limits in brightness temperature of 0.25 K or less are found for furan in 11 galactic sources, and of less than 0.1 K for imidazole in Sgr A and Sgr B2.

131.009 **Association reactions.** D. A. Williams.
Astrophys. Letters, Vol. 10, 17 - 19 (1972).

Although association reactions forming diatomic and small polyatomic molecules occur in the interstellar medium with negligible rates, it is shown here that larger molecules, containing about 10 atoms, may be formed in this way under interstellar conditions. Fairly large molecules have already been identified in the interstellar medium, and it is therefore suggested that any description of the formation of large interstellar molecules must take account of association reactions.

131.010 **Water sources associated with OH emission in the southern Milky Way.**
K. J. Johnston, B. J. Robinson, J. L. Caswell, R. A. Batchelor.
Astrophys. Letters, Vol. 10, 93 - 98 (1972).

A search for H_2O emission at 22.23508 GHz among southern OH emission sources has located five new sources, four of which are associated with H II regions showing intense continuum radiation. Water vapor was detected near 20 per cent of the OH emitters observed in this search.

131.011 **Polarization measurements of 1660 southern OB−stars.** G. Klare, T. Neckel, G. Schnur.
Astron. Astrophys., Suppl. Ser., Vol. 5, 239 - 261 (1972).

For all 1660 OB-stars of the Heidelberg catalogue (Klare

and Szeidl, 1966) polarization data have been measured and are presented in a catalogue. The stars common with Hiltner (1956) have been compared and show a good agreement. Mean errors of the degree and of the direction of polarization are given.

131.012 Comments on a paper by K. Rohlfs: 'On the structure of interstellar matter. I' [Astron. Astrophys., Vol. 12, 43 - 58 (1971)]. W. B. Burton.
Astron. Astrophys., Vol. 16, 158 - 160 (1972). – Letter.

131.013 On analytical and numerical models. A reply to W. B. Burton [Astron. Astrophys., Vol. 16, 158 - 160 (1972)]. K. Rohlfs.
Astron. Astrophys., Vol. 16, 161 - 162 (1972). – Letter.

131.014 Cooling and recombination in interstellar material ionized by supernovae. D. Goldsmith.
Astron. Astrophys., Vol. 16, 286 - 290 (1972).
Detailed calculations of the cooling and recombining interstellar gas, together with a one-dimensional hydrodynamic approximation, model the behavior of interstellar material that has been exposed to a supernova explosion and either partially or totally ionized.

131.015 Evaporation of dirty ice particles surrounding early type stars. III. The size distribution. S. Isobe.
Publ. Astron. Soc. Japan, Vol. 24, 27 - 39 = Tokyo Astron. Obs. Repr. No. 400 (1972).
The size distribution of interstellar grains is determined by a growth mechanism and a destruction process. The dirty ice grain with graphite core is taken as the most probable grain model.

131.016 Radio emission of the line of water vapour from W 49 and Orion nebula.
A. B. Akvilonova, V. I. Ariskin, B. G. Kutuza, R. L. Sorochenko.
Astron. Zhurn. Akad. Nauk SSSR, Vol. 49, 102 - 104 (1972). In Russian. English translation in Soviet Astron. AJ, Vol. 16, No. 1.
The radioline profiles of water vapour from W 49 and Orion nebula at a wavelength of 1.35 cm were obtained with the help of the 22-m radiotelescope of the Physical Institute of the Academy of Sciences. The integral flux density of the observed emission and the luminosity were computed.

131.017 Ultraviolet photometry from the Orbiting Astronomical Observatory. II. Interstellar extinction.
R. C. Bless, B. D. Savage.
Astrophys. Journ., Vol. 171, 293 - 308 (1972).
Interstellar extinction curves over the region $\lambda\lambda 3600-1100$ for 17 stars are presented. The extinction curves generally show a pronounced maximum at $\lambda 2175 \pm 25$, a broad minimum in the region $\lambda\lambda 1800-1350$, and finally a rapid rise to the far-ultraviolet. Large extinction variations from star to star are found, especially in the far-ultraviolet. The data are combined with visual and infrared observations to display the extinction behavior over a range in wavelength of about a factor of 20. The observations appear to require a multicomponent model of the interstellar dust.

131.018 Relative line strengths of intense extreme-ultraviolet lines of Na I and their importance in estimating interstellar abundances. M. W. D. Mansfield, J. P. Connerade.
Astrophys. Journ., Vol. 171, 391 - 392 (1972).

131.019 Interstellar reddening in the North Galactic Polar Cap. I. The dependence on distance from the galactic plane. W. Pfau.
Astron. Nachr., Vol. 293, 195 - 202 (1971/72) = Mitt. Univ.-

Sternw. Jena, No. 112 (1972).
Interstellar reddening within the North Galactic Polar Cap ($b^{II} \geqq 60°$) is determined from published UBV photometry of 38 stars with spectral type earlier than A.

131.020 Eine stellarstatistische Methode zur Bestimmung des Verhältnisses der interstellaren Extinktion zur Verfärbung. S. Rössiger.
Astron. Nachr., Vol. 293, 211 - 219 (1971/72).
After scrutinizing all so far known methods for the determination of the ratio between interstellar extinction and colour excess the paper presents another method based on the comparison of the numbers of stars in an area containing a dark cloud in two different colours. It is tested on several dark clouds.

131.021 A numerical study of the dynamics of H II regions. J. G. Hill, M. C. Marsh.
Monthly Notices Roy. Astron. Soc., Vol. 156, 189 - 206 (1972).
A numerical study is made of the dynamics of H II regions governed by the radiation-hydrodynamic equations of Goldsworthy. The difference equations are explained and applied to plane similarity problems with constant ambient density. The existence of isothermal shocks in the H II region is demonstrated for some of these constant density situations.

131.022 Physical conditions in interstellar hydroxyl and formaldehyde clouds.
R. D. Davies, H. E. Matthews.
Monthly Notices Roy. Astron. Soc., Vol. 156, 253 - 262 (1972).
The physical conditions in interstellar clouds which contain the molecules OH and H_2CO have been derived from an investigation of the H, OH and H_2CO absorption spectra of Cas A, Cyg A and Tau A. The derived parameters of the typical molecule-bearing cloud are listed along with the estimates of the visual obscuration and ultra-violet radiation field.

131.023 On the distribution of OH in the Galaxy. I. Correlation with continuum sources and with formaldehyde.
B. E. Turner.
Astrophys. Journ., Vol. 171, 503 - 518 (1972).
A survey of the Galaxy in the region $337° \leqslant l < 75°$, $-2° \leqslant b \leqslant +2°$ has revealed OH absorption or emission in 180 new sources in a total of 264 directions searched to a limit of 0.7 f.u. The survey is quite strongly selective toward directions of continuum sources. The number of OH and H_2CO clouds is large enough to warrant a statistical study of the association of OH with H_2CO and with continuum sources.

131.024 A critical discussion of Heiles' cloudlets.
A. H. Rots, U. J. Schwarz, H. van Woerden.
Astron. Astrophys., Vol. 16, 344 - 347 (1972).
Heiles' procedures for defining cloudlets and determining their properties are criticized. The conclusions strongly effect the size and mass spectra of interstellar clouds.

131.025 Interference-filter photography of H II regions.
M. G. Smith.
Astron. Astrophys., Vol. 16, 482 - 485 (1972).
"Mono-ionic", interference-filter, image-tube photographs are presented as an efficient tool for planning accurate observations of parameters dependent on line strengths and line profiles in gaseous nebulae. They provide evidence for considerable small-scale variations in excitation conditions within individual H II regions. A larger region of enhanced [N II] emission is found near the position of the radio source, Car II. Comparison of Hα and [N II] regions confirms a recent objection to published temperatures derived from comparison of the linewidths of Hα and [N II] $\lambda 6584$ Å.

131.026 A model of the interstellar medium based on the interstellar calcium and sodium lines.
S. R. Pottasch.
Astron. Astrophys., Vol. 17, 128 - 138 (1972).

Use is made of the Ca^0 to Ca^+ ratio in the interstellar medium to deduce the conditions of electron density and temperature prevailing there. The properties of this medium are further investigated using the observations of the pulsar dispersion measure and the "background" radio recombination line measurements. The problem of the abundances in the interstellar medium is also discussed.

131.027 Beam-maser measurements of the ground-state transition frequencies of OH.
J. J. ter Meulen, A. Dymanus.
Astrophys. Journ., (*Letters*), Vol. 172, L21 - L23 (1972).

Rest frequencies of the four transitions between the $F = 2,1$ hyperfine levels of the $^2\Pi_{3/2}$, $J = 3/2$, Λ-doublet states of OH were determined with a beam-maser spectrometer. The frequencies are accurate to about 100 Hz, and the sum rule is satisfied to within about 100 Hz.

131.028 Predictions on finding the NH and NH$_2$ radicals in interstellar space. B. Kerns, A. B. F. Duncan.
Astrophys. Journ., Vol. 172, 331 - 334 (1972).

Frequencies are calculated for transitions in the NH and NH_2 radicals which are expected to occur in the microwave region usually covered by radioastronomy. NH should be observed in low-temperature sources while NH_2 may be observed in sources whose state temperatures are greater than about $80°K$.

131.029 Survey of molecular lines near the galactic center.
I. 6-centimeter formaldehyde absorption in Sagittarius A, Sagittarius B2, and the galactic plane from $l^{II} = 359°4$ to $l^{II} = 2°2$. N. Z. Scoville, P. M. Solomon, P. Thaddeus.
Astrophys. Journ., Vol. 172, 335 - 353 (1972).

That this line is observed in absorption in all locations indicates a low 6-cm excitation temperature for all molecular clouds. Comparison with 21-cm data indicates that the H_2CO is concentrated in distinct clouds to a much greater extent than atomic hydrogen. There are four dominant clouds occurring at $l^{II} = 0°0$, $0°7$, $0°9$, and $1°7$. In contrast with atomic hydrogen, the strongest formaldehyde features occur at $|V_{LSR}| > 40$ km s^{-1} and are apparently associated with the galactic nucleus. Estimates of the hydrogen density in these clouds indicate a number density in the range of 10^3-10^4 cm^{-3}, and a mass of approximately $5 \times 10^5 M_\odot$.

131.030 A survey of local interstellar hydrogen from OAO-2 observations of Lyman alpha absorption.
B. D. Savage, E. B. Jenkins.
Astrophys. Journ., Vol. 172, 491 - 522 (1972).

The wavelength region near 1216 Å for 69 stars of spectral type B2 or earlier is observed. From the strength of the observed interstellar Lα absorption, column densities of atomic hydrogen were derived over distances averaging 300 pc away from the sun. The OAO data were compared with synthetic ultraviolet spectra, originally derived from earlier higher-resolution rocket observations. An average volume density $n_H = 0.6$ atoms cm^{-3} was derived using all the stars. For stars nearer the sun ($r < 140$ pc) we obtained $n_H = 0.25$ atoms cm^{-3}. There is evidence for a pronounced enrichment of gas toward a number of stars in Scorpius, and a general deficiency seems to exist in the $180° < l^{II} < 270°$ sector. The Lα column densities correlated reasonably well with D-line, H- and K-line, and color-excess measurements for the respective stars.

131.031 Gamma-ray production from p-p reactions in the interstellar medium. D. W. Goldsmith, D. J. Levy.

Bull. American Astron. Soc., Vol. 3, 450 - 451 (1971). – Abstr. AAS.

131.032 H$_2$CO in dust clouds. C. Heiles.
Bull. American Astron. Soc., Vol. 3, 459 (1971). Abstr. AAS.

131.033 Kinematics of dark clouds.
J. M. Greenberg, Y.-K. Minn.
Bull. American Astron. Soc., Vol. 3, 459 (1971). – Abstr. AAS.

131.034 HNCO: The excitation of heavy organic molecules in the interstellar medium.
D. Buhl, L. E. Snyder, J. Edrich.
Bull. American Astron. Soc., Vol. 3, 459 (1971). – Abstr. AAS.

131.035 Observations of CS, HCN, U89.2, and U90.7 in NGC 2264.
B. Zuckerman, B. E. Turner, P. Palmer, M. Morris.
Bull. American Astron. Soc., Vol. 3, 459 (1971). – Abstr. AAS.

131.036 Radiative transport of OH H$_2$O maser emission.
M. M. Litvak.
Bull. American Astron. Soc., Vol. 3, 468 (1971). – Abstr. AAS.

131.037 New microwave H$_2$O sources.
R. H. Rubin, B. E. Turner.
Bull. American Astron. Soc., Vol. 3, 468 (1971). – Abstr. AAS.

131.038 VLBI measurements of the H$_2$O line emission in W49 and Orion A.
J. M. Moran, K. J. Johnston, S. H. Knowles, P. R. Schwartz, G. D. Papadopoulos, B. F. Burke, K. Y. Lo, A. C. Reisz, I. I. Shapiro.
Bull. American Astron. Soc., Vol. 3, 468 (1971). – Abstr. AAS.

131.039 High resolution measurement of the angular size of the water vapor radio sources in W49.
B. F. Burke, L. I. Matveyenko, J. M. Moran, I. G. Moiseyev, S. H. Knowles, B. G. Clark, V. A. Efanov, K. J. Johnston, L. R. Kogan, V. I. Kostenko, K. Y. Lo, D. C. Papa, G. D. Papadopoulos, A. E. E. Rogers, P. R. Schwartz.
Bull. American Astron. Soc., Vol. 3, 468 - 469 (1971). Abstr. AAS.

131.040 Measurement of fractional ionization of interstellar hydrogen toward K3−50.
E. J. Chaisson, L. E. Goad.
Bull. American Astron. Soc., Vol. 3, 471 (1971). – Abstr. AAS.

131.041 The interpretation of diffuse helium recombination lines. M. Jura.
Bull. American Astron. Soc., Vol. 3, 472 (1971). – Abstr. AAS.

131.042 Thermal condensations in cooling interstellar gas.
J. H. Schwarz, R. A. McCray, R. F. Stein.
Bull. American Astron. Soc., Vol. 3, 472 (1971). – Abstr. AAS.

131.043 Pre-main sequence stars. II. Stellar polarization in NGC 2264. M. Breger, H. M. Dyck.
Bull. American Astron. Soc., Vol. 3, 478 (1971). – Abstr. AAS.

131.044 Interstellar motion: Minuet or Rock?

D. G. Wentzel.
Bull. American Astron. Soc., Vol. 3, 497 (1971). – Invited paper.

131.045 Ammonia: Detection of inversion radiation from non-metastable states and new galactic sources.
M. Morris, B. Zuckerman, B. E. Turner, P. Palmer.
Bull. American Astron. Soc., Vol. 3, 499 (1971). – Abstr. AAS.

131.046 Detection of the $1_{11} - 1_{10}$ transition of interstellar formamide.
P. Palmer, C. A. Gottlieb, L. J. Rickard, B. Zuckerman.
Bull. American Astron. Soc., Vol. 3, 499 (1971). – Abstr. AAS.

131.047 On the apparent relationship between high latitude hydrogen cloud structures and the local magnetic field. G. L. Verschuur.
Bull. American Astron. Soc., Vol. 3, 500 (1971). – Abstr. AAS.

131.048 Formation of interstellar molecules on grains in H I clouds. W. D. Watson, E. E. Salpeter.
Bull. American Astron. Soc., Vol. 3, 500 (1971). – Abstr. AAS.

131.049 Amplified spontaneous emission and OH molecules in the interstellar medium.
L. Allen, G. I. Peters.
Nature, Phys. Sci., Vol. 235, 143 - 144 (1972).
Some aspects of the theory of amplified spontaneous emission yield information relevant to the problem of OH emission in the interstellar medium. In particular, problems of dimensions, intensity, saturation and linewidth are discussed in this article.

131.050 Interstellar magnetic field strength and dirty ice grains oriented by paramagnetic relaxation.
G. A. Shah.
Astrophys. Space Sci., Vol. 15, 185 - 194 (1972).
A method of estimating the orientation parameter and the resulting galactic magnetic field consistent with the observed visual ratio of polarization to extinction has been given. The method has been applied to dirty ice grains as an illustration.

131.051 Structure in the interstellar reddening law: 3450–5800 Å. R. K. Honeycutt.
Astron. Journ., Vol. 77, 24 - 28 (1972).
Photoelectric-scanner observations at 20-Å resolution of 11 reddened B supergiants (plus comparison stars) have been used to study the diffuse interstellar absorption bands. In the mean extinction curve we detect absorption features at 5102, 4887, 4760, and 4430 Å. The possible existence of absorption features shortward of 4100 Å is discussed.

131.052 An explanation of the cloudy structure of the interstellar medium. J. G. Hills.
Astron. Astrophys., Vol. 17, 155 - 160 (1972).
The relatively numerous, widely distributed, hot prewhite dwarfs (UV stars) are found to have a profound influence on the interstellar gas. The overlapping Strömgren spheres of these stars result in a gaseous medium in which most of the volume is filled with ionized hydrogen while most of the mass remains un-ionized. The model predicts a monotonic increase in the fraction of the material which is ionized as one progresses towards the galactic center due to the increasing space density of UV stars.

131.053 A comparison between the radial velocities of optical and radio interstellar lines in the southern hemisphere. D. Goniadzki.
Astron. Astrophys., Vol. 17, 378 - 384 (1972).
The velocities of interstellar clouds of the southern hemisphere as determined by optical and radio measurements were studied. Both measurements of each low-velocity (≤ 20 km/s) component refer to the same structure. The optical velocities of high-velocity (> 20 km/s) components do not in general coincide with neutral hydrogen velocities in the direction of the corresponding star.

131.054 Calcium II K-line profiles in front of distant OB stars. J. J. Rickard.
Astron. Astrophys., Vol. 17, 425 - 431 (1972).
A comparison with 21-cm H I line profiles and K-line profiles is made. In the anticenter direction there is good correlation between the shapes of the K-lines and 21-cm lines. The data are consistent with a quiescent interstellar medium showing no peculiar or streaming motions out to approximately 1500 pc from the sun. The high latitude K-lines show many negative velocity features correlated in velocity with the H I gas known as IVC's. The correlation is not as good as at low latitude.

131.055 Interferometer positions of eighteen OH emission sources. E. G. Hardebeck.
Astrophys. Journ., Vol. 172, 583 - 589 (1972).
The two-element interferometer of the Owens Valley Radio Observatory was used to measure the positions of 13 sources in the 1612-MHz line of OH.

131.056 Berkeley survey of high-velocity interstellar neutral hydrogen: II. The section $|b| \geq 15°$. N. H. Dieter.
Astron. Astrophys., Suppl. Ser., Vol. 5, 313 - 368 (1972).
Results of a survey of high-velocity interstellar neutral hydrogen are presented in the form of contour diagrams of antenna temperature as a function of galactic latitude and velocity. This is the second of two sections to the survey and includes observations far from the galactic plane.

131.057 Radio emission at 1400 MHz from galactic H II regions. M. Felli, E. Churchwell.
Astron. Astrophys., Suppl. Ser., Vol. 5, 369 - 432 (1972).
Observations of 130 areas containing 168 optically identified H II regions with a spatial resolution of 10 arc minutes are reported in this paper. Of the 130 regions observed contour maps of 99 regions encompassing 137 Sharpless H II regions are presented here. We were unable to detect radio emission from 31 of the observed regions. In those cases where a radio source could be unambiguously separated from the background emission the integrated flux density is given in addition to the position, maximum brightness temperature, and radio source size.

131.058 Atomic hydrogen observations toward W 3. E. J. Chaisson.
Astron. Astrophys., Vol. 18, 149 - 153 (1972).
Observations of interstellar hydrogen have been made in the direction of the radio nebula W 3. A 21-cm absorption profile is found to be coincident in radial velocity with a well known Cnα recombination line. An unidentified recombination line, if assumed to be an additional Doppler-offset Cnα emission, is shown to have a radial velocity coincident with a second neutral hydrogen absorption feature. Spin temperatures characteristic of three distinct gas condensations are calculated to have values between 35 and 130°K.

131.059 Moleküle im interstellaren Raum. J. Lequeux.
Umschau, 72. Jahrgang, p. 82 - 88 (1972).
During the past few years a considerable number of molecules have been detected in the interstellar space by means of

radio astronomy. Some of them show maser emission. Maybe this interstellar material could be associated with the origin of life.

131.060 Molecules in the interstellar medium.
L. V. Samsonenko.
Uspekhi fiz. nauk, Vol. 105, 363 - 367 (1971). In Russian.
Abstr. in Referativ. Zhurn. 51. Astron., 3.51.688 (1972).

131.061 Les molécules interstellaires. P. Encrenaz.
L'Astronomie, 86e année, p. 105 - 117 (1972).
Presented at the Conférence de la Société Astronomique de France 1971 April 21.

131.062 Interferometric studies of interstellar calcium lines.
L. A. Marschall, L. M. Hobbs.
Astrophys. Journ., Vol. 173, 43 - 62 (1972).
Interferometric, photoelectric scans of the interstellar calcium K-lines in the spectra of 65 stars are presented. The scans were obtained with a PEPSIOS spectrometer having a passband with a full half-intensity width of 1.0 km s^{-1} or 0.013 Å. The fivefold improvement in resolution over that used by Adams reveals numerous line components which correspond very well to those of the interstellar sodium lines, apart from frequent differences in relative intensities.

131.063 Decoupling of magnetic fields in dense clouds with angular momentum. T. Nakano, E. Tademaru.
Astrophys. Journ., Vol. 173, 87 - 101 (1972).
The ionization rate of hydrogen by cosmic rays within dense clouds is calculated in order to investigate the efficiency of ambipolar diffusion in decoupling the magnetic flux from the gas during the collapse of the clouds. The loss of angular momentum via magnetic-dipole radiation of hydromagnetic waves is investigated to estimate the lifetime of clouds whose gravitational, kinetic, and centrifugal forces are in balance. During these "equilibrium" states the cloud may be able to lose sufficient magnetic flux to allow the formation of 1 $M_\odot$ protostars.

131.064 On the circular polarization of HD 226868, NGC 1068, NGC 4151, 3C 273, and VY Canis Majoris. T. Gehrels.
Astrophys. Journ., (*Letters*), Vol. 173, L23 - L25 (1972).
No meaningful amounts of circular polarization were found on NGC 1068, NGC 4151, 3C 273, or on HD 226868 (the star sometimes identified with Cyg X-1). A small amount of circular polarization in the infrared appears to be present in VY CMa.

131.065 The development of research in interstellar absorption, c. 1900–1930. D. Seeley, R. Berendzen.
Journ. History Astron., Vol. 3, 52 - 64 (1972).

131.066 Interstellare Moleküle und der Ursprung des Lebens.
D. Buhl, C. Ponnamperuma.
Sterne, 48. Jahrgang, p. 1 - 10 (1972).

131.067 Processus physiques intéressant les poussières circumstellaires. Masse des grains et accrétion protonique. J.-C. Pecker.
Comptes Rendus Acad. Sci. Paris, Sér. B, Vol. 274, 1001 - 1006 (1972).

131.068 A preliminary classification scheme for interstellar absorbing clouds. S. van den Bergh.
Vistas in astronomy, Vol. 13, (see 003.001), 265 - 277 (1972).

131.069 A possible explanation of the anomalous columnar density of neutral hydrogen in the direction of Orion. W. S. Kovach.
Astrophys. Journ., Vol. 173, 287 - 291 (1972).
The anomalous H^0 columnar density found from Lα and 21-cm measurements in the direction of Orion may be explained on the basis of charged dust grains. We calculate the H^0 columnar density which is required to bring the observations into coincidence and calculate the steady-state number of H^0 atoms on the surface of the grain. Using a range of grain parameters suggested by Hollenbach and Salpeter (1971), we calculate the surface number of H^0 for different grain sizes and nebular conditions.

131.070 The population of high atomic levels at low temperatures. A. K. Dupree.
Astrophys. Journ., Vol. 173, 293 - 300 (1972).
Calculations of b_n coefficients are presented for atomic levels $50 \leq n \leq 250$ under the conditions of low temperature ($T_e = 10°, 20°, 100°, 1000°$K), density ($10^{-3} \leq N_e \leq 10^3$ cm^{-3}), and thermal radiation fields that may be expected in the interstellar medium.

131.071 Abundances of interstellar CH and CH$^+$ radicals. P. Frisch.
Astrophys. Journ., Vol. 173, 301 - 316 (1972).
The spectra of 30 O and B stars were measured for the equivalent widths of the interstellar CH and CH$^+$ lines. Conversion factors are determined between visual estimates of weak line intensities and equivalent widths. Column densities of CH and CH$^+$ are determined with the use of the Strömgren curve of growth. Correlations between line intensity and color excess are searched for. Column densities of CN are determined from equivalent widths measured by Clauser and Thaddeus and by Dunham, and are compared with CH column densities.

131.072 Interstellar formaldehyde. I. The collisional pumping mechanism for anomalous 6-centimeter absorption. P. Thaddeus.
Astrophys. Journ., Vol. 173, 317 - 342 (1972).
"Anomalous" absorption by formaldehyde in diffuse dust clouds—6-cm line absorption in the absence of a continuum source—is generally conceded to result from a pump which maintains the 6-cm excitation temperature below 2°K. A quantum-mechanical treatment is given here of the collisional-pumping process which Townes and Cheung suggest will cool the 6-cm levels by the required amount.

131.073 Structure of the OH/infrared object NML Cygnus.
R. D. Davies, M. R. W. Masheder, R. S. Booth.
Nature, Phys. Sci., Vol. 237, 21 - 24 (1972).
This article describes observations of the OH emitting region around NML Cygnus. The data suggest a model involving a central absorbing cloud ~ 6 X 10^{15} cm in diameter, with the OH emission coming from the outer regions.

131.074 An expanding ring of interstellar gas with center close to the sun.
V. A. Hughes, D. Routledge.
Astron. Journ., Vol. 77, 210 - 214 (1972).
Evidence is given to support the concept of an expanding elliptical ring of dense H I with major and minor axes of 1300 and 560 pc.

131.075 High resolution observations of W 51.
A. H. M. Martin.
Monthly Notices Roy. Astron. Soc., Vol. 157, 31 - 40 (1972).
The northern part of the H II region W 51 has been observed with the Cambridge One-Mile telescope at 2.7 and 5 GHz. Compact components are found whose emission measures agree well with those derived from a non-LTE analysis of recombination line data (Hjellming & Davies). The two brightest condensations coincide with 2 μ infrared sources.

131.076 **Classification of new OH sources.**
D. F. Dickinson, B. E. Turner.
Astrophys. Letters, Vol. 11, 1 - 5 (1972).
　　Main line polarization and satellite line observations have been made on 55 new hydroxyl ion sources. The emission sources can all be described by the OH source categories proposed by Turner.

131.077 **The C^{13} isotope of formaldehyde in Sgr A and Sgr B2.**　J. B. Whiteoak, F. F. Gardner.
Astrophys. Letters, Vol. 11, 15 - 20 (1972).
　　An investigation of the C^{13}H$_2$O and C^{12}H$_2$O absorption near Sgr A and Sgr B2 has been carried out with a beam-width of 4 arc min and filter widths of 33 kHz and 100 kHz.

131.078 **The development of research in interstellar absorption, c. 1900–1930: Part 2.**
D. Seeley, R. Berendzen.
Journ. History Astron., Vol. 3, 75 - 86 (1972).

131.079 **Interaction of hot stars and of the interstellar medium. II. Exciting star and spectra of the bright knot inside the diffuse nebula Sharpless 157.**
M. Chopinet, M. C. Lortet-Zuckermann.
Astron. Astrophys., Vol. 18, 373 - 381 (1972).
　　The bright knot inside the diffuse nebula Sharpless 157 is probably excited mainly by its central star Sh 2–157–1, which has been found to be a late O-type star. The knot has an inhomogeneous brightness distribution. Spectra obtained at the Observatoire de Haute Provence through an image-tube show the ionization structure, a rather low degree of excitation, and emphasize the existence of a bar of high brightness.

131.080 **Galactic shocks in an interstellar medium with two stable phases.**
F. H. Shu, V. Milione, W. Gebel, C. Yuan, D. W. Goldsmith, W. W. Roberts.
Astrophys. Journ., Vol. 173, 557 - 592 (1972).
　　Quasi-steady flows of interstellar gas in a spiral gravitational field are followed for the purpose of investigating galactic shocks and the resultant processes of the formation of stars and interstellar clouds. We model the interstellar medium with two stable phases in which thermal balance is maintained through heating by low-energy cosmic rays. The problem, including transitions between the two phases, is given a general formulation but is solved in an approximation which ignores the difference in fluid velocities of the two phases.

131.081 **Ionization equilibria of calcium and sodium in interstellar clouds.**　R. L. Brown.
Astrophys. Journ., Vol. 173, 593 - 599 (1972).
　　Variations in the ionization equilibria for calcium and sodium as a function of position in interstellar clouds are calculated, including effects due to attenuation of the ultraviolet ionizing radiation by dust grains.

131.082 **Detection of interstellar recombination lines from emitters of intermediate mass.**
E. J. Chaisson, J. H. Black, A. K. Dupree, D. A. Cesarsky.
Astrophys. Journ., (*Letters*), Vol. 173, L131 - L135 (1972).
　　The 18-cm microwave spectra of Orion B and W3A show evidence of an emission feature to the high-frequency side of the carbon recombination line. The new feature originates in a predominantly neutral hydrogen region and can be explained by a superposition of recombination lines from any or all of the following elements: ^{24}Mg, ^{28}Si, ^{32}S, and ^{56}Fe.

131.083 **Polarization observations of the λ4430 diffuse interstellar absorption feature.**　M. F. A'Hearn.
Astron. Journ., Vol. 77, 302 - 305 (1972).
　　Polarization observations have been made of the diffuse interstellar absorption feature at λ4430 using filters to isolate the band and the neighboring continuum in five stars at various galactic longitudes.

131.084 **Profiles of the diffuse interstellar lines.**
G. E. Bromage.
Astrophys. Space Sci., Vol. 15, 426 - 461 (1972).
　　This paper attempts the first rigorous and comprehensive analysis of the classical theory of impurities in homogeneous spheres and coated spheres, as it may be applied to the problem of the diffuse interstellar lines. The theory and computational procedure are described, and the results for spheres and coated spheres are presented. A preliminary extension to small spheroids is discussed.

131.085 **Slipping stream instability of a self-gravitating hydromagnetic gas cloud.**
G. L. Kalra, M. M. Gupta, S. P. Talwar.
Publ. Astron. Soc. Japan, Vol. 24, 261 - 268 (1972).
　　The effect of tangential slippage on the hydromagnetic stability of a self-gravitating interstellar gas cloud is investigated using normal mode analysis.

131.086 **Interstellar motions: Minuet or Rock?**
D. G. Wentzel.
Publ. Astron. Soc. Pacific, Vol. 84, 225 - 239 (1972).
　　This article reviews some of the forces that we now believe control the motions of interstellar gas, and its variations in density and temperature.

131.087 **Polarization of variable stars.**　K. Serkowski.
IAU Colloquium No. 15, (see 012.006), p. 11 - 31 (1972).

131.088 **Polarization of infrared objects.**
A. Kruszewski.
IAU Colloquium No. 15, (see 012.006), p. 32 - 35 (1972).

131.089 **To the explanation of the intrinsic polarization of the light of some red long period variables.**
C. Friedemann.
IAU Colloquium No. 15, (see 012.006), p. 35 (1972). – Abstract.

131.090 **To the explanation of the polarization of the radiation of Mu Cephei.**　C. Friedemann.
IAU Colloquium No. 15, (see 012.006), p. 35 (1972). – Abstract.

131.091 **Polarization of young shell variables.**　M. Breger.
IAU Colloquium No. 15, (see 012.006), p. 40 - 44 (1972).

131.092 **Carbon monoxide observations of dense interstellar clouds.**
A. A. Penzias, P. M. Solomon, K. B. Jefferts, R. W. Wilson.
Astrophys. Journ., (*Letters*), Vol. 174, L43 - L48 (1972).
　　Millimeter observations of the $J = 0$–1 transition of ^{12}C^{16}O have been made in 12 dark interstellar clouds. In four of these we have observed ^{13}C^{16}O lines at approximately one-third the intensity of the ^{12}C^{16}O lines.

131.093 **Infrared emissivities of H$_2$ and HD.**
A. Dalgarno, E. L. Wright.
Astrophys. Journ., (*Letters*), Vol. 174, L49 - L51 (1972).
　　The emissivities in the pure rotational lines of H$_2$ and HD are calculated, and it is suggested that the lines offer a useful means of measuring the interstellar deuterium abundance. The predicted fluxes from the molecular cloud near Sgr B2 seem within the capabilities of present detectors, provided that adequate resolution can be achieved.

131.094   **The influence of the ionized medium on synchrotron emission in interstellar space.**   R. Ramaty.
Astrophys. Journ., Vol. 174, 157 - 163 (1972).

The effect of the ionized gas on synchrotron emission in the interstellar medium is investigated. A detailed calculation of the synchrotron emissivity of cosmic electrons, assumed to have an isotropic pitch-angle distribution in a uniform magnetic field, is made as a function of frequency and observation angle with respect to the field. We treat the theory of synchrotron radiation in a plasma with application to the interstellar medium and compare results with the galactic nonthermal radio background observed in the direction of the anticenter.

131.095   **High-resolution observations of high-velocity neutral hydrogen clouds.**
G. L. Verschuur, T. Cram, R. Giovanelli.
Astrophys. Letters, Vol. 11, 57 - 61 (1972).

The discovery of very small-scale velocity and angular structure in high-velocity neutral hydrogen clouds considerably changes the estimates of the parameters of these clouds and the conditions under which they may be stable entities.

131.096   **Is polyoxymethylene a cosmopolymer?**
R. S. Roche.
Nature, Vol. 235, 217 (1972).

The thermodynamic boundary conditions for the formation of polyoxymethylene from gaseous formaldehyde are derived for the conditions prevailing in the clouds of the interstellar medium. At temperatures below the mean galactic temperature (100°K) polymerisation is thermodynamically allowed for abundance of formaldehyde in the observed range. Polymerisation as a mechanism for grain nucleation is suggested. A central role in chemical evolution is speculatively suggested for formaldehyde. (Note: for an answer to the question "Is polyoxymethylene a cosmopolymer?" see "Occurrence and significance of formaldehyde in the Allende carbonaceous chondrite", I. A. Breger, P. Zubovic, J. C. Chandler, R. S. Clarke, Jr., Nature, Vol. 236, 155 - 158 (1972)).

131.097   **Neutral interstellar hydrogen in the solar neighbourhood.**   F. Macchetto, N. Panagia.
Nature, Phys. Sci., Vol. 235, 56 - 58 (1972).

The interstellar neutral hydrogen density in the solar neighbourhood has been determined from Lyman-α absorption measurements. From the available data only four stars which met some rather stringent requirements were selected. The mean value for the neutral hydrogen density determined from these stars is $N_H = 0.25$ atoms cm^{-3}. The 21-cm results for these stars are also discussed.

131.098   **Why many infrared astronomical sources emit at 100 μm.**
M. Harwit, B. T. Soifer, J. R. Houck, J. L. Pipher.
Nature, Phys. Sci., Vol. 236, 103 - 104 (1972).

Far infrared emission from H II regions seems to be produced by dust grains heated through Ly-α irradiation in a process first discussed by Krishna Swamy and O'Dell for planetary nabulae. Grain temperatures depend weakly on the density of the gas and dust in such a model, and the diffusion of Ly-α further contributes to temperature uniformity. Grains are therefore heated to a temperature in the 40°K range for a wide range of physical conditions. For the galactic center, this model is consistent with the observed emission measure of the extended component of the radio source.

131.099   **Observations of formaldehyde and other molecules in the direction of Bok globules.**
P. Palmer, L. J. Rickard, B. Zuckerman, D. Buhl.
Bull. American Astron. Soc., Vol. 4, 224 - 225 (1972). – Abstr. AAS.

131.100   **Apparent size of OH emission regions.**
P. L. Bender, R. Lang.
Bull. American Astron. Soc., Vol. 4, 225 (1972). – Abstr. AAS.

131.101   **Formation of interstellar molecules via negative ions.**
R. McCray, A. Dalgarno.
Bull. American Astron. Soc., Vol. 4, 225 (1972). – Abstr. AAS.

131.102   **Detection of $J = 2 \rightarrow 1$ rotational transition of interstellar silicon monoxide.**   D. F. Dickinson.
Bull. American Astron. Soc., Vol. 4, 226 (1972). – Abstr. AAS.

131.103   **Observations of the 3.306 mm interstellar emission line and comments on its identification.**
L. E. Snyder, D. Buhl.
Bull. American Astron. Soc., Vol. 4, 227 (1972). – Abstr. AAS.

131.104   **The interstellar reddening law in the ultraviolet deduced from filter photometry obtained by the OAO-II satellite.**   M. Laget.
Bull. American Astron. Soc., Vol. 4, 233 (1972). – Abstr. AAS.

131.105   **Structure in the interstellar reddening law: 3300 - 11,000 Å.**   R. K. Honeycutt.
Bull. American Astron. Soc., Vol. 4, 233 (1972). – Abstr. AAS.

131.106   **Anomalously excited H II regions.**   B. Balick.
Bull. American Astron. Soc., Vol. 4, 235 (1972). Abstr. AAS.

131.107   **The intensity of He II 4686 Å in H II regions and the cosmic-ray flux.**   M. Peimbert, D. Goldsmith.
Bull. American Astron. Soc., Vol. 4, 235 (1972). – Abstr. AAS.

131.108   **Effects of charge exchange reactions on the ultraviolet absorption-line spectra of H I regions.**
R. L. Brown.
Bull. American Astron. Soc., Vol. 4, 242 (1972). – Abstr. AAS.

131.109   **High resolution observations of two very cold H I clouds.**   G. L. Verschuur, G. R. Knapp.
Bull. American Astron. Soc., Vol. 4, 242 (1972). – Abstr. AAS.

131.110   **Possible production of Li, Be and B by low energy cosmic rays.**   J. Audouze.
Bull. American Astron. Soc., Vol. 4, 257 (1972). – Abstr. AAS.

131.111   **Interstellar material ionized by supernovae.**
M. C. Kafatos.
Bull. American Astron. Soc., Vol. 4, 269 (1972). – Abstr. AAS.

131.112   **The wavelength dependence of interstellar polarization in the direction of ζ Ophiuchi.**
T. P. Stecher.
Bull. American Astron. Soc., Vol. 4, 270 (1972). – Abstr. AAS.

131.113   **Interstellar particles and the evolution of the universe.**   Ts. Tsvetkov.
Fiz.-matem. spisanie, Vol. 14, No. 3, p. 187 - 197 (1971). In Bulgarian.

131.114   **Formation of interstellar molecular hydrogen.**
T. J. Lee.
Nature, Phys. Sci., Vol. 237, 99 - 100, 112 (1972).

Combination of hydrogen atoms on ice-coated grains can take place at temperatures above 20 K in the absence of enhanced binding sites.

131.115   **On the intercloud H I-gas.**   U. Mebold.
Astron. Astrophys., Vol. 19, 13 - 26 (1972).

In the present paper some evidence is presented in favour of a two-component interstellar H I-gas and a distribution of the hot part of that gas over a scale comparable to the thickness of our galaxy. About 1300 21-cm line profiles, observed at $b \approx 30°$ and $0° \leqq l \leqq 360°$, have been decomposed into about 2400 gaussian components. Furthermore the density distribution of the hot H I-gas perpendicular to the galactic plane is investigated and presented in an analytical form.

131.116 Interstellar reddening in the north galactic polar cap. II. A cloud model. W. Pfau.
Astron. Nachr., Vol. 293, 275 - 280 = Mitt. Univ.-Sternw. Jena No. 114 (1972).

A statistical treatment of the colour excesses obtained in a previous paper suggests the existence of a system of small interstellar dust clouds (diameter about 1 pc) with large extension (about 1 kpc) from the galactic plane. Other cloud parameters were derived and related observations (interstellar absorption lines, interstellar neutral hydrogen, and polarization data) discussed.

131.117 The absorption feature at 2200 Å in the interstellar reddening curve. K. S. K. Swamy.
Astrophys. Space Sci., Vol. 16, 75 - 80 (1972).

The profile of the absorption feature at 2200 Å has been calculated for model grains of graphite, graphite core-dirty ice mantle and silicate. They are compared with the observed profile obtained by Bless and Savage from a number of early type stars.

131.118 Backscatter of solar resonance radiation – I. H. E. Johnson.
Planet. Space Sci., Vol. 20, 829 - 840 (1972).

A simple model which essentially neglects temperature effects, but which includes gravity, radiation pressure, photoionization and charge exchange, is used to calculate the angular dependence of the intensity of solar Lyman α resonantly scattered from neutral interstellar hydrogen which has penetrated the solar system; the results are then compared with observations. The resonant scattering of He I $\lambda 584$ is also treated.

131.119 The neutral interstellar medium. H. J. Habing.
Mitt. Astron. Ges., No. 31, p. 61 - 69 (1972).
Review article.

131.120 Statistics of absorbing H I concentrations in the galactic disk. W. M. Goss, V. Radhakrishnan.
Mitt. Astron. Ges., No. 31, p. 100 (1972). – Abstract.

131.121 Über den thermischen Zustand der interstellaren Materie. U. Vogel.
Mitt. Astron. Ges., No. 31, p. 201 - 204 (1972).

131.122 Ionisation equilibrium and line intensities for an X-ray heated H I gas.
M. Grewing, M. Walmsley.
Mitt. Astron. Ges., No. 31, p. 204 - 208 (1972).

131.123 Die Dichteverteilung des Zwischenwolken-H I-Gases. U. Mebold.
Mitt. Astron. Ges., No. 31, p. 220 - 224 (1972).

131.124 Observations of sources of maser radio emission with angular resolution 0."0002.
B. F. Burke, K. J. Johnston, V. A. Efanov, B. J. Clark, L. R. Kogan, V. I. Kostenko, K. Y. Lo, L. I. Matveyenko, I. G. Moiseev, J. M. Moran, S. H. Knowles, D. C. Papa, G. D. Papadopoulos, A. E. E. Rogers, P. R. Schwartz.
Astron. Zhurn. Akad. Nauk SSSR, Vol. 49, 465 - 469 (1972). In Russian. English translation in Soviet Astron. AJ, Vol. 16, No. 3.

The sizes of the water vapour radio sources have been measured with a radio interferometer having Simeis–Haystack as baseline. A highest resolution of 0."0002 is obtained.

131.125 The interaction of stars with local dust formations.
V. A. Arshynova, M. M. Dagaev, V. V. Radzievsky.
Astron. Zhurn. Akad. Nauk SSSR, Vol. 49, 524 - 532 (1972). In Russian. English translation in Soviet Astron. AJ, Vol. 16, No. 3.

The re-distribution of the dust density of initially homogeneous cosmic dust clouds by an encounter with hot stars is studied. This density as a function of spherical coordinates is obtained separately for particles moving to the star and for particles having been repulsed from the boundary. It is shown that a deformation of the cloud front approaching the star can be one of the causes of the parabolic losses that are seen at the sides of some nebulae. This problem is solved and the effects of particle collisions and viscous friction of the gas medium are discussed.

131.126 A possible cause of the intensity variations of the interstellar maser. V. S. Strelnitsky.
Astron. Zhurn. Akad. Nauk SSSR, Vol. 49, 649 - 652 (1972). In Russian. English translation in Soviet Astron. AJ, Vol. 16, No. 3.

131.127 Radio recombination lines in H I regions.
I. V. Gosachinsky.
Astron. Zhurn. Akad. Nauk SSSR, Vol. 49, 658 - 661 (1972). In Russian. English translation in Soviet Astron. AJ, Vol. 16, No. 3.

It is shown that the H 157 α radio recombination line, detected in NGC 2024, may be originated in an H I region, only if the source of ionization is ten times more powerful than in ordinary galactic H I regions, according to measurements of the H 271 α recombination line.

131.128 Molecule formation on interstellar grains.
W. D. Watson, E. E. Salpeter.
Astrophys. Journ., Vol. 174, 321 - 340 (1972).

Processes relevant to the formation of molecules (except H_2) on the surface of interstellar dust-grains in H I regions are discussed. The chief purpose of the present paper is to investigate two topics: (i) nonequilibrium evaporation mechanisms for adsorbed particles and (ii) formation of different types of molecules on grain surfaces.

131.129 Time-dependent models of the interstellar gas.
M. Jura, A. Dalgarno.
Astrophys. Journ., Vol. 174, 365 - 376 (1972).

Time-dependent models are presented of the evolution of the interstellar gas following a supernova flash and the results of detailed calculations of the cooling and recombination of the initial hot ionized plasma are described. Two kinds of models reproduce the gross properties of the intercloud regions.

131.130 The variability of the OH source W3.
A. J. Wilson, R. D. Davies, J. Elldèr.
Monthly Notices Roy. Astron. Soc., Vol. 157, 21P - 26P (1972).

Variability has been detected in the 1665 MHz spectrum of the W3 OH source. Observations extending back to 1965 show an average decrease of 8.9 ± 0.9 per cent per year in the -41.7 km s^{-1} feature. Irregular variability on the time scale of a few years is also found. The implications of these observations are discussed.

131.131 Optical depths of neutral hydrogen clouds determined by polarization techniques.
F. F. Gardner, J. B. Whiteoak.
Astrophys. Letters, Vol. 11, 123 - 128 (1972).

Observations have been made of the hydrogen-line-absorption of the linearly-polarized continuum of five radio sources. With the polarization method it is possible to determine at each position the optical depth and the 'expected emission profile', provided that the direction of polarization and the opacity do not both vary across the beam. The results at eleven positions across Centaurus A yield optical depths varying from 0.27 to 0.11, with lower values over the polarized NE component of the central double continuum source.

131.132 A search for neutral hydrogen high-velocity clouds in the directions of six globular clusters.
F. J. Kerr, G. R. Knapp.
Astron. Journ., Vol. 77, 354 - 359 (1972).

A search was carried out for high- and intermediate-velocity H I line emission in the directions of six globular clusters whose spectra show high-velocity interstellar Ca II K-line components. In only one out of the six directions, towards M10, was high-velocity H I emission found. A small hydrogen cloud at a velocity similar to that of the calcium line was found a degree away from the cluster M92.

131.133 Interstellar CN excitation at 2.64 mm.
A. A. Penzias, K. B. Jefferts, R. W. Wilson.
Phys. Rev. Letters, Vol. 28, 772 - 775 (1972).

A sensitive search was made for 2.64-mm line emission from a cloud of CN, whose excitation is known from optical measurements, with essentially a null result. This provides strong support for the proposition that the excitation temperature deduced from the optical CN lines is equal to the temperature of the microwave background.

131.134 The charge-transfer reaction $C^{+2} + He \rightarrow He^+ + C^+$ and its application to the interstellar medium.
R. L. Brown.
Astrophys. Journ., Vol. 174, 511 - 516 (1972).

The rate coefficient for the charge-transfer reaction $C^{+2} + He \rightarrow C^+ + He^+$ is evaluated at thermal energies $T = 10^{2°} - 10^{4°} K$ within the framework of the "orbiting approximation." The effect of this process on the ratio of number densities $n(C^{+2})/n(C^+)$ expected in the interstellar medium is used to show that the C III $\lambda 977$ Å line will be produced preferentially in H I clouds with temperatures $\lesssim 300°K$.

131.135 Interstellar scattering and the sizes of astronomical masers.
R. W. Boyd, M. W. Werner.
Astrophys. Journ., (Letters), Vol. 174, L137 - L140 (1972).

We explore the suggestion that scattering by inhomogeneities in the distribution of electrons causes astronomical masers to appear larger than their actual sizes. It is shown that neither scattering near the source nor scattering in the intervening medium is sufficiently strong to produce this effect.

131.136 Searches for microwave spectral line radiation from some molecules in the interstellar medium.
T. Cato, J. Elldér, B. Höglund, O. E. H. Rydbeck, B. Rönnäng, A. Sume.
Onsala Space Res. Obs., Res. Lab. Electronics, Chalmers Univ. Techn., Gothenburg, Sweden, Res. Rep. No. 109, 22 pp. (1972).

The Onsala 25.6 m radio telescope equipped with travelling wave maser radiometers at the appropriate frequencies has been used in searches for absorption or emission lines from a number of different organic and inorganic molecules in the direction of several radio sources. The molecules include OH, CH, SO_2, HCN, H_2CS, NH_2CHO, and CH_3CHO. All searches gave negative results, with possible exception for H_2CS. Details of the observations are tabulated.

131.137 Circular polarization of Omicron Scorpii: Possible interstellar origin. J. C. Kemp.
Astrophys. Journ., (Letters), Vol. 175, L35 - L37 (1972).

Definite circular polarization has been discovered in light from o Sco. The maximum observed magnitude was at $\lambda 4300$, where $q = -(4.1 \pm 1.0) \times 10^{-4}$. One explanation is a varying grain-alignment direction along the line of sight.

131.138 Interstellar $^{12}C/^{13}C$ ratios toward six stars.
L. M. Hobbs.
Astrophys. Journ., (Letters), Vol. 175, L39 - L41 (1972).

Carbon isotope ratios $a \equiv N(^{12}C)/N(^{13}C)$ are measured for interstellar CH^+ molecules toward six stars, using the $\lambda 4232$ line. Negative results of high sensitivity are obtained for the $^{13}CH^+$ line in all cases, yielding lower limits ranging from $a > 20$ to $a > 77$.

131.139 Detection of silicon monoxide at 87 GHz.
D. F. Dickinson.
Astrophys. Journ., (Letters), Vol. 175, L43 - L46 (1972).

The $J = 2 \rightarrow 1$ rotational transition of SiO at 86,847 MHz has been found in Sgr B2 and Ori A. The column density in Sgr B2 is calculated to be $\sim 4 \times 10^{13}$ cm^{-2}, and the excitation temperature, $\sim 10°K$.

131.140 Interstellar organic chemistry. C. Sagan.
Nature, Vol. 238, 77 - 80 (1972).

131.141 Interstellar molecules. W. Buscombe.
Journ. Astron. Soc. Victoria, Vol. 25, 11 (1972).

131.142 Radio detection of interstellar formaldimine.
P. D. Godfrey, R. D. Brown, B. J. Robinson, M. W. Sinclair.
IAU Circ., No. 2410 (1972).

131.143 Selected topics on the physics of the interstellar medium. G. Münch.
Galactic astronomy, Vol. 1, (see 012.019), 191 - 243 (1970).

131.144 The energy balance of the interstellar medium.
L. Woltjer.
Galactic astronomy, Vol. 2, (see 012.019), 161 - 169 (1970).

131.145 Spectral lines in radioastronomy. G. de Jager.
Monthly Notes Astron. Soc. Southern Africa, Vol. 31, 50 - 51 (1972). – Abstract.

131.146 Des voiles obscurs dans Orion. R. de Terwangne.
Gaz. astron., Antwerpen, No. 3, p. 5 - 11 (1972).

131.147 Cosmic clouds in Orion. R. de Terwangne.
Gaz. astron., Antwerpen, No. 3, p. 12 - 14 (1972).

131.148 Polarization of Be stars at high galactic latitudes.
G. V. Coyne.
Ric. Astron. Specola Vaticana, Castel Gandolfo, Vol. 8, 201 - 210 (1971).

A reconnaissance was made of the polarization in 28 Be stars at galactic latitudes greater than ± 10°. For fifteen of these stars which have polarizations larger than 0.3 % the detailed wavelength dependence of the polarization in 7 colors was studied.

131.149 Interstellar molecules and the origin of life.
D. Buhl, C. Ponnamperuma.
Space Life Sci., Vol. 3, 157 - 164 = National Radio Astron. Obs., Green Bank, Repr. Ser. B, No. 310 (1971).

131.150 Neuere Vorstellungen über den physikalischen Zustand und die Verteilung des interstellaren Mediums.
K. Rohlfs.
Kleinheubacher Berichte, Vol. 14, 85 - 94 = Max-Planck-Inst.

Radioastronomie Bonn, Sonderdruck No. 30 (1971).

This paper gives a review on the observations and the theoretical arguments which, in the last years, have led to a major revision of the generally adopted views of the physical conditions in interstellar space.

131.151 **Interstellar reddening within 200 pc of the sun.**
B. Strömgren.
Quarterly Journ. Roy. Astron. Soc., Vol. 13, 153 - 160 (1972). − Presented at the Woolley symposium (see 012.025).

131.152 **Phase transition in the interstellar medium.**
P. Biermann, R. Kippenhahn, W. Tscharnuter, H. Yorke.
Astron. Astrophys., Vol. 19, 113 - 122 (1972).

Hot interstellar matter heated by cosmic rays is forced out of thermal equilibrium by a spiral arm density wave. This process has been followed by a simple numerical model of thermal evolution. The transition to the cool stable cloud phase may occur within 10^6 years. Thus stars may be formed within approximately 5×10^6 years after the matter has passed through the spiral arm density wave. This agrees with the determination of this time in M 51 by Mathewson et al. (1971).

131.153 **The physical state of nearby interstellar space and the distances and space density of pulsars.**
L. Biermann.
Accad. Nazionale Lincei (Anno 369), Quaderno N. 162, p. 311 - 316 = Separate print Inst. Phys. Astrophys. München (1972).

131.154 **Three investigations of the interstellar medium.**
K. Schocken.
Rep. NASA−TM−X−64577, National Aeronautics and Space Administration, Huntsville, Ala. [Available from NTIS, Springfield, Va.], 30 pp. (1971). − See Phys. Abstr., Vol. 75, No. 20606 (1972).

131.155 **Laboratory work aids interstellar molecule detection.**
National Bureau of Standards, Techn. News Bull., Vol. 56, No. 2, p. 27 (1972).

Reports on the generation and characterization of the thioformaldehyde molecule from its gas phase rotational spectrum using microwave techniques. This molecule has been detected in space.

131.156 **Strong polarization shocks in the interstellar space.**
V. C. Liu, D. Bastos-Netto.
Phys. Letters A, (*Netherlands*), Vol. 39A, 305 - 307 (1972).

The structure of infinitely strong shocks in a slightly ionized gas, suitable for the interstellar HI region, is discussed. The charge-separation field, in closed form, is obtained by using a bi-modal (beam-continuum) distribution and the Mott-Smith approach.

131.157 **Atomic and molecular reactions in space.**
H. S. W. Massey.
From plasma to planet. 21st Nobel symposium 1971, (see 012.028), p. 17 - 37 (1972).

131.158 **Interstellar molecules.** G. R. Carruthers.
Astronaut. Aeronaut. (*USA*), Vol. 9, No. 4, p. 16 - 19, 79 - 80 (1971).

131.159 **Etude des milieux dilués hors équilibre thermodynamique.** J. Bergeron.
Thesis, Univ. Paris. Centre Documentation, C. N. R. S. (1972-04-11), 173 pp. (1972). − See Bull. Signal., Vol. 33, Section 120, No. 7100 (1972).

The broadening of radio recombination lines by electron collisions. See Abstr. 022.001.

RKR Franck−Condon factors for blue and ultraviolet transitions of some molecules of astrophysical interest and some comments on the interstellar abundance of CH, CH$^+$ and SiH$^+$. See Abstr. 022.051.

A low-lying resonance in the spectrum of H$^-$ II. The 2^1P state. See Abstr. 022.085.

Laboratory measurement of the millimeter-wavelength spectrum of formaldehyde. See Abstr. 022.092.

The Parkes survey of 21-centimeter absorption in discrete-source spectra. I. The Parkes hydrogen-line interferometer. See Abstr. 033.001.

Ein Fabry-Perot-Interferometer zur Messung von Radialgeschwindigkeiten an HII-Regionen. See Abstr. 034.058.

Variation of single particle mid-infrared emission spectrum with particle size. See Abstr. 061.011.

Near resonant charge transfer processes at thermal energies. See Abstr. 061.040.

Time dependent radiative transfer. Damping of a temperature fluctuation. II. The two-level atom case. Applications to the interstellar medium. See Abstr. 063.005.

Small molecules in astrophysics: in stellar atmospheres, in comets, and in interstellar space. See Abstr. 064.044.

The collapse of a rotating cloud. See Abstr. 065.076.

The influence of local conditions in the interstellar medium upon star formation. See Abstr. 065.115.

Kollaps einer rotierenden Staubwolke. See Abstr. 066.059.

Origin of organic matter in early solar system−VI. Catalytic synthesis of nitriles, nitrogen bases and porphyrin-like pigments. See Abstr. 105.027.

Mikrowellen-Analogieversuche zur Lichtstreuung an kosmischen Staubpartikeln. See Abstr. 106.017.

Effects of reddening on *UBV* colour transformations. II. See Abstr. 113.003.

Interstellar absorption in the direction of Maffei 1 and 2. See Abstr. 113.015.

Infrared observations of some southern IR−OH sources. See Abstr. 113.017.

Infrared observations of 1612 MHz IR/OH sources. See Abstr. 114.014.

Infrared objects in H II regions. See Abstr. 114.027.

The nature of the Herbig Ae- and Be-type stars. See Abstr. 114.038.

A l-micron search for young stellar objects. See Abstr. 114.039.

Photoelectric measurements of the 6284 Å diffuse interstellar line.   See Abstr. 114.116.

Zur Erklärung der Polarisation der Strahlung von μ Cephei.   See Abstr. 122.007.

The effect of the interstellar gas on the continuum spectrum of 3C 391.   See Abstr. 125.025.

Observations of recombination lines at 16.5 GHz. See Abstr. 132.002.

Internal kinematics of the 30 Doradus nebula in the Large Magellanic Cloud.   See Abstr. 132.008.

High-resolution interferometric observations of the η Carinae nebula.   See Abstr. 132.022.

Pre—main-sequence stars. II. Stellar polarization in NGC 2264 and the nature of circumstellar shells. See Abstr. 132.035.

Low-energy X-rays ruled out as interstellar ionizing mechanism toward K3–50.   See Abstr. 133.005.

Infrared photometry of the H II region Sharpless 266. See Abstr. 133.016.

Interferometry: Effects of irregular propagation media.   See Abstr. 141.142.

Observations of the 21-cm hydrogen-emission line in the direction of 25 pulsars.   See Abstr. 141.529.

Pulse broadening due to multiple scattering in the interstellar medium.   See Abstr. 141.537.

Cosmic-ray source and local interstellar spectra deduced from the isotopes of hydrogen and helium. See Abstr. 143.027.

On the possible role of the interstellar gas in the annual variations of cosmic rays.   See Abstr. 143.031.

H II regions in the Sco OB 1 association. See Abstr. 152.007.

OH observations of the open cluster NGC 2264.

See Abstr. 153.012.

Vergleich der Massenanteile von interstellarem Wasserstoff, Staub und Sternen in offenen Sternhaufen. See Abstr. 153.025.

The high velocity clouds as part of "normal" spiral structure.   See Abstr. 155.015.

Neutral hydrogen self-absorption in a large region toward the galactic center.   See Abstr. 155.025.

Early-type stars and the mean-square electron density in the galactic disk.   See Abstr. 155.042.

A test for relative motions of gas and young stars. See Abstr. 155.043.

Galactic helium abundances. See Abstr. 155.063.

Distribution of neutral hydrogen in the Perseus arm.   See Abstr. 155.066.

Radio frequency absorption line studies of galactic structures.   See Abstr. 155.077.

A study of interstellar reddening in a region in Ophiuchus.   See Abstr. 155.078.

Galactic background at 15 GHz. See Abstr. 155.088.

A survey of the continuum radiation at 820 MHz between declinations –7° and +85°. I. Observations and reductions.   See Abstr. 157.004.

Beobachtungen der H- und He-Radiorekombinationslinien in der Umgebung des galaktischen Zentrums. See Abstr. 157.013.

Errata

131.901   Erratum: 'Radio emission at 1400 MHz from galactic H II regions' [Astron. Astrophys., Suppl. Ser., Vol. 5, 369 - 432 (1972)].
M. Felli, E. Churchwell.
Astron. Astrophys., Suppl. Ser., Vol. 6, 199 - 200 (1972).

132 Emission Nebulae, Reflection Nebulae

132.001 The population of helium triplet states in gaseous nebulae. G. W. F. Drake, R. R. Robbins.
Astrophys. Journ., Vol. 171, 55 - 61 (1972).

Several authors have found that in planetary nebulae the population of helium atoms in the metastable $1s2s\ ^3S$ state calculated by balancing the theoretical rates of formation and and destruction is one or two orders of magnitude greater than that deduced from measured intensity ratios. We reexamine the problem, using updated atomic data and including additional triplet-depopulation mechanisms.

132.002 Observations of recombination lines at 16.5 GHz.
G. D. Papadopoulos, K. Y. Lo, P. Rosenkranz, E. J. Chaisson.
Astrophys. Letters, Vol. 10, 89 - 92 (1972).

Measurements of recombination lines of excited hydrogen in Orion A, W3, W49 and W51 have been made at 16.56 GHz. Assuming local thermodynamic equilibrium, mean electron temperatures are found to be 11000 K for each nebula.

132.003 The nebulosity surrounding the galactic cepheid RS Puppis. R. J. Havlen.
Astron. Astrophys., Vol. 16, 252 - 267 (1972).

The circularly symmetric, ring structured reflection nebula surrounding the 41.4 day period galactic cepheid RS Puppis has been investigated for spatial and time variations in its structure and light intensity. Isodensitracings of several nebular features show no spatial variations but do show variations in light intensity with a period commensurate with that of the cepheid. The nebular arcs are interpreted physically as higher density regions of reflecting material surrounding the cepheid in space. Further, they are incorporated in a more general spherically symmetric model of reflecting shells. By using this model, artificially constructed nebular light and color curves have been fitted to the observations. The formation and evolution of the cepheid-nebula system is discussed.

132.004 Spectrophotometric investigations of filamentary nebulae.
V. F. Esipov, S. A. Kaplan, T. A. Lozinskaya, T. S. Podstrigach.
Astron. Zhurn. Akad. Nauk SSSR, Vol. 49, 105 - 109 (1972). In Russian. English translation in Soviet Astron. AJ, Vol. 16, No. 1.

Spectrograms of separate filaments of supernova remnants S-147 and VRO 42.05.01 were obtained using a contact image converter spectrograph. Intensity ratios I [NII] : I [Hα], I [SII] : I [Hα], I [OIII] : I [Hβ] were measured. The spectrum of optical emission of a shock wave in partly ionized gas was calculated as function of propagation velocity.

132.005 Investigation of several nebulae in Cassiopeia with a Fabry–Perot standard.
V. T. Doroshenko, N. I. Grachev.
Astron. Zhurn. Akad. Nauk SSSR, Vol. 49, 110 - 114 (1972). In Russian. English translation in Soviet Astron. AJ, Vol. 16, No. 1.

A pressure-scanned and a photographic Fabry–Perot interferometer have been used to determine profiles of Hα, Hβ emission lines in a number of nebulae in Cassiopeia. In addition the radial velocities and the halfwidths are presented.

132.006 Observational evidence of collisional excitation in two diffuse nebulae.
R. H. Cromwell, B. T. Lynds.
Astrophys. Journ., Vol. 171, 279 - 284 (1972).

Two reflection nebulae having unusual emission filaments near their edges have been found to have very steep Balmer decrements of the first four Balmer lines. Line ratios of the emission filaments agree well with the theoretical ratios predicted by Parker on the theory of collisional excitation and ionization.

132.007 On the time dependence of emission-line strengths from a photoionized nebula.
J. N. Bahcall, B.-Z. Kozlovsky, E. E. Salpeter.
Astrophys. Journ., Vol. 171, 467 - 482 (1972).

Discussions are given for the time dependence of emission-line intensities from nebulae that are photoionized by a central source with a variable continuum flux. Explicit calculations are carried out for some cases (especially when the nebula can be considered a thin shell) for a temporary increase of continuum flux of simple "square-wave hump" shape. The observational situation is described and further measurements are suggested for a number of objects that are relatively easy to study and whose further investigations seem likely to lead to important conclusions regarding the nebular parameters. These promising objects include QSOs, and N-type, Seyfert, and compact galaxies, as well as some X-ray sources and old novae.

132.008 Internal kinematics of the 30 Doradus nebula in the Large Magellanic Cloud.
M. G. Smith, D. W. Weedman.
Astrophys. Journ., Vol. 172, 307 - 317 = Repr. Arthur J. Dyer Obs., Vanderbilt Univ., *Nashville, Tennessee*, No. 63 (1972).

200 emission-line profiles and radial velocities are observed. Radial-velocity differences of up to 80 km s^{-1} are found between different components of the nebula. The most probable velocity of gas motions is 25 km s^{-1}, almost twice as rapid as in any other LMC H II region. It is concluded that the rapid internal motions are produced by sources within the nebula, probably by the large concentration of Wolf-Rayet stars. The loops and filaments of gas seen in the nebula appear to move coherently.

132.009 VY Canis Majoris. III. Polarization and structure of the nebulosity. G. H. Herbig.
Astrophys. Journ., Vol. 172, 375 - 381 = Contr. Lick Obs., No. 343 (1972).

Direct photographs of VY CMa taken in the 6500 Å region through a Polaroid analyzer show that the small nebula surrounding the star is radially plane-polarized in amounts up to 70 percent. The semistellar nuclei C, D, E, and F in the nebula that were discovered and measured by micrometric observers show the same heavy radial polarization as their nebular background.

132.010 An OH-optical-depth map of the Carina nebula.
H. R. Dickel, J. V. Wall.
Bull. American Astron. Soc., Vol. 3, 459 - 460 (1971). – Abstr. AAS.

132.011 Radio emission from small galactic nebulae.
Y. Terzian, V. Pankonin.
Bull. American Astron. Soc., Vol. 3, 471 (1971). – Abstr. AAS.

132.012 Indirect evidence for the presence of dust in the QSO PHL 5200.
P. D. Noerdlinger, J. D. Scargle, L. J. Caroff.
Bull. American Astron. Soc., Vol. 3, 474 (1971). – Abstr. AAS.

132.013 The He I singlet spectrum in nebulae.
R. R. Robbins.
Bull. American Astron. Soc., Vol. 3, 500 (1971). – Abstr.
AAS.

132.014 A study of Hα and [N II] profiles and ratios in M8 and M42. M. A. Dopita.
Astron. Astrophys., Vol. 17, 165 - 171 (1972).
Temperatures and turbulent motions of these nebulae were determined. The Hα/[N II] brightness ratios were found for 10 positions across the bright core of M8. A lower limit for the abundance of nitrogen has been obtained.

132.015 On the coincidence of OH emission sources and bright knots in H II regions.
H. J. Habing, F. P. Israel, T. de Jong.
Astron. Astrophys., Vol. 17, 329 - 334 (1972).
The H II region NGC 7538 is discussed. The nebula consists of a shell of ionized gas and an associated point source of 0.85 flux units. The point source coincides to within $10''$ with a class I OH emission source, of which an accurate position was recently determined by Hardebeck. Similar coincidences in other H II regions have been reported. A preliminary physical model is suggested.

132.016 Image-tube photography of diffuse nebulae in [S III] λ9532. P. Foukal.
Astrophys. Journ., Vol. 172, 591 - 592 (1972).
Exposures of M17 and the central core of M8 show the potential usefulness of the [S III] λ9532 radiation in the study of emission nebulae. The line is of particular interest in studies of the obscured central region of many of the bright northern objects.

132.017 Radio emission from small galactic nebulae at 606 MHz. Y. Terzian. V. Pankonin.
Astron. Journ., Vol. 77, 115 - 119, 191 - 195 (1972).
A survey for radio emission from small galactic nebulae is presented at 606 MHz. The observations were made with the 1000-ft radio telescope of the National Astronomy and Ionosphere Center (Arecibo Observatory). Of the 85 optical nebulae surveyed 23 have been identified to emit radio radiation. Twentyfive other radio sources have also been detected close to the positions of the optical nebulae, but could not reliably be identified with the optical objects.

132.018 On the temperature and density problem in galactic nebulae. M. Perinotto.
Mem. Soc. Astron. Italiana, Nuova Ser., Vol. 43, 95 - 103 (1972).
Recent computations of transition probabilities and collision strengths are considered in connection with the interpretation of forbidden line emissions in galactic nebulae. The value of using forbidden line intensity ratios for determining the physical conditions in a nebula is outlined. A study of the center of the Orion nebula shows a tendence for the electron density to increase with the ionization potential of the emitting atom.

132.019 Far infrared observations of M 42, NGC 2024 and M 1. I. Furniss, R. E. Jennings, A. F. M. Moorwood.
Nature, Phys. Sci., Vol. 236, 6 - 7 (1972).
Broadband photometric measurements $(40 \mu - 350 \mu)$ of M 42 and NGC 2024 have been made from balloon altitude. Fluxes of 106×10^{-14} watts cm^{-2} and 25×10^{-14} watts cm^{-2} respectively are reported. Observations of M 1 set an upper limit of 2×10^{-14} watts cm^{-2} for the flux in this region, placing restrictions on the shape of its spectrum in the far infrared.

132.020 The line spectra of helium in gaseous nebulae.

M. Brocklehurst.
Monthly Notices Roy. Astron. Soc., Vol. 157, 211 - 227 (1972).
The level populations of neutral helium are calculated for conditions typical of gaseous nebulae. Full collisional redistribution of angular momentum and energy is included, and accurate helium atomic data are used throughout. Intensities of the most important singlet and triplet line series are presented for $n \leqslant 25$.

132.021 Observations of the 3727 Å [O II] doublet in the Orion nebula. J. G. Caplan.
Astron. Astrophys., Vol. 18, 408 - 414 (1972).
We describe a procedure for reducing the photon-counting data obtained from Fabry-Perot scans of the density-sensitive [O II] doublet at 3727 Å in nebulae. Observations are presented for the Orion nebula, obtained with a one-meter siderostat, with a $2'.1$ exploring diaphragm.

132.022 High-resolution interferometric observations of the η Carinae nebula. R. Louise.
Astron. Astrophys., Vol. 18, 475 - 476 (1972). In French.
High-resolution interferometric observations were made on the η Car nebula at the ESO Observatory in the southern hemisphere (La Silla, Chile) during the period of February– March 1970. Line profiles of Hα (λ = 6563 Å) and [N II] (λ = 6584 Å) are derived from these observations at 31 points of the nebula.

132.023 Excitation of nebular spectrum lines. J. B. Kaler.
Astrophys. Journ., Vol. 173, 601 - 609 (1972).
Evidence is presented which supports Seaton's suggestion that permitted lines in gaseous nebulae, other than those of hydrogen and helium, can be produced by direct photoexcitation by the central star rather than by recombination. Observations are presented for three nebulae in which O III lines are found but for which the He II lines are not present. This finding contradicts recombination theory. A strong correlation between the ratio of the strengths of the λ3918 + λ3920 and the λ4267 doublets of C II and the temperature of the exciting star is established.

132.024 Observations of diffuse nebulae in the λ9532 forbidden line of S III. P. Foukal.
Bull. American Astron. Soc., Vol. 4, 233 (1972). – Abstr. AAS.

132.025 An 85α recombination line survey of the Orion nebula.
L. H. Doherty, L. A. Higgs, J. M. MacLeod.
Bull. American Astron. Soc., Vol. 4, 242 (1972). – Abstr. AAS.

132.026 Unusual aspects of IC 410. J. Cuffey.
Bull. American Astron. Soc., Vol. 4, 242 - 243 (1972). – Abstr. AAS.

132.027 The infrared spectrum of NGC 6572: Preliminary results. R. P. Kovar, N. S. Kovar, A. E. Potter, L. Trafton.
Bull. American Astron. Soc., Vol. 4, 269 (1972). – Abstr. AAS.

132.028 The core of Eta Carinae. R. D. Gehrz, E. P. Ney.
Sky Telescope, Vol. 44, 4 - 5 (1972).

132.029 Beobachtungen des Rosetten Nebels bei 6 cm Wellenlänge. W. J. Altenhoff, T. L. Wilson.
Mitt. Astron. Ges., No. 31, p. 100 (1972). – Abstract.

132.030 Weitere Bemerkungen zu der inneren Massenstruktur im Orionnebel. K. Wurm.
Mitt. Astron. Ges., No. 31, p. 201 (1972). – Abstract.

132.031 Spectral investigation of NGC 7635 and BD +60°2522. V. T. Doroshenko.

Astron. Zhurn. Akad. Nauk SSSR, Vol. 49, 494 - 503 (1972). In Russian. English translation in Soviet Astron. AJ, Vol. 16, No. 3.

A quantitative spectral classification of BD +60°2522 was made. The absolute energy distribution in its spectrum was obtained. It is possible to represent the observed anomalous distribution as a sum of the energy distribution of O5- and F5 Ib-stars. An analysis of the physical conditions in the nebula NGC 7635 was carried out. The absolute flux in Hβ for different places of the nebula, the mean electron density and mean electron temperature were determined.

132.032 Helium in the Orion nebula.
A. S. J. Batchelor, M. Brocklehurst.
Astrophys. Letters, Vol. 11, 129 - 133 (1972).

A variable density model of the Orion nebula, consistent with all hydrogen recombination line and radio continuum observations which do not have high angular resolution, has been constructed by Brocklehurst and Seaton (1971, 1972). By combining this model with spatial optical observations of Peimbert and Costero (1969) the total helium abundance can be deduced.

132.033 Hβ photometry of galactic nebulae at $|b| > 20°$.
H. M. Johnson.
Astrophys. Journ., Vol. 174, 591 - 594 (1972).

Hβ is weakly emitted in nebulae at $\langle |b| \rangle = 26°$, and weak Hβ emission may characterize the Galaxy in general at this latitude. Derived $\langle n_e \rangle$ agrees with the value derived from pulsar data by others. The fraction of space in the solar neighborhood filled by photographic nebulae is estimated incidentally. Attention is drawn to the possible analogy of the observed nebulae to fragments of the Gum nebula.

132.034 High-n Balmer transitions in gaseous nebulae.
L. E. Goad, L. Goldberg, J. L. Greenstein.
Astrophys. Journ., Vol. 175, 117 - 125 (1972).

The intensities of the Balmer lines arising from levels with principal quantum number $n \geq 15$, and the intensity of the apparent continuum formed by the overlapping of the Balmer lines, have been measured in the Orion nebula (NGC 1976) and the planetary nebula NGC 7027. These observed intensities have been compared with the predictions of Brocklehurst's recombination theory in order to deduce the electron temperature and density of the nebulae.

132.035 Pre—main-sequence stars. II. Stellar polarization in NGC 2264 and the nature of circumstellar shells.
M. Breger, H. M. Dyck.
Astrophys. Journ., Vol. 175, 127 - 134 (1972).

Stellar polarization between 3000 and 7500 Å has been measured for NGC 2264. Out of 35 stars, four are definitely intrinsically polarized. The mean cluster polarization (of interstellar origin) is found to be very small.

132.036 Observations of 30 Doradus in the infrared.
I. S. Glass.
Nature, Phys. Sci., Vol. 237, 7 - 8 (1972).

Observations of 30 Doradus in the J, H, K and L bands are presented. The flux in the K band is shown to increase with the diameter of the aperture according to the relation $f \propto D^{1.3}$. The distribution with energy bears some relation to that of the galactic centre.

132.037 Optical and radio observations of the Orion nebula.
S. Isobe, N. Kawajiri, T. Ojima, N. Kawano, H. Kurihara.
Tokyo Astron. Bull., Second Ser., No. 218, p. 2549 - 2556 (1972).

Isophotal contours of the Orion nebula are made by radio (7.18 cm) and H_α wavelength observations. Interstellar absorption at the wavelength 6563 Å is obtained by comparison of the radio and the H_α intensities. The values obtained in this paper do not coincide with the values which were obtained by Isobe and Kurihara. The causes of this incoincidence are examined.

132.038 Recombination spectra of gaseous nebulae.
M. J. Seaton.
Comments Atomic Molecular Phys., Vol. 3, No. 2, p. 46 - 52 (1971/72).

Theoretical predictions are compared with H and He lines at optical wavelengths.

132.039 Very long baseline interferometric study of the water-vapor emission regions in Orion A.
G. D. Papadopoulos, B. F. Burke, P. R. Schwartz, K.-Y. Lo, D. C. Papa.
Quarterly Progr. Rep. Res. Lab. Electronics, Mass. Inst. Technology (USA), No. 103, p. 11 - 16 (1971).

The H_2O line emission (rest frequency equal to 22235.08 MHz) regions in W49 and Orion A were measured during June 1970 at three radio astronomical receiving stations. The results on the structure of Orion A are reviewed.

On the interpretation of radio recombination line observations. See Abstr. 022.059.

Moderne Stellar- und Nebel-Photographie. See Abstr. 031.026.

Ein Fabry-Perot-Interferometer zur Messung von Radialgeschwindigkeiten an HII-Regionen. See Abstr. 034.058.

Sternbild Orion — Stätte der Sternentstehung. See Abstr. 065.096.

345-micron ground-based observations of M17, M82, and Venus. See Abstr. 113.002.

Infrared objects in H II regions. See Abstr. 114.027.

The spectrum of Herbig Haro object No. 1. See Abstr. 114.099.

Second catalog of emission-line stars of the Orion population. See Abstr. 114.113.

On the accretion of material by Theta Orionis C. See Abstr. 114.115.

Studies of extremely young clusters. VI. Spectroscopic observations of the ultraviolet-excess stars in the Orion nebula cluster and NGC 2264. See Abstr. 114.122.

Slow flare up in the Pleiades. See Abstr. 122.062.

The Parkes survey of 21-centimeter absorption in discrete-source spectra. III. 21-centimeter absorption measurements on 41 galactic sources north of declination −48°. See Abstr. 131.003.

Radio emission of the line of water vapour from W 49 and Orion nebula. See Abstr. 131.016.

Interference-filter photography of H II regions. See Abstr. 131.025.

VLBI measurements of the H_2O line emission in W49 and Orion A.   See Abstr. 131.038.

Radio emission at 1400 MHz from galactic H II regions.   See Abstr. 131.057.

Interaction of hot stars and of the interstellar medium. II. Exciting star and spectra of the bright knot inside the diffuse nebula Sharpless 157.   See Abstr. 131.079.

Radio recombination lines from planetary nebulae. See Abstr. 133.002.

Spectrophotometric studies of gaseous nebulae. XX. The inhomogeneous, low excitation, planetary NGC 40. See Abstr. 133.006.

Absolute flux density measurements at centimeter wavelengths.   See Abstr. 141.006.

Investigation of the faint nebula identified with the radio source HB-21.   See Abstr. 141.071.

Centimeter-wavelength observations of the radio nebulae NRAO 588/589 and 4C51.12. See Abstr. 141.080.

H II regions in the Sco OB 1 association. See Abstr. 152.007.

Observations of CS, HCN, U89.2, and U90.7 in NGC 2264.   See Abstr. 153.019.

Two new large-diameter galactic regions of faint Hα emission.   See Abstr. 155.068.

133   Planetary Nebulae

133.001   **The ionization structure of planetary nebulae — IX. Luminous filaments.**
D. Van Blerkom, T. T. Arny.
Monthly Notices Roy. Astron. Soc., Vol. 156, 91 - 101 (1972).
Recent observational and theoretical evidence indicates that density condensations which are optically thick to ionizing radiation exist in planetary nebulae. The comet-like structures seen in NGC 7293 are suggested to be optically thick condensations ('heads') and the associated luminous shadows ('tails'). In order to investigate the phenomenon of luminous shadows, a simple model of a planetary nebula is employed.

133.002   **Radio recombination lines from planetary nebulae.**
Y. Terzian, B. Balick.
Astrophys. Letters, Vol. 10, 41 - 47 (1972).
Radio recombination line observations of the H85α line at $\nu = 10522.04$ MHz ($\lambda \sim 3$ cm) of the planetary nebulae NGC 7027, IC 418, and NGC 6210 are presented, together with observations of the Orion nebula, the Rosette nebula, DR 21, and NGC 6888. The H85α line was unambiguously detected from all the above objects except NGC 6210 and NGC 6888, and the observed parameters are presented. LTE electron temperatures are computed, and the line widths are discussed. It is shown that, for NGC 7027 and IC 418, the linewidths can be explained satisfactorily by thermal broadening and nebular expansion. Stark broadening and turbulence seem to be negligible in NGC 7027 and IC 418.

133.003   **Sur un modèle d'éjection de matière dans certaines nébuleuses planétaires.**   R. Louise, S. Roux.
Comptes Rendus Acad. Sci. Paris, Sér. B, Vol. 274, 294 - 297 (1972).
On présente un modèle d'éjection de matière des noyaux des nébuleuses planétaires. Quand la pression de radiation compense exactement la gravitation, l'éjection de matière peut avoir lieu grâce à la rotation de l'étoile. On considère un tel effet pour expliquer la forme pratiquement plate de l'anneau

de NGC 6720 observée par Hua et Louise (1971).

133.004   **Hamburg Schmidt-camera survey of faint planetary nebulae. Cygnus-Perseus region.**
L. Kohoutek.
Astron. Astrophys., Vol. 16, 291 - 300 (1972).
The third part of the Hα Schmidt-camera survey of faint planetary nebulae contains the remaining area in the northern Milky Way $l\ 70° - 146°$, $b \pm 10°$. 32 new planetary nebulae were classified—6 of which are identical with the objects discovered very recently by Kazarjan and Parsamjan—and measured on the Palomar Sky Atlas. In the Appendix three compact small H II regions have been described.

133.005   **Low-energy X-rays ruled out as interstellar ionizing mechanism toward K3−50.**
E. J. Chaisson, L. E. Goad.
Astrophys. Journ., (Letters), Vol. 171, L61 - L65 (1972).
We have detected the H94α and C94α recombination lines ($\lambda 3.8$ cm) originating in an H I region along the line of sight toward the nebula K3−50. For this cloud, soft X-rays (~ 100 eV) are ruled out as the interstellar heating mechanism.

133.006   **Spectrophotometric studies of gaseous nebulae. XX. The inhomogeneous, low excitation, planetary NGC 40.**   L. H. Aller, S. J. Czyzak, E. G. Buerger, P. Lee.
Astrophys. Journ., Vol. 172, 361 - 365 (1972).
Photoelectric and photographic measurements of line intensities, secured at Mount Wilson, Lick, and Kitt Peak Observatories, are interpreted in terms of ionic concentrations, with the aid of two suggested excitational models for the nebula. Large pockets of cool, neutral gas must exist in NGC 40.

133.007   **Infrared studies of galactic nebulae. IV. Continuum and line radiation from planetary nebulae.**
F. C. Gillett, K. M. Merrill, W. A. Stein.
Astrophys. Journ., Vol. 172, 367 - 374 (1972).
Observations are reported of the detection of infrared

radiation from several planetary nebulae not previously known to be radiating at these wavelengths. Broad spectral bandwidth observations indicate that infrared radiation in excess of that expected from atomic processes is a common phenomenon among these objects. The results are discussed in relation to visual and radio-wavelength data.

133.008 Radio recombination lines from planetary nebulae.
Y. Terzian, B. Balick.
Bull. American Astron. Soc., Vol. 3, 471 (1971). – Abstr. AAS.

133.009 The recombination spectrum of the planetary nebula NGC 7027.   J. S. Miller, W. G. Mathews.
Astrophys. Journ., Vol. 172, 593 - 607 = Contr. Lick Obs., No. 327 (1972).
   Detailed measurement of recombination lines and continua of H^+, He^+, and He^{++} at various positions in NGC 7027 show a remarkable agreement with theoretical prediction, contrary to earlier published results. The absence of observable optical stellar flux is consistent with recent theoretical model atmospheres of central stars of planetary nebulae.

133.010 A note to designations of planetary nebulae.
M. Chopinet, M. C. Lortet-Zuckermann.
Astron. Astrophys., Vol. 18, 166 - 167 (1972). – Research note.

133.011 Observations at 2700 MHz of selected planetary nebulae.   L. H. Aller, D. K. Milne.
Australian Journ. Phys., Vol. 25, 91 - 96 (1972).
   Observations at 2700 MHz of 74 planetary nebulae have been obtained with the Parkes 64 m radio telescope. Comparison is made with the optically determined $H\beta$ intensity to obtain extinction coefficients in these directions.

133.012 A new extension of the Helix nebula.
G. Araya, V. M. Blanco, M. G. Smith.
Publ. Astron. Soc. Pacific, Vol. 84, 70 - 71 (1972).
   A west-southwesterly extension to NGC 7293 has been discovered.

133.013 The central star of NGC 1514.   J. L. Greenstein.
Astrophys. Journ., Vol. 173, 367 - 375 (1972).
   Coudé spectra show the presence of stellar He II absorption, making it possible to resolve the stellar continuum into a horizontal-branch A star and a visually subluminous O star. Spectrophotometry of Balmer lines gives the surface gravity of the HB A star and thus its luminosity as M_v = +0.8; the O star has M_v = +2.8. If we adopt the parameters of the nebula from Kohoutek, the ultraviolet luminosity of the O star requires $T \approx 100000°K$, and $M_b = -3.8$, typical of a normal planetary nucleus.

133.014 The emission spectrum of the ion C IV in planetary nebulae.   E. M. Leibowitz.
Monthly Notices Roy. Astron. Soc., Vol. 157, 97 - 113 (1972).
   A theory is developed for the calculation of the energy level population of an ion in planetary nebulae; radiative excitation is taken specifically into account. The theory is applied in the computation of the spectrum of the ion C IV under different excitation conditions and yields measurable differences between a pure recombination spectrum of this ion and a spectrum of radiatively excited C IV.

133.015 Polarization of C IV emission lines in planetary nebulae.   E. M. Leibowitz.
Monthly Notices Roy. Astron. Soc., Vol. 157, 115 - 120 (1972).
   If the ion C IV is radiatively excited in a planetary nebu-

la, three of its emission lines should be polarized. One of the lines, $\lambda 3936$, should be observable and perhaps should also be strong enough to allow the detection of its polarization.

133.016 Infrared photometry of the H II region Sharpless 266.
J. A. Frogel, S. E. Persson, D. E. Kleinmann.
Astrophys. Letters, Vol. 11, 95 - 97 (1972).
   Photometric observations at wavelengths between 1 and $10 \mu m$ of the small H II region Sharpless 266 are presented. Image-tube spectra show that the extended region is of low excitation and that the central object has a bright continuum and strong emission lines.

133.017 Interstellar dust and distances to planetary nebulae.
J. H. Lutz.
Bull. American Astron. Soc., Vol. 4, 234 (1972). – Abstr. AAS.

133.018 Interferometric radio fluxes for planetary nebulae.
J. H. Cahn, R. H. Rubin, B. R. Hermann.
Bull. American Astron. Soc., Vol. 4, 234 - 235 (1972). – Abstr. AAS.

133.019 On the infrared radiation from planetary nebulae.
J. Gürtler.
Astron. Nachr., Vol. 293, 267 - 274 = Mitt. Univ.-Sternw. Jena No. 113 (1972).
   The infrared radiation that has been observed from several planetary nebulae is interpreted as thermal emission of dust particles. The particles are heated either directly by the light of the central stars or indirectly after that light was converted into nebular line radiation. The grains possess a strong emission band in the $11 \mu m$ spectral region. Nebulae that are optically thin for Lyman continuum radiation have a larger infrared excess than optically thick nebulae of the same surface brightness in $H\beta$.

133.020 The evolution of the nuclei of planetary nebulae.
D. J. Faulkner.
Proc. Astron. Soc. Australia, Vol. 2, 72 - 78 (1972).

133.021 Expected infrared spectra from planetary nebulae.
Y. Terzian, D. Sanders.
Astron. Journ., Vol. 77, 350 - 354 (1972).
   Infrared spectra for 16 planetary nebulae and one larger H II region (NGC 6888) have been computed assuming thermal radiation from pure graphite grains surrounding the nebulae. The predicted spectra are compared with the existing infrared observations and general agreement is found. It is indicated that several planetary nebulae should be detectable from ground-based infrared observations at $\lambda 21 \mu m$.

133.022 The peculiar nebula M2-9.
D. A. Allen, J. P. Swings.
Astrophys. Journ., Vol. 174, 583 - 589 (1972).
   M2-9 is a previously unannounced member of the class of objects possessing strong [Fe II] emission lines in their spectra and having prominent infrared continua.

133.023 Étude de la zone neutre des nébuleuses planétaires.
G. Stasinska.
Thesis, 3e Cycle, Spéc. Astrophys., Univ. Paris, 33 pp. (1971). See Bull. Signal., Vol. 33, Section 120, No. 5074 (1972).

The interpretation of total line intensities from optically thin gases. I. A general method.
See Abstr. 022.003.

On the He^+ triplet line intensities.
See Abstr. 022.065.

Nebular photometry with an echelle spectrograph.

See Abstr. 034.004.

Surface fluxes for model atmospheres for the central stars of planetary nebulae. See Abstr. 064.055.

Infrared excesses and forbidden emission lines in early-type stars. See Abstr. 114.007.

The relationship between the pulsation of long-period variables and the ejection of planetary nebulae. See Abstr. 122.024.

The population of helium triplet states in gaseous nebulae. See Abstr. 132.001.

High-n Balmer transitions in gaseous nebulae. See Abstr. 132.034.

Spatial and kinematic parameters of binebulous, centric and annular nebulae. See Abstr. 155.026.

Errata

133.901 Errata: 'The dynamics and thermal stability of planetary nebulae'. [Monthly Notices Roy. Astron. Soc., Vol. 154, 393 - 413 (1971)]. J. H. Hunter, S. Sofia. Monthly Notices Roy. Astron. Soc., Vol. 156, 263 (1972).

134 Crab Nebula

134.001 **Pulsed X-radiation from the Crab nebula.** D. Boclet, G. Brucy, J. Claisse, P. Durouchoux, R. Rocchia. Nature, Phys. Sci., Vol. 235, 69 - 72 (1972).

This communication presents the final results of an analysis of the X-ray emission from the Crab nebula in the 15 to 150 keV energy range.

134.002 **Observation of γ-quanta with energies above 100 MeV from the region of the Crab nebula.** S. A. Volobuev, A. M. Gal'per, V. G. Kirillov-Ugryumov, B. I. Luchkov, Yu. V. Ozerov. Izv. AN SSSR. Ser. fiz., Vol. 35, 2463 - 2465 (1971). In Russian. — Abstr. in Referativ. Zhurn. 51. Astron., 4.51. 623 (1972).

134.003 **Detection of X-ray polarization of the Crab nebula.** R. Novick, M. C. Weisskopf, R. Berthelsdorf, R. Linke, R. S. Wolff. Astrophys. Journ., (*Letters*), Vol. 174, L1 - L8 (1972).

Two different types of X-ray polarimeters were used in a sounding rocket to search for X-ray polarization of the Crab nebula. Polarization was detected at a statistical confidence level of 99.7 percent. If the X-ray polarization is assumed to be independent of energy, the results of this and a previous experiment lead to a polarization of (15.4±5.2) percent at a position angle of $156° \pm 10°$. This results confirms the synchrotron model for X-ray emission from the Crab nebula.

134.004 **Preliminary results for the reddening of the Crab nebula from scanner observations of [S II] lines.** J. S. Miller. Bull. American Astron. Soc., Vol. 4, 233 (1972). — Abstr. AAS.

134.005 **Energy balance of the Crab nebula and the Vela X remnant.** G. Börner, J. M. Cohen.

Astron. Astrophys., Vol. 19, 109 - 112 (1972).

To balance the energy losses from the Crab nebula, energy must be supplied continuously. The short lifetime of the electrons producing the optical and X-ray synchrotron radiation puts a lower limit on the energy output from the Crab. This can be supplied by a rotating neutron star with the period and rate of change of the period of the Crab pulsar. Its moment of inertia and mass are estimated. It is suggested that the Vela pulsar's "glitch" energy is sufficient to maintain the ionization of the Gum nebula.

134.006 **Coronal broadening of the Crab nebula 1969-71. Observations.** R. G. Blesing, P. A. Dennison. Proc. Astron. Soc. Australia, Vol. 2, 84 - 86 (1972).

134.007 **Coronal broadening of the Crab nebula 1969-71. Interpretation.** P. A. Dennison, R. G. Blesing. Proc. Astron. Soc. Australia, Vol. 2, 86 - 88 (1972).

134.008 **The structure of the Crab nebula at 2.7 and 5 GHz — I. The observations.** A. S. Wilson. Monthly Notices Roy. Astron. Soc., Vol. 157, 229 - 253 (1972).

Maps of the brightness and polarization distributions over the Crab nebula have been made with the Cambridge One-Mile telescope at 2.7 and 5 GHz, the resolution being $6'' \times 16''$ arc at the latter frequency. The main results are (i) the basic similarity of the radio nebula to that of the optical continuum, (ii) the absence of any significant variation of spectral index between 0.4 and 5 GHz, (iii) the presence of radio features associated with the bright optical filaments, (iv) complex structure in the distribution of linear polarization, and (v) a general depolarization with increasing wavelength, but stronger depolarization by the bright filaments.

134.009 **Absorption of Crab nebula X-rays.**

R. C. Henry, G. Fritz, J. F. Meekins, T. A. Chubb, H. Friedman.
Astrophys. Journ., Vol. 174, 389 - 397 (1972).

The X-ray absorption is not expected from the filaments in the Crab nebula, but could be due to invisible diffuse matter in the nebula. An alternative explanation is possible if interstellar grains contain a significant amount of matter from supernovae. Upper limits of the average volume densities of elements with X-ray absorption edges in the 0.5−1.5-keV range are given.

Source models with electron diffusion.
See Abstr. 061.049.

Faraday rotation of linearly polarized radio waves from the Crab nebula by the solar corona.
See Abstr. 074.076.

The Chinese guest star of A.D. 1054 and the Crab nebula. See Abstr. 125.013.

Far infrared observations of M 42, NGC 2024 and M 1. See Abstr. 132.019.

Detection of strong interpulses from NP 0532.
See Abstr. 141.501.

Search for pulsed γ-ray emission above 50 MeV from the Crab nebula pulsar NP-0532. See Abstr. 141.507.

The braking index and period-pulse-width distribution of pulsars. See Abstr. 141.509.

Stability of the Crab pulsar.

See Abstr. 141.510.

Results of the 1971 Crab nebula pulsar. Surveillance program at Arecibo Observatory. See Abstr. 141.512.

Very long baseline interferometer observations of NP 0532 at 121.6 MHz. See Abstr. 141.513.

γ-ray absorption for a beam model of NP 0532.
See Abstr. 141.532.

Second speed-up of the Crab pulsar.
See Abstr. 141.533.

Detection of pulsed gamma rays of $\sim 10^{12}$ eV from the pulsar in the Crab nebula. See Abstr. 141.541.

Polarization of the Crab pulsar radiation at low radio frequencies. See Abstr. 141.542.

The height distribution of giant pulses from NP 0532.
See Abstr. 141.544.

Optical timing of the Crab pulsar, NP 0532.
See Abstr. 141.559.

New X-ray pulsar near the Crab nebula.
See Abstr. 142.050.

Veränderliche kosmische Röntgenquellen.
See Abstr. 142.058.

Search for X-rays above 100 keV from the vicinity of NP 0527. See Abstr. 142.072.

Radio Sources, Quasars, Pulsars, X Ray-, Gamma Ray-Sources, Cosmic Radiation

141   Radio Sources, Quasars, Pulsars

Radio Sources, Quasars,

141.001   The relative orientation of radio sources.
M. A. G. Willson.
Monthly Notices Roy. Astron. Soc., Vol. 155, 275 - 282 (1972).

The relative orientation of the major axes of double radio sources is investigated. It is found that sources less than 10° apart tend to lie parallel to each other. It supports the suggestion of Brown that galaxies are orientated non-randomly on a scale of a few hundred megaparsecs.

141.002   The polarization properties of 65 extragalactic sources in the 3C catalogue.   S. Mitton.
Monthly Notices Roy. Astron. Soc., Vol. 155, 373 - 381 (1972).

This paper presents a summary of the polarization properties of 65 sources in the 3C catalogue, based on published material together with newly determined rotation measures. The relation of polarization properties to models of radio sources is discussed, and restrictions can be placed on the magnetic fields and thermal electron densities in these sources.

141.003   The spectral indices of radio sources selected at 1.4 GHz.   M. A. G. Willson.
Monthly Notices Roy. Astron. Soc., Vol. 155, 385 - 394 (1972).

Distributions of spectral indices are estimated for samples of extragalactic radio sources in three ranges of flux density at 1400 MHz. It is concluded that there is a significant tendency for the weakest sources to have steeper spectra.

141.004   Fan-beam surveys of radio sources near declinations +28° and +41°.   M. A. G. Willson.
Monthly Notices Roy. Astron. Soc., Vol. 156, 7 - 49 (1972).

Two strips of sky, of length 9^h and 10^h in right ascension and about 4° wide, have been surveyed at 408 MHz with the Cambridge One-Mile telescope. Narrower strips centred on the same declinations, +28° and +41°, were also surveyed at 1407 MHz. Catalogues of the 320 detected sources are presented. The accurate declinations obtained have been combined with the accurate right ascensions measured in the 4C survey to seek optical identifications for the 4C sources detected. The new flux density measurements have been compared with those of other surveys.

141.005   Fluid dynamic stability of double radio sources.
G. M. Blake.
Monthly Notices Roy. Astron. Soc., Vol. 156, 67 - 89 (1972).

The stability of the radio components of double sources confined by the ram pressure mechanism is discussed, and it is shown that they are unstable to the Rayleigh-Taylor and Kelvin-Helmholtz effects. The non-linear development of the instabilities and the consequences for the evolution of double sources are discussed, and an observational effect suggested.

141.006   Absolute flux density measurements at centimeter wavelengths.   W. J. Medd.
Astrophys. Journ., Vol. 171, 41 - 50 (1972).

Flux density measurements of Cas A, Tau A, Cyg A, and the Orion nebula have been made at 4.5 cm; also, measurements of Cas A and Tau A at 2.2 cm. The results are summarized. The experimental procedure is outlined and the accuracy of the results examined in detail.

141.007   The optical object identified with 3C 455.
H. C. Arp, E. M. Burbidge, C. D. Mackay, P. A. Strittmatter.
Astrophys. Journ., (*Letters*), Vol. 171, L41 - L43 (1972).

It is shown that the optical object identified with 3C 455 is a quasi-stellar object with a redshift $z = 0.543$. The QSO is situated 23″ northeast of the S0 galaxy NGC 7413, which has a redshift of 0.0332.

141.008   The interpretation of variable components in extragalactic radio sources by 'synchro-Compton' theory.
R. D. Blandford, M. J. Rees.
Astrophys. Letters, Vol. 10, 77 - 82 (1972).

Stellar-mass objects, collapsing at a rate $\sim$ 1 per yr, can provide enough energy to power active galactic nuclei and typical QSS's. Such objects may emit their rotational energy as electromagnetic waves at frequencies $\lesssim$ 1 kHz. If these waves impinge onto a surrounding plasma of suitable density they will accelerate relativistic electrons, which then radiate in the radio band. This process can be efficient enough to account for the intensity of the compact components associated with some extragalactic radio sources. The calculated spectrum, time dependence and other properties seem compatible with the radio observations.

141.009   Identification of southern quasi-stellar objects—II.
B. A. Peterson, J. G. Bolton.
Astrophys. Letters, Vol. 10, 105 - 108 (1972).

Eighteen new quasi-stellar objects south of declination −33° have been found from combined radio and optical observations.

141.010   The structure and spectrum of Cygnus A.
D. M. Mills.
Astrophys. Letters, Vol. 10, 109 - 113 (1972).

The observed structure and spectrum of the Cygnus A radio clouds can be explained by confinement of the clouds by the ram pressure of the external medium and depletion of high energy electrons due to synchrotron losses during the early history of the source. Synchrotron self-absorption accounts for the low-frequency turnover in the spectrum for reasonable magnetic field strengths.

141.011   The nature of variable radio structures.
V. N. Kurilchik.
Astrophys. Letters, Vol. 10, 115 - 119 (1972).

A model of variable radio structures as outbursts of clouds of relativistic electrons into stationary magnetic field tubes that expand in a regular manner with distance from nuclei, is considered. Motion and expansion of compact radio sources with apparent super-light velocities and other phenomena, which take place in this model, are discussed.

141.012 Precise optical positions of nine compact radio sources. H. A. Couper.
Astrophys. Letters, Vol. 10, 121 - 124 (1972).

Positions relative to AGK 3 have been measured for nine optical objects identified with radio sources of small angular diameter. Among the best radio determinations the rms difference between optical and radio position is of order 0.6 arc sec in either co-ordinate, which is equal to the standard error of the radio position, thus confirming the optical identification and indicating the coincidence of optical and radio objects.

141.013 Identification and photometry of radiosources of the 3CR catalogue. II. The case of the source 3C 173. G. Wlérick, G. Lelièvre.
Astron. Astrophys., Vol. 16, 53 - 57 (1972). In French.

We have studied the source 3C 173 in the UBV photometric system, on plates taken with the electron camera. This very faint source ($20 \leq B < 22$) has the characteristic properties of a quasar, including a strong UV excess. It belongs to a family of QSS that are radio stable and optically strongly variable.

141.014 Origin of QSOs. J. Terrell.
Nature, Vol. 236, 166 (1972). – Letter.

141.015 Observation optique de la radiosource OJ 287. G. Adam, M. Floquet, J. Marchal, M. Schneider.
Comptes Rendus Acad. Sci. Paris, Sér. B, Vol. 274, 201 - 202 (1972).

Observations d'un objet bleu associé à une radiosource, OJ 287, en photométrie UBV et dans le domaine spectral 3900–4900 Å.

141.016 A Green Bank Sky Survey in search of radio sources at 1400 MHz. II. Spectral properties of the 5C1 and 5C2 sources. J. Maslowski.
Astron. Astrophys., Vol. 16, 197 - 203 (1972).

The results of a 1400 MHz survey of the 5C2 region of the sky, made with the 300-foot radiotelescope of the National Radio Astronomy Observatory, Green Bank, are presented. Forty sources were observed in an area which overlaps 94% of the 5C2 region down to a limiting flux density of 0.09×10^{-26} W m^{-2} Hz^{-1} at 1400 MHz. A comparison of the spectral index distributions between 1400 MHz and 408 MHz for the sources from the 5C1 and 5C2 regions shows that these distributions are significantly different.

141.017 On the radio sources in normal galaxies. V. N. Kurilchik, N. A. Komarov.
Astron. Zhurn. Akad. Nauk SSSR, Vol. 49, 85 - 92 (1972). In Russian. English translation in Soviet Astron. AJ, Vol. 16, No. 1.

The analysis of the radio emission spectra of near spiral galaxies shows that the distribution of the spectral indices has discrete maxima at values $\simeq 0.6$ and $\simeq 1.1$. A classification of the observed spectra was made. A corrected luminosity – surface brightness diagram is presented.

141.018 Photoionization and the emission-line spectra of quasi-stellar objects. K. Davidson.
Astrophys. Journ., Vol. 171, 213 - 231 (1972).

It is found that the elemental abundances cannot be estimated with any assurance; but if "conventional" values are assumed, then certain restrictions may be placed on the ionizing spectra and gas densities. In particular, the ratio between radiation density and gas density is determined. Narrow filaments or sheets of ionized gas seem to be indicated; these may be compressed regions behind shock fronts.

141.019 The spectrum of the quasi-stellar object PHL 957. J. L. Lowrance, D. C. Morton, P. Zucchino, J. B. Oke, M. Schmidt.
Astrophys. Journ., Vol. 171, 233 - 251 (1972).

The emission lines in the Cassegrain spectra have a redshift of $z = 2.69 \pm 0.01$. In addition, the image-tube spectra revealed 64 absorption lines between 3200 and 6800 Å. The absolute energy distribution indicates that this quasar is one of the most luminous known. The coudé spectrum was obtained from 4270 to 4495 Å. It recorded 31 absorption lines including 10 already found. Four absorption-line redshifts at $z = 2.2065, 2.2259, 2.3099,$ and 2.6624 were identified as very probable.

141.020 Search for optical identifications in the 5C2-radio survey. P. Notni, H. Oleak, G. M. Richter.
Astron. Nachr., Vol. 293, 221 - 236 (1971/72).

On plates of the 134/200 cm Schmidt telescope of the Karl-Schwarzschild-Observatory Tautenburg optical objects up to 20^m5 near the position of 5C2-radio-sources were measured astrometrically. For each source two parameters have been derived: the probability for chance identification and the probability for the occurrence of an observed difference between the optical and radio position only by errors of measurements. Both these values enable the number of identifications and its reliability for any selected sample to be determined. Up to 20^m5 about 15 per cent of all radio sources in the 5C2-field are related to optical objects.

141.021 Low frequency, high resolution observations of Virgo A. P. N. Wilkinson, R. J. Peckham.
Monthly Notices Roy. Astron. Soc., Vol. 156, 7P - 11P (1972).

Observations are reported of the angular structure of the central region of Virgo A at 151 and 408 MHz which complement those obtained by synthesis techniques at higher frequencies. The radio counterparts of the jet and counter-jet are shown to have different sub-structures at these low frequencies.

141.022 On the origin of the absorption lines in quasar spectra. R. C. Roeder.
Astrophys. Journ., Vol. 171, 451 - 456 (1972).

The application of Bahcall and Peebles's statistical test for the origin of the absorption lines found in quasar spectra has been extended. The test has been carried out by using several model universes and different values of z_r, the cutoff redshift. Any conclusion about the validity of the cosmological origin of the absorption lines is seen to depend, in the case of multiple redshifts, upon z_c and the cosmological model.

141.023 Results of an initial survey of fine structure in radio sources at 81.5 MHz. S. J. Bell Burnell.
Astron. Astrophys., Vol. 16, 379 - 384 (1972).

Results are presented from an initial survey at 81.5 MHz which used the technique of interplanetary scintillation to measure the angular diameters of radio sources. 163 sources in the region $-10° < \delta < 45°$, $10^h < \alpha < 16^h$ have been studied, and found to contain components with diameters $\lesssim 1\overset{''}{.}0$.

141.024 The detection of radio recombination lines in four radio sources formerly considered supernova remnants. T. L. Wilson, W. J. Altenhoff.
Astron. Astrophys., Vol. 16, 489 - 490 (1972).

Eight sources listed by Reifenstein et al. (1970) as having no radio recombination line emission have been reobserved using the NRAO 140-foot radio telescope. Of the eight, radio recombination lines have been detected in four. Reasons for the failure to detect the line in the earlier survey and consequences of the present results are briefly discussed.

141.025 Optical observations of southern radio sources. E. M. Burbidge, G. R. Burbidge.
Astrophys. Journ., Vol. 172, 37 - 41 (1972).

Redshifts have been measured for 10 radio galaxies, and the spectroscopic features are listed. Direct photographs of some of the objects are described. Spectrograms of four stellar objects suggested as QSO identifications showed all objects to be galactic stars, including one recently suggested as an identification for PKS 1302–49 instead of NGC 4945.

141.026   On the light variation of the quasar 3C273.
I. Jurkevich.
Astrophys. Journ., (Letters), Vol. 172, L29 - L33 (1972).

It is shown that the confidence level with which resonances corresponding to periods in the range of 8–10 and 14–17 years are determined is too low to permit unequivocal claims concerning their reality.

141.027   High-resolution observations of compact radio sources at 13 centimeters. II.
J. J. Broderick, K. I. Kellermann, D. B. Shaffer, D. L. Jauncey.
Astrophys. Journ., Vol. 172, 299 - 305 (1972).

Observations made at 13 cm with a tracking interferometer 25×10^6 wavelengths long have been used to investigate the structure of the compact radio sources. The data show complex structure, so that, in general, it is not possible to discuss specific models with the limited data from this single baseline. It is established, however, that the compact components are confined to a region $\lesssim 1''$ arc.

141.028   Redshifts of twenty radio galaxies.
E. M. Burbidge, P. A. Strittmatter.
Astrophys. Journ., (Letters), Vol. 172, L37 - L40 (1972).

Spectroscopic observations and redshifts of 20 radio galaxies obtained with the Lick 120-inch telescope are presented. Ten of the radio galaxies are from the 3C R catalog, and the remainder are from the 4C, 5C, Ohio, and Parkes catalogs.

141.029   3C279: Evidence for a non-superrelativistic model.
W. A. Dent.
Science, Vol. 175, 1105 - 1106 (1972).

Measurements of the variation of the total flux density of the quasi-stellar radio source 3C279 provide evidence for an alternate model to explain the recently reported apparent source expansion rate of ten times the speed of light.

141.030   Long baseline interferometry at a decametric wavelength.   W. M. Cronyn, W. K. Klemperer, C. L. Rufenach, S. D. Shawhan, J. Basart, T. A. Clark, W. C. Erickson.
Bull. American Astron. Soc., Vol. 3, 438 (1971). –Abstr. AAS.

141.031   The brightness distribution of core-halo radio sources.   R. A. Sramek.
Bull. American Astron. Soc., Vol. 3, 445 - 446 (1971). – Abstr. AAS.

141.032   Observation of differing counts for radio sources of high and low surface brightness.
M. M. Davis, E. B. Fomalont.
Bull. American Astron. Soc., Vol. 3, 446 (1971). – Abstr. AAS.

141.033   Radio source counts at 430 MHz.
D. L. Jauncey, A. E. Niell, J. J. Condon.
Bull. American Astron. Soc., Vol. 3, 446 - 447 (1971). – Abstr. AAS.

141.034   A sample of radio sources complete to 3 f.u. at 430 MHz.   A. E. Niell, D. L. Jauncey.
Bull. American Astron. Soc., Vol. 3, 447 (1971). – Abstr. AAS.

141.035   Spectral index distributions of extragalactic radio sources.   J. J. Condon, D. L. Jauncey, A. E. Niell.
Bull. American Astron. Soc., Vol. 3, 447 (1971). – Abstr. AAS.

141.036   Observations with the Haystack-Goldstone interferometer of phase scintillations due to the solar corona.   C. A. Knight, D. S. Robertson, I. I. Shapiro, A. R. Whitney, A. E. E. Rogers, T. A. Clark, G. E. Marandino, N. R. Vandenberg, R. M. Goldstein.
Bull. American Astron. Soc., Vol. 3, 447 (1971). – Abstr. AAS.

141.037   The flux density of Cassiopeia A at 1440 MHz and its rate of decrease.   J. W. Findlay.
Bull. American Astron. Soc., Vol. 3, 451 (1971). – Abstr. AAS.

141.038   High-accuracy determination of 3 C 273–3 C 279 position difference from long-baseline interferometer fringe phase measurements.
A. R. Whitney, I. I. Shapiro, A. E. E. Rogers, D. S. Robertson, C. A. Knight, T. A. Clark, G. E. Marandino, N. R. Vandenberg, R. M. Goldstein.
Bull. American Astron. Soc., Vol. 3, 465 (1971). – Abstr. AAS.

141.039   High-resolution observations of radio sources near 8 GHz.   J. J. Broderick, B. G. Clark, K. I. Kellermann, W. Cannon, M. H. Cohen, G. H. Purcell, D. B. Shaffer, L. Kogan, V. Kostenko, L. Matveyenko, V. A. Ephanov, I. Moiseev, D. L. Jauncey.
Bull. American Astron. Soc., Vol. 3, 465 (1971). – Abstr. AAS.

141.040   High resolution observations of linear polarization in radio sources.   J. F. C. Wardle.
Bull. American Astron. Soc., Vol. 3, 465 (1971). – Abstr. AAS.

141.041   Circular polarization observations of high-brightness radio sources.   G. A. Seielstad, G. L. Berge.
Bull. American Astron. Soc., Vol. 3, 465 (1971). – Abstr. AAS.

141.042   Interferometer measurements of galactic radio sources toward the galactic center in the 21-cm line.   K. W. Riegel, D. Elliott, J. Scalo.
Bull. American Astron. Soc., Vol. 3, 470 (1971). – Abstr. AAS.

141.043   Rapid differential proper motion in the radio fine structure of 3C279.
J. M. Barnothy, M. F. Barnothy.
Bull. American Astron. Soc., Vol. 3, 472 (1971). – Abstr. AAS.

141.044   A new model for variable radio sources.
R. D. Blandford, M. J. Rees.
Bull. American Astron. Soc., Vol. 3, 472 - 473 (1971). Abstr. AAS.

141.045   Periodicity of 3C-273: A critique.   J. C. Wheeler.
Bull. American Astron. Soc., Vol. 3, 473 (1971). Abstr. AAS.

141.046   Variable quasi-stellar sources.
N. Visvanathan.
Bull. American Astron. Soc., Vol. 3, 473 (1971). – Abstr. AAS.

141.047   The distribution of radiation from relativistically expanding sources.   D. S. De Young.
Bull. American Astron. Soc., Vol. 3, 473 (1971). – Abstr. AAS.

141.048   Density and luminosity evolution of quasars.
M. F. Barnothy, J. M. Barnothy.
Bull. American Astron. Soc., Vol. 3, 473 - 474 (1971). Abstr. AAS.

141.049   Simultaneous, multifrequency observations of

interplanetary scintillations.
G. A. Zeissig, D. C. Backer, R. V. E. Lovelace.
Bull. American Astron. Soc., Vol. 3, 482 (1971). – Abstr.
AAS.

141.050 New evidence for redshift periodicity.
R. G. Lake.
Bull. American Astron. Soc., Vol. 3, 498 (1971). – Abstr.
AAS.

141.051 Probable effects of the terrestrial atmosphere on quasar redshift measurements.
R. C. Roeder, C. C. Dyer.
Nature, Phys. Sci., Vol. 235, 3 - 6 (1972).

There is a significant correlation between the distribution of the published emission-line redshifts of all quasars with two or more lines in their spectra on a scale of 0.1 in redshift and the average number of lines in the spectra. This may be caused by a combination of night sky emission lines and the finite size of the atmospheric observing window, so that no cosmological (or other) significance should be assigned to the distribution.

141.052 Geometrical interpretation of the luminosity of quasars. A. C. Le Floch, J. Lebreton.
Nature, Phys. Sci., Vol. 235, 25 - 26 (1972).

It is suggested that quasars are taking part in an expansion that follows a big bang of their own.

141.053 Optical variability of 20.08.01 Cancri.
W. Wenzel.
Nature, Phys. Sci., Vol. 235, 58 (1972).

141.054 Counts of intense extragalactic radio sources at 1,400 MHz.
A. H. Bridle, M. M. Davis, E. B. Fomalont, J. Lequeux.
Nature, Phys. Sci., Vol. 235, 123 - 126 (1972).

1,400 MHz extragalactic radio source counts indicate that the exponent of the relation between number and flux density is only greater than 1.5 at very low source densities.

141.055 Observations of radio variables at 8000 MHz.
M. A. Stull.
Astron. Journ., Vol. 77, 13 - 16 (1972).

Data are presented on 18 radio sources which show definite or probable variations at 8000 MHz. A number of these have not previously been reported to be radio variables.

141.056 The decrease of flux density of Cassiopeia A and the absolute spectra of Cassiopeia A, Cygnus A and Taurus A. J. W. M. Baars, A. P. Hartsuijker.
Astron. Astrophys., Vol. 17, 172 - 181 (1972).

We find a secular decrease in the flux density of Cas A of $0.90 \pm 0.10\%$ per year. A new analysis is presented of the absolute spectrum of Cassiopeia A using all absolute measurements of up to 1971. The spectra of Cygnus A, Taurus A and Virgo A are also discussed.

141.057 Brightness of the new quasi-stellar source B 2 1215 + 30. G. Klare.
Astron. Astrophys., Vol. 17, 325 (1972).

The brightness of B2 1215 + 30 (= ON 325) is given for 1907 May 7.

141.058 A new complete sample of quasi-stellar radio sources and the determination of the luminosity function.
R. Lynds, D. Wills.
Astrophys. Journ., Vol. 172, 531 - 552 (1972).

New spectroscopic and photometric observations are reported for 19 quasi-stellar objects identified with 4C radio sources in a selected region of the sky. These objects, together with 12 previously studied 4C quasi-stellar radio sources in the same area of sky, form a sample that is complete in radio and optical properties. The combined data for this sample are used to examine the spatial distribution of the objects and to derive their bivariate radio and optical luminosity function at a distance corresponding to a redshift $z = 1.0$. The results are compared with those of a similar study by Schmidt of a complete sample of 3CR quasi-stellar sources.

141.059 Low-mass protogalaxies and absorption lines in quasi-stellar objects. J. Arons.
Astrophys. Journ., Vol. 172, 553 - 562 (1972).

The possibility is explored that the many weak unidentified absorption lines in high-redshift QSO spectra are $L\alpha$ lines of hydrogen. The likely existence of a metagalactic flux of ionizing radiation indicates that hydrogen clouds at high redshifts must be thoroughly ionized. A simple model for the clouds' collapse and for the flux of ionizing radiation is used to compute the distribution of expected $L\alpha$ lines in the spectrum of a QSO as a function of wavelength and H I column density. Crudities in the theory and stronger tests of its validity are also discussed.

141.060 Flux densities of discrete radio sources at 3.5 cm.
A. E. Andrievski, E. E. Spangenberg, I. E. Valtz,
A. G. Gorshkov, V. N. Ivanov, V. K. Konnikova, V. N. Kurilchik, M. G. Larionov, I. G. Moiseev, V. V. Nikitin, V. I. Portman, V. A. Soglasnov.
Izv. Krymskoj Astrofiz. Obs., Vol. 43, 30 - 36 (1971).
In Russian.

The results of measurement of flux density of 130 discrete radio sources at 3.5 cm are given. Observations were made with the 22-meter radiotelescope of the Crimean Observatory in August–September 1968.

141.061 On the distances of quasi-stellar radio sources.
M. A. Arakelyan.
Vestn. AN SSSR, 1971, No. 11, p. 26 - 30. In Russian.
Abstr. in Referativ. Zhurn. 51. Astron., 3.51.816 (1972).

141.062 Beobachtung eines explodierenden Quasars.
P. M. Rapier.
Phys. Blätter, 28. Jahrgang, p. 71 - 73 (1972).

141.063 A plausible energy source and structure for quasi-stellar objects. E. Daltabuit, D. Cox.
Astrophys. Journ., (*Letters*), Vol. 173, L13 - L17 (1972).

If a collision of two large, massive, fast gas clouds occurs, their kinetic energy is converted to radiation in a pair of shock fronts at their interface. The resulting structure is described, and the relevance of this as a radiation source for QSOs is considered.

141.064 Redshifts of southern radio sources.
B. A. Peterson, J. G. Bolton.
Astrophys. Journ., (*Letters*), Vol. 173, L19 - L21 (1972).

Spectroscopic observations are reported for 15 radio sources. Redshifts were measured for nine QSOs and one galaxy.

141.065 Extragalactic radio sources. M. P. Damon.
Spaceflight, Vol. 14, 64 - 67 (1972).

141.066 The distribution of linear polarization in Cassiopeia A at wavelengths of 9.8 and 11.1 cm.
G. S. Downs, A. R. Thompson.
Astron. Journ., Vol. 77, 120 - 133 (1972).

Two series of observations of the brightness distribution of the linearly polarized component of the radiation of Cassiopeia A are described. The first was made at a wavelength of 9.8 cm using a two-element interferometer at Stanford which provided a synthesized beamwidth of $1\rlap{.}'6 \times 2\rlap{.}'7$. The second set of observations was made with the three-element interferometer of the NRAO at a wavelength of 11.1 cm and a beamwidth of $8''.1 \times 9''.3$ was obtained. The results can be explained in terms of a model of the source incorporating Faraday depolarization.

141.067 **Une expérience permettant de confirmer que la vitesse de la lumière reçue de la QSO PKS 2134 + 004 est supérieure à 440, 000 km/sec.** J. Loiseau.
Applied Optics, Vol. 11, 470 - 472 (1972). – Letter.

141.068 **Are quasars associated with bright galaxies?**
J. N. Bahcall, C. F. McKee, N. A. Bahcall.
Astrophys. Letters, Vol. 10, 147 - 152 (1972).
A complete sample of 222 quasi-stellar radio source (QSRS) identifications and 166 radio quiet quasi-stellar object (QSO) identifications has been studied to look for possible correlations with the positions of bright galaxies. The results are consistent with a random correlation between the positions of quasars and bright galaxies for angular separations less than 30 arc min.

141.069 **"Black holes" als Quasarkernmodelle.** K. Fritze.
Sterne, 48. Jahrgang, p. 16 - 19 (1972).

141.070 **The quasars and nuclei of galaxies: a single body or a stellar cluster?**
G. S. Bisnovaty-Kogan, R. A. Sunyaev.
Astron. Zhurn. Akad. Nauk SSSR, Vol. 49, 243 - 252 (1972).
In Russian. English translation in Soviet Astron. AJ, Vol. 16, No. 2.
The stability of a system gas–stars as a model of a quasar or a galaxy nucleus is investigated. The merits and deficiencies of two alternative models: the model of a unique gaseous body and the model of a compact stellar cluster are critically analysed.

141.071 **Investigation of the faint nebula identified with the radio source HB-21.** T. A. Lozinskaya.
Astron. Zhurn. Akad. Nauk SSSR, Vol. 49, 265 - 270 (1972).
In Russian. English translation in Soviet Astron. AJ, Vol. 16, No. 2.
Interferometric observations were carried out of H_α optical emission from the region of radio source HB-21 identified with a supernova remnant. An expansion of the nebulosity with a velocity of $22 \, km/sec \pm 8 \, km/sec$ was discovered.

141.072 **The polarization of radio sources with appreciable redshift.**
P. P. Kronberg, R. G. Conway, J. A. Gilbert.
Monthly Notices Roy. Astron. Soc., Vol. 156, 275 - 282 (1972).
Observations of the linear polarization of quasi-stellar radio sources over a wide range of wavelengths have been combined to investigate the depolarization at long wavelengths. It is suggested that the main cause is Faraday dispersion occurring in or near the source itself. A strong correlation is found between the rate of depolarization and the redshift for 62 quasars.

141.073 **The polarization of radio sources with appreciable red shift.**
P. P. Kronberg, R. G. Conway, J. A. Gilbert.
Journ. Roy. Astron. Soc. Canada, Vol. 66, 66 (1972). – Abstr. Canadian Astron. Soc.

141.074 **A description of the spectrum of the quasi stellar object Ton 202.** C. Barbieri.
Mem. Soc. Astron. Italiana, Nuova Ser., Vol. 43, 157 - 158 (1972). – Letter.

141.075 **Detection of radio emission from GX9+1.**
W. Zaumen, G. T. Murthy, S. Rappaport, R. M. Hjellming, C. M. Wade.
Nature, Vol. 235, 378 - 379 (1972).
A variable point radio source has been detected at 2695 MHz within the position uncertainty of the celestial X-ray source GX9+1. No optical image is visible on the red and blue Palomar Sky Survey prints at the location of the radio source. Searches for point radio emission from GX349+2 and GX340+0 yielded only upper limits of 5×10^{29} Wm^{-2} Hz^{-1} at 2695 MHz.

141.076 **Asymmetries in compact sources at 2,298 MHz.**
A. J. Legg, J. S. Gubbay, D. S. Robertson, G. D. Nicolson, A. T. Moffet.
Nature, Phys. Sci., Vol. 235, 147 - 149 (1972).
This letter compares VLBI observations using two baselines of similar length. The baselines were Woomera, Australia, to Johannesburg and Woomera to Goldstone, California. Differences in fringe visibility were found for a number of the sources observed over both baselines. The results are tabulated.

141.077 **Ohio survey statistics.** J. D. Kraus.
Nature, Phys. Sci., Vol. 236, 4 - 6 (1972).
The number-flux density relation for installment V (3475 sources above 0.18 flux units) of the Ohio survey at 1415 MHz shows no significant departure from that expected for a model universe with integral slope -1.5 corresponding to a uniform, Euclidean universe. Ten to 15 percent of the sources in the Ohio survey (or 1200 to 1800 sources in five installments) may have spectra with flux density enhancement at centimeter wavelengths.

141.078 **Are quasars local or cosmological?**
M. Rowan-Robinson.
Nature, Vol. 236, 112 - 114 (1972).
It is proposed that there are two distinct classes of quasars. In one class the redshift is cosmological, the optical luminosities range from several to several hundred times that of a normal galaxy, and the radio luminosities and dimensions are in the range occupied by the strongest radio-galaxies. In the second class the redshift is mainly intrinsic, the objects are comparatively local (10–100 Mpc), and the optical and radio luminosities are similar to those of radio-emitting Seyfert nuclei. Most of the anomalous properties of quasars, including Arp's associations with peculiar galaxies, are associated with the 'local' quasars, some of which may lie in comparatively nearby clusters.

141.079 **How many variable extragalactic sources at λ3.8 cm?** G. W. Brandie.
Astron. Journ., Vol. 77, 197 - 200 (1972).
Forty-seven sources found in a λ3.8-cm sky survey have been examined for variations in flux density. Over-all, 32% of the sources (15 out of 47) have been found to be variable.

141.080 **Centimeter-wavelength observations of the radio nebulae NRAO 588/589 and 4C51.12.**
R. Butler, V. A. Hughes.
Astron. Journ., Vol. 77, 201 - 206, 255 (1972).
The detailed properties of the two radio nebulae have been obtained from accurate radio observations and from a precise knowledge of distance. Both sources may be classified as average radio nebulosities.

141.081 **Distance estimates for two thermal galactic radio sources.** A. H. Bridle, M. J. L. Kesteven.
Astron. Journ., Vol. 77, 207 - 209 (1972).
Kinematic distances of 3.2 and 4.1 kpc are suggested for the thermal galactic radio sources NRAO 589A and 4C51.12, from observations of their 21-cm absorption profiles.

141.082   The absorption-line QSOs.   C. R. Lynds.
IAU Symposium No. 44, (see 012.005), p. 127 -
138 (1972).

141.083   Diameter of PKS 1514–24 (AP Lib).
S. Ananthakrishnan, A. P. Rao, S. M. Bhandari.
Nature, Phys. Sci., Vol. 235, 167 (1972).
Interplanetary scintillations of the radio source
PKS 1514–24 (AP Lib) were observed at 326.5 MHz using the
Ooty radio telescope in November and December 1971. The
smallest elongation at which observations could be made was,
when the source was 6 degrees away from the sun. The analysis of the data shows that most of the flux is in the point
component. The spectrum extends to 6.5 Hz which puts an upper limit for the compact component of AP Lib at 0.035 arc s.

141.084   The physical properties of the absorption envelopes
of two QSOs.
J. D. Scargle, L. J. Caroff, P. D. Noerdlinger.
IAU Symposium No. 44, (see 012.005), p. 151 - 154 (1972).

141.085   The relation between the optical and centimetric
polarized emission in BL Lac and other QSOs.
T. D. Kinman.
IAU Symposium No. 44, (see 012.005), p. 164 (1972). – Abstract.

141.086   Optical variability of twenty-two quasi-stellar objects.   R. J. Angione, H. J. Smith.
IAU Symposium No. 44, (see 012.005), p. 171 - 178 (1972).

141.087   Can the optical fluctuations of 3C 273 be random?
J. Terrell, K. H. Olsen.
IAU Symposium No. 44, (see 012.005), p. 179 - 186 (1972).

141.088   The angular structures of some compact sources.
H. P. Palmer.
IAU Symposium No. 44, (see 012.005), p. 214 - 215 (1972).

141.089   The radio structure of quasars.
G. K. Miley, G. H. MacDonald.
IAU Symposium No. 44, (see 012.005), p. 216 - 220 (1972).

141.090   Compact radio sources in the nuclei of elliptical
galaxies.   R. D. Ekers.
IAU Symposium No. 44, (see 012.005), p. 222 - 223 (1972).

141.091   Intensity variations in extragalactic radio sources at
13 cm.   G. D. Nicolson.
IAU Symposium No. 44, (see 012.005), p. 224 (1972). – Abstract.

141.092   Observations of variable radio sources at 8.2 mm.
V. A. Efanov, I. G. Moiseev, H. M. Tovmasjan, V. B.
Shteinshleger, V. I. Zagatin.
IAU Symposium No. 44, (see 012.005), p. 225 - 226 (1972).

141.093   Variable radio sources: Comparison of observations
with the adiabatic spherical expansion source model.
E. E. Epstein.
IAU Symposium No. 44, (see 012.005), p. 227 (1972). – Abstract.

141.094   High-resolution observations of variable radio
sources.
A. T. Moffet, J. Gubbay, D. S. Robertson, A. J. Legg.
IAU Symposium No. 44, (see 012.005), p. 228 - 229 (1972).

141.095   Polarization of quasars.   R. G. Conway.
IAU Symposium No. 44, (see 012.005), p. 230 -
231 (1972).

141.096   QSOs and radio galaxies – their spectra and time
variations at radio frequencies.   B. J. Harris.
IAU Symposium No. 44, (see 012.005), p. 232 - 248 (1972).

141.097   Identification optique et photométrie de radio-
sources du catalogue 3CR.
G. Wlérick, G. Lelièvre, P. Véron.
IAU Symposium No. 44, (see 012.005), p. 251 - 257 (1972).

141.098   Optical identifications of compact radio sources.
G. G. Pooley.
IAU Symposium No. 44, (see 012.005), p. 258 (1972). – Ab-
stract.

141.099   Attempts to detect neutral hydrogen in compact ob-
jects.   W. A. Dent.
IAU Symposium No. 44, (see 012.005), p. 259 - 263 = Contr.
Four College Obs., Univ. Mass., Amherst, No. 89 (1972).

141.100   21-cm absorption in BL Lac.   M. H. Cohen.
IAU Symposium No. 44, (see 012.005), p. 271
(1972). – Abstract.

141.101   Theoretical considerations of compact objects.
L. Woltjer.
IAU Symposium No. 44, (see 012.005), p. 277 - 280 (1972).

141.102   The large-scale variations of quasi-stellar objects.
W. H. McCrea.
IAU Symposium No. 44, (see 012.005), p. 283 - 284 (1972).

141.103   A 'single electron' synchrotron radiation model and
the quasi-stellar objects.   D. F. Falla, A. Evans.
IAU Symposium No. 44, (see 012.005), p. 285 - 289 (1972).

141.104   On the evolution of quasars and their remnants.
P. Kafka.
IAU Symposium No. 44, (see 012.005), p. 296 - 299 (1972).

141.105   A matter-antimatter model for quasi-stellar objects.
A. Elvius.
IAU Symposium No. 44, (see 012.005), p. 306 - 310 (1972).

141.106   The N(S) relationship at 1400 MHz.
A. H. Bridle, M. M. Davis.
IAU Symposium No. 44, (see 012.005), p. 437 - 443 (1972).

141.107   Number counts and spectral distribution of radio
sources at centimeter wavelengths.
I. I. K. Pauliny-Toth, K. I. Kellermann, M. M. Davis.
IAU Symposium No. 44, (see 012.005), p. 444 - 452 (1972).

141.108   Statistical properties of QSOs.   A. Braccesi.
IAU Symposium No. 44, (see 012.005), p. 453 -
457 (1972).

141.109   Radio source-counts and cosmology.
M. Rowan-Robinson.
IAU Symposium No. 44, (see 012.005), p. 458 - 463 (1972).

141.110   Redshift distribution of quasi-stellar objects and the
radio source counts.   V. Petrosian.
IAU Symposium No. 44, (see 012.005), p. 464 - 468 (1972).

141.111   The luminosity-volume test for quasi-stellar objects.
M. S. Longair.
IAU Symposium No. 44, (see 012.005), p. 470 (1972).

141.112   Concerning the primordial abundance of helium in
quasi-stellar sources.   J. Silk.
IAU Symposium No. 44, (see 012.005), p. 474 - 477 (1972).

141.113 Radio maps of 31 extragalactic sources at 2.7 and 5.0 GHz.
N. J. B. A. Branson, B. Elsmore, G. G. Pooley, M. Ryle.
Monthly Notices Roy. Astron. Soc., Vol. 156, 377 - 397 (1972).

The Cambridge One-Mile radio telescope has been used to map the structures of 31 extragalactic sources at 2.7 and 5.0 GHz. The maps are presented in this paper together with physical data derived for the various components.

141.114 Identification of 5C4 radio sources.
C. Barbieri, F. Bertola.
Monthly Notices Roy. Astron. Soc., Vol. 156, 399 - 409 (1972).

Two deep-exposure IIIa-J plates obtained with the 48-in. Palomar Schmidt telescope have been used to carry out a re-identification of the 5C4 radio sources in the field of the Coma Cluster of galaxies. Several new candidates, including faint clusters of galaxies, are proposed.

141.115 Observations of radio sources with flat spectra.
B. L. Fanaroff, G. M. Blake.
Monthly Notices Roy. Astron. Soc., Vol. 157, 41 - 53 (1972).

A study has been made of 10 radio sources which appeared from published spectral data to have straight spectra with spectral index $\alpha \sim 0.3$.

141.116 Optical identification of radio-sources selected from the B2 catalogue.
G. Grueff, M. Vigotti.
Astron. Astrophys. Suppl, Vol. 6, 1 - 83 (1972). – Abstr. in Astron. Astrophys., Vol. 18, 494 (1972).

Two samples of radio-sources, statistically complete to 0.25 and 0.90 flux units at 408 MHz in restricted areas, have been selected from the B2 survey to attempt their optical identification on the Palomar Sky Survey. More accurate declinations have been measured for all the 466 radio-sources, using the interferometer of the Owens Valley Radio-Observatory. The accuracy of the final radio-positions is of the order of 6 to 12 arc sec. in R. A., and 2.5 to 20 arc sec in Dec. Identification is suggested for a total of 104 radio-galaxies and 73 quasi-stellar objects. Finding charts and reference star positions are provided for all the fields investigated.

141.117 Low-frequency turnover in the spectrum of 3C 391.
W. Miller Goss.
Astron. Astrophys., Vol. 18, 484 - 486 (1972).

An investigation of 3C 391 has extended the knowledge of the properties of this supernova remnant to 31.4 GHz. The data lend no support to the hypothesis that a discrete H II region is the cause of the low-frequency absorption in the spectrum of 3C 391.

141.118 The NRAO 5-GHz radio source survey. II. The 140-ft "strong", "intermediate", and "deep" source surveys.
I. I. K. Pauliny-Toth, K. I. Kellermann, M. M. Davis, E. B. Fomalont, D. B. Shaffer.
Astron. Journ., Vol. 77, 265 - 284 (1972).

The 140-ft radio telescope at the National Radio Astronomy Observatory has been used at a frequency of 5 GHz to make three separate surveys for discrete sources. The surveys are believed to be essentially complete to 0.6, 0.25, and 0.09 flux unit (f. u.) and cover 0.97, 0.079, and 0.0062 sr, respectively. Positions accurate to a few seconds of arc have been determined with the Owens Valley Radio Observatory interferometer for almost all sources above the completeness levels of 0.6 and 0.25 f. u. Flux densities have been measured at 10.7 GHz and have been used together with the 5-GHz data and previous lower-frequency measurements to determine the spectra of the sources.

141.119 On the mass and distance of the quasi-stellar object 3C 273.
D. F. Falla, A. Evans.
Astrophys. Space Sci., Vol. 15, 395 - 400 (1972).

For the QSO 3C 273 we derive, on the basis of two different theoretical models, expressions for a lower limit to the mass of the QSO, as a function of its distance.

141.120 Electrographic UBV photometry of the jet in 3C 273.
G. E. Kron, H. D. Ables, A. V. Hewitt.
Publ. Astron. Soc. Pacific, Vol. 84, 303 - 305 (1972).

Magnitudes in the UBV system have been determined for the jet in 3C 273 by means of the electronic camera. The precision of the magnitudes is limited by the determination of that portion of the background light contributed by scattering and diffraction from the QSO.

141.121 Relativistic electrons in extragalactic radio sources.
V. N. Kuril'chik.
Uspekhi fiz. nauk, Vol. 105, 770 - 771 (1971). In Russian. Abstr. in Referativ. Zhurn. 51. Astron., 5.51.750 (1972).

141.122 An analysis of the distribution of redshifts of quasars and emission-line objects.
R. G. Lake, R. C. Roeder.
Journ. Roy. Astron. Soc. Canada, Vol. 66, 111 - 119 = Commun. David Dunlap Obs., Richmond Hill, No. 320 (1972).

The distribution of the redshifts of quasars and other emission-line objects has been examined using a new method of power spectrum analysis. Effects with periods of 0.070 and possibly 0.026 are found. The physical significance of the findings is, however, not clear.

141.123 Radiation-pressure-driven mass loss from quasi-stellar objects.
R. F. Mushotzky, P. M. Solomon, P. A. Strittmatter.
Astrophys. Journ., Vol. 174, 7 - 15 (1972).

It is shown that radiation pressure acting on the material in the outer envelopes of QSOs will cause mass outflow unless the mass M of the QSO exceeds a critical value M_s. If the QSOs are at cosmological distances, M_s is of order $10^{10} - 10^{11} M_\odot$, and otherwise scales according to intrinsic luminosity. It is also suggested that under radiative acceleration certain ratios of emission-line to absorption-line redshifts may be preferred as a result of a "line-locking" mechanism.

141.124 Meter-wavelength observations of galactic radio sources and the radiation from the galactic disk.
A. Parrish.
Astrophys. Journ., Vol. 174, 33 - 44 (1972).

The 1000-foot-diameter (305 m) radio telescope at the Arecibo Observatory has been used to observe the continuum emission of galactic sources at low radio frequencies. The following sources were observed: W49, W51, W56, NGC 1499, IC 410, NGC 2175, and NGC 2244. Observations were made at seven frequencies: 53, 65, 74, 89, 111, 196, and 318 MHz. The data presented here have been analyzed by a method based on one previously used by other workers (Wade 1958; Menon 1961; Terzian, Mezger, and Schraml 1968). The present method has been designed to account for the absorption of the background radiation by the H II region, and in doing so to measure the intensity of this radiation.

141.125 Interferometer visibility scintillation.
W. M. Cronyn.
Astrophys. Journ., Vol. 174, 181 - 200 (1972).

The relationship is developed between the spatial structure of random variations in refractive index, the spatial and temporal structure of the complex visibility function, and the intrinsic angular size of a source. The application is primarily to radio scattering in the interplanetary and interstellar media, where the variations in refractive index are proportional to the irregular structure of electron density.

141.126 **3C quasi-stellar objects and bright galaxies.**
C. Hazard, N. Sanitt.
Astrophys. Letters, Vol. 11, 77 - 82 (1972).

The suggestion by Burbidge *et al.* that the correlation between the positions of 3C quasi-stellar objects and bright galaxies provides evidence of a local class of QSO's is examined by the use of other source and galaxy lists. These studies failed to reveal any similar correlation. It is shown the results of the two investigations are difficult to reconcile on the assumption that the 3CR–QSO galaxy correlation has any physical significance.

141.127 **Radio sources in the Ohio Survey and Zwicky compact galaxies.** J. W. Warner.
Astrophys. Letters, Vol. 11, 83 - 86 (1972).

Significant coincidences are found between a sample of radio sources listed by the Ohio Survey and Zwicky's lists of compact galaxies. Positions of 21 Zwicky objects differ from the radio source positions by one standard deviation or less, although coincidences for only 8 of these 21 would be expected by chance. Finding charts, not previously published, are given for 19 objects.

141.128 **Variations optiques de la radiosource OJ 287.**
C. Bertaud, C. Pollas.
Inform. Bull. Variable Stars, (I.A.U. Commission 27), Konkoly Obs., Budapest, No. 621 (1972).

141.129 **Variability of radio sources and spectral type.**
B. H. Andrew, W. J. Medd, G. A. Harvey, J. L. Locke.
Nature, Vol. 236, 445 - 447 (1972).

Five and a half years of observations at the Algonquin Radio Observatory of eighty-four suspected variable sources have shown (i) that between 90% and 100% of sources with centimeter-excess spectra are variable and, (ii) that no source with a normal spectrum shows any evidence of variations.

141.130 **Measurements of the circular polarization of radio sources at 1.4 and 5 GHz.**
J. A. Roberts, J. C. Ribes, J. D. Murray, D. J. Cooke.
Nature, Phys. Sci., Vol. 236, 3 - 4 (1972).

Circular polarization at 1.4 or 5.0 GHz was detected in 9 out of 37 sources studied. When significant polarization was found at both frequencies it had the same sense. The results for self-absorbed sources suggest that the circular polarization is greatest at frequencies in the optically thick region and decreases steadily with increasing frequency.

141.131 **Colours and redshifts of quasars.** S. Goldsmith.
Nature, Phys. Sci., Vol. 236, 122 - 123 (1972).

The correlation between colours and redshifts of 130 quasars is examined. The observations are compared with previous theoretical works. A probable new minimum in U–B at $Z \sim 1.2$ is suggested.

141.132 **Lunar occultation observations of 24 radio sources.**
R. W. Clarke.
Australian Journ. Phys., Vol. 25, 215 - 232 (1972).

Some details of the structure of 24 radio sources have been deduced from measurements of the occultation of radio sources by the moon. A brief description is given of the observing technique and of the method of analysis.

141.133 **Cygnus A, a test of theories of radio sources.**
D. M. Mills.
Bull. American Astron. Soc., Vol. 4, 207 (1972). – Abstr. AAS.

141.134 **Characteristics of radio variables at 2.8 and 4.5 cm.**
W. J. Medd, J. L. Locke, G. A. Harvey, B. H. Andrew.
Bull. American Astron. Soc., Vol. 4, 207 (1972). – Abstr. AAS.

141.135 **A search at 2.8 cm for rapid variations in radio sources.**
J. M. MacLeod, B. H. Andrew, G. A. Harvey, W. J. Medd.
Bull. American Astron. Soc., Vol. 4, 207 (1972). – Abstr. AAS.

141.136 **Rapid expansion at λ3.8 cm in 3C 120.**
D. B. Shaffer, M. H. Cohen, K. I. Kellermann, D. L. Jauncey.
Bull. American Astron. Soc., Vol. 4, 207 - 208 (1972). – Abstr. AAS.

141.137 **Polarization structure of extragalactic sources.**
D. S. De Young, D. E. Hogg.
Bull. American Astron. Soc., Vol. 4, 215 (1972). – Abstr. AAS.

141.138 **Circular polarization of compact sources at 3240 MHz.** E. R. Seaquist.
Bull. American Astron. Soc., Vol. 4, 215 (1972). – Abstr. AAS.

141.139 **Small-pitch-angle synchrotron radiation from some extragalactic radio sources.**
R. I. Epstein, V. Petrosian.
Bull. American Astron. Soc., Vol. 4, 215 (1972). – Abstr. AAS.

141.140 **The accuracy of the NRAO interferometer positions.**
P. Brosche.
Bull. American Astron. Soc., Vol. 4, 215 - 216 (1972). – Abstr. AAS.

141.141 **First results from the bandwidth synthesis interferometer at the University of Texas Radio Astronomy Observatory.** F. N. Bash, J. N. Douglas, F. D. Ghigo, G. F. Moseley, G. W. Torrence.
Bull. American Astron. Soc., Vol. 4, 216 (1972). – Abstr. AAS.

141.142 **Interferometry: Effects of irregular propagation media.** W. M. Cronyn.
Bull. American Astron. Soc., Vol. 4, 216 (1972). – Abstr. AAS.

141.143 **Absorption lines in the spectrum of the quasar Ton 1530.** W. A. Morton, D. C. Morton.
Bull. American Astron. Soc., Vol. 4, 231 (1972). – Abstr. AAS.

141.144 **Absorption-line profiles in the quasar PHL 957.**
D. C. Morton, W. A. Morton.
Bull. American Astron. Soc., Vol. 4, 231 (1972). – Abstr. AAS.

141.145 **A kinematic quasi-stellar source model.**
M. P. Savedoff, B. Nilsen.
Bull. American Astron. Soc., Vol. 4, 231 (1972). – Abstr. AAS.

141.146 **Discovery of new infrared sources associated with W3.** C. G. Wynn-Williams, E. E. Becklin, G. Neugebauer.
Bull. American Astron. Soc., Vol. 4, 232 (1972). – Abstr. AAS.

141.147 **The effect of the interstellar gas on the continuum spectrum of the galactic radio source 3C391.**
M. A. Gordon.
Bull. American Astron. Soc., Vol. 4, 235 (1972). – Abstr. AAS.

141.148 **Statistical association of quasi-stellar radio sources and rich clusters of galaxies.**
J. A. Castro, R. V. Wagoner.
Bull. American Astron. Soc., Vol. 4, 258 (1972). – Abstr. AAS.

141.149 **Linear polarization of radio sources with appreciable redshift.**
P. P. Kronberg, R. G. Conway, J. A. Gilbert.
Bull. American Astron. Soc., Vol. 4, 270 (1972). – Abstr. AAS.

141.150 **Radio sources grouped in the NGC 7331 and Stephan's Quintet area.** H. Arp.
Bull. American Astron. Soc., Vol. 4, 270 - 271 (1972).
Abstr. AAS.

141.151 **Periodicity of 3C 273.** J. C. Wheeler.
Nature, Phys. Sci., Vol. 237, 102 - 104 (1972).
The purpose of this communication is to report the results of an application of the method of the Lebedev group to various versions of the 3C 273 data and to shot noise data. A critical discussion is presented of the claims for periodicity based on this method.

141.152 **On circularly polarized radiation from extragalactic radio sources.** H. Rosenberg.
Astron. Astrophys., Vol. 19, 66 - 70 (1972).
Two mechanisms are investigated, which might explain the recently observed circular polarization in quasi-stellar objects.

141.153 **Interpretation of rotation measures of radio sources.** M. Reinhardt.
Astron. Astrophys., Vol. 19, 104 - 108 (1972).
The rotation measures of 98 extragalactic radio sources with known redshifts are statistically investigated. A significant contribution to the rotation measures by a uniform field was found for quasars, but not for radio galaxies. The additional evidence provided by the rotation and dispersion measures of 19 pulsars suggests that the systematic behaviour of the rotation measures of quasars is to a large extent due to the galactic magnetic field, although an intergalactic contribution cannot be excluded. Estimates of the direction and strength of the uniform component of the galactic magnetic field are given.

141.154 **The counts of radio sources.** M. S. Longair, M. J. Rees.
Comments Astrophys. Space Phys., Vol. 4, 79 - 93 (1972).
We repeat the arguments which show that the counts of radio sources disagree with the predictions of all simple cosmological models in which the number of sources per unit comoving volume is supposed to remain unchanged with epoch ("source-conserving" models), and with those of steady-state theory.

141.155 **Intensitäts- und Polarisationsmessungen bei 3 cm Wellenlänge in Bochum-Sundern.**
E. Hörber, G. Kraus.
Mitt. Astron. Ges., No. 31, p. 100 - 101 (1972). – Abstract.

141.156 **Suche nach Quarkteilchen in Quasarspektren.** M. Iftikharuddin Khan.
Mitt. Astron. Ges., No. 31, p. 193 (1972). – Abstract.

141.157 **The mass and angular momentum losses from spinars.** S. P. S. Anand, M. M. Shara.
Astrophys. Space Sci., Vol. 16, 171 - 178 (1972).
Mestel's stellar wind theory is applied to estimate the mass and angular momentum losses for the recently proposed spinar model of quasars. If a spinar is uniformly rotating and has temperatures in the corotating regions of over a billion degrees, then it is found that all the rotational energy will be lost in 10^3 yr. Hence it is suggested that the temperatures of the corotating regions of spinars must be less than a billion degrees.

141.158 **Observations of linear polarization of radio sources at 7.2 cm.**
H. Tabara, D. Morris, N. Kawajiri, M. Konno.
Publ. Astron. Soc. Japan, Vol. 24, 301 - 308 = Tokyo Astron. Obs. Repr., No. 413 (1972).
The linear polarization of the radiation from thirteen

discrete sources, including seven variable sources, was observed at a frequency of 4.18 GHz.

141.159 **Gli oggetti quasi stellari.** C. Barbieri.
Atti XIII Riunione Soc. Astron. Italiana, Trieste 1969, (see 012.010), p. 141 - 145 (1970).

141.160 **A search for circular polarization in compact sources at 9 mm wavelength.** F. Biraud.
Astron. Astrophys., Vol. 19, 310 - 311 (1972).
No circular polarization, at a level of a few per cents, has been detected at 9 mm from any of 6 compact radio sources.

141.161 **Absorption-line profiles in the quasi-stellar object PHL 957.** D. C. Morton, W. A. Morton.
Astrophys. Journ., Vol. 174, 237 - 252 (1972).
This paper continues the analysis begun by Lowrance, Morton, Zucchino, Oke, and Schmidt on the high-resolution spectrum of PHL 957 ($z_{em} = 2.69$) obtained with the Princeton integrating television system at the coudé focus of the Hale telescope. We give equivalent widths for 28 lines and attempt to match Voigt profiles to most of the features in order to put limits on column densities and Doppler velocities.

141.162 **Absolute measurements of the flux of Cassiopeia A and Taurus A at 1.87 centimeters.**
G. T. Wrixon, J. R. Gott III, A. A. Penzias.
Astrophys. Journ., Vol. 174, 399 - 400 (1972).
Revised values of the flux at 16 GHz of Cas A and Tau A are presented. The presence of a previously reported variation in the flux of Tau A has been attributed to an instrumental effect.

141.163 **Spectroscopic observations of 22 quasi-stellar objects.** E. M. Burbidge, P. A. Strittmatter.
Astrophys. Journ., (Letters), Vol. 174, L57 - L62 (1972).
Spectroscopic observations of 22 QSOs are described, and redshifts of 20 are given. A corrected redshift for 3C 407 is included, and an error in that published for PKS 0226−038 is pointed out. The variability of the large-redshift 5C 2.56 is discussed. Lines of Si II have been found in OI 318.

141.164 **Variability of the optical intensity and plane polarization of OJ 287.** W. L. Williams, A. Rich, P. N. Kupferman, J. A. Ionson, W. A. Hiltner.
Astrophys. Journ., (Letters), Vol. 174, L63 - L64 (1972).
Photometric and plane polarimetric observations of the radio source OJ 287 have been made in the optical region. The photometric data show variations of 0.10 mag with a timescale of 6–7 days superposed on a total decrease of approximately 0.5 mag during 1972 February. The polarimetric data, which exhibit a similar decrease in plane polarization during this same period, suggest the possibility of a periodicity with a timescale of the order of 30 days.

141.165 **Observations of 3C 272.1 at 2.7 and 5.0 GHz.** J. M. Riley.
Monthly Notices Roy. Astron. Soc., Vol. 157, 349 - 357 (1972).
The radio source 3C 272.1, associated with the elliptical galaxy M84 in the Virgo cluster, has been observed at 2.7 and 5.0 GHz with half-power beam widths of $12'' \times 52''$ arc and $6''.5 \times 28''$ arc respectively. The maps obtained show that the source has a spiral structure. Linear polarization of up to 20 per cent in places has been found, consistent with a magnetic field predominantly perpendicular to the axis of the spiral.

141.166 **Accurate positions of radio sources at 408 MHz.** R. W. Hunstead.
Monthly Notices Roy. Astron. Soc., Vol. 157, 367 - 402 (1972).
The Molonglo radio telescope has been used to measure positions and flux densities for 314 small-diameter radio

sources between declinations of +19° and −80°. The mean estimated errors in position are less than 1".5 in right ascension and 2".0 in declination; this accuracy is confirmed by a limited series of comparisons with precise interferometer positions. A large fraction of previously suggested southern identifications is not supported by the more accurate radio positions. Source extensions are found in 25 per cent of the listed sources by application of a sensitive method for detecting the direction of elongation.

141.167　Polarization of radio sources at λ49 cm and λ73 cm.
R. G. Conway, J. A. Gilbert, P. P. Kronberg, R. G. Strom.
Monthly Notices Roy. Astron. Soc., Vol. 157, 443 - 459 (1972).

Measurements are presented of the flux density at λ73 cm and the linear polarization at λ49 and λ73 cm of extragalactic sources, mostly quasars. These measurements establish the degree to which the linear polarization decreases at long wavelengths. Evidence is presented that the depolarization does not occur in the galactic halo.

141.168　Extragalactic radio sources and angular resolution at 25×10^{-6} arc sec.
R. V. E. Lovelace, D. C. Backer.
Astrophys. Letters, Vol. 11, 135 - 138 (1972).

Detection or limits on weak intensity variations of extragalactic radio sources with time scale of about an hour at frequencies 2−20 GHz due to interstellar scintillations could provide information on very small-angle structure, $\lesssim 25 \times 10^{-6}$ arc sec. Over this frequency range the incoherent synchrotron theory of the emission of radio sources suggests minimum sizes in the range $(60-600) \times 10^{-6}$ arc sec and, because of the inverse Compton effect, brightness temperatures $\lesssim 10^{12}$K. Smaller angular structure could exist with higher brightness temperatures.

141.169　The structure of the radio sources PKS 0118+03, PKS 2053−20 and PKS 2057−17.　C. Hazard.
Astrophys. Letters, Vol. 11, 139 - 146 (1972).

This paper describes the detailed structure of three radio sources derived from occultation observations. Two of the sources, PKS 0118+03 and 2053−20, are basically double while PKS 2057−17 is possibly double. PKS 2053−20 and 2057−17 are identified with elliptical galaxies, and PKS 0118+03 is identified with a QSO. In all three cases, however, the radio components appear to be several seconds of arc in extent.

141.170　A search for rapidly varying radio sources.
G. A. Harvey, B. H. Andrew, J. M. MacLeod, W. J. Medd.
Astrophys. Letters, Vol. 11, 147 - 149 (1972).

An attempt to find new, rapidly-varying radio sources was made during the period October−November 1971. Nineteen sources were studied, but no new rapid variables were discovered.

141.171　Fine structure of 3C 225.
V. K. Kapahi, M. N. Joshi, G. Krishna.
Astrophys. Letters, Vol. 11, 155 - 158 (1972).

The radio positions and structures of components of the extragalactic radio source 3C 225 have been derived with a resolution better than 1 arc sec from a lunar occultation observed with the Ooty radio telescope at 326.5 MHz. The two known components of 3C 225, separated by about 6.25 arc min, have been further resolved into simple close doubles with compact subcomponents. The suggested optical identification for one of the two components and the relative orientation of the close doubles support the hypothesis that the two widely-separated components are physically unrelated.

141.172　Polarization measurements of radio sources at 9.55-

mm wavelength.　R. W. Hobbs, J. A. Waak.
Astron. Journ., Vol. 77, 342 - 344 (1972).

The position angle and degree of linear polarization at 31.4 GHz are given for 3C 120, 3C 273, 3C 274, 3C 345, 3C 405, and 3C 454.3. The measurements suggest a high rotation measure (810 rad/m²) for the radio galaxy 3C 274 (Virgo A). Upper limits are set for four other sources, where the uncertainty in the degree of polarization is such that the measured values do not differ significantly from zero.

141.173　Observations of some small-diameter radio sources at 408 MHz.　D. L. Jauncey, R. W. Hunstead.
Astron. Journ., Vol. 77, 345 - 349 (1972).

Accurate positions and flux densities at 408 MHz are given for a selection of 67 radio sources with small-diameter components or peculiar spectra. The mean positional accuracy is ~3" arc in each coordinate. In several cases the catalogued source is shown to be a close double or to have nearby companions. Comparisons with high frequency positions show that few of the sources have significant low frequency extentions.

141.174　Die Entwicklung starker Radioquellen.
G. M. Richter.
Sterne, 48. Jahrgang, p. 76 - 81 (1972).

141.175　Expected density of gravitational-lens quasars.
J. M. Barnothy, M. F. Barnothy.
Astrophys. Journ., Vol. 174, 477 - 482 (1972).

The expected density of gravitational-lens quasars is adequate, or higher than the observed density.

141.176　The flux density of Cassiopeia A at 1440 MHz and its rate of decrease.　J. W. Findlay.
Astrophys. Journ., Vol. 174, 527 - 528 (1972).

The flux density of Cas A at 1440 MHz has been measured over 11 years. The absolute value at epoch 1970.0 and its rate of change with time have been determined.

141.177　A grouping of radio sources in the area of NGC 7331 and Stephan's Quintet.　H. Arp.
Astrophys. Journ., (Letters), Vol. 174, L111 - L114 (1972).

An apparent concentration of radio sources is found in a region of about one square degree of sky centered between the large Sb spiral galaxy NGC 7331 and the smaller, disturbed, and generally higher-redshift neighboring galaxies which make up Stephan's Quintet.

141.178　Search for optical circular polarization in quasars and Seyfert nuclei.　J. D. Landstreet, J. R. P. Angel.
Astrophys. Journ., (Letters), Vol. 174, L127 - L129 (1972).

A search for optical circular polarization has been made in six quasars, the nuclei of three Seyfert galaxies, BL Lac, and OJ 287. With a typical accuracy of 0.1 percent no positive effect is found in any of these objects, several of which have been reported to show circular polarization at radio wavelengths.

141.179　Photoionization models for the emission-line regions of quasi-stellar and related objects.
G. M. MacAlpine.
Astrophys. Journ., Vol. 175, 11 - 30 (1972).

Detailed models are presented for the ionization and thermal structure of a gas which is ionized by ultraviolet radiation with a relatively flat, power-law spectrum. Various physical parameters of the calculations, including spectral slope and intensity of the input radiation, volume filling factor for the gas cloud, atomic number density, and helium abundance, are carefully examined in an effort to better understand their effects upon conditions in the nebula and on the predicted intensities of emission lines. Particular attention is given to 3C 273, but the results are applicable to the investigation of all quasi-stellar and related objects.

141.180 **OJ 287 = VRO 20.08.01.** K. Locher.
IAU Circ., No. 2380 (1972).

141.181 **OJ 287 = VRO 20.08.01** V. P. Tsesevich.
IAU Circ., No. 2389 (1972).

141.182 **Multiple redshifts in QSOs.** J. G. Cohen.
Nature, Vol. 237, 273 (1972).
A statistical procedure for examining the completeness
of the standard line list used in computer searches for absorption redshifts in multiple QSOs is given. It is concluded that the
list of expected absorption lines of Bahcall (Ap. J., Vol. 153,
679, 1968) is essentially complete. Hence more redshifts must
be present than have previously been found.

141.183 **Superlight velocity in quasars and their distances.**
G. M. Richter.
Nature, Phys. Sci., Vol. 237, 71 - 72 (1972).
The paper is a comment to a model by Cavaliere, Morrison
and Sartori which explains the recently observed apparent
superlight velocity in quasars as a geometrical effect (ringlike
flash). It is shown, that with aid of this model the triangulation
distances of quasars can be derived.

141.184 **Radio stars Beta Persei and Beta Lyrae.**
C. M. Wade, R. M. Hjellming.
Nature, Vol. 235, 270 - 271 = National Radio Astron. Obs.,
Green Bank, Repr., Ser. B, No. 313 (1972).
Following the discovery that the radio emission of the
Antares system originates from the B3 V component, observa-
tions of other binary star systems were undertaken with the
NRAO interferometer operating at 2,695 and 8,085 MHz.
Here we report the detection of radio emission from two well
known eclipsing binaries, β Persei and β Lyrae.

141.185 **Position and identification of the Cygnus X-1 radio
source.** C. M. Wade, R. M. Hjellming.
Nature, Vol. 235, 271 = National Radio Astron. Obs., *Green
Bank,* Repr. Ser. B, No. 314 (1972).
The initial measurements of the position of the radio
source associated with the X-ray source Cygnus X-1 were uncertain by 3 to 5 arc s. The results were consistent with two
possible identifications with stars. Here, we report the resolution of the ambiguity by an accurate determination of the
radio position.

141.186 **L'énigme des objets quasi-stellaires. I–V.**
J. Demaret.
Rev. Questions Sci., Vol. 140, 453 - 477; Vol. 141, 73 - 88,
257 - 284, 417 - 436, 514 - 537 = Univ. Liège, Inst. d'Astrophys., Coll. 8°, No. 604 (1969/70). – Review article.

141.187 **Observations at 408 MHz of radio sources from the
4C catalogue. IV. Declination range 20° to 0°.**
R. E. B. Munro.
Australian Journ. Phys., Astrophys. Suppl. No. 22, 51 pp.
(1972).
Radio positions and flux densities measured at 408 MHz
with the Molonglo radio telescope are given for 1392 sources
from the Fourth Cambridge catalogue in the declination range
20° to 0°. The mean spectral indices between 178 and 408
MHz are also considered in the light of these improved flux
densities.

141.188 **The distances to the quasars.** M. Schmidt.
Quarterly Journ. Roy. Astron. Soc., Vol. 13, 297 -
302 (1972). – Presented at the Woolley symposium (see
012.025).

141.189 **Ogromna pomeranja linija ka crvenome kraju u spek-
trima kvazara. [Large dark line in the spectrum of a
quasar.]** R. Danić.
Vasiona, Vol. 20, 35 - 36 (1972).

141.190 **Multiwavelength program for the active sources
3C 120, BL Lac and OJ 287.** E. E. Epstein.
Inform. Bull. Variable Stars, (IAU Commission 27), Konkoly
Obs., Budapest, No. 687 (1972).

141.191 **Quasi-stellar objects.**
K. T. Johansen, J. O. Petersen.
Fys. Tidssk., No. 203, Vol. 69, 57 - 78 (1971). In Danish.

141.192 **Quasi-stellar objects.** K. T. Johansen, J. O.
Petersen.
Fys. Tidssk. (*Denmark*), Vol. 69, No. 4/5, p. 107 - 132 (1971).
In Danish.
Deals with the energy source, construction and theories
evolved to explain the behaviour of quasars.

141.193 **A recent application of detection and estimation
practices in radio and radar astronomy.**
P. M. Rapier.
Spectrosc. Letters (*USA*), Vol. 4, 303 - 311 (1971).
Very-long-baseline-interferometry tests of the exploding
quasar 3C-279 show conclusively the presence of two components that are moving apart at six times the speed of light.
These truly super-light velocities of material objects are shown
to be consistent with a cosmological distance of 3 billion light
years and a sub-light Einsteinian abstract mathematical velocity of 0.95 C.

141.194 **Excitation of a CII line observed in quasars.**
A. R. G. Jackson.
Journ. Phys. B, Atomic and Molecular Phys., Vol. 5, L83 -
L86 (1972). – See Phys. Abstr., Vol. 75, No. 44889 (1972).

141.195 **A gravity generation of electromagnetic radiation
and the luminosity of quasars.** L. Motz.
Nuovo Cimento B, Ser. 11, Vol. 9B, 77 - 82 (1972).
A new mechanism is proposed for the emission of radiation by charged particles in a gravitational field whereby the
gravitational field is coupled directly to the radiation field of
the charge via the principle of equivalence. This leads to a
Larmor formula for the electromagnetic power emitted by a
charge at rest (on the average) in a gravitational field, which
can account for the spectral distribution and the total radiant
energy emitted by quasars. This gravitational mechanism for
the emission of electromagnetic radiation becomes more important than any other mechanism when the radius of the
quasar is near its Schwarzschild radius.

Source models with electron diffusion.
See Abstr. 061.049.

**The induced light pressure under astrophysical con-
ditions.** See Abstr. 063.029.

**Radiation pressure on a test particle in general rela-
tivity.** See Abstr. 066.075.

**Observations of Jupiter, Venus, and source 3C 273
at 2 and 8 mm wavelengths.** See Abstr. 099.071.

**Interplanetary scintillation studies at the Molonglo
Radio Observatory.** See Abstr. 106.015.

Radio flare on β Persei. See Abstr. 121.063.

Radio variations of β Persei and β Lyrae.
See Abstr. 121.091.

Photoelectric observations of BL Lacertae.
See Abstr. 122.020.

UBV observations of FU Ori, SU Tau and OJ 287.
See Abstr. 122.102.

Compact extragalactic nonthermal sources.
See Abstr. 122.133.

Radio observations of two supernova remnants,
HB 21 and IC 443, at 4170 MHz.   See Abstr. 125.017.

The effect of the interstellar gas on the continuum
spectrum of 3C 391.   See Abstr. 125.025.

The Parkes survey of 21-centimeter absorption in
discrete-source spectra. IV. 21-centimeter absorption measurements on low-latitude sources south of declination −46°.
See Abstr. 131.004.

High resolution measurement of the angular size of
the water vapor radio sources in W49.
See Abstr. 131.039.

Observations of sources of maser radio emission with
angular resolution 0.″0002.   See Abstr. 131.124.

Optical depths of neutral hydrogen clouds determined by polarization techniques.   See Abstr. 131.131.

Radio emission from small galactic nebulae at
606 MHz.   See Abstr. 132.017.

The X-ray spectrum of NGC 5128.
See Abstr. 142.003.

The X-ray spectrum of Centaurus A.
See Abstr. 142.019.

On the origin of the magnetic field in the extended
radio and X-ray source in the Coma cluster.
See Abstr. 142.036.

Further radio observations of Cygnus X-1.
See Abstr. 142.069.

Radio identifications of weak extragalactic X-ray
sources.   See Abstr. 142.070.

Possible identification of GX5-1.
See Abstr. 142.071.

Radio detection of Cygnus X-3.
See Abstr. 142.103.

Radio observations of Cygnus X-3.
See Abstr. 142.104.

H II regions in the Sco OB 1 association.
See Abstr. 152.007.

Anzeichen von Aktivität im galaktischen Zentrum.
See Abstr. 155.021.

The formaldehyde absorption of Sagittarius A at
+40 km/sec.   See Abstr. 155.030.

Galactic structure and the apparent size of radio
sources.   See Abstr. 155.032.

The influence of fast particles on the optical emission line spectrum of a nebula: The Balmer decrement.
See Abstr. 158.025.

Limitations on thermal and nonthermal models for
the radiation from extragalactic sources.
See Abstr. 158.027.

A high resolution radio continuum survey of M 51
and NGC 5195 at 1415 MHz.   See Abstr. 158.039.

Radio emission from compact objects.
See Abstr. 158.088.

A probable mechanism of repeated explosions of
compact objects.   See Abstr. 158.094.

Supermassive disks.   See Abstr. 158.095.

The spectral index−luminosity relation for radio
galaxies and quasi-stellar sources.   See Abstr. 158.104.

NGC 1068, 3C 273, and Scorpius X-1: Circular
polarization disputed.   See Abstr. 158.111.

Rapid change in the visibility function of the radio
galaxy 3C 120.   See Abstr. 158.112.

Infrared observations of galaxies and quasi-stellar
objects.   See Abstr. 158.134.

Activity in galaxies and quasars.
See Abstr. 158.142.

The NGC 520 chain of quasars and the "discrepant
redshift" of cluster galaxies.   See Abstr. 158.146.

The luminosity-spectral index relationship for radio
galaxies.   See Abstr. 158.165.

Models for the cosmic evolution of radio galaxies
and quasars.   See Abstr. 162.013.

Cosmological evidence from QSOs and radio galaxies.   See Abstr. 162.028.

Pulsars

search.

141.501 **Detection of strong interpulses from NP 0532.**
J. F. R. Gower, E. Argyle.
Astrophys. Journ., (*Letters*), Vol. 171, L23 - L26 (1972).

Observations of NP 0532 at 146 MHz show that the sporadic bright pulses are associated with the interpulse as well as with the main pulse.

141.502 **Periodicity in the radiofrequency spectrum of the pulsar CP 0328.**
P. A. Sturrock, S. Antiochos, P. Switzer, J. Vallée.
Astrophys. Journ., (*Letters*), Vol. 171, L27 - L30 (1972).

Long-term averaging of a sequence of wide-band radiofrequency spectra of CP 0328 reveals a periodicity not apparent in the original spectra. This may be caused by a mechanism intrinsic to the source, or by a propagation mechanism distinct from ordinary scintillation.

141.503 **An upper mass limit for isolated pulsars.**
J. G. Hills.
Astrophys. Letters, Vol. 10, 27 - 30 (1972).

An isolated neutron star having no energy sources except rotation cannot brake this rotation by the emission of either photons or cosmic ray particles if it is more massive than about 1.7 $M_\odot$. Neutron stars less massive than this limit are likely to release much of their rotational energy by ejections of low-energy cosmic rays.

141.504 **Parameters of 61 pulsars.**
R. N. Manchester, J. H. Taylor.
Astrophys. Letters, Vol. 10, 67 - 70 (1972).

The principal parameters of 61 pulsars are listed in tabular form, together with references to the sources of information.

141.505 **Pulsars, supernova remnants, and the pulsar period-luminosity relation.** G. S. Tsarevsky.
Astrophys. Letters, Vol. 10, 71 - 76 (1972).

The relative space positions of pulsars and supernova remnants (SNR) are compared on the basis of the 'run-away' pulsar hypothesis. The value of the maximum observable separation, 130 pc, obtained from predictions for the upper limit of pulsar velocity and SNR age, is used as the criterion for selecting pulsar-SNR pairs. There is a good correlation between pulsar dispersion measure and supernova distance for the pairs thus selected (20 SNR's and 23 pulsars).

141.506 **Interplanetary scintillation, interstellar scattering of two pulsars.** G. A. Zeissig, R. V. E. Lovelace.
Astron. Astrophys., Vol. 16, 190 - 196 (1972).

The theory of interplanetary scintillations is discussed for the situation where the incident radiation is scattered in angle and amplitude modulated by propagation through the more distant interstellar medium. Observations of interplanetary scintillations of the pulsars CP 0950 and NP 0527 at 318 MHz are described. The variation of the modulation index with solar elongation is derived.

141.507 **Search for pulsed γ-ray emission above 50 MeV from the Crab nebula pulsar NP-0532.**
J. P. Leray, J. Vasseur, J. Paul, B. Parlier, M. Forichon, B. Agrinier, G. Boella, L. Maraschi, A. Treves, L. Buccheri, A. Cuccia, L. Scarsi.
Astron. Astrophys., Vol. 16, 443 - 458 (1972).

A search for pulsed γ-ray emission above 50 MeV from the Crab nebula has been conducted with a balloon borne spark-chamber. The results from six flights performed in 1969 are analyzed in this paper, but only two of the flights were sufficiently accurate in time to provide good data for pulsar re-

141.508 **Pulsar rotation and dispersion measures and the galactic magnetic field.** R. N. Manchester.
Astrophys. Journ., Vol. 172, 43 - 52 (1972).

Observations of pulsar polarization and pulse time of arrival at frequencies between 250 and 500 MHz have been used to determine rotation and dispersion measures for 19 and 21 pulsars, respectively. These measurements have been used to calculate mean line-of-sight components of the magnetic field in the path to the pulsars. The observations are consistent with a relatively uniform field of about 3.5 microgauss directed toward about $l = 90°$ in the local region, but appear to be inconsistent with the helical model for the local field.

141.509 **The braking index and period-pulse-width distribution of pulsars.** D. H. Roberts, P. A. Sturrock.
Astrophys. Journ., Vol. 172, 435 - 441 (1972).

The braking index n and the period-pulse-width distribution of pulsars are reinvestigated by suggesting a new assumption for R_Y, the radius of the neutral points. This model leads to new estimates of the surface magnetic field strength B of pulsars, giving $B = 10^{11.2}$ gauss for the Crab pulsar, and indicates that pulsar magnetospheres are subject to irregular changes associated with irregularities in period.

141.510 **Stability of the Crab pulsar.**
P. Horowitz, C. Papaliolios, N. P. Carleton.
Astrophys. Journ., (*Letters*), Vol. 172, L51 - L54 (1972).

Optical pulses from the Crab pulsar show none of the amplitude or pulse-shape variations associated with the radiofrequency pulsar. Observations designed to detect polar precession were also negative.

141.511 **Short-term variations in pulsars.** D. C. Backer.
Bull. American Astron. Soc., Vol. 3, 452 (1971).
Abstr. AAS.

141.512 **Results of the 1971 Crab nebula pulsar. Surveillance program at Arecibo Observatory.**
J. M. Rankin, D. B. Campbell, C. C. Counselman.
Bull. American Astron. Soc., Vol. 3, 463 (1971). – Abstr. AAS

141.513 **Very long baseline interferometer observations of NP 0532 at 121.6 MHz.** W. C. Erickson,
T. B. H. Kuiper, T. A. Clark, S. H. Knowles, J. J. Broderick.
Bull. American Astron. Soc., Vol. 3, 463 (1971). – Abstr. AAS

141.514 **Average radio spectra of 13 pulsars.**
J. M. Comella.
Bull. American Astron. Soc., Vol. 3, 463 (1971). – Abstr. AAS.

141.515 **Observations of the polarization properties of pulsars at 430 MHz.**
D. B. Campbell, J. M. Rankin, D. C. Backer.
Bull. American Astron. Soc., Vol. 3, 463 (1971). – Abstr. AAS.

141.516 **Observations of pulsar PSR 0833-45.**
P. E. Reichley, G. S. Downs.
Bull. American Astron. Soc., Vol. 3, 463 (1971). – Abstr. AAS.

141.517 **Model of pulsar magnetospheres.**
D. H. Roberts, P. A. Sturrock.
Bull. American Astron. Soc., Vol. 3, 463 - 464 (1971). – Abstr. AAS.

141.518 **Maser emission mechanism from pulsars.** H. Y. Chiu.
Bull. American Astron. Soc., Vol. 3, 464 (1971).
Abstr. AAS.

141.519 **Microquakes and macroquakes in neutron stars.**

D. Pines, J. Shaham.
Nature, Phys. Sci., Vol. 235, 43 - 49 (1972).

Many of the properties of pulsars, particularly those of the Crab nebula and Vela pulsars, can be explained in terms of microquakes and macroquakes.

141.520 **Evidence for hard X-ray pulsations from the Vela pulsar.** F. R. Harnden, Jr., W. N. Johnson III, R. C. Haymes.
Astrophys. Journ., (*Letters*), Vol. 172, L91 - L94 (1972).

A source of hard X-radiation has been detected in the general direction of the Vela X supernova remnant. The data obtained during a 1970 balloon flight also show evidence of periodic hard X-ray pulsations with a frequency near that of the radio pulsar PSR 0833-45.

141.521 **Preliminary results of pulsar CP 1133 observations using Cerenkov light registration from extensive air showers.**
A. A. Stepanian, B. M. Vladimirsky, I. V. Pavlov, V. P. Fomin.
Izv. Krymskoj Astrofiz. Obs., Vol. 43, 42 - 48 (1971).
In Russian.

An installation for the observation of Cerenkov flashes from extensive air showers is described. A choice of essential parameters of the apparatus is discussed. In order to find an optimum field-of-view measurements of the angular size of Cerenkov flash were made. The dependence of a receiving cone upon the minimum energy of the showers registered has been investigated. Pulsar CP 1133 has revealed some flux excess equal to 2.3 standard deviation.

141.522 **Pulsar parameters from timing observations.**
R. N. Manchester, W. L. Peters.
Astrophys. Journ., Vol. 173, 221 - 226 (1972).

Timing observations made over an 18-month interval have been used to compute periods, period derivatives, and positions for 19 pulsars.

141.523 **The structure of pulsar magnetospheres.**
D. H. Roberts, P. A. Sturrock.
Astrophys. Journ., (*Letters*), Vol. 173, L33 - L37 (1972).

Currents drawn from the neutron-star surface may lead to surface heating and hence to evaporation. Substantial accumulation of evaporated gas can occur only in the "force balance" regions where gravitational and centrifugal forces balance. For this model, the braking index $n = 7/3$ and the expected period-pulse-width distribution and period-age distribution agree well with observational data.

141.524 **An unusual polarization feature in PSR 0833—45.**
P. M. McCulloch, P. A. Hamilton, J. G. Ables, M. M. Komesaroff.
Astrophys. Letters, Vol. 10, 163 - 165 (1972).

Polarization measurements of PSR 0833—45 at 400 MHz are reported. An unusual polarization feature, namely a very rapid sweep in position angle during the initial phase of the pulse, was observed.

141.525 **Vier Jahre Pulsarforschung. I.** K.-H. Schmidt.
Sterne, 48. Jahrgang, p. 11 - 15 (1972).

141.526 **Short term intensity variations of pulsars in the frequency range 70—115 MHz. I. Correlation measurements.**
V. N. Brezgunov, V. N. Zlobin, V. A. Udal'tsov.
Astron. Zhurn. Akad. Nauk SSSR, Vol. 49, 279 - 285 (1972).
In Russian. English translation in Soviet Astron. AJ, Vol. 16, No. 2.

A correlation analysis of the short term intensity variations of pulsars CP 0808, 0834, 0950 and 1133 has been made. Periodical variations in all pulsars were detected to be observed

at several frequencies.

141.527 **A new pulsar—supernova association.**
M. I. Large, A. E. Vaughan.
Nature, Phys. Sci., Vol. 236, 117 - 120 (1972).

The pulsar PSR 1154-62 may be associated with a recently discovered radio supernova remnant.

141.528 **On the problem of pulsar distances.**
G. Cavallo, A. Ventura.
Astron. Astrophys., Vol. 18, 287 - 293 (1972).

The justifications for a period-luminosity function for pulsars are examined in detail, and consequences are discussed. Distances to pulsars are derived and these new estimates are compared with recent work on the subject, based on the dispersion measure.

141.529 **Observations of the 21-cm hydrogen-emission line in the direction of 25 pulsars.**
R. Sancisi, M. Klomp.
Astron. Astrophys., Vol. 18, 329 - 334 (1972).

The 25-m Dwingeloo radiotelescope was used to obtain 21-cm emission line profiles in the direction of 25 pulsars. An atlas of the profiles and a table of hydrogen column densities are given.

141.530 **Una possibile causa di emissione gamma da parte di supernovae.** R. F. Femiano.
Mem. Soc. Astron. Italiana, Nuova Ser., Vol. 43, 65 - 68 (1972).

Pulsars, originated in the core of supernovae, can be detected by measuring the γ-ray flux produced by the decay of neutral mesons generated by collisions among the protons, emitted by the pulsar, and the gas in the supernova envelope.

141.531 **Pulsar timing residuals.** F. C. Michel.
Comments Astrophys. Space Phys., Vol. 4, 47 - 52 (1972).

The Crab and Vela pulsars have undergone abrupt, but very small, increases in pulsation period. These changes are superimposed on a general systematic slowing down of the pulsation rate. In addition these pulsars undergo a variable "wander" in period, the physical significance of which remains obscure.

141.532 **γ-ray absorption for a beam model of NP 0532.**
F. W. Stecker. S. Tsuruta.
Nature, Phys. Sci., Vol. 235, 8 - 9 (1972).

A model is presented for estimating the absorption of γ rays in the vicinity of NP 0532 by estimating the effect of beaming as indicated by the pulsations themselves. It is concluded that absorption will not occur below 10 GeV energy even if the γ radiation originates at the surface of the pulsar. This conclusion contradicts previously published results by other authors who concluded that absorption could occur above 1 MeV.

141.533 **Second speed-up of the Crab pulsar.**
E. Lohsen.
Nature, Phys. Sci., Vol. 236, 70 - 71 (1972).

Optical timing of the Crab pulsar revealed a speed-up of $\Delta\nu = 0.0046$ periods per day $= 1.75\times10^{-9}\nu$ at 1971 Oct. 26.2 = JD 2441 250.7, which was apparently not a single, sudden event. The available data indicate that this speed-up was similar to that in September 1969.

141.534 **On the nature of the radio emission of pulsars.**
A. K. Yukhimuk.
Astrofizika, Vol. 7, 611 - 615 (1971). In Russian. English translation in Astrophysics, Vol. 7, No. 4.

The radiation mechanism of a rotating and pulsating

neutron star is examined. It is supposed that the radiation may originate when relativistic electrons move on spiral orbits along the force lines emerging from the magnetic poles of pulsars.

141.535   Improved data for eight Molonglo pulsars.
   A. E. Vaughan, M. I. Large.
Monthly Notices Roy. Astron. Soc., Vol. 156, 25P - 26P (1972).

   The north-south arm of the Molonglo Cross has been used to measure pulsar declinations and periods.

141.536   Discovery of three pulsars.
   A. E. Vaughan, M. I. Large.
Monthly Notices Roy. Astron. Soc., Vol. 156, 27P - 28P (1972).

   The radio source survey in progress at Molonglo now incorporates a concurrent search for pulsars. Brief details of the equipment and of three new pulsars are given.

141.537   Pulse broadening due to multiple scattering in the interstellar medium.   I. P. Williamson.
Monthly Notices Roy. Astron. Soc., Vol. 157, 55 - 71 (1972).

   Pulse broadening is investigated in the limit of strong multiple scattering, in which ray optics is applicable. Two cases are considered: the scattering medium is assumed either to be concentrated in a small fraction of the line of sight near the pulsar, or to extend over the whole line of sight. Analytical solutions for the probability density function are found for these two cases. The relation between this exponential decay time constant and the r. m. s. broadening of the angular size of the source is found, and compared with that expected on the basis of the 'thin-slab' model in which the scattering medium is effectively confined to a plane approximately midway between source and observer.

141.538   Improved data for four pulsars.
   A. G. Lyne, F. G. Smith.
Monthly Notices Roy. Astron. Soc., Vol. 157, 15P - 16P (1972).

   Periods and pulse shapes, measured at 408 MHz, are given for the pulsars PSR 0301 + 19, 1112 + 50, 1541 + 09 and 2154 + 40.

141.539   On the radiation mechanism of pulsars.
   M. Grewing, H. Heintzmann.
IAU Colloquium No. 15, (see 012.006), p. 182 - 184 (1972).

141.540   Corequakes and the Vela pulsar.
   D. Pines, J. Shaham, M. Ruderman.
Nature, Phys. Sci., Vol. 237, 83 - 84, 96 (1972).

   The speedups of the Vela pulsar can be explained by postulating the occurrence of corequakes in a solid core.

141.541   Detection of pulsed gamma rays of $\sim 10^{12}$ eV from the pulsar in the Crab nebula.   J. E. Grindlay.
Astrophys. Journ., (*Letters*), Vol. 174, L9 - L17 (1972).

   The previous evidence we have given for the detection of pulsed γ-rays of $\sim 10^{12}$ eV from NP 0532 has been confirmed by an additional 99 drift scans on the Crab in 1971 November and December. Again, only those extensive air showers detected to be possibly initiated by γ-rays showed a 4.5 σ excess at the phase of the optical interpulse. The ratio of interpulse to primary pulse is $\sim 3.5 : 1$, and the spectrum appears consistent with an extrapolation from the X-ray data.

141.542   Polarization of the Crab pulsar radiation at low radio frequencies.
R. N. Manchester, G. R. Huguenin, J. H. Taylor.
Astrophys. Journ., (*Letters*), Vol. 174, L19 - L23 (1972).

   Pulse shapes and linear polarization parameters for the Crab pulsar have been determined at four frequencies between 110 and 160 MHz. The observations show that the precursor component remains highly polarized at these frequencies, and that additional pulse components, not all polarized at the same position angle, are also present.

141.543   The effect of near fields on the pulsar acceleration of particles.   R. M. Kulsrud.
Astrophys. Journ., (*Letters*), Vol. 174, L25 - L26 (1972).

   It is pointed out that, if the near field is included in the calculation of the acceleration of particles from pulsars, then the electric field can be either bigger or smaller than the magnetic field. This can lead to an appreciable increase in the amount of acceleration previously calculated from the radiation field alone if the phase is such that E exceeds B. E and B are the particle motion in the y- and z-direction respectively.

141.544   The height distribution of giant pulses from NP 0532.
   E. Argyle, J. F. R. Gower.
Bull. American Astron. Soc., Vol. 4, 216 (1972). – Abstr. AAS.

141.545   The interstellar scintillation pattern of PSR 0329+54.
   J. A. Galt, A. G. Lyne.
Bull. American Astron. Soc., Vol. 4, 221 (1972). – Abstr. AAS.

141.546   Maser theory of pulsars.   L. Mertz.
   Bull. American Astron. Soc., Vol. 4, 221 (1972). – Abstr. AAS.

141.547   Geometrical pulsar radiation model.
   Y. Y. Van, A. K. Dunker.
Bull. American Astron. Soc., Vol. 4, 222 (1972). – Abstr. AAS.

141.548   On the tangential injection of relativistic particles into the wave zone of a pulsar.
R. N. Henriksen, D. R. Rayburn.
Bull. American Astron. Soc., Vol. 4, 223 (1972). – Abstr. AAS.

141.549   Ground state binding energies and bound-free cross sections for atoms in pulsar atmospheres.
L. C. Rosen.
Bull. American Astron. Soc., Vol. 4, 271 (1972). – Abstr. AAS.

141.550   Pulsar flux-density spectra.   D. C. Backer.
   Astrophys. Journ., (*Letters*), Vol. 174, L157 - L161 (1972).

   Pulsars 0950+08 and 1133+16 were observed at 1.4 and 5.0 GHz. A break in their equivalent continuum flux-density spectra near 1.0 GHz suggests a critical scale length of 30 cm in the spatial spectrum of coherently emitting charges.

141.551   Pulsar magnetosphere.   J. M. Cohen, A. Rosenblum.
   Astrophys. Space Sci., Vol. 16, 130 - 136 (1972).

   In this paper a self-consistent model of the closed field lines of a pulsar magnetosphere is given. Using this model, it is shown that, close to the star, the assumptions of Goldreich and Julian are justified. Their results are extended to the oblique rotator as well as to stars with magnetic multipoles of arbitrary order and arbitrary orientation.

141.552   Pulsars en hun afstanden.   F. P. Israel.
   Hemel en Dampkring, Vol. 70, 178 - 180 (1972).

141.553   Three years with pulsars.   A. Hewish.
   Mitt. Astron. Ges., No. 31, p. 15 - 21 (1972). –
Karl Schwarzschild lecture 1971.

141.554   Fine structure of spectra of radio emission of pulsars.
   Yu. P. Shitov.
Astron. Zhurn. Akad. Nauk SSSR, Vol. 49, 470 - 482 (1972).
In Russian. English translation in Soviet Astron. AJ, Vol. 16,

No. 3.

The frequency-correlation radii of the fine structure of spectra of seven pulsars (CP 0808, 0834, 0950, 1133, 1919, AP 1237, HP 1507) caused by scattering of their radio emission on irregularities of the interstellar plasma have been measured. The observations have been carried out at the radio astronomy station of the P. N. Lebedev Physical Institute of the USSR in Pushchino, using a twelve-channel radiometer at frequencies 105 and 63 MHz.

141.555 **Stelle di neutroni e pulsars.** F. Pacini.
Atti XIII Riunione Soc. Astron. Italiana, Trieste 1969, (see 012.010), p. 131 - 139 (1970).

141.556 **Injection of relativistic particles into a pulsar wave zone by a near-zone stellar wind.**
R. N. Henriksen, D. R. Rayburn.
Astrophys. Letters, Vol. 11, 107 - 112 (1972).

We show that a relativistic stellar-breeze solution gives a precise description of 'magnetic sling' acceleration of the kind first suggested in the Gold pulsar model. Relativistic particles are produced by corotation at the Dicke-Alfvén point (near the light radius) in a low particle-flux limit, and are subsequently injected into the wave-zone. We tentatively identify the radio precursor pulse of the Crab pulsar NP 0531 with the radiation produced by the corotating charged particles.

141.557 **Pulsar speedups related to metastability of the superfluid neutron-star core.** R. E. Packard.
Phys. Rev. Letters, Vol. 28, 1080 - 1082 (1972).

The sudden speed changes observed in the Vela and Crab pulsars may be caused by transitions between metastable flow states in the superfluid interior of the star.

141.558 **Vier Jahre Pulsarforschung. II.** K.-H. Schmidt.
Sterne, 48. Jahrgang, p. 81 - 86 (1972).

141.559 **Optical timing of the Crab pulsar, NP 0532.**
P. E. Boynton, E. J. Groth, D. P. Hutchinson, G. P. Nanos, Jr., R. B. Partridge, D. T. Wilkinson.
Astrophys. Journ., Vol. 175, 217 - 241 (1972).

Absolute times of arrival of NP 0532 pulses have been measured over a 2-year period. The data are shown to be consistent with a cubic polynomial which describes the secular slowdown, a sudden increase and subsequent exponential decay of the frequency (the glitch of 1969 September 29), and an intrinsic $1/f$ noise component in the frequency.

141.560 **Parameters for two new pulsars.** K. R. Lang.
IAU Circ., No. 2381 (1972).

141.561 **PSR 2154+40.**
J. Sutton, C. Salter, G. Colla, R. Sancisi.
IAU Circ., No. 2386 (1972).

141.562 **Radiation from flares near the magnetic poles of pulsars.** W. M. Glencross.
Nature, Vol. 237, 157 - 158 (1972).

As a neutron star rotates about its axis the open magnetic field lines, which are anchored to the star at one end and to the 'speed of light cylinder' at the other end, becomes twisted. A region in which a high current density flows will exist between these open field lines and the closed lines which co-rotate with the star. If this current is interrupted, the kinetic energy of the charged carriers will be released within a small volume across which a potential difference exceeding 10^{14} V is set up. The way in which this occurs has been described by Alfvén and Carlqvist (Solar Physics, Vol. 1, 220, 1967) as a possible mechanism for solar flares. Electrons and protons accelerated in the large electric field at the instability radiate photons by the synchrotron mechanisms as they spiral in the magnetic field of the

star. The pulsed nature of the emission arises because the emitting region can only be observed during part of each stellar rotation.

141.563 **Classification of pulsar rotating electromagnetic fields.** V. G. Endean.
Nature, Phys. Sci., Vol. 237, 72 - 73 (1972).

It is shown that many important features of the pulsar electromagnetic field which is stationary in the rotating frame may be expressed in terms of the quantity 'scalar potential minus (rotation speed times azimuthal component of vector potential)'. Several field results and a relativistically exact constant of the motion for a charged particle in the field are obtained using this quantity. Implications for the different possible models of the pulsar magnetosphere are discussed.

141.564 **Pulsare – Die seltsamsten Sterne.** E. Obreschkow.
Orion Schaffhausen, 30. Jahrgang, p. 106 - 107 (1972). – Report on a lecture, held by R. Kippenhahn, Konstanz, 1972 January 20.

141.565 **The structure of pulsar magnetospheres.**
R. H. Roberts, P. A. Sturrock.
Inst. Plasma Res., Stanford Univ., Stanford, California. SUIPR Report No. 450, 12 pp. (1972).

Currents drawn from the neutron-star surface may lead to surface heating and hence to evaporation. Substantial accumulation of evaporated gas can occur only in the "force-balance" regions where gravitational and centrifugal forces balance. It is argued that the Y-type neutral points, separating closed and open field lines, must be at the same location. For this model, the braking index n = 7/3 and the expected period-pulse-width distribution and period-age distribution agree well with observational data. A glitch is interpreted as an instability leading to the ejection and/or dumping of accumulated gas.

141.566 **Pulsar theory II. Radiation mechanisms.**
P. Goldreich, F. Pacini, M. J. Rees.
Comments Astrophys. Space Phys., Vol. 4, 23 - 28 (1972).

We outline the main features of some models for the radiation mechanism already proposed by other authors.

141.567 **Long-term intensity variations of pulsars observed at two frequencies.** K. H. Hesse.
Nature, Phys. Sci., Vol. 235, 27 - 29 (1972).

Pulsars are known to exhibit irregular intensity variations with time-scales of weeks and months. For pulsars CP 0808, CP 0834 and CP 1919 the observations have been continued at 81.5 MHz and extended to 151 MHz. A comparison of the daily intensities at the two frequencies has established that they are correlated. It seems therefore that long term variations can be considered to be intrinsic to the pulsar.

141.568 **Characteristics of subpulses of pulsar CP 1133.**
Yu. I. Alekseev, V. V. Vitkevich, V. M. Malofeev.
Nature, Phys. Sci., Vol. 235, 167 - 168 (1972).

We have compared the amplitude and polarization characteristics of the two subpulses in the radio emission from pulsar CP 1133. This comparison of some characteristics of the radio emission in the two subpulses of the pulsar CP 1133 and, in particular, the noticeable differences in their amplitude-time and polarization characteristics lead us to the conclusion that there seem to be two centres of radio emission with highly different parameters. Our results suggest that pulsar CP 1133 at least should be considered as having two active radiating regions which are located side by side. The closeness of these regions having different polarization characteristics suggests that the radio emission is influenced by the local magnetic field of the pulsar and not by the common dipole field as has been supposed.

141.569 The pulsars. F. G. Smith.
Electron. & Power (*GB*), Vol. 18, 57 - 59 (1972).

141.570 Pulsar data and the dispersion relation for light.
J. M. Rawls.
Phys. Rev. D, Particles and Fields, Vol. 5, 487 - 489 (1972).
Recent observations of pulsed γ radiation from NP0532 provide a very accurate check on the dispersion relation obeyed by light. The data are sufficient to rule out the mechanism proposed by Pavlopoulos for breaking Lorentz invariance.

141.571 Pulsars. D. Ter Haar.
Phys. Rep. Phys. Letters, Section C, (*Netherlands*), Vol. 3C, No. 2, p. 57 - 125 (1972).
A survey is given of various experimental data about pulsars. This is followed by a discussion of various attempts to construct theoretical models to account for the observational data.

The physics of pulsars. See Abstr. 003.088.

Time transfer using near-synchronous reception of optical pulsar signals. See Abstr. 035.016.

Results and problems in the investigation of the synchrotron instability. See Abstr. 061.001.

Equations of the transfer of electrons and photons of large energies in magnetic fields. See Abstr. 063.011.

The induced light pressure under astrophysical conditions. See Abstr. 063.029.

Critical core mass for carbon-detonation supernovae. See Abstr. 065.012.

The evolution of growing stellar cores up to carbon ignition: From whence the pulsars? See Abstr. 065.021.

Formation of neutron star spots and its connection with pulsars. I. See Abstr. 065.084.

Carbon ignition in degenerate stellar cores. See Abstr. 065.089.

Electron gas in superstrong magnetic fields: Wigner transition. See Abstr. 065.116.

Relativistic beaming. See Abstr. 066.065.

Recent observations of flare stars and pulsars. See Abstr. 122.023.

The possible line feature in the X-ray background. See Abstr. 125.001.

A model of the interstellar medium based on the interstellar calcium and sodium lines. See Abstr. 131.026.

The physical state of nearby interstellar space and the distances and space density of pulsars. See Abstr. 131.153.

New X-ray pulsar near the Crab nebula. See Abstr. 142.050.

Search for X-rays above 100 keV from the vicinity of NP 0527. See Abstr. 142.072.

Pulsating X-ray sources. See Abstr. 142.102.

On the acceleration of charged particles to cosmic ray energies. See Abstr. 143.059.

Pulsar-produced cosmic rays and the origin of the light elements. See Abstr. 143.065.

Densité spatiale des étoiles Ap. See Abstr. 155.005.

Errata

141.901 Errata: 'Observations of Sgr A at 22.2 GHz with a beamwidth of 1.5 arcminutes' [Bull. American Astron. Soc., Vol. 3, 364 (1971)].
B. G. Leslie, M. L. Meeks, S. Rogers.
Bull. American Astron. Soc., Vol. 3, 502 (1971).

141.902 Errata: 'Hard X-ray spectrum of NP 0532' [Nature, Phys. Sci., Vol. 233, 153 - 155 (1971)].
C. Cavani, F. Frontera, F. Fuligni, D. Brini.
Nature, Phys. Sci., Vol. 235, 40 (1972).

141.903 Erratum: 'Mechanism for the delay of polarization minima in the optical pulsar NP 0532' [Nature, Phys. Sci., Vol. 234, 86 - 87 (1971)]. D. C. Ferguson.
Nature, Phys. Sci., Vol. 235, 80 (1972).

141.904 Erratum: "Evidence for an optically thick component in the galactic radio source Sagittarius B_2" [Astrophys. Journ., (*Letters*), Vol. 165, L87 - L93 (1971)].
R. W. Hobbs, S. B. Modali, S. P. Maran.
Astrophys. Journ., (*Letters*), Vol. 173, L45 (1972).

142 X Ray-, Gamma Ray-Sources

142.001 X-ray spectra of discrete sources in Cygnus.
R. D. Bleach, E. A. Boldt, S. S. Holt, D. A. Schwartz,
P. J. Serlemitsos.
Astrophys. Journ., Vol. 171, 51 - 54 (1972).
X-ray spectral data from a rocket-borne proportional
counter exposure to discrete sources in Cygnus are presented.
The data from Cyg X-1, Cyg X-2, and Cyg X-3 have sufficient
statistical significance to clearly indicate mutually exclusive
spectral forms for the three. Upper limits are presented for
X-ray intensities above 2 keV for Cyg X-4 and Cyg X-5 (Cyg-
nus Loop).

**142.002 Cosmic-ray effects on diffuse gamma-ray measure-
ments.** G. J. Fishman.
Astrophys. Journ., Vol. 171, 163 - 167 (1972).
Calculations and experimental evidence from 600-MeV
proton irradiation indicate that cosmic-ray-induced radioactivity in detectors used to measure the diffuse γ-ray background
produces a significant counting rate in the energy region
around 1 MeV. These counts may be responsible for the observed flattening of the diffuse photon spectrum at this energy.

142.003 The X-ray spectrum of NGC 5128.
M. Lampton, B. Margon, S. Bowyer, W. Mahoney,
K. Anderson.
Astrophys. Journ., (*Letters*), Vol. 171, L45 - L50 (1972).
The X-ray spectrum of NGC 5128 (Cen A) obtained by
sounding rocket and balloon observations is presented in the
range 1–180keV. NGC 5128 may be fitted by a power law
with a photon index $-2 \le n \le -1.45$. Inverse Compton models are presented, and constraints on their parameters are discussed.

**142.004 Spectrum of the cosmic X- and gamma ray back-
ground in the energy range 1 keV–1 MeV.**
K. Kasturirangan , U. R. Rao.
Astrophys. Space Sci., Vol. 15, 161 - 166 (1972).
Available satellite, rocket and balloon observations on
cosmic X- and gamma ray background are critically examined
to understand the spectral characteristics of the radiation. Appropriate corrections have been applied to the balloon observations to account for the multiple Compton scattering of X-rays
in the atmosphere.

**142.005 Some results of the search for point sources of high-
energy gamma rays.**
B. M. Vladimirsky, I. V. Pavlov, A. A. Stepanjan, V. P. Fomin.
Astron. Zhurn. Akad. Nauk SSSR, Vol. 49, 3 - 10 (1972). In
Russian. English translation in Soviet Astron. AJ, Vol. 16, No.1.
The Crimean Astrophysical Observatory installation for
Cerenkov light flashes of cosmic ray air shower detection is
described briefly. Two years data of pulsar CP 1133 are presented. Also the data of different authors are considered. A
flux of high-energy gamma rays in the direction of the galactic
equator in the Cygnus region is revealed. An estimate of the
flux of high-energy quanta generated via inverse Compton
scattering of cosmic ray electrons on the infrared 8°K radiation is made.

**142.006 The Compton effect on thermal electrons in X-ray
sources.** A. F. Illarionov, R. A. Sunyaev.
Astron. Zhurn. Akad. Nauk SSSR, Vol. 49, 58 - 73 (1972). In
Russian. English translation in Soviet Astron. AJ, Vol. 16,
No. 1.
At high optical depth of gas, due to Thomson scattering,
the spectrum of thermal X-ray sources differs strongly from
the spectrum of a hot plasma, optically thin due to free-free

processes. This difference is connected with the exchange of
energy between hot gas and radiation via Compton interaction.

**142.007 Metagalactic gamma rays from relativistic-electron
bremsstrahlung interactions.**
F. W. Stecker, D. L. Morgan, Jr.
Astrophys. Journ., Vol. 171, 201 - 207 (1972).
We present here γ-ray spectra calculated for relativistic-
electron bremsstrahlung interactions at cosmological distances
under the assumption of a single power-law source spectrum
for the electrons. We conclude that such spectra cannot match
the form of the observed cosmic γ-ray spectrum above 1 MeV
as has previously been suggested.

**142.008 Possible observation of high-energy gamma rays
from the Cygnus region.**
M. Niel, G. Vedrenne, R. Bouigue.
Astrophys. Journ., Vol. 171, 529 - 536 (1972).
Results of two high-altitude balloon flights with a spark-
chamber γ-ray telescope have shown the existence, from the
Cygnus region, of a significant γ-ray flux at the 2.5 σ confidence level. Taking into account the angular resolution of our
detector, it seems possible to interpret this flux as associated
either with an extended source or with unresolved sources, in
a zone between $l^{II} \simeq 65°-80°$, $b^{II} \simeq -6°$ to 18°.

**142.009 A probable precursor to the X-ray nova Centaurus
XR-4.** R. D. Belian, J. P. Conner, W. D. Evans.
Astrophys. Journ., (*Letters*), Vol. 171, L87 - L90 (1972).
A short-lived burst of X-rays from a region of the sky
containing Cen XR-4 occurred on 1969 July 7 at 0152 UT, a-
bout 50 hours before the first reported observation of the X-
ray nova outburst of Cen XR-4. The event was observable a-
bove background for a period of about 10 minutes.

142.010 Dynamic spectrum analysis of Cygnus X-1.
M. Oda, M. Wada, M. Matsuoka, S. Miyamoto,
N. Muranaka, Y. Ogawara.
Astrophys. Journ., (*Letters*), Vol. 172, L13 - L16 (1972).
The oscillatory structure of the counting-rate data trains
of Cyg X-1 was studied. It was concluded that the oscillation
lasts typically for several seconds and its frequency drifts with-
in a few seconds repeatedly.

**142.011 Search for pulsed radio emission from Scorpius
X-1 and Cygnus X-1.**
J. H. Taylor, G. R. Huguenin, R. M. Hirsch.
Astrophys. Journ., (*Letters*), Vol. 172, L17 - L19 (1972).
Our results show that Sco X-1 and Cyg X-1 do not radi-
ate detectable amounts of pulsar-like emission.

142.012 The optical polarization of Scorpius X-1.
J. D. Landstreet, J. R. P. Angel.
Astrophys. Journ., Vol. 172, 443 - 446 (1972).
Observations of the wavelength and time dependence of
the optical polarization of Sco X-1 show that the effect is con-
sistent with an interstellar origin.

**142.013 Upper limits to the soft X-ray emission of sources
in Virgo.**
P. Gorenstein, B. Harris, H. Gursky.
Astrophys. Journ., (*Letters*), Vol. 172, L41 - L45 (1972).
The observed region included the center of the Virgo
cluster of galaxies, a source at higher energies which had been
previously identified with M87, and QSO 3C 273. No discrete
sources in Virgo were observed. The upper limit on the

energy flux from sources $\leq 0°5$ is 6.6×10^{-11} ergs cm^{-2} s^{-1} (0.15−0.28 keV) and 5.3×10^{-10} ergs cm^{-2} s^{-1} (0.5−1.2 keV).

142.014 **The low-energy gamma-ray spectrum of Scorpius X-1.** R. C. Haymes, F. R. Harnden, Jr., W. N. Johnson III, H. M. Prichard, H. E. Bosch. Astrophys. Journ., (*Letters*), Vol. 172, L47 - L49 (1972).

A 2-hour balloon altitude observation of Sco X-1, in the 16−930-keV energy band, was conducted from Argentina on 1970 November 25. The 16−41-keV data are consistent with an extension of the low-energy exponential spectrum. In addition, a penetrating component was detected at higher energies. The spectrum is lost in the background at energies greater than 300 keV, where upper limits are established.

142.015 **Limit on line emission in the diffuse X-ray background.** A. Toor, R. E. Price, F. D. Seward. Astrophys. Journ., (*Letters*), Vol. 172, L73 - L75 (1972).

Diffuse X-ray background data from an extended region in the galactic plane and near the South Galactic Pole do not exhibit an enhancement in the spectrum near 7 keV.

142.016 **A new X-ray pulsar near the Crab nebula.** D. Sadeh, M. Meidav, H. Smothers, T. Chubb, H. Friedman. Bull. American Astron. Soc., Vol. 3, 456 (1971). − Abstr. AAS.

142.017 **Hard X-rays from the southern sky.** W. H. G. Lewin, G. R. Ricker, J. E. McClintock, S. Ryckman, M. Gerassimenko. Bull. American Astron. Soc., Vol. 3, 456 (1971). − Abstr. AAS.

142.018 **Galactic X-ray sources in the southern sky.** R. Cruddace, S. Bowyer, M. Lampton, J. Mack, B. Margon. Bull. American Astron. Soc., Vol. 3, 456 (1971). −Abstr. AAS.

142.019 **The X-ray spectrum of Centaurus A.** M. Lampton, B. Margon, S. Bowyer, W. Mahoney, K. A. Anderson. Bull. American Astron. Soc., Vol. 3, 456 (1971). − Abstr. AAS.

142.020 **X-ray observations of the Cygnus region.** J. L. Matteson. Bull. American Astron. Soc., Vol. 3, 456 - 457 (1971). − Abstr. AAS.

142.021 **Correlated transient short-period oscillation in the optical and X-ray flux from Sco X-1.** H. Kestenbaum, J. R. P. Angel, R. Novick, W. J. Cocke. Bull. American Astron. Soc., Vol. 3, 457 (1971). − Abstr. AAS.

142.022 **Coordinated observations of the X-ray star Sco X-1.** W. Forman, E. Kellogg, H. Gursky, H. Tananbaum, R. Giacconi, H. Bradt, G. Moore, W. E. Kunkel, W. A. Hiltner, J. Thomas, B. Warner, P. Vandenbout. Bull. American Astron. Soc., Vol. 3, 457 (1971). − Abstr. AAS.

142.023 **Long-term X-ray observations of Sco X-1 by OSO III.** R. M. Pelling. Bull. American Astron. Soc., Vol. 3, 457 - 458 (1971). − Abstr. AAS.

142.024 **Balloon observations of hard X-rays from GX 349+2.** W. A. Mahoney, K. A. Anderson. Bull. American Astron. Soc., Vol. 3, 458 (1971). − Abstr. AAS.

142.025 **Observations of the Cygnus X region at low radio frequencies.** A. Parrish. Bull. American Astron. Soc., Vol. 3, 469 (1971). − Abstr. AAS.

142.026 **X-ray galaxies.** R. Giacconi, S. Murray, H. Tananbaum, E. Kellogg, H. Gursky. Bull. American Astron. Soc., Vol. 3, 477 (1971). − Abstr. AAS.

142.027 **Extended X-ray sources in the Coma and Perseus clusters.** E. Kellogg, S. Murray, H. Tananbaum, H. Gursky, R. Giacconi, W. Forman. Bull. American Astron. Soc., Vol. 3, 477 (1971). − Abstr. AAS.

142.028 **Diffuse cosmic X-radiation associated with our Galaxy.** E. Boldt, R. D. Bleach, A. Brisken, S. S. Holt, D. A. Schwartz, P. J. Serlemitsos. Bull. American Astron. Soc., Vol. 3, 481 (1971). − Abstr. AAS.

142.029 **On the absorption of gamma rays by photons in pulsars, QSO's and other source objects.** P. D. Guthrie, J. B. Pollack, B. S. P. Shen. Bull. American Astron. Soc., Vol. 3, 498 (1971). − Abstr. AAS.

142.030 **A probable precursor to the X-ray nova Cen XR-4.** R. D. Belian, J. P. Conner, W. D. Evans. Bull. American Astron. Soc., Vol. 3, 498 (1971). − Abstr. AAS.

142.031 **Atmospheric fluorescence as a ground-based method of detecting cosmic X-rays.** J. L. Elliot. Smithsonian Astrophys. Obs., *Cambridge, Mass.*, Special Rep. No. 341, 10 + 150 pp. (1972).

The feasibility of using atmospheric fluorescence as an observational technique for X-ray astronomy has been investigated. The fluorescent light can be detected from a ground-based station, providing a cheap method of observing X-ray sources. Here we have investigated the possibility of detecting only X-ray sources whose flux varies on a time scale of one second or shorter.

142.032 **Production of a diffuse flux of soft X-rays by galactic objects.** J. E. Mack. Nature, Phys. Sci., Vol. 235, 144 - 146, 160, with a correction Vol. 236, 64 (1972).

The spectrum and intensity of galactic sources of soft X-rays are estimated, and several models of possible sources evaluated.

142.033 **Model of Cygnus X-1.** J. C. Jackson. Nature, Phys. Sci., Vol. 236, 39 - 41 (1972).

The hard X-rays from Cygnus X-1 are attributed to inverse Compton scattering of ultraviolet photons from the B star known to be one component of the source. The other component is assumed to be the source of relativistic electrons. This provides a natural explanation of the "anti-eclipse" feature of the X-ray variability.

142.034 **The spectrum of diffuse cosmic X-rays in the 20−125 keV range.** R. K. Manchanda, S. Biswas, P. C. Agrawal, G. S. Gokhale, V. S. Iyengar, P. K. Kunte, B. V. Sreekantan. Astrophys. Space Sci., Vol. 15, 272 - 283 (1972).

Diffuse cosmic X-rays in the energy range 20−125 keV were measured in four balloon flights from Hyderabad, India during 1968−70 using almost identical X-ray telescopes mounted on oriented platforms. The implications of the results are briefly discussed.

142.035 **Neue Entdeckungen in der Röntgen-Astronomie.** T. Schmidt-Kaler. SuW, Vol. 11, 70 - 72 (1972).

142.036 On the origin of the magnetic field in the extended radio and X-ray source in the Coma cluster.
G. C. Perola, M. Reinhardt.
Astron. Astrophys., Vol. 17, 432 - 436 (1972).

The extended X-ray source in the central part of the Coma cluster can be explained by inverse Compton scattering of the relativistic electrons in the extended radio source that coincides with it in position and dimensions. The average magnetic field strength in this source is then 10^{-7} G. Two possibilities of the origin of the field are considered.

142.037 Dissipative processes in neutron-star crusts and the production of blackbody X-ray sources.
R. N. Henriksen, P. A. Feldman, W. Y. Chau.
Astrophys. Journ., Vol. 172, 717 - 728 (1972).

In the first part of this paper we show how a significant fraction of the rotational energy of a neutron star may be converted into heat in the crystalline mantle by the action of a steady wobble. We further show that this process can produce kilovolt surface temperatures and so may imply the existence of blackbody X-ray sources in the soft X-ray band. In the latter part, we consider the actual physical processes involved and demonstrate explicitly that effective blackbody surface temperatures in the kilovolt range can indeed be obtained. We also discuss the importance of plastic flow for large wobble amplitudes, which could lead to small spin-changes and temporarily increased heating of the crust. Application is made to the unusual X-ray source Cen XR-3, and attention is also drawn to the Crab pulsar as a potentially detectable source of this type.

142.038 Evidence for the binary nature of Centaurus X-3 from *Uhuru* X-ray observations.
E. Schreier, R. Levinson, H. Gursky, E. Kellogg, H. Tananbaum, R. Giacconi.
Astrophys. Journ., (*Letters*), Vol. 172, L79 - L89 (1972).

Analysis of data spanning a year of observations of the pulsating X-ray source Cen X-3 from *Uhuru* has revealed the existence of periodic variations in intensity of the source and correlated sinusoidal variations in the period of the 4.8 pulsations. We interpret this effect as due to an occulting binary system. The changes in intensity are then due to occultation of the X-ray source by a large massive companion, and the sinusoidal variations in the period of the 4.8 pulsations are due to Doppler effect.

142.039 Some problems of γ-ray point source search by registration of Cerenkov light from extensive air showers.
A. A. Stepanian, I. V. Pavlov.
Izv. Krymskoj Astrofiz. Obs., Vol. 43, 37 - 41 (1971).
In Russian.

Different methods of high-energy γ-ray source search by registration of Cerenkov light from extensive air showers are considered. The calculation results of the efficiency of different search methods are presented. A preliminary experimental estimate of the influence of atmospheric transparency upon Cerenkov flash registration was made.

142.040 A model for the Centaurus X-3 phenomenon.
G. R. Blumenthal, A. Cavaliere, W. K. Rose, W. H. Tucker.
Astrophys. Journ., Vol. 173, 213 - 220 (1972).

A model for Cen X-3 is presented according to which the X-ray emission comes from an atmosphere heated by shock waves produced by surface pulsations of a white dwarf. This model can explain the luminosity, period, and spectrum of Cen X-3. The way that these quantities vary with pulsation period and amplitude is discussed.

142.041 10 Jahre Röntgen-Astronomie. H. Gursky.
Umschau, 72. Jahrgang, p. 256 - 258 (1972).

Most of the X-ray sources are concentrated along the Milky Way and thus must lie within our own Galaxy. Other X-ray sources have been found to coincide with well known external galaxies such as the radio galaxy M-87 and the quasar 3C273. In addition, a diffuse background of X-rays was discovered which, like the primordial radio background is uniform around the sky.

142.042 Comments on the ionospheric detection of cosmic X-ray phenomena. G. A. Baird, R. J. Francey.
Journ. Geophys. Res., Vol. 77, 1966 - 1970 (1972). — Letter.

142.043 Appearance of λ4650 emission in Cygni X-2.
B. W. Bopp, P. A. Vanden Bout.
Publ. Astron. Soc. Pacific, Vol. 84, 68 - 69 (1972).

The C III-N III λ4650 line has been detected in Cyg X-2. An absorption feature is present to the blue of He II λ4686. Both lines and continuum appear to vary independently in intensity.

142.044 High-energy X-ray sources near the galactic equator between $l^{II} \sim 335°$ and $l^{II} \sim 350°$.
J. E. McClintock, G. R. Ricker, S. G. Ryckman, W. H. G. Lewin.
Astrophys. Journ., (*Letters*), Vol. 173, L57 - L59 (1972).

On 1970 October 15−16 we carried out balloon X-ray observations from Australia. We detected at least two hard X-ray sources (energies > 18 keV) near the galactic equator in the region $\sim 335° < l^{II} < \sim 350°$.

142.045 On the use of long-base time-delay measurements in the study of rapidly varying X-ray stars.
R. Giacconi.
Astrophys. Journ., (*Letters*), Vol. 173, L79 - L81 (1972).

The existence of time variations on the timescale of $10^{-2} - 10^{-3}$ seconds in several of the galactic X-ray sources makes it possible to determine their position by simultaneous measurement from two detectors placed at planetary distances, with a precision comparable to any yet achieved in X-ray astronomy. The significance of a search for even faster time variations in the sources is pointed out.

142.046 Inner bremsstrahlung as a source of X-rays in the steady-state universe. V. Petrosian, R. Ramaty.
Astrophys. Journ., (*Letters*), Vol. 173, L83 - L85 (1972).

We show that because of inner bremsstrahlung from neutron decay, the steady-state universe with neutron creation in diffuse regions is inconsistent with X-ray observations around 100 keV, unless the particle density of the universe is less than about 10^{-7} cm^{-3}.

142.047 Accretion onto a rapidly rotating magnetic white dwarf as a possible model for Cen XR-3.
W. Y. Chau, R. N. Henriksen, P. A. Feldman.
Journ. Roy. Astron. Soc. Canada, Vol. 66, 74 (1972). — Abstr.
Canadian Astron. Soc.

142.048 Cygnus X-1 − a spectroscopic binary with a heavy companion? B. L. Webster, P. Murdin.
Nature, Vol. 235, 37 - 38 (1972).

The B0Ib supergiant, HD 226868, coincident with the Cygnus X-1 source, is a velocity variable, period 5.61 days. A period of 5.6075 days seems present in the hard X-ray intensity. Interpreted as a spectroscopic binary, the star contains an unseen companion with mass at least 2 solar masses, and the minimum in the X-ray intensity occurs when the companion lies between the supergiant and the earth.

142.049 Mechanism for the X-ray pulsations in Cyg X-1.
G. R. Blumenthal, W. H. Tucker.
Nature, Vol. 235, 97 - 98 (1972).

It is suggested that the X-ray pulsations in Cyg X-1 are related to oscillations in a region of high magnetic field which are produced by a giant flare-like event which releases ~ 10^{37} ergs in a region ~ 10^9 cm having a density ~ 10^{16} cm^{-3} and a magnetic field strength ~ 10^6 gauss. These oscillations lead to heating of the plasma in a flux tube which emits thermal radiation in pulses. Synchrotron radiation from high energy particles accelerated in the flare accounts for the hard X-ray emission.

142.050   New X-ray pulsar near the Crab nebula.
D. S. Sadeh, M. Meidav, H. W. Smathers, T. A. Chubb, H. Friedman.
Nature, Vol. 235, 151 - 152 (1972).

On October 17, 1970, the flux of 30–100 keV X-rays from the Crab nebula was measured by a large area balloon-borne proportional counter array. The recorded pulses were subjected to a power spectrum analysis in a search for periodicities secondary to the 33 ms period of CP 0532. A periodicity of 3.827 s was observed, and a single peaked light curve was recorded at this period. The observed X-ray period is 2% greater than the radio period of NP 0527, which was within the instrument field of view.

142.051   Identification of GX3 + 1 from lunar occultations.
A. F. Janes, K. A. Pounds, M. J. Ricketts, A. P. Willmore, L. V. Morrison.
Nature, Vol. 235, 152 - 155 (1972).

The lunar occultation of the galactic X-ray source GX3 + 1 was observed in rocket flights on September 27th and October 24, 1971, and the position of the X-ray source thus obtained about 10 sq arc seconds. This position is compared with a new star plate and the probable identification of the X-ray source with a 16th magnitude star is noted.

142.052   Identification of Cygnus X-1 with HDE 226868.
C. T. Bolton.
Nature, Vol. 235, 271 - 273 (1972).

Radial velocities of HDE 226868 indicate that it is a single-line spectroscopic binary with a period identical to that of the high energy X-ray variations of Cygnus X-1. Hβ at times has a P Cyg profile and HeII λ 4686 has been seen in emission. The Hβ emission indicates that matter is flowing from the B0Ib star to the unseen companion. This companion is too massive to be a normal white dwarf or neutron star.

142.053   Extragalactic origin of low energy gamma rays.
S. V. Damle, R. R. Daniel, G. Joseph, P. J. Lavakare.
Nature, Vol. 235, 319 - 320 (1972).

Measurements have been made on the cosmic gamma rays of energy between 0.25 and 4.2 MeV from a balloon experiment made near the geomagnetic equator using a collimated 7.6 cm NaI(Tl) crystal assembly. The diffuse component of the cosmic gamma rays has been estimated. Using also the upper limits obtained in the same experiment for the gamma ray flux from the galactic center region, it has been possible to show that the diffuse component is of extragalactic origin.

142.054   Discovery of variable circular polarization in the light of the X-ray star Sco X-1.
Yu. N. Gnedin, O. S. Shulov.
Astrofizika, Vol. 7, 529 - 545 (1971). In Russian.
English translation in Astrophysics, Vol. 7, No. 4.

Polarimetric observations of Sco X-1 have been carried out during 14 nights from 1971 June 23 to 1971 July 28. As a result, discovery was made of circular polarization in the yellow-red light. The polarization oscillates around the zero mean level and shows two oscillatory processes.

142.055   Observations of rapid blue variables–VI. Scorpii X-1.
E. L. Robinson, B. Warner.
Monthly Notices Roy. Astron. Soc., Vol. 157, 85 - 95 (1972).

We report photometric observations of Sco X-1 obtained with a view to studying the power spectrum of Sco X-1 during the flare stage, and to search for possible short-lived periodicities at the end of the flares.

142.056   Transition radiation as a source of cosmic X-rays.
R. Ramaty, R. D. Bleach.
Astrophys. Letters, Vol. 11, 35 - 36 (1972).

In order to account for recent X-ray observations below 300 eV by transition radiation, an energy density in interstellar space of about 10 eV cm^{-3} in 10 MeV electrons is required. This seems to rule out transition radiation as an important source of diffuse cosmic X-rays in any energy region.

142.057   Optical candidates for two X-ray sources.
R. J. Brucato, J. Kristian.
Astrophys. Journ., (Letters), Vol. 173, L105 - L107 (1972).

We suggest the bright stars X Per and HD 77581 as possible candidates for the X-ray sources 2U 0352 + 30 and 2U 0900 − 40, respectively. The first is an active, rapidly rotating Be star which is losing mass. The second is similar to BD + 34° 3815, a likely candidate for Cyg X-1, in spectral type and in the possibility that it may be a short-period binary.

142.058   Veränderliche kosmische Röntgenquellen.
J. Trümper.
SuW, Vol. 11, 127 - 132 (1972).

142.059   Some results of a search for point sources of high energy γ-quanta.
A. A. Stepanyan, B. M. Vladimirskij, I. V. Pavlov, V. P. Fomin.
Izv. AN SSSR. Ser. fiz., Vol. 35, 2458 - 2462 (1971). In Russian. – Abstr. in Referativ. Zhurn. 51. Astron., 5.51.596 (1972).

142.060   Variable X-ray sources.
E. M. Kellogg.
IAU Colloquium No. 15, (see 012.006), p. 168 - 172 (1972).

142.061   Variable radio emission from X-ray sources.
L. L. E. Braes, G. K. Miley.
IAU Colloquium No. 15, (see 012.006), p. 173 - 175 (1972).

142.062   The limiting mass of Centaurus X-3.
R. E. Wilson.
Astrophys. Journ., (Letters), Vol. 174, L27 - L30 (1972).

A very low upper limit can be placed on the mass of Cen X-3, which was recently discovered to be a member of an X-ray eclipsing binary system. The principle used is that neither the eclipsing component itself nor any material which may orbit this component can exceed the primary Roche lobe, whose dimensions are determined by the mass ratio. Constraints imposed by the mass limit on possible theoretical models for the X-ray source are briefly discussed.

142.063   A neutron star in Centaurus X-3.
S. Sofia.
Astrophys. Journ., (Letters), Vol. 174, L31 - L33 (1972).

It is shown that the compact object producing the pulsed X-radiation in the binary system Cen X-3 is probably a rotating neutron star of very low mass.

142.064   Shot-noise character of Cygnus X-1 pulsations.
N. J. Terrell, Jr.
Astrophys. Journ., (Letters), Vol. 174, L35 - L41 (1972).

The present analysis was undertaken with the object of

understanding the nature of the puzzling fluctuations of Cyg X-1, and resolving the apparent conflicts between reports. The basic finding is that the X-ray signal gives no evidence of periodicity, but has the same characteristics as shot noise, that is, randomly occurring and overlapping pulses of X-rays.

142.065 A new transient source observed by Uhuru.
T. A. Matilsky, R. Giacconi, H. Gursky, E. M. Kellogg, H. D. Tananbaum.
Astrophys. Journ., (Letters), Vol. 174, L53 - L55 (1972).

A strong X-ray source appeared sometime between 1971 March 25 and 1971 August 17. On August 23, its intensity, as observed by the Uhuru satellite, was about twice that of the Crab nebula, corresponding to about 3.0×10^{-8} ergs cm^{-2} s^{-1} in the range 2–6 keV. On 1971 December 20, its intensity had declined to one-tenth of that value in the same energy range.

142.066 A search for discrete sources of cosmic gamma rays of energy $10^{11} - 10^{12}$ eV.   T. C. Weekes, G. G. Fazio, H. F. Helmken, E. O'Mongain, G. H. Rieke.
Astrophys. Journ., Vol. 174, 165 - 179 (1972).

Twenty-seven celestial sources have been examined for high-energy γ-ray emission in the previously unexplored region of the electromagnetic spectrum from 10^{11} to 10^{12} eV. The principal source investigated was the Crab nebula. Others included supernova remnants, radio galaxies, quasars, magnetic variable stars, pulsars, and X-ray sources, with particular emphasis on M87, M82, 3C 273, and Sgr A.

142.067 Photographic B, V photometry of Sco X-1.
A. J. Wesselink, C. Cesco.
Inform. Bull. Variable Stars, (I.A.U. Commission 27), Konkoly Obs., Budapest, No. 667, 5 pp. (1972).

142.068 Variable stars and X-ray sources.
D. J. MacConnell, A. P. Cowley.
Inform. Bull. Variable Stars, (I.A.U. Commission 27), Konkoly Obs., Budapest, No. 674 (1972).

142.069 Further radio observations of Cygnus X-1.
L. L. E. Braes, G. K. Miley.
Nature, Phys. Sci., Vol. 235, 147 (1972).

New Westerbork observations at 1415 MHz indicate that, apart from a slight decrease in flux after its initial appearance, the radiocounterpart of Cyg X-1 does not show variability. A refinement of the earlier Westerbork position leaves no doubt about the coincidence of the radio source with the spectroscopic binary HDE 226868.

142.070 Radio identifications of weak extragalactic X-ray sources.   A. H. Bridle, P. A. Feldman.
Nature, Phys. Sci., Vol. 235, 168 - 170 (1972).

The UHURU X-ray sources 2 ASE 0227+43 and 2 ASE 1144+19 are identified with the radio sources 3C 66 and 3C 264 respectively. Both radio sources are located in rich clusters of galaxies and have extended components with unusually high spectral indices. The X-ray emission is interpreted as inverse Compton scattering of the 3°K background by relativistic electrons that have escaped from radio galaxies in the clusters.

142.071 Possible identification of GX5-1.
L. L. E. Braes, G. K. Miley, A. A. Schoenmaker.
Nature, Vol. 236, 392 (1972).

Observations at 1415 MHz with the Westerbork synthesis telescope led to the discovery of an unresolved 0.010 f.u. radio source within the X-ray error box of GX5-1. A positive identification must await the detection of variability.

142.072 Search for X-rays above 100 keV from the vicinity of NP 0527.

J. Kurfess, M. Meidav, D. S. Sadeh.
Nature, Vol. 236, 447 - 448 (1972).

A search for pulsed emission above 100 keV from the vicinity of NP 0527 was made from data obtained during two balloon flights. No evidence of a pulsed flux at or near the 3.82 second period reported in 30–100 keV data by Sadeh et al. (Nature, Vol. 235, 151 - 152 (1972)) was observed.

142.073 Identification of Cen X-3.
B. A. Peterson.
Nature, Vol. 236, 449 (1972).

Photometry of the eclipsing binary LR Cen shows that, in March 1972, the secondary minimum of LR Cen nearly coincided with the predicted time of the eclipse of Cen X-3. The similar positions, periods, orbital eccentricities, and phases suggest that LR Cen is the optical counterpart of Cen X-3. However, for this identification to be correct, the period of LR Cen would have to have changed significantly since it was determined in 1932. Observations are now underway to determine the period of LR Cen at the present epoch.

142.074 Hard X-rays from Coma constellation.
R. K. Manchanda, P. C. Agrawal, S. Biswas, G. S. Gokhale, V. S. Iyengar, P. K. Kunte, B. V. Sreekantan.
Nature, Phys. Sci., Vol. 236, 51 - 53 (1972).

In a series of five balloon flights carried out from Hyderabad since April 1968, the region of the sky defined by R.A. = 190 to 330° and δ = −20° to +50° has been scanned for high energy X-ray sources. Evidence for a region of hard X-ray emission (> 20 keV) in Coma constellation, but not coincident with the Coma cluster, has been presented.

142.075 Simultaneous hard X-ray and optical observations of Sco X-1.
M. Matsuoka, M. Fujii, S. Miyamoto, J. Nishimura, M. Oda, Y. Ogawara, S. Ohta, S. Hayakawa, I. Kasahara, F. Makino, Y. Tanaka, P. C. Agrawal, B. V. Sreekantan, Y. Hatanaka, S. S. Rao.
Nature, Phys. Sci., Vol. 236, 53 - 55 (1972).

Simultaneous observations of hard X-ray and optical emissions of Sco X-1 were carried out at Hyderabad, India on May 1, 1971. The correlation coefficient between the X-ray intensity in the energy range 20–30 keV and the B magnitude is about 0.7. The shape of the X-ray spectrum in the energy range 20–40 keV did not appreciably change, whereas the absolute intensity changed with optical flare.

142.076 Intensity and spectrum of the diffuse X-ray background.
R. K. Manchanda, A. Danjo, B. V. Sreekantan.
Nature, Phys. Sci., Vol. 236, 67 - 68 (1972).

The procedure that is generally adopted in all balloon experiments of linear extrapolation of the growth curves determined for large atmospheric depths (> 20 gms/cm²) to obtain the background counting rate of X-ray telescopes corresponding to ceiling altitude is shown to be not valid and to lead to overestimates of the cosmic diffuse X-rays and consequent uncertainty in the spectral results at energies greater than 20 keV.

142.077 Fluctuations in the X-ray intensity of Sco X-1.
R. E. Griffiths, B. A. Cooke.
Nature, Phys. Sci., Vol. 236, 104 - 106 (1972).

Two Skylark rocket observations of Sco X-1 in the energy range 2–14 keV have shown that the source fluctuates on the scale of a few % in amplitude, on a time scale of minutes. The observed fluctuations are consistent with mass motions of the emitting plasma.

142.078 The binary star model for compact galactic X-ray sources: Its place in stellar evolution.   R. P. Kraft.
Bull. American Astron. Soc., Vol. 4, 219 (1972). −Abstr. AAS.

142.079 X-ray observations of Cen X-3.   B. Margon,
S. Bowyer, M. Lampton, R. Cruddace.
Bull. American Astron. Soc., Vol. 4, 220 (1972). – Abstr. AAS.

142.080 UCSD X-ray observations of the Vela region from OSO-7.   M. P. Ulmer, W. A. Baity, W. A. Wheaton, L. E. Peterson.
Bull. American Astron. Soc., Vol. 4, 220 (1972). – Abstr. AAS.

142.081 Hydrogen lines in the spectrum of Sco X-1.
S. Edwards, D. E. Mook, W. A. Hiltner.
Bull. American Astron. Soc., Vol. 4, 227 (1972).

142.082 The less prominent features in the optical spectrum of Sco X-1.
D. Chesley, D. E. Mook, W. A. Hiltner.
Bull. American Astron. Soc., Vol. 4, 227 (1972).

142.083 More photometric properties of Sco X-1.
D. E. Mook, V. Blanco, J. Hesser, W. Kunkel,
B. Lasker.
Bull. American Astron. Soc., Vol. 4, 227 (1972).

142.084 A search for hard X-rays from extragalactic objects.
J. G. Laros, J. L. Matteson, R. M. Pelling.
Bull. American Astron. Soc., Vol. 4, 227 - 228 (1972). – Abstr. AAS.

142.085 Capabilities for celestial nuclear gamma ray monitoring on OSO-7.
E. L. Chupp, D. J. Forrest, P. R. Higbie.
Bull. American Astron. Soc., Vol. 4, 228 (1972). – Abstr. AAS.

142.086 Crustar models for GX 340+0 and other possible blackbody X-ray sources.
P. A. Feldman, R. N. Henriksen, W. Y. Chau.
Bull. American Astron. Soc., Vol. 4, 228 (1972). – Abstr. AAS.

142.087 Results from the X-ray satellite UHURU.
H. Gursky.
Bull. American Astron. Soc., Vol. 4, 255 - 256 (1972). – Abstr. AAS.

142.088 Radio observations of X-ray sources.
R. M. Hjellming.
Bull. American Astron. Soc., Vol. 4, 256 (1972). – Abstr. AAS.

142.089 Theoretical ideas concerning dense objects and X-ray sources.   E. E. Salpeter.
Bull. American Astron. Soc., Vol. 4, 257 (1972). – Abstr. AAS.

142.090 Significance of the small scale isotropy of the cosmic X-ray background radiation.   K. Brecher.
Bull. American Astron. Soc., Vol. 4, 257 - 258 (1972).
Abstr. AAS.

142.091 Measurements of 30–250 keV diffuse cosmic X-rays from OSO-5.   B. R. Dennis, K. J. Frost, A. N. Suri.
Bull. American Astron. Soc., Vol. 4, 259 (1972). – Abstr. AAS.

142.092 Cosmic gamma ray observations above 250 MeV.
K. I. Greisen, B. McBreen, D. Koch, M. F. Campbell,
S. E. Ball, J. P. Delvaille, G. G. Fazio, H. F. Helmken, D. R. Hearn.
Bull. American Astron. Soc., Vol. 4, 259 (1972). – Abstr. AAS.

142.093 Structure and time variability of the X-ray source in the Virgo cluster.   E. Kellogg, H. Tananbaum,
H. Gursky, R. Giacconi, K. Pounds.
Bull. American Astron. Soc., Vol. 4, 260 (1972). – Abstr. AAS.

142.094 Diffuse soft cosmic X-rays.
W. L. Kraushaar, A. N. Bunner, P. L. Coleman, D. McCammon.
Bull. American Astron. Soc., Vol. 4, 260 (1972). – Abstr. AAS.

142.095 The pulsating X-ray source Centaurus X-3.
E. Schreier, H. Tananbaum, E. Kellogg, H. Gursky,
R. Giacconi.
Bull. American Astron. Soc., Vol. 4, 261 (1972). – Abstr. AAS.

142.096 Observations of X-ray sources near the galactic center from 1 to 60 keV from the MIT instrument on OSO-7.   G. Sprott, G. Clark, W. Lewin, H. Schnopper.
Bull. American Astron. Soc., Vol. 4, 261 (1972). – Abstr. AAS.

142.097 Fluctuations of Cygnus X-1.   N. J. Terrell, Jr.
Bull. American Astron. Soc., Vol. 4, 261 (1972).
Abstr. AAS.

142.098 X-ray variability of Cyg X-1.
M. Oda, M. Wada, M. Matsuoka, S. Miyamoto, N. Muranaka, Y. Ogawara.
Bull. American Astron. Soc., Vol. 4, 271 (1972). – Abstr. AAS.

142.099 Anomalous minima in Cen X-3.   B. Margon,
S. Bowyer, M. Lampton, R. Cruddace.
Nature, Phys. Sci., Vol. 237, 104 (1972).
We report an observation made with a rocket-borne argon-methane proportional counter equipped with a 'Mylar' window and sensitive to X-rays in the 0.2 to 10 keV band and compare our results with those of the UHURU satellite observations.

142.100 Extragalactic X-ray sources.
K. Brecher, G. R. Burbidge.
Nature, Vol. 237, 440 - 443 (1972).
The extended X-ray sources associated with the Virgo, Perseus, and Coma clusters of galaxies can best be explained as a result of the Compton scattering of relativistic electrons (generated in the radio sources which are present in each of these clusters) on the microwave background radiation. The same physical process involving microwave infrared or optical photons generated in the nuclei of active galaxies may account for the more compact X-ray sources. Thermal bremsstrahlung of hot gas in clusters of galaxies is a less likely source of X-rays.

142.101 A search for the optical counterpart of Centaurus X-3.   B. A. Peterson.
Proc. Astron. Soc. Australia, Vol. 2, 110 (1972).

142.102 Pulsating X-ray sources.   G. Chanmugam.
Nature, Vol. 237, 449 - 550 (1972).
It has been natural to suggest that the several pulsating X-ray sources with periods ~ 1 s are associated with matter at high densities and in particular with white dwarfs, neutron stars or black holes. Here I suggest that these X-ray sources are stars with central densities intermediate to those of neutron stars and white dwarfs.

142.103 Radio detection of Cygnus X-3.
L. L. E. Braes, G. K. Miley.
Nature, Vol. 237, 506 (1972).
During a continuing search for radio emission from X-ray sources with the Westerbork synthesis telescope, we have detected a highly variable radio source near the X-ray location of Cyg X-3. The observations consisted of four 12 h measurements at a frequency of 1,415 MHz. The strongly variable nature of the source is clear from separate analysis of each of the four 12 h observations. Dates and mean flux densities for these observations are given in a table.

142.104 Radio observations of Cygnus X-3.
R. M. Hjellming, M. Hermann, E. Webster.
Nature, Vol. 237, 507 - 508 (1972).

This letter reports observations of the radio counterpart of Cyg X-3 at 2,695 and 8,085 MHz with the NRAO interferometer at spacings of 2,400, 1,900 and 500 m. Data taken from April 29 to May 3, 1972, confirm the presence of a very variable radio source.

142.105 LR Cen is not Cen X-3. B. A. Peterson.
Nature, Vol. 237, 508 (1972).

The eclipsing variable LR Cen has been suggested as a candidate for the optical counterpart of Cen X-3 on the basis of the near coincidence of the position and period of LR Cen with those of Cen X-3. The difference in periods, however, though small, is significant. I have redetermined the period of LR Cen from my observations of its primary eclipse on March 16, 1972 and from those obtained on April 26, 1972. These observations show that the period of LR Cen is unchanged, and that it is different from the period of Cen X-3.

142.106 Observational evidence relating to a recent theory on the origin of the universal X-ray background.
J. E. Mack, D. E. Robbins.
Astrophys. Space Sci., Vol. 16, 336 - 337 (1972). – Research note.

142.107 Centaurus X-3 and Roche limits of close binary systems. Y. Osaki.
Publ. Astron. Soc. Japan, Vol. 24, 419 - 422 (1972).

The maximum durations of eclipses have been calculated for X-ray eclipsing binaries as a function of mass ratios under the assumption that the X-ray component is a point and the occulting star is confined within the critical Roche lobe of the close binary system. The result is applied to Cen X-3, giving an upper limit of about 0.03 for the mass ratio of the X-ray source to the occulting star. The corresponding mass of the X-ray source is about $0.5 M_\odot$, which favors the white-dwarf interpretation for the X-ray component of Cen X-3.

142.108 The extended X-ray source at M87. E. Kellogg, H. Gursky, H. Tananbaum, R. Giacconi, K. Pounds.
Astrophys. Journ., (Letters), Vol. 174, L65 - L69 (1972).

The strongest X-ray source in the Virgo cluster is extended by about 1° and centered on M87. Its spectrum fits a power law of energy index $- (2.06 \pm 0.10)$. There is no other evidence for X-ray emission from Virgo except for the weak discrete source 2U 1231 + 7. This gives a limit of 2×10^{40} ergs s^{-1} for the X-ray luminosity of an average galaxy in the Virgo cluster.

142.109 Upper limit on the X-ray flux associated with gravitational radiation. G. A. Baird, M. A. Pomerantz.
Phys. Rev. Letters, Vol. 28, 1337 - 1340, with a correction, p. 1545 (1972).

An upper limit in the X-ray flux accompanying pulses of gravitational radiation has been determined with relatively simple balloon-borne apparatus. The minimum detectable flux in space is about 9 orders of magnitude below the energy flux which produces the gravitational signals observed by Weber.

142.110 Galactic gamma-radiation between 200 MeV and 10 GeV. K. Bennett, P. Penengo, G. K. Rochester, T. R. Sanderson, R. K. Sood.
Nature, Vol. 238, 31 - 33 (1972).

It has not yet been established whether most of the γ-ray emission from the Galaxy at energies above 100 MeV radiation is a diffuse line flux from the galactic disk or if it arises from a number of weak point sources, and the production mechanism for each case remains poorly understood. Our experiment was undertaken to study the high energy part of the spectrum be-

tween 200 MeV and 10 GeV, using balloon-borne detectors.

142.111 Distances & absolute luminosities of galactic X-ray sources. F. D. Seward, G. Burginyon, R. Grader, R. Hill, T. Palmieri, J. Stoering.
Bull. American Astron. Soc., Vol. 4, 221 (1972). – Abstr. AAS.

142.112 The X-ray absorption measure of Sco X-1.
A. N. Bunner, P. L. Coleman, W. L. Kraushaar, D. McCammon, F. O. Williamson.
Bull. American Astron. Soc., Vol. 4, 221 (1972). – Abstr. AAS.

142.113 The spectra of ten galactic X-ray sources in the southern sky.
R. Gruddace, S. Bowyer, M. Lampton, J. Mack, B. Margon.
Astrophys. Journ., Vol. 174, 529 - 548 (1972).

Data on 10 galactic X-ray sources located between $l^{II} = 320°$ and $l^{II} = 20°$ were obtained during a rocket flight from Brazil in 1969 June. Detailed spectra of these sources have been compared with bremsstrahlung, blackbody, and power-law models, each including interstellar absorption. A comparison of our results with those of previous investigations provides evidence that five of the sources vary in intensity by a factor of 2 or more, and that three have variable spectra. New or substantially improved positions have been derived for four of the sources observed.

142.114 Long-term temporal variations of the hard X-ray flux from the Centaurus region.
D. A. Schwartz, L. E. Peterson, H. S. Hudson.
Astrophys. Journ., Vol. 174, 549 - 555 (1972).

The University of California at San Diego (UCSD) X-ray telescope aboard the third Orbiting Solar Observatory (OSO-III) observed the Centaurus region daily from 1967 October to 1968 February, and also for five days in 1968 June. For this period, we derive a stable minimum flux of 0.33 ± 0.03 photons $(cm^2 s)^{-1}$ between 7.7 and 38 keV from a persistent hard X-ray source around $l = 305°$.

142.115 On the optical identification of Centaurus X-3.
B. Margon, J. D. Wray.
Astrophys. Journ., (Letters), Vol. 174, L141 - L142 (1972).

We suggest that the emission-line star WRA 795 is the optical counterpart of Cen X-3, the X-ray pulsar which is a member of an occulting binary system.

142.116 Discovery of a periodic pulsating binary X-ray source in Hercules from Uhuru. H. Tananbaum, H. Gursky, E. M. Kellogg, R. Levinson, E. Schreier, R. Giacconi.
Astrophys. Journ., (Letters), Vol. 174, L143 - L149 (1972).

We have discovered a new pulsating X-ray source with a $1^s.24$ period in the constellation Hercules. Analysis of 5 months of data has shown the existence of periodic variations in the intensity of the source and correlated sinusoidal variations in the period of the $1^s.24$ pulsations. As in the case of the pulsating X-ray source Cen X-3, we interpret this effect as due to an occulting binary system.

142.117 Perspective on X-ray sources.
A. Fabian.
New Scient. & Sci. Journ., (GB), Vol. 54, 490 - 491 (1972).

Current theories of the origin of X-ray sources are discussed. Binary stars and galaxy clusters are considered.

142.118 Decametric radio identification of an extragalactic X-ray source.
C. H. Costain, A. H. Bridle, P. A. Feldman.
Astrophys. Journ., (Letters), Vol. 175, L15 - L18 (1972).

A radio source with an unusually high spectral index between 22 and 81.5 MHz is identified with the Uhuru X-ray source 2U 1706+78. It is probably associated with the cluster of galaxies Abell 2256. This identification provides further support for the inverse Compton model of X-ray emission by rich clusters of galaxies.

142.119 Diffuse galactic X-rays from discrete sources.
J. Silk, S. L. Weinberg.
Astrophys. Journ., (Letters), Vol. 175, L29 - L33 (1972).

The galactic ridge is interpreted as a population of intrinsically weak X-ray sources, and calculations are presented of the predicted intensity as a function of scale height and radial gradient. Observations of diffuse galactic X-radiation above 1 keV impose severe constraints on discrete-source interpretations of the galactic-plane excess observed at 0.25 keV.

142.120 Possible identification of X-ray source (IM Normae).
J. Elliot, W. Liller.
IAU Circ., No. 2386 (1972).

142.121 Cyg X-1. V. M. Lyutiy.
IAU Circ., No. 2395 (1972).

142.122 Cen X-3. J. Kristian, R. J. Brucato, J. A. Westphal, I. S. Shklovsky, Yu. N. Efremov, J. L. Elliot, W. Liller.
IAU Circ., No. 2395 (1972).

142.123 Possible optical identification of X-ray source.
W. A. Hiltner, P. Osmer.
IAU Circ., No. 2398 (1972).

142.124 Cen X-3. W. A. Hiltner, B. H. Margon, K. Henize, W. Liller.
IAU Circ., No. 2398 (1972).

142.125 Identifications of X-ray sources. L. L. E. Braes, G. K. Miley, D. J. MacConnell, A. P. Cowley.
IAU Circ., No. 2401 (1972).

142.126 Identifications of X-ray sources.
M. S. Bessell, R. J. Brucato, H. H. Lanning.
IAU Circ., No. 2406 (1972).

142.127 X-ray sources. E. Schreier, R. Giacconi, H. Gursky, E. Kellogg, S. Murray, H. Tananbaum, W. Liller.
IAU Circ., No. 2412 (1972).

142.128 Lunar occultations of X-ray sources.
L. V. Morrison.
IAU Circ., No. 2415 (1972).

142.129 Possible identification of Hercules X-1.
W. Liller.
IAU Circ., No. 2415 (1972).

142.130 Observation of a hard X-ray flare in Cyg X-1.
P. C. Agrawal, G. S. Gokhale, V. S. Iyengar, P. K. Kunte, R. K. Manchanda, B. V. Sreekantan.
Nature, Phys. Sci., Vol. 238, 22 - 23 (1972).

During a balloon flight carried out from Hyderabad on the night of April 6, 1971, we monitored the intensity of Cyg X-1 in the hard X-ray band 20–154 keV for about 3 h. A reanalysis of the data has revealed the occurrence of a "flare" in Cyg X-1 which is similar in characteristics to those recorded in hard X-rays in Sco X-1. We present here details of this flare.

142.131 Photographic observation of Sco X-1.
Y. Hatanaka, S. S. Rao, B. Lokanadham.
Tokyo Astron. Bull., Second Ser., No. 216, p. 2525 - 2527 (1972).

These photographic observations of Sco X-1 were made as a part of simultaneous X-ray and optical observations at Hyderabad, India, in 1971. The present report gives the method of reduction and the results of optical monitoring during several days in April 1971.

142.132 Observation of the diffuse cosmic gamma radiation in the 30–50 MeV region.
H. A. Mayer-Haßelwander, E. Pfeffermann, K. Pinkau, H. Rothermel, M. Sommer.
Max-Planck-Inst. Phys. Astrophys. München, MPI-PAE/Extraterr. 64, 16 pp. (1972).

Results of two balloon flights are presented in which a gamma ray astronomy spark chamber was carried to residual atmospheric pressures of 1.7 g/cm² and 2.2 g/cm², respectively. A diffuse flux of celestial gamma rays was discovered at 30–50 MeV which shows no significant correlation to galactic features and which exceeds a linear interpolation between previously known gamma ray data at 1 MeV and 100 MeV by a factor of between 4 and 10.

142.133 High energy gamma radiation from the region of Cygnus–Cassiopeia.
R. Browning, D. Ramsden, P. J. Wright.
Nature, Phys. Sci., Vol. 235, 128 - 130 (1972).

This communication describes the results obtained with a balloon borne gamma ray telescope which was flown from Palestine, Texas, on September 23/24, 1970. The chief object of this flight was to study the gamma radiation with an energy greater than 100 MeV from the direction of the galactic plane in the region of Cygnus and Cassiopeia. The region of the galactic plane observed during this flight was $80° < l^{II} < 120°$. The locations of the possible point sources of gamma radiation in this area are listed.

142.134 Optical counterpart of Cen X-3.
J. S. Shklovsky, A. M. Cherepashchuk, Yu. N. Efremov.
Nature, Vol. 236, 448 (1972).

We wish to suggest the possibility that the X-ray source Cen X-3 may be identified with the Algol-type variable LR Cen, based on the proximity of their coordinates and, more significantly, on the close coincidence of the orbital periods of the objects (2.08712 days for Cen X-3 and 2.095595 days for LR Cen). These results and the Uhuru data communicated to us by E. Schreier and colleagues lead to the following conclusions: Cen X-3 is a close binary system, its brighter and smaller component being connected with the X-ray source, the minima of the intensity of the X-ray source are connected with absorption or scattering in the atmosphere of the more massive and relatively dark component.

142.135 Search for optical counterpart of GX3+1.
A. W. Rodgers.
Nature, Vol. 237, 273 - 274 (1972).

142.136 Nature of the Cen X-3 system. L. Gratton.
Nature, Vol. 237, 329 - 331 (1972).

In order to explain the X-ray observations of the source Cen X-3 the following model is proposed: the system consists of a central black hole which contains practically the whole mass of the system and a small X-ray source orbiting around it and emitting a pulsed X-ray radiation. A ring or a disk of cold matter is laying inside the orbit of the X source; the radiation from the source evaporates and ionizes the part of the disk facing the source itself. The opaque cloud of plasma thus produced eclipses the X-ray source at the right phase. It seems that a very small mass of the cloud can produce the observed effects.

142.137 The isotropy of the cosmic gamma ray flux between 1 and 6 MeV and its implications for future gamma ray investigations. F. W. Stecker, J. I. Vetle, J. I. Trombka.

Rep. NASA—TM—X—65562, National Aeronautics and Space Administration, Greenbelt, Maryland. [Available from NTIS, Springfield, Va.], 14 pp. (1971).

The alternative hypotheses of galactic and extragalactic origin of the observed cosmic gamma ray flux between 1 and 6 MeV is examined in the light of the most recent spectral information on cosmic rays above 50 MeV.

142.138 **Gamma ray observations of the galactic center and some possible point sources.** C. E. Fichtel, R. C. Hartman, D. A. Kniffen, M. Sommer.
Rep. NASA—TM—X—65564, National Aeronautics and Space Administration, Greenbelt, Maryland. [Available from NTIS, Springfield, Va.], 22 pp. (1971).

A 5 m by 5 m digitized spark chamber gamma ray telescope was flown on three balloon flights to look at the galactic center region, Virgo, and the Crab.

142.139 **Measurement of gamma ray spectra of celestial large energy sources [final Scientific Report, Feb. 1969 – 31 Oct. 1970].** R. C. Haymes.
Rep. AFOSR—TR—71—0548, Rice Univ., Houston, Texas. [Available from NTIS, Springfield, Va], 7 pp. (1971).

Gamma ray spectra were measured in the energy range 25 keV – 975 keV from discrete celestial sources including the Crab nebula, pulsars NP 0532 and PSR 0835—45, the centre of our Galaxy, radiogalaxies Virgo A and Centaurus A, as well as from the galactic sources Cyg XR-1, Sco XR-1 and Cen XR-4. Significant new information on these objects and their energy sources was obtained and is discussed.

142.140 **Rocket astronomy.** H. Friedman.
Rep. Naval Res. Lab., *U. S. A.*, Progr. 1971, p. 1 - 9.

142.141 **The UHURU X-ray instrument.** N. Jagoda, G. Austin, S. Mickiewicz, R. Goddard.
IEEE Trans. Nucl. Sci., Vol. NS—19, 579 - 591 (1972).

On 12 December 1970, the UHURU (SAS-A) X-ray observatory was launched into equatorial orbit with the prime mission of conducting an all-sky survey of astronomical X-ray sources with intensities of 5×10^{-5} Sco X-1 or greater.

Large balloon gathers data on X-ray emission sources.
IEEE Spectrum, Vol. 9, No. 6, p. 89 (1972).

The physics of cosmic X-ray, γ-ray, and particle sources. See Abstr. 003.015.

Preliminary results from the X-ray experiment SL904. See Abstr. 034.006.

Ratio of line intensities in helium-like ions as a density indicator. See Abstr. 062.004.

Continuum radiative transfer in a hot plasma, with application to Scorpius X-1. See Abstr. 062.008.

Can the lumpy distribution of galaxies be detected by X-ray observations? See Abstr. 066.058.

A search for an ionospheric dynamo current effect of the galactic X-ray ionization. See Abstr. 083.016.

Effects of X-ray stars on VLF signal phase. See Abstr. 083.019.

Remarks about X-rays from Eta Carinae. See Abstr. 114.064.

Binary nature of the B supergiant in the error box of the Vela X-ray source. See Abstr. 119.021.

X Persei an X-ray source? See Abstr. 122.060.

Possible identification of X Persei with an X-ray source. See Abstr. 122.061.

Optical polarization of X Per. See Abstr. 122.116.

Is X Persei an X-ray source? See Abstr. 122.117.

X-ray emission from flare stars. See Abstr. 122.126.

On the nature of the Monoceros supernova remnant. See Abstr. 125.005.

A high-sensitivity search for X-rays from supernova remnants in Aquila. See Abstr. 125.014.

X-ray emission from white dwarfs. See Abstr. 126.019.

Gamma-ray production from p-p reactions in the interstellar medium. See Abstr. 131.031.

Detection of X-ray polarization of the Crab nebula. See Abstr. 134.003.

Energy balance of the Crab nebula and the Vela X remnant. See Abstr. 134.005.

Absorption of Crab nebula X-rays. See Abstr. 134.009.

Detection of radio emission from GX9+1. See Abstr. 141.075.

Position and identification of the Cygnus X-1 radio source. See Abstr. 141.185.

Evidence for hard X-ray pulsations from the Vela pulsar. See Abstr. 141.520.

Detection of pulsed gamma rays of $\sim 10^{12}$ eV from the pulsar in the Crab nebula. See Abstr. 141.541.

Gamma-ray observations of the galactic center and some possible point sources. See Abstr. 155.001.

A soft X-ray survey from the galactic center to Vela. See Abstr. 155.006.

Soft X-rays from the vicinity of the North Polar Spur. See Abstr. 155.010.

Origin of the low energy diffuse cosmic X-ray flux. See Abstr. 155.071.

X-ray observations of NGC 5128 (Centaurus A) from Uhuru. See Abstr. 158.029.

M 87 and the X-ray emission from compact objects. See Abstr. 158.089.

NGC 1068, 3C 273, and Scorpius X-1: Circular polarization disputed. See Abstr. 158.111.

X-ray emission from rich clusters of galaxies. See Abstr. 160.020.

Soft X-ray emission from intergalactic gas in the neighbourhood of the Galaxy.   See Abstr. 161.011.

Errata

142.901   Errata: 'Evidence for the binary nature of Centaurus X-3 from *Uhuru* X-ray observations'
[Astrophys. Journ., (*Letters*), Vol. 172, L79 - L89 (1972)].

E. Schreier, R. Levinson, H. Gursky, E. Kellogg, H. Tananbaum, R. Giacconi.
Astrophys. Journ., (*Letters*), Vol. 173, L151 (1972).

142.902   Errata: 'Further observations of the pulsating X-ray source Cygnus X-1 from *Uhuru*' [Astrophys. Journ., (*Letters*), Vol. 170, L21 - L27 (1971)].

E. Schreier, H. Gursky, E. Kellogg, H. Tananbaum, R. Giacconi.
Astrophys. Journ., (*Letters*), Vol. 173, L151 (1972).

143 Cosmic Radiation

143.001 **On collective interactions of nucleons of heavy nuclei of high-energy cosmic rays.**
N. L. Grigorov, D. A. Zhuravlev, V. D. Kozlov, M. A. Kondratieva, I. D. Rapoport, I. A. Savenko.
Dokl. Akad. Nauk SSSR, Ser. Mat. Fiz., Vol. 202, 65 - 66 (1972). In Russian.

143.002 **Interstellar propagation of galactic cosmic-ray nuclei $2 \leq Z \leq 8$ in the energy range 10−1000 MeV per nucleon.** G. M. Mason.
Astrophys. Journ., Vol. 171, 139 - 161 (1972).
The differential kinetic energy per nucleon spectra of galactic cosmic-ray He, Li, Be, B, C, N, and O have been analyzed with the University of Chicago cosmic-ray telescope on board the IMP-5 satellite in 1969−1970. The ratios He/(C+N+O) and (Li+Be+B)/(C+N+O) obtained from these spectra are found. The results are compared with discrepancies existing among other measurements of these ratios, and with predictions of a class of steady-state models of cosmic-ray propagation which assume that Li, Be, and B are absent in cosmic-ray sources.

143.003 **Solar modulation of galactic cosmic radiation.**
U. R. Rao.
Space Sci. Rev., Vol. 12, 719 - 809 (1972).
In this review an attempt is made to present an integrated view of the solar modulation process that causes time variation of cosmic ray particles. After briefly surveying the relevant large and small scale properties of the interplanetary magnetic fields and plasma, the motion of cosmic ray particles in the disordered interplanetary magnetic fields is discussed. The experimentally observed long term variations of different species of cosmic ray particles are summarised and compared with the theoretical predictions from the diffusion−convection model. The effect of the energy losses due to deceleration in the expanding solar wind are clearly brought out. The radial density gradient, the modulation parameter and their long term variation are discussed to understand the dynamics of the modulating region. The cosmic ray anisotropy measurements at different energies are summarised.

143.004 **Cosmic ray isotropy and the origin problem.**
R. Speller, T. Thambyahpillai, H. Elliot.
Nature, Vol. 235, 25 - 29 (1972).
A study of the cosmic ray muon intensity underground in London has revealed small sidereal variations which can be explained in terms of solar motion.

143.005 **Streaming of galactic cosmic rays in the interplanetary magnetic field.**
A. Hashim, M. Bercovitch, J. F. Steljes.
Solar Physics, Vol. 22, 220 - 234 (1972).
The anisotropy observed in galactic cosmic rays can be considered to be the result of radial convective streaming away from the sun with the velocity of the solar wind, and diffusive streaming parallel to the direction of the interplanetary magnetic field (nearly always sunward). This description of the anisotropy is found to be applicable to (a) instances of large 'non-equilibrium' anisotropy observed during the onset, main, or recovery phases of Forbush decreases; (b) enhanced diurnal variations; (c) periods characteristic of the quiet time diurnal variation. These results are analogous to those obtained by McCracken, Rao and Ness.

143.006 **The quiet-time spectra of cosmic-ray electrons of energies between 10 and 200 MeV observed on OGO-5.** J. L'Heureux, C. Y. Fan, P. Meyer.
Astrophys. Journ., Vol. 171, 363 - 376 (1972).
Spectra of cosmic-ray electrons of energies between 10 and 200 MeV were measured over a 1-year period starting 1968 March. Time periods during which no solar-flare events were recorded were selected for the study. It was found that during these quiet periods there were numerous intensity variations of the electron flux. These variations, which are seen only below 25 MeV, do not show marked correlation with any solar or interplanetary-medium parameters. The physical implication of the finding is discussed.

143.007 **Idealized model for the radial gradient of modulated cosmic rays.** J. J. O'Gallagher.
Journ. Geophys. Res., Vol. 77, 513 - 523 (1972).
It is the purpose of this paper to present a simplified model, based on the observed behavior, that provides some basis for a reinterpretation of modulation phenomena at low energies with particular emphasis on the behavior of the radial gradient. This model, although greatly idealized, suggests that in some limit the local interplanetary gradient at low energies may be directly related to the effective radial scale of solar modulation and to the energy dependence of the low-energy interstellar spectrum.

143.008 **Origin of cosmic electrons from several hundreds GeV to muon-poor air-shower energies.**
R. Ramaty, R. E. Lingenfelter.
Bull. American Astron. Soc., Vol. 3, 480 (1971). − Abstr. AAS.

143.009 **Propagation and confinement of ultrarelativistic cosmic rays in the Galaxy.** C. J. Cesarsky.
Bull. American Astron. Soc., Vol. 3, 480 - 481 (1971). Abstr. AAS.

143.010 **On the distortion of stellar anisotropy of galactic cosmic rays by the solar wind.**
A. V. Belov, L. I. Dorman.
Geomagn. Aeronom., Vol. 12, 111 - 112 (1972). In Russian. Brief information.

143.011 **Peculiarities of cosmic ray diffusion in a radially diverging stream of magnetic irregularities.**
A. V. Belov, L. I. Dorman.
Geomagn. Aeronom., Vol. 12, 113 - 114 (1972). In Russian. Brief information.

143.012 **Studies of the chemical composition of cosmic rays with $Z = 3−30$ at high and low energies.**
W. R. Webber, S. V. Damle, J. Kish.
Astrophys. Space Sci., Vol. 15, 245 - 271 (1972).
We have measured the chemical composition of cosmic rays with $Z \geq 2$ over an energy range from ~ 100 MeV/nuc to > 2 GeV/nuc using 2 new large area counter telescopes. A detailed charge spectrum is obtained at both high and low energies. A comparison of solar and cosmic ray abundances reveals certain selective differences, rather than a systematic overabundance of heavy nuclei in cosmic rays, as has been suggested in the past. These differences are discussed in terms of a common nucleosynthesis origin of the two species of particles.

143.013 **Balloon measurements of the energy spectrum of cosmic electrons between 1 and 25 Gev.**
J. A. Earl, D. E. Neely, T. A. Rygg.
Journ. Geophys. Res., Vol. 77, 1087 - 1102 (1972).
This paper presents experimental results on the electron spectrum obtained with the aid of a detector whose response

to both electrons and nuclei was specified in accelerator exposures carried out in support of balloon exposures to primary cosmic rays during three balloon flights made in 1966 and 1967.

143.014 Comment on paper by D. B. Swinson, 'Solar modulation origin of "sidereal" cosmic-ray anisotropies'. [Journ. Geophys. Res., Vol. 76, 4217 - 4223 (1971)]. H. Elliot, T. Thambyahpillai, J. A. Otaola. Journ. Geophys. Res., Vol. 77, 1342 - 1344 (1972), with a reply by D. B. Swinson p. 1345 - 1346.

143.015 Search for antimatter in primary cosmic rays. A. Buffington, L. H. Smith, G. F. Smoot, L. W. Alvarez, M. A. Wahlig. Nature, Vol. 236, 335 - 338 (1972).

A new upper limit on the amount of antimatter in primary cosmic rays has been established. The limits are considerably lower than those for any previous experiment.

143.016 Scattering of high-energy cosmic rays. D. G. Wentzel. Astrophys. Letters, Vol. 10, 167 - 170 (1972).

When cosmic rays stream through highly ionized interstellar regions, they generate hydromagnetic waves that scatter the cosmic rays until the cosmic-ray streaming speed is less than (about) two times the Alfvén speed, independent of the cosmic-ray energy. Confinement to our galaxy of the 100 GeV cosmic rays, at their observed low streaming rates, requires a rather large scattering region, possibly the galactic corona or many nebulae like the Gum nebula.

143.017 Cosmic-ray density distribution normal to the solar equatorial plane and the semi-diurnal anisotropy. P. N. Pathak. Planet. Space Sci., Vol. 20, 533 - 547 (1972).

Distributions of galactic cosmic-ray density normal to the solar equatorial plane are derived by two different methods and the resulting semi-diurnal anisotropy is studied in detail. Following the method suggested by Subramanian and Sarabhai (1967), the semi-diurnal anisotropy arising from these distributions is estimated for neutron monitors situated at Deep River and Huancayo. The implications of the results are discussed and the various assumptions tested.

143.018 Variations of streams of charged particles with various energies from data of automatic interplanetary stations. S. N. Vernov, E. V. Gorchakov, P. P. Ignat'ev, N. G. Galach'ev. Izv. AN SSSR. Ser. fiz., Vol. 35, 2418 - 2422 (1971). In Russian. – Abstr. in Referativ. Zhurn. 62. Issled. kosm. prostranstva, 4.62.226 (1972).

143.019 The energy spectrum of primary cosmic radiation in the range of $10^{11} - 10^{15}$ eV from measurements with Proton 4. V. V. Akimov, N. L. Grigorov, Yu. V. Gubin, V. E. Nesterov, I. D. Rapoport, I. A. Savenko. Izv. AN SSSR. Ser. fiz., Vol. 35, 2434 - 2438 (1971). In Russian. – Abstr. in Referativ. Zhurn. 62. Issled. kosm. prostranstva, 4.62.232 (1972).

143.020 On irregularities in the spectrum of primary cosmic rays in the energy range of 10^{12} eV. V. V. Akimov, N. L. Grigorov, N. A. Mamontova, V. E. Nesterov, I. D. Rapoport, I. A. Savenko. Izv. AN SSSR. Ser. fiz., Vol. 35, 2439 - 2442 (1971). In Russian. – Abstr. in Referativ. Zhurn. 62. Issled. kosm. prostranstva, 4.62.233 (1972).

143.021 Energy spectrum of α-particles of primary cosmic radiation in the range of high energies from meas- urements with Proton. N. L. Grigorov, V. E. Nesterov, V. L. Prokhin, I. D. Rapoport, I. A. Savenko. Izv. AN SSSR. Ser. fiz., Vol. 35, 2443 - 2445 (1971). In Russian. – Abstr. in Referativ. Zhurn. 62. Issled. kosm. prostranstva, 4.62.234 (1972).

143.022 The search for antiprotons in primary cosmic rays. É. A. Bogomolov, N. D. Lubyanaya, V. A. Romanov, M. G. Totubalina. Izv. AN SSSR. Ser. fiz., Vol. 35, 2448 - 2452 (1971). In Russian. – Abstr. in Referativ. Zhurn. 51. Astron., 4.51.647 (1972).

143.023 Transient north-south anisotropies in the cosmic ray intensity. J. B. Mercer, D. N. H. Barker, W. K. Griffiths, C. J. Hatton. Planet. Space Sci., Vol. 20, 721 - 729 (1972).

Using data from four polar neutron monitors, a search has been made for north–south anisotropies in the cosmic radiation during the period August 1966 – November 1967 (solar rotations 1820–1837). The existence of anisotropies with large N–S components is discussed in relation to the study of anisotropies in the ecliptic plane.

143.024 A cosmic ray density gradient perpendicular to the ecliptic plane. A. Hashim, M. Bercovitch. Planet. Space Sci., Vol. 20, 791 - 801 (1972).

We have made a detailed study of the effect of the particle drift in the diurnal variation data for 1967 and 1968 from a variety of cosmic ray monitors. The results are compared with estimates of the gradient from annual wave measurements.

143.025 Cosmic-ray electrons. H. C. van de Hulst. Quarterly Journ. Roy. Astron. Soc., Vol. 13, 10 - 24 (1972). – The George Darwin lecture delivered on 1970 November 13.

143.026 Extragalactic cosmic rays and the production of light elements. J. A. de Freitas Pacheco. Astron. Astrophys., Vol. 18, 450 - 452 (1972).

It is shown that the galactic diffusion coefficient D calculated previously (Pacheco, 1971) for relativistic cosmic rays remains basically unchanged irrespective of whether the primaries are of galactic or extragalactic origin.

143.027 Cosmic-ray source and local interstellar spectra deduced from the isotopes of hydrogen and helium. G. M. Comstock, K. C. Hsieh, J. A. Simpson. Astrophys. Journ., Vol. 173, 691 - 709 (1972).

From the experimental information now available on the quartet of isotopes ^1H, ^2H, ^3He, and ^4He, and recognizing that, except for small contributions from C, N, and O nuclei, a closed generic relationship exists within the quartet, one can obtain a self-consistent model for cosmic-ray hydrogen and helium propagation from the sources in the Galaxy to the orbit of earth.

143.028 The chemical composition of cosmic rays – I. L. H. Aller. Sky Telescope, Vol. 43, 285 - 287 (1972).

143.029 The chemical composition of cosmic rays – II. L. H. Aller. Sky Telescope, Vol. 43, 362 - 363 (1972).

143.030 Cosmic-ray increases at the Antarctic Base of General Belgrano. H. S. Ghielmetti. Inform. Bull. Southern Hemisph., No. 19, p. 32 - 33 (1971).

143.031 On the possible role of the interstellar gas in the

annual variations of cosmic rays. E. Ya. Gidalevich.
Izv. AN SSSR. Ser. fiz., Vol. 35, 2479 - 2482 (1971). In
Russian. – Abstr. in Referativ. Zhurn. 51. Astron., 5.51.390
(1972).

**143.032 A region of the effective 11-year modulation of
galactic cosmic rays (from meteorite data).**
A. K. Lavrukhina.
Izv. AN SSSR. Ser. fiz., Vol. 35, 2498 - 2502 (1971). In
Russian. – Abstr. in Referativ. Zhurn. 51. Astron., 5.51.393
(1972).

**143.033 Diffusion and stochastic variations of galactic cos-
mic rays.**
A. G. Zusmanovich, E. V. Kolomeets, Ya. E. Shvartsman.
Izv. AN SSSR. Ser. fiz., Vol. 35, 2525 - 2529 (1971). In
Russian. – Abstr. in Referativ. Zhurn. 51. Astron., 5.51.394
(1972).

**143.034 Study of the three-dimensional anisotropy of cos-
mic rays of high and low energies.**
L. I. Dorman, E. A. Eroshenko, O. I. Inozemtseva.
Izv. AN SSSR. Ser. fiz., Vol. 35, 2503 - 2507 (1971). In
Russian. – Abstr. in Referativ. Zhurn. 51. Astron., 5.51.446
(1972).

**143.035 The effects of the solar active regions on the propa-
gation of galactic cosmic rays.** A. M. Altukhov,
G. F. Krymskij, A. I. Kuz'min, N. P. Chirkov, P. A. Krivoshap-
kin, G. V. Skripin, I. A. Transkij, V. P. Mamrukova.
Izv. AN SSSR. Ser. fiz., Vol. 35, 2508 - 2518 (1971). In
Russian. – Abstr. in Referativ. Zhurn. 51. Astron., 5.51.447
(1972).

**143.036 The 11-year solar modulation of the intensity of
cosmic rays in the stratosphere.**
G. A. Bazilevskaya, A. N. Kvashnin, A. K. Pankratov, A. K.
Svirzhevskaya, Yu. I. Stozhkov, A. N. Charakhch'yan, T. N.
Charakhch'yan.
Izv. AN SSSR. Ser. fiz., Vol. 35, 2483 - 2487 (1971). In
Russian. – Abstr. in Referativ. Zhurn. 51. Astron., 5.51.464
(1972).

**143.037 The effect of ionization losses on the lifetime of
cosmic rays in the Galaxy.** P. Velinov.
Izv. AN SSSR. Ser. fiz., Vol. 35, 2466 - 2471 (1971). In
Russian. – Abstr. in Referativ. Zhurn. 51. Astron., 5.51.655
(1972).

**143.038 Multifarious temporal variations of low-energy rela-
tivistic cosmic-ray electrons.**
F. B. McDonald, T. L. Cline, G. M. Simnett.
Journ. Geophys. Res., Vol. 77, 2213 - 2231 (1972).
A detailed examination is made of the intensity variations of 3- to 12-MeV interplanetary electrons. The data are
from the Goddard cosmic-ray experiment on the Imp satellites and cover the period from just before the last solar minimum through the onset of the present solar maximum (i.e.,
from December 1963 through August 1969). A morphology
for the intensity changes is tentatively proposed that includes
solar-flare-associated events, solar corotating increases, Forbush decreases, quiet-time increases, and the long-term 11-year
variation.

**143.039 Consequences of "scaling" for high energy cosmic
rays.** J. Wdowczyk, A. W. Wolfendale.
Nature, Phys. Sci., Vol. 236, 29 - 30 (1972).
A preliminary analysis of the effect of adopting of the
"scaling" hypothesis for high energy cosmic rays interactions
is given. It is shown that if scaling is applicable then the effec-

tive mass of the cosmic rays primaries must increase with increase of their energy.

**143.040 Solar modulation and the chemical composition of
the cosmic radiation.** C. J. Waddington.
Nature, Vol. 236, 391 - 392 (1972).
It now appears that except at very low energies the effects of solar modulation of cosmic ray intensities can be
treated as an adiabatic deceleration of the individual particles.
Thus a particle detected with an energy E in the inner solar
system, traversed interstellar space with an energy $E+\Delta E$.
Hence nuclei of fragmentation origin were formed at a higher
energy than that at which they are observed. Consequently if
the fragmentation process is energy dependent the abundances
of the fragmentation nuclei detected at E will depend on the
magnitude of ΔE. Since ΔE varies during the solar cycle so
will the abundances. Consideration of the semi-empirical fragmentation cross-sections suggest that for observations in the
200–500 MeV/n range, manganese and chromium should
show the most pronounced variations during a solar cycle.
Presently published data seem inadequate to verify this effect,
but currently available detectors should be adequate. A failure
to observe such composition variations would imply either the
cross-sections are grossly incorrect or the model of solar modulation is wrong.

**143.041 The acceleration of cosmic rays in supernova rem-
nants.** J. P. Ostriker, R. M. Kulsrud, J. E. Gunn.
Bull. American Astron. Soc., Vol. 4, 260 (1972). – Abstr. AAS.

143.042 Propagation of Z > 75 cosmic rays.
D. N. Schramm.
Bull. American Astron. Soc., Vol. 4, 261 (1972). – Abstr. AAS.

143.043 Scattering of high-energy cosmic rays.
D. G. Wentzel.
Bull. American Astron. Soc., Vol. 4, 261 - 262 (1972).
Abstr. AAS.

**143.044 Solar-diurnal, sporadic and annual variations of
cosmic rays.** G. A. Gonchar, A. G. Zusmanovich,
E. V. Kolomeets, R. A. Chumbalova, Yu. A. Shakhova.
Izv. AN SSSR. Ser. fiz., Vol. 35, 2519 - 2524 (1971).
In Russian. – Abstr. in Referativ. Zhurn. 51. Astron., 6.51.
557 (1972).

**143.045 On the irregularity in the spectrum of primary
cosmic rays in the energy range of 10^{12} eV.**
V. V. Akimov, N. L. Grigorov, N. A. Mamontova, V. E. Nesterov, I. D. Rapoport, I. A. Savenko.
Izv. AN SSSR. Ser. fiz., Vol. 35, 2439 - 2442 (1971).
In Russian. – Abstr. in Referativ. Zhurn. 51. Astron., 6.51.
766 (1972).

**143.046 The energetic spectrum of primary cosmic rays in
the range of 10^{11} - 10^{15} eV according to measurements on satellite Proton 4.** V. V. Akimov, N. L. Grigorov,
Yu. V. Gubin, V. E. Nesterov, I. D. Rapoport, I. A. Savenko.
Izv. AN SSSR. Ser. fiz., Vol. 35, 2434 - 2438 (1971).
In Russian. – Abstr. in Referativ. Zhurn. 51. Astron., 6.51.
767 (1972).

**143.047 The energetic spectrum of α-particles of primary
cosmic rays in the range of high energies according
to measurements on Proton satellites.** N. L. Grigorov,
V. E. Nesterov, V. L. Prokhin, I. D. Rapoport, I. A. Savenko.
Izv. AN SSSR. Ser. fiz., Vol. 35, 2443 - 2445 (1971).
In Russian. – Abstr. in Referativ. Zhurn. 51. Astron., 6.51.
768 (1972).

143.048 Solar modulation of galactic particles. Numerical integration of the steady Fokker-Planck equation.
N. J. Martinić.
Acta Univ. Upsaliensis, Uppsala Diss. Fac. Sci. 4. [Distributed by Almqvist & Wiksell, Stockholm, Sweden. Price skr. 25.00], 4 + 42 pp (1972).

Galactic particles, in our case protons and electrons, are modulated by the solar atmosphere. From the numerical integration of the steady, spherically symmetric Fokker-Planck equation, which governs this phenomenon as a first approximation, we show that the galactic flux measured at the earth presents the effect of: a) convection, i.e. the sweeping-out of the galactic particles at the convective outward velocity of the solar wind (∼400 km/sec at steady conditions), creating a gradient of these particles; b) diffusion, i.e. the dynamical equilibrium between the gradient of galactic particles and the flux of particles which tends to neutralise this gradient; c) the adiabatic deceleration, i.e. the 'cooling-down' effect of expansion of the bulk of galactic particles tied to the magnetic inhomogeneities frozen in the solar plasma.

143.049 Low frequency radio emission from extensive air showers. H. R. Allan.
Nature, Vol. 237, 384 - 385 (1972).

The observed radio emission from extensive air showers at frequencies between 30 and 100 MHz is in good agreement with theory. At lower frequencies the situation is much less satisfactory. The calculation presented here is intended to set a generous upper limit to the typical field strength per unit bandwidth based on the least controversial elements of the theory. Because the claimed experimental values are greater even than this upper limit, I suggest that further checks are needed.

143.050 North-south asymmetry of cosmic ray intensity increase before magnetic storms.
N. S. Kaminer, V. P. Kashevarov, A. E. Kuzmicheva, L. I. Dorman.
Geomagn. Aeronom., Vol. 12, 385 - 391 (1972). In Russian.

143.051 Some aspects of measuring the differential spectrum of cosmic rays.
A. A. Kolchin, V. V. Lebedev, G. P. Skrebtzov.
Geomagn. Aeronom., Vol. 12, 392 - 397 (1972). In Russian.

143.052 27-daily variations of the anisotropy of cosmic rays. L. Kh. Shatashvili, A. K. Pankratov, A. A. Stepanyan.
Geomagn. Aeronom., Vol. 12, 538 - 541 (1972). In Russian.
Brief information.

143.053 Radial gradients and anisotropies due to galactic cosmic rays. I. H. Urch, L. J. Gleeson.
Astrophys. Space Sci., Vol. 16, 55 - 74 (1972).

In this paper we will present computations of the anisotropies and radial gradients of galactic cosmic ray particles which have penetrated into the solar cavity. Particular attention is given to particles with kinetic energies less than 200 MeV/nucleon where no adequate theoretical analysis has hitherto been available.

143.054 On the chemical composition of cosmic rays.
V. S. Ptuskin.
Kosm. Issled., Vol. 10, 351 - 357 (1972). In Russian.

143.055 Angular distribution and energy spectrum of cosmic ray single neutrons with energies higher than 5×10^{11} eV at mountain altitudes.
Kh. P. Babayan, N. G. Boyadjian, E. A. Mamidjanian.
Izv. Akad. Nauk Armyan. SSR, Fizika, Vol. 6, 168 - 173

(1971). In Russian.

The results of the study of single neutrons with energies higher than 5×10^{11} eV at mountain altitudes by means of an ionization calorimeter supplemented with proportional counters are given. The angular distribution and the integral energy spectrum of single neutrons are obtained.

143.056 Hochenergie-Astrophysik. W. K. H. Schmidt.
Umschau, 72. Jahrgang, p. 377 - 382 (1972).

143.057 Aspects of high energy astrophysics. J. R. Prescott.
Proc. Astron. Soc. Australia, Vol. 2, 70 - 71 (1972).

143.058 Observations at 6 MHz of radio pulses from extensive air showers. D. G. Felgate, T. J. Stubbs.
Proc. Astron. Soc. Australia, Vol. 2, 90 - 91 (1972).

143.059 On the acceleration of charged particles to cosmic ray energies. M. Grewing, H. Heintzmann.
Mitt. Astron. Ges., No. 31, p. 91 - 94 (1972).

143.060 Dust grain origin of cosmic ray air showers.
S. Hayakawa.
Astrophys. Space Sci., Vol. 16, 238 - 240 (1972).

It is suggested that cosmic rays of energies as high as 10^{20} eV consist of dust grains of relativistic energies. Such dust grains as typical in interstellar space are accelerated first by a strong radiation pressure of luminous, compact galaxies and then by magnetic processes.

143.061 Galactic cosmic ray modulation by the interplanetary medium (including the problem of the outer boundary). L. I. Dorman.
Astrophys. Space Sci. Library, Vol. 29, Part II, (see 012.013), p. 67 - 91 (1972).

143.062 Low-energy cosmic rays in interplanetary space.
S. N. Vernov, G. P. Lyubimov.
Astrophys. Space Sci. Library, Vol. 29, Part II, (see 012.013), p. 92 - 109 (1972).

143.063 Extragalactic cosmic rays.
K. Brecher, G. R. Burbidge.
Astrophys. Journ., Vol. 174, 253 - 291 (1972).

Two related questions are examined: Are there cosmic rays outside galaxies? Do the cosmic rays we detect locally originate outside our Galaxy? We consider the possible sources of extragalactic cosmic rays, their propagation, and their confinement in several kinds of extragalactic systems: clusters of galaxies, superclusters, and throughout metagalactic space. We develop no theory for the spectral shape and composition of the cosmic rays at their source, but constraints imposed by changes in composition during propagation are considered in detail.

143.064 Constant-crossed-field acceleration, a mechanism for the generation of cosmic rays by strong low-frequency electromagnetic waves. M. Grewing, H. Heintzmann.
Phys. Rev. Letters, Vol. 28, 381 - 384 (1972).

Magnetized stars contracting to neutron star dimensions or collapsing to a black hole will emit large quantities of electromagnetic radiation. It is shown that under favorable circumstances cosmic rays can be generated with energies in excess of 10^{21} eV.

143.065 Pulsar-produced cosmic rays and the origin of the light elements. K. Sitte.
Nuovo Cimento B, Ser. 11, Vol. 7B, 110 - 118 (1972).

Particles accelerated in the vicinity of pulsars must traverse the expanding shell of the supernova before emerging as cosmic rays. This passage results in a considerable intensity reduction and in the production of light nuclei by fragmentation of heavy nuclei of the shell gas.

143.066   Production and propagation of particles with A > 81 in the Galaxy.
T. B. Kaiser, J. R. Wayland, G. Gloeckler.
Phys. Rev. D, Particles and Fields, Vol. 5, 307 - 313 (1972).

A detailed mathematical model is developed to describe the transformation of the charge composition of superheavy (A > 81) cosmic rays by spallation on interstellar hydrogen in the Galaxy.

143.067   Three-dimensional cosmic ray anisotropy in interplanetary space. I. Formulation of cosmic ray daily variation produced by axis-symmetric anisotropy.
K. Nagashima.
Rep. Ionosph. & Space Res. Japan, Vol. 25, 189 - 211 (1971).
See Phys. Abstr., Vol. 75, No. 33588 (1972).

143.068   Three-dimensional cosmic ray anisotropy in interplanetary space. II. General expression of annual modulation of daily variation by frequency modulation method and its application to the modulation due to earth's revolution around sun.   K. Nagashima, H. Ueno.
Rep. Ionosph. & Space Res. Japan, Vol. 25, 212 - 241 (1971).

143.069   The solar semi-diurnal anisotropy of the cosmic radiation.   Z. Fujii.
Rep. Ionosph. & Space Res. Japan, Vol. 25, 242 - 270 (1971).
See Phys. Abstr., Vol. 75, No. 33768 (1972).

143.070   An analysis of the solar semi-diurnal variation of the cosmic radiation.   S. Mori, S.-I. Yasue, M. Ichinose.
Rep. Ionosph. & Space Res. Japan, Vol. 25, 271 - 284 (1971).
See Phys. Abstr., Vol. 75, No. 33769 (1972).

143.071   Antiproton spectrum in cosmic rays.   M. C. Chen.
Nuovo Cimento B, Ser. 11, Vol. 8B, 343 - 357 (1972).

The steady-state antiproton spectrum in cosmic rays as a result of the collision between primary cosmic protons and interstellar hydrogen is calculated by means of the statistical model of two fire-balls in multiple-particle production. Results based on Fermi's original one-fire-ball model are also presented for comparison.

143.072   The anisotropy of cosmic rays of galactic origin above 10^{17} eV.
S. K. Karakula, J. L. Osborne, E. Roberts, W. Tkaczyk.
Journ. Phys. A, General Phys., Vol. 5, 904 - 915 (1972).

By following the trajectories of protons in the Galaxy the authors predict the detailed form of the anisotropy of ultrahigh energy cosmic rays of galactic origin for given magnetic fields and source distributions. Three specific magnetic field models based on interpretations of the observational data are used and they consider sources uniformly distributed throughout the galactic disc or spiral arms or concentrated at the galactic centre. Data on the arrival directions of extensive air showers are then compared with the predicted distributions for these models, and, assuming that metagalactic cosmic rays are isotropic, upper limits are given for the fraction of cosmic rays of galactic origin.

143.073   The cosmic-ray spectral modulation above 2 GV during the descending phase of solar cycle number 19. I. A comprehensive treatment of the neutron monitor data from the worldwide station network and latitude surveys.
F. Bachelet, N. Iucci, G. Villoresi, B. Sporre.
Nuovo Cimento B, Ser. 11, Vol. 7B, 17 - 33 (1972).

143.074   The cosmic-ray spectral modulation above 2 GV during the descending phase of solar cycle number 19. II. A study of large Forbush decreases.
F. Bachelet, N. Iucci, G. Villoresi.
Nuovo Cimento B, Ser. 11, Vol. 7B, 34 - 44 (1972).

143.075   The cosmic-ray spectral modulation above 2 GV during the descending phase of solar cycle number 19. III. A study of the long-term variation.
F. Bachelet, N. Iucci, G. Villoresi.
Nuovo Cimento B, Ser. 11, Vol. 7B, 45 - 61 (1972).

The physics of cosmic X-ray, γ-ray, and particle sources.   See Abstr. 003.015.

Cosmic rays and nuclear interactions at high energies.   See Abstr. 003.121.

Progress in elementary particle and cosmic ray physics, Vol. 10.   See Abstr. 003.142.

Cherenkov and transient radiation of uniformly moving charge in random inhomogeneous medium.   See Abstr. 063.030.

Forbush decreases in the flux of galactic cosmic rays and associated VLF night-time propagation phenomena.   See Abstr. 083.017.

Interaction of solar and galactic cosmic-ray particles with the moon.   See Abstr. 094.031.

Meteor variation of cosmic ray intensity for type I and II meteor streams.   See Abstr. 104.005.

The Os-Pt-Hg abundance peak in Ap stars and the problem of very heavy cosmic rays.   See Abstr. 114.041.

Acceleration of cosmic rays in supernova remnants.   See Abstr. 125.028.

Interstellar motions: Minuet or Rock?   See Abstr. 131.086.

The intensity of He II 4686 Å in H II regions and the cosmic-ray flux.   See Abstr. 131.107.

Possible production of Li, Be and B by low energy cosmic rays.   See Abstr. 131.110.

Stellar Systems

151 Kinematics and Dynamics of Stellar Systems

151.001 Some elementary applications of the virial theorem to stellar dynamics.
S. Chandrasekhar, D. D. Elbert.
Monthly Notices Roy. Astron. Soc., Vol. 155, 435 - 447 (1972).

The dynamical evolution of spherical and spheroidal systems of mass points is examined with the aid of the scalar and the tensor forms of the virial theorem.

151.002 Collective oscillations of a spherically symmetric system of rotating masses.
A. M. Fridman, I. G. Shukhman.
Dokl. Akad. Nauk SSSR, Ser. Mat. Fiz., Vol. 202, 67 - 70 (1972). In Russian.

151.003 A numerical hydrodynamic study of coalescence in head-on collisions of identical stars.
F. G. P. Seidl, A. G. W. Cameron.
Astrophys. Space Scil, Vol. 15, 44 - 128 (1972).

A two-dimensional hydrodynamic code has been developed for numerical studies of stellar collisions. The motivation for the study has been the suggestion by Colgate that collisions among stars in a dense galactic core can lead to growth of stellar masses by coalescence and thus to an enhanced rate of supernova activity. The specific results reported here refer to head-on collisions between identical polytropes of index 3 having solar mass and radius.

151.004 The evolution of a multi-phase space density collisionless one-dimensional stellar system.
S. Cuperman, A. Harten.
Astron. Astrophys., Vol. 16, 13 - 20 (1972).

The evolution in time of collisionless one-dimensional stellar systems is investigated for the case of a four phase space density configuration. The investigation is carried out by following the motion of the boundary curves enclosing the four phase space regions, each of which is characterized by a different phase space density. A partial qualitative agreement with Lynden-Bell's prediction that the velocity dispersions should be inversely proportional to the initial phase space densities is found.

151.005 Modèle tridimensionnel et instationnaire de système stellaire auto-gravitant. Applications aux amas globulaires et aux galaxies. J.-P. Petit.
Comptes Rendus Acad. Sci. Paris, Sér. B, Vol. 274, 373 - 376 (1972).

On cherche une solution de Schwarzschild à l'équation de Vlasov telle que l'ellipsoïde des vitesses ait des propriétés spéciales. Une solution analytique tridimensionnelle apparaît qui semble assez cohérente vis-à-vis des données d'observation sur les amas globulaires et les galaxies.

151.006 Numerical study of a four-dimensional mapping. C. Froeschlé.
Astron. Astrophys., Vol. 16, 172 - 189 (1972).

The study of dynamical systems with three degrees of freedom can be reduced to the study of a four-dimensional mapping using the method of surface of section. Therefore a mapping T of a four-dimensional space into itself has been

studied using three numerical methods of a non-graphical nature: 1) The divergence of two initially close orbits. 2) The variation of the largest eigenvalue of the linear tangential mapping of T^1. 3) A local fitting of the points in the four-dimensional "surface of section" by a quadratic surface.

151.007 Oscillation and overstability of density waves in a rotating disk-like star-gas system. S. Kato.
Publ. Astron. Soc. Japan, Vol. 24, 61 - 86, with an addendum, p. 293 (1972).

A possible mechanism of maintenance of density waves in spiral galaxies is suggested. It is shown that small amplitude tightly-wound density waves in a rotating disk-like star-gas system can become overstable if two sub-systems of stars and gases interact by the processes of birth and disintegration of stars. Attention is paid in particular to the cases where the disk consists mainly of stars, and the gas temperature is low in the sense that the sound velocity in the gas is small compared with the dispersion velocity of stars.

151.008 The problem of three bodies with nonstationary Newton-Gook interaction and its application to motions in the metagalaxy. T. B. Omarov.
Astron. Zhurn. Akad. Nauk SSSR, Vol. 49, 191 - 196 (1972). In Russian. English translation in Soviet Astron. AJ, Vol. 16, No. 1.

The summarized statement of the celestial-mechanical problem of two bodies is solved by quadrature when that problem is considered against a background of uniformly distributed and isotropically expanding gravitating matter of the Einstein—de Sitter world.

151.009 The structure of a quasi-stationary spherical system of stars of different masses. I. Mihaila.
Astron. Zhurn. Akad. Nauk SSSR, Vol. 49, 197 - 203 (1972). In Russian. English translation in Soviet Astron. AJ, Vol. 16, No. 1.

A model of a quasi-stationary spherical system of stars of different masses with isotropic velocity distribution is obtained. It is assumed that the potential is produced by stars of average mass and that the stars of other masses constitute a small addition. The density, the velocity of the centroid and the velocity dispersion are determined for stars of different masses.

151.010 Polytropic and isothermal plane-symmetric configurations. E. R. Harrison, R. G. Lake.
Astrophys. Journ., Vol. 171, 323 - 330 = Contr. Five College Obs., Univ. Mass., *Amherst*, No. 111 (1972).

In this paper we idealize rotating disk systems and assume that density and pressure in the equilibrium state depend locally only on the perpendicular distance to the equatorial plane. We show that the hydrostatic equation, with the aid of the mass equation, can be integrated directly to give a pressure-mass relation. We then assume that density and pressure are related by a polytropic equation of state and derive in closed form the solution for all plane-symmetric polytropes of $-1 < n \leq \infty$.

151.011 Nonlinear condensations in self-gravitating media. I. A numerical approach.

J.-L. Tassoul, G. Pellieux.
Astrophys. Journ., Vol. 171, 485 - 493 (1972).

We investigate, within the framework of Newtonian mechanics, the evolution of finite-amplitude fluctuations in extended self-gravitating media. Pressure effects are fully taken into account.

151.012 Nonlinear condensations in self-gravitating media. II. An analytical approach.
J.-L. Tassoul, M. Tassoul.
Astrophys. Journ., Vol. 171, 495 - 501 (1972).

This paper deals with the nonlinear gravitational instability of infinite homogeneous media. Second-order asymptotic expansions are constructed to describe small- but finite-amplitude motions.

151.013 Bias of virial-theorem estimates of mass. D. Sher.
Astrophys. Journ., Vol. 171, 537 - 538 (1972).

It is shown that the usual procedure for estimating the mass of a cluster by means of the virial theorem leads to an overestimate of up to 30 percent because the velocity dispersion varies with position in the cluster.

151.014 Virial mass determinations of bound and unstable groups of galaxies. S. J. Aarseth, W. C. Saslaw.
Astrophys. Journ., Vol. 172, 17 - 35 (1972).

Virial theorem masses of small N-body configurations for three different ensembles representing groups of 8, 16, or 32 galaxies are computed. Dynamical evolution, projection effects, and the effects of incomplete data are used to simulate actual observations. The validity of theoretical assumptions is tested against dynamically consistent configurations of one N-body system with 500 particles. In the second part, we examine the observational implications of the hypothesis that groups of galaxies are unbound and expanding. Evolution tracks of virial masses have also been computed for several types of mass-loss mechanisms and different rates. Finally, we discuss several methods for distinguishing between the hidden-mass hypothesis and the expansion hypothesis.

151.015 Numerical experiments concerning the random force in stellar systems. L. Cohen, A. Ahmad.
Bull. American Astron. Soc., Vol. 3, 441 (1971). – Abstr. AAS.

151.016 Cylindrical explosions in differentially rotating gas systems. R. H. Sanders, K. H. Prendergast.
Bull. American Astron. Soc., Vol. 3, 441 - 442 (1971). – Abstr. AAS.

151.017 The stability problem of oscillations along the axis of symmetry in a galaxy. III. Lyapunov's linear theory and a generalization of the results. P. Andrle.
Bull. Astron. Inst. Czechoslovakia, Vol. 23, 40 - 47 (1972).

The stability problem of oscillations along the axis of symmetry in a galaxy was studied by determining the first order perturbations in previous papers (Andrle, 1969, 1970). Those results are generalized by using Lyapunov's qualitative method.

151.018 A galactic formation model based on post-big bang fragmentation. W. K. Brown.
Astrophys. Space Sci., Vol. 15, 293 - 306 (1972).

A new model of galactic formation is presented. A primeval distribution of angular momentum is derived, which rests on the postulated presence of mass flows at the early stage of the post-big bang universal expansion, when fragmentation occurs. The shape of any particular fragment and its orientation with respect to the mass flow predestine the morphology of the galaxy that will be produced in the subsequent collapse. A fragment-to-disk mapping transform allows examination of the galactic disk mass distributions that result from the postulates

of the model.

151.019 On the statistics of stellar dynamical systems. R. H. Miller.
Astrophys. Journ., Vol. 172, 685 - 688 (1972).

The phase space is divided into cells with at most one occupant per cell by the pair correlation recently reported. The exclusion leads to a statistics similar to the Fermi-Dirac statistics. When values appropriate to the Galaxy are inserted into the empirical forms suggested by the numerical experiments, no degeneracy is indicated.

151.020 On the damping of pre-galactic vortical motions. G. V. Chibisov.
Astron. Zhurn. Akad. Nauk SSSR, Vol. 49, 286 - 293 (1972). In Russian. English translation in Soviet Astron. AJ, Vol. 16, No. 2.

The damping of vortical motions which are the source of pre-galactic turbulence is calculated by an exact solution of the kinetic equation taking into account the effects of radiative viscosity and friction in a "hot" expanding universe for arbitrary optical depth.

151.021 On the direction of trailing in spiral galaxies. V. L. Polyachenko, A. M. Fridman.
Astron. Zhurn. Akad. Nauk SSSR, Vol. 49, 367 - 370 (1972). In Russian. English translation in Soviet Astron. AJ, Vol. 16, No. 2.

The paper deals with the problem of the direction of trailing in spiral galaxies. Results of theoretical papers, obtained hitherto, contradict the data of observations. As particular examples, a solid-body rotating disc and one with large central mass are cosidered.

151.022 The solution of the problem of the third integral of motion. I. T. A. Agekjan.
Astron. Zhurn. Akad. Nauk SSSR, Vol. 49, 371 - 377 (1972). In Russian. English translation in Soviet Astron. AJ, Vol. 16, No. 2.

The exact analytical form of the third integral of motion is found for the general case of a potential field with axial symmetry.

151.023 Dynamics of a rotating gaseous ellipsoid in an expanding universe. H. Nariai, M. Fujimoto.
Progr. Theor. Phys., *Japan*, Vol. 47, 105 - 117 (1972).

In order to assess the importance of the coupling of vorticity with the cosmic expansion in the process of galaxy formation, dynamical equations for a rotating gaseous ellipsoid with uniform density embedded in an expanding universe are derived on the basis of our Newtonian hydrodynamical equations rewritten so as to be useful in a rotating frame of reference relative to the cosmological one. Numerical integrations of the equations are carried out on gaseous ellipsoids of the initial density $\rho = 1.2\rho_B$, $1.1\rho_B$ and $1.05\rho_B$, where ρ_B is the background density of the expanding universe.

151.024 Newtonian hydrodynamics in an expanding universe in terms of the scalar-tensor theory. H. Nariai.
Progr. Theor. Phys., *Japan*, Vol. 47, 118 - 133 (1972).

In view of a possible relevance of the scalar-tensor theory of gravity to various cosmological and astrophysical problems, Newtonian hydrodynamics suitable at the matter dominant stage of the Brans-Dicke universe is formulated and a formalism for dealing with various non-linear effects of cosmic fluid being important during the formation of galaxies is developed in comparison with the previous one based on the general relativistic cosmology.

151.025 Stability of a uniform, non-rotating spheroid collapsing under its own gravitation. S. A. E. G. Falle.

Monthly Notices Roy. Astron. Soc., Vol. 156, 265 - 273 (1972).

The stability of a uniform, non-rotating, oblate gaseous spheroid collapsing under its own gravitation is investigated using a linearized stability analysis. The collapse is unstable in the sense that the density of a subcondensation grows relative to the mean density.

151.026   Collisional processes in stellar systems.
I. H. Gilbert.
Gravitational N-body problem. IAU Colloquium No. 10, (see 012.004), p. 5 - 12 (1972) = 06.151.050.

151.027   Polarization clouds and dynamical friction.
A. J. Kalnajs.
Gravitational N-body problem. IAU Colloquium No. 10, (see 012.004), p. 13 - 17 (1972) = 06.151.031.

151.028   A certain discontinuous Markov process in stellar dynamics.   W. Tscharnuter.
Gravitational N-body problem. IAU Colloquium No. 10, (see 012.004), p. 18 - 21 (1972) = 06.151.051.

151.029   Relaxation times in strictly disk systems.
G. B. Rybicki.
Gravitational N-body problem. IAU Colloquium No. 10, (see 012.004), p. 22 - 26 (1972). = 06.151.052.

151.030   Numerical experiments on the N-body problem.
S. J. Aarseth.
Gravitational N-body problem. IAU Colloquium No. 10, (see 012.004), p. 29 - 43 (1972) = 06.042.045.

151.031   Monte Carlo models of star clusters.   M. Hénon.
Gravitational N-body problem. IAU Colloquium No. 10, (see 012.004), p. 44 - 59 (1972) = 06.151.032.

151.032   A fluid-dynamical method for computing the evolution of star clusters.   R. B. Larson.
Gravitational N-body problem. IAU Colloquium No. 10, (see 012.004), p. 60 - 61 (1972).

151.033   On the lifetimes of galactic clusters.   R. Wielen.
Gravitational N-body problem. IAU Colloquium No. 10, (see 012.004), p. 62 - 70 (1972) = 06.151.033.

151.034   Disruption of star clusters through passing interstellar clouds investigated by numerical experiments.
P. Bouvier, G. Janin.
Gravitational N-body problem. IAU Colloquium No. 10, (see 012.004), p. 71 - 72 (1972).

151.035   Numerical experiments on the escape from non-isolated clusters and the formation of multiple stars.
A. Hayli.
Gravitational N-body problem. IAU Colloquium No. 10, (see 012.004), p. 73 - 87 (1972) = 06.151.034.

151.036   Binary evolution in stellar systems.
S. J. Aarseth.
Gravitational N-body problem. IAU Colloquium No. 10, (see 012.004), p. 88 - 98 (1972) = 06.151.035.

151.037   On the reproducibility of run-away stars formed in collapsing clusters.   C. Allen, A. Poveda.
Gravitational N-body problem. IAU Colloquium No. 10, (see 012.004), p. 114 - 123 (1972) = 06.151.036.

151.038   Numerical experiments on pair correlations and on 'thermodynamics'.   R. H. Miller.
Gravitational N-body problem. IAU Colloquium No. 10, (see 012.004), p. 124 - 130 (1972).

151.039   A numerical experiment on relaxation times in stellar dynamics.   M. Lecar, C. Cruz-González.
Gravitational N-body problem. IAU Colloquium No. 10, (see 012.004), p. 131 - 135 (1972) = 06.151.037.

151.040   Recent developments of integrating the gravitational problem of N-bodies.
V. Szebehely, D. G. Bettis.
Gravitational N-body problem. IAU Colloquium No. 10, (see 012.004), p. 136 - 147 (1972) = 06.151.038.

151.041   The use of integrals in numerical integrations of the N-body problem.   P. E. Nacozy.
Gravitational N-body problem. IAU Colloquium No. 10, (see 012.004), p. 153 - 164 (1972) = 06.042.047.

151.042   Collisionless stellar dynamics.   G. Contopoulos.
Gravitational N-body problem. IAU Colloquium No. 10, (see 012.004), p. 169 - 178 (1972) = 06.151.039.

151.043   On the origin and permanence of galactic spirals.
C. C. Lin.
Gravitational N-body problem. IAU Colloquium No. 10, (see 012.004), p. 179 (1972). − Abstract.

151.044   The hose-pipe instability in stellar systems.
R. M. Kulsrud, J. W. K. Mark, A. Caruso.
Gravitational N-body problem. IAU Colloquium No. 10, (see 012.004), p. 180 - 183 (1972) = 06.151.053.

151.045   On the stability of an encounterless self-gravitating constant density system.   S. Goldstein.
Gravitational N-body problem. IAU Colloquium No. 10, (see 012.004), p. 184 - 193 (1972) = 06.151.040.

151.046   Exact statistical mechanics of a one-dimensional self-gravitating system.   G. B. Rybicki.
Gravitational N-body problem. IAU Colloquium No. 10, (see 012.004), p. 194 - 210 (1972) = 06.151.054.

151.047   Numerical experiments in collisionless systems.
R. H. Miller.
Gravitational N-body problem. IAU Colloquium No. 10, (see 012.004), p. 213 - 230 (1972) = 06.151.055.

151.048   Dynamics of plane stellar systems.   F. Hohl.
Gravitational N-body problem. IAU Colloquium No. 10, (see 012.004), p. 231 - 249 (1972) = 06.151.056.

151.049   Numerical experiments in spiral structure.
W. J. Quirk.
Gravitational N-body problem. IAU Colloquium No. 10, (see 012.004), p. 250 - 253 (1972).

151.050   On the number of isolating integrals in systems with three degrees of freedom.   C. Froeschle.
Gravitational N-body problem. IAU Colloquium No. 10, (see 012.004), p. 254 - 261 (1972) = 06.151.057.

151.051   Numerical experiments on Lynden-Bell's statistics.
M. Lecar, L. Cohen.
Gravitational N-body problem. IAU Colloquium No. 10, (see 012.004), p. 262 - 275 (1972) = 06.151.041.

151.052   A phase-space boundary integration of the Vlasov equation for collisionless one-dimensional stellar systems.   S. Cuperman, A. Harten, M. Lecar.
Gravitational N-body problem. IAU Colloquium No. 10, (see 012.004), p. 276 - 289 (1972) = 06.151.042.

151.053 **The collective relaxation of two-phase-space-density collisionless one-dimensional selfgravitating systems.**
S. Cuperman, A. Harten, M. Lecar.
Gravitational N-body problem. IAU Colloquium No. 10, (see 012.004), p. 290 - 310 (1972) = 06.151.043.

151.054 **Numerical experiments with a one-dimensional gravitational system by an Euler-type method.**
G. Janin.
Gravitational N-body problem. IAU Colloquium No. 10, (see 012.004), p. 311 (1972). — Abstract.

151.055 *N-body problem and gas dynamics in one dimension.*
F. Nahon.
Gravitational N-body problem. IAU Colloquium No. 10, (see 012.004), p. 312 (1972). — Abstract.

151.056 **Stability properties for encounterless self-gravitational stellar gas and plasma.**
M. R. Feix, J. P. Doremus, G. Baumann.
Gravitational N-body problem. IAU Colloquium No. 10, (see 012.004), p. 347 - 364 (1972) = 06.151.044.

151.057 **Direct integration methods of the N-body problem.**
S. J. Aarseth.
Gravitational N-body problem. IAU Colloquium No. 10, (see 012.004), p. 373 - 387 (1972) = 06.042.048.

151.058 **Treatment of close approaches in the numerical integration of the gravitational problem of N bodies.**
D. G. Bettis, V. Szebehely.
Gravitational N-body problem. IAU Colloquium No. 10, (see 012.004), p. 388 - 405 (1972) = 06.042.049.

151.059 **The Monte Carlo method.**   M. Hénon.
Gravitational N-body problem. IAU Colloquium No. 10, (see 012.004), p. 406 - 422 (1972) = 06.151.058.

151.060 **The fluid-dynamical method.**   R. B. Larson.
Gravitational N-body problem. IAU Colloquium No. 10, (see 012.004), p. 423 - 427 (1972).

151.061 **The model of spherical concentric shells.**
G. Janin.
Gravitational N-body problem. IAU Colloquium No. 10, (see 012.004), p. 428 - 430 (1972).

151.062 **Integration methods where force is obtained from the smoothed gravitational field.**   F. Hohl.
Gravitational N-body problem. IAU Colloquium No. 10, (see 012.004), p. 431 - 441 (1972) = 06.151.059.

151.063 **Blast waves in rotating media.**   L. F. Rossner.
Nature, Phys. Sci., Vol. 235, 68 - 69 (1972).
The relevance of a blast wave model to the flow of gas from the center of the Galaxy is re-examined.

151.064 **The velocity variation of a star as a purely discontinuous random process. II. Massive stars in clusters.**   V. S. Kaliberda, I. V. Petrovskaya.
Astrofizika, Vol. 7, 663 - 670 (1971). In Russian.
English translation in Astrophysics, Vol. 7, No. 4.
The method of solving the second Kolmogorov-Feller equation developed earlier is applied to the investigation of the evolution of the velocity distribution function for the group of stars with masses $m_1 = 2\bar{m}$, where $\bar{m}$ is the average mass of a cluster star. The escape rate of stars considered and the amount of energy taken away by the dissipated stars in different moments of time are also found.

151.065 **On the generating mechanism of spiral structure.**

D. Lynden-Bell, A. J. Kalnajs.
Monthly Notices Roy. Astron. Soc., Vol. 157, 1 - 30 (1972).
We show that the presence of waves lowers the angular momentum in the inner parts and increases it in the outer parts. The role of spiral structure is to carry angular momentum from the inner parts to the outer parts so that the waves may grow. We consider bars in galaxies to be standing waves which have grown enough to orient and trap the major axes of orbits with two lobes so that they lie along the bar.

151.066 **The water-bag model in a cylindrical two-dimensional stellar system.**
J. P. Doremus, G. Baumann, M. R. Feix.
Astrophys. Letters, Vol. 11, 37 - 40 (1972).
For a rotating two-dimensional rod stellar system with cylindrical symmetry, the 'water-bag' model is equivalent, in the steady state, to a polytrope of index $n = 1$ ($\gamma = 2$). We show that for this water-bag model the minimum of the total energy for a given mass and a given total angular momentum corresponds to the steady state, with uniform rotation, of the collisionless Boltzmann equation.

151.067 **The damping of the galactic density waves by their induced shocks.**   A. J. Kalnajs.
Astrophys. Letters, Vol. 11, 41 - 43 (1972).
The shock induced in the gaseous component of a galaxy by a tightly wrapped density wave always damps the wave. For shocks of the strength proposed by Roberts the damping times can be shorter than one galactic year.

151.068 **Random gravitational encounters and the evolution of spherical systems. III. Halo.**
L. Spitzer, Jr., S. L. Shapiro.
Astrophys. Journ., Vol. 173, 529 - 547 (1972).
Accumulation of stars in the halo of a spherical stellar system, together with escape of stars from the system, result from stellar encounters within a dense central core; the resultant halo structure is studied both analytically and numerically for an isolated system. On the assumption that conditions in the central core remain constant with time, a simple steady-state solution is obtained for the outermost halo, or "fringe," defined as the region from which stars may escape after a few additional orbits through the central core of the system.

151.069 **The ejection of massive objects from galactic nuclei: Interactions between the massive object and the galactic gas.**   W. C. Saslaw, D. S. De Young.
Astrophys. Letters, Vol. 11, 87 - 93 (1972).
A black-body or a spinar type model for a massive object can produce a long luminous trail when the wake is heated by cosmic rays or when the ionization front acts as a catalyst for star formation. The results are applied to Markarian 205.

151.070 **Dichtewellen — eine Erklärung der Spiralstruktur?**
D. Wiedemann.
Orion Schaffhausen, 30. Jahrgang, p. 6 - 10 (1972).

151.071 **Motions of the stars and gas in spiral arms.**
R. M. Humphreys.
Bull. American Astron. Soc., Vol. 4, 263 (1972). — Abstr. AAS.

151.072 **Patterns of waves in galactic disks.**   C. Hunter.
Bull. American Astron. Soc., Vol. 4, 263 (1972). Abstr. AAS.

151.073 **A new treatment of stellar encounters in the computation of N-body systems.**   G. Janin, D. G. Bettis.
Bull. American Astron. Soc., Vol. 4, 264 (1972). — Abstr. AAS.

151.074 **On the permanence of galactic spirals.**
C. C. Lin, S. I. Feldman.

Bull. American Astron. Soc., Vol. 4, 264 (1972). – Abstr. AAS.

151.075 Construction of a static self-consistent model of a disk galaxy in a computer. R. H. Miller.
Bull. American Astron. Soc., Vol. 4, 264 (1972). – Abstr. AAS.

151.076 Evidence for density wave streaming in three spiral galaxies. D. H. Rogstad, G. S. Shostak.
Bull. American Astron. Soc., Vol. 4, 265 (1972). – Abstr. AAS.

151.077 The effect of masses on the evolution of triple gravitational systems. V. Szebehely.
Bull. American Astron. Soc., Vol. 4, 267 (1972). – Abstr. AAS.

151.078 The collapse of a proto-galaxy.
A. E. Wright, K. A. Innanen.
Bull. American Astron. Soc., Vol. 4, 267 - 268 (1972).
Abstr. AAS.

151.079 The stability of small dynamical systems.
J. K. Estes.
Bull. American Astron. Soc., Vol. 4, 270 (1972). – Abstr. AAS.

151.080 The dynamics of spiral structure. Lecture notes.
G. Contopoulos.
Astronomy Program and Center for Theoretical Physics, Univ. Maryland. Preprint Library, Dep. Phys. Astron., Univ. Maryland, College Park, Maryland. 3 + 118 pp. (1972).

151.081 The gravitational field of flat galaxies.
M. Clutton-Brock.
Astrophys. Space Sci., Vol. 16, 101 - 119 (1972).
A set of bi-orthogonal pairs of functions is derived which is especially suited for the calculation of the gravitational field of flat galaxies. This field calculation can be used for a relatively cheap and simple computer simulation of galaxies.

151.082 Vers une explication de la structure spirale.
G. Courtès.
L'Astronomie, 86ᵉ année, p. 318 - 321 (1972).

151.083 Computational models of gravitationally interacting galaxies. A. E. Wright.
Monthly Notices Roy. Astron. Soc., Vol. 157, 309 - 333 (1972).
Numerical computations of the dynamics of interacting binary galactic systems are presented. It is concluded that the tails and bridges frequently observed can be explained in the majority of cases as a result solely of the gravitational interactions within the system.

151.084 Modelle zur Kinematik von Sternhaufen.
R. Dvorak, E. Göbel.
Mitt. Astron. Ges., No. 31, p. 123 - 124 (1972).

151.085 Auflösungszeiten von Sternhaufen.
W. Tscharnuter.
Mitt. Astron. Ges., No. 31, p. 143 (1972). – Abstract.

151.086 A numerical experiment in the problem of accretion.
V. I. Aleshin, S. A. Kaplan.
Astron. Zhurn. Akad. Nauk SSSR, Vol. 49, 489 - 493 (1972).
In Russian. English translation in Soviet Astron. AJ, Vol. 16, No. 3.
By the method of numerical experiment it is possible to calculate the gas accretion in the proto-cluster of protostars with taking into account the dependence of mass growth on the changing velocity of protostars.

151.087 Changes in the gravitational energy of galaxies during collisions. K. S. Sastry.
Astrophys. Space Sci., Vol. 16, 284 - 294 (1972).
The problem of the change in gravitational energy of a colliding galaxy due to tidal effects is considered. The change in the internal energy, the mass of escaping matter and the change in the mean radius of the test galaxy have been estimated for a relative velocity of 1000 km s⁻¹ for three distances of closest approach for the following four cases: (a) both galaxies centrally concentrated, (b) both galaxies homogeneous, (c) test galaxy centrally concentrated, field galaxy homogeneous, and (d) test galaxy homogeneous, field galaxy centrally concentrated. Some implications of the results are discussed.

151.088 Random gravitational encounters and the evolution of spherical systems. IV. Isolated systems of identical stars. L. Spitzer, Jr., T. X. Thuan.
Astrophys. Journ., Vol. 175, 31 - 61 (1972).
Models of isolated stellar systems with stars all of the same mass have been recomputed, allowing for the exact dependence of the velocity diffusion coefficients on stellar velocity, and extending to later evolutionary phases than before. These models permit a comparison of escape rates with a "local" theory, where the velocity distribution function at any point is arbitrarily assumed to be the same as that obtained from the Fokker-Planck equation for a homogeneous system, with a rectangular profile for the potential $\phi(r)$, and where ξ_e, the overall probability of escape per star per interval of time, is computed by averaging over the entire system the local escape probability at each point.

151.089 The equilibria and oscillations of a family of uniformly rotating stellar disks. A. J. Kalnajs.
Astrophys. Journ., Vol. 175, 63 - 76 (1972).
We investigate the stability of a family of self-consistent uniformly rotating collisionless stellar disks by solving the normal mode problem. Every member of the family is unstable. From this family we generate new models that can be either stable or unstable. We give two examples of each.

151.090 Irregular spiral structure—causes and effects.
J. Rickard.
Galactic astronomy, Vol. 1, (see 012.019), 331 - 334 (1970).

151.091 Theory of spiral structure. C. C. Lin.
Galactic astronomy, Vol. 2, (see 012.019), 1 - 93 (1970).

151.092 Shock formation and star formation in galactic spirals. W. W. Roberts.
Galactic astronomy, Vol. 2, (see 012.019), 201 - 214 (1970).

151.093 A blast wave model. A. B. Solinger.
Galactic astronomy, Vol. 2, (see 012.019), 215 - 232 (1970).

151.094 On the comparison between the density-wave theory and observations. C. Yuan.
Galactic astronomy, Vol. 2, (see 012.019), 233 - 250 (1970).

151.095 An exact solution for a collisionless flat galactic model. S. Aoki.
Celestial mechanics. Symposium Tokyo, (see 012.020), p. 7 - 14 (1972). In Japanese.

151.096 Models of disk-like stellar systems. M. Miyamoto.
Celestial mechanics. Symposium Tokyo, (see 012.020), p. 15 - 20 (1972). In Japanese.

151.097 Cascade dynamics. I. Collective phenomena in an abstract dynamical system. Y. Aizawa.
Celestial mechanics. Symposium Tokyo, (see 012.020), p. 50 - 56 (1972). In Japanese.

151.098   **Variations of stellar orbits in a collapsing galaxy.**
N. Owaki.
Celestial mechanics. Symposium Tokyo, (see 012.020), p.
102 - 107 (1972). In Japanese.

151.099   **Modified exponential models of stellar systems.**
J. Einasto, L. Einasto.
Tartu Astron. Obs. Teated, No. 36, p. 3 - 45 (1972).
  Standard descriptive functions have been calculated for
models of stellar systems with a modified exponential density
function. The following functions are tabulated: the gravitational potential and its radial and vertical gradients , the surface
density, the velocity dispersion perpendicular to the fundamental plane and the logarithmic gradient of the circular velocity function. Graphs are given for the determination of parameters of the model from the observed surface and the integrated brightness.

151.100   **Experimental studies of the numerical stability of
the gravitational n-body problem.**
R. H. Miller.
Journ. Comput. Phys., Vol. 8, 449 - 464 (1971).
  The gravitational n-body problem is remarkably unstable
numerically, in spite of the fact that near constancy of the
energy and angular momentum makes it appear that the calculation should be reliable. In an attempt to understand why
this should be so, the properties of numerical solutions were
explored through a perturbation technique in which equations
of motion for the differences between two systems are integrated numerically. Implications of the results concerning the
utility of computational n-body problem are discussed.

151.101   **Partial iterative refinements.**   R. H. Miller.
Journ. Comput. Phys., Vol. 8, 464 - 471 (1971).
  A few well-defined parameters may be iteratively refined
by a reduced Newton–Raphson process on the generalized inverse. The method has been used in a gravitational n-body integration to control the usual ten first integrals of motion. The
discussion is based on that example to show how the details
may be carried through.

151.102   **Evolution of a stationary disk of stars.**
F. Hohl.
Journ. Comput. Phys., Vol. 9, 10 - 25 (1972).
  An improved potential solver for calculating the gravitational potential of isolated disk galaxies is presented. The potential solver is used to investigate the evolution of initially
stationary axisymmetric disks of stars for various values of the
initial velocity dispersion.

151.103   **R_0 from exponential disk models.**   A. Toomre.
Quarterly Journ. Roy. Astron. Soc., Vol. 13, 241 -
242 (1972). – Presented at the Woolley symposium (see
012.025).

151.104   **The problem of mass deficiency of galaxy agglom-
erations.**   P. Droz-Vincent.
Fiz. Szemle, Vol. 22, No. 1, p. 21 - 24 (1972). In Hungarian.
See Phys. Abstr., Vol. 75, No. 41059 (1972).

151.105   **Dynamics of self-gravitating systems. Structure of
galaxies.**   C. C. Lin.
Stud. in Math., Vol. 7, 116 - 149 (1971). – See Bull. Signal.,
Vol. 33, Section 120, No. 7094 (1972).

  **Ergodicity and scattering of orbits in the negative
curvature zone of classical dynamical systems.**
See Abstr. 042.057.

  **Non-linear theory of gravitational instability in the
expanding universe. III.**   See Abstr. 061.014.

  **A Lagrangian derivation of the action-conservation
theorem for density waves.**   See Abstr. 062.033.

  **Applications de la théorie cinétique des gaz à la
physique des plasmas et à la dynamique des galaxies.**
See Abstr. 062.047.

  **Galactic evolution. I. Differential effects of radia-
tion pressure around stars of different spectral types on neutral atoms and dust grains.**   See Abstr. 065.069.

  **The problem of n bodies with variable gravitational
constant and some dynamical features of large-scale cosmic
systems.**   See Abstr. 066.012.

  Velocity effects in triple stellar systems.
See Abstr. 117.009.

  **The quasars and nuclei of galaxies: a single body or
a stellar cluster.**   See Abstr. 141.070.

  **Application of the density-wave theory of spiral
structure: Shock formation along the Perseus arm.**
See Abstr. 155.034.

  Sind galaktische Dichtewellen beobachtbar?
See Abstr. 155.064.

  Collapse in galactic nuclei.   See Abstr. 158.116.

  Tidal interactions of galaxies.
See Abstr. 158.151.

  **An observational approach to the problem of spiral
structure. II.**   See Abstr. 159.008.

152 Stellar Associations

152.001 **Distances of bright stars in stellar ring 274.**
L. N. Kolesnik.
Astron. Astrophys., Vol. 16, 155 - 157 (1972). In German.
BV-photometry and spectral types of 110 stars in the field around stellar ring 274 have been used to determine the distances of bright stars in the ring. Distances have been calculated for 7 stars that might be members of the stellar ring. It is likely, that 4 high-luminosity stars are at the same distance. However stars of spectral types A0–F2 are evidently background stars.

152.002 **Spectroscopic observations of stars in two star rings.**
G. W. Preston.
Publ. Astron. Soc. Pacific, Vol. 84, 25 - 27 (1972).
Spectral types of 27 stars in or near star rings 373 and 519 of Isserstedt (1968) indicate that these rings are chance configurations of field stars.

152.003 **Infrared and optical observations of a young stellar group surrounding BD+40°4124.**
K. M. Strom, S. E. Strom, M. Breger, A. L. Brooke, J. Yost, G. Grasdalen, L. Carrasco.
Astrophys. Journ., (*Letters*), Vol. 173, L65 - L70 (1972).
Observations of a group of stars surrounding the Herbig Be-type star BD+40°4124 suggest that this object is the brightest member of a relatively small stellar group. The age of this group is extremely small, probably less than 5×10^5 years.

152.004 **Investigation of the T-associations in the Cygnus region.** K. D. Avulov.
Tsirk. Astron. Inst., *Tashkent*, No. 29 (376), p. 12 - 20 (1971). In Russian.

152.005 **Luminosity functions for 16 associations in the Large Magellanic Cloud.** P. B. Lucke.
Bull. American Astron. Soc., Vol. 4, 223 (1972). – Abstr. AAS.

152.006 **A near-infrared survey of Cygnus OB2.**
P. J. Jarecke.
Bull. American Astron. Soc., Vol. 4, 233 (1972). – Abstr. AAS.

152.007 **H II regions in the Sco OB 1 association.**
A. Laval.
Astron. Astrophys., Vol. 19, 82 - 91 (1972). In French.
Two bright nebulae are related to the Sco OB 1 association. RCW 113 is situated in the southern part, IC 4628 in the northern part. The purpose of this study is to know whether the two observed H II regions originated at the same time from the same initial gas cloud. Optical and radio-astronomical data have been used. Radio sources of high electron density have been found in IC 4628.

152.008 **Photometrische und spektroskopische Untersuchung eines Sternrings in Monoceros.**
J. Isserstedt, T. Schmidt-Kaler.
Mitt. Astron. Ges., No. 31, p. 132 (1972). – Abstract.

152.009 **Qualitatives Sternringmodell.**
W. Haupt, W. Schlosser.
Mitt. Astron. Ges., No. 31, p. 133 (1972). – Abstract.

152.010 **Sternringe und Sternhöhlen als Zufallsphänomene.**
J. Meurers.
Mitt. Astron. Ges., No. 31, p. 133 - 138 (1972).

152.011 **Erste Ergebnisse einer EB-Untersuchung im Hydra Ring.** F. V. Prochazka.
Mitt. Astron. Ges., No. 31, p. 138 - 142 (1972).

152.012 **A review of problems concerning associations and star clusters.** G. Larsson-Leander.
Centre de Données Stellaires, Inform. Bull. No. 3, (see 002.020), p. 17 - 18 (1972).

152.013 **I Scorpii and NGC 6231.** L. A. Milone.
Bol. Inst. Mat., Astron., Fis., Univ. Nacional Córdoba, Vol. 3, No. 2, p. 25 - 34 (1971).
The aim of this paper is to discuss the distance of the association and the properties of interstellar absorption in the region around, and so to obtain the absolute magnitude of several Of and WR stars, and other very bright ones belonging to it. Incidentally, approximate values for its age are derived.

152.014 **Cinématique et évolution dans l'association Sco OB 1.**
A. Laval.
Thesis, Sci. Phys., Univ. Provence. Centre Documentation, C.N.R.S. (1971–10–26), 95 pp. (1971). – See Bull. Signal., Vol. 33, Section 120, No. 5069 (1972).

The positions and proper motions of 127 stars in the region of WX Centauri. See Abstr. 112.011.

Seven new spectroscopic binaries in Cepheus.
See Abstr. 119.004.

Über das Verhalten der Flare-Sterne in den Plejaden und in einigen anderen Assoziationen.
See Abstr. 122.145.

OB star distribution in Puppis.
See Abstr. 155.024.

153 Galactic Clusters

153.001 Membership of NGC 6940 and IC-1805.
W. L. Sanders.
Astron. Astrophys., Vol. 16, 58 - 59 (1972).
Membership probabilities for stars in the field of NGC 6940 and revised probabilities for those in the field of IC-1805 are presented.

153.002 The open cluster NGC 6811. U. Lindoff.
Astron. Astrophys., Vol. 16, 315 - 321 (1972).
Magnitudes and colours in the UBV-system have been determined for 377 stars in the open cluster NGC 6811 based on a combination of photoelectric and photographic photometry. Spectral classes for 40 relatively brighter stars have been determined from objective prism-plates.

153.003 Circumstellar shells in the young cluster NGC 2264. II. Infrared and further optical observations.
S. E. Strom, K. M. Strom, A. L. Brooke, J. Bregman, J. Yost.
Astrophys. Journ., Vol. 171, 267 278 (1972).
Flux measurements at 1.6, 2.2, and 3.4 μ for 42 stars in the young cluster NGC 2264 suggest that a significant number of these stars are surrounded by circumstellar material which produces observable infrared excesses. Hydrogen line profiles are used to estimate surface gravities and thereby to predict visual luminosities for a number of pre–main-sequence A stars. In addition to the observed infrared excesses, infrared deficiencies are observed for a number of stars.

153.004 The Pleiades lower main sequence. B. F. Jones.
Astrophys. Journ., (Letters), Vol. 171, L57 - L60 = Contr. Lick Obs., No. 357 (1972).
It is shown that the faint members of the Pleiades fall almost entirely below the zero-age main sequence. The most likely explanation is that these stars are surrounded by circumstellar shells having predominantly gray extinction properties.

153.005 Pekuliare A-Sterne in offenen Sternhaufen. I. Photoelektrische UBV-Photometrie und MK-Klassifikation von Sternen im Gebiet von NGC 7039.
E. Želwanowa, W. Schöneich.
Astron. Nachr., Vol. 293, 155 - 161 (1971/72) = Mitt. Astrophys. Obs. Potsdam, No. 137 (1972).
Photoelectric UBV magnitudes were obtained for 34 stars in the region of the galactic cluster NGC 7039. MK spectral types for 175 BD stars in a field of $3°.5 \times 3°.5$ around the cluster were also obtained. In this field and in a neighbouring field 8 new peculiar stars were discovered. The cluster does not contain Ap stars.

153.006 Pekuliare A-Sterne in offenen Sternhaufen. II. Resultate der Suche nach pekuliaren A-Sternen im Gebiet des Haufens Tr 2 (Cr 29). E. Želwanowa.
Astron. Nachr., Vol. 293, 163 - 166 (1971/72) = Mitt. Astrophys. Obs. Potsdam, No. 138 (1972).
In a region of $4°.3 \times 4°.3$ around the open cluster Tr 2 a search for peculiar A stars was carried out on objective prism plates. 6 objects were discovered.

153.007 Pre-main-sequence stars. I. Light variability, shells, and pulsation in NGC 2264. M. Breger.
Astrophys. Journ., Vol. 171, 539 - 548 (1972).
A light-variability study has been undertaken in the young cluster NGC 2264. Among the nonpulsating pre-main-sequence A and F stars, only about 25 percent have shown short-period light variability or light variations since 1953. There exists a good correlation between light variability and (other) shell indicators. The variability data confirm previous

suggestions that the fainter A/F stars are surrounded by thick shells. Two δ Scuti variables with periods of 3 hours were detected. Arguments for cluster membership are given. If cluster membership is confirmed, then they would be the first known cepheid-like pre-main-sequence pulsators.

153.008 Lithium abundances of stars in open clusters.
R. R. Zappala.
Astrophys. Journ., Vol. 172, 57 - 74 = Contr. Lick Obs., No. 342 (1972).
Image-intensifier instrumentation was used at the coudé focus of the Lick 120-inch reflector to obtain spectrograms of F to K main-sequence stars in the Hyades, Pleiades, and Praesepe clusters, as well as stars still contracting toward the main sequence in NGC 2264. The abundance of lithium was calculated from the measured equivalent widths of Li I $\lambda6707$, a grid of model atmospheres, and an empirical correction for saturation effects.

153.009 Proper motions, membership, and stellar content of the old cluster NGC 188.
A. R. Upgren, W. S. Mesrobian, S. J. Kerridge.
Astron. Journ., Vol. 77, 74 - 84 (1972).
Proper motions and probabilities of membership have been determined for 228 stars in the field of NGC 188 from plates covering a 40-yr interval taken with the 30-inch Thaw Refractor of the Allegheny Observatory. The survey is complete to a limiting blue magnitude of 15.6. An eight-parameter solution was made in which the motions are represented by a bivariate frequency function with circular and elliptical models for the motions of the cluster members and the field stars, respectively. A 67-star reference frame provides a stable solution in which a clear distinction between members and field stars is made.

153.010 Membership of the open cluster IC 4665.
W. L. Sanders, W. F. van Altena.
Astron. Astrophys., Vol. 17, 193 - 200 (1972).
Probabilities of membership, based on relative proper motions, for 275 stars in the field of IC 4665 are given. The cluster proper motion dispersion (m.e.) of $0''.0009$ yields 19 probable members in the limited field studied. The main sequences is believed to terminate at $M_V \cong + 3.5$, or F6V.

153.011 The color excesses and metallicities of the open clusters NGC 2360 and NGC 3680.
R. D. McClure.
Astrophys. Journ., Vol. 172, 615 - 622 (1972).
Intermediate-bandpass photometry on the David Dunlap Observatory (DDO) system is reported for the giant stars in the intermediate to old open clusters NGC 2360 and NGC 3680. Color excesses and cyanogen-band strengths corrected for surface gravity were determined. The metallicity inferred for both clusters is similar to solar abundance.

153.012 OH observations of the open cluster NGC 2264.
K. W. Riegel, R. M. Crutcher.
Astrophys. Journ., (Letters), Vol. 172, L107 - L110 (1972).
The young open cluster NGC 2264 has been observed as a source of 1667- and 1665-MHz OH line emission. Further observations of the cluster and several nearby points in the sky, in the main lines and in the 1612- and 1720-MHz satellite lines, show that the OH emission is from interstellar dust clouds.

153.013 NGC 2516 and the Pleiades group. O. J. Eggen.
Astrophys. Journ., Vol. 173, 63 - 86 (1972).
UBV observations to visual magnitude 16 are given for

stars in NGC 2516 together with new UBV observations of members of IC 2602 and 2391 and intermediate-band results for the members of the ζ Sculptoris cluster. The clusters of the Pleiades group are found to represent a considerable age spread $(1-8 \times 10^7$ years), and this is confirmed by both the luminous blue and red group members unattached to clusters. The distribution of group members in the $(M_{bol}, \log T_e)$-plane is understandable in terms of evolutionary tracks for stellar models of masses $5-15 M_\odot$.

153.014 Mean absolute magnitudes and colour indices of the concentrations of red giants on colour−magnitude diagrams of open clusters. A. E. Vasilevsky.

Astron. Zhurn. Akad. Nauk SSSR, Vol. 49, 378 - 388 (1972). In Russian. English translation in Soviet Astron. AJ, Vol. 16, No. 2.

Relations age−mean absolute magnitude and age−mean colour index for red giants of open clusters forming concentrations on the colour−magnitude diagram, were determined on the basis of published photometric data. For this purpose the zero-age main sequence, distance moduli and ages for 32 clusters were found.

153.015 The Alpha Persei cluster. N. M. Artiukhina.

Astron. Zhurn. Akad. Nauk SSSR, Vol. 49, 389 - 396 (1972). In Russian. English translation in Soviet Astron. AJ, Vol. 16, No. 2.

On the basis of proper motions, curves of the radial distribution of the apparent and spatial density for possible cluster members brighter than $12^m.0_{pg}$ are constructed and discussed.

153.016 An investigation of field stars near NGC 2168 (M 35). K. M. Cudworth.

Astron. Astrophys., Vol. 18, 318 - 324 = Lick Obs. Bull. No. 621 (1972).

The field stars which were measured in a previous study of the open cluster NGC 2168 were investigated with regard to their secular parallaxes and their luminosity function. The main sequence stars were segregated by means of proper motion dispersions and approximate distances were determined for these objects, including a correction for interstellar extinction. The motion of NGC 2168 relative to "stars of zero parallax" was determined.

153.017 A new photometric investigation of NGC 1662. S. M. Hassan.

Mem. Soc. Astron. Italiana, Nuova Ser., Vol. 43, 117 - 122 (1972).

Three colour photometric observations in the UBV system have been secured with the 74"-reflector of Kottamia Observatory (U.A.E.) for improving the distance determination of NGC 1662. A discussion of the membership and a determination of the age of the cluster, using theoretical evolutionary tracks, are also included.

153.018 The reddening, distance modulus, chemical composition and age of the galactic cluster NGC 752. R. A. Bell.

Monthly Notices Roy. Astron. Soc., Vol. 157, 147 - 156 (1972).

The $uvby\beta$ colours of the stars in the galactic cluster NGC 752 have been analysed using theoretical colours computed from synthetic stellar spectra. The points discussed include (I) the reddening, (II) the distance modulus, (III) the chemical abundances of the cluster stars, (IV) the comparison of the cluster M_v, $\log T_{eff}$ and c_1, $b-y$ diagrams with corresponding diagrams obtained from evolutionary tracks, and (V) the comparison of theoretical and observed $U-B$ and $u-b$ colours.

153.019 Observations of CS, HCN, U89.2, and U90.7 in NGC 2264.

B. Zuckerman, M. Morris, P. Palmer, B. E. Turner.
Astrophys. Journ., (Letters), Vol. 173, L125 - L129 (1972).

Because of the narrow line widths HCN hyperfine structure in this cluster is resolved. Improved rest frequencies for the unidentified lines U89.2 (X-ogen) and U90.7 ("HCN") have been obtained.

153.020 The determination of reddening and ultraviolet excess of clusters from UBV observations of red giants.

F. D. A. Hartwick, R. D. McClure.
Publ. Astron. Soc. Pacific, Vol. 84, 288 - 291 (1972).

It is empirically verified that the use of the commonly assumed color excess ratio of 0.72 to find the reddening of clusters from UBV observations of giant stars gives erroneous results. A procedure is outlined for obtaining correct intrinsic colors of giant stars given an independent estimate of the reddening from blue stars, or for obtaining the reddening if the ultraviolet excess for the giants is assumed.

153.021 Photographic photometry of the cluster NGC 6866. W. Sutantyo, B. Hidajat.

Inform. Bull. Southern Hemisph., No. 19, p. 31 (1971).

Nine Hamburg Schmidt plates taken with the filter-emulsion combinations which match the UBV photometric system have been measured for 410 stars in the neighborhood of NGC 6866. The distance modulus of the cluster has been determined by fitting the zero-age main sequence to the observed C-M diagram.

153.022 Membership of the open cluster NGC 6819. W. L. Sanders.

Astron. Astrophys., Vol. 19, 155 - 158 (1972).

Probabilities of membership, based on relative proper motions, for 189 stars in the field of NGC 6819 are given. The cluster proper motion dispersion (m.e.) of 0''.0012 yields 88 probable members.

153.023 Die mittleren Geschwindigkeiten der Sterne in offenen Sternhaufen. W. Lohmann.

Astron. Nachr., Vol. 293, 259 - 265 = Mitt. Astron. Rechen-Inst., Heidelberg, Ser. B (1972).

According to the definition of the "Umwandlungsmasse" = "transformation mass" for the computation of the masses of star clusters from star counts and after its determination from an appropriate observational material the masses of 39 open star clusters and the mean velocities of their stars are derived. Finally relations to escape velocities are discussed.

153.024 Zur Kinematik der offenen Sternhaufen. H. van Schewick.

Mitt. Astron. Ges., No. 31, p. 119 - 122 (1972).

153.025 Vergleich der Massenanteile von interstellarem Wasserstoff, Staub und Sternen in offenen Sternhaufen. R. Schwartz.

Mitt. Astron. Ges., No. 31, p. 218 - 220 (1972).

153.026 Photometric study of the open cluster NGC 2232. J. J. Clariá.

Astron. Astrophys., Vol. 19, 303 - 309 (1972).

Three color photoelectric photometry is presented for 43 stars in the vicinity of the southern galactic cluster NGC 2232. Photoelectric measurements of Hβ line intensity of 22 early-type stars are also presented. Independent analyses of UBV and Hβ data confirm that NGC 2232, with a minimum membership of 19 stars, is located at a distance of 360 pc. A nuclear age of 20×10^6 years was obtained for the cluster.

153.027 A photometric study of the open cluster NGC 2477.

F. D. A. Hartwick, J. E. Hesser, R. D. McClure.
Astrophys. Journ., Vol. 174, 557 - 572 (1972).

The results of a photometric study of NGC 2477 based on the combined results of photoelectric and photographic *UBV* photometry and intermediate-band photometry on the David Dunlap Observatory (DDO) system are presented.

153.028 The H-R diagram of the open cluster IC 2602.
H. A. Abt, W. W. Morgan.
Astrophys. Journ., (*Letters*), Vol. 174, L131 - L135 (1972).

Revised spectral types have been determined for White-oak's cluster members of types A0 and earlier. The resulting H-R diagram shows a rather narrow main sequence and a previously known silicon star, HD 92664. The most remarkable feature of the cluster is the spectrum of its brightest star, θ Car: lines of N III are abnormally strong, while C III is abnormally weak.

153.029 Three-colour photometry of NGC 7245 and NGC 7226. S. Karaali.
Publ. Istanbul Univ. Obs., No. 92, 5 pp. (1971).

Three-colour photometry of the clusters NGC 7245 and NGC 7226 have been studied in UBV system and their distances determined with the "fitting method" have been compared with those found in RGU system.

153.030 Note on the mass spectrum of Hyades cluster.
T. Yokoo.
Mem. Osaka Kyoiku Univ. (*Japan*), Vol. 19, 1 - 6 (1970).
See Bull. Signal., Vol. 33, Section 120, No. 3949 (1972).

On the problem of the identification of cluster stars. See Abstr. 041.037.

Pulsational instability of stars during deuterium burning. See Abstr. 065.099.

UBV photographic photometry of stars in the region $AR_{1950}: 17^h03^m - 17^h41^m Decl_{1950}: -28.8°$ to $-33.4°$. III. The catalogue and identification maps of open star clusters: NGC 6405, NGC 6383, "NGC 6374", Av 2, NGC 6416 and H alpha emission regions: Gum 67 (Av 3), Gum 68 (Av 2). See Abstr. 113.010.

An observation of variation of colour excesses of hot stars with time. See Abstr. 113.013.

A highly reddened star near NGC 6231. See Abstr. 113.034.

Equivalent width data for several Am stars in clusters and comparison standards. See Abstr. 114.008.

Studies of extremely young clusters. VI. Spectroscopic observations of the ultraviolet-excess stars in the Orion nebula cluster and in NGC 2264. See Abstr. 114.122.

The mass-luminosity relationship of the Hyades and other stars. See Abstr. 115.011.

The redetermination of the zero age main sequence. See Abstr. 115.018.

IC 4665, a cluster of binaries. See Abstr. 119.002.

Spectroscopic binaries in the open cluster NGC 2516. See Abstr. 119.003.

The Hyades spectroscopic binary HD 27149. See Abstr. 119.018.

Slow flare up in the Pleiades. See Abstr. 122.062.

New flare stars in the Pleiades region. See Abstr. 122.080.

Über das Verhalten der Flare-Sterne in den Plejaden und in einigen anderen Assoziationen. See Abstr. 122.145.

New flares in the Pleiades. See Abstr. 122.157.

New variable star in the open cluster NGC 7128. See Abstr. 123.027.

Pre-main-sequence stars. II. Stellar polarization in NGC 2264 and the nature of circumstellar shells. See Abstr. 132.035.

Auflösungszeit von Sternhaufen. See Abstr. 151.005.

A review of problems concerning associations and star clusters. See Abstr. 152.012.

On the z-motion of some young clusters in the Galaxy. See Abstr. 155.004.

OB star distribution in Puppis. See Abstr. 155.024.

Remarks on Sculptor type systems. See Abstr. 158.171.

154   Globular Clusters

154.001   The globular clusters NGC 6362 and NGC 6752.
G. Alcaino.
Astron. Astrophys., Vol. 16, 220 - 229 (1972).

A *BV* photographic investigation on the globular clusters NGC 6362 and NGC 6752 was carried out at the Cerro Tololo Inter-American Observatory (CTIO) with the 1.5 meter reflector and the plate material was reduced with the iris-photometer of the European Southern Observatory. For NGC 6362, 235 stars were calibrated photographically from a photoelectric sequence of 61 stars, for NGC 6752, 339 stars were calibrated photographically from 71 stars previously obtained by the author at CTIO.

154.002   The luminosity function for globular clusters. I. Observations on M5 and M13.
M. Simoda, K. Tanikawa.
Publ. Astron. Soc. Japan, Vol. 24, 1 - 12 (1972).

Luminosity functions for M5 and M13 have been obtained down to the visual magnitude 21 and 19, respectively, using plates by the Kitt Peak 84-inch and Palomar 200-inch reflectors. Our results are in essential agreement with the result for M13 by Simoda and Kimura (1968).

154.003   The luminosity function for globular clusters. II. Evolutionary interpretation for M5, M13, and M92.
M. Simoda.
Publ. Astron. Soc. Japan, Vol. 24, 13 - 25 (1972).

The luminosity functions for M5 and M13 obtained in paper I (Simoda and Tanikawa 1972) and for M92 by Hartwick (1970) are compared with the results of evolution theory.

154.004   The globular clusters and the distance scale.
B. V. Kukarkin, R. M. Rusev.
Astron. Zhurn. Akad. Nauk SSSR, Vol. 49, 121 - 133 (1972). In Russian. English translation in Soviet Astron. AJ, Vol. 16, No. 1.

The mean apparent V and B magnitudes of the five (V 5 and B 5) and twenty-five (V 25 and B 25) brightest stars in 37 globular clusters were derived on the base of colour-magnitude diagrams. The apparent distance moduli for 38 globular clusters were derived. The apparent distance moduli as functions of V 5, B 5, V 25, B 25 and also of the absolute integral magnitude and index of metallicity of the globular clusters were studied.

154.005   On Hartwick's two-dimensional classification of globular clusters.   A. V. Mironov.
Astron. Zhurn. Akad. Nauk SSSR, Vol. 49, 134 - 136 (1972). In Russian. English translation in Soviet Astron. AJ, Vol. 16, No. 1.

154.006   Scanner abundance studies. V. Globular clusters in M31.   H. Spinrad, F. Schweizer.
Astrophys. Journ., Vol. 171, 403 - 410 (1972).

Photoelectric observations of eight luminous globular clusters in M31 confirm and extend van den Bergh's conclusions about the wide range of metallic-line strength in their integrated spectra. It appears very likely that the M31 globular clusters with the strongest lines, H87, B282, and MII, are super-metal-rich. The presence of hot stars is required to explain their strong ultraviolet and blue continua.

154.007   Gravities of blue horizontal-branch stars.
A. G. D. Philip.
Astrophys. Journ., (*Letters*), Vol. 171, L51 - L55 (1972).

In this preliminary study, four-color observations have been obtained of blue horizontal-branch stars in the globular clusters M3, M4, M13, M55, and M92. Surface gravities computed for 35 stars of populations I and II by means of photoelectric scans or high-dispersion spectra have been used to calibrate δc_1 against $\delta \log g$.

154.008   The globular cluster NGC 4833.   J. Menzies.
Monthly Notices Roy. Astron. Soc., Vol. 156, 207 - 221 (1972).

Photoelectrically calibrated photographic photometry of about 800 stars in NGC 4833 is presented. The colour magnitude and two-colour diagrams suggest the cluster belongs to the extremely metal deficient class of globular clusters like M92. The ratio of giants to horizontal branch stars is about unity suggesting a high helium abundance. The giant branch appears to have a gap at about the same magnitude relative to the horizontal branch as does M15. Eight new suspected variables have been found.

154.009   Globular clusters NGC 6304 and NGC 6401.
A. Terzan, B. Rutily.
Astron. Astrophys., Vol. 16, 408 - 416 (1972).   In French.

The photometric study of the plates obtained in 1968 and 1970 at the newtonian focus of the 80-cm, 193-cm telescopes of the Haute Provence Observatory, and with the Schmidt telescope (48-inch) of Mount Palomar enabled us: 1) to make sequences of red and infrared magnitudes about the globular cluster NGC 6304, 2) to establish a new sequence m_r in the field of the globular cluster NGC 6401, 3) to reveal 102 new variable stars around the cluster NGC 6401, 4) to measure the extreme values m_1 and m_2 of these 102 new variable stars, to have a general view of their average brightness and amplitudes.

154.010   Gravities of blue horizontal-branch stars.
A. G. D. Philip.
Bull. American Astron. Soc., Vol. 3, 479 - 480 (1971). Abstr. AAS.

154.011   UV - bright stars in globular clusters.
R. J. Zinn, E. B. Newell, J. B. Gibson.
Bull. American Astron. Soc., Vol. 3, 480 (1971). – Abstr. AAS.

154.012   Catalogue of photometric characteristics and proper motions of stars in the neighbourhood of the globular clusters M3 and M5. Proper motions and space velocities of the clusters.   L. V. Zhukov.
Trudy Glav. Astron. Obs. Pulkovo, Ser. 2, Vol. 78, 160 - 214 (1971). In Russian.

The B magnitudes for 873 stars in a wide neighbourhood of the globular cluster M5 were determined in a system close to the H. Arp magnitude system. The B magnitudes for 499 stars in the neighbourhood of the globular cluster M3 were determined from four plates taken with the normal astrograph, and V magnitudes of 400 stars in the neighbourhood of the same cluster from plates taken with the 26″ refractor. The B and V magnitudes were obtained in a system close to the magnitude system of the Kukarkins. The B−V were determined for 289 stars. The relative proper motions of the globular clusters M3 and M5 were determined using the stars with V and B−V magnitudes determined reliably. The relative proper motions of M3 and M5 obtained are compared with the results of other authors. Probable reasons of the difference between the results are discussed. The absolute proper motions of M3 and M5 and their space velocities V have been obtained. The catalogue of photometric and kinematic characteristics of 764 stars of M3 and 1141 stars of M5 are given in the appendix.

154.013 **A possible explanation of the gaps on the giant branch of M15.**
P. Demarque, J. G. Mengel, A. V. Sweigart.
Astrophys. Journ., (*Letters*), Vol. 173, L27 - L32 (1972).

It is conjectured that fast rotating cores in the red giants of M15 might explain the possible existence of gaps on the giant branch of the cluster. In going outward through the transition layer between the rotating core and the nonrotating envelope the temperature gradient $|dT/dr|$ will increase abruptly. As the hydrogen-burning shell traverses this transition layer, a sudden increase in the luminosity might therefore be expected. A numerical simulation of this phenomenon has confirmed this prediction. Some implications of this hypothesis on the evolution of Population II stars are briefly discussed.

154.014 **RR Lyrae variables in NGC 5897.**
A. Wehlau, H. Sawyer Hogg, N. Potts.
Journ. Roy. Astron. Soc. Canada, Vol. 66, 72 (1972). – Abstr. Canadian Astron. Soc.

154.015 **Clustering properties of the luminosity function in galactic globular clusters.**
F. Caputo, V. Castellani.
Mem. Soc. Astron. Italiana, Nuova Ser., Vol. 43, 161 - 165 (1972). – Letter.

154.016 **The Palomar Sky Survey and space distribution of globular clusters stars.**
V. Castellani, F. D'Antona, R. De Amicis, F. Smriglio.
Mem. Soc. Astron. Italiana, Nuova Ser., Vol. 43, 167 - 172 (1972). – Letter.

154.017 **A search for UV-bright stars in 27 globular clusters.**
R. J. Zinn, E. B. Newell, J. B. Gibson.
Astron. Astrophys., Vol. 18, 390 - 402 (1972).

Twenty-seven globular clusters have been searched for stars that lie to the left of the giant branch and above the horizontal branch in the HR-diagram. We present the results of our survey in order to provide a finding list for future investigations of post-horizontal branch evolution.

154.018 **Pseudo color-magnitude diagrams for three southern hemispheric globular star clusters.**
R. E. White, T. T. Kraft.
Publ. Astron. Soc. Pacific, Vol. 84, 298 - 302 (1972).

The initial results of a photometric investigation concerning suspected metal-rich globular star clusters are presented in the form of pseudo color-magnitude (CM) diagrams, i.e., graphical displays of the photographic plate measurements of apparent brightness. The globular star clusters reported upon here are: NGC 3201, NGC 5927, and NGC 6388. Judging from the shapes of their pseudo-CM diagrams, the first cluster is definitely not metal-rich, but the remaining two clusters are.

154.019 **Helium content of M92 deduced from the subgiant luminosity function.** J. Faulkner.
Nature, Phys. Sci., Vol. 235, 27 (1972).

It is shown that application of theoretical luminosity functions (due to Simoda and Iben) to Hartwick's observations of M92 yields a high helium content for this cluster, in agreement with Hartwick's preference based on earlier stellar models. A rough estimate for the uncertainty yields a value $Y \sim 033 \pm 0.05$.

154.020 **NGC 6352: A globular cluster with normal metal abundance?** F. D. A. Hartwick, J. E. Hesser.
Bull. American Astron. Soc., Vol. 4, 241 (1972). – Abstr. AAS.

154.021 **The colour-magnitude diagram of the globular cluster NGC 6981.** R. J. Dickens.

Monthly Notices Roy. Astron. Soc., Vol. 157, 281 - 297 (1972).

UBV photometry for stars in the globular cluster NGC 6981 = M72 is presented. The colour-magnitude diagram shows a horizontal branch populated on both sides of the variable star region. The two-colour ($U-B/B-V$) diagram for field stars is used to derive a reddening of $E(B-V) = 0.07 \pm 0.01$ mag, and the two-colour diagram for cluster giants used to derive an ultra-violet excess of $\delta(U-B) = 0.15 \pm 0.05$ mag at $(B-V)_0 = 1.0$. The latter value shows that the cluster is relatively metal rich.

154.022 **The colour-magnitude diagram of the globular cluster NGC 7099.** R. J. Dickens.
Monthly Notices Roy. Astron. Soc., Vol. 157, 299 - 307 (1972).

UBV photometry of the globular cluster NGC 7099 is reported. The ultra-violet excess of $\delta(U-B) = 0.33$ mag shows that NGC 7099 is an extremely metal-poor cluster. The colour-magnitude diagram shows a strong blue population of the horizontal branch, with very few RR Lyrae variables and stars on the red side of the instability region. A reddening of $E(B-V) = 0.06$ mag is derived from the two-colour diagram for field stars.

154.023 **(U−B)-colours and U-magnitudes of 63 RR Lyrae variables of the globular cluster NGC 5139 (ω Cen).**
Mitt. Astron. Ges., No. 31, p. 168 - 170 (1972).

154.024 **Bright red giants of the globular clusters M3, M5 and M13.** Z. I. Kadla, N. Spasova.
Astron. Zhurn. Akad. Nauk SSSR, Vol. 49, 504 - 513 (1972). In Russian. English translation in Soviet Astron. AJ, Vol. 16, No. 3.

All the members of M3, M5 and M13 belonging to the brightest part of the giant branch $\Delta V = V - V_5 \approx 1.30$, where V_5 is the mean magnitude of the five brightest members, have been found by means of measured proper motions and V magnitudes. Some statistical results on these objects are reported.

154.025 **The globular cluster M92.**
O. A. Melnikov, Z. I. Kadla.
Astron. Zhurn. Akad. Nauk SSSR, Vol. 49, 661 - 663 (1972). In Russian. English translation in Soviet Astron. AJ, Vol. 16, No. 3.

Additional observational data provide the evidence that the giants with strong CN- and CH-bands and strong metallic lines assumed to be members of the globular cluster M92 are in reality field stars.

154.026 **The abnormally metal-rich globular cluster NGC 6352.** F. D. A. Hartwick, J. E. Hesser.
Astrophys. Journ., Vol. 175, 77 - 88 (1972).

A color-magnitude (C-M) diagram has been constructed for the globular cluster NGC 6352 from photoelectric UBV measurements to $V \approx 19$ for 63 stars and from photographic B, V measurements of 397 stars on five or more plate pairs. The C-M diagram, which has been found to be comparable to those of 47 Tucanae and M71, confirms the integrated photometry of van den Bergh and indicates that NGC 6352 is indeed metal rich. A blue "straggler" sequence is also identified.

154.027 **Luminosity functions of globular clusters.**
I. R. King, C. P. Wilson.
The evolution of population II stars, (see 012.024), p. 29 - 30 (1972).

154.028 **Distributions and masses of horizontal-branch stars in clusters.** I. R. King.
The evolution of population II stars, (see 012.024), p. 31 - 33 (1972).

154.029 Photometry of blue horizontal-branch stars in globular clusters.   A. G. D. Philip.
The evolution of population II stars, (see 012.024), p. 35 - 40 (1972).

154.030 Intermediate band photometry of red stars in globular clusters.   W. Osborn.
The evolution of population II stars, (see 012.024), p. 43 (1972).

154.031 Asymptotic and red giant branch stars in globular clusters.   S. E. Strom.
The evolution of population II stars, (see 012.024), p. 51 - 53 (1972).

154.032 Comments on blue horizontal-branch stars.   E. B. Newell.
The evolution of population II stars, (see 012.024), p. 55 - 56 (1972).

154.033 On the two Oosterhoff period-groups of globular clusters.   T. S. van Albada, N. H. Baker.
The evolution of population II stars, (see 012.024), p. 193 - 201 (1972).

On the problem of the identification of cluster stars. See Abstr. 041.037.

Oscillatory thermal instabilities at the onset of helium shell burning.   See Abstr. 065.030.

Mixed models for blue horizontal branch stars. See Abstr. 065.113.

Narrow- and broad-band photometry of red stars.

VII. Luminosities and temperatures for halo-population red stars of high luminosity.   See Abstr. 113.007.

The spectra and colors of two blue horizontal-branch stars in Omega Centauri.   See Abstr. 114.016.

Uncertainties in the theoretical determination of blue edges and estimates of Y and $L/L_{\odot}$ for RR Lyrae stars in globular clusters.   See Abstr. 122.025.

The RR Lyrae stars in Baade's field near NGC 6522. See Abstr. 122.132.

A search for neutral hydrogen high-velocity clouds in the directions of six globular clusters. See Abstr. 131.132.

A review of problems concerning associations and star clusters.   See Abstr. 152.012.

The identity and aliases of the stellar system Terzan 5.   See Abstr. 155.062.

IRC -20385, the nearest galaxy or the richest globular cluster?   See Abstr. 158.035.

10-color photometry of elliptical galaxies and globular clusters.   See Abstr. 158.138.

Electronographic photometry of stars in the globular clusters of the Magellanic Clouds.   See Abstr. 159.010.

Electronographic photometry of star clusters in the Magellanic Clouds — III. The colour-magnitude diagram of NGC 2257.   See Abstr. 159.011.

155 Structure and Evolution of the Galaxy

155.001 **Gamma-ray observations of the galactic center and some possible point sources.**
C. E. Fichtel, R. C. Hartman, D. A. Kniffen, M. Sommer.
Astrophys. Journ., Vol. 171, 31 - 40 (1972).

A 0.5 × 0.5 m digitized spark-chamber γ-ray telescope was flown on three balloon flights to look at the galactic center region, Virgo, and the Crab nebula. An excess flux above atmospheric background of over 4 standard deviations was found for γ-rays exceeding 100 MeV coming from the galactic center region, but there was no statistically significant excess in the 50−100-MeV interval. No positive evidence was found for radiation from any of the sources M 87, 3C 273, or the Crab nebula.

155.002 **Neutral hydrogen in an interior region of the Galaxy; the longitude interval 22° to 42°. III. The Scutum arm.** W. W. Shane.
Astron. Astrophys., Vol. 16, 118 - 148 (1972).

Neutral hydrogen observations described in preceding papers have been used to construct a schematic map of the Galaxy in the region where the Scutum arm is seen tangentially, between 4.0 and 6.5 kpc from the galactic center. For this purpose, a kinematic model has been adopted in which a radial expansion is superimposed upon a velocity field derived from the first-order density-wave theory.

155.003 **Further evidence of explosive events in the galactic nucleus.**
R. H. Sanders, G. T. Wrixon, A. A. Penzias.
Astron. Astrophys., Vol. 16, 322 - 326 (1972).

A survey of high velocity hydrogen has been carried out in a region near the galactic center ($358° < l^{II} < 12°; −12° < b^{II} < 0°$) using the 20-ft. horn reflector at Crawford Hill. The purpose of this survey was to observe high velocity hydrogen possibly associated with the galactic center and to map the excent of this hydrogen. We conclude that the observations are consistent with a picture of recurring explosive events in the galactic nucleus as well as a more continual form of mass expulsion.

155.004 **On the z-motion of some young clusters in the Galaxy.** S. Isobe.
Publ. Astron. Soc. Japan, Vol. 24, 41 - 48 = Tokyo Astron. Obs. Repr. No. 401 (1972).

Space velocities of 12 open clusters are calculated. The large z-motions of NGC 457, NGC 581, Tr 1, and IC 1805 (de Vegt, Gail, and Gehlich 1968) are presumably reduced by giving corrections of the systematic error of the FK4 system and the regional error of the preliminary AGK3 catalog to the proper motion data used.

155.005 **Densité spatiale des étoiles Ap.** P. Renson.
Publ. Astron. Soc. Japan, Vol. 24, 149 - 151 (1972).

The number density of Ap stars is found to be near $4 × 10^{-5} pc^{-3}$. This seems to invalidate Imoto and Kanai's (1971) statistical argument against the origin of pulsars from Ap stars.

155.006 **A soft X-ray survey from the galactic center to Vela.**
R. W. Hill, G. Burginyon, R. J. Grader, T. M. Palmieri, F. D. Seward, J. P. Stoering.
Astrophys. Journ., Vol. 171, 519 - 528 (1972).

A 640 cm² proportional-counter system, sensitive from 0.2 to 12 keV, was used to survey the galactic plane from Sagittarius to Vela. Spatial resolution was 1°, at best. The soft X-ray sky is dominated by a region of strong emission in the vicinity of Vela X and has an appearance quite different from the sky viewed at photon energies of a few kilovolts. Many sources were located in one dimension in the direction of the galactic center. Locations were derived for two strong sources; GX 292 + 0 or Cen 3, and GX 322 + 0 or Cir 1. Another weaker source, GX 296 − 2 or Cen 5, was located in a region of area 0.6 square degrees. A definite source of soft X-rays, which seems to coincide with the object η Car, was seen. Spectra are derived for all sources, and numbers are given for interstellar material in the line of sight.

155.007 **Detectability of ionized neon at the galactic center.**
M. Kaufman.
Astron. Astrophys., Vol. 16, 361 - 368 (1972).

Infrared measurements of the galactic nucleus by Low et al. (1969) are interpreted in terms of a [Ne II] emission line at 12.8 μ superimposed on a continuum. With allowance for the absorption of resonantly trapped λ 1.08 μ photons by dust particles mixed with the ionized gas, the postulated [Ne II] emission models can be made consistent with the upper limit obtained by Spinrad et al. (1971) for the equivalent width of He I λ 1.08 μ.

155.008 **On a blue filament in the southern Milky Way.**
W. Schlosser.
Astron. Astrophys., Vol. 16, 486 - 488 (1972). In German.

One of the most striking features on wide-angle photographs of the southern Milky Way in the ultraviolet (3200 - 3800 Å) is a thin filament, which stretches from about $l = 325°$ to $l = 345°$. Evidence is accumulating that this filament displays the optically most active part of the spiral arm−I.

155.009 **The spectrum of low-energy gamma radiation from the galactic-center region.**
W. N. Johnson III, F. R. Harnden, Jr., R. C. Haymes.
Astrophys. Journ., (Letters), Vol. 172, L1 - L7 (1972).

A balloon-altitude observation was conducted 1970 November 25, of the galactic-center region, at energies between 23 and 930 keV. The radiation detected from the 24° FWHM area of Sagittarius during the 7-hour observation has a differential photon number spectrum that may be approximated by a power law with a spectral index of 2.37 ± 0.05.

155.010 **Soft X-rays from the vicinity of the North Polar Spur.**
A. N. Bunner, P. L. Coleman, W. L. Kraushaar, D. McCammon.
Astrophys. Journ., (Letters), Vol. 172, L67 - L72 (1972).

Soft X-rays in the region $E < 284$ eV and $500 < E < 1000$ eV have been detected from the vicinity of the North Polar Spur.

155.011 **Is the Orion arm a major arm?**
C. Yuan, L. Liebovitch.
Bull. American Astron. Soc., Vol. 3, 440 (1971). − Abstr. AAS.

155.012 **The distribution and kinematics of luminous stars in the Carina spiral feature.** R. M. Humphreys.
Bull. American Astron. Soc., Vol. 3, 440 - 441 (1971). − Abstr. AAS.

155.013 **UBV photometry in the nuclear bulge of the Galaxy.** S. van den Bergh.
Bull. American Astron. Soc., Vol. 3, 441 (1971). − Abstr. AAS.

155.014 **New mass models of the galactic system.**
K. A. Innanen,
Bull. American Astron. Soc., Vol. 3, 441 (1971). − Abstr. AAS

155.015 **The high velocity clouds as part of "normal" spiral structure.** G. L. Verschuur.
Bull. American Astron. Soc., Vol. 3, 469 (1971). − Abstr. AAS.

155.016 **A kinematic method for the derivation of the galactic distribution of hydrogen.** W. B. Burton.
Bull. American Astron. Soc., Vol. 3, 470 (1971). − Abstr. AAS.

155.017 **A loop-like structure in the neutral hydrogen gas near $l = 130°$, $b = +35°$.** C. R. Tolbert.
Bull. American Astron. Soc., Vol. 3, 470 (1971). − Abstr. AAS.

155.018 **Far ultraviolet background radiation.** R. C. Henry.
Bull. American Astron. Soc., Vol. 3, 477 (1971). − Abstr. AAS.

155.019 **An upper limit on the line emission in the diffuse background.**
A. Toor, R. E. Price, F. D. Seward.
Bull. American Astron. Soc., Vol. 3, 481 (1971). − Abstr. AAS.

155.020 **Rocket infrared observations of the galactic center region.** M. O. Harwit, J. R. Houck, J. L. Pipher, B. T. Soifer.
Bull. American Astron. Soc., Vol. 3, 499 (1971). − Abstr. AAS.

155.021 **Anzeichen von Aktivität im galaktischen Zentrum.** F. Gondolatsch.
SuW, Vol. 11, 64 - 67 (1972).

155.022 **The local linearized velocity field in the presence of a spiral density wave.** K. Rohlfs.
Astron. Astrophys., Vol. 17, 246 - 252 (1972).
The streaming motions caused by spiral density waves effect the radial velocity field in the solar neighbourhood in two ways: i) The local Oort constant A shows a periodic variation across the spiral arm, the minimum is adopted in the arm centre for trailing spirals. ii) The nodes of the sin 2l variation of radial velocity is turned by a deviation angle ϵ.

155.023 **An objective-prism study of galactic structure in Cassiopeia.** N. Martin.
Astron. Astrophys., Vol. 17, 253 - 266 (1972). In French.
Our catalogue gives radial velocities of 358 stars and distances of 254 stars. The distances of O−B stars have two maxima, one situated at about 2.6 kpc, the other about 4.0 kpc. Radial velocities are compatible with Oort's theory. The effect of quality is however very important.

155.024 **OB star distribution in Puppis.** R. J. Havlen.
Astron. Astrophys., Vol. 17, 413 - 424 (1972).
UBV and Hβ photometry have been completed for all of the OB stars of the Tonantzintla surveys between $l^{II} = 240°$ and $250°$ in Puppis. The existence of the two associations Pup OB 1 and Pup OB 2 is confirmed. The cepheid AQ Pup is a likely candidate for membership in Pup OB 1. The galactic cluster NGC 2467 is found to have some members in common with Pup OB 2 and is probably more distant than previously indicated. In general the OB star distribution correlates well with the H I distribution found by Hindman and Kerr at $l^{II} = 245°$.

155.025 **Neutral hydrogen self-absorption in a large region toward the galactic center.**
K. W. Riegel, R. M. Crutcher.
Astron. Astrophys., Vol. 18, 55 - 69 (1972).
A large region of the galactic plane, covering the longitude range 345° to 25° and the latitude range ±6° has been surveyed in the 21-cm spectral line. Most of the area surveyed is found to contain a very prominent H I self-absorbing cloud.

155.026 **Spatial and kinematic parameters of binebulous, centric and annular nebulae.** W. E. Greig.
Astron. Astrophys., Vol. 18, 70 - 78 (1972).
Paper I (Greig, 1971) showed that the agreement between the morphology and certain line ratios was 95 per cent. Herein new and unpublished data are shown to agree with the predictions of Paper I. It is shown that nearly half of the local planetary nebulae have circular orbits and are likely intermediate Population I.

155.027 **An expanding and rotating ring of gas 2.4 kpc from the galactic center.** R. H. Sanders, G. T. Wrixon.
Astron. Astrophys., Vol. 18, 92 - 96 (1972).
Observations made in the direction of the galactic center at negative velocities have shown that the large feature found by van der Kruit (1970) south of the plane at positive longitudes, extends also to negative longitudes with its velocity becoming more negative in the sense of galactic rotation. A kinematic model that we fit to the expanding arm at + 135 km/s also describes quite well the velocity distribution with longitude of the van der Kruit feature.

155.028 **On the vertex deviation.** M. Mayor.
Astron. Astrophys., Vol. 18, 97 - 105 (1972).
The kinematic properties of different stellar samples are discussed. We seek in particular the more plausible interpretation of the deviation of the vertex: reflection of initial conditions, or non-axisymmetrical perturbation of the potential.

155.029 **The density wave model of the inner parts of the Galaxy.** B. Basu, A. K. Roy.
Bull. Astron. Inst. Czechoslovakia, Vol. 23, 117 - 123 (1972).
The density wave model of the spiral structure has been extended to the inner parts of the Galaxy. Angular velocities from Schmidt's model were used beyond 3 kpc from the center. Inside of that distance, lower values were adopted, less steeply rising towards the center. Other parameters have been calculated over the entire inner region of the Galaxy. These values have been used to calculate the values of the basic density, the amplitudes of the density and potential perturbation, the amplitudes of the radial and tangential velocity perturbation, and the radial velocity dispersion, in gas and stars separately, over the entire Galaxy.

155.030 **The formaldehyde absorption of Sagittarius A at +40 km/sec.** F. F. Gardner, J. B. Whiteoak.
Astrophys. Letters, Vol. 10, 171 - 174 (1972).
An investigation of CH_2O near the galactic center has been carried out with a beamwidth of 4.2 arc min and a frequency resolution of 33 kHz (2 km/sec). The +40 km/sec absorption feature is associated with a single continuum component of Sgr A. The positional change with velocity is consistent with rotation about an axis perpendicular to the plane of the Galaxy.

155.031 **Small-scale structure of the Milky Way.** T. B. Pyatunina.
Astrofiz. Issled. Izv. Spets. Astrofiz. Obs., Vol. 3, 124 - 127 (1971). In Russian.
From observations at 3.95 cm the upper limit to fluctuations of radio brightness of the Milky Way has been obtained. The upper limit to the power of galactic radio sources, and the upper limit to the r.m.s. electron density fluctuations are found. A search for young H II regions supports the existing theoretical estimates of their spatial density.

155.032 **Galactic structure and the apparent size of radio sources.** A. C. S. Readhead, A. Hewish.
Nature, Vol. 236, 440 - 443 (1972).

Observations of radio sources at 81.5 MHz show that interstellar scattering is more important than some pulsar data suggest. Low frequency measurements also provide a new means for studying the distribution of ionized gas in the Galaxy.

155.033 **A new contribution to the study of the Carina spiral feature.**
J. H. Bigay, R. Garnier, Y. P. Georgelin, Y. M. Georgelin.
Astron. Astrophys., Vol. 18, 301 - 309 (1972). In French.

A $4°.5 \times 4°.5$ region in Carina has been attentively studied. With Perot-Fabry rings, we have obtained radial velocities of 10 H II regions; non circular motions are found only in the outer edge of the spiral feature. A comparison with radio observations is made. From spectroscopic and photoelectric observations we have obtained distances of 29 exciting stars or early type stars; these observations proved that at this longitude we are looking tangentially along a spiral arm between 2 and 5 kpc.

155.034 **Application of the density-wave theory of spiral structure: Shock formation along the Perseus arm.**
W. W. Roberts, Jr.
Astrophys. Journ., Vol. 173, 259 - 283 (1972).

In light of the observational data, a model based on the density-wave and galactic-shock-wave concepts is proposed for the Perseus arm. The model considered is the two-armed spiral shock model that delineates the large-scale spiral structure and describes the large-scale motion of the interstellar gas, and the young optical objects born out of the gas, over the galactic disk. In this model, the Perseus arm is visualized to consist of a galactic shock wave embedded in a background density wave.

155.035 **The mass of neutral atomic hydrogen in the Galaxy.**
S. J. Goldstein, Jr.
Astrophys. Journ., Vol. 173, 285 - 286 (1972).

Twenty-one-centimeter observations of the galactic poles lead to a mass of optically thin neutral atomic hydrogen in the Galaxy in the range 2.3 to $2.9 \times 10^9 M_\odot$ on the assumption that the hydrogen is a uniform disk with a radius of 15 kpc.

155.036 **Observations of the diffuse galactic light.**
F. E. Roach, L. L. Smith, J. Pfleiderer, C. Batishko, K. Batishko.
Astrophys. Journ., Vol. 173, 343 - 352 (1972).

Systematic observations during five nights were made with a field of 137 arc seconds at a fixed declination of $20°\,42'$ through one arm of the Milky Way over a range of right ascension from 257° to 336°. Corrections for zodiacal light, airglow, and integrated starlight leave a residual which is interpreted as the diffuse galactic light. The residual, plotted against galactic latitude, suggests that the diffuse light in the sample studied emanates from several discrete sources.

155.037 **The theory for a galactic halo density function.**
K. A. Innanen, A. G. Ryman.
Journ. Roy. Astron. Soc. Canada, Vol. 66, 67 (1972). – Abstr. Canadian Astron. Soc.

155.038 **Outer spiral structure and high velocity clouds in the Galaxy.** R. D. Davies.
Nature, Vol. 237, 88 - 91 (1972).

New observations of high velocity clouds show that these may be interpreted in terms of an extension of the outermost spiral arm, confirming and extending the model of Habing.

155.039 **X-rays from a galactic wind.**
J. A. Burke, F. D. A. Hartwick.
Nature, Phys. Sci., Vol. 236, 4 (1972).

The low energy diffuse X-rays may originate in a Milky Way galactic wind. Fluxes predicted by galactic wind models are in reasonable accord with the observed values.

155.040 **The extension of Loop III into southern latitude.**
J. Milogradov-Turin.
Monthly Notices Roy. Astron. Soc., Vol. 157, 1P - 4P (1972).

The galactic radio feature, Loop III, is prominent only north of the galactic plane. A new 38 MHz survey has revealed features south of the plane which appear to be continuations of the loop.

155.041 **The distribution of mass in the galactic nucleus.**
R. H. Sanders, T. Lowinger.
Astron. Journ., Vol. 77, 292 - 297 (1972).

The near-infrared observations of the region near the galactic center by Becklin and Neugebauer are used to derive a mass model for the galactic nucleus which has the form of an inhomogeneous spheroidal distribution. This spheroid representing the galactic nucleus is combined with the Schmidt model of the galactic disk and the resulting rotation curve is calculated. This rotation curve is shown to be quite similar to the observed neutral-hydrogen rotation curve in the inner region of the Galaxy.

155.042 **Early-type stars and the mean-square electron density in the galactic disk.**
D. O. Richstone, K. Davidson.
Astron. Journ., Vol. 77, 298 - 301 (1972).

An assessment is made of the total supply of ionizing photons from hot stars, within several hundred parsecs of the sun. The resulting photoionized H II regions probably contribute an average measure of the order of 5 pc cm^{-6}, through the galactic disk.

155.043 **A test for relative motions of gas and young stars.**
E. Moore.
Publ. Astron. Soc. Pacific, Vol. 84, 273 - 280 (1972).

If the gas in H II regions and the associated young stars have large-scale motions relative to each other in the galactic plane, there should be radial-velocity differences which vary systematically with galactic longitude. An analysis with 152 stars in 53 H II regions shows that stars and gas appear to move together at nearly zero relative velocity, with a standard error of estimate of 15 km/sec. The error is reduced to 8 km/sec if analysis is restricted to H II regions with three or more stars.

155.044 **The stellar population in the nuclear bulge of the Galaxy.** S. van den Bergh.
Publ. Astron. Soc. Pacific, Vol. 84, 306 - 313 (1972).

New observations in the low-absorption window that is centered on NGC 6522 yield a foreground reddening $E_{B-V} = 0.45$ and a distance of 9.4 ± 2.0 kpc to the center of the Galaxy. In a color-magnitude diagram the red giants of the nuclear bulge are concentrated in a zone that lies between the giant branch of M3 and the lower envelope to the giant region occupied by field stars near the sun. The lower (metal-rich) part of this region is much more strongly populated than is the upper (metal-poor) part of this zone. It is concluded that the dominant stellar population in the nuclear bulge is metal rich.

155.045 **The distribution of variable stars in the Galaxy.**
L. Plaut.
IAU Colloquium No. 15, (see 012.006), p. 317 - 326 (1972).

155.046 **Radio pulses from the direction of the galactic centre.**
V. A. Hughes, D. S. Retallack.
Bull. American Astron. Soc., Vol. 4, 216 (1972). – Abstr. AAS.

155.047 **Observations of the soft diffuse X-ray flux near the galactic plane, $\ell^{II} = 150°$ to $220°$.** P. L. Coleman, A. N. Bunner, W. L. Kraushaar, D. McCammon, F. O. Williamson
Bull. American Astron. Soc., Vol. 4, 220 (1972). – Abstr. AAS.

155.048 **UCSD X-ray observations in the galactic plane from OSO-7.** L. E. Peterson, W. A. Baity, W. A. Wheaton, M. Ulmer.
Bull. American Astron. Soc., Vol. 4, 220 - 221 (1972). – Abstr. AAS.

155.049 **Descriptive functions of the Galaxy.** J. Einasto, L. Einasto.
Tartu Astron. Obs. Teated, No. 36, p. 46 - 54 (1972).

155.050 **An attempt to estimate the galactic mass density in the vicinity of the sun.** M. Jõeveer.
Tartu Astron. Obs. Teated, No. 37, p. 3 - 13 (1972).
A method is proposed for the determination of the galactic gravitational acceleration across the galactic plane from a comparison of the distances perpendicular to the galactic plane and the motions of young stars with their ages. From statistics of a small sample of nearby late B-type stars it follows that the galactic mass density near the sun is 0.085 $M_\odot/pc^3$.

155.051 **Evidence for a radial composition gradient in the disk of the Galaxy.** K. A. Janes, R. D. McClure.
Bull. American Astron. Soc., Vol. 4, 241 - 242 (1972). – Abstr. AAS.

155.052 **The spectrum of low-energy gamma radiation from the galactic center region.**
W. N. Johnson III, F. R. Harnden, Jr., R. C. Haymes.
Bull. American Astron. Soc., Vol. 4, 259 (1972). – Abstr. AAS.

155.053 **A survey of the Galaxy using UHURU.** S. Murray, H. Tananbaum, T. Matilsky, E. Kellogg, H. Gursky, R. Giacconi.
Bull. American Astron. Soc., Vol. 4, 260 (1972). – Abstr. AAS.

155.054 **Kinematics of the intermediate age galactic population.** A. A. Blaauw, C. D. Garmany, P. A. Ianna, C. R. Tolbert.
Bull. American Astron. Soc., Vol. 4, 262 (1972). – Abstr. AAS.

155.055 **A dynamical model with a "3 kpc arm".** W. L. Peters, W. W. Roberts, Jr.
Bull. American Astron. Soc., Vol. 4, 265 (1972). – Abstr. AAS.

155.056 **Shock formation along the Perseus spiral arm, and related dynamical phenomena.**
W. W. Roberts, Jr.
Bull. American Astron. Soc., Vol. 4, 265 (1972). – Abstr. AAS.

155.057 **Motions in the galactic center and the "3-kpc arm".** S. C. Simonson III, G. L. Mader.
Bull. American Astron. Soc., Vol. 4, 266 (1972). – Abstr. AAS.

155.058 **Velocity structures in hydrogen profiles.** M. A. Tuve, S. Lundsager.
Bull. American Astron. Soc., Vol. 4, 267 (1972). – Abstr. AAS.

155.059 **On the spiral structure of the Milky Way system.** C. Yuan, L. S. Liebovitch.
Bull. American Astron. Soc., Vol. 4, 268 (1972). – Abstr. AAS.

155.060 **An empirical determination of the star density in the galactic halo with three-colour-photometric methods (IV, SA 71).** R. P. Fenkart, R. Wagner.
Astron. Astrophys., Vol. 19, 1 - 7 (1972).
With the same three-colour-photometric methods as in our earlier halo-papers (Becker, 1956; Fenkart, 1967 and 1968) we separated the disc- and halo-populations in the direction of SA 71 (b = −35°) and determined their density-gradients for different intervals in absolute magnitude.

155.061 **On the kinematic distribution of galactic neutral hydrogen.** W. B. Burton.
Astron. Astrophys., Vol. 19, 51 - 65 (1972).
The shape of 21-cm line profiles is very sensitive to small variations in the streaming motions (Burton, 1971). A method is described in this paper for producing maps of the kinematic structure of at least the outer portions of the Galaxy.

155.062 **The identity and aliases of the stellar system Terzan 5.** I. R. King.
Astron. Astrophys., Vol. 19, 166 (1972).
The same rich stellar system has been reported independently under at least three names. This note attempts to dispel some of the confusion.

155.063 **Galactic helium abundances.** E. Churchwell, P. G. Mezger.
Mitt. Astron. Ges., No. 31, p. 99 (1972). – Abstract.

155.064 **Sind galaktische Dichtewellen beobachtbar?** K. Rohlfs.
Mitt. Astron. Ges., No. 31, p. 143 - 147 (1972).

155.065 **Zur galaktischen Struktur bei $230° < l^{II} < 355°$ auf Grund lichtelektrischer UBV-Hβ-Photometrie von 55 südlichen offenen Sternhaufen.** N. Vogt, A. F. J. Moffat.
Mitt. Astron. Ges., No. 31, p. 224 - 226 (1972).

155.066 **Distribution of neutral hydrogen in the Perseus arm.** J. F. W. Perry, H. L. Helfer.
Astrophys. Journ., Vol. 174, 341 - 360 (1972).
A statistical analysis is applied to 21-cm data to derive models of the density distribution of interstellar gas in small volumes of spiral arms. We analyzed many small volumes within four large segments of the Perseus arm and one segment of an interarm feature, each observed with the 140-foot telescope and 400-channel receiver at NRAO. The resulting cloud distributions are presented.

155.067 **X-ray emission from the galactic disk.** R. D. Bleach, E. A. Boldt, S. S. Holt, D. A. Schwartz, P. J. Serlemitsos.
Astrophys. Journ., (Letters), Vol. 174, L101 - L106 (1972).
A search was made for a diffuse component of X-rays > 1.5 keV associated with an interarm region of the Galaxy at galactic longitudes in the vicinity of 60°. A statistically significant excess associated with a narrow disk component was detected. The angular extent of this component has a most probable value of 2° and may be as large as 7° at 90 percent confidence. Several possible emission models are evaluated, with the most likely candidate being a population of unresolvable low-luminosity discrete sources.

155.068 **Two new large-diameter galactic regions of faint Hα emission.** G. Courtès, J. P. Sivan.
Astrophys. Letters, Vol. 11, 159 - 162 (1972).
In order to detect the limit of the general emission of the Hα line in the Milky Way, a photographic survey has been undertaken with a 55° field camera using a very narrow interference filter centred on Hα. Two of the plates reveal two new large-diameter regions of faint emission at high galactic latitudes. In one of these pictures, the whole Cygnus complex is recorded.

155.069 **The gravitational acceleration perpendicular to the galactic plane.** E. Gould, P. O. Vandervoort.

Astron. Journ., Vol. 77, 360 - 365 (1972).

A new method for the determination of the gravitational acceleration perpendicular to the galactic plane is formulated in terms of a set of virial equations governing the perpendicular structure of a stellar subsystem of the Galaxy. When applied to a sample of A-type stars in the general direction of the North Galactic Pole, the method yields values in the range $0.19-0.28\,M_\odot\,pc^{-3}$ for the density of matter in the solar neighborhood. The apparent discrepancy between this result and the conventional value $0.15\,M_\odot\,pc^{-3}$ is found to be marginal when the statistics of the sample is considered.

155.070 **A new upper limit to the local population II density.** D. Weistrop.
Astron. Journ., Vol. 77, 366 - 373 (1972).

The nature of faint stars at high galactic latitudes is investigated through U, B, V photometry of several thousand stars between $V = 12$ and 18 near the North Galactic Pole. Several models for the luminosity function and density distribution of population II field stars are assumed. Assuming also that all stars bluer than $(B-V)_0 = 0.5$ and fainter than $V = 17$ are members of population II, an upper limit to the local population II density is calculated. A value of $1.1 \times 10^{-4}\,(M/L)$ solar masses per pc^3 is found, where (M/L) is the mass-luminosity ratio of population II field stars. The result is significantly lower than local densities computed from the number of nearby stars suspected of halo membership. Possible reasons for the discrepancy are discussed.

155.071 **Origin of the low energy diffuse cosmic X-ray flux.** J. L. Culhane, A. C. Fabian.
Nature, Vol. 237, 379 - 381 (1972).

While only the gross features of the low energy diffuse flux have been observed, it seems that a galactic component of diffuse soft X-rays is required. Forces of this flux should be distributed about the galactic plane with a scale height similar to that of the hydrogen.

155.072 **Polarization of the diffuse galactic light.** J. G. Sparrow, E. P. Ney.
Astrophys. Journ., Vol. 174, 717 - 720 (1972).

Polarization measurements made from the satellite OSO-5 show that the polarized intensity in the direction of the Scutum arm of the Galaxy is different in intensity and direction of the polarization from that observed due to the zodiacal light. The results are interpreted in terms of a model in which the galactic starlight is scattered by interstellar dust.

155.073 **Observation of the diffuse cosmic gamma radiation in the 30–50-MeV region.**
H. A. Mayer-Hasselwander, E. Pfeffermann, K. Pinkau, H. Rothermel, M. Sommer.
Astrophys. Journ., (*Letters*), Vol. 175, L23 - L28 (1972).

Results of two balloon flights are presented in which a γ-ray astronomy spark chamber was carried to residual atmospheric pressures of 1.7 and 2.2 g cm⁻², respectively. A diffuse flux of celestial γ-rays was discovered at 30–50 MeV which shows no significant correlation with galactic features and which exceeds a linear interpolation between previously known γ-ray data at 1 and 100 MeV by a factor of between 4 and 10.

155.074 **Basic problems on the structure and dynamics of our Galaxy.** B. J. Bok.
Galactic astronomy, Vol. 1, (see 012.019), 2 - 119 (1970).

155.075 **Spiral structure of galaxies.** J. H. Oort.
Galactic astronomy, Vol. 1, (see 012.019), 121 - 145 (1970).

155.076 **Box orbits of nearby stars.** R. Woolley.

Galactic astronomy, Vol. 2, (see 012.019), 95 - 159 (1970).

155.077 **Radio frequency absorption line studies of galactic structures.** T. K. Menon.
Galactic astronomy, Vol. 2, (see 012.019), 251 - 300 (1970).

155.078 **A study of interstellar reddening in a region in Ophiuchus.** C. Schalén.
Rep. Obs. Lund, No. 3, 55 pp. (1972).

This paper contains a study of a region in Ophiuchus centred at $l^{II} = 33°$, $b^{II} = +5°5$. The investigation is based on photometric and spectral plates from Warner and Swasey Observatory, Cleveland. The two-colour diagram of the whole region is given. It seems probable that the reddening law is normal. No part of the region is free from interstellar extinction. The relations between colour excess and distance modulus have been derived for two areas with different amount of reddening. The distribution of the absorbing matter is discussed.

155.079 **Comments on systematic and didactic approach to galactic research.** H. Eelsalu.
Tartu Astron. Obs. Teated, No. 37, p. 14 - 23 (1972).

155.080 **Kinematics and the distance scale.** O. J. Eggen.
Quarterly Journ. Roy. Astron. Soc., Vol. 13, 152 (1972). – Presented at the Woolley symposium (see 012.025).

155.081 **Spiral and halo structure of our galaxy on the basis of optical distance determinations.** W. Becker.
Quarterly Journ. Roy. Astron. Soc., Vol. 13, 226 - 240 (1972). – Presented at the Woolley symposium (see 012.025).

155.082 **The uses of early-type stars for galactic distances.** A. D. Thackeray.
Quarterly Journ. Roy. Astron. Soc., Vol. 13, 243 - 251 (1972). – Presented at the Woolley symposium (see 012.025).

155.083 **Galactic rotation constants.** S. V. M. Clube.
Quarterly Journ. Roy. Astron. Soc., Vol. 13, 252 - 257 (1972). – Presented at the Woolley symposium (see 012.025).

155.084 **V_0 from Magellanic Cloud orbits.** A. Toomre.
Quarterly Journ. Roy. Astron. Soc., Vol. 13, 266 - 270 (1972). – Presented at the Woolley symposium (see 012.025).

155.085 **General discussion of galactic constants.**
Quarterly Journ. Roy. Astron. Soc., Vol. 13, 271 - 273 (1972). – Presented at the Woolley symposium (see 012.025).

155.086 **Direct observations of infrared and submillimeter background radiation.** M. Harwit.
Bull. American Astron. Soc., Vol. 4, 256 (1972). –Abstr. AAS.

155.087 **Kinematics of gas near the galactic center.**
R. H. Sanders, N. Z. Scoville, E. A. Spiegel.
Comments Astrophys. Space Phys., Vol. 4, 15 - 21 (1972).

We wish to summarize recent developments in the observational studies of the flow of gas outward from the central regions of our own Galaxy and mention some possible kinematic conclusions that may tentatively be drawn from them.

155.088 **Galactic background at 15 GHz.**
H. Hirabayashi, H. Yokoi, M. Morimoto.
Nature, Phys. Sci., Vol. 237, 54 - 56 (1972).

Galactic background radiation was observed at 15 GHz (2 cm) using the 7 m antenna of KDD for several points along the Milky Way. Observed brightness is typically 0.03 K. Comparing the brightness with those at lower frequencies it was found that the radiation is mainly thermal at this frequency. The brightness of the thermal radiation is then compared with the strength of the recombination lines, which gives an electron temperature of the diffuse interstellar gas of about 3000 K.

155.089 **The expansion of the Galaxy.**
G. B. Field, M. J. Rees, D. W. Sciama.
E.S.R.O., S. P., No. 52, (see 012.030), p. 29 - 32 (1971).

155.090 **The equilibrium and stability of the gaseous compo-
nent of the Galaxy.**   S. A. Kellman.
Thesis, Univ. California, Berkeley. [Available from Univ. Microfilms, Ann Arbor, Mich., U.S.A. Order No. 71−15812], 90 pp. (1970).
   The distribution of gas satisfying both the hydrostatic equilibrium and Poisson conditions with distance above the galactic plane is derived and compared with Schmidt's observations of the gas density at the galactic tangential points.

155.091 **A study of neutral hydrogen gas motion within
spiral arms and in the local region.**
R. H. Harten.
Dissertation Fac. Graduate School, Univ. Maryland. 120 pp. (1971).
   A study of the gas motion in all the main spiral features shows the presence of systematic velocity gradients in the direction perpendicular to the galactic plane. These velocities vary smoothly over a large extent in longitude indicating the cause to be a large scale effect. There appears to be a strong correlation between the local magnetic field and the velocity distribution of the local gas.-*DKM*

The discovery of our Galaxy.
See Abstr. 003.139.

**Remeasurement of the rest frequency of the 36-
centimeter radio line of methanol.**   See Abstr. 022.070.

**Precessional corrections, apex and galactic rotation
derived from proper motions by a maximum likelihood
method.**   See Abstr. 043.002.

Spatial inhomogeneities of nucleosynthesis.
See Abstr. 061.042.

**Galactic evolution. I. Differential effects of radia-
tion pressure around stars of different spectral types on neutral
atoms and dust grains.**   See Abstr. 065.069.

**Radio pulses from the direction of the galactic cen-
tre.**   See Abstr. 066.020.

Gravitational radiation: the theoretical aspect.
See Abstr. 066.122.

The statistical analysis of stellar kinematics.
See Abstr. 111.003.

New kinematical data for bright southern OB stars.
See Abstr. 112.001.

**Kinematics of faint M stars near the north galactic
pole, and the mass density in the solar neighbourhood.**
See Abstr. 112.013.

**Faint O−B2 stars in the Vela, Carina, Centaurus,
and Crux sections of the southern Milky Way.**
See Abstr. 113.016.

**Integrated U-V color index of the southern Milky
Way.**   See Abstr. 113.030.

The absolute magnitudes of the barium stars.
See Abstr. 114.118.

**A comparison of the luminosities of Perseus-arm
stars in the Hγ and MK systems.**   See Abstr. 115.003.

**Masses, radii and luminosities of RR Lyrae variable
stars.**   See Abstr. 122.029.

Soft X-ray background and flare stars.
See Abstr. 122.115.

**Observational aspects of RR Lyrae and W Virginis
stars: Some conundrums of stellar populations and galactic
distribution.**   See Abstr. 122.147.

**A study of galactic supernova remnants: I. Distances,
radio luminosity function and galactic distribution.**
See Abstr. 125.012.

**The Parkes survey of 21-centimeter absorption in
discrete-source spectra. V. Note on the statistics of absorbing
H I concentrations in the galactic disk.**   See Abstr. 131.005.

**Polarization measurements of 1660 southern OB−
stars.**   See Abstr. 131.011.

**Comments on a paper by K. Rohlfs: 'On the struc-
ture of interstellar matter. I'** [Astron. Astrophys., Vol. 12, 43 - 58 (1971)].   See Abstr. 131.012.

**On analytical and numerical models. A reply to W.
B. Burton** [Astron. Astrophys., Vol. 16, 158 - 160 (1972)].
See Abstr. 131.013.

**Interstellar reddening in the North Galactic Polar
Cap. I. The dependence on distance from the galactic plane.**
See Abstr. 131.019.

**On the distribution of OH in the Galaxy. I. Correla-
tion with continuum sources and with formaldehyde.**
See Abstr. 131.023.

**Survey of molecular lines near the galactic center.
I. 6-centimeter formaldehyde absorption in Sagittarius A,
Sagittarius B2, and the galactic plane from $l^{II} = 359°.4$ to
$l^{II} = 2°.2$.**   See Abstr. 131.029.

**An explanation of the cloudy structure of the inter-
stellar medium.**   See Abstr. 131.052.

**Berkeley survey of high-velocity interstellar neutral
hydrogen: II. The section $|b| \geq 15°$.**   See Abstr. 131.056.

**Radio emission at 1400 MHz from galactic H II
regions.**   See Abstr. 131.057.

Interstellar motions: Minuet or Rock?
See Abstr. 131.086.

**Why many infrared astronomical sources emit at
100 μm.**   See Abstr. 131.098.

**Interstellar reddening in the north galactic polar cap.
II. A cloud model.**   See Abstr. 131.116.

Die Dichteverteilung des Zwischenwolken-H I-Gases.
See Abstr. 131.123.

Meter-wavelength observations of galactic radio sources and the radiation from the galactic disk. See Abstr. 141.124.

Diffuse cosmic X-radiation associated with our Galaxy. See Abstr. 142.028.

Limit on line emission in the diffuse X-ray background. See Abstr. 142.015.

Intensity and spectrum of the diffuse X-ray background. See Abstr. 142.076.

Significance of the small scale isotropy of the cosmic X-ray background radiation. See Abstr. 142.090.

Observations of X-ray sources near the galactic center from 1 to 60 keV from the MIT instrument on OSO-7. See Abstr. 142.096.

Diffuse galactic X-rays from discrete sources. See Abstr. 142.119.

High energy gamma radiation from the region of Cygnus–Cassiopeia. See Abstr. 142.133.

Gamma ray observations of the galactic center and some possible point sources. See Abstr. 142.138.

Blast waves in rotating media. See Abstr. 151.063.

Irregular spiral structure—causes and effects. See Abstr. 151.090.

Theory of spiral structure. See Abstr. 151.091.

A systematic search for high-velocity hydrogen outside the galactic plane. I. See Abstr. 157.001.

Berkeley survey of high-velocity interstellar neutral hydrogen. I. The section $b = -15°$ to $+15°$, $l = 10°$ to $250°$. See Abstr. 157.002.

A survey of linear polarization at 1415 MHz. I. Method of reduction and results for the North Polar Spur. See Abstr. 157.003.

A systematic search for high-velocity hydrogen outside the galactic plane. II. See Abstr. 157.005.

A survey of positive velocity neutral hydrogen above the galactic plane between $l = 252°$ and $l = 322°$. See Abstr. 157.008.

New theory for giant loops. See Abstr. 157.009.

High velocity hydrogen apparently associated with continuum jets near the galactic center. See Abstr. 157.010.

An unusual high-velocity hydrogen feature. See Abstr. 157.011.

Beobachtungen der H- und He-Radiorekombinationslinien in der Umgebung des galaktischen Zentrums. See Abstr. 157.013.

The distortion of the galactic plane emission in the anticentre region. See Abstr. 157.014.

Galaktische und metagalaktische Hintergrundstrahlung. See Abstr. 162.047.

156 Galactic Magnetic Field

156.001 **The origin and form of the galactic magnetic field. I. Parker's dynamo model.** J. H. Piddington. Cosmic Electrodynamics, Vol. 3, 60 - 70 (1972).

Parker's dynamo theory of the galactic magnetic field is examined with the conclusion that it fails for a number of reasons.

156.002 **Generation of a large-scale magnetic field of the Galaxy. II.** S. I. Vainshtein, A. A. Ruzmaikin. Astron. Zhurn. Akad. Nauk SSSR, Vol. 49, 449 - 452 (1972). In Russian. English translation in Soviet Astron. AJ, Vol. 16, No. 2. – Short note.

156.003 **A dynamic equation for stochastic magnetic field lines in the Galaxy.** F. C. Jones. Rep. NASA–TM–X–65514, National Aeronautics and Space Administration, Greenbelt, Maryland. [Available from NTIS, Springfield, Va.], 34 pp. (1971). – See Phys. Abstr., Vol. 75, No. 13927 (1972).

Interpretation of rotation measures of radio sources. See Abstr. 141.153.

Pulsar rotation and dispersion measures and the galactic magnetic field. See Abstr. 141.508.

Errata

156.901 **Erratum: "A dynamic equation for stochastic magnetic field lines in the Galaxy"** [Astrophys. Journ., Vol. 169, 477 - 485 (1971)]. F. C. Jones. Astrophys. Journ., Vol. 172, 525 (1972).

157 Galactic Radio Radiation

157.001 A systematic search for high-velocity hydrogen outside the galactic plane. I.
J. van Kuilenburg.
Astron. Astrophys., Suppl. Ser., Vol. 5, 1 - 19 (1972).

A survey is presented of neutral hydrogen with radial velocities between -270 and -60 kms^{-1} at positive declinations and radial velocities between $+60$ and $+270$ kms^{-1} at negative declinations. The driftscans on which the survey is based were separated by $2°5$ in declination. The region $-15° < b^{II} < 15°$ is not included in this survey.

157.002 Berkeley survey of high-velocity interstellar neutral hydrogen. I. The section $b = -15°$ to $+15°$, $l = 10°$ to 250°. N. H. Dieter.
Astron. Astrophys., Suppl. Ser., Vol. 5, 21 - 80 (1972).

Results of a survey of high-velocity interstellar neutral hydrogen are presented in the form of contour diagrams of antenna temperature as a function of galactic latitude and velocity. This is the first of two sections to the survey and includes observations near the galactic plane.

157.003 A survey of linear polarization at 1415 MHz.
I. Method of reduction and results for the North Polar Spur. T. A. T. Spoelstra.
Astron. Astrophys., Suppl. Ser., Vol. 5, 205 - 238 (1972).

The reduction of observations of linear polarization of the galactic background radiation in the direction of the North Polar Spur at 1415 MHz is discussed. The antenna pattern shows some asymmetry. It may be assumed that systematic effects that could be introduced by the observational method can be neglected. The most important effect to correct for is the azimuth- and elevation-dependent spurious polarization. The results are presented in a table.

157.004 A survey of the continuum radiation at 820 MHz between declinations $-7°$ and $+85°$. I. Observations and reductions. E. M. Berkhuijsen.
Astron. Astrophys., Suppl. Ser., Vol. 5, 263 - 311 (1972).

Contour maps of absolute brightness temperature at 820 MHz are presented being the result of a fully sampled continuum survey with the Dwingeloo telescope. The resolution of the map is $1°2$. Correction for radiation into the antenna sidelobes has been made. Polarization percentages are given in a table.

157.005 A systematic search for high-velocity hydrogen outside the galactic plane. II. J. v. Kuilenburg.
Astron. Astrophys., Vol. 16, 276 - 281 (1972).

A survey is presented of neutral hydrogen with radial velocities between -270 and -60 km/s and $+60$ and $+270$ km/s.

157.006 New measurements of the cosmic noise background spectrum down to 200 kHz.
L. W. Brown, J. K. Alexander.
Bull. American Astron. Soc., Vol. 3, 469 (1971). – Abstr. AAS.

157.007 RAE-1 satellite observations of the southern Milky Way at 3.93 MHz. J. K. Alexander, L. W. Brown.
Bull. American Astron. Soc., Vol. 3, 469 - 470 (1971).
Abstr. AAS.

157.008 A survey of positive velocity neutral hydrogen above the galactic plane between $l = 252°$ and $l = 322°$.
P. Wannier, G. T. Wrixon, R. W. Wilson.
Astron. Astrophys., Vol. 18, 224 - 231 (1972).

A survey is presented of positive velocity hydrogen in the range $+18$ to $+334$ km/s occupying the region of the sky between $l = 252°$ and $322°$ and $b = +10°$ and $+30°$. A number of high velocity features discovered by other observers are seen as are several others previously undetected high velocity complexes. The possibility is discussed that these are high latitude extensions of spiral arms lying far outside our own arm.

157.009 New theory for giant loops.
J. C. Brandt, S. P. Maran.
Nature, Vol. 235, 38 - 39 (1972).

The galactic radio spurs, including the North Polar Spur and the Cetus Arc, are ancient fossil Strömgren spheres (FSS), in a more advanced evolutionary state than that of the Gum nebula. Thus, they are not supernova remnants, but rather are artifacts of the ionization of the interstellar medium by supernovae. FSS expand into their cooler surroundings and the associated hydromagnetic shocks compress the ambient cosmic ray gas and generate the observed radio emission by the van der Laan mechanism.

157.010 High velocity hydrogen apparently associated with continuum jets near the galactic center.
R. H. Sanders, G. T. Wrixon.
Astron. Astrophys., Vol. 18, 467 - 470 (1972).

Observations at 21-cm of the distribution of high velocity hydrogen near the center of the Galaxy are presented. Two hypotheses for the origin of the features reported are discussed.

157.011 An unusual high-velocity hydrogen feature.
P. Wannier, G. T. Wrixon.
Astrophys. Journ., (*Letters*), Vol. 173, L119 - L123 (1972).

21-cm observations have been made of an unusual high-velocity feature in the region of the south galactic pole. Along a single narrow cloud some 60° in length we see a systematic variation of the radial velocity in the range -400 km s$^{-1} < V_r < -60$ km s^{-1}. The velocity behavior is suggestive of rigid motion.

157.012 Galactic radio emission in the 21-cm line and the continuum. G. Westerhout.
Galactic astronomy, Vol. 1, (see 012.019), 147 - 190 (1970).

157.013 Beobachtungen der H- und He-Radiorekombinationslinien in der Umgebung des galaktischen Zentrums.
P. G. Mezger, E. Churchwell.
Mitt. Astron. Ges., No. 31, p. 100 (1972). – Abstract.

157.014 The distortion of the galactic plane emission in the anticentre region. J. Milogradov-Turin.
Publ. Dep. Astron. Univ. Beograd, No. 3, p. 25 - 28 (1971).

An attempt to interpret the radio continuum brightness in the anticentre region is made taking as the basis the right - handed helical model of the local magnetic field. It is shown that the shape of the contours of brightness distribution can be well explained by this model and the assumption that part of the radiation comes from the neighbouring region of the Perseus arm.

Das Spektrum der Radiostrahlung.
See Abstr. 033.025.

The influence of the ionized medium on synchrotron emission in interstellar space. See Abstr. 131.094.

Galactic background at 15 GHz.
See Abstr. 155.088.

158 Single and Multiple Galaxies

158.001 High resolution observations of neutral hydrogen in M33–I. The hydrogen distribution.
M. C. H. Wright, P. J. Warner, J. E. Baldwin.
Monthly Notices Roy. Astron. Soc., Vol. 155, 337 - 356 (1972).

Observations of the galaxy M33 at 21-cm wavelength have enabled the neutral hydrogen to be mapped with an angular resolution of 1.5×3.0 and a velocity resolution of 39 km s^{-1}. The distribution of neutral hydrogen is roughly uniform out to a distance of 5 kpc from the nucleus and it is approximately coextensive with the visible disk of the galaxy. H I features are associated with the stellar spiral structure. Two H I extensions at the ends of the major axis may be due to warping of the hydrogen plane.

158.002 Inclination and absorption effects on the apparent diameters, optical luminosities and neutral hydrogen radiation of galaxies – I. Optical and 21-cm line data.
J. Heidmann, N. Heidmann, G. de Vaucouleurs.
Mem. Roy. Astron. Soc., Vol. 75, 85 - 104 (1972). – Abstr. in Monthly Notices Roy. Astron. Soc., Vol. 155, 505 (1972).

The observed optical luminosities and apparent diameters of galaxies, and their 21-cm H I line emission are subject to systematic effects depending on inclination and on galactic and internal absorption. In order to study these effects we have considered (1) all major sets of homogeneous optical data and derived standard photometric parameters for 64 galaxies of all types, of which 36 are previously unpublished results of the Mt Stromlo—McDonald program, and (2) data on 21-cm line emission in a uniform system for 92 spiral and irregular galaxies. The combination of optical and radio data eliminates the danger of circular arguments and resolves several important questions on systematic effects in optical data.

158.003 Inclination and absorption effects on the apparent diameters, optical luminosities and neutral hydrogen radiation of galaxies – II. Empirical properties.
J. Heidmann, N. Heidmann, G. de Vaucouleurs.
Mem. Roy. Astron. Soc., Vol. 75, 105 - 119 (1972). – Abstr. in Monthly Notices Roy. Astron. Soc., Vol. 155, 506 (1972).

The data on optical diameters and luminosities and neutral hydrogen emission of 92 galaxies presented in Paper I are empirically analyzed to estimate the optical effects of inclination and absorption in spirals.

158.004 Inclination and absorption effects on the apparent diameters, optical luminosities and neutral hydrogen radiation of galaxies – III. Theory and applications.
J. Heidmann, N. Heidmann, G. de Vaucouleurs.
Mem. Roy. Astron. Soc., Vol. 75, 121 - 141 (1972). – Abstr. in Monthly Notices Roy. Astron. Soc., Vol. 155, 507 - 508 (1972).

In this third paper we present a theoretical analysis and some applications of the effects of absorption on apparent luminosities and diameters discussed in Paper II.

158.005 Emission-line intensities and *UBV* magnitudes for twenty-three Markarian galaxies. D. W. Weedman.
Astrophys. Journ., Vol. 171, 5 - 12 = Arthur J. Dyer Obs. Vanderbilt Univ., Nashville, Repr. No. 62 (1972).

Intensities for the [O III] λλ4959, 5007 emission lines were measured with a photoelectric spectrum scanner to calibrate spectrophotometry for 23 representative objects from Markarian's list of galaxies with strong ultraviolet continua. The 23 objects include 10 galaxies that resemble the Seyfert galaxies, two galaxies with strong but narrow nuclear emission lines, and 11 diffuse galaxies. *UBV* magnitudes are given for 24 Markarian galaxies.

158.006 On the integral-sign galaxy MCG 12-7-28.
H. B. Richer, S. Sharpless, B. I. Olson.
Astrophys. Journ., Vol. 171, 13 - 16 (1972).

Some properties of a peculiar galaxy with the apparent shape of an integral sign are discussed.

158.007 Search for faint companions to M31.
S. van den Bergh.
Astrophys. Journ., (*Letters*), Vol. 171, L31 - L33 (1972).

Three faint objects that are probably Sculptor-like dwarf galaxies have been found near the Andromeda galaxy.

158.008 Variations of the hydrogen lines emitted by the nucleus of IC 450. M.-H. J. Ulrich.
Astrophys. Journ., (*Letters*), Vol. 171, L35 - L36 (1972).

The component of the Balmer lines blueshifted by 3000 km s^{-1} from the systemic velocity of IC 450 and observed in 1969 and 1970 by Khachikian and Weedman is undetectable on spectra we have taken in 1971 February and October.

158.009 Gas motions in the center of the Seyfert galaxy NGC 7469. M.-H. J. Ulrich.
Astrophys. Journ., (*Letters*), Vol. 171, L37 - L40 (1972).

Image-tube spectra of the nucleus of NGC 7469, taken with a dispersion of 28 Å mm^{-1}, show that the line [N II] λ6583 and the narrow core of Hα are resolved into two components 3.5 Å wide and separated by 3.8 Å. The observations are interpreted as evidence for the presence of two gas clouds partly overlapping along the line of sight.

158.010 On the ejection of gas from active nuclei of galaxies.
I. S. Shklovsky.
Astrophys. Letters, Vol. 10, 5 - 7 (1972).

The distribution of the filaments of gas in the 'counter-jet' of the radio galaxy NGC 4486 shows that the motion of gas is governed by regular outer magnetic fields parallel to the rotation axis of the nucleus of this galaxy. Such magnetic fields may originate from an interaction of magnetized clouds of relativistic particles with the interstellar medium. These clouds appear to be ejected in the direction of the rotation axis of the nucleus. The 'fan jet' in NGC 4486 is explained by an activity of the equatorial regions of the nucleus. A similar activity is responsible for the halo of NGC 4486. This halo must be a comparatively flat system that has an elliptical outline due to projection effects.

158.011 The X-ray flux variations from M 87.
G. Cavallo, A. Messina.
Astrophys. Letters, Vol. 10, 61 - 65 (1972).

The hypothesis that the X-ray flux variations of M 87 are due to supernova outbursts occurring in the Virgo cluster is shown to be consistent with the observations. Intrinsic X-ray luminosity, total energy emitted in X-rays, and outburst rate can then be expressed as functions of the single parameter τ, the X-ray emission decay time. Reasonable values are obtained for $\tau \simeq 15$ days and $\tau \simeq 30$ days. The last value would suggest that mostly Type I supernovae are responsible for this X-ray emission. It is unlikely that supernovae make a substantial contribution to the diffuse X-ray background.

158.012 A spectroscopic study of the irregular galaxy NGC 3067. I. J. Danziger, F. R. Chromey.
Astrophys. Letters, Vol. 10, 99 - 104 (1972).

A spectroscopic analysis of the type II irregular galaxy NGC 3067 shows it to have an interstellar gas component of low density. Rotation curves give a mass of $1.8 \times 10^{10} M_\odot$ in the inner one third of the visible galaxy. A mass density of $1.5 M_\odot$ pc^{-3} is higher than average but the mass-to-luminosity ratio 6.8, and the absolute visual magnitude -19.2 are consistent with mean values quoted for type II irregular galaxies.

158.013 Non-velocity redshifts in galaxies.
G. de Vaucouleurs, A. de Vaucouleurs.
Nature, Vol. 236, 166 (1972). – Letter.

158.014 The mass-luminosity ratio of spiral galaxies.
B. M. Lewis.
Astron. Astrophys., Vol. 16, 165 - 171 (1972).
The rise in the mass-luminosity ratio in the outer parts of spiral galaxies can be explained if most of the mass is in the form of neutral hydrogen, with a temperature of 1 to $10°$K. A model of NGC 300 with 87% of the mass in the form of a disk of cold gas is shown to provide an adequate explanation of the 21 cm observations.

158.015 The nonstellar continuum of the Seyfert galaxy NGC 1068. N. Kaneko.
Publ. Astron. Soc. Japan, Vol. 24, 145 - 148 (1972).
We show that the starlike nucleus in NGC 1068 is an emitter of a nonstellar continuum. The result reveals that the continuous spectrum agrees with the energy distribution of a hot black body.

158.016 Galaxies with complex blue nuclei.
B. Vorontsov-Velyaminov, G. Zaitseva, V. Lyutyj.
Astron. Zhurn. Akad. Nauk SSSR, Vol. 49, 93 - 96 (1972). In Russian. English translation in Soviet Astron. AJ, Vol. 16, No. 1.
We stress the cosmogonic significance of cases when inside a region composed of cold stars in the inner part of the flat component there is a nuclear region composed of hot blue stars. Some parameters of the nuclei are evaluated and a list of galaxies with complex blue nuclei is presented.

158.017 On the absorption of light in galaxies.
B. I. Fessenko.
Astron. Zhurn. Akad. Nauk SSSR, Vol. 49, 97 - 101 (1972). In Russian. English translation in Soviet Astron. AJ, Vol. 16, No. 1.
Four statistical methods for estimating the light absorption in galaxies with known inclination classes are considered.

158.018 The peculiar extragalactic system NGC 6438.
E. M. Burbidge, G. R. Burbidge.
Astrophys. Journ., Vol. 171, 253 - 255 (1972).
The peculiar interacting system NGC 6438 has been investigated at the Cerro Tololo Inter-American Observatory. It is shown that the irregular component and the S0 component have about the same redshift velocity of +2400 km s^{-1}, so that the large velocity differences originally reported by Sersic are not confirmed.

158.019 The stellar content of the nuclei of nearby galaxies. II. A note on NGC 4594.
H. Spinrad, B. J. Taylor.
Astrophys. Journ., Vol. 397 - 401 (1972).
Scans of the nuclear regions of the giant Sa galaxy NGC 4594 have been obtained on the Spinrad-Taylor photometric system. A synthesis of the galaxy spectrum in terms of stellar constituents yields a population model similar to, but containing somewhat redder stars than the nuclei of M 31 and M 81.

158.020 Helligkeitsschätzungen von 7 Markarian-Galaxien auf photographischen Himmelsaufnahmen.
G. Jackisch.
Astron. Nachr., Vol. 293, 175 - 178 (1971/72).
Seven Markarian galaxies were investigated for optic variability on plates taken with the Sonneberg astrographs. The variability of the galaxy no. I was discovered and the significance confirmed. No changes in brightness, larger than the mean error of ±0.10 mag could be found for the other galaxies.

158.021 On the distribution of angular momenta of galaxies.
M. Reinhardt.
Monthly Notices Roy. Astron. Soc., Vol. 156, 151 - 163 (1972).
Available observations of position angles and axial ratios of galaxies are discussed. There are statistically significant departures from the distributions corresponding to a random distribution of angular momenta, and there is evidence that these departures cannot be ascribed to selection effects. Connections between the distribution of angular momenta in aggregations of galaxies and of the aggregations as a whole are indicated. Methods in preparing catalogues of position angles and axial ratios are discussed.

158.022 Photometry of compact galaxies.
B. S. P. Shen, P. D. Usher, J. W. Barrett.
Astrophys. Journ., Vol. 171, 457 - 461 (1972).
Photometric histories of the N galaxies 3C 390.3 and PKS 0521−36 have been determined from plates of the Harvard collection. Four other compact galaxies (Markarian 9, I Zw 92, II Zw 136, III Zw 77) show no evidence of historical variability beyond the uncertainties of the photometry.

158.023 The color change of a population model based on the M32 galaxy over the four-billion-year look-back time to 3C 295. H. Spinrad.
Astrophys. Journ., Vol. 171, 463 - 466 (1972).
The systematic change with age of the stellar content of the Spinrad-Taylor M32 model is used to simulate the change in the color of the integrated light of a giant E galaxy over the look-back time of 4×10^9 years, appropriate for the distant radio galaxy 3C 295.

158.024 The spectra of two galaxies near PKS 2251 + 11.
L. B. Robinson, E. J. Wampler.
Astrophys. Journ., (Letters), Vol. 171, L83 - L86 = Contr. Lick Obs., No. 359 (1972).
Redshifts for two galaxies in the group associated with the quasi-stellar source PKS 2251 + 11 have been determined from the measurements of the H and K absorption lines of Ca II.

158.025 The influence of fast particles on the optical emission line spectrum of a nebula: The Balmer decrement. S. M. V. Aldrovandi, D. Péquignot.
Astron. Astrophys., Vol. 17, 88 - 96 (1972).
We study the effect of a monoenergetic flux of relativistic electrons on the relative intensity of the Balmer lines of a nebula. This effect is first evaluated taking only into account its influence through excitation of the gas, without solving ionization and thermal balances, so that the electronic temperature and the ionization degree are given parameters. Secondly, for a pure hydrogen gas, assuming that the relativistic electrons are the only source of energy, we solve ionization and thermal balances.

158.026 The case for correlated optical and radio variability of the Seyfert galaxy 3C 120. P. D. Usher.
Astrophys. Journ., (Letters), Vol. 172, L25 - L27 (1972).
The history of the optical variability of 3C 120 suggests that the occurrence of an optical event at epoch 1967.5 ± 0.25 is the precursor to the radio event II documented by Pauliny-Toth and Kellermann. Delay times for radio variability

following optical variability are predicted.

158.027 Limitations on thermal and nonthermal models for the radiation from extragalactic sources.
T. W. Jones, P. J. Kellogg.
Astrophys. Journ., Vol. 172, 283 - 298 (1972).

Both thermal and nonthermal models for the radiation from QSOs and Seyfert galaxies are discussed in terms of the size limitations imposed by the models. It is concluded that if daily fluctuations in the luminosity of Seyfert galaxies are confirmed and do represent a maximum size for the sources, then neither dust models nor ordinary synchrotron models can explain the radiation. Colgate's proposal involving the production of infrared radiation by a nonthermal plasma is considered, and a more detailed theory is derived which indicates that the process cannot produce infrared radiation.

158.028 Identification of the nucleus of NGC 5128.
W. E. Kunkel, H. V. Bradt.
Bull. American Astron. Soc., Vol. 3, 444 (1971). – Abstr. AAS.

158.029 X-ray observations of NGC 5128 (Centaurus A) from Uhuru.
H. Tananbaum, E. Kellogg, H. Gursky, R. Giacconi.
Bull. American Astron. Soc., Vol. 3, 444 (1971). – Abstr. AAS.

158.030 High resolution photographs of the nuclei of M 31 and M 32 obtained by Stratoscope II.
R. E. Danielson, E. S. Light, M. Schwarzschild, M. G. Tomasko.
Bull. American Astron. Soc., Vol. 3, 445 (1971). – Abstr. AAS.

158.031 New meter-wavelength radio observations of M 31.
J. M. Durdin, Y. Terzian.
Bull. American Astron. Soc., Vol. 3, 445 (1971). – Abstr. AAS.

158.032 Mapping the distribution of polarized radiation of radio galaxies at 2695 and 8085 MHz.
E. Fomalont, M. C. H. Wright.
Bull. American Astron. Soc., Vol. 3, 446 (1971). – Abstr. AAS.

158.033 Hot Seyfert winds. A. M. Wolfe.
Bull. American Astron. Soc., Vol. 3, 474 (1971).
Abstr. AAS.

158.034 Line profiles in the elliptical galaxies NGC 1889, 3115, 4473, and 4494.
D. C. Morton, R. A. Chevalier.
Bull. American Astron. Soc., Vol. 3, 476 (1971). – Abstr. AAS.

158.035 IRC-20385, the nearest galaxy or the richest globular cluster? E. Becklin, G. Neugebauer, H. Spinrad, I. King, H. E. Smith. R. Stone.
Bull. American Astron. Soc., Vol. 3, 498 (1971). – Abstr. AAS.

158.036 Radial velocities of galaxies.
G. Chincarini, H. J. Rood.
Astron. Journ., Vol. 77, 4 - 8 (1972).

We present a list of radial velocities of galaxies derived from spectrograms obtained primarily in 1970 with the Carnegie image tube Cassegrain spectrograph attached to the KPNO 84-inch telescope. The text describes our reduction procedures. The sample of galaxies is non-homogeneous, representing parts of various special investigations on groups and clusters of galaxies.

158.037 A neutral hydrogen study of the galaxy NGC 253.
W. Huchtmeier.
Astron. Astrophys., Vol. 17, 207 - 214 (1972).

Observations of the 21-cm line of the galaxy NGC 253

yield a H I mass of $7.8 \times 10^9 \, M_\odot$ or nearly 6% of the total mass of $1.3 \times 10^{11} \, M_\odot$ for a distance of 3.4 Mpc. The integrated H I-contours indicate an asymmetry in the distribution of neutral hydrogen. Parameters of the H I-distribution and the rotation curve are deduced by a model-fitting procedure.

158.038 21-cm line study of the peculiar galaxy NGC 5253.
L. Bottinelli, L. Gouguenheim, J. Heidmann.
Astron. Astrophys., Vol. 17, 445 - 449 (1972).

The properties of the H I gas content are characteristic of an early-type galaxy, most probably lenticular. This type is consistent with the large-scale stellar distribution.

158.039 A high resolution radio continuum survey of M 51 and NGC 5195 at 1415 MHz.
D. S. Mathewson, P. C. van der Kruit, W. N. Brouw.
Astron. Astrophys., Vol. 17, 468 - 486 (1972).

The most striking feature of the radio map is the clear delineation of two radio spiral arms which lie along the inside edges of the bright optical arms and coincident with the dust lanes. The nuclei of M 51 and NGC 5195 are strong radio sources and intrinsically much more intense than the galactic centre. Several discrete sources are resolved from the spiral arm emission and all are identified with H II regions. One of them is a supernova remnant candidate as its emission is too strong to be thermal in origin and it is linearly polarized. The interaction between M 51 and NGC 5195 is evident in the radio picture.

158.040 Wolf-Rayet stars in M 33.
J. D. Wray, G. J. Corso.
Astrophys. Journ., Vol. 172, 577 - 582 (1972).

Twenty-five objects in M33 have been identified as Wolf-Rayet stars on the basis of narrow-band interference-filter photography.

158.041 The hydrogen lines in Markarian 6.
T. F. Adams.
Astrophys. Journ., (Letters), Vol. 172, L101 - L103 (1972).

Results of new observations are presented showing that the second component in the hydrogen lines in Markarian 6 was still present in 1971 April and November. This is in conflict with Ulrich's report that the second component had disappeared. An attempt is made to relate the N galaxy 3C 390.3 and the Seyfert galaxy NGC 5548 to Markarian 6.

158.042 Correlation between the intrinsic radiopower and the spectral index of radiogalaxies.
M. P. Véron, P. Véron, A. Witzel.
Astron. Astrophys., Vol. 18, 82 - 91 (1972).

It was found by several authors that there is a statistical trend for radiogalaxies with a high monochromatic intrinsic power to have a steeper spectral index. It is shown that, if we restrict our sample to "elliptical radiogalaxies", that is elliptical galaxies identified with a radiosource with two or more components, none of them being associated with the nucleus of the galaxy, there is a good correlation between the two above mentioned parameters.

158.043 21-cm line study of the heavily obscured galaxy IC 10.
L. Bottinelli, L. Gouguenheim, J. Heidmann.
Astron. Astrophys., Vol. 18, 121 - 125 (1972).

The heavily obscured galaxy IC 10 has been observed in the 21-cm line of neutral hydrogen and the neutral hydrogen diameter is measured along the East-West direction. Using a statistical relation between the H I diameter and the optical one, an apparent optical diameter of about 17' is deduced. The distance of IC 10 is estimated to be (2.9 ± 1.5) Mpc. A latitude-velocity diagram indicates that the galaxies in the UMa-Cam cloud and in our Local Group may constitute a chain or an elongated complex of galaxies.

158.044 Disk and ring structure in the universe.
S.-S. Huang.
Sky Telescope, Vol. 43, 225 - 228 (1972).

158.045 Kompakte Galaxien. J. Hoffmann.
Umschau, 72. Jahrgang, p. 7 - 11 (1972).

There are three teams of astronomers investigating compact galaxies: Zwicky, USA; Richter, GDR; Markarian, USSR. The most significant findings are: absolute brightness of 84 % of the objects is -20^M0, the diameters vary between 0.2 and 10 kiloparsec.

158.046 Multicolour photometry of the double system NGC 5194/5195.
I. I. Pronik, K. K. Chuvaev.
Izv. Krymskoj Astrofiz. Obs., Vol. 43, 101 - 112 (1971). In Russian.

The prime-focus image-converter on the 2.6-meter Shajn telescope has been used to take photographs of the double system NGC 5194/5195 through 6–9 filters. The results are presented in detail.

158.047 Energy of galaxies.
Aviatsiya i kosmonavtika (NRB), Vol. 13, No. 6, p. 1 - 3 (1971). In Bulgarian.

158.048 Exploding galaxies. B. Ilieva.
Kosmos (NRB), 1971, No. 8, p. 6 - 8. In Bulgarian.

158.049 The spectra of 80 galaxies in Markarian's second list and the space density of the Markarian galaxies.
W. L. W. Sargent.
Astrophys. Journ., Vol. 173, 7 - 24 (1972).

Low-dispersion spectroscopic observations have been made of 80 objects in Markarian's second list of galaxies with an abnormally strong ultraviolet continuum. Redshifts have been measured for 68 objects which include one bright quasi-stellar object (Markarian 132) and 62 emission-line galaxies. These include four galaxies of the Seyfert type (Nos. 79, 106, 124 and 176) and six members of a class of dwarf galaxies with strong, high-excitation, emission lines described by Sargent and Searle. Markarian's first three lists contain 302 objects, distributed over 2000 square degrees of the sky. We show that the lists are reasonably complete down to m_p = 15.5 while at m_p = 16.5 they are incomplete by a factor of about 5. Redshifts obtained by various authors have been used to derive absolute magnitudes M_p for 61 out of 70 galaxies in list I and for 94 out of 130 in list II. Down to m_p = 15.5 redshifts are available for 109 out of 125 galaxies in these two lists. The data have been used to derive the space density $\Phi(M_p)$, in units of Mpc^{-2} per unit magnitude interval, for Markarian galaxies with $-13 \geq M_p \geq -22$. The results are given in a table and are then compared with estimates of $\Phi(M_p)$ for field galaxies. We discuss some consequences of these estimates of space density.

158.050 Inferences from the composition of two dwarf blue galaxies. L. Searle, W. L. W. Sargent.
Astrophys. Journ., Vol. 173, 25 - 33 (1972).

The emission spectra of two dwarf compact galaxies, I Zw 18 and II Zw 40, which were earlier described as "isolated extragalactic H II regions," have been analyzed. It is shown that most of the mass in I Zw 18 and II Zw 40 is probably in the form of interstellar hydrogen gas. The observed colors are used with the composition data to infer that the present rate of star formation exceeds the past average rate. It is shown that most of the mass in I Zw 18 and II Zw 40 is probably in the form of interstellar hydrogen gas. The observed colors are used with the composition data to infer that the present rate of star formation exceeds the past average rate. It is argued that the galaxies are either young (in the sense that most of their star formation has occurred in recent times) or that the star formation in them occurs in intense bursts which are separated by long quiescent periods.

158.051 Optical study of nearby galaxies. H. D. Ables.
Publ. United States Naval Obs., *Washington,* Second Ser., Vol. 20, Part 4, 126 pp. (1971).

The present study was undertaken to derive optical parameters for several galaxies which have been studied in detail by radio astronomers. This study presents the results of an isophotometric investigation of six nearby galaxies, IC 356, NGC 6946, NGC 1569, IC 342, IC 1613, and A 1009.

158.052 Gaseous ejections in NGC 4486 and problems of galactic nuclei activity. I. S. Shklovsky.
Astron. Zhurn. Akad. Nauk SSSR, Vol. 49, 233 - 242 (1972). In Russian. English translation in Soviet Astron. AJ, Vol. 16, No. 2.

The question about the nature of the "conter-jet" and the so-called "fan-jet" in NGC 4486 is discussed. The former is generated by ejection of substance in the direction of the axis of rotation of some compact body (nucleus). The "fan-jet" was formed during an explosion in the equatorial region of a nucleus when the duration of the explosion was about $^1/_3$ of the rotation period. The question is discussed about the evolution of planetary nebulae generated in NGC 4486 with small specific momentum ($\sim$ 3 M$_\odot$ per year).

158.053 Seyfert galaxies. E. A. Dibai.
Priroda, No. 1.72, p. 11 - 19 (1972). In Russian.

158.054 An attempt at detecting "protogalaxies" from fluctuations in the radio brightness of the Metagalaxy.
T. B. Pyatunina.
Astrofiz. Issled. Izv. Spets. Astrofiz. Obs., Vol. 3, 128 - 134 (1971). In Russian.

From observations made at 3.95 cm the upper limit to the fluctuations in the radio brightness of the Metagalaxy is estimated. A comparison is made of the observations with the fluctuations in the radio brightness of the Metagalaxy calculated on the basis of several hypotheses concerning the formation of galaxies. An estimate is also obtained of the upper limit to the fluctuations in the radio brightness of the polarized relict radiation at 6.6 cm.

158.055 Distribution of gas and dust in M 81.
L. P. Connolly, P. Z. Mantarakis, L. A. Thompson.
Publ. Astron. Soc. Pacific, Vol. 84, 61 - 63 (1972).

A new determination of M 81's inclination angle (58° ± 2°) allows us to obtain a face-on view of the galaxy in both dust and gas components. A logarithmic spiral is fitted to the dust diagram, and statistics on the gas distribution are presented.

158.056 Blue condensations associated with elliptical and S0 galaxies. A. Stockton.
Astrophys. Journ., Vol. 173, 247 - 256 (1972).

Spectroscopic observations have been obtained for a number of systems for which one or more faint blue knots are found in the vicinity of an elliptical or S0 galaxy. Of particular interest are NGC 3561 and IC 1182, which have luminous bridges connecting the galaxies and the blue knots.

158.057 Further observations of Maffei 2 at 4.5 and 2.8 centimeters. M. B. Bell, E. R. Seaquist.
Astrophys. Journ., Vol. 173, 257 - 258 (1972).

We report here further measurements of the angular size, flux density, and position of Maffei 2 at 4.5 and 2.8 cm.

158.058 **The nucleus of M82.**
P. P. Kronberg, C. J. Pritchet, S. van den Bergh.
Astrophys. Journ., (*Letters*), Vol. 173, L47 - L50 (1972).
Radio observations with the NRAO interferometer have been combined with optical observations obtained with the 200-inch Hale telescope to study the structure of the nuclear region of M82.

158.059 **A first approximation to the effect of evolution on q_0.**
B. M. Tinsley.
Astrophys. Journ., (*Letters*), Vol. 173, L93 - L97 (1972).
Because the spectral energy distributions of nearby giant elliptical galaxies are inconsistent with a very steep initial luminosity function, their rate of evolution must be such as to move the magnitude-redshift relation to an apparent value of q_0 at least 0.5 greater than its true value.

158.060 **The luminosity−spectral index relationship for radio galaxies.**   J. M. MacLeod, L. H. Doherty.
Journ. Roy. Astron. Soc. Canada, Vol. 66, 66 (1972). − Abstr. Canadian Astron. Soc.

158.061 **Line profiles in the Seyfert galaxy NGC 1068.**
J. R. Auman, J. Eilek, T. J. Ulrych, G. A. H. Walker.
Journ. Roy. Astron. Soc. Canada, Vol. 66, 71 - 72 (1972).
Abstr. Canadian Astron. Soc.

158.062 **Effect of infalling matter on the heavy element content of a galaxy.**   R. B. Larson.
Nature, Phys. Sci., Vol. 236, 7 - 8 (1972).
It is shown that if the interstellar medium in a galaxy is replenished by a continuing inflow of intergalactic matter, its heavy element content tends to approach a constant limiting value which is independent of both the infall rate and the early history of the galaxy and depends only on stellar evolution processes.

158.063 **Galaxies with ultraviolet continuum. IV.**
B. E. Markarian, V. A. Lipovetsky.
Astrofizika, Vol. 7, 511 - 519 (1971). In Russian.
English translation in Astrophysics, Vol. 7, No. 4.
The fourth list of galaxies with ultraviolet continuum containing 99 objects is given.

158.064 **Morphology of some Markarian galaxies.**
A. T. Kalloghlian.
Astrofizika, Vol. 7, 521 - 528 (1971). In Russian.
English translation in Astrophysics, Vol. 7, No. 4.
A morphological study of 30 Markarian galaxies on plates obtained with the 2-meter telescope of the Tautenburg Observatory has been carried out.

158.065 **Stellar populations in galaxies.**   S. van den Bergh.
IAU Symposium No. 44, (see 012.005), p. 1 - 11 (1972).

158.066 **The gaseous content of galaxies.**   M. S. Roberts.
IAU Symposium No. 44, (see 012.005), p. 12 - 36 (1972).

158.067 **Structural and kinematic properties of populations of the Andromeda galaxy.**   J. Einasto.
IAU Symposium No. 44, (see 012.005), p. 37 - 45 (1972).

158.068 **The population I content of the elliptical companions of M31.**   P. W. Hodge.
IAU Symposium No. 44, (see 012.005), p. 46 - 47 (1972).

158.069 **Gas in the nucleus and disk of M31.**
V. C. Rubin, W. K. Ford, Jr.
IAU Symposium No. 44, (see 012.005), p. 49 - 55 (1972).

158.070 **Distribution of dust and H II regions in spiral galaxies.**   B. T. Lynds.
IAU Symposium No. 44, (see 012.005), p. 56 - 61 (1972).

158.071 **Spiral arm patches in Sc and SBc galaxies.**
I. Pronik, K. Chuvaev.
IAU Symposium No. 44, (see 012.005), p. 62 - 65 (1972).

158.072 **Observations of H II regions in Sc galaxies.**
L. Searle.
IAU Symposium No. 44, (see 012.005), p. 66 (1972). − Abstract.

158.073 **The neutral hydrogen distribution in spiral and irregular galaxies.**   R. D. Davies.
IAU Symposium No. 44, (see 012.005), p. 67 - 73 (1972).

158.074 **Absorption by neutral hydrogen in the irregular galaxy M82.**   M. Guélin, L. Weliachew.
IAU Symposium No. 44, (see 012.005), p. 74 (1972). − Abstract.

158.075 **Thermal radio emission from normal galaxies.**
Y. Terzian.
IAU Symposium No. 44, (see 012.005), p. 75 - 81 (1972).

158.076 **Mass-luminosity ratios and sizes of giant elliptical galaxies.**   I. R. King, R. Minkowski.
IAU Symposium No. 44, (see 012.005), p. 87 - 88 (1972).

158.077 **On the mass-to-light ratios for double galaxies.**
R. J. Dickens, J. V. Peach.
IAU Symposium No. 44, (see 012.005), p. 89 - 92 (1972).

158.078 **Do galaxies evolve along the R−ω sequence?**
W. C. Saslaw.
IAU Symposium No. 44, (see 012.005), p. 93 - 96 (1972).

158.079 **Classification of compact objects: QSS, QSOs, N-type and compact galaxies, Seyfert and galactic nuclei.**   W. W. Morgan.
IAU Symposium No. 44, (see 012.005), p. 97 - 103 (1972).

158.080 **Astrophysical statistics of 745 compact galaxies near the galactic north pole.**
L. Richter, N. B. Richter, P. Schneller.
IAU Symposium No. 44, (see 012.005), p. 104 - 108 (1972).

158.081 **Optical spectra of compact objects.**
E. M. Burbidge.
IAU Symposium No. 44, (see 012.005), p. 109 - 126 (1972).

158.082 **Spectral energy distributions of nuclei of peculiar galaxies.**   J. B. Oke.
IAU Symposium No. 44, (see 012.005), p. 139 - 150 (1972).

158.083 **The Balmer lines in the Seyfert galaxies NGC 5548 and NGC 4151.**   R. Weymann, R. Cromwell.
IAU Symposium No. 44, (see 012.005), p. 155 - 159 (1972).

158.084 **Physical conditions in Seyfert type galaxies Markarian 9, 10 and 42.**   E. Ye. Khachikian.
IAU Symposium No. 44, (see 012.005), p. 160 - 162 (1972).

158.085 **Infrared radiation from compact objects.**
G. Neugebauer.
IAU Symposium No. 44, (see 012.005), p. 163 (1972). − Abstract.

158.086 **Activity in the nuclei of Seyfert galaxies in the visual and infrared.**   A. G. Pacholczyk.

IAU Symposium No. 44, (see 012.005), p. 165 - 170 (1972).

158.087 **A study of the continua of the nuclei of galaxies.**
D. Alloin, Y. Andrillat, S. Souffrin.
IAU Symposium No. 44, (see 012.005), p. 188 - 189 (1972).

158.088 **Radio emission from compact objects.**
K. I. Kellermann.
IAU Symposium No. 44, (see 012.005), p. 190 - 213 (1972).

158.089 **M87 and the X-ray emission from compact objects.**
M. S. Longair.
IAU Symposium No. 44, (see 012.005), p. 249 - 250 (1972).

158.090 **Neutral hydrogen in compact galaxies.**
J. Heidmann.
IAU Symposium No. 44, (see 012.005), p. 264 - 265 (1972).

158.091 **Seyfert galaxies.** B. M. Lewis.
IAU Symposium No. 44, (see 012.005), p. 267 - 268 (1972).

158.092 **Radio observations of neutral hydrogen in four Seyfert galaxies.**
R. J. Allen, B. F. Darchy, R. Lauqué.
IAU Symposium No. 44, (see 012.005), p. 269 (1972).

158.093 **The law of momentum conservation and some problems of metagalactic astronomy.** I. S. Shklovsky.
IAU Symposium No. 44, (see 012.005), p. 272 - 276 (1972).

158.094 **A probable mechanism of repeated explosions of compact objects.** L. M. Ozernoy.
IAU Symposium No. 44, (see 012.005), p. 290 - 295 (1972).

158.095 **Supermassive disks.**
R. V. Wagoner, E. E. Salpeter.
IAU Symposium No. 44, (see 012.005), p. 300 - 305 (1972).

158.096 **Matter-antimatter annihilation as an energy source in Seyfert galaxies.**
G. Steigman, P. A. Strittmatter.
IAU Symposium No. 44, (see 012.005), p. 311 (1972).

158.097 **Transient annular structures in exploding galaxies.**
J. L. Sérsic.
IAU Symposium No. 44, (see 012.005), p. 313 (1972). – Abstract.

158.098 **Cosmological information from galaxies and radio galaxies.** J. V. Peach.
IAU Symposium No. 44, (see 012.005), p. 314 - 340 (1972).

158.099 **The radial velocities of galaxies near NGC 7331.**
C. R. Lynds.
IAU Symposium No. 44, (see 012.005), p. 376 - 379 (1972).

158.100 **Ejection of small compact galaxies from larger galaxies.** H. C. Arp.
IAU Symposium No. 44, (see 012.005), p. 380 - 392 (1972).

158.101 **The diameter-redshift relation.** W. A. Baum.
IAU Symposium No. 44, (see 012.005), p. 393 - 396 (1972).

158.102 **Galactic and metagalactic magnetic fields.**
K. Brecher.
IAU Symposium No. 44, (see 012.005), p. 520 - 525 (1972).

158.103 **The Seyfert galaxies NGC 1068 and NGC 1566.**
M. G. Smith, D. W. Weedman, H. Spinrad.
Astrophys. Letters, Vol. 11, 21 - 26 (1972).
Photometric, photographic and spectroscopic observations are presented to compare the Seyfert galaxies NGC 1068 and 1566, which are at nearly identical distances and have similar apparent diameters.

158.104 **The spectral index–luminosity relation for radio galaxies and quasi-stellar sources.**
A. H. Bridle, M. J. L. Kesteven, B. Guindon.
Astrophys. Letters, Vol. 11, 27 - 30 (1972).
Data on the spectra of radio galaxies and quasi-stellar sources between 100 MHz and 7 GHz are used to examine the spectral index-luminosity relation.

158.105 **Concentration indices of galaxies.**
C. W. Fraser.
Observatory, Vol. 92, 51 - 54 (1972).
Results of a major programme on surface photometry of galaxies indicate that the concentration indices depend upon morphological type. Attention is drawn to the results of the central surface brightness of the exponential component of spirals, which do not confirm an earlier assertion that this brightness has a constant value throughout the entire spiral sequence.

158.106 **Adriaan van Maanen and internal motions in spiral nebulae: A historical review.** N. S. Hetherington.
Quarterly Journ. Roy. Astron. Soc., Vol. 13, 25 - 39 (1972).

158.107 **mm-observations of the exploding galaxy M82.**
G. Feix.
Astron. Astrophys., Vol. 18, 481 - 483 (1972).
M82 has been mapped at 81 GHz by a resolution of 1.2'. The loop stability of solar prominences has been compared with galactic prominences of M82.

158.108 **The redshift-distance relation. I. Angular diameter of first-ranked cluster galaxies as a function of redshift: The aperture correction to magnitudes.** A. Sandage.
Astrophys. Journ., Vol. 173, 485 - 499 (1972).
Micrometric estimates of the apparent angular diameters of first-ranked ellipticals in clusters, obtained from a homogeneous series of plates, are tightly correlated with redshift. The growth curve of magnitude with measuring aperture is obtained from new photoelectric data. The aperture correction follows from this formulation as a function of z and q_0 alone for first-ranked ellipticals. A method for differential correction of the aperture function is developed that takes account of cosmic differences in intrinsic diameters and different q_0 values.

158.109 **The emission-line velocity field in M82.**
H. M. Heckathorn.
Astrophys. Journ., Vol. 173, 501 - 527 (1972).
A split-spectrum study of the apparent velocity field in the peculiar galaxy M82 (NGC 3034) has been completed. Velocity curves were measured from the emission lines of $H\alpha$, [N II], and [S II] on 20 spectra obtained by using fiber-optic image intensifiers, and a contour map of the emission-line velocity field has been constructed. The observations are interpreted from two standpoints.

158.110 **The absorption-line Seyfert galaxy Markarian 231.**
T. F. Adams, D. W. Weedman.
Astrophys. Journ., (*Letters*), Vol. 173, L109 - L111 (1972).
The spectral features of this galaxy are reported in detail.

158.111 **NGC 1068, 3C 273, and Scorpius X-1: Circular polarization disputed.**
J. C. Kemp, R. D. Wolstencroft, J. B. Swedlund.
Astrophys. Journ., (*Letters*), Vol. 173, L113 - L118 (1972).

Null results to within 0.05 percent were obtained for the circular polarization of NGC 1068 and 3C 273, and similarly for Sco X-1 except possibly on one occasion.

158.112 Rapid change in the visibility function of the radio galaxy 3C 120.
D. B. Shaffer, M. H. Cohen, D. L. Jauncey, K. I. Kellermann.
Astrophys. Journ., (*Letters*), Vol. 173, L147 - L150 (1972).

Second-epoch observations of the radio galaxy 3C 120 with the Goldstack interferometer show apparent rapid changes in structure similar to those observed for the quasars 3C 273 and 3C 279. Our earlier observations of 3C 84 and VRO 42.22.01 are confirmed.

158.113 Extremely compact galaxy CGCG 1622 + 4112.
H. Karoji, K. Kodaira.
Publ. Astron. Soc. Japan, Vol. 24, 239 - 246 (1972).

Spectra of the extremely compact galaxy CGCG 1622 + 4112 were obtained with an image-intensifier spectrograph ($\lambda\lambda 3900 - 6600$ Å, 220 Å mm^{-1}). By taking spectra of comparison stars with the same instrument, we were able to analyze the spectroscopic data quantitatively.

158.114 Active galaxies with radio trails in clusters.
G. K. Miley, G. C. Perola, P. C. van der Kruit, H. van der Laan.
Nature, Vol. 237, 269 - 272 (1972).

New radio data and a critical examination of the interacting galaxies hypothesis to explain complex radio sources in clusters lead us to reject a causal relation between sources in clusters. Instead, the "head-tail" radio galaxies are proposed to represent radio trails of galaxies, active in their own right, along trajectories through a dense intergalactic medium.

158.115 An observational approach to the problem of spiral structure. III.
M. E. Dixon, V. L. Ford, J. W. Robertson.
Astrophys. Journ., Vol. 174, 17 - 26 (1972).

The possibility of using the techniques of surface photometry to study the mechanism which produces spiral structure is examined, and a method is proposed which employs surface color ($U - B$) as an age parameter and ($B - V$) as an indicator of interstellar absorption. The method is tested by comparing surface colors of selected regions in the major arm of the LMC with an age parameter f based on the H-R diagrams of the regions.

158.116 Collapse in galactic nuclei. W. G. Mathews.
Astrophys. Journ., Vol. 174, 101 - 108 = Contr. Lick Obs., No. 351 (1972).

A time-dependent stability analysis for gas collapsing into galactic centers suggests that coherent objects much more massive than ordinary stars are likely to form.

158.117 Spiraalnevels en hun kernen. P. C. van der Kruit.
Hemel en Dampkring, Vol. 70, 71 - 85 (1972).

158.118 Is NGC 404 a variable object? E. H. Geyer.
Inform. Bull. Variable Stars, (I.A.U. Commission 27), Konkoly Obs., Budapest, No. 614, 3pp. (1972).

158.119 On the variability of NGC 404.
B. V. Kukarkin, N. E. Kurochkin.
Inform. Bull. Variable Stars, (I.A.U. Commission 27), Konkoly Obs., Budapest, No. 636 (1972).

158.120 Remark concerning NGC 404. L. Meinunger.
Inform. Bull. Variable Stars, (I.A.U. Commission 27), Konkoly Obs., Budapest, No. 638 (1972).

158.121 Note on the galaxy M51 and the galaxy NGC 404 (also M33). S. Gaposhkin.
Inform. Bull. Variable Stars, (I.A.U. Commission 27), Konkoly Obs., Budapest, No. 648 (1972).

158.122 Radio maps of Maffei 2. R. Love.
Nature, Phys. Sci., Vol. 235, 53 - 56 (1972).

Observations of Maffei 2 in the radio continuum at 1419.0 MHz with an angular resolution of 49".5 by 57".5, and in the neutral hydrogen line with an angular resolution of 3'.1 × 3'.6 and a velocity resolution of 39 km s^{-1} are reported. A central core and surrounding disc emission are found in the continuum. The HI observations lead to the conclusion that it is most likely an Sb spiral at a distance of 4 Mpc. The resulting neutral hydrogen mass is then 1.0×10^9 solar masses.

158.123 The redshifts of 27 radio galaxies.
J. B. Whiteoak.
Australian Journ. Phys., Vol. 25, 233 - 235 (1972). – Short communication.

158.124 The optical spectrum of the Seyfert galaxy 3C 120.
G. A. Shields, J. B. Oke, W. L. W. Sargent.
Bull. American Astron. Soc., Vol. 4, 208 (1972). – Abstr. AAS.

158.125 The dynamics of multiple radio galaxy explosions.
W. A. Christiansen.
Bull. American Astron. Soc., Vol. 4, 208 (1972). – Abstr. AAS.

158.126 New observations of Markarian galaxies.
D. W. Weedman.
Bull. American Astron. Soc., Vol. 4, 213 (1972). – Abstr. AAS.

158.127 Violent mass motions in Markarian 78.
T. F. Adams.
Bull. American Astron. Soc., Vol. 4, 213 (1972). – Abstr. AAS.

158.128 Spectra of extragalactic rings.
J. C. Theys, E. A. Spiegel, J. Toomre.
Bull. American Astron. Soc., Vol. 4, 213 - 214 (1972). – Abstr. AAS.

158.129 Stellar velocity fields in galaxies: NGC 5866.
S. M. Simkin.
Bull. American Astron. Soc., Vol. 4, 214 (1972). – Abstr. AAS.

158.130 Stellar motions in the nuclear disk of M31.
V. C. Rubin, W. K. Ford, Jr.
Bull. American Astron. Soc., Vol. 4, 214 (1972). – Abstr. AAS.

158.131 Internal kinematics of two compact galaxies.
R. W. O'Connell, R. P. Kraft.
Bull. American Astron. Soc., Vol. 4, 214 (1972). – Abstr. AAS.

158.132 Model of the encounter between NGC 5194 and 5195. A. Toomre, J. Toomre.
Bull. American Astron. Soc., Vol. 4, 214 - 215 (1972). – Abstr. AAS.

158.133 A search for circular polarization in extragalactic objects. K. H. Nordsieck.
Bull. American Astron. Soc., Vol. 4, 223 (1972). – Abstr. AAS.

158.134 Infrared observations of galaxies and quasi-stellar objects. F. Low, G. Rieke.
Bull. American Astron. Soc., Vol. 4, 223 - 224 (1972). – Abstr. AAS.

158.135 Infrared observations of the nucleus of NGC 253.
E. E. Becklin, G. Neugebauer.
Bull. American Astron. Soc., Vol. 4, 224 (1972). – Abstr. AAS.

158.136 Photometry of Messier 81.
J. C. Brandt, J. K. Kalinowski, R. G. Roosen.
Bull. American Astron. Soc., Vol. 4, 224 (1972). – Abstr. AAS.

158.137 A UBV photoelectric study of M 51.
M. S. Burkhead, R. K. Honeycutt.
Bull. American Astron. Soc., Vol. 4, 224 (1972). – Abstr. AAS.

158.138 10-color photometry of elliptical galaxies and globular clusters. S. M. Faber.
Bull. American Astron. Soc., Vol. 4, 224 (1972). – Abstr. AAS.

158.139 The Wing-Ford band at 9910 Å as a population indicator in M31. A. E. Whitford.
Bull. American Astron. Soc., Vol. 4, 230 (1972). – Abstr. AAS.

158.140 Evidence for short-lived phenomena in galaxies from real-time image orthicon surveys.
J. A. Hynek, J. R. Dunlap, R. J. Altizer.
Bull. American Astron. Soc., Vol. 4, 230 - 231 (1972). – Abstr. AAS.

158.141 Emission line profiles in the spectrum of the Seyfert galaxy NGC 1068. J. Eilek, J. R. Auman, T. J. Ulrych, G. A. H. Walker, L. Kuhi.
Bull. American Astron. Soc., Vol. 4, 231 (1972). – Abstr. AAS.

158.142 Activity in galaxies and quasars. P. A. Sturrock.
Bull. American Astron. Soc., Vol. 4, 232 (1972).
Abstr. AAS.

158.143 Collapse in galactic nuclei. W. G. Mathews.
Bull. American Astron. Soc., Vol. 4, 232 (1972).
Abstr. AAS.

158.144 An optical survey of southern galaxies with bright nuclei. M. G. Smith.
Bull. American Astron. Soc., Vol. 4, 237 - 238 (1972). – Abstr. AAS.

158.145 Scd galaxies: Variation of global properties as determined from 21-cm observations.
G. S. Shostak, D. H. Rogstad.
Bull. American Astron. Soc., Vol. 4, 238 (1972). – Abstr. AAS.

158.146 The NGC 520 chain of quasars and the "discrepant redshift" of cluster galaxies.
M. F. Barnothy, J. M. Barnothy.
Bull. American Astron. Soc., Vol. 4, 238 - 239 (1972). – Abstr. AAS.

158.147 Selection effects and the distribution of orientations of galaxies. D. L. Hawley, P. J. E. Peebles.
Bull. American Astron. Soc., Vol. 4, 239 (1972). – Abstr. AAS.

158.148 Galaxy distributions in the Zwicky catalogue.
M. G. Hauser, J. N. Bahcall.
Bull. American Astron. Soc., Vol. 4, 239 (1972). – Abstr. AAS.

158.149 The gas content of galaxies. W. J. Quirk.
Bull. American Astron. Soc., Vol. 4, 265 (1972).
Abstr. AAS.

158.150 Evidence for the density wave nature of the spiral structure in M51. R. B. Tully.
Bull. American Astron. Soc., Vol. 4, 267 (1972). – Abstr. AAS.

158.151 Tidal interaction of galaxies.
N. N. Kozlov, R. A. Siuniaev, T. M. Eneev.
Dokl. Akad. Nauk SSSR, Ser. Mat. Fiz., Vol. 204, 579 - 582 (1972). In Russian.

158.152 Some current studies of galaxies. P. W. Hodge.
Sky Telescope, Vol. 44, 23 - 27 (1972).

158.153 Un groupe Maffei de galaxies. J. Heidmann.
L' Astronomie, 86ᵉ année, p. 257 - 259 (1972).

158.154 1400-MHz survey of bright galaxies. J. Pfleiderer.
Mitt. Astron. Ges., No. 31, p. 180 - 182 (1972).

158.155 An irregular galaxy of the Local System.
Yu. N. Efremov.
Zemlya i Vselennaya, 1972, No. 3, p. 26 - 28. In Russian.

158.156 Turbulent plasma "piles" in the nuclei of galaxies.
S. A. Kaplan, V. N. Tsytovich.
Astron. Zhurn. Akad. Nauk SSSR, Vol. 49, 647 - 649 (1972). In Russian. English translation in Soviet Astron. AJ, Vol. 16, No. 3.
    The hypothesis is proposed that turbulent plasma piles can exist in the nuclei of galaxies and it is shown that the maximum of emissivity of these piles can be in the infrared region of spectra.

158.157 Ricerche extragalattiche in Italia. F. Bertola.
Atti XIII Riunione Soc. Astron. Italiana, Trieste 1969, (see 012.010), p. 125 - 128 (1970).

158.158 Polarization and velocity field in the galaxy M 82.
A. Elvius.
Astron. Astrophys., Vol. 19, 193 - 196 (1972).
    The model of M 82 tentatively suggested by Sandage (1971) on the basis of polarization measures in the Hα emission line is discussed. It is pointed out that the assumption of scattering of Hα rather than recombination in the filaments implies a new interpretation of the observed radial velocities.

158.159 The radio-continuum radiation from Messier 33.
Y. Terzian, V. Pankonin.
Astrophys. Journ., Vol. 174, 293 - 299 (1972).
    Radio observations of the spiral galaxy Messier 33 are presented at 318 and 606 MHz. These observations were performed with the Arecibo Observatory 1000-foot radio telescope. The radio spectrum of M33 is discussed, and it is shown that its spectral index is in the range −0.17 to −0.20. This somewhat flat spectrum suggests a mixture of thermal and nonthermal radiation.

158.160 Dust near the center of M31.
H. M. Johnson, M. M. Hanna.
Astrophys. Journ., (Letters), Vol. 174, L71 (1972).
    Specially processed copies of a plate show the dust.

158.161 Radial variations in the strengths of absorption features in the spectra of two early-type galaxies.
G. A. Welch, W. T. Forrester.
Astron. Journ., Vol. 77, 333 - 341 (1972).
    A spectrophotometric study based on image-tube spectrograms has been conducted of selected absorption features in the galaxies NGC 3115 (E7/S0) and NGC 4472 (E2). Significant radial variation – in the sense of decreasing strength with distance – occurs in the violet cyanogen bands at λ3883 and λ4216, the G band, the Mg I + MgH blend at λ5175, and the "D" line of neutral sodium at λ5892. A qualitatively consistent interpretation is that the relative abundance of metallic elements and of late dwarf stars is high in the nuclei of these galaxies.

158.162 The profiles of the hydrogen lines in NGC 3516 and NGC 5548. M.-H. J. Ulrich.
Astrophys. Journ., Vol. 174, 483 - 488 (1972).
    Profiles of the Hα line emitted by the nucleus of

NGC 3516 and NGC 5548 have been obtained from image-tube spectra with a dispersion of 28 Å mm^{-1}. Both Hα profiles are asymmetrical, the short-wavelength side being the more intense. The observed profiles are compared to an electron scattering profile and to the profile caused by an opaque sphere surrounded by an expanding envelope.

158.163   **Velocity dispersions in galaxies. I. The E7 galaxy
          NGC 7332.**   D. C. Morton, R. A. Chevalier.
Astrophys. Journ., Vol. 174, 489 - 498 (1972).

A coudé spectrum of the E7 galaxy NGC 7332 with 0.9 Å resolution from 4186 to 4364 Å was obtained with the Princeton SEC vidicon television camera and the Hale telescope. Comparisons with spectra of G and K giant stars, numerically broadened for various Maxwellian velocity distributions, give a dispersion velocity in the line of sight of 160 ± 20 km s^{-1} with the best fit at G8 III.

158.164   **The optically variable galaxy V 395 Herculis.**
          H. E. Bond.
Astrophys. Journ., (*Letters*), Vol. 174, L163 = Contr. Louisiana State Univ. Obs., *Baton Rouge,* No. 66 (1972).

The object V 395 Herculis appears to be a compact galaxy rather than a galactic star. Extremely rapid optical variations have been reported for this object. It is not known to be a radio source.

158.165   **The luminosity-spectral index relationship for radio
          galaxies.**   J. M. MacLeod, L. H. Doherty.
Nature, Vol. 238, 88 - 89 (1972).

The relationship between spectral index and luminosity for the 3CR extragalactic sources suggests that sources with high radio luminosity have, on the average, steeper spectra than do the low luminosity sources. We show here that even for radio galaxies this relationship holds only for a particular subgroup in the 3CR catalogue. For those galaxies with power law spectra there is a clear relationship between luminosity and spectral index, but for those galaxies which do not have simple power law spectra, luminosity is independent of spectral index.

158.166   **An unusual variable object near β And.**
          E. H. Geyer.
IAU Circ., No. 2380 (1972).

158.167   **Object near β And = NGC 404.**   C. B. Stephenson.
IAU Circ., No. 2382 (1972).

158.168   **A jet in NGC 5253.**
          J. L. Sérsic, G. Carranza, M. Pastoriza.
IAU Circ., No. 2413 (1972).

158.169   **Descriptive functions of the Andromeda galaxy.**
          J. Einasto, U. Rümmel.
Tartu Astron. Obs. Teated, No. 36, p. 55 - 63 (1972).

A new model of the Andromeda galaxy, M31, is proposed. Tables of functions characterizing the gravitational field of M31 are given.

158.170   **Morfología y fotometría de galaxias australes.**
          H. A. Dottori.
Bol. Inst. Mat., Astron., Fis., Univ. Nacional Córdoba, Vol. 3, No. 2, p. 5 - 23 (1971).

158.171   **Remarks on Sculptor type systems.**
          S. van Agt.
The evolution of population II stars, (see 012.024), p. 97 - 102 (1972).

158.172   **The structure of Seyfert nuclei.**
          M. J. Rees, W. L. W. Sargent.
Comments Astrophys. Space Phys., Vol. 4, 7 - 14 (1972).

The physical conditions of the gas in Seyfert nuclei are probably exceedingly complex and inhomogeneous. However, all the available spectral data even for NGC 4151, which is the best-studied of these objects, may be explicable in terms of a two component model: a comparatively tenuous region $\gtrsim$ 25 pc in size and a compact region which emits the broad wings. Although the spectral information on the broad wings is very sparse and imprecise, we shall discuss the possible properties of the latter region, and the cause of the line broadening.

158.173   **Radio survey of the spiral nebulae M31 and M33.**
          J. H. Spencer.
Quarterly Progr. Rep. Res. Lab. Electronics, Mass. Inst. Technology (*USA*), No. 103, p. 16 - 20 (1971).

Examines the ionized hydrogen in these two galaxies.

Extragalactic astronomy.   See Abstr. 003.137.

Evolution of elements in the galaxies.
See Abstr. 061.047.

The induced light pressure under astrophysical conditions.   See Abstr. 063.029.

345-micron ground-based observations of M17, M82, and Venus.   See Abstr. 113.002.

Classical cepheids: Cornerstone to extragalactic distances?   See Abstr. 122.152.

On the estimate of frequency of novae in the Andromeda nebula and the Galaxy.   See Abstr. 124.002.

Novae in M 31 discovered and observed at Asiago from 1963 to 1970.   See Abstr. 124.006.

On the radio sources in normal galaxies.
See Abstr. 141.017.

Low frequency, high resolution observations of Virgo A.   See Abstr. 141.021.

Redshifts of twenty radio galaxies.
See Abstr. 141.028.

Extragalactic radio sources.   See Abstr. 141.065.

Are quasars associated with bright galaxies?
See Abstr. 141.068.

The quasars and nuclei of galaxies: a single body or a stellar cluster?   See Abstr. 141.070.

The absorption-line QSOs.   See Abstr. 141.082.

The physical properties of the absorption envelopes of two QSOs.   See Abstr. 141.084.

The angular structures of some compact sources.
See Abstr. 141.088.

Compact radio sources in the nuclei of elliptical galaxies.   See Abstr. 141.090.

QSOs and radio galaxies – their spectra and time variations at radio frequencies.   See Abstr. 141.096.

Attempts to detect neutral hydrogen in compact objects.   See Abstr. 141.099.

Theoretical considerations of compact objects.

See Abstr. 141.101.

3C quasi-stellar objects and bright galaxies.
See Abstr. 141.126.

Radio sources in the Ohio Survey and Zwicky compact galaxies.   See Abstr. 141.127.

Radio sources grouped in the NGC 7331 and Stephan's Quintet area.   See Abstr. 141.150.

Observations of 3C 272.1 at 2.7 and 5.0 GHz.
See Abstr. 141.165.

A grouping of radio sources in the area of NGC 7331 and Stephan's Quintet.   See Abstr. 141.177.

Search for optical circular polarization in quasars and Seyfert nuclei.   See Abstr. 141.178.

X-ray galaxies.   See Abstr. 142.026.

Radio identifications of weak extragalactic X-ray sources.   See Abstr. 142.070.

The extended X-ray source at M87.
See Abstr. 142.108.

Virial mass determinations of bound and unstable groups of galaxies.   See Abstr. 151.014.

The ejection of massive objects from galactic nuclei: Interactions between the massive object and the galactic gas.
See Abstr. 151.069.

The gravitational field of flat galaxies.
See Abstr. 151.081.

Theory of spiral structure.
See Abstr. 151.091.

Scanner abundance studies. V. Globular clusters in M31.   See Abstr. 154.006.

The identity and aliases of the stellar system Terzan 5.   See Abstr. 155.062.

Spiral structure of galaxies.
See Abstr. 155.075.

Infall of matter in galaxies.   See Abstr. 161.001.

Models for the cosmic evolution of radio galaxies and quasars.   See Abstr. 162.013.

Cosmological evidence from QSOs and radio galaxies.   See Abstr. 162.028.

Errata

158.903 Errata: 'Some unusual southern hemisphere objects.'
[Astron. Journ., Vol. 76, 775 - 776 (1971)].
P. K. Lü.
Astron. Journ., Vol. 77, 254 (1972).

158.902 Erratum: "On the interpretation of the Hα profile of the Seyfert galaxy NGC 5548" [Astrophys. Journ., Vol. 169, 449 - 453 (1971)].   K. S. Anderson.
Astrophys. Journ., Vol. 172, 785 (1972).

158.901 Erratum: 'Photometry of the outer corona of M87'
[Astrophys. Letters, Vol. 4, 17 - 22 (1969)].
G. de Vaucouleurs.
Astrophys. Letters, Vol. 10, 145 (1972). − See 01.158.065.

159   Magellanic Clouds

159.001   Kinematic study of ionized hydrogen in the Large Magellanic Cloud.   M. F. Chériguene, G. Monnet.
Astron. Astrophys., Vol. 16, 28 - 37 (1972). In French.

The radial velocities of 120 H II regions in the Large Magellanic Cloud have been obtained by the interference method using the Hα line, with a mean dispersion of 20 Å mm^{-1}. 1200 points have been measured on 63 plates. A comparison with the former published velocities (H II regions or neutral hydrogen) shows a systematic difference. A rotation curve is drawn by assuming a negligible expansion.

159.002   A photoelectric Hα surface photometry of the Small Magellanic Cloud.   T. Schmidt.
Astron. Astrophys., Vol. 16, 95 - 102 (1972).

A photoelectric Hα + [N II] surface photometry of the Small Magellanic Cloud has been undertaken by means of the Heidelberg 50 cm Cassegrain-reflector at the Boyden Observatory, South Africa.

159.003   Mini clusters in the Large Magellanic Cloud.   S. I. Gaposhkin.
Bull. American Astron. Soc., Vol. 3, 444 - 445 (1971). – Abstr. AAS.

159.004   4 color and Hβ photometry of the brightest stars in the Magellanic Clouds.   P. S. Osmer.
Bull. American Astron. Soc., Vol. 3, 453 (1971). – Abstr. AAS.

159.005   Color-magnitude diagrams of five faint clusters of the Large Magellanic Cloud.   P. W. Hodge.
Smithsonian Astrophys. Obs., *Cambridge, Mass.,* Special Rep. No. 337, 4 + 18 pp. (1971).

Color-magnitude diagrams of five faint clusters in the Large Magellanic Cloud are obtained by astrophotometry of ADH Schmidt plates centered on three faint photoelectric sequences. Two of the clusters are old, with well-established giant branches that are at least 2 mag brighter than their main sequences, while the other three are young, with bright, sparsely populated main sequences and several giant members.

159.006   Comparison of the Large Magellanic Cloud catalogs of Sanduleak and Fehrenbach-Duflot.   N. Sanduleak.
Astron. Astrophys., Vol. 17, 326 - 328 (1972).

A comparison results in the identification of probable high-velocity stars in the foreground of the LMC and a group of stars that might include horizontal-branch stars or possibly "bridge" stars located between our Galaxy and the Large Cloud.

159.007   Die Magellanschen Wolken.   R. X. McGee.
Umschau, 72. Jahrgang, p. 209 - 213 (1972).

Recent radioastronomical observations indicate that the Large Cloud of Magellan appears to be a barred spiral with two large arms and two smaller arms in different planes. Recent radio observations of the Small Cloud lead inevitably to the conclusion that the Small Cloud may be a barred spiral of much the same type as the Large Cloud.

159.008   An observational approach to the problem of spiral structure. II.   M. E. Dixon, V. L. Ford.
Astrophys. Journ., Vol. 173, 35 - 42 (1972).

The spatial distributions of "old" and "young" stars in the vicinity of the major arm of the Large Magellanic Cloud are compared in order to determine whether or not the major arm is a wavelike feature which propagates smoothly through the interstellar gases in the azimuthal direction.

159.009   Photometry of bright stars in the Large Magellanic Cloud.   J. Dachs.
Astron. Astrophys., Vol. 18, 271 - 286 (1972).

Photoelectric photometry in the *UBV* system down to a limiting visual magnitude of 14$^\mathrm{m}$9 is presented for 68 stars in two fields in the northern part of the Large Magellanic Cloud (LMC). The first field is located around the triple association NGC 1871/1869/1873, the second one in the surroundings of the stellar aggregate NGC 2034.

159.010   Electronographic photometry of stars in the globular clusters of the Magellanic Clouds.   M. F. Walker.
IAU Symposium No. 44, (see 012.005), p. 48 (1972). – Abstract.

159.011   Electronographic photometry of star clusters in the Magellanic Clouds – III. The colour-magnitude diagram of NGC 2257.   M. F. Walker.
Monthly Notices Roy. Astron. Soc., Vol. 156, 459 - 469 (1972).

Electronographic magnitudes and colours of stars in the cluster NGC 2257 in the Large Magellanic Cloud have been measured on electrographs taken with a Spectracon image-converter. The colour-magnitude diagram has been derived to $V = 22.3$.

159.012   Identification charts for clusters in the Small Magellanic Cloud.   E. M. Lindsay, J. McFarland.
Irish Astron. Journ., Vol. 10, 128 (1972).

159.013   Dust clouds in the LMC.   P. W. Hodge.
Bull. American Astron. Soc., Vol. 4, 223 (1972). Abstr. AAS.

159.014   Hα emission stars in the Large Magellanic Cloud.   B. Bohannan.
Bull. American Astron. Soc., Vol. 4, 242 (1972). – Abstr. AAS.

159.015   An X-ray observation of the Large Magellanic Cloud.   I. R. Tuohy, J. R. Harries, A. J. Broderick, K. B. Fenton, A. P. J. Luyendyk.
Proc. Astron. Soc. Australia, Vol. 2, 111 - 112 (1972).

159.016   Die Magellanschen Wolken.   T. Schmidt.
Mitt. Astron. Ges., No. 31, p. 71 - 86 (1972).
Review article.

159.017   Zum Magnetfeld der Magellanschen Wolken.   W. Deinzer, T. Schmidt.
Mitt. Astron. Ges., No. 31, p. 177 - 178 (1972).

159.018   UBVRI-Photometrie in der Kleinen Magellanschen Wolke.   G. Schnur, G. Klare, T. Neckel, J. Pelayo, T. Vives.
Mitt. Astron. Ges., No. 31, p. 179 - 180 (1972).

159.019   UBV-Photometrie heller Sterne in der Großen Magellanschen Wolke.   J. Dachs.
Mitt. Astron. Ges., No. 31, p. 180 (1972). – Abstract.

159.020   A more precise definition of structural peculiarities of the instability strip of classical cepheids from Gascoigne's photoelectric data in the Small Magellanic Cloud.   N. N. Yakimova.

Astron. Zhurn. Akad. Nauk SSSR, Vol. 49, 514 - 523 (1972). In Russian. English translation in Soviet Astron. AJ, Vol. 16, No. 3.

A comparison of photometric characteristics in the BV system (colour, brightness, amplitudes of brightness and colour) of about 30 classical cepheids from peripheric regions of the SMC is carried out. The basis of this comparison is the photoelectric data of Gascoigne.

159.021 **The Magellanic Clouds.** A. D. Thackeray.
Monthly Notes Astron. Soc. Southern Africa, Vol. 31, 47 (1972). – Abstract.

159.022 **Spectrographic and photometric observations of supergiants and foreground stars, in the direction of the Large Magellanic Cloud.**
A. Ardeberg, J.-P. Brunet, E. Maurice, L. Prévot.
Astron. Astrophys., Suppl. Ser., Vol. 6, 249 - 309 (1972).

UBV photometry and slit spectra of intermediate dispersion (74 Å mm^{-1}) have been obtained at La Silla for 409 stars belonging to the Large Magellanic Cloud (Part 1 of catalogue), for 42 possible LMC members (Part 2 of catalogue) and for 132 galactic foreground stars (Part 3 of catalogue) among which 19 have high radial velocities. The stars have been selected at the Marseille Observatory from plates taken with the objective-prism radial velocity astrograph of the European Southern Observatory. V magnitudes, $(B-V)$ and $(U-B)$ colour indices, MK spectral types and radial velocities are presented. Remarks are given concerning the positions, spectral types, magnitudes and colours.

159.023 **The Large Magellanic Cloud: Its topography of 1830 variable stars.** S. I. Gaposhkin.
SAO, *Cambridge, Mass.,* Special Report No. 310, 16 + 61 + 111 + A5 pp. (1970).

This new "topographic" study of the Large Magellanic Cloud (LMC) and its 1830 variables is based on 1, 200,000 photographic and on several thousand photovisual observations obtained with the author's method within 3 years on 2000 Harvard plates covering the years 1893 to 1961. The findings are presented in detail.

159.024 **Étude du Petit Nuage de Magellan par la technique du prisme-objectif de Fehrenbach.** A. Florsch.
Publ. Obs. Astron. Strasbourg, Vol. 2, Fasc. 1, 6 + 134 pp. (1972). – Diss. Univ. Strasbourg.

159.025 **The field RR Lyrae stars in the Large Magellanic Cloud.** J. A. Graham.
The evolution of population II stars, (see 012.024), p. 187 - 192 (1972).

159.026 **The Magellanic Clouds and the distance scale.** S. C. B. Gascoigne.
Quarterly Journ. Roy. Astron. Soc., Vol. 13, 274 - 281 (1972). – Presented at the Woolley symposium (see 012.025).

159.027 **Nitrogen deficiency in the Small Magellanic Cloud.**

N. Sanduleak, D. J. MacConnell, P. S. Hoover.
Nature, Vol. 237, 28 - 29 (1972).

We have made observations of the planetary nebulae in the Magellanic Clouds. The spectra of the planetaries in the H$_\alpha$ region were obtained on unwidened objective-prism plates taken with the University of Michigan Curtis Schmidt telescope at the Cerro Tololo Inter-American Observatory in Chile. In the Small Cloud only one of 32 objects has a detectable 6584 Å line, and in the others the line must be very weak and may be absent. By contrast, approximately 40% of the LMC objects definitely show the nitrogen line. From these results we conclude that nitrogen must be deficient in the Small Cloud planetary nebulae.

The Parkes survey of 21-centimeter absorption in discrete-source spectra. I. The Parkes hydrogen-line interferometer. See Abstr. 033.001.

Envelope opacities and the cepheid loops at 5 $M_\odot$. See Abstr. 065.081.

The strength of the O I λ7774 line in the brightest stars in the Magellanic Clouds. See Abstr. 114.017.

Photometry of long period variables in the Magellanic Clouds. See Abstr. 122.075.

Cepheid variables in Large and Small Magellanic Clouds. See Abstr. 122.137.

Nova Mensae 1970b in the Large Magellanic Cloud. See Abstr. 124.100.

Internal kinematics of the 30 Doradus nebula in the Large Magellanic Cloud. See Abstr. 132.008.

Observations of 30 Doradus in the infrared. See Abstr. 132.036.

Luminosity functions for 16 associations in the Large Magellanic Cloud. See Abstr. 152.005.

V_0 from Magellanic Cloud orbits. See Abstr. 155.084.

An observational approach to the problem of spiral structure. III. See Abstr. 158.115.

Errata

159.901 **Errata: "The brightest stars in the Magellanic clouds"** [Monthly Notices Roy. Astron. Soc., Vol. 121, 337 - 385 (1960)].
M. W. Feast, A. D. Thackeray, A. J. Wesselink.
Monthly Notices Roy. Astron. Soc., Vol. 155, 383 (1972).

160   Clusters of Galaxies

160.001   Redshifts versus apparent magnitudes for clusters of galaxies.   G. C. McVittie.
Monthly Notices Roy. Astron. Soc., Vol. 155, 425 - 434 (1972).

An attempt is made to reconstitute the apparent magnitudes (V_c) of the brightest members of clusters of galaxies from the available information. The redshifts of the clusters being denoted by z, it is argued that the only empirical, and statistically significant, conclusion is that $V_c - 5 \log z$ is constant for 39 clusters with redshifts in the range $0 < z < 0.461$. The main purpose of the paper is to describe a theoretical method for finding the parameters of zero-pressure uniform models of the universe when the empirical conclusion is accepted.

160.002   Additional scrutiny of a proposed test for cluster expansion.   H. J. Rood.
Astrophys. Journ., Vol. 171, 1 - 3 (1972).

If a centrally concentrated cluster is expanding because a force has accelerated the galaxies outward, its velocity dispersion does not decrease from the cluster center to its borders, but first increases from the center to a "turnover radius".

160.003   The unlikeliness that molecular hydrogen stabilizes clusters of galaxies or closes the universe.
T. W. Noonan.
Astrophys. Journ., Vol. 171, 209 - 211 (1972).

Photodissociation by ultraviolet radiation prevents the existence of intergalactic molecular hydrogen in sufficient amounts either to stabilize clusters of galaxies or to close the universe. Molecular hydrogen in the coronas of elliptical galaxies is also excluded.

160.004   On the tests for superclustering of Abell clusters.   W. Fullerton, P. Hoover.
Astrophys. Journ., Vol. 172, 9 - 16 (1972).

Abell's catalog of rich clusters of galaxies has been examined for evidence of second-order clustering in three dimensions. We have carried out further tests in order to show that systematic errors in the catalog are most probably the reasons why the serial correlation coefficient is large. The two-dimensional distribution of clusters on the sky has also been reanalyzed with the χ^2 test in order to examine the angular scale of apparent superclustering as a function of distance. We find no correlation with distance.

160.005   Detailed color index distribution within galaxies of the Virgo cluster.   J. D. Wray.
Bull. American Astron. Soc., Vol. 3, 477 (1971). – Abstr. AAS.

160.006   Redshifts of galaxies near the Hercules cluster.   L. P. Bautz.
Bull. American Astron. Soc., Vol. 3, 477 - 478 (1971). Abstr. AAS.

160.007   Redshifts and evolutionary effects for galaxies in distant clusters.   J. B. Oke.
Bull. American Astron. Soc., Vol. 3, 478 (1971). – Abstr. AAS.

160.008   The unlikeliness that molecular hydrogen stabilizes clusters of galaxies or closes the universe.
T. W. Noonan.
Bull. American Astron. Soc., Vol. 3, 478 (1971). – Abstr. AAS.

160.009   Re-examination of evidence for superclustering of rich clusters.   P. Hoover, W. Fullerton.
Bull. American Astron. Soc., Vol. 3, 498 (1971). – Abstr. AAS.

160.010   Form types for clusters of galaxies.   L. P. Bautz.
Astron. Journ., Vol. 77, 1 - 3 (1972).

Rich clusters of galaxies have been classified by the relative contrast between the brightest member and the typical bright galaxy population of the cluster. This parameter is not correlated with richness. A list of 111 newly classified Abell clusters is presented.

160.011   The cluster of galaxies Abell 1413.
T. W. Noonan.
Astron. Journ., Vol. 77, 9 - 12 (1972).

The cluster Abell 1413 at a redshift of 0.143 shows an effect which may be interpreted as a luminosity segregation, indicating that this property probably persists over look-back times of order 0.14 of the inverse of the Hubble constant. The cluster has the same richness as the Coma cluster but only about 60 % as the radius, unless what appears to be the brightest member is really a foreground galaxy.

160.012   Intergalactic clouds in the Coma cluster?
D. Goldsmith, J. Silk.
Astrophys. Journ., Vol. 172, 563 - 575 (1972).

We present a theoretical discussion of the stability of the intergalactic medium within the Coma cluster of galaxies, and conclude that a narrow range of parameters may exist over which intergalactic clouds can account for the virial discrepancy in the Coma cluster. The recent discovery of X-radiation from the central region of the Coma cluster is interpreted as thermal emission from the (hotter) intercloud medium.

160.013   Absence of Lyman-alpha emission from the Coma cluster of galaxies.   R. C. Henry.
Astrophys. Journ., (*Letters*), Vol. 172, L97 - L100 (1972).

The redshifted Lα flux from the Coma cluster of galaxies is determined to be less than 6000 photons cm^{-2} s^{-1}, leading to severe new restrictions on hot ionized gas models for the gravitational binding of the cluster.

160.014   A suggested concept for galaxy-cluster size for use in the angular-size cosmological test.
T. W. Noonan.
Astron. Journ., Vol. 77, 134 - 137 (1972).

The mean distance of cluster members from the line of sight through the cluster center is suggested as a suitable radius for clusters of galaxies in using the cosmological test of angular size versus redshift.

160.015   Analysis of X-ray background fluctuations.
A. C. Fabian.
Nature, Phys. Sci., Vol. 237, 19 - 21 (1972).

An analysis of the X-ray background similar to the $\log N - \log S$ plots familiar from radio work shows that an origin in clusters of galaxies cannot be ruled out.

160.016   Problems concerning the extragalactic distance scale.   G. O. Abell.
IAU Symposium No. 44, (see 012.005), p. 341 - 352 (1972).

160.017   The velocity-distance relation and the Hubble constant for nearby groups of galaxies.
G. de Vaucouleurs.
IAU Symposium No. 44, (see 012.005), p. 353 - 366 (1972).

160.018 Rapid evolution of galactic nuclei. W. G. Tifft.
IAU Symposium No. 44, (see 012.005), p. 367 - 375 (1972).

160.019 Analysis of the magnitude-redshift relation including possible effects of evolution.
J.-E. Solheim, B. M. Tinsley.
IAU Symposium No. 44, (see 012.005), p. 397 - 400 (1972).

160.020 X-ray emission from rich clusters of galaxies.
H. Gursky, A. Solinger, E. M. Kellogg, S. Murray, H. Tananbaum, R. Giacconi, A. Cavaliere.
Astrophys. Journ., (*Letters*), Vol. 173, L99 - L104 (1972).

The *Uhuru* catalog has revealed a significant number of rich clusters associated with X-ray sources. Combined with earlier observations of X-ray emission from the Virgo, Coma, and Perseus clusters, this suggests that most, if not all, rich clusters include an X-ray emission region of large size and of net luminosity $10^{43} - 10^{44}$ ergs s^{-1}.

160.021 Intergalactic extinction in the local supercluster. I. Selective extinction.
G. de Vaucouleurs, A. de Vaucouleurs, H. G. Corwin, Jr.
Astron. Journ., Vol. 77, 285 - 287 (1972).

An analysis of 566 color residuals of 262 normal galaxies observed at McDonald Observatory between 1960 and 1968 indicates that galaxies close to the supergalactic plane have a significant color excess averaging +0.03 mag in the center sector (northern galactic hemisphere) and +0.01 mag in the anticenter sector.

160.022 A note on mass loss during collisions between galaxies and the formation of giant systems.
J. S. Gallagher III, J. P. Ostriker.
Astron. Journ., Vol. 77, 288 - 291 (1972).

Using models based on the density distribution observed in the E 1 galaxy NGC 3379, the mass loss is computed for hyperbolic collisions on an approximate but conservative basis. Binary collisions are then summed to show that large numbers of stars will be evaporated from elliptical members of dense clusters of galaxies.

160.023 Intergalactic ionized hydrogen in nearby groups of galaxies. P. Chamaraux, T. Montmerle, M. Tadokoro.
Astrophys. Space Sci., Vol. 15, 383 - 394 (1972).

Stability of the nearest 14 groups of galaxies is investigated by means of Jacobi's criterion. The apparent instability is tentatively explained by presence of intergalactic ionized hydrogen in each group. Physical parameters of the gas are derived by means of the general modified form of the virial theorem and through the assumption of the equipartition among several modes of energy.

160.024 Redshift, morphology, and integrated magnitude relationships in the Coma cluster. W. G. Tifft.
Bull. American Astron. Soc., Vol. 4, 238 (1972). – Abstr. AAS.

160.025 Concerning the forms of the velocity - distance relation clusters. L. P. Bautz, G. O. Abell.
Bull. American Astron. Soc., Vol. 4, 239 (1972). – Abstr. AAS.

160.026 Absence of Lyman alpha emission from the Coma cluster of galaxies. R. C. Henry.
Bull. American Astron. Soc., Vol. 4, 239 (1972). – Abstr. AAS.

160.027 Spectrum of soft X-rays from the Virgo cluster.
R. C. Catura, P. C. Fisher, H. M. Johnson, A. J. Meyerott.
Bull. American Astron. Soc., Vol. 4, 258 (1972). – Abstr. AAS.

160.028 Clusters of galaxies and the N-body problem.
D. G. Saari.
Bull. American Astron. Soc., Vol. 4, 265 (1972). – Abstr. AAS.

160.029 The predominant motion in spherical systems of galaxies.
I. Karachentsev, W. Zonn, A. Shcherbanovsky.
Astrophys. Letters, Vol. 11, 151 - 154 (1972).

Statistical study of kinematics of some clusters and groups of galaxies supports the hypothesis of their instability. The study is based on the examination of regression curves of radial velocities of members of a cluster on the projections (on the celestial sphere) of their distances from the center of a cluster.

160.030 Redshifts for galaxies in and near Abell 2147.
L. P. Bautz.
Astron. Journ., Vol. 77, 331 - 332, 401 - 402 (1972).

Sixteen galaxies southwest of the Hercules cluster were observed to determine the true spatial arrangement of two apparent arcs of galaxies. Twelve of them, including the peculiar system Arp 324, are projected against the rich cluster Abell 2147. Four are away from the main body of the cluster. All but two of the group have redshifts comparable to galaxies in the Hercules cluster. A 2147 is at roughly the same distance as the Hercules cluster.

160.031 Density data and emission measure for a model of the Coma cluster. I. R. King.
Astrophys. Journ., (*Letters*), Vol. 174, L123 - L125 (1972).

The density, the square of the density, the velocity dispersion, and the escape velocity are given numerically, in space and in projection, for a model fitted to the Coma cluster of galaxies. The total emission measure is calculated. Lyman-α observations of higher sensitivity will be needed to confirm or dismiss the hypothesis that the "missing mass" is hot gas.

Statistical association of quasi-stellar radio sources and rich clusters of galaxies. See Abstr. 141.148.

Upper limits to the soft X-ray emission of sources in Virgo. See Abstr. 142.013.

Extended X-ray sources in the Coma and Perseus clusters. See Abstr. 142.027.

On the origin of the magnetic field in the extended radio and X-ray source in the Coma cluster. See Abstr. 142.036.

Structure and time variability of the X-ray source in the Virgo cluster. See Abstr. 142.093.

Extragalactic X-ray sources. See Abstr. 142.100.

The extended X-ray source at M87. See Abstr. 142.108.

Far ultraviolet background radiation. See Abstr. 155.018.

Inclination and absorption effects on the apparent diameters, optical luminosities and neutral hydrogen radiation of galaxies–III. Theory and applications. See Abstr. 158.004.

Radial velocities of galaxies. See Abstr. 158.036.

The law of momentum conservation and some problems of metagalactic astronomy. See Abstr. 158.093.

The redshift-distance relation. I. Angular diameter of first-ranked cluster galaxies as a function of redshift: The aperture correction to magnitudes.   See Abstr. 158.108.

Active galaxies with radio trails in clusters. See Abstr. 158.114.

The NGC 520 chain of quasars and the "discrepant redshift" of cluster galaxies.   See Abstr. 158.146.

Selection effects and the distribution of orientations of galaxies.   See Abstr. 158.147.

Intergalactic clouds in the Coma cluster? See Abstr. 161.005.

Possible evidence for an intergalactic medium in clusters of galaxies.   See Abstr. 161.007.

Intergalactic reddening of galaxies. See Abstr. 161.012.

161   Intergalactic Matter

161.001   **Infall of matter in galaxies.**   R. B. Larson.
Nature, Vol. 236, 21 - 23 (1972).
If the formation of galaxies leaves behind uncondensed intergalactic matter, this material will continue to fall into galaxies and form new stars. Oort's estimate for the infall rate in our Galaxy is consistent with theoretical expectations and with the present star formation rate, suggesting that the presence of a Pop. I component in galaxies is a result of the continuing infall of intergalactic matter. Many properties of extragalactic systems, including spiral structure and the existence of active nuclei, may be explainable as a result of the accretion of intergalactic matter.

161.002   **Intergalactic medium – A review.**   G. B. Field.
Bull. American Astron. Soc., Vol. 3, 437 (1971).
Abstr. AAS.

161.003   **Photoionized intergalactic hydrogen.**
J. Arons, D. W. Wingert.
Bull. American Astron. Soc., Vol. 3, 437 (1971). – Abstr. AAS.

161.004   **Limits on the heavy element abundance of the intergalactic medium from an X-ray spectrum of 3C 273.**
S. Bowyer, B. Margon, M. Lampton.
Bull. American Astron. Soc., Vol. 3, 437 (1971). – Abstr. AAS.

161.005   **Intergalactic clouds in the Coma cluster?**
D. W. Goldsmith, J. Silk.
Bull. American Astron. Soc., Vol. 3, 437 - 438 (1971). – Abstr. AAS.

161.006   **Limits on a clumpy intergalactic gas.**
J. Silk, J. Tarter.
Bull. American Astron. Soc. , Vol. 3, 438 (1971). – Abstr. AAS.

161.007   **Possible evidence for an intergalactic medium in clusters of galaxies.**   D. S. De Young.
Astrophys. Journ., (Letters), Vol. 173, L7 - L11 (1972).
An initial list of 64 simple double radio sources is examined to determine if any morphological differences exist between those sources that lie within clusters of galaxies and those that lie without. After elimination of possible selection effects, the remaining list of 34 sources is investigated for differences in component separation and ratio of component size to separation. The mean component separation for those sources not in clusters of galaxies is slightly more than twice the mean separation for sources in clusters, and the mean size-to-separation ratio for sources outside clusters is about 0.6 the mean of those inside. It is argued that this may be due to the presence of a significant amount of intergalactic gas in clusters of galaxies.

161.008   **Intergalactic matter and radiation.**
G. R. Burbidge.
IAU Symposium No. 44, (see 012.005), p. 492 - 517 (1972).

161.009   **Relativistic bremsstrahlung as a source of γ-ray background flux.**
R. G. Cruddace, J. Silk, H. Reeves.
Bull. American Astron. Soc., Vol. 4, 258 - 259 (1972).
Abstr. AAS.

161.010   **Metagalactic γ-rays from relativistic bremsstrahlung interactions.**   F. W. Stecker, D. L. Morgan, Jr.
Bull. American Astron. Soc., Vol. 4, 261 (1972). – Abstr. AAS.

161.011   **Soft X-ray emission from intergalactic gas in the neighbourhood of the Galaxy.**
R. Hunt, D. W. Sciama.
Monthly Notices Roy. Astron. Soc., Vol. 157, 335 - 348 (1972).
The gravitational influence of our Galaxy on a hypothetical local group plasma, which was calculated in a previous paper, is here shown to give rise to a flux of soft X-rays which may already have been detected. If the emitting gas contains a small admixture of heavy elements expelled by the Galaxy, soft X-ray lines might be detectable. For example, if the relative abundance of oxygen $\sim 10^{-2}$ of its solar value, the combined equivalent width of the 21.6, 21.8 and 22.1 Å lines of O VII would ~ 10 Å for an emitting temperature $\sim 2 \times 10^6 \, ^\circ$K.

161.012   **Intergalactic reddening of galaxies.**   B. Takase.
Publ. Astron. Soc. Japan, Vol. 24, 295 - 299 = Tokyo Astron. Obs. Repr., No. 412 (1972).
An analysis of the corrected colors C_0 in the Reference Catalogue of Bright Galaxies (de Vaucouleurs and de Vaucouleurs 1964) indicates that the lower the supergalactic latitude of galaxies, the redder their reduced color. This fact suggests the presence of intergalactic absorbing matter concentrated toward the supergalactic plane.

161.013   **Some remarks on intergalactic magnetic fields.**
M. J. Rees, M. Reinhardt.
Astron. Astrophys., Vol. 19, 189 - 192 (1972).

Observational upper limits to the strength of intergalactic magnetic fields are discussed. The role such fields might play in the early stages of the hot model of the universe, and in galaxy formation after the recombination era, is investigated.

161.014   **Limits on intergalactic helium from the 3C 273 X-ray spectrum.**
B. Margon, S. Bowyer, M. Lampton.
Astrophys. Journ., Vol. 174, 471 - 475 (1972).

An X-ray spectrum of the quasi-stellar object 3C 273 in the interval 0.25–10 keV has been obtained by sounding-rocket observations. An upper limit on the X-ray optical depth to 3C 273 has been calculated from the data, permitting upper limits to be set on the absolute abundance of helium in the intergalactic medium.

161.015   **Enrichment of intergalactic matter.**
J. Silk, R. S. Siluk.
Astrophys. Journ., Vol. 175, 1 - 10 (1972).

The primordial gas out of which the Galaxy condensed may have been significantly enriched in heavy elements. A specific mechanism of enrichment is described, in which quasi-stellar sources eject enriched matter into the intergalactic medium. Expressions are given for the degree of enrichment Z of the intergalactic gas as a function of redshift z, and we show that our hypothesis implies that the present density of intergalactic gas must be at least a factor 3 larger than the mean density in galaxies at the present epoch. A variation of Z with z is predicted.

161.016   **A possible origin of the high-velocity neutral hydrogen gases.**   T. L. Chow, M. P. Savedoff.
Nuovo Cimento B, Ser. 11, Vol. 8B, 130 - 142 (1972).

Using the modified characteristic method, the consequences when high-momentum extra-galactic clouds collide with galactic gases are calculated. Both variable density and radiation loss are included hydrodynamically. The results show that most of the high-velocity neutral hydrogen gases may result from recombination after the galactic post-shock gas radiates its surplus energy.

Étude des milieux dilués hors équilibre thermodynamique.   See Abstr. 131.159.

Effect of infalling matter on the heavy element content of a galaxy.   See Abstr. 158.062.

Intergalactic clouds in the Coma cluster?
See Abstr. 160.012.

Intergalactic extinction in the local supercluster. I. Selective extinction.   See Abstr. 160.021.

Errata

161.901   **Erratum: 'Photographic detection of 'intergalactic' matter in the Coma cluster'** [Astrophys. Journ., (*Letters*), Vol. 169, L3 - L5 (1971)].
G. A. Welch, G. N. Sastry.
Astrophys. Journ., (*Letters*), Vol. 171, L81 (1972).

162 Structure and Evolution of the Universe, Cosmology

162.001 Cosmological models in a conformally invariant gravitational theory—I. The Friedmann models.
F. Hoyle, J. V. Narlikar.
Monthly Notices Roy. Astron. Soc., Vol. 155, 305 - 321 (1972).

The present paper discusses the formulation of the Friedmann cosmological models in terms of a conformally invariant gravitational theory. This theory is Machian in the sense that the mass of a particle arises from the interaction of the particle with a mass field generated by other particles. The three Friedmann cases $k = 0, \pm 1$ are discussed in detail from this point of view.

162.002 Cosmological models in a conformally invariant gravitational theory—II. A new model.
F. Hoyle, J. V. Narlikar.
Monthly Notices Roy. Astron. Soc., Vol. 155, 323 - 335 (1972).

It was seen in a previous paper how the mass of a particle can be determined in terms of a mass field $m(X)$. A dimensionless coupling λ constant is introduced, the individual particle mass being $\lambda m(X)$. Since the mass field can be explicitly calculated in terms of cosmic time τ and L^{-3}, the cosmic particle density, the numerical value of λ can be determined by relating $\lambda m(X)$ to empirically known particle masses. The model resulting from the dependence of λ on τ involves continuous creation of matter.

162.003 On the redshift-magnitude relation in hierarchical cosmologies. M. J. Haggerty, J. R. Wertz.
Monthly Notices Roy. Astron. Soc., Vol. 155, 495 - 503 (1972).

The usual redshift-magnitude relation is valid when the expansion rate of the universe, though time-dependent, is uniform. Modifications due to larger-scale clustering are suggested. They are based on a model of light propagation in which light moves at a constant velocity with respect to the mean motion of the matter through which it passes. Specific results are obtained which could serve as a potential test of big-bang hierarchical cosmological models.

162.004 A critique of Rees's theory of primordial gravitational radiation. J. C. Jackson.
Monthly Notices Roy. Astron. Soc., Vol. 156, 1P - 5P (1972). Short communication.

162.005 Some conformally flat cosmological models of axial symmetry.
N. P. Bondarenko, P. K. Kobushkin.
Dokl. Akad. Nauk SSSR, Ser. Mat. Fiz., Vol. 202, 558 - 559 (1972). In Russian.

162.006 Dynamic motions in the early universe.
A. D. Chernin.
Astrophys. Letters, Vol. 10, 125 - 128 (1972).

Dynamic motions on space scales approaching the Friedmann radius can generate rotating and non-rotating condensations in protogalactic matter. The condensations separate from the expanding metagalactic medium when the mean density of the universe becomes comparable to the present interior density of galaxies.

162.007 Damping of adiabatic perturbations in an expanding universe. G. V. Chibisov.
Astron. Zhurn. Akad. Nauk SSSR, Vol. 49, 74 - 84 (1972). In Russian. English translation in Soviet Astron. AJ, Vol. 16, No. 1.

The damping of adiabatic density perturbations by radiative viscosity and heat conductivity is considered with taking into account the real duration of plasma recombination in a hot expanding universe.

162.008 Dissipative effects in the expansion of the universe. I. R. A. Matzner, C. W. Misner.
Astrophys. Journ., Vol. 171, 415 - 432 (1972).

We consider dissipative processes in anisotropic homogeneous world models, and show that dissipation reduces the anisotropy. A discussion is given of the viscosity approximation and its range of applicability. Examples are presented which have been calculated by the use of a simple approximation to the collision-time method, using the cross-section appropriate to weak-interaction neutrino scattering.

162.009 Dissipative effects in the expansion of the universe. II. A multicomponent model for neutrino dissipation of anisotropy in the early universe. R. A. Matzner.
Astrophys. Journ., Vol. 171, 433 - 448 (1972).

A detailed multicomponent model is developed for dissipative processes in Euclidean homogeneous cosmological models. These processes involve neutrinos which might have long mean free times t_ν, in interaction with other constituents (electrons, photons) which are thermalized by electromagnetic interactions, and whose weak interactions produce thermal neutrinos with mean rate t_e^{-1}. The method used is a mean-free-time approximation generalized to the different rates of destruction and production t_ν^{-1} and t_e^{-1}.

162.010 The effect of temperature variations on the primeval production of helium. F. M. Ipavich.
Astrophys. Journ., Vol. 171, 449 - 450 (1972).

Random temperature fluctuations in the primeval fireball are shown to lead to an increase in the cosmic helium abundance relative to the abundance in the absence of such fluctuations. This conclusion is opposite to that reached by Silk and Shapiro.

162.011 A spherically symmetric inhomogeneous cosmological model. C. T. J. Dodson.
Astrophys. Journ., Vol. 172, 1 - 8 (1972).

The model derives from a diagonal metric tensor with components $g_{\alpha\alpha} = R^2(t)e^{-\lambda r}, \alpha = 1, 2, 3, g_{44} = -1$, where $\lambda \geq 0$ is constant. The case $\lambda = 0$ reduces to the Einstein—de Sitter model. When $\lambda \neq 0$, an evolutionary bounded universe is obtained with a center at $r = 0$ and a plausible distribution of matter. The time elapsed since an initial singularity is identically H^{-1}, the Hubble age.

162.012 Limits on the local deviation of the universe from a homogeneous model.
A. Sandage, G. A. Tammann, E. Hardy.
Astrophys. Journ., Vol. 172, 253 - 263 (1972).

Modern data exist which permit us to set more stringent limits. The tests are of two kinds. (1) The redshift-magnitude relation defined locally ($0.003 \leq z < 0.03$), using only well-defined standard candles (first-ranked cluster galaxies), puts limits on perturbations to the velocity field; this in turn restricts the permitted values of the density contrast. (2) Direct counts of galaxies in depth put limits on θ in $N(r) \propto r^{-\theta}$, where θ is the "hierarchical thinning rate" defined by Wertz (1971) and by Haggerty and Wertz (1971).

162.013 Models for the cosmic evolution of radio galaxies and quasars. J. N. Bahcall.
Astrophys. Journ., Vol. 172, 265 - 282 (1972).

The subject of this paper is the evolution in cosmic time of radio galaxies and quasars. The paper consists of four main parts: (1) a mathematical formulation; (2) a discussion of "ultraconservative" models for radio-galaxy evolution; (3) a discussion of "symmetric" models for the coupled evolution of quasars and radio galaxies; and (4) an application of the luminosity-volume test to an essentially complete sample of radio galaxies. Our major purpose is to exhibit with the aid of simple models the observational distinctions between classes of explanations for radio-source evolution.

162.014 Quantum cosmologies in the quaternion formulation of general relativity theory. K. C. Jacobs.
Bull. American Astron. Soc., Vol. 3, 480 (1971). – Abstr. AAS.

162.015 Physical conditions and evolution in the hadron era. P. Mészáros.
Nature, Phys. Sci., Vol. 235, 50 - 52 (1972).

In this article the hadronic equation of state is discussed in the light of recent developments concerning an ultimate temperature. The value of the adiabatic exponent γ is determined and found to be less than $4/3$. This is applied to Friedman universes to evaluate the effects on the expansion rate and on the growth of fluctuations during the hadron era.

162.016 Covariant chronogeometry and extreme distances. I. I. Segal.
Astron. Astrophys., Vol. 18, 143 - 148 (1972).

A space-time model is proposed which is locally identical with Minkowski space in its geometrical and causality features, but globally distinct. Its physical implications are effectively identical with those of Minkowski space at classical distances, but are materially different as regards cosmology and the classification of elementary particles.

162.017 Conditions for the rapid growth of perturbations in an expanding universe. W. C. Saslaw.
Astrophys. Journ., Vol. 173, 1 - 5 (1972).

The relation, originally derived by Bonnor, which governs the growth of spherical perturbations in an isotropic universe is shown to apply also to expanding anisotropic world models which may rotate. New solutions of this equation are found in which perturbations can grow exponentially.

162.018 Rotational perturbations in anisotropic cosmology. V. A. Ruban, A. D. Chernin.
Astron. Zhurn. Akad. Nauk SSSR, Vol. 49, 447 - 449 (1972). In Russian. English translation in Soviet Astron. AJ, Vol. 16, No. 2. – Short note.

162.019 Comments on the insensitivity of $\langle V/V_m \rangle$ to random shuffling of the redshifts in the luminosity-volume test. R. Carswell, R. Weymann.
Monthly Notices Roy. Astron. Soc., Vol. 156, 19P - 23P (1972).

Several artificial luminosity functions are considered in an Einstein–de Sitter cosmological model and the values of $\langle V/V_m \rangle$ for the 'actual' and 'shuffled' luminosity functions are computed. The relative insensitivity of $\langle V/V_m \rangle$ to the complete shuffling of the redshifts is shown to be a very general property.

162.020 Cosmology and the arrow of time. D. Layzer.
Vistas in astronomy, Vol. 13, (see 003.001), 279 - 287 (1972).

162.021 Sulla termodinamica e la cosmologia. G. Wataghin.
Atti Accad. Nazionale Lincei, Rend. Sci. fis., mat., nat., Ser. 8, Vol. 50, 725 - 729 (1971).

A model of an expanding universe filled with neutrino-pairs is described. The origin of the weakly interacting phases of the neutrinos and the photons in the actual universe is analyzed, and the role of rapid cooling processes of hadrons and of photons at a certain "leptonic" epoch is pointed out.

162.022 Cosmological models with two fluids. I. Robertson–Walker metric. C. B. G. McIntosh.
Australian Journ. Phys., Vol. 25, 75 - 82 (1972).

Exact solutions in terms of elementary functions are given for flat, homogeneous and isotropic, relativistic cosmological models which contain two fluids, each with an equation of state of given special form. Special cases are analogous to models with a Robertson–Walker metric with $k = \pm 1$ and to anisotropic models of the type discussed by Jacobs respectively.

162.023 Cosmological models with two fluids. II. Conformal and conformally flat metrics. C. B. G. McIntosh, J. M. Foyster.
Australian Journ. Phys., Vol. 25, 83 - 89 (1972).

Similar solutions to those in Part I are given for two-fluid cosmological models when the Robertson–Walker metric in its usual form is replaced by "conformal" and "conformally flat" forms.

162.024 Light out of darkness vs order out of chaos. P. J. E. Peebles.
Comments Astrophys. Space Phys., Vol. 4, 53 - 58 (1972).

To trace the evolution of the universe back in time we need a physical theory, and for purposes of discussion let us adopt general relativity theory (with $\Lambda = 0$). We also need to know what the universe is like now, and again for the purposes of discussion let us take the expanding Friedmann model perturbed by irregularities on scales up to 30 Mpc, say. There are many initial states that could end up like this, and the task is to decide which of these states is most reasonable.

162.025 The observational status of cosmological helium. L. Searle, W. L. W. Sargent.
Comments Astrophys. Space Phys., Vol. 4, 59 - 64 (1972).

In this article we shall review the observations which relate to the helium content of pre-galactic matter. We shall show that there is strong evidence that the bulk of the helium observed in the absorption spectra of young, hot stars in our Galaxy and in the emission spectra of H II regions in our Galaxy and other, nearby, galaxies was already present when these objects formed.

162.026 A combined cosmological test. J.-E. Solheim.
IAU Symposium No. 44, (see 012.005), p. 401 - 403 (1972).

162.027 Notes on cosmology. J. Pachner.
IAU Symposium No. 44, (see 012.005), p. 404 - 406 (1972).

162.028 Cosmological evidence from QSOs and radio galaxies. M. J. Rees.
IAU Symposium No. 44, (see 012.005), p. 407 - 436 (1972).

162.029 Selection of cosmological models using QSOs. R. C. Roeder.
IAU Symposium No. 44, (see 012.005), p. 471 - 473 (1972).

162.030 Ghost images in a universe with inhomogeneous mass distribution. J. M. Barnothy, M. F. Barnothy.
IAU Symposium No. 44, (see 012.005), p. 478 - 491 (1972).

162.031 General review of cosmological theories.

F. Hoyle.
IAU Symposium No. 44, (see 012.005), p. 526 - 531 (1972).

162.032 On Faraday rotation in cosmology. R. R. Burman.
Publ. Astron. Soc. Japan, Vol. 24, 291 - 292 (1972).
A comparison is made of Faraday rotation by a cosmic magnetic field in the steady-state and Einstein-de Sitter cosmologies.

162.033 Test of the expanding universe postulate.
M. J. Geller, P. J. E. Peebles.
Astrophys. Journ., Vol. 174, 1 - 5 (1972).
The effect of an expansion of the universe, as opposed to a tired-light effect for the cosmological redshift, may be apparent in the available observational data.

162.034 The electromagnetic radiation of the universe.
R. A. Syunyaev.
Priroda, No. 4.72, p. 69 - 81 (1972). In Russian.

162.035 Time direction of information propagation and cosmology. D. T. Pegg.
Australian Journ. Phys., Vol. 25, 207 - 214 (1972).
Conventional electrodynamics is used to obtain a condition which, if satisfied, allows information to be received from the past only, and ensures that the retarded potential is the only consistent solution.

162.036 Inner bremsstrahlung as a source of X-rays in the steady state universe. V. Petrosian, R. Ramaty.
Bull. American Astron. Soc., Vol. 4, 243 - 244 (1972). – Abstr. AAS.

162.037 Uniform model universes containing matter and blackbody radiation. T. L. May, G. C. McVittie.
Bull. American Astron. Soc., Vol. 4, 244 (1972). – Abstr. AAS.

162.038 Yukawa range and cosmological constant, the frontiers of a hierarchical universe.
J. M. Barnothy, M. F. Barnothy.
Bull. American Astron. Soc., Vol. 4, 244 - 245 (1972). – Abstr. AAS.

162.039 Atomic matter-antimatter rearrangement and annihilation cross sections and cosmological applications.
D. L. Morgan, Jr.
Bull. American Astron. Soc., Vol. 4, 260 (1972). – Abstr. AAS.

162.040 The cosmological annihilation in the Omnes model and the observed γ-ray spectrum.
J. L. Puget, F. W. Stecker.
Bull. American Astron. Soc., Vol. 4, 261 (1972). – Abstr. AAS.

162.041 On galaxy formation from primeval universal turbulence. N. Dallaporta, F. Lucchin.
Astron. Astrophys., Vol. 19, 123 - 134 (1972).
The problem of galaxy separation from the continuous matter background due to density fluctuations is reconsidered in the frame of the universal turbulence theory of primeval matter, following the general outline to the problem due to Ozernoy *et al.* In common with these authors, the principal omission consists in the neglect of the dissipative effects occurring in the turbulence.

162.042 The symmetry of matter and antimatter in universe.
B. Krygier, J. Krempec.
Urania Kraków, Vol. 43, 21 - 24 (1972). In Polish.

162.043 Über die Möglichkeit der Entstehung von Galaxien aus Materiequellen. P. von der Osten-Sacken.

Mitt. Astron. Ges., No. 31, p. 193 - 195 (1972).

162.044 The past of the universe and helium.
A. G. Doroshkevich, I. D. Novikov, R. A. Syunyaev.
Zemlya i Vselennaya, 1972, No. 3, p. 21 - 25. In Russian.

162.045 Particle creation in isotropic cosmologies.
L. Parker.
Phys. Rev. Letters, Vol. 28, 705 - 708, with a correction p. 1497 (1972).
The simplest covariant generalization of the scalar wave equation leads to significant pion creation and annihilation processes near an isotropic Friedmann-type singularity. Estimates for a plausible initial state yield pion creation of the same order of magnitude as obtained by Zeldovich near an anisotropic Kasner-type singularity.

162.046 Hydrogen-antihydrogen interactions.
B. R. Junker, J. N. Bardsley.
Phys. Rev. Letters, Vol. 28, 1227 - 1229 (1972).
The interaction potential between hydrogen and antihydrogen is obtained by a variational calculation.

162.047 Galaktische und metagalaktische Hintergrundstrahlung. G. Dautcourt.
Astron. Nachr., Vol. 293, 281 - 283 (1972).
The ratio f of the intensity of metagalactic background radiation emitted from galactic cores to the radiation flux coming from the center of our Galaxy is studied as a function of cosmological model parameters. It is seen that some cosmological information may thus also be obtained in cases where an absolute calibration of the radiation flux is uncertain, as in the case of gravitational radiation.

162.048 Hot universe, cosmic rays of ultrahigh energy and absolute reference system.
H. Sato, T. Tati.
Progr. Theor. Phys. Japan, Vol. 47, 1788 - 1790 (1972). Letter.

162.049 The distance-redshift relation for universes with no intergalactic medium.
C. C. Dyer, R. C. Roeder.
Astrophys. Journ., (*Letters*), Vol. 174, L115 - L117 (1972).
The distance-redshift relation is derived for model universes in which there is negligible intergalactic matter and in which the line of sight to a distant object does not pass close to intervening galaxies. When fitted to observations, this relation yields a higher value of q_0 than does a homogeneous model.

162.050 The magnitude-redshift relation in Brans-Dicke cosmology. B. M. Tinsley.
Astrophys. Journ., (*Letters*), Vol. 174, L119 - L121 (1972).
The magnitude-redshift relation expected in a flat Brans-Dicke world is discussed. Effects of stellar evolution, both in the conventional manner and as caused by the time-variation of the gravitational constant, are included.

162.051 Radio testing of cosmological models. Part I. Methods of observation. M. Urbanik.
Postępy Astron., Vol. 20, 3 - 22 (1972). In Polish.
In this article, that constitutes a part of a review of radio testing methods of cosmological models, are presented the basic observational methods that enable to form complete samples of radio sources. Such samples constitute basic data for the investigation of the universe by radio sources counts.

162.052 Isotropic models of radiative universe.
J. Krempeć, B. Krygier.
Postępy Astron., Vol. 20, 67 - 77 (1972). In Polish.

Solutions of the field equations and the equations of the geodetic lines of the light rays are given for both closed and open isotropic radiative universes. The light rays move on circular orbits in a closed world and on hyperbolic orbits in an open one. For a semi-closed world it is not possible to tie up the external and internal metrics. This suggests that these worlds are not real.

162.053 Curved space cosmological bounds on the time variation of G.   R. E. Morganstern.
Nature, Phys. Sci., Vol. 237, 70 - 71 (1972).

A differential expression arising from the Brans-Dicke cosmological equations is used to show that the time-variation of the gravitational constant, $\dot{\Lambda}_0$, decreases with pressure for all curvatures. Moreover, the explicit curvature contribution to $\dot{\Lambda}_0$ is shown to be the same order of magnitude as the flat space contribution and carries the sign of the spatial curvature ($k=\pm 1.0$). Thus, whereas for $k=+1$, $\dot{\Lambda}_0$ is at most an order of magnitude larger than its flat space value, for $k=-1$, $\dot{\Lambda}_0$ may be quite small.

162.054 Inhomogeneous world models and high-frequency
approximation.   E. Saar.
Izv. Akad. Nauk Ehstonskoj SSR, Vol. 20, (Fiz., Mat., 1971, No. 4), 420 - 425 = Tartu Astron. Obs. Teated, No. 38.

Cosmological models, homogeneous only on the average, are studied by using the high-frequency approximation. It is shown that the density inhomogeneities give raise to an additional energy-momentum tensor, realizing the feedback between the evolution of the structure and the overall dynamics of the universe.

162.055 Distances to galaxies: The Hubble constant, the
Friedmann time, and the edge of the world.
A. Sandage.
Quarterly Journ. Roy. Astron. Soc., Vol. 13, 282 - 296 (1972). – Presented at the Woolley symposium (see 012.025).

162.056 The first fact of evolutionary cosmology.
D. D. Clayton.
Comments Astrophys. Space Phys., Vol. 4, 29 - 33 (1972).

162.057 Global and non-global problems in cosmology.
G. F. R. Ellis, D. W. Sciama.
General relativity, (see 003.017), p. 35 - 59 (1972).

162.058 On the transparency of extragalactic space to very-
high-energy photons. II.   S. A. Bonometto,
F. Lucchin.
Nuovo Cimento Lettere, Ser. 2, Vol. 2, 1299 - 1304 (1971).

162.059 Periodic model of the evolving universe.
R. G. Zaikov.
Comptes Rendus Acad. Bulg. Sci., Vol. 24, 1301 - 1313 (1971).

162.060 A photon rest mass and electromagnetic dispersion
in cosmology.   R. R. Burman.
Phys. Letters A, Vol. 38A, No. 2, p. 96 (1972).

The effect of a photon rest mass on electromagnetic dispersion is calculated for the steady-state and Einstein–de Sitter cosmologies.

162.061 A static or an expanding universe?   V. A. Firsoff.
Astron. & Space (GB), Vol. 1, 220 - 226 (1971).

162.062 Point sources in nonhomogeneous universe.
F. de Felice.
Nuovo Cimento B, Ser. 11, Vol. 7B, 333 - 340 (1972).

Use is made of the linear perturbation theory to calculate the luminosity of a point source in a nonhomogeneous Friedmann universe.

162.063 Variational principles and spatially-homogeneous
universes, including rotation.
M. A. H. MacCallum, A. H. Taub.
Commun. Math. Phys., Vol. 25, 173 - 189 (1972).

The validity of imposing spatial homogeneity on the variations in the usual action principle for Einstein's equations is studied.

162.064 Universal magnetic fields.   E. N. Parker.
American Scientist, Vol. 59, 578 - 585 (1971).
See Phys. Abstr., Vol. 75, No. 26581 (1972).

162.065 Homogeneous rotating universe with flat space.
M. Demianski, L. P. Grishchuk.
Commun. Math. Phys., Vol. 25, 233 - 244 (1972).

The homogeneous, anisotropic cosmological model is considered. It satisfies three physically reasonable conditions: it is space homogeneous, possesses flat space like sections and is filled with expanding, rotating and shearing matter. The asymptotic solution is presented and general properties are discussed. The question how the rotation influences the behaviour of matter near the singularity is investigated.

162.066 Velocity-dominated singularities in irrotational
hydrodynamic cosmological models.
E. P. T. Liang.
Journ. Math. Phys., New York, Vol. 13, 386 - 393 (1972).
See Phys. Abstr., Vol. 75, No. 26592 (1972).

162.067 Current problems in cosmology.   S. Weinberg.
Proc. AIP Conference, No. 2, p. 247 - 255 (1971).

Discusses problems in the early evolution of the universe. This presents problems in general relativity, hydrodynamics, statistical mechanics, and elementary particle physics. In order to resolve the problems farther observational evidence is required particularly of the microwave neutrino and graviton backgrounds and also the helium abundance.

162.068 The matter-antimatter phase transition: A deriva-
tion of the critical temperature.
R. Aldrovandi, S. Caser.
Nuclear Phys. B, (Netherlands), Vol. 38B, 593 - 599 (1972).

A new estimate of the temperature of the matter-antimatter phase transition in the black-body radiation is given using the theoretical nucleon-antinucleon phase-shifts computed in a previous work.

162.069 The possible role of elementary particle physics in
cosmology.   R. Omnes.
Phys. Rep. Phys. Letters, Section C (Netherlands), Vol. 3C, No. 1, p. 1 - 56 (1972).

This is a review which is concerned with the following topics: (a) the thermodynamical behaviour of a system of particles at ultrahigh temperatures, (b) the astrophysical consequences of these effects in the framework of an evolutionary isotropic universe and their relevance to the problems of the origin of matter and galaxies.

162.070 A non-singular universe with torsion.
W. Kopczynski.
Phys. Letters A, (Netherlands), Vol. 39A, 219 - 220 (1972).

The problem of a spherically-symmetric gravitational field produced by spinning dust is solved in the framework of the Einstein-Cartan theory. There exists a two-parameter family of world-models without singularities of the Friedmann type.

162.071 Matter – antimatter hydrodynamics: the coalescence
effect.   R. Omnes.
Rep. Rept-71/25, Lab. Phys. Theor. and Hautes Energies, Orsay, France. [Available from NTIS, Springfield, Va.], 33 pp.

(1971). – See Phys. Abstr., Vol. 75, No. 44820 (1972).

162.072 **Relict radiation.** P. Andrle.
Kozmos, Vol. 3, 35 - 36 (1972). In Czech.
Review article.

162.073 **Stability of homogeneous universes.** S. Bonanos.
Commun. Math. Phys., Vol. 26, 259 - 270 (1972).
The stability of a class of homogeneous cosmological models is investigated. It is shown that the perturbation problem for six such universes can be reduced to a system of ordinary differential equations.

162.074 **Closed rotating cosmologies containing matter described by the kinetic theory. Entropy production in the collision time approximation.** R. A. Matzner.
Journ. Math. Phys., *New York*, Vol. 13, 931 - 937 (1972).
A collision time approximation for the production of partly collisional particles ('neutrinos') from a collision dominated fluid ('electrons and photons') within closed anisotropic Bianchi type IX cosmologies is discussed.

162.075 **Non-homologous expansion in cosmology.**
A. D. Chernin, B. S. Moros, Yu. V. Vandakurov.
Phys. Letters A, (*Netherlands*), Vol. 39A, 233 - 234 (1972).
A simple model suggested shows that cosmic expansion, which is non-homologous at an initial stage, can become, in principle, homologous and isotropic asymptotically in the limit of infinitely large time.

162.076 **The oscillatory regime near the singularity in Bianchi-type IX universes.** M. P. Ryan, Jr.
Ann. Physics, Vol. 70, 301 - 322 (1972).
The region near the initial singularity of the most general case of Bianchi-type IX universes is investigated.

162.077 **Cosmological models with non-zero pressure.**
J. Kulhanek, G. Szamosi.
Relativity and gravitation, Haifa 1969, (see 012.029), p. 221 - 224 (1971).

162.078 **Cosmological implications of the microscopic CP violation.** Y. Ne'Eman, Y. Achiman.
Relativity and gravitation, Haifa 1969, (see 012.029), p. 259 - 264 (1971).

162.079 **Some notes on cosmology.** J. Pachner.
Relativity and gravitation, Haifa 1969, (see 012.029), p. 265 - 268 (1971).

162.080 **The recent renaissance of observational cosmology.**
D. W. Sciama.
Relativity and gravitation, Haifa 1969, (see 012.029), p. 283 - 304 (1971).

162.081 **A cosmological theory of gravitation.**
S. J. Prokhovnik.
Relativity and gravitation, Haifa 1969, (see 012.029), p. 275 - 282 (1971).

162.082 **Neutrino processes in the lepton era of the universe and hot big bang cosmology.** D. Ray.
Journ. Phys. A, General Phys., Vol. 5, L 17 - L 18 (1972).

162.083 **Antimatière et astronomie.** J. Vandermeulen.
Rev. Questions Sci. Belgique, Vol. 142, 455 - 487 (1971). – See Bull. Signal., Vol. 33, Section 120, No. 1230 (1972).

162.084 **Spatially homogeneous rotating world models.**
I. Ozsváth.

Journ. Math. Phys., *New York*, Vol. 12, 1078 - 1082 (1971).
See Bull. Signal., Vol. 33, Section 120, No. 3682 (1972).

162.085 **Velocity-dominated singularities in irrotational dust cosmologies.** D. Eardley, E. Liang, R. Sachs.
Journ. Math. Phys., *New York*, Vol. 13, 99 - 107 (1972).

162.086 **The cosmological significance of observations at far-infrared and millimetre wavelengths.** M. J. Rees.
E. S.R.O., S. P., No. 52, (see 012.030), p. 23 - 28 (1971).

162.087 **Relativistic cosmology and space platforms.**
R. Ruffini, J. A. Wheeler.
E. S. R. O., S. P., No. 52, (see 012.030), p. 45 - 174 (1971).

Gravitation and universe. See Abstr. 003.048.

Physical cosmology. See Abstr. 003.106.

The origin of the universe. See Abstr. 003.107.

Hamiltonian cosmology. See Abstr. 003.116.

Edwin Hubble and a relativistic, expanding model of the universe. See Abstr. 004.008.

On the fragmentation of a contracting hydrogen cloud in an expanding universe. See Abstr. 065.010.

Sternmaterie.III. Das "schwarze Loch" und das expandierende Weltall. See Abstr. 065.105.

Origin of the cosmic microwave background radiation in a chaotic universe. See Abstr. 066.071.

Closed cosmological solutions to Einstein's field equations. See Abstr. 066.128.

Gravitation theory and oscillating universe. See Abstr. 066.129.

The possible line feature in the X-ray background. See Abstr. 125.001.

Interstellar particles and the evolution of the universe. See Abstr. 131.113.

Radio source-counts and cosmology. See Abstr. 141.109.

Redshift distribution of quasi-stellar objects and the radio source counts. See Abstr. 141.110.

Metagalactic gamma rays from relativistic-electron bremsstrahlung interactions. See Abstr. 142.007.

Inner bremsstrahlung as a source of X-rays in the steady-state universe. See Abstr. 142.046.

Observational evidence relating to a recent theory on the origin of the universal X-ray background. See Abstr. 142.106.

A galactic formation model based on post-big bang fragmentation. See Abstr. 151.018.

Dynamics of a rotating gaseous ellipsoid in an expanding universe. See Abstr. 151. 023.

Newtonian hydrodynamics in an expanding universe

in terms of the scalar-tensor theory.   See Abstr. 151.024.

Cosmological information from galaxies and radio galaxies.   See Abstr. 158.098.

The unlikeliness that molecular hydrogen stabilizes clusters of galaxies or closes the universe.
See Abstr. 160.003.

A suggested concept for galaxy-cluster size for use in the angular-size cosmological test.
See Abstr. 160.014.

Analysis of X-ray background fluctuations.
See Abstr. 160.015.

The velocity-distance relation and the Hubble constant for nearby groups of galaxies.   See Abstr. 160.017.

Some remarks on intergalactic magnetic fields.
See Abstr. 161.013.

Enrichment of intergalactic matter.
See Abstr. 161.015.

Errata

162.901   Errata: 'A complete redshift-magnitude formula.'
          [Astron. Journ., Vol. 76, 751 - 755 (1971)].
S. E. Kaufman.
Astron. Journ., Vol. 77, 254 (1972).

Author Index

AVERY, R. W.
022.080
AVGYEJEV, JU. F.
094.259
AVULOV, K. D.
152.004
AXEL, L.
099.033
AXFORD, W. I.
074.059
078.025
102.007
AXISA, F.
077.030
AYRES, T. R.
072.008
AZCARRAGA, A.
082.088
BAARS, J. W. M.
033.036
141.056
BABAYAN, KH. P.
143.055
BABU, G. S. D.
116.901
BACHELET, F.
143.073 .074 .075
BACKER, D. C.
074.067
141.049 .168 .511 .515
.550
BACON, M. E.
022.035
BADALYAN, O. G.
071.008
BADESCO, R.
081.027
BAGDASARYAN, M. B.
078.015
BAGGALEY, W. J.
104.036
BAGLIN, A.
065.098
BAHCALL, J. N.
061.043
080.037 .038
132.007
141.068
158.148
162.013
BAHCALL, N. A.
141.068
BAHNER, K.
031.028
034.082
122.131
BAHNG, J. D. R.
124.103
BAHR, J. L.
022.019
BAILEY, A. D.
079.101
BAILEY, G. J.
083.035
BAILEY, J. M.
042.018
BAILEY, N. G.
094.010
BAIRD, G. A.
142.042 .109

BAITY, W. A.
142.080
155.048
BAIZE, P.
118.014
BAKER, D.
053.005 .014
BAKER, N.
115.017
BAKER, N. H.
154.033
BAKHAREV, A. M.
103.105
BAKHCHIVANDZHI, V. E.
031.010
BAKKE, J. C.
034.131
BAKOS, G. A.
118.011
121.036
BALAMORE, D. S.
064.036
BALAZS, L. G.
122.157
BALDINELLI, L.
009.025
BALDINI, R.
034.031
BALDWIN, J. E.
158.001
BALDWIN, M. E.
010.001
BALDWIN, R.
033.043
BALDWIN, R. B.
105.059
BALICK, B.
125.101
131.106
133.002 .008
BALL, S. E.
034.138
142.092
BALLARIO, M. C.
085.002
BALLEVRE, J.
034.041
BALLY, M.
124.102
BALODIS, JA. C.
055.007
BAME, S. J.
073.050
074.034
BANDERMANN, L. W.
098.058
BANDYOPADHYAY, P.
022.118
BANFI, V.
091.025
BANFIELD, R. M.
113.040
BANNISTER, T. C.
053.013
BANSAL, B. M.
094.009 .062
BAPPU, M. K. V.
075.001
BARABASHOV, N. P.
031.040

BARANNE, A.
034.080 .081
BARATTA, G. B.
123.014
BARBANIS, B.
008.110
BARBARO, G.
065.066
BARBARROJA, R.
075.015
BARBIER, M.
119.019
BARBIERI, C.
113.032
141.074 .114 .159
BARBON, R.
008.107
082.052
125.107
BARDEEN, J. M.
066.038
BARDSLEY, J. N.
162.046
BARKAT, Z.
061.010
064.009
065.021 .052 .058
BARKER, D. N. H.
143.023
BARKER, F. C.
022.053
BARLAI, K.
122.101
BARLETTI, R.
012.023
083.028
BARLOW, B. C.
081.007
BARNES, C. A.
065.141
BARNES, C. W.
080.032
BARNES, J. A.
044.029
BARNES, J. V.
113.001
BARNES, T. G.
099.020
BARNES III, T. G.
122.028
BARNOTHY, J. M.
141.043 .048 .175
158.146
162.030 .038
BARNOTHY, M. F.
141.043 .048 .175
158.146
162.030 .038
BAROCAS, V.
082.084
BARRA, A. L. DE LA
115.012
BARRETT, A. H.
114.048 .093
BARRETT, J. W.
158.022
BARRICELLI, N. AALL
107.008
BARROW, C. H.
099.004

BENJAMIN JR., F. S.
003.025
BEN'KOVA, N. P.
083.059
BENNETT, D. J.
084.231
BENNETT, J. E.
082.026
BENNETT, K.
142.110
BENNETT, L.
094.009
BENNETT JR., C. L.
033.062
BERAN, D. W.
082.031
BERCOVITCH, M.
143.005 .024
BERDOT, J. L.
105.035
BERENDZEN, R.
014.005
131.065 .078
BEREZINSKIJ, V. S.
011.013
BERGE, G. L.
141.041
BERGEAT, J.
121.009
BERGER, M. J.
082.025
BERGER, P. S.
074.001
BERGER, X.
052.032
BERGERON, J.
131.159
BERGEY, J. A.
034.128
BERGFRIED, D. E.
033.059
BERGH, C. DE
099.021
BERGH, S. VAN DEN
122.061
131.068
155.013 .044
158.007 .058 .065
BERKHUIJSEN, E. M.
157.004
BERN, K.
113.026
BERNACCA, P. L.
114.081 .108
BERNFELD, D.
031.045
BERNOT, M.
075.010
BERRY, R.
034.013
036.008
BERTAUD, C.
123.042
125.102
141.128
BERTEL, L.
082.002
BERTHEL, R. O.
022.081
BERTHELSDORF, R.
134.003

BERTIAU, F. C.
113.044
BERTOLA, F.
012.021 .022 .023
034.100
141.114
158.157
BERTOTTI, B.
066.099
BERTSCH, D. L.
071.034
078.001
BESPROZVANNAYA, A. S.
083.061
BESSELL, M. S.
122.058
142.126
BEST, A.
084.253
BETHE, H. A.
065.143
BETTIS, D. G.
021.007
151.040 .058 .073
BEYER, M.
102.012
121.002
122.006
BEYNON, W. J. G.
083.042
BEZCHASTNOV, I. M.
009.024
BHANDARI, S. M.
141.083
BHARGAVA, B. N.
084.216
BHATIA, P. K.
062.032
BHATNAGAR, A.
073.058
BHATNAGAR, K. B.
042.067
BHATNAGAR, V. P.
074.053
BHATT, T. R.
123.007
BHATTACHARYYA, J. C.
099.058
BHONSLE, R. V.
033.039
BHOWMIK, G.
061.041
BIBRING, J. P.
094.023
BIDELMAN, W. P.
008.041
041.052
121.043
BIEMANN, K.
091.029
097.015
BIERMANN, K.-R.
005.009
BIERMANN, L.
008.099
064.044
102.022
107.011
131.153
BIERMANN, P.
065.087 .108

BIERMANN, P.
066.059
117.002 .017 .028
131.152
BIGAY, J. H.
113.004
155.033
BIGGS, E. S.
003.170
BIGNENS, P.
009.013
BILLAUD, G.
112.014
BILLINGS, D. E.
074.016
BINDER, A. B.
097.014 .065 .097 .105
099.016
101.001
BINGHAM, D. K.
084.268 .269
BINGHAM, J.
122.104
BINNENDIJK, L.
121.025 .026
BINNS, W. R.
034.141
BIONDI, M. A.
083.037
BIRAUD, F.
141.160
BIRD, J.
079.103
BIRKEBAK, R. C.
094.040 .121 .256 .271
BIRNBAUM, M.
003.038
BIRULIN, A. I.
082.068
BIRYUKOV, YU. L.
063.032
BISCHOF, W.
121.058 .090
BISCHOFF, M.
002.901
041.035
BISHOFF, K.
083.056
BISKUP, M.
003.026
BISNOVATYI-KOGAN, G. S.
065.082
BISNOVATY-KOGAN, G. S.
062.029
064.028
065.049
125.011 .023
141.070
BISWAS, S.
034.142
142.034 .074
BIVAS, M.
046.028
BJORKHOLM, P.
094.020
BLAAUW, A.
008.064
012.018
111.006
BLAAUW, A. A.
155.054

GETCHELL, B. C.
046.023
GEYER, E. H.
121.035
154.023
158.118 .166
GEZARI, D. Y.
113.002
115.009
GHAZI, A.
097.067
GHETU, I.
098.009
GHEZLOUN, A.
041.044
098.026
GHIELMETTI, H. S.
009.009
143.030
GHIGO, F. D.
141.141
GHOSE, S.
094.087
GIACAGLIA, G. E. O.
042.032 .033 .047
081.016
GIACCONI, R.
142.022 .026 .027 .038
     .045 .065 .093 .095
     .108 .116 .127 .901
     .902
155.053
158.029
160.020
GIANNONE, P.
065.112
117.005 .032
GIANNUZZI, M. A.
008.122
117.005 .032
GIBB, F. G. F.
094.137
GIBBONS, J. H.
065.142
GIBSON, E. G.
080.023
GIBSON, E. K.
094.009
GIBSON, J.
113.001
GIBSON, J. B.
103.102 .115
154.011 .017
GIBSON, U. T.
103.102 .115
GICLAS, H.
098.019
GICLAS, H. L.
103.114 .115 .127
112.015
GIDALEVICH, E. YA.
143.031
GIERASCH, P. J.
097.108
GIESE, R.-H.
008.024
061.026
106.016
GIETZEN, J. W.
034.067

GILBERT, I. H.
151.026
GILBERT, J. A.
141.072 .073 .149 .167
GILBERT, J. C.
002.035
GILL, J. R.
012.027
GILLARD, R.
044.055
GILLETT, F. C.
133.007
GILLETTE, R.
051.002
GILLUM, D. E.
094.008 .066
GILMORE, A. C.
103.014 .115
GILVARRY, J. J.
066.070
GIMMESTAD, G. G.
022.038
GINDILIS, L. M.
011.022 .035
GINZBURG, V. L.
061.050
066.029
GIOVANELLI, R.
131.095
GIOVANELLI, R. G.
073.005
GIRNSTEIN, H. G.
033.028
GIULI, R. T.
107.013
114.030
GIVER, L. P.
093.019
GLAGOLEVSKY, J. V.
116.004 .005
GLASS, B. P.
094.149
GLASS, I.
114.130
GLASS, I. S.
132.036
GLASS, N. W.
084.025
GLASSER, M. L.
065.116
GLAZHEVSKA, A.
084.404
GLEDHILL, J. A.
099.056
GLEESON, L. J.
143.053
GLEISSBERG, W.
008.054
GLENCROSS, W. M.
141.562
GLENN, W. H.
079.005 .007
GLESKE, I. U.
034.130
GLICKER, S.
131.007
GLIESE, W.
111.002 .005
GLIKSON, A. Y.
081.007

GLOECKLER, G.
143.066
GLUSHKO, V. N.
082.116 .122 .126
GNEDIN, YU. N.
142.054
GNEZDILOV, A. A.
077.046
GOAD, L. E.
131.040
132.034
133.005
GODDARD, R.
142.141
GODFREY, P. D.
131.142
GODISOV, N. P.
021.006
GODOLI, G.
008.037
072.035
GOEBEL, C. J.
066.010
GOEBEL, E.
151.084
GOERTZ, C.
099.034
GOETTIG, C.
123.030
GOETZ, A. F. H.
094.088
GOETZ, W.
122.145 .146
GOGOSHEV, M. M.
014.002
GOJSA, N. I.
085.005
GOKHALE, G. S.
142.034 .074 .130
GOKHBERG, M. B.
084.254
GOLD, T.
051.001
GOLDBACH, C.
062.014 .026 .044
GOLDBERG, B. A.
122.121
GOLDBERG, J. N.
066.136
GOLDBERG, L.
008.033
012.011
074.005
132.034
GOLDMAN, I.
066.129
GOL'DOVSKIJ, D. YU.
094.140
GOLDREICH, P.
063.040
094.134
125.006
141.566
GOLDSMITH, D.
131.014 .107
160.012
GOLDSMITH, D. W.
131.031 .080
161.005
GOLDSMITH, S.
022.071 .096

LOWRANCE, J. L.
034.024
141.019
LOWREY, B. E.
054.003
LOZINSKAYA, T. A.
132.004
141.071
LUBYANAYA, N. D.
143.022
LUC, P.
022.030
LUCAS, J. W.
003.089
LUCCHETTI, S. C.
123.053
LUCCHIN, F.
162.041 .058
LUCHKOV, B. I.
134.002
LUCIA, F. C. DE
022.114
LUCK, J. MCK.
118.023
LUCKE, P. B.
152.005
LUCY, L. B.
064.036
121.023
LUDWIG, A. C.
033.066
LUE, P. K.
158.903
LUEBBERS, G. L.
003.090
LUEST, R.
008.099
LUEST, RH.
103.106
LUGOVENKO, V. N.
084.242
LUKAS, R.
103.005
119.022
123.001 .032 .037
LUKATSKAYA, F. I.
113.005
LUKIN, D. S.
093.023
LUKINA, N. YU.
005.001
LUMME, K.
100.006
LUNDQUIST, C. A.
081.016
LUNDSAGER, S.
155.058
LUNEL, M.
121.009
LUPISHKO, D. F.
097.101
LUTENKO, V. F.
084.038
LUTZ, J. H.
113.037
133.017
LUTZ, T. E.
113.037
LUYENDYK, A. P. J.
159.015

LUYTEN, W. J.
112.016 .017 .018
113.046
117.036
L'VOV, B. V.
022.044
L'VOV, V. G.
046.014
LYKOUDIS, P. S.
072.009
LYNCH, M. A.
099.011 .024
LYNDEN-BELL, D.
012.025
111.001
151.065
LYNDS, B. T.
132.006
158.070
LYNDS, C. R.
141.082
158.099
LYNDS, R.
141.058
LYNE, A. G.
141.538 .545
LYNGA, G.
034.107
LYNGSTAD, E.
077.002
LYSOV, V. P.
093.025
LYTTLETON, R. A.
102.006
LYUBIMOV, G. P.
078.017
143.062
LYUTIY, V. M.
142.121
LYUTYJ, V.
158.016
MAANDERS, E. J.
033.057
MABUCHI, H.
105.047
MACALPINE, G. M.
141.179
MACAU, J. P.
034.110
MACAU-HERCOT, D.
022.106
MACCAGNI, D.
078.012
MACCALLUM, M. A. H.
162.063
MACCHETTO, F.
131.097
MACCONNELL, D. J.
114.118
142.068 .125
159.027
MACDONALD, G. H.
141.089
MACDONALD, G. J. F.
094.193
MACHADO, M. E.
073.043 .045
MACK, D. A.
031.012
MACK, J.
142.018 .113

MACK, J. E.
142.032 .106
MACKAY, C. D.
141.007
MACKIE, J. B.
010.024
MACKLIN, R. L.
065.142
MACLENNAN, C. G.
078.009
MACLEOD, J. M.
132.025
141.135 .170
158.060 .165
MACRAE, D. A.
008.119
032.010
114.063
MACRIS, C. J.
012.011
073.001 .071 .080
MACVEY, J. W.
051.004
MADDEN, T. R.
094.122
MADER, G. L.
155.057
MADORE, K.
115.015
122.084
MAEDA, M.
033.052
MAEHLUM, B. N.
034.011
083.018
MAEYAMA, Y.
003.091
MAFFEI, P.
002.026
MAGEE, N. H.
065.038 .039
MAGNANT, F.
074.045
MAHADEVAN, P.
079.101
MAHANTA, M. N.
066.121
MAHLE, S. H.
082.099
MAHONEY, W.
142.003 .019
MAHONEY, W. A.
142.024
MAIER, E. J.
079.101
084.021
MAILER, N.
003.092
MAITZEN, H. M.
116.002 .016
MAJDEN, E. P.
031.009
MAJERNIK, V.
066.039
MAJUMDAR, S. K.
064.053
MAKALKIN, A. B.
099.025 .035
MAKAROV, A. N.
099.030

MCFADDEN, W. H.
094.164
MCFARLAND, J.
159.012
MCGEE, J. D.
034.088
MCGEE, R. X.
159.007
MCGILL, G. E.
097.087
MCGOVERN, W. E.
099.067
MCILVENNA, J. F.
033.061
MCILWAIN, C. E.
034.146
MCINNES, B.
012.023
MCINNES, P. A.
033.043
MCINTOSH, B. A.
104.035
MCINTOSH, C. B. G.
162.022 .023
MCINTOSH, P. S.
036.003
073.048
MCKAY, D. S.
094.009
MCKAY, G. A.
094.043
MCKEE, C. F.
062.043
141.068
MCKEE, C. R.
125.015 .021
MCKEITH, C. D.
114.136
MCKELLAR, A. R. W.
091.031
MCKENZIE, D. L.
034.129
076.029
MCKINNEY, W.
079.101
MCL. MARSH, L.
066.108
MCMULLAN, D.
034.096
MCNALL, J. F.
034.022 .026 .027
MCNAMARA, A. G.
079.101
MCNAMARA, L. F.
022.068
MCPHERRON, R. L.
084.201
MCVITTIE, G. C.
160.001
162.037
MCWILLIAM, S. F.
045.018
MEABURN, J.
034.001
MEADOWS, A. J.
003.169
105.037
MEBAGISHVILI, I. I.
083.064
MEBOLD, U.
131.115 .123

MECHTLY, E. A.
079.101
MEDD, W. J.
123.004
141.006 .129 .134 .135
       .170
MEDNIKOVA, N. V.
083.050
MEDRANO, R.
094.146
MEDVEDEVA, G. I.
003.154
MEDVEDEVA, L. I.
112.005
MEED, W. J.
141.134
MEEKINS, J. F.
073.016
134.009
MEEKS, M. L.
033.049
141.901
MEEUS, J.
047.019
054.022
055.001
091.014
094.233
096.006 .009 .011
097.049
098.901
099.022 .049
101.003
103.102
104.025
MEFFROY, J.
042.045
MEGESSIER, C.
113.035
MEIDAV, M.
142.016 .050 .072
MEIER, R. R.
082.028
MEILLER, V.
099.063
MEINIG, M.
044.023
MEINUNGER, L.
122.143
123.046
158.120
MEISSINGER, H. F.
051.034
MEISTER, J.
094.146
MELCHIOR, P.
003.094
041.019 .020
MELCHIORRI, F.
032.055
MEL'NIKOV, O. A.
005.001
154.025
MELSON, W.
094.158
MELTON, L. A.
082.017
MENA, L.
041.041
MENDE, S. B.
084.004

MENDELL, R. B.
079.101
MENDELL, W. W.
094.110
MENDIS, D. A.
102.007
MENDOZA V, E. E.
122.901
MENG, C.-I.
084.234
106.023
MENGEL, J. G.
065.020 .120
154.013
MENNESSIER, M. O.
043.002
MENON, T. K.
155.077
MENTALL, J. E.
022.115
131.007
MENZEL, D. H.
003.095 .096
051.023
074.097
MENZIES, J.
154.008
MERCER, J. B.
143.023
MERCIER, J. P.
078.012
MEREDITH, R. E.
022.033
MERGENTALER, J.
076.042
MERLIN, P.
114.119
MERMILLIOD, J. C.
041.037
113.039
MERRILL, K. M.
122.133
133.007
MERTS, A. L.
065.038 .039
MERTZ, L.
141.546
MESHKOVA, T. S.
041.010 .012
MESROBIAN, W. S.
118.024
153.009
MESSELL, K.
099.062
MESSINA, A.
158.011
MESSMER, J.
066.067
MESTEL, L.
065.075
MESZAROS, P.
162.015
METZ, W. D.
034.014
061.004
MEULEN, J. J. TER
131.027
MEUNIER, R.
094.023
MEURERS, J.
008.158

MIZUNO, S.
073.006
MIZUTANI, H.
094.225
MNATSAKANIAN, R. G.
125.019
MOCHNACKI, S. W.
121.001 .003
MODALI, S. B.
082.108
141.904
MODI, V. J.
042.011
MODISETTE, J. L.
074.066
114.030
MOELLER, C.
003.099
MOFFAT, A. F. J.
116.002
155.065
MOFFET, A. T.
141.076 .094
MOFFETT, R. J.
083.035
MOFFETT, T. J.
122.003
MOGRO-CAMPERO, A.
078.002
084.410
MOHAN, C.
065.011
MOHAN, S. N.
052.003
MOISEEV, I.
141.039
MOISEEV, I. G.
077.018 .019 .038
131.124
141.060 .092
MOISEYEV, I. G.
099.071
131.039
MOLCHANOV, V. A.
082.126
MOLNAR, M. R.
114.013 .086
116.010
MOLTON, P. M.
099.043
MOMCHEV
099.050
MONAGHAN, J. J.
022.079
MONIN, A. S.
003.100
093.032
MONJES, J. A.
094.099
MONNET, G.
034.092
159.001
MONTGOMERY, D. R.
097.009
MONTLE, R. E.
123.020
MONTMERLE, T.
065.146
160.023
MOOK, D. E.
142.081 .082 .083

MOORCROFT, D. R.
083.006
084.005
MOORE, C. B.
094.009
MOORE, E.
155.043
MOORE, E. P.
032.001
MOORE, G.
142.022
MOORE, H. J.
097.023
MOORE, J. G.
079.104
MOORE, P.
003.101
096.012
123.034 .039
MOORE, R. L.
073.020 .037
MOORE, W. H.
021.007
MOORE JR., J. T.
097.027
MOORE-SITTERLY, C.
114.057
MOORWOOD, A. F. M.
132.019
MOOS, H. W.
114.032 .114
MORAN, J. M.
131.038 .039 .124
MORANDO, B.
005.011
097.090
MORANZINO, C.
032.012
MOREL, P. J.
118.003
MORENO, G.
074.094
MORENO, H.
113.003
MORGAN, B. L.
034.088
MORGAN, J. W.
094.005 .067 .902
105.011
MORGAN, T. E.
114.033 .096
MORGAN, W. W.
114.045
153.028
158.079
MORGAN JR., D. L.
142.007
161.010
162.039
MORGANSTERN, R. E.
162.053
MORI, S.
143.070
MORIMOTO, M.
155.088
MORIN, F. J.
094.112
MORISI, F.
123.023
MOROS, B. S.
162.075

MOROZ, V. I.
097.092
MOROZHENKO, A. V.
103.100
MOROZOVA, G. V.
042.025
MORRIS, C. S.
123.034
MORRIS, D.
141.158
MORRIS, E. C.
097.014
MORRIS, M.
131.035 .045
153.019
MORRISON, B. L.
099.063
MORRISON, D.
099.040
MORRISON, D. A.
094.009 .261
MORRISON, J. A.
094.033
MORRISON, L. V.
096.007
142.051 .128
MORRISON, P.
011.007
MORSE, D. L.
062.035
MORTON, D. C.
034.087
141.019 .143 .144 .161
158.034 .163
MORTON, W. A.
141.143 .144 .161
MOSELEY, G. F.
141.141
MOSELEY, J. T.
083.004
MOSIER, S. R.
084.016
MOSKALEVA, G. V.
103.100
MOSS, D. L.
065.064
117.033
MOSS, F. J.
081.007
MOSSMAN, D. J.
105.044
MOTTONI, G. DE
032.015
091.020
MOTZ, L.
141.195
MOUTSOULAS, M.
002.028
MOUTSOULAS, M. D.
094.189
MOZER, F. S.
084.406
099.042
MRKOS, A.
095.005
103.103 .107 .116 .129
MUCKE, H.
011.023
MUCKE, R.
011.023

VASHKOV'YAK, M. A.
054.015
VASIL'EV, E. M.
074.071
VASIL'EV, V. G.
046.015
VASILEVSKY, A. E.
153.014
VASILIEV, K. N.
083.014
VASILIEVA, E. K.
072.031
VASILYEV, O. B.
021.005 .006
080.012 .020
082.044
VASILYEV, V. P.
099.028
VASSEUR, J.
141.507
VASSILYEVA, G. Y.
072.041
VASYLIUNAS, V. M.
084.274
VAUCLAIR, G.
121.011
VAUCLAIR, S.
065.067
114.117
VAUCOULEURS, A. DE
158.013
160.021
VAUCOULEURS, G. DE
003.096
097.027 .054
158.002 .003 .004 .013
.901
160.017 .021
VAUGHAN, A. E.
141.527 .535 .536
VAUGHAN, L. M.
103.107
VAUGHN, L. M.
103.107 .108
VAZIAGA, M. J.
064.057
VEDRENNE, G.
142.008
VEER, F. VAN'T
121.009
VEGT, C. DE
041.003
VEHRENBERG, H.
098.002
VEIO, F. N.
003.133
VELGHE, A. G.
114.129
VELINOV, P.
143.037
VELTMANN, UE. I.
099.036
VENKATESWARAN, S. V.
076.003
079.101
VENO, S.
064.056
VENTURA, A.
141.528
VERDET, J. P.
097.091

VERDET, J. P.
099.021 .901
VERHULST, F.
042.003
VERNAZZA, J. E.
073.008
VERNIANI, F.
003.168
VERNOV, S. N.
063.006
078.016
084.407
106.011
143.018 .062
VERNOVA, L. V.
083.038
VERON, M. P.
158.042
VERON, P.
141.097
158.042
VERSCHUUR, G. L.
131.047 .095 .109
155.015
VESECKY, J. F.
079.101
VESELY, C. D.
098.023 .043
103.114
VETLE, J. I.
142.137
VETTER, U.
081.023
VEVERKA, J.
097.027 .069
098.001 .039 .040
VICENTE, R. O.
044.044
045.009
VICKERS, D. G.
066.076
VIDAL, N.
065.044
VIDIAKIN, V. V.
042.046
VIESTAVKIN, A. N.
077.019
VIGIER, J.-P.
066.004 .036 .077
VIGOTTI, M.
141.116
VILAIN, C.
066.094
VILLAMEDIANA, J. F.
031.005
VILLANTE, U.
084.901
VILLORESI, G.
143.073 .074 .075
VINCENT, M.
008.021
VINNIKOV, E. M.
035.005
VINTI, J. P.
052.005
VIOTTI, R.
012.021 .023
123.014
VIRDEFORS, B.
113.026

VIRGO, D.
094.142 .165 .901
VIRSKAJA, N. F.
081.012
VISHNIAC, W.
003.134
VISHNIAC, W. V.
097.020
VISHVESHWARA, C. V.
066.111
VISVANATHAN, N.
034.034
141.046
VITKEVICH, V. V.
033.018
106.018 .020
141.568
VITRICHENKO, E. A.
119.009 .010
VIVES, T.
159.018
VLADIMIROVA, G. V.
083.063
VLADIMIRSKIJ, B. M.
142.059
VLADIMIRSKY, B. M.
141.521
142.005
VLASCEANU, V. I.
098.009
VLASOV, M. N.
082.011 .081
VLASOV, V. I.
106.018
VOGEL, U.
131.121
VOGT, N.
155.065
VOGT, R. W.
077.004
VOGT, S. S.
034.048
VOIGT, H. H.
008.059
VOISE, W.
003.135
VOJTA, YA.
083.038
VOLKOFF, I.
003.136
VOLKOV, V. P.
011.003
VOLLAND, H.
082.065 .069
VOLOBUEV, S. A.
134.002
VOLOGDIN, A. G.
083.047
VOLOVIK, V. D.
099.028
VOLYNSKIJ, B. A.
053.012
061.028
VOLZ, F. E.
082.007
VONDJIDIS, A.
003.034
VONDRAK, J.
044.048
VOORDES, H. R.
121.070

Subject Index

ABSOLUTE MAGNITUDES
115.000 .007
ABSOLUTE MAGNITUDES
OB STARS
114.071
ABSOLUTE MAGNITUDES
OF STARS
152.013
ABSOLUTE MAGNITUDES
SPECTROSCOPIC BINARIES
119.001
ABSOLUTE MAGNITUDES
WOLF RAYET STARS
152.013
ABSORPTION
GALAXIES
158.017
ABSORPTION
INTERGALACTIC MATTER
161.012
ABSORPTION
INTERSTELLAR MATTER
113.015
131.078 .083
ABSORPTION
VENUS ATMOSPHERE
093.034
ACCESSORIES
012.003
ACCRETION
MOON
094.225
ACCRETION
NEUTRON STARS
065.006
ACCRETION
PROTOCLUSTERS
151.086
ACHONDRITES
105.011
ACOUSTIC WAVES
SOLAR ATMOSPHERE
080.004
AIRGLOW
082.000

AIRGLOW
MARS ATMOSPHERE
097.038
AI VELORUM STARS
122.034
ALBEDO
MARS
097.101
ALBEDO
MOON
094.183
ALFVEN WAVES
CHROMOSPHERE
073.024
ALFVEN WAVES
SOLAR WIND
074.099
ALGOL
RADIO RADIATION
121.063
ALGOL SYSTEMS
064.048
ALMANACS
047.000
ANDROMEDA NEBULA
158.007 .160 .169
.173
ANDROMEDA NEBULA
GLOBULAR CLUSTERS
154.006
ANDROMEDA NEBULA
NOVAE
124.002 .006
ANTIMATTER
143.015
162.046 .068 .071
.083
ARCTURUS
UV SPECTRA
114.114
ARTIFICIAL SATELLITES
054.000
ARTIFICIAL SATELLITES
OBSERVATIONS
055.000

ASSOCIATIONS
010.000
ASSOCIATIONS STELLAR
152.000
A STARS
ATMOSPHERES
114.009
A STARS
LUMINOSITIES
115.019
A STARS
MODELS
065.098
A STARS
PECULIAR
113.035
114.041 .051 .054
.055 .111 .117
122.037
153.005 .006
155.005
A STARS
PHOTOMETRY
113.001
A STARS
PROPER MOTIONS
112.008
A STARS
ROTATION
116.008
ASTEROIDS
LIGHT CURVES
098.024
ASTEROIDS
MASSES
098.012
ASTEROIDS
ORBITS
098.027
ASTRODYNAMICS
052.000
ASTROLABE OBSERVATIONS
112.014
ASTROMETRY
PHOTOGRAPHIC
041.003